Cavicchi, Thomas J.

Fundamentals of electrical engineering.

$72.00

| DATE | | | |
|---|---|---|---|
| | | | |
| | | | |
| | | | |
| | | | |
| | | | |
| | | | |
| | | | |
| | | | |
| | | | |
| | | | |
| | | | |
| | | | |

DANVILLE
PUBLIC
LIBRARY

# Fundamentals of Electrical Engineering

## Principles and Applications

**THOMAS J. CAVICCHI**

*Department of Engineering*
*Grove City College, Pennsylvania*

PRENTICE HALL Englewood Cliffs, NJ 07632

Library of Congress Cataloging-in-Publication Data

Cavicchi, Thomas J.
Fundamentals of electrical engineering : principles and applications / Thomas J. Cavicchi.
p. cm.
ISBN 0-13-059601-9
1. Electric engineering. I. Title.
TK146.C235 1993
621.3--dc20 92-30761
CIP

Acquisitions Editor: Alan Apt
Development Editor: Sondra Chavez
Production Editor: Joe Scordato
Marketing Manager: Tom McElwee
Copy Editor: Barbara Zeiders
Cover Designer: Bruce Kenselaar
Prepress Buyer: Linda Behrens
Manufacturing Buyer: Dave Dickey
Supplements Editor: Alice Dworkin
Editorial Assistant: Shirley McGuire

A Paramount Communications Company
Englewood Cliffs, New Jersey 07632



Printed in the United States of America

10 9 8 7 6 5 4 3 2

ISBN 0-13-059601-9

PRENTICE-HALL INTERNATIONAL (UK) LIMITED, *London*
PRENTICE-HALL OF AUSTRALIA PTY. LIMITED, *Sydney*
PRENTICE-HALL CANADA INC., *Toronto*
PRENTICE-HALL HISPANOAMERICANA, S.A., *Mexico*
PRENTICE-HALL OF INDIA PRIVATE LIMITED, *New Delhi*
PRENTICE-HALL OF JAPAN, INC., *Tokyo*
SIMON & SCHUSTER ASIA PTE. LTD., *Singapore*
EDITORA PRENTICE-HALL DO BRASIL, LTDA., *Rio de Janeiro*

*This book is dedicated to my immediate family—a family of teachers, literally and/or effectively: my father, Richard H. Cavicchi (mechanical engineering/mathematics/physics); my mother, Mary Anne Cavicchi (political science); my sister, Elizabeth M. Cavicchi (physics and art); her husband, Alva L. Couch (computer science); my brother, Richard E. Cavicchi (physics); his wife, Clare L. Cavicchi (historic preservation); and their baby daughter, Violet K. Cavicchi (about the miracle of brand-new life).*

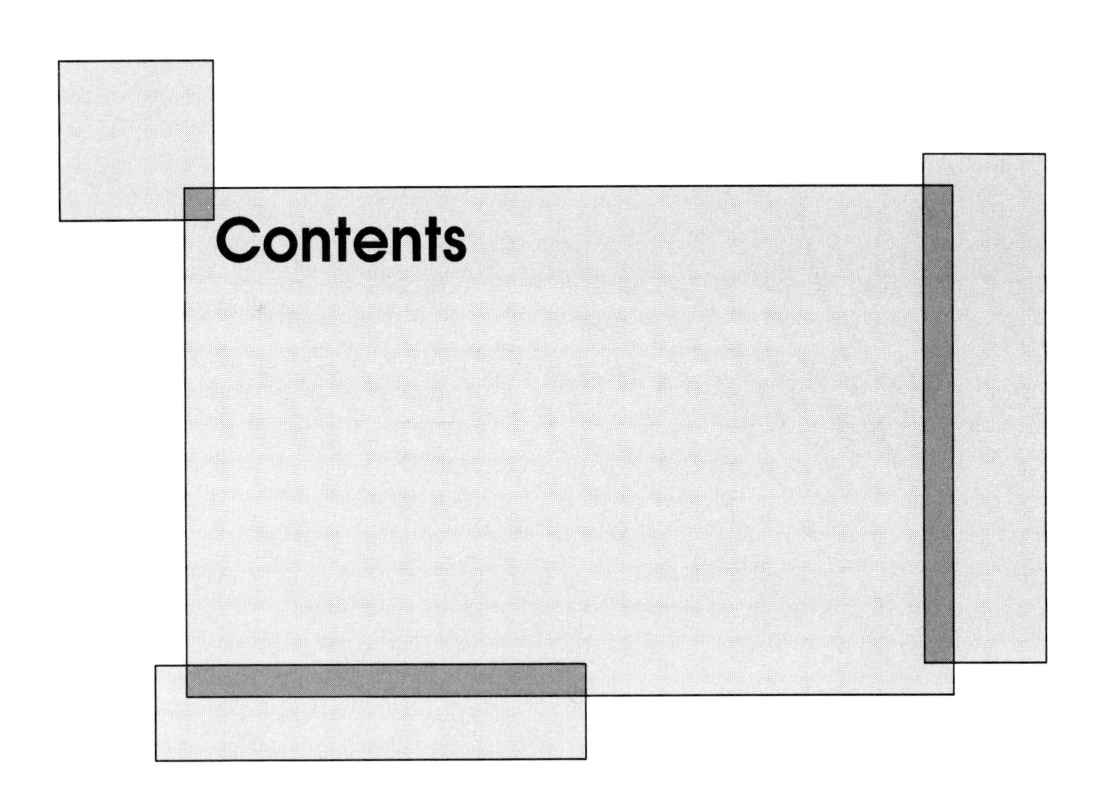
Contents

* Indicates advanced subsection

* Indicates advanced subsection

* Indicates advanced subsection

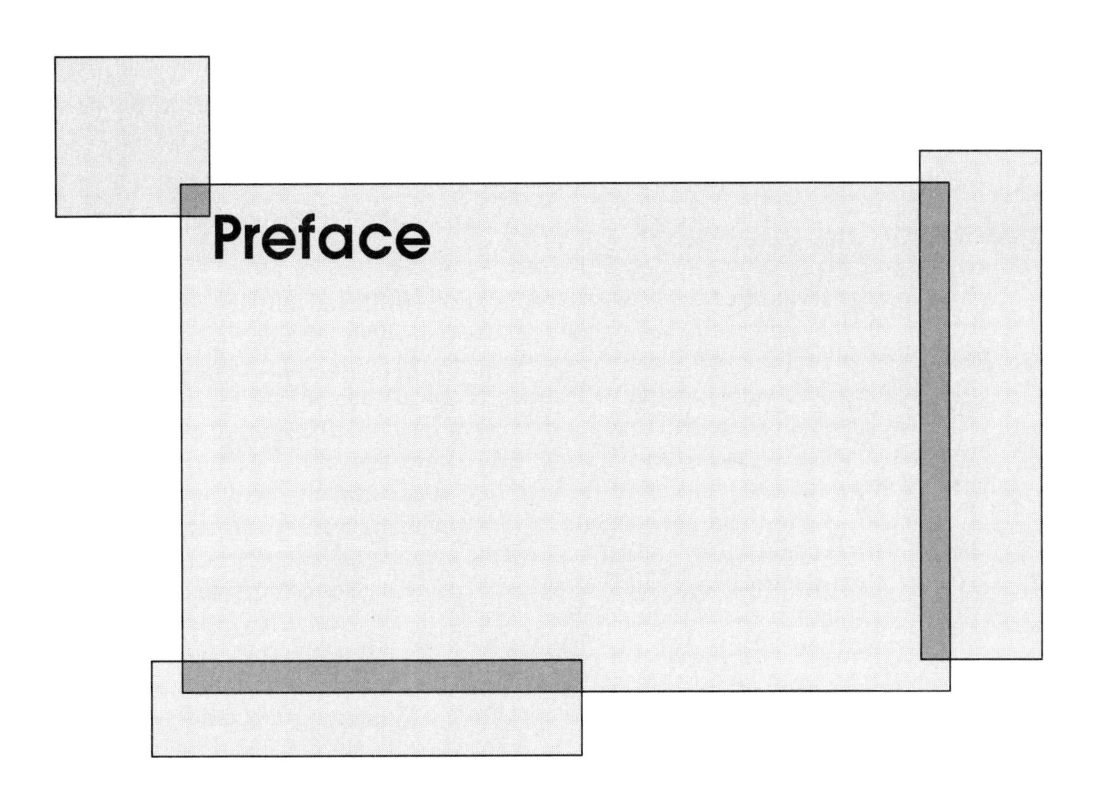

# Preface

*Fundamentals of Electrical Engineering* focuses on the three main areas of electrical engineering (EE) most vital to all engineers: analog electronics, digital electronics, and electromechanical machinery. It builds toward these with a strong foundation of fundamental physical laws and circuit analysis techniques. Student interest is retained by using physical reasoning and numerous everyday application examples. This book gives top priority to demonstrating relevance of the subject, a key element in teaching EE to non-electrical engineering (non-EE) students.

In this era of VCRs, personal computers, and fax machines, there has been an explosion of information processing, storage, and transmission. Computers are used daily by nearly all engineers. This is also an era of environmental awareness. Concerns about how to convert and transmit energy (and transport people) efficiently demand knowledge about electromechanical machinery as a first step. Just as the world is becoming one gigantic international network, engineering is becoming one huge interdisciplinary field of study. More than ever, engineers from all branches are becoming teammates with electrical engineers and consequently must learn the language. The primary purpose of this book is to aid in this task and go beyond nomenclature to understanding.

Although this book was written mainly for non-EE students, it is also hoped that it will be of value to undergraduate EE students. We will take a fresh look at concepts and devices both old and new, simple and complex. Teaching electrical engineering to non-EE students can be particularly challenging and rewarding, because these students take such a fresh look at the field. None of the assumptions an electrical engineer takes for granted are accepted without convincing justification. Many of the questions answered in this book were raised by my non-EE students. Photocopy versions of most of this text have withstood and been enhanced by the test of repeated classroom use for several years. The result is a friendly and conversational, yet thorough and precise approach.

## ORGANIZATION AND SUBSTANCE

The only prerequisites for using this book are freshman physics and calculus; otherwise, the material is self-contained. *Fundamentals of Electrical Engineering* contains five parts: (1) Physical Laws, (2) Electric Circuits, (3) Analog Electronics, (4) Digital Electronics, and (5) Electromechanical Machines. The Introduction, Chapter 1, sets the stage and tone for the rest of the book. Chapters 2 and 3 present the fundamental physical laws behind EE. These laws are introduced using an historical perspective, and often through actual experiments by which they were discovered. A thorough introduction to electromagnetic laws is necessary for the non-electrical engineering student to grasp the meaning of such abstract concepts as voltage, flux, electric power, and voltage sources, as well as comprehend the operation and characteristics of inductors and capacitors. Both electricity and magnetism are first introduced through the discussion of forces, with which non-electrical engineers can easily identify. Moreover, interesting examples such as how a common household battery works, how long you can use a Walkman, camcorder, or smoke alarm without changing batteries, what is the cost of energy for batteries vs. wall-outlet ac, and what automobile battery ratings mean, bring an otherwise formal set of principles to everyday life.

Concise summaries of these laws and a detailed tabular comparison of them are provided for courses that wish to briefly cover or omit electromagnetism. A clear transition from electromagnetics to circuit theory is provided in Chapter 4. Interesting and occasionally amusing examples include an automotive lighting system, and what it takes to blow a kitchen fuse. Chapter 5 introduces the frequency domain from a practical point of view: cleaning up a noisy transducer signal by passive linear filtering. In addition, basic ac circuit analysis is made easier by a concise review of complex arithmetic. Chapter 6 provides an understandable discussion of complex power in physical terms. The discussions on power distribution systems and grounding should clear away a lot of the mystery surrounding these systems.

Part 3 begins with motivation for why we need electronics, followed by the principles by which semiconductor devices achieve these goals. Detailed physical explanations of the operation of diodes and transistors are given in terms a non-EE can understand. It is essential to connect these concepts to the physical world if the student is going to "buy" the simplified circuit models used here and in digital electronics. Yet a simple question such as "Why do we *need* holes, not just electrons, to make a practical diode?" is generally left unanswered in textbooks. We answer this and many other intriguing questions in device physics. Applications such as an AM transistor radio and an electronic flash unit show that these devices and electronic circuit analysis really do have relevance in day-to-day living. In Chapter 8 an extensive potpourri of op amp application examples shows the versatility of this essential signal-processing building block. As a reward for previous studies, the reader is treated in Chapter 9 to a host of transducer and signal conditioning applications in mostly nonmathematical terms. Industrial transducers such as the strain gage, liquid crystal display, and smoke/pollution detectors are complemented by home entertainment examples, including audio speakers, microphones, and tape recording heads.

Digital electronics is discussed in chapters 10 through 12. Again, motivation is provided by immediate application to engaging examples. For instance, we consider the use of logic to determine who makes the NFL playoffs. Modern TV remote control on/off switching is introduced as an application of the toggle flip-flop, and a few of the binary signal transmission details involved in TV remote control are examined. The automatic washing machine is studied as an example of state machine design. We investigate what happens when a number is typed into a personal computer. Case studies on the control of microwave ovens and automotive engines by microprocessors provide stimulating material for reading and discussion.

It can well be argued that today's non-electrical engineers may be more likely to encounter electromechanical machines than most EEs. Thus it is essential that a solid presentation of machine principles be included. Basic ideas are conveyed through a first-principles approach in Chapters 13 through 15. Extensive application discussions show how the formal, mathematical ideas actually influence machine selection decisions. These are often the most popular chapters of all with my students.

## PEDAGOGICAL AIDS

The greatest pedagogical aid is accurate and lucid explanations. There are 150 examples that demonstrate how the principles are used in applications. The number of examples per chapter varies to reflect the character of the material (quantitative vs. expository). Substantial end-of-chapter summaries are provided which review and reinterpret the material, as well as show where the text is going. The Appendix to Chapter 4 contains a list of pitfalls for the beginning student to avoid. This list was compiled from mistakes commonly made by non-EE students during exams.

There are over 820 end-of-chapter problems that serve to build skills and evaluate student abilities. Many of these problems involve real-world applications that will interest non-EE students. Some examples are: power drawn by household appliances, lightning-bolt charge, what happens to voltage when you unplug a power cord, magnetic speaker, magnetron in microwave oven, compact disk $R_{eq}$ compared with self-cleaning oven $R_{eq}$, D-cell flashlight, commercial space heater circuit with fan motor, tuning filtering for TV station reception, three-speed oscillating fan, dc long-distance power transmission, automatic nighttime floodlight switch, audio amplifier tone control filtering and multi-stage analysis, voltage multiplier, removal of mother's heartbeat from fetal heartbeat signal, refrigerator thermostat, ultrasonic medical imaging system position transducers, bicycle tachometer, ultrasonic hyperthermia cancer treatment temperature measurement, pH measurement, doppler bloodflow meter, logic used to determine whether a meeting of Chairs of the Board will fail, digital odometer for car, on/off control of microwave oven, and many more. Many problems have multiple parts; they can be broken into two or more shorter problems. The problems sets are broken into two classifications: easy to moderately difficult, and advanced-level.

Finally, a small number of "advanced" sections are marked with an asterisk in the Table of Contents and by a star in the left margin of the text, and are printed in smaller type. These may be omitted without loss of continuity.

An instructor's manual is available to instructors which contains solutions for all of the end-of-chapter problems, as well as supplementary materials that for brevity were excised from the main text. Detailed coverage of the superposition circuit solution method, op amp input and output impedance calculations, direct memory access, introductions to assembly language and C, and V-curves for synchronous machines are located in the instructor's manual. Other miscellaneous materials such as a course syllabus, a lecture outline and overhead transparency masters are also included.

## USAGE OF THIS BOOK

Typical courses are one-semester or two-semester introductions to electrical engineering. Two-semester treatments can leisurely cover the entire book. An effective schedule might be Parts 1, 2, and 5 (fundamental laws, dc and ac circuit analysis, machines) the first semester, and Parts 3 and 4 (analog and digital electronics) the second semester.

Difficult choices concerning what to leave out must be made for one-semester courses. Instructors wishing to maintain depth could largely omit Part 4 (digital electronics) or Part 5 (machines), and cover the remaining four parts in moderate detail. Other compromises can be made, such as covering only dc or only ac machines, or omitting the difficult topic of microprocessors, while still covering the essentials of both Parts 4 and 5. Survey courses can focus on introductory materials in all parts. Usually an instructor naturally sees certain areas as most vital and interesting, and will spend more time with them. Indeed, prioritizing and selecting from a profusion of topics was a major task in writing this book.

Comments and/or errata may greatly help to improve future editions of this text. The e-mail address of the author is: Cavicchi@gcc.edu

To receive an updated errata file, include the following in your message: send errata for fundamentals of ee

## ACKNOWLEDGMENTS

I have had valuable discussions regarding this work with many members of the University of Akron Department of Electrical Engineering and others. These include James Grover, Fadi Sibai, and Alva L. Couch on digital electronics and microprocessors; Gordon Danielson, Richard E. Cavicchi, Richard Nemer, Greg Lewis, Michael Radanovich, and C. F. Chen on electronic devices; Mark Viola, Nathan Ida, and Elizabeth M. Cavicchi on electromagnetics; Milton Kult on a variety of aspects of circuit theory; and Donald Zinger, Malik Elbuluk, and Robert Grumbach on electromechanical machinery. For development of homework sets, I have been fortunate to receive help from students Michael Radanovich, Xiaoqing

Chen, and David Boll, and from Professor Tom B. Stenis of Texas Tech University. I have received helpful suggestions from those who reviewed early versions of the manuscript for this book: Tom B. Stenis, Joseph A. Coppola of Syracuse University, and Zoran Gajic of Rutgers University. However, I assume all responsibility for the technical accuracy of this book. Most of all, I have received much moral support from my department head, Dr. C. S. Chen, and my Prentice Hall editors, Elizabeth Kaster, Alan Apt, Sondra Chavez and Joe Scordato.

*Thomas J. Cavicchi*

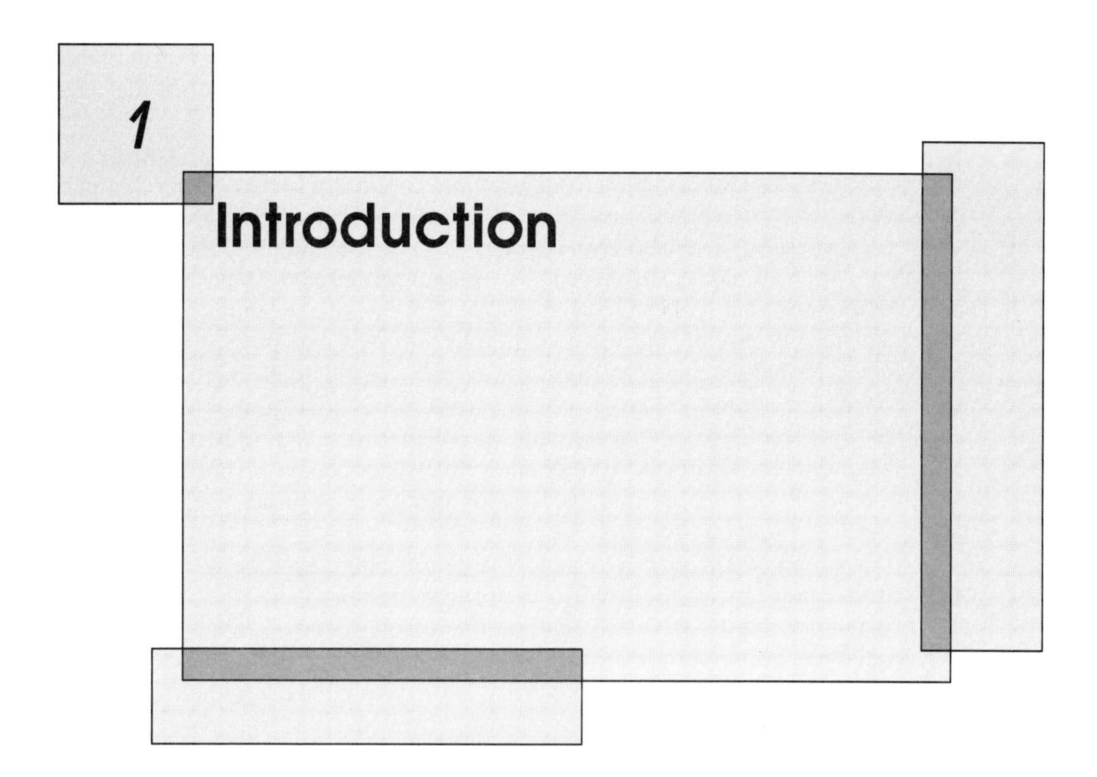


## 1.1 WHY STUDY ELECTRICAL ENGINEERING?

Why on earth would a person who is not planning to become an electrical engineer need to learn about electrical engineering? Of what practical use is circuit analysis? It is an undeniable fact that nearly all aspects of the average American's lifestyle in the twenty-first century will depend on the results of electrical engineering. So on a personal level, the first question may fairly be reworded, "Am I curious about my world?"

Think about it. It is 6:30 A.M. My peace is rudely interrupted by an irritating radio ad. After turning off the radio and noting from the digital clock that it is now 6:45, I get up and turn on an electric light. After a nice hot shower (electrically heated in my case), I use my electric shaver and get dressed. Then I throw some bread in the toaster and (carefully) cook my eggs in the microwave oven. All this time, my new electronically controlled heat pump is keeping my home cozy.

I glance at my digital watch and it is getting late. My major exercise for the day is about to begin: A sprint from the kitchen to the garage. Fortunately, I just got a new battery, so the car starts right up. Little do I know that the ignition timing, fuel injection, suspension, transmission, and braking are all under computer control. But I *am* aware 15 minutes later that the electronic display panel is telling me that the gas tank is just about empty. That will have to wait.

Into my cubicle I go (needless to say, *these* lights were turned on long ago), only to find a pile of FAXs needing immediate attention. I write up some memos on my personal computer, get a laser printer hardcopy, make copies of that in the copy room for other staffers, and FAX back my response. There are also several phone messages on my answering machine that need attention. Some of these I will take care of on the way home with my new cellular phone.

Now, however, I am called to the floor. Over there are some electrical engineers (EEs) who need information from me about conditions for maximum chemical reaction rates of certain liquid waste materials resulting from our manufacturing. (In this story, I am a modern chemical engineer.) Someone had an idea about using ultraviolet rays to break down these toxic organic wastes into water and harmless acidic compounds. The EEs have been assigned as part of the team to develop the required electronic monitoring and control for the three-stage process. Eventually, the process is to be fully automated. In fact, because of the public relations benefits and tax incentives, our company is giving several of us time to work on this.

Back in the office, I continue working on the presentation for Thursday. I justify the budget for my project using some spreadsheet-generated projections, and present some results on pie and bar graphs using my graphics package. Color hardcopies, of course, go to my bosses.

After a long and productive day I drive home, singing all the way along with the great jazz musician, Jelly Roll Morton, on my on-board cassette deck. On the way, I remember that I need gas. I read on the digital gas pump that while before, the gas tank was empty, now my wallet will be. Thus, instead of going to the record store to buy a new CD, I go to the library and borrow it—using their computer catalog to see what else they own and using my bar-code library card to check out the CD. I also check out a VCR tape of an old classic mystery.

Once home, I turn on the news with my remote control "zapper," with which I victoriously mute the commercials. Simultaneously, my new computer-controlled washing machine does the laundry. After the news, I spend 20 minutes on the exercycle with digital readout and computer-generated difficulty settings. I try out my new three-party phone option to boast about my new baby niece. Then I make microwave popcorn and settle down with the CD. All in a day's work.

Although the scenario above may not be conducive to intellectual and spiritual growth, it does represent the lifestyle of an increasing fraction of the American engineering work force. Every aspect of the day involved electric or electronic equipment. Most engineers are by nature curious. Reading this book will help begin to satisfy and stimulate that curiosity. Hopefully, it will prompt further study beyond what can be done in a single introductory volume.

In this book we describe some of the important ideas behind the electric-powered products mentioned in our story, as well as many others. It can, however, only scratch the surface of the complexity of modern electronics. Any introductory science course must start with the fundamental physical laws. These laws comprise experimental results quantified and generalized by mathematical models. From the laws, devices and machines can be designed that do useful things in our lives, at home and at work.

If you follow through this entire book, you will be well prepared to communicate with electrical engineers in situations such as the one the chemical engineer faced in the story above. Having a non-electrical engineering background, you bring a unique perspective to an applications situation that the electrical engineer does not have. By learning about electrical engineering concepts and devices, you will be able more readily to use this perspective in intelligent ways that can bring you success, fame, and international awards—or at least pass the "Fundamentals of Engineering" exam.

## 1.2 TYPES AND LEVELS OF STUDY

As any basic EE textbook will tell you in its introduction, the applications of electrical engineering are far and wide. No such endless litany will be reproduced here. However, there are a few main categories into which electric or electronic circuits may be placed. Fundamentally, there are *information* (*signal*) *processing* circuits and *energy processing* circuits.

Information processing circuits generate, combine, and transmit information, or convert it into other useful forms. For example, measurements are taken using *sensors*. The electrical representation of these measurements may be transmitted over wires and then be converted to forms suitable for a computer to store and process. The computer may then further convert this information into pictorial form using graphics programs within the computer. Also, appropriate controls may be generated to change the behavior of the measured physical system. This control

Back in 1893, Tesla (see inset) dreamed of the long-distance transmission of information and power; both of these essential tasks are still being refined today. He also evidently dreamed of seating himself amidst his magical laboratory-lightning; wisely, he chose to realize this dream by trick photography. *Sources:* Courtesy of the Burndy Library, Norwalk, CT. (main photo). Smithsonian Institution photo No. 52223 (inset).

information is again converted into a form suitable to drive an *actuator* that physically affects the system. Much more will be said about these circuits throughout this book.

Energy processing systems store, transmit, and/or convert energy from one form to another. For example, a power line transmits electric power from the power plant to the user's building. A light bulb converts low-frequency or constant electric energy into heat and high-frequency radiated energy that we recognize as light. Electric heaters are similar to light bulbs but emphasize the generation of material heating (via phonons) much more than visible light (photons). A motor converts electric energy into mechanical energy, and a generator does the reverse. Naturally, once the electric energy is converted to rotational energy, further mechanical systems can be used to convert the rotations into pumping action, linear motion, lifting, pressing, and so on.

As is the case in many branches of science, there is a wide range of levels of study within electrical engineering. We may consider the overall system configuration and operation, of which electric circuits may be only a portion. Even within this categorization, we could consider a pocket AM radio a complete system or the entire national telecommunications network as a single system.

At the other extreme is the atomic, quantum mechanical description of electronic devices. Although this level survives the most exacting scrutiny, it is impractical except to the designer of electronic devices. The circuit level is a compromise: Models of circuit elements or subsystems are used for representing the terminal behavior of the device.

Such *models* are chosen to be sufficiently accurate to reliably reflect the behavior of the device in a particular application. Usually, the more accurate the model, the more complicated it is. We choose the simplest model that still incorporates the important physical behavior observed. If the measured quantities are determined to an accuracy of the order of $10^{-3}$ units and the model for the processing circuit is accurate to within 10 units, a better model is needed. If the model is accurate to within $10^{-9}$ units, perhaps a simpler model would suffice.

The set of models used in this book is the pragmatic, self-consistent set commonly used by electrical engineers. Depending on the application, the tolerances for values are typically 10% to less than 1%. Usually, this precision is sufficient for overall prediction of behavior. It might be concluded that a system can be no more accurate than its highest-tolerance device. However, this may not be true when the values of some components turn out to be more critical than others in determining behavior. The only way to know for certain is to perform error calculations, which can be a major undertaking in circuit design.

Our goal in the following pair of chapters (Chapters 2 and 3), however, is merely to introduce the major electric and magnetic quantities and relations that are responsible for observed macroscopic behavior of all the devices considered in this book. Having this strong foundation to call upon, there should be no "surprises" later, except possibly in the form of refinements.

Keep your chin up when you see a vector or integral sign in the following pages. Take comfort in the fact that these "founding fathers" equations in our practical usage will almost always be simplified to scalar quantities in algebraic equations. *Without having seen* the defining integral forms, the physical meanings of the laws are lost and "knowing" the equations becomes rather a memorization game. *Having seen them* and the vectors involved provides a true picture and physical insight. Such insight can, for example, enable you to predict not only that the motor of an electric car will spin when you turn it on, but whether the car will go forward or backward. We all know the importance of being certain about this!

Let us now begin this fascinating journey that will help us understand the modern world around us. Maybe in the process we shall gain an appreciation of the miracle of human genius, the limitations and barriers the world imposes on it, and where we as individuals should stand in light of where this technological revolution is pushing us.

# Part One: PHYSICAL LAWS

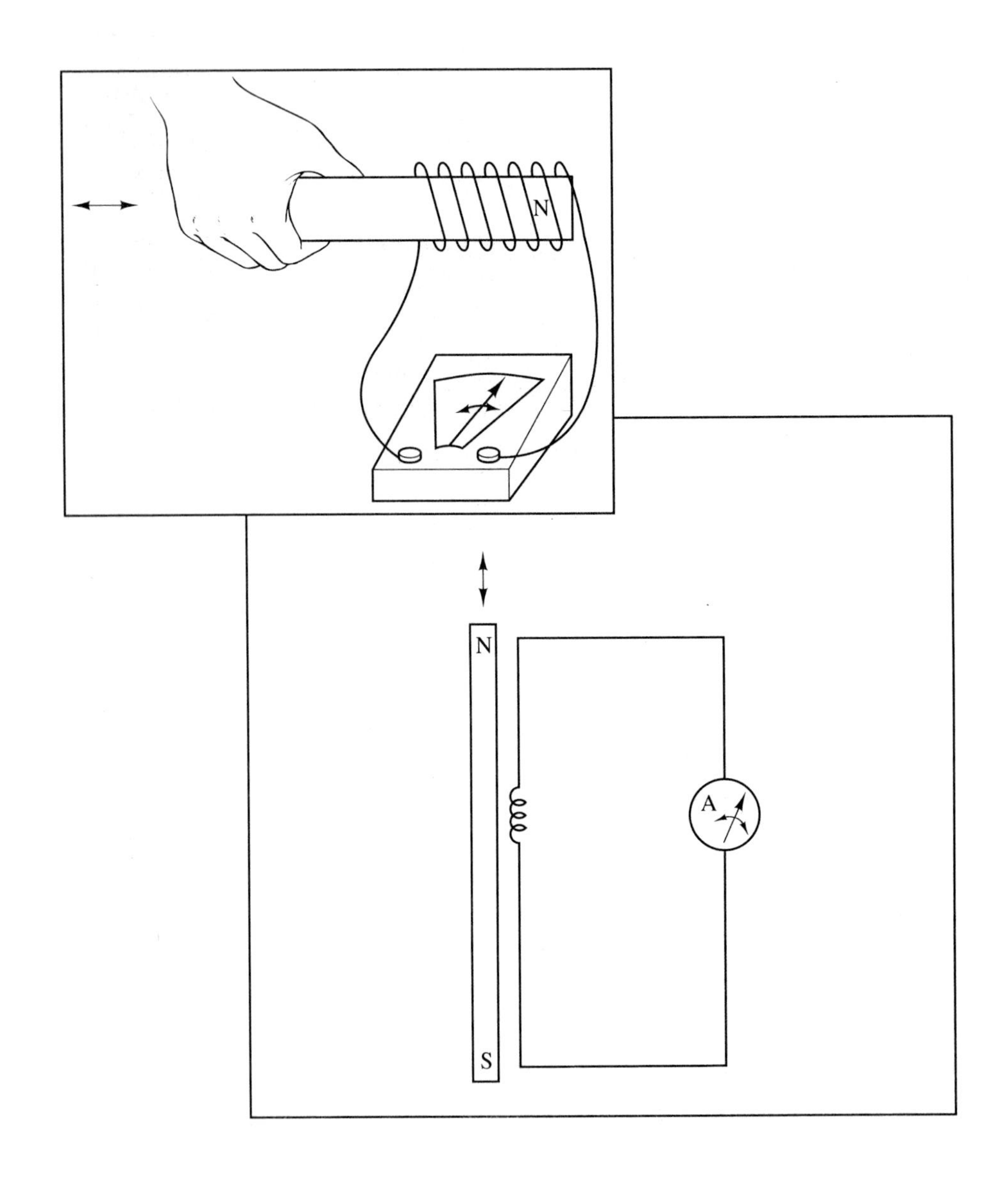

# 2

# Physical Laws I: Electricity

## 2.1 INTRODUCTION

About 2600 years ago, a Greek philosopher named Thales discovered that when a soft, yellow stone called amber was rubbed with cloth, it could pick up feathers, straw, and other light objects. (Today's most familiar example of this electrification is "static cling" between clothes.) This was a strange kind of force, because it could be felt at a distance. The only other known force like this was gravity. In 1570, William Gilbert of England named this phenomenon *electricity*, after *elektron*, the Greek word for amber. But it was not until more than two hundred years later that quantitative studies of electricity began.

## 2.2 ELECTRIC CHARGE AND COULOMB'S ELECTRIC FORCE

Meanwhile, it soon was discovered that there were two "kinds" of electricity: An electrified object attracted some things but repelled other objects electrically charged the same way it was. In 1746, Benjamin Franklin proposed calling these "positive" and "negative" electric charges. This property of matter called *electric charge* is ultimately responsible for all of the effects to be described in this book. We shall have much more to say about it as we go along.

Slightly earlier, in 1729, Steven Gray of England formalized the common idea that "electricity" (actually, electric charge) could be transmitted from one object to another via some materials but not via others. Because of the analogy with heat, the

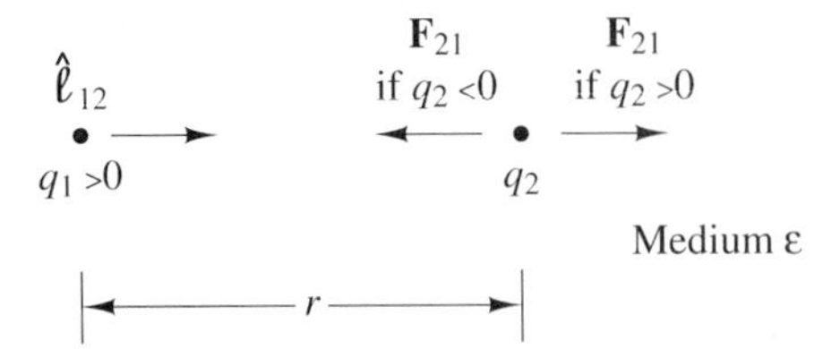


**Figure 2.1** Coulomb force between two charges.

former were called *conductors* and the latter *insulators*. Today, every material is characterized in this sense by its electrical conductivity $\sigma$.

Finally, in 1785 Charles Augustine de Coulomb performed an experiment in which small pith balls were electrically charged. One ball was held in place while the other was on the end of the lever arm of a torsion (torque-measuring) balance—a device he invented. He found that the force between the electric charges was proportional to the amounts of charge on each ball and inversely proportional to the square of the distance between them (see Fig. 2.1). The force was repulsive if the charges were alike and attractive if the charges were opposite. In today's notation, *Coulomb's law* has the form

$$\mathbf{F}_{21} = \frac{q_1 q_2}{4\pi\epsilon r^2}\hat{\boldsymbol{\ell}}_{12} \tag{2.1}$$

where $\mathbf{F}_{21}$ is the force vector on charge 2 due to charge 1 in N (newtons, after Sir Isaac), $q_1$ and $q_2$ are the two charges in C (coulombs, after Charles), $r$ is the distance in m (meters) between the charges, $\hat{\boldsymbol{\ell}}_{12}$ is the unit vector pointing from charge 1 to charge 2, and $\epsilon$ is the "fudge factor" needed to make the numerical value of the right-hand side newtons. As a point of reference, the smallest denominations of charge in the real world are *electrons*, each having a charge of $-e = -1.602 \cdot 10^{-19}$ C. Thus in 1 C there are $6.24 \cdot 10^{18}$ electrons.

The last parameter defined above, $\epsilon$, is named *permittivity*. It is the only parameter in Eq. (2.1) that characterizes the medium between the charges. In a vacuum, as well as in air, $\epsilon = \epsilon_0 = 8.85 \cdot 10^{-12}$ $C^2/(N \cdot m^2)$ or F/m, where F is farads (after Michael Faraday), defined in Sec. 2.7. More will be said about this strange number when we study magnetic fields.

Permittivity is different from conductivity, another electrical parameter of material that was mentioned above and is discussed further in Sec. 2.9. Recall that conductivity indicates how easily electric charge can be made to pass through the material. For example, a good conductor will transfer a given amount of charge very quickly, while a poor conductor exposed to the same electric force will take a much longer time to transfer the same amount of charge.

Permittivity, contrastingly, indicates how readily the molecules of material become aligned under the force due to a given external charge. The electric force stretches each molecule so that one end becomes positively charged and the other negatively charged. This process is called *polarization*. Because all internal polarized molecules have both positively and negatively charged ends, the material is called a *dielectric* (double-electric). The dielectric is an insulator, because while the electrons of the atoms can be displaced by an electric field, they are bound to the atom. In a conductor, the electron is free to move from atom to atom.

If polarization happens easily, there will be a large uncanceled charge on the surface of the material. Refer to Fig. 2.2. On the surfaces nearest the external charge,

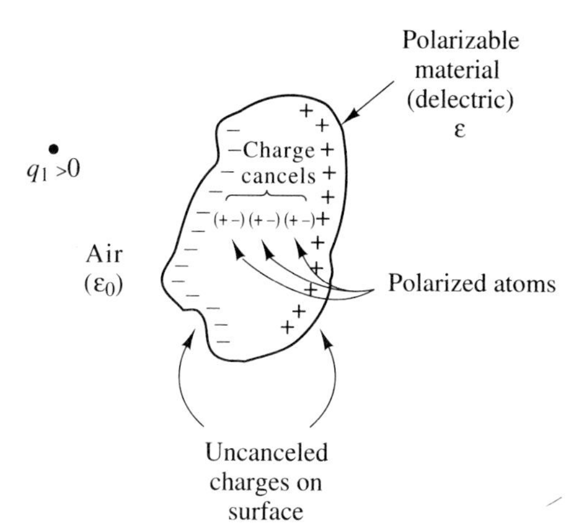


**Figure 2.2** Dielectric in the presence of a point charge $q_1$.

this induced surface charge will be of type opposite that of the external charge due to the attraction between opposite charges. On the surfaces facing away from the external charge, the induced surface charge will be the same as that of the external charge. Within the material, the adjacent polarized molecules cancel each other's charge; the interior material thus behaves like uncharged (electrically insulating) material.

An easily polarizable material is called a good dielectric and has high permittivity (because it readily *permits* polarization). If the medium in which the charges in Eq. (2.1) are immersed is a strong dielectric, the effects of each charge are reduced. This is true because the charge on the surface of the dielectric partly cancels the charges $q_1$ and $q_2$. Therefore, if $\epsilon$ is large, the force $\mathbf{F}_{21}$ is reduced in magnitude, as the presence of $\epsilon$ in the denominator of Eq. (2.1) indicates. This will be shown explicitly in Fig. 2.7.

To avoid having to remember the value of $\epsilon_0$ when classifying dielectrics, the relative permittivity has been defined as $\epsilon_r = \epsilon/\epsilon_0$ for a material having permittivity $\epsilon$. Thus a strongly dielectric (polarizable) material will have $\epsilon_r \gg 1$, while for all dielectrics, $\epsilon_r > 1$. For metals, air, and vacuum (*free space*), $\epsilon_r = 1$. Values of $\epsilon_r$ for other materials of interest in circuit components include 2 to 8 for quartz, Bakelite, paper, and mica, 10 for aluminum trioxide, 81 for distilled water, and 25 to 10,000 for tantalum oxides and titanate (titanium dioxide) porcelains.

Incidentally, the static electricity effects originally observed with amber were due to charge rubbing off the cloth onto the amber as a result of the frictional contact during rubbing. Once charged, the amber could attract other objects according to Eq. (2.1).

**Example 2.1**

A negative point charge $q_1$ of magnitude 2 μC (i.e., 2 microcoulombs, or $2 \cdot 10^{-6}$ C) is placed in a vacuum. Find the force on a point charge $q_2 = 1$ pC (i.e., 1 picocoulomb, or $10^{-12}$ C) a distance 1 mm from $q_1$.

**Solution** From Eq. (2.1) we have

$$\mathbf{F}_{21} = \frac{q_1 q_2}{4\pi\epsilon_0 r^2}\hat{\boldsymbol{\ell}}_{12} = \frac{2 \cdot 10^{-6} \cdot 10^{-12}}{4\pi \cdot 8.85 \cdot 10^{-12} \cdot 10^{-6}}\hat{\boldsymbol{\ell}}_{12}$$
$$= 0.018\hat{\boldsymbol{\ell}}_{12} \text{ newtons.}$$

Before concluding this section, a practical application well known to environmental engineers is the electrostatic precipitator. In these systems, polluted gas to be purified (particles removed) is directed to a chamber bounded by two electrodes having a huge electric potential difference between them (and thus a strong electric field—see Secs. 2.3 and 2.6 for precise definitions of electric field and potential).

The field is made strong enough to partly ionize the fine particles in the air without sparking, a condition known as corona discharge. Once ionized (charged), the particles experience Coulomb's force and are driven to the collector electrode (at ground potential), at which point they fall into a dust hopper. These purifiers require little power, do not significantly reduce the air pressure in the supplying conduit, and are 99% effective.

## 2.3 THE ELECTRIC FIELD

Because the force in Eq. (2.1) acts at a distance, and exists and varies throughout space, $\mathbf{F}_{21}$ may be termed a *force field*. Notice that if $q_2$ is considered a "test" charge, it experiences the force $\mathbf{F}_{21}$ at a given location due to $q_1$. The force on the test charge is proportional to its own charge, $q_2$. It is convenient to define a field quantity that is independent of the value of the test charge $q_2$. This can be done by defining the *electric field intensity vector*

$$\mathbf{E} = \frac{\mathbf{F}}{q}, \tag{2.2}$$

where in this discussion $q = q_2$. The units of $\mathbf{E}$ are N/C, although we shall see below that the more common units for $\mathbf{E}$ are volts/meter, with volts (V) defined below. This electric field is equivalently the force vector that a unit charge (1 C) would experience in the presence of $q_1$. From Eqs. (2.1) and (2.2),

$$\mathbf{E}_{21} = \frac{q_1}{4\pi\epsilon r^2}\hat{\boldsymbol{\ell}}_{12} \tag{2.3}$$

is the electric field at location 2 (an observation point) due to charge 1, where now $r$ is the distance and $\hat{\boldsymbol{\ell}}_{12}$ the unit vector from charge $q_1$ to our observation point. Because this direction is always radial, we may write $\hat{\boldsymbol{\ell}}_{12} = \hat{\boldsymbol{\ell}}_r$, where $\hat{\boldsymbol{\ell}}_r$ is the unit vector in the outward, radial direction originating from $q_1$. Note that division of Eq. (2.1) by $q_2$ in Eq. (2.2) indeed normalizes the effect of $q_1$ at the location of test charge $q_2$ (removes from the field due to $q_1$ the dependence on the test charge $q_2$).

**Example 2.2**

Under conditions of Example 2.1, find the electric field due to $q_1$ at $q_2$.

**Solution** From Eq. (2.2), $\mathbf{E}_1 = \mathbf{F}_{21}/q_2 = (0.018/10^{-12})\hat{\boldsymbol{\ell}}_{12} = 1.8\text{-}10^{10}\,\hat{\boldsymbol{\ell}}_{12}$ V/m.

The astute reader will question why the force on the test charge is due only to the electric field that would exist were the test charge absent. Clearly, the electric field from the test charge dramatically influences the total electric field at the test charge—it is infinite there! Our only answer is that ignoring this component of the field produces observed results. Indeed, what would be the direction of this infinite force? In this case, the appropriate unit vector is $\hat{\boldsymbol{\ell}}_{22}$, which is undefined!

## 2.4 LINES OF ELECTRIC FLUX

Pictorially, we may represent **E** by *flux lines* with arrows. The terminology *flux* stems historically from the old idea that electricity was a "fluid," continually flowing out of charges just like water sprayed out from an omnidirectional sprinkler head. In this context, "flux" means the time rate of transfer of fluid across a given surface. The magnitude and direction of the "velocity" of this flow at any point is represented by the electric field at that point. From Eq. (2.3) we conclude that the flux lines emanating from a positive charge point radially outward, and those from a negative charge, radially inward. The fluid analogy is still helpful today if we remove the idea of a "time rate" from the discussion. We may instead think of the flux lines as possible paths on which a point charge would travel under the influence of this electric field.

Because the magnitude of **E** is proportional to the charge producing it, the total number of flux lines emanating from a point charge is chosen to be proportional to the amount of charge producing that flux. The spatial *density* of lines at any point in space indicates the magnitude of **E** at that point, and the arrows indicate the direction. Figure 2.3 illustrates Eq. (2.3), showing several sample flux lines as they pass through a spherical surface $S_1$. For clarity, each arrow tip was chosen to be drawn at the point of intersection of the line with $S_1$. The lines become sparser far from $q_1$ just as $1/r^2$ in Eq. (2.3) does, and infinitely dense at the point charge. See the problems for consideration of the limits of applicability of the Coulomb electric field, Eq (2.3).

An amazing fact can now be appreciated. Refer to Fig. 2.4, which shows an additional, larger concentric spherical surface $S_2$ also having $q_1$ at its center. The number of lines passing through $S_2$ equals the number passing through $S_1$. The flux lines are uniformly distributed because of the spherical symmetry of the point

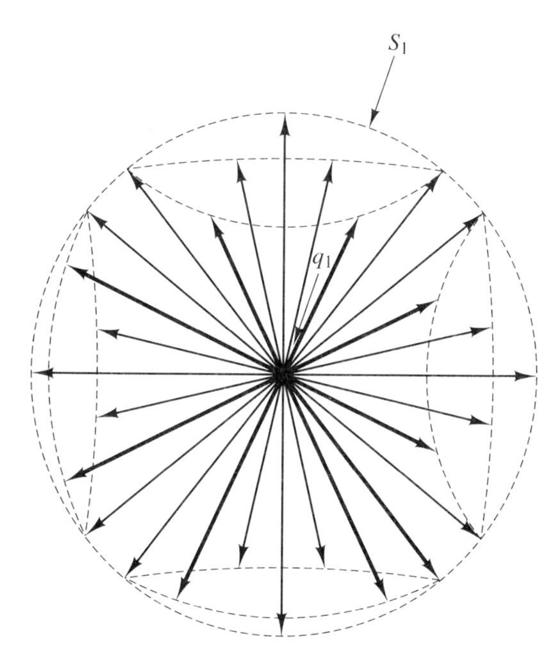


**Figure 2.3** Electric flux lines from a charge through a spherical surface centered on the charge.

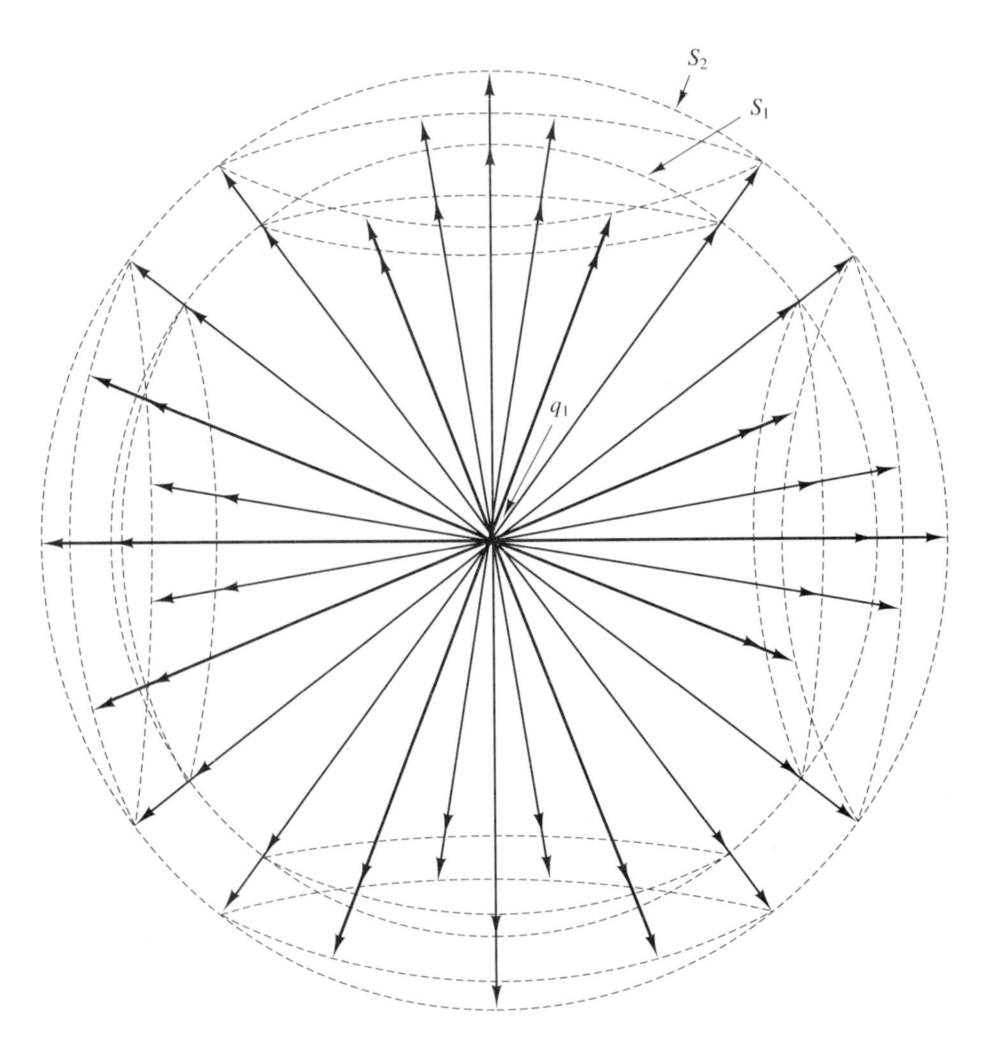


**Figure 2.4** The number of flux lines passing through one sphere equals that through a concentric sphere (charge at center).

charge. Therefore, the total number of lines passing through a spherical surface of *any* size having $q$ at its center is the same and is equal to the density of lines at the surface times the surface area.

The density of lines is proportional to the magnitude of the electric field, by definition. Recall that the surface area of a spherical shell of radius $r$ is $A = 4\pi r^2$. Therefore, the total number of lines is proportional to the product of $E$ and $A$:

$$\text{number of flux lines proportional to } EA = \frac{q_1}{4\pi\epsilon r^2} 4\pi r^2 = \frac{q_1}{\epsilon}. \tag{2.4}$$

Here and elsewhere, the symbol of a vector quantity such as $E$ shown in italic type rather than boldface implies the magnitude of the vector.

This relation was generalized early in the nineteenth century by Karl Friedrich Gauss of Germany. Again, the analogy with fluid flow is used, where fluid mass

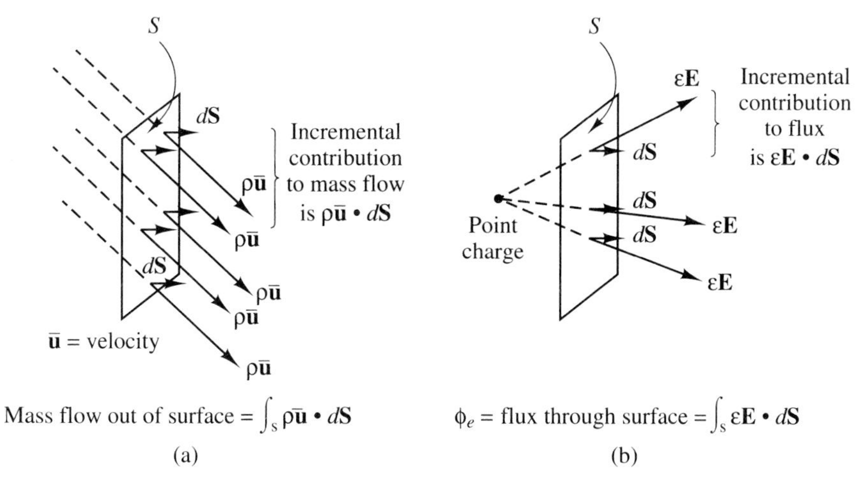


**Figure 2.5** (a) Mass flow through an aperture (open surface); (b) electric flux through an aperture.

density times the velocity vector is analogous to $\epsilon\mathbf{E}$. Referring to Fig. 2.5, the total *flow* or electric flux $\phi_e$ out of a surface is the integral of the component of the "mass density times velocity" perpendicular to the surface (see Fig. 2.5a; in our case, the perpendicular component of $\epsilon\mathbf{E}$, as in Fig. 2.5b):

$$\int_S \epsilon\mathbf{E}\cdot d\mathbf{S} = \phi_e. \tag{2.5}$$

In Eq. (2.5), "$\cdot$" is the dot product between vectors, and the magnitude of the vector $d\mathbf{S}$ is an incremental surface area element of $S$, while the direction of $d\mathbf{S}$ is the defined outward perpendicular ("normal") to that incremental surface element of $S$. The dot product is needed because to the degree that $\mathbf{E}$ is not perpendicular to the surface, the effective area through which "flow" can proceed along $\mathbf{E}$ is reduced. See Fig. 2.5a for an example of this reduction in effective area.

If the surface is closed (Fig. 2.6), the analogy with fluid mechanics can again be used, where now conservation of mass in a control volume is relevant. The total flow out of the surface must be equal to that provided by all "sources" within the surface. Hence

$$\oint_S \epsilon\mathbf{E}\cdot d\mathbf{S} = q_1, \tag{2.6}$$

where the circle around the integral sign indicates a closed surface and where $q_1 = \phi_e$ for a closed surface. That is, the total electric flux piercing the surface is equal to the enclosed charge. Equation (2.6) is known as *Gauss's law*, in honor of its discoverer.

Note that flux lines emanating from any charge outside $S$ provide no net contribution to the integral in Eq. (2.6). Any flux line going out of $S$ also passes in (a negative outgoing contribution) somewhere else on $S$, unless the charge producing that line is within $S$. The electric field associated with Coulomb's law, Eq. (2.3), is actually a special case of the more general Gauss's law, Eq. (2.6) (see the problems). Thus Eq. (2.6) is consistent with what we already know.

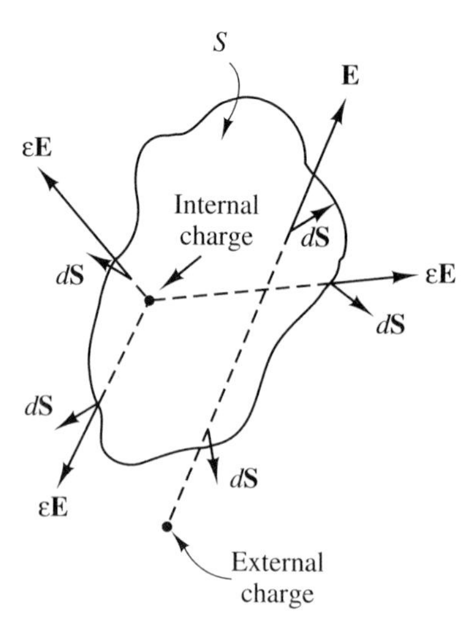


**Figure 2.6** External charge makes no net contribution to outgoing flux, whereas internal charge does.

Now note that Eq. (2.3) can be rewritten as

$$\mathbf{E}_{21} = \frac{q_1 + q_b}{4\pi\epsilon_0 r^2}\,\hat{\boldsymbol{\ell}}_{12}, \tag{2.7}$$

where

$$q_b = -q_1\left(1 - \frac{1}{\epsilon_r}\right) \tag{2.8}$$

is the effective induced bound surface charge in the dielectric. The charges making up $q_b$ are called *bound* because they are not free to move as are, for example, free conduction electrons. Instead, these are the net charges on the surface molecules of the dielectric material and are immobile, due to molecular forces. The surface of the dielectric is in this case that which envelopes $q_1$ (see Fig. 2.7). The word "effective" above means that $q_1 + q_b$ would create the same field in free space ($\epsilon = \epsilon_0$) that $q_1$ does in dielectric material with $\epsilon = \epsilon_r\epsilon_0$. Notice that because $\epsilon_r > 1$, $q_b$ always has the opposite sign of $q_1$, so indeed the dielectric reduces the electric field. That is, the charge $q_1$ "looks like less" when immersed in a dielectric.

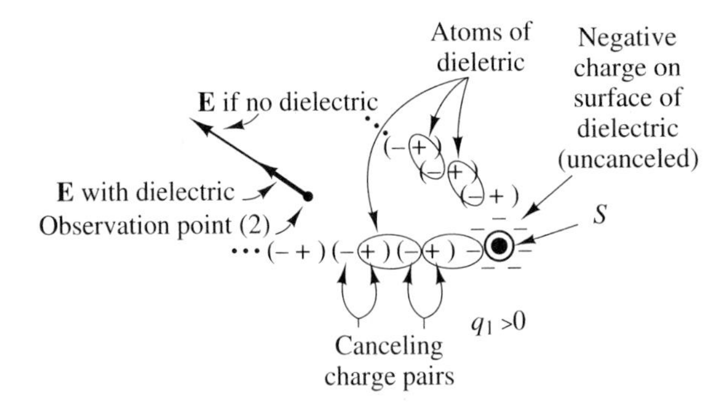


**Figure 2.7** Induced bound charge on the surface $S$ of a dielectric enveloping a positive free charge.

## 2.5 DECOMPOSITION OF THE ELECTRIC FIELD INTENSITY VECTOR

In Eqs. (2.7) and (2.8) we decomposed the net charge responsible for the observed electric field into two components: the free charge $q_1$ and the bound charge $q_b$. We may equivalently decompose the observed electric field itself into two components: a field **D** due only to free charges and a field **P** due only to bound charges. Using Eqs. (2.3), (2.7), and (2.8), write

$$\mathbf{E} = \frac{q_1}{4\pi\epsilon r^2}\hat{\boldsymbol{\ell}}_{12} \tag{2.9a}$$

$$= \left(\frac{q_1}{4\pi\epsilon_0 r^2} + \frac{q_b}{4\pi\epsilon_0 r^2}\right)\hat{\boldsymbol{\ell}}_{12} \tag{2.9b}$$

$$= \left[\frac{q_1}{4\pi\epsilon_0 r^2} - \frac{q_1(\epsilon_r - 1)}{4\pi\epsilon_0 \epsilon_r r^2}\right]\hat{\boldsymbol{\ell}}_{12} \tag{2.9c}$$

$$= \frac{1}{\epsilon_0}(\mathbf{D} - \mathbf{P}), \tag{2.9d}$$

where

$$\mathbf{D} = \epsilon_0\left[\frac{q_1}{4\pi\epsilon_0 r^2}\hat{\boldsymbol{\ell}}_{12}\right] = \epsilon_0 \begin{pmatrix}\text{component of electric field}\\ \text{due only to free charges}\end{pmatrix} \tag{2.10}$$

and because $\epsilon_0\epsilon_r = \epsilon$,

$$\mathbf{P} = \epsilon_0\left[\frac{q_1(\epsilon_r - 1)}{4\pi\epsilon r^2}\hat{\boldsymbol{\ell}}_{12}\right] = -\epsilon_0 \begin{pmatrix}\text{component of electric field}\\ \text{due only to bound charges}\end{pmatrix} \tag{2.11}$$

It is important to recognize that **D** is also equal to $\epsilon$ times the *total* electric field [just simplify Eq. (2.10)]:

$$\mathbf{D} = \frac{q_1}{4\pi r^2}\hat{\boldsymbol{\ell}}_{12} \tag{2.12a}$$

$$= \epsilon\mathbf{E}. \tag{2.12b}$$

From Eq. (2.12a) we see that **D** in this case is independent of the medium $\epsilon$; it has nothing to do with bound charge or polarization. All those effects are contained in the *electric polarization vector* **P**. [Strictly, only the component of **D** normal to the interface between media is independent of the media, so other components of **D** actually depend on bound as well as free charges. Because of the spherical symmetry of this problem, **D** has only a normal component to the interface between the point charge and the dielectric medium; so in this case, **D** is completely independent of the medium. In general, Eqs. (2.12b) and (2.13) below define **D** and are always true.] Also, because $\epsilon_r \geq 1$, Eq. (2.11) shows that **P** has the same direction as **E** (as does **D**). However, from Eq. (2.9d) this means that the effect of **P** is to reduce the field strength **E** from what it would be were $\epsilon_r = 1$. This conclusion agrees with that of our earlier discussion in Sec. 2.2 of the effect of $q_b$: The field within a strong dielectric is reduced. In the same way that Eq. (2.12b) was obtained, from Eq. (2.11) we find that $\mathbf{P} = (\epsilon - \epsilon_0)\mathbf{E}$.

Finally, we substitute Eq. (2.12b) into Gauss's law, Eq. (2.6), to obtain its most commonly stated form,

$$\oint_S \mathbf{D} \cdot d\mathbf{S} = q_{\text{enc}}, \tag{2.13}$$

where in the discussion above, the charge $q_{\text{enc}}$ enclosed within the closed surface $S$ is $q_1$. Because of the direct relation between $\mathbf{D}$ and electric flux ($q_{\text{enc}}$) without any conversion constants, $\mathbf{D}$ is called the *electric flux density vector.* It must be stressed that $q_{\text{enc}}$ is the *free*, not the bound charge within the surface $S$. [Using the representation Eq. (2.7) rather than Eq. (2.3) for $\mathbf{E}$, we *could* instead write Eq. (2.6) as $\oint_S \epsilon_0 \mathbf{E} \cdot d\mathbf{S} = q_1 + q_b$, but this is not usually done.] It might be said that $\mathbf{D}$ "converts" any medium to yield $q_{\text{free}}$ upon its integration over a closed surface.

## 2.6 ELECTRIC POTENTIAL (VOLTAGE)

Given that the electric (Coulomb) force exists and can be measured, it follows that it should be able to do some work. After all, the work $W$ in J (joules, after James Prescott Joule, a British physicist of the nineteenth century whose law is described below) done in acting *against* a force $\mathbf{F}$ from point $a$ to point $b$ is

$$W_{ab} = -\int_a^b \mathbf{F} \cdot d\boldsymbol{\ell}, \tag{2.14}$$

where $d\boldsymbol{\ell}$ is the path increment and direction and "•" is again the dot product (above, "against" is the reason for the minus sign; "by" would give no minus sign). Therefore, we may define the work to move a unit positive charge from point $a$ to point $b$ against Coulomb's force as

$$V_{ab} = -\int_a^b \mathbf{E} \cdot d\boldsymbol{\ell}, \tag{2.15}$$

where $V_{ab}$ is called the change in electric potential, or the *voltage* between points $a$ and $b$. The minus sign reflects the fact that the potential of the positive charge falls in the direction that $\mathbf{E}$ points. The units of voltage are J/C, or V (volts, after Alessandro Volta, discussed below). We may accordingly define absolute potentials $V_a$ and $V_b$ with respect to an arbitrarily assigned common zero absolute potential, analogous to elevations defined with respect to sea level. Only the difference between potentials is important and is calculated in Eq. (2.15). The rise in potential from $a$ to $b$, $V_b - V_a$, is the work done to move a unit charge from $a$ to $b$, namely $V_{ab}$; that is, $V_{ab} = V_b - V_a$.

An important point of interest is that the Coulomb force is conservative. Recall that a conservative force is one for which the work done in moving from point $a$ to point $b$ is independent of the path taken. The adjective "conservative" is used because energy is conserved on a round trip; for example, on the way out you must work against the force, but on the way back it pushes you home. In equation form, where either $\mathbf{E}$ or $\mathbf{F}$ may be used,

$$\oint_C \mathbf{E} \cdot d\boldsymbol{\ell} = 0, \tag{2.16}$$

where the circle around the integral sign indicates that the contour $C$ is a closed path.

The reason the electric force is conservative is that $\mathbf{F}_{21} = -\mathbf{F}_{12}$ (due to the fact that $\hat{\ell}_{12} = -\hat{\ell}_{21}$). An example of a nonconservative force is friction, for which $\mathbf{F}_{21} = \mathbf{F}_{12}$; the one pushing fights against the force both ways. Another is a *voltage source*, which involves a non-Coulomb electric force that we discuss in Sec. 2.8.

We now more formally define absolute potential. Because the electric force is conservative, we may without error choose a path in Eq. (2.15) that goes first to a third reference point $R$. This point $R$ is defined to have absolute potential $V_R$ of zero volts: $V_R = 0$. Thus. $V_a \equiv -V_{aR} = -(0 - V_a)$ and $V_b \equiv -V_{bR}$ are the *absolute potentials* of points $a$ and $b$. Consequently, $V_{ab}$ may be decomposed as

$$V_{ab} = V_{aR} + V_{Rb} = V_{aR} - V_{bR} = V_b - V_a. \tag{2.17}$$

Defining absolute potentials, always with respect to an assumed *zero absolute potential* reference point, is common practice in circuit analysis. Such a point is informally referred to as a *ground*, a subject we treat in detail in Sec. 6.8 and other later sections.

**Example 2.3**

Find the voltage between a point $r_1 = 1$ m from charge $q_1 = +1$ μC and another point $r_2 = 2$ m from $q_1$. Does your answer depend on whether these points lie on the same radial ray from $q_1$?

**Solution** Define the point of zero absolute potential to be infinitely far from $q_1$, because the electric field and force there are zero. Then the absolute potential $V(r)$ at any point a distance $r$ from $q_1$ is, by Eq. (2.15) with $\mathbf{E}\cdot d\boldsymbol{\ell} = E\hat{\ell}_{12}\cdot d\boldsymbol{\ell} = -Ed\ell$ ($\mathbf{E}$ is 180° in opposition to the motion of the charge along our path) and $d\ell = -dr'$ ($r'$ decreases along our path),

$$\begin{aligned} V(r) &= V_{\infty r} = -\int_{\infty}^{r} \frac{q_1}{4\pi\epsilon_0 r'^2} dr' \\ &= \frac{q_1}{4\pi\epsilon_0 r'}\Bigg|_{\infty}^{r} = \frac{q_1}{4\pi\epsilon_0}\left(\frac{1}{r} - \frac{1}{\infty}\right) = \frac{q_1}{4\pi\epsilon_0 r}, \end{aligned} \tag{2.18}$$

which is positive because it takes work to move a (positive) unit charge toward the positive charge $q_1$.

Thus the requested voltage is

$$\begin{aligned} V_{21} &= V_1 - V_2 = V(r_1) - V(r_2) \\ &= \frac{q_1}{4\pi\epsilon_0}\left(\frac{1}{r_1} - \frac{1}{r_2}\right) = \frac{10^{-6}}{4\pi \cdot 8.85 \cdot 10^{-12}}\left(\frac{1}{1} - \frac{1}{2}\right) \approx 4500 \text{ V}, \end{aligned}$$

which is positive because $r_2 < r_1$; a positive unit charge to be pushed closer to $q_1$ (also positive) requires us to do positive work against the repulsive Coulomb force. The answer does not depend on whether these points lie on the same radial ray from $q_1$, for all points on a spherical shell of radius $r$ centered on $q_1$ are at the same potential. Therefore, it takes no work to move from one point to another on such an *equipotential surface.*

### 2.6.1 An Analogy with Gravity

To close this section, it is extremely interesting that all the electric field quantities have close analogies with gravitation field quantities. Thus you can often use your intuition about the force of gravity and gravitational potential energy to help you understand the electric field and voltage. For example, the force $\mathbf{F}_{21}$ in newtons on a mass of $m_2$ kilograms a distance $r$ meters from another mass of $m_1$ kilograms is

$$\mathbf{F}_{21} = -G\frac{m_1 m_2}{r^2}\,\hat{\boldsymbol{\ell}}_{12}, \tag{2.19}$$

where $G = 6.673 \cdot 10^{-11}\,\mathrm{N \cdot m^2/kg^2}$, $\hat{\boldsymbol{\ell}}_{12}$ is the unit vector pointing from mass 1 to mass 2, and the minus sign indicates that the gravitational force is always attractive (unlike charge, mass is always positive). Note that the gravitational force is independent of the medium between the masses, unlike the electric force, which depends on $\epsilon$ of the medium between the charges. Coulomb himself noted the similarity of Eq. (2.19) (discovered by Newton) to his own law.

Furthermore, the gravitational field strength (force per unit mass) is defined analogously to $\mathbf{E}$ in Eq. (2.2) as

$$\mathbf{g} = \frac{\mathbf{F}}{m}, \tag{2.20}$$

where of course here $\mathbf{F}$ is the gravitational (not electric) force. Near the surface of the earth, $r \approx R_e$, the radius of the earth, and $m_1 = M_e$, the mass of the earth. Consequently, $\mathbf{F}_{21} = -gm_2\hat{\boldsymbol{\ell}}_{12}$, where $g = 9.8\,\mathrm{m/s^2}$ is the magnitude of $\mathbf{g}$ for the earth and $-\hat{\boldsymbol{\ell}}_{12}$ points "down." Finally, the work done in hoisting a mass $m$ from ground up to a height $h$ is

$$W = -\int_0^h -mg\hat{\boldsymbol{\ell}}_{12}\cdot d\boldsymbol{\ell} = mgh, \tag{2.21}$$

or the potential energy per unit mass (gravitational potential) is $W/m = gh$. The gravitational potential energy per unit mass is analogous to the electrical potential energy per unit charge, which is called voltage or electrical potential ($V = W/q = Ed$). The unity of nature is amazing!

## 2.7 CONDUCTORS AND CAPACITORS

So far we have been dealing with fictitious *point* charges, each of which takes up zero space. Practical charged objects are not points; they are macroscopic. But if this is true, how was Coulomb able to predict correctly the forces between the charged spheres, having finite size? The answer is that he was making use of Gauss's law [Eq. (2.6) or (2.13)] without knowing it. The surface charge on the spherical pith balls was fairly uniform. Thus the field at any point outside the balls was identical to that produced by an equivalent point charge equal to the total surface charge, located at the center of the ball.

Rather than continuing to discuss pith balls or point charges, we now move on to the more practical process of charging two initially uncharged sizable metal plates. These plates both have area $A$ and are separated a distance $d$ by a dielectric of permittivity $\epsilon$ (see Fig. 2.8). The method of charging the plates is by removal of electrons from one plate and depositing those electrons on the other. This is a very common task in electric circuits. A voltage source such as a battery (see Sec. 2.8)

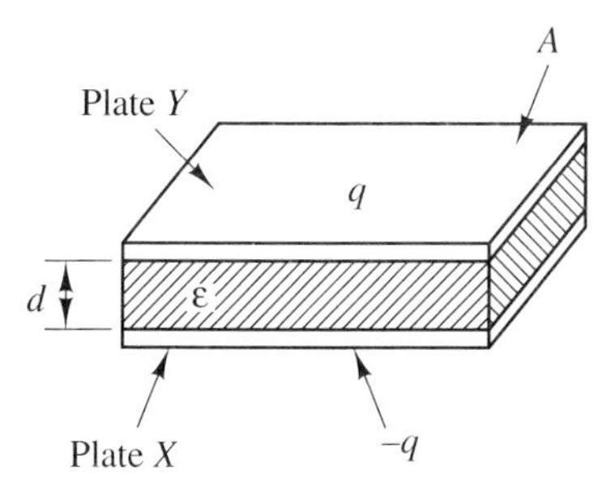


**Figure 2.8** Parallel-plate capacitor.

does the pulling of electrons off one plate and pushing them onto the second plate, via wires connected to the plates.

The charge storage system, consisting of the plates and dielectric, is called a *capacitor.* It was originally called a condenser because it was a dense reservoir (condenser) of electricity (electric charge), whose electric density was controllable by the applied voltage. The analogy was drawn between voltage across the plates and pressure applied to a balloon: In both cases the amount inside (charge in capacitor, air in balloon) was variable, determined by the applied "pressure."

The only mobile charges in crystalline metal are (negatively charged) electrons; the positively charged nuclei are spatially fixed in the crystal lattice. We assume that the metal plates are *perfect conductors.* This means that under even the slightest push from an electric field, the conduction electrons will almost instantaneously spatially redistribute themselves until the net force on every electron is zero.

The only exception is forces on surface charges (electrons) that are directed perpendicular to the surface because these cannot be equilibrated, except possibly by the electrons being removed from the conductor. Removal of electrons from conductors is generally not necessary in electric circuits because of the connecting wires. We can conclude that the electric field within and on the surface of a perfect conductor is zero, except for surface fields that are perpendicular to the surface.

By Eq. (2.15), it therefore takes zero energy to move an electron around within a perfect conductor (which is continually in a state of instantaneous equilibrium). Thus at all points on the surface of or within a perfect conductor, the potential energy is the same. Although metals are not *perfect* conductors, they are so good that the perfect conductor model describes observed behavior very well over short distances, especially for isolated metal objects. The simplest and most common circuit element that is often assumed to be a perfect conductor is wire. However, no material, even wire, is a perfect conductor, though it may be approximately modeled as one.

In our example, suppose that the lower plate $X$ has charge $-q$ and the upper plate $Y$ has charge $+q$. Then the free electrons in plate $X$ bunch up on the surface of plate $X$ nearest plate $Y$ because of the Coulomb force attraction to the positive charge on $Y$. Similarly, the free electrons in plate $Y$ bunch up on the side farthest from plate $X$. The charge distributes itself so that all forces except those perpendicular to the surface of the plate are canceled by Coulomb forces of the surface electrons.

What does the electric field look like? The answer is easily determined if we ignore the complicated field patterns occurring near the edges of the plates. Rigorous analysis shows that these *fringing fields* are often negligible for practical capacitors.

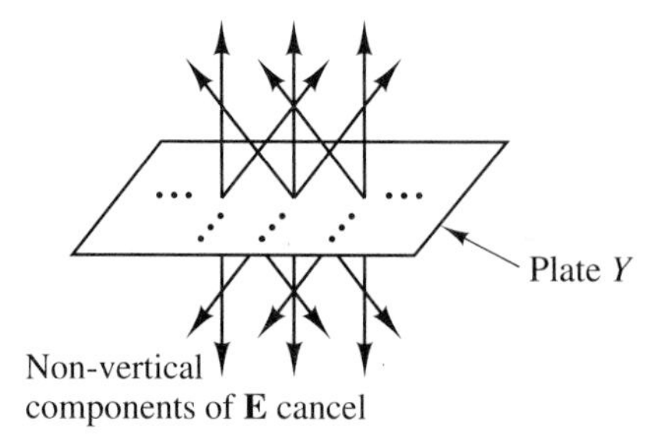


**Figure 2.9** Nonvertical components of the electric field surrounding a capacitor plate cancel.

Away from the edges of the plates, the charges are uniformly distributed on the charged surfaces. We shall assume that they are uniformly distributed over the entire charged surface of the plate.

First, let us find the field due to a single plate in free space having positive charge $q$ (i.e., plate $Y$ alone). In Fig. 2.9 the surface charge is assumed to be composed of arrays of pointlike charges, each contributing an electric field given by Eq. (2.3). Notice, however, that any off-vertical component of the field due to one charge is canceled by that of a neighboring charge. Therefore, the total field is strictly vertical (again, assuming no fringing at the edges).

Now consider both plates (Fig. 2.10). The electric field due to the negatively charged plate (plate $X$ alone) is identical to that of plate $Y$, except that **E** points in the opposite direction. From Fig. 2.10 it is seen that everywhere outside the plates, the sum of the two electric fields is zero, while between the plates they add. Thus, between the plates, there is a vertical electric field pointing down from $Y$ to $X$.

With this knowledge of field direction, we may apply Gauss's law, Eq. (2.6), to find the value of $E$ (the magnitude of **E**) between the plates. Take the closed surface $S$ to be a box having upper and lower surfaces parallel to and on either side of one of the plates (say, the upper plate $Y$), as shown in Fig. 2.11. The only side of $S$ having nonzero **E** passing through it is the lower surface, between the plates. Because this field is vertical, it is parallel with the normal of the bottom surface of $S$. Therefore, the integrated dot product in Eq. (2.6) is just the uniform magnitude of $\epsilon\mathbf{E}$, $\epsilon E$, times the upper plate area $A$:

$$\epsilon EA = q,$$

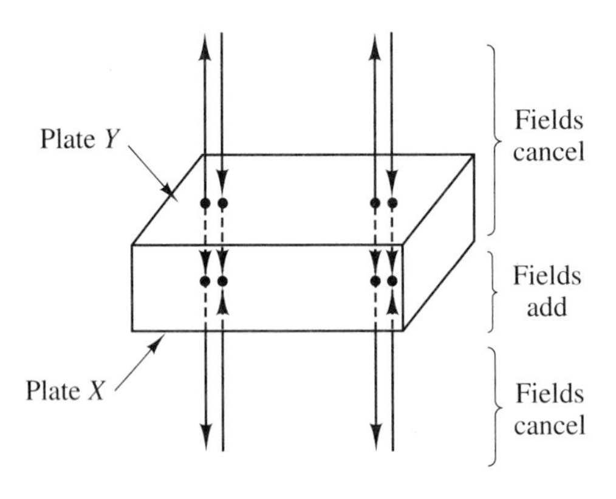


**Figure 2.10** Demonstration that the electric field of a capacitor is zero everywhere except between the plates.

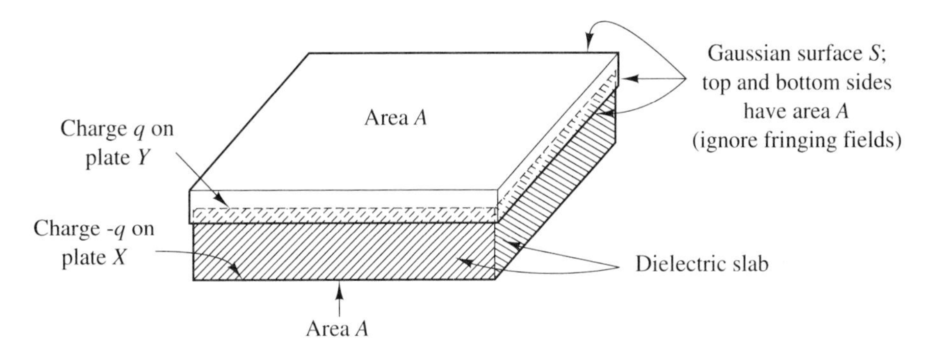


**Figure 2.11** Gaussian surface for calculating the electric field between the plates of a capacitor.

or

$$E = \frac{q}{\epsilon A}. \tag{2.22}$$

The difference in potential between the plates when they have equal and opposite charges $q$ and $-q$ can be calculated from Eq. (2.15). In Eq. (2.15), let $a$ be the lower plate $X$ ($x = 0$) and $b$ be the upper plate $Y$ ($x = d$). Let the path be (upward) vertical, so that the dot product of $\mathbf{E}$ (down) and $d\boldsymbol{\ell}$ (up) is just $-E\, d\boldsymbol{\ell}$. Then

$$V_Y - V_X = V_{XY} = Ed = \frac{d}{\epsilon A} q. \tag{2.23}$$

Note that $V_{XY}$ is the energy it takes to move a positive unit charge from the surface of plate $X$ to the surface of plate $Y$.

Let us build up the charge on these plates incrementally by moving tiny charges $dq$ from one plate to the other repeatedly, and in the process building $q$ up from zero to $Q$. With the result Eq. (2.23) in hand, the required energy for each $dq$ transfer is $dq\, V_{XY}(q)$, and thus the total energy required to build up charge $Q$ on one plate and $-Q$ on the other is

$$\begin{aligned} \text{energy} &\approx \int_0^Q V_{XY}(q)\, dq = \frac{d}{\epsilon A} \int_0^Q q\, dq \\ &= \frac{1}{2} \frac{d}{\epsilon A} Q^2. \end{aligned} \tag{2.24}$$

To complete the example, let us use some realistic numerical values for the parameters and see what is the required energy. Suppose that we wish to transfer $Q = -10^{-8}$ C $= -10$ nC of charge from plate $Y$ to plate $X$. The area of the plates is 1 cm$^2$ = $10^{-4}$ m$^2$. The distance between the plates is 0.1 mm = $10^{-4}$ m, and the relative permittivity of the dielectric is 100. The required energy is therefore approximately (neglecting fringing effects)

$$\text{energy} \approx 0.5 \cdot \frac{10^{-4}}{100 \cdot 8.85 \cdot 10^{-12} \cdot 10^{-4}} \cdot (-10^{-8})^2 = 56 \text{ nJ}. \tag{2.25}$$

By Eq. (2.23), the voltage between the plates is

$$V_Y - V_X = V_{XY} = \frac{10^{-4}}{100 \cdot 8.85 \cdot 10^{-12} \cdot 10^{-4}}(10^{-8}) \approx \frac{1}{10^{-9}}(10^{-8}) = 10 \text{ V}. \quad (2.26)$$

The relation in Eq. (2.23) between the voltage across two conducting parallel plates and the charge stored on them can be generalized for any system of charged conductors. For example, the corresponding relation for two charged metal spheres is considered in the problems. Also, long wires exhibit the behavior of capacitors even when this is not intended, a fact important to keep in mind whenever sending information along wires.

The relation between voltage and charge always ends up being that the voltage between conductors is proportional to the charge on them. Alternatively, if the potential difference between two conductors is held fixed at $V$ volts, there will be a charge $Q$ coulombs on the conductors ($+Q$ on one, $-Q$ on the other) proportional to $V$. The proportionality constant $C$ is called *capacitance*:

$$Q = CV. \quad (2.27)$$

The name "capacitance" stems from the simple analogy between a capacitor and a tank. In this analogy, $Q$ is the water and $V$ is the water level in the tank. The capacity of the tank when the tank is filled up to a certain level is of course related to the geometric parameters, such as the length and width of the cross section at all levels below the surface. Similarly, geometric parameters as well as permittivity determine capacitance. For the two parallel plates, comparison of Eq. (2.27) with Eq. (2.23) shows that

$$C = \frac{\epsilon A}{d}. \quad (2.28)$$

An intuitive explanation of Eq. (2.28) is as follows. The greater the area of the plates, the more charge can be stored on them for a given potential difference between the plates (and so the larger $C$ becomes). From Eq. (2.23) we see that the larger the distance $d$ between the plates, the more rapidly the potential grows with $q$ during charging. Therefore, the larger the distance $d$, the smaller the amount of charge on the plates at a given potential (and therefore, a smaller $C$). From Eq. (2.22) and discussions in Sec. 2.2, we know that the larger $\epsilon$ is, the smaller will be the electric field between the plates for a given charge on the plates. Thus the amount of work to move a given amount of charges is reduced, so that the amount of charge displaced for a given applied voltage (and therefore $C$) is increased.

The units of capacitance are $C^2/J$, or farad (F) after Michael Faraday of England, who in the 1830s pioneered the ideas of charge on the surface of conductors and the dependence of capacitance on permittivity. Therefore, from Eq. (2.28) we see that the units of $\epsilon$ are also F/m, as noted in Sec. 2.2. Capacitors of practical size and materials usually have capacitance less than $10^{-3}$ F; therefore, μF, nF, or pF appear most commonly in practice. For example, the third equality in Eq. (2.26) was written in the form $V = Q/C$, where $C = 10^{-9}$ F = 1 nF.

As a final note, substitution of Eqs. (2.28) and (2.27) into Eq. (2.24) gives

$$\begin{array}{l}\text{energy stored in the electric} \\ \text{field within a capacitor}\end{array} = \frac{Q^2}{2C} = \frac{QV}{2} = \frac{CV^2}{2}, \quad (2.29)$$

which is the familiar formula $\frac{1}{2}CV^2$ encountered in freshman physics courses.

**Example 2.4**

What is the energy stored in the electric field between the plates of a 1-μF capacitor charged to 10 V? How much charge is on the capacitor plates?

**Solution** By Eq. (2.29), energy stored $= \frac{1}{2} \cdot 10^{-6} \cdot 10^2 = 50$ μJ. The charge on the plates, by Eq. (2.27), is $Q = 10^{-6} \cdot 10 = 10$ μC. By comparison, the energy stored in a 200-μF capacitor charged to 400 V in a camera electronic flash unit (discussed fully in Sec. 7.7.4) is $\frac{1}{2} \cdot 2 \cdot 10^{-4} \cdot (400^2) = 16$ J, easily lethal in its high-potential form.

## 2.8 THE ELECTRIC BATTERY

All the electric theory above may have been interesting to discover, but it remained of little use until continuous flows of charge called *electric currents* were made possible. For example, Otto von Guericke of Germany produced some electric sparks way back in 1650, but these did not make for a good reading lamp. (Perhaps it was the similarity between these electric sparks and gunpowder explosions from a charged firearm that explains using "charge" to name the electrical property of matter.)

The real breakthrough came in 1800 when Alessandro Volta of Italy invented the battery, a stack of zinc, paper, and copper layers immersed in acid. Each zinc/paper/copper sandwich is called a *voltaic cell.* Volta's battery, called the "voltaic pile," which was a series-connected set (pile or battery—analogous to the *batt*ery of army troops used to fight *batt*les) of voltaic cells, made continuous electric currents possible. It took only another 80 years for Edison to be granted his electric light patent!

The essential idea in a common household battery is that chemical energy can be used to produce charge separation, similar to that described in Example 2.4 and other discussion in Sec. 2.7. The difference is that this charge separation is maintained continually even when charges are allowed to move under its influence in ways that normally would tend to diminish it. When charges are separated, potential energy exists due to the electric (Coulomb) force. If the charges are allowed to move under the influence of this electric force, then in the process they can do work such as lighting a bulb, playing a transistor radio, and so on. In the case of a battery, the energy required to separate the charges comes from the chemical reactions which produce the separation of charge.

Without getting into a detailed discussion of battery structure and chemistry, the essential ideas may still be understood. A basic flashlight battery has an outer zinc electrode and an inner carbon electrode, immersed in a paste of pyrolucite (i.e., manganese dioxide, $MnO_2$), carbon (C), ammonium chloride ($NH_4Cl$), and water.

The zinc electrode (*anode*) when immersed in the paste oxidizes (i.e., its apparent charge increases) to $Zn^{2+}$ and $2e^-$, where $e^-$ is an electron. Over at the carbon electrode (*cathode*), the pyrolucite reduces (i.e., its apparent charge decreases) to hydrous manganous oxide [MnO(OH)]. The latter reaction requires electrons. These are provided by the electrons remotely generated over at the anode when an external electric circuit (such as a light bulb) is connected across the terminals. That is, for the chemical reaction to proceed, the electrons must pass from their origin at the anode, through the connected electric circuit, and finally into the carbon cathode. They cannot pass through the paste because it is an insulator.

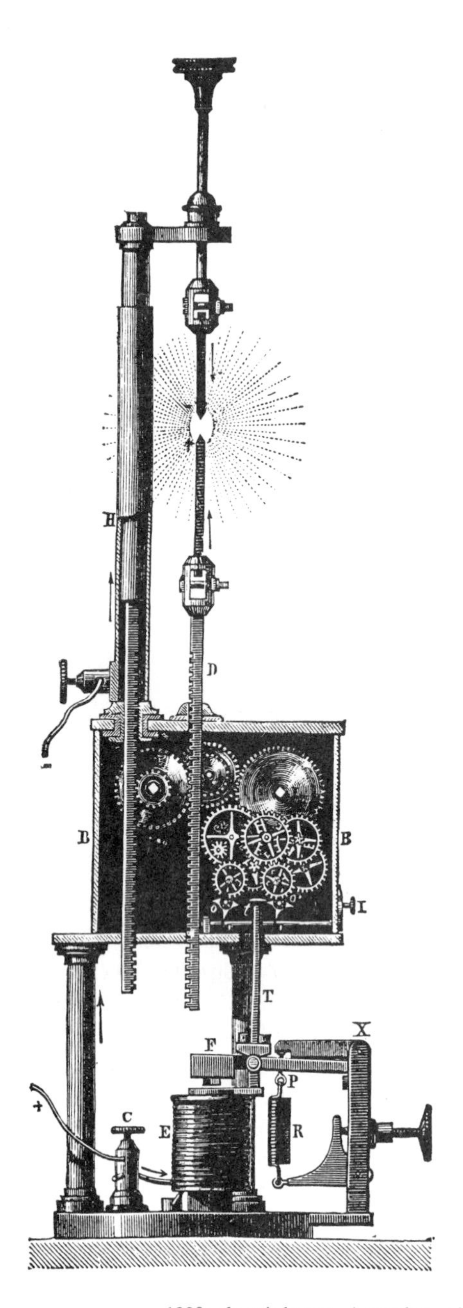


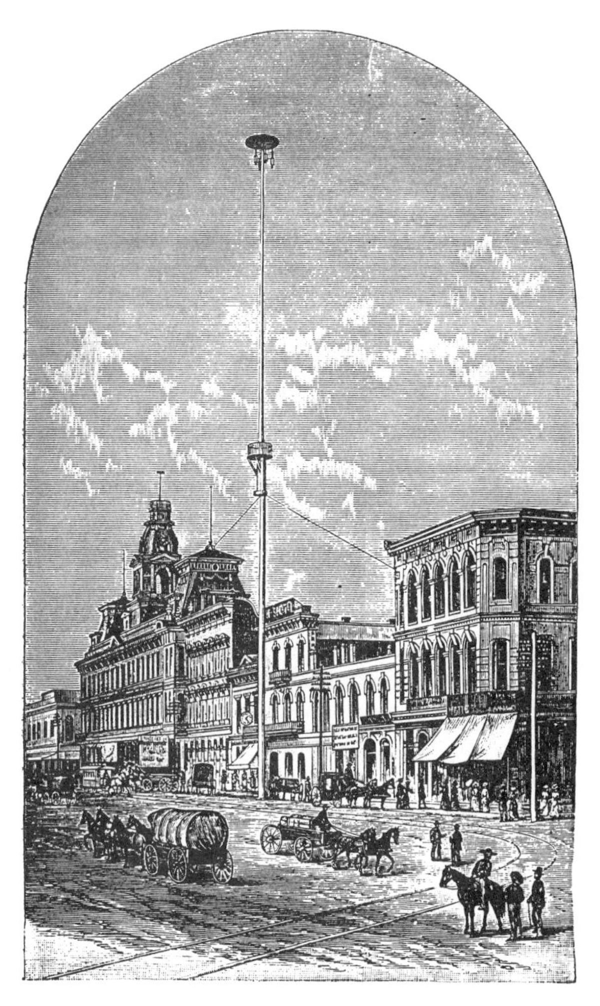

1880s electric lamps. A spark was initiated by bringing the carbon pencil tips close together (see lamp on left). As the pencils wore down and the gap widened, the decreased current relaxed the magnetic force of the electromagnet, which in turn through clockwork narrowed the gap. These lamps reportedly lasted for "hours" before failing! The first public lighting systems such as the one shown on the right were based on exactly the same arc light mechanisms, and would have frightened even Douglas Fairbanks.

The reason that the reaction occurs (provided that the circuit is connected) is that the reaction lowers the energy of the system. This difference in energy is used to do electrical work in the external electric circuit. Thus chemical potential energy is converted into electric energy. In the last part of this book we shall investigate how mechanical energy is converted into electric energy. Most high-power applications are best served by the latter method, while low-power and/or portable circuits can use the former.

The most interesting point here is that by chemical means, a *charge separation* has been accomplished, and we know from above that this means that electric energy is available. The fact that there is plenty of zinc and paste in the cell allows the charge separation to be maintained over a long period of time even with the connected circuit consuming energy. Chemical energy is converted to electrical energy continuously. Without the circuit connected, the reaction ceases until the battery is used again.

Another interesting point is that independent of the load current or how long the battery is used (within limits), the voltage across a battery remains relatively constant. This is because the electric potential generated in the cell is directly related to the chemical reaction energy. This energy depends primarily on the type of reaction taking place—not on how long or at what rate it has been occurring. For example, the chemical reactions of the flashlight D cell discussed above maintain a constant voltage of 1.5 V. A 9-V transistor radio battery is merely six of these 1.5-V cells in series [6(1.5) = 9]. Incidentally, the D cell is a single cell, so technically it should not be called a battery (of cells).

A useful model for describing the behavior of batteries and other voltage sources is as follows. They can be modeled as having a non-Coulomb electric field $\mathbf{E}_{\text{V.S}}$ that is nonzero only within the source. As mentioned in Sec. 2.6, this non-Coulomb force is nonconservative. This is because of its spatial limitation. For example, the integral of $\mathbf{E}_{\text{V.S.}}\cdot d\boldsymbol{\ell}$ taken along one path between the two voltage source terminals and lying within the source gives a nonzero value, called the *electromotive force* (emf) and denoted $\mathcal{V}$. The integral of $\mathbf{E}_{\text{V.S.}}\cdot d\boldsymbol{\ell}$ using a path taken outside the source is zero, not $\mathcal{V}$, because outside, $\mathbf{E}_{\text{V.S.}} = 0$.

Let a closed contour start at one terminal of the voltage source, pass through the source to the other terminal, and go back to the first terminal by a path outside the source. From our discussion we know that the integral of $\mathbf{E}_{\text{V.S.}}$ around this closed contour is $\mathcal{V} + 0 = \mathcal{V}$. Because the total electric field $\mathbf{E}$ is the sum of $\mathbf{E}_{\text{V.S.}}$ and the Coulomb field of the separated charges $\mathbf{E}_{\text{COUL}}$, and because $\mathbf{E}_{\text{COUL}}$ is conservative [Eq. (2.16) applies], the closed-contour integral of $\mathbf{E}\cdot d\boldsymbol{\ell}$ is equal to the emf $\mathcal{V}$. In general, when there are several voltage sources of emf $\mathcal{V}_n$, the right-hand side of Eq. (2.16) is not zero, but (including time dependence for generality) rather

$$\oint_C \mathbf{E}(t)\cdot d\boldsymbol{\ell} = \sum_n \mathcal{V}_n. \tag{2.30}$$

The nonconservative electric field $\mathbf{E}_{\text{V.S.}}$ is produced by the chemical energy in a battery (or other energy source in other voltage source types). It causes the charge separation described above. In equilibrium, with no circuit connected, the resulting Coulomb field exactly cancels $\mathbf{E}_{\text{V.S.}}$ within the voltage source.

When a circuit is connected, charge flows in the circuit under the Coulomb force, tending to decrease the amount of charge separated and therefore the Coulomb force itself. However, the non-Coulomb force within the voltage source remains because it is independently produced (via chemical reaction, etc.); it is now partly uncanceled. This uncanceled field effectively moves charge within the voltage source continually, thereby restoring the charge separation and thus maintaining the voltage across the terminals. We discuss voltage sources further in Sec. 2.10.

Even though this description does not accurately reflect the physical current flow process within a chemical battery, it is a self-consistent and equivalent model that can be used to account correctly for the terminal behavior of any type of voltage source.

(*Caution*: "Anode" in all cases other than batteries refers to the *higher* rather than the lower potential terminal; and vice versa for the cathode. Prove this to yourself by consulting a dictionary, where the two given meanings for "anode" are opposite! However, the root meaning of "anode" is "way up," while that of "cathode" is "way down." Thus the battery usage of these terms is exceptional. This anomaly probably stems from the similar and relevant terms in chemistry, "*an*ion" and "*cat*ion." A *cat*ion is a positive ion that moves toward the battery *cat*hode, and an *an*ion is a negative ion that moves toward the battery *an*ode. All these terms were devised by Faraday.)

## 2.9 ELECTRIC CURRENT AND RESISTANCE

The battery was the first of many types of instruments that maintain a constant voltage between their terminals. This voltage is maintained even when charge is allowed to flow through a circuit connected between the terminals.

Previously, the only time that charge would flow was when charged objects were brought into contact. Upon contact, a charge transfer would occur that would bring about an equilibrium almost immediately. In this equilibrium there was zero net charge on either object and thus zero potential difference. The only possible electrical energy storage and transfer were of this type, quantified in Eq. (2.24). From numerical values such as in Eq. (2.25), the minimal industrial and societal impact of this electricity is clear.

When a continuous resupply of charge to one terminal maintained at higher potential than the other is possible, significant work can be done. The energy required to maintain that potential can come from chemical, mechanical, or solar energy in the form of batteries, generators, or solar cells. Such a device that maintains a potential difference $V$ volts between its terminals is said to be a voltage source of $V$ volts.

We know what will happen if an insulator such as air is placed between the terminals: Nothing. We also know what happens if a near-perfect conductor such as a short piece of wire is placed across the terminals: The battery dies or explodes, or a fuse blows. Neither of these extremes is very useful. What about something in between? Available materials span a range of conductivity from the most highly insulating materials such as mica to metals such as copper, whose conductivity is $10^{25}$ times greater!

Suppose that the dielectric sandwiched between the plates of the capacitor in Fig. 2.8 is replaced by a material with moderate conductivity and, for the sake of

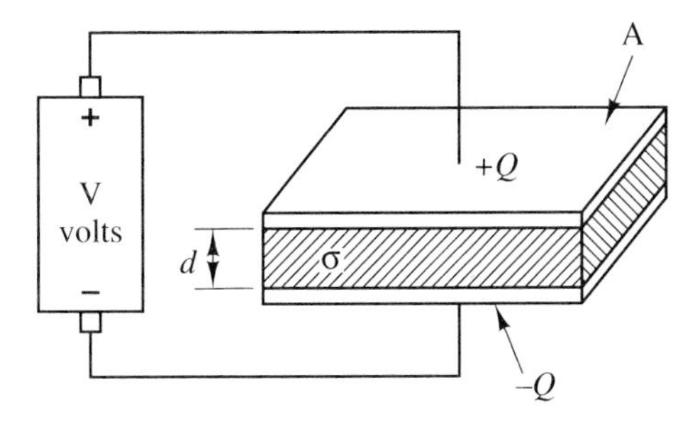


**Figure 2.12** Conductor connected to a voltage source.

specificity, $\epsilon_r = 1$. Let there be a voltage source of voltage $V$ volts whose terminals are connected to the plates, as shown in Fig. 2.12. Then according to Eq. (2.23) and the surrounding discussion, there will be an electric field pointing down in between the plates, uniform over the area $A$.

By Eq. (2.2), there will be an upward force on electrons on the lower plate of magnitude $F = qE$. When the "sandwich meat" was an insulating dielectric, no charge would flow once the capacitor attained a charge $Q = CV$ on its plates. That is, the free electrons remain on the surface of the metal plates, unable to traverse the insulating dielectric. But if we have a conducting sandwich, charge is relatively free to move under this electric force $F$.

An electron in a vacuum exposed to an electric force will accelerate according to $F = ma$, where $m$ is the electron mass and $a$ is its acceleration. In a solid, however, the crystal lattice of atoms interferes with this acceleration. The modern theory describing this interaction is highly advanced (quantum mechanics) but can be summarized simplistically as follows.

As an electron travels through the lattice, it is continually *scattered* from its normal path by the electric forces of the positive metallic ions. Each time such a *collision* occurs, it loses essentially all its momentum, and energy is transferred to the crystal. This energy, which was initially provided by the voltage source now manifests itself in increased thermal motion of the metal atoms—it is converted to heat.

After each collision, the electron again accelerates under the electric field provided by the voltage source. However, because of all these collisions, it never attains high speed in the direction of **E**. Of course, due to thermal agitation the electron undergoes very high speed *randomly directed* motion. But it is the *directed* drift due to **E** that is of practical use because it is directed and under our control via the voltage source, and can do real work or transmit elsewhere the battery energy in electrical form.

Assuming complete stoppage upon collision, the average velocity $U$ that the electron does attain is $U = a\,\Delta t$, where $\Delta t$ is one-half the mean time between collisions (the one-half in order to give the average). The average velocity of electrons is therefore proportional to $E$ because $a$ is. This is the observed behavior of electron flow in metals: The mean velocity is proportional to the applied electric field. Therefore, a constant-in-time voltage source produces a constant flow of electrons in proportion to $E$.

For a variety of reasons (including differences in materials and measurement considerations), the mean velocity is not the common variable describing charge flow. Rather, the electric current (denoted $I$) is defined as the time rate of charge flow through any cross section of a conductor through which the charge flows. The current $I$ is proportionately related to the average drift velocity $U = dx/dt$ as follows:

$$
\begin{aligned}
I &= \frac{dq}{dt} = \lim_{\Delta t \to 0} \frac{\Delta q}{\Delta t} = \lim_{\Delta t \to 0} \frac{\Delta q}{\Delta \text{vol.}} \frac{\Delta \text{vol.}}{\Delta t} = \lim_{\Delta t \to 0} \frac{\Delta q}{\Delta \text{vol.}} \lim_{\Delta t \to 0} \frac{A\,\Delta x}{\Delta t} \\
&= (Nn_e e)(AU) \qquad (2.31) \\
&= \left(\frac{\text{charge}}{\text{volume}}\right) \cdot \left(\frac{\text{volume of charges flowing by}}{\text{unit time}}\right),
\end{aligned}
$$

where $N$ is the number of atoms per unit volume, $n_e$ the number of free electrons per atom, $e$ the charge of an electron, and $A$ the cross-sectional area of the conductor.

Above, we concluded that $U$ is proportional to the applied voltage $V$:

$$
\begin{aligned}
U &= a\,\Delta t \\
&= \frac{F}{m}\Delta t \\
&= \frac{eE}{m}\Delta t \qquad (2.32) \\
&= \frac{eV}{md}\Delta t,
\end{aligned}
$$

so that from Eq. (2.31),

$$
\begin{aligned}
I &= Nn_e eAU \\
&= \frac{e^2\,\Delta t\, Nn_e}{m}\frac{A}{d}V, \qquad (2.33)
\end{aligned}
$$

or

$$
I = \left(\frac{\sigma A}{d}\right)V, \qquad (2.34)
$$

where $\sigma = e^2\,\Delta t\, Nn_e/m$ is called the *conductivity* of the material. The reciprocal of conductivity is called *resistivity* $\rho$: $\rho = 1/\sigma$. Both $\rho$ and $\sigma$ are used in characterizing the conductivity of materials.

Equation (2.34) is known as *Ohm's law,* after Georg Ohm of Germany, who discovered it experimentally in 1827. The reciprocal of the quantity in parentheses is called the *resistance* $R$ of the chunk of material (sandwiched between the plates in Fig. 2.12) through which current flows. The form of Ohm's law used universally and for the remainder of this book is

$$
\begin{aligned}
\text{current} &= \frac{\text{voltage}}{\text{resistance}} \\
I &= \frac{V}{R}, \qquad (2.35)
\end{aligned}
$$

where

$$R = \rho \frac{d}{A}. \tag{2.36}$$

The units of current are C/s, or amperes (A) after André-Marie Ampère of France, who in the 1820s made great discoveries concerning currents that we shall consider below. An ampere is the current required to flow in two parallel wires 1 m apart in order for there to exist a force between them of $2 \cdot 10^{-7}$ N on each meter of wire. This force, which is different from Coulomb's force, is discussed in Sec. 3.2.

Resistance has the units V/A, or ohms (Ω, after the similarity between the sound of "omega" and that of "ohm"), after Georg Ohm. Precisely, a chunk of material has a resistance of 1 Ω if a current of 1 A flows through it when a potential of 1 V is placed across it.

**Example 2.5**

A carbon resistor is connected to a 9-V battery. The current is measured by a current-measuring device called an ammeter (yes, after André) to be 1 mA. What is the value of the resistance?

**Solution** By Ohm's law, Eq. (2.35), $R = V/I = 9/0.001 = 9000\ \Omega = 9\ \mathrm{k}\Omega$, where k represents 1000.

**Example 2.6**

Suppose that in Example 2.5 the measurements of the resistor are: length = 1.5 cm, area = circle of radius 1 mm.

**(a)** What is the resistivity of the material constituting the resistor?

**(b)** If we stacked 10 of these identical resistors end to end, what would be the resistance?

**Solution**

**(a)** By Eq. (2.36), $\rho = RA/d = 9000[\pi(10^{-3})^2]/0.015 = 1.88\ \Omega \cdot \mathrm{m}$.

**(b)** $R = \rho\ (10d)/A = 90\ \mathrm{k}\Omega$. Notice that stacking 10 resistors end to end is equivalent to one resistor of the same material 10 times as long as the otherwise identical original resistor.

It may be asked, just how fast do electrons actually move through a wire? Typical values for copper are $N = 10^{29}$ atoms/m$^3$ and $n_e = 1$ electron/atom. Suppose that 1 A is flowing through a copper wire of cross-sectional area 1 mm$^2$ = $10^{-6}$ m$^2$. Then from Eq. (2.33),

$$U = \frac{I}{Nn_e eA} = \frac{1}{10^{29} \cdot 1.6 \cdot 10^{-19} \cdot 10^{-6}} = 0.063\ \mathrm{mm/s},$$

which is very slow.

How can it be, then, that when a light switch is turned on way across the room (from the light) that the light goes on instantly? The answer is clear when we think of the Boston marathon. When the start of the race is called, the runners begin moving slowly. Yet the person first in line knows the race has begun as soon as does the runner in back, who is perhaps one-fourth of a mile or more behind. Even a deaf runner would get the message instantly from the pushing and shoving, which moves through the crowd much faster than individual participants run.

The same phenomenon holds true in electric circuits. When the light switch is turned on, an electromagnetic wave field (discussed in Sec. 3.10.4) travels at the speed of light through the wire. Thus, essentially instantaneously, there is an electric field acting on all electrons to give them a push at nearly the same instant. This wave is not dependent on matter traveling at the speed of light ($6.7 \cdot 10^8$ miles/h) any more than the sound wave at the marathon depends on runners running at the speed of sound (742 miles/h)! In the first case, the electrons move at 0.063 mm/s = $1.4 \cdot 10^{-4}$ mile/h, and in the other case the people are running at less than 12 miles/h.

**Example 2.7**

The newest method of determining body composition is the "whole body" bioelectrical impedance method. Essentially, a controlled current is passed into one hand of the patient and out of the corresponding ankle, by means of electrodes. In practice, sinusoidal currents are usually used (50 kHz), but this modifies the results minimally, so we may consider the current here as constant in time.

We may multiply the numerator and denominator of Eq. (2.36) by $d$ to obtain $R = \rho d^2/\text{vol.}$, where vol. is the volume of the object, in this case the human body and $d$ is now essentially the height of the person. This may be solved for the volume as $\text{vol.} = \rho d^2/R$. An empirical relation between $R$ and total body water (TBW) has been proposed: TBW (kg) $= 1.825 + 0.585 d^2/R$, where $d$ is in centimeters. The fat-free mass (FFM) is then FFM = TBW/0.7194, so that the percent body fat is, in terms of total weight ($W$), % fat $= 100\% \cdot (W - \text{FFM})/W$.

What is the percent body fat of a 157-lb person 5 ft 7 in. tall whose measured resistance is 370 Ω?

**Solution** We can express % fat in terms of weight in pounds, height in inches, and resistance in Ω as

$$\%\ \text{fat} = 100\% \cdot \left[ \frac{(W\ \text{lb}/(2.2 \cdot \text{lb/kg}) - \dfrac{1.825 + 0.585 \dfrac{\left(\dfrac{d}{0.39\ \text{in./cm}}\right)^2}{R}}{0.7194}}{W/2.2\ \text{kg}} \right]$$

$$= 100\% \left[ 1 - \frac{2.537 + 5.346 \dfrac{d^2}{R}}{0.454 W} \right]$$

which for $W = 157$ lb, $d = 67$ in., and $R = 370$ Ω comes out to 5.4%, a healthy person indeed.

### 2.9.1 Conservation of Charge

Now that we have defined electric currents, we can add a few postscripts to our discussion of the charging of capacitors. In Sec. 2.7 it was mentioned that the capacitor was charged using wires connected to the plates. Thus charge moved through the wires at a particular rate that probably varied with time. An instantaneous current $i(t)$ is defined as the rate at which charge passes through a cross section such as that of a wire at time $t$, $dq_{xsec}(t)/dt$:

$$i(t) = \frac{dq_{xsec}(t)}{dt}. \tag{2.37}$$

This definition of current is also valid for any other situation involving the flow of electric charge.

A basic principle of the conservation of charge may be articulated regarding the capacitor-charging situation, as depicted in Fig. 2.13. The statement is that the sum of charges that pass through any closed surface is equal to the charge enclosed within the surface. Expressing the charge conservation law in the form of time rates of charge movement, the sum of all the currents $i_n(t)$ flowing into a closed surface $S$ is equal to the rate at which charge is deposited within that surface, $dq_{enc}/dt$, where $q_{enc}$ is the charge enclosed by the surface:

$$\sum_{n,S} i_n(t) = \frac{dq_{enc}(t)}{dt}. \tag{2.38}$$

For surface $S_1$ in Fig. 2.13, Eq. (2.38) takes the form $i(t) = dq(t)/dt$, where $q_{enc}(t) = q(t)$ is the charge on the upper capacitor plate. For $S_2$, Eq. (2.38) becomes $i(t) - i(t) = \frac{d}{dt}[q(t) - q(t)] = 0$ and for $S_3$ we write $i(t) - i(t) = 0$. The law of charge conservation, Eq. (2.38), may seem trivial and obvious, but it is one of the two most fundamental laws in basic circuit analysis. It again is valid for any situation involving the flow of electric charge.

One final point on instantaneous current and the capacitor. If we differentiate Eq. (2.27) with respect to time, we obtain $dq(t)/dt = C\,dv(t)/dt$, or

$$i(t) = C\frac{dv(t)}{dt}. \tag{2.39}$$

In circuit analysis, Eq. (2.39) is actually more commonly used than Eq. (2.27) because Eq. (2.39) involves current and voltage, the two primary electric circuit variables. We shall encounter Eq. (2.39) repeatedly in this book.

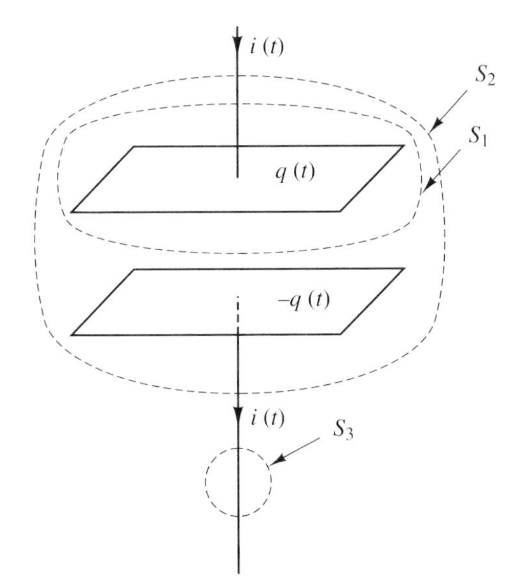


**Figure 2.13** Three surfaces for demonstration of conservation of charge.

## 2.10 ELECTRIC POWER

The electric force, when acting on charge through a distance, does work. If, instead, we overcome this force by pushing charge against it, we do work. The electric field itself is the seat of stored energy [see Eq. (2.29)]. It is common knowledge that power is the time rate at which work is done or energy is stored: $p = dw/dt$, where $p$ is the power and $w$ is the energy.

The units of power are joules/second, or watts (W), after the Scottish inventor James Watt, who greatly improved the steam engine in the 1770s. Watt came up with a unit of power that anyone of his day could understand: the power of a horse, or horsepower. Even today the horsepower is a common unit for designating mechanical power: 1 hp = 746 W. (A struggling horse could make about seven light bulbs glow!)

The electric potential as defined in Eq. (2.15) is the work required to move a unit charge (1 C) from point $a$ to point $b$. This definition can be rephrased to say that when 1 C of positive charge drops 1 V, 1 J of work is done by the electric field on the charge. More precisely, $v = \lim_{\Delta q \to 0} \left( \frac{\Delta w}{\Delta q} \right) = \partial w / \partial q$, where $\partial$ signifies partial differentiation, so that the charge being forced is too small to significantly affect the environment in ways that would make $v$ dependent on the test charge $\Delta q$. Therefore, if $q$ coulombs drop $v$ volts, it is not hard to guess that $q \cdot v$ joules of work is done by the electric field:

$$w = qv. \tag{2.40}$$

(The use of lowercase letters is common in the context of time-varying variables, which we shall be considering below.)

This formula should not be surprising; it was already encountered as the energy stored in the electric field within a capacitor in Eq. (2.29). The only difference was the $\frac{1}{2}$ factor, which was due to the fact that as work was done (the capacitor charged), both $q$ and $v$ were increasing. That is, the field being pushed against [and therefore $V(q)$] was zero when the first charge was being transferred (when the accumulated charge on the plates was $q = 0$). We pushed against the maximum force for which $V = V(Q)$ at only the end, when $q = Q$. The mathematics accounted for this by giving the $\frac{1}{2}$ factor.

We define $p(t)$ as the power absorbed by a circuit element at any time $t$. The circuit element is any device having metal electrodes with which to connect it electrically to the rest of a circuit (see Fig. 2.14). Let us suppose that we have a *passive circuit element* connected to a battery or other voltage source. A passive element is one that can absorb, store, and return electric (or magnetic) energy, but it cannot be considered a source.

A *source* or *active element* is an element that can continually deliver electric energy to other circuit elements because it in turn receives energy from other mechanical, chemical, solar, or even other electric sources. The electric energy source therefore receives energy and converts it into electrical energy in a form appropriate for the rest of a circuit. Hence *energy converter* is actually a better name than *energy source*, but they are often called the latter and in fact serve as sources of voltage for an electric circuit.

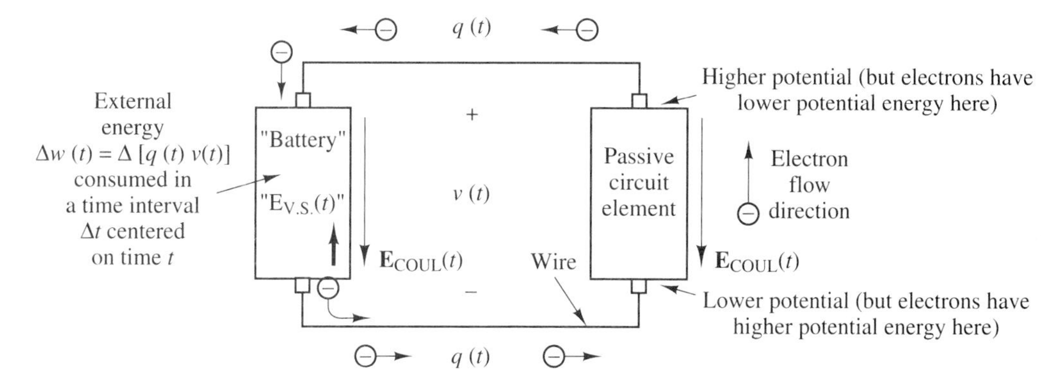


**Figure 2.14** Passive element connected to a voltage source. Depiction of the electric fields and charge carriers involved.

As described previously, it is as though within the voltage source there were an electric field "$\mathbf{E}_{\text{V.S.}}(t)$" continually pushing the charges against the Coulomb electric field $\mathbf{E}_{\text{COUL}}(t)$. Of course, in a battery the charges are not actually pushed by an internal electric field but rather, released at the higher potential terminal and absorbed at the other due to the chemical reaction—hence the quotes on "$\mathbf{E}_{\text{V.S.}}(t)$". In other voltage sources such as electromechanical generators, there may be an actual externally generated electric field $\mathbf{E}_{\text{V.S.}}(t)$ within the voltage source. The importance of $\mathbf{E}_{\text{V.S.}}(t)$ or its equivalent is that, by whatever means, it produces the usable terminal voltage $v(t)$.

At time $t$, one terminal of the passive element is $v(t)$ volts higher than the other, under the influence of the voltage source. For a chemical battery, $v(t)$ would be a constant, $V$ volts. We shall eventually encounter time-varying voltage sources; our discussion here will therefore be generalized to include them. The voltage source is effectively drawing negatively charged electrons from the higher potential electrode of the passive element where they have low potential energy, and returning the same amount of electrons to the other electrode of the passive element at a lower potential, where the electrons have higher potential energy.

*Note: Because the charge of electrons is negative, an electric field tends to push them in the direction opposite that of* $\mathbf{E}$. The electric field points from regions of higher potential to those of lower potential. Thus in an electric field, electrons tend to move from regions of lower potential to those of higher potential. Conventionally, we have defined voltage $V$ to describe the potential energy of positive, not negative charges. Furthermore, another convention is that *the polarity of a current $I$ or $i(t)$ always refers to the direction that positive charge would flow.* Therefore, if a current $I$ is shown pointing in one direction, we know that the true charge flow in a conductor, which is electrons (negative charge), is in the *opposite* direction (see Fig. 2.15).

For the voltage source to set up its charge separation [and therefore electric field and voltage $v(t)$] and force charge $q(t)$ through the passive circuit element during a time interval $\Delta t$ centered on time $t$ requires an amount of energy $\Delta w_{\text{ext}}(t) = \Delta[q(t)v(t)]$. This energy is the total requirement, including both the energy involved in forcing electrons through the circuit element and the energy in setting up the electric field $\mathbf{E}_{\text{V.S.}}$ within the source to provide this force. It is therefore

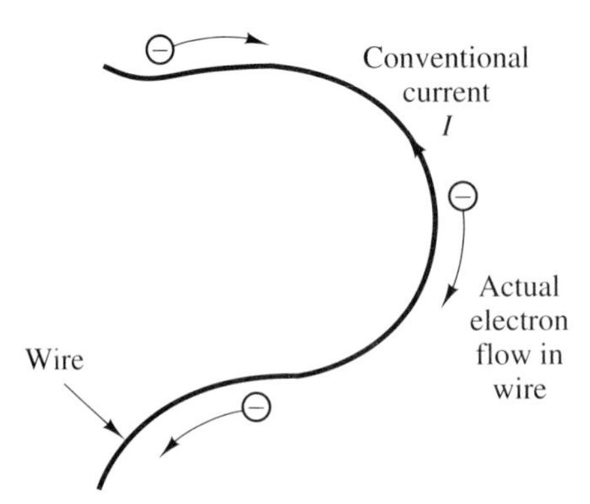


**Figure 2.15** Actual electron flow opposes conventional current direction.

the energy that the voltage source draws from the external energy source driving it (chemical, mechanical, or electrical)—hence the subscript "ext." Notice that as $q(t)$ and $v(t)$ vary with time, so does $\Delta w_{\text{ext}}(t)$. Expressing the energy this way will help us find the instantaneous power $p(t)$ into any circuit element.

The power that the source draws from the external energy source driving it is $p_{\text{ext}}(t) = dw_{\text{ext}}(t)/dt$. The passive element may absorb energy in a variety of ways: It may dissipate energy in the form of resistive heating as discussed previously, according to Joule's law (given below), it may store energy in electric (capacitor) or magnetic fields (inductor, discussed in Sec. 3.10.2), or absorb energy in a combination of these ways. (In the case of the capacitor, the energy stored in the electric field is in addition to that stored in the electric field of the voltage source, $\mathbf{E}_{\text{V.S.}}$.)

Yet for all of these types of energy absorption, the voltage source provides energy to the passive element in the same form: effectively pushing electrons against the potential difference $v(t)$. (Physically, the electrons leaving the voltage source may not be the same as those entering it, but the effect is the same as if they were.) To arrive at a complete energy balance, we compute the total demand the voltage source places on its supplying source (chemical, mechanical, etc.). This power demand may be expressed in terms of $q(t)$ and $v(t)$ as

$$p_{\text{ext}}(t) = \lim_{\Delta t \to 0} \frac{\Delta w_{\text{ext}}(t)}{\Delta t} = \frac{dw_{\text{ext}}(t)}{dt} = \frac{d\,[q(t)\,v(t)]}{dt}. \tag{2.41}$$

By the derivative of a product,

$$p_{\text{ext}}(t) = q(t)\frac{dv(t)}{dt} + v(t)\frac{dq(t)}{dt}. \tag{2.42}$$

What do these terms mean? Recall that the derivative of the product is equivalent to the total derivative:

$$p_{\text{ext}}(t) = \left.\frac{\partial w_{\text{ext}}(t)}{\partial v(t)}\right|_{\substack{q(t)\text{ held}\\ \text{constant}}} \frac{dv(t)}{dt} + \left.\frac{\partial w_{\text{ext}}(t)}{\partial q(t)}\right|_{\substack{v(t)\text{held}\\ \text{constant}}} \frac{dq(t)}{dt}, \tag{2.43}$$

where $\partial$ signifies partial differentiation.

The meaning of the first term is as follows. From $w_{\text{ext}} = qv$, we know that when $q(t)$ is held constant, $\partial w_{\text{ext}}(t)/\partial v(t) = q(t)$. [Thus the first term in Eq. (2.43) is equal to the first term in Eq. (2.42).] The statement "$q(t)$ held constant" means that this is energy *not associated* with the motion of charges through the passive element. It does not mean that $q(t)$ must be constant; it is just an identification of energy not associated with changes in $q(t)$.

This term is nonzero only if the voltage is changing with time [$dv(t)/dt \neq 0$]. If the voltage is constant, the energy stored in the electric field is constant, requiring zero power to maintain, in the absence of current. The term $q(t) \cdot dv(t)/dt$ thus represents the power involved in changing the electric field within the voltage source, thereby increasing the energy stored in the electric field $\mathbf{E}_{\text{V.S.}}$. It is not the power required to drive actual currents through the passive load element. [Of course, to raise or lower $v(t)$ *does* require charge motion, but this is not the same charge as $q(t)$, that going into the passive element, so it is not counted in $dq(t)/dt$.]

Now consider the second term. From $w_{\text{ext}} = qv$ we know that when $v(t)$ is held constant, $\partial w_{\text{ext}}(t)/\partial q(t) = v(t)$. The statement "$v(t)$ held constant" means that we are considering the work done as charge moves (is pushed) against a given defined electric field associated with the voltage $v(t)$. Note that this does *not* imply that $v(t)$ must be constant in time any more than the first term requires the charge $q(t)$ to remain constant. The holding of $v(t)$ "constant" is only for the purpose of distinguishing the two types of power; $v(t)$ may change for calculation of $p_{\text{ext}}(t)$ for the next instant. This work done is precisely the kind of electric work that effectively is provided by the voltage source to the passive element. Therefore, the power $p(t)$ consumed by the passive circuit element is this term alone, or

$$p(t) = p_{\text{ext}}(t) - q(t)\frac{dv(t)}{dt} = v(t)\frac{dq(t)}{dt}. \tag{2.44}$$

The derivative $dq(t)/dt$ is the rate at which negative charge is leaving one terminal of the passive circuit element at the higher potential and reentering its other terminal at the lower potential. Again, this corresponds to a *rise* in potential energy of the electrons. Thus $v(t)\ dq(t)/dt$ is the rate of work externally performed by the electric field set up in the voltage source, under force of the chemical reaction if, for example, the voltage source is a battery.

As there is only one path through which the charge can flow, this $dq(t)/dt$ is the current $i(t)$ flowing through both the passive element and the terminals of the voltage source. Therefore, the commonly encountered expression for power absorbed by a circuit element is

$$p(t) = v(t)i(t), \tag{2.45}$$

where $v(t)$ is the voltage across the circuit element and $i(t)$ is the electric current flowing through it at time $t$.

In more complicated circuits where several circuit elements are connected together in a network, the same result may be used. To compute the power into the network, the additional elements may merely be viewed together as one equivalent (more complicated) element (see Chapter 4). Also, the power $p(t)$ absorbed by an individual element of the network calculated as $v(t)i(t)$ is the contribution to the total demand on the source made by this element. In Sec. 4.4 we deal more precisely with power sign conventions, voltage polarities, and current directions.

The reader should also be forewarned that the power absorbed or supplied by a voltage source is nevertheless designated by $v(t)i(t)$. After all, the term $q(t)\,dv(t)/dt$ is either zero (for constant voltage sources) or may be very small compared with $v(t)i(t)$. Furthermore, omitting $q(t)\;dv(t)/dt$ allows an internally consistent energy balance to be written for the circuit without consideration of the external supply to the voltage source(s) for the circuit. It is also possible that the voltage source itself may be absorbing power from other sources within the circuit.

**Example 2.8**

A circuit element has a constant voltage of 4 V across it and a constant current of 20 mA flowing through it. What is the instantaneous power into this element?

**Solution** By Eq. (2.45), $p(t) = 4 \cdot 0.02 = 0.08$ W, which is also independent of time.

Because of Ohm's law, Eq. (2.35), the power dissipated by a resistor as expressed in Eq. (2.45) is

$$p(t) = v(t)i(t) \tag{2.46a}$$

$$= \frac{v(t)v(t)}{R} = \frac{v^2(t)}{R} \tag{2.46b}$$

$$= i(t)Ri(t) = i^2(t)R. \tag{2.46c}$$

The last expression, Eq. (2.46c), which says that the power dissipated in a conductor is proportional to the square of the current flowing through it, was observed experimentally by James Prescott Joule in 1841 and is therefore known as *Joule's law.*

This idea of resistive heating has been put to good use by mechanical engineers for electric boilers. These boilers, which consume a few kilowatts to tens of megawatts require no chimneys or vents, no combustion air, and no fuel storage areas. They supply primary power for many varieties of heating/ventilating/air conditioning systems. For example, some defroster elements in refrigerators use them. To obtain proper control, the dependence of resistivity on temperature (which can be strong) must be known and accounted for.

Systems use either special heating elements as the dissipating member (shielded by metal from the water), or the water itself. In the latter case, stepped-up voltages from 1 to 10 kV or more are applied to two electrodes in contact with the water. Given that the resistivity of water is $10^{-4}\ \Omega \cdot$ cm and the heat conductivity of water (see Example 2.12), it would be an interesting calculation to find how long it would take to heat a given volume of water, assuming a practical geometry.

### 2.10.1 Practical Examples

Now that we have discussed the battery, current, and power, several simple types of calculation can be made concerning electric appliances around the house. The following examples literally "bring home" and extend some of the ideas discussed in this chapter.

**Example 2.9**

**(a)** We may consider the meaning of a commonly used battery rating: ampere · hours. Recall that modern clock radios have a "backup" 9-V battery for maintaining the clock function should the power go out. During an outage, the clock display turns off, but the settings of time, alarm, and any other control feature settings on the unit are maintained/updated. Normally, the battery is disconnected, so there is zero drain on it; during an outage, the current is 3.4 mA. A new 9-V battery has the capacity to deliver 1 mA for 565 h; that is, it has a storage capacity of 565 mA · h. How long must a power outage be in effect for the clock to lose its memory?

**(b)** With both the clock display and the radio going, the current is 0.1 A. How long would the 9-V battery last under this condition were it used for this?

**Solution**

**(a)** Very simply, 565 mA · h/3.4 mA = 166 h, or roughly 1 week. That is sure to be sufficient for all but the worst disasters. Note that the actual energy stored in the battery is A · h times the voltage (9 V). Thus the stored energy is 0.565 A · h · 9 V · 3600 s/h = 18.3 kJ.

**(b)** Now the value is 0.565 A · h/0.1 A = 5.6 h. In practice, the battery would discharge even faster than a normal calculation would predict due to the performance degradation caused by such heavy loading (for a 9-V battery). This is why other types of battery are used for portable stereos. For example, a 20-W boom box draws a maximum current of 2.2 A at 9 V. We obtained 2.2 A by $p = vi$, where $p = 20$ W and $v = 9$ V. A 9-V battery would lose all its energy in much less than 15 minutes. With six D cells in series, each of which has a capacity of 14,250 mA · h (and therefore the same mA · h for the set of six, because all the current must pass through each series-connected cell), the playing time is 14.25 A · h/2.2 A ≈ 7 h. Note that the energy stored in one fresh D cell is 14.25 A · h · 1.5 V · 3600 s/h = 77 kJ, 4.2 times that stored in a 9-V battery.

**Example 2.10**

A smoke alarm draws 6 μA from a 9-V battery. How often must the battery be replaced?

**Solution** In theory, the alkaline battery would last 565 mA · h/0.006 mA = $9.4 \cdot 10^4$ h ≈ 11 years. Better have it checked every year to be sure, though. The self-drain alone due to internal leackage current in the battery will discharge the battery in 4 years or so. Thus the smoke alarm presents very little drain on the battery compared with self-drain. Also, testing the alarm or having it actually "go off" requires 10.7 mA. The beeping could therefore theoretically continue for 565/10.7 = 53 h, or about 2 days. In practice, the battery would probably give out much sooner (e.g., in most situations the battery would not be factory fresh).

**Example 2.11**

A 6-V Walkman radio draws, on the average, 40 mA. How long can we play the Walkman, given that the capacity of the AA series-connected cells is 2450 mA · h? What is the power consumed by this radio?

**Solution** A maximum of 2450 mA · h/40 mA = 61 h, but again in practice the value will be significantly less. The energy in a fresh AA cell is 2.45 A · h · 1.5 V · 3600 s/h = 13.2 kJ, making the total energy stored in the 4AA cell battery 52.8 kJ. The power consumed by the radio is $p = vi = 6 \cdot 0.04 = 0.24$ W.

We close this chapter with two simple examples of power/energy calculation around the house.

**Example 2.12**

(**a**) The power company charges 11 cents/kW · h. How much energy does it take to heat 1 gallon of water 90°F if we are told that the price of doing this is 2.5 cents?

(**b**) To do equivalent amounts of dishwashing by hand requires 3 gallons of hot water, 60 times/month; an automatic washer uses 10 gallons of water but needs to be run only 30 times/month. What is the cost of each, and which is cheaper?

**Solution**

(**a**) Dividing the cost per gallon by the unit energy cost gives the energy to heat 1 gallon of water as 2.5 cents/(11 cents/kW · h) = 0.23 kW · h. We can convert this to joules: 0.23 kW · h · 3600 s/h = 830 kJ. (Compare this with the energy in a battery or cell, calculated above.) Also, 1°F = 5/9°C. The energy required to heat 1 gallon of water 1°F is therefore 1 gallon · 3.78 liters/gallon · 1000 milliliters/liter · 1 $cm^3$/milliliter · 1 gram/$cm^3$ · 1 calorie/(gram · °C for water) · 4.184 J/calorie · 5/9°C/1°F = 8786 J/°F. To heat 1 gallon 90°F therefore requires 790 kJ. It would seem that the power company is underestimating the cost of heating 1 gallon, because it is unlikely that the heater operates at 790/830 = 95% efficiency.

(**b**) For dishwashing by hand, the cost per month is 3 gallons/wash · 60 washes/month · $0.025/gallon = $4.50/month. For the automatic washer, the cost is 10 gallons/wash · 30 washes/month · $0.025/gallon = $7.50—1.7 times as expensive, not including the investment and maintenance cost of the dishwasher.

**Example 2.13**

(**a**) What is the price per A · h at 120 V, assuming purely resistive loads? Use 11 cents/kW · h.

(**b**) Compare the cost/kW · h of electricity from a power company with that for typical batteries.

**Solution**

(**a**) Assuming 120-V line voltage, 1 kW involves a current $i = p/v =$ 1000 W/120 V = 8.3 A, so that 1 kW · h = 8.3 A · h at 120 V. Thus the cost of 1 A · h is (11 cents/kW · h)/(1 kW · h/8.3 A · h) = 1.3 cents/A · h.

(**b**) At this writing, the list price of a 9-V battery is $3.19, not including tax, which we will ignore here. From above, we know that such a battery has a capacity of 18.3 kJ = 18.3 kW · s · 1 h/3600 s = 5.1 W · h. Thus the cost per kW · h is $3.19/5.1 W · h · 1000 W/kW = $625/kW · h, 5700 times as expensive as house current! D cells are somewhat cheaper. The cost per cell is $1.60. Recalling that the D-cell capacity is 77 kJ = 77 kW · s · 1 h/3600 s = 21.4 W · h, the cost per kW · h is $1.60/21.3 W · h · 1000 W/kW = $75/kW · h, still 680 times as expensive as house current. C cells have capacity 7100 mA · h and are usually the same price per cell, so they are roughly twice as expensive per kW · h as D cells; however, they are smaller. Penlight (AA) cells have 2450 mA · h and cost $1.30 apiece, yielding $353/kW · h.

A typical 12-V automobile battery costs about $60. Its capacity is about 40 A · h, or roughly 1.7 MJ. This means that the cost per kW · h is $60/(40 · 12) · 1000 = $125/kW · h, a little more than the C cell. (Of course, this is an unfair comparison because the battery is recharged many times for far less than $60.) To give an example of how severe loading effects on batteries can be, one rating for a car battery is the *cold-start*, or *cold-cranking*, *amperes*. It is the current that the battery can deliver for 30 s at 0°F while maintaining a voltage of at least 7.2 V at the battery terminals. (At cold temperatures, the oil is thick and strains the oil pump, which thus demands twice the starting current as that at 80°F. Simultaneously, the chemical reactions in the battery are reduced to 40% of normal. On the other hand, the latter fact means that the self-discharge is much slower, so the shelf life is longer at cold temperatures.) A typical value of cold-start current is 500 A, which times 30 s is 500 · 30/3600 = 4.2 A · h. The battery "dies" after one-tenth of its capacity at normal loads.

The other rating commonly used today is *reserve capacity:* The number of minutes the car can run (at 80°F, maintaining at least 10.5 V) if the charging system fails and a basic set of extras such as lights, heater, and windshield wipers are operating (at 25 A). Modern batteries will typically last 90 minutes. Assuming that the ignition/computer control takes about 1 to 4 A (primarily the ignition coil), the lighting system 20 A, and that the other accessories consume another 2 to 3 A, this gives 25 A · 1.5 h = 38 A · h, similar to the foregoing figure of 40 A · h, as expected.

It must be concluded that the only reason batteries are used is because of their portability and convenience—certainly not their cheapness.

## 2.11 SUMMARY

In this first major chapter, the fundamental laws of electricity have been presented in some detail. Electricity is one of two related phenomena due to electric charge; the other is magnetism, the subject of Chapter 3. The electric force which governs electric phenomena is known as the Coulomb force. It is directed along the lines of the electric force field set up by a distribution of charge. In particular, the direction of the electric force between two charges is along the line joining them.

The strength of this force depends not only on this charge distribution but also the medium in which the force acts. A material that is easily polarized under the influence of an external charge distribution tends to decrease the magnitude of this force, compared with its magnitude in a vacuum. Because the force also depends on the charge on which the force is acting, it is convenient to normalize this force to that charge; the result is known as the electric field.

In regions away from the charge distribution responsible for the electric field, we may quantify its intensity by evaluating its flux density: the number of lines per unit area at the location in question. In fact, the closed-surface integral of the flux density vector is equal to the enclosed charge causing the electric field. The total electric field is decomposable into this flux density and the effect of the medium, represented by the polarization vector.

The electric potential between two points is defined as the work required to move a unit charge from one point to the other. For all points within perfect conductors, this work is zero. But for practical conductors, there is always some resistance to such movement. In the latter case, it is found that the oppositional forces (and therefore the potential energy difference between the ends of the conductor) increase in proportion to the rate at which charge is moving through the conductor, similar to viscosity. This is known as Ohm's law. The potential difference is called voltage, the rate at which charge moves through the conductor is called current, and the proportionality constant is called electrical resistance. In metallic conductors, the current is carried by electrons, which have negative charge. Because conventional current indicates the direction of positive charge flow, actual electron flow is in the opposite direction from conventional current.

The driving force behind the continuous current is a voltage source, such as a battery. Normally, when charge is allowed to flow, the charge distribution responsible for the charge motion is depleted, so that the current soon stops. Alternatively, a charge not part of the charge distribution making up the field would have zero work done on it when traveling around a circuit. Because resistance is like friction—nonconservative—charge could not be made to circulate continually through practical conductors. The voltage source is the solution: another nonconservative force field that overcomes the frictional forces in practical conductors. The mathematical result is that the integral of the electric field around a closed loop is zero unless there are sources of voltage (emfs), in which case the integral is equal to the sum of the emfs on that loop.

A capacitor is a very useful device for achieving certain kinds of circuit behavior, which will be introduced throughout the remainder of this book. It consists of two parallel plates, between which is a slab of dielectric. When a voltage is applied, charge in proportion to the voltage accumulates on the plates, causing an electric field in the dielectric. The proportionality constant is called capacitance. Electric energy is actually stored in the resulting electric field.

The rate of accumulation of charge on the plates of a capacitor is equal to the current on the wires connected to them. This is known as charge conservation. Without plates that allow charge buildup, this conservation law may be restated to say that the sum of all currents flowing into a circuit junction is zero. This law and the other concerning the integral of the electric field around a closed loop form the fundamental pair of relations known as Kirchhoff's laws, which are discussed in detail in Sec. 4.5.2.

It was shown in full detail that the power absorbed by any circuit element is equal to the product of the voltage across and current through that element. This is the power drain on any electric energy source connected to the element. The electric energy source itself presents a drain on chemical or mechanical energy sources driving it. Some of this energy is used in setting up the nonconservative electric field (emf).

This chapter closed with several examples concerning batteries, power, and energy costs of household appliances. These show the real-world application of the principles discussed previously. In Chapter 3 we consider the magnetic field, which is equally useful for other devices such as motors, generators, radio, and many of the other modern conveniences that we enjoy.

## PROBLEMS

**2.1.** Suppose that we have three collinear charges (all in one line): first $q_A = 2$ C, then to the right 2 m, $q > 0$, and then 3 m to the right of $q$, $q_B = 4.5$ C. Assume that the medium is homogeneous air ($\epsilon = \epsilon_0$). What is the net force on $q$? (Give both magnitude and direction.) Denote "to the right" by "$+y$ direction."

**2.2.** Two point charges are located a distance $d$ apart. The resulting electric field due to the point charges is zero at a location that is finite distances away from both the charges and not on the line segment joining them. What must be true about the signs of the charges? Where can the total field be zero?

**2.3.** Two charges of equal sign and magnitude are placed on metal conducting spheres, as shown in Fig. P2.3. Charge $q_1$ is fixed, while the sphere holding $q_2$ is suspended by a thread from point $A$. If $q_1 = q_2 = 1$ μC and $\theta = 40°$, how much does the sphere holding $q_2$ weigh?

**2.4.** The electric field strength $E$ due to an infinite line of charge is given by $E = \lambda/(2\pi r\epsilon_0)$, where $\lambda$ is the charge per unit length and $r$ is the radial distance from the wire.

**(a)** Show that this formula for $E$ still has the units of N/C.

**(b)** What is the attractive force on a charge $q_1$ of 1 μC when 1 mm from the line of charge if $\lambda =$ nC/m?

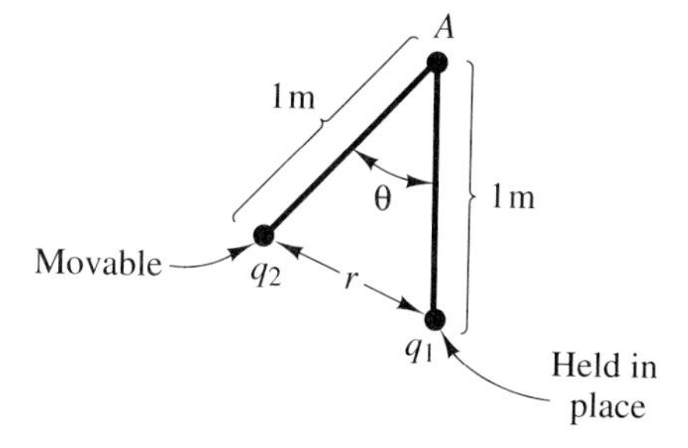

**Figure P2.3**

**2.5.** What is the torque $\boldsymbol{\tau}$ exerted by a uniform electric field $\mathbf{E}$ on an electric dipole consisting of two charges, $+q$ and $-q$, separated by a distance vector $\mathbf{l}$, and at an angle $\theta$ as shown in Fig. P2.5? If $\mathbf{E}$ and $\mathbf{l}$ are both in the plane of the paper, in what direction is $\boldsymbol{\tau}$? If $\theta = 90°$, $q = 1\ \mu C$, $l = 1$ cm, and $E = 100$ V/cm, what is the magnitude of this force?

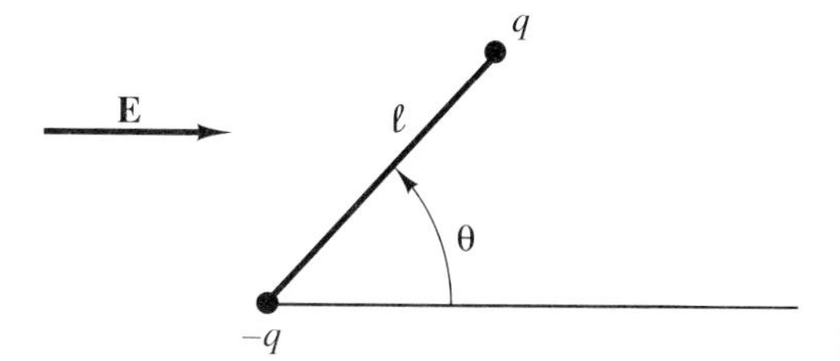

**Figure P2.5**

**2.6.** An irregularly shaped conductor has a charge of $q$. What is the shape of the surfaces of constant $E$ far from the conductor?

**2.7.** Two point charges $q_1$ and $q_2$ are 1 m apart. What is the total electric flux passing through a sphere of radius 0.6 m whose center is:
(**a**) On the midpoint of the line segment joining the charges.
(**b**) At the point $q_1$.
(**c**) At a point 0.3 m from charge $q_2$.
(**d**) Greater than 1 m from both charges.

**2.8.** Derive Coulomb's law for point charges by applying Gauss's law, integrating over a spherical surface. For what kinds of charge distributions (besides point charges) will Coulomb's law be valid?

**2.9.** A point charge $q = 10^{-4}$ C is embedded in a dielectric with relative permittivity $\epsilon_r = 3$. Find the magnitudes of the vectors $\mathbf{E}$, $\mathbf{D}$, and $\mathbf{P}$ at a distance of 10 cm from the charge.

**2.10.** (**a**) In Fig. P2.10, find the voltage $V_{AC}$ between points $C$ and $A$. Which point has the higher potential? Try to provide a physical reason.
(**b**) Find the voltage between points $B$ and $A$, $V_{AB}$. Provide your reasoning.

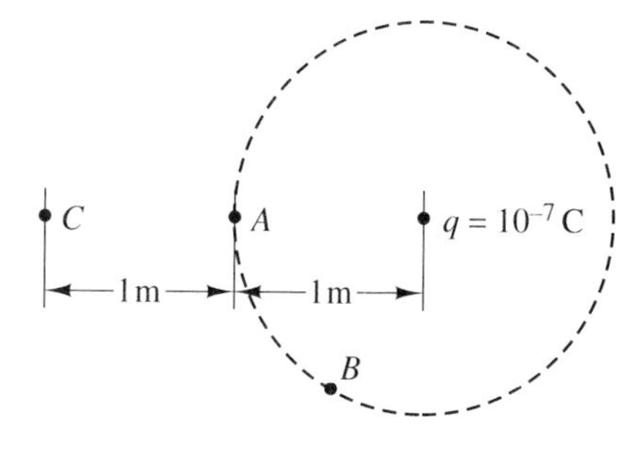

**Figure P2.10**

**2.11.** Suppose that a negative charge $q_1$ of magnitude 1 $\mu$C is placed in a vacuum. Find:

**(a)** The electric field at a point 1 m from $q_1$.

**(b)** The work done in bringing an electron from a point very far away ("infinitely" far away) from $q_1$ to a point 0.5 m away.

**(c)** The magnitude of the force on the electron at 0.5 m.

**2.12.** Given that the charge-to-mass ratio of an electron is $e/m \approx 10^{11}$ C/kg and that the acceleration of gravity is $\approx$ 10 m/s$^2$, find the required voltage between two metal plates that will balance the force of gravity on an electron. What does this say about gravitational vs. electric forces at the subatomic level?

**2.13.** Two equal and opposite point charges are located a distance $d$ apart, as shown in Fig. P2.13.

**(a)** What is the potential difference between points $A$ and $B$? Can you determine it without resorting to integration? What is the electric field at points $A$ and $B$?

**(b)** What if both charges are equal and of value $+q$? Answer qualitatively and then quantitatively.

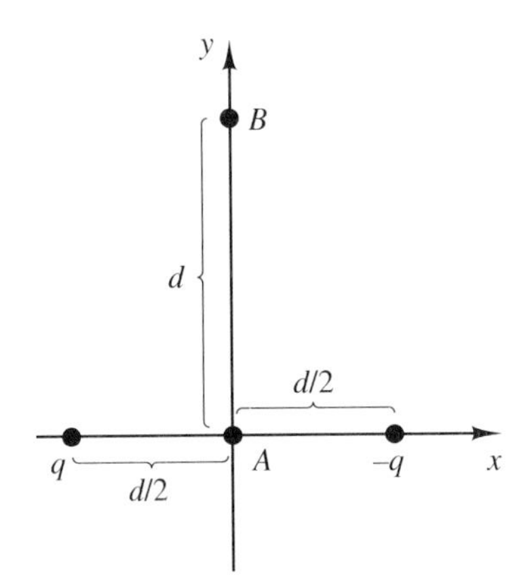


**Figure P2.13**

**2.14.** Consider again the infinite line of charge of Problem 2.4. In terms of $\lambda$, what is the potential difference between a point 1 m from the line and another point 2 m from the line?

**2.15.** **(a)** If **E** is zero throughout a region, must the potential $V$ also be zero throughout the region? Why or why not?

**(b)** If the potential $V$ is zero throughout a region, must **E** also be zero throughout the region? Why or why not?

**2.16.** A capacitor is made from a roll of aluminum foil and paper. The roll of foil measures 75 ft by 1 ft; the paper is $\frac{1}{150}$ in. thick and has a relative permittivity of $\epsilon_r = 3.5$.

**(a)** What will be the capacitance of the resulting capacitor if the two plates are made by cutting the long side of the unrolled foil rectangle in half, and sandwiching the paper between the two halves?

**(b)** How long would the roll of foil have to be to have a capacitance of 1 F?

**2.17.** A 1-$\mu$F capacitor is plugged into a wall outlet (120 V ac—actually a 60-Hz sinusoidal voltage of amplitude $120\sqrt{2} = 169.7$ V). It is left there for a few seconds, and then is removed. What will be the charge on the capacitor when removed? Assume that the capacitor is fully charged to its equilibrium value in a time very short compared with $\frac{1}{60}$ Hz, a realistic assumption. **Do not try this experiment on your own—you could get seriously shocked.**

**2.18.** It takes $10^{-6}$ J to fully charge a capacitor from a 100-V source. What is the capacitance of the capacitor? If the capacitor is a parallel-plate capacitor with $d = 0.1$ mm and a dielectric for which $\epsilon_r = 5$, what is the area of the plates?

**2.19.** A cylindrical capacitor is made from two coaxial cylinders of radii $a$ and $b$, length $L$, and dielectric constant $\epsilon_r$. Using Gauss's law, the electric field between the cylinders when the there are charges $q$ and $-q$ on, respectively, the inner and outer conductors is radially outward directed and has magnitude $E = q/(2\pi\epsilon_0 rL)$. Find the capacitance of this system. How does your answer compare with the capacitance of a parallel-plate capacitor with area equal to that of the inner cylinder? Or equal to that of the outer cylinder?

**2.20.** Suppose that a $\frac{1}{2}$-mile by $\frac{1}{2}$-mile cloud 10,000 ft high undergoes a lightning bolt discharge. The current is found to have the time dependence shown in Fig. P2.20. The peak voltage is 200 million volts.

(**a**) What charge was on the cloud before the lightning bolt? Use the graph to obtain a rough estimate.

(**b**) What is the capacitance of the cloud–earth system? (Use $Q = CV$.)

(**c**) Repeat part (**b**) by using $C = \epsilon_0 A/d$. Do your answers agree?

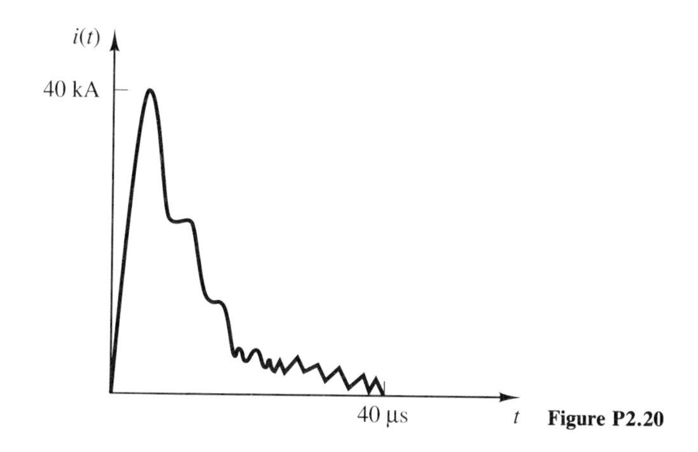


**Figure P2.20**

**2.21.** A capacitor is constructed on a silicon chip by forming a rectangle of aluminum 100 μm by 75 μm on top of an insulating silicon dioxide ($SiO_2$) dielectric having thickness $T_{ox} = 20$ nm (see Fig. P2.21). The capacitor has a capacitance of 13 pF. What is the relative permittivity of the $SiO_2$? (*Note:* Below the $SiO_2$ is a silicon base, which we may consider to be the other capacitor plate, having an effective area equal to that of the aluminum plate.)

**2.22.** Suppose that a battery charger supplies a 2-A constant current to a car battery for 3 h. With a voltmeter, we measure the battery voltage to rise with time according to the formula $(10 + t/2)$ V, where $t$ is in hours. Thus every 2 h it gains a full extra volt.

(**a**) Find how much charge was delivered to the battery in 3 h.

(**b**) What is the terminal voltage of the battery after this time?

(**c**) How much energy was added to the auto storage battery during this 3 h of charging?

(**d**) How much money did it cost to charge the battery for these 3 h, neglecting losses in the battery charger and assuming that home electricity costs 5 cents/kWh?

**2.23.** A plastic trough measures 1 m by 10 cm by 10 cm deep and is filled with water (conductivity $\sigma = 2\cdot 10^{-4}\ \Omega^{-1}\ \text{m}^{-1}$). A 100-V battery has one terminal connected to a

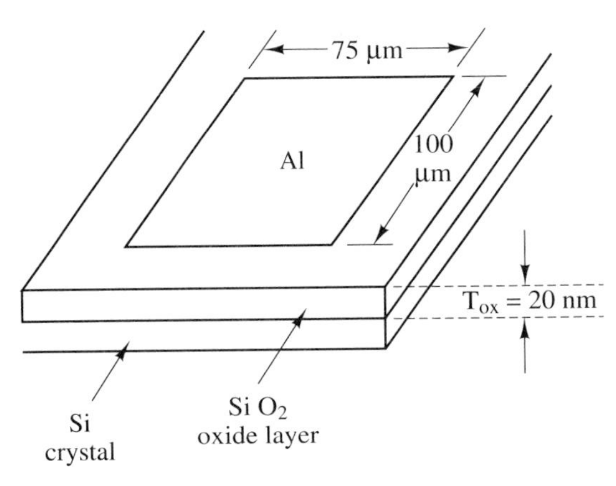


Figure P2.21

plate immersed in the far-right end of the trough, and the other terminal is connected to a similar facing plate at the far-left end.

**(a)** What is the potential difference between the two wires inserted into the water a distance $d$ apart along the direction perpendicular to the end plates ("axial")?

**(b)** What is the resistance of the water in the trough between the plates?

**(c)** Given that a typical digital voltmeter has an input resistance of 10 MΩ, what do you think you would measure if you connected the wires in part **(a)** to the dc voltage inputs of the voltmeter? Answer qualitatively.

**2.24.** When the current $i(t)$ in Fig. P2.24a is applied to an electric network via a current source, the voltage waveform across the network is as shown in Fig. P2.24b. Find mathematical expressions for the instantaneous power $p(t)$ and the energy $w(t)$ delivered to the network for $t \geq 0$. Complete the table below and sketch $p(t)$ and $w(t)$. Recall that for $t \geq t_0$, the accumulated energy into the network is $w(t) = w(t_0) + \int_{t_0}^{t} p(t')dt'$. Assume that $i(t < 0) = v(t < 0) = 0$.

| Interval (s) | Current, $i(t)$ (A) | Voltage, $v(t)$ (V) | Power, $p(t)$ | Energy, $w(t)$ |
|---|---|---|---|---|
| $t \leq 0$ | 0 | — | 0 | 0 |
| $0 \leq t \leq 0.5$ | $2t$ | 3 | | |
| $0.5 \leq t \leq 1.0$ | $2 - 2t$ | 3 | | |
| $1.0 \leq t \leq 2.0$ | 0 | $6 - 3t$ | | |
| $2.0 \leq t$ | 0 | 0 | | |

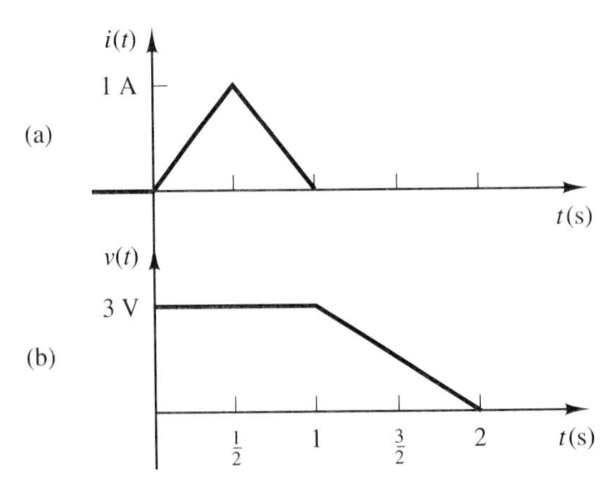


Figure P2.24

**2.25.** A camcorder draws 9.5 W from a 12-V 1500-mA · h battery pack. How long will the battery last?

**2.26.** Give the currents average drawn by the following household devices, assuming that the wall voltage is 120 V dc. (For these devices, the results will be approximately correct also for the more accurate assumption of ac.)

**(a)** $\frac{1}{4}$-W night-light.

**(b)** 1100-W laundry hand iron.

**(c)** 150-W floodlight.

**(d)** 1500-W hair dryer.

**2.27.** Give the average power drawn by the following household devices, assuming that the wall voltage is 120 V dc. (For these devices, the results will be approximately correct also for the more accurate assumption of ac.)

**(a)** 4.2-A vacuum cleaner.

**(b)** 1.2-A 20-in. screen color TV set.

**(c)** 2.6-A chain saw.

**(d)** 3.5-A edge trimmer.

**(e)** 0.5-A three-speed oscillating fan.

**2.28.** A fully charged auto battery, rated at 30 A · h, supplies current to a starter motor. The starter draws 300 A for 5 s each time the engine is cranked. How many times can the engine be cranked before the battery dies?

**2.29.** A pack of four 75-W incandescent lamps costs \$2.29, and each lasts 750 h. An equivalent fluorescent lamp draws only 18 W and lasts 10,000 h but costs \$21.95.

**(a)** Assuming that wall electricity costs 8 cents/kWh for the entire period, what is the electricity cost to supply 10,000 h of light for each type of lamp?

**(b)** Including the cost of bulbs, which costs less, and by how much?

**2.30.** A 100-W 120-V light bulb has a resistance of 9.5 Ω at room temperature.

**(a)** What is its resistance $R_a$ when the filament is heated at lighting temperature?

**(b)** Suppose that the resistance rises from 9.5 Ω to the value found in part (a) in 50 ms with a linear dependence on time. For simplicity, assume that the wall voltage is constant-in-time[dc; in fact, it is sinusoidal (ac)]. How much energy is dissipated by the light bulb during this 50-ms interval?

**(c)** What is the average power (total energy divided by total time) over this 50-ms interval?

**(d)** What are the maximum and minimum values of instantaneous power?

**2.31.** Estimate the cost per year of a leaky faucet that drips once per second (there are about 12,600 drops in a gallon). The electric cost per °F to heat 1 gallon of water is 0.028 cents. Assume that we heat the water to 140°F from the ground temperature of 50°F.

**2.32.** An electronic flash unit stores sufficient charge on a 200-μF capacitor to bring its voltage up to 400 V. This large voltage is needed to discharge the xenon flashbulb upon closing the shutter. The flash unit uses two AA (1.5-V) batteries. Assume that the flash circuit has an efficiency of 10%.

**(a)** For typical AA batteries, how many flash pictures can we take before the batteries go dead? How many 36-picture rolls of film is this?

**(b)** How much charge is stored on the capacitor when fully charged?

**(c)** Also give the stored energy in joules.

**2.33.** **(a)** The terminal voltage of an automobile battery declines as a result of discharging; it can be modeled by the formula $V_T = 11.8\text{ V} + (\%\text{ charged}) \cdot 0.008\text{ V}$. Thus when there is insufficient charge to use for operating the car, the voltage is 11.8 V. What is the terminal voltage of the fully charged battery?

**(b)** Permanent damage to the battery can result if the battery is less than 75% charged.

If we were to determine the state of charge by a terminal voltage measurement, when should we become worried?

**(c)** The electrolyte in an auto battery is a mixture of water and sulfuric acid, the latter of which has a specific gravity of 1.84, where specific gravity is the ratio of mass density of a substance to that of water. If the specific gravity of a completely discharged battery is 1.12, what are the percentages of water and acid in that case? If the specific gravity of the mixture rises by 0.0014 for each percent that the charge is increased, what is the composition of the water–acid mixture at full charge?

**2.34.** Find the spatial energy densities ($W \cdot h/cm^3$) of the following batteries:

**(a)** A 6-V rectangular lantern battery, 3 in. by 3.75 in., with a capacity of 7.5 A · h.

**(b)** A 1.5-V cylindrical AA battery, $r = \frac{9}{32}$ in., $d = 2$ in., with a capacity of 1.2 A · h.

**(c)** A 1.5-V cylindrical watch battery, $r = 0.23$ in., $d = 0.19$ in., with a capacity of 0.19 A·h. (*Note:* 1 $in^3 = 16.39\ cm^3$.)

**2.35.** Why is it unwise to connect together different types of commercially available batteries (e.g., D and AA)?

**2.36.** An electric oven draws 25 A at 120 V.

**(a)** What power does it dissipate?

**(b)** At 11 cents/kWh, how much does it cost to bake a cake (45 min)?

## ADVANCED PROBLEMS

**2.37.** Is it possible for a finite set of point charges to produce an electric field that is zero everywhere? Why or why not? Comment on the limitations of validity of Coulomb's electric field relation, Eq. (2.3).

**2.38.** In Chapter 7 we shall study electronic devices. One electronic device, called the diode, has a charged region within its essentially silicon crystal structure. Although the charges (ions) are actually immobile, they are not due to polarization and are uncanceled. Thus for Gauss's law they can be considered "free" charge. Suppose that this "depletion" (charged) region contains $N_a = 10^{16}$ positively charged ions per $cm^3$, each having charge $+e$ (where $e$ is the magnitude of the charge on an electron). The region to the left of the depletion region is neutral ($E = 0$), and the electric field is assumed to have only an $x$ component elsewhere (see Fig. P2.38). If this depletion region is 100 nm thick, what is the field $E_{max}$ at the right edge of the charged region? (*Hints:* For silicon, $\epsilon_r = 11.7$. Also, by Gauss's law, if we take a surface $S$ with one side where $E = 0$ and the other side a distance $x$ within the charged region, the enclosed charge increases in linear proportion to $x$.)

**2.39.** A spherical charged solid material immersed in a vacuum has a uniform charge density of $\rho = 10^{-10}\ C/m^3$ and a permittivity $\epsilon = 10\epsilon_0$. The sphere has a radius of $R = 1$ cm (see Fig. P2.39).

**(a)** Plot the electric field magnitude as a function of radius. *Hints:*

**(i)** $\int_S \epsilon \mathbf{E} \cdot d\mathbf{S} = q_{enc} = \int_V \rho\, dV$.

**(ii)** $\mathbf{E}$ is radially directed outward and is spherically symmetric.

**(iii)** The volume of a sphere is $4\pi r^3/3$ and the area of a spherical shell is $4\pi r^2$.

**(iv)** Consider three cases: $r < R(\rho = 10^{-10}\ C/m^3,\ \epsilon_r = 10)$

$r = R$[to make plotting $E(r)$ easy].

$r > R(\rho = 0$—careful, $q_{enc} \neq 0!$, $\epsilon_r = 1)$.

**(v)** For convenience, use $\epsilon_o \approx 10^{-11}$ F/m.

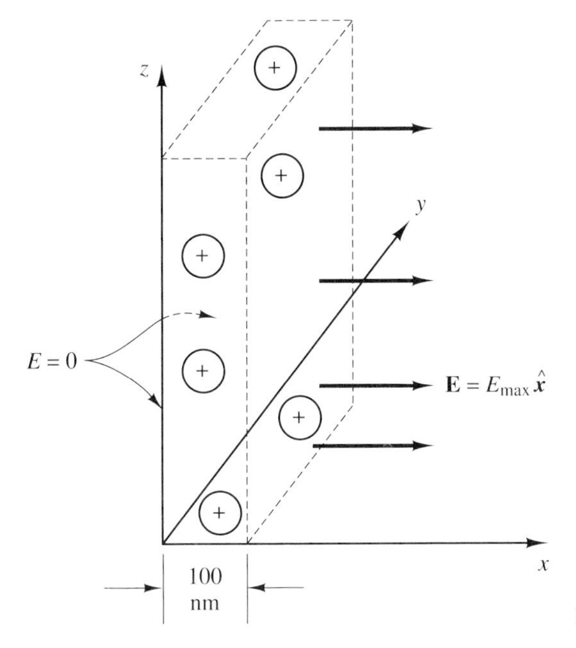


**Figure P2.38**

(**b**) Plot the electric potential $V(r)$, assuming that $V(r = \infty) = 0$. (*Hints:* $V_{ab}(r) = -\int_a^b \mathbf{E}(\mathbf{r}) \cdot d\mathbf{r}$. **E** is radically directed and spherically symmetric, so $\mathbf{E}(\mathbf{r}) \cdot d\mathbf{r} = |\mathbf{E}(r)|\, dr$. Also:
For $r > R$, $a = -\infty$, $b = r$.
For $r = R$, $a = -\infty$, $b = R$.
For $r < R$, use $a = R$ and $b = r$; add this to $V(R)$.
[What is $V(0)$?] Always use the appropriate expression for electric field in part (a) for the given integration interval!)

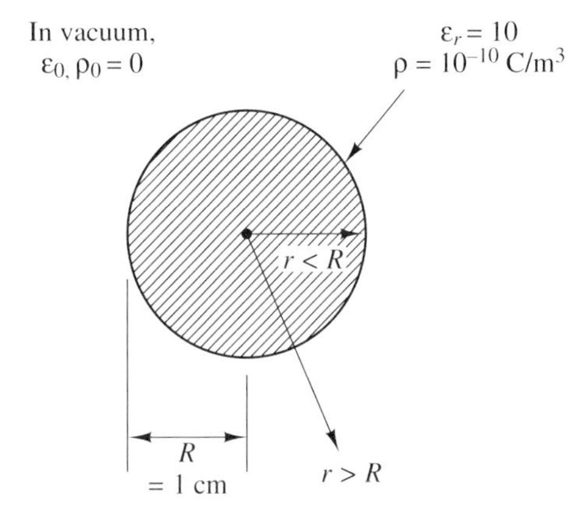


**Figure P2.39**

**2.40.** Suppose that a brave (or stupid) person decides to test personally the laws we have studied. He goes to an noninsulated 760,000-V power line, gets on a well-insulated ladder, and hangs from the copper wire. Suppose that he still is fine. Then he lets go and falls to the ground, $d = 10$ ft below. He is still fine. Why was he not shocked when he hit the ground? What was his potential when he was in midair (height $y$) with respect to ground? What circuit model might be appropriate for this situation?

**2.41.** A capacitor is constructed from two concentric spheres, as shown in Fig. P2.41. The inner sphere has radius 1 cm and the outer sphere has radius 5 cm. What is the capacitance of this capacitor? What is the maximum electric field magnitude when the capacitor is charged to 100 V? [*Hint:* Using Gauss's law, integrate electric flux over a spherical surface to find the electric field a distance $r$ from the center of the capacitor spheres. Then determine the potential difference in terms of the charge $Q$ by integrating $E(r)$ radially from $r = 1$ cm to $r = 5$ cm.]

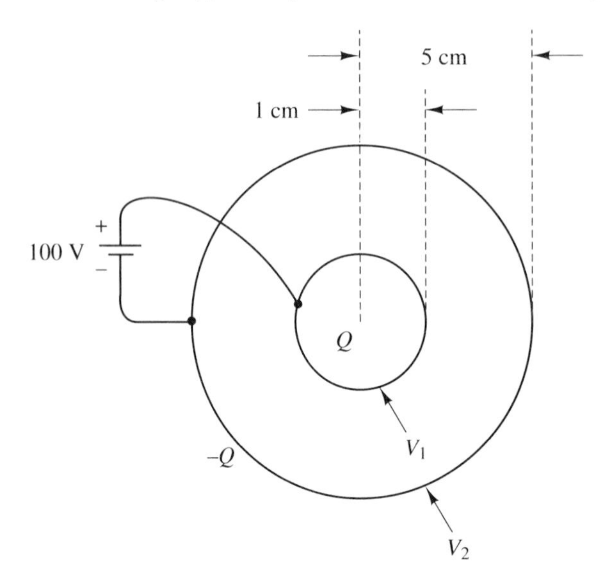


**Figure P2.41**

**2.42.** When we plug wires into an ac wall outlet, it is dangerous to touch them because they have a 120-V potential difference (which, incidentally, is sinusoidally oscillating). To simplify this discussion, suppose that it is a constant 120 V. Let the wire ends away from the wall be unconnected to anything, and consider a typical 6-ft power cord. When we unplug the cord from the outlet, suddenly it is no longer dangerous to touch the wires. Where did the voltage "go"? Or is there still 120 V across it? Could we measure it with a voltmeter? How can we model this situation properly? (*Hint:* Think about capacitors and storage of energy. How much charge and energy are stored, and how?)

**2.43.** Concerning Problem 2.42, suppose that the ends are now connected, through an on/off switch, to an electrical load. If the switch is off, is the situation the same as in Problem 2.42? What if the switch is on?

**2.44.** Consider the process of charging two initially uncharged metal spheres of radius $a$ whose centers are a distance $R$ apart by removal of electrons from one ball and depositing those electrons on the other. The spheres are immersed in a dielectric having permittivity $\epsilon$. A voltage source such as a battery does the pulling of electrons off of one plate and pushing of them onto the second sphere, via wires connected to the spheres.

**(a)** Describe the electron distribution on the spheres—where on each sphere are electrons "bunched up"?

**(b)** The difference in potential between the spheres when they have equal and opposite charges $q$ and $-q$ can be calculated by a method outside the scope of this book called the method of images. The simple result is that this potential difference, $V_{12}$, is

$$V_{12} = \frac{q}{2\pi\epsilon_0 aX}, \tag{P2.1}$$

where the constant $X \approx 1 + a/R + (a/R)^2$, assuming that $a \ll R$. Note that $V_{12}$ is the energy it takes to move a unit charge from the surface of one sphere to the surface of the other.

With the result Eq. (P2.1) in hand, no integration of electric fields over distances is required to find the potential; that has already been done for us. Let us build up the charge on these spheres incrementally by moving tiny charges $dq$ from one sphere to the other repeatedly, and in the process building $q$ up from zero to $Q$.

Using Eq. (P2.1), and noting that the required energy for each $dq$ transfer is $dq \cdot V_{12}(q)$, find the total energy required to build up charge $Q$ on one sphere and $-Q$ on the other.

**(c)** Suppose that we wish to transfer $Q = 10^{-4}$ C of charge from a metal ball of radius $a = 1$ cm $= 0.01$ m to an identical ball whose center is $R = 10$ cm $= 0.1$ m away from that of the first ball. Thus $a/R = 0.1$. Find the number of joules required to do this charge transfer. Also find the capacitance of this capacitor, using Eq. (P2.1). Repeat for $R = 2$ cm $= 0.02$ m so that $a/R = 0.5$.

**2.45.** How would the results of Problem 2.29 change if it were assumed that one uses the lamp only 2 h 17 min per day and the interest rate is 6%? The interest would increase the real cost to you, because if you had not bought the light, that money would be in the bank earning interest. Assume a 3% annual inflation rate.

3

# Physical Laws II: Magnetism and Electromagnetism

## 3.1 INTRODUCTION

Electrical engineering is doubly fascinating. Not only is there an invisible electric field that can be used to light and heat our rooms at night, but there is also the invisible magnetic field, without which devices as diverse as electric motors, radio/TV broadcasting, audio speakers, audiocassette recorders and VCRs, and refrigerator magnets would be impossible.

Magnetic and electric fields have much in common, which is understandable, given that they both arise from electric charge. The iron-filing-patterns demonstration that everyone sees in elementary school has been repeated again and again for over 2000 years. Titas Lucretius Carus did it in the year 56 B.C.

In those days there were no electromagnets, only "permanent" magnets found in the ground, called *lodestone.* This word evidently stems from the capability of compasses (invented by the Chinese in 1000–3000 B.C.) made from it to guide or "lead" (in old English, "lode") the traveler. Because the lodestone was found in Magnesia, an ancient Turkish city, it was also called a magnet.

Just as was the case with electricity, not much was done with magnetism until the nineteenth century. The long delay is really due to lack of quantitative models until then that would allow electric and magnetic behavior to be understood, predicted, and ultimately created and controlled using designed apparatus.

A significant factor extending this delay was the discouragement of scientists, who in the middle ages were sometimes labeled as black magicians and punished for or prevented from carrying out their research. In a sense those blind persecutors unknowingly had a bit of the truth. Our knowledge of electromagnetism has (along with other technologies) brought onto the earthly scene nuclear weapons, much of our pollution, excessive consumption, and many other ills facing us today. However, at this point the genie is out of the bottle. It is up to us to understand physical phenomena (and human nature) even better to try to solve the problems that have arisen with and from technology.

## 3.2 THE MAGNETIC FORCE

### 3.2.1 Coulomb's Magnetic Force

It had long been known that an end of a piece of lodestone would attract one end and repel the other end of a second piece. The two ends, or poles, were labeled north (N) and south (S) with reference to the magnetic poles of the earth. (Originally, the northerly pull on a pole of the compass needle was thought to come from *Polar*is, the guiding North Star.) Coulomb showed experimentally that these poles attract and repel according to the same type of law as holds for the force between electric charges, Eq. (2.1):

$$\mathbf{F}_{21} = \frac{\mu_0 m_1 m_2}{4\pi r^2}\,\hat{\boldsymbol{\ell}}_{12}, \tag{3.1}$$

where $\mathbf{F}_{21}$ is the force vector on magnetic pole 2 due to pole 1 in newtons, $m_1$ and $m_2$ are the two magnetic pole strengths, $r$ is the distance in meters between the poles, $\hat{\boldsymbol{\ell}}_{12}$ is the unit vector pointing from pole 1 to pole 2, and $\mu_0$ is the medium parameter, needed to make the numerical value of the right-hand side newtons. Much more will be said below about $\mu$, the *permeability* of the medium (in this case air or free space, so $\mu = \mu_0$), which in a sense is a measure of how "permeable" the substance is to magnetic fields.

It must be stressed from the outset that unlike electric charge, magnetic poles *always* occur in N-S pairs (dipoles). There are no isolated magnetic poles (monopoles, or "magnetic charges"), contrary to the implication of Eq. (3.1). If a bar magnet is broken, two dipole (N-S) magnets result. Thus the appropriate picture for Coulomb's experimental result [Eq. (3.1)] is as shown in Fig. 3.1, where the contributions from the distant poles are ignored. In addition, as with electric charges, the idea of a "point" magnetic pole is an idealization or mathematical convenience. Rather than try to impose a patched-up analog to each and every electric field equation, scientists soon discarded the monopole idea implicit in Eq. (3.1) for the

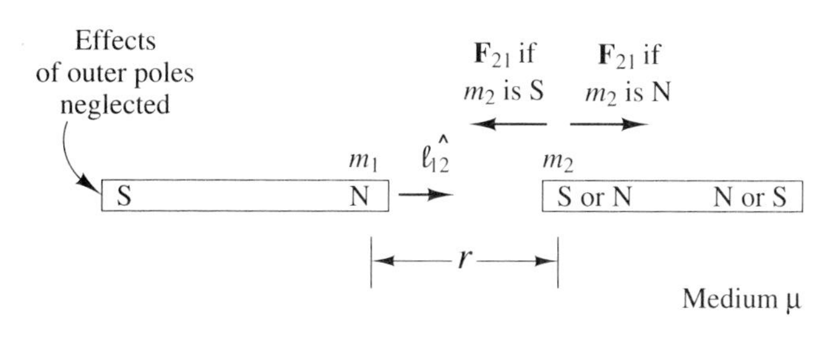


**Figure 3.1** Coulomb's force between magnetic poles. The forces shown are those acting on the right magnet (2) due to the left magnet (1).

much more realistic dipole model. In particular, this new model successfully explains the forces Coulomb observed in Eq. (3.1), without requiring monopoles.

If iron filings are placed on a paper held above a bar magnet, a pattern such as in Fig. 3.2a appears, when the paper is shaken to facilitate the magnetic force in moving the filings around. What is the meaning of the direction of the arrows in Fig. 3.2? They indicate the strength and direction of the *magnetic flux density vector* **B**. Specifically, recall that the direction of the flux lines of **E** indicates the direction of the force that **E** exerts on a unit positive charge at that point in space. Similarly, we can imagine that if a magnetic monopole or charge *did* exist, the direction of the magnetic flux lines would indicate the direction of the force that **B** exerts on a unit positive magnetic point charge (point north pole) at that point in space.

A similar meaning can be ascribed to these lines for their action on tiny dipoles, such as the iron filings. Even if free to move, a dipole obviously will not travel along **B** lines, because the force on one pole is in a direction opposite that of the force on the other pole. But if the poles of the dipole were not originally aligned with **B**, a torque in proportion to $B$ (the magnitude of **B**) would tend to align the dipole with **B**.

The iron filings are slender dipoles that align themselves due to this torque. When aligned, they themselves actually contribute to the **B** field. The bar magnet, itself a large dipole, is the source of **B** because its millions of constituent dipoles are mostly aligned. That alignment is inherent if the bar is lodestone, and can be induced in bars of similar metals like that of the iron filings (*ferromagnetic* material) using current-carrying coils wrapped around the bar as discussed below. In the latter case, part of the alignment remains even when the coil is removed, resulting in a "permanent" magnet; this is known as *residual magnetism.*

This pattern from the magnetic dipole is identical to that of the lines of uniform electric potential surrounding an electric dipole, rotated 90° with respect to the orientation of the magnetic dipole. Far from the dipole, it is similar to the pattern of the electric field surrounding an electric dipole having the same orientation as the magnetic dipole (see Fig. 3.2b).

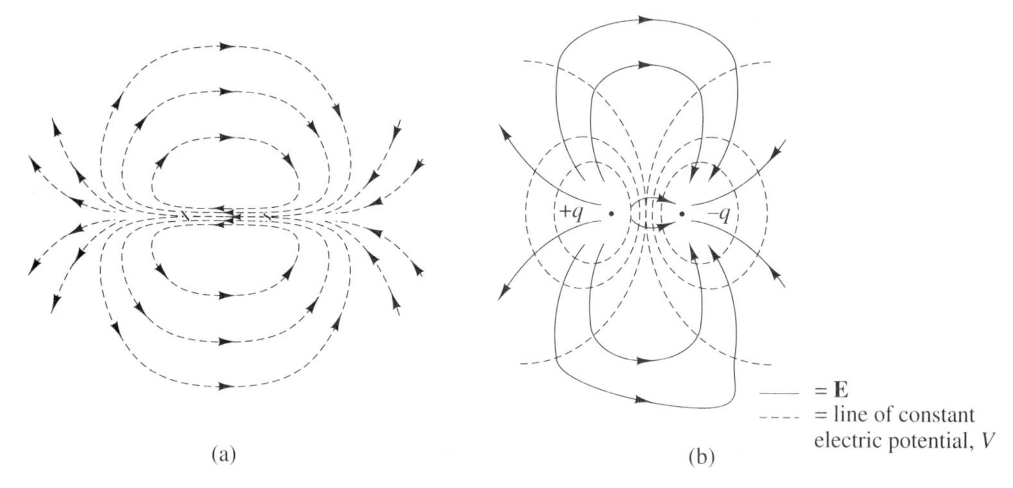


**Figure 3.2** (a) Magnetic field/flux lines surrounding a bar magnet (magnetic dipole); (b) electric field lines and equipotentials surrounding an electric dipole.

### 3.2.2 Ampère's Magnetic Force

In 1819, a Danish professor named Hans Christian Oersted was giving a "basic EE" lecture when he made a startling discovery. He was showing how steady currents created heat in resistive conductors. When the circuit was connected and disconnected, he noticed that a compass which happened to be lying on his demonstration table deflected one way and then the other. By accident, he had discovered *electromagnetic induction.* That is, magnetic effects could be produced by an electric circuit. (A less romantic but perhaps more plausible account has Oersted intentionally studying the electric effects on the compass, inspired by the effects that lightning was known to have on compasses.)

Soon after that, Oersted discovered that a wire in which electric currents were flowing moved (and thus experienced a force) when a bar magnet was brought near it (see Fig. 3.3a). There is a magnetic field due to the bar magnet, the magnetic flux density vector **B**. He found that the force was proportional to both the current and the magnetic flux density, and also perpendicular to both. In today's notation, the force $\mathbf{F}_{/\mathrm{ul}}$ per unit length (hence the subscript "/ul") on a wire carrying a current $I$ in a magnetic field **B** (here, **B** being due to the bar magnet) is

$$\mathbf{F}_{/\mathrm{ul}} = I\hat{\ell} \times \mathbf{B}, \tag{3.2}$$

Oersted discovers electromagnetism while lecturing. Note voltaic pile and compass on the laboratory table. He could not foresee how others would soon build on this discovery to transform the classical era into modern times. *Source:* Courtesy of the Burndy Library, Norwalk, CT.

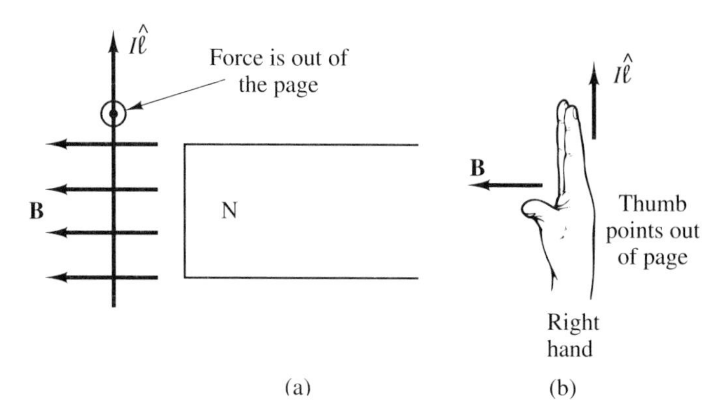


**Figure 3.3** (a) Force on a current-carrying wire in the presence of the magnetic field of a bar magnet; (b) right-hand rule for determination of the direction of the magnetic force in (a).

where $\times$ is the symbol for the vector cross product and $\hat{\ell}$ is a unit vector pointing in the direction of the current. Figure 3.3a illustrates Eq. (3.2) for $I$ near the magnetic pole, in which case **B** is essentially uniform and perpendicular to $\hat{\ell}$ for the section of the wire directly opposite the pole. The magnitude $F_{/\text{ul}}$ of the force per unit length is thus $F_{/\text{ul}} = IB$. Thus the force is proportional to both the current and the magnetic field, and is perpendicular to both.

Unlike the electric force, the magnetic force exerted is not in the same direction as the field but is perpendicular to both **B** and the current direction $\hat{\ell}$. As with all cross products, the direction of the force is determined by the right-hand rule (see Fig. 3.3b). Point the fingers of your right hand in the direction of $I\hat{\ell}$. Orient your hand so that when viewed from your thumb, a counterclockwise motion (180° or less) of your fingers points in the direction of **B**. Your thumb then points in the direction of $\mathbf{F}_{/\text{ul}}$. In Fig. 3.3 the direction of $\mathbf{F}_{/\text{ul}}$ is out of the page.

André-Marie Ampère in 1820 took this one step further: He found that two parallel wires carrying currents repelled or attracted each other with no bar magnet present! (see Fig. 3.4). In today's notation, the force per unit length of wire is

$$\mathbf{F}_{21/\text{ul}} = \frac{\mu_0 I_1 I_2}{2\pi r}\hat{\ell}_{21}, \tag{3.3}$$

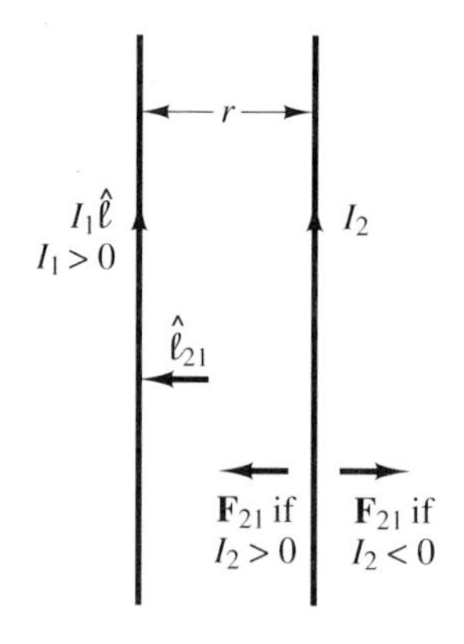


**Figure 3.4** Magnetic force between current-carrying wires.

where $\mathbf{F}_{21/ul}$ is the force per unit length of wire in newtons on wire 2 (and $\mathbf{F}_{12/ul} = -\mathbf{F}_{21/ul}$), $I_1$ and $I_2$ are the two currents in amperes (whose polarities are both defined with respect to the same direction), $r$ is the distance in meters that the wires are held apart, $\hat{\boldsymbol{\ell}}_{21}$ is the unit vector pointing from wire 2 to wire 1, and $\mu_0$ is the permeability in $N/A^2$ of the medium between the wires (typically, air or other material for which $\mu \approx \mu_0$). For practical limitations on Eq. (3.3), see the problems.

Note that if both currents flow in the same direction ($I_1$ and $I_2$ have the same sign), the wires attract each other. If the directions of the currents are opposite, they repel each other. This is the reverse of the situation of Coulomb's force, where like charges repel.

As presented in Eqs. (3.1) and (3.3), usually $\mu = \mu_0$ because typically the wires or bar magnets are immersed in air (free space). Moreover, essentially all materials except ferromagnetic metals have $\mu \approx \mu_0$. Thus even if, for example, the wires are covered with insulation, use of $\mu_0$ is appropriate. The value of $\mu_0$ has been assigned as $4\pi \cdot 10^{-7}$ $N/A^2$ (see below).

Just as with permittivity, the relative permeability $\mu_r$ of any medium is defined as $\mu_r = \mu/\mu_0$. The value of $\mu_r$ in magnetic materials ranges from 1000 to 1,000,000, depending on the base material and alloy. Unlike the electric case, the operational situations for which a single, constant value of $\mu_r$ for a magnetic medium can be defined are quite restricted. This point is discussed further in Sec. 3.8.

**Example 3.1**

Suppose that the force between two wires a distance $r = 1$ m apart, and both containing the same current $I$, is measured with a *very* accurate device to be $2.00 \cdot 10^{-7}$ N/m. What current is flowing through the wires?

**Solution** By Eq. (3.3), the current is $I = (F_{21/ul} 2\pi r/\mu_0)^{1/2} = [2 \cdot 10^{-7} 2\pi \cdot 1/(4\pi \cdot 10^{-7})]^{1/2} = 1$ A. Actually, the International System of Units (SI) uses this example to *define* the ampere, as noted in Sec. 2.9 when we introduced the ampere. The C (coulomb) is actually in turn defined as 1 A · s.

[What about the value of $\mu_0$? In the cgs system, $\mu_0$ was defined to be 1. The abampere (10 A) was then defined as the current in two wires held in air 1 cm apart that is necessary to produce 1 dyne $= 10^{-5}$ N per centimeter between them. Thus in mks, $\mu_0$ had to be $10^{-5}$ N/dyne $\cdot\, 10^{-2}$ abampere$^2$/A$^2 = 10^{-7}$ N/A$^2$; the cm units cancel. The $4\pi$ is merely due to a subsequent rescaling of all electromagnetic equations that minimizes the occurrences of "$4\pi$" in the other unit definitions.

The theory of electromagnetic waves shows that $1/(\epsilon_0\mu_0)^{1/2} = c$, the speed of light (see Sec. 3.10.4). With $\mu_0$ assigned and $c$ measurable as $3 \cdot 10^8$ m/s, $\epsilon_0$ had to have the value $8.85 \cdot 10^{-12}$ F/m. One may also measure $\epsilon_0$ via Eq. (2.1) and obtain the same number. Determining the rationale behind unit definitions can be a very complicated and confusing enterprise!]

Comparing Eq. (3.2) with Eq. (3.3), both of which were observed experimentally, we may hypothesize that in Eq. (3.3) each wire serves as the magnetic field source for the other, that is, as **B** in Eq. (3.2). Thus Eqs. (3.2) and (3.3) are different expressions for the same magnetic force. In Eq. (3.3) the cross product has already been taken to obtain the direction of the force, $\hat{\boldsymbol{\ell}}_{21}$.

Suppose that we view $I_2\hat{\boldsymbol{\ell}}$ as the $I\hat{\boldsymbol{\ell}}$ in Eq. (3.2) and wish to determine the

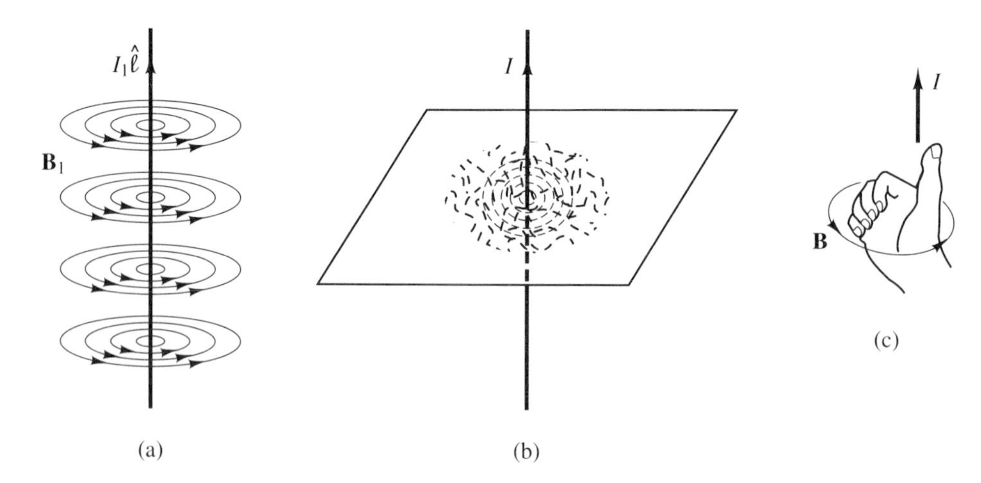


**Figure 3.5** (a) Magnetic field about a current-carrying wire; (b) iron filings pattern for this field; (c) right-hand rule for determination of the direction of the magnetic field in (a) and (b), a straight current-carrying wire.

magnetic field $\mathbf{B}_1$ due to $I_1$ [the $\mathbf{B}$ in Eq. (3.2)]. Let $I_1 > 0$, $I_2 > 0$. Comparing the magnitudes of Eqs. (3.2) and (3.3), it is clear that the magnitude of $\mathbf{B}_1$ is $\mu_0 I_1/(2\pi r)$.

Let us determine the direction of $\mathbf{B}_1$. Looking at Fig. 3.4, we ask: What unit direction vector when crossed with $\hat{\boldsymbol{\ell}}$ gives the known direction of the actual force on wire 2 in Eq. (3.3), namely $\hat{\boldsymbol{\ell}}_{21}$? The right-hand rule provides the answer: into the page. If we moved the wire containing $I_2$ over to the other side of $I_1$, the answer would be "out of the page." Moving $I_2$ all around $I_1$, we conclude that the direction of $\mathbf{B}_1$ must, at every point, be tangent to the circle passing through that point, whose center and normal are the wire containing $I_1$ (see Fig. 3.5a).

Notice that the density of lines indicates the strength of $\mathbf{B}$. Because $\mathbf{B}$ falls with $1/r$, the density of lines is greatest near the wire. Dropping the subscript 1 for generality and designating the tangential (*azimuthal*) angle by $\hat{\boldsymbol{\ell}}_\phi$, we have

$$\mathbf{B} = \frac{\mu_0 I}{2\pi r}\,\hat{\boldsymbol{\ell}}_\phi. \tag{3.4}$$

Indeed, if one pierced a piece of paper with the wire and dropped iron filings on it, the pattern would appear as shown in Fig. 3.5b, in agreement with Eq. (3.4). Recall that tiny dipoles such as iron filings align themselves with $\mathbf{B}$. The degree of alignment (markedness of the iron filings pattern) would decline far from the wire, as $1/r$. Note that if you point the thumb of your right hand in the direction of $I_1$, the direction of $\mathbf{B}_1$ is that in which your fingers point when curled (see Fig. 3.5c).

## 3.3 MAGNETIC FIELD INTENSITY VECTOR AND MAGNETIC POTENTIAL

In Sec. 3.7 we decompose $\mathbf{B}$ into two parts, similar to the way in which $\mathbf{E}$ was decomposed. Only one of the two parts includes the effects of macroscopic electric currents such as those that flow in wires (as opposed to ferromagnetic sources), and is mainly independent of the medium. (The reason for saying "mainly" here rather than "completely" will be given in Sec. 3.7, when we fully discuss the decomposition of **B**.) In analogy with **D**, where essentially we removed $\epsilon$ from Coulomb's law [see Eq. (2.12a)], we remove $\mu = \mu_0$ from Eq. (3.4) and define the *magnetic field intensity vector* **H** for a wire:

$$\mathbf{H} = \frac{I}{2\pi r}\hat{\boldsymbol{\ell}}_\phi. \tag{3.5}$$

We see that the units of **H** are A/m. In this case we obviously have

$$\mathbf{H} = \frac{\mathbf{B}}{\mu_0}. \tag{3.6}$$

Equation (3.6) always holds in air; that is, $\mathbf{B} = \mu\mathbf{H}$ with the substitution $\mu = \mu_0$. In ferromagnetic materials ($\mu \gg \mu_0$), the linear relation $\mathbf{B} = \mu\mathbf{H}$ does not hold true for certain operational situations, while for others it approximately holds true. In magnetic materials the true relation between **B** and **H** is determined experimentally by plotting a *magnetization curve* ($B$ vs. $H$), which we examine in Sec. 3.8. However, the definition that **H** is the component in the decomposition of **B** including "macroscopic" currents such as in wires *always* holds.

Equation (3.5) will be of help below in finding the magnetic field due to a coil of wire, which is a basic building block in the construction of transformers, motors, and generators. Ampère saw a way to generalize Eq. (3.5) in much the same way that Gauss's law, Eq. (2.6), generalizes Coulomb's law, Eq. (2.3). We may rewrite the magnitudes of the vectors in Eq. (3.5) as

$$H \cdot (2\pi r) = I. \tag{3.7}$$

Notice that **H** is along $\hat{\boldsymbol{\ell}}_\phi$ and that $H$ is uniform all around any circle whose center and normal are the current-carrying wire. Because $2\pi r$ in Eq. (3.7) is the circumference of one such circle, it is at least plausible to write

$$\oint_C \mathbf{H} \cdot d\boldsymbol{\ell} = I, \tag{3.8}$$

where again the circle around the integral sign indicates a closed-loop integration path $C$.

Compare Eq. (3.8) with Eq. (2.16), both of which are called *circulation integrals,* or the "circulation of **E**" (or **H**). Recall that for voltage sources [see Eq. (2.30)], the right-hand side of Eq. (2.16) is not zero, but rather, the sum of electromotive forces $\mathcal{V}_n$ (or sources of electric potential) around the loop. This is due to the nonconservative fields $\mathbf{E}_{\text{V.S.}}$ in each source. Similarly, the "force" associated with **H** is not conservative because the round-trip "work" done is not equal to zero, but rather to $\mathcal{F} = I$, where $\mathcal{F}$ is called a *magnetomotive force* (mmf), or source of magnetic potential. Actually, $\mathcal{F}$ is quite analogous to the source of electric potential $\mathcal{V}$ (a voltage source).

The fact that nonconservative fields exist does not mean that circuit concepts involving potential cannot be used for electric or magnetic devices. For example, one may still speak of a rise or fall of magnetic potential $\mathcal{F}_{ab}$ along sections of a chosen path that encircles $I$ once, where even for the non-source magnetic potential drop we maintain the script notation (unlike $V_{ab}$) to avoid confusion with a force $F$. As long as the path is defined, the rise or fall of potential along it can be determined unambiguously. This scalar magnetic potential $\mathcal{F}_{ab}$ can be defined similarly to the scalar electric potential $V_{ab}$ in Eq. (2.15) as

$$\mathcal{F}_{ab} = -\int_a^b \mathbf{H} \cdot d\boldsymbol{\ell}. \tag{3.9}$$

Note that if $I = 0$ [or if the contour $C$ in Eq. (3.8) does not include $I$], then **H** is conservative.

In Sec. 4.5 we obtain *very* simple electric circuit relations based on Eq. (2.30). Also, when we begin our study of electromechanical machines in Chapter 13, we shall use equally simple rules (in Sec. 13.2) for magnetic circuit analysis based on Eq. (3.8).

Why the quotes around "force" and "work" in the discussion of Eq. (3.8) above? Let us examine for a moment the physical meaning of Eq. (3.8). To do this, we first derive the units of **B**.

From a purely dimensional standpoint, comparison of Eq. (3.1) with Eq. (3.3) indicates that the units of a monopole $m$ are $\mathrm{A \cdot m}$. From Eq. (3.3), the units of $\mu$ are $\mathrm{N/A^2}$. With these identifications, it is experimentally verifiable that Eq. (3.1) may be rewritten as $\mathbf{F}_{21} = m_2\mathbf{B}_{21}$; that is, $\mathbf{B}_{21}$ would be the force per unit monopole at location 2 due to pole 1, if the monopole (at location 1) actually existed in isolation. Thus **B** has units of $\mathrm{N/(A \cdot m)}$, or tesla (T), after Nicola Tesla. Tesla (see photo on page 3), a Yugoslavian-American, was the inventor of the induction and synchronous motors, pioneered ac power systems including three-phase voltages, and presented early work on the fluorescent light.

Now if both sides of Eq. (3.8) are multiplied by $\mu$, the units of both sides would be N/A, which when multiplied by the monopole would give $\mathrm{N \cdot m}$ = joules. In situations for which $\mathbf{B} = \mu\mathbf{H}$, the left-hand side is the integral of $\mathbf{B} \cdot d\hat{\boldsymbol{\ell}}$. Thus the left-hand side of Eq. (3.8) multiplied by $\mu$ (the line integral of $\mathbf{B} \cdot d\hat{\boldsymbol{\ell}}$) gives the work required to push a unit monopole once around the closed path $C$ enclosing a wire containing current $I$.

The right-hand side indicates that this work is equal to $\mu I$. Clearly, then, the line integral in Eq. (3.8) when multiplied by $\mu$ is directly and physically analogous to electric potential. However, as monopoles do not exist, Eq. (3.8), even if multiplied by $\mu$, does not represent actual work, or potential energy. The practical magnetic force that can do actual work appears in Eqs. (3.2) and (3.3).

Even if monopoles did exist, Eq. (3.8) without multiplication by $\mu$ dimensionally does not represent potential energy. Nevertheless, as noted above for the case of one loop, conventionally Eq. (3.8) is said to define a *source* of magnetic potential or magnetomotive force, $\mathcal{F}$, as

$$\mathcal{F} = \oint_C \mathbf{H} \cdot d\boldsymbol{\ell} = NI, \tag{3.10}$$

where the new parameter $N$ generalizes Eq. (3.8) and is the number of wires of current containing $I$ around which the contour $C$ travels. Of course, $\mathcal{F}$ is also the circulation of **H**, as mentioned previously.

The parameter $N$ takes on great importance when coils of wire are involved, in which each loop of the coil contains the same current $I$. Clearly, when $N$ is large, $H$ also becomes large, which means that $B$ is large, which means that the magnetic force in Eq. (3.2) can be large. This is a key to understanding motors, which we discuss at the end of this book. The units of $\mathcal{F}$ and the scalar magnetic potential $\mathcal{F}_{ab}$ are A · t (ampere-turns).

Equation (3.10) is known as *Ampère's circuital law,* because Ampère recognized it first. It is of tremendous importance in the study of electromechanical machines and all "magnetic circuits." It holds true and can be used to determine **H** even in ferromagnetic materials for cases where $\mathbf{B} \neq \mu\mathbf{H}$. Thus Eq. (3.10) constrains **H** just as Eq. (2.13) constrains **D**.

Like the contour of Eq. (2.30), the contour of integration in Eqs. (3.8) and (3.10) need not be a circle. Also, the wires containing $I$ need not lie at its center; the contour need only be closed and enclose the wires for Eq. (3.10) to apply. Finally, most generally, $NI$ would be replaced by the sum of the currents of all wires enclosed by $C$, when not all wires have the same current $I$.

For mathematically inclined readers, Eq. (3.10) under these most general assumptions can easily be proved by using the Cauchy integral formula. In this context the current wires can be thought of as simple singularities (e.g., note the $1/r$ in Eq. (3.5)!). The contour integral of an appropriate function of a complex variable can be derived whose real part is the circulation of **H**, Eq. (3.10), and imaginary part is the *flux* of **H** flowing out normal to the contour.

For this situation the result of complex integration is purely real. The Cauchy integral formula thus produces two results simultaneously: Eq. (3.10) and the fact discussed below (see Eq. (3.13)) that the flux of the magnetic field out of a closed contour (or surface) is zero.

An analogy can also be made between magnetostatics (our present discussion) and fluid mechanics. For example, **H** is analogous to fluid velocity, and $I$ is a vortex. If $I = 0$, the "fluid" is irrotational. It should also be noted that the assumption that Eq. (3.5) is exactly true is valid only for $\mu = \mu_0$ (no magnetic materials present), while Eq. (3.8) is always valid (see Sec. 3.7 and the end of Sec. 3.8).

James Clerk Maxwell, some of whose work in the mid-nineteenth century we discuss in Sec. 3.10.3, used such an analogy to develop the theory of electromagnetism. Well aware that fluid mechanics at that time was already fairly well developed, he took advantage of this to make tremendous discoveries in electromagnetism.

## 3.4 THE SOLENOID (ELECTROMAGNET)

Ampère also noticed that when one current-carrying wire was wrapped in a coil, it behaved like a bar magnet. For example, it can attract bar magnet poles according to Eq. (3.1) and attract wires according to Eq. (3.2). The ends of the coil produce a magnetic field just like that of the bar magnet (see Fig. 3.6).

One may understand why this is true by looking again at Fig. 3.5a, the field from a straight wire. Because of the effects of neighboring loops and parts of the same loop, the field about one loop of the coil is no longer circularly symmetric around the wire, as in Fig. 3.5a. To see how they combine, we consider the field plots

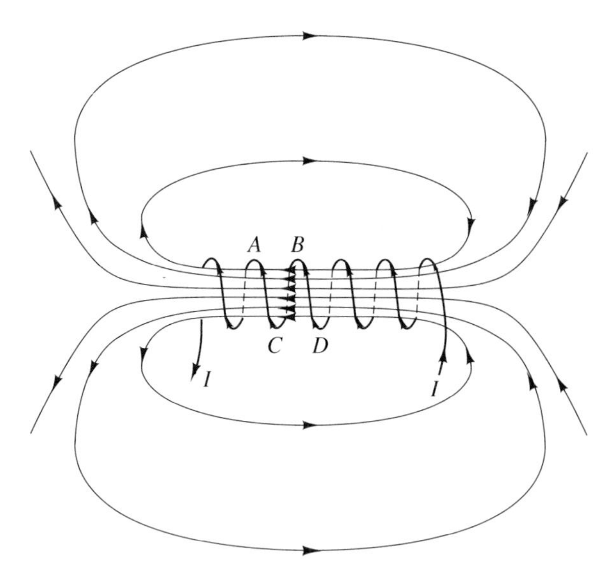


**Figure 3.6** Magnetic field of a solenoid.

in Fig. 3.7. Each of these can be obtained by performing iron filing experiments, or mathematically by adding the two or more contributing vector components at each point in space.

In Fig. 3.7a are shown the individual head-on field patterns of wires at locations $A$ and $B$ in Fig. 3.6, which for the moment we view as straight wires. These individual component fields are just as in Fig. 3.5a, but their sum is the distorted picture given in Fig. 3.7b. The currents in Fig. 3.7a and b are in the same direction: into the page. Figure 3.7c and d are the corresponding diagrams if the currents are in opposite directions, such as wires at locations $A$ and $C$ in Fig. 3.6. In Fig. 3.7e is shown the net field due to the combined effects in Fig. 3.7b and d for wires at locations $A$, $B$, $C$, and $D$ in Fig. 3.6.

Notice that all the left-to-right lines are confined to (squeezed into) the interior of the coil in Figs. 2.22e and 2.21 because the field lines cannot cross. (The field lines of no vector field can cross because if they did, two different directions for the field would be specified at the same location, which is impossible.) Outside the coil, the **B** lines spread out as shown.

Incidentally, the direction of **B** *within* the single current loop represented by Fig. 3.7d can again be determined by an application of the right-hand rule. Curl your fingers in the direction of the current flow. Your thumb will point in the direction of **B**. This last right-hand rule, illustrated in Fig. 3.8, is extremely helpful in determining the direction of **B** within and through coils such as the solenoid, discussed in Example 3.2. It is used repeatedly in the study of motors and generators.

---

**Figure 3.7** (a) Magnetic field about two current-carrying wires having currents in the same direction: the individual contributions from the two wires; (b) resulting field due to the two components in (a); (c) magnetic field about two current-carrying wires having currents in opposite directions: the individual contributions from the two wires; (d) resulting field due to the two components in (c); (e) magnetic field about a three-turn coil.

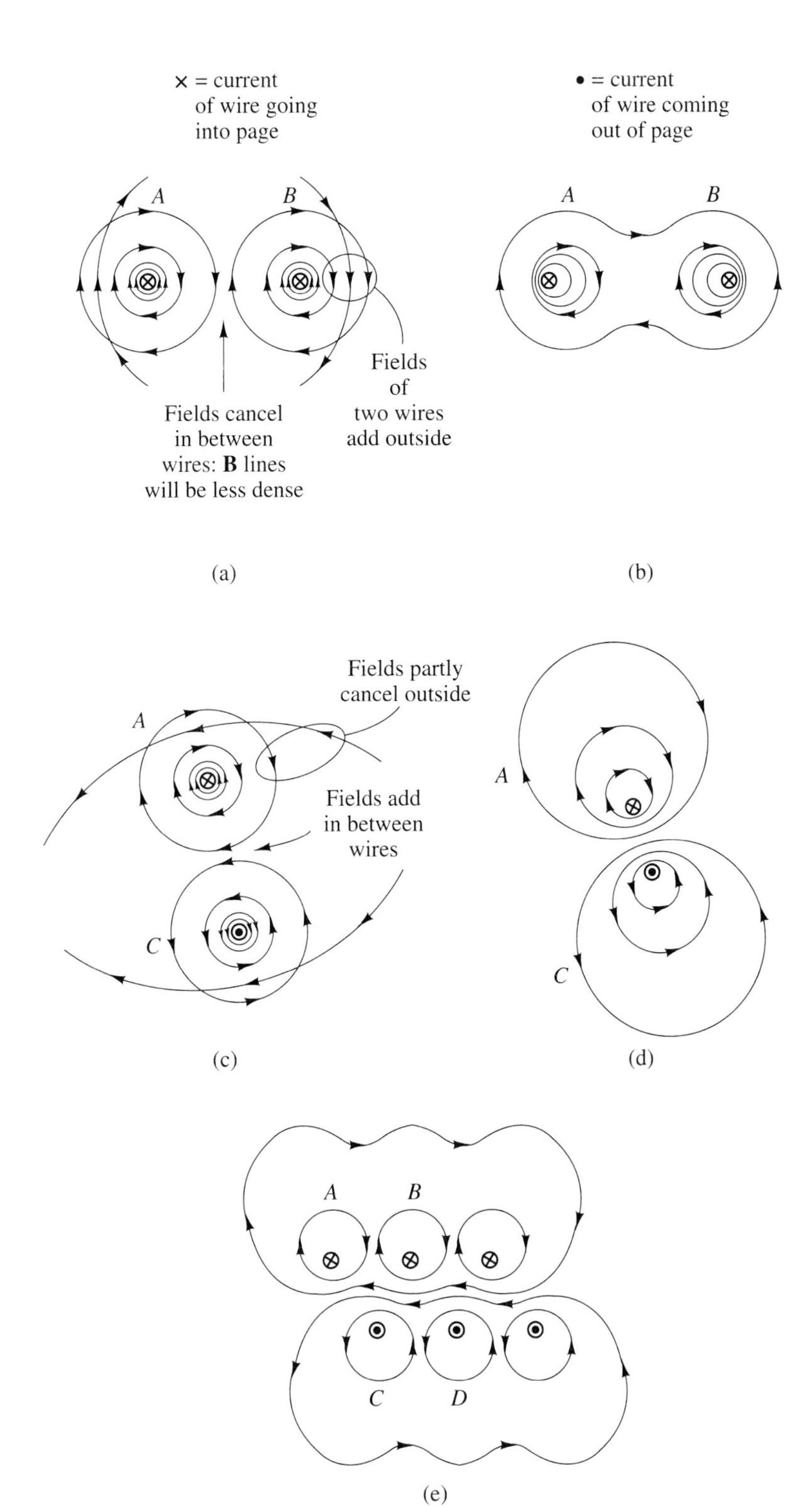
× = current
of wire going
into page
• = current
of wire coming
out of page
A
B
Fields
of
two wires
add outside
Fields cancel
in between
wires: B lines
will be less dense
(a)
(b)
Fields partly
cancel outside
Fields add
in between
wires
A
C
(c)
(d)
A
B
C
D
(e)

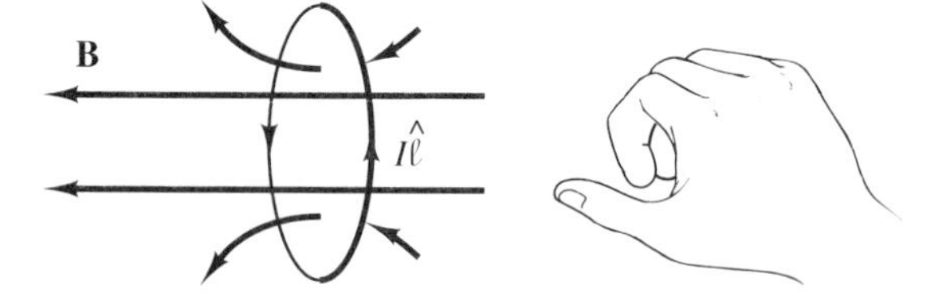


**Figure 3.8** Right-hand rule for determination of the direction of the magnetic field within a current-carrying loop.

From the **B** field that we have just constructed from addition of fields due to wires having currents in opposite directions (Fig. 3.7d), we have as a bonus obtained the magnetic field of a single loop of wire. This is because in the plane pictured in Fig. 3.7d, the contributions from other parts of the loop cancel each other. Therefore, merely rotate the field distribution in Fig. 3.7d about the horizontal axis in the page to imagine the three-dimensional **B** field due to a loop. The ideal solenoid is essentially an extended-width loop (like a section of pipe) and will be explored fully in Examples 3.2 and 3.3. (The origin of *solenoid* is the Greek word meaning "pipe.")

Before doing so, it is helpful to reconsider for a moment the nonconservative nature of **B** (or **H**) in Eq. (3.10). Figure 3.9 shows the current loop of Fig. 3.8 along with two points $a$ and $b$ whose magnetic potential difference we seek. Notice that if path $C_1$ is taken, $\mathbf{H}\cdot d\boldsymbol{\ell}$ is always positive, while along the path $C_2$, $\mathbf{H}\cdot d\boldsymbol{\ell}$ is negative. Therefore, the integral in Eq. (3.9) depends on the path taken, and thus $\mathcal{F}_{ab}$ is not well defined.

As noted above, no ambiguity exists if we clearly designate the path to be taken and restrict the contour to go all the way around [pass through the current loop(s)] only one time before closing on itself. In this context the potential $\mathcal{F}_{ab}$ *is* well defined with respect to the specified path, even though **H** is nonconservative.

**Example 3.2**

Compute the magnetic field intensity **H** and the flux density **B** within an ideal solenoid (also called an *inductor*), shown in Fig. 3.10. This solenoid is $d = 10$ cm long, has $N = 100$ turns, and has a radius of $a = 1$ cm. The current in the solenoid is $I_s = 1$ A. The coil is wound so that the direction of **H** (and thus **B**) is from right to left. Also, compute the force on the nearby wire shown in Fig. 3.10, in which a current $I_w = 10$ A flows. Assume that **B** at the wire is roughly equal to **B** within the coil for the 2-cm region opposite the diameter of the coil and falls off rapidly from there on out. This assumption is valid if the wire is very close to the solenoid.

**Solution** An ideal solenoid means that the adjacent coils are packed extremely close together. When this is true, the fields from all the loops produce a field that from far away looks like that from one loop—shown in Fig. 3.7d. Near the outer surface of the

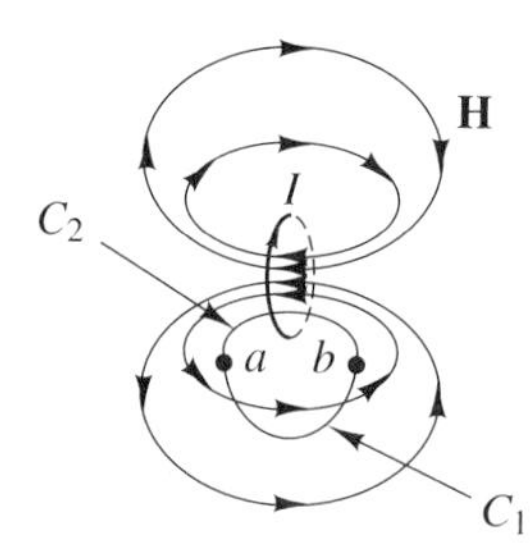


**Figure 3.9** Demonstration that the magnetic field is nonconservative; magnetic potential depends on path.

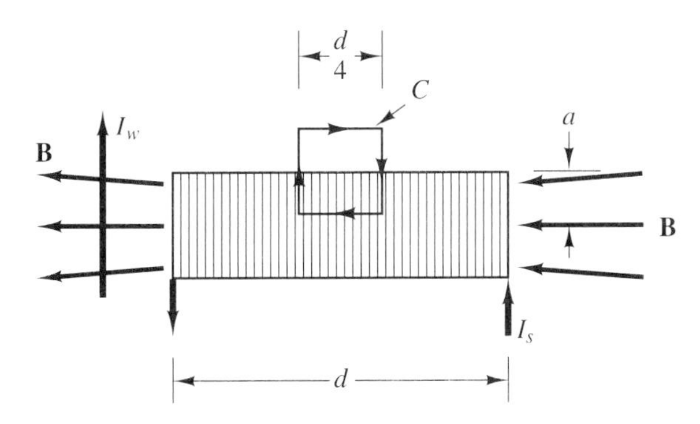


**Figure 3.10** Solenoid and nearby wire.

solenoid, the field is very weak, due to the spreading outside the loop, and is assumed to be zero for an ideal solenoid. Therefore, we can use the picture of Fig. 3.10 in conjunction with Ampère's circuital law, Eq. (3.10), to find **H**. Because we neglect fringing effects, the radius of the coil does not enter our calculation.

In Fig. 3.10 a typical contour is shown, whose top and bottom sides are chosen for convenience to be $d/4$. The magnetic field is assumed to be zero for the portion of $C$ outside the solenoid. Contributions to the circulation integral of **H** in Eq. (3.10) due to the portions of the vertical sides within the coil are zero, because there **H** is perpendicular to the contour, so $\mathbf{H} \cdot d\boldsymbol{\ell} = 0$. Therefore, the magnitude of **H** within the solenoid is

$$H = \frac{(N/4)\, I_s}{d/4} = \frac{N I_s}{d} = 100\left(\frac{1}{0.1}\right) = 1000 \text{ A} \cdot \text{t/m}.$$

Because the medium within the coil is air, $B = \mu_0 H = 4\pi \cdot 10^{-7} \cdot 10^3 = 0.0013$ T.

The force on the wire is out of the page (see Fig. 3.3) and by Eq. (3.2) has magnitude $F = I_w B = 10 \cdot 0.0013 = 0.013$ N per meter. We assume that 2 cm of the wire experiences this force and we neglect the force on other portions of the wire due to the spreading of **B**. Thus the force on the wire is roughly $0.013 \cdot 0.02 = 2.6 \cdot 10^{-4}$ N—very small.

**Example 3.3**

In Example 3.2 suppose that a cast steel core having $\mu_r = 1000$ is placed within the solenoid, and fills it. Now what is the force on the external wire? Assume that for this operating condition, $\mathbf{B} = \mu \mathbf{H}$.

**Solution** Again we can assume that **B** at the wire in front of the solenoid is approximately equal to that within the solenoid, even with the steel core in place. Unlike **E** but just like **D** in electrostatics, **B** lines are continuous from medium to medium. Rather, it is $H = B/\mu$ that changes abruptly between media, having differing $\mu$ such as from the steel core to the surrounding air.

(Note that both flux densities **D** and **B** can change from medium to medium. For example, **B** is shown to spread out somewhat outside the solenoid in Fig. 3.10. For the case where the core is steel, the direction change can be sudden. More precisely, it is the component of **B** or **D** normal to the surface between adjacent media that is continuous. This means that every flux line incident on a boundary continues on the other side, although its direction may change. For this reason it is more convenient to plot flux lines of **B** or **D** than those of **H** or **E**. More on these considerations in Sec. 3.7.)

Thus to find the force on the wire when the solenoid has a steel core, we merely multiply the force found above by $\mu_r$ to obtain the new force, which is $2.6 \cdot 10^{-4} \cdot 1000 =$

0.26 N. Although this is still small, it is definitely detectable. If a bundle of wires (e.g., a side of a coil having 100 turns) each carrying a current of 10 A is brought near the solenoid, the force on the bundle (coil side) is now 26 N = 5.9 lb—quite sizable. The crucial role magnetic materials and coils play in motors should now be clear.

### 3.4.1 Magnetic Dipoles versus Current Loops

From the results, Eqs. (3.3) and (3.10), the fact that a solenoid produces a dipole **B** field, and other observations, Ampère correctly concluded that *all* magnetic forces are actually due to electric charge in motion. This motion could be electric current in wires, or it could be electrons spinning about their axes of rotation in ferromagnetic material. In the latter case, if the spinning directions of most electrons agree, there is an overall effect equivalent to a current on the surface of the magnet (see Fig. 3.11).

In practice, the viewpoints of magnetic materials as being composed of either magnetic dipoles or of current loops are both compatible and equivalent. The engineer is free to choose whichever viewpoint seems most convenient for the application.

## 3.5 ELECTRICITY VERSUS MAGNETISM

The conclusion of Ampère described above was quantified as a result of the studies of Leigh Page of Yale University in 1912. He used the results of special relativity to identify the fundamental magnetic force between charges in relative motion as

$$\mathbf{F}_{21} = q_2 \mathbf{u}_2 \times \frac{\mu q_1 \mathbf{u}_1 \times \hat{\boldsymbol{\ell}}_{12}}{4\pi r^2} \tag{3.11a}$$

$$= q_2 \mathbf{u}_2 \times \mathbf{B}_{21}, \tag{3.11b}$$

where $\mathbf{u}_1$ and $\mathbf{u}_2$ are the velocities of charges $q_1$ and $q_2$, respectively, and the other parameters are as defined previously.

Relation (3.11b) is nothing other than the microscopic representation of

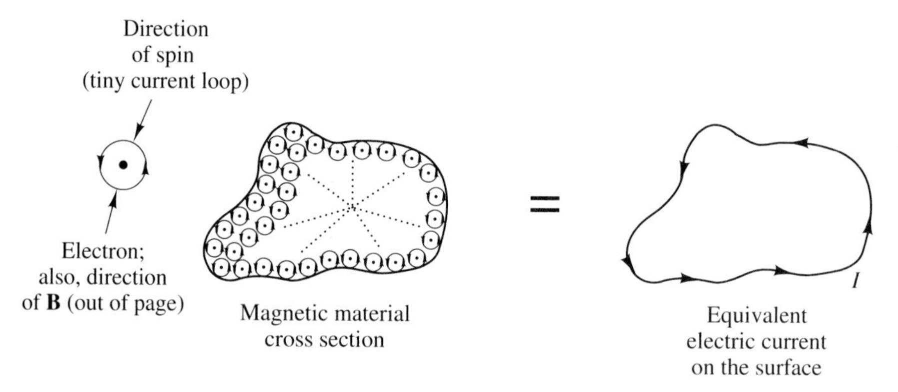


**Figure 3.11** Cross section of magnetic material showing individual electron spin components, and the equivalent surface loop current around the perimeter.

Eq. (3.2) ($\mathbf{F}_{/\text{ul}} = I\hat{\ell} \times \mathbf{B}$). One may also draw a close analogy between Eqs. (3.11a) and (2.1) (Coulomb's force). Both forces between charges are proportional to the product of the charges and inversely proportional to the square of the distance between them. A clear analogy between Eqs. (3.11b) and (2.2) ($\mathbf{F}_{21} = q_2\mathbf{E}_1$) also exists.

We see that the only differences between electric and magnetic forces are (1) that the electric force always exists wherever charges are, whereas the magnetic force must have the charges be in relative motion to exist (and its magnitude depends on these velocities), and (2) that the direction of the electric force is along the line joining the charges, whereas that of the magnetic force is perpendicular to both the charge velocity $\mathbf{u}_2$ and the magnetic field $\mathbf{B}_{21}$.

In general, a charge $q$ moving with velocity $\mathbf{u}$ will experience both electric and magnetic forces. Using Eqs. (2.2) and (3.11), this may be expressed as

$$\mathbf{F} = q[\mathbf{E} + \mathbf{u} \times \mathbf{B}]. \tag{3.12}$$

This total force, valid whatever the sources of $\mathbf{E}$ and $\mathbf{B}$ actually are, is called the *Lorentz force* after its discoverer Hendrik Antoon Lorentz, a Dutch turn-of-the-century pioneer in special relativity and its electromagnetic implications.

Incidentally, from the point of view of the charge $q$ in Eq. (3.12), $\mathbf{u} = 0$. The velocity $\mathbf{u}$ is with respect to the reference frame in which the force is measured or observed. Yet the charge must experience the same net force that we say it has by writing Eq. (3.12) from our frame of reference! In fact, the charge would classify what we are calling the $\mathbf{u} \times \mathbf{B}$ force as an additional electric force on it. This electric field is due to the time-varying magnetic flux the charge perceives as it cuts across the flux lines of $\mathbf{B}$, which occurs even if $\mathbf{B}$ is spatially uniform. Further insight on this point will be gained in Sec. 3.10, in which we consider time-varying electromagnetics.

To further contrast electricity versus magnetism, consider the following fundamental difference between dielectric and magnetic media. When a dielectric object is polarized, the bound charges on its surface generate a field normal to the object surface. The fields of all the internally polarized molecules cancel, and the depth of the object does not substantially influence the field intensity outside the object.

This is not so with a magnetized piece of magnetic material. All the aligned dipoles produce magnetic fields that *add up*. Equivalently, all the surface current "loops" from all cross sections such as the one in Fig. 3.11 go the same way, so that their magnetic fields add up. This fact is one reason why the size of a motor must be increased to handle larger load torque (and thus produce a larger $B$). The width of the object influences the intensity similarly in both electrically polarized and magnetized objects: It merely determines the spatial extent of significant field intensity and the particular pattern of the total field.

The distinction from dielectrics just raised is a major reason why high-power motors and generators are based on magnetic rather than electric forces. The magnetic energy/unit volume is proportional to $B^2$, the square of the magnitude of $\mathbf{B}$; the electric energy density is proportional to $E^2$. Only where energy density is high over large volumes can practical amounts of work be done. Because of the presence of the buildup property that occurs in magnetic materials and its absence in dielectrics, magnetic energy is the better choice for energy conversion systems.

## 3.6 LINES OF MAGNETIC FLUX

Another essential difference between electricity and magnetism is as follows. A current not only has magnitude but also has a direction. That is, coupled with its direction, it is a vector. An electric charge has only magnitude, no direction; it is a scalar. Currents (vectors) are the sources of magnetic fields, while charges (scalars) are the sources of electric fields. (We omit for the moment the effects of time-varying fields. In that case, a time-varying electric/magnetic field serves as a vector source of the magnetic/electric field.)

Recall that flux lines of **E** and **D** begin and end on charges, which are scalar sources. Flux lines of **B** arising from vector sources, such as currents (charges having nonzero velocities), encircle the source and have no beginnings or ends. This is evident in Figs. 3.2 and 3.5 through 3.7. The lines shown there are the flux lines, or magnetic field (**B**) lines. Like electric field flux lines, magnetic flux lines do have direction. For a magnetic dipole, they come out of the north pole of the source, around, and back into the source at its south pole. For a line current, they encircle the wire according to a right-hand rule. However, they always close on themselves. Unlike electric flux lines, no charge (in this case magnetic) exists on which magnetic flux lines can originate.

Why is the fact that the flux lines close on themselves significant? One very important consequence is that unlike electric fields, the total flux of **B** out of *any* closed surface is zero. One may verify this by looking at any of the **B** fields in Figs. 3.2 through 3.10. Any closed surface placed anywhere in the **B** field has an equal number of lines entering the surface as leaving it. In equation form, if $S$ is a closed surface,

$$\oint_S \mathbf{B} \cdot d\mathbf{S} = 0. \tag{3.13}$$

Comparing with Eq. (2.13), we see that Eq. (3.13) is equivalent to saying that there is no magnetic charge (point source of magnetic flux). The magnetic flux $\phi_m$ through any open (unclosed) surface $S$ is defined by

$$\int_S \mathbf{B} \cdot d\mathbf{S} = \phi_m. \tag{3.14}$$

The practical significance of magnetic flux $\phi_m$ will be disclosed in Sec. 3.10, where it is directly related to magnetically induced voltage. Its units are webers ($\text{Wb} = \text{T} \cdot \text{m}^2 = \text{N} \cdot \text{m/A} = \text{J/A}$) after Wilhelm Eduard Weber, a German experimentalist who in the mid-nineteenth century furthered the understanding of magnetic materials.

It is believed that Gauss formulated Eq. (3.13), which parallels the electric field counterpart he derived, Eq. (2.13). One is reminded how Coulomb discovered both Eqs. (2.1) and (3.1), the inverse-square electric and magnetic force laws. However, Coulomb's equations were experimentally based, while Gauss's were theoretically derived.

The practical significance of Eq. (3.13) is that the fluxes through all open surfaces having the same defining perimeter contour are equal. In Fig. 3.12 the defining perimeter contour is $C$; Eq. (3.13) implies that

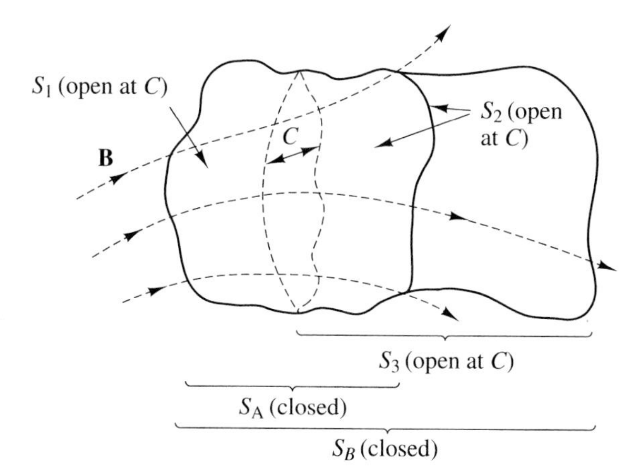


**Figure 3.12** The fluxes through all open surfaces limited by the same contour $C$ are equal.

$$\oint_{S_A} \mathbf{B} \cdot d\mathbf{S} = 0, \tag{3.15}$$

so that

$$-\int_{S_1} \mathbf{B} \cdot d\mathbf{S} = \int_{S_2} \mathbf{B} \cdot d\mathbf{S} = \phi_m, \tag{3.16}$$

where $\phi_m$ is the left-to-right flux in Fig. 3.12. Similarly,

$$\oint_{S_B} \mathbf{B} \cdot d\mathbf{S} = 0, \tag{3.17}$$

so that, making use of Eq. (3.16), we have

$$\int_{S_2} \mathbf{B} \cdot d\mathbf{S} = \int_{S_3} \mathbf{B} \cdot d\mathbf{S} = \phi_m. \tag{3.18}$$

Notice that nothing has been said about the medium in all this. For example, $S_2$ may be in a medium having $\mu_2$ and $S_3$ may be in an adjacent medium with $\mu_3 \neq \mu_2$.

In words, Eq. (3.18) states that in a simple *magnetic circuit* defined by a cross-sectional boundary $C$, the flux through any surface normal to the circuit path is the same everywhere in the circuit (see Fig. 3.13). We shall use this fact frequently when we deal with magnetic devices (transformers, iron-core electromagnets, etc.) later in the book. Notice that if somewhere along the circuit the cross-sectional area changes (e.g., increases), the flux density **B** will change (decrease) even though the total flux through the varying cross-sectional areas will remain constant.

## 3.7 DECOMPOSITION OF THE MAGNETIC FLUX DENSITY VECTOR

In one sense, **B** and **D** are analogous in that they both are referred to as flux densities, due to the fact that their normal components are continuous across boundaries between media. But in a more important sense, **B** and **E** are the analogous vectors because they are behind the observable forces $\mathbf{F} = I\hat{\ell} \times \mathbf{B}$ and $\mathbf{F} = q\mathbf{E}$.

Just as was the case with the electric field **E**, the magnetic field (flux density) **B** can be decomposed into two components: One entirely dependent on the medium

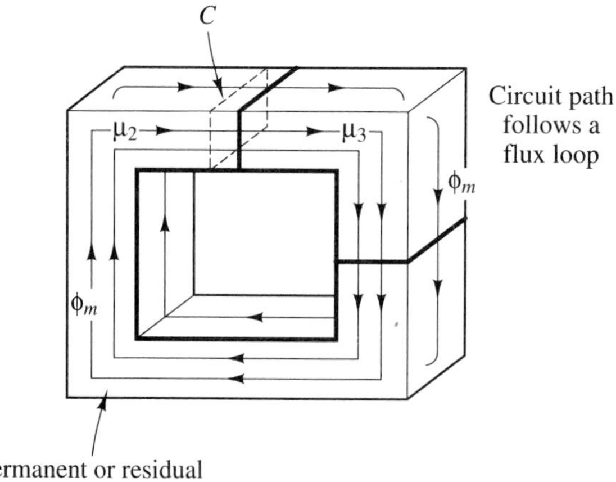


**Figure 3.13** Inhomogeneous magnetic circuit; the magnetic flux is the same in all sections.

and the other one including non-medium-based sources. In the case of the electric field, **E** was decomposed into the polarization **P**, which was due exclusively to bound charges occurring only in dielectric media, and the flux density **D**, which included the effects of free charges that are foreign to the insulating dielectric medium.

The magnetic flux density vector **B** can be decomposed similarly. However, the situation is more complicated in that while $\mathbf{D} = \epsilon\mathbf{E}$ for constant $\epsilon$ always holds true, $\mathbf{B} = \mu\mathbf{H}$ for constant $\mu$ holds only approximately, and then under only certain operating conditions.

Consider Fig. 3.14, which shows a piece of permanently magnetized ferromagnetic material with a current-carrying coil wrapped around it. That is, we have a solenoid with a ferromagnetic core, as studied above. The contribution of coil current to **B** is analogous to that of free charges to **E**. The contribution of the permanent magnet to **B** is analogous to that of the dielectric to **E**.

The conventional decomposition of **B**, the total magnetic field, is as follows. It is a simple mathematical model designed to conveniently represent observed behavior. Within the magnetic material, and only there, exists a medium source of **B** called the *magnetization vector* **M**. Whatever is left over is **H**. The formal statement is

$$\mathbf{H} = \frac{\mathbf{B}}{\mu_0} - \mathbf{M}, \tag{3.19}$$

which is rewritten in a form analogous to Eq. (2.9d) as

$$\mathbf{B} = \mu_0(\mathbf{H} + \mathbf{M}). \tag{3.20}$$

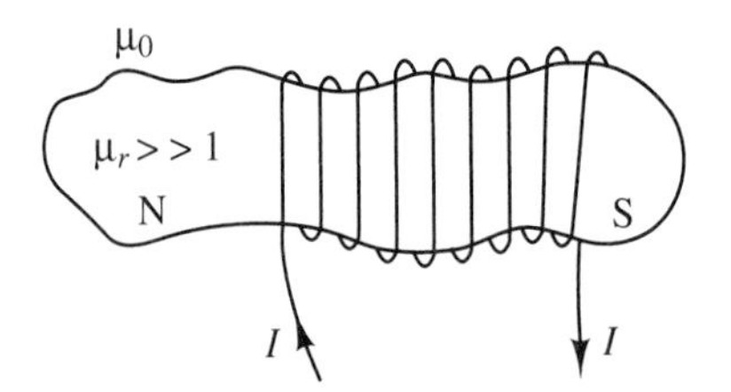


**Figure 3.14** Piece of magnetic material wrapped by a current-carrying coil.

Notice that for $\mathbf{M} = 0$ (free space), Eq. (3.19) reduces to Eq. (3.6), as expected. If $\mathbf{B} = \mu\mathbf{H}$ holds in a magnetic medium, evidently $\mathbf{M} = \chi\mathbf{H}$, where $\chi$ is called the *magnetic susceptibility*. This name is used because if **H** is due primarily to externally supplied currents, $\chi$ indicates how easily the substance is magnetized.

As noted previously, all the **B** lines in one material continue into the next because they have no beginnings or ends; this is experimentally verifiable. The lines of **M** begin and end on the boundaries of the magnetic material; this is a convenient definition of **M**.

This latter fact, along with Eq. (3.19), implies that some of the lines of **H** might also begin and end on the boundary of the magnetic material in order that **B** lines have no beginnings or ends. This is in fact the case. See Fig. 3.15, where for clarity the coil is not shown because $I = 0$. Thus even though $I = 0$, $H \neq 0$. No, there is not a mistake in Fig. 3.15; **H** does point opposite to **B** within the magnetic material! Outside the material is air in which $\mathbf{M} = 0$ so that $\mathbf{B} = \mu_0\mathbf{H}$; thus outside, **H** is in the same direction as **B**.

In principle, the value of **H** within the magnetic material when $I = 0$ can be calculated from Eq. (3.10). Its particular value depends on the geometrical structure of the magnetic object. If in Eq. (3.10) one takes a path partly through the magnetic material and partly outside it, the integral of $\mathbf{H} \cdot d\boldsymbol{\ell}$ around it must be zero because there is no enclosed current. However, **H** must be nonzero outside the material (where $\mu = \mu_0$) if we are to have $\mathbf{B} = \mu_0\mathbf{H}$ there. Thus, within the magnetic material, **H** must have direction *opposite* that of **B** so that the portions of the integral of $\mathbf{H} \cdot d\boldsymbol{\ell}$ from within and outside the material sum to zero (see Fig. 3.16). Satisfaction of Eq. (3.10) then dictates that **H** within the material be nonzero, and in the opposite direction. This implies, strangely enough, that in this case of no current in the coil, "$\mu$" $\equiv \mathbf{B}/\mathbf{H}$ is negative!

In words, as predicted from Eq. (3.19), **H** lines, as opposed to **B** lines, *can* have beginnings and ends: They begin at "north" poles of media interfaces and end on

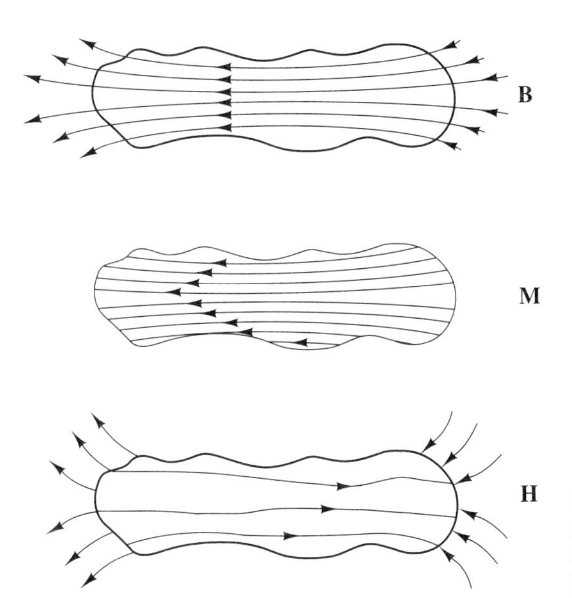


**Figure 3.15** Magnetic flux density, magnetization, and magnetic field intensity in the ferromagnetic object of Fig. 3.14 with $I = 0$.

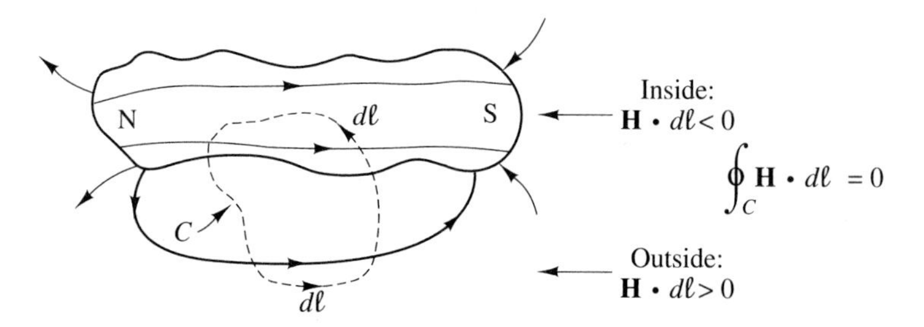


**Figure 3.16** Demonstration by Ampère's law that the magnetic field intensity within magnetized material is nonzero even if there is no externally supplied excitation current.

"south" poles of media interfaces. This is true only for the component of **H** due to magnetic materials; the lines of the component of **H** due to electric currents $I$ (such as in Fig. 3.9) close on themselves, as do all lines of **B**.

Even though **H** was defined in Eq. (3.5) to be proportional to $I$ in free space, we are now forced to conclude that in magnetic materials, **H** is partly composed of ferromagnetic effects. This is because, for whatever reason, the theorists wished to have what from the point of view of our discussion is the somewhat arbitrary requirement that $\mathbf{B} = \mu_0\mathbf{H}$ always hold in free space even when magnetic materials are contributing to **B**. If it were not for this requirement, an "**H**" could be defined that was zero everywhere whenever no macroscopic currents are present. This would result in a neater and more intuitive decomposition of **B**.

## 3.8 THE *B–H* CURVE

Let us consider for a moment the process of creating a "permanent" magnet from an initially unmagnetized ferromagnetic material. We refer again to Fig. 3.14, which is a solenoid with a magnetic material core, which we equivalently called an *electromagnet.*

The component of **H** due to $I$ was calculated in Example 3.2. It can be in one of two directions within the magnetic material: left to right or right to left. Consequently, it is simplest to deal only with the magnitudes of **B** and **H**, but allowing negative "magnitudes" to indicate left to right as opposed to right to left. In the following discussion we concern ourselves only with $B$ and $H$ within the magnetic material.

One can always write $B = \mu H$ if $\mu$ is allowed to be a function of $H$: $\mu = \mu(H)$. In reality, the permeability $\mu$ is approximately independent of $H$ only for a certain range of values of $H$. This range is known as the linear region of the $B$–$H$ curve for the given medium. This is the range where one may speak of a $\mu_r$ for a given material, such as we did before in mentioning that it could be 1000 or more.

For permanently magnetized materials, the linear region does not extend down to $H = 0$. We have already seen how $\mu$ can be negative for very small $H$ (zero applied current) if the material is permanently magnetized. For large values of $H$, $\mu$ declines so that the $B$–$H$ curve flattens; this range of $H$ is called the *saturation region.*

A typical $B$–$H$ curve for an initially unmagnetized ferromagnetic material is

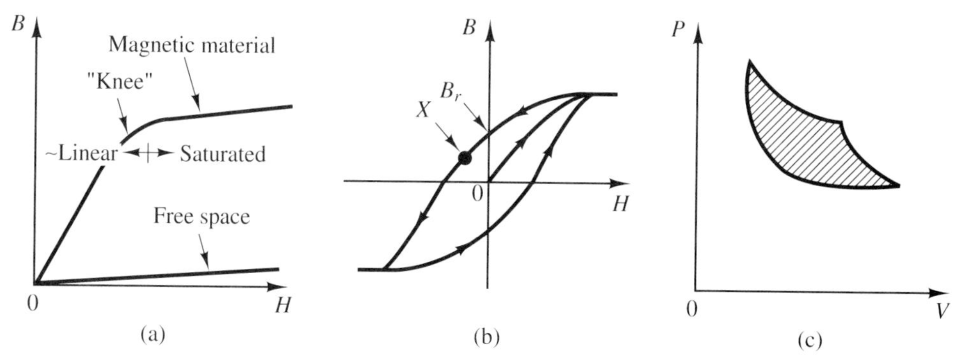


**Figure 3.17** (a) Magnetization curves for free space and a typical magnetic material; (b) *B–H* curve of a typical ferromagnetic material; (c) analogous *P–V* curve of the Carnot cycle.

shown in Fig. 3.17a. Shown for comparison in Fig. 3.17a is the *B–H* relation for free space ($B = \mu_0 H$); it is the lower curve. Actually, the true slope ($\mu_0$) of the free-space curve is far smaller in relation to that of the *B–H* curve for a typical magnetic material (e.g., 1/1000) but was magnified for visibility. Notice that the linear region of free space extends indefinitely; there is no saturation region for free space. Also, it is experimentally verifiable that the slope of the *B–H* curve in the saturation region is asymptotically equal to that of the free-space curve: $\mu_0$.

There will be a different *B–H* curve for each type of magnetic material. Therefore, both the threshold value of *B* that roughly divides the linear and saturated regions and the permeability value within the linear region depend on the material. Typical values are shown in Table 3.1.

Why does saturation occur? Quantum theory has an explanation. Originally, with no applied magnetic field, the spins of the electrons are aligned within little volumes (*domains*) of the magnetic material. This is because electron alignment on a small scale (domain) decreases the overall energy of the material, while large-scale alignment is slightly higher. (In true lodestone, the large-scale alignment *is* favored.) Thus the field directions of the magnetic domains are randomly oriented. The random orientation produces a net *B* of zero.

If a nonzero magnetic field strength *H* is applied by means of the solenoid with $I \neq 0$, some of the domains become aligned, producing a nonzero net *B*. The larger *H* is, the more domains align and thus the larger *B* is. But eventually, the vast majority of domains are aligned, so that further increases in *H* will not substantially increase the number of aligned domains and hence *B*. Therefore, the *B–H* curve flattens.

**TABLE 3.1**

| Material | Linear region | $\mu_r$ in linear region |
|---|---|---|
| Sheet steel | $B < 1$ T | 5700 |
| Cast steel | $B < 1$ T | 1300 |
| Cast iron | $B < \frac{1}{2}$ T | 400 |

If $H$ continues to increase (by increasing $I$), the only contribution to $B$ comes from the solenoid field alone, and none from the magnetization. Thus for very large $I$, the magnetic material only provides the saturation $B$ added onto a linear $B$–$H$ curve with slope $\mu_0$.

What happens if the current is turned off? Most of the domains become randomly oriented due to thermal effects. That is, once aligned, it takes energy to disalign the domains again even though disalignment results in a lower final energy. Energy sufficient to disalign some of the domains comes from thermal energy. However, not all the domains disalign. Thus there is a nonzero residual magnetic flux density $B_r$ for $H = 0$. It is this $B_r$ that is observed in permanent magnets magnetized by applied currents. To make $B$ become zero again, an $H$ of direction opposite that previously applied must now be applied, called the coercive magnetic field intensity $H_c$.

We have already discussed the fact that in a magnetized material with zero current, **H** points in the direction opposite that of **B**; that is, $H$ is negative. Thus the relevant point on the $B$–$H$ curve for zero current and a permanent magnet is in the second quadrant (e.g., point $X$ in Fig. 3.17b). The $B$–$H$ curve is defined by the type of material, not the geometrical structure. The point on this curve (value of $H$) that applies when $I = 0$ is determined by geometrical parameters, and can be calculated using Ampère's circuital law, Eq. (3.10).

The process is symmetrical for a magnetic field intensity in the opposite direction, that is, for an $H$ of opposite polarity. Together, then, a complete $B$–$H$ curve can be drawn (Fig. 3.17b), which is seen to be of a *hysteresis* or irreversible or state-dependent form. That is, the correct $B$–$H$ curve branch is dictated by which $B$ saturation most recently occurred (either positive or negative)—the present status of the magnetic domains. This is the meaning of the arrows: the arrows on a given curve point from the maximum value that had to be most recently attained in order now to be on that branch of the curve (or, in the case of the initial magnetization curve, from zero).

*Note: Hysteresis* is from the Greek word meaning "delay." Note from the $B$–$H$ curve that indeed if we begin at one saturation point and decrease $H$ in magnitude, $B$ remains at near the saturation level for awhile before finally also coming back down. Thus $B$ lags behind $H$, hence the name "hysteresis."

It can be shown from James Clerk Maxwell's electromagnetic equations and the electromagnetic energy flow equation (which is beyond the scope of this course) that energy is lost in the form of heat during the cyclical process. This energy is called *hysteresis heating*. If written in the form of an energy volume density, it is equal to the area between the two $B$–$H$ curve branches; $dW = H\,dB$.

Hysteresis heating is associated with the friction in continually bringing about realignment of the magnetic domains as $H$ and thus $B$ are changed, and limits the power that the magnetic machine or device can safely handle. Clearly, this type of loss is minimal for materials having narrow hysteresis loops. Such materials are sought and used in high-powered magnetic devices such as transformers, motors, and generators.

A good permanent magnet will have a wide hysteresis curve, so that its residual magnetic field is larger (closer to its saturation magnetic field). Permanent magnets are not continually cycled on the hysteresis curve, so loss discussions are irrelevant.

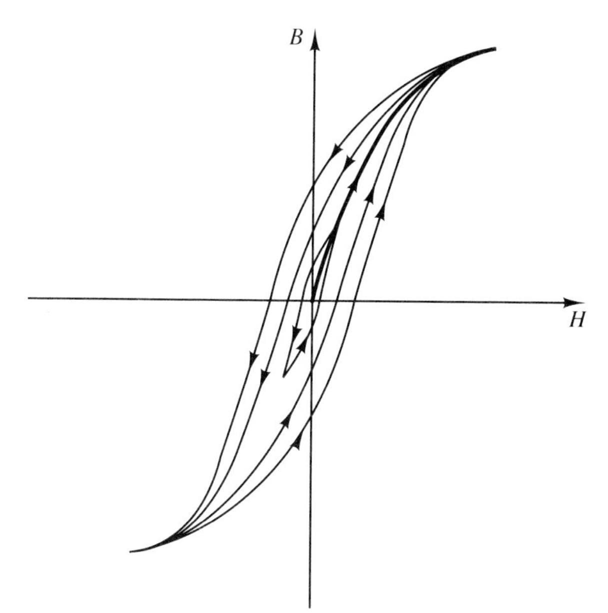


**Figure 3.18** Multiple hysteresis loops depending on farthest point reached on magnetization curve before decreasing $|H|$.

The hysteresis plot and mathematical form of energy loss should remind the mechanical engineering student of the $P$–$V$ diagram for a Carnot cycle, where $P$ is the pressure, $V$ is the volume, and the net work done by the gas during one cycle is $dW = P\,dV$ and equals the area between the two branches of the $P$–$V$ curve (Fig. 3.17c). On a more pedestrian level (excuse the pun), if one runs a circuital course that includes a hill, one is exhausted from the run but is no higher after one complete cycle than at the start! In all these cases, nonconservative forces are literally "at work."

One final complexity is that there is not a single $B$–$H$ curve, but rather a family of curves for a particular medium. Figure 3.18 shows a few of these curves. The particular return path depends on how far up the magnetization curve we most recently went (how high we let $I$ become) before reducing $I$. Going all the way to saturation determines the widest hysteresis loop.

If a material is chosen whose widest loop is narrow, one may roughly speak of a single $B$–$H$ curve, such as the magnetization curve in Fig. 3.17a, that approximately applies whether or not the material has been previously magnetized. In this case, below saturation we may roughly say that $B = \mu H$ with $\mu$ independent of $H$. In so doing, we neglect both the finite width of the $B$–$H$ curve and the effect that for zero applied current, $H$ opposes $B$. Effectively, we may also say that we ignore the component of $H$ due to magnetization. Under this assumption, $H$ is proportional to the applied current, as was assumed in Example 3.3, involving the solenoid with the steel core.

## 3.9 MAGNETIC RELUCTANCE

Before leaving magnetostatics (the study of constant-in-time magnetism), we introduce one further quantity that is sometimes helpful in magnetic circuit calculations such as those we will consider in our study of electromechanical machinery. In a magnetic circuit such as shown in Fig. 3.13, it is usually reasonable to assume that within the magnetic core, $B$ is uniform over any cross section, such as that bounded by curve $C$ in Fig. 3.13. Thus if at a given location in the core the cross-sectional area is $A$, then Eq. (3.9) reduces to $BA = \phi_m$. Along any section of the core in Fig. 3.13 the flux is constant, which would be true even if $A$ (and therefore $B$) varied along the section.

Suppose we take a path in Eq. (3.9) that goes against the direction of $\mathbf{H}$ in Fig. 3.13, so that the result (when multiplied by $\mu$) would be positive work done on a monopole, if monopoles existed. *If* conditions are such that it can be assumed that $B$ and $H$ in a magnetic material are proportional ($B = \mu H$), Eq. (3.9) may be rewritten for such a section as

$$\mathcal{F}_{ab} = -\int_a^b \mathbf{H}\cdot d\boldsymbol{\ell} = \int_a^b \frac{B}{\mu}\,d\ell = \phi_m \int_a^b \frac{d\ell}{\mu A} = \phi_m \mathcal{R}_{ab}, \tag{3.21}$$

where

$$\mathcal{R}_{ab} = \int_a^b \frac{d\ell}{\mu A} = \frac{d}{\mu A} \tag{3.22}$$

is called the *reluctance* of the section $ab$ of the magnetic circuit over which the potential difference $\mathcal{F}_{ab}$ is calculated and $d = b - a$ for a simple uniform magnetic circuit section. If integrated all the way around, the result would also be equal to $NI$, if a coil of $N$ turns carrying current $I$ were wrapped around the core. In that case $\mathcal{R}$ would be the reluctance of the entire magnetic circuit. The units of reluctance are $\text{A}\cdot\text{t/Wb}$ (where $t$ signifies turns). The inverse of reluctance is $1/\mathcal{R} = \mathcal{P}$, which is called *permeance*.

An analogy can be drawn between permeance and capacitance. Let us rewrite the derivation of capacitance using Eqs. (2.15), (2.12b), and (2.5) as follows:

$$V_{ab} = -\int_a^b \mathbf{E}\cdot d\boldsymbol{\ell} = \int_a^b \frac{D}{\epsilon}\,d\ell = \phi_e \int_a^b \frac{d\ell}{\epsilon A} = \frac{\phi_e}{C_{ab}} = \frac{Q}{C_{ab}}, \tag{3.23}$$

where

$$\frac{1}{C_{ab}} = \int_a^b \frac{d\ell}{\epsilon A} = \frac{d}{\epsilon A}, \tag{3.24}$$

which, although just a restatement of Eq. (2.28), has a striking similarity to Eq. (3.22).

The analogy is that just as $\phi_e\ (= Q) = CV$, $\phi_m = \mathcal{P}\mathcal{F}$, where $C = \epsilon A/d$ for a capacitive element and $\mathcal{P} = \mu A/d$ for a simple magnetic circuit element. Note that just as $\phi_m$ is an open-surface flux, so is the flux $\phi_e$ that is actually used in Eq. (3.23); $\phi_e$ happens to equal the closed-surface flux $Q$ because $\mathbf{D} = 0$ everywhere but on the open surface portion (plate face inside the capacitor). Furthermore, $C$ and $\mathcal{P}$ are both proportionality constants between a potential and a flux. Thus permeance is a sort of magnetic capacitance or capacity of the magnetic device. It is an indicator of the magnetic energy $\frac{1}{2}\mathcal{P}\mathcal{F}^2$ stored in a magnetic device having a specified magnetic potential $\mathcal{F}$ across it, just as $C$ is an indicator of the electric energy $\frac{1}{2}CV^2$ stored in a capacitor having a potential $V$ across it.

The analogy just made is unfortunately not the one generally made in magnetic circuit theory discussions, even though it is physically the correct one. Instead, an analogy between Eqs. (3.21)–(3.22) and Ohm's law in a conductor, Eqs. (2.35)–(2.36), is given.

To show this analogy we proceed as follows. Recall that in the derivation of average conduction electron velocity in Eq. (2.32) we replaced $E$ by $V/d$. Thus a generalization of Eq. (2.34) may be obtained by restoring $V/d$ to $E$, yielding $I = \sigma AE$. Because the direction of $I$ is the same as that of $E$, this can be written $I\hat{\ell} = \sigma A\mathbf{E}$. Also, no matter how the cross section $A$ and the conductivity $\sigma$ of the conducting material may vary between the "terminals" of the slab, the current $I$ through all cross sections will be equal. Consequently, for a piece of conducting material such as in Fig. 2.12, Eq. (2.15) can be written as

$$V_{ab} = -\int_a^b \mathbf{E}\cdot d\boldsymbol{\ell} = \int_a^b \frac{I}{\sigma A}\,d\ell = I\int_a^b \frac{d\ell}{\sigma A} = IR_{ab}, \tag{3.25}$$

where

$$R_{ab} = \int_a^b \frac{d\ell}{\sigma A} = \frac{d}{\sigma A}, \tag{3.26}$$

the latter equality in Eq. (3.26) true for uniform conductors. Because of the mathematical similarity between Eqs. (3.25)–(3.26) and Eqs. (3.21)–(3.22), relation (3.21) is often called the "Ohm's law" of magnetic circuits, even though this is not the correct physical analogy, despite the mathematical similarity. After all, $I$ and $Q$ are not analogous; it is $\phi_m$ and $Q = \phi_e$ that are. As far as mathematical form, Eqs. (3.23)–(3.24) are equally similar to Eqs. (3.21)–(3.22). Furthermore, resistance is a dissipative quantity, while both reluctance and capacitance are nondissipative parameters quantifying, respectively, the inverse of magnetic capacity and the electric capacity of a structure.

## 3.10 TIME-VARYING ELECTROMAGNETICS

### 3.10.1 Faraday's Law

For the most part in this chapter, we have been considering electric and magnetic fields that are constant in time. Even when we allowed the voltage and current to vary in defining the instantaneous power, we tacitly assumed that variations were sufficiently slow that the static electric equations still applied. For example, the "quasistatic" assumption behind the definition of instantaneous power holds up to very high frequencies.

However, certain other electromagnetic effects have been observed that depend entirely on time-varying electric or magnetic fields for their existence. For instance, Michael Faraday in 1831 made the following discovery (see Fig. 3.19a). Two coils $a$ and $b$ were placed on an iron ring. The joined ends of coil $b$ were run over a distant compass needle, while the ends of coil $a$ were connected to a battery or disconnected by means of a switch. When the battery was connected, the needle momentarily deflected one way; when the battery was disconnected, it deflected the other way. It did not remain deflected when the battery was left either on or off, but

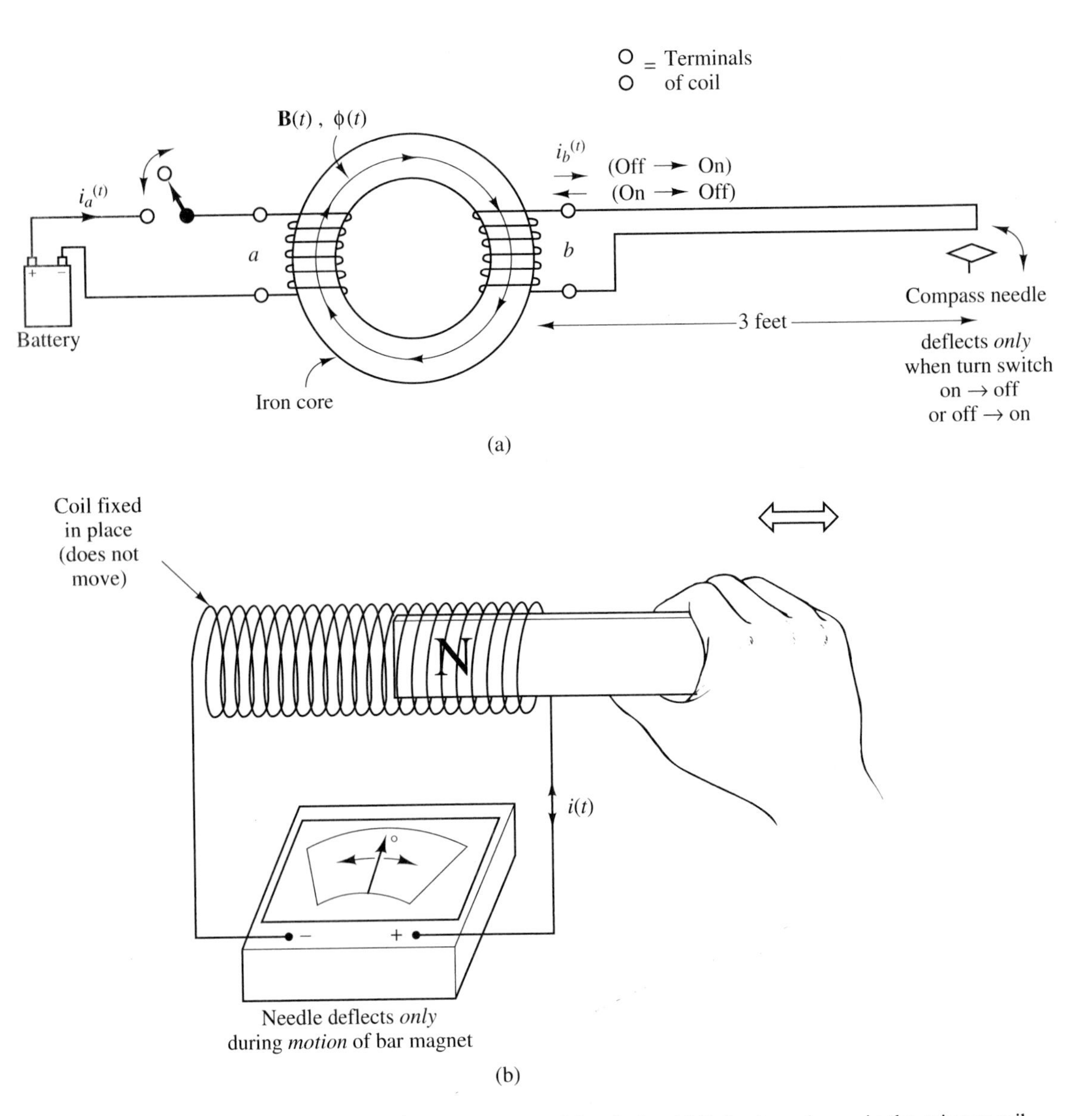


**Figure 3.19** (a) Induction of an emf [producing $i_b(t)$] due to a change in the primary coil current (and therefore magnetic field). One of Faraday's experiments. (b) Movement of bar magnet field produces emf in coil and thus a response on the ammeter. Another of Faraday's experiments.

returned to its original position. If the iron core was replaced by a nonmagnetic (wooden) core, the same effect was observed, although it was much weaker.

Obviously, a current $i_b(t)$ flowed in coil $b$, in order for there to have been a deflection of the distant compass needle (recall Oersted's discovery). This meant that there was an induced electric field in coil $b$ forcing its electrons to move. This electric field existed only when the applied magnetic field due to current $i_a(t)$ in coil *a was changing*. This was the first "transformer": coil $a$ is called the primary, $b$ is called

The actual "Faraday ring" (quite humble beginnings for what has now become a key element in transmitting efficiently gigawatts of power over hundreds of miles.) For comparison, a modern power transformer is shown below. *Sources:* IEEE Center for the History of Electrical Engineering (top photo); Virginia Transformer (bottom photo).

the secondary, and the iron core couples the magnetic field between the two coils. We shall study the practical use of transformers in Secs. 6.5 through 6.7.

The same effect could be observed when the primary and iron core were replaced by a moving bar magnet (see Fig. 3.19b). When the bar magnet was rapidly moved into (or out of) the coil, an ammeter deflected one way (or the other), indicating an instantaneous current in the coil. If the bar magnet was left motionless either within or outside the coil, the ammeter registered zero current.

In both experiments, an electric field was undeniably produced by a changing magnetic field. This electric field, when integrated along the coil path, naturally produced a potential difference (voltage) between the terminals of the coil. According to Ohm's law, $i(t) = v(t)/R$, the observed currents therefore flowed in the resistive wires. Recall that in the first experiment, a long wire (run over the compass needle) was connected across the coil terminals, while in the second experiment an ammeter was placed across the terminals. In either case, a potential difference generated between the coil terminals produced currents in the connected circuit.

Faraday was more an experimentalist than a theoretician. Franz Neumann of Germany in 1845 formulated the mathematical rule behind Faraday's discovery. An *electromotive force* (the contour integral of a nonconservative electric field) is observable at the terminals of a coil exposed to a changing magnetic field. If the coil has $N$ turns and the magnetic flux through the surface of the loops of the coil is $\phi_m(t)$, the instantaneous emf, $\mathcal{V}(t)$, is

$$\mathcal{V}(t) = N\frac{d\phi_m(t)}{dt}. \tag{3.27}$$

Relation (3.27) is known as *Faraday's law*. Like the battery, the emf in Eq. (3.27) is due to a nonconservative electric field. Consequently, for a general path in which there might be batteries or other voltage sources as well as time-varying magnetic fields, we may generalize Eq. (2.30) as follows:

$$\oint_C \mathbf{E}(t)\cdot d\boldsymbol{\ell} = \sum_n \mathcal{V}_n + N_n\frac{d\phi_{m,n}}{dt}, \tag{3.28}$$

where the $\mathcal{V}_n$ are all the voltage sources around the loop $C$ and $N_n\phi_{m,n}$ are the magnetic fluxes passing through the contour $C$ due to all the inductors around the loop. Equation (3.28) with $d\phi_m/dt = 0$ will be shown to yield one of the two fundamental electric circuit equations in Sec. 4.5.2 [the other being charge conservation, which we introduced in Eq. (2.38)].

**Example 3.4**

A bar magnet is known to have a magnetic flux density of 0.13 T. Suppose that initially the bar magnet is within the coil (as in Fig. 3.19b). If we suddenly pull it out and very far away from the coil in $\frac{1}{2}$ s, what is the average voltage observed at the coil terminals? The coil has 100 turns of radius 2 cm. Assume that the flux decreases in a linear manner with time, for simplicity.

**Solution** The initial flux $\phi_m(0) = \int_S \mathbf{B}\cdot d\mathbf{S} = BA = 0.13\cdot\pi(0.02)^2 = 1.6\cdot 10^{-4}$ Wb. The final flux is zero. This change of flux occurs in a time of $\frac{1}{2}$ s. The average emf (voltage) generated at the terminals of the coil is, by Eq. (3.27),

$$\mathcal{V}_{\text{avg}} = N\frac{\Delta\phi_m}{\Delta t} = \frac{100\cdot 1.6\cdot 10^{-4}}{0.5} \approx 0.03\ \text{V}.$$

Near the end of the course we will study electric generation by Faraday's law. From this example it should be apparent why generators are rotated at very high speeds and with very many turns of wire: Doing so will increase the emf to more practical values than 0.03 V.

How may the polarity of the induced voltage be determined? Heinrich Lenz in 1834 concisely stated the rule known as *Lenz's law:* The polarity will be such that the induced currents produce a magnetic field that opposes the *change* in the main magnetic field. Why is this so? It is because of conservation of energy. The electric field induced within the coil sets up (via currents) a magnetic field whose pole nearest the bar magnet will repel *movement* of the bar magnet. Of course, this means a pole of the same sign as that of the nearest of $\Delta B$, according to Eq. (3.1). If it were otherwise, we could have perpetual motion! For the case of Fig. 3.19a, we would have indefinitely increasing currents—perpetually increasing usable electric energy.

**Example 3.5**

As the bar magnet is pulled out, does current $i(t)$ in the rightmost conductor Fig. 3.19b (connected to the + terminal of the ammeter) point up or down?

**Solution** When the bar magnet is removed, the right-to-left flux passing through the coil due to the north pole of the bar magnet decreases. Consequently, $i(t)$ will have polarity so as to oppose this change. In this case, the induced $B$ field therefore points right to left. Notice that the induced $B$ points in this case *with* the bar magnet $B$ field. The direction of the external $B$ does *not* determine the direction of the induced $B$. It is *change* in the external $B$ that is opposed by the induced $B$, and determines its direction. If we point our right-hand thumb left, our fingers curl in the direction of current. According to the winding direction shown in the diagram, this means that $i(t)$ points down.

To close this subsection, consider the following amusing analogy involving Lenz's law. Consider two voters: one conservative and the other liberal. The conservative says, "The country is going bleeding-hearted liberal! I will fix that!" and votes Republican. The liberal says, "World War III will occur if I do not do something now!" and votes Democratic. In each case, the voter is using a form of Faraday's law: Vote $= -d\phi/dt_{\text{assumed}}$, where here $\phi$ represents the liberal/conservative state of the nation.

Both think they are following Lenz's law: Oppose the change (which to them is undesirable). The problem is that they have a biased or assumed sign for $d\phi/dt$. Lenz's law is vote $= -d\phi/dt_{\text{actual}}$; Lenz is very perceptive! We, however, do not know which voter has the correct polarity. It is like defining the polarity of the voltage across a resistor. In reality, we may or may not have guessed right about which terminal is actually at higher potential (see Sec. 4.4). Lenz's law uses the *actual numerical value* of $d\phi/dt$ to determine the induced voltage polarity.

### 3.10.2 Self-induction and Inductance

Not only can a $d\phi_m/dt$ from an external source (other coil $a$ or bar magnet in Fig. 3.19) produce an electric field within a coil $b$, but the current flowing through the coil $b$ itself also can, whenever connected to a circuit in which its current varies with time. This is true even in the absence of any external $d\phi_m/dt$. Joseph Henry of Albany Academy (in New York) discovered this around 1832. Henry actually independently discovered "Faraday's" law, Eq. (3.27), slightly earlier than did Faraday in England, but Henry was slow to publish and lost the glory. But the discovery of *self-induction,* in which current within a (coiled) wire can produce an emf within the same wire, is uniquely Henry's.

We may find the emf of self-induction in terms of the rate of change of the current $i(t)$ flowing through the coil as follows. In Example 3.2 we found that within a current-carrying coil (solenoid), $H \approx NI/d$, which can be written for time-varying currents as $H(t) \approx Ni(t)/d$. For a medium in which it is valid to assume that $B = \mu H$, we may write $B(t) = \mu Ni(t)/d$. From Eq. (3.14), the flux passing through the coil, assuming that $B(t)$ is uniform over the cross-sectional area $A$ of the coil, is $\phi_m(t) = B(t)A = \mu ANi(t)/d$. Consequently, from Eq. (3.27) we have

$$\begin{aligned} \mathcal{V}_{\text{self}}(t) &= N\frac{d\phi_m(t)}{dt} = N\frac{\mu AN}{d}\frac{di(t)}{dt} \\ &= L\frac{di(t)}{dt}, \end{aligned} \tag{3.29}$$

where for a coil,

$$L = \frac{\mu AN^2}{d}. \tag{3.30}$$

The parameter $L$ is called *inductance.* Its units are $\text{V} \cdot \text{s/A} = \text{N} \cdot \text{m/A}^2 = \text{J/A}^2$, which are called henrys (H) after Joseph Henry. Any circuit or circuit element has an associated inductance, not necessarily expressible in the form of Eq. (3.30) but always relating a self-induced voltage to the rate of change of current flowing through the inductive element. However, $L$ can be made intentionally large by using a coil, as in Eq. (3.30) with large $N$ and/or $\mu$ (e.g., iron core) and/or $A$, wound tightly (small $d$).

The self-induced voltage $\mathcal{V}_{\text{self}}(t)$ is equal to the instantaneous voltage $v(t)$ across the coil. The polarity of the self-induced voltage is again determined by Lenz's law. We may consider the current in the coil analogous to the angular velocity of a spinning bicycle wheel, inductance analogous to rotational inertia, and the induced voltage is analogous to inertial torque. The larger the inductance (rotational inertia), the more resistant is the coil (wheel) to changes in its current (angular velocity), that is, the greater the oppositional emf (inertial torque).

### 3.10.3 Maxwell's Equations

James Clerk Maxwell was born in Scotland the year of Faraday's discovery of the magnetically induced emf. By the time he was 35, he had fully described light as an electromagnetic phenomenon. Unlike most of his predecessors, Maxwell predominantly made mathematical arguments rather than appeals to experiment. In particular, he made the remarkable hypothesis that a time-varying electric field produces a magnetic field. This was based on a detailed hydrodynamic model of the propagation of electricity and magnetism in space. The effect was not experimentally noticeable because the time variations were too slow to produce a detectable $B$. This fact makes his proposal (later verified in every instance) all the more ingenious.

We can see the need for this additional source of magnetic fields by considering application of Ampère's circuital law, Eq. (3.10), to the wire–capacitor connection in Fig. 3.20a. A current $i_A(t)$ flows through the wire into the (circular) capacitor plates as shown because of a time-varying voltage source connected across the wire–capacitor system (for simplicity, it is not shown).

Because no current passes through surface $S_1$ defined by contour $C$, Eq. (3.10) indicates that the integral of $\mathbf{H} \cdot d\boldsymbol{\ell}$ is zero. The circular symmetry of this configuration thus implies that $\mathbf{H}$ is zero on $C$. Yet iron filings or other experiments would indicate that $\mathbf{H}$ is definitely not zero, but rather as shown in the figure—similar to

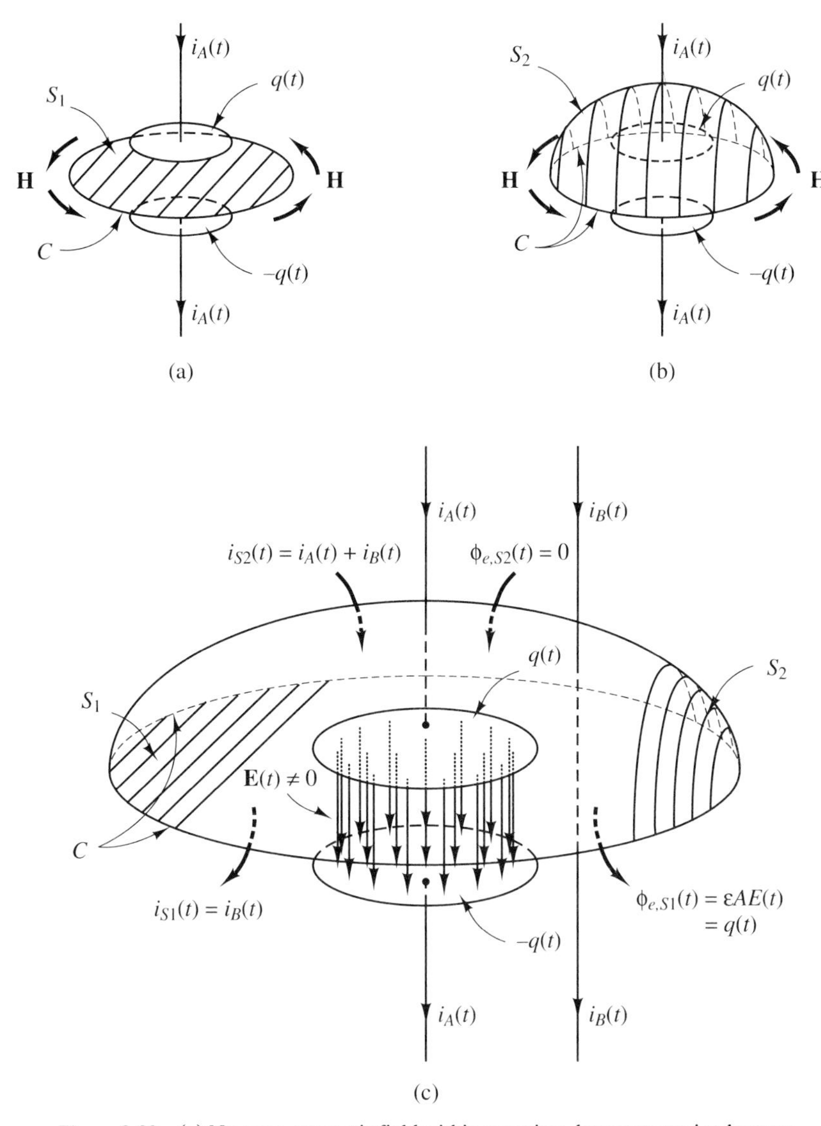


**Figure 3.20** (a) Nonzero magnetic field within capacitor demonstrates inadequacy of Ampère's law. (b) If the surface in Ampère's law is stretched to include the wire, it yields the correct result. (c) Maxwell's solution of the problem: the time-varying electric field between capacitor plates causes an additional source of magnetic field.

that around a current-carrying wire. Furthermore, if the surface bounded by $C$ is stretched over the capacitor plate to form $S_2$ as shown in Fig. 3.20b, Eq. (3.10) yields the correct nonzero value for $H$. In fact, a detailed mathematical study of this problem (involving Stokes's theorem), which Maxwell himself performed, concludes that the answer should be independent of the surface as long as it is bounded by $C$. The surface need not be in the plane of $C$; for that matter, $C$ itself need not be constrained to lie in a plane!

Maxwell solved this problem as follows. Refer to Fig. 3.20c, in which the only changes are that both surfaces $S_1$ and $S_2$ are shown at once, and a current-carrying wire $i_B(t)$ with no capacitor is added that passes through both surfaces. Let us call the total current passing through surfaces $S_1$ and $S_2$, respectively, $i_{S1}(t)$ and $i_{S2}(t)$. Clearly, $i_{S1}(t) = i_B(t)$ and $i_{S2}(t) = i_A(t) + i_B(t)$. Also, let us call the total electric flux [see Eq. (2.5)] passing through surfaces $S_1$ and $S_2$, respectively, $\phi_{e,S1}(t)$ and $\phi_{e,S2}(t)$. Using Eqs. (2.5) and (2.22), we see that $\phi_{e,S1}(t) = \epsilon AE(t) = q(t)$, where $A$ is the area of the circular capacitor plates, while $\phi_{e,S2}(t) = 0$ because the field outside the plates (and thus crossing $S_2$) is zero.

By conservation of charge, Eq. (2.38), the total current passing through $S_2$ minus that passing through $S_1$ equals the rate of increase of charge $q(t)$ on the upper capacitor plate: $i_{S2}(t) - i_{S1}(t) = dq(t)/dt$. The sum of surfaces $S_1$ and $S_2$ can be thought of as $S$ in Gauss's law, Eq. (2.6). The integral of $\epsilon\mathbf{E}(t)\cdot d\mathbf{s}$ on $S_2$ equals $-\phi_{e,S2}(t)$ as defined in Fig. 3.20c, because $d\mathbf{s}$ is an *outward* normal. Thus Gauss's law, Eq. (2.6), can be written in our terminology as $\phi_{e,S1}(t) - \phi_{e,S2}(t) = q(t)$, or

$$\frac{d}{dt}[\phi_{e,S1}(t)] - \frac{d}{dt}[\phi_{e,S2}(t)] = \frac{d}{dt}[q(t)] = i_{S2}(t) - i_{S1}(t), \tag{3.31a}$$

or

$$\frac{d}{dt}[\phi_{e,S1}(t)] + i_{S1}(t) = \frac{d}{dt}[\phi_{e,S2}(t)] + i_{S2}(t). \tag{3.31b}$$

Now we have something that does not depend on the surface chosen: $d\phi_e(t)/dt + i(t)$. If this is used instead of just $i(t)$, the result is consistent with experiment. Recall that the units of $\phi_e$ are those of electric charge, coulombs, so $d\phi_e(t)/dt$ has the units of current. This additional *equivalent current* term $d\phi_e(t)/dt$ is also essential for the recognition that light and radio waves are electromagnetic phenomena, as we shall show below. The form of Ampère's circuital law as amended by Maxwell is strikingly similar to Faraday's law with emfs, as in Eq. (3.28):

$$\oint_C \mathbf{H}(t)\cdot d\boldsymbol{\ell} = \sum_n I_n + \frac{d\phi_{e,n}}{dt}, \tag{3.32}$$

where the $\phi_{e,n}$ are the electric fluxes due to all capacitors whose middles are encircled by $C$, or are other changing electric fluxes enclosed by $C$.

### ☆ 3.10.4 The Electromagnetic Wave

Without electromagnetic waves, you could not be reading this page, for there would be no light. Neither could you listen to a radio, watch TV, use a cellular phone, cook with your microwave oven, get a suntan, or behold a mountain view. For that matter, we would not be

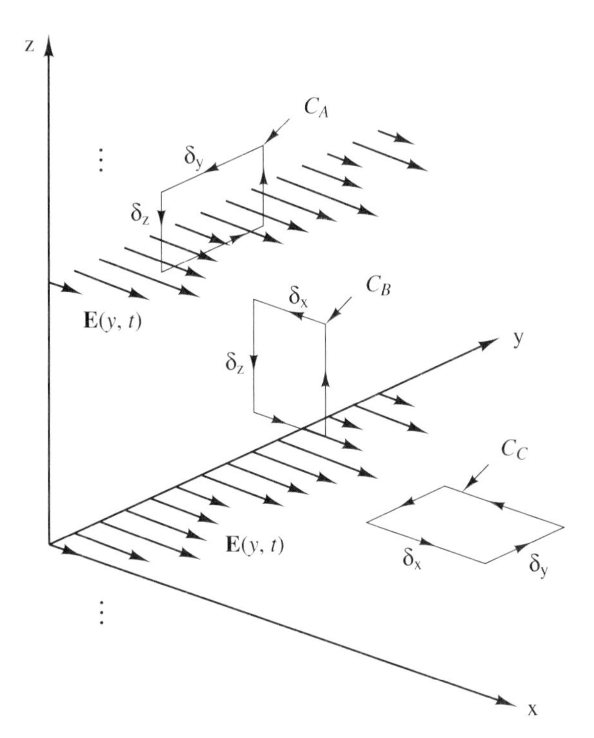


**Figure 3.21** Contours and surfaces required in the derivation of the wave equation from the integral form of Maxwell's equations.

alive, for there would be no means for energy from the sun to reach us on earth. So our life depends on Maxwell's equations. Let's see how.*

Let us consider only time-varying electric and magnetic fields (not constant-in-time fields) for only these make up electromagnetic waves. For simplicity, we consider an electric field $\mathbf{E} = E_x(y,t)\hat{\mathbf{x}}$; that is, an electric field which has a nonzero component only in the $x$ direction $\hat{\mathbf{x}}$, and which varies only with the $y$ coordinate and time $t$ (see Fig. 3.21). This electric field is to be caused by a time-varying magnetic field $\mathbf{B}(t)$, and the medium is free space. We shall now show that $\mathbf{E}(y,t)$ satisfies a wave equation, by calculating circulations on the three contours $C_A$, $C_B$, and $C_C$. These contours all have sides of lengths $\delta x$, $\delta y$, and/or $\delta z$, as shown in Fig. 3.21.

Because $\mathbf{E}(y,t)$ is perpendicular to all sides of $C_A$, its circulation around $C_A$ is zero. Similarly, because (1) $\mathbf{E}(y,t)$ is perpendicular to the left and right sides of $C_B$ and (2) its value on the top side equals that on the bottom, the circulation of $\mathbf{E}(y,t)$ is also zero on $C_B$. These two circulations yield no information, but $C_C$ will. Again, $\mathbf{E}(y,t)$ is perpendicular to the sides along $\hat{\mathbf{y}}$ of $C_C$, but its values along the other two sides are both nonzero *and different:* $E_x(y_1,t)$ on one side and $E_x(y_1,t) + \Delta_y E_x(y_1,t)$ on the other, where $\Delta_y$ represents a variation between $y = y_1$ and $y = y_1 + \delta y$ of the variable it precedes (here, $E_x$). Consequently, for infinitesimal $\delta x$ and $\delta y$,

$$\oint_{C_C} \mathbf{E}\cdot d\boldsymbol{\ell} = E_x(y_1,t)\,\delta x - [E_x(y_1,t) + \Delta_y E_x(y_1,t)]\,\delta x \tag{3.33a}$$

$$= -\Delta_y E_x(y_1,t)\,\delta x \tag{3.33b}$$

*The following is an adaptation, with all steps shown, of the derivation in *The Electromagnetic Field in Its Engineering Aspects,* by G. W. Carter (American Elsevier Publishers, New York, 1967), pp. 267–270.

$$= -\frac{\partial E_x(y_1,t)}{\partial y}\,\delta y\,\delta x \tag{3.33c}$$

$$= -\frac{\partial \phi_{m,C_C}(t)}{\partial t} \tag{3.33d}$$

$$= -\frac{\partial B_z(y_1,t)}{\partial t}\,\delta x\,\delta y, \tag{3.33e}$$

where in Eq. (3.33c) the chain rule was used, in Eq. (3.33d) Faraday's law, Eq. (3.28), with no emfs was used (with $\phi_{m,C_C}$ the magnetic flux through $C_C$), and Eq. (3.33e) follows from Eq. (3.14). Note that the time-varying $\mathbf{B}(t)$ producing $\mathbf{E}(t)$ must point in the $\hat{\mathbf{z}}$ direction in order to produce the flux $\phi_{m,C_C}$ flowing through the surface bounded by $C_C$ (see Fig. 3.21).

We now calculate the circulations of $\mathbf{B}(t)$ using Maxwell's corrected version of Ampère's circuital law, Eq. (3.32), multiplied by $\mu_0$. Note that there are no currents in this case, so the circulation of $\mathbf{B}(t)$ is just $\mu_0 d\phi_e/dt$, where $\phi_e$ is the electric flux through the surface bounded by the circulation contour. Because we have assumed that there is only an $x$ component of $\mathbf{E}$, only $C_A$ has nonzero electric flux through it; the other contours open to $z$ or $y$ directions, through which $\mathbf{E}$ lines and thus $\phi_e$ do not pass. Furthermore, neither field varies with $x$ or $z$: $\mathbf{B} = B_z(y,t)$. Thus we proceed in identical fashion to that above:

$$\oint_{C_A} \mathbf{B}(t)\cdot d\boldsymbol{\ell} = B_z(y_1,t)\,\delta x - [B_z(y_1,t) + \Delta_y B_z(y_1,t)]\,\delta x \tag{3.34a}$$

$$= -\Delta_y B_z(y_1,t)\,\delta x \tag{3.34b}$$

$$= -\frac{\partial B_z(y_1,t)}{\partial y}\,\delta y\,\delta x \tag{3.34c}$$

$$= -\frac{\mu_0 \partial \phi_{e,C_A}(t)}{\partial t} \tag{3.34d}$$

$$= -\frac{\mu_0 \epsilon_0 \partial E_x(y_1,t)}{\partial t}\,\delta x\,\delta y, \tag{3.34e}$$

where in Eq. (3.34c) the chain rule was used, in Eq. (3.34d) Ampère's circuital law as amended by Maxwell equation (3.32) with no currents was used (with $\phi_{e,C_A}$ the electric flux through $C_A$), and Eq. (3.34e) follows from Eq. (2.5).

Now the fun part. Differentiate the equality between Eqs. (3.34c) and (3.34e) with respect to $y$ and divide by $\delta x\,\delta y$ to obtain

$$\frac{\partial^2 E_x(y_1,t)}{\partial y^2} = -\frac{\partial^2 B_z(y_1,t)}{\partial y\,\partial t} \tag{3.35}$$

and differentiate the equality between Eqs. (3.33c) and (3.33e) with respect to $t$ to obtain

$$\frac{\mu_0 \epsilon_0 \partial^2 E_x(y_1,t)}{\partial t^2} = -\frac{\partial^2 B_z(y_1,t)}{\partial y\,\partial t} \tag{3.36}$$

Now just equate the left-hand sides of Eqs. (3.35) and (3.36), noting that the right-hand sides are identical:

$$\begin{aligned} \frac{\partial^2 E_x(y_1,t)}{\partial y^2} &= \frac{\mu_0 \epsilon_0 \partial^2 E_x(y_1,t)}{\partial t^2} \\ &= \frac{1}{c_0^2}\frac{\partial^2 E_x(y_1,t)}{\partial t^2} \end{aligned} \tag{3.37}$$

where $c_0 = 1/(\mu_0\epsilon_0)^{1/2}$ clearly has the dimensions of distance/time (by examination of the dimensions in Eq. (3.37)), and in fact is the speed of light, $2.998 \cdot 10^8$ m/s! Recall that this was mentioned back when $\mu_0$ was defined.

From a dynamical perspective, Eq. (3.28) implies that when there is a time-varying **B** (and therefore a time-varying $\phi_m$), there will be a nonzero **E**; Eq. (3.32) implies that when **E** varies with time, there will be a nonzero $\mathbf{H}(t)$ [and therefore $\mathbf{B}(t)$]. So as one field varies, it causes the other to exist. In this manner, the electromagnetic disturbance spreads out or "propagates" throughout space. Note that the time-varying **E** in Eq. (3.34e) is associated with a spatial variation in **B** [Eq. (3.34c)]. Together, these time and space relations cause a pulse or oscillation in **E** or **B** to travel at the speed of light $c_0$ along the $y$ direction to other regions.

In fact, the general solution of the wave equation (3.37) is $f(y - c_0t) + g(y + c_0t)$, where $f(\cdot)$ and $g(\cdot)$ are arbitrary functions. Suppose that $f(w)$ is a pulse function, having a maximum value at $w = 0$. This maximum is the crest of a wave when we replace $w$ with $w = y - c_0t$ and the field is independent of the $x$ and $z$ directions. For every value of $t$, the crest of the wave appears at a different value of $y$: $y = c_0t$—which increases with time $t$. Thus the function $f(y - c_0t)$ is a wave traveling in the $+y$ direction as $t$ increases, while similarly $g(y + c_0t)$ is another wave traveling in the opposite direction. We now show that indeed $f(y - c_0t) + g(y + c_0t)$ satisfies Eq. (3.37) by substituting it for $E_x(y,t)$:

$$\frac{\partial[f(y - c_0t) + g(y + c_0t)]}{\partial y} = f'(y - c_0t) + g'(y + c_0t), \tag{3.38a}$$

so that

$$\frac{\partial^2[f(y - c_0t) + g(y + c_0t)]}{\partial y^2} = f''(y - c_0t) + g''(y + c_0t). \tag{3.38b}$$

Also,

$$\frac{\partial[f(y - c_0t) + g(y + c_0t)]}{\partial t} = c_0[-f'(y - c_0t) + g'(y + c_0t)], \tag{3.38c}$$

so that as in the right-hand side of Eq. (3.37),

$$\frac{1}{c_0^2}\frac{\partial^2[f(y - c_0t) + g(y + c_0t)]}{\partial t^2} = f''(y - c_0t) + g''(y + c_0t), \tag{3.38d}$$

which equals Eq. (3.38b), thus proving that the wave equation is satisfied by $f(y - c_0t) + g(y + c_0t)$.

Unlike sound waves, which rely on molecules bumping each other to propagate, electromagnetic waves such as light can travel in a vacuum ("free space"). They in fact can transport energy from place to place, as is done by sunlight and radio waves. Although this may be hard to imagine, it is true, as is evidenced by the transfer of energy from the sun, through outer space (essentially a vacuum), to us on earth.

There were many precursors to Maxwell's discovery of electromagnetic waves in the early 1860s. In a crude experiment, Henry observed electromagnetic wave propagation and even compared it with light back in the late 1830s. Gustav Kirchhoff correctly hypothesized in 1857 that electricity propagated along wire at the speed of light. It took Maxwell to formulate wave theory authoritatively in his *Treatise on Electricity and Magnetism* of 1873. Fourteen years later, Maxwell's theory was proved true by the experiments of Heinrich Hertz of Germany.

The study of electromagnetic waves is a highly intricate and mathematical enterprise. For example, for practical problem solving, "divergence" and "curl" operators are used to express Maxwell's equations in a pointwise form, more convenient in many cases than the integral forms we have considered. We shall not further pursue waves in a quantitative manner because such an approach lies outside the scope of this book. However, with the understanding gained from this chapter, you have all the tools (except for differential vector operators such as the "del") required to take on such a study.

Before closing, a few practical words. Heat transfer engineers should recognize that radiation heat transfer is merely the propagation of electromagnetic waves! The predominant wavelengths are infrared: 0.3 to 25 $\mu$m. Thermal imaging systems depend on this electromagnetic radiation, which all objects emit, for their operation (e.g., "night vision" systems).

An application of electromagnetic waves familiar to civil engineers is the geodimeter (geodetic distance meter). It is an electromagnetic distance measuring system; it determines the time for electromagnetic energy to travel from one end of the line of measurement to the other end and back. For long-distance measurements, a sinusoid (see Chapter 5) of, for example, frequency 30 MHz is generated at the transmitter. The phase relation between the internal signal and the measured reflected signal indicates a fraction of a wavelength distance (at this frequency, the wavelength $\lambda = c/f = 3 \cdot 10^8/3 \cdot 10^6 = 200$ m). Specifically, if $\theta$ is the phase difference between the internal and reflected waves, the distance is $d = c(\theta/[2\pi]) \cdot (1/f)/2 = \lambda\theta/[4\pi]$. Larger unit distances are made doing the same procedure at different frequencies. For short-range measurements, simpler infrared detection systems are used.

Also, many environmental engineers routinely use remote sensing equipment to monitor the environment over large areas. An airborne electromagnetic source emits a wide range of frequencies, and the frequency content of the reflected signal provides information about the chemical/geometrical nature of the terrain. For example, radar scanning provides relief displacement mapping.

## 3.11 SUMMARY

In this first pair of chapters, we have introduced all of the physical and mathematical laws governing electric and magnetic fields. These laws provide the foundation of essentially all electric, electronic, and magnetic circuits that we use in our daily lives.

In particular, in this chapter we have considered the laws governing magnetic phenomena. Just as electric charges attract and repel each other, magnetic "poles" also do. Unlike electric charge, which can be isolated as positive ions and negative ions or electrons, there are no isolated magnetic monopoles. Originally, the only magnetic forces observed were those between inherently magnetized materials such as lodestone. The usefulness of magnetism in science and engineering became manifest when it was discovered that electric currents could produce magnetic effects.

These magnetic forces were quantified by a magnetic field. Just as there are electric forces between charges or on a charge in the presence of an electric field, there are also magnetic forces between current-carrying wires or on a current-carrying wire in the presence of a magnetic field. The direction of the magnetic force in the former case (of parallel wires) is along the line perpendicular to both wires which joins them. In the latter case, it is that of the vector cross product between the magnetic field and the current direction.

The theory that magnetic fields are always caused by charge motion of various types unifies the explanation of various magnetic effects. Thus both electric and magnetic fields and forces are all caused and experienced by electric charge. In combination, the net force on a moving electron due to electric and magnetic forces is called the Lorentz force.

Whereas a polarizable material tends to cancel the electric field within it, a ferromagnetic material tends to accentuate the magnetic field within it. Consequently, iron is commonly used to maximize magnetic-based effects. Also, solenoidal coils can mimic bar magnets. The more densely the coils are arranged, the greater the magnetic field strength for a given coil current.

Just as the integral of the electric field around a closed loop is equal to the sum of emfs along the loop, the integral of the magnetic field around a closed loop is equal to the sum of currents in wires enclosed by the loop. Thus currents are the source of nonconservative magnetic fields, just as are emfs in the electric case. In both cases it is these nonconservative fields that may be set up that are responsible for practical, large-scale electric and magnetic behavior.

Although it does not represent work because of the nonexistence of monopoles, the magnetic potential is defined analogously to the electric potential. It is useful in solving magnetic circuit problems in the same way as electric potential is for analyzing electric circuits. Magnetic circuits come up when we study transformers and electromechanical machines later in the book.

Again in analogy with the electric case, the magnetic field may be decomposed into a medium-based component called the magnetization vector and another called the magnetic field intensity, which includes the effects of current-carrying wires. In a sense, the magnetic field intensity represents an excitation, and the magnetization represents the medium response.

The situation is complicated by the physical means of this response, which involves the alignment of "magnetic domains" over frictional opposition. The result is that the total magnetic field "lags behind" changes in excitation, made evident in the hysteresis $B$–$H$ curve. It is the hysteresis phenomenon that causes residual magnetism and thus allows us to make permanent magnets out of electromagnets. Furthermore, eventually all domains are aligned so that further increase in excitation does not significantly increase the total magnetic field; this is called saturation.

Just as electric potential is proportional to electric flux (charge) in a capacitor, magnetic potential is proportional to magnetic flux in a section of magnetic field-containing material. The name of the latter proportionality constant is permeance. Its reciprocal, called reluctance, is often stated as being analogous to electrical resistance.

Having described all the basic electric and magnetic variables and concepts, we considered the interaction of electricity and magnetism. This interaction occurs only when the fields (and therefore fluxes) vary with time. The more obvious interaction was that a time-varying magnetic field could produce nonconservative electric fields. For example, moving a bar magnet through a coil produces a current in that coil if it is connected in a closed circuit. Again, these nonconservative fields are ideal for doing real work; for example, driving electricity through our home appliances.

Moreover, the self-emf in an inductor is a useful signal-processing tool in modern electronic circuits. The parameter describing the self-inductance of a solenoid or other circuit element is called inductance. It is an indicator both of self-emf magnitude and of energy stored within the magnetic field of the coil.

Less obvious to the early pioneers was how a time-varying electric field could create magnetic effects. It took the genius of Maxwell to realize this, and add this nonconservative field to Ampère's original law. Without this term, not easily observable at low frequencies used by those experimentalists, electromagnetic waves such as light would be impossible! We closed our discussion by showing how the doubly coupled electromagnetic field satisfies a wave equation.

In closing this pair of chapters, we may now write all of these fundamental laws of electromagnetics together, where the equation number refers to its number of appearance in this and the previous chapters and where the first group of equations is known as Maxwell's equations:

### *Basic Electromagnetic Field Equations*

Faraday's law with emfs:

$$\oint_C \mathbf{E}(t) \cdot d\boldsymbol{\ell} = \sum_n \mathcal{V}_n + N_n \frac{d\phi_{m,n}}{dt} \tag{3.28}$$

specializing to

$$v(t) = L\, di(t)/dt \text{ for a coil, with} \tag{3.29}$$

$$L = \frac{\mu A N^2}{d} \tag{3.30}$$

Ampère's circuital law as amended by Maxwell:

$$\oint_C \mathbf{H}(t) \cdot d\boldsymbol{\ell} = \sum_n I_n + \frac{d\phi_{e,n}}{dt} \tag{3.32}$$

Gauss's law for **D**:

$$\oint_S \mathbf{D} \cdot d\mathbf{S} = q_{\text{enc}} \tag{2.13}$$

Gauss's law for **B**:

$$\oint_S \mathbf{B} \cdot d\mathbf{S} = 0 \tag{3.13}$$

### *Force Equations*

Coulomb's electric force on a charge:

$$\mathbf{F}_{21} = \frac{q_1 q_2}{4\pi\epsilon r^2} \hat{\boldsymbol{\ell}}_{12} = q_2 \mathbf{E}_1 \qquad (2.1), (2.2)$$

Ampère's magnetic force on a current-carrying wire:

$$\mathbf{F}_{21/\text{ul}} = \frac{\mu_0 I_1 I_2}{2\pi r} \hat{\boldsymbol{\ell}}_{21} = I_2 \hat{\boldsymbol{\ell}} \times \mathbf{B}_1 \qquad (3.3), (3.2)$$

Lorentz's electromagnetic force on a charge:

$$\mathbf{F} = q[\mathbf{E} + \mathbf{u} \times \mathbf{B}] \tag{3.12}$$

### *Field Decompositions/Material Relations*

Electric field decomposition:

$$\mathbf{E} = \frac{\mathbf{D} - \mathbf{P}}{\epsilon_0} \tag{2.9d}$$

Material relations for electric fields:

$$\mathbf{D} = \epsilon \mathbf{E} \tag{2.12b}$$

$\mathbf{P} = (\epsilon - \epsilon_0)\mathbf{E}$ [see paragraph preceding (2.13)]

Magnetic field decomposition:

$$\mathbf{B} = \mu_0(\mathbf{H} + \mathbf{M}) \tag{3.20}$$

Material relations for magnetic fields:

$\mathbf{B} = \mu\mathbf{H}$ (under linear assumption; $B$–$H$ curve always valid) [see (3.6)ff]

$\mathbf{M} = \chi\mathbf{H}$ (under linear assumption) [see paragraph after (3.20)]

### *Potential Definitions*

Electric potential:

$$V_{ab} = -\int_a^b \mathbf{E}\cdot d\boldsymbol{\ell} \qquad (2.15)$$

Magnetic potential:

$$\mathcal{F}_{ab} = -\int_a^b \mathbf{H}\cdot d\boldsymbol{\ell} \qquad (3.9)$$

### *Flux Definitions*

Electric flux:

$$\int_S \mathbf{D}\cdot d\mathbf{S} = \phi_e \qquad (2.5)$$

Magnetic flux:

$$\int_S \mathbf{B}\cdot d\mathbf{S} = \phi_m \qquad (3.14)$$

### *Flux–Potential Relations*

Electric—proportionality constant is capacitance:

$$Q = CV \xrightarrow[d/dt]{} i(t) = C\frac{dv(t)}{dt} \qquad (2.27), (2.39)$$

with

$$C = \frac{\epsilon A}{d} \qquad (2.28)$$

Magnetic—proportionality constant is permeance = 1/reluctance:

$$\phi_m = \mathcal{P}\mathcal{F} = \frac{1}{\mathcal{R}}\mathcal{F}$$ [see (3.21) ff., paragraph following (3.24)]

with

$$\mathcal{R} = \frac{d}{\mu A} \qquad (3.22)$$

### *Energy Storage*

Energy in electric field of a capacitor:

$$W_e = \tfrac{1}{2}CV^2 \qquad (2.29)$$

Energy in magnetic field of an inductor:

$$W_m = \tfrac{1}{2}L\,i^2 \qquad \text{(problem)}$$

***Electric Current Definition as the Time Rate of Charge Flowing through a Cross Section***

$$i(t) = \frac{dq_{xsec}(t)}{dt} \tag{2.37}$$

***Conservation of Charge for a Closed Surface S***

$$\sum_{n,S} i_n(t) = \frac{dq_{enc}(t)}{dt} \tag{2.38}$$

***Ohm's Law***

$$I = \frac{V}{R} \tag{2.35}$$

with

$$R = \frac{\rho d}{A} \tag{2.36}$$

***Power Absorbed by a Circuit Element***

$$p(t) = v(t)i(t) \qquad (2.45), (2.46)$$

Fortunately, basic circuit analysis requires only a small subset of these laws and only scalar quantities. The most important definitions and relations for electric circuit analysis shall be the electric potential between two points [Eq. (2.15)], electric current [Eq. (2.37)], the circulation of **E** [Eq. (2.30), which is Eq. (3.28) for the case $d\phi_m/dt = 0$], charge conservation [Eq. (2.38)] with $dq(t)/dt = 0$ [this and the preceding relation are called Kirchhoff's laws], Ohm's law [Eq. (2.35)], the power absorbed by a circuit element [Eq. (2.46); b and c valid for resistors], and the current–voltage relations for capacitors [Eq. (2.39)] and inductors [Eq. (3.29ff)].

When we study electronics, we shall return to Gauss's law [Eq. (2.13)] and Coulomb's electric force [Eq. (2.1)] in addition to the laws above. In our study of transducers, the parametric computations of resistance [Eq. (2.36)], capacitance [Eq. (2.28)], and inductance [Eq. (3.30)] will be of renewed interest. For the discussion of machines, Faraday's law [Eq. (3.27), which is Eq. (3.28) for the case $\mathcal{V}_n = 0$], Ampère's magnetic force [Eqs. (3.2) and (3.3)], Ampère's original circuital law [Eq. (3.10), which is Eq. (3.32) for the case $d\phi_e/dt = 0$], the relation between $B$ and $H$ (such as $B = \mu H$ under a linear assumption), the parametric computation of reluctance [Eq. (3.22)], the magnetic potential [Eq. (3.9)], magnetic flux [Eq. (3.14)], and the magnetic flux–potential proportionality [Eq. (3.21)].

As you can see, nearly all of the equations we have stated in this chapter have practical significance in one or more areas of electrical engineering that we study in this book. Typically, in our usage, simplified forms of these equations which avoid calculus are sufficient. Occasionally, other equations not mentioned in the list above are either tacitly applied in those we use or have special uses outside the scope of our study. It is now time to begin *applying* these equations in ways that do great things such as turn on lights, filter audio signals, run computers, and turn motors.

## PROBLEMS

**3.1.** A 10-cm bar magnet can be modeled as having a magnetic monopole of strength $m$ on one end and one of strength $-m$ on the other end. If two such magnets are placed on the $x$ axis so that the two $+m$ ends are 10 cm apart and the two $-m$ ends are 30 cm apart, what is the force on one magnet due to the other?

**3.2.** What would happen to the poles of a bar magnet were the bar bent into a circle (ring) and the ends joined?

**3.3.** A section of wire of length 1 m carries a current of 1 A. The wire makes an angle of 30° with respect to a uniform magnetic field $B$ of magnitude 1 T. What are the magnitude and direction of the force on the wire? Draw a figure indicating this direction, along with the current direction and $\mathbf{B}$.

**3.4.** Suppose that you are given a very light but strongly magnetized scrap of iron. If all you have is a piece of cloth and you are outside on a sunny day, how can you determine which is the north pole and which is the south pole of the magnetized scrap?

**3.5.** In what respect is the planarian worm similar to a bar magnet?

**3.6.** Four small compasses are placed about 1 ft away from each other on the ground: one toward the north, one toward the east, one toward the south, and one toward the west. Then a very strong bar magnet is placed in the middle, with its N pole toward the west and its S pole toward the east. Draw a figure showing the resulting orientations of the compasses. Assume that the compasses do not affect each other.

**3.7.** Suppose that a straight wire carrying 0.5 A is placed in "the plane of the paper, pointing down" and is subjected to a magnetic field pointing "into the page." The spatial extent of the magnetic field is such that exactly 1 m of the wire is exposed to this $B = 0.3$ T field; for the rest of the wire, $B = 0$. Find the magnitude and direction of the force on the wire.

**3.8.** Two long, parallel wires are 1 m apart. The wire on the left has $I_1 = 1$ A pointing "up" and the wire on the right has current $I_2 = 2$ A pointing "down" (the currents are thus in opposite directions). A third wire carrying $I_3 = 1$ A is now placed parallel to the first two wires.

**(a)** On what locus of positions must the third wire be placed so that the net force on it due to the first two wires is zero? On what locus are the two forces on the third wire of the same magnitude? Are these two loci the same?

**(b)** In order that the total force on the third wire be zero, does it matter which direction ("up" or "down") the 1-A current in the third wire flows? Why or why not?

**3.9.** A permanent-magnet speaker is shown in Fig. P3.9, which shows a cross-section (the outer leg actually goes all around the center leg). Suppose that at this moment $i(t) > 0$. What are the magnitude and direction of the force on the coil? What if $i(t) < 0$? If a paper cone is attached to the coil, can you now envision how a speaker works?

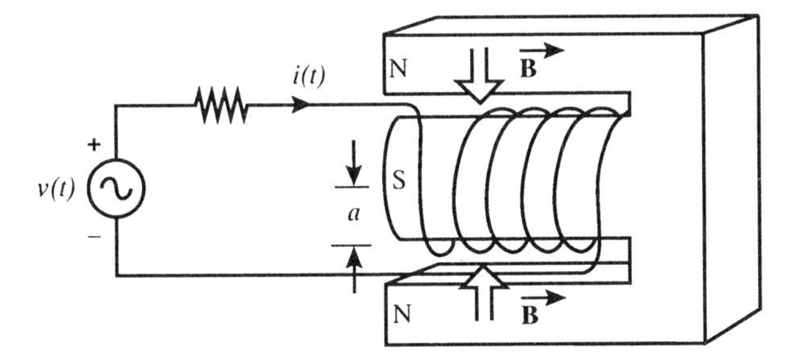


**Figure P3.9**

**3.10.** Why is it common practice to twist together two signal wires in a signal-processing applications circuit? Are the wires likely to carry parallel or antiparallel currents?

**3.11.** Suppose that we have a long, straight wire carrying current $I_1$. A circular loop carrying current $I_2$ and having radius $r$ is laid flat on the wire (so that the straight wire is in the plane of the disk enclosed by the loop), and the straight wire passes through the center of the disk. What is the force on the current loop? Supposing the wire and loop to be in the plane of a page viewed from above, $I_1$ is "up" on the page and $I_2$ is counterclockwise.

**3.12.** (**a**) Suppose that we have a straight wire in a plane, carrying current $i$. Perpendicular to this plane is an applied, uniform magnetic field that is nonzero only over a square $2r \times 2r$, and the wire passes through the center of that region. The current goes from left to right, and the $B$ field is directed "out of the page." What is the force on the wire (both direction and magnitude)?

(**b**) Someone squeezes (forms) a semicircle of radius $r$ into the wire while maintaining the wire in the same plane. The center of the circle of which the semicircle is half is the center of the region of nonzero $B$. Calculate the force on the wire now; compare with your result in part (a).

**3.13.** A wattmeter (a device used for measuring electrical power) is constructed by suspending two wires 1 mm apart over a length of 3 m, as shown in Fig. P3.13. A force balance determines the force between the wires, from which the power can be calculated. Suppose that the balance measures a force of $6 \cdot 10^{-5}$ N between the wires. What is the power consumed by the load? (*Hint:* You do not need to know the value of voltage source $V$.)

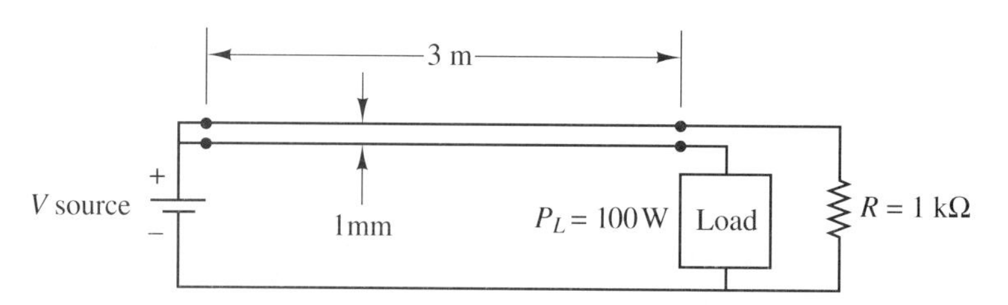


**Figure P3.13**

**3.14.** Two long, straight wires cross at right angles in a plane, such that one is along the $\hat{y}$ axis and the other is along the $\hat{x}$ axis. If each wire carries a current $I$ (the first, $I_1$, in the $x$ direction and the second, $I_2 = I_1 = I$, in the $y$ direction), where in the plane is the total $B$ field from the two wires equal to zero?

**3.15.** Repeat Problem 3.14 for the case in which the currents $I_1$ and $I_2$ are not necessarily equal.

**3.16.** Current $I$ flows in the thin cylindrical metal shell in Fig. P3.16. What is $|\mathbf{B}|$ inside and outside the shell? Your answer should be in terms of $\mu_0$, $r$, and $I$.

**3.17.** Derive a formula for $B$ within a long, straight wire of radius $a$, assuming that the current is uniformly distributed over the wire cross section and that within the wire, $\mu = \mu_0$. Is the field continuous at the surface of the wire? [*Hint:* Multiply both sides of Eq. (3.8) by $\mu_0$.]

**3.18.** A solenoid is made from 30 m of wire. The wire is 2.5 mm thick (diameter = 2.5 mm). What will be the resulting self-inductance if (**a**) the wire is wrapped tightly around a cylinder of radius 1 cm, or (**b**) the wire is wrapped tightly around a cylinder of radius 2 cm? (In both cases, the cylinder is removed after wrapping.)

**3.19.** An electric field of 1000 V/m and a magnetic field of 0.1 T are at right angles to each other. An electron moving at right angles to both fields experiences zero net force. What is its speed?

**3.20.** A cylindrical solenoid is measured to have a **B** field of 0.01 T with no core present.

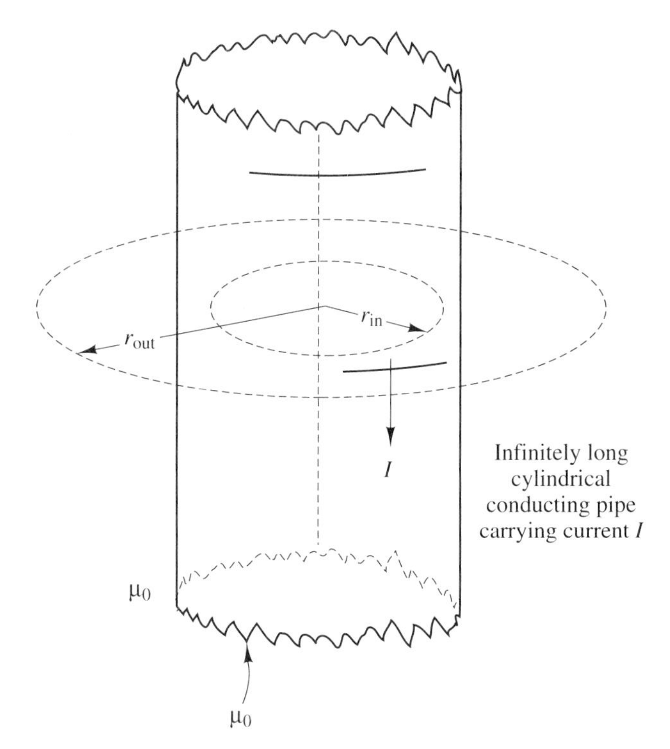


**Figure P3.16**

If the solenoid is 5 cm long, find the relationship between the number of turns $N$ in the coil and the current $I$ in the coil. Use this relationship to give one practical possibility for the values of $N$ and $I$.

**3.21.** An electron having initial velocity of $3 \cdot 10^7$ m/s is injected into a chamber having a transverse magnetic field of 0.05 T (see Fig. P3.21).

**(a)** What is the radius of its circular path? These circular paths can be observed in a device known as a bubble chamber. The same principle is used in the cyclotron,

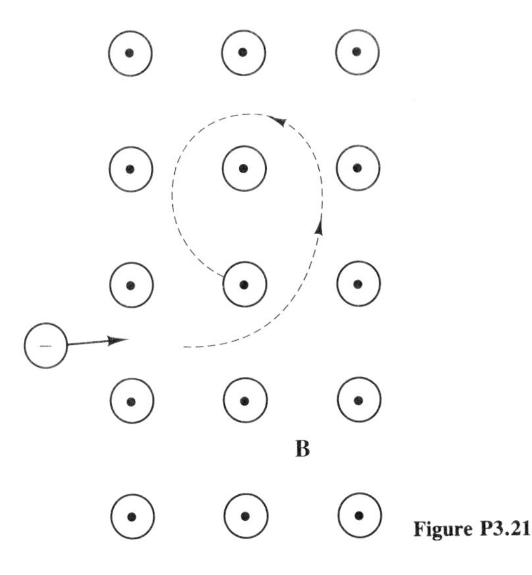


**Figure P3.21**

an instrument that accelerates charged particles to great speeds in order to study the effects of subatomic collisions.

**(b)** Show that the net work done on the charge in one orbit is zero.

**3.22.** An electron gun fires an electron at $u = 3.52 \cdot 10^4$ m/s in a direction parallel to the current in a long, straight wire, 1 m away from it. Under the influence of the magnetic field of the wire, the electron is observed to spiral around the wire, always staying 1 m away from it. Is the electron going in the direction of the current, or in the opposite direction? What is the current in the wire? Recall that centrifugal force of a rotating object is $a = mu^2/r$.

**3.23.** A speed separator is shown in Fig. P3.23. The electric field **E** points to the right and has a magnitude of 100 N/C. The magnetic field **B** is perpendicular to the page. A stream of electrons moving at various speeds but all with velocities in the $+\hat{y}$ direction enters the chamber at point 1. If we want the electrons leaving the chamber at the tiny hole at point 2 to have a speed of $u = 3 \cdot 10^7$ m/s, what should the direction and magnitude of **B** be? (For only one value of **B** will the electrons be on-track at the exit point; and in this case, all electrons with speeds other than $3 \cdot 10^7$ m/s will be forced left or right, off-target.)

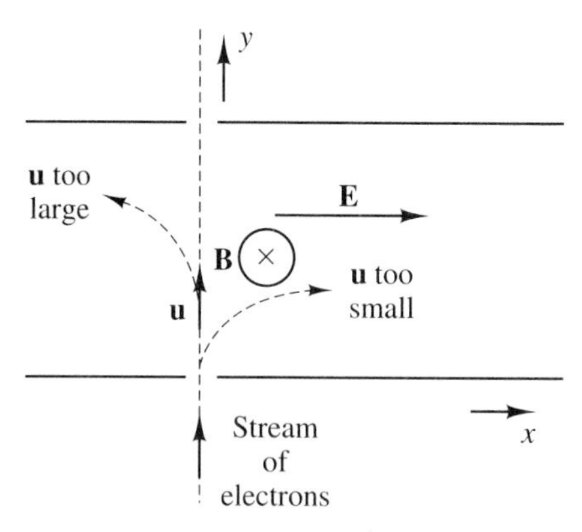


**Figure P3.23**

**3.24.** Use the expression for magnetic force given in Eq. (3.11) to determine that the force is attractive between two parallel current-carrying wires whose currents are in the same direction.

**3.25.** A square loop of wire is 1 m on each side, centered in the $xy$ plane at the origin. In this region, $\mathbf{B} = [x^2 + 0.1\ \text{m}]\hat{z}$ (T). Find the total flux through the square.

**3.26.** A solenoid is 5 cm long, has 500 turns, is cylindrical with a radius of 0.5 cm, and has a core with $\mu_r = 2500$. A current of 10 mA is in the solenoid coil. Find:

**(a)** $H$ within the coil.

**(b)** $B$ within the coil.

**(c)** The reluctance of the core.

**3.27.** In terms of our study of electricity and magnetism, describe the operation of a simple automotive ignition system; refer to Fig. P3.27.

**3.28.** A patient undergoing magnetic resonance imaging for diagnostic purposes is placed in a powerful constant magnetic field built up over a period of several minutes. In another instance, however, a person is accidentally exposed to a powerful sinusoidally varying magnetic field from a very large motor and is injured. Why the different results?

**3.29.** Which *cannot* change instantaneously in an inductor: the voltage across it, or the current through it? Why?

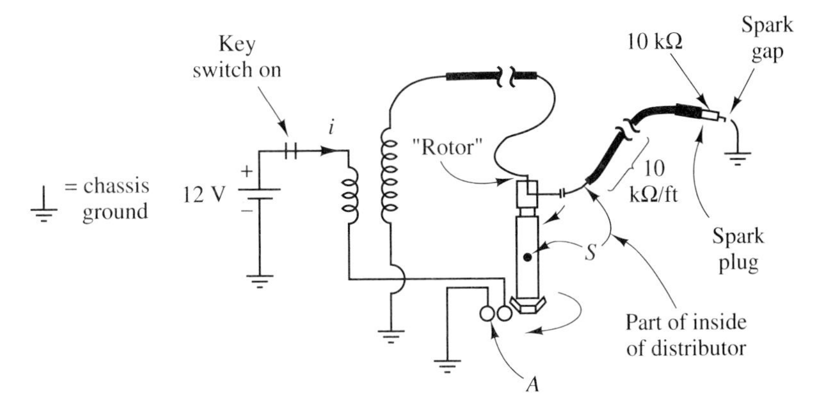


**Figure P3.27**

**3.30.** Derive the energy stored in a current-carrying inductor, $\frac{1}{2}Li^2$ using relevant equations in this chapter. (*Hint:* Use the fact that energy is the integral of power, which is $v \cdot i$ of the inductor.)

**3.31.** Consider again an electric cord plugged into a wall socket. Suppose that the switch was on before pulling the plug. Upon pulling the plug, we notice a spark. Why does this spark occur?

**3.32.** For the coil shown in Fig. P3.32 in which the magnetic field is $\mathbf{B}(t) = 0.1 \cos(300t)\hat{y}$ (T), determine:

(**a**) The polarity (positive or negative) of the variable $v(t)$ for time $t$ such that $0 < t < \pi/600$ s (use Lenz's law).

(**b**) The polarity of $v(t)$ for the time interval $\pi/600 < t < \pi/300$ s.

(**c**) $v(t)$ for all time using the values of parameters given in the diagram.

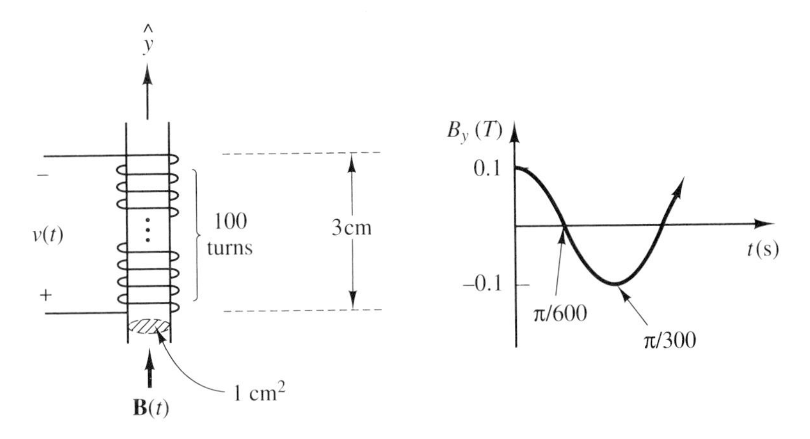


**Figure P3.32**

**3.33.** A uniform magnetic field $B = B_0 = 0.5$ T field is perpendicular to a plane and is restricted to a rectangle; outside the rectangle, $B = 0$. A wire loop in the shape of an isosceles right triangle is within the rectangle, with its hypotenuse flush on the right edge. At $t = 0$, someone begins pulling the loop out of the nonzero $B$ region, at constant velocity of $u = 10$ cm/s and perpendicularly to the hypotenuse edge (to the right). The distance from the hypotenuse to the opposite vertex of the triangle ("width" of the triangle) is $d = 10$ cm.

**(a)** What is the induced emf in the loop, both in symbols and by substituting in the given parameter values?

**(b)** Suppose that the loop has a resistance of 0.1 Ω. What is the current in the loop?

**(c)** In what direction does this current flow (clockwise or counterclockwise, when viewed from above)?

**(d)** For how long is there a nonzero emf?

**3.34.** Suppose that in Problem 3.33, the triangular loop is replaced by a circular loop of radius $r$. What is the emf generated, as a function of time?

**3.35.** A rectangular loop of width $w$ is pulled out of a region of constant **B**, similar to Problem 3.33. Compare the rate of work required to pull the loop at a steady velocity $u$ with the power dissipated by the current in the loop, assuming that the loop has resistance $R$. Neglecting friction, what is the efficiency of generating power from mechanical power in this case? The results in this problem we will come to again in more practical settings when we discuss machines in Chapters 13 through 15.

**3.36.** What is the speed of electromagnetic waves (e.g., light) in water, which has a relative dielectric constant of $\epsilon_r = 78$?

## ADVANCED PROBLEMS

**3.37.** The equation for the magnetic field magnitude of a long, straight current-carrying wire is $B = \mu_0 I/(2\pi r)$. For $r = 0$, this gives $B = \infty$. Does that really happen with actual wires? Comment on the practical limitations of, for example, Eq. (3.3).

**3.38.** A toroid having radius $R = 1$ m is constructed of an iron ring ($\mu_r = 2000$) having a circular cross section of radius $a = 1$ cm. A current $I = 2$ A is passed through the $N = 1000$ turns wrapped uniformly around the toroid. Assuming that the $B$ field is negligible outside the iron, and directed around the toroidal axis, use Ampère's circuital law [Eq. (3.10)] to find the magnitude of the magnetic flux density $B$ and flux $\phi$ in the iron. Try two ways: **(a)** path just like $C$ in Fig. 3.10, and **(b)** path all the way around the circular iron core; you should get the same answer either way.

**3.39.** An alternative form of Eq. (3.11) can be used to find the magnetic field due to a very long (ideally, infinite-length) wire. Refer to Fig. P3.39. A charge $q_1$ moving at velocity $\mathbf{u}_1$ is indistinguishable from a current element of length $d\mathbf{l}_1$ having a strength $I_1$: $q_1\mathbf{u}_1 = I_1 d\mathbf{l}_1$. Consequently, $\mathbf{B}_{21}$ in Eq. (3.11) may be rewritten as $\mathbf{B}_{21} = \mu I_1 d\mathbf{l}_1 x \hat{l}_{12}/(4\pi r^2)$. The current element can be thought of as a tiny portion of the entire wire, so that the contribution to the total $B$ field due to that element, $\mathbf{B}_{21}$, may simply be called $d\mathbf{B}$. In this form, the equation just cited for $d\mathbf{B}$ is known as the law of Biot and Savart (1820), named after Jean Biot and Felix Savart of France. Use this law of Biot and Savart to find the total field at any point a distance $a$ from an infinitely long vertical wire carrying current $I_1$ along the **z** axis ($d\mathbf{l} = dz\hat{z}$). [*Hint:*

$$\int \frac{dz}{(z^2 + a^2)^{3/2}} = \frac{z}{a^2(z^2 + a^2)^{1/2}}.$$

You should get the result in Eq. (3.4).]

**3.40.** Due to symmetry, the **B** field found in Problem 3.39 [and given by Eq. (3.4)] will be the same for all points on any vertical line—in particular, along a vertical current-carrying wire. Suppose that this second wire carries a current $I_2$ and passes through

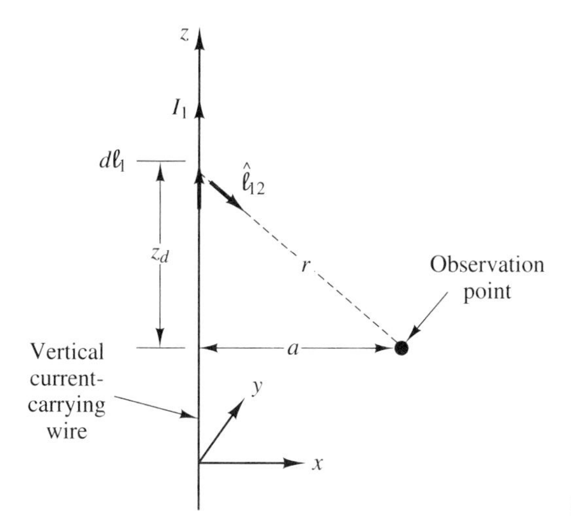


**Figure P3.39**

the point $(a, 0, 0)$, where the $(x, y)$ origin is defined to be at the wire $I_1$. Using Eq. (3.2), $\mathbf{F}_{ul} = I\mathbf{l} \times \mathbf{B}$, find the magnitude and direction of this force in terms of $I_1$, $I_2$, $a$, and $\mu_0$. Your answer should agree with the experimentally determined equation, Eq. (3.3) (Ampère's law).

**3.41.** In Fig. P3.39, suppose that the wire is only of finite length. That is, only a portion of current-carrying wire $2d$ long is unshielded and producing a magnetic field to outside observers. The rest of the circuit providing the wire with current is enclosed with shielding so that outside observers are not exposed to any fields from that portion of the circuit. Specifically, suppose that the wire extends only from $-d$ to $+d$ on the $z$ axis.

**(a)** What is the magnetic field at the observation point $(a, 0, 0)$?

**(b)** Show that your answer reduces to $B = \mu_0 I/(2\pi r)$ as $d \rightarrow \infty$.

**(c)** How small must $d$ be before there is a $100x\%$ (i.e., $0 < x < 1$) error in assuming an infinitely long wire? Finally, consider the example case of 10% error ($x = 0.1$).

**3.42.** **(a)** A scientist injects a positive ion with velocity **u** into a constant-in-time, uniform (according to him) magnetic field **B** and observes it being deflected by a Lorentz force **F** (see Fig. P3.42a). What is the magnitude $F$ and the direction of **F**?

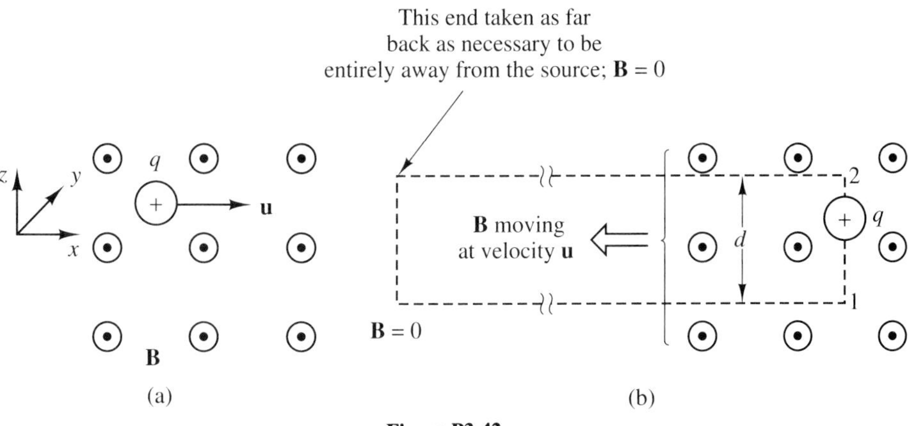


**Figure P3.42**

(**b**) At the same time, a tiny imaginary scientist sitting atop the stationary (according to him) ion sees a magnetic field-generating apparatus moving backward past him. He observes a force due to the electric field induced by the changing magnetic flux inside the loop shown in Fig. P3.42b. Assuming that $\mathbf{E} \cdot \mathbf{d} = 0$ everywhere except between points 1 and 2, compute the magnitude and direction of this force. (Let $d$ be arbitrarily small.) This result should agree with that in part (a) because of the principle of relativity.

**3.43.** Is it possible for a material to have a relative dielectric constant $\epsilon_r$ less than 1? Why or why not? How does your answer compare with the corresponding situation regarding magnetic fields?

**3.44.** An electron is thermally liberated from the cathode of a Hull magnetron, used in microwave ovens (see Sec. 12.7.2) and radar transmitters. When free, it is accelerated toward the positive anode (see Fig. P3.44). Before it reaches the anode, however, it is deflected by a magnetic field. The electron then passes through a uniform electric field having magnitude 6000 V/m at the microwave "resonant" cavity. Because the electric field opposes the motion of the electron, the electron gives up kinetic energy to the microwave cavity.

(**a**) How much energy is delivered to the microwave cavity?

(**b**) Suppose that the voltage between the main anode and cathode is 300 V. Assume that before deflection, the electron essentially is accelerated through all 300 V. Estimate the fraction of the total kinetic energy acquired in traversing the 300 V that is lost in imparting energy to (by deceleration in) the microwave cavity. (This is a grossly simplified description of a real microwave system. For example, the 6000-V/m electric field is in practice oscillating at the "resonant frequency" and is caused by previously accelerated electrons. Swirling clouds of electrons rather than individual electrons are accelerated.)

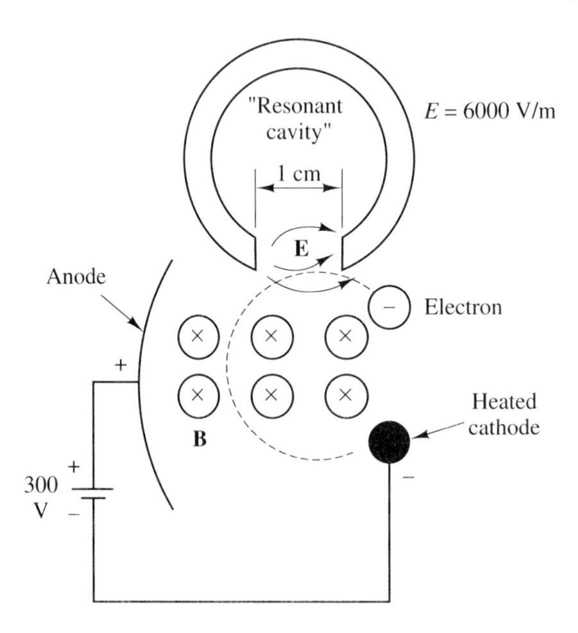


**Figure P3.44**

**3.45.** Is it physically possible to have the following magnetic field: $\mathbf{B} = (r^2 - r)\hat{r}$, where $\hat{r}$ in spherical coordinates is the radial unit vector? [*Hint:* Consider satisfaction of Gauss's law, Eq. (3.13).]

# Part Two:
# ELECTRIC CIRCUITS

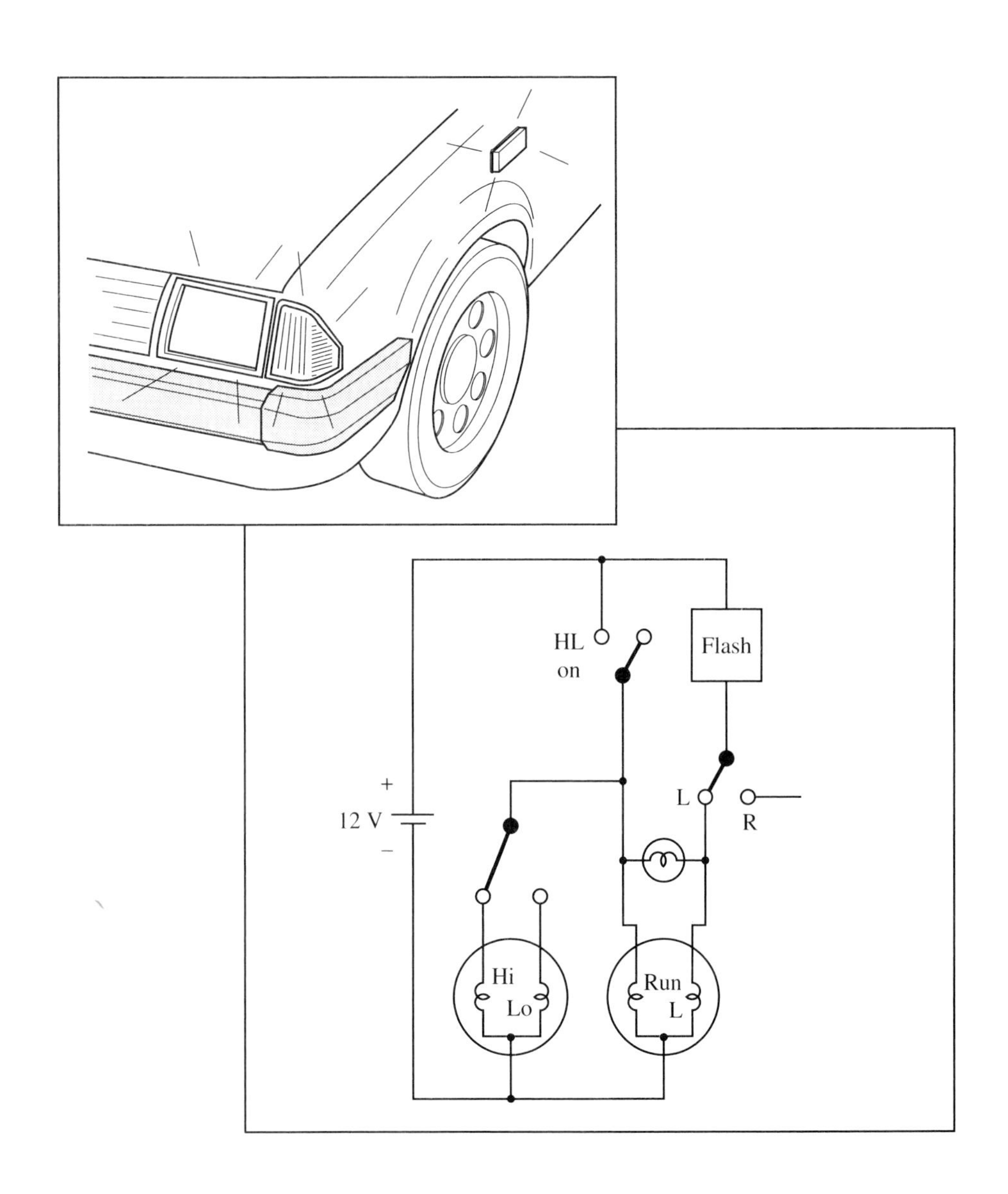

# 4

# Electric Circuits I: Basics

## 4.1 INTRODUCTION

In Chapters 2 and 3, the fundamental electromagnetic quantities and relations were presented. We found that all electromagnetic phenomena derive from the property of matter known as electric charge. That is perhaps one reason why the subject of this book is known as "electrical engineering" rather than "electromagnetic engineering."

Additionally, everyone knows that in practice, the transmission of energy to our homes is done by means of supplying an electrical potential or voltage. In theory, the power company could supply magnetic potential by running to our homes "wires" of ferromagnetic material rather than copper, transmitting magnetic flux rather than electric charge flow. The user could tap energy off the line by wrapping a coil around the magnetic power line at his or her home, just like the secondary coil $b$ in Fig. 3.19. To its credit, this method would eliminate $i^2R$ resistive line losses; no electric currents, only $\phi_m$ would be in the lines.

However, the efficiency and cheapness of using copper cables to transmit electric energy far exceeds that of the equivalent magnetic approach. Magnetic materials are expensive, lossy, and inconvenient for power transmission. In general, the disadvantages of using magnetic circuits compared with electric circuits are so severe that they are used only when there is no alternative: In electromechanical machines, transformers, radio receivers, and filters/oscillators. Moreover, we shall see in Sec. 8.8.4 that the last item above is presently being replaced by electronically synthesized equivalents.

These are just a few of the reasons why we shall usually focus on electric field effects. In this chapter in particular, we deal exclusively with electric potentials and the currents they produce in simple devices. That is, we shall be studying electric circuits, not magnetic circuits.

## 4.2 REVIEW OF SOME RESULTS FROM CHAPTER 2

In Chapter 2 we defined the electric forces and fields. Because our focus is now on electric circuits as explained in the introduction, let us briefly review that portion of material. This material is also reviewed in the summary of Chapter 2. We also briefly mention the solenoid and Faraday's law, from Chapter 3. The material in Chapter 3 is reviewed in its summary, Sec. 3.11. Also, a complete table of electromagnetic relations is presented near the end of that summary.

The fundamental electric force is the Coulomb force $\mathbf{F}_{21} = q_1 q_2 \hat{\ell}_{12}/(4\pi\epsilon r^2)$ ($F$ in newtons) between electric charges $q_1$ and $q_2$ [in coulombs (C)] separated a distance $r$ meters in a medium having permittivity $\epsilon$ [farads/meter (F/m)]. The electric field $\mathbf{E}$ [in volts/meter (V/m)] is the Coulomb force per unit charge, $\mathbf{E} = \mathbf{F}/q$. In a dielectric medium (with permittivity $\epsilon$), the surface effects of polarization tend to reduce the electric field strength of a given charge distribution from what it would be in free space ($\epsilon = \epsilon_0$). Dimensionless "relative" permittivity of a medium is defined as $\epsilon_r = \epsilon/\epsilon_0$.

As a measure of field strength, we defined *lines of electric flux* $\phi_e$ (in coulombs) whose density is reflected in the magnitude of the electric field vector, $E$. The total flux emanating from a closed surface is, by Gauss's law, equal to the electric charge that surface encloses. The electric field may be decomposed into a component $\mathbf{P}$, including only polarization effects (bound charge), and another, $\mathbf{D}$, incorporating the effect of free charge (both $\mathbf{D}$ and $\mathbf{P}$ in C/m$^2$). This decomposition is $\mathbf{E} = (\mathbf{D} - \mathbf{P})/\epsilon_0$, where $\mathbf{D} = \epsilon\mathbf{E}$ and $\mathbf{P} = (\epsilon - \epsilon_0)\mathbf{E}$.

The energy required to move a unit charge (1 C) from point $a$ to point $b$ is the electric potential difference or voltage $V_{ab} = -\int_a^b \mathbf{E}\cdot d\boldsymbol{\ell}$ (because work $= -\int_a^b \mathbf{F}\cdot d\boldsymbol{\ell}$). The units of $V_{ab}$ are volts (V = J/C). It is often helpful to define "absolute" potentials in circuit analysis (analogous to mountain elevations), with one potential arbitrarily assigned zero absolute potential (analogous to sea level). We then have $V_{ab} = V_b - V_a$, where $V_a$ and $V_b$ are the absolute potentials at points $a$ and $b$ with respect to the "common" or zero absolute potential.

An important circuit element is the capacitor, which has two charged metal parallel plates of area $A$, separated by a dielectric slab (permittivity $\epsilon$) of thickness $d$. In the charged capacitor, the electric field exists only between the plates and is uniform and perpendicular to them. As a result of the electric field occurring when the plates have charges $q$ and $-q$, respectively, there is a voltage between the plates. This voltage is $V = q/C$, where $C = \epsilon A/d$ [in farads (F)] is called the *capacitance.* When charged to a voltage $V$, an energy of $\frac{1}{2}CV^2$ is stored in the electric field between the plates.

A voltage source is a device that maintains a potential difference between its terminals, independent of charge flow that would normally tend to decrease it. Whereas the Coulomb electric field due to free charges is conservative (net work around a closed path is zero), the voltage source does net work on a charge as it traverses a closed path passing through the voltage source and a connected circuit. This work per unit charge is called the *electromotive force* (emf, in V). In equation form,

$$\oint_C \mathbf{E}(t) \cdot d\boldsymbol{\ell} = \sum_n \mathcal{V}_n, \tag{4.1}$$

where $\mathcal{V}_n$ are all of the voltage sources (or other emfs, such as inductor self-emfs) through which the closed path $C$ (i.e., circuit loop) passes. This equation is the basis for one of the two fundamental relations of circuit analysis, which are known as Kirchhoff's laws.

The existence of voltage sources allows the continual flow of charge because the nonconservative electric field continually restores the charge separation that would normally diminish as a result of charge flow due to Coulomb's force. This continual flow of charge is known as electric current $i$ [in amperes (A)]. Quantitatively, $i = dq_{\text{xsec}}/dt$, where $q_{\text{xsec}}$ is the charge flowing across the cross section of the conductor at which $i$ is calculated. It has been determined experimentally that the current $i$ that flows through most conductors is proportional to the potential difference $v$ (voltage) between their ends. This is Ohm's law, written as $i = v/R$, where $R$ is called the resistance [in ohms ($\Omega$)] of the conducting object. For a uniform slab of material, $R$ may be calculated as $R = \rho d/A$, where $\rho$ is the resistivity in $\Omega \cdot \text{m}$, $d$ is the slab thickness in meters, and $A$ is the area of the slab in $\text{m}^2$.

Conventional current flow is the direction that positive charge would move. In simple electric conductors, the charge flow is done by electrons, which have negative charge. It is helpful to keep in mind that the direction of current specified in a circuit diagram is opposite that of the motion of the electron charge carriers.

The quantitative statement of charge conservation is most conveniently given in terms of rate of charge flow and charge accumulation. The sum of the currents flowing into any closed surface $S$ is equal to the rate of increase of the charge enclosed by $S$:

$$\sum_{n,S} i_n(t) = \frac{dq_{\text{enc}}(t)}{dt}. \tag{4.2}$$

This relation forms the basis of the other of Kirchhoff's laws, which are so essential in circuit analysis.

With current defined, it may be stated that the current flowing into a capacitor is equal to the time rate of change of the voltage across it: $i = C\,dv/dt$, which is the time derivative of $q = Cv$.

The solenoid (inductor) is another important electric circuit element. Rather than storing energy in an electric field (as does a capacitor), it stores energy in a magnetic field. It is a coil of $N$ turns of wire, within which there may be a core of magnetic material or just air.

Faraday's law implies that there will be a voltage developed across the terminals of the coil that is proportional to the time rate of change of the current flowing in the solenoid. Mathematically, $v = L\,di/dt$. The parameter $L$ is called the inductance of the solenoid [in henrys (H)] and is equal to $\mu AN^2/d$, where $\mu$ ($\text{N/A}^2$) is a medium parameter called the *permeability* (of the core), $A$ is the area of the loops, and $d$ is the length of the solenoid. Similarly to the case of a capacitor, the magnetic energy stored in the magnetic field of an inductor is $\frac{1}{2}Li^2$.

Notice how the circuit element parameters $R$, $C$, and $L$ embody both the material parameters ($\rho$, $\epsilon$, and $\mu$) and the geometrical parameters ($N$, $A$, and $d$) of the respective circuit elements. The medium parameter for a given circuit element is chosen by the designer and is typically uniform throughout the element. The geometrical parameters actually represent the results of integrating field quantities over distances or surfaces that define the spatial extent of the circuit element. See, for examples, Eqs. (3.21) through (3.26) in which $d$ is the result of integration along a path taken through the element. By this means, known as *lumped elements*, we may avoid performing integrals when doing simple circuit analysis; the element parameter embodies the result of integration through the homogeneous circuit element.

The power absorbed by a circuit element is $p = vi$ [in watts (W)]. This expression was shown to be valid for time-varying voltage and current (denoted by lowercase letters), resulting in an "instantaneous power" into a circuit element. Because $v = iR$, we have for the instantaneous power absorbed by a resistor $p = vi = i^2R = v^2/R$.

## 4.3 ELEMENTARY CIRCUIT ELEMENTS

### 4.3.1 Ideal and Practical Passive Elements

What is an electric circuit? We already studied the simple circuit shown in Fig. 2.14 for the purpose of deriving the power absorbed by a passive circuit element. In that circuit a voltage source was connected across the terminals of the element. In general, any number and configuration of elements, passive and/or active, connected together so as to include one or more closed paths constitutes an electric circuit. Recall that an electric circuit element is any device having two or more metal terminals with which to connect it to other devices (usually, by wires or printed circuit lines). Also, an active element is one that can, over time, continually deliver energy to the rest of the circuit; passive elements can only absorb or store energy. Thus the time average of $p_{abs}$ is $\geq 0$ for passive elements, but can have any value for active elements.

The whole idea of circuit analysis is to characterize these elements using the simplest model that still includes the important factors causing the observed electrical behavior. The three most basic passive circuit parameters we shall consider are resistance $R$, capacitance $C$, and inductance $L$. The symbols for these parameters are given in Fig. 4.1. The squiggle of the resistance symbol represents the hindrance or impedance of current caused by such a device, as discussed in detail in Chapter 2. In fact, we use the term *impedance* in Chapter 6 as a generalization of resistance. The parallel bars of the capacitance symbol signify the parallel plates of a capacitor, and the curls of the inductance symbol represent the turns of the inductor coil.

If we have a piece of partially conducting material attached to two wire leads, we call it a *resistor*. A typical resistor is shown in Fig. 4.2a. They are fabricated to have particular values and tolerances, because of their utility as building blocks for

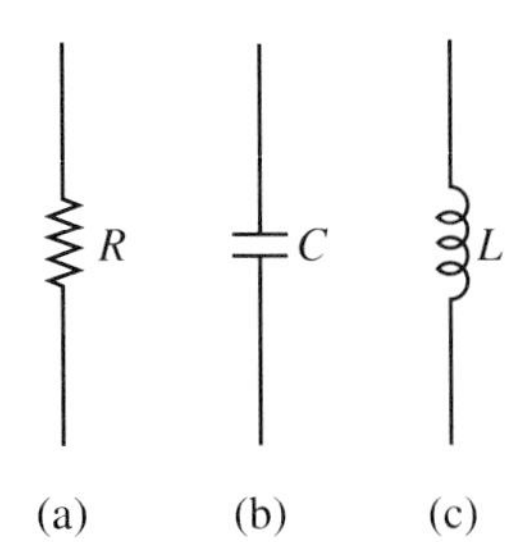

**Figure 4.1** Symbols for (a) resistance, (b) capacitance, and (c) inductance.

achieving a designed circuit behavior. A resistor usually has the color brown, and its value can be identified by the color bands painted on it, according to Table 4.1.

An ideal resistor has only resistance $R$. The ideal resistor model is shown in Fig. 4.2b. If we required a more exact model of the resistor shown in Fig. 4.2a, the model shown in Fig. 4.2c or a portion of it could be used. The metal terminals of the resistor form a sort of "parallel-plate" set, resulting in a capacitance $C_R$, and any current-carrying component exhibits at least a small inductance, here labeled $L_R$. Because the capacitance $C_R$ and inductance $L_R$ are very small, their effects may usually be ignored in circuit calculations.

The importance of $C_R$ and $L_R$ relative to $R$ depends on the rate of change of the voltage $v(t)$ across and current $i(t)$ through the resistor. In Chapter 5 we shall find that the effects of $C_R$ and $L_R$ in Fig. 4.2c increase with the frequency of oscillations in $v(t)$ and $i(t)$ if, in fact, they have such high-frequency content. In this chapter we are concerned primarily with constant-in-time voltages and currents. In this case $C_R$ and $L_R$ produce negligible effects, and the model of Fig. 4.2b is sufficiently accurate.

If we attach parallel plates to the resistor terminals and replace the conducting material of the resistor with a dielectric, we obtain the capacitor, appearing physically as shown in Fig. 4.3a. It may be tubular or planar and comes in a variety of colors; modern capacitors typically have their value and voltage limitation numerically printed on the device. In the case of tubular capacitors, the "plates" are rolled up in a cylinder.

The simplest model for this device is pure capacitance, as shown in Fig. 4.3b. Note that because $i = C\,dv/dt$ for an ideal capacitor, for constant-in-time voltages and currents $i = C \cdot 0 = 0$. Thus an ideal capacitor is an open circuit for constant-in-time excitation. However, keep in mind that this does *not* imply that the voltage across a capacitor is zero for constant-in-time excitation; it may in fact be very large.

Again, because the dielectric has conductivity and inductance (albeit very

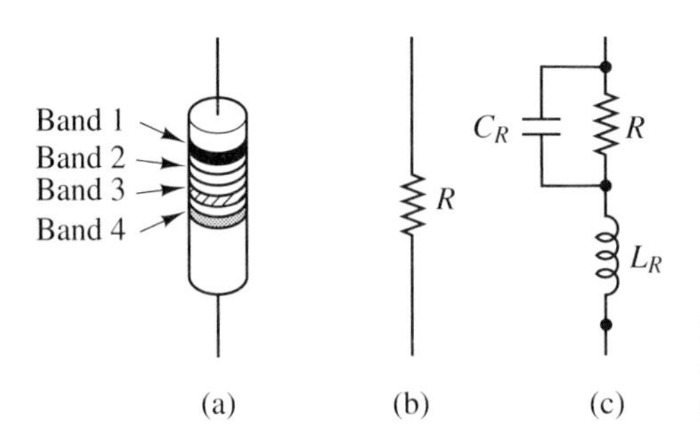

**Figure 4.2** (a) Physical resistor; (b) ideal resistor model; (c) practical resistor model.

**TABLE 4.1** RESISTOR COLOR CODE

| | Digits | Multiplier | Resistance if band 1 = yellow, band 2 = violet |
|---|---|---|---|
| Black | 0 | $10^0 = 1$ | $47 \cdot 10^0 = 47\ \Omega$ |
| Brown | 1 | $10^1 = 10$ | $47 \cdot 10^1 = 470\ \Omega$ |
| Red | 2 | $10^2 = 100$ | $47 \cdot 10^2 = 4.7\ \text{k}\Omega$ |
| Orange | 3 | $10^3 = 1000$ | $47 \cdot 10^3 = 47\ \text{k}\Omega$ |
| Yellow | 4 | $10^4 = 10{,}000$ | · 470 kΩ |
| Green | 5 | $10^5 = 100{,}000$ | · 4.7 MΩ |
| Blue | 6 | $10^6 = 1{,}000{,}000$ | · 47 MΩ |
| Violet | 7 | $10^7 = 10{,}000{,}000$ | 470 MΩ |
| Gray | 8 | | ↑ order of magnitude is the same as these examples for all values of bands 1 and 2. |
| White | 9 | | |

1 2 3 4
Bands

Tolerances: Brown = ±1%, red = ±2%, gold = ±5%, silver = ±10%, no color = ±20%.

Bands 1 and 2 are the significant digits, band 3 is the multiplier, and band 4 is the tolerance. Band 1 is identified as being the closest band to an end of the resistor, or the edge band that is not silver or gold. Only certain standard values of resistance are available, spanning the whole range with tolerances. There are also ±1% resistors that either give the numerical value in Ω directly, use a numerical four-digit code similar to the one above but without colors (also, no tolerance digit, but an extra value numerical digit for precision), or five bands with brown for tolerance. Note that the order of pure colors follows "ROY G BIV" (without indigo), which is the order of increasing light-wave frequency. Also, the power of 10 in the "multiplier" is equal to the corresponding "digit" value.

small), a model involving $C$, $R_C$, and $L_C$ must be used for maximum accuracy. For example, over very long periods of time leakage of charge from one plate to the other occurs, due to the nonzero conductivity of the dielectric. (Recall that one way of looking at electrical conductivity of material is as a measure of how long it takes to transfer a given amount of charge through a section of that material.) In such cases, a resistance in parallel with the capacitance would improve the accuracy of the model. However, the model in Fig. 4.3b is sufficient for our purposes.

Finally, if the conducting material of the resistor is instead replaced by a wire coil, the predominant behavior of the device is inductance. The inductor is shown in Fig. 4.4a, and the ideal model is given in Fig. 4.4b. Note that because $v = L\,di/dt$ for an ideal inductor, for constant-in-time voltages and currents $v = L \cdot 0 = 0$. Thus an ideal inductor is a short circuit for constant-in-time excitation. However, keep in mind that this does *not* imply that the current in an inductor is zero for constant-in-time excitation; it may in fact be very large.

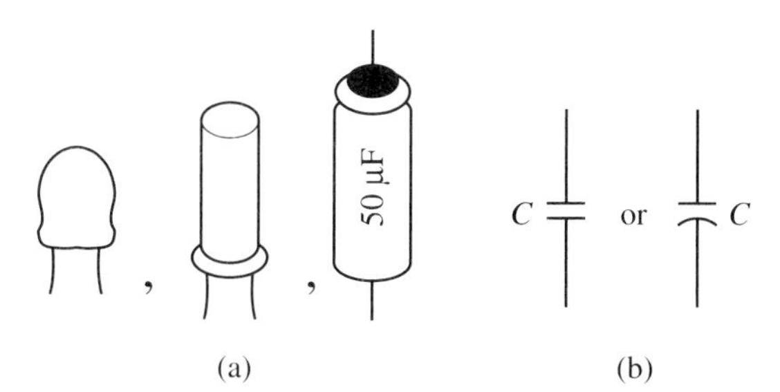


**Figure 4.3** (a) Typical physical capacitors; (b) ideal capacitor model.

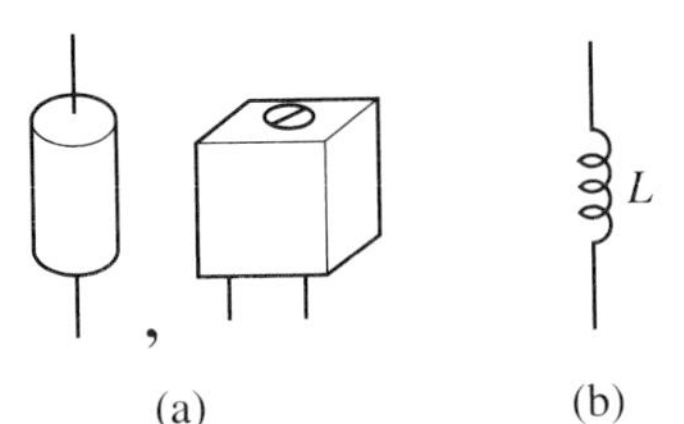

**Figure 4.4** (a) Typical physical inductors; (b) ideal inductor model.

Inductors come in a variety of shapes and sizes. Unfortunately, their inductance value is often not given on the device and consequently usually must be checked experimentally. Frequently, their value is adjustable by turning a screwed-in core lug.

Resistance of the wire forming the inductor coil is an important factor in accounting for the nonideal behavior of an inductor, as is the capacitance between turns of the coil. For constant-in-time and low-frequency voltages and currents, the resistance $R_L$ dominates the device behavior. For very high frequencies, the capacitance $C_L$ dominates the overall behavior, while for a specific moderate range, $L$ dominates.

Of the three basic passive components, inductors tend to be by far the most nonideal in practice. For our purposes we shall model the inductor as ideal. However, for all three devices $R$, $C$, and $L$, it is extremely important to be aware of the possibility that what ostensibly is one type of device in practice may strongly exhibit qualities of one or more of the other types.

One further passive element we have already discussed is wire. A scrap of insulated wire is shown in Fig. 4.5a. It is modeled as having zero resistance and is symbolized with a line (Fig. 4.5b). Recall that the wire is modeled as a perfect conductor, so that the voltage *across it* is zero (by Ohm's law, $v = iR = i \cdot 0 = 0$). That is, *everywhere along an ideal wire, the electric potential is the same (zero potential variation along it)*. Thus if one point on the wire is at $V_1$ volts with respect to a common "0 V" terminal, *every* point on that wire is at potential $V_1$ volts. Long wires used for transmitting data typically must be modeled as having resistance, capacitance, and inductance distributed along them.

### 4.3.2 Ideal Active Elements

We have also already introduced, in Chapter 2, an active element: the ideal voltage source. Such a device maintains a voltage $V$ across its terminals, irrespective of the current drawn by a load placed across it. Recall that a voltage is an electric potential

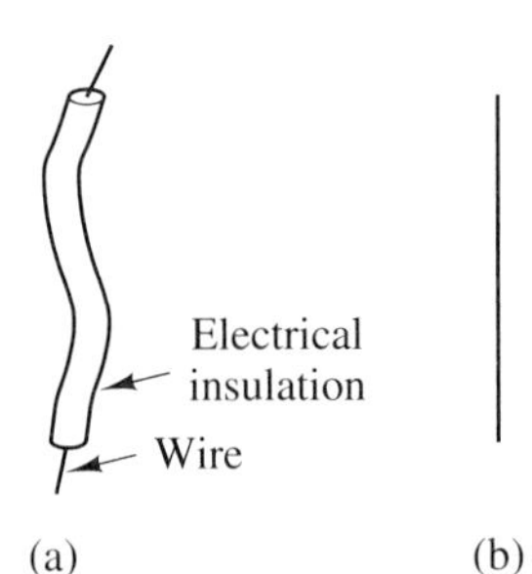

**Figure 4.5** (a) Typical physical wire; (b) ideal wire model.

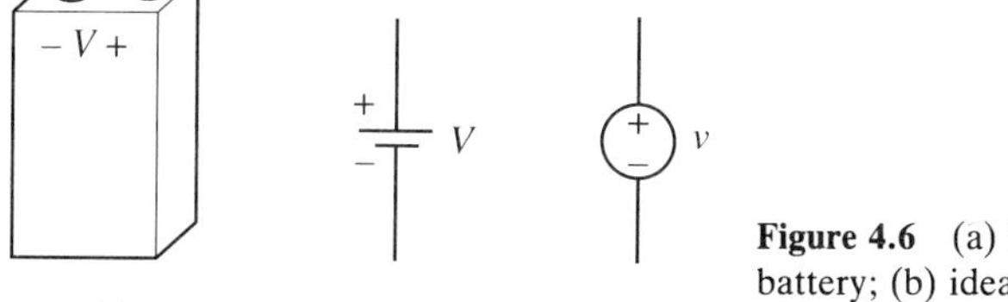


**Figure 4.6** (a) Typical physical 9-V battery; (b) ideal battery symbol; (c) ideal general voltage source symbol.

energy difference between two points: In dropping 1 volt in electric potential, 1 coulomb of charge loses 1 joule of electric potential energy.

An example voltage source, a battery, is shown in Fig. 4.6a. The symbol for an ideal battery or other constant-in-time voltage source is given in Fig. 4.6b. Another symbol commonly used to represent a voltage source is shown in Fig. 4.6c; it is used for both constant-in-time and time-varying voltage sources. In cases where the sign of the voltage alternates with time, the "+" sign in Fig. 4.6c means that whenever the actual potential distribution is such that $v > 0$, the terminal marked "+" is actually the one at higher potential, and is the lower potential terminal at other times. (This convention is true for *all* labeled alternating polarity voltages, and a similar convention concerns alternating-sign currents.) No voltage source is ideal; we shall in Sec. 4.9 consider a more accurate model involving an ideal voltage source in series with a *source resistance* $R_s$.

As noted previously, in electric circuit analysis we need consider only voltage and current as variables. Each element has a particular voltage across it and current within it. For simple passive elements, those two variables are not independent: They are related by the so-called *v–i relation* of the element. For active elements and more complicated passive semiconductor elements, one may be essentially independent of the other. The $v$–$i$ relations for the elements basic to circuit analysis are

$$i = \frac{v}{R} \qquad \text{[resistor]} \tag{4.3a}$$

$$i = C\frac{dv}{dt} \qquad \text{[capacitor]} \tag{4.3b}$$

$$L\frac{di}{dt} = v \qquad \text{[inductor]} \tag{4.3c}$$

$$v = 0,\ R = 0,\ \text{independent of } i \qquad \text{[wire]} \tag{4.3d}$$

$$v = V,\ \text{independent of } i \qquad \text{[independent voltage source]} \tag{4.3e}$$

$$v = v(v_x \text{ or } i_x) \qquad \text{[dependent voltage source]} \tag{4.3f}$$

$$i = I,\ \text{independent of } v \qquad \text{[independent current source]} \tag{4.3g}$$

$$i = i(v_x \text{ or } i_x) \qquad \text{[dependent current source],} \tag{4.3h}$$

where Eqs. (4.3f)–(4.3h) will be discussed momentarily.

When $i = f(v)$, where $f(\cdot)$ is any function, the $i$–$v$ relation may be plotted on a graph known as an *i–v characteristic*. For example, Eq. (4.3a) for the resistor has an $i$–$v$ characteristic shown in Fig. 4.7a; it is a straight line through the origin, with

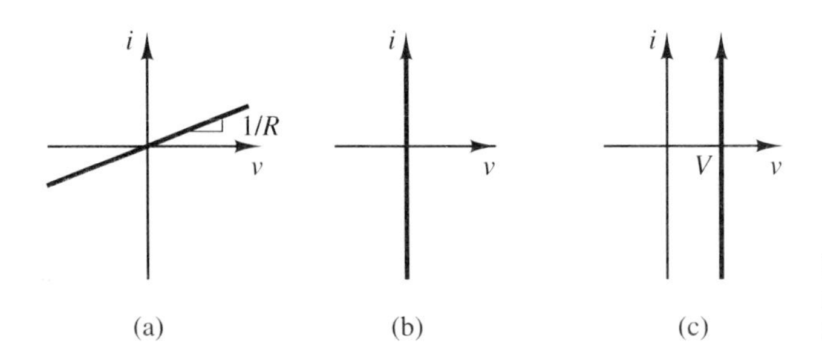


**Figure 4.7** (a) The $i$–$v$ plot for ideal resistor; (b) $i$–$v$ plot for ideal wire; (c) ideal $i$–$v$ plot for voltage source.

slope $1/R$. The $i$–$v$ relation (4.3d) for the wire is shown in Fig. 4.7b; no matter what current $i$ flows in it, the wire has zero volts across it.

A similar characteristic to that of the wire exists for the $i$–$v$ relation (4.3e) for an independent ideal voltage source and is shown in Fig. 4.7c for a voltage source for which $v = V > 0$. The one difference is that the vertical line is shifted over to the source voltage $V$. Thus a wire is indistinguishable from a *zero-volt voltage source.* If the voltage $V$ varies with time, the vertical line in Fig. 4.7c will shift correspondingly back and forth with time.

The reason that the adjective "ideal" is used is that for such a voltage source, the terminal voltage is maintained at $V$ irrespective of the current that may pass through it. Of course, there are limitations to such an idealization in practice, which we consider in Sec. 4.9. Nevertheless, the ideal model is extremely useful as a component in the practical model. The adjective "independent" means that the value of the voltage maintained is independent of any other voltage or current elsewhere in the connected circuit. There are also ideal voltage sources whose voltage is maintained at a value that is a function of a voltage or current elsewhere in the circuit [$v_x$ or $i_x$ in Eq. (4.3f)], again irrespective of the current through the voltage source. Naturally, such a source is called a *dependent source* because it depends on another voltage or current.

Dependent voltage sources are quite useful in modeling amplifiers or other active circuits. A primary example is the integrated circuit called the operational amplifier, to which Chapter 8 is entirely devoted. You cannot go to your supply store and buy a dependent voltage source as you can a battery. Rather, the dependent source is an accurate model of the behavior of certain devices or networks as they function in a particular circuit. For most cases to be encountered in practice, the type of dependence of a dependent source on $v_x$ or $i_x$ is linear proportionality. The symbol for a dependent source and its typical placement in a voltage-boosting (amplification) circuit are shown in Fig. 4.8, where for linear proportionality, $\alpha$ is a constant.

Just as there are methods for generating voltages that are independent of the current, there are methods for generating currents that are independent of the voltage. These active elements are called, not surprisingly, *current sources.* Often, a transducer, a "stage" or module in an electronic circuit, or a circuit component (such as a transistor) exhibits the ability to maintain a current within it regardless of the voltage across it. The symbol for a current source and its $i$–$v$ characteristic are shown in Fig. 4.9.

Just as there are both independent and dependent voltage sources, there are independent and dependent current sources analogously defined. The symbol for a

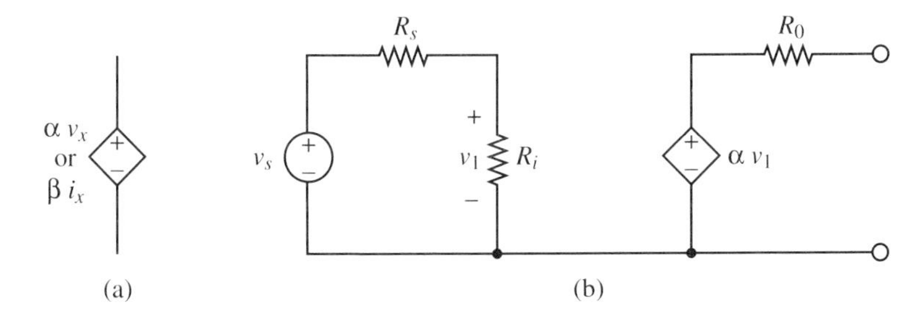


**Figure 4.8** (a) Dependent voltage source; (b) typical placement in an electronic circuit diagram.

dependent current source and its typical placement in a transistor amplifier is shown in Fig. 4.10 (where $\beta$ is a constant). Do not be concerned about the unfamiliar circuit elements in Fig. 4.10b; the intention is merely to illustrate that they have real-world applications. We shall have more to say about dependent sources later in this chapter and in Chapters 7 through 9 (on electronics).

We cannot draw $i$–$v$ characteristics of capacitors or inductors because the time rate of change of one of the variables $v$ or $i$ is involved. In Chapter 5 we shall see that for sinusoidal excitation, we can draw (complex-amplitude) linear $i$–$v$ characteristics for these elements, analogous to Ohm's law for conductors.

## 4.4 POLARITY CONVENTIONS

### 4.4.1 Amplitude Polarity

Polarity for voltage sources is indicated by the "+" symbol for the terminal of higher potential, and "−" for the terminal of lower potential. Also, for batteries, the larger bar is used for representing the higher-potential terminal (and thus is redundant, except in indicating that this is a constant-in-time voltage source and not some other type of element). For current sources, the polarity is indicated by the direction of the arrow, which points in the direction of the flow of positive-charge current (again recall that actual electron flow is in the opposite direction because electron charge is negative).

Perhaps the most important concept to keep in mind when performing circuit calculations is the following rule concerning polarity: *Every voltage and current variable used in analyzing a circuit must be assigned a polarity on the circuit diagram.*

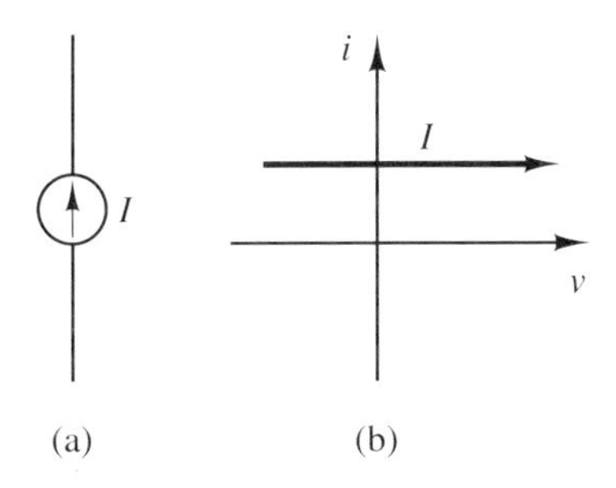


**Figure 4.9** (a) Symbol for ideal current source; (b) $i$–$v$ plot for ideal current source.

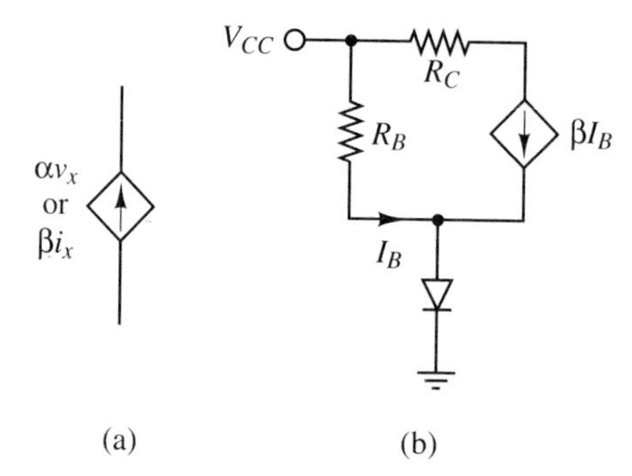


**Figure 4.10** (a) Dependent current source; (b) typical placement in an electronic circuit diagram.

*If the value of the variable is known, the polarity is naturally chosen to agree with the true polarity. If the variable is unknown, a polarity must be arbitrarily assigned. If this assigned ("guessed") polarity turns out to be opposite the actual polarity, the value of the variable obtained from calculations will be a negative number.* The importance of this rule cannot be overstated, because violating it will produce a wrong answer. Some examples will illustrate the use of the polarity rule.

**Example 4.1**

In Fig. 4.11a a 5-V voltage source is shown. Two people who do not know this value have assigned variables $v_1$ and $v_2$ with polarities as shown in Fig. 4.11b and c, respectively. What are the actual numerical values of $v_1$ and $v_2$?

**Solution** From the given information, we know that in fact, the upper terminal is 5 V *higher* than the lower terminal. Person 1 has correctly assumed this, so $v_1 = 5$ V. Person 2 has guessed wrong, so $v_2 = -5$ V.

There is absolutely no shame in guessing wrong; it is impossible to know for certain beforehand what the true polarity is, especially for complicated circuits. (However, experience will "increase your percentage.") The only thing to be sure not to do is forget to assign and label the polarities of every variable you will be using in the analysis! Also, use those polarities consistently throughout the analysis.

**Example 4.2**

In Fig. 4.12a a −3-A current source is shown. Two people who do not know this value have assigned variables $i_1$ and $i_2$ with assumed positive polarities (directions) as shown in Fig. 4.12b and c, respectively. What are the correct values of $i_1$ and $i_2$?

**Solution** From the information given, we know that, in fact, the true positive-charge equivalent current is going "down" from the upper to the lower terminal. Person 1 has "guessed wrong" as far as assumed positive direction, so $i_1 = -3$ A. Person 2 has assumed the correct positive polarity of this current, so $i_2 = 3$ A.

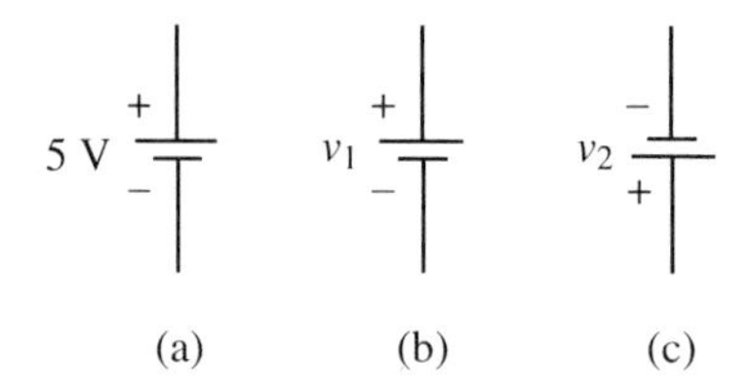


**Figure 4.11** (a) Actual voltage source in Example 4.1; (b), (c) polarity assumptions made by two people.

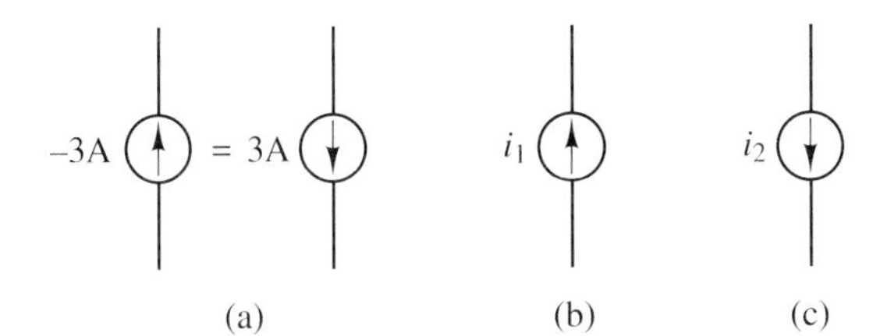


**Figure 4.12** (a) Actual current source in Example 4.2; (b), (c) polarity assumptions made by two people.

### 4.4.2 Power Polarity

What polarity convention is assumed in the $i$–$v$ relations (4.3)? The convention for all passive elements is that in Eqs. (4.3), positive-charge current [$i$ in Eqs. (4.3)] flows into the assumed higher-potential terminal in the definition of the voltage $v$. Thus the passive element $i$–$v$ relations in Eqs. (4.3) correspond to the assumed polarity sets shown in Fig. 4.13.

The reason for this is the associated *power convention,* now to be discussed, which is *valid for all circuit elements, both passive and active.* No, this is not an arms control summit but a means of energy bookkeeping. The rule is that the power *absorbed* by a circuit element is $vi$ if and only if current $i$ is drawn (i.e., symbolically defined to be) flowing into the terminal at the assumed higher potential, $v$ volts higher than the other terminal. If this product $vi$ comes out to be a positive number when the actual numerical values are substituted, the element is absorbing electrical power $vi$; otherwise, it is said to be providing, releasing, or generating power. Conversely, if $i$ is drawn flowing into the lower potential terminal, the power absorbed by the element is $-vi$, which again may end up being positive (absorbing) or negative (providing) (see Fig. 4.14).

Actually, this rule has a firm physical basis. Continually forcing positive charges into the terminal at higher potential (and receiving them back from the other terminal at a lower potential) requires the source to be doing work on the charges. This amounts to the source losing power to the element. That is, the element is absorbing power from the connected electrical energy source. This is a property of all practical passive elements ($R$, $C$, $L$, practical wire): on the time average, they dissipate power due to conduction losses. (Note that *ideal* $C$ and $L$ absorb zero power over a time average, and *ideal* wire absorbs zero instantaneous power.)

Conversely, if the current enters the lower potential terminal of the element and leaves at a higher potential, the element itself must be doing work on the charge. Hence, in this situation, the element is providing, releasing, or generating power for the rest of the circuit. This is characteristic of active elements such as voltage and

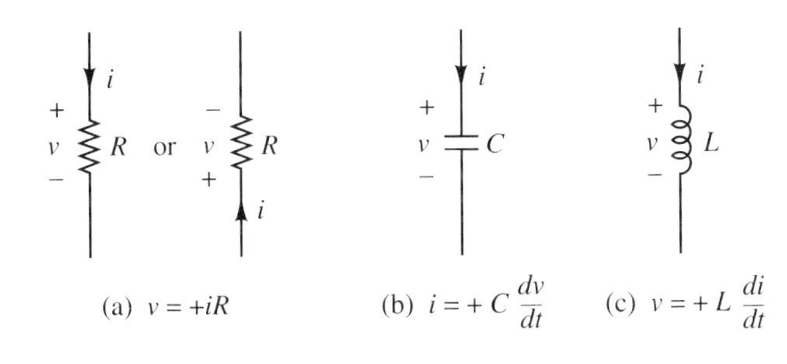


**Figure 4.13** Polarity conventions for (a) resistor, (b) capacitor, and (c) inductor.

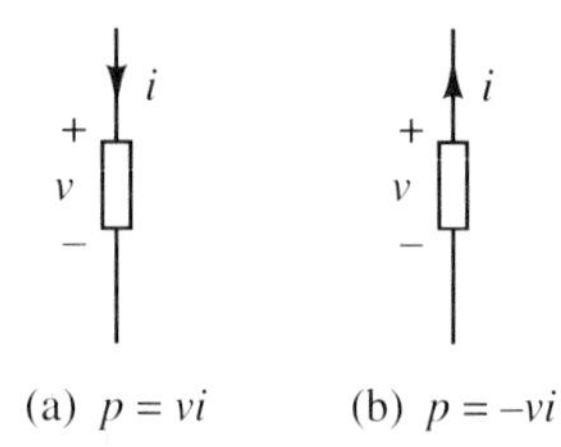


**Figure 4.14** Power conventions for a circuit element.

current sources. However, in the problems at the end of the Chapter, it is made clear that when opposed by other sources in a multisource circuit, active sources also can absorb power.

If you substitute "mass" for "charge," "velocity" for "current," and "mechanical potential" for (electric) "potential," the statements above are familiar rules for power determination in mechanical systems. An important logical conclusion of these laws is that for a *resistor*, which *always* dissipates energy, the *actual value* of positive-charge current *always* flows into the terminal at *actual* higher potential.

As far as *symbolic* voltage and current (i.e., variable polarity definitions you choose) for resistors, *if the arrow of the defined current i points into the "+" terminal of the symbolic defined voltage v, we may write $v = iR$ and $p = vi$*; otherwise, we must write $v = -iR$ and $p = -vi$. A little practice will make all of this second nature. Whichever way you choose to symbolize current and voltage, if you find that the power absorbed by a resistor calculates to a negative value, there must be a mistake somewhere!

**Example 4.3**

A resistor is found (either by ammeter measurements or circuit analysis) to have a current +2 A flowing from left to right in Fig. 4.15. Joe, Mary, and Bob have labeled the voltage across the resistor as $v_1$, $v_2$, and $v_3$ as shown in Fig. 4.13a, b, and c, respectively. What are the values of $v_1$, $v_2$, and $v_3$? In each case, what is the power absorbed by the resistor?

**Solution** Consideration of the rule stated above concerning Ohm's law, we see that $v_1 = -4 \cdot 2 = -8$ V and $v_2 = +4 \cdot 2 = 8$ V. As for Bob, to bed for him with no dessert, because he forgot to label the polarity of $v_3$—consequently, the numerical value of $v_3$ *cannot be determined.* The power into the resistor is, from Fig. 4.15a, $-v_1 i = -(-8) \cdot 2 = 16$ W, because $i$ is shown going into the (symbolic) lower potential terminal. This power is also equal to $+v_2 i = 8 \cdot 2 = 16$ W because in that case (Fig. 4.15b), $i$ is shown going into the (symbolic) higher potential terminal. Again, Bob in Fig. 4.15c has left the problem indeterminate.

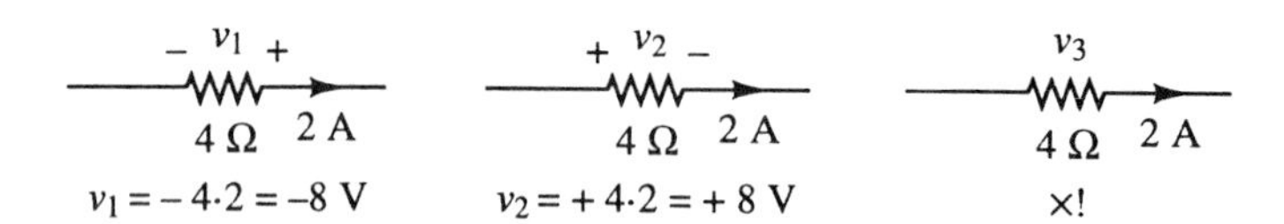


**Figure 4.15** Variable assignments for resistor voltage made by three people.

## 4.5 FROM ELECTROMAGNETICS TO CIRCUIT THEORY

### 4.5.1 Quasistatic Approximation

With the elements and polarity conventions defined above, we may begin to study the basic principles and techniques of circuit analysis. All of the physical laws governing circuit behavior were explained in Chapters 2 and 3, and were reviewed in Sec. 4.2.

These laws apply without modification except for a few simplifications. For example, except when magnetic fields are intentionally created in inductors, their value and rate of change are very small. Hence, in Eq. (4.1) we did not include the additional $d\phi_m/dt$ for the overall loop $C$ that results from Faraday's law, as given in Eq. (3.28). (For inductors along the loop, we classify $N_n\, d\phi_{m,n}/dt$ as an emf $\mathcal{V}_n = L\, di/dt$.) Also, it takes extremely rapidly varying electric fields for the term $d\phi_e/dt$ in Eq. (3.32) to become significant in anything other than capacitors. As a result, electromagnetic waves are very weak in elementary circuits; they are obviously extremely important in radio circuits, which are based on electromagnetic wave propagation.

A detailed study of electromagnetic wave theory shows that the power radiated in the form of electromagnetic waves (and thus lost from the circuit into space) is proportional to $(d/\lambda)^2$, where $d$ is the length of the element carrying the oscillating current that causes the wave and $\lambda$ is the wavelength of the wave. The wavelength of any wave is equal to $c/f$, where $c$ is the propagation speed of the wave and $f$ is its frequency in Hz (hertz, after Heinrich Hertz, who experimentally verified Maxwell's wave theory).

For the 60-Hz wall outlet frequency, the wavelength of an electromagnetic wave is $\lambda = 5{,}000{,}000$ m! When squared in the denominator of $(d/\lambda)^2$, this power is quite negligible compared with other power terms in the energy balance. The wavelength of a typical UHF TV wave is roughly 1 m. At this frequency, practical-sized antennas may be used to emit high-power electromagnetic waves efficiently, while simultaneously circuit theory may be used for the receiver electronic components, which are much smaller than 1 m in dimension. That these statements are true is very fortunate for those who like to watch TV.

The domain of electromagnetics on which we here focus, in which waves are insignificant, is called the *quasistatic regime*. Under this assumption, electromagnetic waves are not a significant component of the total electric and magnetic fields. Essentially, $d\phi_e/dt$ in Eqs. (3.32) and (3.34d) and/or $d\phi_m/dt$ in Eqs. (3.28) and (3.33d) are negligible, so that electric and magnetic fields are uncoupled, as far as the generation of waves is concerned.

The exception to the assumption $d\phi_e/dt \approx 0$ is between capacitor plates, where, however, $d\phi_m/dt$ is assumed to be zero; waves are thus still insignificant. Similarly, $d\phi_m/dt$ is assumed to be zero everywhere in a circuit except through a coil—in which $d\phi_e/dt$ is assumed to be zero. Without the fields strongly doubly coupled, waves are an insignificant component of the total fields. We thus see how circuit theory is actually an approximation of field theory, which we shall find much easier to work with—no vectors, only scalars.

### 4.5.2 Kirchhoff's Laws

**Kirchhoff's voltage law (KVL).** Gustav Kirchhoff in 1845 proposed two fundamental electric circuit laws. Let us now obtain his voltage law, based on Eq. (4.1). To do so, first recall that potential is the integral of $\mathbf{E}\cdot d\boldsymbol{\ell}$. The voltage across each circuit element located along the path $C$ in Eq. (4.1) therefore represents a portion of the closed-loop integration of $-\mathbf{E}\cdot d\boldsymbol{\ell}$. Equivalently, if we break up the path $C$ in Eq. (4.1) into sections, one for each circuit element, we obtain

$$\sum_n \int_{C_n} \mathbf{E}\cdot d\boldsymbol{\ell} = \sum_n \mathcal{V}_n. \tag{4.4}$$

The emfs $\mathcal{V}_n$ in Eq. (4.4) include the self-emfs of inductors through which the closed loop may pass; see Eq. (3.29), by which we equivalently incorporate the term $N_n\, d\phi_{m,n}/dt$ of Eq. (3.32).

If we take the integral clockwise around the circuit, then $+\mathcal{V}_n$ in Eq. (4.4) has symbolic polarity defined so that on our path we hit the minus terminal of $\mathcal{V}_n$ first. This is because the result of integration for $\mathcal{V}_n > 0$ is positive for this convention. This means that the voltage source (via $\mathbf{E}$) is doing positive net work *on* a positive charge if the charge follows the integration path.

Equation (4.4) can be simplified by considering the circuit in Fig. 4.16, where we assume that $V > 0$. Because the wires have zero resistance and because the potential of the upper terminal of the voltage source is $V$ volts higher than its lower terminal, it must be true that terminal $b$ of the resistor is $V$ volts higher than terminal $a$. Thus in Eq. (4.1) if we take a path around the circuit of Fig. 4.16, the result of the closed-loop integral of $\mathbf{E}\cdot d\boldsymbol{\ell}$ is equal to $\mathcal{V} = V$. Now in Eq. (4.4), the integral of $\mathbf{E}\cdot d\boldsymbol{\ell}$ from $b$ to $a$ is just $V_b - V_a$ (see Sec. 4.2 or 2.6). Except for the voltage source, the integral of $\mathbf{E}\cdot d\boldsymbol{\ell}$ around the remainder of the path (the wires) is approximately zero, because $E$ in a near-perfect conductor such as wire is extremely small.

The voltage across $R$ is $V_1 = V_b - V_a$. Consequently, Eq. (4.4) for this case is $V_1 = V$, or $-V + V_1 = 0$. Using the latter form, if we begin at $a$ and go clockwise, $-V\ (< 0)$ is the *drop* in voltage as we go by the emf (i.e., a rise in potential), and $V_1\ (> 0)$ is the *drop* in voltage as we go by the resistor.

Suppose that there are several passive and/or active elements around a closed circuit loop, each with a particular potential $V_n(t)$ across its terminals, equal to either the integral of $-\mathbf{E}\cdot d\boldsymbol{\ell}$ through, or the emf of the element. Generalizing the reasoning applied to Fig. 4.16, and allowing time-dependent voltages, Eq. (4.4) reduces to *Kirchhoff's voltage law (KVL)*:

$$\sum_n v_n(t) = 0, \tag{4.5}$$

which says that *the sum of the voltage drops around any closed loop is zero.*

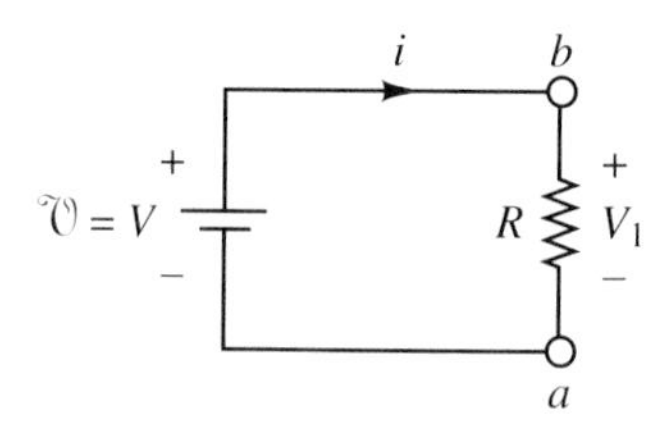


**Figure 4.16** Circuit for Example 4.4.

The easiest way to avoid sign errors regarding KVL is as follows. Define voltage polarities however you like for all unknown voltages; voltage source polarities or other known voltage polarities are of course fixed. When going around a loop (either clockwise or counterclockwise), sum the voltage drops around the loop to zero as in Eq. (4.5). This may be accomplished for each element *by using the voltage polarity sign of the element (active or passive) which you first encounter in passing by that element.*

**Example 4.4**

Write KVL, Eq. (4.5), for the circuit of Fig. 4.16, and determine the current $i$ flowing in the resistor and voltage source.

**Solution** Taking the loop clockwise from $a$, we first hit the "$-$" sign of the voltage source. Therefore, the first term of Eq. (4.5) is $-V$. Moving on to $b$, we first hit the "$+$" sign of $V_1$, so we write $+V_1$ to obtain $-V + V_1 = 0$, which of course is the equation we used in deriving Eq. (4.5).

There is only one path through which current can flow in Fig. 4.16, so the current in the voltage source equals that in the resistor, which by Ohm's law is $i = V_1/R = V/R$. With these calculations, we have actually solved our first circuits problem!

**Example 4.5**

I wish to play my large 20-W 9-V boom box down on the beach, but my D cells are dead. I decide to run it from the car battery (12 V), using some old speaker wire in my trunk. The wire is copper, American Wire Gage (AWG) 25; this wire has a resistance of 32 Ω per 1000 ft. Refer to Fig. 4.17.

**(a)** What is the equivalent resistance of my boom box?

**(b)** What current does the boom box normally draw from the D cells?

**(c)** If the beach is 60 ft from the car, what voltage will be provided for my boom box, and what will the current be?

**(d)** Suppose that we used a cigarette lighter adapter for connecting the wire to the car. Would we blow a car fuse?

**Solution**

**(a)** From the point of view of the D cells, the entire boom box running at the given operating point is indistinguishable from a single equivalent resistance $R_{eq} = R_L$, which would draw the same load current. This idea is illustrated in Fig. 4.17a. We know the power absorbed by the boom box under normal operation: $P = 20\text{ W} = V^2/R_{eq}$, so that $R_{eq} = V^2/P = (9\text{ V})^2/20\text{ W} = 4.0\ \Omega$.

**(b)** From Ohm's law, $I = V/R_{eq} = 9\text{ V}/4.0\ \Omega = 2.25\text{ A}$. Actually, the 20-W figure is the maximum load (cassette player playing at full volume); ordinarily, the power and current would be significantly less.

**(c)** The situation is illustrated in Fig. 4.17b. The equivalent circuit is shown, where $R_{out}$ and $R_{in}$ are the resistances of the wires from the car to the boom box and back, respectively. The value of each is $60\text{ ft} \cdot 32\ \Omega/1000\text{ ft} = 1.9\ \Omega$. Because there is only one path for current to flow, the same current $I_1$ flows in each of the series-connected components. Whenever elements are connected end to end so that they carry the same current, they are said to be connected in series, and one calls the combination a *series circuit*. By KVL, $-12\text{ V} + V_{R\text{out}} + V_{RL} + V_{R\text{in}} = 0$. By Ohm's law,

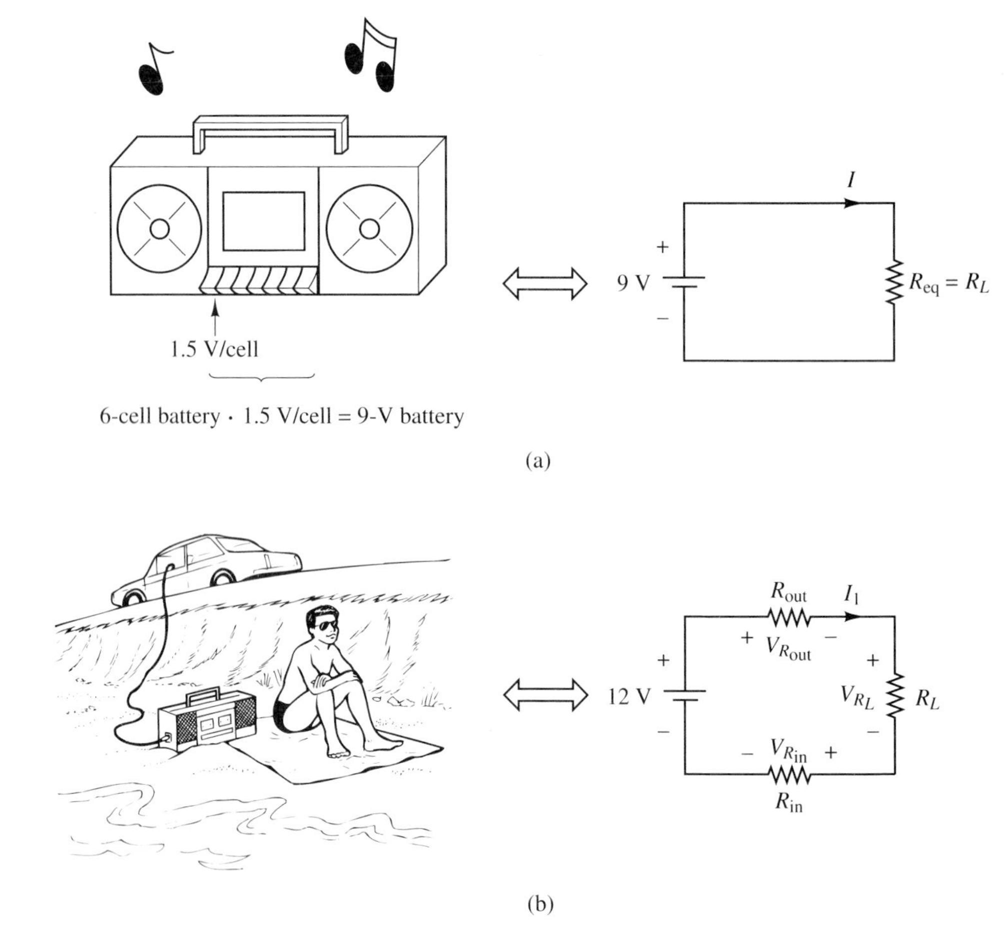


**Figure 4.17** (a) Simple resistive model of boom box; (b) circuit model for extended power cord to boom box.

$V_{Rout} = V_{Rin} = I_1R_{out} = I_1R_{in} = 1.9I_1$, while from Ohm's law and (a), $V_{RL} = I_1R_L = 4.0I_1$. Thus KVL can be written $(1.9 + 1.9 + 4.0)I_1 = 12$, or $I_1 = 12/7.8 = 1.54$ A. This is the current through our boom box when connected via the wires to the car battery. By Ohm's law, the voltage across the boom box is $1.54\text{ A} \cdot 4.0\ \Omega \approx 6.2$ V. Thus the wires caused a 5.8-V drop from the car battery, a significant drop reflecting energy dissipation in the wires. In practice, depending on the particular circuit requirements and actual wire, this procedure may or may not provide sufficient voltage with which to run the stereo (at minimum, the radio should work). If we used a long section of household power cord wire (e.g., AWG 16, with 4 Ω/1000 ft), we would have $R_{out} = 0.24\ \Omega$, so $I_1 = 12\text{ V}/(0.24 + 0.24 + 4.0) = 2.7$ A, so that the voltage across the boom box would be 10.7 V; this would work well.

**(d)** The current drawn from the car is 1.54 A or 2.7 A in the two scenarios of wire choice above. The rating of the fuse for the circuit supplying the cigarette lighter is typically 20 A. There should be no problem, as long as we do not short the wires at the connections.

**Example 4.6**

(a) A Walkman draws 30 mA from a 6-V source at a low volume setting. What are the power consumption and equivalent resistance of the Walkman?

(b) If I really "crank" it (i.e., turn up the volume), the current rises to 56 mA. What is the power and equivalent resistance when fully "cranked"?

**Solution**

(a) Again, $R = V/I = 6\ \text{V}/0.03\ \text{A} = 200\ \Omega$. $P = I^2R = (0.03)^2 \cdot 200 = 0.18$ W. (We could also use the method of Chapter 2: $p = vi = 6\ \text{V} \cdot 0.03 = 0.18$ W.) It does not take much input power to operate a radio at sufficient volume to drive a headset.

(b) $R = 6\ \text{V}/0.056 = 107\ \Omega$. Thus the equivalent resistance of the Walkman is strongly a function of the volume control setting. $P = (0.056)^2 \cdot 107 = 0.34$ W; still quite low power, especially compared with the boom box, which powers large speakers and a cassette player. Note that if the LED FM stereo indicator is on, the current in both parts (a) and (b) increases by 8 mA.

**Kirchhoff's current law (KCL).** The other fundamental circuit law by Kirchhoff allows us to analyze circuits in which not all elements have the same current. It is just a restatement of charge conservation, Eq. (4.2), with one simplification. If we consider only *nodes*, which are connections between two or more circuit elements, there is no possibility of charge storage, as there is with a capacitor (see Fig. 2.13). We may think of a node as a junction of water pipes in which water flows through but cannot be stored, and a capacitor as a water tank that *can* store water.[1] Consequently, the right-hand side of Eq. (4.2) is zero, giving *Kirchhoff's current law (KCL)*:

$$\sum_n i_n(t) = 0, \tag{4.6}$$

which says that the sum of the currents $i_n$ leaving (or equivalently, entering) any node is zero. Use of "leaving" rather than "entering" is arbitrary because the right-hand side is now zero; "leaving" tends to be more convenient in calculations and usually will be used in this book.

**Example 4.7**

Given the circuit fragment in Fig. 4.18a, determine the current $I_1$ by using KCL, Eq. (4.6).

**Solution** Remember that a node is an equipotential. Everywhere on the top wire the potential is the same, so it is all one node, as drawn in Fig. 4.18b. By Eq. (4.6), KCL for the upper node is $-1\ \text{A} + 2\ \text{A} + I_1 - 4\ \text{A} = 0$, or $I_1 = 3$ A. Notice that in a circuit diagram, a node may be "spread out" as in Fig. 4.18a. But this is just a way of drawing a diagram neatly, for the diagram of Fig. 4.18b is electrically equivalent to that of Fig. 4.18a. Although drawing curved lines may be helpful to envision nodes, it is

[1] [One *could* view the node as one plate of a capacitor, the other being a conductor at reference potential. However, in $C = \epsilon A/d$, $A$ is extremely small (the node area), $d$ between the plates is very large, so that $C$ and therefore $q = Cv$ is extremely small. Nevertheless, all conductors have capacitance, even if it is negligibly small.]

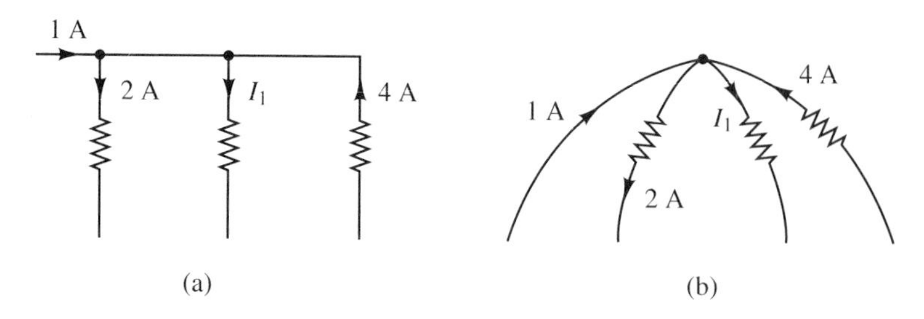


**Figure 4.18** Circuit fragment for Example 4.7.

impractical for large circuits; in practice, straight lines and right angles are used in drawing diagrams whenever feasible.

**Example 4.8**

(**a**) Each branch of a household wiring system (120 V) is connected to the main lines through a fuse or circuit breaker. The rating of a kitchen circuit breaker is 20 A. In this kitchen I have a microwave oven (1500 W), a mixer (175 W), a blender (375 W), a toaster (800 W), an electric light (100 W), a night light ($\frac{1}{4}$ W), and a can opener (1.5 A)—all of these going at once in this busy kitchen (see Fig. 4.19a). Do I blow the circuit breaker?

(**b**) What if I now turn on the coffee maker (900 W)?

**Solution**

(**a**) For this calculation we ignore any inductive effects of the motors in, for example, the blender. These devices are all connected in parallel (see Fig. 4.19b). Whenever elements are connected one across the other so that they share the same pair of nodes and thus have the same voltage across their terminals, they are said to be connected in parallel. One calls such a combination a *parallel circuit.* Because $p = vi$, $i = p/v$, which for $v = 120$ V (ac, 60 Hz) is for the devices above: 12.5 A, 1.5 A, 3.1 A, 6.7 A, 0.8 A, 0.002 A, and the can opener was given as drawing 1.5 A. However, the overhead light is, by requirement of the National Electrical Code® (NEC) on a separate circuit. Furthermore, the kitchen outlets are on two separate 20-A branches (also an NEC requirement). Suppose that the devices happen to be distributed over the branches as shown in Fig. 4.19b, dictated by where we plugged the appliances in (see Fig. 4.19a). Then by Kirchhoff's current law, these currents add to a total load of 19.2 A for one branch and 6.1 A for the other. The first load is straining its circuit breaker, but it should work; breakers are designed to carry 110% rated current indefinitely. It is not recommended that you try this overloading in your own kitchen—chances are, you will blow the breaker or fuse and may encounter possible electric hazards.

(**b**) The coffee maker draws 900/120 = 7.5 A. When added to the 19.2 A of the other appliances on that branch (see Fig. 4.19b), the result is 26.7 A—the breaker will probably "trip" in about a minute! Was the coffee maker the culprit? Not really. If we had plugged it in on the other branch, nothing would have blown. It pays to know a little house wiring!

For the majority of this chapter we shall be considering circuit behavior for constant-in-time voltage and current sources (i.e., constant-in-time excitation). In this case, the direction of current flow in every component does not change with time.

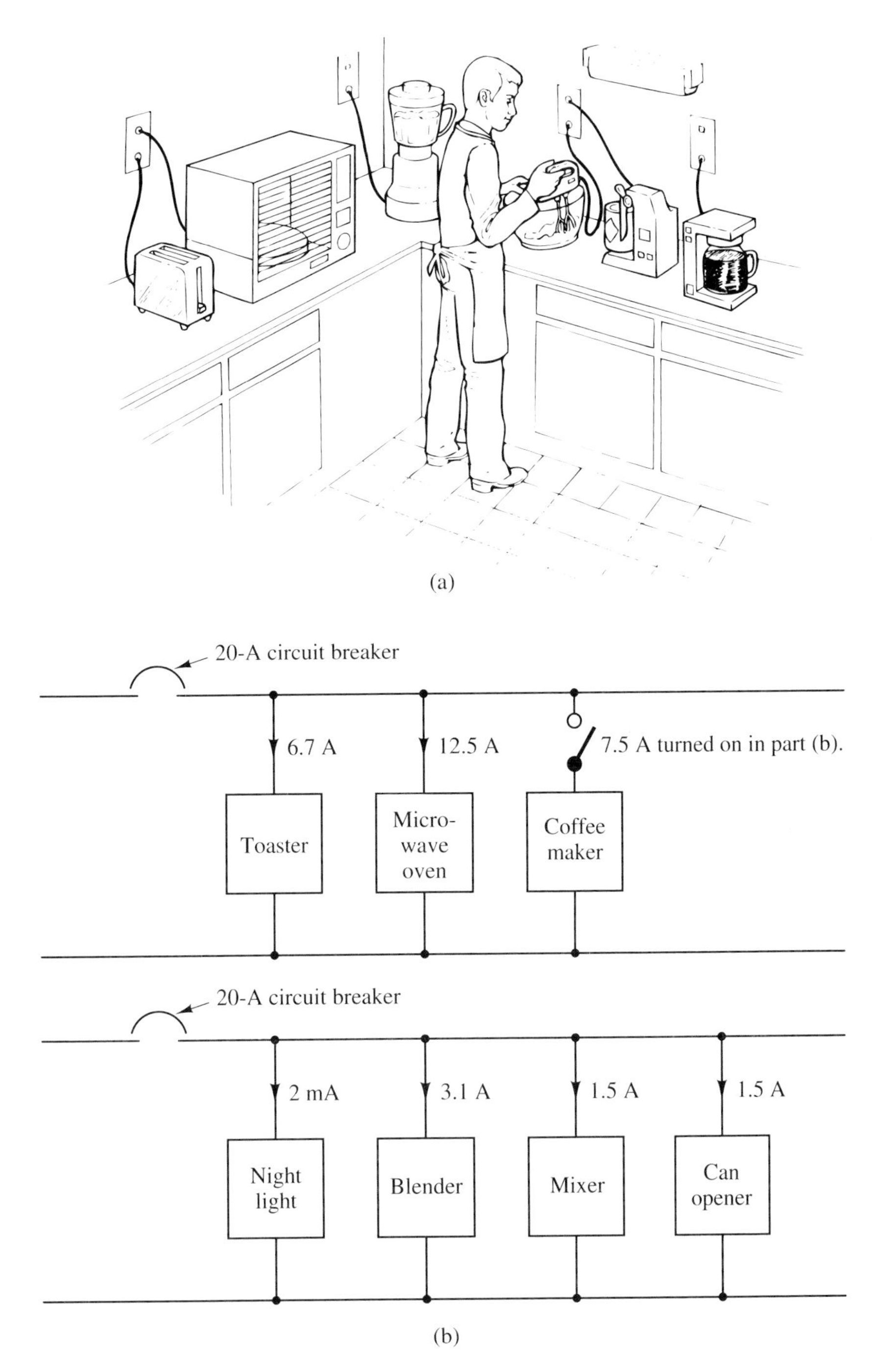


**Figure 4.19** (a) Physical kitchen appliance setup; (b) parallel circuit representation.

For example, in Fig. 4.16, $i$ flows clockwise. Thus we may speak of the *direction* of each *current* for all time. This cannot be said when the direction of currents periodically *alternates* in time, as is the case with electricity from our 60-Hz wall outlets.

For these reasons, constant-in-time currents are called *direct currents* (dc), while currents from the wall outlet are called *alternating currents* (ac). We shall have much more to say about alternating currents in Chapters 5 and 6. For now, we restrict our attention primarily to circuit analysis when the sources of voltage and/or current are constant in time, that is, dc analysis. Those examples we consider here using 120-V ac line voltages are simple examples where the differences between using dc and ac analysis are small and/or are ignored for simplicity.

We conventionally use capital letters for dc voltage and current variable symbols ($V$, $I$), and lowercase letters for time-varying voltages and currents ($v$, $i$). We also use lowercase letters when either constant-in-time or time-varying voltages and currents would satisfy the stated conditions and conclusions.

## 4.6 NODE VOLTAGES

Consider the circuit in Fig. 4.20. Suppose that the values of $v_B$, $R_S$, $R_L$, and $i_C$ are known and we wish to determine the voltage $v_L$ across the load resistor $R_L$. In examining Fig. 4.20, the first thing the reader is tempted to proclaim is that the polarity of $v_L$ is missing. However, this is not true. The voltage $v_L$ is called a *node voltage*, which is the voltage (potential difference) between the node so labeled and the 0-V *reference node* (shown in Fig. 4.20 with a ground symbol on the bottom node).

A node voltage such as $v_L$ refers to the absolute potential of a node (node $L$ in Fig. 4.20), with respect to the 0-V reference node. The polarity is *implied* to be as follows: If the node voltage is a positive number, that node is at higher potential than the 0-V reference; otherwise, it is less than or equal to 0 V. This polarity convention for specification of a node voltage is opposed to that for the specification of the voltage across a particular element, in which case + and − symbols *must* be labeled to avoid ambiguity.

Although in Fig. 4.20 only one element comes between the node $L$ and ground, namely $R_L$ (or the current source), a node at which a node voltage is defined can in practice be several (any number of) elements away from the ground node. For example, notice that although $v_L$ is the voltage of a node connected to resistor $R_S$, it is *not* the voltage across $R_S$—that voltage is $v_S$. Rather, $v_L$ is the voltage across the series combination of $R_S$ and $v_B$. Clearly understanding the distinction between

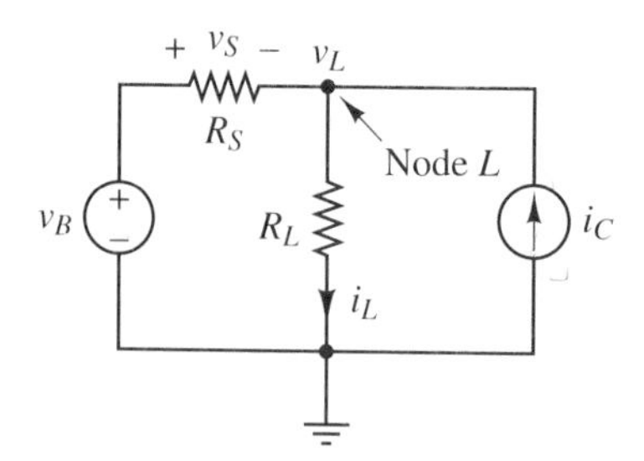


**Figure 4.20** Example circuit demonstrating a node and node voltage.

the two representations—voltage across an element and node voltage—is absolutely essential in order to perform circuit analysis successfully. Also, node voltages are the foundation of an efficient circuit solution method introduced in Sec. 4.12.

To solve the stated problem, we may now proceed as follows. We note that the voltage across $R_L$ in this case *is* equal to the node voltage $v_L$, because the terminals of $R_L$ are connected to the node $L$ (with node voltage $v_L$) and to the 0-V reference node. KVL may be written $-v_B + v_S + v_L = 0$, or $v_L = v_B - v_S$. We may now use KCL and Ohm's law at node $L$: $-v_S/R_S + v_L/R_L - i_C = 0$, or $v_S = R_S(v_L/R_L - i_C)$. Thus substitution of $v_S$ into the expression for $v_L$ gives

$$v_L = v_B - R_S\left(\frac{v_L}{R_L} - i_C\right).$$

## 4.7 EQUIVALENT PASSIVE COMPONENTS

It is very common that several passive components of similar type may be connected together in series or parallel networks. When overall voltages and currents are important rather than those associated with individual components of the network, a single circuit element equivalent to the network may be determined. Doing so may substantially simplify the analysis.

A simple example would be to calculate the total current consumed by a chain of Christmas lights. If the *equivalent resistance* of the chain $R_{eq}$ is known, the total current $i$ is just $i = v_{app}/R_{eq}$, where $v_{app}$ is the voltage applied to the chain. After reading this section, you will be able to verify the following results: If the $N$ lights are series connected, $R_{eq} = NR$; if they are parallel connected, $R_{eq} = R/N$.

As we have seen, many common household items may be modeled from the outside world as being equivalent resistances. Therefore, it is both simple and helpful to discuss the formation of equivalent elements using the resistor as an example passive component type.

It was noted in Sec. 2.9 that 10 identical resistors placed end to end were equivalent to a single resistor having 10 times the resistance of one of those placed end to end. Effectively, the $d$ in $R = \rho d/A$ was multiplied by 10, thereby multiplying $R$ by 10. We may generalize this concept by referring to Fig. 4.21a, which shows two resistors $R_1$ and $R_2$ connected in series.

The voltage across $R_1$ is $v_1$, and that across $R_2$ is $v_2$. Thus the voltage across the

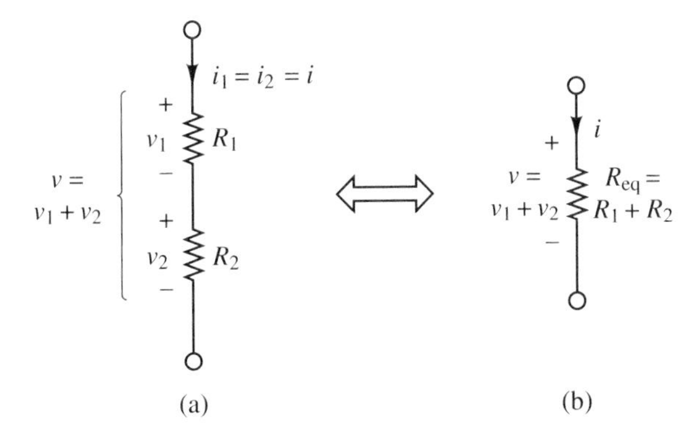


**Figure 4.21** (a) Two series-connected resistances; (b) their circuit equivalent.

series combination of $R_1$ and $R_2$ is $v = v_1 + v_2$. The same current flows in both $R_1$ and $R_2$, for there is only one path in which current can flow. By Ohm's law, this current is $i = v_1/R_1 = v_2/R_2$. Therefore, $v_1 = iR_1$ and $v_2 = iR_2$, so that the voltage across the network is $v = i(R_1 + R_2)$.

To the outside world, there is no difference between having these two resistors connected in series and a single resistor of resistance $R_{eq} = v/i = R_1 + R_2$. A current $i$ flowing in either the series combination of $R_1$ and $R_2$ or the equivalent resistance $R_{eq}$ has a voltage $i(R_1 + R_2)$ across it (see Fig. 4.21b). Simple repetition of this idea results in the conclusion that *any number of resistances in series add: The equivalent resistance of a chain of resistors connected in series is equal to the sum of the component resistances forming the chain.*

$$R_{eq} = \sum_n R_n, \qquad R_n \text{ connected in series.} \tag{4.7}$$

It is clear from Eq. (4.7) that the equivalent resistance of series-connected resistors is larger than any of the contributing resistances.

**Example 4.9**

Suppose in Fig. 4.21a that $R_1 = 1\ \Omega$ and $R_2 = 10\ \text{k}\Omega$. What is $R_{eq}$?

**Solution** The equivalent resistance is $R_{eq} = R_1 + R_2 = 1 + 10{,}000 = 10.001\ \text{k}\Omega$. For typical tolerances, $R_1$ makes essentially no contribution to the total resistance. In general, if one of the resistances $R_n$ in Eq. (4.7) is orders of magnitude greater than the sum of the others connected in series with it, the equivalent resistance will be roughly equal to that largest resistance $R_n$.

**Example 4.10**

Show how equivalent resistances simplify the boom-box example, Example 4.5.

**Solution** Notice that in Fig. 4.17b, $R_{out}$, $R_L$, and $R_{in}$ are connected in series. In this case, only a slight simplification can be made: Combine the two wire resistances $R_{eq} = R_{out} + R_{in} = 1.9 + 1.9 = 3.8\ \Omega$. It is *not* desirable here to combine $R_L$ with $R_{out}$ and $R_{in}$, because we would have no way of determining the portion of the total 12 V across the boom box—the whole point of the analysis! The general principle is never to eliminate ("equivalent-away") an element whose individual current or voltage you are interested in finding. The simplified circuit is shown in Fig. 4.22.

It may be argued that $R_{out}$ and $R_{in}$ cannot be combined, because they are separated by $R_L$. However, as far as circuit operation is concerned, it makes no difference where in the chain the boom box is put. In part (c) of that example, if the two wires were connected in series with the boom box, with the other end of the boom box connected directly to the cigarette lighter, the voltage would still be just

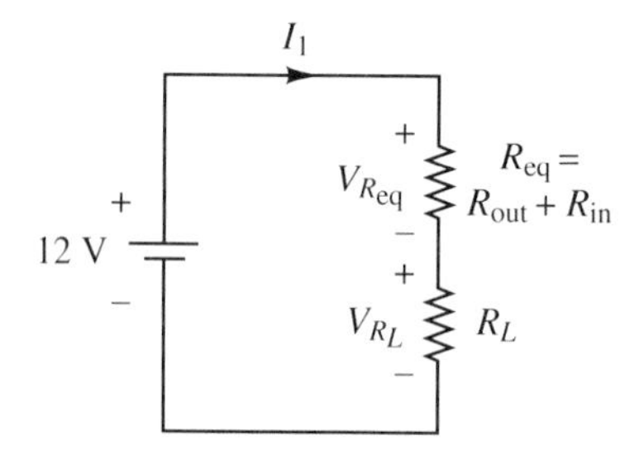


**Figure 4.22** Simplification of circuit of Example 4.5 when the resistances of two power cords are combined.

as weak across the stereo as it was in the example, when connected between the far ends of the two wires as in Fig. 4.17b. We may always mentally or pictorially "slide" a resistor to a convenient place anywhere along *the same branch of series-connected elements*.

By KVL, we have $-12\text{ V} + V_{R\text{eq}} + V_{RL} = 0$; by Ohm's law, this gives $I_1 = 12\text{ V}/(R_{\text{eq}} + R_L) = 12/(3.8 + 4.0) = 1.54\text{ A}$; finally, again by Ohm's law, $V_{R\text{eq}} = 3.8 \cdot I_1 \approx 5.8\text{ V}$, $V_{RL} = 4.0 \cdot I_1 = 6.2\text{ V}$, as before.

What about resistors in parallel? To be precise, elements are connected in parallel if *both of the terminals of one element are connected to those of the other; that is, they share the same pair of nodes*. See Fig. 4.23a for an example and Fig. 4.23b for a counterexample.

In the latter example (Fig. 4.23b), $R_1$ is *not* in parallel with $R_2$ because $R_3$ causes the lower terminals of $R_1$ and $R_2$ not to be connected. Parallel-connected elements have the same voltage across them (and series-connected elements have the same current through them). In Fig. 4.23b, $R_1$ and $R_2$ in general have different voltages and currents. Note, however, that the *series combination* of $R_1$ and $R_3$ *is* in parallel with $R_2$.

The equivalent resistance for parallel-connected resistors will be that which, having the same voltage $v$ placed across it, has the same current $i$ through it as the parallel network. Consider Fig. 4.23a, which shows three resistances connected in parallel. If the resistors were identical and equal to $R$, we would expect that $R_{\text{eq}} = R/3$ because the area $A$ in $R = \rho d/A$ has been increased threefold. For general resistor values, we proceed as follows. Because the resistors are connected in parallel, the voltage across them is the same: $v_1 = v_2 = v_3 = v$. By KCL and Ohm's law, $i = i_1 + i_2 + i_3 = v(1/R_1 + 1/R_2 + 1/R_3)$, so that $R_{\text{eq}} = v/i = 1/(1/R_1 + 1/R_2 + 1/R_3)$. The general rule for any number of resistances in parallel is that *the equivalent resistance equals the reciprocal of the sum of the reciprocals of the parallel-connected resistors.*

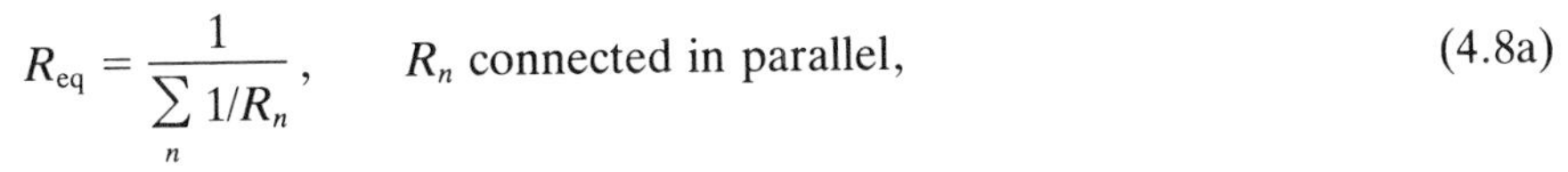

$$R_{\text{eq}} = \frac{1}{\sum_n 1/R_n}, \qquad R_n \text{ connected in parallel,} \tag{4.8a}$$

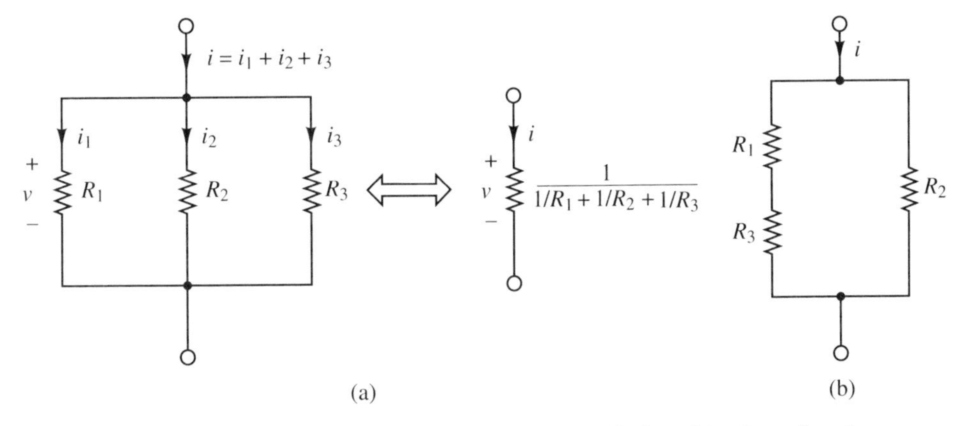


**Figure 4.23** (a) Three resistors in parallel, and their equivalent; (b) resistors $R_1$ and $R_2$ are *not* in parallel.

which for the special case of two resistors reduces to the product of the two resistances divided by their sum:

$$R_{\text{eq}} = \frac{R_1 R_2}{R_1 + R_2}, \qquad R_1 \text{ and } R_2 \text{ connected in parallel.} \tag{4.8b}$$

**Example 4.11**

Suppose in Fig. 4.23a that $R_3 = \infty$; that is, $R_3$ is removed from the circuit (*open circuited*). What is $R_{\text{eq}}$?

**Solution** With $R_3 = \infty$; $R_{\text{eq}}=1/(1/R_1 + 1/R_2) = R_1R_2/(R_1 + R_2)$, which is the formula for any two resistors connected in parallel.

In fact, it is easy to see in Eq. (4.8) that if one of the $R_n$ is very large compared with the others, it will make essentially no contribution to $R_{\text{eq}}$. Contrarily, if one of the $R_n$ is very small compared with the others, its $1/R_n$ value dominates the denominator, so that $R_{\text{eq}} \approx 1/(1/R_n) = R_n$ for $R_n$ small. In particular, if two resistors are connected in parallel and have greatly differing values, the equivalent resistance is roughly equal to the *smaller* of the contributing resistances. For instance, 10 Ω in parallel with 1 kΩ is $10 \cdot 1000/1010 \approx 9.9$ Ω (roughly, 10 Ω). And if one of the resistances is 0 Ω (*short circuited*), the resistance of the parallel combination is also 0 Ω.

Another observation from Eq. (4.8) is that the equivalent resistance of parallel-connected resistors is smaller than any of the contributing resistances. This is because they all increase the denominator of Eq. (4.8), while one of them alone would yield just $1/(1/R_n) = R_n$.

The above procedure can also be applied to determine the equivalent capacitance of capacitors in series and parallel connections, as well as equivalent inductance of inductors connected in series and parallel. The only difference is that we substitute $i = C\, dv/dt$ or $v = L\, di/dt$ for Ohm's law. Note that for $y = A\, dx/dt$, the mean value theorem tells us that $x(t) = 1/A \int_{-\infty}^{t} y(t')\, dt'$, where we assume, as we always can in practice, that $x(-\infty) = 0$.

To find the equivalent parameter for series-connected elements, we always express $v$ in terms of $i$, using the integral for the case of the capacitor. For parallel-connected elements, we express $i$ in terms of $v$. The results are that inductors follow the same rule as resistors, while capacitors follow the reverse:

$$L_{\text{eq}} = \sum_n L_n, \qquad L_n \text{ connected in series} \tag{4.9}$$

$$L_{\text{eq}} = \frac{1}{\sum_n 1/L_n}, \qquad L_n \text{ connected in parallel} \tag{4.10}$$

$$C_{\text{eq}} = \frac{1}{\sum_n 1/C_n}, \qquad C_n \text{ connected in series} \tag{4.11}$$

$$C_{\text{eq}} = \sum_n C_n, \qquad C_n \text{ connected in parallel.} \tag{4.12}$$

Proofs of these results are left as exercises for the interested reader.

Writing $v = Ri$, $v = L\,di/dt$, $v = (1/C)\int_{-\infty}^{t} i(t')\,dt'$, it should be apparent why capacitance is "backwards": Its *reciprocal* is analogous to $R$ and $L$. For resistors, capacitors, and inductors, we may use Eqs. (4.7) through (4.12) to determine series–parallel combinations by applying successively the individual rules. We may not, however, combine mixed types of elements. For example, you can never add capacitance to inductance or resistance!

**Example 4.12**

Find the equivalent components for the element configurations shown in Fig. 4.24.

**Solution**

(**a**) First combine the two parallel resistances, which are both $R_B$: $R_B R_B/(R_B + R_B) = R_B/2$. This is a general result: If two *equal* resistors (or inductors) are connected in parallel, the equivalent resistance (or inductance) is one-half that of the constituent elements. Of course, by doing so, we are doubling the area $A$ in $R = \rho d/A$. The total $R_{\text{eq}}$ is thus $R_A + R_B/2$.

(**b**) Because capacitors in parallel add, we have 0.02 μF + 0.1 μF + 3 μF = 3.12 μF.

(**c**) At first the combination appears formidable, but when redrawn as shown, it

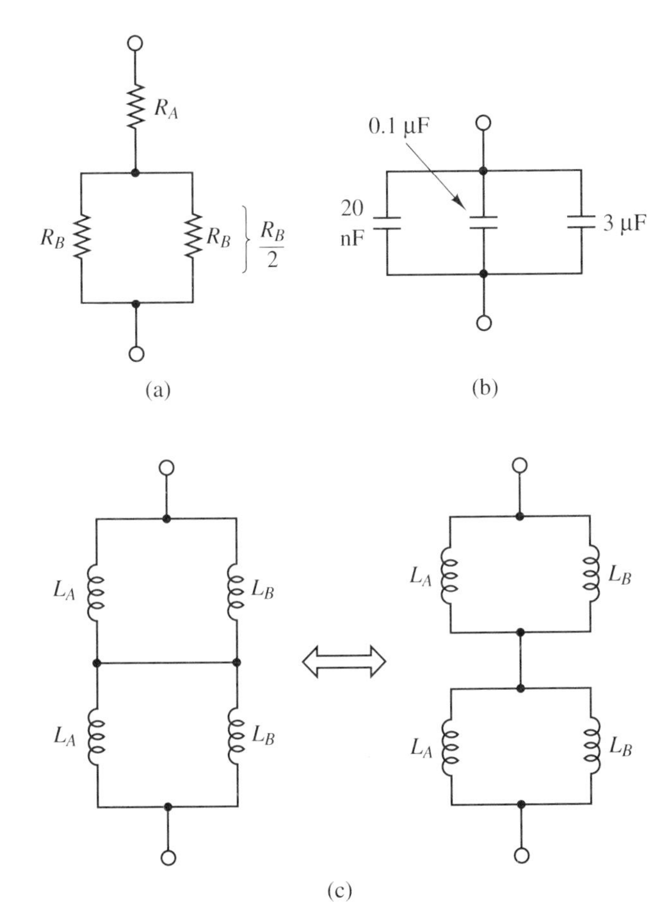


**Figure 4.24** Passive networks whose equivalents are determined in Example 4.12.

is clearly two parallel-connected inductors in series with two other parallel-connected inductors. Thus $L_{eq} = 2L_A L_B/(L_A + L_B)$. Note that if $L_A = L_B$, then $L_{eq} = L_A$.

As is true throughout the study of this or any other engineering discipline, it is most informative to carry out calculations using symbols rather than numbers. That way, when the numbers change, the same resulting relation can be reused with the new numbers. If numbers are used throughout the analysis but are subsequently changed because of a new design or application situation, the entire derivation must be repeated. However, numbers are also intuitive, involve less algebra, and are ultimately used in any practical situation. So we present a mixture in this book.

**Example 4.13**

The circuit in Fig. 4.25 is known as a *Wheatstone bridge.* It was named after Charles Wheatstone of England, who made many inventions in telegraphy and electromechanical machinery, although the invention of this circuit is credited to Samuel H. Christie's work in the 1840s. In this circuit, an element whose resistance $R_x$ is unknown is connected to the bridge configuration as shown, in order to be measured.

One of the other resistors, $R_v$, is adjustable. It is known as a *potentiometer* because it "meters" or supplies in a measured amount the voltage (potential) available across its terminals due to its variable resistance. Alternatively, it is sometimes called a *rheostat* because "rheo" is Greek for current and "stat" means a stabilizing or controlling device (e.g., thermostat); by changing its resistance it also regulates current through the branch in which it is connected. The volume control knob on an audio system is a potentiometer. In the Wheatstone bridge, $R_v$ is adjusted until the ammeter registers zero current. Determine $R_x$ under this condition. (Do not confuse the ammeter with a current source.)

**Solution** We assume that the values of $R_1$ and $R_2$ are known and that $R_v$ has been calibrated so that once we adjust it we can read its value in Ω. When the ammeter (which in practice always has nonzero resistance) reads zero current (adjusted to null), this means that nodes $A$ and $B$ are at the same potential and would be, even were the connecting ammeter removed: $V_A = V_B$. If they were not at the same potential, an

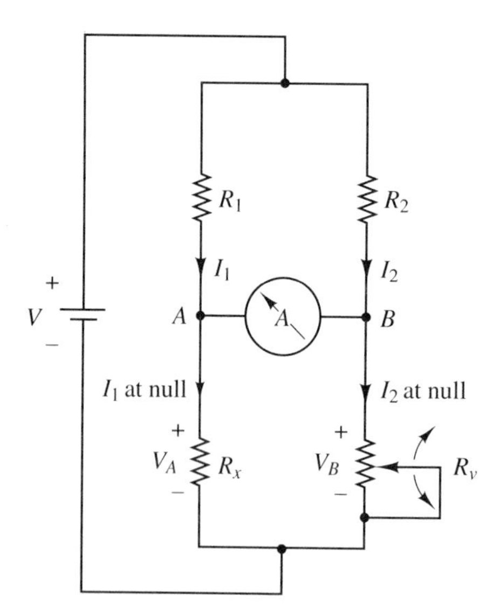


**Figure 4.25** Wheatstone bridge for measurement of $R_x$.

electric field would exist between the nodes that would cause a current in the ammeter. Consequently, by Ohm's law, $I_1R_x = I_2R_v$. Similarly, $I_1R_1 = I_2R_2$. Dividing the latter equation by the former gives $R_1/R_x = R_2/R_v$ or $R_x = R_v\,(R_1/R_2)$.

Notice that the result is independent of the applied voltage $V$, a good feature because therefore $V$ (typically imprecise and liable to vary over time) need not be calibrated. To be able to measure large $R_x$ despite the fact that $R_v$ has a limited range, $R_2$ can be switched to one of a fixed set of alternate values to match the order of magnitude of $R_x$. Bridge circuits with such range switches are commonly used in electrical measuring instruments.

**Example 4.14**

As an exercise in using KVL and KCL as well as a dependent source, solve for the voltages $V_1$ and $V_3$ in the circuit of Fig. 4.26.

**Solution** Notice that for clarity nodes 1 and 2 have been labeled above resistors $R_1$ and $R_2$. Also, $V_1$ and $V_2$ are in this case equal to the node voltages of nodes 1 and 2 because $V_1$ and $V_2$ "stretch all the way" from the corresponding node to the bottom node, which would be the reference node. We shall formalize these comments in Sec. 4.12. For now, let us just write KCL for node 1 to obtain

$$\frac{V_1}{R_1} - \alpha V_3 + I_S = 0.$$

For node 2, KCL reads

$$-I_S + \frac{V_2}{R_2} + \frac{V_3}{R_3} = 0.$$

To simplify this, write KVL for the rightmost leg, completing the loop through $R_2$ to obtain

$$-V_S + V_3 - V_2 = 0,$$

or

$$V_2 = V_3 - V_S.$$

Thus KCL for node 2 can be rewritten

$$-I_S + V_3\left(\frac{1}{R_2} + \frac{1}{R_3}\right) - \frac{V_S}{R_2} = 0,$$

or

$$V_3 = \frac{R_3(V_S + I_S R_2)}{R_2 + R_3}.$$

Note that $V_3$ was found without solving for $V_1$ simultaneously. However, $V_1$ depends on $V_3$ by KCL for node 1, which may now be solved for $V_1$ as

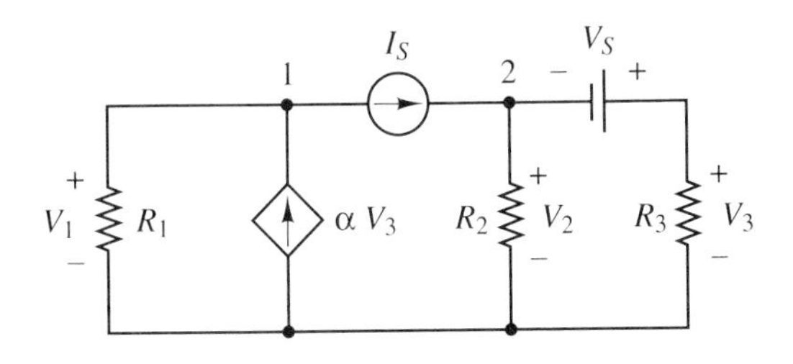


**Figure 4.26** Circuit for Example 4.14.

$$V_1 = R_1(\alpha V_3 - I_S)$$
$$= R_1\left[\frac{\alpha R_3(V_S + I_S R_2)}{R_2 + R_3} - I_S\right].$$

This type of analysis is very common in practice: use of both KVL and KCL, each where appropriate. Where there are current sources and/or many parallel-connected elements, KCL is more convenient. When voltage sources and/or series-connected elements occur, KVL is convenient.

Incidentally, the units of $\alpha$ must be A/V = $\Omega^{-1}$, or those of conductance. Playfully reversing the letters of "ohm" resulted in the name "mho" = $\Omega^{-1}$ = ℧ for conductance. Another name for the same unit is Siemens (S). Keep in mind, however, that this does not mean that the dependent source is a resistor. After all, the voltage and current related by $\alpha$ are those of different devices. In fact, dependent sources are often models of active circuits.

## 4.8 VOLTAGE AND CURRENT DIVISION

### 4.8.1 Voltage Division

We are now in the process of building up a set of tools of use in the design and analysis of electric and electronic circuits. Such tools make life easier. Recall how the circuit of Fig. 4.17b of Example 4.5 was simplified to that in Fig. 4.22 in Example 4.10 by means of an equivalent resistance. However, we still had to use KVL to find the voltage $V_{RL}$. Let us study this circuit once more to obtain one further simplification, in the generic form shown in Fig. 4.27.

A voltage source $v_S$ is placed across the series combination of $R_1$ and $R_2$, resulting in respective voltages $v_1$ and $v_2$ across $R_1$ and $R_2$. The same current flows in all elements; call that current $i$. It is desired to determine $v_1$ and $v_2$ in terms of $v_S$, $R_1$, and $R_2$. We solved this in Example 4.10 for a particular case; here $i = v_S/(R_1 + R_2)$, so that

Same

$$v_1 = R_1 i = \frac{R_1}{R_1 + R_2} v_S, \tag{4.13a}$$

and similarly,

Same

$$v_2 = \frac{R_2}{R_1 + R_2} v_S \quad \text{voltage divider for two resistances connected in series.} \tag{4.13b}$$

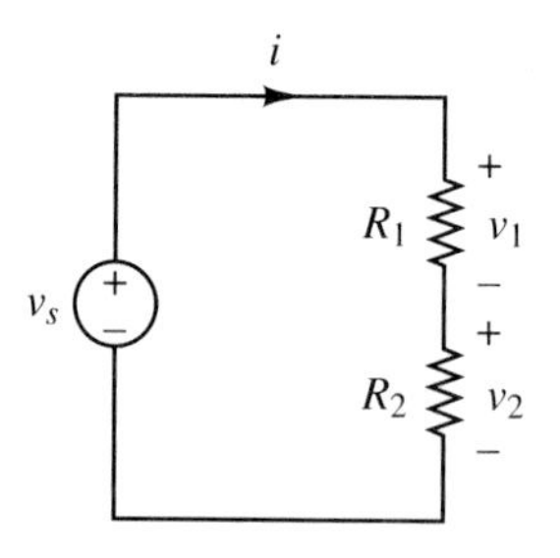


**Figure 4.27** Voltage divider circuit.

Note that the numerator resistor is the *same* as the one whose voltage is calculated. The source voltage $v_S$ is said to be *divided* between $v_1$ and $v_2$. Of course, $v_1 + v_2 = [(R_1 + R_2)/(R_1 + R_2)]v_S = v_S$. Note that the ratios $R_1/(R_1 + R_2)$ and $R_2/(R_1 + R_2)$ are both positive numbers less than 1: *fractions.*

With Eq. (4.13) we may re-solve Examples 4.5 and 4.10 very easily. The desired voltage across $R_L$ is just $12\text{ V} \cdot R_L/(R_L + R_{eq}) = 12 \cdot 4.0/(3.8 + 4.0) = 6.2\text{ V}$ as before. This is the simplest possible solution. There is no point in rederiving Eq. (4.13) every time that result is used; hence it is helpful to memorize it so that it can be used whenever convenient.

In general, the same procedure may be used to show that if many resistors are connected in series and a voltage source is applied across the combination, the voltage $v_m$ across one of the resistors $R_m$ is

$$v_m = \frac{R_m}{\sum_n R_n} v_S \qquad \text{general voltage divider for resistances connected in series.} \tag{4.14}$$

[Note the use of lowercase letters, indicating that Eqs. (4.13) and (4.14) are also valid for time-varying voltages.]

*Caution:* Equations (4.13) and (4.14) can be used *only* if all the resistors are in series, with no other branches emanating from the string. For example, in Fig. 4.20 we cannot find $v_L$ using voltage division because the presence of the current source $i_c$ invalidates the claim that $R_S$ and $R_L$ are in series, which otherwise would be true.

**Example 4.15**

Comment on voltage division [Fig. 4.27, Eqs. (4.13)] for the cases (a) $R_1 \gg R_2$, (b) $R_1 = R_2$, and (c) $R_1 \ll R_2$ where "$\gg$" and "$\ll$" mean "much greater/less than."

**Solution**

(**a**) From Eq. (4.13) we find that if $R_1 \gg R_2$, then $v_1 \gg v_2$. Furthermore, as $R_1 \to \infty$, $v_1 \to v_S$ while $v_2 \to 0\text{ V}$. This shows that when one of the resistors is far greater than the other, essentially all of the voltage appears across the larger resistor; the voltage across the other is comparatively negligible.

(**b**) From Eq. (4.13), if $R_1 = R_2$, then $v_1 = v_2 = v_S/2$. That is, if the resistors are equal, the voltage is distributed equally between components $R_1$ and $R_2$.

(**c**) Noting the mirror symmetry with part (a), we find that if $R_1 \ll R_2$, then $v_1 \ll v_2$. Also, as $R_2 \to \infty$, $v_1 \to 0\text{ V}$ while $v_2 \to v_S$.

### 4.8.2 Current Division

An analogous helpful rule can be obtained for finding the currents through resistors connected in parallel. Again we consider the case of two resistors, this time connected in parallel and powered by a parallel current source (see Fig. 4.28). The voltages across all elements are equal; call that voltage $v$. It is desired to determine $i_1$ and $i_2$ in terms of $i_s$, $R_1$, and $R_2$. We considered this situation for three resistors in Fig. 4.23, but did not end up calculating the individual currents. In Fig. 4.23, the current $i$ was equivalent to $i_s$ in Fig. 4.28. From that analysis [see Eq. (4.8b)] we may write that $v = i_s R_{eq}$, where $R_{eq} = 1/(1/R_1 + 1/R_2) = R_1R_2/(R_1 + R_2)$. Therefore,

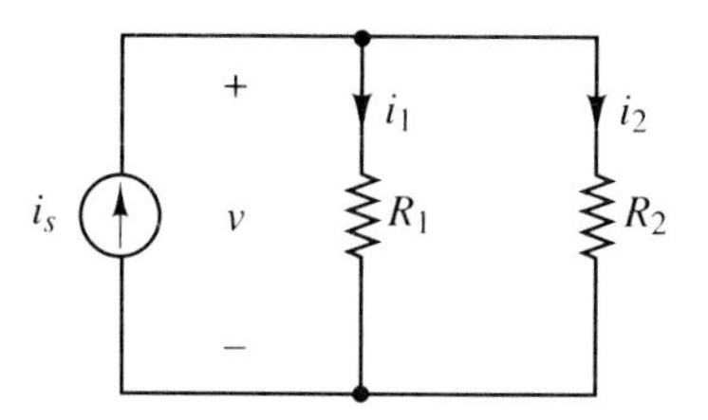


**Figure 4.28** Current divider circuit.

Different

$$i_1 = \frac{v}{R_1} = \frac{R_2}{R_1 + R_2} i_s, \tag{4.15a}$$

and similarly,

Different

$$i_2 = \frac{R_1}{R_1 + R_2} i_s \quad \text{current divider for two resistances connected in parallel.} \tag{4.15b}$$

Note that the numerator resistor is *the other resistor* from the one whose current is calculated. The source current $i_s$ is said to be *divided* between $i_1$ and $i_2$. Of course, $i_1 + i_2 = \{(R_1 + R_2)/(R_1 + R_2)\}\, i_s = i_s$. Again, there is no point in rederiving Eq. (4.15) every time that result is used; hence it is helpful to memorize it so that it can be used whenever convenient.

In general, the same procedure [using Eq. (4.8a)] may be used to show that if many resistors are connected in parallel and a current source is applied across the combination, the current $i_m$ in one of the resistors $R_m$ is

$$i_m = \frac{1}{R_m \sum_n \left(\frac{1}{R_n}\right)} i_s \quad \text{general current divider for resistances connected in parallel.} \tag{4.16}$$

*Caution:* Equations (4.15) and (4.16) can be used *only* if all the resistors are in parallel, with no other elements appearing in any of the branches. For example, in Fig. 4.20 we cannot find $i_L$ using current division because the presence of the voltage source $v_B$ invalidates the claim that $R_s$ and $R_L$ are in parallel, which otherwise would be true.

**Example 4.16**

Comment on current division [Fig. 4.28, Eqs. (4.15)] for the cases (a) $R_1 \gg R_2$, (b) $R_1 = R_2$, and (c) $R_1 \ll R_2$.

**Solution**

(**a**) From Eq. (4.15), we find that if $R_1 \gg R_2$, then $i_1 \ll i_2$. Furthermore, as $R_1 \to \infty$, $i_1 \to 0$ A while $i_2 \to i_S$. This shows that when one of the resistors is far smaller than the other, essentially all of the current flows in the smaller resistor; the current in the other is comparatively negligible.

(**b**) From Eq. (4.15), if $R_1 = R_2$, then $i_1 = i_2 = i_S/2$. That is, if the resistors are equal, the current is distributed equally between components $R_1$ and $R_2$.

(**c**) Noting the mirror symmetry with part (a), we find that if $R_1 \ll R_2$, then $i_1 \gg i_2$. Furthermore, as $R_2 \to \infty$, $i_1 \to i_S$ while $i_2 \to 0$ A.

### 4.8.3 Practical Examples

To illustrate the real-world application of voltage and current division, we consider the following examples from the lab and at home.

**Example 4.17**

Suppose that it is desired to examine on an oscilloscope a pulsing waveform from the high-voltage section of an electronic circuit. (An oscilloscope is an instrument with a video screen used to visually examine and measure time-varying voltages.) We model the pulsing voltage to be measured as a voltage source $v(t)$ (see Fig. 4.29). The problem is that these pulses are hundreds of volts, suspected to reach 400 V at the peaks, but the inputs of the oscilloscope can handle only voltages of up to around 20 V. Choose the resistor $R_2$ to make a reading of 20 V for the input voltage to the 'scope $v_{in}(t)$ correspond to $v(t) = 400$ V. Resistors come in $\frac{1}{4}$ W, $\frac{1}{2}$ W, and higher average power capabilities. What type should we choose for $R_2$?

**Solution** Although it would first appear that the oscilloscope violates the assumption of a voltage divider for the otherwise series connection of the 1-kΩ resistor and $R_2$, its effect on the voltage $v_{in}(t)$ is negligible. This is because the input resistance of the scope is 1 MΩ, which in parallel with the 1-kΩ resistor is, by Eq. (4.8b), 999 Ω—indistinguishable from 1 kΩ within typical 5 to 20% tolerances. Thus by measuring $v_{in}(t)$ we do not significantly affect the rest of the circuit, an essential requirement of all measurement equipment. Therefore, we can solve Eq. (4.13b) for $R_2$ to obtain $(R_1 + R_2)v_{in}(t) = R_1 v(t)$ or $R_2 = R_1[v(t)/v_{in}(t) - 1]$. With $R_1 = 1$ kΩ, $v_{in}(t) = 20$ V, and $v(t) = 400$ V, this becomes $R_2 = 1000 \cdot (400/20 - 1) = 19$ kΩ.

The current $i$ through both resistors is 400 V/20 kΩ = 20 mA. The peak power absorbed by $R_2$ is $i^2 R_2 = 0.02^2 \cdot 19000 = 7.6$ W. Naturally, in practice it is always economical to use the smallest wattage resistor that can handle the current it will have to carry, because it is the cheapest and physically the smallest. Because most of the time the voltage is far from the peak, probably a 1-W or even $\frac{1}{2}$-W resistor would probably suffice. A $\frac{1}{4}$-W might become too hot or even burn over time; it definitely would burn if the 400 V were constant. To be certain, we would have to compute an *average* current $i_{avg}$ and then use $i_{avg}^2 R_2$; we will have much more to say about average power dissipation in Chapter 6. Notice that the power dissipated in the 1-kΩ resistor is only one-nineteenth of that absorbed by $R_2$, the 19-kΩ resistor.

**Example 4.18**

The circuit of Fig. 4.30 is proposed for an electric stove, where the resistors are heating elements. Although the two (equal) resistances $R_B$ are in parallel, it is desired that all three elements dissipate the same energy, that is, provide the same heating effect. Find the relation between $R_B$ and $R_A$ to satisfy this requirement.

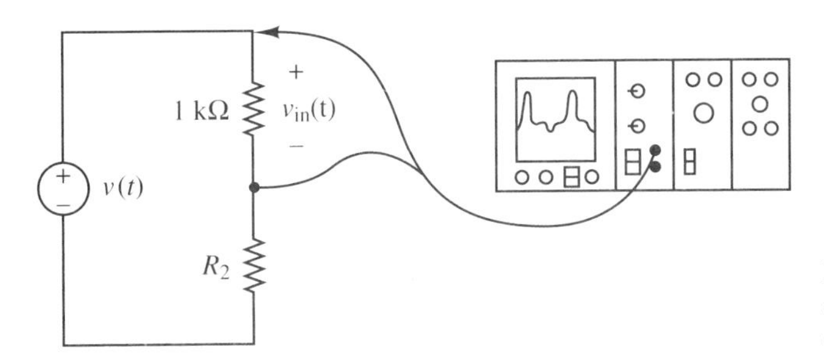


**Figure 4.29** Use of voltage divider to display high-voltage pulse waveform on oscilloscope.

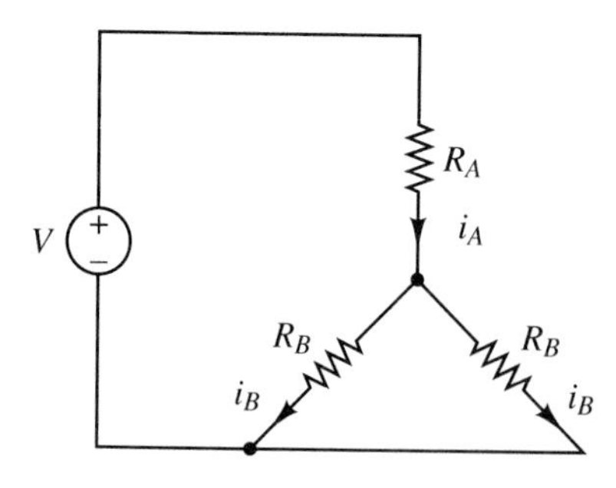


**Figure 4.30** Heating element circuit.

**Solution** Because the two lower resistors are equal and connected in parallel, they both have the same current $i_B$. To have all elements generate the same heat, we must have $i_B^2 R_B = i_A^2 R_A$. By the current divider, we know that the current $i_B$ is one-half of the total current $i_A$. Thus we must have $i_B^2 R_B = (2i_B)^2 R_A$, or $R_B = 4R_A$.

**Example 4.19**

A circuit that we all use is the lighting system in an automobile. A basic circuit for one side (say, the left side) of the car is shown in Fig. 4.31. It leaves out such features as hazard flashing, "burned-out headlamp" indicator, "lights are on" buzzer when the door opens, and so on. It also leaves out the light network signaling that the car is backing up. Describe its operation, in particular that of the sidemarker lights. That part of this example will provide some additional practice with voltage dividers.

**Solution** First, notice the bubble at the top marked "12 V." This is an example of a node voltage. In particular, it is the 12-V car battery positive potential, with respect to ground. *Ground,* or reference, is shown at the bottom of the schematic diagram and is connected to the automobile chassis.

Consider the headlight network, on the left side. With the headlights switch turned on, the dashboard headlight indicator (dashboard light) has 12 V across it. This and the other dash indicator lights take roughly $\frac{1}{4}$ A, which means that their resistance when lighted is 12 V/0.25 A = 48 Ω. Actually, in practice this light would be attached to a potentiometer for dimming the dashboard light. The headlight itself has a low and a high beam, selected by the high/low beam switch. When the high beam is used, another dashboard indicator signifies this. The main headlight takes 3 A on low beam and 4 A on high beam. Therefore, the resistance when lighted is 4 Ω and 3 Ω, respectively.

Whenever the headlight switch is on, all front and rear small *run* lights are also on, as well as the main headlights. [Oddly enough, the run lights are usually called *parking* lights, even though the usual reason they would be on when the car is parked is by mistake! This is as inexplicable as the names for places we drive (parkway) and places we park (driveway).] These run lights (20 Ω when lighted) are one of two filaments in a bulb; the other filament (6 Ω when lighted) serves as brake indication and/or turn signal functions. In practice, the run lights may be independently selected without turning on the main headlights by an additional headlight switch position.

The reason the words "when lighted" are used is that the resistance when lighted is roughly 10 times the resistance at room temperature for all these bulbs. That is, Ohm's law does not hold for light bulbs over the wide range of applied voltage that includes the transition from not lighted to lighted. This is because the resistance of the filament increases strongly with temperature. Contrarily, carbon resistors (Fig. 4.2a) are designed to have a resistance as fixed as possible, and to be operated at fixed (room) temperature.

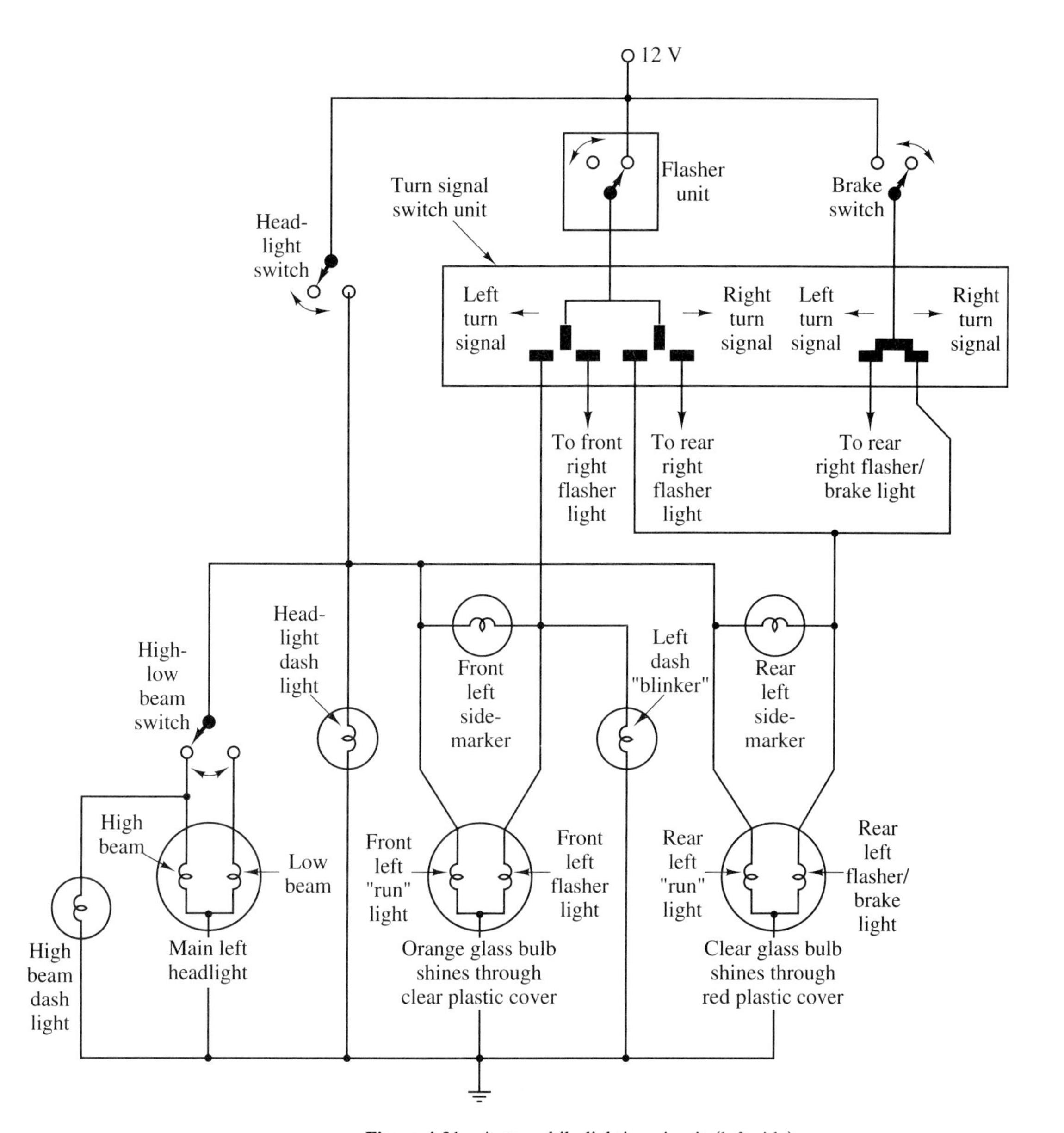


**Figure 4.31** Automobile lighting circuit (left side).

The more interesting part of this commonly used lighting circuit is the turn signal/brake light system. It is desired that if braking and turn signaling simultaneously, the side corresponding to the direction the car is turning will blink, while the other side maintains the braking signal. This is true in the back; in the front, braking is never to be indicated, while turning is always signaled. The circuit of Fig. 4.31 accomplishes this, as follows.

Examine the turn signal switch unit. The left section is the part that controls the turn signal flashing, and the right part controls the brake lights. Normally, all flashers are off; the sliding contact when in the middle does not reach either of the flasher

Although this 1907 advertisement concerns the spark coil, it equally well could portray the contemporary excitement about headlights on automobiles. The earliest turn signals came later, and were optional equipment.

contacts. When slid to the left, contact is made only with the left flashers (front and rear), as required, and vice versa. If the brake switch is on and neither turn signal is selected, both left and right brake lights are on in the rear. This is because the braking sliding contact switch normally does reach (contacts) both of the flasher/brake light contacts. When slid to the left to indicate a left turn (the sliding is simultaneous with the other slide switches), contact is broken with the right contact, which is connected to the left flasher/brake light. Hence the left bulbs flash, while the right rear bulb still indicates braking (the front right flasher bulb is then totally disconnected).

When signaling a turn, a bimetallic, self-heated switch turns on to heat and turns off by different bending of the two heated metals, which eventually breaks the connection. This process takes place periodically, causing the connected lamps to flash. A separate flasher unit is used for "hazard" lights, not shown in Fig. 4.31.

Finally, consider the *sidemarker* lights (which appear near all four corners of the *sides* of the car). If the headlights are on, close examination of Fig. 4.31 shows that 12 V is applied to the left terminals of the sidemarkers, while the right terminals go to ground through the flasher lights. Equivalent circuits are shown in Fig. 4.32a.

When cold, the sidemarker has resistance 4.7 $\Omega$, while the flasher light has resistance 0.6 $\Omega$. The flasher in parallel with the dashboard "blinker" (4.7 $\Omega$) is roughly 0.53 $\Omega$. By the voltage divider relation Eq. (4.13b), the voltage across the sidemarker is $12 \cdot 4.7/(4.7 + 0.53) = 10.8$ V, while that across the flasher is only $12 - 10.8 = 1.2$ V. The 10.8 V is enough to make the sidemarker light and heat up, causing its resistance to increase to 44 $\Omega$, while the flasher bulb remains cool. Thus a new voltage divider holds when the sidemarker becomes heated: Its voltage is now $12 \cdot 44/(44 + 0.53) = 11.9$ V, and now there is only 0.1 V across the flasher bulb.

Therefore, when the headlights are on, the sidemarkers normally are also on. If

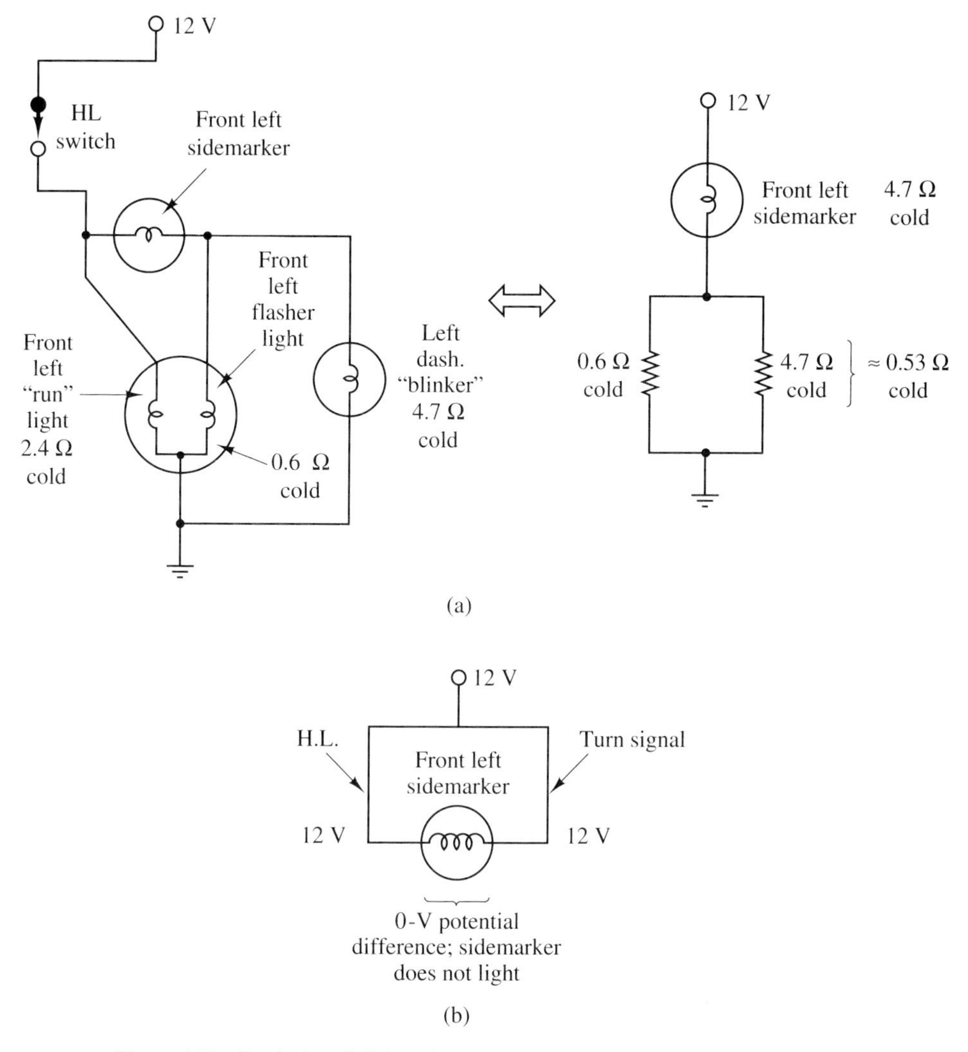


**Figure 4.32** Equivalent lighting circuits when headlights are on and (a) no turn signal is selected, and (b) turn signal is selected, during a flash.

the headlights are on and the turn signal for that side is selected, both sides of the sidemarker are then at roughly 12 V, and the sidemarker turns off (see Fig. 4.32b). Of course, this is true only during a flash; during the "off" time, the sidemarker stays on as when the turn signal is not selected. Thus when the headlights are on, the sidemarker flashes oppositely the turn signal flasher; when one is on, the other is off, and vice versa.

When the headlights are off, and the turn signal for that side is selected, in Fig. 4.31 the right terminal of the sidemarker is now at 12 V during a flash (0 V otherwise). Now the sidemarker is grounded through the headlight bulb. Note that one of either the high or the low beam is always selected at the high–low beam switch even when the main headlights switch is turned off, so there is always a low-resistance path to ground. (There are also much higher-resistance paths to ground through the run lights.) The headlight has roughly the same resistance (0.3 to 0.6 Ω cold) as the flasher. Therefore, the same argument as that given above applies, resulting in diagrams similar to those in Fig. 4.32a, with the turn signal (flasher) light replaced by the headlight. In this case the sidemarker is on and off in synchrony with the flasher light, unlike before when the headlights were on. This is because their "hot" end terminals are now connected to each other. This example should "drive home" the importance of voltage dividers in everyday life.

The various currents during lighting of the various bulbs are measured to be as follows: headlight: 3-A low beam, 4.1-A high beam, flasher filament 2.0 A, run filament 0.6 A, dashboard indicators, and sidemarkers both 0.24 A. Considering that both left and right sides operate simultaneously except turn signals, we may calculate the maximum total current to be (using high-beam and during a turn signal flash) $2 \cdot 4.1 + 2 \cdot 2.0 + 4 \cdot 0.6 = 14.6$ A. Inclusion of dashboard and other "courtesy" bulbs will increase this further. All of the numbers in this example are actual measurements made by the author on his car.

## 4.9 PRACTICAL ACTIVE ELEMENTS AND SOURCE TRANSFORMATION

We have already noted that a real inductor is more accurately modeled as an ideal inductor $L$ in series with a loss resistance $R_L$. Similarly, a wire can be modeled as an ideal conductor in series with a loss resistance, and a resistor can be modeled as a pure resistance in parallel with a capacitance and/or inductance. Real devices exhibit resistance, capacitance, and inductance, but are designed to emphasize the ideal characteristics of one type of device as much as possible.

The situation of nonideal circuit elements is analogous to the description of a typical person: Nobody is perfect. While it is nice and simple to characterize a person as perfect (and many would argue we were designed to behave that way), it is more accurate to include the person's faults while retaining those attributes the person shares with the ideal person (see Fig. 4.33a). The differences from the ideal person are seen primarily under "load" conditions, unless the person is consistently "malfunctioning." Usually, it is when a person is irritated by everyday or excessive stresses ("full load") that the bad qualities exhibit themselves; under "open-circuit" (*no-load*) conditions, he or she can be an angel! We know that deep down they *are* angels, but we can observe only "terminal characteristics" under "loading conditions."

The same holds true with electric circuit components: not only for passive elements such as the inductor in Fig. 4.33b, but also for active elements such as voltage and current sources. For example, the effect of a low-resistance (severe) load on a battery is that the terminal voltage available to the load is substantially reduced from the terminal voltage available at no load. This effect may be incorporated into the model of a practical battery as an ideal voltage source $V_S$ in series with an *internal source resistance* $R_S$; see Fig. 4.33c, where $V_S = 12$ V.

From the voltage divider relation (4.13b), the voltage across a load $R_L$ placed across the battery terminals is $V_S R_L/(R_L + R_S)$. If we know the internal resistance of the source, $R_S$, we can conveniently predict the load voltage for any load placed across the voltage source, within the limits of validity of the model of Fig. 4.33c. Writing the relation for $v_L$ for a general source $v_S$ as $v_L = v_S x/(1 + x)$ where $x = R_L/R_S$, the load voltage versus $R_L/R_S$ is as shown in Fig. 4.34. Note the difference from the curve for an ideal source (for which $v_L$ is independent of $R_L$).

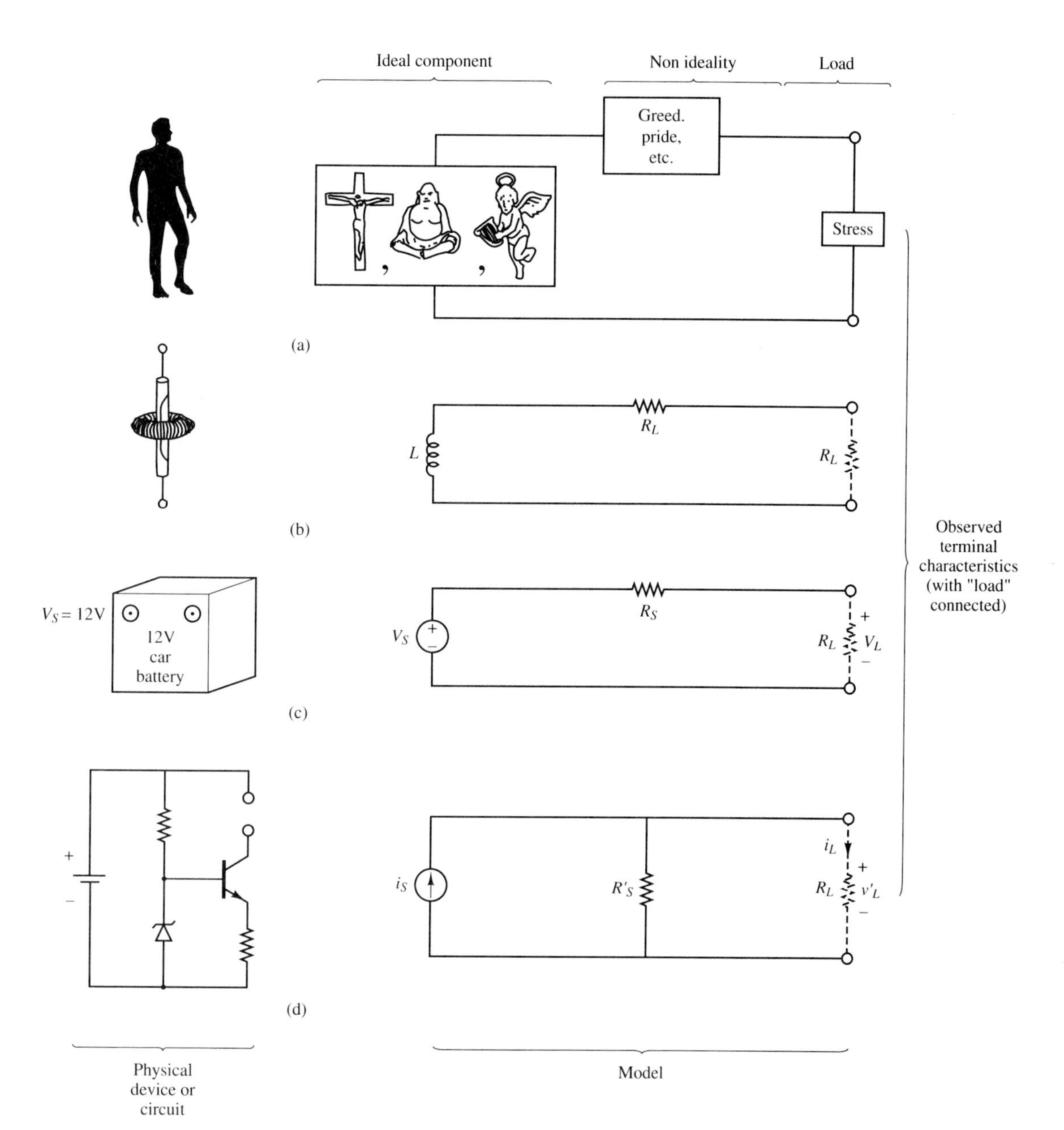


**Figure 4.33** Physical devices and models: (a) person; (b) inductor; (c) voltage source (battery); (d) current source.

Previously, we have ignored $R_S$ and its effects. The following example illustrates the error in doing so.

**Example 4.20**

**(a)** An auto battery has an internal resistance of 50 mΩ. The open-circuit voltage of "12-V" batteries is actually slightly higher, nominally 12.6 V. If the open-circuit voltage of a real battery is measured to be 12.5 V, what is the voltage appearing across

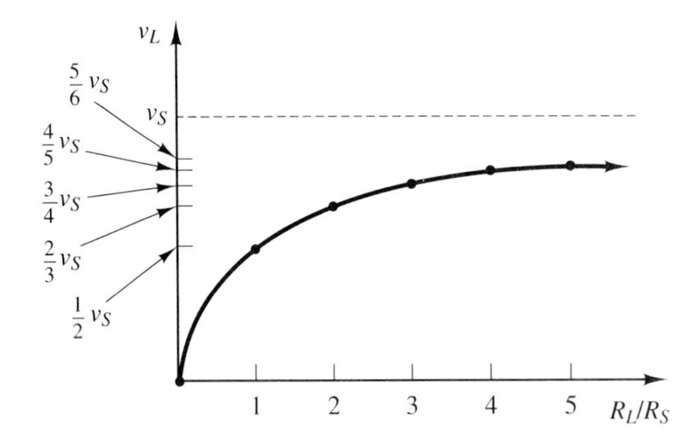


**Figure 4.34** Load voltage $v_L$ versus $R_L/R_s$ for a practical voltage source of voltage $v_s$ and internal resistance $R_s$. Dashed curve is for an ideal source.

the headlights (the terminal voltage of the battery) when on? Assume that only the high-beam headlights and run lights are on.

(**b**) If the left blinker is now turned on, what will be the terminal voltage?

**Solution**

(**a**) The open-circuit voltage is the voltage appearing across the battery with no lights turned on—nothing connected (ignoring *parasitic drains*). The parallel combination of two high-beam headlights is $3/2 = 1.5\ \Omega$. The parallel combination of the four "run" lights is $20/4 = 5\ \Omega$; thus the resistance the lights present to the battery is $1.5\ \Omega || 5\ \Omega = 1.5 \cdot 5/(1.5 + 5) = 1.2\ \Omega$, where the symbol "||" means "in parallel with." By the voltage divider relation (4.13b), the *terminal voltage* is $V_T = 12.5\ \text{V} \cdot 1.2\ \Omega/(1.2\ \Omega + 0.05) = 12.0\ \text{V}$.

(**b**) During the time that the blinkers are on, we now have $6\ \Omega || 6\ \Omega = 3\ \Omega$ in parallel with the $1.2\ \Omega$ above, which comes out to $3 || 1.2 = 0.9\ \Omega$. Now the terminal voltage is $V_T = 12.5 \cdot 0.9/(0.9 + 0.05) = 11.8$ V. Actually, the blinkers are on only roughly half the time, so we would expect to measure with a dc voltmeter the average between 12.0 V (no blinkers) and 11.8 V (with blinkers). This is 11.9 V, which is precisely what has been measured with a dc voltmeter, with all other parameters measured as having the values above. Given that such a heavy load resulted in only a loss of 0.6 V in terminal voltage, we see that the auto battery is an amazingly robust voltage source.

Just as the model of a practical voltage source is an ideal one in series with an internal resistance $R_S$, the model of a practical current source is a current source in parallel with $R_S'$. Having the source resistance in series with the ideal voltage source modeled the fact that for less-than-infinite load resistances $R_L$, the full open-circuit voltage is not available to the load. Similarly, for greater-than-zero $R_L$, we find that the full source current $i_S$ of a current source is not available to the load. Having $R_S'$ in parallel with the current source "draws off" part of $i_S$ via a current divider, just as $R_S$ reduces the available voltage from a battery via a voltage divider.

The model for a practical voltage source was shown in Fig. 4.33c, which is just Fig. 4.27 with $R_1 = R_S$ and $R_2 = R_L$. Similarly, the practical current source is obtainable from Fig. 4.28 with $R_1 = R_S'$ and $R_2 = R_L$, and is shown in Fig. 4.33d. By the current divider relation (4.15b), the load current $i_L$ is $i_L = i_S R_S'/(R_S' + R_L)$, which can be rewritten as $i_L = i_S/(1 + x)$, where now $x = R_L/R_S'$. The load current $i_L$ versus $R_L/R_S'$ is shown in Fig. 4.35.

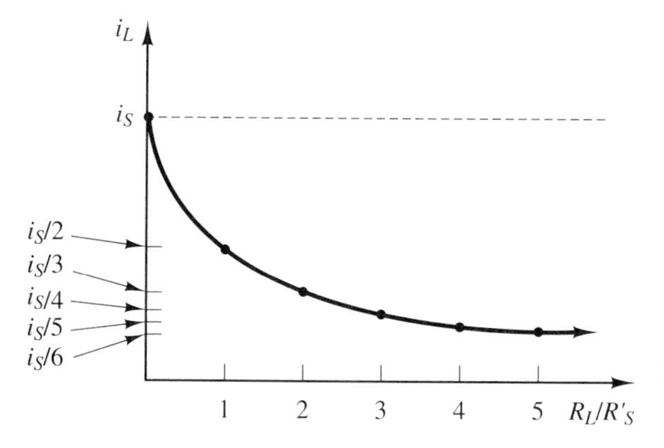


**Figure 4.35** Load current $i_L$ versus $R_L/R_s$ for a practical current source of voltage $v_s$ and internal resistance $R_s'$. Dashed curve is for an ideal source.

Although one cannot go out and buy a current source "battery," these do exist in electronic circuits. In addition, we shall now find that a practical voltage source can be transformed into an equivalent practical currents source, or vice versa. Doing this can be convenient in various circuit analysis situations.

This "source transformation" idea is a result of the fact that a practical voltage source such as that shown in Fig. 4.33c is equivalent to a practical current source as shown in Fig. 4.33d *provided* that $i_S = v_S/R_S$ and $R_S' = R_S$. They are equivalent *from the point of view of* $R_L$ if when any load $R_L$ is connected to either practical source, the same voltage appears across $R_L$ in both cases (and by Ohm's law, the same current flows within it).

The proof of this assertion is simple. Suppose that the same load $R_L$ is placed across both sources. What is the voltage across $R_L$ in each case? For the voltage source we use the voltage divider equation (4.13b) to obtain

$$v_L = \frac{v_S R_L}{R_S + R_L}, \tag{4.17}$$

where $v_L$ is the voltage across the load $R_L$, as shown in Fig. 4.33c. The voltage $v_L'$ in Fig. 4.33d is $i_L R_L$, so that

$$v_L' = \frac{i_S R_S' R_L}{R_S' + R_L}. \tag{4.18}$$

We require that $v_L' = v_L$ for all $R_L$.

There are two unknowns: $R_S'$ and $i_S$, which can be determined by two equations. These equations may be obtained by setting equal the right-hand sides of Eqs. (4.17) and (4.18) for any two values of $R_L$, $R_{L,1}$, and $R_{L,2}$:

$$\frac{v_S}{(R_S + R_{L,1})} = \frac{i_S R_S'}{R_S' + R_{L,1}}$$

and

$$\frac{v_S}{R_S + R_{L,2}} = \frac{i_S R_S'}{R_S' + R_{L,2}}.$$

Two simple values to use are $R_{L,1} = 0$ and general $R_{L,2}$. For $R_{L,1} = 0$ we obtain $i_S = v_S/R_S$. Substituting this into the $R_{L,2}$ equation gives, after dividing both sides by $v_S/(R_S + R_{L,2})$,

$$1 = \frac{R_S'(R_S + R_{L,2})}{R_S(R_S' + R_{L,2})} = \frac{1 + R_{L,2}/R_S}{1 + R_{L,2}/R_S'},$$

which gives $R_S' = R_S$.

Consequently, if in the analysis of a circuit we find that a voltage source $v_S$ appearing in the schematic diagram is inconvenient whereas a current source would be convenient, we merely replace the [voltage source, source resistance] series combination with the equivalent [current source, source resistance] parallel combination. The current source value is just $i_S = v_S/R_S$, and the source resistances are equal. Similarly, to convert an unwanted current source $i_S$ into a voltage source, do the reverse, letting $v_S = i_S R_S$. The results are summarized in Fig. 4.36.

The source transformation comes in handy quite frequently in the design and analysis of electric and electronic circuits. Note, however, that only if a source resistance appears explicitly in series with the voltage source to be transformed (or in parallel with the current source to be transformed) can the source transformation be performed.

**Example 4.21**

**(a)** In the circuit of Fig. 4.37a, convert the current source to a voltage source.

**(b)** Find the current $i_4$ through the 4-Ω resistor.

**Solution**

**(a)** The first step is to recognize that we indeed have a current source in parallel with a source resistance, namely 4 Ω. This is made evident by redrawing Fig. 4.37a in the form of Fig. 4.37b. Notice that no potential or current has changed as a result of this redrawing; the top and bottom nodes still are attached to the same elements in the same way. Finally, use of the source transformation idea yields Fig. 4.37c, where the equivalent voltage is 3 A · 4 Ω = 12 V. Notice the polarity, which must be drawn to be consistent with those in Fig. 4.36. We shall simplify this circuit further in Sec. 4.11.

**(b)** Although the source resistance of the current source in Fig. 4.37c obviously has the same resistance as the 4-Ω resistor in Fig. 4.37a, it does *not* in general carry the same current. Thus, to solve for the current $i_4$ in Fig. 4.37a, we *must not* source-transform away the 4 Ω-resistor. We may, however, source-transform the voltage source, as that will not affect $i_4$ but will simplify the solution. The result is shown in Fig. 4.37d. Finally, recall that just as voltage sources in series add (by KVL), current sources in parallel add (by KCL). Consequently, Fig. 4.37d may be redrawn as in Fig. 4.37e. Note the polarity of the 2.5-A source: It opposes $i_4$, so $i_4$ will be negative. We now have a simple current divider, so $i_4 = -2.5\text{ A} \cdot 2/(2 + 4) = -0.833$ A. *Lesson: Never eliminate the variable whose value you wish to find!*

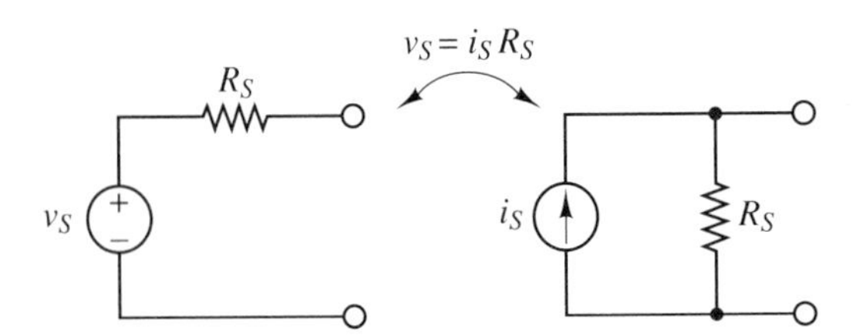


**Figure 4.36** Source transformation.

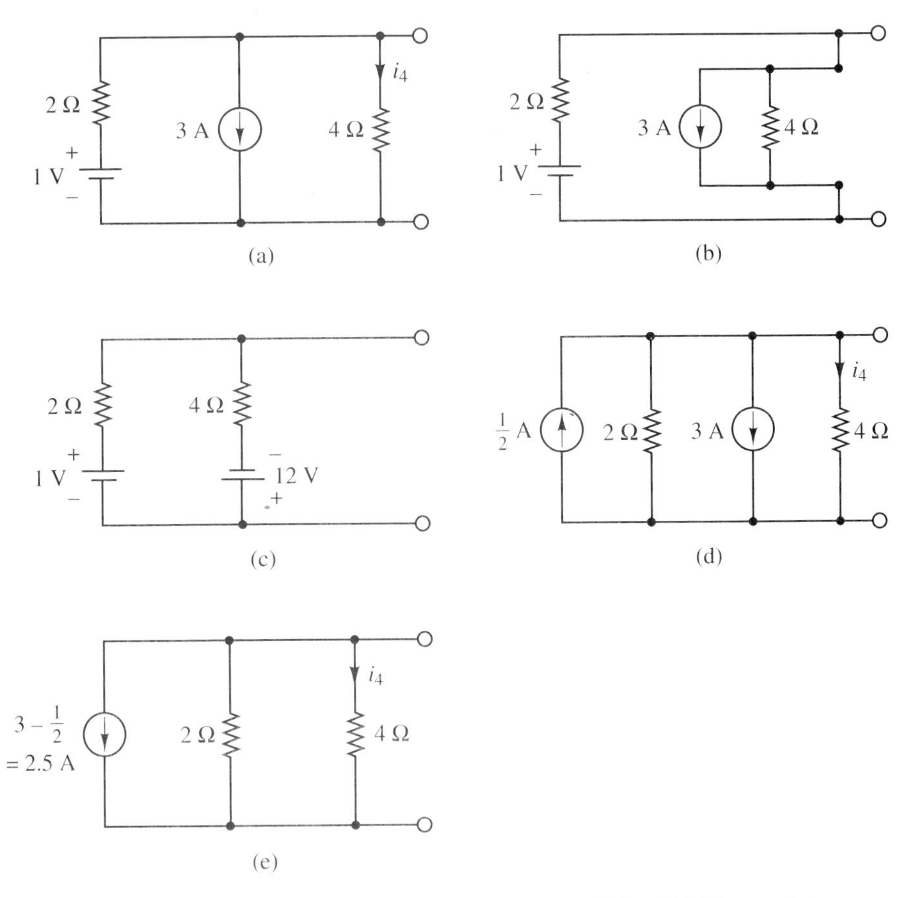


**Figure 4.37** (a) Circuit of Example 4.21; (b) redrawing of (a) to highlight practical current source; (c) after source transformation; (d) alternate analysis retaining original 4-Ω resistor; (e) simplification of (d).

**Example 4.22**

Find the current source equivalent to the automobile battery in Example 4.20, which had $R_S = 50\ \text{m}\Omega$ and $v_S = V_B = 12.6$ V.

**Solution** Quite simply, $i_S = I_S = 12.6/0.05 = 252$ A and $R_S = 50\ \text{m}\Omega$; a hefty current source. Note, however, that because $R_S$ is so small, it draws most of the current when the source is driving loads having resistances even as small as those of headlights or other high-current, low-resistance loads. In this sense, the current is wasted on $R_S$, so the car battery is not considered a good current source. However, it must be stressed that the characteristics of automobile batteries are dynamic; they change with environmental and loading conditions. For example, $R_S$ can increase substantially in cold weather.

In general, a good voltage source (here, the car battery) is not a good current source, and a good current source is not a good voltage source. This is because a good voltage source has a small $R_S$ while a good current source has a large $R_S$, where "good" means that nearly all of the ideal voltage or current is available to a wide range of loads $R_L$.

Alternatively, we say that a voltage source drives high-impedance loads well, and a current source drives low-impedance loads well, because in both cases the actual load (terminal) voltage or current is nearly equal to the ideal value of the source. In fact, when a signal source such as a detecting transducer has a low internal resistance, it is considered to be (and is modeled by) a voltage source. If it has a large internal resistance, it is considered to be a current source.

## 4.10 NONLINEAR ELEMENTS AND THE LOAD LINE

In the automotive lighting example, Example 4.19, it was noted that the resistance of light bulbs changes with temperature and therefore with current strength. Ohm's law does not hold over the wide range of currents from "cold" to "hot" (lighted). For the sidemarker bulb, the $i$–$v$ characteristic was determined experimentally to be that in Fig. 4.38.

Although Ohm's law does not hold globally for the sidemarker, it does hold *locally*. That is, if we focus on a particular point on the curve and consider only *small deviations* from that point, the curve is practically linear. The slope is equal to that of the tangent to the curve at that point, as shown: $i \neq v/R$, but $di = dv/R(v)$, where now $R$ is a function of the applied voltage $v$. This generalization of Ohm's law holds for essentially any passive conducting device if the excursions from the *operating point* are small enough. The concept of a relation being highly nonlinear on a large scale but essentially linear on a small scale is central in electronic design.

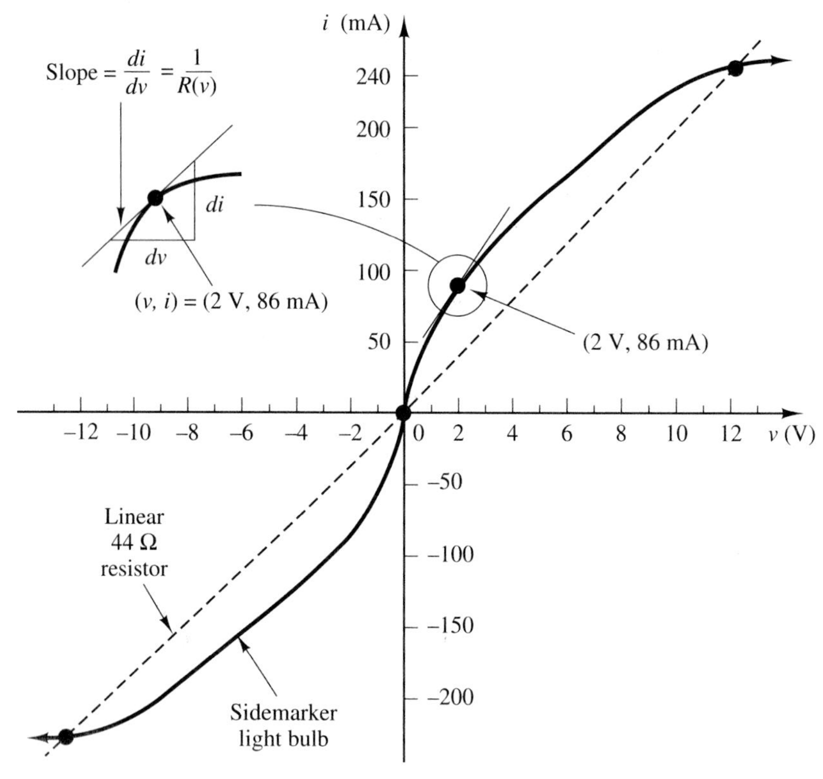


**Figure 4.38** Nonlinear $i$–$v$ characteristic of a sidemarker automobile light bulb.

The sidemarker was measured to have a resistance of 4.7 Ω cold and 44 Ω when lighted. The cold resistance was measured using a digital multimeter, an instrument used for measuring voltage, current, and resistance. The multimeter measures resistance by applying a very accurate, internally generated 1 mA current source to the resistance, and simultaneously measuring the voltage across the resistance (which equals that across the 1-mA current source placed in parallel with the resistor). At this very small applied current of 1 mA (light bulb cold), the slope of the *i–v* curve must be 1/(4.7 Ω), corresponding to a voltage of only 0.0047 V—too small to read from the curve in Fig. 4.38.

For an applied voltage of 1 V, the curve is already highly nonlinear because the light filament is already heating up; from the graph we find that the "resistance" is already 17 Ω. When we said that the resistance of the sidemarker was 44 Ω at 12 V, we meant that it behaved the same as the 44-Ω resistor *for that operating voltage only*. That operating point is the point of intersection of the light bulb *i–v* curve with the straight-line *i–v* characteristic of a linear 44-Ω resistor, which is also shown (dashed) in Fig. 4.38.

Because the two circuit elements (sidemarker and resistor) whose *i–v* characteristics are shown in Fig. 4.38 are both passive, if they alone were connected in parallel, there would be 0 V across them and 0 A through them. That is, there is nothing to counteract the dissipation in order to maintain a current. The operating point of such a "circuit" is just the origin, where incidentally, the two curves also intersect. But if one of the elements connected in parallel is active, there *can* be

The fascination with electric lighting was epitomized at the 1893 World's Fair in Chicago with the impressive arrays of electric lights. Note the Ferris wheel and in-flight balloon in background.

nonzero voltage and current. As we argued concerning Fig. 4.16, the voltage and current of two parallel-connected devices must both be the same.

If the resistor in Fig. 4.16 is replaced by the sidemarker light bulb and the voltage source is a 12-V car battery, we have the circuit shown in Fig. 4.39a, with $i$–$v$ characteristics in Fig. 4.39b. The intersection of the $i$–$v$ curves is the *operating point* of the circuit, because it is the point at which necessarily the voltage and current of both parallel-connected devices are identical. That is, by KVL they must have the same voltage because they are connected in parallel, and by KCL they must have the same current because there is only one current path. The two $i$–$v$ characteristics indicate the sets of possible operating points of each device; upon parallel connection of the elements, the operating points must match, which occurs at the intersection of the two $i$–$v$ curves.

The operating point for the circuit in Fig. 4.39a is (12 V, 0.24 A), where the current 0.24 A was noted in Example 4.19. The $i$–$v$ characteristic of the bulb is said to be a *load line* on the voltage source $i$–$v$ plot, because the bulb acts as a load on the source.

When a device is nonlinear and does not follow a simple $i$–$v$ curve, as is the case for the light bulb, the load line is the only way to predict the operating point

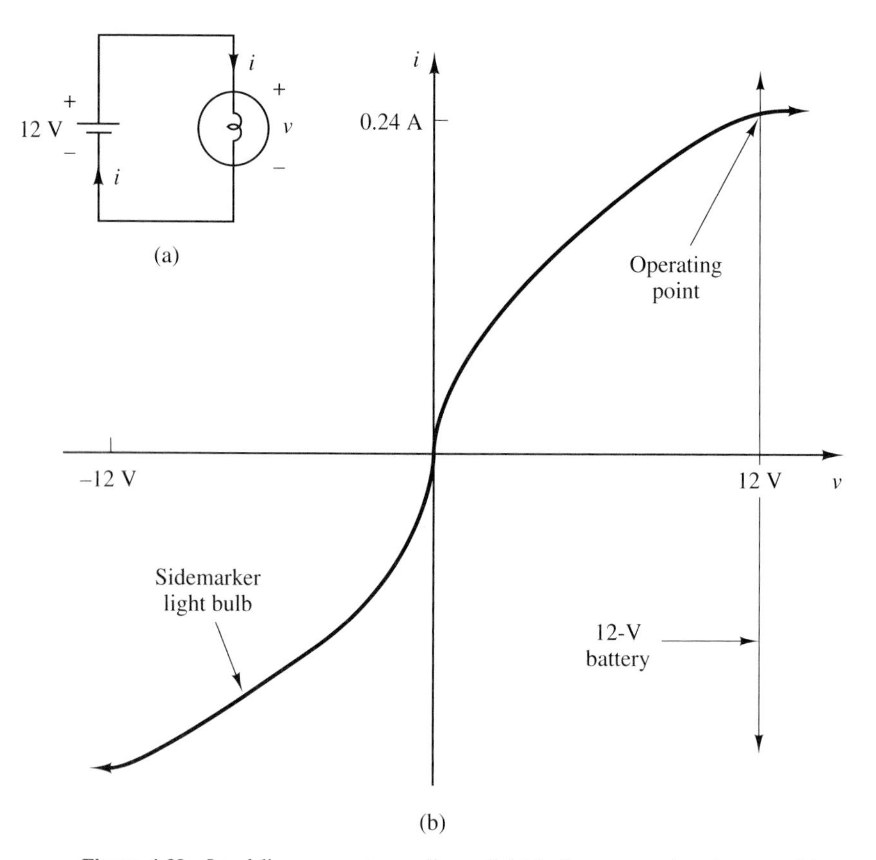


**Figure 4.39** Load line concept—nonlinear light bulb connected to battery: (a) circuit; (b) simultaneous $i$–$v$ characteristics showing operating point.

of a device in-circuit. Note that the source (active part of the circuit) need not be a voltage source; it could have any shaped $i$–$v$ curve. This sort of graphical analysis can be very useful in electronic circuit/device design and analysis, where many $i$–$v$ curves are highly nonlinear. It is a general circuit analysis concept valuable not only for its utility in analysis but also conceptually for being able visually to distinguish a load from a source. The load line concept is also used in Sec. 4.11.

## 4.11 EQUIVALENT ACTIVE COMPONENTS: THÉVENIN AND NORTON EQUIVALENTS

The load line concept just developed can be used to derive the parameters of a simple two-element network that can be made equivalent to any network involving only resistors and voltage and/or current sources. It works for both independent and dependent sources. An extension of the theorem that works for circuits, including capacitors and inductors, when the sources are sinusoidal is discussed in Chapter 5. This procedure is helpful when we encounter a complicated network whose terminal characteristics are all that is important to us, and not the individual currents and voltages of the elements within that network.

What is the simple two-element equivalent? A practical voltage source (ideal voltage source in series with a source resistance). Such an equivalent can be found for any pair of terminals in the network. This result is known as *Thévenin's theorem,* after M. L. Thévenin of France, who proposed the idea in a paper published in 1888. The equivalent network is known as the *Thévenin equivalent.*

The bottom-line assumption of this theorem is that the network whose equivalent we seek is "linear." This means that any current or voltage within the network is expressible as a sum of *constant multiples* of the sources. Therefore, in general, the currents and voltages are linear combinations of each other, with constant additive terms also possible. In particular, the voltage $v$ and current $i$ at any pair of terminals are related by an $i$–$v$ characteristic having the linear form $v = v_{eq} - iR_{eq}$, where $v_{eq}$ and $R_{eq}$ are constants with respect to $i$ (not necessarily with respect to time).

We may proceed to determine the parameters of the model, $v_{eq}$ and $R_{eq}$, as follows. The circuit diagram of the equivalent model with a load connected is shown in Fig. 4.40a, and its $i$–$v$ characteristic (with $v_{eq} > 0$) is shown in Fig. 4.40b. By KVL, and by setting $v = v_L$ and $i = i_L$,

$$v_L = v_{eq} - i_L R_{eq}. \tag{4.19}$$

Thus if $i_L = 0$, then $v_L = v_{oc} = v_{eq}$; this is the open-circuit (no load) condition, which, recalling Ohm's law, occurs for $R_L = \infty$. That is, if we can measure or calculate the open-circuit voltage of the original network $v_{oc}$, we have found one of the two model parameters of the original network, $v_{eq} = v_{oc}$.

We can rearrange Eq. (4.19) to give

$$i_L = \frac{v_{eq} - v_L}{R_{eq}}. \tag{4.20}$$

Thus if $v_L = 0$, then $i_L = i_{sc} = v_{eq}/R_{eq} = v_{oc}/R_{eq}$; this is the short-circuit (infinite load) condition, occurring for $R_L = 0$. Therefore if we connect a wire across the terminals of the original network and measure or calculate the current in that wire,

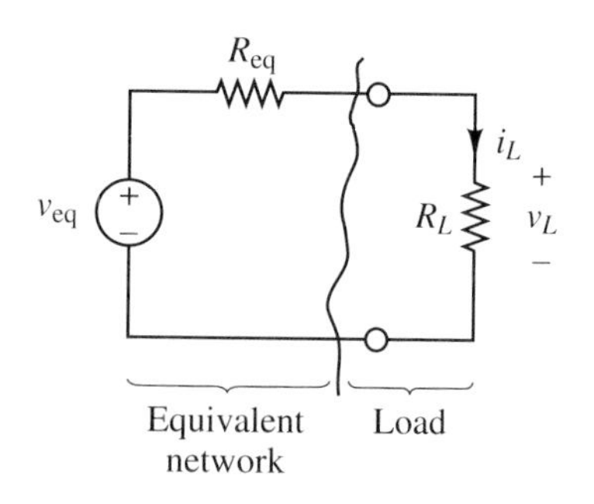


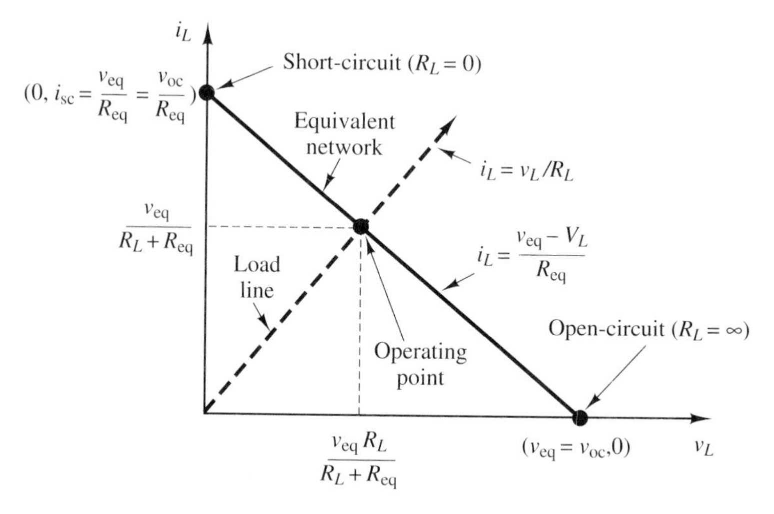


**Figure 4.40** (a) Thévenin equivalent circuit, with load; (b) load line showing ways of determining equivalent circuit parameters.

$i_{sc}$, we then obtain another equation involving the other model parameter of the equivalent network, $R_{eq}$. Consequently, we may in principle then solve for $R_{eq}$ using

$$R_{eq} = \frac{v_{oc}}{i_{sc}}. \tag{4.21}$$

Furthermore, these extreme points determine the intercepts of the $i$–$v$ curve of the practical voltage source, as shown in Fig. 4.40b. Compare this with the $i$–$v$ curve of an ideal voltage source of voltage $v_{eq}$, a vertical line passing through $(v_{eq}, 0)$; the $i$-intercept, $v_{eq}/R_{eq}$, goes up to infinity (vertical line) if $R_{eq} = 0$ (ideal source).

The $i$–$v$ characteristic of a load having a particular value of $R_L$ (between 0 and $\infty$) is superimposed on that of the practical voltage source $i$–$v$ curve in Fig. 4.40b; the former is the load line, with slope $1/R_L$. By varying $R_L$, the operating point (intersection of the two curves) ranges from the short-circuit operating point ($i_L$ axis) to the open-circuit operating point ($v_L$ axis). By choosing any two values of $R_L$, the $i$–$v$ characteristic of the practical source is determined completely. That is, $v_{eq}$ and $R_{eq}$ can be determined.

It is always safe to let $R_L = \infty$ be one of the load values for determination of $v_{eq}$ and $R_{eq}$, for this represents no load. The measured terminal voltage is then $v_{eq} = v_{oc}$. Also, when just analyzing a circuit on paper, we may set $R_L = 0$, find $i_{sc}$ by circuit analysis, obtain $R_{eq} = v_{oc}/i_{sc}$, and thus have the model completely specified.

In measurement situations, however, it is better not to short-circuit a voltage source. That is, experimentally it may be unsafe or unreliable to try to obtain $R_{eq}$ using $R_L = 0$ (via $i_{sc}$). Instead, to determine $R_{eq}$, a load value $R_L = R_{L,1}$ typically to be encountered in actual operation could be connected, and the terminal voltage $v_{L,1}$ measured. By Ohm's law, Eq. (4.19) with $v_L = v_{L,1}$ and $R_L = R_{L,1}$ can be rewritten

$$v_{L,1} = v_{eq} - \frac{v_{L,1} R_{eq}}{R_{L,1}}, \tag{4.22}$$

which after substitution of $v_{eq} = v_{oc}$ is solved for $R_{eq}$ to give

$$R_{eq} = \left(\frac{v_{oc}}{v_{L,1}} - 1\right) R_{L,1}. \tag{4.23}$$

This is actually how the internal resistance of the auto battery of 50 m$\Omega$ used in Example 4.20 was originally obtained. The open-circuit voltage was measured, yielding $v_{eq} = v_{oc} = 12.5$ V, and the terminal voltage at known $R_{L,1} = 1.2\ \Omega$ (load) was measured ($v_{L,1} = 12.0$ V) to determine $R_{eq}$ by Eq. (4.23). [Note that Eq. (4.23) is the identical formula to the one we obtained for $R_2$ in Example 4.17.]

Notice that by a source transformation, we may alternatively represent the network by a practical current source (an ideal current source in parallel with a source resistance). The source resistance is $R_{eq}$ and the current source value must be $v_{eq}/R_{eq} = i_{sc}$. This model is called the *Norton equivalent,* after the American engineer Edward L. Norton, who realized this in the 1920s. The form of the Thévenin equivalent circuit that is valid whenever independent sources are present is summarized in Fig. 4.41. The Thévenin and Norton relationship is summarized in Fig. 4.42.

It should be noted that if there are no independent sources in the network whose equivalent we seek, we will find that $v_{oc} = 0$ and $i_{sc} = 0$, making $R_{eq}$ indeterminate in Eq. (4.21). This is because there is no driving force to make the dependent source function. With no independent sources, the network thus appears to the outside world as a passive device—only a resistance. The only way a terminal voltage could be generated is for an external source to supply it. For example, a test current source of, say, 1 A (for convenience on paper; 1 mA on the lab bench) may be

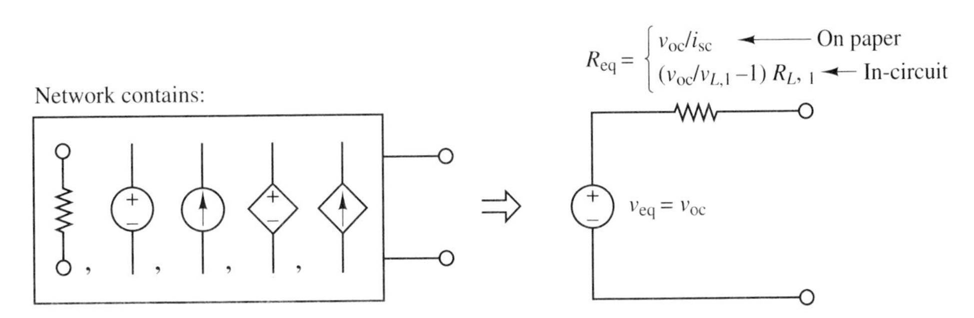


**Figure 4.41** Thévenin's theorem for circuits with independent sources.

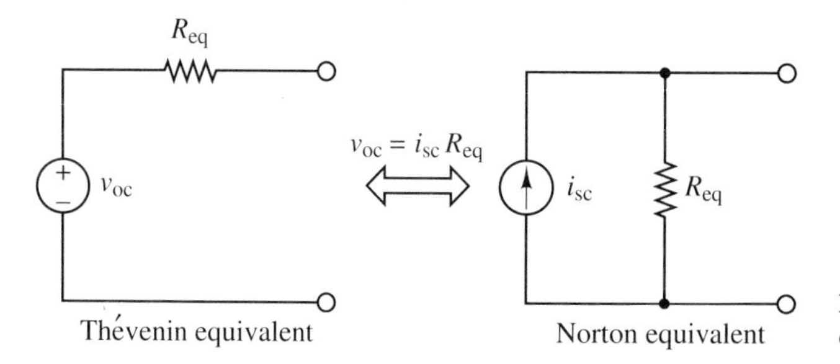


**Figure 4.42** Thévenin and Norton equivalent circuits.

applied, and the resulting terminal voltage $v_t$ can be measured or calculated (see Fig. 4.41). Then, noting that $v_{eq} = 0$, Eq. (4.20) may be solved for $R_{eq}$ (or just use Ohm's law) to yield

$$R_{eq} = \frac{v_t}{1\text{ A}}, \tag{4.24}$$

where in this case the sign of $i_L$ is negative because the 1-A test current flows *into* the network; normally, it would flow out. Thus in this case the complete equivalent network is just a resistor of resistance $R_{eq}$. This situation is summarized in Fig. 4.43. As mentioned previously, this is precisely the technique a multimeter uses to measure resistance of a passive network (no independent sources, or all independent sources turned off).

**Example 4.23**

In Example 4.21 it was promised that the circuit of Fig. 4.37c would be further simplified in this section. Do so by finding Thévenin's equivalent circuit (see Fig. 4.44).

**Solution** All we need are the open-circuit voltage $v_{oc}$ and the short-circuit current $i_{sc}$; then $v_{eq} = v_{oc}$ and $R_{eq} = v_{oc}/i_{sc}$. In this example, the excitation is dc, so we will use uppercase letters. To find $V_{oc}$, refer to Fig. 4.44a. First note that $I_{2,oc} = I_{4,oc} = I$ because there is only one path for the current. The open-circuit voltage is found by KVL: For the rightmost loop,

$$12\text{ V} - I \cdot 4\ \Omega + V_{oc} = 0,$$

so $V_{oc} = 4 \cdot I - 12$ V. By KVL for the leftmost loop,

$$-1\text{ V} + I \cdot 2\ \Omega + I \cdot 4\ \Omega - 12\text{ V} = 0,$$

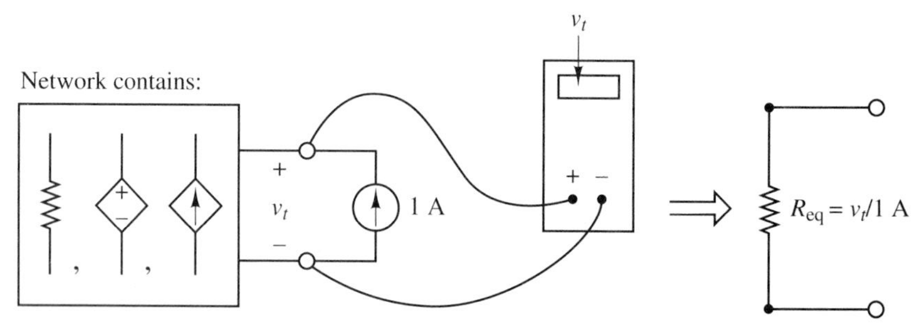


**Figure 4.43** Thévenin's theorem for networks containing only dependent sources (and resistors).

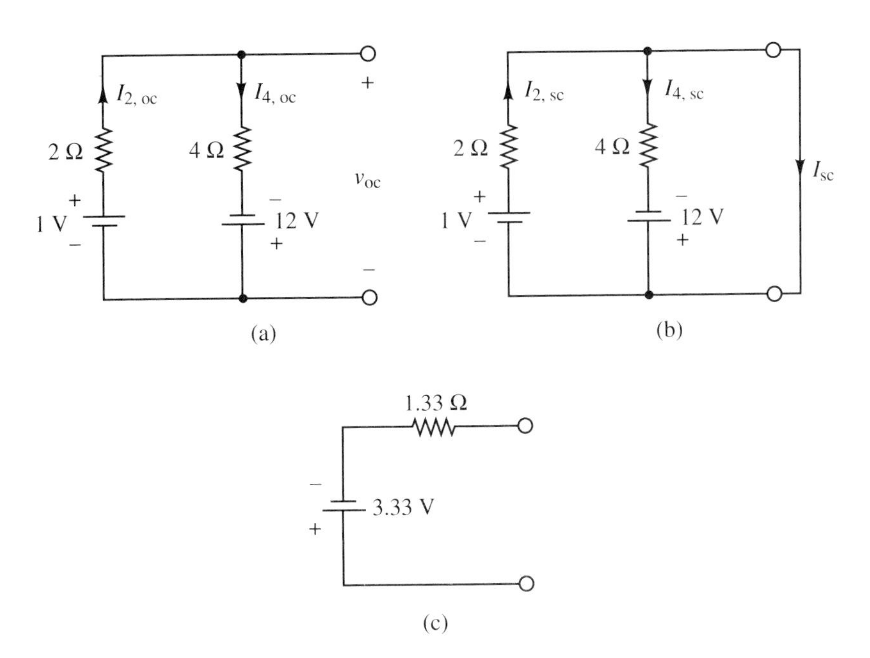


**Figure 4.44** Further simplification of circuit from Example 4.21: (a) determination of open-circuit voltage; (b) determination of short-circuit current; (c) resulting Thévenin equivalent.

or $I = 13/6 = 2.17$ A, so $V_{\text{eq}} = V_{\text{oc}} = 4 \cdot 2.17 - 12 = -3.33$ V. The short-circuit current may be found by drawing Fig. 4.44b. (*Note:* Do *not* use any values of current or voltage found for the open-circuited condition in the short-circuit-condition circuit.) By KCL at either the top or bottom node, $I_{\text{sc}} = I_{2,\,\text{sc}} - I_{4,\,\text{sc}}$, so we need $I_{2,\,\text{sc}}$ and $I_{4,\,\text{sc}}$. Be sure to notice that $I_{4,\,\text{sc}} \neq I_{4,\,\text{oc}}$ and $I_{2,\,\text{sc}} \neq I_{2,\,\text{oc}}$! This is because the circuit connections are different; in addition, $I_{2,\,\text{sc}} \neq I_{4,\,\text{sc}}$ because of the short circuit. Now KVL for the rightmost loop is

$$12\text{ V} - I_{4,\,\text{oc}} \cdot 4\ \Omega = 0,$$

or $I_{4,\,\text{oc}} = 12/4 = 3$ A. Also, from the outer loop, KVL gives

$$-1\text{ V} + I_{2,\,\text{sc}} \cdot 2\ \Omega = 0,$$

so $I_{2,\,\text{sc}} = 0.5$ A. Now from KCL, $I_{\text{sc}} = 0.5 - 3 = -2.5$ A. Finally, $R_{\text{eq}} = V_{\text{oc}}/I_{\text{sc}} = -3.33/-2.5 = 1.33\ \Omega$. Thus the Thévenin equivalent circuit is as shown in Fig. 4.44c.

It should be mentioned that there is a shortcut for finding $R_{\text{eq}}$ when the network consists only of independent sources and resistors, as in Example 4.23. Recall that for any resistance–source combination, the equivalent $i$–$v$ characteristic is given by Eq. (4.19). If $v_{\text{eq}} = 0$, then $v_L = i_L R_{\text{eq}}$, and the equivalent circuit reduces to the pure resistance $R_{\text{eq}}$.

We can guarantee $v_{\text{eq}} = 0$ by turning off all independent voltage and current sources. Turning off a voltage source is equivalent to replacing it by a short circuit (wire) because a 0-V voltage source is indistinguishable from a short circuit (see Fig. 4.7b). Similarly, turning off a current source is equivalent to replacing it by an open circuit because a 0-A current source is indistinguishable from an open circuit.

If all of the sources are independent, then $R_{eq}$ is just the series–parallel combination of the resistive network left over when all the sources are turned off. In Example 4.23, both of the sources are independent voltage sources, to be replaced by short circuits. Doing so leaves us with the 2-Ω and 4-Ω resistors in parallel, yielding $R_{eq} = 2 \cdot 4/(2 + 4) = 1.33\ \Omega$, as found above.

**Example 4.24**

Find the Thévenin equivalent circuit of the network shown in Fig. 4.45.

**Solution** First we find $V_{oc}$, the output voltage under open-circuit conditions, that is, with no load connected to the output terminals. In Fig. 4.45a, we let the current in the 2-Ω resistor be $I_{2,oc}$, where "oc" again means "open circuit." Under no load, the current in the 1-Ω resistor is $2\text{ A} + I_{2,oc}$. This allows us to write KVL around the 1-V, 1-Ω, and 2-Ω loop:

$$-1\text{ V} + (2\text{ A} + I_{2,oc}) \cdot 1\ \Omega + I_{2,oc} \cdot 2\ \Omega = 0,$$

which yields $I_{2,oc} = -\frac{1}{3}$ A. This gives $V_{eq} = V_{oc} = (-\frac{1}{3}) \cdot 2 = -\frac{2}{3}$ V. To find $R_{eq}$, we turn off both the independent sources and redraw the circuit as shown in Fig. 4.45b. We see that $R_{eq} = 3\ \Omega + 1\ \Omega || 2\ \Omega \approx 3.67\ \Omega$.

The assertion that with the independent sources off, the equivalent network reduces to a resistor implies that if we applied a test current of value $i_t$, the measured terminal voltage $v_t$ would be $v_t = i_t R_{eq}$. Recalling this fact indicates why one may not "turn off" dependent sources: Even if all the independent sources are turned off, the dependent sources will still respond to the test current $i_t$, thereby affecting the value of the terminal voltage $v_t = i_t R_{eq}$ and thus $R_{eq} = v_t/i_t$.

To understand *fully* what all this means requires knowledge of the modes of transistor modeling—large- versus small-signal models. The dependent source in a small-signal circuit diagram may tacitly rely on a power source that is not shown in

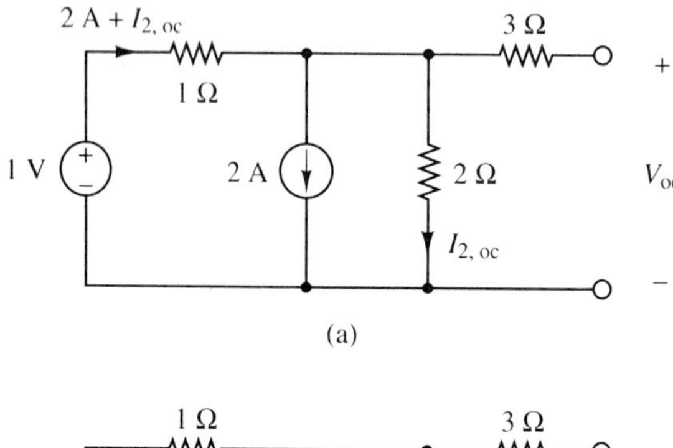


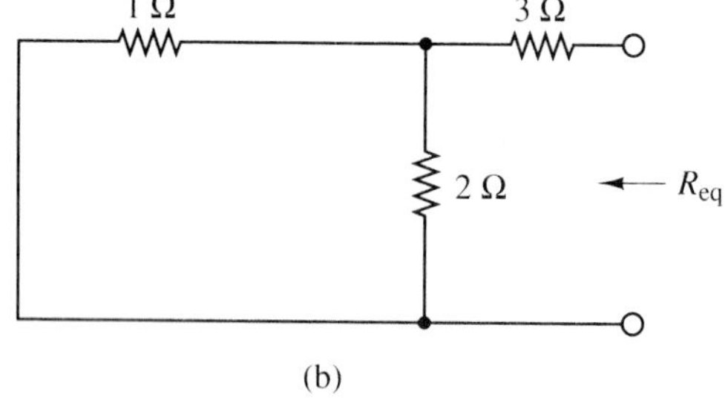


**Figure 4.45** (a) Circuit for Example 4.24; (b) same circuit with both independent sources turned off for calculation of $R_{eq}$.

the diagram; see Chapter 7 for details. The result is, however, simple: No shortcut for dependent sources. If there exist both independent and dependent sources, we may use Eq. (4.21) to find $R_{eq}$ by first solving for $v_{oc}$ and $i_{sc}$. If all the sources are dependent, we must apply $i_t$ and calculate or measure $v_t$ to obtain $R_{eq} = v_t/i_t$.

### 4.11.1 Condition for Maximum Power Transfer

A handy little spin-off from Thévenin's theorem is the following corollary: Maximum power is transferred from a source network to a load if the load resistance is equal to the Thévenin equivalent resistance $R_{eq}$. We can make so bold a statement as this because anything that is true about the Thévenin equivalent and a load (Fig. 4.40) is also true about any network having that same equivalent which is driving the same load. Thus all we need to show is that $R_L = R_{eq}$ is the condition for $v_{eq}$ in Fig. 4.40 to deliver maximum power to $R_L$.

In Fig. 4.40, $v_L = v_{eq}R_L/(R_L + R_{eq})$ and $i_L = v_{eq}/(R_L + R_{eq})$, so that $p_L = v_L i_L = v_{eq}^2 R_L/(R_L + R_{eq})^2$. To maximize $p_L$, set the derivative with respect to $R_L$ equal to zero:

$$\frac{dp_L}{dR_L} = \frac{v_{eq}^2[(R_L + R_{eq})^2 - R_L \cdot 2(R_L + R_{eq})]}{(R_L + R_{eq})^4}$$

$$= \frac{v_{eq}^2(R_{eq}^2 - R_L^2)}{(R_{eq}^2 + R_L^2)^4} = 0,$$

or

$$R_L = R_{eq}$$

for maximum source power transferred to the load. Looking at the expression for $p_L$ above, this makes sense. For $R_L \ll R_{eq}$, $p_L$ is roughly proportional to $R_L$, because the denominator of $p_L$ is approximately $R_{eq}^2$ (thus for very small $R_L$, $p_L$ increases with $R_L$). For $R_L \gg R_{eq}$, $p_L$ depends on $R_L$ roughly according to $1/R_L$ (thus for very large $R_L$, $p_L$ decreases with $R_L$). Incidentally, substitution of $R_L = R_{eq}$ into the expression for $p_L$ gives $p_{L,\max} = v_{eq}^2/(4R_{eq})$.

Note that this analysis assumes that $R_{eq}$ is fixed, and we seek to choose the best possible $R_L$. If, instead, we could vary $R_{eq}$, we would choose it to be zero for maximum power to $R_L$, because $R_{eq} = 0$ minimizes the denominator of $p_L$. Physically, making $R_{eq}$ small minimizes the power dissipated in $R_{eq}$.

Setting $R_L = R_{eq}$ is called *matching* the load to the source. However, in most practical signal processing applications, it is not until the final stage of amplification in which a high-power-absorbing load must be driven that load–source matching is desirable. In earlier signal processing stages, issues such as noise immunity and distortion minimization take precedence over maximum power transfer. Consequently, usually people design circuits to have very high "input impedance," that is, circuits that from the point of view of the source are effectively a very large $R_L$ value. This is done to avoid loading down the source and thus maximize $v_L$, as opposed to setting $R_L = R_{eq}$, where $R_{eq}$ is the equivalent source resistance.

Also, one should not confuse power transfer with power gain; maximum power gain of an amplifier may occur for low "power transfer to the load." We return to these issues in Chapter 7.

**Example 4.25**

One case in which maximum power transfer *is* desired is in the final amplification stage of an audio amplifier, where a speaker must be driven. A 3-Ω speaker is connected to a tone generator having an output resistance of 600 Ω. Assume that the frequency of the signal generator is such that we may consider the speaker to be well modeled as pure resistance, which for certain low-frequency tone ranges is a realistic assumption. Compare the power that would be delivered to a matched load with that actually delivered to the 3-Ω speaker. If the open-circuit voltage amplitude of the signal generator output is 5 V, what is the terminal voltage amplitude when the speaker is connected?

**Solution** We effectively have been told that $R_{eq} = 600\ \Omega$ for the signal generator; thus the value of $R_L$ yielding maximum power to $R_L$ would be 600 Ω—far from the 3 Ω of our speaker. The ratio of $p_L$ at any $R_L$ to $p_{L,\max}$ is $[v_{eq}^2 R_L/(R_L + R_{eq})^2]/[v_{eq}^2/(4R_{eq})] = 4R_L R_{eq}/(R_L + R_{eq})^2$. Substituting in the given values gives only $4 \cdot 3 \cdot 600/(603)^2 \approx 2\%$ of the power that would be transferred to the load were it a 600-Ω speaker instead of a 3-Ω speaker—very poor matching! If $v_{eq} = v_{oc} = 5$ V, then with the 3-Ω speaker connected, $v_L = v_{eq}R_L/(R_L + R_{eq}) = 5 \cdot 3/603 = 0.025$ V (which in fact was measured at the frequencies in the vicinity of 400 Hz for this speaker). For a 600-Ω speaker $v_L$ would be $5/2 = 2.5$ V—100 times that for this poor matching situation.

We shall see in Sec. 6.5 that a transformer can be used to "convert" the 3-Ω load to a 600-Ω load, thereby transferring maximum power to the 3-Ω resistor; old radios usually had such a transformer just before the speaker to provide such matching.

## 4.12 NETWORK THEOREMS

### 4.12.1 Introduction

At this point we have all the tools required for carrying out the analysis of resistive circuits, with dependent and/or independent voltage and/or current sources. It is, however, occasionally helpful to have a systematic method of circuit solution, especially if the number of variables becomes large. In such cases, computer aided design packages such as SPICE (Simulation Program with Integrated Circuit Emphasis) can be used to relieve us from algebraic nightmares and the inevitable resulting errors.

Just in case SPICE is not available to you or you do not have time to learn it, the methods of this section are sufficiently general to handle as much complexity as you are likely to encounter in the realm of linear (nonelectronic) electric circuits. Actually, for most circuits, a few KCL node equations, a few KVL loop equations, and an ounce of common sense are sufficient, without resorting to the matrix methods of this section.

Nevertheless, these methods are sometimes of use; they produce a matrix equation for the given circuit, either for a set of node voltages or a set of mesh currents. Clearly, a computer routine that solves linear equations would be essential for large circuits. For just a few variables, however, Cramer's method is convenient with only a hand calculator. Thus we now state it:

**Cramer's method.** Suppose that we have a set of linear equations, for the sake of specificity, for a set of node voltages $v_i$. Assume that these linear equations are of the form

$$\sum_n G_{m,n} v_n = i_m, \tag{4.25a}$$

or in matrix form,

$$\boldsymbol{G}\mathbf{v} = \mathbf{i}, \tag{4.25b}$$

where the components of $\boldsymbol{G}$ are $G_{m,n}$, those of $\mathbf{v}$ are $v_n$, and those of $\mathbf{i}$ are $i_m$. In Eq. (4.25a), the $i_m$ and $G_{m,n}$ are known values. The $G_{m,n}$ have units $\Omega^{-1}$ and thus are conductances, while $i_m$ have units of A. The $G_{m,n}$ are constant coefficients; $i_m$ could depend on time, in which case there would be a different solution of Eq. (4.25) for each instant of time. We shall restrict our attention to constant-in-time $\mathbf{i}$ and $\mathbf{v}$. Cramer's method, the proof of which can be found in any linear algebra textbook, says that the solution of Eq. (4.25b) is

$$v_m = \frac{|\boldsymbol{G}_m|}{|\boldsymbol{G}|}, \tag{4.26}$$

where the matrix $\boldsymbol{G}_m$ is the same as $\boldsymbol{G}$, except that the $m$th column is replaced by $\mathbf{i}$ [the right-hand side of Eq. (4.25b)].

### 4.12.2 Nodal Analysis

The goal of nodal analysis and mesh analysis in Sec. 4.12.3 is to generate a matrix equation of the form in Eq. (4.25). This first method, nodal analysis, is actually used by the SPICE package. The steps are simple, where "independent node" means involving at least one element not in the other node equations:

1. Assign node voltage variables $v_m$ to each node $m$, except for one designated a 0-V reference. For convenience, choose the reference to be a node connecting many branches.
2. Convert any practical voltage sources whose voltages or currents you do not care to find (if any) into practical current sources by source transformation. Ideal voltage sources naturally cannot be converted. Around all remaining voltage sources and the nodes they connect, draw dashed circles called *supernodes.*
3. Write KCL for all independent nodes except those within dashed circles, using Ohm's law whenever needed. Relate pairs of nodes within dashed circles by the voltage source connecting them.
4. Collect terms and put in the form of Eq. (4.25); solve. Use Ohm's law, KVL and/or KCL, to obtain the desired currents or voltages originally sought.

Independent and dependent sources are handled identically. The method is illustrated with the following example.

**Example 4.26**

In the circuit of Fig. 4.46, determine $V_1$ and $I_5$ using the nodal analysis method.

**Solution**

*Step 1.* This step is already performed in Fig. 4.46a. Note that $V_1$ is both the voltage across the 1-$\Omega$ resistor and the node voltage for the node between the 1-$\Omega$

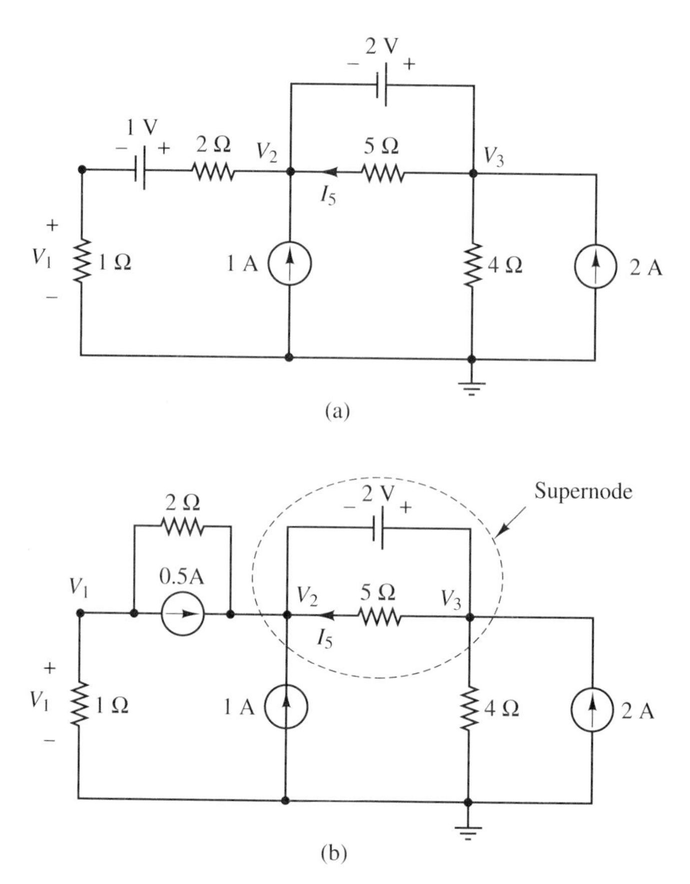


**Figure 4.46** (a) Circuit for Example 4.26, with node voltages shown; (b) circuit (a) redrawn with supernode and after source transformation.

resistor and 1-V battery. Also, a node need not be drawn between the 1-V battery and the 2-Ω resistor because it will not generate an independent equation. In fact, in step 2 we convert the 1 V–2 Ω series combination to a practical current source. In nodal analysis, we always express all contributions to KCL in terms of the node voltages, not the previously defined voltages across individual elements.

*Step 2.* The current source just mentioned is 1 V/2 Ω = 0.5 A. The 2-V source has been modeled as "ideal" (no internal resistance shown) and thus cannot be transformed. We thus encircle it, as well as the 5-Ω resistor, with a supernode dashed circle. Because the current through the 5-Ω resistor would be added and subtracted when writing KCL for the supernode if it were not included in the supernode (thus producing no effect), we include it in the supernode and reduce the required algebra. These results are shown in Fig. 4.46b.

*Step 3.* We sum the currents leaving each node to zero, so that for all Ohm's law terms, the node voltage of the node for which we are writing KVL is *positive* [i.e., is $V_a$ rather than $V_b$ in $(V_a - V_b)/R$]. For the node having node voltage $V_1$, KCL reads

$$\frac{V_1 - 0}{1\ \Omega} + 0.5\ \text{A} + \frac{V_1 - V_2}{2\ \Omega} = 0,$$

or

$$1.5V_1 - 0.5V_2 = -0.5 \text{ A},$$

or

$$3V_1 - V_2 = -1 \text{ A}.$$

For the supernode, having node voltages $V_2$ and $V_3$, KCL reads

$$-0.5 \text{ A} + \frac{V_2 - V_1}{2\ \Omega} - 1 \text{ A} + \frac{V_3 - 0}{4\ \Omega} - 2 \text{ A} = 0,$$

or

$$-0.5V_1 + 0.5V_2 + 0.25V_3 = 3.5 \text{ A},$$

or

$$2V_1 - 2V_2 - V_3 = -14 \text{ A}.$$

By KCL, $V_3 = V_2 + 2$ V or, formally dividing by 1 $\Omega$ to convert to currents, $V_3 = V_2 + 2$ V, so that division by 1 $\Omega$ to convert to current, $-V_2 + V_3 = 2$ A.

*Step 4.* Putting the results above into matrix form, we have

$$\begin{bmatrix} 3 & -1 & 0 \\ 2 & -2 & -1 \\ 0 & -1 & 1 \end{bmatrix} \begin{bmatrix} V_1 \\ V_2 \\ V_3 \end{bmatrix} = \begin{bmatrix} -1 \\ -14 \\ 2 \end{bmatrix}.$$

The determinant of $\boldsymbol{G}$ is $3 \cdot (-2 - 1) + 2 = -7$. Therefore,

$$V_1 = \frac{\begin{vmatrix} -1 & -1 & 0 \\ -14 & -2 & -1 \\ 2 & -1 & 1 \end{vmatrix}}{(-7)} = \frac{-(-2 - 1) + (-14 + 2)}{-7} = 1.286 \text{ V}$$

$$V_2 = \frac{\begin{vmatrix} 3 & -1 & 0 \\ 2 & -14 & -1 \\ 0 & 2 & 1 \end{vmatrix}}{(-7)} = \frac{3(-14 + 2) + 2}{-7} = 4.857 \text{ V}$$

$$V_3 = \frac{\begin{vmatrix} 3 & -1 & -1 \\ 2 & -2 & -14 \\ 0 & -1 & 2 \end{vmatrix}}{(-7)} = \frac{3(-4 - 14) + 4 - (-2)}{-7} = 6.857 \text{ V}.$$

Finally, $V_1 = 1.286$ V and $I_5 = (V_3 - V_2)/5\ \Omega = 0.4$ A.

### 4.12.3 Mesh Analysis

Mesh analysis is the dual of nodal analysis. Instead of defining and solving for node voltages, we define and solve for loop currents. Instead of obtaining an equation of the form Eq. (4.25), we obtain an equation of the form

$$\sum_n R_{m,n} i_n = v_m, \tag{4.27a}$$

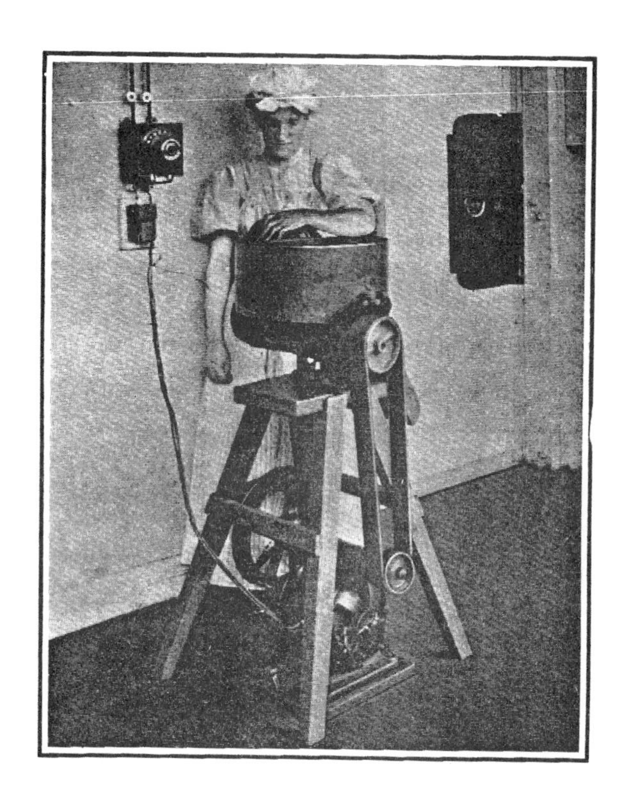

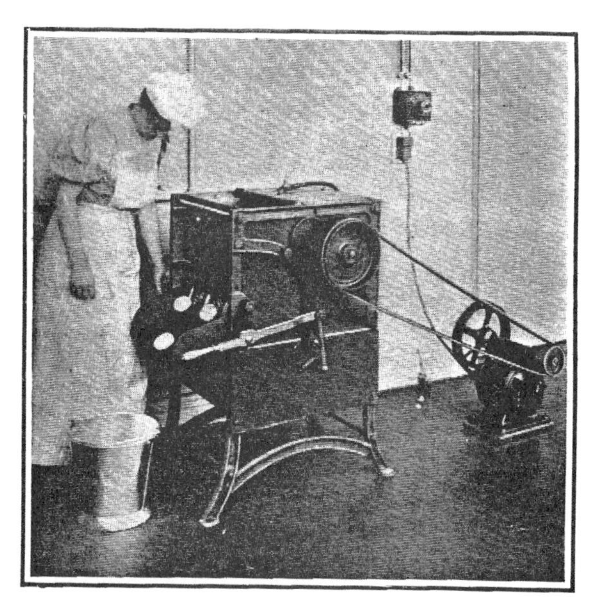

"Some novel uses of electricity," or in modern parlance, food processors. Above is an Electric Machine for Chopping Cabbage, and below is an Electric Potato-Paring Machine. Note the vitality of the latter machine, as evidenced by the dented pail. These 1906 models are no longer on the market; we must settle for what is now available. *Source: Scientific American* (April 28, 1906).

or in matrix form,

$$\boldsymbol{R}\mathbf{i} = \mathbf{v}. \tag{4.27b}$$

The steps are all analogous to those of nodal analysis, where "independent mesh" means involving at least one element not in the other mesh equations:

1. Assign clockwise mesh current variables $i_m$ to each mesh $m$.
2. Convert any practical current sources whose voltages or currents you do not care to find (if any) into practical voltage sources by source transformation. Ideal current sources naturally cannot be converted.
3. Write KVL for all independent meshes except those with current sources, using Ohm's law whenever needed. Relate mesh currents of adjacent meshes that are separated by a current source, using the current source connecting them.
4. Collect terms and put in the form of Eq. (4.27); solve. Use Ohm's law, KVL, and/or KCL to obtain the desired currents or voltages originally sought.

**Example 4.27**

In the circuit of Fig. 4.46, determine $V_1$ and $I_5$ using the mesh analysis method.

**Solution**

*Step 1.* We first redraw the circuit in Fig. 4.47, showing the mesh currents.

*Step 2.* We cannot convert the 1-A source because it has been modeled as ideal (no internal resistance shown). We convert the 2 A–4 Ω practical current source to a $2\text{ A}\cdot 4\ \Omega = 8\text{ V}$ practical voltage source through which $I_2$ now flows. Thus the mesh current $I_4$ is eliminated. (By the way, $I_4 = 2\text{ A}$, the rightmost current source.) Note that doing so does not eliminate any variable we seek; if it did, we could not perform the source transform.

*Step 3.* We take our loops clockwise (along the mesh currents) so that all *RI* terms from loop currents within the mesh are added rather than subtracted in the KVL loop equations. (*RI* terms from loop currents outside the mesh, if any, are subtracted.) For the lower, outside loop, KVL reads (mentally inserting the 8 V–4 Ω source-transformed voltage source on the right):

$$I_1 \cdot 1\ \Omega - 1\text{ V} + I_1 \cdot 2\ \Omega + (I_2 - I_3) \cdot 5\ \Omega + I_2 \cdot 4\ \Omega + 8\text{ V} = 0,$$

or

$$3I_1 + 9I_2 - 5I_3 = -7\text{ V}.$$

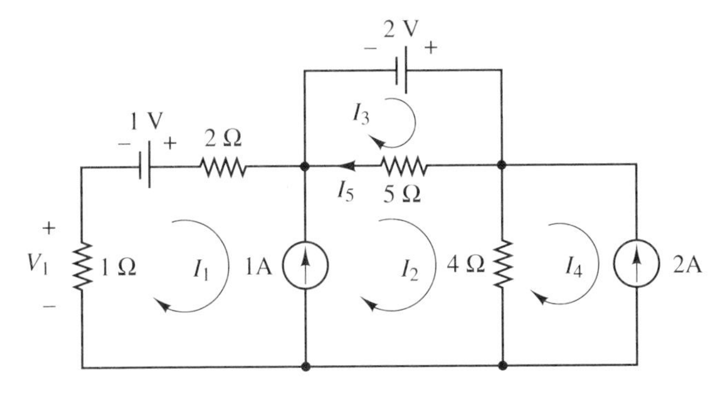


**Figure 4.47** Circuit for Example 4.27, with mesh currents shown.

Incidentally, the distinction between loop currents (e.g., $I_2$) and current in an element (e.g., $I_5$) is analogous to the difference between node voltages and voltage across an element. They are just different representations of current or voltage that are well suited to a particular solution method. In mesh analysis, we always express all contributions to KVL in terms of the mesh currents, not the previously defined currents in individual elements. Note that the right-to-left current in the 5-Ω resistor ($I_5$) is $I_3 - I_2$. Thus KVL for the top mesh is

$$-2\text{ V} + (I_3 - I_2)\cdot 5\ \Omega = 0,$$

or

$$5I_2 - 5I_3 = -2\text{ V}.$$

By KCL at the 1-A current source, we have $1\text{ A} = I_2 - I_1$ or (formally multiplying by 1 Ω to convert to a voltage),

$$I_1 - I_2 = -1\text{ V}.$$

*Step 4.* Putting the results above into matrix form yields

$$\begin{bmatrix} 3 & 9 & -5 \\ 0 & 5 & -5 \\ 1 & -1 & 0 \end{bmatrix}\begin{bmatrix} I_1 \\ I_2 \\ I_3 \end{bmatrix} = \begin{bmatrix} -7 \\ -2 \\ -1 \end{bmatrix}.$$

The determinant of $\boldsymbol{R}$ is $3(-5) - 9(5) - 5(-5) = -35$. Therefore,

$$I_1 = \frac{\begin{vmatrix} -7 & 9 & -5 \\ -2 & 5 & -5 \\ -1 & -1 & 0 \end{vmatrix}}{(-35)} = \frac{-7(-5) - 9(-5) - 5(2+5)}{-35} = 1.286\text{ A}$$

$$I_2 = \frac{\begin{vmatrix} 3 & -7 & -5 \\ 0 & -2 & -5 \\ 1 & -1 & 0 \end{vmatrix}}{(-35)} = \frac{3(-5) + 7(5) - 5(2)}{-35} = -0.286\text{ A}$$

$$I_3 = \frac{\begin{vmatrix} 3 & 9 & -7 \\ 0 & 5 & -2 \\ 1 & -1 & -1 \end{vmatrix}}{(-35)} = \frac{3(-5-2) - 9(2) - 7(-5)}{-35} = 0.114\text{ A}.$$

Finally, $V_1 = -I_1 \cdot 1\ \Omega = 1.286$ V and $I_5 = I_3 - I_2 = 0.4$ A, in agreement with Example 4.26.

One word of caution is necessary. Although dependent sources are handled identically, in practice two things must be kept in mind. The variable on which the source depends (the *control variable*) must *not* be eliminated in step 2 of either method by source transformation, because the dependent source value could then no longer be determined. Also, in forming the matrix equation, the control variable must be expressed in terms of the node voltages (for nodal analysis) or the mesh currents (mesh analysis).

Which method is better to use in practice: nodal or mesh analysis? Essentially, the answer is the same as that which would be given to the question "KCL or KVL?".

If there are many parallel branches and current sources, nodal analysis (KCL-based) is easier. If there are many series-connected elements and voltage sources, mesh analysis (KVL-based) is easier.

Formally, one may compare the numbers of independent equations generated by the two methods. Let $N_N$ be the number of nodes connecting two or more elements, $N_B$ be the number of single-element branches connecting any two nodes, $N_V$ be the number of independent and dependent voltage sources, and $N_I$ be the number of independent and dependent current sources.

In nodal analysis, potentially there is one node equation for every node except the reference node: $N_N - 1$. However, all voltage sources directly relate the node voltages at their two terminals, thereby reducing the number of node voltage equations by one. (This relation is very simple and in practice may be "done in your head," thereby eliminating one variable from the outset.) From these points it is very easy to conclude that the number of node equations $N_{\text{node}} = N_N - 1 - N_V$.

In mesh analysis, potentially there is one mesh equation for every mesh. Many branches and few nodes means that many of the branches are in parallel—there are many meshes. This is because all branches must be connected to something; if there are only a few nodes, many of the elements will have to be connected to the same nodes—in parallel. Contrarily, if for the same number of branches there are many nodes, it must mean that many of the branches are "used up" on series connections. As far as the number of meshes goes, these many-branch series connections are like one long branch; thus many nodes decreases the number of meshes.

Quantitatively, graph theory shows that the number of meshes is $N_B - (N_N - 1)$. Try it out and you will find that it always works for "planar" circuits, which are those which can be drawn so that no element crosses over another. The number of independent KVL equations is further reduced by the number of current sources, because each directly relates its two adjacent mesh currents via KCL. Consequently, the number of independent mesh equations is $N_{\text{mesh}} = N_B - N_N + 1 - N_I$.

Thus to minimize the effort, use nodal analysis if $N_{\text{node}} < N_{\text{mesh}}$; otherwise, use mesh analysis. In Examples 4.25 and 4.26, $N_N = 5$, $N_B = 8$, $N_V = 2$, and $N_I = 2$. Therefore, $N_{\text{node}} = 5 - 1 - 2 = 2$ and $N_{\text{mesh}} = 8 - 5 + 1 - 2 = 2$. In Example 4.25 we did not eliminate $V_2$ or $V_3$ in our head, and in Example 4.26 we did not eliminate $I_1$ or $I_2$ in our head, so in both cases we ended up with three equations. Clearly, for that example mesh and loop analysis are of equal difficulty; in other circuits, $N_{\text{node}}$ and $N_{\text{mesh}}$ may greatly differ.

**Example 4.28**

Calculate $N_{\text{node}}$ and $N_{\text{mesh}}$ for the circuits of (a) Fig. 4.20 and (b) Fig. 4.26.

**Solution**

**(a)** $N_N = 3$, $N_B = 4$, $N_V = 1$, $N_I = 1$, so that $N_{\text{node}} = 3 - 1 - 1 = 1$ while $N_{\text{mesh}} = 4 - 3 + 1 - 1 = 1$, so again $N_{\text{node}} = N_{\text{mesh}}$.

**(b)** $N_N = 4$, $N_B = 6$, $N_V = 1$, $N_I = 2$, so that $N_{\text{node}} = 4 - 1 - 2 = 1$ while $N_{\text{mesh}} = 6 - 4 + 1 - 1 = 2$. In this case, as we might have guessed from the parallel appearance of the circuit, $N_{\text{node}} < N_{\text{mesh}}$; again, in more complicated circuits, the differences could be more dramatic.

## 4.13 TRANSIENT ANALYSIS

### 4.13.1 First-Order Capacitive Circuits

When the voltage $v_B(t)$ applied to a capacitor through a resistor $R$ changes suddenly from 0 V to $V_B$ upon the closing of a switch (see Fig. 4.48), we know from Sec. 2.7 that the capacitor $C$ will charge through $R$ until its voltage $v_C$ becomes equal to $V_B$. This is because until then the potential energy of electrons is higher at the battery (− terminal) than at the capacitor (− terminal). Thus, eventually, a charge $Q = CV_B$ must reside on the capacitor plates so that the potential energy of electrons on the two sides are equal: $v_C = V_B$. Until then, charge progressively accumulates on the capacitor plates.

How long does it take for the charge $Q$ to be transferred to the capacitor plates? We know from Chapter 2 that conductance (and thus its reciprocal, resistance) can be viewed as the time it takes to transfer a given amount of charge through the material. Therefore, we expect that the time for $C$ to charge up must depend on $R$, and increase with $R$. But the question of how long it takes has a surprising answer: forever. This is explained as follows. First of all, in a resistor, $i = v_R/R$. While $v_C$ increases, $v_R = V_B - v_C$ decreases, and hence the current $i$ decreases. That is, the rate of charging, $dq/dt = i$, decreases. When $v_C$ becomes extremely close to $V_B$, the rate of charging, $i = (V_B - v_C)/R$, becomes extremely slow. Eventually, when $v_C$ is infinitesimally close to $V_B$, $dq/dt \approx 0$; therefore, $v_C$ can never quite reach $V_B$.

In practice, however, $R$ *does* indicate how long it takes for *the bulk* of the charging to occur. To precisely determine how it indicates this requires solving a first order differential equation. We assume that before $t = 0$, $i = 0$ and $v_B = 0$. Later we shall focus on the case that $v_C(t = 0) = 0$, but for generality we will begin by assuming a general initial condition $v_C(0)$ for $v_C(t)$ at $t = 0$. The current $i$ in $R$ equals the current in $C$. Noting that $i = C dv_C/dt$, we may write KVL for this $RC$ circuit (Fig. 4.48) as

$$RC\frac{dv_C(t)}{dt} + v_C(t) = v_B(t), \tag{4.28a}$$

or

$$\frac{dv_C(t)}{dt} + \frac{1}{RC}v_C(t) = \frac{1}{RC}v_B(t). \tag{4.28b}$$

As is well known from an introductory study of differential equations, the *integrating factor* $\exp\left(\int_0^t dt/RC\right)$ provides a formal solution. This is because for any $v_C(t)$, the derivative of a product combined with the derivative of an exponential shows that

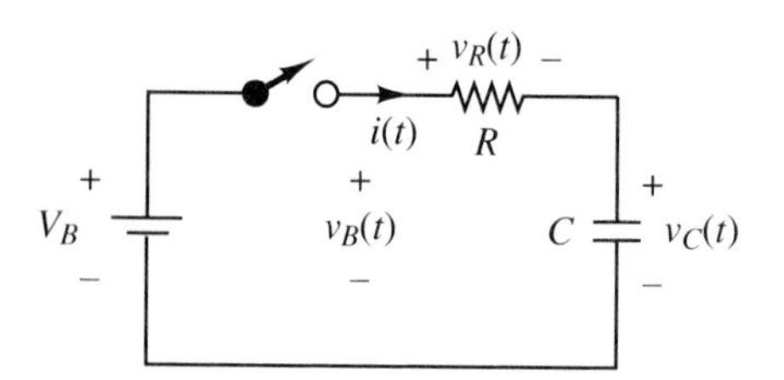


**Figure 4.48** $RC$ circuit, with battery $V_B$ connected at $t = 0$.

$$\frac{d}{dt}[e^{\int dt/RC}v_C(t)] = e^{\int dt/RC}\left[\frac{dv_C(t)}{dt} + \frac{v_C(t)}{RC}\right]. \tag{4.29}$$

Thus if both sides of Eq. (4.28b) are multiplied by exp ($\int dt/RC$) = exp ($t/RC$), the left-hand side of the result may be replaced by the left-hand side of Eq. (4.29), yielding

$$\frac{d}{dt}[e^{t/RC}v_C(t)] = e^{t/RC}\frac{v_B(t)}{RC}. \tag{4.30}$$

We can now easily solve for $v_C(t)$ by integrating both sides of Eq. (4.30) with respect to time from $-\infty$ to $t$. Of course, the integration provides the answer only to within a constant $K$, for differentiating an added constant will contribute zero to the differential equation and thus still satisfy Eq. (4.30). After performing the integration, multiplication of both sides of the result by exp $(-t/\tau_C)$, where

$$\tau_C = RC \tag{4.31}$$

is called the time constant, allows us to solve for $v_C(t)$:

$$v_C(t) = e^{-t/\tau_C}\left[\frac{1}{\tau_C}\int_{-\infty}^{t} e^{t'/\tau_C}v_B(t')\,dt' + K\right] \tag{4.32a}$$

$$= \frac{1}{\tau_C}\int_{-\infty}^{t} v_B(t')e^{-(t-t')/\tau_C}\,dt' + Ke^{-t/\tau_C} \tag{4.32b}$$

$$= v_{C,f}(t) + v_{C,n}(t). \tag{4.32c}$$

In Eq. (4.32c), $v_{C,f}(t)$ represents the *forced response,* that is, the component of $v_C(t)$ due entirely to the effects of the forcing function $v_B(t)$. Also, $v_{C,n}(t)$ is the *natural response,* that is, whatever amount of the type of response that would occur in the absence of $v_B(t)$ which is necessary for $v_C(t)$ to be continuous at $t = 0$ [$v_C(0^-) = v_C(0^+)$; the reason for this continuity will be discussed in detail below]. This component exists if there is a mismatch between the initial condition on $v_C(t)$ and the forced response $v_{C,f}(0)$, which is not always the case. Actually, $v_{C,n}(t)$ arises only because some or all of the charge on the capacitor at a particular time ($t = 0$, for example) originates from a source *other* than $v_B(t)$. The magnitude $K$ of $v_{C,n}(t)$ depends on $v_B(t)$ and the external source of charge contributing to $v_C(0)$, but its shape, exp$(-t/\tau_C)$, depends on neither, only on the $RC$ circuit configuration. In particular, we will find for the step input case that $K = 0$ if $v_C(0) = 0$; that is, no sources of charge other than $v_B(t)$ have previously been charging the capacitor.

We can write $K$ explicitly in terms of $v_C(0)$ and $v_B(t)$ merely by evaluating Eq. (4.32b) at $t = 0$:

$$v_C(0) = \frac{1}{\tau_C}\int_{-\infty}^{0} v_B(t')\,e^{-(0-t')/\tau_C}\,dt' + K, \tag{4.33a}$$

or

$$K = v_C(0) - \frac{1}{\tau_C}\int_{-\infty}^{0} v_B(t')\,e^{t'/\tau_C}\,dt', \tag{4.33b}$$

which allows us to rewrite Eq. (4.32b) as

$$v_C(t) = \frac{1}{\tau_C}\int_0^t v_B(t')\,e^{-(t-t')/\tau_C}\,dt' + v_C(0)\,e^{-t/\tau_C}. \qquad \text{[general } v_B(t) \text{ and } v_C(0)\text{]}. \tag{4.34}$$

Thus we see that it does not matter exactly what form $v_B(t)$ has for $t < 0$; only its result manifest at $t = 0$ in $v_C(0)$ affects $v_C(t)$ for $t > 0$. Note that all Eqs. (4.32) through (4.34) are valid for *any* $v_B(t)$—not just a battery turned on at $t = 0$. Also, the same idea can be used to splice together results for cases where there are discontinuities in $v_B(t)$: at each discontinuity time $t_m$, there will be a different $K_m$ chosen to satisfy $v_C(t_m^-) = v_C(t_m^+)$.

Now, in our study of what happens after the switch in Fig. 4.48 is closed, we have $v_B(t) = V_B$ for $t > 0$ and $v_B(t) = 0$ for $t < 0$ (a step input). This causes Eq. (4.34) to reduce to

$$v_C(t) = \frac{V_B}{\tau_C}\int_0^t e^{-(t-t')/\tau_C}\,dt' + v_C(0)\,e^{-t/\tau_C} \tag{4.35a}$$

$$= V_B(1 - e^{-t/\tau_C}) + v_C(0)\,e^{-t/\tau_C} \tag{4.35b}$$

$$= [v_C(0) - V_B]\,e^{-t/\tau_C} + V_B \qquad [v_B(t) \text{ a step, general } v_C(0)]. \tag{4.35c}$$

Equation (4.35c) states first of all that the final value of $v_C(t)$ is $v_C(\infty) = V_B$. It says further that for finite $t$, the amount that $v_C(t)$ is below $V_B$ is the difference between the final value $V_B$ and the initial value $v_C(0)$, weighted by the decaying factor $\exp(-t/\tau_C)$.

If $v_C(0) = 0$, $v_C(t)$ becomes

$$v_C(t) = V_B(1 - e^{-t/\tau_C}) \qquad [\text{valid } \textit{only} \text{ for } v_B(t) \text{ a step, } v_C(0) = 0]. \tag{4.36}$$

Notice that in Eq. (4.36), $v_C(0) = 0$ and $v_C(\infty) = V_B$, as we argued above. In Eq. (4.36), the *time constant* $\tau_C$ is a measure of how quickly the *transient* term $-V_B \exp(-t/\tau_C)$ dies away, leaving only the *steady-state* term, $V_B$. The smaller $\tau_C$ is, the more quickly $V_B \exp(-t/\tau_C)$ decreases with time, so the faster $v_C(t)$ approaches $V_B$ (and $0.63V_B$); see Fig. 4.49, which shows $v_C(t)$ for two values of $\tau_C$. Because $v_C$ never reaches $V_B$, the $0.63V_B$ point is used to determine the speed of the response, which depends only on $RC$. This is in agreement with our previous physical arguments above.

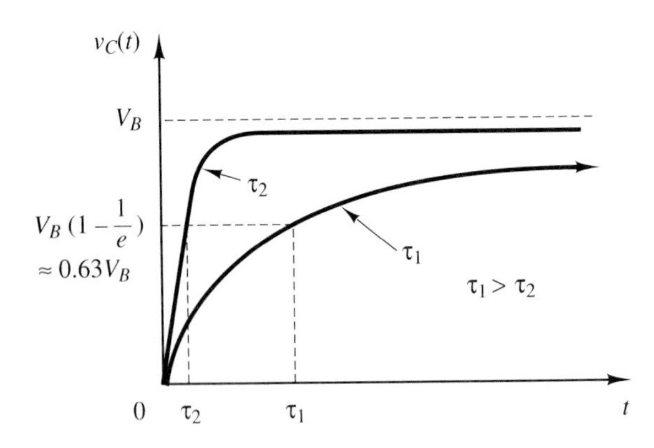


**Figure 4.49** Time dependence of the voltage of the charging capacitor in Fig. 4.48 for two values of $\tau_C$.

Concerning the current, we know that as $t$ approaches infinity it approaches zero. Mathematically, the current may be computed from Eq. (4.35c) as

$$i(t) = C\frac{dv_C(t)}{dt} \tag{4.37a}$$

$$= \frac{V_B - v_C(0)}{R}e^{-t/\tau_C} \qquad [v_B(t) \text{ a step, general } v_C(0)], \tag{4.37b}$$

which for $v_C(0) = 0$ reduces to

$$i(t) = \frac{V_B}{R}e^{-t/\tau_C} \qquad [\text{valid } only \text{ for } v_B(t) \text{ a step, } v_C(0) = 0], \tag{4.38}$$

where we used the fact that $C/\tau_C = C/(RC) = 1/R$. Equations (4.37b) and (4.38) indicate that the smaller $\tau_C$ is, the faster $i(t)$ dies out (due to the negative exponential); see Fig. 4.50, which shows $i(t)$ for two values of $\tau_C$, assuming that $v_C(0) = 0$. If it is $C$ that is varied in order to change $\tau_C$ ($R$ fixed), the initial current is the same for all $\tau_C$ (see Fig. 4.50a). However, if it is $R$ that is varied ($C$ fixed), the initial current, $i(0) = V_B/R$, increases for smaller $\tau_C$ (and $R$) (see Fig. 4.50b).

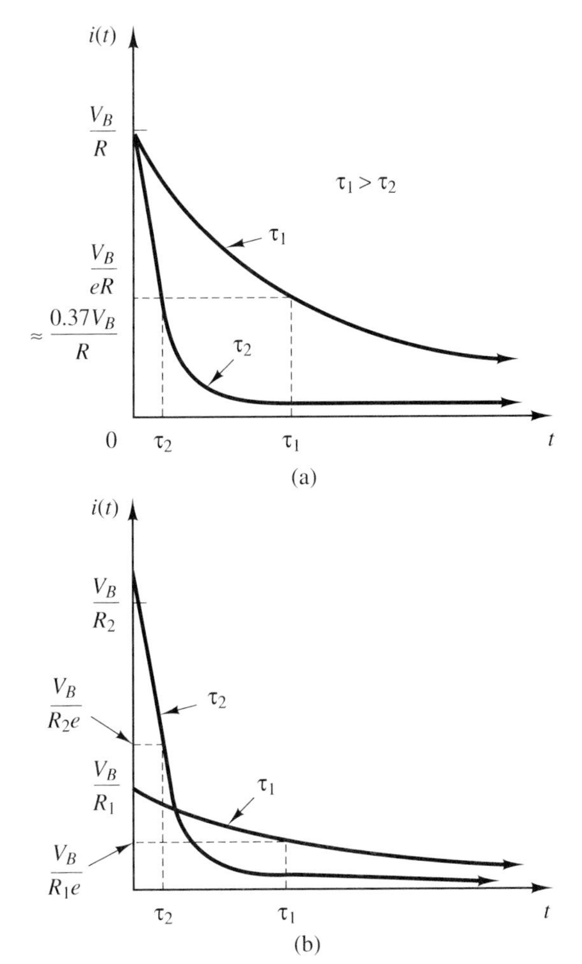


**Figure 4.50** Time dependence of the current of the charging capacitor in Fig. 4.48 for two values of $\tau_C$: (a) $R$ fixed, $C$ varies; (b) $R$ varies, $C$ fixed.

Notice that while $i(t)$ in Fig. 4.50 changes suddenly at $t = 0$, $v_C(t)$ never does. If $v_C(t)$ were to change suddenly, its derivative with respect to time and thus $i(t)$ would become infinite—a physical impossibility. Therefore, the voltage of a capacitor can never change suddenly. This is analogous again to the water in a tank: One may suddenly turn on the faucet [analogous to $i(t)$], but the water level [analogous to $v_C(t)$] will never *suddenly* jump to a new value.

The description above is equivalent to saying that a capacitor behaves like a voltage source for short periods of time. In fact, when performing time-varying circuit analysis, it is often convenient to assume for $t$ just after the sudden change that the capacitor is an ideal voltage source, having voltage equal to the initial voltage on the capacitor. (In the case $v_C(0^-) = 0$, a 0-V source—a short circuit. Contrast this with dc, where the capacitor behaves like an *open* circuit!) For short periods of time, the assumption that $v_C$ maintains its initial value is valid. No matter how fast we dump water into Lake Erie, its level is going to stay roughly the same for quite awhile; it is a very large "capacitor."

What are the practical implications of all this? There are many, but two important ones will now be mentioned. If one has used the ac wall voltage to generate a dc voltage by means of "rectification" (discussed in Chapter 7), the output voltage from the rectifier will be ripply. Due to the fact that the capacitor voltage cannot change suddenly, it can be used at the output to smooth out the ripples to make a more constant (better) output voltage. A large capacitor will do the best job, because $\tau_C = RC$ is then large, making the capacitor (output) voltage sluggish.

In the case of the rectifier, the sluggishness of $v_C$ is used to advantage. But in other signal processing applications, it can be a nuisance. For example, in digital circuits it is common to have a *square wave* or *clock:* a series of high/low pulses with square corners. This clock signal is used for making things happen in a computer. If it must travel along a wire such as an oscilloscope probe (see Fig. 4.51a), the capacitance of the wire can interfere with transmission, smoothing out the square wave, as shown in Fig. 4.51b.

In Fig. 4.51a the parameters $R_s$ and $C_s$ are the source resistance and capacitance (yes, one can model a rapidly time-varying signal generator as having a capacitance). A simplified model of the probe would have effective resistance $R_p$ and capacitance $C_p$. In Fig. 4.51b, the square-wave signal is shown along with the output voltage at the far end of the probe, for two values of $C_p$: $C_{p1} \ll C_{p2}$.

Unfortunately, the situation is yet more complicated, because the inductance of the probe significantly affects the response to a square wave. Consequently, the actual shape one would observe is more like an S curve than an exponential, and we cannot compute a simple $RC$ time constant for the probe system. Nevertheless, the combined effects of capacitance and inductance do distort and smooth the original square wave. A typical value for the time for the output voltage of a section of coaxial cable to drop to $1/e$ of its maximum value is 40 $\mu$s or less.

### 4.13.2 First-Order Inductive Circuits

Like the capacitor, an inductor has a similar resistance to change. This time, however, it is the current that cannot change suddenly, while the voltage can. Referring to Fig. 4.52, the switch is again closed at $t = 0$. Again by KVL, and noting that $v = L\,di(t)/dt$,

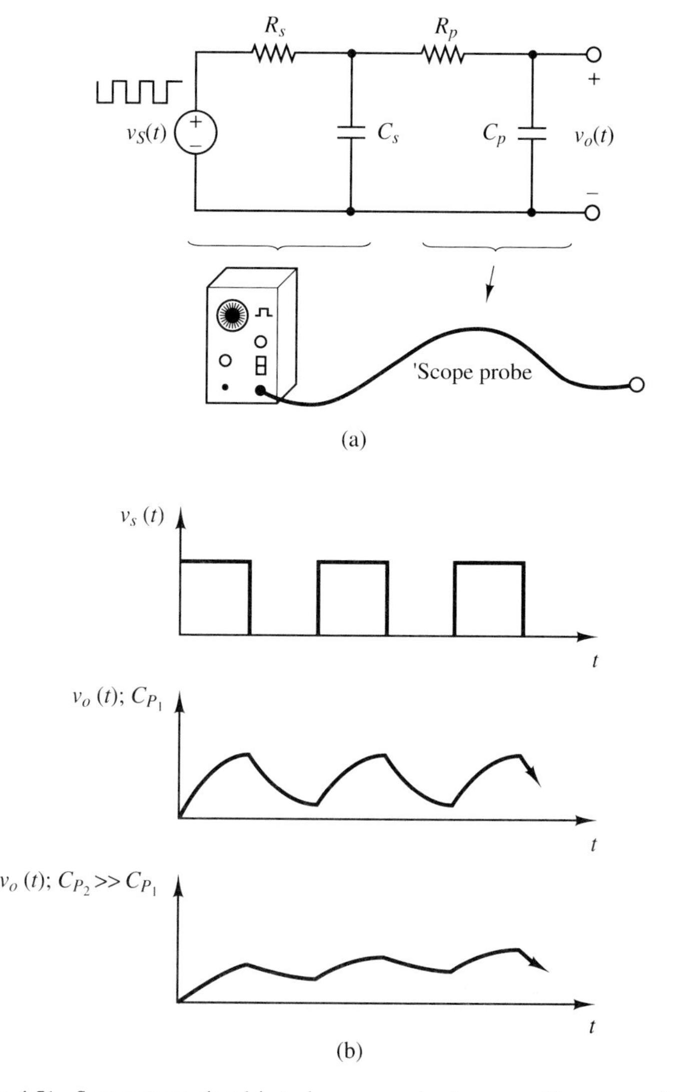


**Figure 4.51** Square-wave signal is to be measured using an oscilloscope probe: (a) physical situation and simplified modeling circuit; (b) square-wave source signal and output voltage assuming the simplified model in (a) for two values of probe capacitance: $C_{p1} \ll C_{p2}$.

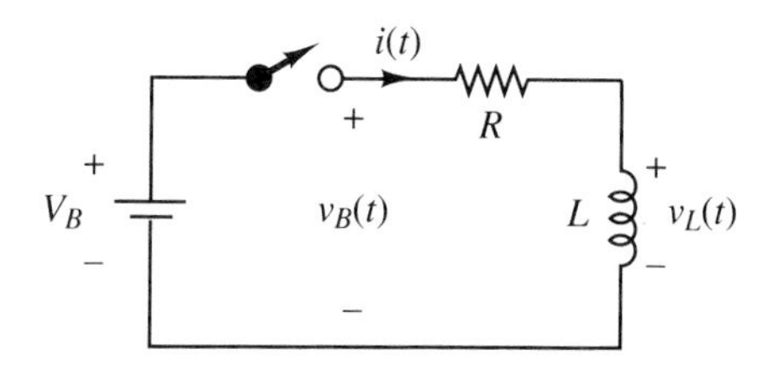


**Figure 4.52** *RL* circuit, with battery $V_B$ connected at $t = 0$.

$$i(t)R + L\frac{di(t)}{dt} = v_B(t), \tag{4.39a}$$

or

$$\frac{di(t)}{dt} + \frac{R}{L}i(t) = \frac{v_B(t)}{L}. \tag{4.39b}$$

Equation (4.39a) is of the same form as Eq. (4.28b); the only differences are the variable and the coefficients. Specifically, in Eq. (4.28b) replace $v_C(t)$ by $i(t)$, $RC$ on the left-hand side by $L/R$, and $RC$ on the right-hand side by $L$. Following identical steps to those in Eqs. (4.29) through (4.36) gives, for $v_B(t)$ a step and $i(0) = 0$,

$$i(t) = \frac{V_B}{R}(1 - e^{-t/\tau_L}), \tag{4.40}$$

where

$$\tau_L = \frac{L}{R}. \tag{4.41}$$

From Eq. (4.40) we may determine the voltage across the inductor $v_L(t) = L\, di(t)/dt$ to be

$$v_L(t) = V_B e^{-t/t_L}. \tag{4.42}$$

The inductor voltage is of the same form as the capacitor current in Eq. (4.38) and Fig. 4.50, while the inductor current has the same form as the capacitor voltage in Eq. (4.32b) and Fig. 4.49. It is left as an exercise to sketch the inductor voltage and current waveforms.

It is again helpful to consider the practical implications of the transient behavior of an inductor. Because $v_L(t) = L\, di(t)/dt$, a sudden change in $i(t)$ would require an infinite voltage $v_L(t)$, which in practice is impossible. If current is flowing in an inductor and the circuit is broken, $di(t)/dt$ is enormous, making $v_L(t)$ huge. This very large voltage can cause sparking and is one reason why large motors are not suddenly disconnected. It is also the principle behind spark generation in an automotive spark plug.

Normally, the current in an inductor changes gradually, like the voltage across a capacitor. In particular, in *step response* problems such as the one depicted in Fig. 4.52, the inductor at $t = 0$ acts like a current source, with current equal to the initial value of inductor current (in this particular case, 0 A). The situation is analogous to that of a bicycle wheel. When spinning, we can immediately put on the brakes or push to pedal harder, but the bicycle does not immediately respond to our desire.

**Example 4.29**

Consider the $RC$ network shown in Fig. 4.48. Given that $R = 10\ \Omega$, $C = 2\ \mu\text{F}$, $V_B = 25$ V, $v_C(t = 0^-) = -5$ V, and that the switch in Fig. 4.48 closes at $t = 0$,

**(a)** Find the time constant $\tau_C$.

**(b)** Determine equations for $i(t)$, $v_C(t)$, and $v_R(t)$ for $t > 0$.

**(c)** Sketch $i(t)$ and $v_C(t)$ versus time.

**(d)** Find $v_C(t_0)$, $v_R(t_0)$, and $i(t_0)$, where $t_0 = 1.5\ \tau_C$.

**Solution**

(**a**) $\tau_C = 10\ \Omega \cdot 2\ \mu\text{F} = 20\ \mu\text{s}$.

(**b**) From Eq. (4.37b), $i(t) = [(25\ \text{V} - (-5\ \text{V}))/10]\ \exp\ (-t/2 \cdot 10^{-5}) = 3\ \exp\ (-50{,}000t)$ A, where $t$ is in seconds. Furthermore, from Eq. (4.35c), $v_C(t) = (-5 - 25)\ \exp\ (-50{,}000t) + 25\ \text{V} = -30\ \exp\ (-50{,}000t) + 25$ V. Finally, $v_R(t) = V_B - v_C(t) = [V_B - v_C(0)]\ \exp\ (-t/\tau_C) = [25 - (-5)]\ \exp\ (-50{,}000t)\ \text{V} = 30\ \exp\ (-50{,}000t)$ V.

(**c**) See Fig. 4.53. Note an interesting fact: If we draw straight lines through the initial conditions on $i(t)$ and $v_C(t)$ having slope equal to those of $i(t)$ and $v_C(t)$ at $t = 0^+$, we reach the final values [$i(\infty) = 0$ A and $v_C(\infty) = V_B = 25$ V] in *exactly one time constant* ($t = \tau_C$)! This effect is generally true for first-order transients. We note from Eq. (4.37b) that this straight line is, at $t = \tau_C$, $i_{\text{str}}(\tau_C) = i(0^+) + [di(t)/dt](0^+)\tau_C = [V_B - v_C(0)]/R + \{[V_B - v_C(0)]/R\}\ (-1/\tau_C)\tau_C = 0 = i(\infty)$ and for the capacitor voltage [see Eq. (4.35c)] it is $v_{C,\text{str}}(\tau_C) = v_C(0^+) + [dv_C(t)/dt](0^+)\tau_C = v_C(0^+) + [v_C(0) - V_B](-1/\tau_C)\ \tau_C = V_B = v_C(\infty)$. Thus we have an additional meaning for the time constant: It is the time it would take the variable to reach its final value if, rather than changing exponentially, it were to change linearly with time according to its actual initial slope.

(**d**) From the solution of part (b), $v_C(1.5\tau_C) = -30\ \exp\ (-1.5) + 25 = 18.3$ V, $v_R(t) = 30\ \exp\ (-1.5) = 6.7\ \text{V}\ [= V_B - v_C(1.5\tau_C)]$, and $i(1.5\tau_C) = 3\ \exp\ (-1.5) \approx 0.67$ A.

### 4.13.3 Second-Order Circuits

In reality, as noted in Sec. 4.3.1, real inductors and capacitors also have, respectively, capacitance and inductance. Furthermore, any complicated circuit is likely to have an equivalent inductance, capacitance, and resistance. This point was

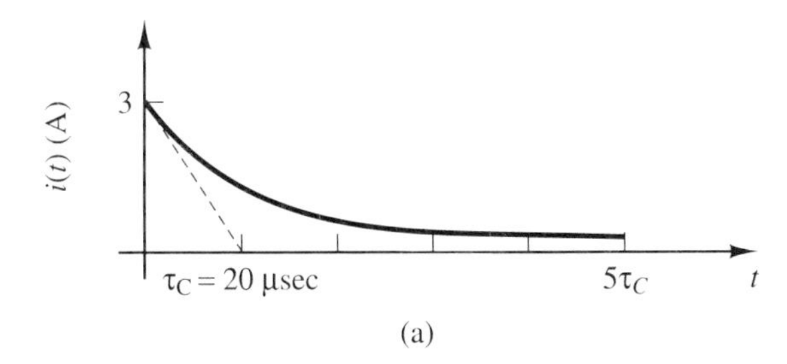


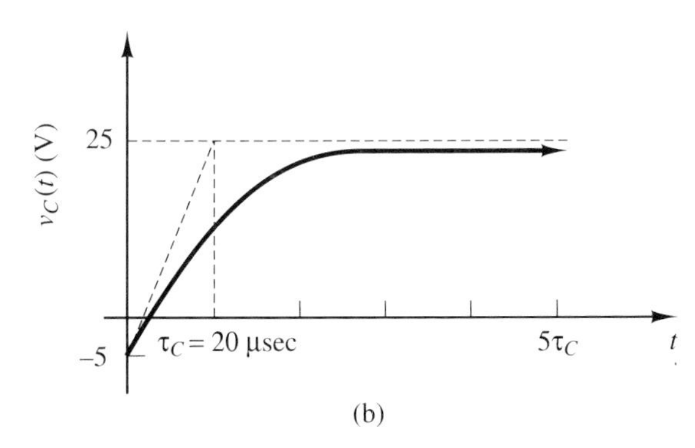


**Figure 4.53** For the solution of Example 4.29: (a) $i(t)$; (b) $v_C(t)$.

already noted in Sec. 4.13.1, where the coaxial cable voltage waveform did not have a simple exponential time dependence, but rather a smoother, sometimes "ringing" appearance.

If either capacitance or inductance greatly dominates, the exponential response will occur in the manner of Secs. 4.13.1 and 4.13.2. When both capacitance and inductance are present (*LC* circuits), there is the possibility for *oscillation*. The oscillation is between energy storage in electric and magnetic fields. This is similar to the mechanical situation in which energy oscillates between mechanical potential energy and kinetic energy, such as in a pendulum or other vibrating system.

Just as in such a mechanical system a second-order differential equation describes the dynamics ($F = ma + kx = md^2/dx^2 + kx$), a second-order differential equation describes *LC* circuits. A damping term is additionally present in the differential equation if friction is present in the mechanical system or resistance is present in the *LC* circuit. A detailed mathematical study of *LC*/*LRC* circuits would have to be reserved for a second course thoroughly devoted to electric circuits.

These second-order systems are useful in oscillator circuits such as are found in radio tuners, and in Chapter 7 we shall return to them. Also, whenever "ringing" is observed in a measured waveform, it is probably due to an *LC* combination somewhere. Often this ringing is undesirable; the *L* and *C* may be unintentional, as in the example of an oscilloscope probe. A detailed study of a parallel *LRC* combination shows that the frequency of this ringing is $\{1/(LC) - [1/(2RC)]^2\}^{1/2}/2\pi$. We shall encounter the so-called *resonant frequency* $1/[2\pi(LC)^{1/2}]$ again in Chapter 5.

## 4.14 SUMMARY

In this extensive chapter we have introduced the fundamental concepts of electric circuit analysis. First we reviewed the material of Chapters 2 and 3 that is most directly relevant to basic circuit analysis. The primary passive circuit elements: resistance, capacitance, and inductance, were reviewed along with their circuit symbols and $i$–$v$ characteristics. To provide electrical power to these elements, voltage and current sources, as well as their symbols and $i$–$v$ plots, were introduced. Dependent sources are a special type of active element that are determined by a voltage or current elsewhere in the circuit.

When performing analysis on a circuit, every variable must have its polarity or direction assigned; otherwise, its value can not be determined. You may flip a coin to choose the symbolic polarity, but then stick with it for all the analysis. In accordance with energy conservation, power is absorbed if positive current flows into the terminal of an element at higher potential.

Kirchhoff's laws were obtained from previously derived electromagnetic equations under the quasistatic approximation. The sum of the voltages around any loop is zero, as is the sum of currents into any node. When currents and voltages are constant in time, they are known as dc currents and voltages.

As an alternative to specification of the voltage across an individual circuit element, one may specify the potential of any particular point in the circuit with respect to a reference point. These potentials "in the circuit" rather than "across an element" are called node voltages because they represent the potential of a node. Similarly, loop currents may be defined that are not currents in a particular element but rather around a given loop. Methods of circuit solution based on node voltages and loop currents yield matrix systems for the unknown circuit variables.

The idea of an equivalent network is exceedingly useful in circuit analysis and design. Series–parallel combinations of passive elements (all of the same type) may be replaced by a single equivalent element. Resistive circuits powered by voltage and/or current sources can be replaced by a "practical voltage source," that is, a voltage source in series with a source resistance. By the source transformation idea, this Thévenin equivalent may be replaced by a practical Norton current source. Determination of the active equivalent allows us to match a load to the source for maximum power transfer to the load, if desired.

These practical sources reflect the fact that a load placed on any source causes the terminal current or voltage to be reduced from that at no load. The deviation from ideal is modeled by the presence of the source resistance for both voltage and current sources (see Fig. 4.54a). By the voltage divider or current divider rules, the amount of degradation is easily computable for any connected load.

A good voltage source has small $R_s$ ($R_{s1}$ in Fig. 4.54b), and a good current source has large $R_s$ ($R_{s3}$). Typically, if you are presented with a source having $R_s$ on the order of several ohms, it is usually best to model it as a voltage source; if $R_s$ is kilohms or more, it acts like a current source. Of course, it is really the comparison between $R_s$ and the load $R_L$ the source is driving that truly determines which model is more appropriate. A source transformation easily takes you from one model to the other.

The load line is a graphical technique for circuit solution when the circuit elements have nonlinear $i$–$v$ characteristics. Where the two $i$–$v$ curves intersect indicates the operating point of the circuit: Any two elements or networks connected in parallel must simultaneously have the same voltage and current.

The model of an ideal capacitor at dc is an open circuit, while that of an ideal inductor

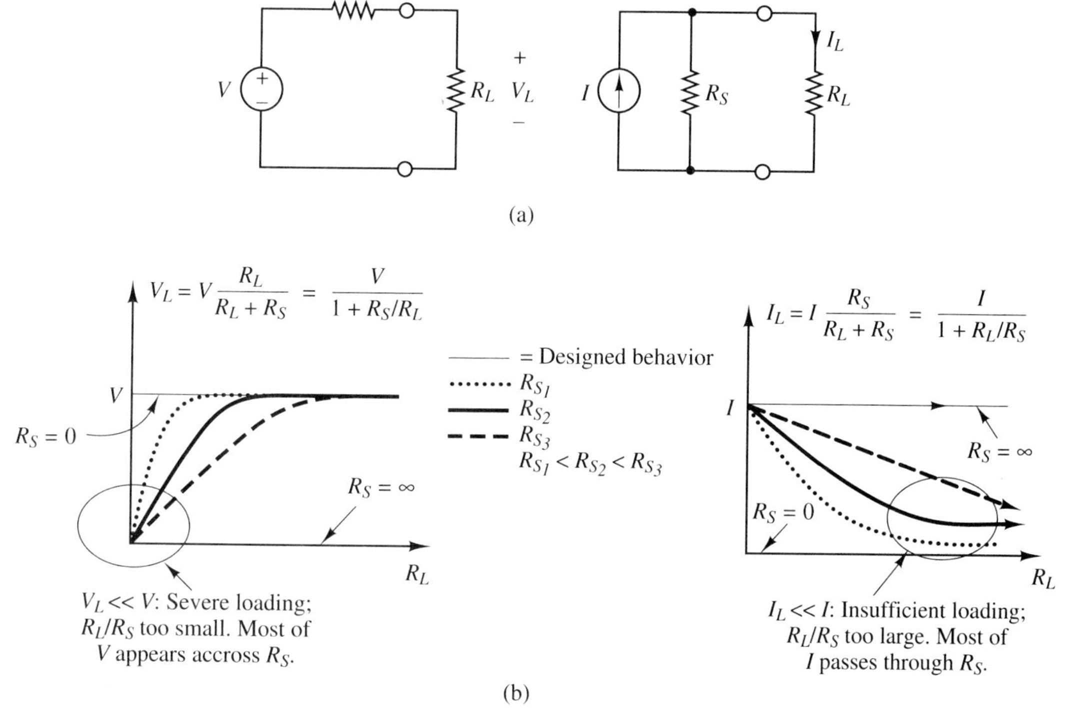


**Figure 4.54** Practical sources: (a) circuits; (b) output characteristics.

at dc is a short circuit. This is because no current flows in a capacitor unless its voltage is changing, and no voltage appears across an inductor unless its current is changing.

But if a sudden change in applied voltage (presented through a series resistor) occurs, a gradual buildup of electric or magnetic energy is stored in the device. As a result, the terminal voltage or current changes exponentially. During the period immediately following the sudden change in applied voltage, the capacitor behaves like a constant voltage source (rather than like an open circuit at dc), because its voltage cannot change suddenly. The value of this modeling voltage source is the capacitor voltage before the change. Analogously, the inductor behaves like a constant current source (rather than like a short circuit at dc) because its current cannot change suddenly. The value of this modeling current source is the inductor current before the change.

Circuits having capacitance or inductance but not both are describable by a first-order differential equation. The exponential changes are characterized by the time constant parameter, which is $RC$ or $L/R$. If the circuit has both capacitance and inductance, it satisfies a second-order differential equation, and may oscillate at the frequency $\{1/(LC) - [1/(2RC)]^2\}^{1/2}/2\pi$.

Our next two chapters take a look at another important type of circuit excitation: sinusoidal. We shall find many advantages to this form of excitation, which plays such a dominant role in both signal and power transmission/processing systems.

# Appendix

## Some Common Pitfalls

- Never add a current to a voltage. Checking units will prevent this from happening.
- Never add a resistance to a capacitance or inductor. The $i$–$v$ relations for these devices are different.
- In Fig. A4.1, $R_1$ and $R_2$ are *not* in parallel, because of the voltage source.

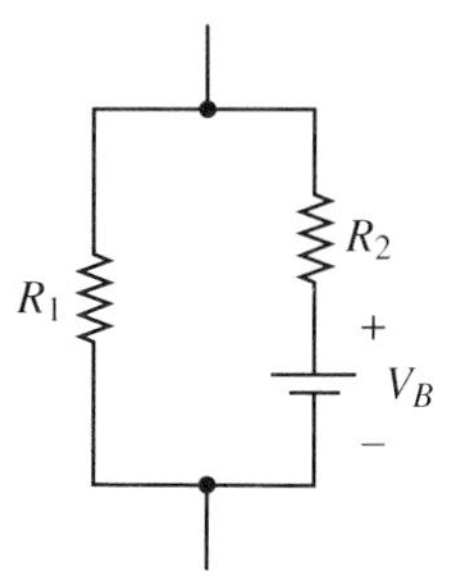


**Figure A4.1** $R_1$ is not in parallel with $R_2$.

- When given a situation such as that in Fig. A4.2, don't give the *politician's* answer "$v_L = i_L R_L$"—$i_L$ is another unknown! Instead, say "$v_L = [R_L/(R_L + R_s)]v_s$." That is, whenever possible answer directly, in terms of *known* quantities (in Fig. A4.2, $v_s$, $R_L$, $R_s$) instead of "passing the buck" and answering in terms of other *unknown* quantities (in Fig. A4.2, $i_L$).

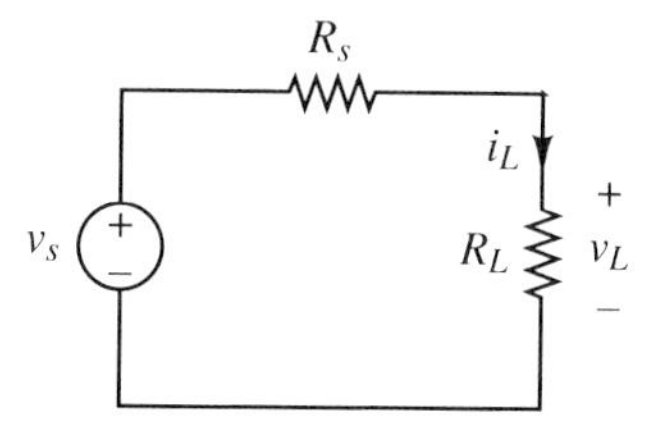


**Figure A4.2** If asked for $v_L$, recognize the voltage divider and give the direct answer $v_L = [R_L/(R_L + R_s)]v_s$, not $v_L = i_L R_L$.

- In Fig. A4.3a, if you just write "$v_R$" in your analysis without defining its polarity in the figure as in Fig. A4.3b, no one else will know whether you mean the polarity in Fig. A4.3b or its opposite. You are also much more likely to make a mistake in your analysis. The same goes for current. If you just say "$i$," is it clockwise or counterclockwise? No one knows, and an error is very likely to result. Indicate the current direction, as in Fig. A4.3b. In the circuit diagram, fully define, including polarity/direction, *every* variable you use in analysis.

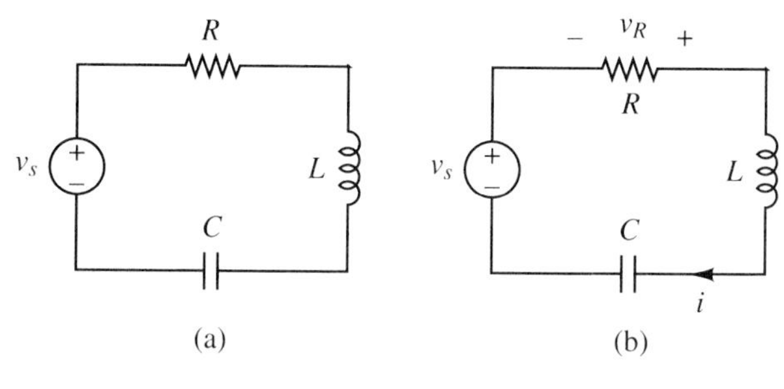


**Figure A4.3** (a) A given circuit without defined voltage and current symbols. For analysis, it must be amended as shown in (b) or errors are very likely to result.

- $V = IR$, *not* $I = VR$ or $V = I/R$.
- Capacitors in *parallel* add; resistors and inductors in *series* add. Otherwise, the equivalent is the reciprocal of the sum of reciprocals of the individual elements.
- Never say "voltage through" an element or circuit. Voltage is a potential difference *across* the terminals of the element.
- Never say "current across" an element; say "current through" or better, "current in."
- The voltage across a current source is *not* zero. When writing KVL, it is tempting to "skip over" a current source, thereby assuming zero volts across it. A new voltage variable must be defined across the current source, or KCL written instead of KVL.
- If it is required to turn off a voltage source, replace it by a short circuit. If it is required to turn off a current source, replace it by an open circuit. This is important when calculating Thévenin equivalent resistances. Do not mix these up. If you recall their physical meanings, you probably won't.
- If finding $v_{oc}$ and $i_{sc}$ for Thévenin or Norton equivalent circuits, do not use voltages or currents found for the open-circuited circuit (used for finding $v_{oc}$) in the short-circuited circuit (used for finding $i_{sc}$)—these are different circuit configurations.
- Never apply Ohm's law to a voltage source, a current source, or anything other than a resistor.

- If finding the equivalent resistance $R_{eq}$ seen by $R_L$, then $R_L$ should not appear in the expression for $R_{eq}$.
- If, in Fig. A4.4, you are asked to find $i_1$, do not combine $R_1$ and $R_2$ in parallel. That would eliminate from the circuit the very variable that you seek.

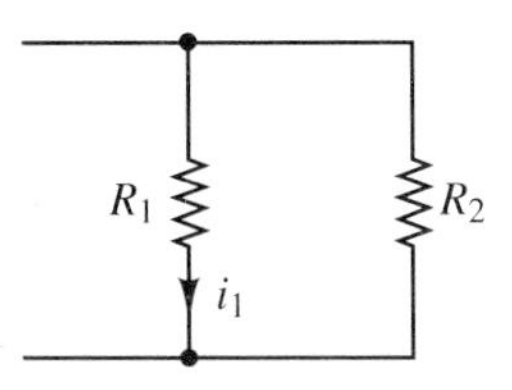


**Figure A4.4** Do not combine $R_1 || R_2$ if $i_1$ is sought.

- In Fig. A4.5, KCL is written as follows:

$$-I + \frac{V_A}{R_1} + \frac{V_A - (V_B + V_{R3})}{R_2} = 0.$$

Do not forget $V_{R3}$, tempting as it is to write "$(V_A - V_B)/R_2$" for that term.

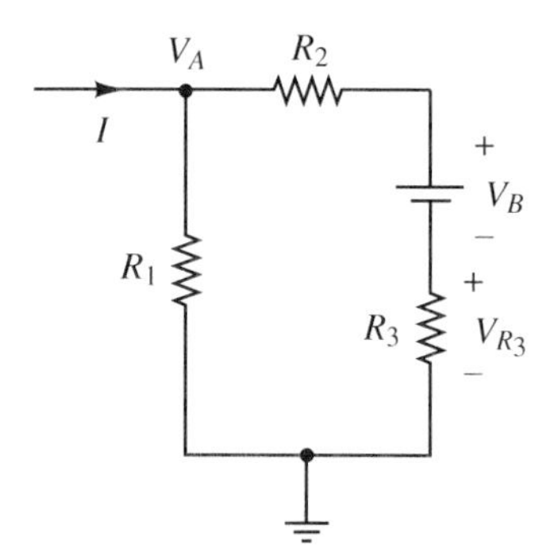


**Figure A4.5** Watch out for "hidden" voltage drops such as $V_{R3}$.

- Never turn off a dependent source.
- When writing a voltage divider, $V_1 = [R_1/(R_1 + R_2)]V$, do not forget to "tack on" the $V$. Similarly, when writing a current divider, $I_1 = [R_2/(R_1 + R_2)]I$, do not forget the factor $I$.
- In Fig. A4.6, when writing KVL, $V_1 = (I_1 - I_2)R_1$, do not forget the "$-I_2$" that is hiding in the lower loop. In general, take enough time to guarantee yourself that you have not missed any terms in a KVL or KCL equation.

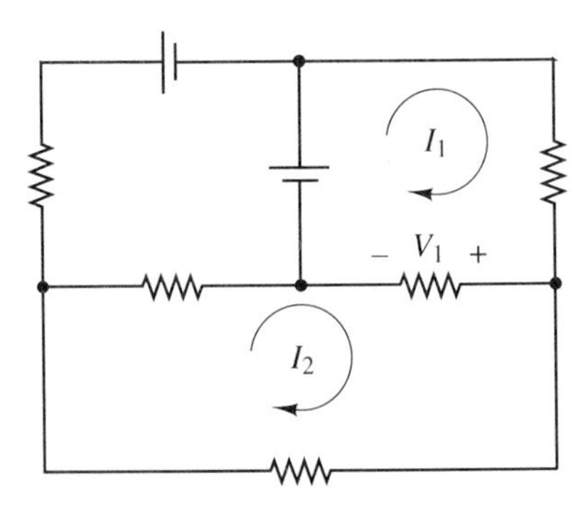


**Figure A4.6** It is easy to miss one of the contributing currents in loop analysis.

- Suppose in Fig. A4.7 that it is required to find $v_{sr}$, $v_{pq}$, and $v_{qs}$. Do not write KVL for this loop in order to solve for the current—the one and only current in this circuit is *given* as 1 A!

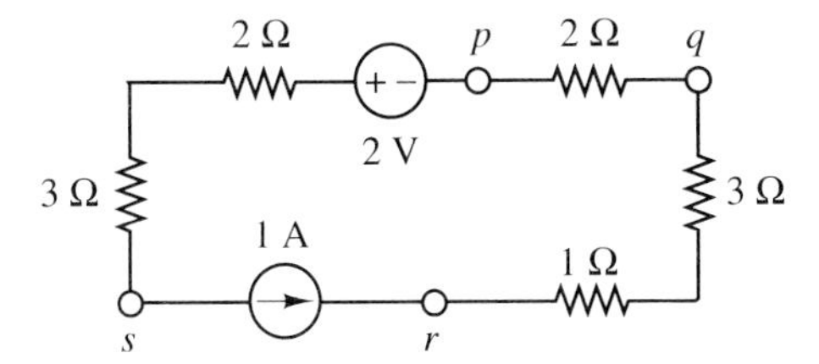


**Figure A4.7** If the current is given, as in this example (1 A), recognize the fact and use it to your advantage; don't try to solve for it.

- In Fig. A4.8a, $I$ is *not* equal to 2 V/30 Ω, because the voltage across the 30-Ω resistor is not in general 2 V. Similarly, in Fig. A4.8b, the current $I_2$ is *not* equal to $AV_1/R_2$, for the same reason. In using $I = V/R$, $V$ must be the voltage across $R$, not across a neighboring component such as a voltage source!

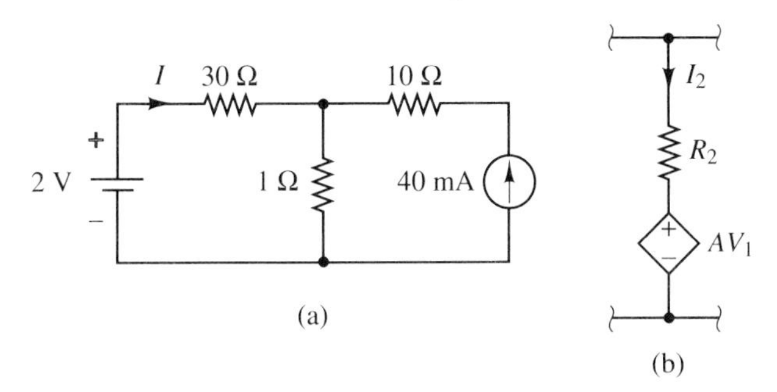


**Figure A4.8** (a) $I \neq 2$ V/30 Ω; (b) $I_2 \neq AV_1/R_2$.

## PROBLEMS

**4.1.** In Fig. P4.1, find $I_1$ and $I_2$.

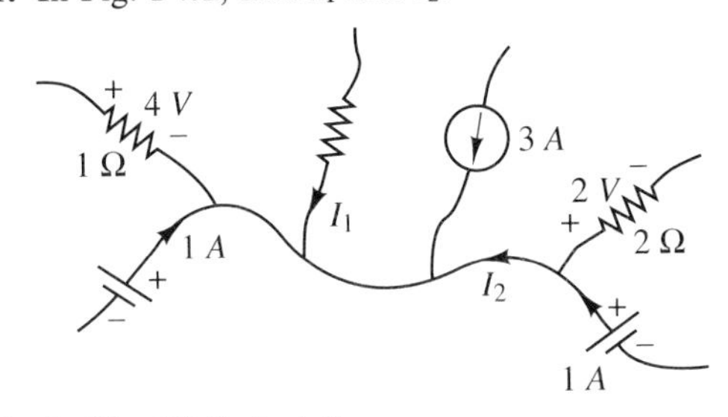


**Figure P4.1**

**4.2.** In Fig. P4.2, find $R_{eq}$.

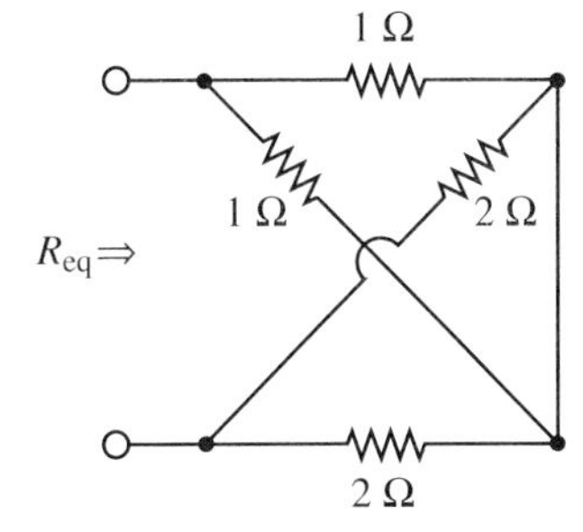


**Figure P4.2**

**4.3.** In Fig. P4.3, find $R_{eq}$.

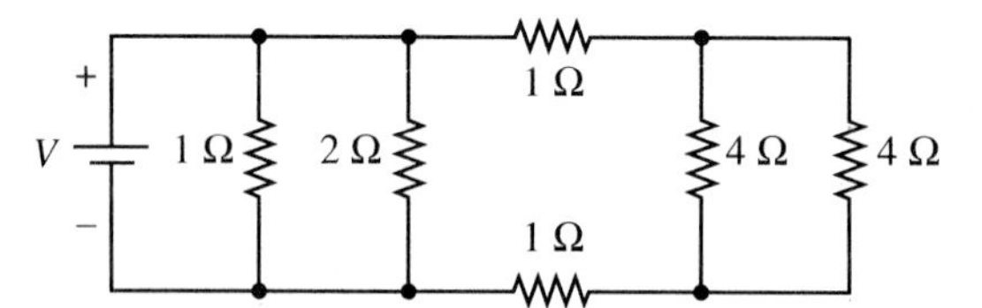


**Figure P4.3**

**4.4.** Find $I_2$ and $I_3$ using KCL and Ohm's law in Fig. P4.4.

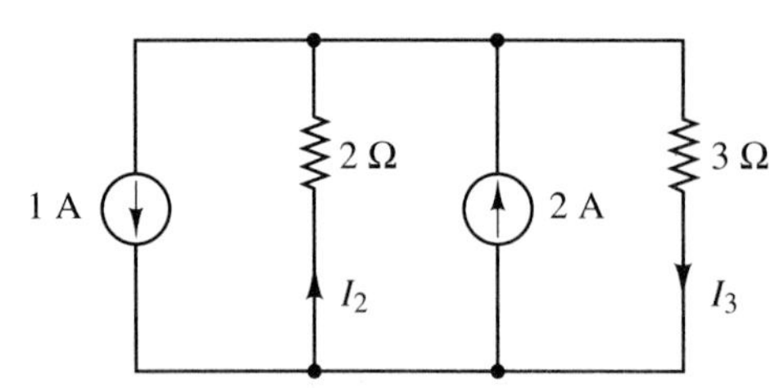


**Figure P4.4**

**4.5.** In Fig. P4.5, find $R_{eq}$.

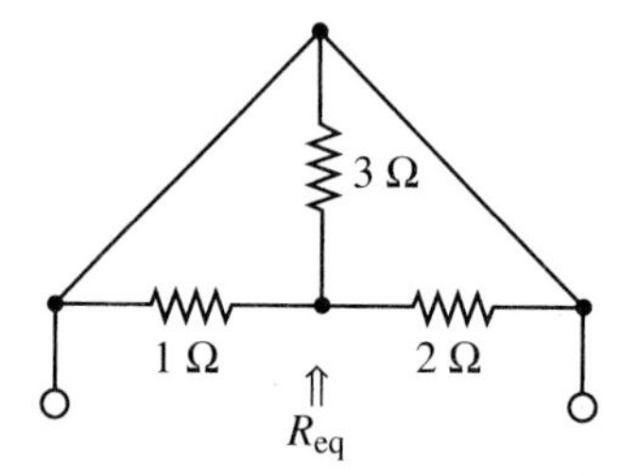


**Figure P4.5**

**4.6.** Replace the hypotenuse wire of the left right triangle in Fig. P4.5 by a 4-Ω resistor. Now find $R_{eq}$. (*Hint:* Redraw the circuit to look more familiar.)

**4.7.** Suppose three capacitors $C_1 = 1$ μF, $C_2 = 2$ μF, and $C_3 = 3$ μF are connected in parallel with each other and a 200 V constant voltage source. Find $C_{eq}$, $q_{total}$, $q_1$, $q_2$, and $q_3$ (the charge on $C_{eq}$ and the charge on each capacitor).

**4.8.** **(a)** Suppose an inductor $L_1 = 1$ mH is in series with the parallel combination of two other inductors $L_2 = 2$ mH and $L_3 = 3$ mH. Find $L_{eq}$ seen by the source.

**(b)** A voltage source $v(t)$ is connected across this series-parallel combination of inductors. Find the voltage of self-induction $v_2$ across coil $L_2$ (+ side near the + terminal of $v(t)$) when the current $i_1$ in coil 1 is changing at the rate of 3 A/s. Let $i_1$, $i_2$, and $i_3$ be currents in the respective inductors directed away from the + terminal of the source $v(t)$.

**4.9.** Draw the circuit diagram of a standard two-D-cell flashlight. In which direction does positive current flow through the batteries: toward the light or toward the back? How and where does the current return?

**4.10.** Find the power absorbed by each element in Fig. P4.10.

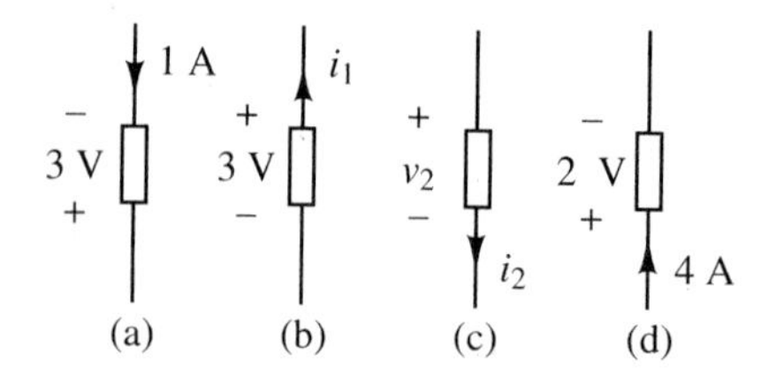


**Figure P4.10**

**4.11.** **(a)** Find the power absorbed by each element in Fig. P4.11. Which are absorbing energy, and which are delivering energy to the rest of the circuit?

**(b)** What passive element is the $20v_1$ source in this circuit equivalent to? Provide the value of that element.

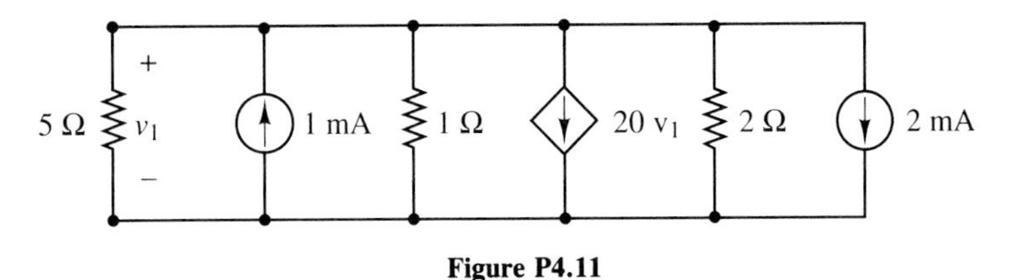


**Figure P4.11**

**4.12.** For each of the circuits in Fig. P4.12, find the indicated unknown ($v_x$ or $i_x$), by using voltage or current dividers.

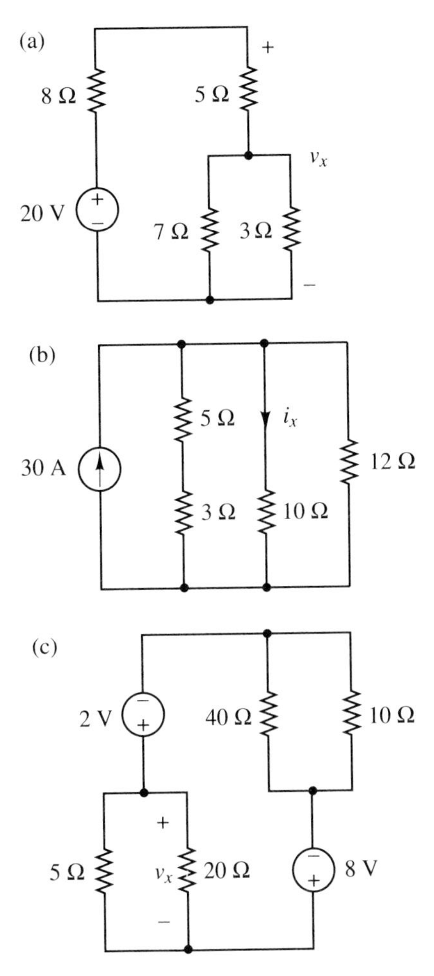


**Figure P4.12**

**4.13.** For the circuit in Fig. P4.13:

**(a)** Find $I_o$ by circuit reduction, that is, by replacing series and parallel resistor combinations with equivalent resistances.

(b) Find $I_1$ by using a current divider
(c) Find $V_1$ by using a voltage divider
(d) Find the power dissipated in the 3-Ω resistor.

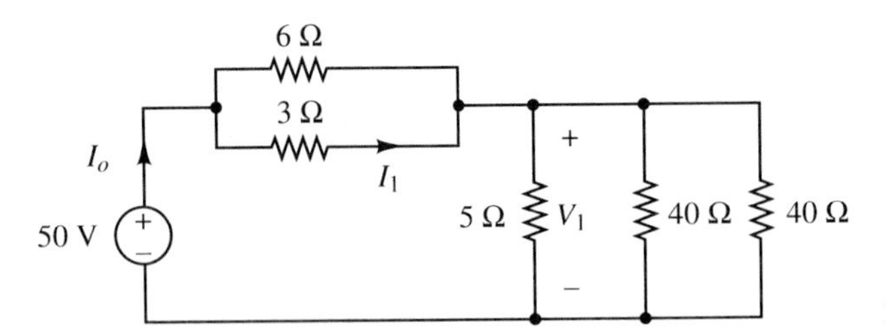


**Figure P4.13**

**4.14.** For the circuit in Fig. 4.40a, suppose that $v_s = 10$ V, $R_S = 100$ Ω, and $R_L = 400$ Ω. Find the power in the load $P_{RL}$ and the percent efficiency, $100\% \cdot P_{RL}/P_{vs}$, where $P_{vs}$ is the power available from the source $v_s$.

**4.15.** In Fig. P4.15, what are $R_1$ and $R_2$? Be careful with signs.

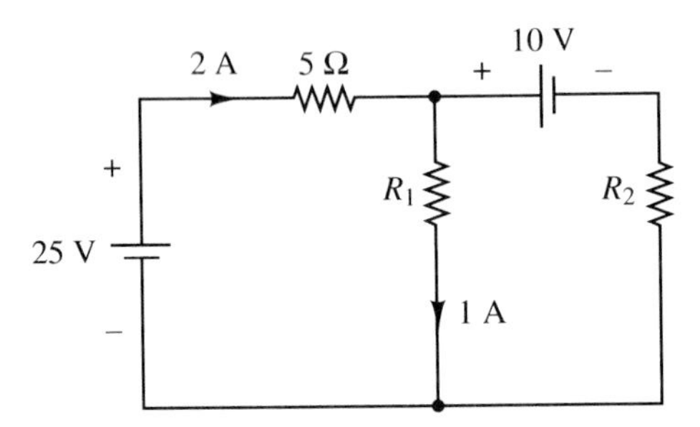


**Figure P4.15**

**4.16.** Find the resistances $R_1$ and $R_2$ in the circuit in Fig. P4.16. Also, find the power into the current and voltage sources. Does each absorb or generate electric power?

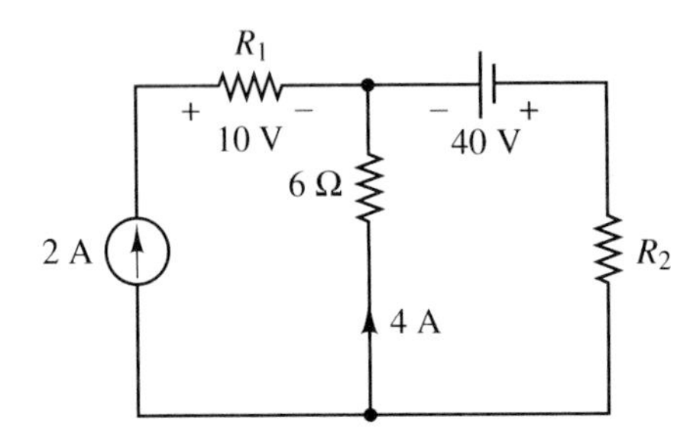


**Figure P4.16**

**4.17.** Compare the equivalent resistance of a compact disk player (which draws 12 W) with that of a self-cleaning oven (which draws 4000 W).

**4.18.** Is the resistance of an accurate ammeter very high or very low? Why? Repeat for a voltmeter rather than an ammeter.

**4.19.** Capacitors and inductors behave either like short circuits, open circuits, current sources, or voltage sources at particular instants in time. State what (i) capacitors and (ii) inductors act like (and why) under the following conditions:

(a) Initially disconnected and with zero energy stored in them, power is turned on; answer for the moment power is turned on.

(b) Initially, with nonzero energy stored in them, they are suddenly connected to a circuit that in the steady state will result in a change of energy in the capacitor or inductor; answer for the moment the capacitor or inductor is connected to the new circuit. This part also holds for being connected to the same circuit when the power supply or signal source suddenly changes value.

(c) In any constant-in-time excited circuit long after all transient effects have disappeared (the steady state). Note which (current or voltage) is zero in the steady state for capacitors and for inductors.

**4.20.** Based on your answers in Problem 4.19 or on information in the chapter concerning how capacitors and inductors behave in the steady state, determine the steady-state values $I_1$, $I_2$, $I_3$, $V_1$, $V_2$, and $V_3$ in Fig. P4.20.

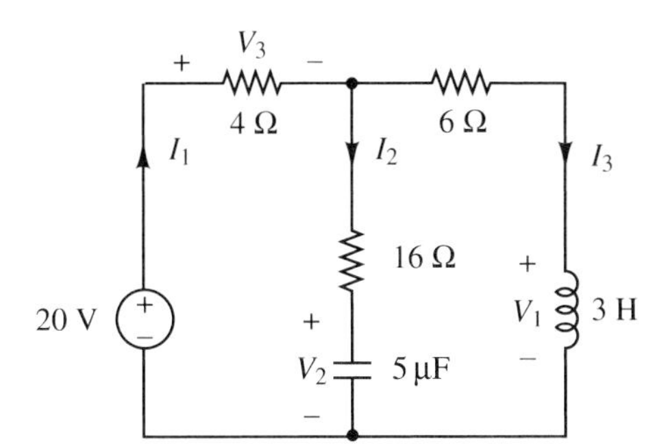


**Figure P4.20**

**4.21.** For the circuit of Fig. P4.21, find the steady-state values of $i_1$, $i_2$, $i_3$, $v_1$, $v_2$, and $v_3$ and the energy stored in each inductor and capacitor.

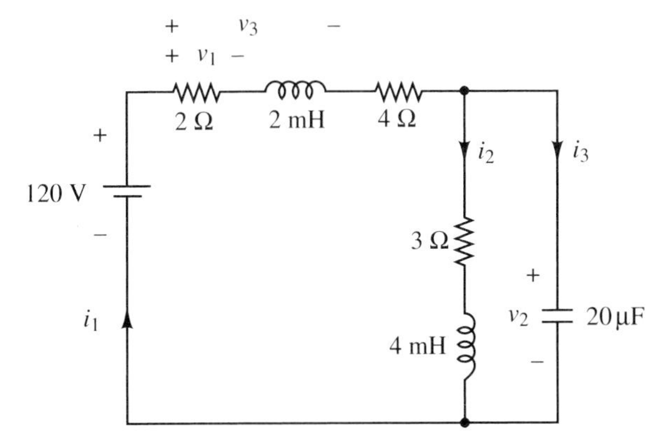


**Figure P4.21**

**4.22.** Design a simple space heater circuit that has the following features: Whenever the heating element is on, a fan runs; both require 120 V (ac). Whenever the temperature reaches the one desired, the circuit is automatically turned off by a thermostat switch. [The thermostat operates on the bending of dissimilar metals upon heating. The amount of bending (and thus, the temperature) required to throw the switch depends on the setting of a rotary knob.] The simple circuit you will obtain is an actual example of a commercial space heater.

**4.23.** For the circuit in Fig. P4.23, find $I_1$, $I_2$, $I_3$, $I_4$, and $V_3$ in terms of $V_A$, $V_B$, and the resistance values. Simplify where possible.

**4.24.** In Fig. P4.24, determine the reading of a digital voltmeter when connected (**a**) red to $m$, black to $r$; (**b**) red to $q$, black to $r$; (**c**) red to $q$, black to $m$; (**d**) red to $p$, black to $n$.

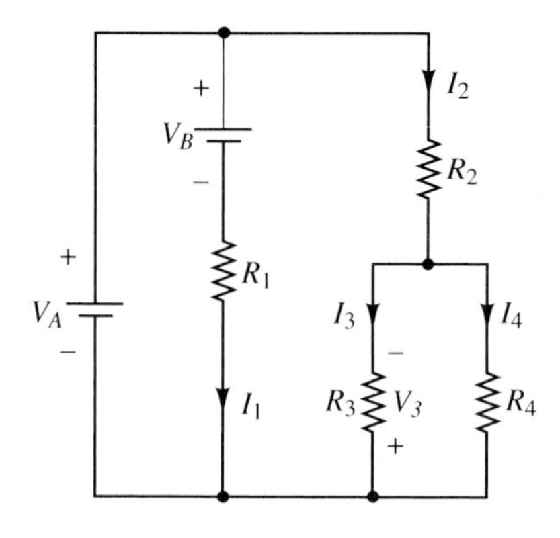


**Figure P4.23**

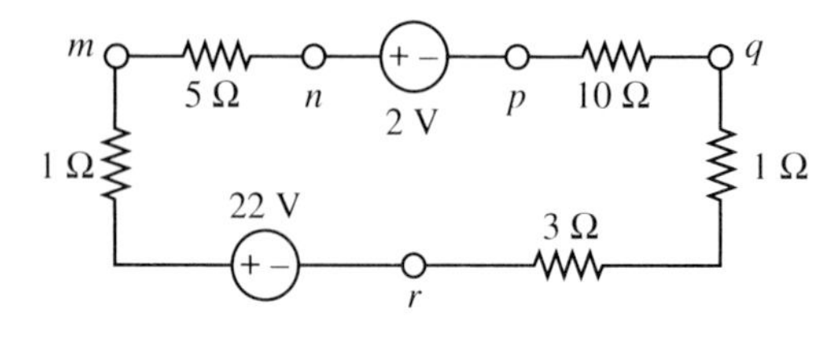


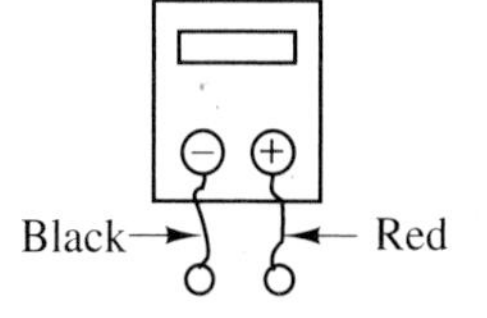


**Figure P4.24**

**4.25.** In the circuit in Fig. P4.25, determine **(a)** $v_{sr}$, **(b)** $v_{pq}$, and **(c)** $v_{qs}$. **(d)** What power does the 1-A source absorb? Is it supplying energy to the rest of the circuit, or receiving it? Why?

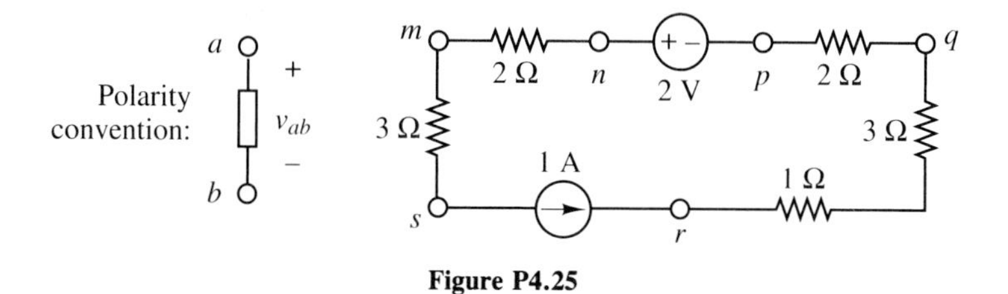


**Figure P4.25**

**4.26.** For the circuit in Fig. P4.26, find $V_{ab}$, $V_{fd}$, $V_{bc}$, and the power into the 2-A current source. (Is it supplying or receiving energy?) Use the polarity convention given in Fig. P4.25.

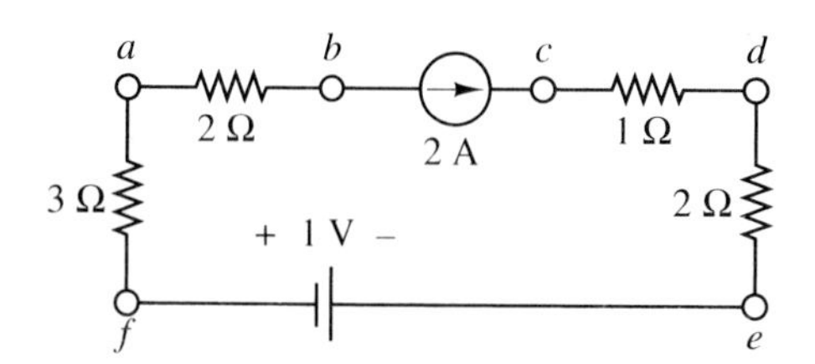


**Figure P4.26**

**4.27.** Resistors $R_A$, $R_B$, and $R_C$ are all connected in series with a voltage source of constant-in-time voltage V. Let $I$ be the current directed out of the + terminal of the source $V$, and $V_A$ and $V_B$ be respective voltages across $R_A$ and $R_B$ with + sides nearest the + side of $V$. Find:

(**a**) The total resistance across the voltage source.

(**b**) The current $I$.

(**c**) The voltage $V_B$.

(**d**) The power absorbed by $R_C$.

(**e**) Suppose that $R_A = 2\ \Omega$, $R_B = 1\ \Omega$, and $R_C = 6\ \Omega$. Find $V$ such that $V_A = 4$ V.

**4.28.** For the circuit in Fig. P4.28:

(**a**) Find the power into element $a$; is it absorbing or generating power?

(**b**) If the power into element $b$ is $-24$ W, indicate on the diagram the actual voltage, including polarity, across element $b$.

(**c**) If element $c$ is a 3-Ω resistor, what is the total power into the three-element network?

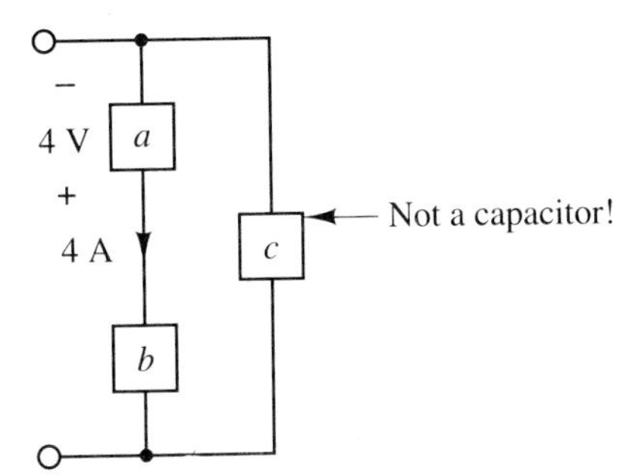


**Figure P4.28**

**4.29.** (**a**) To determine the $i$–$v$ characteristic of the circuit fragment shown in Fig. P4.29, a variable external source $v_T$ is applied to the terminals, and the resulting terminal current $i_T$ is measured. Predict the current $i_T$ that would be measured were this experiment carried out. Your answer should be in terms of $v_T$, $I_S$, $R_1$, and $R_2$ (not in terms of $v_1$). Simplify your expression as much as possible.

(**b**) Draw the $i_T$–$v_T$ characteristic, with all axes and pertinent points fully labeled (e.g., intercepts) and indicating the value of slopes of any lines. Assume that $I_S > 0$.

(**c**) Suppose that the source $v_T$ were replaced by a resistor $R_L$. Determine the operating point ($v_T$, $i_T$), where now the voltage $v_T$ is due only to the action of the current source $I_S$.

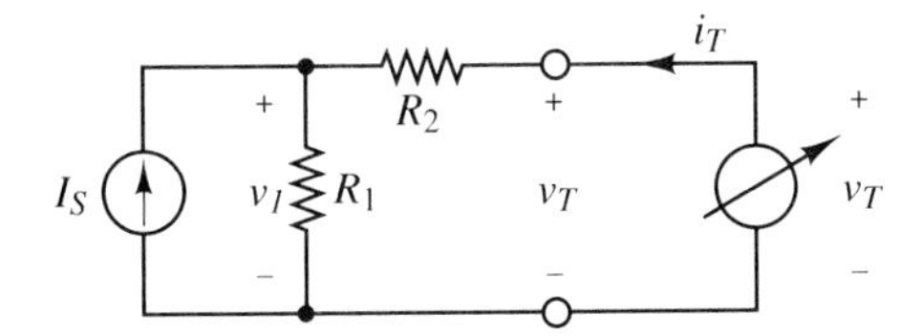


**Figure P4.29**

**4.30.** Suppose that a 35-V constant-in-time voltage source is connected across the following network: a 5 Ω resistor in series with the parallel combination of a 10 Ω and a 4 Ω resistor. Find the current $i$ in the 4 Ω resistor directed away from the + terminal of the voltage source and the voltage across the 10 Ω resistor, where the + side is nearer the + side of the voltage source. Use voltage and/or current dividers.

**4.31.** Design a current divider that has an equivalent resistance of 50 Ω and divides the current in a ratio of 3:1.

**4.32.** Write down all the equations needed for loop analysis of the circuit in Fig. P4.32. From your solution, give the magnitude and direction of the current in every resistor in the circuit.

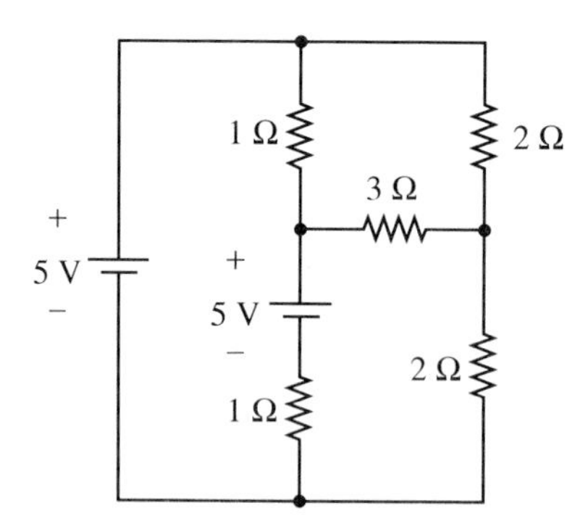


**Figure P4.32**

**4.33.** For the circuit in Fig. P4.33, (**a**) use mesh (loop) analysis to solve for the current in the 200 V-source (provde polarity), and (**b**) find $V_I$.

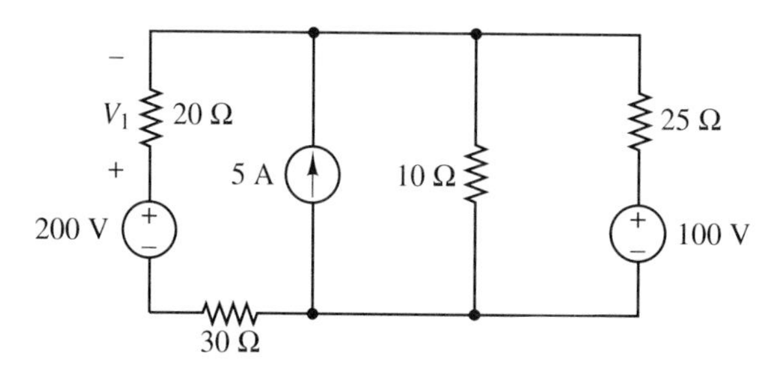


**Figure P4.33**

**4.34.** In one sense, a current-dependent voltage source is like a resistance, but in one very important respect it is different. How is this dependent source a "resistance/source hybrid"? Differentiate between the two, and describe the dependent source from the viewpoint of a source element. Note that there is another major distinction concerning power. Repeat for a voltage-dependent current source. (*Hint:* Think of the dependent source as a generalization of resistance, where the resistance is a "self-dependent" dependent source.)

**4.35.** Find $I$ in Fig. P4.35 using source transformations.

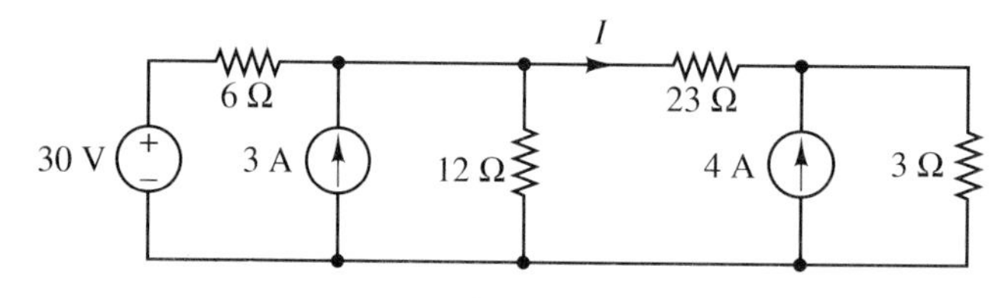


**Figure P4.35**

**4.36.** In Fig. P4.36, use mesh analysis to solve for the current $I_1$. Ignore the voltage $V_x$ in this problem.

**4.37.** In Fig. P4.36, use nodal analysis to find $V_x$.

**4.38.** (**a**) Find $v_{oc}$ in the circuit in Fig. P4.38 using mesh analysis.

(**b**) Find the equivalent resistance seen from terminals $A$ and $B$.

(**c**) Draw and fully label the Norton equivalent circuit for this network, using your results from parts (a) and (b).

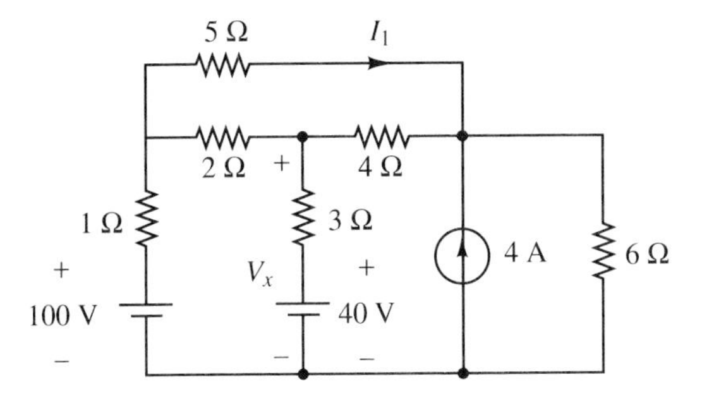


Figure P4.36

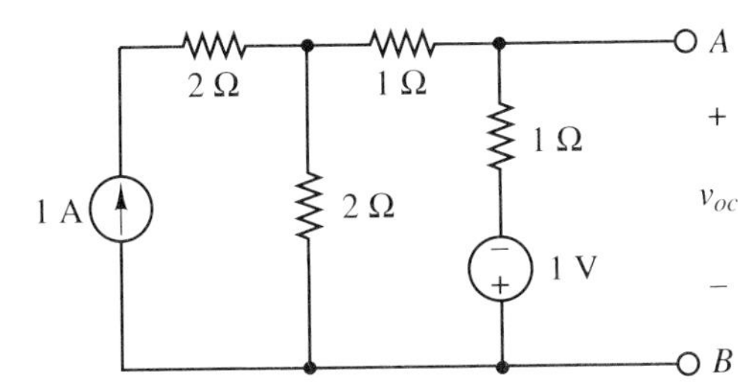


Figure P4.38

**4.39.** For the circuit in Fig. P4.39, use loop analysis to obtain a matrix equation $\boldsymbol{R}\mathbf{I} = \mathbf{V}$, where $\mathbf{I}$ is the vector for loop currents, $\boldsymbol{R}$ is a matrix, and $\mathbf{V}$ is a vector both of whose elements you must determine. Assume that $R_1, R_2, R_3, V_A, V_B, I_A$, and $I_B$ are all known parameters. Solve the matrix equation for the loop currents. Using these, determine the magnitude and direction of the currents in resistors $R_1$, $R_2$, and $R_3$, and the current in the source $V_B$.

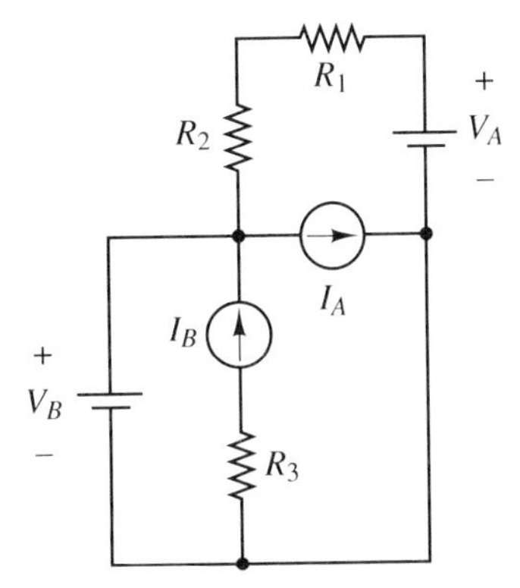


Figure P4.39

**4.40.** Using nodal analysis with Cramer's rule, find the current in each resistor in the circuit in Fig. P4.40. In your answer, be sure to define the direction of each resistor current.

**4.41.** In Fig. P4.41, use nodal analysis to obtain two equations for the two node voltages. Write the nodal matrix equation and solve for the node voltages and $i_x$. Indicate the units of the matrix and vector elements.

**4.42.** The whole idea of Thévenin equivalent circuits is that as far as $R_L$ is concerned, a complicated network $A$ (see Fig. P4.42) is indistinguishable from the simple Thévenin network $B$. When loading considerations and calculations are being made, it is far easier to deal with the entirely equivalent network $B$, once its parameters are found,

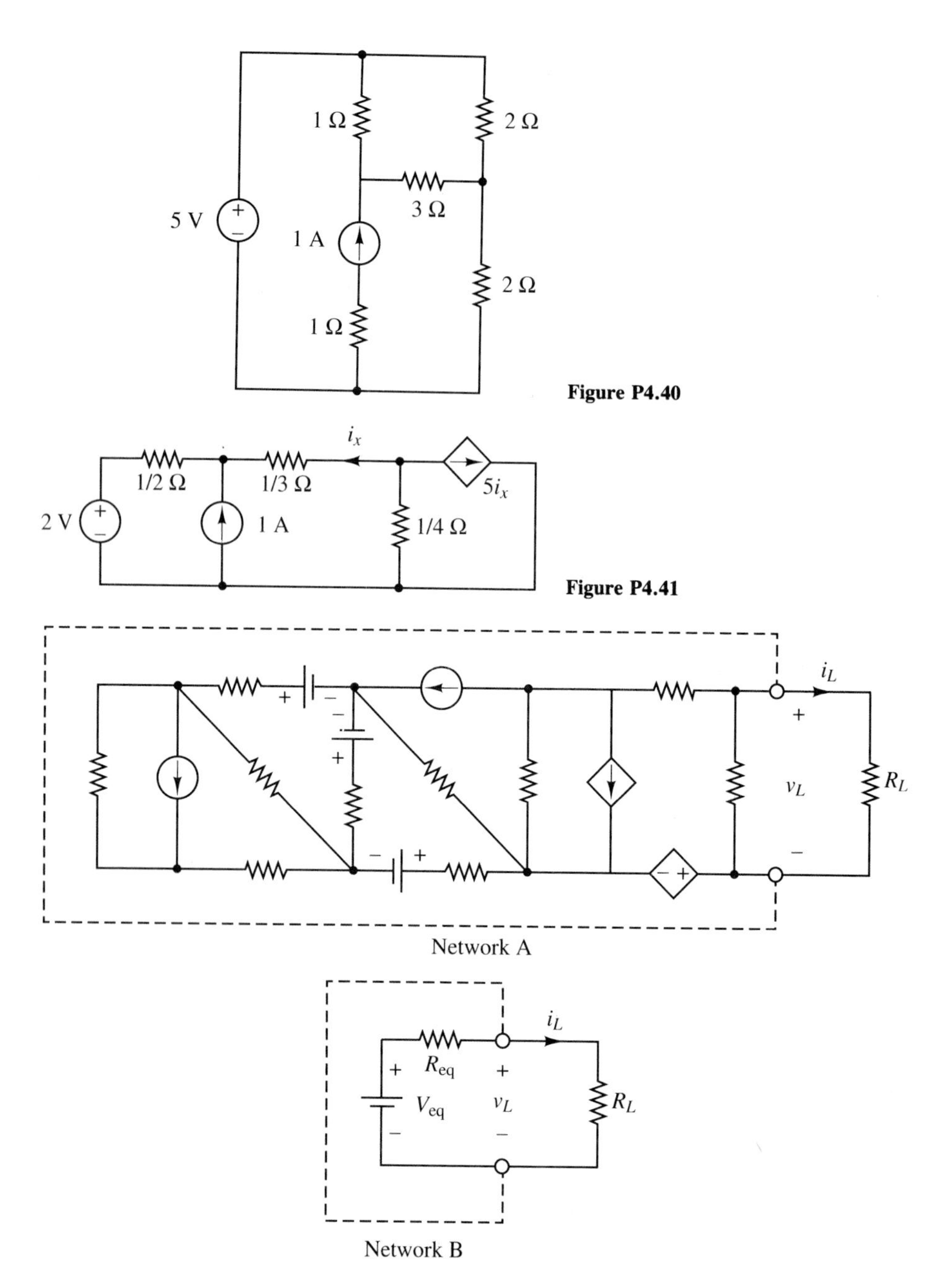


Figure P4.40

Figure P4.41

Figure P4.42

than with $A$. Argue logically exactly how measurements of $v_{oc}$ and $i_{sc}$ made on network $A$ are used to determine $v_{eq}$ and $R_{eq}$ for network $B$. (*Hint:* Use symbology such as $v_{oc,A}$, $v_{oc,B}$, $i_{sc,A}$, and $i_{sc}, B$.)

**4.43.** Why in Fig. 4.44b do we draw the $i_{sc}$ pointing down?

**4.44.** For the circuit shown in Fig. P4.44:

(**a**) Find the Thévenin equivalent circuit looking into terminals $a$ and $b$.

(b) A 5-Ω resistor is placed across the terminals. Find the power dissipated in the 5-Ω load resistor.

(c) What value of load resistance $R_L$ should be placed across terminals $a$ and $b$ so that maximum power is delivered?

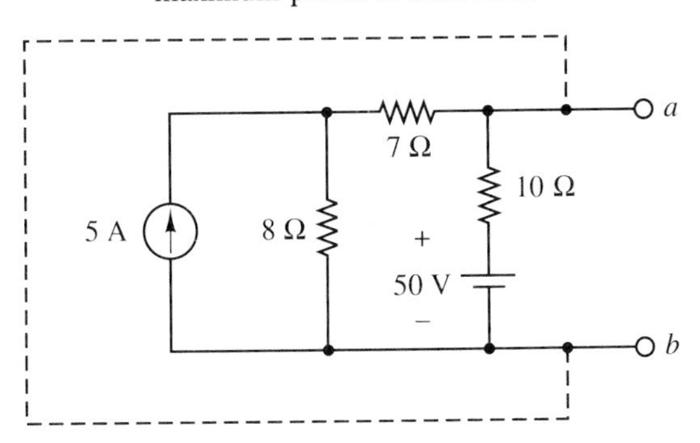


**Figure P4.44**

**4.45.** Find the voltage gain $A_v = v_3/v_1$ of the circuit in Fig. P4.45 using nodal analysis.

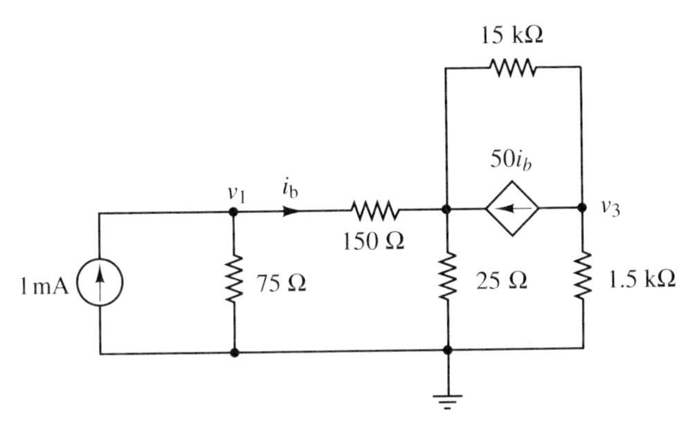


**Figure P4.45**

**4.46.** The circuit in Fig. P4.46 is a realistic model of an integrated-circuit electronic amplifier, which we study in detail in Chapter 7.

(a) Find the nodel analysis matrix equation; do not invert. Consider $v_s$, $A$, and all the resistors as known quantities. Express your answer (matrix equation) in terms of these. (*Hint:* There are only two required node voltages, and they are already defined in Fig. P4.46.)

(b) Suppose that $R_1 = \infty$, $R_S \ll R_F$, $A/R_o \gg 1/R_F$, and $R_o \ll$ (both $R_F$ and $R_L$). Find $v_1$ and $v_o$ by simplifying the matrix you found in part (a), then solving it by Cramer's method. Simplify.

(c) As $A \to \infty$, what values do $v_1$ and $v_o$ in part (b) approach?

(d) Let $R_S = 30\ \Omega$, $R_F = 2000\ \Omega$, $R_1 = 10{,}000\ \Omega$, $R_o = 30\ \Omega$, $R_L = 8\ \Omega$, and $A = 3000$. Using the matrix equation in part (a), what is the voltage gain, $v_o/v_s$, that is, the output voltage $v_o$ divided by the input voltage $v_s$? Compare this result with the approximate results for $v_o/v_s$ obtained in parts (b) and (c). Are they all close?

(e) Using the component values given in part (d), find the Thévenin equivalent seen by $R_L$. (Do *not* turn off $Av_1$. Find $R_{\text{eq}} = v_{\text{oc}}/i_{\text{sc}}$.)

**4.47.** In the circuit in Fig. P4.47, use nodal analysis to find $I_2$.

**4.48.** (a) Find $v$ in Fig. P4.48 by superposition. (*Hint:* Think "dividers.")

(b) Find $i_x$ in Fig. P4.41 by superposition.

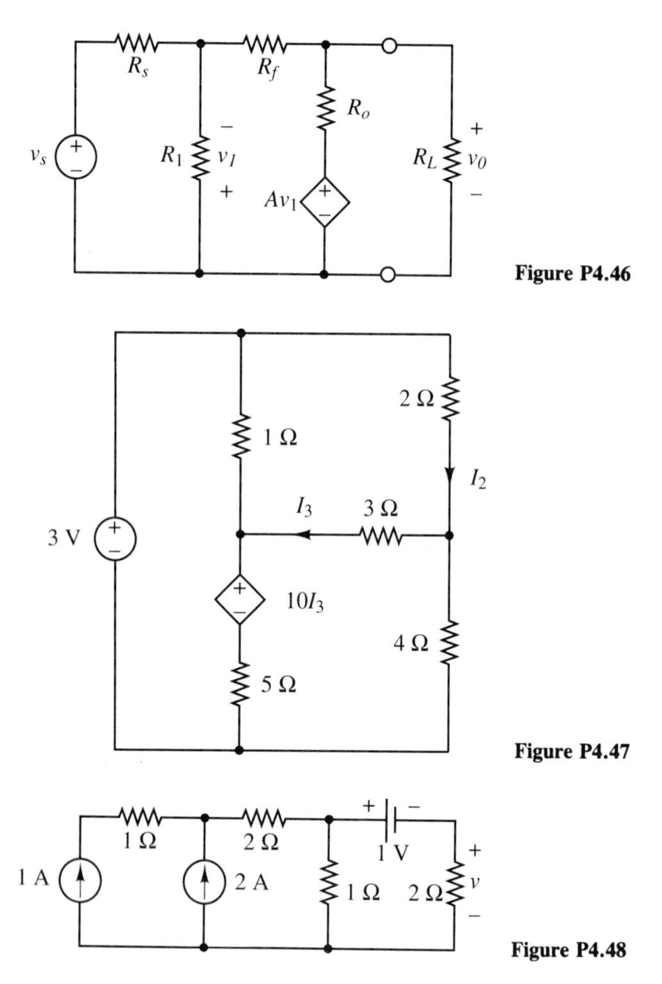


Figure P4.46

Figure P4.47

Figure P4.48

**4.49.** In the circuit of Fig. P4.49, find the current $i_{90}$ in the 90-Ω resistor by using superposition.

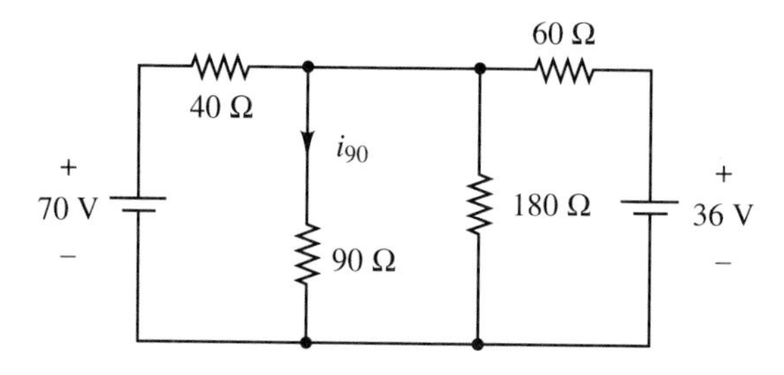


Figure P4.49

**4.50.** In the circuit in Fig. P4.50, solve for $I_L$ by **(a)** loop analysis, **(b)** node analysis, and **(c)** superposition. Which method was easiest, and why?

**4.51.** For the circuit in Fig. 4.40, draw the load voltage, the load current, and the load power as functions of $R_L$. Draw all on the same axes, to show how the power is maximized at $R_L = R_S$. Note points of inflection, if any, for each curve. (Do $v_L$ and $i_L$ change concavity "up or down" for any value of $R_L$?) In particular, prove that $R_L = R_{eq}$ is

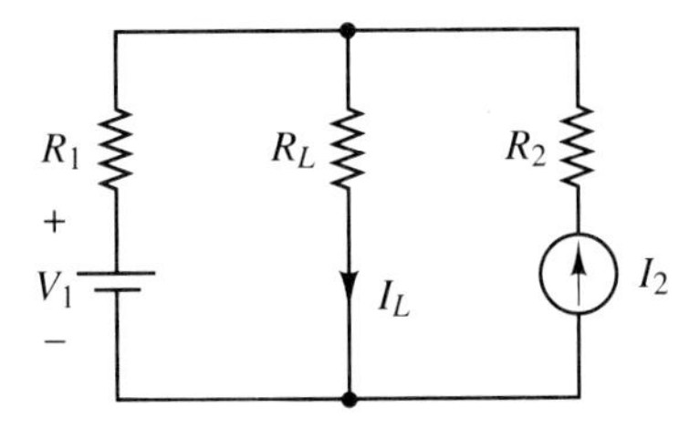

Figure P4.50

a maximum, not a minimum, by showing the concavity of $p_L$ at $R_L = R_{eq}$. For what value of $R_L$ does $p_L$ change concavity?

**4.52.** In the circuit of Fig. P4.52, for what value of $R_L$ is maximum power delivered to it? What is that power? (*Hint:* Use the Thévenin equivalent.)

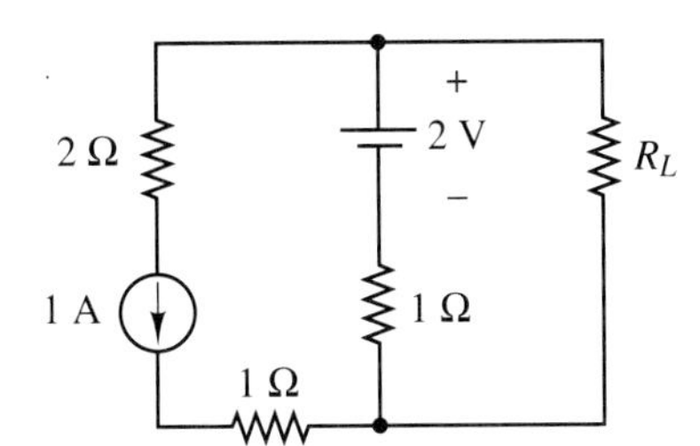

Figure P4.52

**4.53.** (**a**) Find the Thévenin equivalent of the circuit in Fig. P4.53. (Use mesh analysis to help you.)
(**b**) What is the Norton equivalent of this circuit?

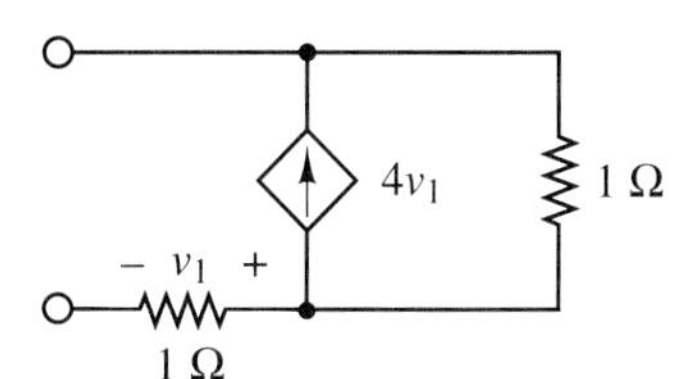

Figure P4.53

**4.54.** Suppose that we had a device with the $i$–$v$ characteristic shown in Fig. P4.54, and we connected it across a practical voltage source of voltage $V_S$ and internal resistance $R_S$.
(**a**) Sketch the $i$–$v$ characteristic of the source and the device on the same set of axes. For the source, here define $i$ to be coming out of the terminal at higher potential. Label all critical points and give the slope.
(**b**) Determine the operating point $(i, v)$ if
(i) $V_S = 10$ V, $R_S = 10\ \Omega$
(ii) $V_S = 10$ V, $R_S = 100\ \Omega$
(iii) $V_S = 0.5$ V, $R_S = 10\ \Omega$

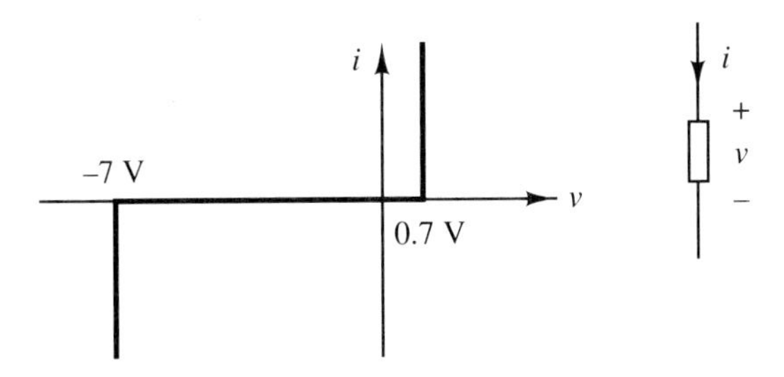

Figure P4.54

(iv) $V_S = -1$ V, $R_S = 10\ \Omega$
(v) $V_S = -10$ V, $R_S = 100\ \Omega$
We shall see in Chapter 7 that the device whose $i$–$v$ characteristic is given in Fig. P4.54 is a solid-state rectifier (diode).

**4.55.** Consider the circuit in Fig. P4.55. Where convenient, denote two resistors $R_A$ and $R_B$ in parallel as $R_A \| R_B$.

**(a)** What units does the coefficient 2 have in the dependent source $2v_x$?

**(b)** Find $R_L$ such that maximum power is delivered to it, and find that maximum load power. (*Hint:* First find the Thévenin equivalent circuit the load sees.)

**(c)** Using your work above, give both the Thévenin and Norton equivalent networks seen by $R_L$.

**(d)** Using (for example) node analysis, calculate for the condition of maximum power transfer examined in parts (b) and (c) the power lost in $R_1$, $R_2$, and $R_3$ as well as the power delivered by the source $V_1$.

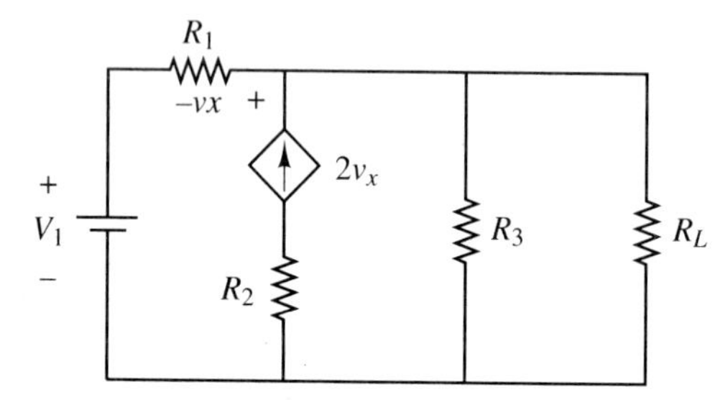


**Figure P4.55**

**4.56.** Find the Thévenin equivalent of the circuit in Fig. P4.56. Draw the complete Thévenin equivalent circuit diagram with all parameters determined. (*Hint:* Use loop analysis to find $V_{oc}$. Assume that $R_1$, $R_2$, $R_3$, $I_A$, and $V_A$ are all known.)

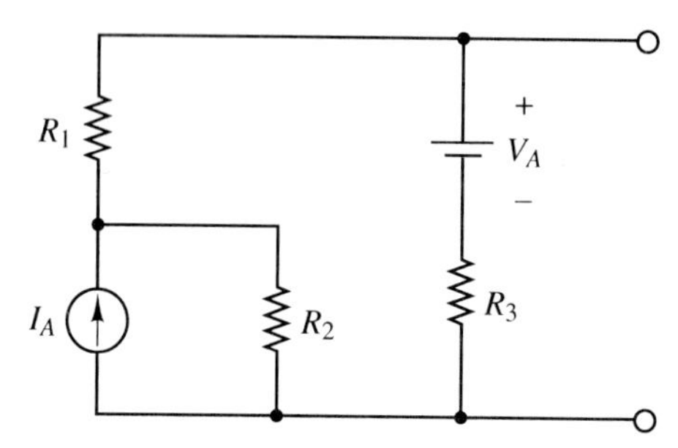


**Figure P4.56**

**4.57.** Find the Thévenin equivalent of the circuit in Fig. P4.57. Draw the complete Thévenin equivalent circuit diagram with all parameters determined. (*Hint:* Apply a 1-A test source and use $R_{eq} = v_T/1$ A; solve for $v_T$ by nodal analysis. Assume that $R_1$ and $R_2$ are known.)

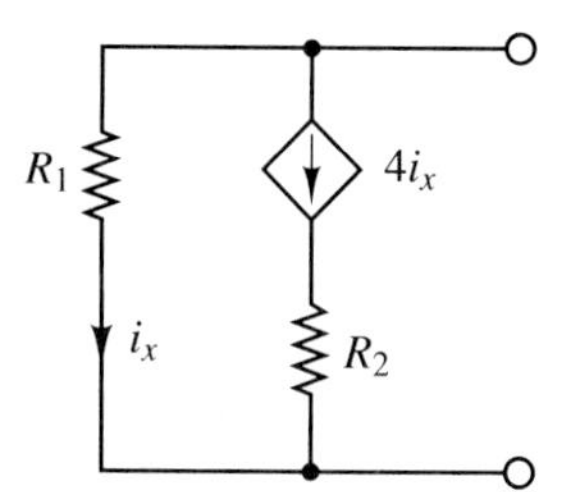


**Figure P4.57**

**4.58.** **(a)** For the circuit in Fig. P4.58, find the nodal analysis matrix equation; do not invert. Consider $I_o$, $A$, and all the resistors as known quantities. Express your answer (matrix equation) in terms of these.

**(b)** Now let $R_1 = \infty$ for this part *only*. Now solve for the output voltage $V_o$ using the simpler matrix equation by Cramer's method. Simplify your result as much as possible; explain why your result does or does not make sense.

**(c)** Now assume that $R_1 = 1\ \text{k}\Omega$, $R_2 = 2\ \text{k}\Omega$, $R_3 = 3\ \text{k}\Omega$, $R_L = 4\ \text{k}\Omega$, $I_o = 10$ mA, $A = 100$. Find $I_1$ and $V_o$.

**(d)** Using the component values in part (c), find the Norton equivalent circuit seen by $R_L$.

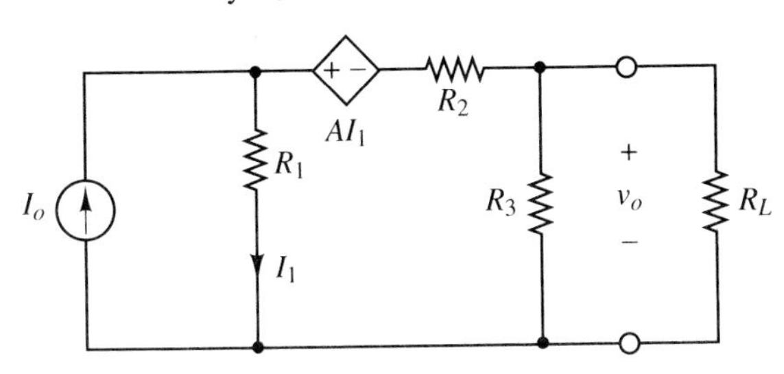


**Figure P4.58**

**4.59.** Use a source transformation and KVL to aid you in finding the Thévenin equivalent circuit of the network shown in Fig. P4.59. Recall that in this case you must find *both* $V_{oc}$ and $I_{sc}$.

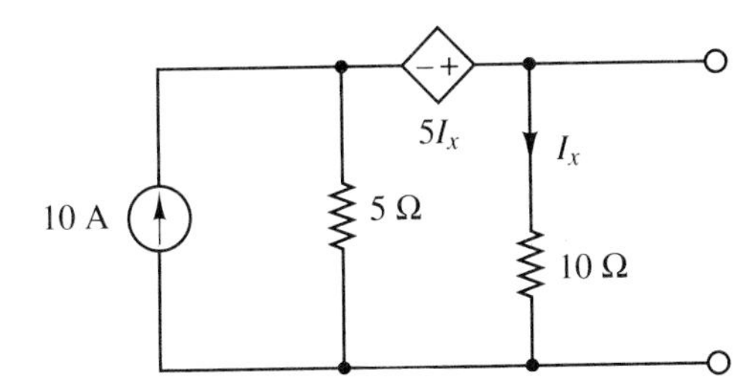


**Figure P4.59**

**4.60.** Find the Thévenin equivalent of the network in Fig. P4.60 as seen by $R_L$.

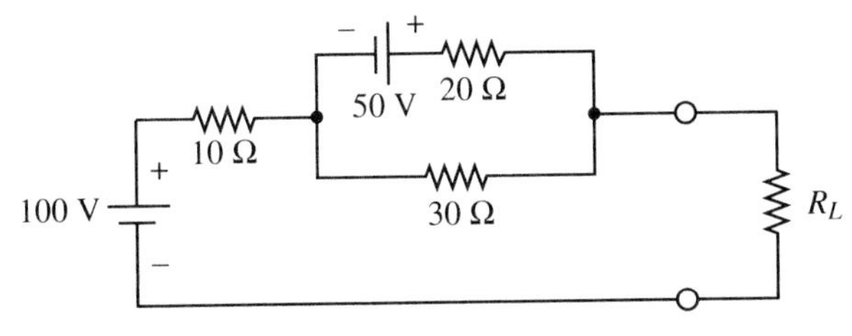


**Figure P4.60**

**4.61.** **(a)** Find the Thévenin equivalent of the circuit in Fig. P4.61, as seen from the terminals $A$ and $B$. [*Hints:* (i) There are no independent sources in this circuit, which should tell you something about and how to find $R_{th}$. (ii) Use is made of KVL, KCL, and Ohm's law in finding $R_{th}$.]

**(b)** Find the Norton equivalent.

**4.62.** For the circuit in Fig. P4.62:

**(a)** Find the Thévenin equivalent seen by the 2-Ω resistor.

**(b)** Find the Norton equivalent seen by the 3-Ω resistor.

**4.63.** Compare an inductor with coil resistance included (series $RL$ combination) driven by

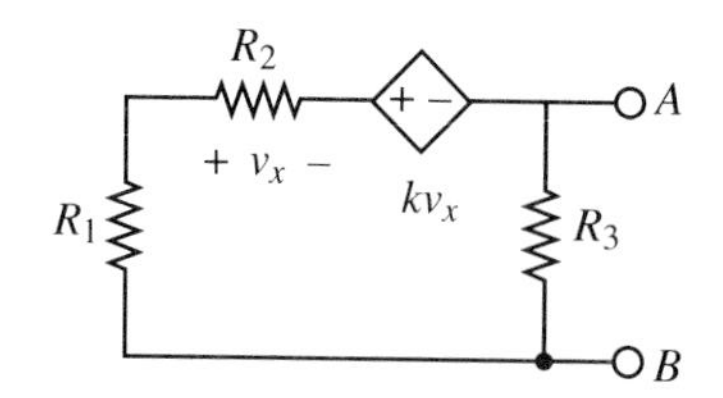


Figure P4.61

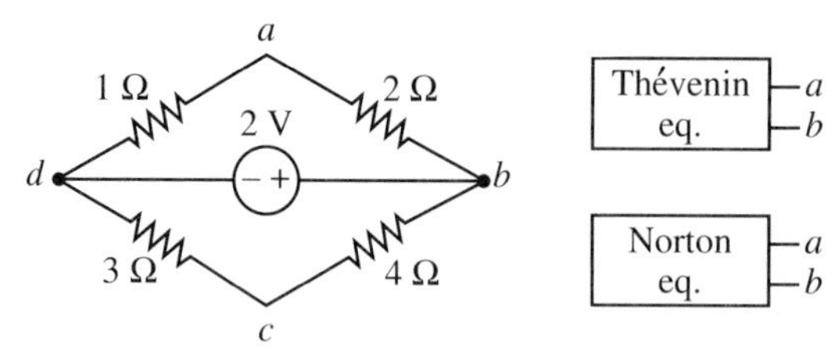


Figure P4.62

a voltage source with the motion of a mass subject to a force, where friction is involved. Write the corresponding differential equations. Which variables and parameters are analogous? Describe in as much detail as possible.

**4.64.** Suppose a voltage source $v_s(t)$ connected the parallel combination of $R = 5\ \Omega$, $C = 200\ \mu\text{F}$, and $L = 200$ mH. The respective currents flowing away from the $+$ terminal of $v_s$ are $i_R(t)$, $i_c(t)$, and $i_L(t)$. For $v_s$ having time dependence as shown in Fig. P4.64:

(**a**) Find $i_R(t_o)$, $i_C(t_o^-)$, $i_C(t_o^+)$, where $t_o = 1\ s$, assuming that $i_L(0) = 0$.

(**b**) Find the energy stored in $L$ and $C$ at $t_o$.

(**c**) Find the power supplied to $R$ at $t = 2\ s$.

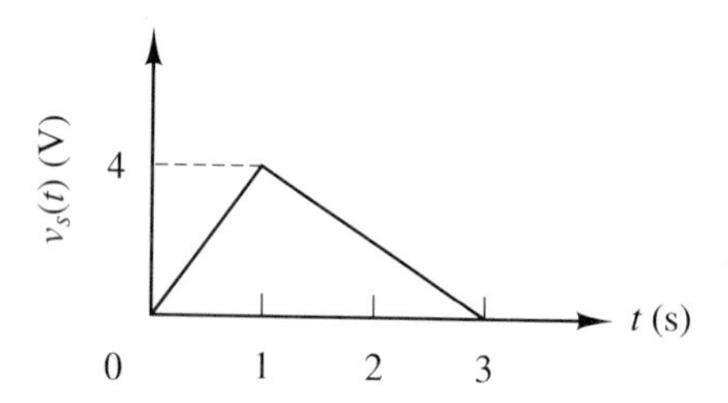


Figure P4.64

**4.65.** Suppose a voltage $v(t)$ is connected across the series combination of $R = 2\ \Omega$, $L =$ 10 H, and $C = 0.02$ F. The respective voltages with $+$ terminals nearer the $+$ terminal of $v$ are $v_R(t)$, $v_L(t)$, and $v_C(t)$, and the current $i(t)$ is that flowing out of the $+$ terminal of $v(t)$. For the current waveform shown in Fig. P4.65, find sketch $v_R(t)$, $v_L(t)$, $v_C(t)$, and $p_R(t)$, assuming $v_C(0) = 0$. Provide the maximum and minimum values of your curves.

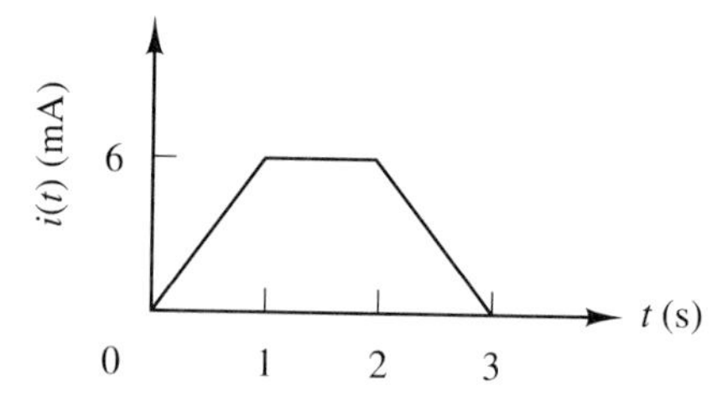


Figure P4.65

**4.66.** Using Eqs. (4.28) through (4.36) for the first-order capacitive circuit as a model, derive Eqs. (4.40) and (4.41) for the current in a first-order inductive circuit. Note that the analog to Eq. (4.34) should be

$$i(t) = \frac{1}{L}\int_0^t v_B(t')\, e^{-(t-t')/\tau_L}\, dt' + i(0)\, e^{-t/\tau_L}. \qquad \text{(P4.1)}$$

**4.67.** Compare Eq. (4.34) with Eq. (P4.1) dimensionally. What are the relations analogous to Eqs. (4.37b) and (4.38) for the first-order inductive circuit?

**4.68.** Plot the inductor voltage and current waveforms for the circuit in Fig. 4.50; fully label your axes and all important points on them. In both cases, draw curves on the same axes for two values of $\tau_L$, one larger than the other. For the current, make two graphs: one where $L$ changes for the two values of $\tau_L$ and $R$ is fixed, and the other for $L$ fixed and $R$ varying. Assume that $i(0) = 0$.

**4.69.** For the circuit in Fig. 4.48, suppose that $V_B = 10$ V, $R = 10$ kΩ, and $C = 100$ μF. Assume that the capacitor is initially uncharged. Find numerical values of $v_C(t)$, $v_R(t)$, and $i(t)$ when **(a)** $t = 1$ s and **(b)** $t = 10$ s.

**4.70.** Repeat Problem 4.69 for **(a)** $t = \tau_C$ and **(b)** $t = 5\tau_C$.

**4.71.** Consider the $RC$ network shown in Fig. 4.48. Given that $R = 2$ kΩ, $C = 10$ μF, $V_B = 100$ V, $v_C(t = 0^-) = 20$ V, and that the switch in Fig. 4.48 closes at $t = 0$:

**(a)** Find the time constant $\tau_C$.

**(b)** Determine equations for $i(t)$ and $v_C(t)$ for $t > 0$.

**(c)** Sketch $i(t)$ and $v_C(t)$ versus time.

**(d)** Find $i(t = 1.6\tau)$.

**4.72.** Suppose a constant-in-time voltage source $V_B = 120$ V is connected, via a switch, to $R_1 = 10$ kΩ in series with the parallel combination of $C = 10$ μF and $R_2 = 110$ Ω. Suppose that the switch $S$ has been closed a long time. Let $v_R$ be the voltage across $R_2$, with + terminal nearer the + terminal of $V_B$ and $i_R$ be the current in $R_2$ flowing away from the + terminal of $V_B$.

**(a)** Find the steady-state value of $v_R$.

**(b)** Now let the switch $S$ be opened at $t = 0$. Find expressions for and sketch $v_R(t)$ and $i_R(t)$, and give the time constant for this circuit.

**4.73.** Suppose that $C = 2$ μF is connected, via a switch, to $R = 10$ Ω. Let $v_C$ and $v_R$ be defined to have common − terminals, and let $i$ be the current flowing into the + side of $v_R$. Suppose that $v_C(0^-) = 25$ V, and the switch closes at $t = 0$.

**(a)** Find and sketch $v_R(t)$ for $t > 0$.

**(b)** Find $i(0^+)$ and $i(2\tau_C)$.

## ADVANCED PROBLEMS

**4.74.** To conserve energy, a college student turns on a 1500-W electric heater. For simplicity, assume that at his wall outlet, the constant voltage of 120 V is available. Two 75-W lights and a 12-W clock-radio draw power from the same circuit.

**(a)** How much current is drawn off the line?

**(b)** All of this current passes through a 15-A fuse having resistance 3 mΩ. How much power is dissipated in the fuse?

**(c)** If the heating due to this power dissipation in the fuse causes its current rating to degrade by 700 mA/h, how long will it be before the student is left cold and in the dark?

**4.75.** A component designated "X" has an $i$–$v$ characteristic that is a circle centered on (0 V, 0 A) and passing through the points (0 V, 1 A) and (1 V, 0 A). Its terminals are connected across a 1 Ω resistor, such that $i$ has polarity defined such that it flows into the designated (symbolic) + side of $v$. What are the possible operating points of this circuit? Show the graphical solution.

**4.76.** A simplified model of a long transmission line is the infinite-length network shown in Fig. P4.76. There is a small amount of conduction between the wires (modeled by the $100R$ resistances), distributed along the wires, and also a series resistance for each unit length of wire, symbolized by the $R$ resistances. Compute the equivalent resistance $R_{eq}$ looking into the terminals of one end of this "infinitely long" transmission line. (*Hint:* Adding one repeated element does not change the value of $R_{eq}$, so the rest of the line, looking just like $R_{eq}$, can in fact be replaced by $R_{eq}$. This makes the problem very easy! Also note that $R_{eq}$ must be positive.)

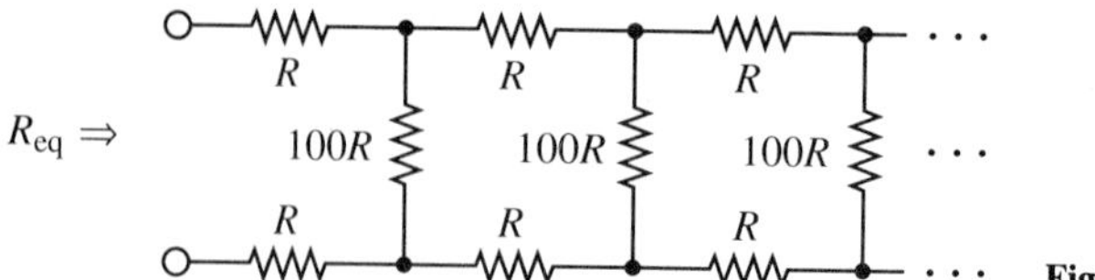


**Figure P4.76**

**4.77.** Find the equivalent resistance $R_{eq}$ of the resistive network shown in Fig. P4.77. (*Hint:* $R_{eq}$ *cannot* be found by series–parallel combinations.)

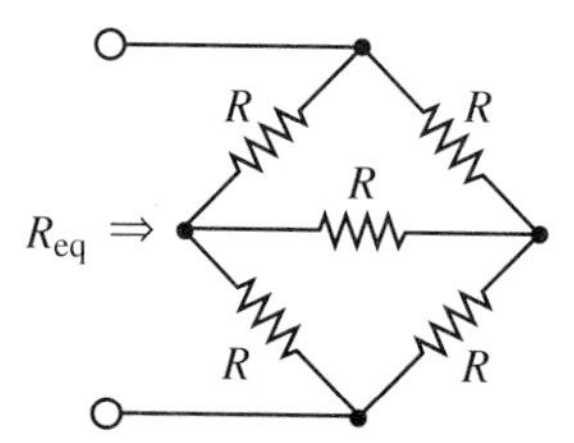


**Figure P4.77**

**4.78.** Find the Thévenin equivalent of the integrated-circuit electronic amplifier described in Problem 4.46 seen by $R_L$ for the polarity of $v_1$ defined oppositely to that in Fig. P4.46. Do not at first substitute any numbers for the resistances. Then substitute the values from Problem 4.46, part (d). What strange result do you obtain? Can you comment on this result?

**4.79.** A node can be thought of as part of a capacitor with respect to ground if in the actual circuit it runs along parallel with the ground wire. If this is the case, cannot charge be stored on that node? If so, does KCL (net charge into node is zero) still hold?

**4.80.** Find the Norton equivalent seen by $R_L$ in Fig. P4.80.

**4.81.** For the circuit in Fig. P4.81:

**(a)** Find the differential equation satisfied by the loop current $i_2(t)$.

**(b)** Let $v_g = 8e^{-2t}$ V, $i_1(0^+) = 2$ A, and $i_2(0^+) = 9$ A. Find $di_2(t)/dt$ at $t = 0^+$.

**(c)** For the values given in part (b), show that $i_2 = 3e^{-t} + 4e^{-2t} + 2e^{-6t}$ A is the correct solution by substituting it into the differential equation for $i_2(t)$ found in part (a).

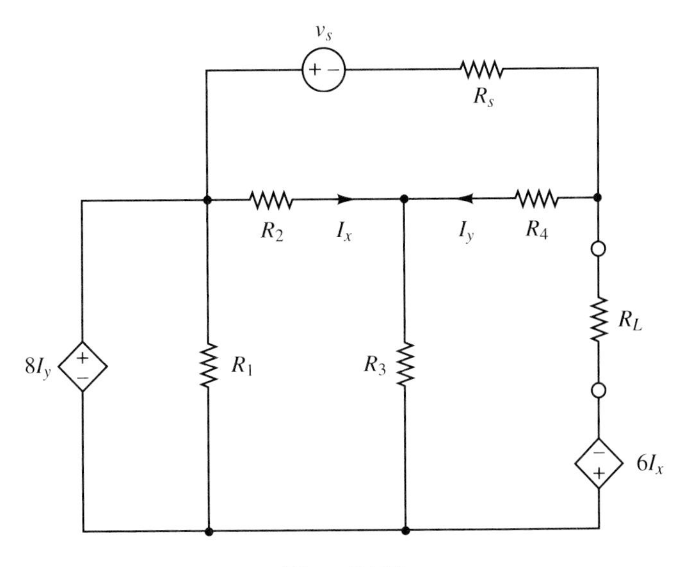


**Figure P4.80**

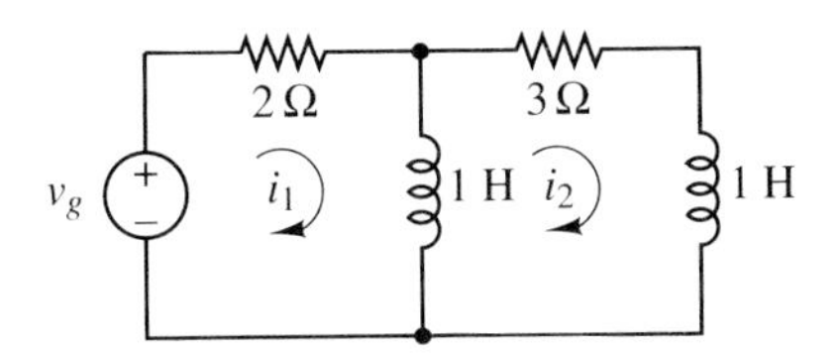


**Figure P4.81**

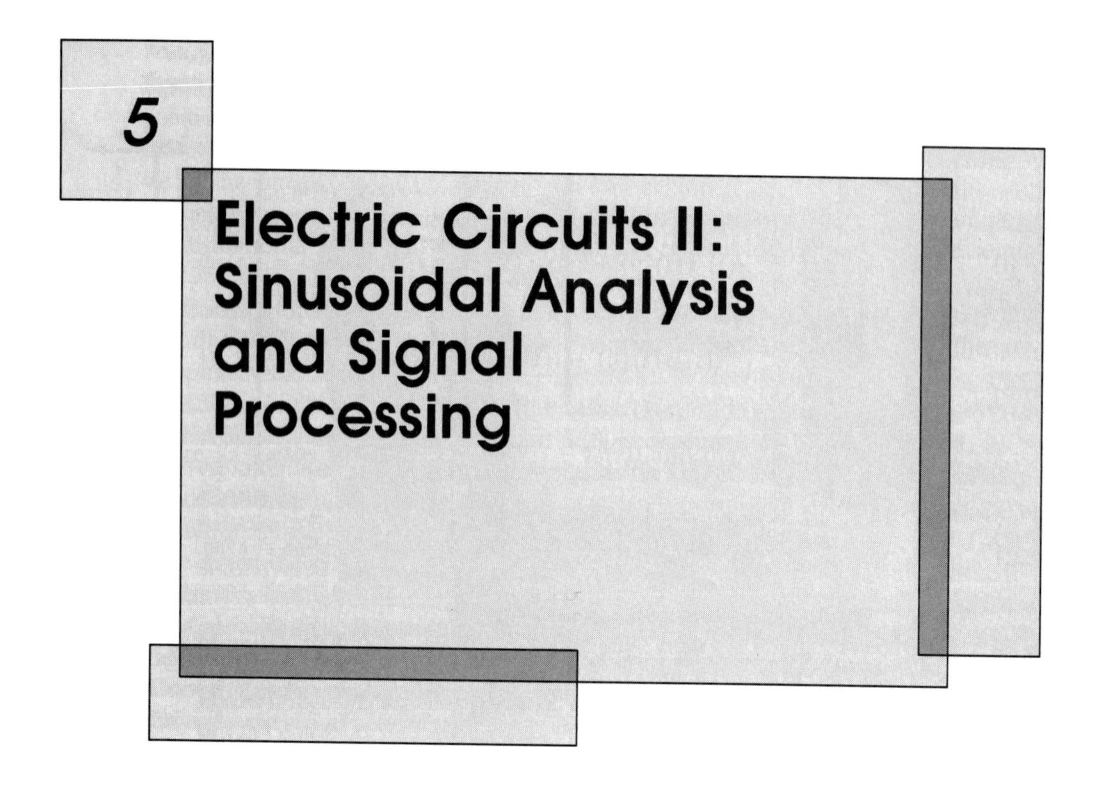

5
Electric Circuits II: Sinusoidal Analysis and Signal Processing


## 5.1 INTRODUCTION

In Chapter 4, current and voltage were usually constant in time. Because such current flows in the same direction all the time, it is called *direct current.* However, it was shown that when a step voltage is supplied to an inductive or capacitive circuit, the resulting current and voltage change with time. These time-varying functions describe the dynamics of a device or network.

In the study of transducers (see Chapter 9) it is precisely the time variation of the quantity to be measured that gives value to the transducer, for if the measured value were constant, there would be no need for continual measurements. These time-varying functions describe the state of some physical process or event—hence they are called signals (the root word being "sign"). Some obvious examples of types of information a signal may possess are numerical, verbal, and musical information.

Alternatively, the time-varying function can be a vehicle for transporting power rather than information. In this case, an advantage of using time-varying waveforms is the ability to use transformers to step up/down voltages and currents. We shall see that transformers operate only for time-varying voltages and currents. The ability to scale voltages up and down is important because efficient transmission voltages are very high, whereas safe voltages for users are relatively quite low.

Often the purpose of a device or network is to modify a signal or to extract from it a few important parameters. The network (or system) performs the same operation on *any* signal presented to it as input. That operation, describable in various forms, is then sufficient to characterize the system. The differential equation is the form most directly obtainable from the circuit physics, but least easy to work with. An easier form, the impulse response, involves the solution of the differential equation. And easiest to use and often to obtain (using sinusoidal steady-state analysis) is the *frequency response.*

The differential equation modeling a simple energy storage circuit and its solution were discussed in Chapter 4. In this chapter the corresponding impulse response and frequency response descriptions will be developed in simple ways. Sinusoidal steady-state (monofrequency) analysis, the understanding of which is prerequisite to the calculation and use of the frequency response, is given careful consideration. The advantages of using the frequency response for basic analysis will be demonstrated by examples, including low- and high-pass filters.

An appendix derives *convolution* as a consequence of linear shift-invariant systems. It is upon this foundation that filtering theory is based. Shifting emphasis from signal processing to power transport, in Chapter 6 we deal with power in sinusoidal waveforms, power factor correction, three-phase systems, and power distribution systems.

## 5.2 SINUSOIDS AND FOURIER ANALYSIS

### 5.2.1 The Sinusoid

A sinusoid is a waveform whose time dependence can be written as either the sine or the cosine function. For example, *any* real sinusoid can be represented by

$$v(t) = V\cos(\omega t + \theta) \tag{5.1}$$

(see Fig. 5.1). The symbol $V$ represents the amplitude (maximum value) of the sinusoid. The radial frequency of the sinusoid is designated $\omega = 2\pi f = 2\pi/T$, where $f$ is the frequency in hertz ($s^{-1}$ or cycles/s) and $T$ is the time period of one cycle in seconds [i.e., $v(t + T) = v(t)$]. Finally, $\theta$ is the *phase angle* (usually abbreviated "phase"), which is related to $v(t = 0)$ by $v(0) = V\cos(\theta)$. Thus three parameters are sufficient to specify a sinusoid: its magnitude, frequency, and phase angle.

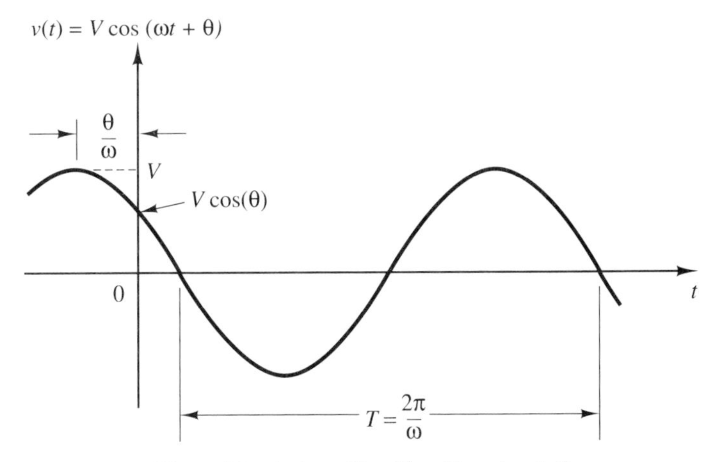


**Figure 5.1** A sinusoid, $v(t) = V\cos(\omega t + \theta)$.

Before proceeding, it may be asked: Why are we studying sinusoidal signals or ac (alternating current)? For power transmission, the convenience of time-varying voltages and currents is well known. Transformers (discussed in Chapter 6) step voltage up for economical transmission and down for safe usage. Transformers are based on Faraday's law, Eq. (3.27), which requires time-varying voltages; they do not work for dc. But why must the time dependence be *sinusoidal?* Sinusoidal waveforms are often most efficient and convenient to generate. Also, in power conversion a sinusoidal current in the field windings of an ac motor creates a rotating magnetic field that makes smooth motor rotation possible.

### 5.2.2 Fourier Series and Integrals

Perhaps the most fundamental advantage of sinusoidal time dependence arises in the area of signal processing. The analysis of the passage of sinusoids through linear circuit systems is easy. Moreover, the determination of the effects of a system on sinusoids is useful in characterizing the effects of the system on *general* signals.

Specifically, Jean B. de Fourier in 1822 showed that essentially any waveform (signal) can be decomposed into an infinite (or finite) sum or integral of appropriately weighted sinusoids of equally spaced frequencies. The specification of the weights (some of which may be zero) for all these *harmonic frequencies* is as complete a representation of the signal as the original specification of the value of the signal at every instant in time.

Furthermore, Fourier has provided the necessary transformation integrals (or sums) with which to convert from one representation to the other. For example, a periodic function $f(t)$ having period $T$ [i.e., $f(t + T) = f(t)$ for all time $t$] can be exactly constructed from knowledge of the Fourier series coefficients $a_n$ and $b_n$ of the sines and cosines composing it:

$$f(t) = \frac{a_0}{2} + \sum_{n=1}^{\infty} [a_n \cos(n\omega_0 t) + b_n \sin(n\omega_0 t)], \tag{5.2}$$

where

$$a_n = \frac{2}{T}\int_{-T/2}^{T/2} f(t) \cos(n\omega_0 t)\, dt$$

and

$$b_n = \frac{2}{T}\int_{-T/2}^{T/2} f(t) \sin(n\omega_0 t)\, dt, \tag{5.3}$$

in which $\omega_0 = 2\pi/T$ is called the *fundamental frequency*. It can be seen from Eq. (5.2) that for periodic functions, the constituent components all have frequencies that are multiples of $\omega_0$ apart (harmonics of $\omega_0$).

Finally, if we know the decomposition of the general signal into sinusoids and we know the effect of a system on sinusoids of all frequencies, we can determine the effect of the linear system on the general signal by superposition. That is, the output will be the sum of the system outputs for each frequency, scaled by the corresponding coefficient $a_n$ or $b_n$.

**Example 5.1**

Expand the periodic pulse train in Fig. 5.2 over a sum of sines and cosines.

**Solution** For this example, $\omega_0 = 2\pi/8 = \pi/4$ rad/s. For convenience, use the zero-centered period for this analysis; any one period will give the same result. Noting that the function is zero from $t = -4$ to $-2$ and from $t = 2$ to 4, Eqs. (5.3) give

$$a_n = \frac{2}{8}\int_{-2}^{2} 1 \cos\frac{n\pi t}{4}\,dt$$
$$= \frac{1}{n\pi}\left(\sin\frac{n\pi}{2} - \sin\frac{-n\pi}{2}\right) = \frac{2\sin(n\pi/2)}{n\pi}.$$

Because for $n = 0$ the above yields 0/0, substitute the zero for $n$ *before* integration to obtain

$$a_0 = \frac{1}{4}\int_{-2}^{2} 1\cos(0)\,dt = 1.$$

The sine coefficients are

$$b_n = \frac{1}{4}\int_{-2}^{2} \sin\frac{n\pi t}{4}\,dt$$
$$= \frac{-1}{n\pi}\left(\cos\frac{n\pi}{2} - \cos\frac{-n\pi}{2}\right) = 0.$$

Because $f(t)$ is an even function, it makes sense that all the sine coefficients $b_n$ are zero because $\sin(n\omega_0 t)$ is an odd function of $t$. Finally, putting the results together allows the sinusoidal representation of the pulse train $f(t)$:

$$f(t) = \frac{1}{2} + 2\sum_{n=1}^{\infty}\frac{\sin(n\pi/2)}{n\pi}\cos\frac{n\pi t}{4}.$$

Notice that the frequency spacing of sinusoidal components of $f(t)$ in Eq. (5.2) is $\omega_0 = 2\pi/T$. This means that the periodic function $f(t)$ can be represented exactly by *only* the sinusoids of frequencies $n\omega_0$, *n an integer*; no frequencies such as $(n + \frac{1}{2})\,\omega_0$, for example, contribute.

Aperiodic (not infinitely repeating) functions may be considered as periodic functions having period $T = \infty$, so that the repetitions "never occur." This makes the frequency spacing $\omega_0 = 2\pi/\infty = 0$! So aperiodic functions, unlike periodic functions, may contain components having frequencies $\omega$ of *any* value; the spacing between allowed frequencies is zero. Consequently, for aperiodic functions $f(t)$ the series (sum) expansion becomes an integral expansion. The appropriate form is then the Fourier integral or Fourier transform

$$f(t) = \frac{1}{\pi}\int_{0}^{\infty}[a(\omega)\cos(\omega t) + b(\omega)\sin(\omega t)]\,d\omega, \tag{5.4}$$

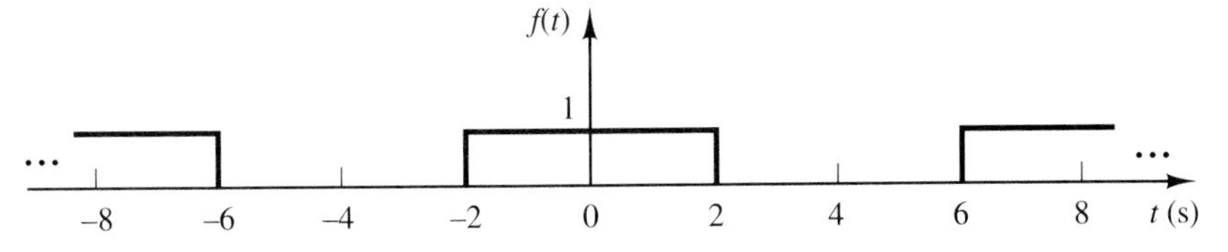


**Figure 5.2** Periodic pulse train used in Example 5.1.

where

$$a(\omega) = \int_{-\infty}^{\infty} f(t) \cos(\omega t)\, dt$$

and (5.5)

$$b(\omega) = \int_{-\infty}^{\infty} f(t) \sin(\omega t)\, dt$$

for any real-valued, aperiodic $f(t)$ for which the integrals in Eq. (5.5) exist.

**Example 5.2**

Expand the aperiodic function in Fig. 5.3 over an integral of sines and cosines.

**Solution** Using Eqs. (5.5), we obtain

$$a(\omega) = \int_{-1}^{1} 1 \cos(\omega t)\, dt = \frac{2 \sin(\omega)}{\omega}$$

and

$$b(\omega) = \int_{-1}^{1} 1 \sin(\omega t)\, dt = 0,$$

again, because $f(t)$ is an even function of $t$. Thus

$$f(t) = \frac{2}{\pi} \int_{0}^{\infty} \frac{\sin(\omega)}{\omega} \cos(\omega t)\, d\omega.$$

### 5.2.3 Further Motivation for Using Sinusoidal Analysis

This is all elegant and interesting mathematics, but of what use is it in this course? The answer lies in the characterization of signals and *linear shift-invariant* (LSI) systems. A linear system is essentially one for which doubling the input doubles the output. A shift-invariant system is one that if given the same input as before, but now shifted, it produces the same output as before, but now shifted. For further details and examples concerning these definitions, see the appendix.

The important goal in mind here is the ability to determine the effect of some electric/electronic system on an input. It is also shown in the appendix that the output of an LSI system subjected to a general input $x(t)$ is a weighted [by $x(\cdot)$] sum of time-reversed outputs due to a special set of inputs: shifted impulses. The output of an LSI system having as its input an unshifted impulse (i.e., an impulse occurring at $t = 0$) is called the *impulse response* $h(t)$.

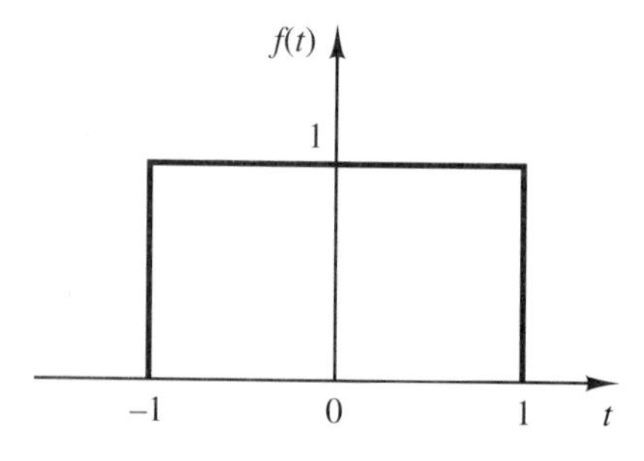


**Figure 5.3** Aperiodic pulse used in Example 5.2.

Two men who made essential contributions to sinusoidal analysis. Fourier (left) showed that we can use sinusoidal analysis to facilitate the study of all signals, while Steinmetz (right) in 1893 discovered the phasor approach to sinusoidal analysis, saying, "Analysis of the complex plane is very well worked out, hence by reducing the electrical problems to the analysis of complex quantities they are brought within the scope of a known and well-understood science." *Source:* Smithsonian Institution photo No. 56822 (left photo).

Consider calculation of the output $y(t)$ of an LSI system with impulse response $h(t)$ due to an input $x(t)$. It will be shown in Sec. 5.5.3 that Fourier's integrals allow us to avoid the tricky convolution integral in Eq. (A5.8e) of the appendix. Instead, all we must do is multiply the two coefficient functions of frequency ($n\omega_0$ or $\omega$) that represent $h(t)$ and $x(t)$, respectively. These resulting coefficients are, for example, the $a_n$ and $b_n$ of $y(t)$.

Not only do the Fourier expansions simplify the input–output calculations of a system as described above, but they also help identify the pertinent features of a signal $x(t)$. For example, a noise component may be mainly in one frequency range while the signal is in another. Such knowledge aids in designing an LSI system to eliminate the unwanted noise. The system $h(t)$ is then designed such that its frequency coefficients multiply the signal components by one and the noise components by zero. This process is known as *linear filtering.* There is no way to accomplish this without considering the sinusoidal expansions of $x(t)$.

Fortunately, the form of the expansion in Eqs. (5.4) and (5.5) may be simplified. Oddly enough, the key to the simplification is in the use of *complex exponentials.* In order to understand the meaning of complex exponentials and how to use them, it is helpful to take a step back and consider a simpler problem.

Suppose that two sinusoidal voltage sources, having the same frequency $\omega$ but differing magnitudes and phases, are connected in series (see Fig. 5.4). By KVL, the total voltage $v_3(t)$ will be the sum of the two. But what is the form of the sum? Specifically, is there a simple, single-term expression for

$$v_3(t) = v_1(t) + v_2(t), \tag{5.6}$$

where

$$\begin{aligned} v_1(t) &= V_1 \cos(\omega t + \theta_1) \\ v_2(t) &= V_2 \cos(\omega t + \theta_2), \end{aligned} \tag{5.7}$$

and if so, what is it?

The solution can be found in a straightforward (though laborious) way without resorting to complex exponentials. It consists of the following steps:

**1.** Expand $v_1(t)$ and $v_2(t)$ using the identity for the cosine of a sum:

$$\begin{aligned} v_1(t) &= V_1 [\cos(\omega t) \cos(\theta_1) - \sin(\omega t) \sin(\theta_1)] \\ v_2(t) &= V_2 [\cos(\omega t) \cos(\theta_2) - \sin(\omega t) \sin(\theta_2)] \end{aligned} \tag{5.8}$$

**2.** Group and sum the coefficients of $\cos(\omega t)$ and $-\sin(\omega t)$ in $v_1(t) + v_2(t)$:

$$\begin{aligned} v_3(t) = [V_1 \cos(\theta_1) + V_2 \cos(\theta_2)] \cos(\omega t) \\ - [V_1 \sin(\theta_1) + V_2 \sin(\theta_2)] \sin(\omega t). \end{aligned} \tag{5.9}$$

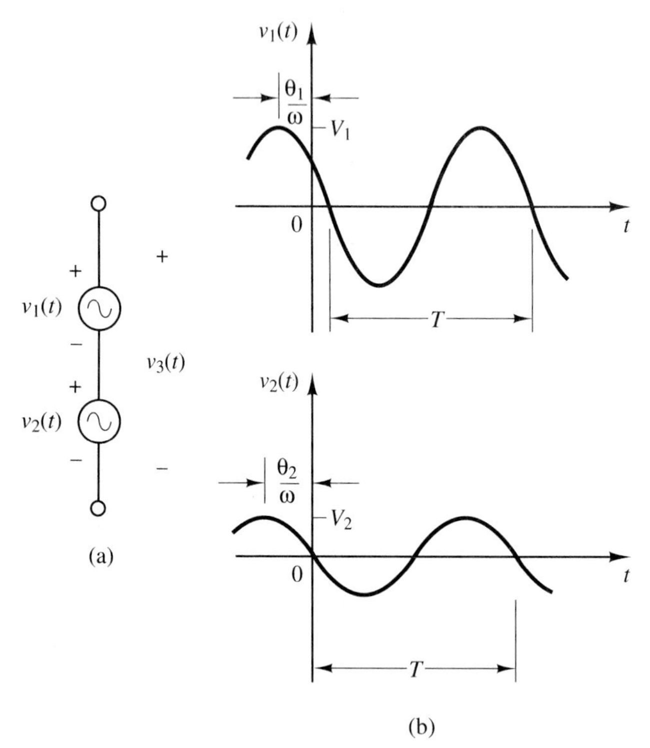


**Figure 5.4** Two voltages of different magnitudes and phase angles: (a) voltages connected in series; (b) individual waveforms.

**3.** Consider the bracketed coefficients in Eq. (5.9) as rectangular coordinates in a plane whose horizontal and vertical axes are, respectively, the coefficients of $\cos(\omega t)$ and $-\sin(\omega t)$ (see Fig. 5.5). [This is sensible because $\cos(\omega t)$ and $\sin(\omega t)$ are orthogonal functions, so their coefficients can be thought of as coordinates along orthogonal axes.]

**4.** In order to recognize the result as a sinusoid of frequency $\omega$ (see step 5), express these rectangular coordinates in polar form: $(V, \theta)$. This can be accomplished by noting that the rectangular coordinates can now also be written in terms of the polar form parameters as $[V\cos(\theta), V\sin(\theta)]$ so that

$$V\cos(\theta) = V_1\cos(\theta_1) + V_2\cos(\theta_2) \tag{5.10a}$$

$$V\sin(\theta) = V_1\sin(\theta_1) + V_2\sin(\theta_2). \tag{5.10b}$$

Square both sides of Eqs. (5.10a) and (5.10b) and add to obtain

$$V = [V_1^2 + V_2^2 + 2V_1V_2\cos(\theta_2 - \theta_1)]^{1/2}, \tag{5.11}$$

and divide Eq. (5.10b) by Eq. (5.10a) to obtain

$$\theta = \tan^{-1}\left[\frac{V_1\sin(\theta_1) + V_2\sin(\theta_2)}{V_1\cos(\theta_1) + V_2\cos(\theta_2)}\right]. \tag{5.12}$$

**5.** To obtain a form matching that of the original sinusoids, substitute the new polar form expression for the coefficients into the original sum in Eq. (5.9):

$$\begin{aligned} v_3(t) &= V\cos(\theta)\cos(\omega t) - V\sin(\theta)\sin(\omega t) \\ &= V\cos(\omega t + \theta), \end{aligned} \tag{5.13}$$

by the cosine of a sum identity. Thus the sum of two sinusoids of the same frequency but differing magnitudes and phases is itself a sinusoid of the same frequency, having magnitude $V$ and phase $\theta$ as found above in step 4 and given in Eqs. (5.11) and (5.12).

Not only was this a tedious and painful enterprise, but it is hard to remember because it involves some subtle recognitions. If instead of working explicitly with the time functions (in the *time domain*), equivalent steps are performed using frequency functions such as $a(\omega)$ and $b(\omega)$ in Eq. (5.5) (in the *frequency domain*), the algebra becomes easy. In fact, it is equally easy for any number of sinusoids of the same frequency to be added.

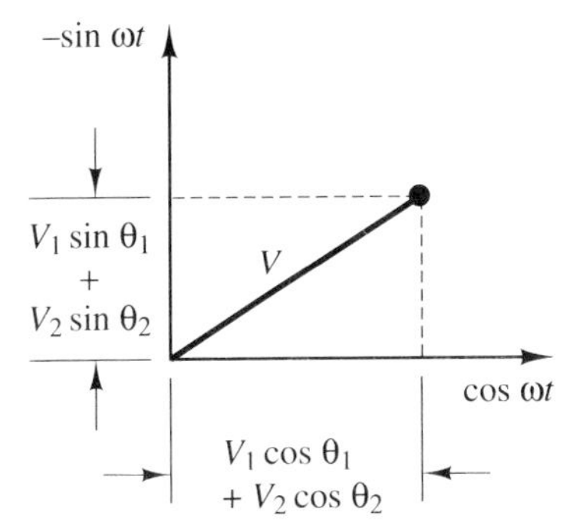


**Figure 5.5** Graphical construction for adding two sinusoids of different magnitudes and phases. The orthogonal axes represent coefficients of the orthogonal functions $\cos(\omega t)$ and $-\sin(\omega t)$.

Another huge bonus of learning to work in the frequency domain is that the differential equations for circuits involving linear energy storage elements (inductors and capacitors) all become simple algebraic equations. This result, valid for a single frequency $\omega$, brings about a generalized version of Ohm's law.

This new Ohm's law, also valid only in the frequency domain at one particular frequency, applies to all linear, passive circuit elements! Thus we may perform series and parallel combinations, voltage and current dividers, mesh and node analysis, and so on. Finally, simpler expressions for the Fourier expansions in terms of complex exponentials will unify the entire discussion. They will show exactly how the single-frequency analysis fits into the larger picture introduced above of multifrequency signals and linear filtering.

## 5.3 COMPLEX NUMBERS AND EXPONENTIALS

To reap all the rewards promised above, it is essential to have a basic understanding of and ability to manipulate complex numbers. As everyone knows, there is no real-valued solution of the equation $x^2 = -1$. But if the symbol $j$ is chosen to designate the square root of $-1$, then clearly $x = \pm j$. (In electrical engineering, we use the symbol $j$ rather than $i$ for $\sqrt{-1}$ because the symbol $i$ is reserved for current.)

In fact, the solutions of *all* algebraic equations (not only $x^2 = -1$) belong to the following set of ordered pairs: real numbers, and $j$ scaled by real numbers (called *imaginary* numbers). These ordered pairs are called *complex numbers.*

For example, a complex number $\overline{Z} = x + jy$, where $x$ and $y$ are real numbers has *real part* $x$ and *imaginary part* $y$. (The overbar, here over $Z$, signifies that the number or variable below it is complex.) The ordered pair $(x, y)$ can be displayed conveniently as the position of a point on a plane defined by a pair of orthogonal axes (see Fig. 5.6).

For performing addition and subtraction, the rectangular form $x + jy$ is most convenient. Consider two complex numbers $\overline{Z}_1 = x_1 + jy_1$ and $\overline{Z}_2 = x_2 + jy_2$. Then

$$\begin{aligned} &\text{Addition:} && \overline{Z}_1 + \overline{Z}_2 = x_1 + x_2 + j(y_1 + y_2) \\ &\text{Subtraction:} && \overline{Z}_1 - \overline{Z}_2 = x_1 - x_2 + j(y_1 - y_2). \end{aligned} \tag{5.14}$$

**Example 5.3**

Let $\overline{Z}_1 = 3 + j4$ and $\overline{Z}_2 = 7 - j$. What are (a) $\overline{Z}_1 + \overline{Z}_2$ and (b) $\overline{Z}_1 - \overline{Z}_2$?

**Solution**

**(a)** $\overline{Z}_1 + \overline{Z}_2 = 3 + 7 + j(4 - 1) = 10 + j3.$

**(b)** $\overline{Z}_1 - \overline{Z}_2 = 3 - 7 + j(4 + 1) = -4 + j5.$

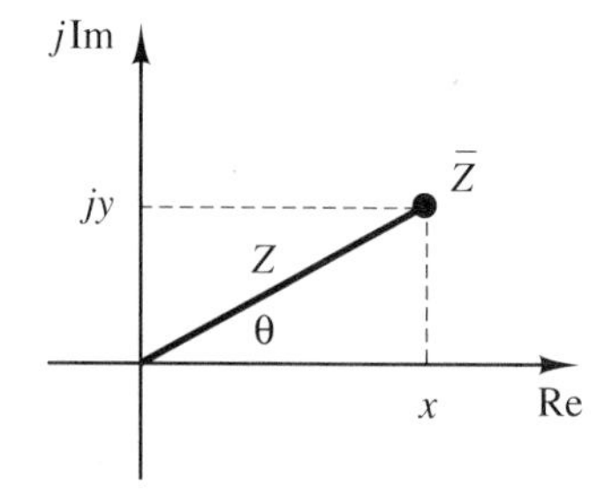


**Figure 5.6** Complex number on the complex plane; both polar and rectangular coordinates are shown.

Multiplication and division are also straightforward but can best be demonstrated after introduction of the polar form. To introduce the polar form, let us examine the square root of $-1$ in more detail. In the complex plane, the number $j$ has magnitude one and is oriented along the positive vertical axis. It can be viewed as a unit "vector" of magnitude 1, rotated counterclockwise 90° from the positive horizontal axis (see Fig. 5.7). The magnitude referred to above is just the radial coordinate in a polar coordinates description, while the 90° represents the polar angle. Hence

$$\begin{aligned} j &= 1 \cos(90°) + j1 \sin(90°) \\ &= 1\underline{/90°}, \end{aligned} \tag{5.15}$$

where the last expression is a very convenient and common way of representing the polar form (magnitude $\underline{/\text{phase angle}}$).

Similarly, a general complex number $\overline{Z} = x + jy$ can be written in polar form as

$$\overline{Z} = Z\underline{/\theta}, \tag{5.16a}$$

where

$$Z = (x^2 + y^2)^{1/2} \tag{5.16b}$$

$$\theta = \tan^{-1}\frac{y}{x}, \tag{5.16c}$$

$Z$ and $\theta$ being, respectively, the magnitude and phase angle of $\overline{Z}$. (Figure 5.6 also shows $Z$ and $\theta$.) Conversely,

$$\begin{aligned} x &= Z \cos(\theta) = \text{Re}\{\overline{Z}\} \\ y &= Z \sin(\theta) = \text{Im}\{\overline{Z}\}, \end{aligned} \tag{5.17}$$

where $\text{Re}\{\cdot\}$ and $\text{Im}\{\cdot\}$ are, respectively, operators that select the real and imaginary part of their complex argument. Note that for two complex numbers $\overline{Z}_1$ and $\overline{Z}_2$,

$$\begin{aligned} \text{Re}\{\overline{Z}_1\} + \text{Re}\{\overline{Z}_2\} &= x_1 + x_2 \\ &= \text{Re}\{\overline{Z}_1 + \overline{Z}_2\}; \end{aligned} \tag{5.18}$$

similarly for the imaginary operator.

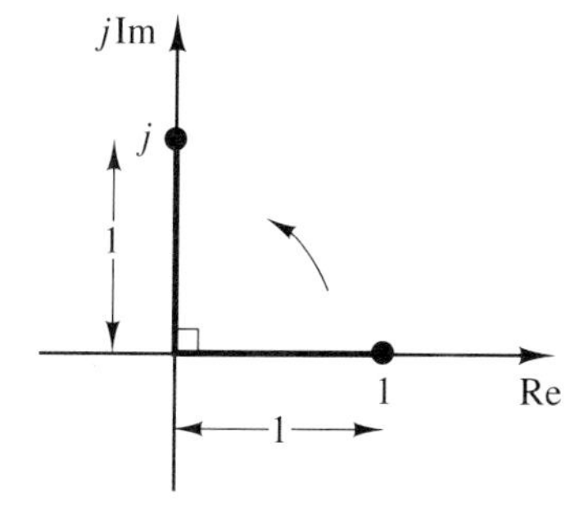


**Figure 5.7** Graphical demonstration that $j = 1\underline{/90°}$.

**Example 5.4**

Find (a) the rectangular form of $\bar{Z} = 2\,\angle 120°$ and (b) the polar form of $\bar{Z} = 3 + j4$.

**Solution**

(**a**) By Eq. (5.17), $x = 2\cos(120°) = -1$ and $y = 2\sin(120°) = \sqrt{3}$. Therefore, $\bar{Z} = -1 + j\sqrt{3}$.

(**b**) By Eqs. (5.16), $Z = (3^2 + 4^2)^{1/2} = 5$ and $\theta = \tan^{-1}(4/3) \approx 53.1°$. Thus $\bar{Z} = 5\,\angle 53.1°$.

From comparison of the Taylor expansions (developed by Brook Taylor in 1712) for the sine, cosine, and exponential, Leonhard Euler found in 1748 that

$$e^{j\theta} = \cos(\theta) + j\sin(\theta). \tag{5.19}$$

This relation is known as *Euler's formula.* The left side, $e^{j\theta}$, is called a *complex exponential,* for obvious reasons. It can be visualized on the complex plane as a unit vector rotated counterclockwise an angle $\theta$ from the positive $x$ axis ("real axis"; see Fig. 5.8). That is, the convention is that for $\theta > 0$, we rotate counterclockwise, and for $\theta < 0$, we rotate clockwise. If $\theta = \omega t$, the right side consists of temporal sinusoids for its real and imaginary parts. Therefore, for $\theta = \omega t$, the complex exponential may be termed a *complex sinusoid.*

Suppose that Euler's formula [Eq. (5.19)] is multiplied by the radial coordinate $\bar{Z}$ of the complex number $\bar{Z}$. Then the left side, called the *exponential form,* is seen as just a complete, compact way of expressing the polar form of any complex number $\bar{Z}$:

$$\begin{aligned}\bar{Z} &= Ze^{j\theta}\\ &= Z\,\angle\theta\\ &= Z\cos(\theta) + jZ\sin(\theta).\end{aligned} \tag{5.20}$$

But more than that, the exponential form follows all the rules for multiplication, division, and exponentiation that are so simple for the exponential function.

Multiplication: $$\bar{Z}_1\bar{Z}_2 = Z_1Z_2e^{j(\theta_1+\theta_2)} \tag{5.21}$$

Division: $$\frac{\bar{Z}_1}{\bar{Z}_2} = \frac{Z_1}{Z_2}e^{j(\theta_1-\theta_2)} \tag{5.22}$$

Integer power: $$\bar{Z}^n = Z^n e^{jn\theta} \tag{5.23}$$

Fractional power: $$\bar{Z}^{1/n} = Z^{(1/n)}e^{j(\theta+2\pi k)/n}, \qquad 0 \le k \le n-1, \tag{5.24}$$

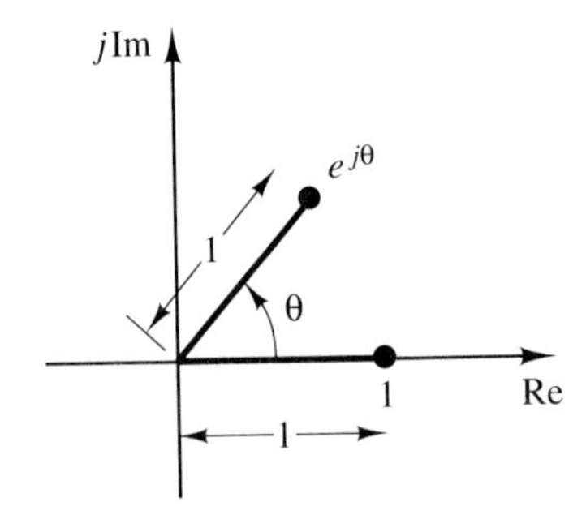


**Figure 5.8** Complex exponential on the complex plane.

where $k$ is an integer. The $2\pi k$ in Eq. (5.24) is necessary because there are $n$ distinct $n$th roots. When any of the roots in Eq. (5.24) is raised to the $n$th power, it is easy to see that $\overline{Z}$ results because $e^{j2\pi k} = \cos(2\pi k) + j\sin(2\pi k) = 1$ for all integers $k$. Without the $2\pi k$, only one of the $n$ roots would be found. These roots are found at radius $Z^{1/n}$ and at angles evenly spaced at intervals of $2\pi k/n$, beginning at angle $\theta/n$, in the complex plane.

Note that multiplication or division of a complex number $\overline{Z}_1$ by another complex number $\overline{Z}_2$ results in two modifications of $\overline{Z}_1$: magnitude scaling (by $Z_2$ for multiplication or $1/Z_2$ for division) and rotation through the angle $\theta_2$ (counterclockwise for multiplication or clockwise for division). Integer powers are just an extension of multiplication: magnitude scaling (from $Z$ to $Z^n$) and rotation (from $\theta$ to $n\theta$).

Some useful simple polar forms that often come in handy and are easily determined from the complex plane and Euler's formula are

$$\begin{aligned} 1 &= e^{j0} \\ j &= e^{j\pi/2} \\ -1 &= e^{j\pi} \\ -j &= e^{j3\pi/2} = e^{-j\pi/2}. \end{aligned} \tag{5.25}$$

(Usually, people express the phase in radians when using the exponential form and in degrees when using the $Z\,\underline{/\theta}$ form.) Another highly useful property concerning $j$ is easily found by dividing $j^2 = -1$ by $-j$:

$$\frac{1}{j} = -j. \tag{5.26}$$

As a simple exercise, verify that $j^2 = -1$ by writing $j^2 = j \cdot j$ and using Euler's formula [Eq. (5.19)].

Suppose that we are given the rectangular forms $\overline{Z}_1 = x_1 + jy_1$ and $\overline{Z}_2 = x_2 + jy_2$. The product $\overline{Z}_1 \cdot \overline{Z}_2$ may be found in rectangular form directly, as follows:

$$\begin{aligned} \overline{Z}_1\overline{Z}_2 &= (x_1 + jy_1)(x_2 + jy_2) \\ &= x_1x_2 + jx_1y_2 + jy_1x_2 + j^2y_1y_2 \\ &= x_1x_2 - y_1y_2 + j(x_1y_2 + x_2y_1). \end{aligned} \tag{5.27}$$

**Example 5.5**

Let $\overline{Z}_1 = 1 + j$ and $\overline{Z}_2 = 2\,\underline{/90^\circ}$. Find (a) $\overline{Z}_1\overline{Z}_2$, (b) $\overline{Z}_2/\overline{Z}_1$, and (c) $\overline{Z}_1^2$ and $\overline{Z}_2^2$.

**Solution**

(a) Put $\overline{Z}_1$ in polar form: $\overline{Z}_1 = \sqrt{2}\,\underline{/45^\circ}$. Then $\overline{Z}_1\overline{Z}_2 = 2\sqrt{2}\,\underline{/135^\circ}$.

(b) $\overline{Z}_2/\overline{Z}_1 = (2/\sqrt{2})\,\underline{/45^\circ} = \sqrt{2}\,\underline{/45^\circ}\ (= \overline{Z}_1)$.

(c) $\overline{Z}_1^2 = (\sqrt{2})^2 e^{j2\pi/4} = 2e^{j\pi/2} = j2\ (= \overline{Z}_2)$ and $\overline{Z}_2^2 = 2^2 e^{j2(\pi/2)} = 4e^{j\pi} = -4$.

**Example 5.6**

Find the three cube roots of $-8 = 8e^{j\pi}$.

**Solution** From Eq. (5.24), the roots are found as

$$(-8)^{1/3} = 8^{1/3} e^{j(\pi + 2\pi k)/3}, \qquad 0 \le k \le 2, \tag{5.28}$$

where $-1 = e^{j\pi}$ was used. These three roots can be written

$$\begin{aligned} k = 0&: \quad 2e^{j\pi/3} \\ k = 1&: \quad 2e^{j(\pi/3 + 2\pi/3)} = 2e^{j\pi} = -2 \qquad \text{(the familiar, real-valued root)} \\ k = 2&: \quad 2e^{j(\pi/3 + 4\pi/3)} = 2e^{j5\pi/3}. \end{aligned} \tag{5.29}$$

Note that for $k = 3$, the root is $2e^{j7\pi/3} = 2e^{j\pi/3}$, which equals the $k = 0$ root. Thus there are only $n = 3$ distinct roots. The three roots are shown on the complex plane in Fig. 5.9.

One more useful concept is the *complex conjugate*. A meaning of the word "conjugate" is "having features in common but opposite in some particular." The features in common between a complex number $\overline{Z}$ and its complex conjugate $\overline{Z}^*$ are the real part and the absolute value of the imaginary part, and also the magnitude $|\overline{Z}| = |\overline{Z}^*| = Z$. The "opposite in some particular" is the sign of the imaginary part; thus if $\overline{Z} = x + jy$, then $\overline{Z}^* = x - jy$. In fact, to obtain the complex conjugate of *any* algebraic expression containing complex numbers, just replace $j$ wherever it occurs by $-j$ and the result will be the complex conjugate.

One convenient expression involving the complex conjugate is that for the magnitude of a complex number $\overline{Z}$:

$$\begin{aligned} \overline{Z}\,\overline{Z}^* &= (x + jy)(x - jy) \\ &= x^2 + y^2 \\ &= Z^2, \end{aligned} \tag{5.30}$$

or

$$Z = (\overline{Z}\,\overline{Z}^*)^{1/2}. \tag{5.31}$$

**Example 5.7**

Without explicitly converting numerator and denominator to polar form, find $Z$ if $\overline{Z} = (1 + j2)/(3 + j4)$.

**Solution** Using Eq. (5.31) yields

$$Z = \left(\frac{1 + j2}{3 + j4}\cdot\frac{1 - j2}{3 - j4}\right)^{1/2} = \left(\frac{1 + 4}{9 + 16}\right)^{1/2} = \frac{1}{\sqrt{5}} \approx 0.45.$$

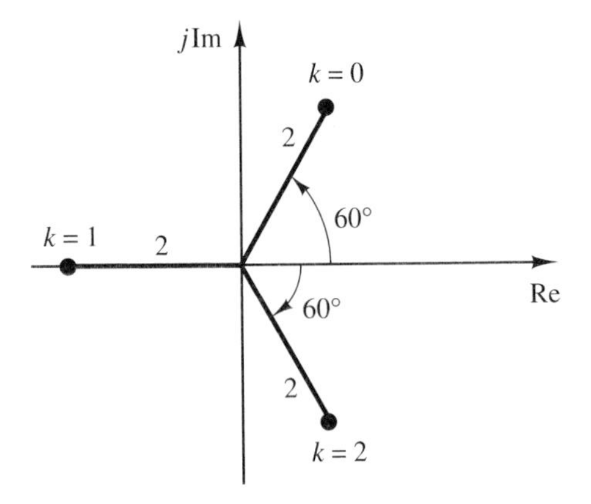


**Figure 5.9** The three roots of $-8$ on the complex plane.

Note that in polar form the magnitudes of $\overline{Z}$ and $\overline{Z}^*$ are identical, and the absolute value of the phase of $\overline{Z}^*$ equals that of $\overline{Z}$; the only difference is that the signs of the phases of $\overline{Z}$ and $\overline{Z}^*$ are opposite; $\overline{V} = V \cdot e_{-j\theta}$.

Also notable is that any complex number $\overline{Z}$ plus its complex conjugate equals twice the real part of $\overline{Z}$:

$$\begin{aligned} \overline{Z} + \overline{Z}^* &= x + jy + x - jy \\ &= 2x \\ &= 2\ \mathrm{Re}\{\overline{Z}\}. \end{aligned} \tag{5.32}$$

The sum of a complex number and its conjugate is always a real number! Similarly,

$$\overline{Z} - \overline{Z}^* = j2\ \mathrm{Im}\{\overline{Z}\}. \tag{5.33}$$

An additional interesting and useful identity results if the complex conjugate of Euler's formula [Eq. (5.19)] is added to the unmodified Euler's formula. The result is

$$\cos(\theta) = \frac{e^{j\theta} + e^{-j\theta}}{2}. \tag{5.34a}$$

Similarly, if the conjugate of Euler's formula is subtracted from Eq. (5.19), we find that

$$\sin(\theta) = \frac{e^{j\theta} - e^{-j\theta}}{2j}. \tag{5.34b}$$

Yet another case where the complex conjugate commonly comes into play is in the division of complex numbers for cases where the rectangular forms of the operands are most easily available and where the result is desired in rectangular form.

**Example 5.8**

Find the rectangular form of $(3 + j)/(2 - j)$.

**Solution** Multiply numerator and denominator by the complex conjugate of the denominator, $2 + j$:

$$\frac{(3 + j)(2 + j)}{(2 - j)(2 + j)} = \frac{1}{5}(5 + j5) = 1 + j.$$

Here are two simple rules to remember for numerical calculations: (1) usually the polar form is more convenient for carrying out multiplication and division (see the problems for the relatively clumsy general rectangular form formula for division as used in Example 5.8); and (2) addition and subtraction are always carried out using the rectangular form.

## 5.4 PHASORS, IMPEDANCE, AND THE FREQUENCY DOMAIN

### 5.4.1 The Phasor

The brief review above on complex numbers and exponentials covers all the necessary concepts for understanding the complex analysis carried out in this book. In applying these ideas to sinusoidal analysis, one more tool is needed: The *phasor,* a

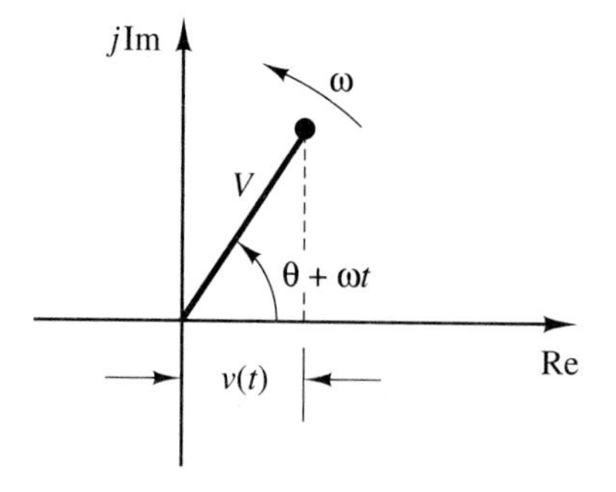


**Figure 5.10** Depiction of sinusoidal $v(t)$ as the real part of a complex sinusoidal function.

concept developed by Charles Steinmetz around the turn of the century. Just as $Z\cos(\theta) = \text{Re}\{\overline{Z}\} = \text{Re}\{Ze^{j\theta}\}$, a real sinusoid $v(t)$ can be considered the real part of a complex exponential:

$$\begin{aligned} v(t) &= V\cos(\omega t + \theta) \\ &= \text{Re}\{Ve^{j(\omega t + \theta)}\}. \end{aligned} \tag{5.35}$$

This is a remarkable fact: A sinusoidal waveform can be viewed as the real part of a complex sinusoidal time function. See Fig. 5.10 for an illustration of Eq. (5.35).

Furthermore, the complex exponential $Ve^{j(\omega t + \theta)}$ can be viewed in the complex plane as a *phasor*. The phasor has magnitude $V$, is initially (at $t = 0$) at an angle $\theta$ from the positive real axis, and rotates counterclockwise at the rate of once every $2\pi/\omega$ seconds (see Fig. 5.11).

Although intuitively *phasor* should refer to a rotating vector which indicates the phase at any time during rotation, the convention is that the term "phasor" refers to everything *except* the $e^{j\omega t}$ factor. For example, for the sinusoid above, the phasor $\overline{V}$ would be

$$\overline{V} = Ve^{j\theta}. \tag{5.36}$$

That is, $\overline{V}$ is the (complex) amplitude of the complex sinusoid $Ve^{j(\omega t+\theta)}$.

Equivalently, the phasor is the entire complex sinusoid evaluated at $t = 0$, because $e^{j\omega\cdot 0} = 1$. Finally, then, the sinusoidal waveform expressed in terms of the phasor $\overline{V}$ is simply

$$v(t) = \text{Re}\{\overline{V}e^{j\omega t}\}. \tag{5.37}$$

A phasor actually represents the complex amplitude (with magnitude and phase parameters) of the component of a signal having frequency $\omega$. Because in the

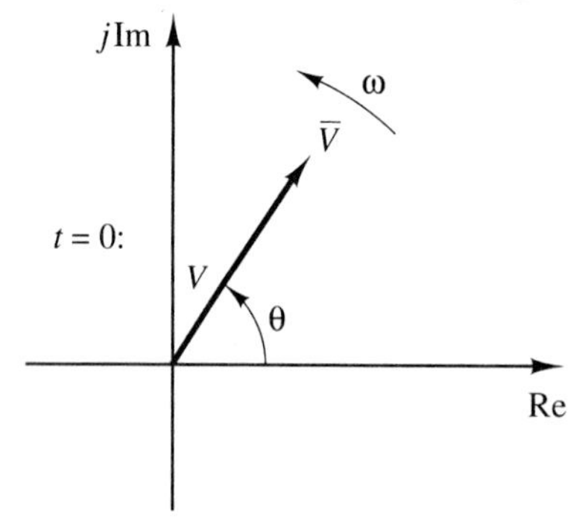


**Figure 5.11** A phasor in the complex voltage plane rotates with angular frequency $\omega$. Shown is its position at $t = 0$.

present analysis all signals are assumed to have the same frequency, there is no need to subscript or otherwise label the phasor with $\omega$. In general, a distinct complex amplitude (phasor) would have to be specified for *every* value of $\omega$, and be so labeled. Those phasors $Ve^{j\theta}$ at each frequency $\omega$ are directly related to Fourier's coefficients $a(\omega)$ and $b(\omega)$ at each $\omega$. The collection of all those complex amplitudes, called the frequency response, is discussed in Sec. 5.5.

### 5.4.2 Addition and Differentiation of Sinusoids Using Phasors

We are now in a position to solve, using the frequency domain, the problem of the sum of two sinusoids having the same frequency but differing magnitudes and phases. Recalling various results from this and the section on complex numbers (Sec. 5.3), we write

$$\begin{aligned}
v_3(t) &= V_1 \cos(\omega t + \theta_1) + V_2 \cos(\omega t + \theta_2) \qquad \text{[using Eqs. (5.6) and (5.7)]} \\
&= \text{Re}\{V_1 e^{j(\omega t + \theta_1)}\} + \text{Re}\{V_2 e^{j(\omega t + \theta_2)}\} \qquad \text{[using Eq. (5.35)]} \\
&= \text{Re}\{(V_1 e^{j\theta_1} + V_2 e^{j\theta_2})\, e^{j\omega t}\} \qquad \text{[using Eq. (5.18)]} \\
&= \text{Re}\{\overline{V}_3 e^{j\omega t}\} \qquad \text{[see Eqs. (5.39) and (5.40) below]} \\
&= \text{Re}\{[V_3 \cos(\theta_3) + jV_3 \sin(\theta_3)][\cos(\omega t) + j \sin(\omega t)]\} \\
&\qquad \text{[see Eqs. (5.20) and (5.19)]} \\
&= V_3[\cos(\theta_3)\cos(\omega t) - \sin(\theta_3)\sin(\omega t)] \qquad \text{[see Eq. (5.27)]} \\
&= V_3 \cos(\omega t + \theta_3) \qquad \text{(cosine of sum identity),}
\end{aligned} \tag{5.38}$$

where

$$\begin{aligned}
\overline{V}_3 &= V_3 e^{j\theta_3} \\
&= \overline{V}_1 + \overline{V}_2 \\
&= V_1 \cos(\theta_1) + V_2 \cos(\theta_2) + j[V_1 \sin(\theta_1) + V_2 \sin(\theta_2)] \\
&= [(V_1 \cos(\theta_1) + V_2 \cos(\theta_2))^2 + (V_1 \sin(\theta_1) + V_2 \sin(\theta_2))^2]^{1/2} \\
&\qquad \cdot e^{j \tan^{-1}\left\{\frac{V_1 \sin(\theta_1) + V_2 \sin(\theta_2)}{V_1 \cos(\theta_1) + V_2 \cos(\theta_2)}\right\}} \\
&= [V_1^2 + V_2^2 + 2 \cdot V_1 V_2 \cdot \cos(\theta_1 - \theta_2)]^{1/2} \cdot e^{j \tan^{-1}\left\{\frac{V_1 \sin(\theta_1) + V_2 \sin(\theta_2)}{V_1 \cos(\theta_1) + V_2 \cos(\theta_2)}\right\}}
\end{aligned} \tag{5.39}$$

and

$$\begin{aligned}
\overline{V}_1 &= V_1 e^{j\theta_1} \\
\overline{V}_2 &= V_2 e^{j\theta_2}.
\end{aligned} \tag{5.40}$$

where by comparing the first and last expressions in Eq. (5.39), we obtain explicit expressions for $V_3$ and $\theta_3$.

Actually, the procedure can be greatly streamlined down to the following three simple steps, using the definitions given above where necessary:

1. Transform to the frequency domain by defining the phasors $\overline{V}_1$ and $\overline{V}_2$ [Eq. (5.40)].
2. Add the phasors (using rectangular coordinates) to obtain $\overline{V}_3$ [Eq. (5.39)] (or graphically perform the "vector" sum in the complex plane by parallelogram addition, as shown in Fig. 5.12).
3. From the polar form of $\overline{V}_3$, extract $V_3$ and $\theta_3$. The time-domain result can be written down immediately as $v_3(t) = V_3 \cos(\omega t + \theta_3)$.

As is evident, a tremendous amount of labor and thought is saved. The ability to add many such sinusoids is crucial for ac circuit analysis in, for example, the satisfaction of Kirchhoff's laws. More generally, the advantages of working in the frequency domain are exploited in numerous areas of both electrical and nonelectrical engineering.

**Example 5.9**

Find $V$ and $\theta$ in

$$v(t) = \cos(\omega t) + 3\cos\left(\omega t + \frac{\pi}{4}\right) + \frac{1}{2}\sin(\omega t) = V\cos(\omega t + \theta).$$

**Solution** Following the simple steps above:

**1.** $\overline{V}_1 = 1 \angle 0°$ and $\overline{V}_2 = 3 \angle 45°$. Because $\sin(\omega t) = \cos(\omega t - \pi/2)$, $\overline{V}_3 = \frac{1}{2} \angle -90°$.

**2.** $\overline{V} = \overline{V}_1 + \overline{V}_2 + \overline{V}_3 = 1 + 3/\sqrt{2} + j3/\sqrt{2} - j1/2 \approx 3.12 + j1.62 \approx 3.52 \angle 27.4°$.

**3.** $V = 3.52$ and $\theta = 27.4° = 0.478$ rad so that $v(t) = 3.52\cos(\omega t + 0.478)$.

Another advantage of ac circuit analysis was hinted at in Sec. 5.2.3: Linear differential equations in the time domain become linear algebraic equations in the frequency domain. The key to the effect is an extraordinary property of the exponential function: $d/dx(e^{\alpha x}) = \alpha e^{\alpha x}$. In words, the form of the exponential function remains unchanged under differentiation. It is merely multiplied by the coefficient, in the exponent, of the differentiation variable. For ac circuit analysis, $\alpha$ is $j\omega$ and $x$ is time $t$. Thus

$$\frac{d}{dt}(e^{j\omega t}) = j\omega e^{j\omega t}. \tag{5.41}$$

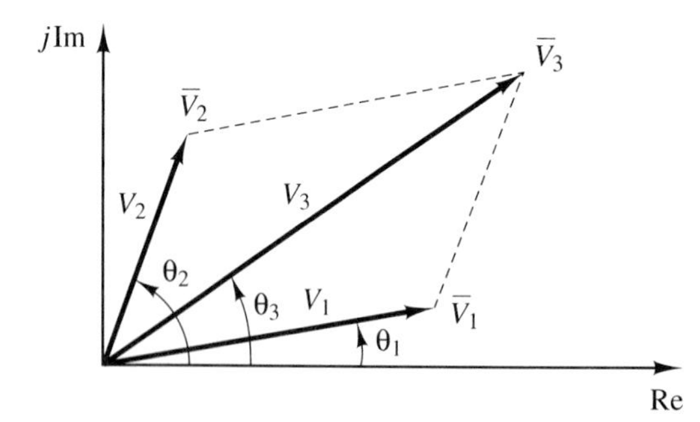


**Figure 5.12** Phasor diagram for the addition of three sinusoids having common frequency and different amplitudes and phases.

So differentiation is simply multiplication by $j\omega$ in the frequency domain. Contrarily, differentiation of real-valued sinusoids in the time domain results in a change of functional form (cosine to $-$sine or sine to cosine). (The exponential form is also invariant to indefinite integration.) The process is summarized in Fig. 5.13.

Use of transforms from the given domain to one more convenient is a common process in mathematics and engineering. One of the simplest transforms that was once useful to scientists for arithmetic was the logarithm. In the days before calculators, multiplication was a very tedious operation. But because $\log(xy) = \log(x) + \log(y)$ [a consequence of the facts $10^u \cdot 10^v = 10^{(u+v)}$ with $u = \log(x)$, $v = \log(y)$, and $\log[10^w] = w$ where $w = u + v$], multiplication could be transformed into addition.

The product $xy$ was sought. First the logs of $x$ and $y$ were found using tables of the logarithm, which were then widely available. Now in the *log domain,* multiplication becomes just addition, so one would add the two logs and obtain $\log(xy)$. Exponentiation of the result (inverse transformation) using the table in reverse would yield the desired product, $xy$. Thus the log domain avoids not differentiation (as does the frequency domain), but rather, multiplication.

There are also other ramifications. For example, when combined with the result in Sec. 5.2.3 [Eqs. (5.6), (5.7), and (5.13)] that the sum of sinusoids having equal frequencies is itself a sinusoid of the same frequency, the differentiation property allows the following conclusion to be made.

If a circuit containing only linear elements is excited by a sinusoidal source of frequency $\omega$, *all* voltages and currents within the circuit will also be sinusoidal of frequency $\omega$! A more fundamental result behind this conclusion is that the solution of a linear differential equation has the same functional form as the driving function plus all of its derivatives up to the order of the differential equation. Notice, however, that all these derivatives have the same form of time dependence for complex sinusoids. Moreover, because in the frequency domain all terms have the same form, $\bar{A}e^{j\omega t}$, they are far easier to group and factor. In general, when the complex exponential form is used, the various phase and magnitude differences that may exist between various voltages and currents are merely differing complex coefficients of $e^{j\omega t}$. The need for continual use of trigonometric identities is eliminated.

The condition of sinusoidal excitation with only one frequency present is known as the sinusoidal steady state. "Steady state" indicates that all transient

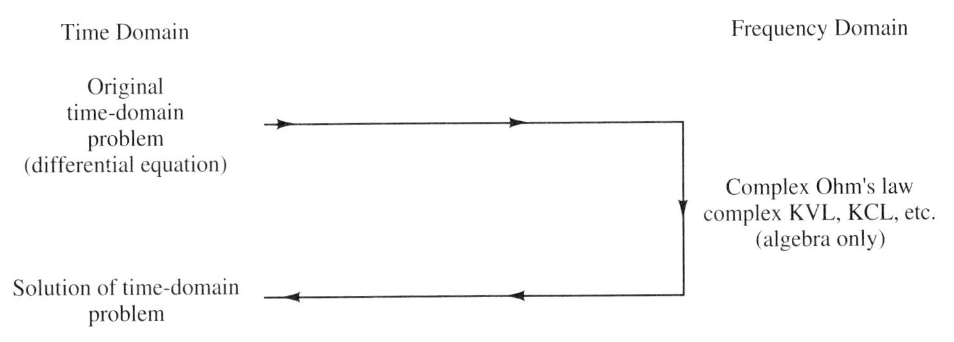


**Figure 5.13** Frequency-domain solution of a differential equation of a sinusoidally excited circuit.

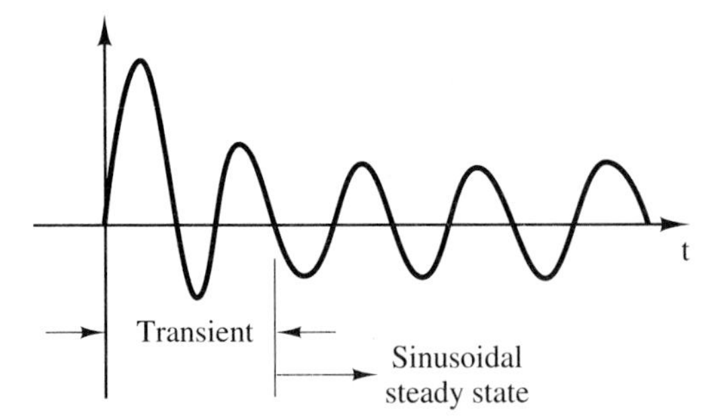


**Figure 5.14** Transient waveform, showing transition to the sinusoidal steady state.

(decaying exponential) voltages and currents are assumed to have long since died away, as illustrated in Fig. 5.14. In real circuits, this need not be very long, only five or so time constants (often milliseconds). The practical application of steady-state sinusoidal analysis permeates all electric and electronic circuits in which there are time-varying voltages and currents.

### 5.4.3 Generalized Ohm's Law for Sinusoidal Excitation

In Sec. 5.2.3 it was promised that a complex Ohm's law valid at any one particular frequency for all linear, passive circuit elements could be obtained in the frequency domain. This fact is a direct consequence of the differentiation property [Eq. (5.41)]. Consider the time-domain current–voltage relation for a capacitor:

$$i(t) = C\frac{dv(t)}{dt}. \tag{5.42}$$

Suppose that $v(t)$ is sinusoidal. Then in the frequency domain it has a phasor representation $\overline{V}$, and $v(t) = \text{Re}\{\overline{V}e^{j\omega t}\}$. Substituting $v(t)$ above gives

$$i(t) = C\frac{d}{dt}\text{Re}\{\overline{V}e^{j\omega t}\}. \tag{5.43}$$

Can the time derivative be brought inside the Re{·} operator and thereby allow use of the property $d/dt\{e^{j\omega t}\} = j\omega e^{j\omega t}$? To find out, compare $d/dt(\text{Re}\{\overline{V}e^{j\omega t}\})$ with $\text{Re}\{d/dt(\overline{V}e^{j\omega t})\}$:

$$\begin{aligned}\frac{d}{dt}(\text{Re}\{\overline{V}e^{j\omega t}\}) &= \frac{d}{dt}[\text{Re}\{\overline{V}\}\cos(\omega t) - \text{Im}\{\overline{V}\}\sin(\omega t)] \\ &\qquad \text{[using Eqs. (5.19), (5.27)]} \\ &= \omega[-\text{Re}\{\overline{V}\}\sin(\omega t) - \text{Im}\{\overline{V}\}\cos(\omega t)],\end{aligned} \tag{5.44}$$

while

$$\begin{aligned}\text{Re}\left\{\frac{d}{dt}(\overline{V}e^{j\omega t})\right\} &= \text{Re}\{j\omega\overline{V}e^{j\omega t}\} \\ &= \text{Re}\{\omega(j\,\text{Re}\{\overline{V}\} - \text{Im}\{\overline{V}\})[\cos(\omega t) + j\sin(\omega t)]\} \\ &= \omega[-\text{Re}\{\overline{V}\}\sin(\omega t) - \text{Im}\{\overline{V}\}\cos(\omega t)] \\ &= \frac{d}{dt}(\text{Re}\{\overline{V}e^{j\omega t}\}).\end{aligned} \tag{5.45}$$

So, yes, Re{·} and $d/dt$ can be exchanged so that $i(t)$ can be rewritten as

$$\begin{aligned} i(t) &= C\,\mathrm{Re}\left\{\frac{d}{dt}(\bar{V}e^{j\omega t})\right\} \\ &= \mathrm{Re}\{j\omega C\bar{V}e^{j\omega t}\}. \end{aligned} \tag{5.46}$$

But $i(t) = \mathrm{Re}\{\bar{I}e^{j\omega t}\}$, so that substituting the phasor representation of $i(t)$ into Eq. (5.46) gives

$$\mathrm{Re}\{\bar{I}e^{j\omega t}\} = \mathrm{Re}\{j\omega C\bar{V}e^{j\omega t}\} \tag{5.47}$$

or,

$$\mathcal{R}\mathrm{e}\{(\bar{I} - j\omega C\bar{V})e^{j\omega t}\} = 0, \tag{5.48}$$

The factor $\exp\{j\omega t\}$ is never zero, and it rotates in the complex plane. It is thus impossible for Eq. (5.48) to be satisfied FOR ALL TIME unless the coefficient of $\exp\{j\omega t\}$ is zero. Otherwise, it could be satisfied (the real part of the quantity in brackets in Eq. (5.48) equal to zero) at only two instants of time every period. (This same reasoning can also be used to prove Eqs. (5.38)–(5.40).) Thus

$$\bar{I} = j\omega C\bar{V}. \tag{5.49}$$

In summary, suppose that it is desired to find the current $i(t)$ through the capacitor given the sinusoidal voltage $v(t)$ across it (see Fig. 5.15a). Instead of solving a differential equation as in Fig. 5.15a, all that needs to be done is (see Fig. 5.15b) (1) determine the phasor $\bar{V}$ representing $v(t)$, (2) multiply $\bar{V}$ by $j\omega C$ to obtain $\bar{I}$, and (3) multiply $\bar{I}$ by $e^{j\omega t}$ and take the real part to obtain $i(t)$.

Note that Eq. (5.49) can be rewritten as

$$\begin{aligned} \bar{V} &= \frac{1}{j\omega C}\bar{I} \\ &= \frac{-j}{\omega C}\bar{I}. \end{aligned} \tag{5.50}$$

The current–voltage relation for a resistor $R$ is $v(t) = Ri(t)$. In the frequency domain Ohm's law takes the similar form $\bar{V} = R\bar{I}$. Comparing with Eq. (5.50) shows that in the frequency domain, the only difference between the capacitor and resistor current–voltage relations is that $R$ has been replaced by $-j/(\omega C)$.

**Example 5.10**

A capacitance of 0.01 μF has across it a 60-Hz voltage of magnitude 1 V and zero phase; that is, $v_C(t) = 1\cos(377t)$ V (where $\omega = 2\pi f = 377$ rad/s). What is the current $i(t)$ through the capacitor?

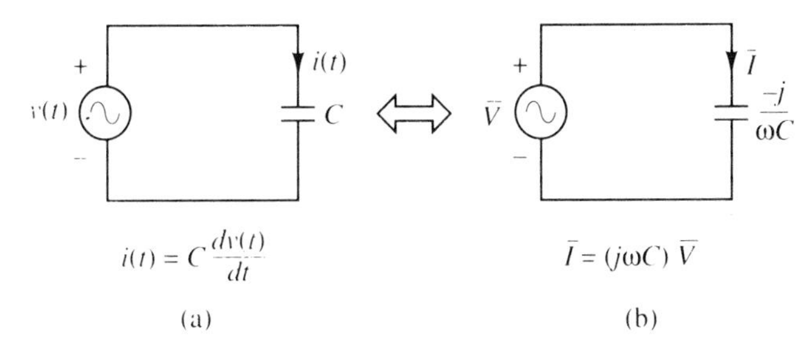


**Figure 5.15** Contrast between analysis of a sinusoidally excited capacitive circuit in (a) the time domain and (b) the frequency domain.

**Solution** Solve the problem in the frequency domain using Eq. (5.49). $\bar{V} = 1\text{ V}\,\angle 0°$, which gives $\bar{I} = j377(0.01 \cdot 10^{-6})\,\angle 0° = 3.77\ \mu\text{A}\,\angle 90°$. Thus

$$i(t) = 3.77 \cdot 10^{-6} \cos(377t + \pi/2) = -3.77 \sin(377t)\ \mu\text{A}.$$

*Caution:* Avoid the common error of equating $i(t)$ with $\bar{I}$; note that $i(t)$ depends on $t$, whereas $\bar{I}$ does not. Also, $i(t)$ is real-valued, while $\bar{I}$ is complex-valued. That is, do *not* write $i(t) = 3.77\ \mu\text{A}\,\angle 90°$.

What about inductors? The current–voltage relation in the time domain is

$$v(t) = L\frac{di(t)}{dt}. \tag{5.51}$$

Using analysis similar to that for the capacitor, we quickly obtain

$$\bar{V} = j\omega L\bar{I}. \tag{5.52}$$

Again the form is identical to that for a resistor, where $R$ has now been replaced by $j\omega L$.

These results can be unified by introducing the concept of a *complex impedance* $\bar{Z}$. The bar over $Z$ indicates that $\bar{Z}$ is complex; however, $\bar{Z}$ is *not* a phasor and does not rotate in time. Rather, it is a complex constant relating current and voltage phasors. For a resistor, $\bar{Z}$ happens to have zero imaginary part and is just the real number $R$ $(=Re^{j0})$. The resistor thus introduces zero phase shift, so the voltage and current for a resistor are said to be in phase.

Figure 5.16, called a *phasor diagram,* shows, on the complex plane at a given time $t > 0$, the voltage and current phasors of a resistor. Because the phasors shown represent different quantities (voltage and current), its primary use is in the display of phase relations between the phasors; magnitudes of different types of quantities cannot be compared. Thus for the resistor, the current and voltage phasors lie on the same ray from the origin. *Both* phasors, of course, rotate counterclockwise at angular frequency $\omega$ as time progresses.

Next, consider the capacitor. The impedance of an ideal capacitor has no real part and is the negative imaginary number $-j/(\omega C)$. The negative real number $-1/(\omega C)$ is called the *reactance* $X_C$ of the capacitor at frequency $\omega$. Notice that $|X_C|$ is a decreasing function of $\omega$. Finally, the inductor impedance is the positive imaginary number $j\omega L$. The positive real number $\omega L$ is called the reactance $X_L$ of the inductor at frequency $\omega$. Notice that $X_L$ is an increasing function of $\omega$. Phasor diagrams for capacitors and inductors are presented below when the phase relationships are established.

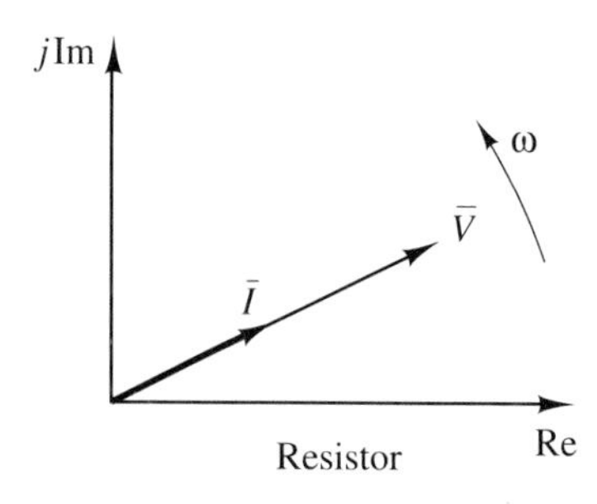


**Figure 5.16** Phasor diagram of voltage across and current in a resistor.

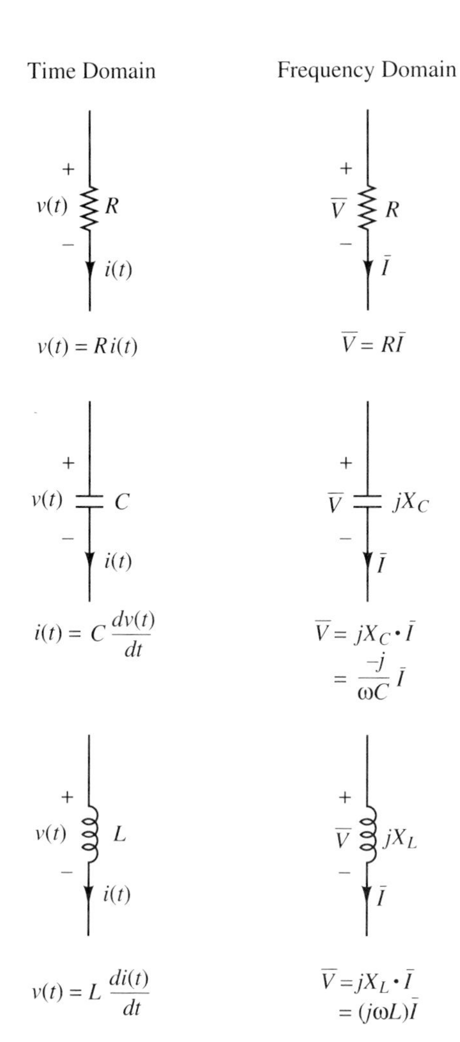


**Figure 5.17** Time-domain (left) and frequency-domain (right) $i$–$v$ relations for the three basic passive circuit element types. Note that all frequency-domain relations are proportionalities (complex Ohm's law).

The results above are summarized in Fig. 5.17. This figure shows the process of transforming $v(t)$ and $i(t)$ into frequency-domain variables (phasors), and $R$, $L$, and $C$ into frequency-domain constants. As noted in Sec. 5.2.3, with this new general Ohm's law, parallel and series combinations of inductors, capacitors, and resistors can all be handled as easily as so many resistors. The net impedance $\overline{Z}$ merely becomes complex with both nonzero real and imaginary parts. The real part is always called the resistive component or *resistance,* while the imaginary part is called the reactive component or *reactance*:

$$\overline{Z} = R + jX. \tag{5.53}$$

Because impedances in series add, the rectangular form of $\overline{Z}$ in Eq. (5.53) corresponds to a resistance $R$ connected *in series* with a reactance $X$.

### 5.4.4 Frequency-Domain Circuit Analysis

In analyzing linear circuits in the sinusoidal steady state, we need not bother writing down the differential equation. Instead, the complex Ohm's law

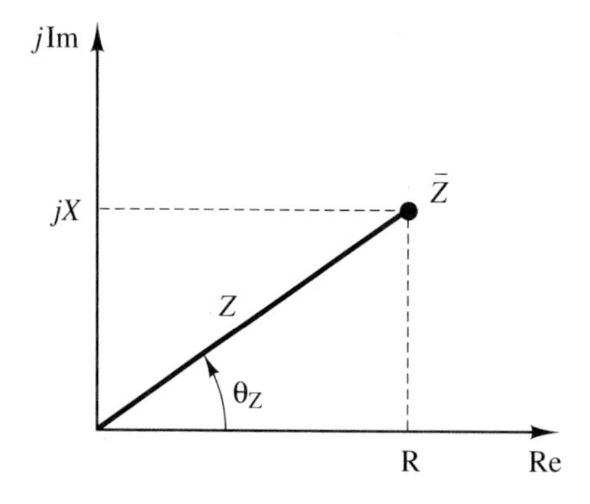


**Figure 5.18** Complex impedance $\overline{Z}$ shown on the complex impedance plane. Both polar and rectangular forms are indicated.

$$\overline{V} = \overline{I}\,\overline{Z} \tag{5.54}$$

is all that is needed for sinusoidal analysis. Alternatively, we may write

$$\overline{Z} = \frac{\overline{V}}{\overline{I}} = Z\underline{/\theta_z}\,, \tag{5.55}$$

where, by division of complex numbers [Eq. (5.22)] and use of the polar form [Eqs. (5.16)] applied to Eq. (5.53),

$$Z = \frac{V}{I} = (R^2 + X^2)^{1/2} \tag{5.56a}$$

and

$$\theta_Z = \theta_v - \theta_i = \tan^{-1}\left(\frac{X}{R}\right). \tag{5.56b}$$

The complex impedance is shown on the complex plane in Fig. 5.18. Again, $\overline{Z}$ is not a phasor, but rather, just a complex number. Some examples should clarify the methods of using these results in sinusoidal circuit analysis.

**Example 5.11**

For the circuit in Fig. 5.19a, find $i(t)$ and express it in the form $i(t) = I\cos(\omega t + \theta_i)$. The voltage source has time dependence $v_s(t) = V_s\cos(\omega t + \theta_s)$.

**Solution** A circuit diagram showing phasors and complex impedance is given in Fig. 5.19b. The phasor representation of the voltage source is $\overline{V}_s = V_s e^{j\theta_s}$.

The voltage across the inductor has phasor representation $\overline{V}_L$, which by the complex Ohm's law is $j\omega L\overline{I}$, while that across the resistor is $\overline{V}_R = R\overline{I}$. By KVL, $\overline{V}_s = \overline{V}_R + \overline{V}_L = (R + j\omega L)\overline{I}$. Therefore,

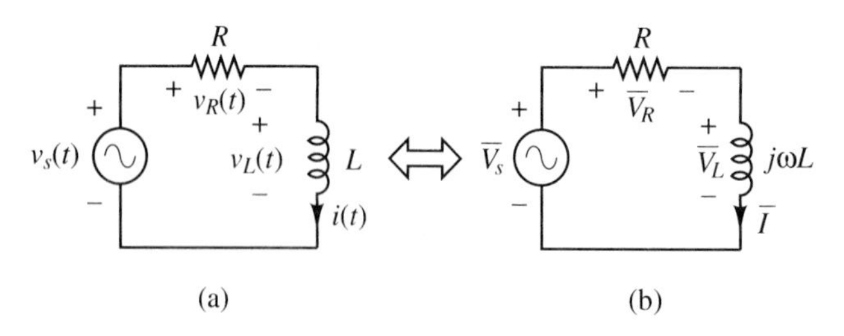


**Figure 5.19** Circuit for Example 5.11 in (a) the time domain and (b) the frequency domain.

$$\bar{I} = \frac{1}{R + j\omega L}\bar{V}_s$$
$$= \frac{V_s}{R + j\omega L}e^{j\theta_s}. \tag{5.57}$$

Expressing the denominator in polar form gives

$$\bar{I} = \frac{V_s}{[R^2 + (\omega L)^2]^{1/2}}e^{j[\theta_s - \tan^{-1}(\omega L/R)]}. \tag{5.58}$$

Alternatively, the total impedance seen by the source is the series combination of $R$ and $L$. Defining $\bar{Z}$ as the total impedance, we have

$$\bar{Z} = R + j\omega L$$
$$= Ze^{j\theta_Z}, \tag{5.59}$$

where

$$Z = [R^2 + (\omega L)^2]^{1/2} \tag{5.60a}$$

and

$$\theta_Z = \tan^{-1}\left(\frac{\omega L}{R}\right). \tag{5.60b}$$

The current phasor $\bar{I}$ is then easily seen to be

$$\bar{I} = Ie^{j\theta_i}$$
$$= \frac{\bar{V}_s}{\bar{Z}}, \tag{5.61}$$

the same result as in Eq. (5.58), where

$$I = \frac{V_s}{Z} \tag{5.62a}$$

and

$$\theta_i = \theta_s - \theta_Z. \tag{5.62b}$$

Finally, $I$ and $\theta_i$ are substituted into $i(t) = I\cos(\omega t + \theta_i)$ to obtain the desired time-domain waveform, which amounts to a transformation from the frequency domain back to the time domain. [Now we see why $j$ rather than $i$ is used for $\sqrt{-1}$: Eq. (5.61) would read $\bar{I} = Ie^{i\theta_i}$ and $\bar{V} = Ve^{i\theta_v}$—which would all be very confusing because the $i$ preceding $\theta$ has nothing to do with current, whereas the $i$ subscript in $\bar{I}$ has everything to do with current!]

**Example 5.12**

To get a feel for how simple the foregoing calculations really are, a numerical example of the same problem solved in Example 5.11 is now presented. Let $v_s(t) = 10\cos(377t)$ V, $L = 0.1$ H, and $R = 100\ \Omega$. Find $i(t)$.

**Solution**

$$\bar{Z} = 100 + j377(0.1) = 100 + j37.7\ \Omega \approx 107\ \Omega\ \underline{/20.7^\circ}.$$
$$\bar{V}_s = 10\text{ V}\ \underline{/0^\circ}.$$

Therefore,

$$\bar{I} = \frac{\bar{V}_s}{\bar{Z}} = \frac{10 \angle 0^\circ}{107 \angle 20.7^\circ} = 0.093 \angle -20.7^\circ = 0.093 \text{ A} \angle -0.361 \text{ rad.}$$

Finally, these results are expressed in the time domain as

$$i(t) = 93 \cos(377t - 0.361) \text{ mA.}$$

**Example 5.13**

Now suppose in Example 5.11 that the voltage phasor across the inductor $\bar{V}_L$ is desired (again see Fig. 5.19b).

**Solution** $\bar{V}_L$ can be found by a complex voltage divider:

$$\begin{aligned}\bar{V}_L &= \frac{j\omega L}{R + j\omega L}\bar{V}_s \\ &= \frac{j\omega L\,(R - j\omega L)}{R^2 + (\omega L)^2}\bar{V}_s \\ &= \frac{\omega L e^{j[\pi/2 - \tan^{-1}(\omega L/R)]}}{[R^2 + (\omega L)^2]^{1/2}}\bar{V}_s.\end{aligned} \tag{5.63}$$

Or, by taking advantage of the current phasor found in Eq. (5.58) and using the complex Ohm's law [Eq. (5.54)], we obtain

$$\begin{aligned}\bar{V}_L &= \bar{Z}_L \bar{I} \\ &= j\omega L\left(\frac{V_s}{Z}\right)e^{j(\theta_s - \theta_z)}.\end{aligned} \tag{5.64}$$

Noting that

$$\begin{aligned}\bar{V}_L &= j\omega L \bar{I} \\ &= \omega L \bar{I} e^{j\pi/2},\end{aligned} \tag{5.65}$$

the inductor voltage phasor $\bar{V}_L$ is seen to be at an angle 90° counterclockwise from the current phasor $\bar{I}$. Recall that the direction of rotation for phasors is counterclockwise as time progresses. It can thus be concluded that for an inductor the current lags behind the voltage by 90°, consistent with the fact that the current in an inductor cannot change suddenly. The phasor diagram for current and voltage in an inductor, given in Fig. 5.20, should now be understandable.

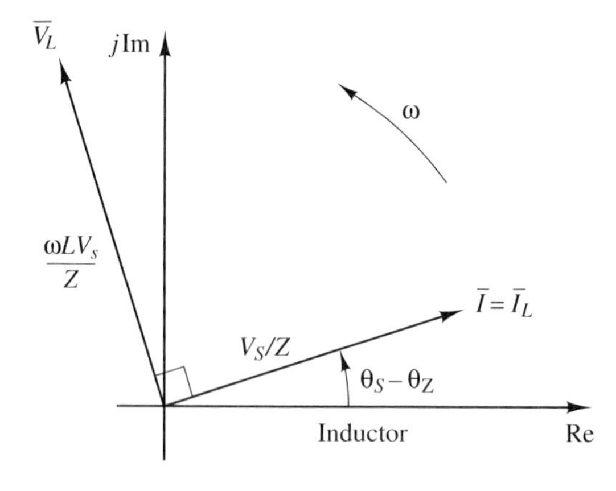


**Figure 5.20** Phasor diagram of voltage across and current in an inductor in the context of the $RL$ circuit in Fig. 5.19.

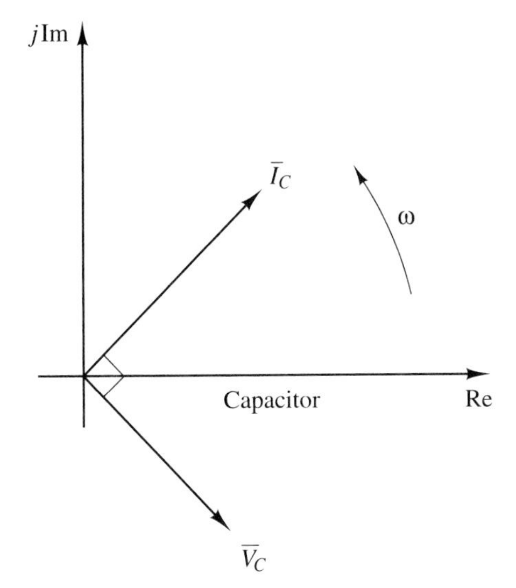


**Figure 5.21** Phasor diagram of voltage across and current in a capacitor.

The reverse situation is true for a capacitor. Recalling Eq. (5.49), we have

$$\overline{I}_C = j\omega C\overline{V}_C, \tag{5.66}$$

so that in the case of a capacitor, the current leads the voltage by 90° (see Fig. 5.21).

A fun and easy way to remember the phase relations for capacitors and inductors is to think of "ELI the ICE man." Here $E$ symbolizes voltage, $I$ current, $L$ an inductor, and $C$ a capacitor. Because $I$ comes after $E$ in "ELI," we are reminded that the current in an inductor lags behind the voltage (by 90°). Similarly, because $I$ precedes $E$ in "ICE," we are reminded that the current in a capacitor leads the voltage (by 90°). In fact, impedances that are inductive are called *lagging* and capacitive impedances are called *leading,* the convention being that "lag" and "lead" refer to the *current* with respect to the voltage.

Recall that $X_L > 0$ and $X_C < 0$. In general, an impedance with positive or negative net reactance is called, respectively, inductive or capacitive. To illustrate this point, consider the series combination of a resistor, an inductor, and a capacitor shown in Fig. 5.22. The equivalent impedance is

$$\overline{Z}_{\text{eq}} = R + j\left(\omega L - \frac{1}{\omega C}\right). \tag{5.67}$$

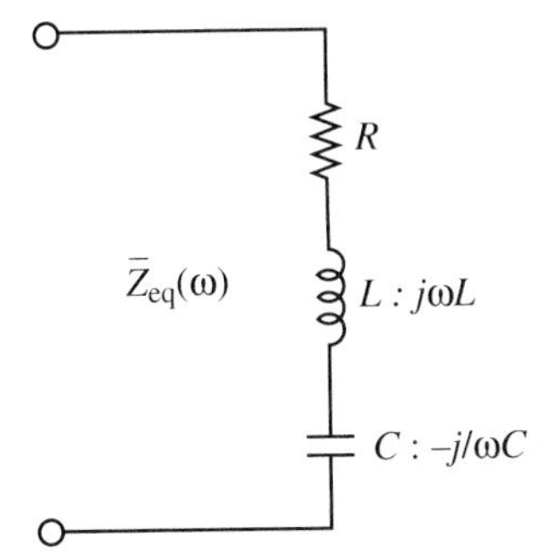


**Figure 5.22** Series *RLC* combination shown as an equivalent impedance $Z_{\text{eq}}(\omega)$.

Figure 5.23a and b shows for two cases the three component impedances and their vectorial addition, which is the net impedance $\overline{Z}_{eq}$. In Fig. 5.23a the inductive reactance is larger than the magnitude of the capacitive reactance, while the reverse is true in Fig. 5.23b. Consequently, $\overline{Z}_{eq}$ is inductive in Fig. 5.23a [because $X_{eq} = \omega L - 1/(\omega C) > 0$] and capacitive in Fig. 5.23b (because $X_{eq} < 0$), even though in both cases capacitors and inductors are both present. We are reminded of our discussion in Sec. 4.3.1, where a single circuit element had aspects of all three types of passive component.

To illustrate these ideas further, suppose that the phase of $v_{eq}(t)$, the voltage across the entire series impedance in Fig. 5.22, is taken to be zero. Figure 5.24a shows the relation between the current phasor and the various voltage phasors for the situation of Fig. 5.23a ($\overline{Z}_{eq}$ is inductive), and the corresponding voltage time waveforms are shown in Fig. 5.24b. Note that $\overline{V}_R$ and $\overline{I}_R$ are aligned, $\overline{I}$ lags $\overline{V}_L$, and $\overline{I}$ leads $\overline{V}_C$.

The vectorial sum of $\overline{V}_R$, $\overline{V}_L$, and $\overline{V}_C$ is equal to $\overline{V}_{eq}$. Also note that because we have (arbitrarily) taken $\overline{V}_{eq}$ along the real axis, $\overline{V}_R$ does *not* lie on the real axis, but instead, lies at an angle $\theta_R$ behind $\overline{V}_{eq}$.

Because in this example $\theta_R \approx -20°$, the first crest of $v_R(t)$ appears about $20/360 = 1/18$ of a cycle after that of $v_{eq}(t)$ (which is located at $t = 0$ in Fig. 5.24b). The voltage $v_R(t)$ is said to lag $v_{eq}(t)$ by 20°. Similarly, because $\theta_C \approx -110°$, the first crest of $v_C(t)$ appears about one-third of a cycle after that of $v_{eq}(t)$, and because $\theta_L \approx +70°$, the first crest of $v_L(t)$ appears about one-fifth of a cycle *before* that of $v_{eq}(t)$. The voltage $v_L(t)$ is said to lead $v_{eq}(t)$ by 70°.

Table 5.1 summarizes some of the important component frequency-domain relations. With the generalization of Ohm's law to the complex case, which is valid in the frequency domain, there are further bonuses. For example, we can now apply voltage and current division, loop and node analysis, and Thévenin and Norton equivalent circuits to the design and analysis of linear circuits in the sinusoidal steady state.

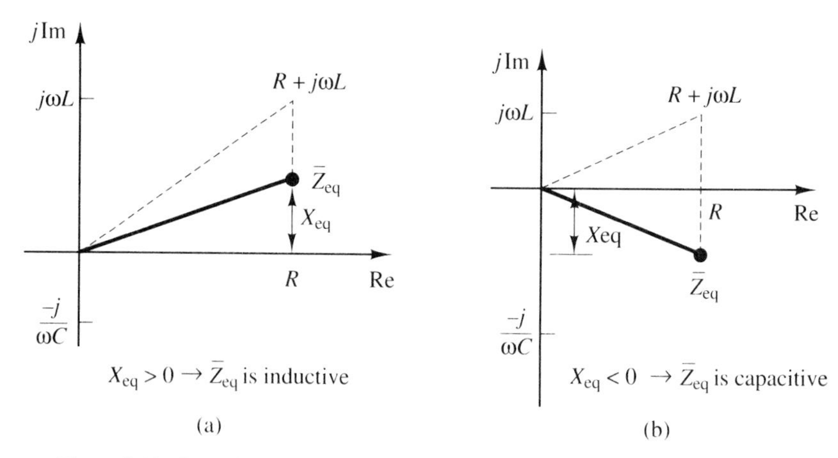


**Figure 5.23** Impedance diagrams for series *RLC* combination when the combination $Z_{eq}$ is (a) inductive and (b) capacitive.

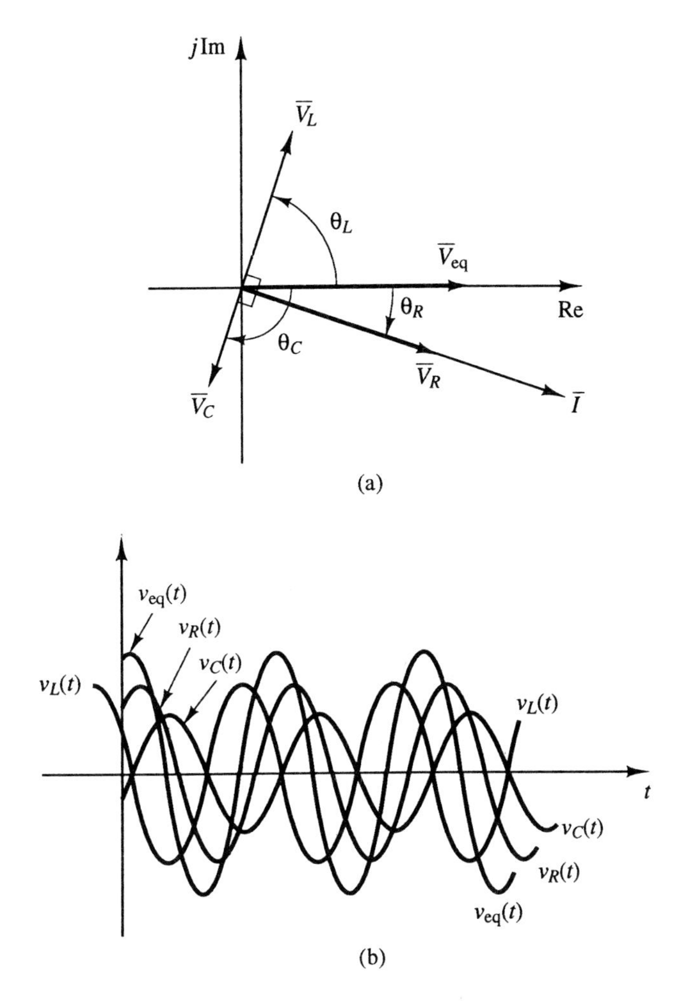


**Figure 5.24** (a) Phasor diagram for series $RLC$ combination when it is inductive; (b) associated time waveforms.

**TABLE 5.1** IMPEDANCE AND PHASE RELATIONS OF THE THREE BASIC PASSIVE CIRCUIT ELEMENTS

| Component | Impedance | $\overline{I}$–$\overline{V}$ phase relation |
|---|---|---|
| Resistor | $R$ | $\overline{I}$ and $\overline{V}$ are in phase |
| Inductor | $jX_L$ where $X_L = \omega L$ | $\overline{I}$ lags $\overline{V}$ by 90° (ELI) |
| Capacitor | $jX_C$ where $X_C = -1/(\omega C)$ | $\overline{I}$ leads $\overline{V}$ by 90° (ICE) |

Always: $\overline{V} = \overline{I}\,\overline{Z}$. If $i(t) = I\cos(\omega t)$, then $v(t) = IZ\cos(\omega t + \theta_Z)$.
This is because $\overline{V} = Ve^{j\theta_v} = Ie^{j\theta_i}Ze^{j\theta_Z} = IZe^{j(\theta_i + \theta_Z)}$.

## 5.5 FREQUENCY RESPONSE AND FILTERING

### 5.5.1 Introduction

As well as being a convenient tool for monofrequency voltage and current calculations, the frequency domain has other uses. For example, selection of individual sinusoidal components from the complicated waveform produced by an antenna is the basis for radio and television communication. Audiophiles are always trying to pin down the "cutoff frequencies" of a given piece of stereo equipment on the market. In biomedical applications such as the artificial heart, the frequencies of the sinusoidal components having the largest magnitudes can provide diagnostic information. Speech synthesis or recognition depends on analysis of the "frequency content" of the sounds of interest. Economic and weather forecasting (equally unreliable for long-term prediction) can use knowledge about periodicities that may exist in the relevant data.

In this section a simple description of the frequency response is presented. It is followed by a real-life application of the frequency response that a non-electrical engineer could expect to encounter on the job. As we have seen for the phasor method, it is far more convenient mathematically to work with complex sinusoids $e^{j\omega t}$ than with sines and cosines. For the same reasons, it is highly advantageous to reformulate Fourier's expansions in terms of $e^{j\omega t}$ rather than the sines and cosines of Eqs. (5.2) through (5.5).

In Sec. 5.5.2, it is shown that simplified expressions for Fourier series and Fourier integrals may be obtained by use of complex exponentials. The results are that the Fourier series becomes

$$f(t) = \sum_{n=-\infty}^{\infty} \bar{F}_n e^{jn\omega_0 t} \tag{5.68}$$

$$\bar{F}_n = \frac{1}{T}\int_{-T/2}^{T/2} f(t)\, e^{-jn\omega_0 t}\, dt, \tag{5.69}$$

and the Fourier transform becomes

$$f(t) = \frac{1}{2\pi}\int_{-\infty}^{\infty} \bar{F}(\omega)\, e^{j\omega t}\, d\omega \tag{5.70}$$

$$\bar{F}(\omega) = \int_{-\infty}^{\infty} f(t)\, e^{-j\omega t}\, dt. \tag{5.71}$$

Also, an example of using each of these complex exponential expansions is presented.

### ☆ 5.5.2 Fourier Expansions over Complex Exponentials

Just after the discussion of Fourier series and integrals, it was stated that the formulas could be simplified by the introduction of complex exponentials. This will now be demonstrated first for the Fourier series of periodic functions $f(t)$. Recalling Eqs. (5.34), the Fourier series expression for $f(t)$ presented in Eq. (5.2) can be rewritten

$$f(t) = \frac{a_0}{2} + \frac{1}{2}\left[\sum_{n=1}^{\infty} a_n(e^{jn\omega_0 t} + e^{-jn\omega_0 t}) + \frac{b_n(e^{jn\omega_0 t} - e^{-jn\omega_0 t})}{j}\right] \tag{5.72a}$$

$$= \frac{a_0}{2} + \frac{1}{2}\left[\sum_{n=1}^{\infty} (a_n - jb_n)e^{jn\omega_0 t} + \sum_{n=1}^{\infty} (a_n + jb_n)e^{-jn\omega_0 t}\right]. \tag{5.72b}$$

Let $n' = -n$ in the second sum of Eq. (5.72b). That sum can then be expressed as

$$\frac{1}{2}\sum_{n'=-\infty}^{-1} (a_{-n'} + jb_{-n'})e^{jn'\omega_0 t}. \tag{5.73}$$

But from the formulas for $a_n$ and $b_n$ in Eqs. (5.3) and the fact that cosine is an even function while sine is an odd function,

$$a_{-n} = a_n \quad \text{and} \quad b_{-n} = -b_n, \tag{5.74}$$

which is true whether or not $f(t)$ is real-valued. Changing the symbol $n'$ to $n$, the entire expression for $f(t)$ can be written as

$$f(t) = \sum_{n=-\infty}^{\infty} \tfrac{1}{2}(a_n - jb_n)e^{jn\omega_0 t} \qquad (b_0 = 0). \tag{5.75}$$

Define the complex coefficient

$$\bar{F}_n = \tfrac{1}{2}(a_n - jb_n). \tag{5.76}$$

Then

$$f(t) = \sum_{n=-\infty}^{\infty} \bar{F}_n e^{jn\omega_0 t}, \tag{5.77}$$

which is the complex Fourier series. Note that if (and *only* if) $f(t)$ is real valued, then Eqs. (5.74) translate to $\bar{F}_{-n} = \bar{F}_n^*$. Furthermore, from the formulas for $a_n$ and $b_n$ in Eq. (5.3), the definition of $\bar{F}_n$ in Eq. (5.76), and Euler's formula [Eq. (5.19)], we have

$$\bar{F}_n = \frac{1}{T}\int_{-T/2}^{T/2} f(t)\, e^{-jn\omega_0 t}\, dt. \tag{5.78}$$

Equations (5.77) and (5.78) are the complex Fourier series pair.

**Example 5.14**

Repeat Example 5.1 but expand $f(t)$ over complex sinusoids using Eqs. (5.77) and (5.78).

**Solution**

$$\bar{F}_n = \frac{1}{8}\int_{-2}^{2} e^{-jn\pi t/4}\, dt = \frac{1}{j2n\pi}(e^{jn\pi t/2} - e^{-jn\pi t/2})$$

$$= \frac{\sin(n\pi/2)}{n\pi},$$

where the last step used Eq. (5.34b). For $n = 0$ use

$$\bar{F}_0 = \frac{1}{8}\int_{-2}^{2} 1\, dt = \frac{1}{2}.$$

Finally,

$$f(t) = \frac{1}{2} + \sum_{n=-\infty}^{\infty} \frac{\sin(n\pi/2)}{n\pi} e^{jn\pi t/4},$$

which took a lot less effort than did Example 5.1. Note that because $\sin(-n\pi/2)/(-n\pi) = \sin(n\pi/2)/(n\pi)$, use of Eq. (5.34a) in the result above gives

$$f(t) = \frac{1}{2} + \sum_{n=1}^{\infty} \frac{\sin(n\pi/2)}{n\pi} 2 \cos\frac{n\pi t}{4},$$

which is the same result as that obtained in Example 5.1.

Exactly the same procedure can be used to find the complex exponential Fourier integral. Using Eqs. (5.34), the Fourier integral expression for an aperiodic time function $f(t)$ in Eq. (5.4) is

$$f(t) = \frac{1}{2\pi}\int_0^{\infty}\left[a(\omega)(e^{j\omega t} + e^{-j\omega t}) + \frac{b(\omega)(e^{j\omega t} - e^{-j\omega t})}{j}\right] d\omega \tag{5.79a}$$

$$= \frac{1}{2\pi}\left\{\int_0^{\infty} [a(\omega) - jb(\omega)]\, e^{j\omega t}\, d\omega + \int_0^{\infty} [a(\omega) + jb(\omega)]\, e^{-j\omega t}\, d\omega\right\}. \tag{5.79b}$$

In the second integral in Eq. (5.79b), let $\omega' = -\omega$. Then it can be written

$$-\int_0^{-\infty} [a(-\omega') + jb(-\omega')]\, e^{j\omega' t}\, d\omega' = \int_{-\infty}^{0} [a(\omega') - jb(\omega')]\, e^{j\omega' t}\, d\omega', \tag{5.80}$$

because, from Eqs. (5.5), for any $f(t)$ (complex or real),

$$a(-\omega) = a(\omega) \quad \text{and} \quad b(-\omega) = -b(\omega). \tag{5.81}$$

Changing the symbol $\omega'$ to $\omega$, the entire expression for $f(t)$ becomes

$$f(t) = \frac{1}{2\pi}\int_{-\infty}^{\infty} [a(\omega) - jb(\omega)]\, e^{j\omega t}\, d\omega. \tag{5.82}$$

Defining the complex coefficient

$$\overline{F}(\omega) = a(\omega) - jb(\omega) \tag{5.83}$$

gives

$$f(t) = \frac{1}{2\pi}\int_{-\infty}^{\infty} \overline{F}(\omega)\, e^{j\omega t}\, d\omega, \tag{5.84}$$

which is called the *inverse Fourier transform.* Note that if (and *only* if) $f(t)$ is real valued, then Eqs. (5.81) translate to $\overline{F}(-\omega) = \overline{F}^*(\omega)$. Furthermore, from the formulas for $a(\omega)$ and $b(\omega)$ in Eq. (5.5) and the definition of $\overline{F}(\omega)$ in Eq. (5.83),

$$\overline{F}(\omega) = \int_{-\infty}^{\infty} f(t)[\cos(\omega t) - j\sin(\omega t)]\, dt, \tag{5.85}$$

which by Euler's formula is

$$\overline{F}(\omega) = \int_{-\infty}^{\infty} f(t)\, e^{-j\omega t}\, dt, \tag{5.86}$$

and is called the *Fourier transform* of $f(t)$. Equations (5.84) and (5.86) are the complex Fourier transform pair.

What has been gained? A neat, compact expansion of the periodic (or aperiodic) time function $f(t)$ over a sum (or integral) of complex exponentials. Often at this point it is objected that $n$ (or $\omega$) is allowed to take on negative values. After all, there is no such thing as a negative frequency. But it should be remembered that complex exponentials are just mathematically convenient ways of representing real sinusoids. For example, from Eq. (5.34a), $\cos(\omega t)$ may be expressed as $\cos(\omega t) = (e^{j\omega t} - e^{-j\omega t})/2$. Thus $\cos(\cdot)$, which is real-valued, is now conveniently expressed in terms of complex exponentials (one with "negative" frequency $-\omega$), with all their desirable properties.

Now consider the terms of the complex expansion of $f(t)$ for $f(t)$ periodic [or aperiodic]. Supposing that $f(t)$ is real-valued, (1) $\bar{F}_{-n} = \bar{F}_n^*$ [or $\bar{F}(-\omega) = \bar{F}^*(-\omega)$] and (2) $e^{-jn\omega_0} = (e^{jn\omega_0})^*$ [or $e^{-j\omega t} = (e^{j\omega t})^*$]. As a result, the positive and corresponding negative frequency terms in the expansions of $f(t)$ are complex conjugate pairs. Consequently, the sum of each positive frequency term and the corresponding negative frequency term in the expansion of $f(t)$ is always a real number, by Eq. (5.32). And that real number represents the component of $f(t)$ having frequency $n\omega_0$ [or $\omega$].

A simple instance of this was already demonstrated in Example 5.14 for periodic $f(t)$. The complex expansion was shown to be equal to the purely real expansion of Example 5.1. An example for an aperiodic function is now considered.

**Example 5.15**

Repeat Example 5.2 but expand $f(t)$ over complex sinusoids using Eqs. (5.84) and (5.86).

**Solution**

$$\bar{F}(\omega) = \int_{-1}^{1} e^{-j\omega t}\,dt = \frac{1}{-j\omega}(e^{-j\omega} - e^{j\omega}) = \frac{2\sin(\omega)}{\omega},$$

so that

$$f(t) = \frac{1}{2\pi}\int_{-\infty}^{\infty} \frac{2\sin(\omega)}{\omega} e^{j\omega t}\,d\omega.$$

By Eq. (5.34a), this integral can be written as an integral over real-valued functions:

$$f(t) = \frac{2}{\pi}\int_{0}^{\infty} \frac{\sin(\omega)}{\omega}\cos(\omega t)\,d\omega,$$

which is in exact agreement with the result of Example 5.2.

### 5.5.3 Linear Filtering: Practical Examples

How do all the results above about complex exponential Fourier expansions fit in with the desired *frequency response?* It is shown in the appendix that the output $y(t)$ of a linear shift-invariant system is equal to the convolution of the input signal $x(t)$ with the system impulse response $h(t)$:

$$y(t) = \int_{-\infty}^{\infty} x(t-\tau)\,h(\tau)\,d\tau. \tag{5.87}$$

Suppose that $x(t) = e^{j\omega t}$, a complex exponential. Then

$$y(t) = \int_{-\infty}^{\infty} e^{j\omega(t-\tau)}\,h(\tau)\,d\tau \tag{5.88a}$$

$$= e^{j\omega t}\int_{-\infty}^{\infty} h(\tau)\,e^{-j\omega\tau}\,d\tau \tag{5.88b}$$

$$= x(t)\,\bar{H}(\omega), \tag{5.88c}$$

where from Eq. (5.88b) the frequency response $\bar{H}(\omega)$ is the Fourier transform of the impulse response $h(t)$, exactly as defined in Eq. (5.86)! So yet another wonderful property of the complex exponential has been found: It is an *eigenfunction* of LSI systems. That is, when a complex exponential $x(t)$ is given as input to the system, the output is just a time-independent constant, $\overline{H}(\omega)$, times the input time function $x(t)$.

Of course, in practice $x(t)$ will be real-valued, for example, $\cos(\omega t)$. The real-valued output $y(t)$ will then be [using Eqs. (5.34a) and (5.88)]

$$y(t) = \frac{e^{j\omega t}\overline{H}(\omega) + e^{-j\omega t}\overline{H}(-\omega)}{2} \tag{5.89a}$$

$$= \frac{e^{j\omega t}\overline{H}(\omega) + e^{-j\omega t}\overline{H}^*(\omega)}{2}, \tag{5.89b}$$

$$= \frac{e^{j\omega t}\overline{H}(\omega) + [e^{j\omega t}\overline{H}(\omega)]^*}{2}, \tag{5.89c}$$

where Eq. (5.89b) follows because $h(t)$ is assumed to be real valued, so that $\overline{H}(-\omega) = \overline{H}^*(\omega)$. Noting that the two terms in Eq. (5.89c) are a complex conjugate pair, we have

$$y(t) = \text{Re}\{e^{j\omega t}\overline{H}(\omega)\}, \tag{5.90}$$

where the two in Eq. (5.32) cancels the $\frac{1}{2}$ in Eq. (5.89c). One important fact stands out from all this. Because $e^{j\omega t}$ has magnitude 1, the magnitude of $\overline{H}(\omega)$, $H(\omega)$, indicates the level of output for a unity amplitude sinusoidal input of frequency $\omega$. Thus if $\overline{H}(\omega)$ is specified for all $\omega$, we know what the system will do to *any* input. This is because by Fourier's theorem, Eqs. (5.2) and (5.4), the input is decomposable into a sum (integral) of functions such as $e^{j\omega t}$ for various values of $\omega$.

**Example 5.16**

Suppose that $x(t) = 2\cos(377t)$ is sent into a linear system having frequency response $\overline{H}(\omega) = j\omega/10$. What is $y(t)$?

**Solution** By Eq. (5.90),

$$y(t) = 2\,\text{Re}\left\{e^{j377t}\frac{j377}{10}\right\}$$

$$= 75.4\,\text{Re}\{e^{j(377t + \pi/2)}\}$$

$$= 75.4\cos\left(377t + \frac{\pi}{2}\right) = -75.4\sin(377t).$$

So because the input $x(t)$ was just a single sinusoid of radial frequency 377 rad/s, it was scaled by the factor $H(377) = 377/10$ and shifted in phase by 90°. We note that this particular $h(t)$ is a differentiator $(j\omega)$ with constant scaling (1/10). Indeed, (1/10) $dx(t)/dt = -[377(2)/10]\sin(377t) = -75.4\sin(377t)$. We shall return to differentiator circuits in Sec. 8.8.4.

In fact, the magnitude of the frequency response, $H(\omega)$ [or often 20 $\log_{10}\{H(\omega)\}$], is the function usually plotted versus $\omega$ to indicate to what degree the given system passes each frequency in the spectrum. That is the plot the audio buffs are squinting at. In their case the music is the input, the recording/playback system is the system $H(\omega)$, and the output is the signal sent to the speakers.

A system designed such that $H(\omega)$ is relatively large for low frequencies and small for high frequencies is called a *low-pass filter*. Similarly, there are high-pass

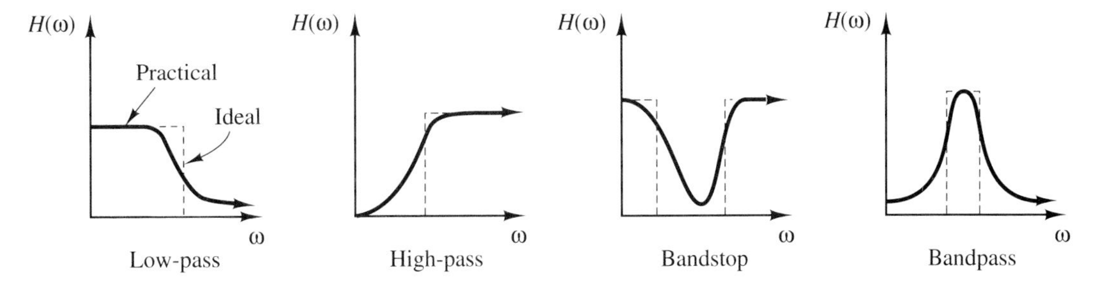


**Figure 5.25** Frequency responses of various filter types.

filters, bandstop filters, and bandpass filters. Examples of these are shown in Fig. 5.25. The smooth curves are practical filter responses approximating the square, dashed ideal responses.

An illustrative example should help clarify some of the foregoing ideas. Suppose that $v_s(t)$ is the voltage output from a transducer measuring the pressure in a cyclical process such as a pump or an engine (so the pressure is periodic; see Fig. 5.26a). This particular signal happens to be the sum of a sinusoid of frequency $\omega_1$ and another weaker signal at $2\omega_1$. But somehow, between the transducer and the amplifier, some high-frequency noise has contaminated the signal, as shown in Fig. 5.26b.

How can the signal be "cleaned up"? An easy way is to place $v_s(t)$ across the series $RL$ combination as in the $RL$ circuit we studied previously (see Fig. 5.19). Consider the output voltage to be $v_R(t)$; that is, look at $v_R(t)$ instead of the noisy $v_s(t)$. Using a complex voltage divider, valid for any particular single frequency $\omega$, the voltage phasor across the resistor is found to be

$$\begin{aligned} \overline{V}_R &= \frac{R}{R + j\omega L}\overline{V}_s \\ &= \frac{R}{[R^2 + (\omega L)^2]^{1/2}} V_s e^{j(\theta_s - \theta_Z)}. \end{aligned} \tag{5.91}$$

Although originally, such expressions were derived for analysis of the monofrequency sinusoidal steady state, there is no reason why such evaluations cannot be made individually for *all* values of $\omega$. Because $e^{j(\theta_s - \theta_Z)}$ has unity magnitude, the term $R/[R^2 + (\omega L)^2]^{1/2}$ can be viewed as $H(\omega)$. Therefore, we could write for the voltage magnitudes in Eq. (5.91),

$$V_R = H(\omega)\, V_s. \tag{5.92}$$

It is clear that $H(\omega)$ decreases with $\omega$ because $\omega$ appears only in the denominator. $H(\omega)$ can be rewritten in a more illuminating form:

$$\begin{aligned} H(\omega) &= \frac{R}{[R^2 + (\omega L)^2]^{1/2}} \\ &= \frac{1}{[1 + (\omega L/R)^2]^{1/2}} \\ &= \frac{1}{[1 + (\omega/\omega_0)^2]^{1/2}}, \end{aligned} \tag{5.93}$$

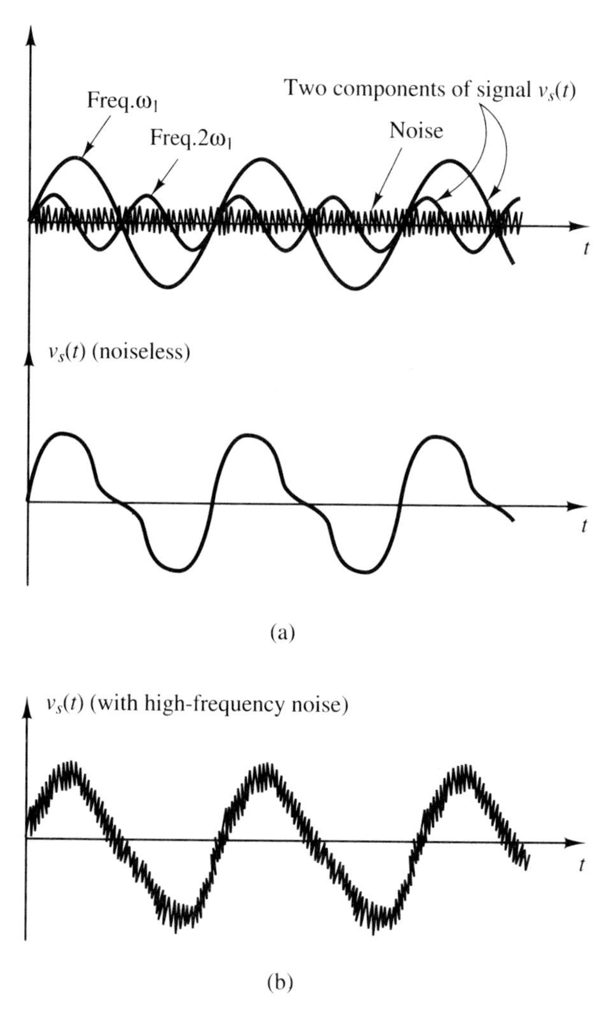


**Figure 5.26** (a) Components of the noisy version of the sum of two sinusoids, and the uncorrupted sum; (b) total (measured) noisy waveform.

where $\omega_0 = R/L$ and is called the *breakpoint* or *cutoff* angular frequency. Notice that $H(0) = 1$, $H(\omega_0) = 1/\sqrt{2}$, and $H(\infty) = 0$. Power is proportional to the square of signal quantities, such as those proportional to $H(\omega)$. The frequency $\omega = \omega_0$ therefore corresponds to the half-power transmission frequency (relative to the maximum value of 1 at $\omega = 0$). A plot of $H(\omega)$ might appear as shown in Fig. 5.27. Again the dashed response is the ideal low-pass filter.

Of course, not only does $h(t)$ have a frequency-dependent spectrum (a Fourier transform), but so does $v_s(t)$. Its spectrum, $\overline{V}_s(\omega)$, would not however, be termed a frequency response because $v_s(t)$ is the *input* signal, not the impulse response. In accordance with the stated assumptions about the noisy input signal $v_s(t)$, its spectrum might appear as in Fig. 5.28.

But what is the spectrum of $v_R(t)$? To find out, the surprisingly simple result

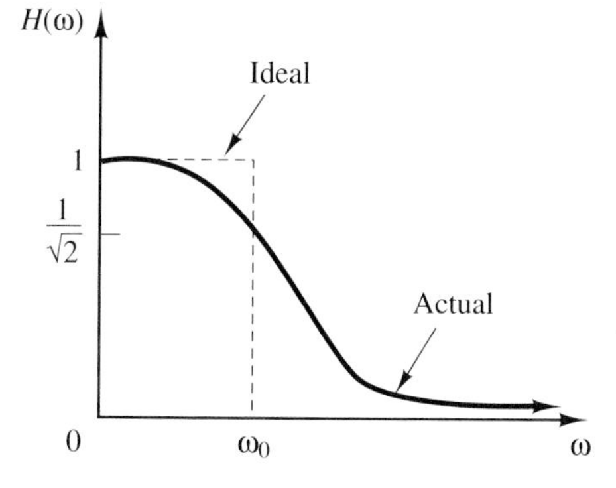


**Figure 5.27** Frequency response of a low-pass filter.

for the *general* case will now easily be shown. Previous notation will be used, such as in Eq. (5.87) for input, impulse response, and output waveforms.

Again recall the fact, shown in the appendix, that the output of an LSI system $y(t)$ is the convolution of the input $x(t)$ with the impulse response of the system $h(t)$. The Fourier transform of the output $y(t)$ is therefore equal to

$$\begin{aligned}
\overline{Y}(\omega) &= \int_{-\infty}^{\infty}\int_{-\infty}^{\infty} h(\tau)x(t-\tau)\,d\tau\, e^{-j\omega t}\,dt \qquad &&\text{[using Eq. (5.87) in Eq. (5.86)]} \\
&= \int_{-\infty}^{\infty} h(\tau)\int_{-\infty}^{\infty} x(t-\tau)\,e^{-j\omega t}\,dt\,d\tau \qquad &&\text{(reverse order of integration).}
\end{aligned} \tag{5.94}$$

Let $u = t - \tau$. Then $t = u + \tau$ and $dt = du$, so that

$$\begin{aligned}
\overline{Y}(\omega) &= \int_{-\infty}^{\infty} h(\tau)\int_{-\infty}^{\infty} x(u)\,e^{-j\omega(u+\tau)}\,du\,d\tau \\
&= \int_{-\infty}^{\infty} h(\tau)\,e^{-j\omega\tau}\,d\tau \int_{-\infty}^{\infty} x(u)\,e^{-j\omega u}\,du \\
&= \overline{H}(\omega)\,\overline{X}(\omega).
\end{aligned} \tag{5.95}$$

That is, convolution in the time domain is merely multiplication in the frequency domain! This result, known as the *convolution theorem*, is omnipresent in signal processing applications.

Thus yet *another* benefit of working in the frequency domain is evident. If the Fourier transforms of the input signal and of the impulse response are known, the output time function is simply the inverse Fourier transform of the product of

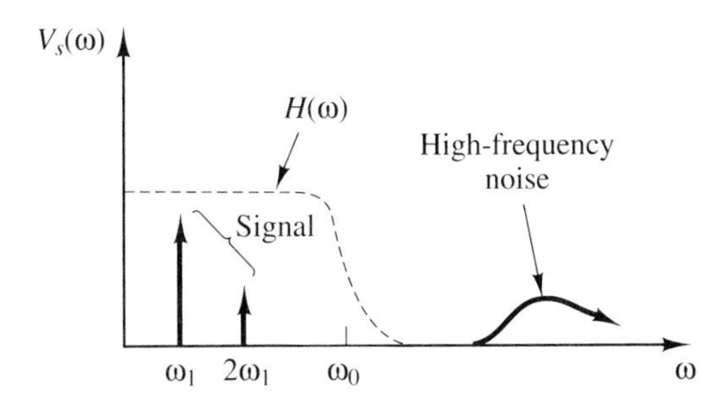


**Figure 5.28** Spectrum of noisy measured signal, before filtering, with $H(\omega)$ of Fig. 5.27 superimposed.

the two Fourier transforms. This procedure is often conceptually and/or computationally easier than "convolving" in the time domain (as done in Example A5.4 in the appendix). Note that the general result, Eq. (5.95), was already heuristically found to hold for a particular example, in magnitude form in Eq. (5.92).

Now the output spectrum in the noisy signal example can be obtained by inspection: Just multiply point by point the two curves in Figs. 5.27 and 5.28. The result is shown in Fig. 5.29a and the corresponding time waveform is shown in Fig. 5.29b. As is evident from both the output spectrum and output waveform, the noise is greatly reduced while the signal is passed essentially without change (distortion). The problem has been solved successfully.

Not only that, but what appeared to be a rather boring and useless *RL* circuit can actually be put to a good purpose. And the narrow-minded sinusoidal steady-state complex Ohm's law analysis has given way to a full-frequency-range characterization of the circuit. Specifically, an understanding has been gained of the filtering action of $R$ and $L$ upon *any* input signal, not just a single-frequency input. The *RL* circuit acts in this example as a low-pass filter. We here see the value of leaving $\omega$ as a variable until the end in ac circuit analysis.

The technique is also general and can be applied to any linear circuit. Suppose now that instead of high-frequency noise, some heavy machinery in the factory is causing a low-frequency rumble in the measured signal, as shown in Fig. 5.30. Replace the inductor of the previous circuit by a capacitor (see Fig. 5.31) and look again at $v_R(t)$.

In the frequency domain at frequency $\omega$,

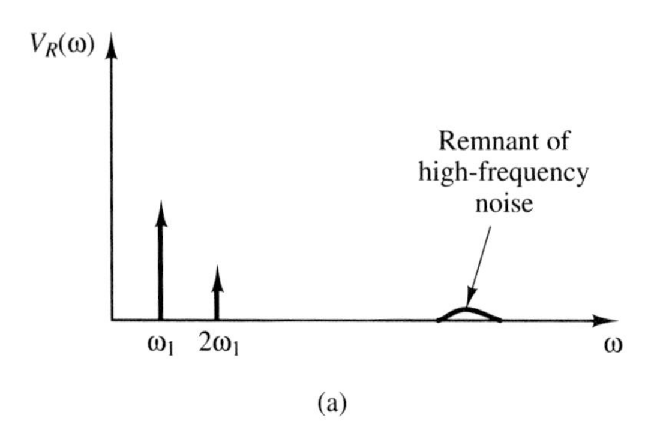


(a)

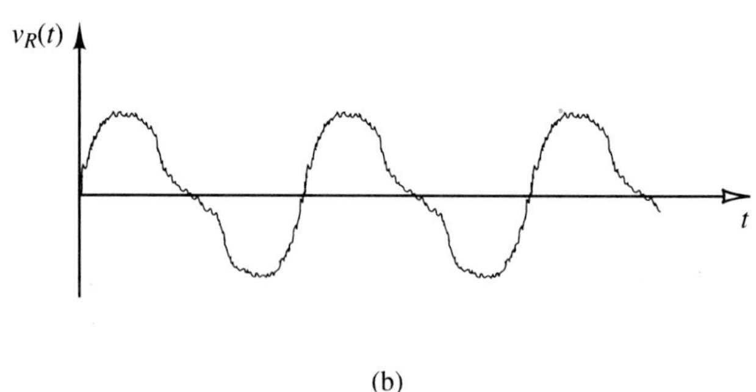


(b)

**Figure 5.29** Output spectrum (a) and time waveform (b) after filtering. The two sinusoids remain strong and undistorted, while the noise has largely been eliminated.

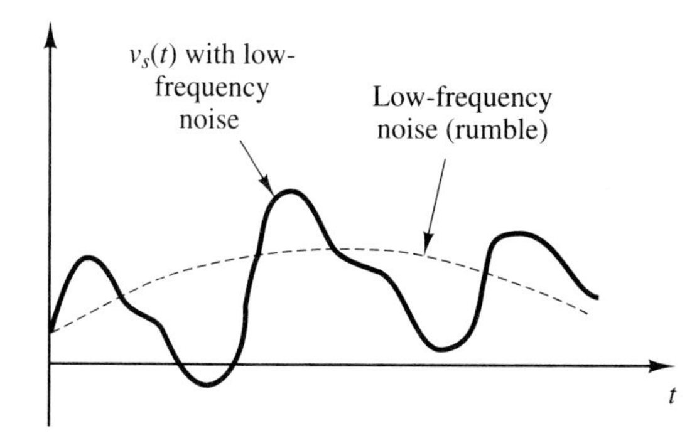


**Figure 5.30** Measured signal composed of two sinusoids plus a low-frequency "rumble" noise.

$$\begin{aligned}\frac{\bar{V}_R}{\bar{V}_s} &= \frac{R}{R - j/(\omega C)} \\ &= \frac{R}{\{R^2 + [1/(\omega C)]^2\}^{1/2}} e^{-j\theta_Z} \\ &= \frac{1}{\{1 + [1/(\omega RC)]^2\}^{1/2}} e^{-j\theta_Z},\end{aligned} \tag{5.96}$$

where $\theta_Z = -\tan^{-1}[1/(\omega RC)]$. Therefore, the magnitude of the frequency response of this filter is

$$\begin{aligned}H(\omega) = \frac{V_R}{V_s} &= \frac{1}{\{1 + [1/(\omega RC)]^2\}^{1/2}} \\ &= \frac{1}{[1 + (\omega_0/\omega)^2]^{1/2}},\end{aligned} \tag{5.97}$$

where $\omega_0 = 1/(RC)$ is the cutoff frequency, for which $H(\omega) = 1/\sqrt{2}$. A plot of $H(\omega)$ (ideal high-pass filter is dashed) appears in Fig. 5.32. Interestingly, $\omega_0 = 1/\tau_C$, where $\tau_C$ is the $RC$ time constant characterizing transient behavior ($\omega_0$ characterizes steady-state behavior).

Clearly, $H(\omega)$ increases, with $\omega$, up to unity as $\omega$ approaches infinity and is thus called a *high-pass filter,* with cutoff frequency $\omega_0$. The unwanted part of $v_s(t)$ is the low-frequency rumble, assumed to have frequency below that of the desired signals. The spectrum $V_s(\omega)$, is shown in Fig. 5.33. Notice that in this case the noise includes a dc ($\omega = 0$) component, which caused the offset in Fig. 5.30. In this case $\omega_0 = 1/(RC)$ can be designed to lie in between the noise and signal frequency ranges (see Fig. 5.33). The result will be that the rumble is attenuated, while the signal is passed. So $v_R(t)$ will appear similar to the noiseless $v_s(t)$ in Fig. 5.26a. And all this is true because of the simple convolution theorem result, Eq. (5.95).

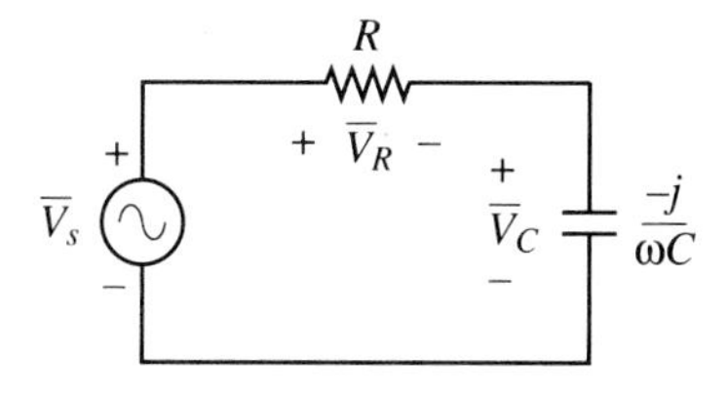


**Figure 5.31** Simple high-pass filter circuit to eliminate low-frequency noise.

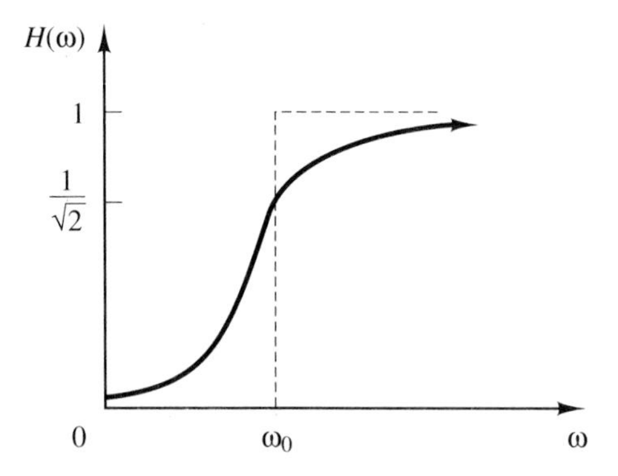


**Figure 5.32** Frequency response of high-pass filter.

It should be noted that the *RL* circuit can also be used as a high-pass filter if $v_L(t)$ is taken to be the output voltage instead of $v_R(t)$. By taking the magnitude of both sides of Eq. (5.63), we have

$$\begin{aligned} \frac{V_L}{V_s} &= \frac{\omega L}{[R^2 + (\omega L)^2]^{1/2}} \\ &= \frac{1}{\{1 + [R/(\omega L)]^2\}^{1/2}} \\ &= \frac{1}{[1 + (\omega_0/\omega)^2]^{1/2}}, \end{aligned} \tag{5.98}$$

where $\omega_0 = R/L$ as in Eq. (5.93). This magnitude frequency response has a form identical to that of the *RC* circuit in which $v_R(t)$ was taken as the output, so it is a high-pass filter. Similarly, the voltage across the capacitor in the *RC* circuit is a low-pass version of the input signal $v_s(t)$.

In conclusion, the input–output behavior of a system must be determined from the frequency response for the particular connection configuration rather than merely by what type of components the circuit contains.

**Example 5.17**

Consider the *RLC* circuit in Fig. 5.34. If the voltage $v_0(t)$ across the resistor is taken to be the output voltage and $v_s(t)$ the input voltage, what is $\overline{H}(\omega)$? Plot $H(\omega)$. What are the frequencies of maximum-power and half-power?

**Solution** The net complex impedance is $\overline{Z} = R + j\,[\omega L - 1/(\omega C)]$. Thus $\overline{H}(\omega)$ can be found by a complex voltage divider of $\overline{V}_o/\overline{V}_s$:

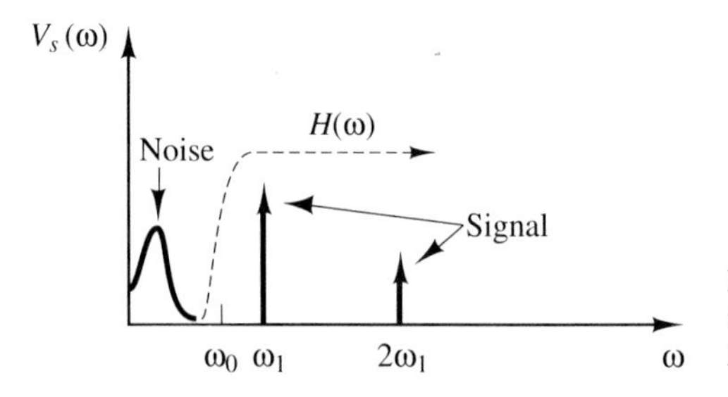


**Figure 5.33** Spectrum of noisy measured signal, before filtering, with $H(\omega)$ of Fig. 5.32 superimposed.

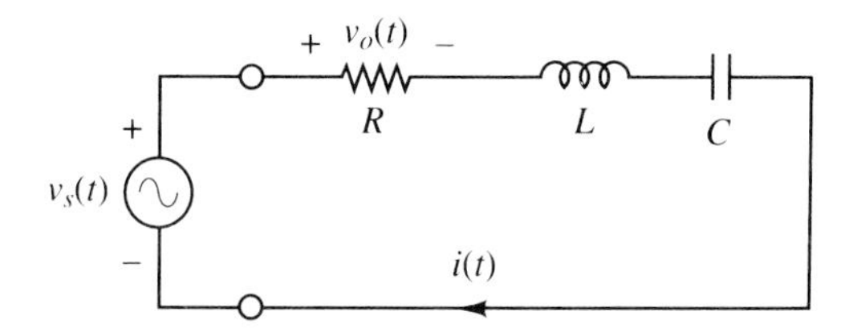


**Figure 5.34** Series *RLC* for Example 5.17.

$$\overline{H}(\omega) = \frac{\overline{V}_o}{\overline{V}_s} = \frac{R}{R + j[\omega L - 1/(\omega C)]}$$
$$= \frac{R}{\{R^2 + [\omega L - 1/(\omega C)]^2\}^{1/2}} e^{-j\tan^{-1}\{[\omega L - 1/(\omega C)]/R\}}, \tag{5.99}$$

so that

$$H(\omega) = \frac{1}{\{1 + [(\omega^2 LC - 1)/(\omega RC)]^2\}^{1/2}}$$
$$= \frac{1}{(1 + \{[\omega L - 1/(\omega C)]/R\}^2)^{1/2}}. \tag{5.100}$$

Substitution of $\omega = 0$ and $\omega = \infty$ into $H(\omega)$ shows that $H(0) = H(\infty) = 0$. The maximum value of $H(\omega)$ occurs if the second term in the denominator vanishes: If $\omega L = 1/(\omega C)$, or $\omega = \omega_0 = 1/(LC)^{1/2}$. This frequency $\omega_0$, for which $X_{\text{net}} = \omega L - 1/(\omega C) = 0$ and for which $H(\omega)$ is maximum, is called the *resonant frequency*. What are the half-power frequencies? By inspection of the expression for $H(\omega)$ in Eq. (5.100), $H(\omega)$ will be $1/\sqrt{2}$ when the second term in the denominator is unity:

$$\omega L - \frac{1}{\omega C} = \pm R \tag{5.101}$$

or

$$\omega^2 \pm \frac{\omega R}{L} - \frac{1}{LC} = 0. \tag{5.102}$$

Solving this quadratic equation for $\omega$ and using $1/(LC) = \omega_0^2$ gives four solutions:

$$\omega_{1-4} = \pm\frac{R}{2L} \pm \left[\left(\frac{R}{2L}\right)^2 + \omega_0^2\right]^{1/2}. \tag{5.103}$$

The positive-valued frequency solutions are

$$\omega_{1,2} = \pm\alpha + (\alpha^2 + \omega_0^2)^{1/2}, \tag{5.104}$$

where $\alpha = R/(2L)$ is called the *damping coefficient.* Notice that $\omega_1$ and $\omega_2$ are *not* centered on $\omega_0$; this is reflected in the asymmetry of the plot of $H(\omega)$ in Fig. 5.35. The *bandwidth* of $H(\omega)$ is $\text{BW} = \omega_2 - \omega_1 = R/L = 2\alpha$. The bandwidth indicates for what band of frequencies more than half the input power is transmitted to the output.

The bandwidth of a system is an extremely important parameter in communication systems of all types. In audio systems it implies the fidelity of reproduction, in digital data channels it indicates at what rate digital information may be sent, and in vibration systems (such as ultrasonic transducers) it indicates the sharpness of pulses that can be produced.

One further parameter of a reactive circuit is the *quality factor,* $Q(\omega)$, defined as

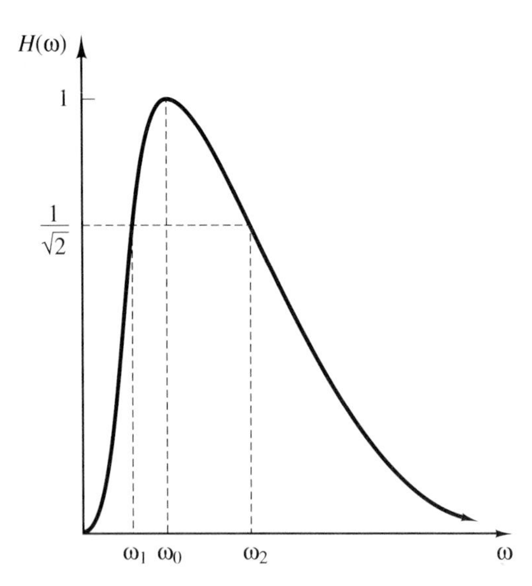


**Figure 5.35** Frequency response of series *RLC* circuit.

$$Q(\omega) = 2\pi \frac{\text{peak energy stored per cycle}}{\text{energy dissipated per cycle}}. \tag{5.105}$$

It is left to the problems to show that for the *series RLC* circuit Fig. 5.34, $Q \equiv Q\ (\omega = \omega_0)$ = resonant frequency/bandwidth = $\omega_0/\mathrm{BW} = (L/C)^{1/2}/R$ = resonant reactance/resistance = $X_0/R$, where $X_0 = X_{L,0} = \omega_0 L = -X_{C,0} = 1/(\omega_0 C) = (L/C)^{1/2}$, and that for a lossy inductor $R + j\omega L$ ($RL$ in Fig. 5.19), $Q(\omega) = \omega/\omega_0 = X_L/R = \omega L/R$, where here $\omega_0 = R/L$ = low-pass $RL$ filter cutoff frequency [and thus $Q(\omega_0) = 1$ for an inductor]. Also left for homework is a practical application of a related (parallel) *RLC* network: a simple TV tuner resonator.

### 5.5.4 A Mechanical Analogy

An analogy can, in fact, be made with mechanical systems, the following of which is one example. Consider the dashpot, spring, and mass $m$ in Fig. 5.36. A cranking force of magnitude $F$ is applied to the mass such that its vertical component oscillates in time with radial frequency $\omega$. The motion of the piston in the dashpot is governed by viscous forces proportional (by the constant $c$) to velocity $u = dx/dt$, and the motion of the spring is governed by the restoring force $F$, which is proportional (by the constant $k$) to displacement $x$. The differential equation for the position is

$$\frac{d^2x}{dt^2} + \frac{c}{m}\frac{dx}{dt} + \frac{k}{m}x = \frac{F}{m}\cos(\omega t). \tag{5.106}$$

The constant $m/c$ is a time constant $\tau$, and the constant $(k/m)^{1/2}$ is a resonant frequency $\omega_0$. This is entirely analogous to the series *RLC* circuit of Example 5.17 (see Fig. 5.34) with $v_s(t) = V_o \cos(\omega t)$, where the differential equation for the charge $q(t)$ is, by KVL, substitution of $i = dq/dt$, and rearrangement:

$$\frac{d^2q}{dt^2} + \frac{R}{L}\frac{dq}{dt} + \frac{1}{LC}q = \frac{V_o}{L}\cos(\omega t). \tag{5.107}$$

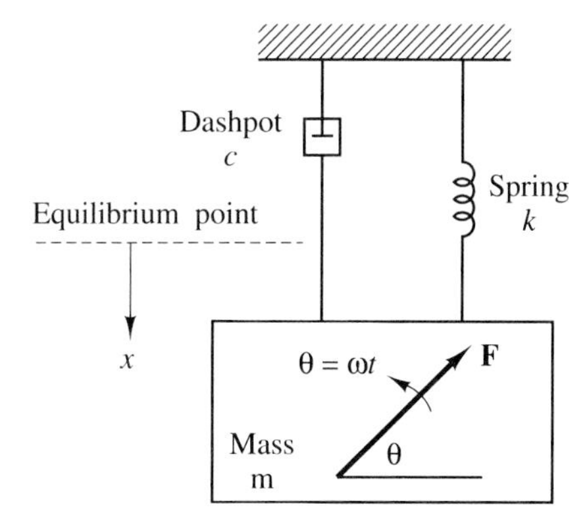


**Figure 5.36** Mechanical analog of the series *RLC* circuit.

In Eq. (5.107), the time constant is $L/R$ and the resonant frequency is $1/(LC)^{1/2}$. In fact, we may form a table of analogous quantities for this problem as shown in Table 5.2.

The solutions of the electrical problem, both transient and steady state, are well documented and can be applied to the mechanical case. A very detailed and interesting discussion of these ideas can be found in *Shock and Vibration Handbook* by C. M. Harris (McGraw-Hill, New York, 1988). It develops, for mechanical systems, analogs of Kirchhoff's laws, series–parallel combinations, superposition, Thévenin and Norton equivalents, and complex impedance for sinusoidal analysis.

### 5.5.5 Example of Low-Pass Filtering

As a final note, it is interesting to make connections with an example used in the appendix to illustrate convolution. Recall that in Example A5.4 of the appendix, the square input $x(t)$ was "smoothed" by the sine-section filter $h(t)$. This effect can now be easily understood. The Fourier transform of $h(t)$ is the system frequency response $H(\omega)$. Although it requires some work to find the exact form of $H(\omega)$, the result is that $H(\omega)$ decreases with $\omega$—that is, it acts as a low-pass filter.

**TABLE 5.2** ANALOGY BETWEEN MECHANICAL AND ELECTRICAL SECOND-ORDER VIBRATION SYSTEMS

| Mechanical system | Electric/magnetic system |
|---|---|
| Mass, $m$ | Inductance, $L$ |
| Damping coefficient, $c$ | Resistance, $R$ |
| Spring constant, $k$ | Reciprocal of capacitance, $1/C$ |
| Displacement, $x$ | Electric charge, $q$ |
| Velocity, $u = dx/dt$ | Current, $i = dq/dt$ |
| Exciting force, $F\cos(\omega t)$ | Exciting voltage, $V\cos(\omega t)$ |
| Kinetic energy, $(1/2)mu^2$ | Magnetic energy, $(1/2)Li^2$ |
| Oscillator potential energy, $(1/2)kx^2$ | Electric energy, $(1/2)Cv_c^2 = (1/2)q^2/C$ |
| Power, $\mathbf{F}\cdot\mathbf{u}$ | Power, $vi$ |
| Viscous time constant, $\tau = c/m$ | Inductive time constant, $\tau = L/R$ |
| Resonant frequency, $\omega_0 = (k/m)^{1/2}$ | Resonant frequency, $\omega_0 = 1/(LC)^{1/2}$ |
| Mass flux, $\overline{F}/\overline{U}$ | Impedance, $\overline{V}/\overline{I}$ |

From Example 5.15, the Fourier transform of the square signal $x(t)$ is similar to $\sin(\omega)/\omega$, which has nonzero components theoretically out to $\omega = \infty$. But they are attenuated by the low-pass $H(\omega)$ [see Eq. (5.95)], so the sharp corner in $x(t)$ containing all the high frequencies disappears in the output $y(t)$.

## 5.6 SUMMARY

The topic of this chapter has been the analysis of sinusoids and sinusoidal decompositions of signals. Sinusoids take a special place in electrical engineering for a variety of reasons. Perhaps most important is the fact that if the input to a linear shift-invariant system is a sinusoid, the output will also be a sinusoid of the same frequency but altered amplitude and phase. This allows an extremely useful system characterization, the frequency response. It should be noted that the vast majority of circuits are well modeled by LSI systems. More precisely, at least portions of or operational modes of a circuit that processes signals are usually well modeled by LSI systems.

The simplified analysis for sinusoids is usable for general signals because of Fourier decomposition. Each sinusoidal component may be treated with the simplified analysis. After necessary computations, the results are brought together and the resulting waveform is synthesized with Fourier's expansions. Without the frequency domain, something as simple as addition of sinusoids of the same frequency but differing magnitudes and phases is a difficult, tedious task.

In the frequency domain, such an operation is as easy as complex number addition or equivalently, two-dimensional vector addition. Once the sum is found, merely multiply it by $e^{j\omega t}$ and take the real part to obtain the time-domain waveform:

$$A_1 \cos(\omega t + \theta_1) + A_2 \cos(\omega t + \theta_2) = \text{Re}\{(\bar{A}_1 + \bar{A}_2)e^{j\omega t}\}$$
$$= |\bar{A}_1 + \bar{A}_2| \cos(\omega t + \angle\{\bar{A}_1 + \bar{A}_2\}).$$

Another prime example of the simplification available in the frequency domain is the equivalence of differentiation in the time domain to multiplication by $j\omega$ in the frequency domain. This allows problems involving inductors and capacitors, which have differential equation relationships between current and voltage, to be solved using only algebraic equations. For example, $d^3/dt^3$ becomes merely multiplication of the phasor by $(j\omega)^3$.

A complex form of Ohm's law is valid for any single frequency, where in general, resistance is replaced by complex impedance and voltage and current time waveforms are replaced by phasor representations. The only price to be paid is the necessity of introducing complex number calculations. Now an entire network can be characterized by a single complex impedance. The network is inductive if the voltage phasor is counter-clockwise from (in time, ahead of) the current phasor, and capacitive if the reverse is true. It is resistive if the two are in phase.

By introduction of the complex forms of Fourier expansions, we were able to define the frequency response of an LSI system. It represents the degree to which an LSI system will pass a sinusoidal input signal and alter its phase. The frequency response is a means both for characterization of existing systems and design of new ones.

Examples of low- and high-pass filters illustrated the fundamental idea of noise reduction by linear filtering. Filters can be as simple as a capacitor or inductor voltage divider or, as we shall see in Chapter 8, highly sophisticated integrated circuits.

The multiplication-by-frequency-response result was derived from the convolutional form of the input–output relation of an LSI system (which in turn is derived in the appendix).

It is important to remember that it is not the component types alone that determine the filtering behavior of a system, but the configuration and input–output signals chosen. The same circuit can be either a low-pass or a high-pass filter, depending on which voltage is defined to be the output. It is the frequency response for the specific configuration that will make clear the filtering behavior.

In Chapter 6 we complete our basic studies of sinusoidal analysis by considering sinusoidal power. While the ideas of this chapter are essential to the understanding and design of signal processing circuitry, those of the next (built upon some of the concepts in this chapter) are crucial in the design and application of electrical energy conversion equipment. Both areas are omnipresent in laboratories and factories.

# Appendix
# The Impulse Response and Convolution

Consider again the $RC$ circuit in Fig. 4.48 for the case in which $K = 0$ [no mismatch between actual initial value $v_C(0)$ and the contribution to $v_C(0)$ made by $v_B(t < 0)$]. See the end of this appendix for the case of general $K$ and also for further important comments on Eq. (A5.1). Recall that the voltage across the capacitor $v_C(t)$ was a smoothed version of the sharp step input battery voltage $v_B(t)$. The expression for $v_C(t)$ was found to be [see Eq. (4.32b); note that in this appendix we use the symbol $\tau$ rather than $t'$, as is conventionally done—do *not* confuse the time integration variable $\tau$ with the time constant $\tau_C = RC$]

$$v_C(t) = \frac{1}{RC}\int_{-\infty}^{t} v_B(\tau)e^{-(t-\tau)/RC}\,d\tau, \tag{A5.1}$$

which can be rewritten as

$$v_C(t) = \int_{-\infty}^{t} v_B(\tau)h(t-\tau)\,d\tau, \tag{A5.2}$$

where

$$h(t) = \frac{1}{RC}e^{-t/RC} \qquad \text{for } t > 0 \qquad \text{(zero otherwise)} \tag{A5.3}$$

is called the *impulse response.* The reason for this name is easy to see and will be shown after introduction of the delta (impulse) function.

An impulse function $\delta(t)$ (also called a *Dirac delta function*) is a pulse occurring at $t = 0$ that is infinitely large in magnitude and infinitesimally narrow in width:

$$\delta(t) = \begin{cases} 0 & t \neq 0 \\ \infty & t = 0 \end{cases} \qquad \text{and} \qquad \int_{-\infty}^{\infty} \delta(t)\,dt = 1. \tag{A5.4}$$

The delta function is shown in Fig. A5.1. It is analogous to the momentum impulse during a collision often studied in freshman physics.

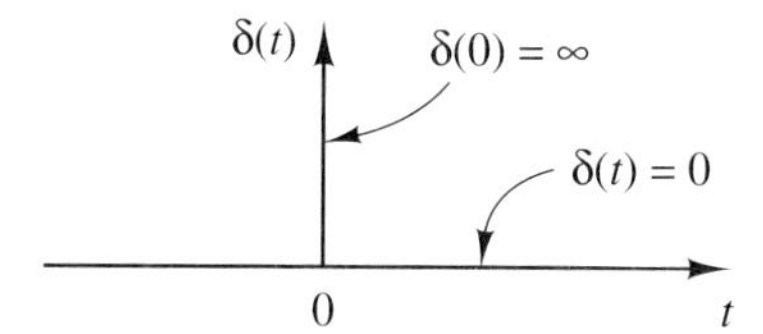


**Figure A5.1** Dirac delta function.

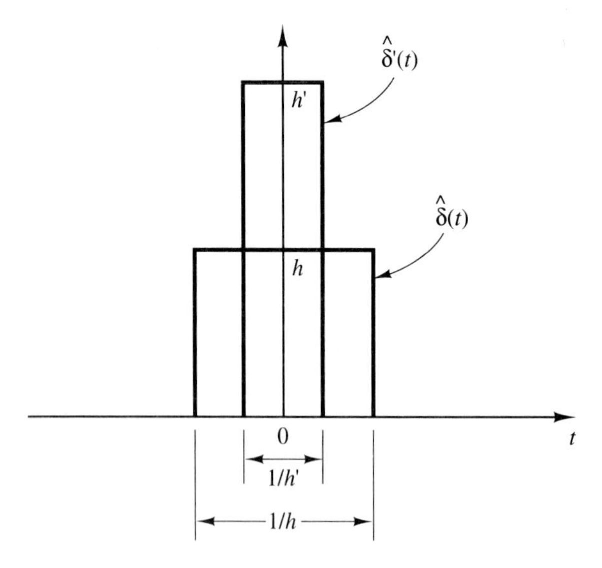


**Figure A5.2** Two examples in a sequence of finite functions approaching a Dirac delta function.

**Example A5.1**

An example of how an approximation of a delta function may arise in practice is a sharp pulse (see Fig. A5.2). As $h \to \infty$, the pulse $\hat{\delta}(t)$ approaches $\infty$ at $t = 0$ and zero for all other $t$. Its area (integral over $t$), though, stays equal to 1 because the width is the inverse of the height.

As a result of Eqs. (A5.4), the impulse function has the sampling property

$$\begin{aligned}\int_a^b \delta(\tau) f(t-\tau)\, d\tau &= f(t)\int_a^b \delta(\tau)\, d\tau \\ &= f(t) \qquad \text{for } a < 0 < b \qquad \text{(zero otherwise)}\end{aligned} \tag{A5.5}$$

for all $f(t)$, where the first equality holds because $\delta(\tau)$ is zero except for $\tau = 0$. Equation (A5.5) says that out of all $\tau$ from $-\infty$ to $\infty$, the delta function in the integral selects only $f(\tau = t)$ as the result: It samples $f(\tau)$. The factors forming the integrand of Eq. (A5.5) are illustrated in Fig. A5.3.

The first integrand, $\delta(\tau)$, is nonzero only at $\tau = 0$, where it is infinite. Now

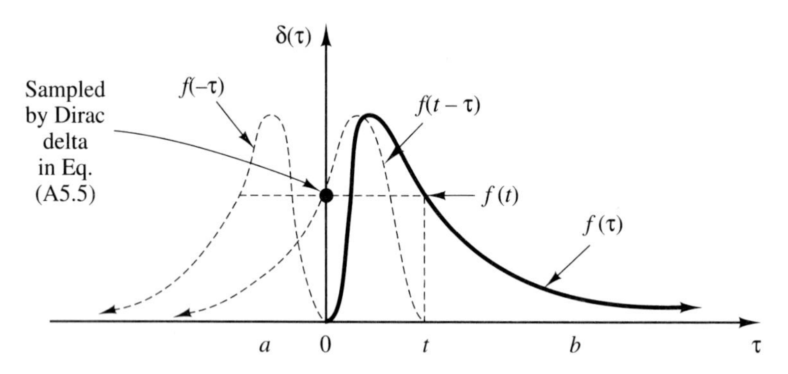


**Figure A5.3** Factors in the integrand of the sampling integral Eq. (A5.5).

suppose that $f(\tau)$ has the shape indicated. Then $f(-\tau)$ has the inverted appearance, as shown. Shifting $f(-\tau)$ to the right by $t$ gives the other factor of the integrand, $f(t-\tau)$. The only place where $f(t-\tau)$ is multiplied by anything other than zero is at $\tau = 0$, where $\delta(\tau)$ is infinite. The result is $f(t-0) = f(t)$.

**Example A5.2**

Evaluate $I = \int_{-\infty}^{\infty} \delta(\tau) \cos(\omega\tau)\, d\tau$.

**Solution** Because $\delta(\tau)$ is zero except at $\tau = 0$,

$$I = \int_{-\infty}^{\infty} \delta(\tau) \cos(0)\, d\tau = \cos(0) \int_{-\infty}^{\infty} \delta(\tau)\, d\tau = \cos(0) = 1.$$

Again the sampling property has been observed. In this example, the $t$ in Eq. (A5.5) was zero, so the result was $\cos[\omega(t = 0)] = \cos(0)$.

Now suppose that the input $v_B(t)$ was set to $\delta(t)$ in Eq. (A5.2) above. The result for $t > 0$ would be

$$v_C(t) = \int_{-\infty}^{t} \delta(\tau) h(t-\tau)\, d\tau = h(t). \tag{A5.6}$$

Thus $h(t)$ is the output or response ($v_C(t)$) when the input is an impulse ($v_B(t) = \delta(t)$); that is, $h(t)$ is the impulse response.

It will now be shown that the form for the output in Eq. (A5.2) is actually applicable to all linear shift-invariant (LSI) systems. Formally, a *linear operator* (system) $\mathscr{L}\{\cdot\}$ is one for which

$$\mathscr{L}\{a_1 x_1(t) + a_2 x_2(t)\} = a_1 \mathscr{L}\{x_1(t)\} + a_2 \mathscr{L}\{x_2(t)\}, \tag{A5.7}$$

where $a_1$ and $a_2$ are constants and $x_1(t)$ and $x_2(t)$ are inputs to the system (see Fig. A5.4). A *shift-invariant system* is merely one for which a shifted input will result in a shifted (but otherwise identical) output compared with that for the unshifted input.

**Example A5.3**

Suppose that $\mathscr{L}\{x(t)\} = \int_{-\infty}^{t} x(\tau)\, d\tau$. (a) Is $\mathscr{L}$ a linear system? (b) Is $\mathscr{L}$ shift-invariant?

**Solution**

(**a**) Testing this $\mathscr{L}\{\cdot\}$ to see whether Eq. (A5.7) is satisfied, we obtain

$$\begin{aligned}\mathscr{L}\{a_1 x_1(t) + a_2 x_2(t)\} &= \int_{-\infty}^{t} [a_1 x_1(\tau) + a_2 x_2(\tau)]\, d\tau \\ &= a_1 \int_{-\infty}^{t} x_1(\tau)\, d\tau + a_2 \int_{-\infty}^{t} x_2(\tau)\, d\tau \\ &= a_1 \mathscr{L}\{x_1(t)\} + a_2 \mathscr{L}\{x_2(t)\}.\end{aligned}$$

So yes, $\mathscr{L}$ is a linear system.

(**b**) To determine whether $\mathscr{L}$ is shift-invariant, set $y(t) = \mathscr{L}\{x(t)\}$. Consider

$$\mathscr{L}\{x(t-t')\} = \int_{-\infty}^{t} x(\tau - t')\, d\tau.$$

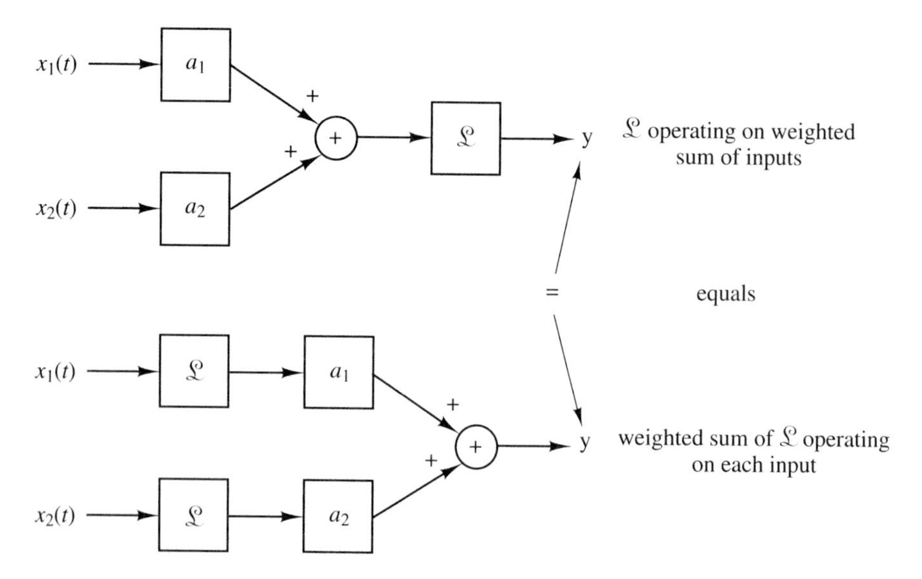


**Figure A5.4** Graphical illustration of the meaning of a linear operator (system).

Now let $u = \tau - t'$. Then $d\tau = du$, and if $\tau = t$, then $u = t - t'$. Thus

$$\mathscr{L}\{x(t - t')\} = \int_{-\infty}^{t-t'} x(u)\, du = y(t - t').$$

So yes, $\mathscr{L}$ is also shift-invariant.

Suppose that $x(t)$ is the input to an LSI system $\mathscr{L}\{\cdot\}$ and $y(t)$ is the output. Then (see below for explanation of steps)

$$y(t) = \mathscr{L}\{x(t)\} = \mathscr{L}\left\{\int_{-\infty}^{\infty} \delta(\tau)x(t - \tau)\, d\tau\right\} \tag{A5.8a}$$

$$= \mathscr{L}\left\{\int_{-\infty}^{\infty} x(\tau)\delta(t - \tau)\, d\tau\right\} \tag{A5.8b}$$

$$= \int_{-\infty}^{\infty} x(\tau)\mathscr{L}\{\delta(t - \tau)\}\, d\tau \tag{A5.8c}$$

$$= \int_{-\infty}^{\infty} x(\tau)h(t - \tau)\, d\tau \tag{A5.8d}$$

$$= \int_{-\infty}^{\infty} x(t - \tau)h(\tau)\, d\tau, \tag{A5.8e}$$

where (a) is obtained from $\mathscr{L}\{x(t)\}$ using Eq. (A5.5) "in reverse," (b) is just a change of integration variables ($\tau_{\text{new}} = t - \tau_{\text{old}}$), (c) is due to linearity [Eq. (A5.7)], (d) is just the definition of the impulse response (and implies shift invariance), and (e) is again due merely to a change of integration variables (again, $\tau_{\text{new}} = t - \tau_{\text{old}}$).

The expression Eq. (A5.8d) reduces to the form of Eq. (A5.2) if $h(t < 0) = 0$, for in that case $h(t - \tau) = 0$ if $t - \tau < 0$; that is, the integrand is zero for $\tau > t$. A system for which $h(t < 0) = 0$ is called a causal system, because that condition means that outputs do not depend on future inputs [$v_B(\tau > t)$ in Eq. (A5.2)]. Physically realizable systems are generally causal.

It has now been shown that LSI systems can be characterized by a *convolutional* integral for the input–output relation. One meaning of the word "convolution" is "folded together with one part upon another." Above, either $x(t)$ [Eq. (A5.8e)] or $h(t)$ [Eq. (A5.8d)] is seen to be shifted and time reversed with respect to the integration variable while the other is not, and the integral of the resulting product is taken.

Figure A5.5 depicts the process. The input $x(t)$ is shown in Fig. A5.5a, $h(t)$ appears in Fig. A5.5b, and Fig. A5.5c shows the factors forming the integrand of Eq. (A5.8e). In Fig. A5.5c, $x(t)$ is shown time-reversed and shifting to the right as $t$ increases. As later and later values of $x(t)$ come in (as the shift to the right progresses), they first excite the initial peak of $h(\tau)$ and later work their way down to the last, dying-away ripples in $h(\tau)$. The output is the linear sum of the point-by-point products of the time-reversed, shifted input and the impulse response. Note that because $h(\tau < 0) = 0$, future values of $x(t)$ ($\tau > t$) are multiplied by zero.

Alternatively, we could reason from Eq. (A5.8d) as follows. The output $y(t)$ at any particular time is a weighted [according to $x(\tau)$] sum of shifted, time-reversed impulse responses. (Because the system is shift-invariant and by definition of the impulse response, this is the same as saying the following. The output is a weighted [by $x(\tau)$] superposition of (time-reversed) outputs due to a special set of inputs: shifted impulses at $\tau = t$.) A contributing impulse response begins for each instant that $x(\tau)$ is nonzero, is weighted by $x(\tau)$, and lasts for exactly the duration of $h(t)$. Because $h(\cdot)$ is time-reversed in Eq. (A5.8d), the contribution from the $x(\tau)$-weighted impulse response comes from later and later values of $h(\cdot)$ as $t$ increases.

The impulse response form is more convenient than the differential equation form because the output $y(t)$ is now isolated on the left-hand side of Eq. (A5.8). Of course, the impulse response convolution integral is precisely the farthest we can proceed in the time-domain solution of the differential equation without knowing the input $x(t)$.

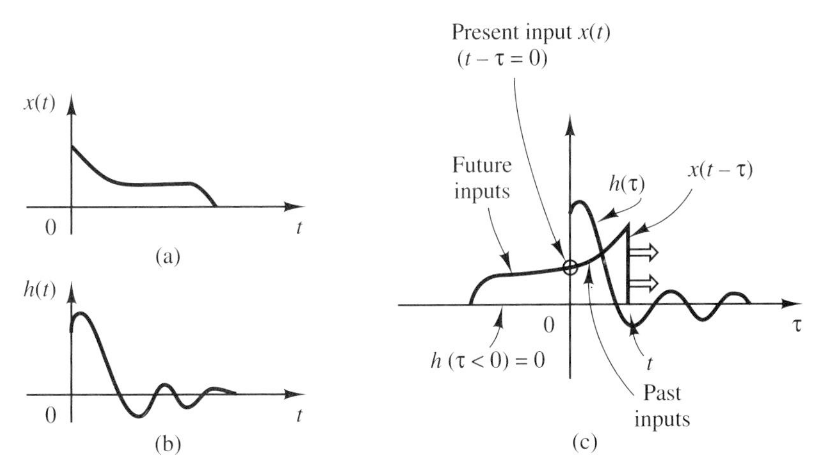


**Figure A5.5** (a) Input $x(t)$ to a linear system; (b) system impulse response $h(t)$; (c) factors in the convolution integral, Eq. (A5.8e).

**Example A5.4**

Convolve the two functions $x(t)$ and $h(t)$ in Fig. A5.6a, where $h(t) = \sin(\pi t)$ for $0 < t < 1$ and zero otherwise. That is, find

$$y(t) = \int_{-\infty}^{\infty} x(t - \tau)h(\tau)\, d\tau.$$

**Solution** The graphical method is easiest. Just time-reverse $x(t)$ to get $x(-\tau)$ and then shift that to the right by $t$. Graph on the $\tau$ axis along with $h(\tau)$ (the other factor of the integrand) and integrate the product. Figure A5.6b through e shows the time-reversed and shifted $x(t - \tau)$ and $h(\tau)$ for four distinct time intervals. In Fig. A5.6b, $t < 0$. Because there is no overlap between $x(t - \tau)$ and $h(\tau)$, $y(t) = 0$. [So $h(t)$ is causal,

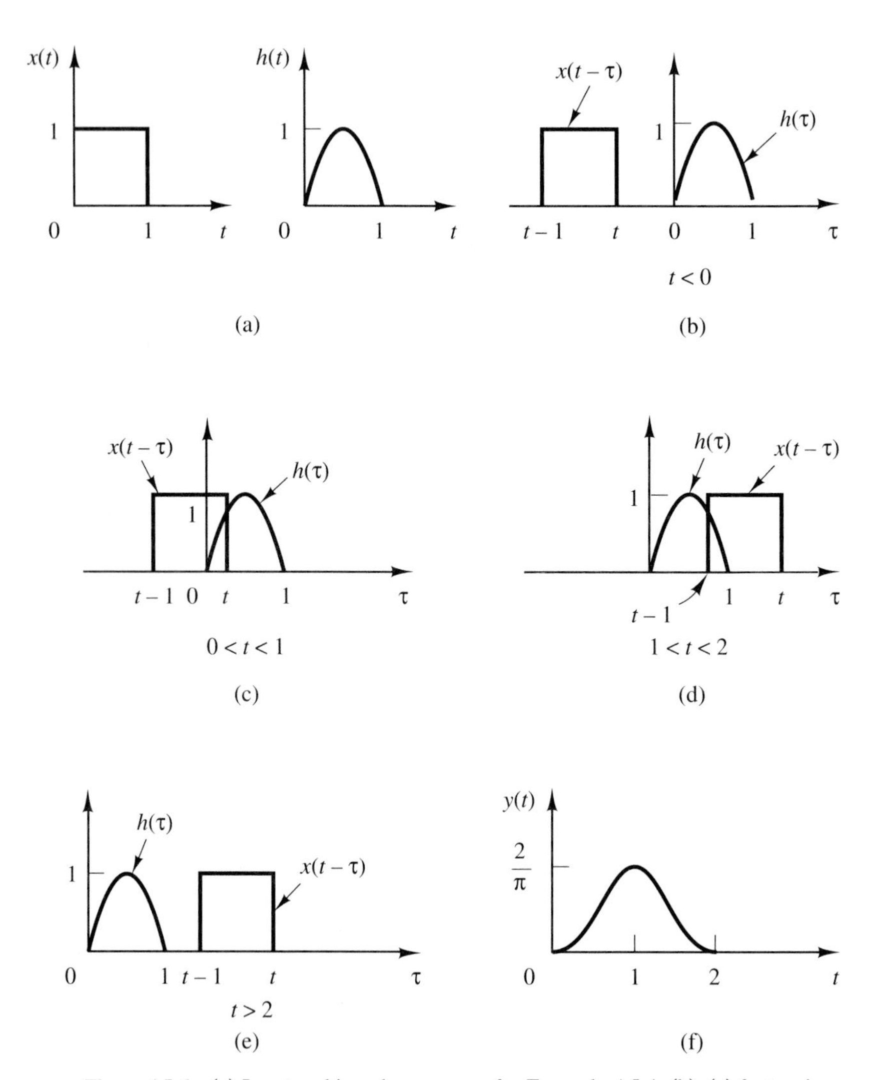


**Figure A5.6** (a) Input and impulse response for Example A5.4; (b)–(e) factors in the convolution integral for (b) $t < 0$, (c) $0 < t < 1$, (d) $1 < t < 2$, and (e) $t > 2$; (f) output $y(t)$.

which can also be inferred from Fig. A5.6a, where $h(t < 0) = 0$.] In Fig. A5.6c, $0 < t < 1$, in which case the overlap begins at $\tau = 0$ and ends at $\tau = t$:

$$y(t) = \int_0^t \sin(\pi t)\, dt = -\frac{1}{\pi}[\cos(\pi t) - \cos(0)] = \frac{1}{\pi}[1 - \cos(\pi t)].$$

The limits of integration in Fig. A5.6d for $1 < t < 2$ are $\tau = t - 1$ to $\tau = 1$, so that

$$y(t) = \int_{t-1}^{1} \sin(\pi t)\, dt = -\frac{1}{\pi}\{\cos(\pi) - \cos[\pi(t-1)]\}$$

$$= \frac{1}{\pi}[1 - \cos(\pi t)].$$

Finally, for $t > 2$ there is again no overlap, so $y(t) = 0$. Piecing together $y(t)$ for the four ranges gives the plot in Fig. A5.6f.

Notice that the duration of $y(t)$ is equal to the duration of $x(t)$ plus that of $h(t)$; this is true in general. In light of the paragraph on Eq. (A5.8d), this makes sense. The last instant that $x(t)$ is nonzero, the last "new" impulse response is "sent out" from the system. It, and therefore $y(t)$, will terminate at the end of the duration of this last $h(\cdot)$. So an effect of the filter $h(t)$ is to extend the duration of $y(t)$ beyond the duration of $x(t)$. Another important point evident in Fig. A5.6f is that for this particular filter $h(t)$, the output $y(t)$ is a *smoothed* version of the sharp-cornered square input $x(t)$. This point is pursued in Sec. 5.5.5.

To conclude this appendix, we consider briefly the case of general $v_C(0)$ [not $v_C(0) = 0$, as assumed above]. The question is raised: What do we do with the extra term $K \exp(-t/\tau_C)$ in Eq. (4.32b)? The answer is that we just have to add it on at the end. For example, assuming that we know $v_C(0)$ [which is necessary to know in order to find $v_C(t)$ in any case], we can use Eq. (4.34), replacing the first integral by the form in Eq. (A5.1), with the lower limit increased to zero.

The convolution in Eq. (A5.1) produces *both* forced and natural responses resulting from a startup of $v_B(t)$. The only thing it does not include is a contribution to $v_C(0)$ caused by a charging of the capacitor by a source *other than* $v_B(t)$. There is no way to find an $h(t)$ that when convolved with any and all $v_B(t)$ will produce the same independently chosen initial condition $v_C(0)$, part of which has *nothing at all to do with* $v_B(t)$.

## PROBLEMS

**5.1.** A sinusoidal voltage $v(t)$ has successive maxima of 2.0 V at $t = 1.0$ s and $t = 3.0$ s. Write down the numerical expression for $v(t)$. What is $v(0)$?

**5.2.** A sinusoidal voltage $v(t) = A\cos(\omega t + \theta)$ has a period of 0.1 ms, a phase angle of $-30°$, and is equal to 1 V at $t = 0$. Write down the numerical expression for $v(t)$.

**5.3.** A sinusoidal current $i(t)$ has a frequency of 1.2 kHz, an amplitude of 0.5 A, and has the value 0.2 A at $t = 0$.

(**a**) Find $i(t)$ for all $t$.

(**b**) At what time nearest $t = 0$ does $i(t)$ reach a maximum? (*Note:* The peak occurs *before* $t = 0$.)

**5.4.** If $v_1(t) = A\cos(\omega t)$ and $v_2(t) = B\cos(\omega t + \pi)$, find $v_1(t) + v_2(t)$, $v_1(t) - v_2(t)$, $v_1(t)v_2(t)$, and $v_1(t)/v_2(t)$. Which of these oscillates the fastest?

**5.5.** Equations (5.2) give the Fourier series expansion of a general periodic time function. Supposing that $f(t)$ represents the voltage across a circuit element. What is the physical significance of $a_0/2$? Repeat if $f(t)$ represents the power into the circuit element.

**5.6.** Suppose that $f_0(t) = -1$ for $-2 < t < 0$, 1 for $0 < t < 2$, and 0 otherwise (i.e., one negative pulse followed immediately by one positive pulse). Expand over a Fourier series $f(t) = f_0(t + nT)$ where $n$ is any integer and $T = 8$ [$f(t)$ is the "periodic extension" of $f_0(t)$ with intervals of zero between repetitions]. Compare your results with those of Example 5.1.

**5.7.** For an aperiodic function of time $f(t)$, the Fourier series expansion becomes an integral expansion and the $a_n$ and $b_n$ become $a(\omega)$ and $b(\omega)$, as shown in Eq. (5.4). Suppose that $b(\omega) = 0$ and $a(\omega) = 1$ for $0 \le \omega \le 2$ (and zero otherwise).
- **(a)** Find $f(t)$ corresponding to the given $a(\omega)$ and $b(\omega)$.
- **(b)** Is $f(t)$ an even or an odd function of $t$? What is the maximum value of $f(t)$, and for what time does it occur? What are $f(\infty)$ and $f(-\infty)$? (*Hint:* $\lim_{x\to 0}[\sin(x)/x]=1$.)

**5.8.** A periodic function $f(t)$ having period $T = 2\pi$ is expanded in a Fourier series, resulting in the coefficients $a_0 = 0, a_1 = \frac{3}{4}, a_2 = 0, a_3 = \frac{1}{4}, a_{n>3} = 0$, and $b_1 = \frac{3}{4}, b_2 = 0$, $b_3 = -\frac{1}{4}, b_{n>3} = 0$. Find $f(t)$; simplify and express as a simple function of $\sin(t)$ and $\cos(t)$. [*Hint:* recall the trigonometric identity for $\sin^3(t)$ and $\cos^3(t)$.]

**5.9.** With $f(t)$ as defined in Fig. 5.2 (Example 5.1), define $g(t)$ as follows: $g(t) = f(t)$ for $-4 < t < 4$ and zero otherwise.
- **(a)** Find the Fourier coefficient functions of frequency $a(\omega)$ and $b(\omega)$ defined in Eq. (5.5).
- **(b)** Compare $a(\omega)$ and $b(\omega)$ with $a_n$ and $b_n$ in Example 5.1; determine the relation between them.

**5.10.** If $\overline{Z}_1 = 1 + j$ and $\overline{Z}_2 = -1 + j$, find $\overline{Z}_1 + \overline{Z}_2$, $\overline{Z}_1 - \overline{Z}_2$, $Z_1$, $\angle\overline{Z}_1$, $Z_2$, $\angle\overline{Z}_2$, $\overline{Z}_1 \cdot \overline{Z}_2$, $\overline{Z}_1/\overline{Z}_2$, $\overline{Z}_1^*$, $\overline{Z}_2^*$, $\overline{Z}_1^{1/2}$, and $\overline{Z}_2^{1/2}$. Where possible, express all answers in both polar and rectangular forms.

**5.11.** Repeat Problem 5.10 if $\overline{Z}_1 = j$ and $\overline{Z}_2 = 3 - j$.

**5.12.** Reduce $(1 + 3j)/(4 - j) + (2 - j)/(5j)$ to the form $\overline{Z} = x + jy$.

**5.13.** By equating real and imaginary parts, solve for the real numbers $x$ and $y$ in the equation $(2 - 3j)^2 - 3\overline{Z} = \overline{Z}^*$, where $\overline{Z} = x + jy$.

**5.14.** By multiplying the numerator and denominator by $\overline{Z}_2^*$ find $\overline{Z}_1/\overline{Z}_2$ in rectangular form in terms of $x_1$, $x_2$, $y_1$, and $y_2$.

**5.15.**
- **(a)** For the complex numbers $\overline{Z}_1$, $\overline{Z}_2$, $\overline{Z}_3$, and $\overline{Z}_4$ shown in Fig. P5.15, provide the appropriate angles $\theta_1$, $\theta_2$, $\theta_3$, and $\theta_4$ in terms of the positive numbers $x$ and $y$.
- **(b)** Most calculators for $\tan^{-1}(\cdot)$ produce an angle between $-\pi/2$ and $\pi/2$, yet clearly this is not appropriate for $\overline{Z}_2$ and $\overline{Z}_4$. Suppose that the angles the calculator shows are, respectively, $\theta_A$ and $\theta_B$. What is $\theta_2$ in terms of $\theta_A$ and what is $\theta_1$ in terms of $\theta_B$?

**5.16.** Solve $\overline{Z}^4 + 4 = 0$ for all solutions $\overline{Z}$, and factor $\overline{Z}^4 + 4$ into two quadratic factors having only real coefficients of the powers of $\overline{Z}$.

**5.17.** Taylor's series for expanding a function $f(x)$ over powers of $x$ (i.e., MacLaurin's series) is

$$f(x) = f(0) + xf'(0) + x^2f''(0)/2! + x^3f'''(0)/3! + \cdots + x^n f^{(n)}(0)/n! + \cdots$$

Find the MacLaurin series for $\cos(\theta)$, $\sin(\theta)$, and $\exp(j\theta)$. Using these, compare $\cos(\theta) + j\sin(\theta$, with $\exp(j\theta)$. By finding them equal, conclude that Euler's formula [Eq. (5.19)] is valid.

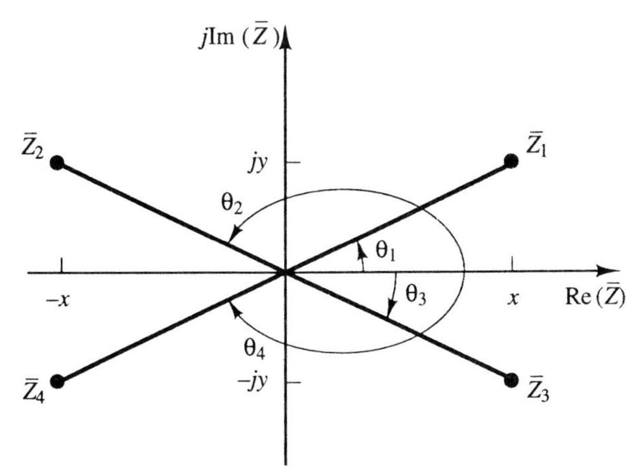


**Figure P5.15**

**5.18.** We showed in Sec. 5.4.3 that

$$d/dt\,\{\text{Re}\,(\bar{V}\exp[j\omega t])\} = \text{Re}\,\{d/dt\,(\bar{V}\exp[j\omega t])\}.$$

Show that, also,

$$d/dt\,\{\text{Im}\,(\bar{V}\exp[j\omega t])\} = \text{Im}\,\{d/dt\,(\bar{V}\exp[j\omega t])\}.$$

**5.19.** Give the cosine-based phasors representing the following sinusoids.
(**a**) $f_1(t) = 10\cos(\omega t + 0.2)$
(**b**) $f_2(t) = 3\cos(\omega t)$
(**c**) $f_3(t) = 5\sin(\omega t)$
(**d**) $f_4(t) = 9\sin(\omega t + 1.5)$

**5.20.** Find the time signals (sinusoids with frequency $\omega$) corresponding to the following phasors (assume that the phasors are cosine based).
(**a**) $\bar{G}_1 = 0.4\exp(j30°)$
(**b**) $\bar{G}_2 = 2\exp(-j0.2\text{ rad})$
(**c**) $\bar{G}_3 = -\exp(j)$

**5.21.** The voltage appearing across a load satisfies the following differential equation: $3dv(t)/dt + 400v(t) = \cos(500t)$. Find the steady-state solution of this equation using phasor arithmetic.

**5.22.** Repeat Problem 5.21 for $v(t)$ that satisfies the differential equation $d^2v(t)/dt^2 + 7000\,dv(t)/dt + 30{,}000\,v(t) = 300\cos(200t)$.

**5.23.** Suppose that two current sources $i_1(t)$ and $i_2(t)$ are connected in parallel, and both across a resistor $R = 1\text{ k}\Omega$. Both are defined to have the same symbolic direction into $R$, and let $v(t)$ be the voltage across $R$ such that the currents flow into the symbolic + side of $v$. Let $i_1(t) = 4\cos(\omega t + 45°)$ A and $i_2(t) = 3\cos(\omega t - 45°)$ A.
(**a**) Find the phasors for the currents $i_1(t)$ and $i_2(t)$.
(**b**) Find the phasors for $i(t)$ and $v(t)$.
(**c**) Draw a phasor diagram showing all the phasors above.
(**d**) Find the voltage $v(t)$ across the load resistor.

**5.24.** A voltage $v(t) = 4\cos(2\pi\cdot 200t - 10°)$ V is applied to (i) a resistor $R = 1\ \Omega$, (ii) a capacitor $C = 500\ \mu\text{F}$, and (iii) an inductor $L = 30$ mH.
(**a**) Find the corresponding phasors for the currents in the resistor, capacitor, and the inductor.
(**b**) Draw all these current phasors in one phasor diagram. Do your results agree with

the phase relations for each element concerning current vs. voltage that were given in the chapter?

**5.25.** Suppose that a sinusoidal voltage source $e(t)$ is connected across the parallel combination of two impedances $\overline{Z}_1$ and $\overline{Z}_2$. The current in $e(t)$ directed out of its + side is $i_{TOT}(t)$, and the currents $i_1(t)$ and $i_2(t)$ in, respectively $\overline{Z}_1$ and $\overline{Z}_2$ are both flowing away from the + side of $e(t)$. Suppose also that we are given that

$$i_1(t) = 3\cos(377t + 80°)\text{ A}$$

$$i_2(t) = 3\cos(377t - 40°)\text{ A}$$

$$e(t) = 6\cos(377t)\text{ V}$$

**(a)** Find $i_{TOT}(t) = I_{TOT}\cos(377t + \theta_{iTOT})$. Express your result in this single-sinusoid form; do not leave in terms of the variables $i_1(t)$ and $i_2(t)$.

**(b)** Find $\overline{Z}_{eq}$ seen by $e(t)$.

**5.26.** A resistance $R = 10\ \Omega$ and an inductance $L = 5$ mH are in series. The resistor voltage is $v_R(t) = 5.0\cos(377t - 45°)$ (V).

**(a)** Obtain the time-dependent voltage across the series combination $v_T(t)$ by using phasor addition. Draw the voltage phasor diagram, fully labeled.

**(b)** Find $\overline{Z}_{eq}$ of the series combination; express in polar form.

**5.27.** Given the circuit fragment in Fig. P5.27:

**(a)** Find $i_1(t)$. Be careful with your final angle determinations!

**(b)** Suppose that $\overline{Z} = \frac{1}{2} - j\frac{1}{4}\ \Omega$. What is $v_o(t)$?

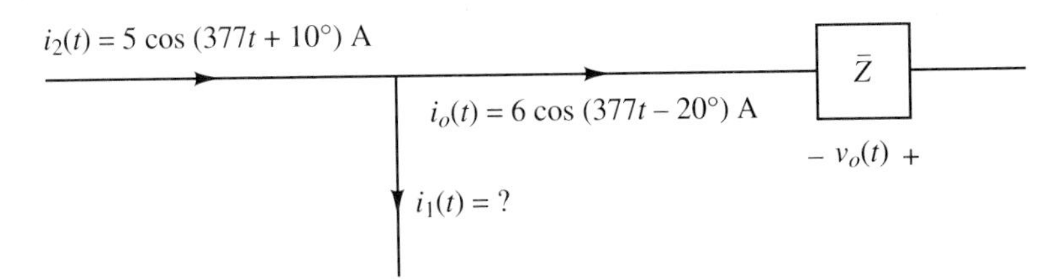


**Figure P5.27**

**5.28.** A parallel circuit of $R$ and $C$ has an applied voltage $v(t) = V\cos(\omega t + \theta_v)$. Find the current $i(t)$ passing through the voltage source in terms of $R$, $C$, $\omega$, $V$, and $\theta_v$. [Let the polarity of $i(t)$ be defined as coming out of the + terminal of $v(t)$.] Simplify as much as possible.

**5.29.** **(a)** In Fig. P5.29, find $\overline{Z}_L$ as seen by the voltage source (in terms of $R$, $L$, $C$, and $\omega$). Express in both polar and rectangular forms.

**(b)** With $\omega = 1000$ rad/s, $R = 1000\ \Omega$, $L = 0.5$ H, $C = 1\ \mu$F, express $\overline{Z}_L$ in polar and in rectangular form.

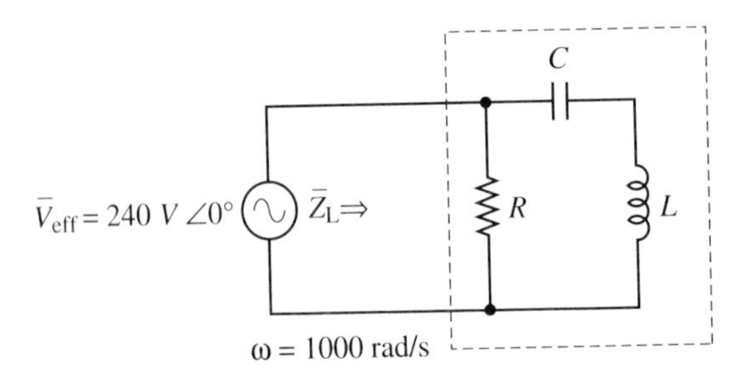


**Figure P5.29** (Disregard "eff" in this figure; it is used in a Chap. 6 problem that refers to this figure).

**5.30.** Suppose that in Fig. P5.29 we reverse the positions of $R$ and $C$, and let $\overline{V}_{\text{eff}} = 100\text{ V}\angle 0°$.

**(a)** Find, in both rectangular and polar forms, the impedance $\overline{Z}_L$ in terms of $R$, $L$, $C$, and $\omega$.

**(b)** Now let $\omega = 10$ rad/s, $L = 0.1$ H, $C = 0.01$ F, $R = 1\ \Omega$. What is $Z_L$ in both polar and rectangular forms?

**5.31.** A current source $i_s(t) = 3\cos(2\pi \text{ft})$ A is connected across the parallel combination of: (i) the series combination ($\overline{Z}_1$) of a 0.70362 H inductor and a 10 μF capacitor and (ii) a resistor $\overline{Z}_2 = 200\ \Omega$. The currents $i_1(t)$ and $i_2(t)$ for the two load branches are directed along the defined direction of $i_s(t)$.

**(a)** Find $i_1(t)$ and $i_2(t)$ for $f = 60$ Hz.

**(b)** Find $i_1(t)$ and $i_2(t)$ for $f = 50$ Hz.

**(c)** Find the equivalent impedance seen by the source.

**5.32.** A current source $i(t) = 5\cos(300t + 30°)$ A is applied to unknown devices that are either resistors, capacitors, or inductors. We measure the voltages below; what are the components to which we apply $i(t)$, including their values?

**(a)** $v(t) = 10\cos(300t + 30°)$ V

**(b)** $v(t) = 0.2\cos(300t + 120°)$ V

**(c)** $v(t) = 4\cos(300t + 60°)$ V

**(d)** $v(t) = 6\sin(300t + 30°)$ V

**5.33.** A sinusoidal voltage $v(t) = 9\cos(1000t)$ V is applied to the network shown in Fig. P5.33, where $R = 300\ \Omega$, $L = 75$ mH, and $C = 10$ μF.

**(a)** Determine $i_R(t)$ and $v_R(t)$. (*Hint:* It may be convenient to use complex mesh analysis.)

**(b)** Draw the phasor diagram for $\overline{V}$ and $\overline{V}_R$. What is the strange result you obtain in this problem regarding the magnitude of $\overline{V}_R$ compared with the magnitude of $\overline{V}$?

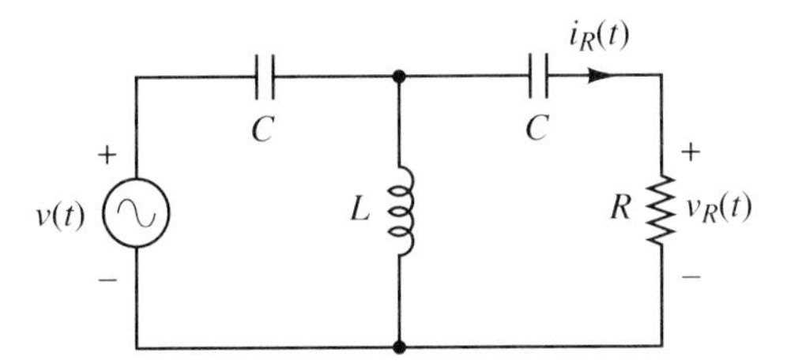


**Figure P5.33**

**5.34.** **(a)** For the network in Fig. P5.33, find the total impedance $\overline{Z}_{\text{eq}}$.

**(b)** Draw the equivalent series $RC$ and $RL$ circuit, and determine the value of $C_{\text{eq}}$ or $L_{\text{eq}}$ involved.

**5.35.** In Fig. P5.35, which models a transistor amplifier driving an inductive load, find the output voltage $v_o(t)$ when the input voltage $v_s(t) = 10\cos(100t)$ V.

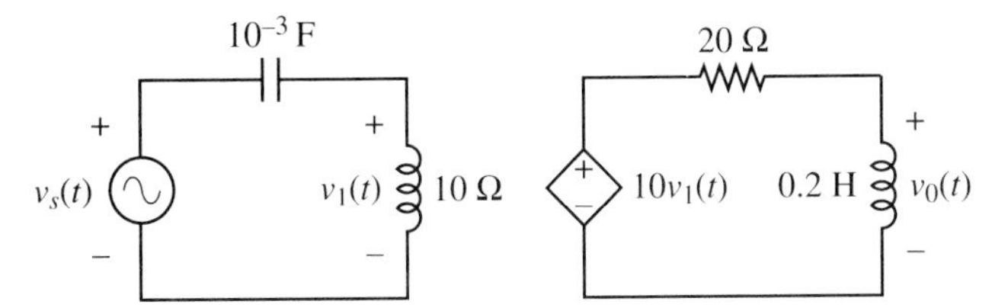


**Figure P5.35**

**5.36.** For the circuit in Fig. 5.19a, let $R = 10\ \Omega$, $L = 100$ mH, and $v_s(t) = 6\cos(200t) + 4\sin(300t)$ V. Using superposition (i.e., by considering the response due to each sinusoid separately), determine the current $i(t)$ through the inductor.

**5.37.** (a) Find the complex Thévenin equivalent (with equivalent resistance and equivalent reactance) of the circuit in Fig. P5.37 which is valid at frequency $\omega$.

(b) At what frequency $\omega_0$ is the Thévenin equivalent a phasor voltage source in series with a pure resistance? Find this special Thévenin equivalent.

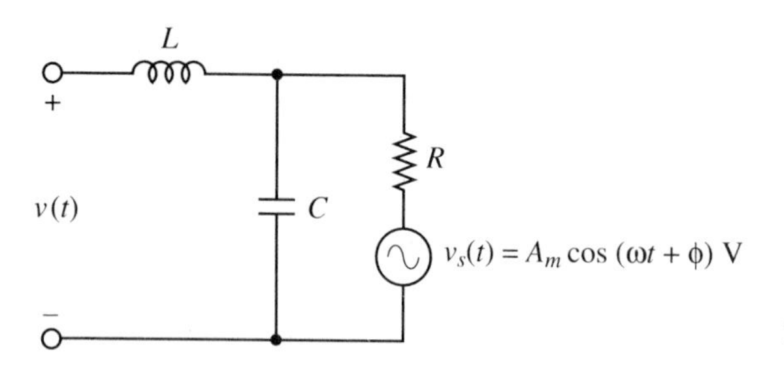


Figure P5.37

**5.38.** (a) Find the equivalent complex impedance $\overline{Z}_{eq}$ of the following network: resistor $R_C$ in parallel with the series combination of resistor $R_A$, inductor $L$, capacitor $C$, and resistor $R_B$. Express your result in polar form. For what frequency (rad/s) does $\underline{/\overline{Z}_{eq}} = 0°$? Does this "resonance" condition result in a maximum or minimum voltage across $\overline{Z}_{eq}$ for a given input sinusoidal current magnitude at the specified frequency? (*Hint:* Compare the magnitude of $\overline{Z}_{eq}$ for nonresonance with resonance frequencies.)

(b) Simplify your polar form expression for $\underline{/\overline{Z}_{eq}}$ for the case $R_A = R_B = R_C/2 = R/2$.

**5.39.** A series $LC$ combination is excited with a sinusoidal voltage at frequency $\omega$.

(a) Find the frequency $\omega = \omega_0$ when the equivalent impedance is zero.

(b) For what frequencies is the overall impedance capacitive, and for what range is it inductive?

**5.40.** A periodic time function $f(t)$ is shown in Fig. P5.40. Find the complex exponential Fourier series expansion for $f(t)$.

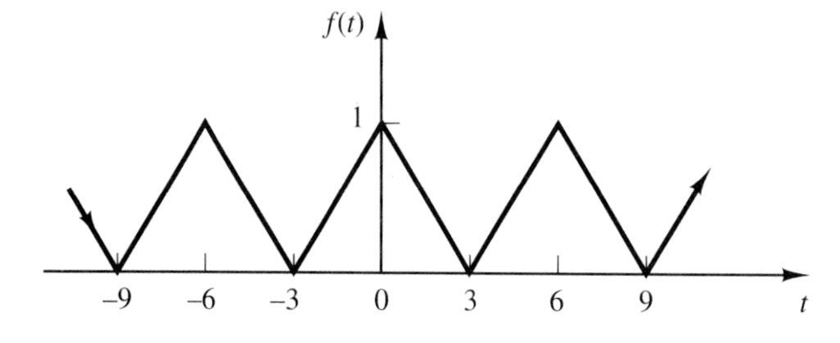


Figure P5.40

**5.41.** Let $g(t)$ be one period of $f(t)$ in Problem 5.40; that is, $g(t) = f(t)$ for $|t| \leq 3$ and 0 otherwise. Find the Fourier transform of $g(t)$.

**5.42.** In terms of $\overline{F}(\omega)$, the Fourier transform of $f(t)$, find the Fourier transform of $g(t)$, $G(\omega)$, when:

(a) $g(t) = f(at)$, where $a$ is a constant.

(b) $g(t) = f(t - t_0)$ where $t_0$ is a constant.

**5.43.** It is desired to pass all frequencies above 27 Hz from one stage of an audio amplifier to the next, but all dc is to be blocked. (We have to change over at *some* frequency, and there is not much signal of interest to the ears below 27 Hz.) We model the input of the second amplifier stage in Fig. P5.43 as having an input resistance of 3 kΩ. What coupling capacitance should we use? (In Fig. P5.43, an amplifier is represented by a triangle.)

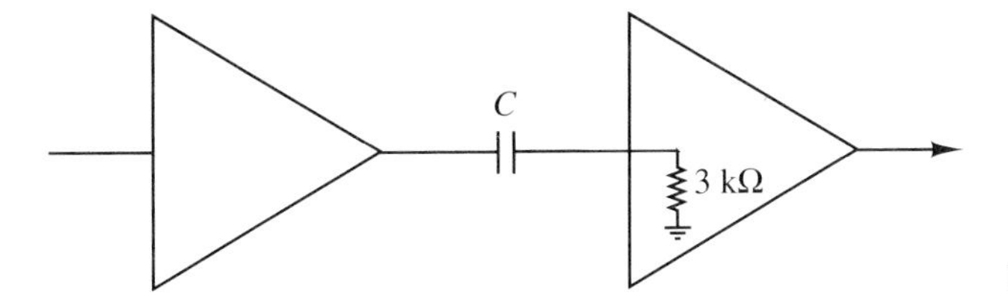


**Figure P5.43**

**5.44.** A voltage $v_i(t)$ is connected across the series combination of resistor $R$ and capacitor $C$ (an $RC$ filter). The output voltage $v_o(t)$ is the voltage across the capacitor, where $v_i(t)$ and $v_o(t)$ have common $-$ terminals.

(**a**) Determine the frequency response of this filter, $\overline{V}_o/\overline{V}_i$.

(**b**) Draw the magnitude of the frequency response as a function of $\omega$.

(**c**) Determine the half-power cutoff frequency. Is this a low- or a high-pass filter?

**5.45.** The tuner network of an AM radio (which we discuss in much more detail in Sec. 7.7.3), is shown in Fig. P5.45.

(**a**) Find the resonant frequency of the tuner and the resonant equivalent impedance seen by the antenna. Is this a minimum or a maximum impedance? What should it be for the given application?

(**b**) The frequency range for AM broadcasting is 520 to 1610 kHz. If $L = 80\ \mu$H, what range must $C$ be adjustable over to have the resonant frequency extend over the entire AM band?

(**c**) Sketch the magnitude frequency response of the circuit [input is $i_s(t)$ and output is $v_o(t)$] and determine the quality factor at resonance when $R = 32$ kΩ, $L = 80$ $\mu$H, and $\omega_0 = 2\pi \cdot 800$ kHz.

(**d**) Find the bandwidth BW in hertz for general values of $R$, $L$, and $C$. To do so, you will have two quadratic equations for $\omega$; take the two positive roots for $\omega_1$, $\omega_2$ at which $H(\omega_1) = H(\omega_2) = H(\omega_0)/\sqrt{2}$. Then BW $= |\omega_2 - \omega_1|$. What is the BW for the values in part (c)?

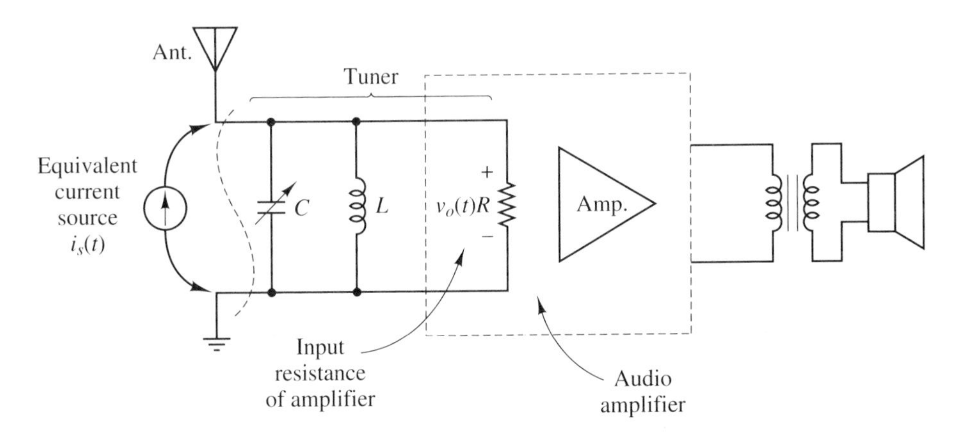


**Figure P5.45**

**5.46.** For the circuit shown in Fig. P5.46:

**(a)** Find the resonant frequency $\omega_0$.

**(b)** Find the quality factor $Q(\omega_0) = R/X_0$ where $X_0 = \omega_o L = 1/(\omega_0 C)$ (a result proved in Problem 5.53).

**(c)** If $i_1(t) = 2 \cos(\omega_0 t)$ mA and $i_2(t) = 2 \cos(0.9\omega_0 t)$ mA, find $v_o(t)$.

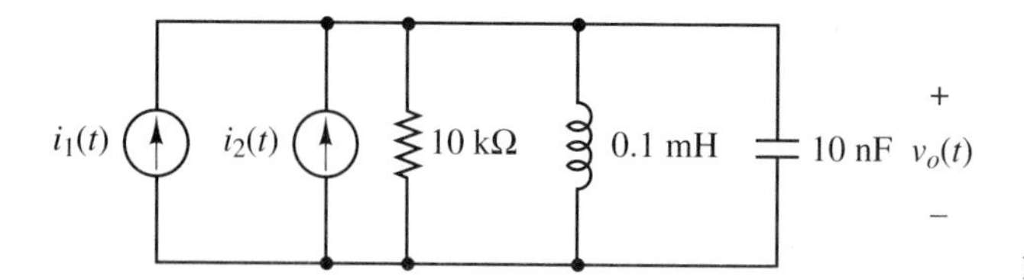


**Figure P5.46**

**5.47.** Suppose that a series $RLC$ circuit is driven by $v_1(t) = 10 \cos(\omega t)$ V, where $\omega$ is a variable that can be made to range from $\omega = 0$ to very large values. Such a voltage source is called a sine-wave signal generator. If $L = 1$ mH, $C = 1$ nF, and $R = 10\ \Omega$:

**(a)** Sketch $H(\omega)$, the magnitude frequency response, where the output voltage is taken to be that across the resistor. [*Note:* To make a good plot, it may help you to also do parts (b), (c), and (e) at this time.]

**(b)** Find the value of $\omega$ for which $H(\omega)$ achieves its maximum.

**(c)** Determine that maximum value of $H(\omega)$.

**(d)** Find the capacitor voltage phasor magnitude at that frequency.

**(e)** Find the bandwidth of this filter in kHz.

**5.48.** $RC$ filters such as that shown in Fig. P5.48 are commonly used in audio tone control circuitry.

**(a)** Determine the magnitude frequency response of the circuit, $H(\omega) = |\bar{V}_o/\bar{V}_i|$.

**(b)** Sketch $H(\omega)$ for $R \neq 0$. What happens if $R = 0$?

**(c)** Is this a treble control (low-pass type, passing dc up to a maximum frequency) or a bass control (high-pass type, passing from a minimum frequency to a high frequency)?

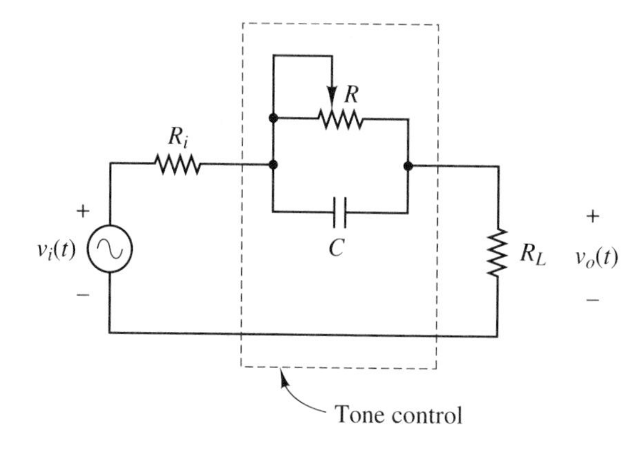


**Figure P5.48**

**5.49.** Another tone control (see Problem 5.48) is shown in Fig. P5.49.

**(a)** Determine the magnitude frequency response of the circuit, $H(\omega) = |\bar{V}_o/\bar{V}_i|$.

**(b)** Sketch $H(\omega)$ for $R \neq \infty$. What happens if $R = \infty$?

**(c)** Is this a treble control (low-pass type, passing dc up to a maximum frequency) or a bass control (high-pass type, passing from a minimum frequency to a high frequency)?

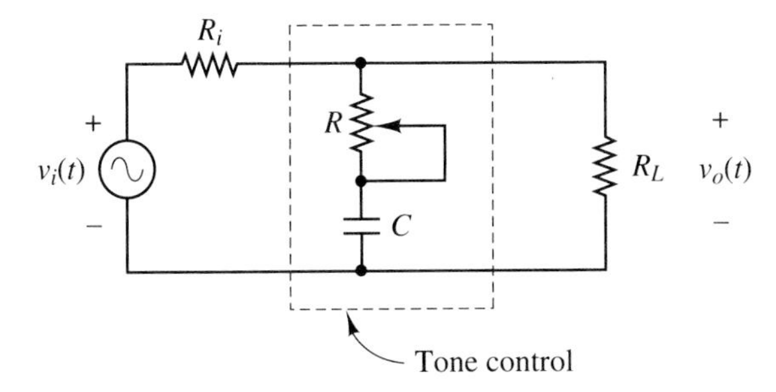


**Figure P5.49**

## ADVANCED PROBLEMS

**5.50.** An important theorem in electronics, known as Miller's theorem, allows us to conveniently break up a circuit into two portions that can then be treated separately. We begin with the circuit in Fig. P5.50a, which shows an impedance joining two networks, and convert it to two disconnected networks as shown in Fig. P5.50b. Define the gain of Fig. P5.50a as $\overline{A}_v = \overline{V}_2/\overline{V}_1$. Find $\overline{Z}_1$ and $\overline{Z}_2$ of Fig. P5.50b in terms of $\overline{Z}$ of Fig. P5.50a and $\overline{A}_v$ so that all corresponding voltages and currents in the two networks are equal. For large $|\overline{A}_v|$, which magnitude impedance is much larger than the other: $Z_1$ or $Z_2$? What happens to the model as $A_v \rightarrow \infty$?

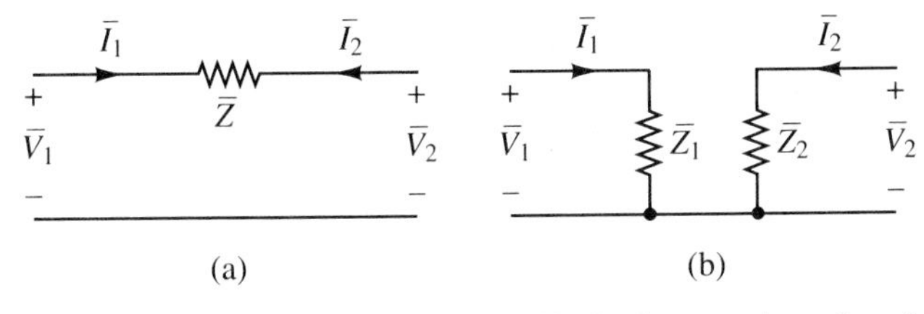


**Figure P5.50**

**5.51.** A typical use of Miller's theorem introduced in Problem 5.50 is in the simplification of a bipolar transistor amplifier circuit model. (We will study transistor amplifiers in Chapter 7.) At high frequencies, the circuit model shown in Fig. P5.51 accurately describes the amplifier. Assume that $C_b$, $C_{bc}$, $g_m$, and $R_0$ are known circuit parameters.

**(a)** Find $\overline{A}_v = \overline{V}_0/\overline{V}_{be}$ at frequency $\omega$. (*Hint:* Just look at the dependent source and output resistor and this is very easy!)

**(b)** Transform the circuit using Miller's theorem to one looking more like Fig. P5.50(b). Rember to use $-(j/\omega C)$ for the impedance of a capacitor, and express your "$\overline{Z}_1$" and "$\overline{Z}_2$" as $-j/(\omega C_1)$ and $-j/(\omega C_2)$. $C_1$ is called the "Miller capacitance."

**(c)** Find the total input-side capacitance $C_t$ (parallel combination of $C_b$ and the Miller capacitance $C_m$).

**(d)** Suppose that $g_m = 0.5\ \Omega^{-1}$, $R_0 = 400\ \Omega$, $C_b = 100$ pF, $C_{bc} = 5$ pF, and $r_b = 1$ k$\Omega$. Notice that for high frequencies the input impedance will be small, so that there will not be much input voltage $v_{be}$ or thus output voltage $v_o$. That is, we have a

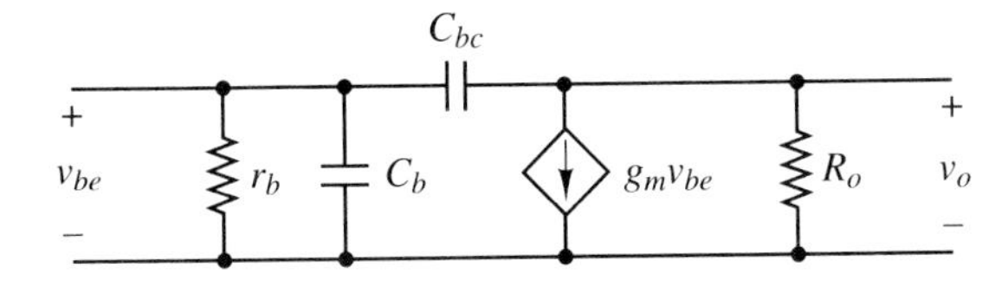


**Figure P5.51**

low-pass filter. Find the Miller capacitance $C_m$ (in pF) and the cutoff frequency $f_0 = \omega_0/2\pi = 1/(2\pi r_b C_t)$ (in kHz).

**5.52.** **(a)** In Eq. (5.105), the quality factor $Q(\omega)$ was defined. Using this definition, prove that the results claimed after that equation to hold for the series $RLC$ ciruit are true, that is, that $Q \equiv Q(\omega = \omega_0)$ = resonant frequency/bandwidth $= \omega_0/\text{BW} = (L/C)^{1/2}/R$ = resonant reactance/resistance $= X_0/R$ where $X_0 = X_{L,0} = \omega_0 L = X_{C,0} = 1/(\omega_0 C)$. In the process, also determine $Q(\omega)$ (at any frequency) of a series $RLC$ circuit. (*Hints:* The energy dissipated in one voltage–current cycle in the resistor of the series $RLC$ circuit is $I_{\max}^2 R\pi/\omega$, where $I_{\max}$ is the current amplitude at frequency $\omega$ in the series $RLC$ circuit. We shall prove this in a Chapter 6 homework problem, using information in Chapter 6. For $\omega < \omega_0$, the series $RLC$ impedance is net capacitive [$X_{\text{net}} = -1/(\omega C) + \omega L < 0$ because $|-1/(\omega C)| > \omega L$], so the maximum stored energy will be $\frac{1}{2}CV_C^2$; that is, more energy will be stored in the cpacitor than in the inductor. For $\omega > \omega_0$ the impedance is net inductive, so the maximum energy will be the $\frac{1}{2}LI^2$ store in the inductor.)

**(b)** For a lossy inductor $R + j\omega L$ ($RL$ in Fig. 5.19), show that $Q(\omega) = \omega/\omega_0 = X_L/R = \omega L/R$, where here $\omega_0 = R/L$ = low-pass $RL$ filter cutoff frequency [and thus $Q(\omega_0) = 1$ for an inductor]. Again the energy dissipated per cycle is $I_{\max}^2 R\pi/\omega$.

**5.53.** Results analogous to those for series $RLC$ combinations stated at the end of Sec. 5.5.3 exist for parallel-connected $RLC$ combinations. Show that for a parallel-connected $RLC$ combination ($R$, $L$, and $C$ connected in parallel, and energized by a parallel-connected oscillating current source), $Q_T(\omega_0) = R/X_0$ where $X_0 = \omega_0 L = 1/(\omega_0 C)$, where $\omega_0 = 1/(LC)^{1/2}$. [*Hint:* The energy dissipated in one voltage–current cycle is $V_{\max}{}^2\pi/(\omega R)$, where $V_{\max}$ is the voltage amplitude of the parallel combination at frequency $\omega$. In this case, for $\omega < \omega_0$, the parallel $RLC$ impedance is net inductive (less impedance takes more current), so the maximum stored energy will be $\frac{1}{2}LI_L^2$; that is, more energy will be stored in the inductor than in the capacitor. For $\omega > \omega_0$, the maximum energy will be the $\frac{1}{2}CV^2$ stored in the capacitor.]

**5.54.** A parallel model for a lossy inductor equivalent to the series model $\overline{Z}_L = R_S + j\omega L_S$ can be made (see Fig. P5.54): $Z_L = R_P \| j\omega L_P$.

**(a)** Find $R_S$ and $L_S$ in terms of $R_P$, $L_P$, and $Q_P(\omega) = R_P/(\omega L_P)$.

**(b)** Find $R_P$ and $L_P$ in terms of $R_S$, $L_S$, and $Q_S(\omega) = \omega L_S/R_S$. [*Hint:* Use intermediate results from part (a) and divide the two equations.]

**(c)** Show that the $Q_L(\omega)$ of either model of the inductor is the same: $Q_L(\omega) = Q_P(\omega) = R_P/(\omega L_P) = Q_S(\omega) = \omega L_S/R_S$. Physically, we know this must be true because $Q(\omega)$ is the ratio of stored to dissipated energies for the same device, so its value should be independent of which exact representation model is used. Note how both expressions for $Q_T$ ($RLC$ network; see Problem 5.53) and $Q_L$ ($RL$ network; this problem) involve a ratio which is the inverse of that of the corresponding series connection (parallel networks have the form $R/X$, while series networks have the form $X/R$).

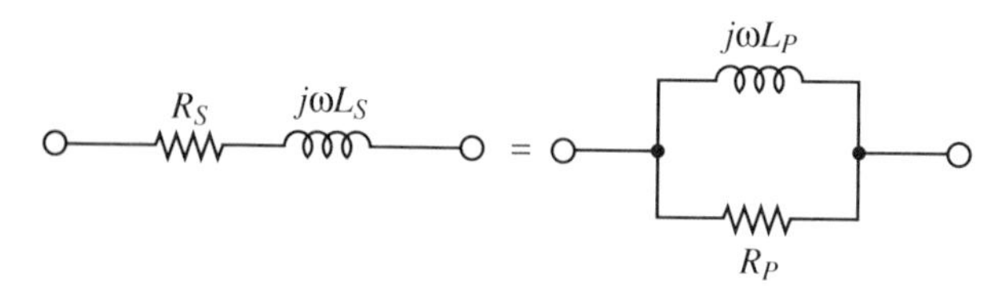


**Figure P5.54**

**5.55.** Suppose that we wish to design a parallel *RLC* filter bank for reception of TV channel 4. Channel 4 has a center frequency of 69 MHz and its bandwidth (like those of all TV stations) is 6 MHz. We wish to design a resonator whose resonant frequency $\omega_0$ is at $(2\pi \text{ rad/Hz}) \cdot 69$ MHz and whose bandwidth is also 6 MHz (see Fig. P5.55(a)). That way, given that all station signals come in at once, our resonator responds only to the desired channel 4. In practice, a different bank can be used for each channel, switched by the channel selector. Suppose that the inductors available to us all have $Q_L(\omega_0) = 80$. An additional parallel resistance $R$ models the load that the resonator will drive. For this example, let it be $R = 500\ \Omega$. In the following, make use of the results involving $Q(\cdot)$ from Problems 5.53 and 5.54 concerning the parallel *RLC* circuit and refer to Figs. P5.54 and P5.55.

**(a)** What is the $Q$ required for the entire *RLC* network?

**(b)** What values of $L$ and $C$ are required? (*Hint:* You will have to solve for the inductor parallel resistance $R_P$.)

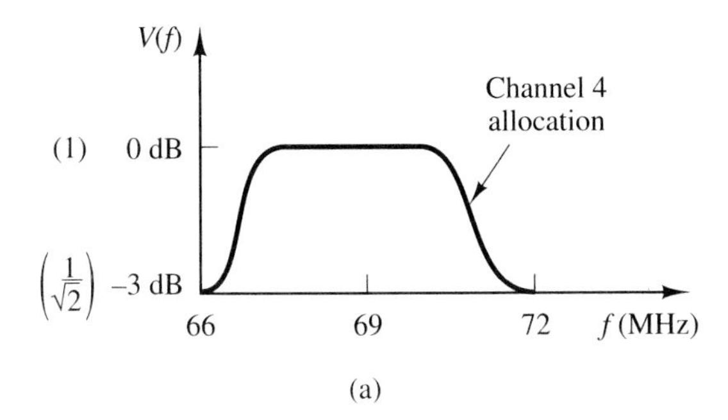


(a)

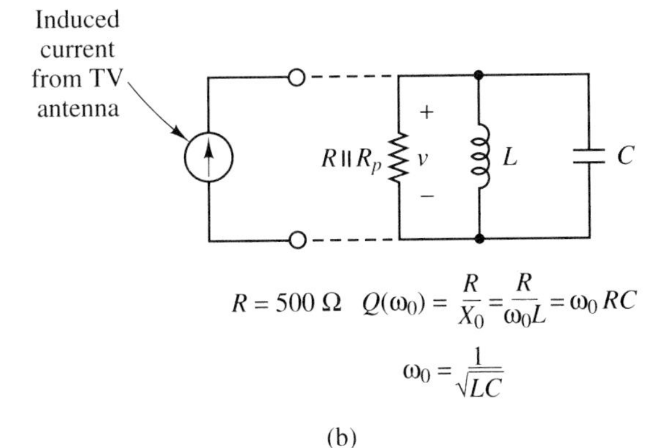


$$R = 500\ \Omega \quad Q(\omega_0) = \frac{R}{X_0} = \frac{R}{\omega_0 L} = \omega_0 RC$$

$$\omega_0 = \frac{1}{\sqrt{LC}}$$

(b)

**Figure P5.55**

**5.56.** Suppose that we have a parallel combination of a 10-k$\Omega$ resistor, a 0.1-mH inductor, and a 1-$\mu$F capacitor.

**(a)** Find the resonant frequency.

**(b)** Suppose that the network is driven by a current source $i_S(t)$ at the resonant frequency. Find the current phasor (defined in same direction as $\bar{I}_S$) through the inductor in terms of $I_S$, the magnitude of $\bar{I}_S$.

**5.57.** Determine which of the following systems are linear and shift invariant. ($a$, $b$, and $c$ are constants.)

**(a)** $L\{x(t)\} = x(t)$

**(b)** $L\{x(t)\} = ax(t) + b$

(c) $L\{x(t)\} = x^2(t)$
(d) $L\{x(t)\} = x(t)/(t + 1)$
(e) $L\{(t)\} = c$

**5.58.** Calculate the convolution $g(t) = f_1(t) * f_2(t)$ where $f_1(t)$ and $f_2(t)$ are as shown in Fig. P5.58.

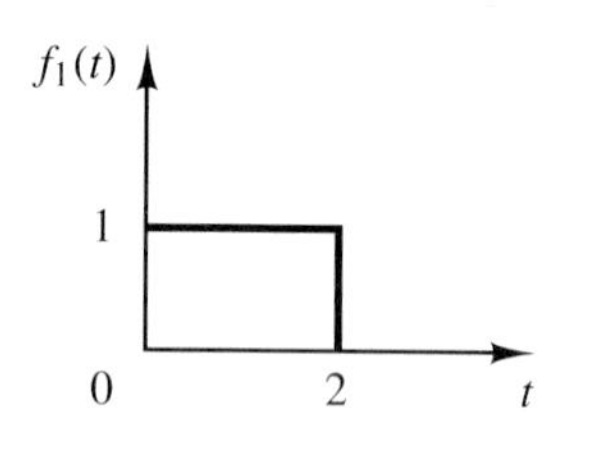


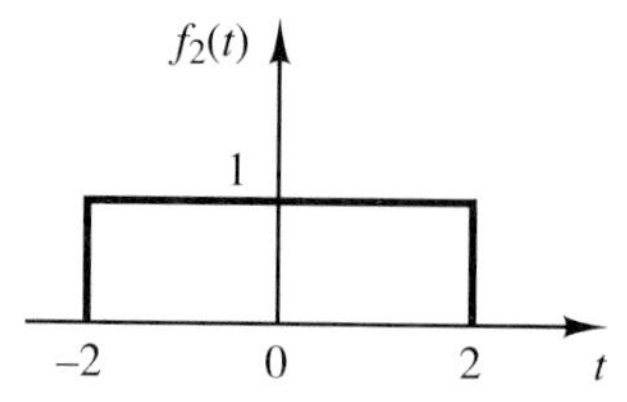


**Figure P5.58**

**5.59.** Suppose that we input $x(t) = A \sin(\omega_0 t)$ into a linear system with frequency response $\overline{H}(\omega_0)$ (whose impulse response is real-valued). What is the output $y(t)$? [*Hint:* Use Eqs. (5.89) as a model.]

**5.60.** Using the Dirac delta sampling property, evaluate the following integrals:

(a) $\displaystyle\int_{-\infty}^{\infty} \exp(-t^2/2)\, \delta(t)\, dt$

(b) $\displaystyle\int_{-\infty}^{\infty} \cos(\omega_0 t)\, \delta(t - 2)\, dt$

(c) $\displaystyle\int_{-\infty}^{\infty} \frac{1 + \omega^2}{1 + \omega^4}\, \delta(\omega - 1)\, d\omega$

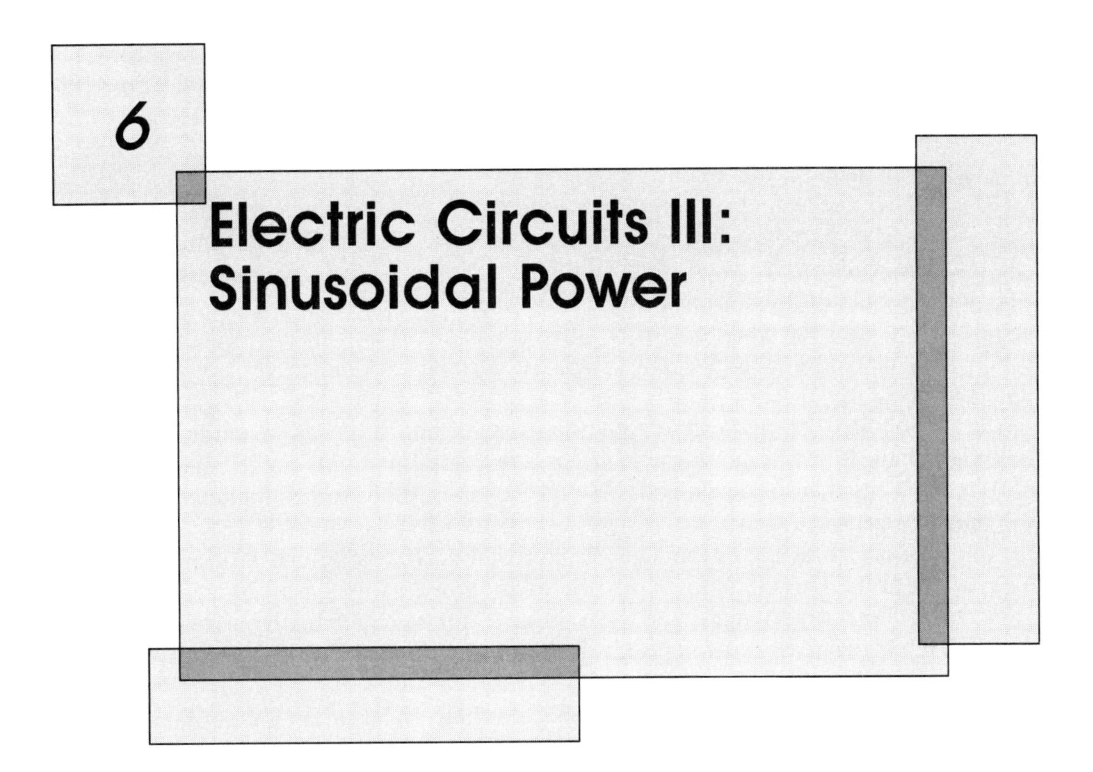

# 6 Electric Circuits III: Sinusoidal Power

## 6.1 INTRODUCTION

Up to this point, our analysis of sinusoidal waveforms has been confined to voltage, current, and impedance calculations. These are the quantities of greatest immediate interest in signal processing circuitry, for they contain and modify the actual signal content. In another branch of electrical engineering called electrical energy conversion, an additional quantity is of tremendous importance: the average power consumed by the circuit or network. This area includes motors, lighting, heaters, radiating radio antennas, and so on, each of which may have large amplitude sinusoidal waveforms.

In these applications it is the average power that is potentially useful to the user to drive the desired loads. However, the user must pay for both the average power consumed by the load and $I^2R_{tl}$ losses due to resistance $R_{tl}$ in the power transmission lines. Additionally, large currents drawn by the user increase the required power transmission equipment ratings and therefore their costs. These factors can be significant for the high-power circuitry involved in industrial energy transfer equipment. They are generally not important in signal processing equipment, excepting cases such as high-power radiating antennas.

It will be shown in this chapter that the most efficient delivery of power (e.g., lowest $I^2R_{tl}$ losses for delivery of a given average power) occurs for cases where the voltage and current waveforms are in phase. This important fact greatly influences the design and implementation of high-power circuitry. If the phase angle between current and voltage cannot economically be made zero, we at least attempt to minimize it.

In signal processing, the instantaneous value of a voltage or current contains the very information of interest. By contrast, in energy conversion applications it is the effective value (a kind of average value, defined below) which will determine how bright the light glows, how fast the motor spins, how much warmth the heater radiates; usually, there is no "signal."

Now the average power delivered to the load, expressible in terms of effective values, is of paramount interest. In reactive loads, there is also the power of energy storage, called reactive power. As this contributes to the $I^2R_{tl}$ discussed above, it must be quantified. Just as complex impedance was defined as a convenient mathematical tool for dealing with resistive (dissipative) and reactive (energy-storing) circuit elements, complex power will be defined to handle average and reactive powers.

We shall consider in Sec. 6.5 the transformer, a device we studied briefly in Sec. 3.10. First the ideal transformer is discussed in some detail, and then more realistic models are developed. Finally, the applications of this fundamental device are reviewed.

Because of the advantages of improved performance and efficiency, electric power in the United States is distributed in three phases, each 120° apart from the others. So various circuits and issues in three-phase circuit analysis are considered. We next take up the area of power distribution systems, by which we receive electric power to do useful work.

Finally, the absolutely crucial matter of electrical safety is considered briefly in Sec. 6.8. Facts about shocking, preventive measures, and the important design technique of grounding, are introduced. Before doing anything around high-power equipment, be sure to read this material and consult experienced personnel.

## 6.2 EFFECTIVE VALUES

Before proceeding to discuss sinusoidal power, the concept of *effective value* should be explained. The effective value of some time-varying voltage or current is that dc voltage or current value which would dissipate the same power in a resistor $R$ as the time-varying voltage or current would dissipate on the time average in $R$. To the resistor, this dc value and the original time-varying function are, on the average, *effectively* the same, hence the name "effective." As an example, consider the simple current source $i(t) = I\cos(\omega t)$ connected to resistor $R$ in Fig. 6.1. The instantaneous power dissipated in $R$ is

$$\begin{aligned} p(t) &= i^2(t)R \\ &= I^2R\cos^2(\omega t). \end{aligned} \tag{6.1}$$

The period $T$ of $i(t)$ is equal to $2\pi/\omega$. The time-average power dissipated in $R$ is

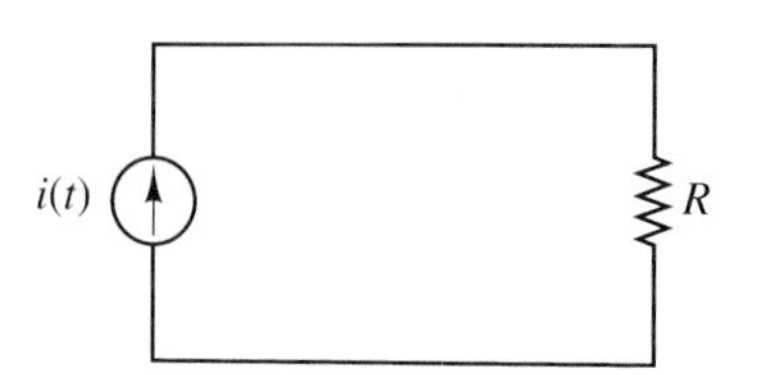


**Figure 6.1** Circuit useful for describing the meaning of effective value.

$$
\begin{aligned}
P &= \frac{I^2R}{T}\int_0^T \cos^2(\omega t)\,dt \\
&= \frac{1}{2}\left(\frac{I^2R}{T}\right)\int_0^T [1+\cos(2\omega t)]\,dt \\
&= \frac{I^2R}{2} \\
&= \left(\frac{I}{\sqrt{2}}\right)^2 R \\
&= I_{eff}^2 R,
\end{aligned} \tag{6.2}
$$

where $I_{\text{eff}} = I/\sqrt{2}$. It is justifiable to call $I/\sqrt{2}$ the effective current for the following reason. The average power $P$ is $(I/\sqrt{2})^2R = I_{\text{eff}}^2R$. But that is the same expression as that for the (constant) power dissipated in resistor $R$, through which passes a dc current of value $I_{\text{eff}} = I/\sqrt{2}$.

What operations led to the conclusion that $I_{\text{eff}} = I/\sqrt{2}$? First the time-varying waveform was squared [Eq. (6.1)], then the time average was computed [Eq. (6.2)], and finally, the square root was taken to obtain $I_{\text{eff}}$. So the effective value $I_{\text{eff}}$ is the *root* of the *mean* of the *square* of $i(t)$. Consequently, the effective value is also referred to as the root-mean-square (rms) value. The rms operation thus arises from average power calculations and can be written in general as

$$I_{\text{eff}} = \left[\frac{1}{T}\int_0^T i^2(t)\,dt\right]^{1/2}. \tag{6.3}$$

For any *sinusoidal* waveform (voltage or current), the effective (rms) value is the amplitude of the sinusoid divided by $\sqrt{2}$. (An example would be the rms voltage, 110 V, found in household electric outlets. The maximum of the sinusoidal voltage is actually $110\text{ V}\cdot\sqrt{2} \approx 156\text{ V}$.) So again, the effective (rms) value is the dc source having the same dissipative *effect* (i.e., the same capability of supplying resistive power) as the time-varying waveform, on the average. Because of its significance, for the remainder of this chapter all voltage and current amplitudes are expressed in terms of the corresponding effective value. Thus for the sinusoidal case, the subject of this chapter, the amplitudes are recoverable, if desired, from the effective values by

$$V = \sqrt{2}V_{\text{eff}} \tag{6.4a}$$

and

$$I = \sqrt{2}I_{\text{eff}}. \tag{6.4b}$$

Every type of periodic waveform has a particular form for the effective value in terms of the maximum value of the waveform, not necessarily involving $\sqrt{2}$. But all waveform types having a given maximum value can be compared in terms of their respective effective values. So the effective (equivalent dc) value is a standard for comparison.

**Example 6.1**

Both the pulse train $v_1(t)$ in Fig. 6.2 and the sinusoid $v_2(t) = V\sin(2\pi t)$ have zero average value. What are their respective effective values?

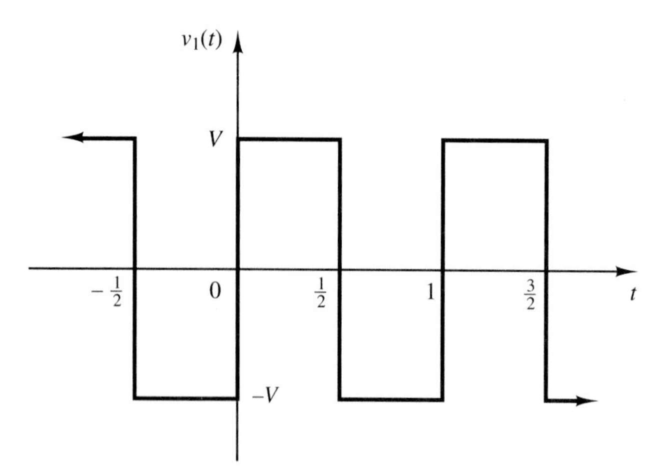


**Figure 6.2** Pulse train for Example 6.1.

**Solution** The effective value of $v_1(t)$ is (using Fig. 6.2)

$$V_{1,\text{eff}} = \left[\frac{1}{1}\int_0^1 v_1^2(t)\,dt\right]^{1/2} = (V^2 \cdot 1)^{1/2} = V.$$

From Eq. (6.4a), $V_{2,\text{eff}} = V/\sqrt{2}$. So the pulse train has a larger effective value for the same amplitude $V$. Note that squaring $v_1(t)$ (which occurs in the calculation of power) makes it indistinguishable from a dc voltage of $V$ volts. Therefore, the effective value of $v_1(t)$ equals its maximum value.

In the frequency domain, the voltage and current phasors $\overline{V}$ and $\overline{I}$ are replaced by effective voltage and current phasors $\overline{V}_{\text{eff}}$ and $\overline{I}_{\text{eff}}$. Thus

$$\overline{V}_{\text{eff}} = V_{\text{eff}}\angle\underline{\theta_v} = V_{\text{eff}}\,e^{j\theta v}$$

and (6.5)

$$\overline{I}_{\text{eff}} = I_{\text{eff}}\angle\underline{\theta_i} = I_{\text{eff}}\,e^{j\theta i}$$

**Example 6.2**

For the waveforms of Example 6.1, determine their average values over their first half-period (during which they are both positive). Obtain their *form factors,* which is the ratio of the effective value to the half-period average value.

**Solution** The average value of any function $f(t)$ on the interval from $t = 0$ to $t = T$ is the integral of $f(t)$ from 0 to $T$ divided by $T$. For both waveforms, we take $T = \frac{1}{2}$. The integral of $V \sin(2\pi t)/T$ is $-V \cos(2\pi t)/(2\pi \cdot \frac{1}{2})$, which when evaluated at the limits $\frac{1}{2}$ and 0 gives $2V/\pi$. The form factor is therefore $(V/\sqrt{2})/(2V/\pi) = \pi/(2\sqrt{2}) = 1.11$. For the pulse train, the average value from 0 to $\frac{1}{2}$ is just $V$, which equals the effective value. Therefore, the pulse train has a form factor of unity.

## 6.3 POWER OF A SINUSOIDAL EXCITATION

### 6.3.1 Instantaneous Sinusoidal Power

As derived in Sec. 2.10, the instantaneous power into a circuit element or network having a voltage $v(t)$ across it and a current $i(t)$ through it is

$$p(t) = v(t)i(t). \tag{6.6}$$

If the voltage and current waveforms are sinusoids of frequency $\omega$ and respective effective magnitudes $V_{\text{eff}}$ and $I_{\text{eff}}$ and phases $\theta_v$ and $\theta_i$, the instantaneous power becomes [using Eqs. (6.4)]

$$p(t) = 2V_{\text{eff}}I_{\text{eff}} \cos(\omega t + \theta_v) \cos(\omega t + \theta_i). \tag{6.7}$$

### 6.3.2 Average Sinusoidal Power

The instantaneous power $p(t)$ in Eq. (6.7) can be put in a very illuminating form by using a trigonometric identity. Recall the identity involving the cosine of a sum and of a difference for angles $A$ and $B$:

$$\begin{aligned} \cos(A + B) &= \cos(A)\cos(B) - \sin(A)\sin(B) \\ \cos(A - B) &= \cos(A)\cos(B) + \sin(A)\sin(B). \end{aligned} \tag{6.8}$$

Adding the two formulas together and rearranging gives

$$\cos(A)\cos(B) = \tfrac{1}{2}[\cos(A + B) + \cos(A - B)]. \tag{6.9}$$

Now identify $A = \omega t + \theta_v$ and $B = \omega t + \theta_i$. Then the instantaneous power in Eq. (6.7) can be written

$$p(t) = V_{\text{eff}}I_{\text{eff}}[\cos(2\omega t + \theta_v + \theta_i) + \cos(\theta_v - \theta_i)]. \tag{6.10}$$

Notice that the factor of 2 in Eq. (6.7) arising from the product of the current and voltage magnitudes in terms of the effective values cancels the $\frac{1}{2}$ factor from the trigonometric identity.

The first term in Eq. (6.10) is merely a sinusoid oscillating in time every $2\pi/(2\omega) = \pi/\omega$ seconds, that is, twice as frequent as either $v(t)$ or $i(t)$. Its time-average value is zero, as is true for *any* temporal sinusoid. On the other hand, the second term is *constant* in time; the time-average value of that term is, of course, equal to that constant. Consequently, the time-average value of $p(t)$, conventionally given the symbol $P$ and called the average power, is just

$$P = V_{\text{eff}}I_{\text{eff}} \cos(\theta_v - \theta_i). \tag{6.11}$$

The term $\cos(\theta_v - \theta_i)$ is called the *power factor* (often abbreviated PF). The minus sign in front of $\theta_i$ comes directly from that in the trigonometric identity in Eq. (6.9).

The quantity $P$ is the average rate at which the network or device consumes energy. Such energy may provide mechanical work as is true for motors, drive other circuit elements as is true for lighting and electronic circuits, or may merely be dissipated as heat through resistances as occurs in all circuits. The important point is that the energy flow is one-way; it is never returned to the source, but rather, is converted to other types of energy, such as heat or mechanical work. Clearly, it is $P$ that is potentially useful to the consumer.

Note that for any passive element, the resistive component of $\overline{Z}$ is nonnegative. Thus $|\theta_v - \theta_i| = |\theta_Z| \leq \pi/2$, so by Eq. (6.11), $P \geq 0$. It is the line current $I_{\text{eff}}$ in Eq. (6.11) that is responsible for the $I^2R_{\text{tl}}$ losses mentioned previously. For a given value of usable load power $P$ and line voltage $V_{\text{eff}}$, the current $I_{\text{eff}}$ can be minimized if the phase angles are such that the cosine term is maximized [by Eq. (6.11)]. The cosine

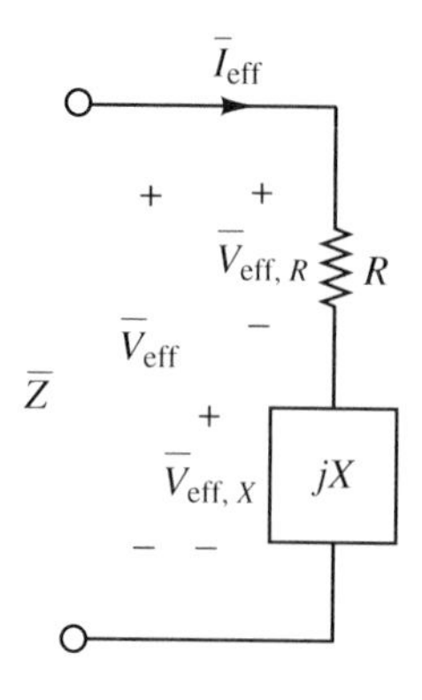


**Figure 6.3** General complex impedance represented in series form: a resistance $R$ in series with a reactance $X$.

term (PF) is maximum for $\theta_v = \theta_i$, in which case it is unity. It is less than 1 if $\theta_v$ is not equal to $\theta_i$, that is, if the complex impedance of the network had a nonzero reactive component. For in that case, the impedance angle $\theta_Z = \theta_v - \theta_i$ is nonzero.

Call the equivalent complex impedance of the power-consuming network or device $\bar{Z} = R + jX$ (see Fig. 6.3). That is, $\bar{Z}$ is made up of an equivalent resistance $R$ in series with an equivalent reactance $X$. What is the time-average power absorbed by only the resistance if the total voltage across $\bar{Z}$ is sinusoidal with effective value $\bar{V}_{\text{eff}}$? First note that the current through $R$ is the same as that through $X$; its effective value is $\bar{I}_{\text{eff}}$. The development leading to Eq. (6.11) holds for *any* component or network, including $R$. So the average power absorbed by $R$ will be the product of $I_{\text{eff}}$, the effective voltage magnitude *across* $R$, $V_{\text{eff},R}$, and the cosine of the difference in phase angles of $\bar{V}_{\text{eff},R}$ and $\bar{I}_{\text{eff}}$. But in a resistor the voltage and current are in phase; hence the cosine term is 1. $V_{\text{eff},R}$ is just the magnitude of a complex voltage divider:

$$\begin{aligned} V_{\text{eff},R} &= \left|\frac{R}{R + jX}\right| V_{\text{eff}} \\ &= V_{\text{eff}} \frac{R}{(R^2 + X^2)^{1/2}} \\ &= V_{\text{eff}} \cos(\theta_Z) \qquad \text{(by Fig. 5.18)} \\ &= V_{\text{eff}} \cos(\theta_v - \theta_i) \qquad \text{[by Eq. (5.56b)].} \end{aligned} \tag{6.12}$$

Consequently, because $\theta_R = 0$ and thus $\cos(\theta_R) = 1$, the average power absorbed by $R$ is

$$\begin{aligned} V_{\text{eff},R} I_{\text{eff}} \cdot 1 &= V_{\text{eff}} I_{\text{eff}} \cos(\theta_v - \theta_i) \cdot 1 \\ &= P, \end{aligned} \tag{6.13}$$

the average power absorbed by the *total* impedance $\bar{Z}$! So the average power $P$ into any network or device due to sinusoidal excitation is equal to the average power absorbed by the resistive component of the complex impedance of the network. And, of course, because $V_{\text{eff},R} = I_{\text{eff}} R$, it follows from Eq. (6.13) that the average power into any linear network is equal to $I_{\text{eff}}^2 R$, where $R$ is the resistive component of the impedance of the network.

### 6.3.3 Reactive Sinusoidal Power

What about the average power absorbed by the reactive component? Following the same procedure as above, first note that the phase difference between the voltage and current in a reactance is $\pm 90°$ (see Table 5.1). Consequently, the cosine of that phase difference is zero. Therefore, the average power absorbed by a reactance is zero, by Eq. (6.11). But there are still nonzero instantaneous voltages across and currents through these capacitors or inductors.

Although the time-average power absorbed is zero, reactances can contribute to the total line current, resulting in increased $I^2R_{tl}$ losses and power transmission equipment ratings for which the user must pay. So it is worthwhile to get a handle on the magnitude of the *instantaneous* power into the reactance.

From Eq. (6.10), the instantaneous power into a reactance is

$$p_X(t) = V_{\text{eff},X} I_{\text{eff}} \cos(2\omega t + \theta_{v,X} + \theta_i), \tag{6.14}$$

where $V_{\text{eff},X}$ and $\theta_{v,X}$ are, respectively, the effective magnitude and phase of the voltage across $X$. Notice that the second (constant-in-time) cosine term in Eq. (6.10) is zero because $|\theta_{v,X} - \theta_i| = 90°$. Thus the instantaneous power into $X$ is a sinusoid of radial frequency $2\omega$ and thus has zero average value.

A sinusoidal function periodically goes negative. So Eq. (6.14) indicates that the power into the reactance is sometimes negative. That is, the flow of power is two-way; sometimes power flows out of $X$ back to the source. The energy sent back to the source comes from the energy stored in the electric or magnetic field associated with the energy storage device $X$. This sinusoidal power term also exists for the resistor. However, the constant second term in Eq. (6.10), being unity, prevents the total instantaneous power into the resistance from ever actually becoming negative.

As an illustration of the ideas above, consider the plots of voltage $v(t) = \sqrt{2}\, V_{\text{eff}} \cos(\omega t)$, current $i(t) = \sqrt{2}\, I_{\text{eff}} \cos(\omega t + \theta)$, and power $p(t) = v(t)i(t)$ shown in Fig. 6.4. From Eq. (6.10) it can be predicted that $p(t)$ oscillates at radial frequency $2\omega$: $V_{\text{eff}} I_{\text{eff}} \cos(2\omega t + \theta)$, and will have an average value $V_{\text{eff}} I_{\text{eff}} \cos(\theta)$. In Fig. 6.4a, $\theta = 0$, which represents the purely resistive case (no reactance). Because whenever the voltage is negative the current is also, $p(t)$ is always nonnegative.

Figure 6.4b shows the case $\theta = 45°$, a capacitive and resistive case. Notice that between times $t_1$ and $t_2$ and between $t_3$ and $t_4$ for example, $p(t)$ is negative because $v(t)$ and $i(t)$ have opposite sign. During this time, energy is returned from the stored electric field to the source. However, for all devices such as this one that have a resistive component, the *average* value of $p(t)$, $I_{\text{eff}}^2 R$, will be positive; this is seen in Fig. 6.4b in that the waveform $p(t)$ is centered above, not on or below the time axis.

Finally, Fig. 6.4c shows the case $\theta = 90°$: a pure reactance (capacitive—no resistive component). As expected, the power has zero average value [$VI \cos(90°) = 0$].

We can express the magnitude of $p_X(t)$ in Eq. (6.14) in a form similar to that of Eq. (6.11). The effective voltage across $X$ can be expressed in terms of $V_{\text{eff}}$, the effective voltage across the total impedance $\bar{Z} = R + jX$, by again taking the magnitude of a complex voltage divider:

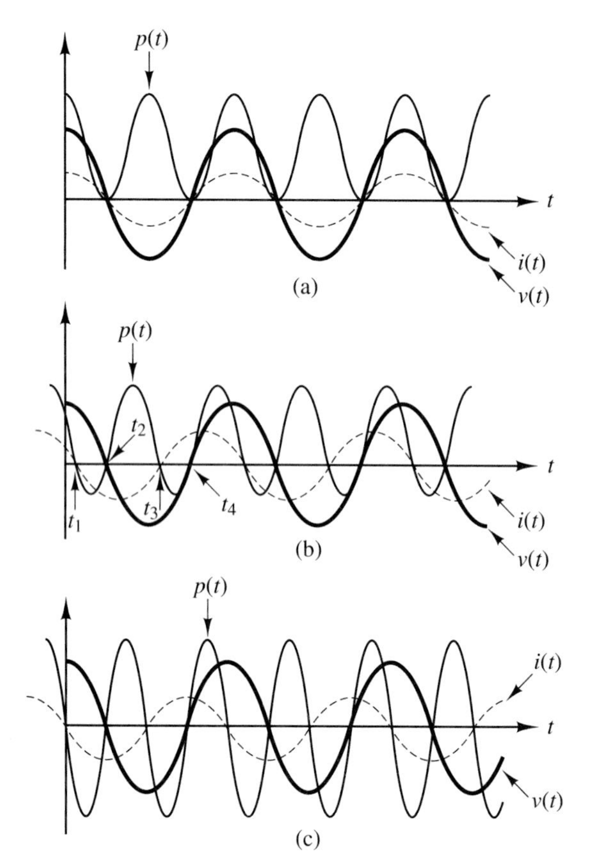


**Figure 6.4** Voltage, current, and power waveforms for (a) a purely resistive impedance, (b) a capacitive/resistive impedance, and (c) a purely capacitive impedance.

$$V_{\text{eff},X} = \left|\frac{jX}{R + jX}\right| V_{\text{eff}} \tag{6.15a}$$

$$= V_{\text{eff}} \frac{X}{(R^2 + X^2)^{1/2}} \tag{6.15b}$$

$$= V_{\text{eff}} \sin(\theta_Z) \qquad \text{(by Fig. 5.18)} \tag{6.15c}$$

$$= V_{\text{eff}} \sin(\theta_v - \theta_i) \qquad \text{[by Eq. (5.56b)]}, \tag{6.15d}$$

where in Eq. (6.15b) we decided to retain the sign of $X$, instead of using $|X|$. The *magnitude* (with sign of $X$) of the instantaneous power into $X$ is called the *reactive power* $Q$. Expressed in terms of the effective values of the sinusoidal excitation $V_{\text{eff}}$ and resulting effective current $I_{\text{eff}}$, $Q$ is [from the magnitude of Eq. (6.14)]

$$\begin{aligned} Q &= V_{\text{eff},X} I_{\text{eff}} \\ &= V_{\text{eff}} I_{\text{eff}} \sin(\theta_v - \theta_i) \qquad \text{[by Eq. (6.15)]}. \end{aligned} \tag{6.16}$$

Physically, $Q$ represents the amplitude of oscillations of the rate at which energy is continually exchanged between the power source and the electric (or magnetic) field associated with a capacitor (or inductor). It also retains the sign of $X$ (via $\theta_v - \theta_i$) in order to distinguish inductive from capacitive reactive power. The energy delivered to $X$ is always quickly returned: The energy exchange is a cyclical process of frequency $2\omega$. And because $V_{\text{eff},X} = I_{\text{eff}}X$, it follows from Eq. (6.16) that the reactive power $Q$ into any linear network is equal to $I_{\text{eff}}^2 X$, where $X$ is the reactive component of the impedance of the network. Note that in Fig. 6.4c, $Q$ was equal to the magnitude of the total instantaneous power ($P = 0$).

It may help to compare the effects of a reactance on current and power. For a given voltage placed across an impedance $\overline{Z}$, reactance $X$ hinders *current* passing through $\overline{Z}$ (via the cyclical storage and release of energy in an electric or magnetic field). Mathematically, this is seen as an increase of $Z = (R^2 + X^2)^{1/2}$ in $I = V/Z$. Roughly speaking, for a given line current and line voltage, reactance diverts *power* from its useful resistive form to those cyclical energy exchanges to and from the electric and/or magnetic field. Mathematically, this is seen as a decrease in $\cos(\theta_v - \theta_i)$ in $P = V_{\text{eff}} I_{\text{eff}} \cos(\theta_v - \theta_i)$.

Because $Q$ is proportional to the *sine* of $\theta_v - \theta_i$ [Eq. (6.16)], it is also proportional to the *sign* of $\theta_v - \theta_i$. Therefore, because $\theta_i < \theta_v$ for an inductor, $Q$ is positive for an inductive reactance. By similar reasoning, $Q$ is negative for a capacitive reactance.

### 6.3.4 Decomposition of Sinusoidal Power

Notice that $P$ is an average power, while $Q$ is a maximum magnitude power (with sign)—they are different types of quantities as Fig. 6.5a summarizes pictorially. Nevertheless, they are typically combined as the perpendicular legs of a right triangle, known as the *power triangle* (see Fig. 6.5b).

The hypotenuse of the triangle, called the *apparent power* $S$, has length $V_{\text{eff}} I_{\text{eff}}$ and is at an angle $\theta_v - \theta_i$ from the horizontal axis. Mathematically, $S$ is the magnitude of the oscillatory component of the total instantaneous power into the impedance. We denote in Fig. 6.5 the operation of extracting the magnitude of a sinusoidal function by "[ ]"—this is *not* an absolute value operator. The oscillatory component is just the instantaneous power into $\overline{Z}$ minus the time-average power into $\overline{Z}$. In Fig. 6.5 we denote the time-averaging operation by "< >." $S$ is called apparent power because it is the power that rms current and voltage measurements would indicate is being supplied to the load. Some of the power is, however, "lost" in reactive power.

The horizontal component $P$ is called the power or the *average (resistive) power.* (The name "power," although commonly used, is liable to cause confusion because of its lack of a distinguishing modifier.) The vertical component $Q$ is, again, appropriately called the reactive power.

Although the units of $S$, $P$, and $Q$ are all volts times amperes, the following convention exists to help distinguish the three quantities. The units of $S$ are called volt-amperes (V · A), those of $P$ are called watts (W), and those of $Q$ are called volt-amperes-reactive (VAR).

Furthermore, as noted briefly in Sec. 6.3.3,

$$\begin{aligned} Q &= V_{\text{eff},X} I_{\text{eff}} \\ &= (I_{\text{eff}} X) I_{\text{eff}} \\ &= I_{\text{eff}}^2 X \\ &= V_{\text{eff}}^2 \frac{X}{Z^2}, \end{aligned} \tag{6.17}$$

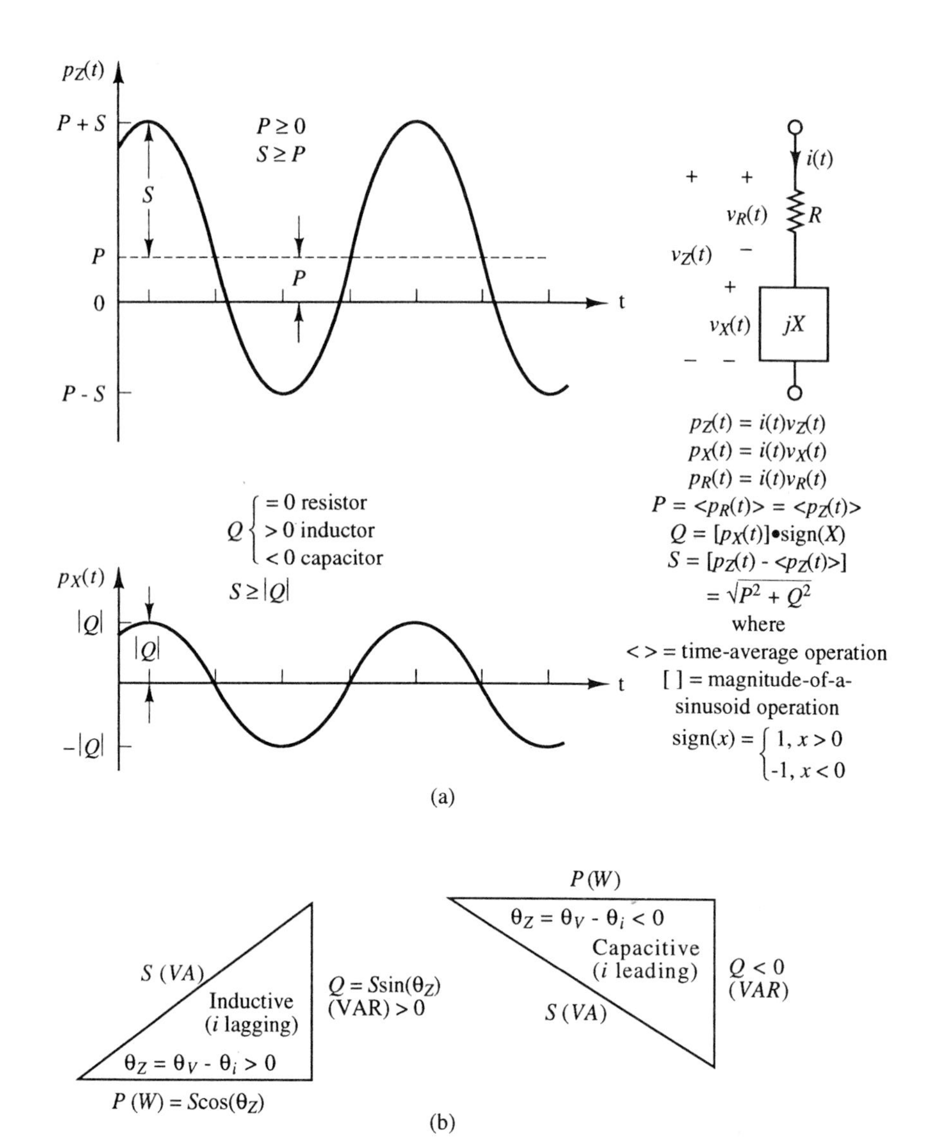


**Figure 6.5** Power relationships for a general impedance $\overline{Z}$ (resistance $R$ in a series with reactance $X$): (a) instantaneous power into impedance and into reactance plotted with $P$, $Q$, and $S$ indicated; (b) power triangle for inductive and for capacitive impedances.

where $Z^2 = R^2 + X^2$. To Eqs. (6.17) may be added the following expression for $P$, obtained from Fig. 6.5, Eq. (6.13), and Ohm's law:

$$\begin{aligned} P &= V_{\text{eff},R} I_{\text{eff}} \\ &= (I_{\text{eff}} R) I_{\text{eff}} \\ &= I_{\text{eff}}^2 R \\ &= V_{\text{eff}}^2 \frac{R}{Z^2}, \end{aligned} \tag{6.18}$$

and $S$ can be written

$$\begin{aligned} S &= V_{\text{eff},} I_{\text{eff}} \\ &= (I_{\text{eff}} Z) I_{\text{eff}} \\ &= I_{\text{eff}}^2 Z \\ &= \frac{V_{\text{eff}}^2}{Z}. \end{aligned} \tag{6.19}$$

**Example 6.3**

Suppose a network consists of a resistor $R = 200\ \Omega$ in series with the parallel combination of an inductor $L = 0.5\text{H}$ and a capacitor $C = 10\mu F$. It is driven by a 60-Hz, 10-V effective value voltage source. Calculate $P$, $Q$, and $S$ into the $RLC$ load.

**Solution** The net reactance is

$$X = \frac{-\omega L/(\omega C)}{\omega L - 1/(\omega C)} = \frac{\omega L}{1 - \omega^2 LC}$$

$$= \frac{377(0.5)}{1 - (377)^2(0.5)10^{-5}} = 651\ \Omega.$$

Thus $Z = [(200)^2 + (651)^2]^{1/2} = 681\ \Omega$. The effective current is $I_{\text{eff}} = V_{\text{eff}}/Z = 10/681 = 14.7$ mA. Now using Eqs. (6.17) through (6.19), $P = (0.0147)^2(200) = 0.043$ W, $Q = (0.0147)^2(651) = 0.14$ VAR, and $S = (0.0147)^2(681) = 0.15$ VA. If we thought we could get a handle on (that is, a rough estimate of) the usable (resistive) power for this load by multiplying the terminal rms voltage and current (i.e., by calculating the apparent power), we would be completely wrong. Nearly all of the apparent power goes into reactive power and not to $R$.

A practical application of the effects of reactive power should be familiar to polymer scientists: *Dielectric heating*. Polymers such as polyester have $\epsilon_r \approx 5$ to 10. When sandwiched between two metal plates, the polymer acts like the dielectric slab of a capacitor. When a strong sinusoidal electric field is applied to the plates, the polymer heats up. This behavior is used for preheating the polymer for compression and transfer molding. Preforms can be heated to over 100°C in less than a minute this way. The required molding temperature is reduced, and the production process is made more efficient with fewer discarded specimens.

Let us consider briefly this phenomenon of dielectric heating. If for simplicity we consider a capacitance as being purely reactive, then $Z = |X|$ in Eq. (6.17), so that $Q = V_{\text{eff}}^2/|X|$. Recall that this is the power involved in charging and discharging the capacitor, with the resulting stored and returned energy in/from the capacitor. Within the dielectric slab, recall that the electric dipoles continually realign themselves with the externally applied alternating polarity electric field (voltage). In so doing, friction converts some of the reactive energy into heat. This fraction of $Q$ is called the *loss factor* "tan (δ)." Clearly, the equation for heating power $P_h$ is $P_h = Q \tan(\delta) = \omega CV^2 \tan(\delta)$. If the loss factor is high, the substance is easy to heat. The loss factor is a strong function of frequency and is substantially an empirically defined factor. In fact, $\tan(\delta)$ increases with $\omega$, so $P_h$ increases strongly with $\omega$. For low frequencies (<1 MHz), $\tan(\delta)$ is negligible, making $P_h$ negligible. The frequencies used to heat polymers are in the range 20 to 50 MHz or more.

An additional quantity present in nearly all descriptions of sinusoidal power is the *complex power* $\bar{S}$. It may easily be constructed from the power triangle in Fig. 6.5 as

$$\begin{aligned}\bar{S} &= P + jQ \qquad \text{(rectangular form)} \\ &= V_{\text{eff}} I_{\text{eff}} [\cos(\theta_v - \theta_i) + j\sin(\theta_v - \theta_i)] \qquad \text{[Eqs. (6.11) and (6.16)]} \\ &= S\angle\theta = Se^{j\theta} \qquad \text{(polar form)},\end{aligned} \tag{6.20}$$

where $S = V_{\text{eff}} I_{\text{eff}}$ is, again, called the apparent power and $\theta = \theta_v - \theta_i$. Consequently, $P = S\cos(\theta)$ and $Q = S\sin(\theta)$. Everything here has a name. The cosine factor $\cos(\theta)$ is (again) called the power factor, and $\sin(\theta)$ is called the *reactive factor*.

Finally, $\bar{S}$ may be expressed in terms of the effective voltage and current phasors, as all the necessary elements are in Eq. (6.20):

$$\begin{aligned}\bar{S} &= V_{\text{eff}} I_{\text{eff}} e^{j(\theta_v - \theta_i)} \\ &= (V_{\text{eff}} e^{j\theta_v})(I_{\text{eff}} e^{-j\theta_i}) \\ &= \bar{V}_{\text{eff}} \bar{I}^*_{\text{eff}}.\end{aligned} \tag{6.21}$$

The complex power $\bar{S}$ is similar to complex impedance $\bar{Z}$ in that it is often expressed in terms of phasors, yet is not itself a phasor and does not rotate—it is merely a complex constant. In fact, $\bar{S}$ can be conveniently expressed in terms of $\bar{Z}$ and the magnitude of one phasor using the complex Ohm's law $\bar{V} = \bar{I}\bar{Z}$:

$$\begin{aligned}\bar{S} &= \bar{V}_{\text{eff}} \bar{I}^*_{\text{eff}} \\ &= \bar{V}_{\text{eff}} \left(\frac{\bar{V}_{\text{eff}}}{\bar{Z}}\right)^* \\ &= \frac{V^2_{\text{eff}}}{\bar{Z}^*} \qquad \text{[by Eq. (5.31)]},\end{aligned} \tag{6.22}$$

or

$$\begin{aligned}\bar{S} &= (\bar{I}_{\text{eff}} \bar{Z})\bar{I}^* \\ &= I^2_{\text{eff}} \bar{Z}.\end{aligned} \tag{6.23}$$

Thus $\bar{S}$ has the same phase angle as does $\bar{Z}$. Equations (6.22) and (6.23) are the sinusoidal steady-state expressions for power that are analogous to, respectively, $P = V^2/R$ and $P = I^2R$, which apply to constant-in-time excitation of resistive circuit elements.

Although dimensionally the same, the real ($P$) and imaginary ($Q$) components of $\bar{S}$ are different types of quantities ($P$ is an average value while $Q$ is a maximum value). Thus no physical significance can be directly ascribed to the complex number $\bar{S}$ except as follows. Its magnitude is the apparent power discussed above, and its phase angle indicates what fraction of the apparent power is lost to reactance and, by its sign, which type of reactance (inductive or capacitive).

In any case, Eqs. (6.11) and (6.16) certainly suggest the construction of $\bar{S}$. Actually, $\bar{S}$ is a mathematical construct to conveniently keep track of the trigonometry relations inherent between $P$, $Q$, and the all-important $I_{\text{eff}}$. ($\bar{S}$ helps us "mind our $P$s and $Q$s.")

A virtue of $\bar{Z}$ is that it allows us to write $\bar{V} = \bar{I}\bar{Z}$ for sinusoidal excitation, just as $V = IR$ is written for constant-in-time excitation. A virtue of $\bar{S}$ is that it allows us to write $\bar{S} = \bar{V}\bar{I}^*$, just as $P = VI$ is written for constant-in-time excitation. The only drawback is that the physical interpretations of $\bar{Z}$ and of $\bar{S}$ are not obvious; they are abstract and hard to visualize.

Remember that the total power into several elements is the sum of the individual powers no matter what the connection configuration, because of energy conservation. This also is true for so-called complex power $\bar{S}$ "drawn" by each component. (But do not add powers from individual sources in superposition problems in which only one source is turned on at a time, because power is not a linear function of the signals!)

A word of caution. Often the power factor $\cos(\theta)$ rather than $\theta$ is given in design problems. Because $\cos(-\theta) = \cos(\theta)$, the sign of $\theta$ must be indicated in addition to the power factor in order to determine $\theta$ without ambiguity. Recall the fact that $\theta = \theta_v - \theta_i$ and the convention for the terms "lead" and "lag" (which always describe the *current* relative to the voltage). Therefore, a leading power factor corresponds to $\theta < 0$ (i.e., $\theta_i > \theta_v$; capacitive) and a lagging power factor corresponds to $\theta > 0$ (i.e., $\theta_i < \theta_v$; inductive) (see Table 5.1). So whenever specifying a power factor, we must always append the designation "lagging" or "leading," whichever is appropriate.

**Example 6.4**

Consider the following situation, of great relevance to all news-lovers. A TV set rated at 110 W is turned on for the nightly News Hour (1 hour long).

**(a)** How much energy was expended in our getting the news? (Answer in MJ and kW · h.)

**(b)** Suppose that the electric utility charges 11 cents/kW · h. How much does it cost to see the News Hour?

**(c)** What is the power factor of the TV, and what is the apparent power into it? The current and voltage waveforms are as shown in Fig. 6.6.

**Solution**

**(a)** The power rating is $P = \text{Re}\{\bar{S}\} = 110$ W. The energy is just the time the set is run times the power, or 1 h · 3600 s/h · 110 W = 0.4 MJ = 0.11 kW · h.

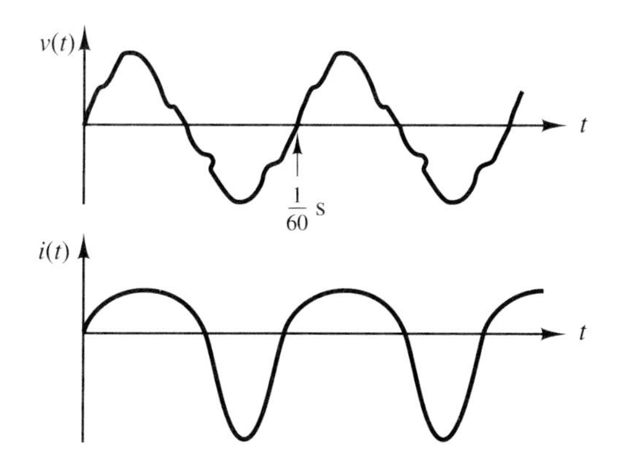


**Figure 6.6** Measured voltage and current waveforms at the input to a TV set. The power factor is essentially 1.

(**b**) 0.11 kW · h · 11 cents/kW · h = 1.2 cents.

(**c**) Although there is distortion in the current waveform (due to the generation of multifrequency signals as a result of rectifier switching), overall the voltage and current waveforms are essentially perfectly in phase. Thus the TV presents an approximately resistive load to the line. Clearly, the power factor is essentially unity, and the apparent power $S$ is practically equal to the average power $P$, 110 W. Unity power factor can be assumed for most electronic appliances. When sizable motors are involved, however, it may decrease, which brings us to our next topic: power factor correction.

## 6.4 POWER FACTOR CORRECTION

To show an important application of the theory above, the problem of *power factor correction* for heavy, inductive loads will now be examined. This issue has already been described: A given average load power $P$ into a load $\bar{Z}_1 = R + jX_1$ is required for a fixed effective value of line voltage $V_{\text{eff}}$. It is desired to minimize $I_{\text{eff}}$ in order to minimize $I_{\text{eff}}^2 R_{\text{tl}}$ losses in the power transmission lines, where $R_{\text{tl}}$ is the resistance of the transmission line.

Power companies also object to unnecessarily large line currents for reasons other than the line losses and will charge accordingly. For a fixed generated voltage, the power factor will dictate the kVA (apparent power) requirement. The size and cost of all the transmission equipment depends on this, or equivalently, the maximum current that the various machine windings must be able to handle and the required generator cooling costs. Although for low-power customers having a high power factor, the main cost is energy, other customers with low power factors and/or high peak loads increase the kVA requirements of the power distribution system.

Power factor correction is not an issue for the homeowner, whose power factor is high and power requirements small. But industrial customers find it economical to minimize the current magnitude while still supplying the load-absorbing average power $P$. Because $P = V_{\text{eff}} I_{\text{eff}} \cos(\theta)$, where $\theta = \theta_v - \theta_i$, we may solve for the effective current magnitude:

$$I_{\text{eff}} = \frac{P}{V_{\text{eff}} \cos(\theta)}, \tag{6.24}$$

so that in order to minimize $I_{\text{eff}}$ while maintaining constant $P$, the only option is to increase the power factor $\cos(\theta)$. As mentioned before, this can be done by trying to make $\theta_v$ as close as possible to $\theta_i$. This amounts to minimizing the net reactance that causes $\theta_v$ and $\theta_i$ to be out of phase. Recall that $\theta = \theta_Z$ [Eq. (5.56b)]. Suppose that $\bar{Z}_1$ is inductive, so that $\theta > 0$. Then the power triangle showing $\bar{S}$, $P$, and $Q$ on the complex plane appears as in Fig. 6.7a.

Note that we cannot simply add a negative reactance $X_C$ to inductive load $X_1$ to make the total load purely real because $X_C$ is to be placed in *parallel* with $\bar{Z}_1$. The capacitors are placed in parallel to maintain the same line voltage across the load; the load power and load behavior must *not* be affected. For example, the expression for the net reactance after addition of $X_C$ in parallel involves $R$. However, a fair amount of algebra can be used in conjunction with Eqs. (6.17) and (6.18) for the net impedance before ($\bar{Z} = \bar{Z}_1$) and after ($\bar{Z} = \bar{Z}_1 \,||\, jX_C$) introduction of $X_C$ to show the same results now to be worked out using powers. In practice, it is more

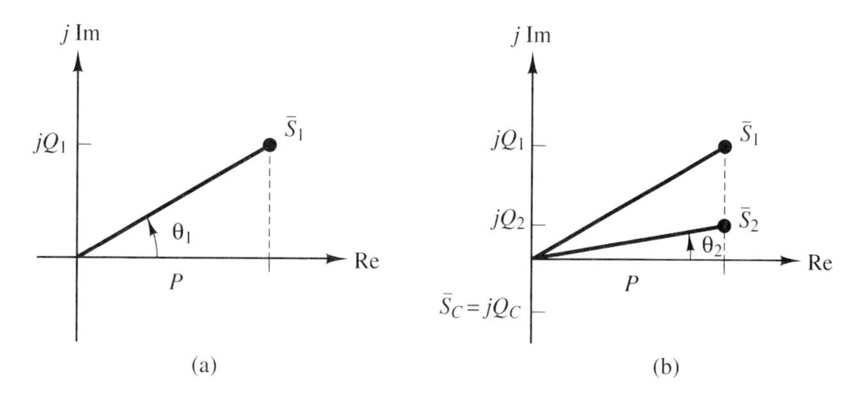


**Figure 6.7** Improvement of power factor: (a) before addition of corrective capacitance; (b) after correction ($\bar{S}_2$).

convenient to use the power diagram instead because powers of elements always add, whether connected in series or in parallel.

By placing a capacitive reactance $jX_C$ in parallel with the load $\bar{Z} = R + jX_1$, a negative $Q$ (imaginary part $jQ_C$) is added to the original $\bar{S} = P + jQ_1$, as shown in Fig. 6.7b; note that indeed $\theta$ has been reduced. (The subscripts 1 and 2 refer, respectively, to before and after addition of the parallel capacitors.)

A cost benefit is obtained because the usable average power has not at all been decreased, yet the line current is reduced by the increase in $\cos(\theta)$, thereby reducing $I^2R_{\text{tl}}$ costs on the power line. Physically, the reactive power required by the inductive load is locally cyclically supplied (and returned to) the capacitor instead of all the way back to the power company. Improving the power factor also reduces the required kVA rating of the power transmission equipment (and thus its size and cost), or rather, increases the capability of the same equipment.

Of course there *is* a cost: installment of the capacitors. The price of capacitors used for power factor improvement is typically proportional to the $Q$ they must support; the larger the corrective $Q$, the more capacitors must be placed in parallel to handle the reactive power.

**Example 6.5**

A factory has a 10-kW inductive load with power factor 0.75, operating at 60 Hz (see Fig. 6.8a). To reduce the line current magnitude and therefore power bills, the manager decides to try to raise the power factor to 0.95 (see Fig. 6.8b). What value of capacitance should be placed in parallel with the load to accomplish this? The effective line voltage is 220 V.

**Solution** First note that "inductive load" means that $i(t)$ lags $v(t)$, "lagging," so $\theta = \theta_v - \theta_i > 0$. Therefore, $\bar{S} = S\underline{/+\cos^{-1}(PF)}$. ["Leading" would imply $\bar{S} = S\underline{/-\cos^{-1}(PF)}$.] The 0.95 desired PF is assumed also "lagging"; otherwise, "leading" would have been specified. Use the line voltage $V_{\text{eff}}$ as the reference; that is, let its phasor representation be $V_{\text{eff}}\underline{/0^\circ}$. Again let the subscripts 1 and 2 designate, respectively, the situation before and after addition of the parallel capacitance. The average power after modification, $P_2$, must equal the average power before, $P_1$; so call the common value $P$. Then

$$P = S_1 \cos(\theta_1). \tag{6.25}$$

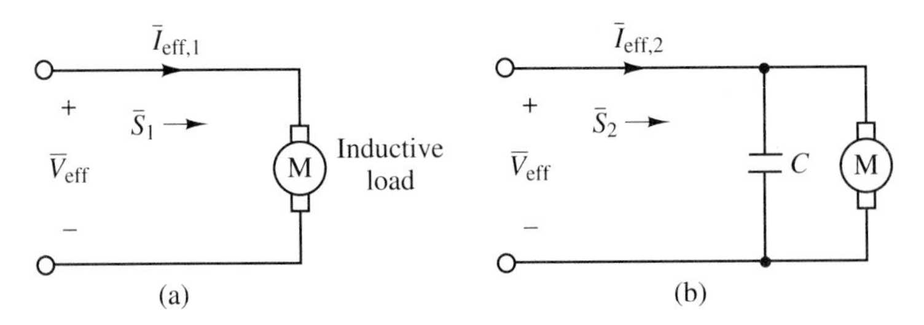


**Figure 6.8** Circuit diagrams for power factor correction: (a) before and (b) after addition of corrective capacitance.

Solving for the apparent power $S_1$ gives

$$S_1 = \frac{P}{\cos(\theta_1)}, \tag{6.26}$$

or, substituting in given values and adhering to the convention for units of apparent power,

$$S_1 = \frac{10{,}000}{0.75} = 13{,}330 \text{ VA}. \tag{6.27}$$

The complex power $\bar{S}_1$ is therefore

$$\begin{aligned}\bar{S}_1 &= S_1\underline{/\theta_1} = 13{,}330 \underline{/\cos^{-1}(0.75)} \\ &= 13{,}330 \text{ VA} \underline{/41.4^\circ} \\ &= 10{,}000 \text{ W} + j8815 \text{ VAR}.\end{aligned} \tag{6.28}$$

To improve the power factor to 0.95 and still have $P_2 = P$, the new complex power will have to be

$$\begin{aligned}\bar{S}_2 &= \frac{10{,}000}{0.95}\underline{/\cos^{-1}(0.95)} \\ &= 10{,}530 \text{ VA} \underline{/18.2^\circ} \\ &= 10{,}000 \text{ W} + j3289 \text{ VAR}.\end{aligned} \tag{6.29}$$

Consequently, the needed additional $Q$ to add to $\bar{S}_1$ is

$$Q_2 - Q_1 = 3289 - 8815 = -5526 \text{ VAR}. \tag{6.30}$$

Recall the fact that because the needed change in $Q$ is negative, capacitance is needed, not inductance.

The required value of capacitance can easily be determined as follows. The complex power into the capacitor is clearly

$$\bar{S}_C = j(Q_2 - Q_1) = -j5526 \text{ VAR}. \tag{6.31}$$

The complex impedance of the capacitor is

$$\bar{Z}_C = \frac{-j}{\omega C} = \frac{-j}{377C}, \tag{6.32}$$

because the frequency is 60 Hz. But by Eq. (6.22),

$$\bar{S} = \frac{V_{eff}^2}{\bar{Z}^*}. \tag{6.33}$$

So in this case

$$-j5526 = \frac{(220)^2}{+j/(377C)}, \tag{6.34}$$

or

$$C = \frac{5526}{377 \cdot (220)^2} = 303\ \mu\text{F}. \tag{6.35}$$

For those who like general formulas, the general solution of this problem is the expression

$$C = \frac{Q_1 - Q_2}{\omega V_{\text{eff}}^2}, \tag{6.36a}$$

where

$$\begin{aligned} Q_{1,2} &= \frac{P}{\text{PF}_{1,2}} \sin[\cos^{-1}(\text{PF}_{1,2})] \\ &= P\left(\frac{1}{\text{PF}_{1,2}^2} - 1\right)^{1/2}, \end{aligned} \tag{6.36b}$$

which is true for lagging or unity power factors before and after correction.

A valid question might be: Why not just increase the power factor to unity to achieve the optimal situation? As mentioned above, the cost of these high-power capacitors is proportional to the reactive power $Q$ they must be able to handle.

First recall that $I_{\text{eff}}$ is the power transmission line current. To observe the effect of diminishing returns in improving PF beyond, say, 0.95, compare the percent change in $I_{\text{eff}}^2 R_{\text{tl}}$ with the change in corrective $Q$ in going from PF = 0.95 to PF = 1.0. At PF = 0.75, the effective current was

$$I_{\text{eff},1} = \frac{S_1}{V_{\text{eff}}} = \frac{13{,}330}{220} = 60.6\ \text{A}. \tag{6.37}$$

At PF = 0.95,

$$I_{\text{eff},2} = \frac{S_2}{V_{\text{eff}}} = \frac{10{,}530}{220} = 47.9\ \text{A}. \tag{6.38}$$

At PF = 1.00, $S_3$ would be equal to $P$, so that

$$I_{\text{eff},3} = \frac{S_3}{V_{\text{eff}}} = \frac{10{,}000}{220} = 45.5\ \text{A}. \tag{6.39}$$

Consider first the effect of adding the 303 μF that raised PF from 0.75 to 0.95.

The percent change in $I_{\text{eff}}$ in the change from PF = 0.75 to 0.95 was $100\% \cdot (47.9 - 60.6)/60.6 = -21\%$. Thus $I_{\text{eff}}^2 R_{\text{tl}}$ was reduced by $-[(47.9^2 - 60.6^2)/60.6^2] \cdot 100\% = [1 - (0.75/0.95)^2]100\% = 38\%$. The percent change in $I_{\text{eff}}$ in the change from PF = 0.95 to 1.00 would be $100\% \cdot (45.5 - 47.9)/47.9 = -5\%$. This means that only a $-[(45.5^2 - 47.9^2)/47.9^2] \cdot 100\% = [1 - (0.95/1)^2] \cdot 100\% = 9.8\%$ further reduction in $I_{\text{eff}}^2 R_{\text{tl}}$. Yet the percent change in capacitive $Q$ needed to change PF from 0.95 to 1.00 would be $100\% \cdot [3289/(3289 + 5526)] = 37\%$. Thus a huge expenditure on additional parallel capacitance (37% increase) would have to be paid to decrease $I_{\text{eff}}^2 R_{\text{tl}}$ by only 10%. Often a PF near 0.95 is found to be a near optimal trade-off. The situation could be improved further if all transmission lines were made of room-temperature superconductors coming from room-temperature fusion power plants!

## 6.5 THE TRANSFORMER

In our discussion of Faraday's law in Sec. 3.10, we found that an emf is induced in a coil when the magnetic flux $\phi_m$ passing through it changes with time. For brevity, in this section we omit the subscript "$m$." This $d\phi/dt$ could come from the magnetic field produced by another source coil, a phenomenon known as *mutual inductance.* This mutual or cross-inductance takes place in a transformer, a multicoil device we introduced in Sec. 3.10.1. The reasons for the name "transformer" will soon be evident. Alternatively, all or part of the $d\phi/dt$ can also be due to the current flowing within the same coil, if it is time varying. This is called *self-induction.* Both stand-alone inductors and windings on transformers experience self-induction.

Transformers are typically operated under sinusoidal excitation, the standard type of excitation for power transmission. High-voltage pulse-generating circuits such as those found in automotive ignition systems and photoflash circuits are significant exceptions. In those examples, sudden current changes (unlike the smooth sinusoidal variations) are arranged to produce the desired high-voltage pulses.

To understand transformer behavior, it is helpful to recall single-frequency (sinusoidal) excitation of an inductor. In Sec. 5.5.3 we saw how an inductor could act as a filter of a general signal. Now, however, we focus on the case in which only one frequency is present. We studied this case before; in Sec. 5.4.4 the phasor diagram of an inductor was given (Fig. 5.20) in which the current lagged behind the voltage by 90°. That discussion concerned single-frequency excitation, of interest here.

We focus our initial analysis on the ideal transformer, for which there is zero resistance in the windings, zero heating losses in the core (negligible hysteresis—a very narrow, linear $B$–$H$ curve), and all of the flux remains within the core. This last assumption implies that the fluxes through the primary and secondary coils are equal. Also, we shall at first assume a finite $\mu$ and then let $\mu$ go to infinity, which leads to a reasonably accurate and very simple model for transformer behavior.

To determine the behavior of a transformer merely involves application of Faraday's law to each coil. A transformer and its circuit symbol are given in Fig. 6.9. This figure shows a construction facilitating understanding of transformer operation. In practice, the coils are usually wound on top of each other to minimize *leakage,* a phenomenon in which a fraction of the flux from one coil does not pass through the other.

Suppose that on the primary (coil 1 in Fig. 6.9) a voltage $v_1(t)$ is applied via a practical sinusoidal voltage source $v_s(t)$, resulting in a current flowing in the primary. With no load connected (switch open), this current, $i_1(t) = i_{1,\mathrm{nl}}(t)$, establishes the flux in the core and is called the *magnetizing current.*

We wish to determine first the secondary voltage $v_2(t)$ when the secondary is open-circuited (no load) as indicated in Fig. 6.9. The switch is open, so $i_2(t) = 0$. Therefore, the *self-induction* voltage across the secondary winding and the *mutual-*

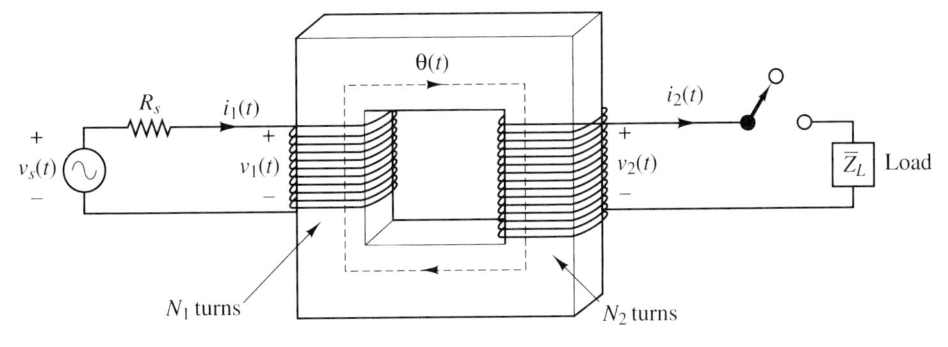


(a)

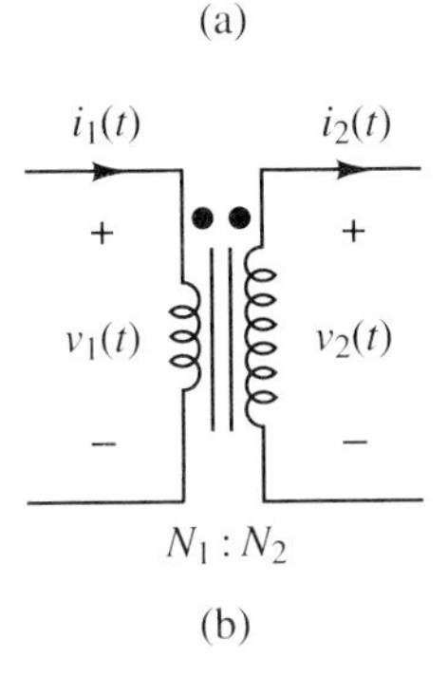


(b)

**Figure 6.9** Transformer: (a) physical device in-circuit; (b) circuit symbol of a transformer.

*induction* voltage across the primary due to the secondary are both zero. By Faraday's law, the self-induction voltage across the primary is $v_1(t) = N_1\, d\phi(t)/dt$, where $N_1$ is the number of turns on the primary. The mutual-induction voltage on the secondary due to the primary flux is $v_2(t) = N_2\, d\phi(t)/dt$, where $N_2$ is the number of turns on the secondary coil.

Use Lenz's law to check the polarity of these voltages in Fig. 6.9, in which at this moment $i_1(t) > 0$ and increasing. Lenz's law will indicate the direction of the current. Recall that because the voltage developed is an emf (a voltage source), positive current flows out of the higher-potential terminal (in this case, the upper terminal of $v_1$). You should find that at this moment, $v_1(t)$ and $v_2(t)$ are both positive.

The dots in the circuit symbol of Fig. 6.9b indicate a pair of terminals whose potentials will be in phase. That is, when the dotted terminal of the primary is higher/lower than the other primary terminal, the same relationship will hold for the dotted secondary terminal. Placement of both dots on either the top or on the bottom agrees with the transformer shown in Fig. 6.9a.

Let the *turns ratio* $a$ of the transformer be defined as

$$a \equiv \frac{N_1}{N_2}. \tag{6.40}$$

Elimination of $d\phi(t)/dt$ from the foregoing Faraday's law equations for $v_1(t)$ and $v_2(t)$ gives

$$v_2(t) = \frac{N_2}{N_1} v_1(t) = \frac{v_1(t)}{a}. \qquad (6.41)$$

Thus if $N_2 > N_1$, the $v_2(t)$ waveform is an amplified version of $v_1(t)$, by the reciprocal of the turns ratio. Similarly, for $N_2 < N_1$, the secondary voltage will be smaller than the primary voltage by the factor $1/a$. Clearly, Eq. (6.41) implies that the phasors are similarly related: $\overline{V}_2 = \overline{V}_1/a$. If the voltage is amplified ($a < 1$), the transformer is called *step-up*; if the voltage is attenuated ($a > 1$), it is called *step-down*. There are also $a = 1$ transformers used for isolation purposes (described below).

If we assume that the voltage source internal resistance $R_s$ is very small (which in practice is a very good approximation in power transmission applications), then $v_1(t) \approx v_s(t)$ whether or not a load is connected across the secondary coil. We shall assume this in the following. Consequently, we need not subscript $v_1(t)$ with "nl" for no load. We will find, however, that $i_1(t)$ changes greatly when the load is connected. Thus let $i_{1,\text{nl}}(t)$ designate the primary current at no load (the magnetizing current) and $\overline{I}_{1,\text{nl}}$ the corresponding current phasor.

Because $\mu$ of the core is very large, so is the primary winding inductance $L_1 = \mu A N_1^2/d$, where $A$ is the core cross section and $d$ is the mean length around the core. By the complex Ohm's law, the current $\overline{I}_{1,\text{nl}} = \overline{V}_1/(j\omega L_1)$ will therefore be small relative to a typical load current, a fact we use below.

Note that the relation above for the current indicates that the current in the primary takes on the value such that the flux it produces causes a self-emf equal to the applied voltage. This approximation ignores both magnetic and electric losses (the eddy current and hysteresis iosses typically dominating over electrical winding losses).

We may obtain an estimate of the flux and magnetic field in the core from $v_1(t) = N_1 d\phi(t)/dt$. Both $v_1(t)$ and $\phi(t)$ are sinusoidal, so they have phasor representations $\overline{V}_1$ and $\overline{\Phi}$, related by $\overline{V}_1 = j\omega N_1 \overline{\Phi}$, where here we assume effective values. If the cross section of the core is $A$, the effective value of the flux is $\Phi = V_1/(\omega N_1)$ and $B = \Phi/A$.

**Example 6.6**

A power distribution transformer (run at 60 Hz) has $N_1 = 1800$, a core 3 in. by 4 in. having a mean path through the core of 2 ft, and is designed to step down 4600 V to 230 V (effective = rms values).

**(a)** Find the number of turns on the secondary and the turns ratio.

**(b)** Find the (rms) flux and magnetic field in the transformer.

**(c)** If the relative permeability of the steel core is 5000, what magnetization (no-load) current is required (rms)? Assume that we operate just below the knee of the magnetization curve, so that the $B$–$H$ curve is linear.

**Solution**

**(a)** From the phasor form of Eq. (6.41), 230 V = $(N_2/1800) \cdot 4600$ V, or $N_2 = 90$ turns, or a turns ratio of $a = 1800/90 = 20$.

**(b)** From the discussion above, $\Phi = 4600/(377 \cdot 1800) = 6.8$ mWb. The effective magnetic field is thus 6.8 mWb/(3 in. · 0.0254 m/in. · 4 in. · 0.0254 m/in.) = 0.88 T. We see that the linear $B$–$H$ assumption is approximately valid.

(c) From Eq. (3.10), $(B/\mu)d = N_1 I_{1,\text{nl}}$, or $[0.88/(4\pi \cdot 10^{-7} \cdot 5000)] \cdot 2 \cdot 12 \cdot 0.0254 = 1800 I_1$, or $I_{1,\text{nl}} = 47$ mA. For the typical multiampere currents flowing in such a power transformer, this magnetizing current is negligible.

Suppose now that a load having impedance $\bar{Z}_L$ is connected across the terminals of the secondary coil. The voltage induced on the secondary due to the primary flux, $v_2(t)$, is an emf (a voltage source) for the load. By the complex Ohm's law, the load current $\bar{I}_2 = \bar{V}_2/\bar{Z}_L$ differs in phase from $\bar{V}_2$ by the load impedance angle $\theta_Z$.

This sinusoidal current $i_2(t)$ produces a flux in the core that tends to oppose the original $d\phi/dt$, by Lenz's law. With $d\phi/dt$ reduced, the total emf of the primary, $v_1(t)$, is momentarily reduced. (The flux through the primary coil now has both a positive self-induced and a negative cross-induced "mutual" component.) Therefore, by KVL the voltage across $R_s$ will be increased.

By Ohm's law for $R_s$, the current in the primary coil will instantly be increased by an amount designated $\bar{I}_{1,L}$, resulting in an increased flux sufficient to restore equilibrium. After all, as $\bar{I}_1$ increases, so will $\bar{V}_1 = j\omega L_1 \bar{I}_1$, where $L_1$ is the primary inductance. If we assume that $R_s$ is very small compared with $j\omega L_1$, we must in equilibrium have $\bar{V}_1 \approx \bar{V}_s$; this condition determines the additional primary current $\bar{I}_{1,L}$. All of these events actually happen simultaneously, so $\bar{I}_{1,L}$ is in phase with $\bar{I}_2$.

All this while, the magnetizing current $\bar{I}_{1,\text{nl}}$ must be supplied as a part of the KVL balance. Thus the total primary current with load is $\bar{I}_1 = \bar{I}_{1,\text{nl}} + \bar{I}_{1,L}$. The complete phasor diagram is shown in Fig. 6.10.

By Eqs. (3.10) and (3.21) (which in Chapter 13 we call "$K\mathcal{F}L$"), the total reluctance of the core times the flux is equal to the sum of the mmfs, $N_1 i_1(t) - N_2 i_2(t)$, where the minus sign reflects the fact that $i_1$ and $i_2$ go opposite ways around the dashed path in Fig. 6.10a. If we now take $\mu$ to be infinity, the core reluctance is zero and therefore the sum of mmfs is zero. This does *not* imply that the flux in the core is zero or even small, but rather is equivalent to saying that $i_{1,\text{nl}}(t) \approx 0$. Having $N_1 i_1(t) - N_2 i_2(t) = 0$ results in $i_1(t)$ and $i_2(t)$ being in phase and proportional:

$$i_2(t) = \frac{N_1}{N_2} i_1(t) = a i_1(t), \tag{6.42}$$

which in the frequency domain reads $\bar{I}_2 = a\bar{I}_1$. The important thing to note is that if $N_2 > N_1$, $i_2(t)$ is an *attenuated* version of $i_1(t)$ (opposite the rule for voltage). For $N_2 < N_1$, the reverse is true. We see that whenever voltage is amplified, current is attenuated, and vice versa.

This fact has a major impact on the input–output power relation of an ideal transformer. The complex power delivered to the load is, by Eqs. (6.41) and (6.42),

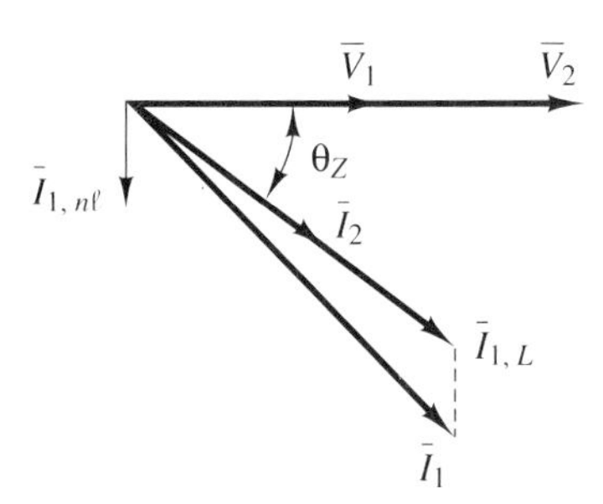


**Figure 6.10** Phasor diagram for a transformer with load.

$$\bar{S}_L = \bar{V}_2 \bar{I}_2^*$$
$$= \left(\frac{N_2}{N_1}\bar{V}_1\right)\left(\frac{N_1}{N_2}\bar{I}_1^*\right) \tag{6.43}$$
$$= \bar{V}_1 \bar{I}_1^*,$$

which is precisely the power supplied at the input of the transformer! Naturally, this is only approximately true for a real transformer; the output power is always slightly less than the input power (typically, by only 2% or less) in real transformers, due to core and winding losses. However, the essential point is that the transformer provides zero power gain even though it does provide voltage gain or current gain (but not both simultaneously). We shall return to this fact in Chapter 7.

By now, the reason this device is called a transformer should be evident. It transforms (via the turns ratio $a$) the amplitude of voltages and currents from its input to its output while leaving the shape of the input waveform unchanged in the output. There is another transformation performed by a transformer: impedance. If the load impedance is $\bar{Z}_L$, the input impedance of the transformer–load combination is

$$\bar{Z}_{\text{in}} = \frac{\bar{V}_1}{\bar{I}_1}$$
$$= \left(\frac{N_1}{N_2}\right)^2 \frac{\bar{V}_2}{\bar{I}_2} \tag{6.44}$$
$$= \left(\frac{N_1}{N_2}\right)^2 \bar{Z}_L = a^2 \bar{Z}_L.$$

Thus we have the remarkable result that an output load $\bar{Z}_L$ looks like a load of impedance $a^2\bar{Z}_L$ to the input. This is one practical use for transformers: converting a load from low to high impedance, or vice versa. For a step-up transformer ($a < 1$), the load impedance is reduced, and vice versa.

Also, recall the fact that the secondary voltage and current are proportional to the input voltage and current for an ideal transformer. That is, there is no distorting phase shift at any frequency. Thus it may be used for multifrequency signals such as appear in audio circuits.

**Example 6.7**

Recall Example 4.24, where we tried to match a 3-Ω speaker to a 600-Ω output impedance amplifier. The results were terrible—only 2% of the load power that could be obtained if in fact the load were impedance matched to the source. Find the turns ratio required to match the load to the source. Draw the diagrams for circuits with the speaker directly connected to the amplifier and with the impedance transformer to achieve maximum power transfer. For simplicity, we assume here that the Thévenin equivalent of the amplifier is a sinusoidal voltage source and has effective voltage $V_{\text{eq}}$ equal to 5 V. As in Example 4.24, we assume that the equivalent source resistance is 600 Ω and the speaker is a pure resistance of 3 Ω. Assume an ideal transformer.

**Solution** The systems without and with the impedance transformer are shown, respectively, in Fig. 6.11a and b. Setting $Z_{\text{in}} = 600\ \Omega$ and $Z_L = 3\ \Omega$, we obtain from Eq. (6.44) $a = N_1/N_2 \approx 14$. This is a voltage step-down transformer that will result in an average load power of $V_{\text{eq}}^2/(4Z_{\text{eq}}) = 5^2/(4 \cdot 600) = 10$ mW instead of $V_{\text{eq}}^2 Z_L/(Z_L + Z_{\text{eq}})^2 =$

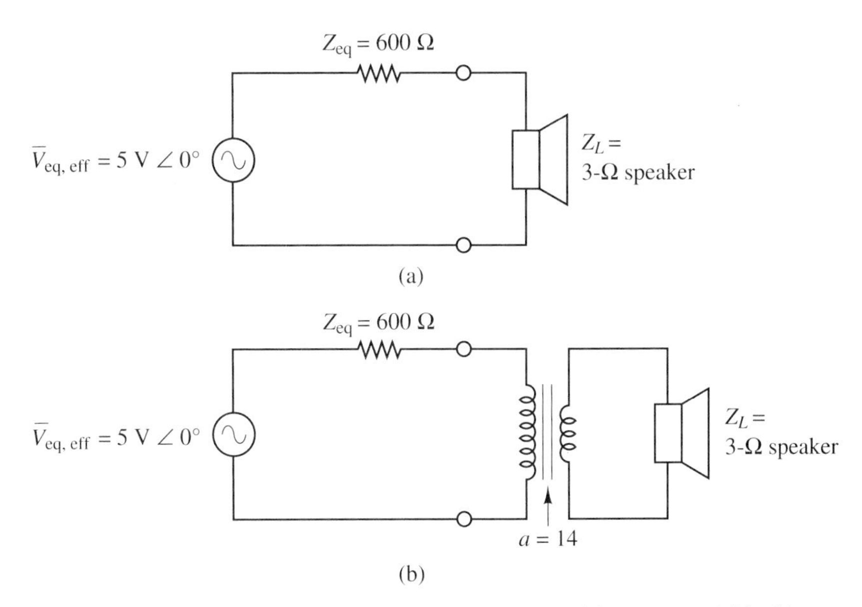


**Figure 6.11** Speaker connections to the audio amplifier (a) without and (b) with impedance-matching transformer.

$5^2 \cdot 3/(3 + 600)^2 = 0.21$ mW for the 3-Ω speaker alone. As noted in Example 4.24, without the impedance transformer we obtain only $0.21/10 \approx 2\%$ of the power available to a matched load connected to the same source. Before the days of low-impedance output amplifier stages, this technique was common in audio circuits.

An interesting but involved design question is how to choose $N_1$ ($N_2$ then being fixed at $N_2 = N_1/a$). A fundamental consideration is material characteristics. A given wire insulation and thickness can safely carry a certain maximum current (causing severe heating, which can break down winding insulation), specified by the given load operating at required voltage. A larger flux will generate the required voltage for fewer turns; hence it is desirable to operate just below the knee of the $B$–$H$ curve. There is a trade-off between core size and number of turns. The result of analysis is that for a given voltage, a higher-power transformer will have a larger core and fewer, thicker windings carrying heavier current; a lower-power transformer will have a smaller core and more thin wire carrying less current.

Practical transformers violate to various extents all of the simplifying assumptions of the ideal transformer. However, modern transformers are quite efficient, so our analysis fairly well approximates the actual behavior. Furthermore, the ideal transformer may be used as part of a practical transformer model. For example, to account for resistive losses, series resistances can be added to a model involving an ideal transformer, as we shall see below. Usually, the excitation current can be neglected. More significant are the leakage and associated effects of finite $\mu$.

To account for both leakage and winding losses, the theory of mutual inductance can be applied. Because the magnetic paths for fluxes produced by either coil and passing through both coils are identical (both around the core), the reluctance $\mathcal{R}_1 = d_1/(\mu_1 A_1) = \mathcal{R}_2 = \mathcal{R} = d/(\mu A)$, where $d = d_1 = d_2$ is the common mean path length of the core, $A = A_1 = A_2$ is the uniform cross-sectional area of the core, and $\mu = \mu_1 = \mu_2$ is the uniform permeability of the core. The flux due to $i_1(t)$ is $\phi_1(t) = N_1 i_1(t)/\mathcal{R}$, and because self-inductance is $L = N^2/\mathcal{R}$, $\phi_1(t) = L_1 i_1(t)/N_1$; similarly, $\phi_2(t) = L_2 i_2(t)/N_2$. Keep in mind below that for $i_1 > 0$ and $i_2 > 0$, $\phi_1$ is clockwise while $\phi_2$ is counterclockwise.

Assuming that a fraction $K < 1$ of the flux $\phi_2(t)$ reaches the primary (typically, $K$ is at least 0.99 for iron-core transformers), then KVL with Faraday's law reads

$$\begin{aligned} v_1(t) &= N_1 \frac{d}{dt}\{\phi_1(t) - K\phi_2(t)\} + R_1 i_1(t) \\ &= N_1 \frac{d}{dt}\left\{\frac{L_1 i_1(t)}{N_1} - \frac{KL_2 i_2(t)}{N_2}\right\} + R_1 i_1(t) \\ &= L_1 \frac{di_1(t)}{dt} - M\frac{di_2(t)}{dt} + R_1 i_1(t), \end{aligned} \tag{6.45}$$

where $M = K(N_1/N_2)L_2 = K(L_1L_2)^{1/2}$ because $N_1/N_2 = (L_1/L_2)^{1/2}$, and where $R_1$ is the resistance of the primary winding. For no leakage ($K = 1$), $M = (L_1L_2)^{1/2}$.

Similarly, because of the symmetry of the reception/transmission of flux processes (reciprocity), the same fraction $K$ of $\phi_1$ reaches the secondary, so that by the same reasoning

$$v_2(t) = -L_2 \frac{di_2(t)}{dt} + M\frac{di_1(t)}{dt} - R_2 i_2(t), \tag{6.46}$$

where $R_2$ is the resistance of the secondary winding. With the exception of inclusion of winding resistance, this is the approach taken in elementary transformer presentations. However, it does not provide a convenient, simple circuit model with parameters that are easily measured.

A more profitable approach is as follows. Instead of working with the fluxes $\phi_1$ and $\phi_2$ produced by the two windings and the leakage factor $K$, decompose the flux through each coil into that clockwise flux common to both coils ($\phi$) and the individual leakage components $\phi_{l1}$ for the primary and $\phi_{l2}$ for the secondary which do not reach the other coil. Thus the flux through coil 1 is $\phi_1 - K\phi_2 = \phi + \phi_{l1}$ and that through coil 2 is $\phi_2 - K\phi_1 = -\phi + \phi_{l2}$. Thus Eq. (6.45) is now written as

$$\begin{aligned} v_1(t) &= N_1 \frac{d}{dt}\{\phi(t) + \phi_{l1}(t)\} + R_1 i_1(t) \\ &= N_1 \frac{d\phi(t)}{dt} + L_{l1}\frac{di_1(t)}{dt} + R_1 i_1(t) \end{aligned} \tag{6.47}$$

and Eq. (6.46) becomes

$$v_2(t) = N_2 \frac{d\phi(t)}{dt} - L_{l2}\frac{di_2(t)}{dt} - R_2 i_2(t), \tag{6.48}$$

where $L_{l1}$ and $L_{l2}$ are defined as the leakage inductances of the primary and secondary windings. [*Note:* Opposite the situation for the primary, the current $i_2(t)$ is defined as flowing *out* of the assumed positive voltage terminal. Thus $-di_2(t)/dt$ makes a positive contribution to the secondary voltage.]

Because $\phi$ is common to primary and secondary, it is the flux equivalently presented to an ideal transformer. Hence $e_1 = N_1\, d\phi(t)/dt$ and $e_2 = N_2 d\phi(t)/dt$ are,

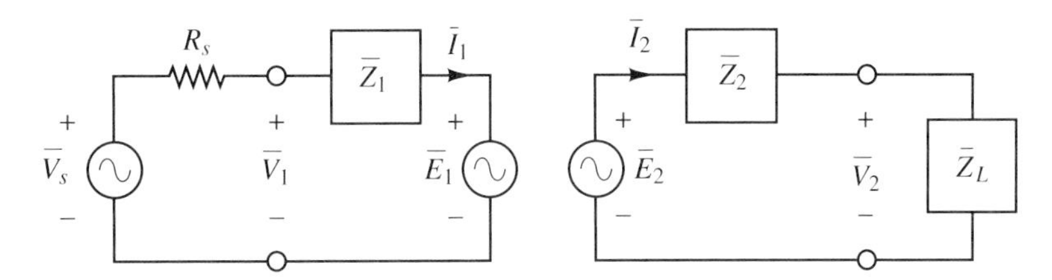


**Figure 6.12** Simplified equivalent circuit of a practical transformer.

respectively, the net primary and secondary voltages on the ideal transformer portion of this model. Writing Eqs. (6.47) and (6.48) in the frequency domain,

$$\overline{V}_1 = \overline{E}_1 + (R_1 + jX_{l1})\overline{I}_1 \tag{6.49a}$$

$$\overline{V}_2 = \overline{E}_2 - (R_2 + jX_{l2})\overline{I}_2, \tag{6.49b}$$

where $X_{l1} = \omega L_{l1}$ and $X_{l2} = \omega L_{l2}$ are the primary and secondary leakage reactances. The circuit diagram model, including a load $\overline{Z}_L$ on the secondary, is now as shown in Fig. 6.12, where $\overline{Z}_1 = R_1 + jX_{l1}$ and $\overline{Z}_2 = R_2 + jX_{l2}$.

The important combinations of parameter values can now be determined as follows, resulting in a yet simpler model sufficient for most application/design purposes. Short circuit the secondary and measure the values of the voltage $V$, current $I$, and average power $P$ at the primary. (The average power is measured by a wattmeter, discussed in Sec. 6.6.3.) Then $R_{\text{tr}} = P/I^2$ and $X_{\text{tr}} = [(V/I)^2 - R_{\text{tr}}^2]^{1/2}$ and $\overline{Z}_{\text{tr}} = R_{\text{tr}} + jX_{\text{tr}}$. Now the circuit model, amended with a load $\overline{Z}_L$ and *viewed from the primary* (*hence* $\overline{Z}_L$ *is multiplied by* $a^2$), is as shown in Fig. 6.13. A similar equivalent circuit may be derived from the point of view of the secondary. When viewed from one side, all that remains of the transformer is $\overline{Z}_{\text{tr}}$ and the factor $a^2$ multiplying $\overline{Z}_L$.

In terms of the circuit model of Fig. 6.12, these equivalent values have the following meanings. With the secondary short-circuited, $V_2$ in Eq. (6.49b) is zero. Again let $a = N_1/N_2$, so that $\overline{E}_2 = \overline{E}_1/a$ and thus from Eq. (6.49b), $\overline{E}_1 = a(R_2 + jX_{l2})\overline{I}_2$. As in the ideal model, we assume that to a good approximation (merely neglecting the excitation current, $\overline{I}_{1,\text{nl}}$), $\overline{I}_2 = (N_1/N_2)\overline{I}_1 = a\overline{I}_1$, so that $\overline{E}_1 = a^2(R_2 + jX_{l2})\overline{I}_1$. When substituted into Eq. (6.49a), we obtain

$$\overline{V}_1 = [R_1 + a^2R_2 + j(X_{l1} + a^2 X_{l2})]\overline{I}_1, \tag{6.50}$$

so that

$$R_{\text{tr}} = R_1 + a^2R_2 \tag{6.51a}$$

$$X_{\text{tr}} = X_{l1} + a^2X_{l2}. \tag{6.51b}$$

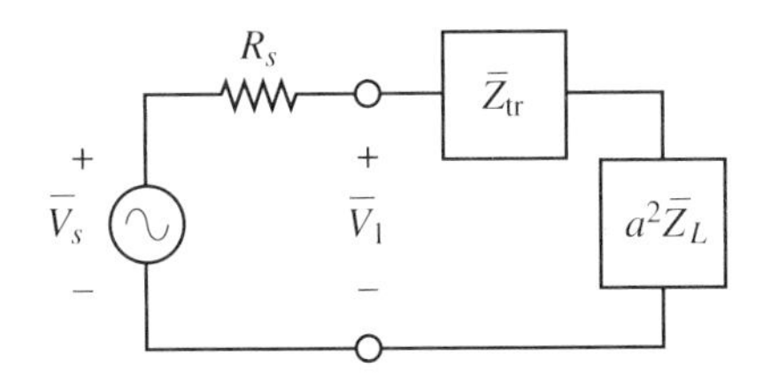


**Figure 6.13** Simplest equivalent circuit of a transformer and load, as seen from the primary side (from the voltage source).

From the simple and workable model that has resulted in Fig. 6.13, we see how much more useful the leakage reactance model is than the mutual inductance model.

**Example 6.8**

With the secondary of a 7.92 kV:240 V transformer short-circuited, we measure $V_{1,sc} = 600$ V, $I_{1,sc} = 3$ A, and average power $P_{sc} = 300$ W. With a finite load $\overline{Z}_L = 1\ \Omega + j0.5\ \Omega$ connected, (a) find the net impedance seen by the source, (b) determine the power factor seen by the source, and (c) compare these values with those we would obtain were we to neglect leakage.

**Solution** First note that $a = 7920/240 = 33$.

(**a**) $R_{tr} = P/I^2 = 300/3^2 = 33\ \Omega$, $X_{tr} = [(V/I)^2 - R_{tr}^2]^{1/2} = [(600/3)^2 - 33^2]^{1/2} = 197\ \Omega$, and $\overline{Z}_{tr} = R_{tr} + jX_{tr} = 33\ \Omega + j197\ \Omega$. Thus from Fig. 6.14 we see that with the load connected, the total impedance is $\overline{Z}_{eq} = \overline{Z}_{tr} + a^2\overline{Z}_L = 33 + 33^2 + j(197 + 33^2 \cdot 0.5) = 1122 + j742 = 1345\ \Omega\underline{/33.5^\circ}$.

(**b**) PF $= \cos(33.5^\circ) = 0.83$.

(**c**) Now $\overline{Z}_{eq} = a^2\overline{Z}_L = 1089 + j545 = 1218\ \Omega\underline{/26.6^\circ}$ and $PF = \cos(26.6^\circ) = 0.89$ lagging. The magnitude differs from that in part (a) by 9.5%, and the power factor is off by 6% from that in part (b). Thus the leakage in this case is significant but not devastating.

Let us finally consider a few common applications of transformers. The most familiar use for transformers is in power distribution. Step-up transformers ($N_2 \gg N_1$) boost voltages to maximize transmission efficiency and back down to safe levels ($N_2 \ll N_1$) at the user's residence, as discussed in Sec. 6.7.

Another use is for circuit isolation. Notice that only the *difference* in potential between the secondary terminals is specified. The absolute potential floats, which makes for a safer circuit. That way, if you happen to touch a line wire to a wire in the load circuit, nothing will happen. The absolute potential of the load will merely ride on the line wire potential instead of causing a disastrous short circuit.

The application of the transformer as an impedance modifier was already mentioned. A typical example occurs when the output impedance of an audio amplifier cannot be matched to the speaker it drives. To achieve maximum power transfer to the speaker requires impedance matching, as noted in Sec. 4.11.1. This can easily be accomplished for the unmatched amplifier–speaker pair by placing a suitable transformer between the amplifier output and the speaker.

Finally, the *autotransformer* is a variation on the transformer that is efficient for small voltage step-up or step-down or can provide an inexpensive means for realizing an adjustable ac voltage power supply. The autotransformer, as the name suggests, generates its output voltage in the same physical coil as the one to which the input is applied. We no longer speak of a primary and secondary, but rather of a common winding $N_c$ and a series winding $N_s$ (see Fig. 6.14a). Because the output voltage $\overline{V}_2$ is developed across $N_c + N_s$, its voltage will be larger than the input voltage $\overline{V}_1$ by the turns ratio $(N_c + N_s)/N_c = 1/a'$; the current is reduced by the factor $a'$.

The autotransformer is efficient in electric power systems for overcoming small line voltage drops. The advantage is that the apparent input–output power (assuming a lossless transformer) is greater than that actually in the windings by the factor

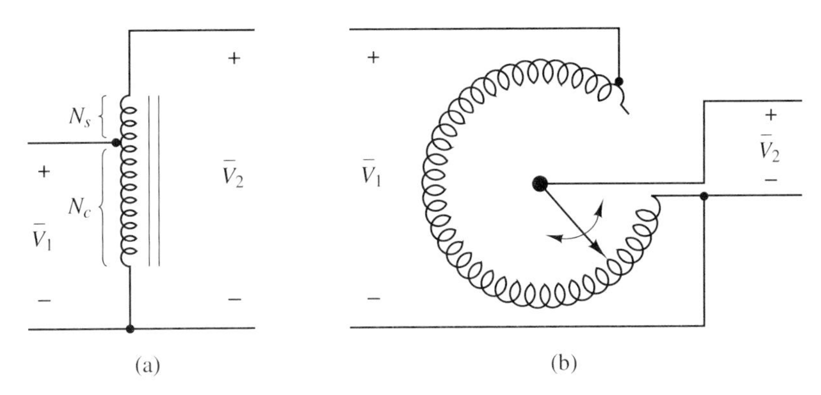


**Figure 6.14** Autotransformer: (a) fixed turns ratio; (b) adjustable turns ratio (variac).

$(N_s + N_c)/N_s$. This is shown later in a problem. Therefore, for the same apparent power, the transformer maximum power rating can be reduced; the transformer can be made to handle only $N_s/(N_s + N_c)$ times the actual input–output apparent power.

If the common coil windings are exposed to allow an output metal wiper to pass over them, the output voltage can be adjusted (see Fig. 6.14b). This arrangement is a very convenient and inexpensive way of obtaining a variable ac voltage source from the wall voltage, with high current capability. The only price to be paid by having the single coil is a loss of isolation between the output and line voltages. It is known as a *variac,* a contraction of "variable ac."

## 6.6 THREE-PHASE CIRCUIT ANALYSIS

### 6.6.1 Introduction

When we consider the large-scale distribution of power, efficiency is clearly of paramount importance. Discovered long ago was the fact that if three voltage sources of the same amplitude but 120° apart in phase are used instead of a single ac source, transmission efficiency is improved. Additionally, a main consumer of this transmitted power is electric motors. Three-phase power-motor systems also have higher efficiency than single-phase systems. Even construction of three-phase systems may be cheaper than that of equivalent single-phase systems, as building three lower-power systems may be cheaper than one system that must handle greater power.

Another advantage is that the total instantaneous power into a balanced three-phase load is constant. This means that the torque developed in a three-phase motor is approximately constant in time, which means smooth operation. Finally, the ability of three-phase systems to establish a rotating magnetic field electrically is of prime importance for ac machines (see Chapter 15).

This section is devoted to a presentation of some simple mathematics of three-phase circuits and phasors. The constant power into a balanced three-phase load is derived. Then various three-phase circuit connections are considered, along with the line-phase relations for current and voltage. A device for measuring the average power into a load, called a wattmeter, is discussed next. Finally, relations between "wye" and "delta" loads are provided.

### 6.6.2 Basic Configurations

First consider three sinusoidal currents or voltages of equal magnitude and frequency $\omega$ but differing from each other in phase by $120° = 2\pi/3$. Such voltages or currents are called *balanced* voltages or currents. It will now be shown that their sum is zero. The sum of three such functions is

$$\begin{aligned} &A\left[\cos(\omega t) + \cos\left(\omega t - \frac{2\pi}{3}\right) + \cos\left(\omega t - \frac{4\pi}{3}\right)\right] \\ &= A\left[\cos(\omega t) + \cos(\omega t)\cos\left(\frac{2\pi}{3}\right) + \sin(\omega t)\sin\left(\frac{2\pi}{3}\right)\right. \\ &\quad \left. + \cos(\omega t)\cos\left(\frac{4\pi}{3}\right) + \sin(\omega t)\sin\left(\frac{4\pi}{3}\right)\right] \\ &= A\left[\left(1 - \frac{1}{2} - \frac{1}{2}\right)\cos(\omega t) + \left(\frac{\sqrt{3}}{2} - \frac{\sqrt{3}}{2}\right)\sin(\omega t)\right] = 0. \end{aligned} \tag{6.52}$$

We can also visualize this graphically by considering the three sinusoids of equal magnitude and frequency $\omega$ but differing from each other in phase by 120° as shown in Fig. 6.15. At any instant (such as $t_1$ or $t_2$), mentally adding up the waveforms does indeed result in zero, no matter what value of $t$ we choose.

To calculate the total instantaneous power in the three phases, assume that in each phase the voltage leads the current by $\theta$. This would occur, for example, if each voltage were applied to identical impedances $\overline{Z} = Z\angle\theta$; the set of three impedances is in this case called a *balanced load.* The individual instantaneous phase powers are

$$p_A(t) = v_A(t)i_A(t) = VI\cos(\omega t)\cos(\omega t - \theta) \tag{6.53a}$$

$$p_B(t) = v_B(t)i_B(t) = VI\cos\left(\omega t - \frac{2\pi}{3}\right)\cos\left(\omega t - \frac{2\pi}{3} - \theta\right) \tag{6.53b}$$

$$p_C(t) = v_C(t)i_C(t) = VI\cos\left(\omega t - \frac{4\pi}{3}\right)\cos\left(\omega t - \frac{4\pi}{3} - \theta\right). \tag{6.53c}$$

In the notation of Eq. (6.5), define $\theta_v$ and $\theta_i$ as follows:

| Phase | $\theta_v$ | $\theta_i$ |
|---|---|---|
| $A$ | $0$ | $-\theta$ |
| $B$ | $-2\pi/3$ | $-2\pi/3 - \theta$ |
| $C$ | $-4\pi/3$ | $-4\pi/3 - \theta$ |

Now the resulting instantaneous power for each phase can immediately be written from Eq. (6.10). The total power will be the sum of the individual phase powers (by conservation of energy). Thus

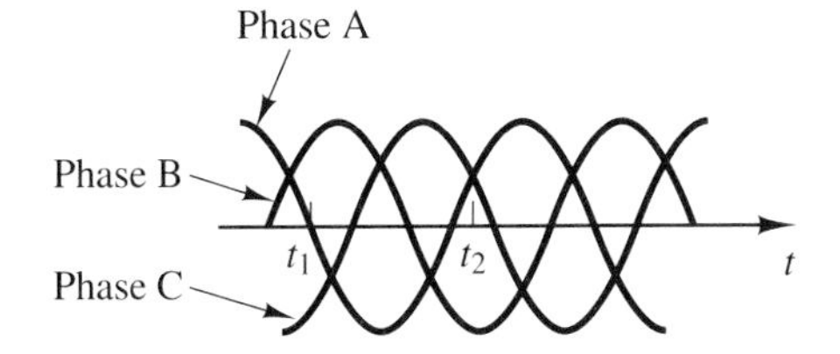


**Figure 6.15** Balanced sinusoidal functions, that is, all having the same magnitude and frequency, but differing in phase by 120°.

$$
\begin{aligned}
P_{\text{tot}}(t) &= p_A(t) + p_B(t) + p_C(t) \\
&= (\text{VI}/2)\Big[\cos(2\omega t - \theta) + \cos(\theta) \\
&\quad + \cos\left(2\omega t - \frac{4\pi}{3} - \theta\right) + \cos(\theta) \\
&\quad + \cos\left(2\omega t - \frac{8\pi}{3} - \theta\right) + \cos(\theta)\Big].
\end{aligned}
\tag{6.54}
$$

Recalling that $8\pi/3$ radians = 480° modulo 360° is 120°, it is clear that the three time-dependent terms are balanced functions (equal magnitude and frequency, and 120° apart in phase). Therefore, they sum to exactly zero. All that is left in Eq. (6.54) are the three identical $\cos(\theta)$ terms. The total instantaneous power is therefore

$$
\begin{aligned}
p_{\text{tot}}(t) &= 3 \cdot \frac{VI}{2}\cos(\theta) \\
&= 3V_{\text{phase}} I_{\text{phase}} \cos(\theta),
\end{aligned}
\tag{6.55}
$$

where $V_{\text{phase}}$ and $I_{\text{phase}}$ are the effective phase voltage and current. That is, the instantaneous total power $p_{\text{tot}}(t)$ is, for all $t$, just three times the *average* power in a single phase [see Eq. (6.11)]!

The total instantaneous power has thus been shown to be a *constant* in time. Thus the average power is equal to the instantaneous power. For balanced voltages, currents, and motor winding impedances, this result means that any three-phase motor running at constant speed experiences a total instantaneous torque that is constant in time. Such a torque is far preferable to a pulsating torque and is an asset of three-phase ac motors. It also means that three-phase generators under constant torque from the driving turbines will produce constant electric power.

Therefore, compared with single-phase, which requires a double-frequency pulsating driving torque, three-phase is again more desirable. The double-frequency opposing torque creates shaft vibration and consequent losses and eventual destruction of the shaft. Three-phase is a smooth operation.

Recall that the relation between the voltage and current phasors of the three inductive-load phases is that shown in Fig. 6.16. Those phasor relations exactly describe the situation where the voltages are placed across a balanced load some distance away (see Fig. 6.17). The load could be either an actual machine or, more commonly, a three-phase transformer that in turn services single- and three-phase equipment at more moderate voltages (the intervening step-down transformer is here omitted for brevity).

In Fig. 6.17 there is no common reference. None of the voltage–current

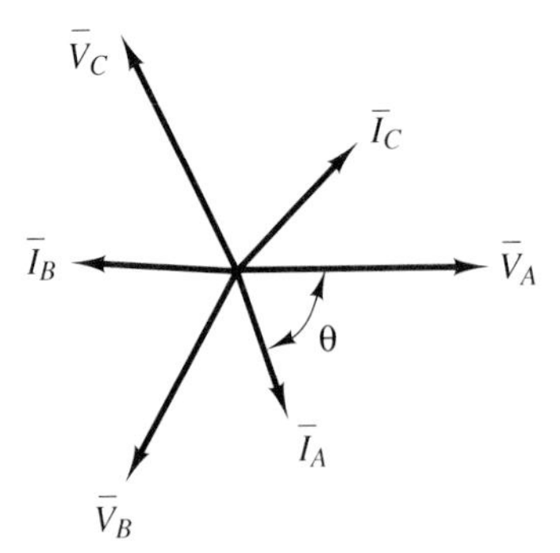


**Figure 6.16** Phasor diagram for a three-phase balanced load.

relations there would be in any way affected if the three negative wires were fused together to make a common reference. Figure 6.18 would result. Both the source (left end) and the load (right end) are now connected in what is called a Y or *wye connection* (owing to its shape). By KCL, the sum of the current phasors $\overline{I}_A$, $\overline{I}_B$, and $\overline{I}_C$ flows into the common minus wire. But these three currents by assumption are of equal magnitude and are 120° apart, so the total current is zero. Thus for balanced loads, no current flows in this fourth "return" or "neutral" wire, and it can be eliminated (see Fig. 6.19).

(In practice, the fourth wire is retained because of imperfect balance, but it can be a much thinner wire because its current is small and ideally zero. Remember that this single thin wire is to be compared with Fig. 6.17, where three thick wires were required. This is one of the big advantages of three-phase systems.)

In Fig. 6.19 the phases and lines are labeled where we now again assume perfect balance. Quantities associated with individual phases are called phase quantities, and those associated with lines, line quantities. We follow here the same convention used in Chapter 2 to define potentials: $V_{ab} = V_b - V_a$. Therefore, the phase *voltages* are $\overline{V}_A$, $\overline{V}_B$, and $\overline{V}_C$, while the *line voltages* are $\overline{V}_{BA}$, $\overline{V}_{CB}$, and $\overline{V}_{AC}$, as shown. Notice that the line currents in this circuit are equal to the phase currents. How are the line and phase voltages related? Consider $\overline{V}_{AB}$, as vectorially constructed in Fig. 6.20. From Fig. 6.19 it is seen that $\overline{V}_{BA} = \overline{V}_A - \overline{V}_B$, where $|\overline{V}_A| = |\overline{V}_B| = V_P$. By the law of cosines,

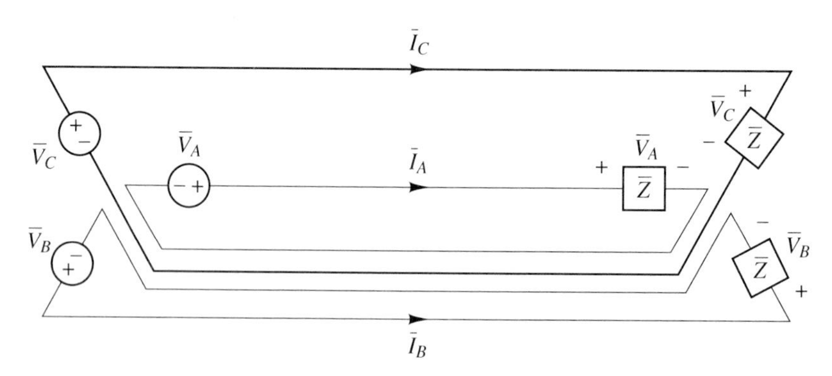


**Figure 6.17** Three-phase transmission line with balanced sources and loads.

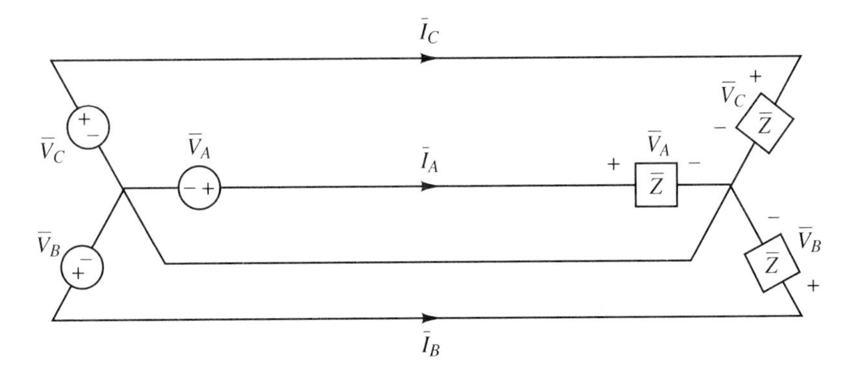


**Figure 6.18** The three negative wires of Fig. 6.17 are fused together into one to reduce wiring requirements.

$$|\bar{V}_{BA}|^2 = V_P^2 + V_P^2 - 2V_P^2 \cos(120°)$$
$$= 3V_P^2, \tag{6.56}$$

so

$$|\bar{V}_{BA}| = \sqrt{3}\,V_P, \tag{6.57}$$

or

$$V_{\text{line}} = \sqrt{3}\,V_{\text{phase}} \tag{6.58}$$

for a wye connection.

Now Eq. (6.55) for the total power can be rewritten in terms of line quantities as

$$p_{\text{tot}}(t) = 3\,V_{\text{phase}}\,I_{\text{phase}}\cos(\theta)$$
$$= 3\cdot\frac{V_{\text{line}}}{\sqrt{3}}\,I_{\text{line}}\cos(\theta) \tag{6.59}$$
$$= \sqrt{3}\,V_{\text{line}}\,I_{\text{line}}\cos(\theta).$$

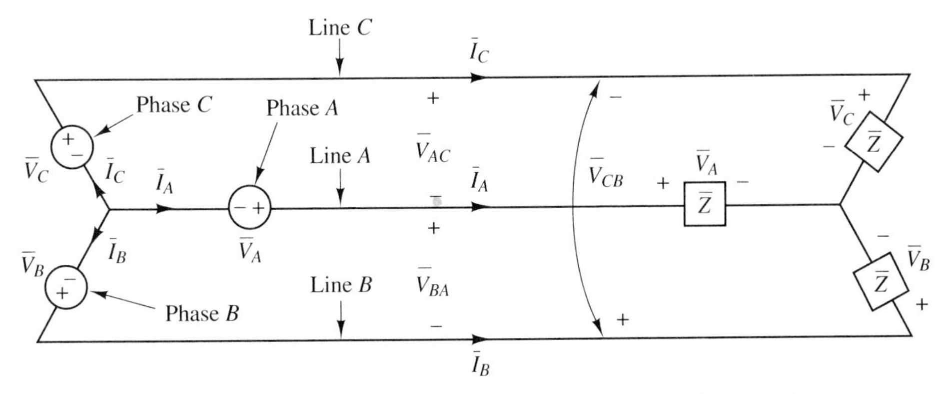


**Figure 6.19** For a balanced system, the current in the fourth wire is zero, so it may be eliminated.

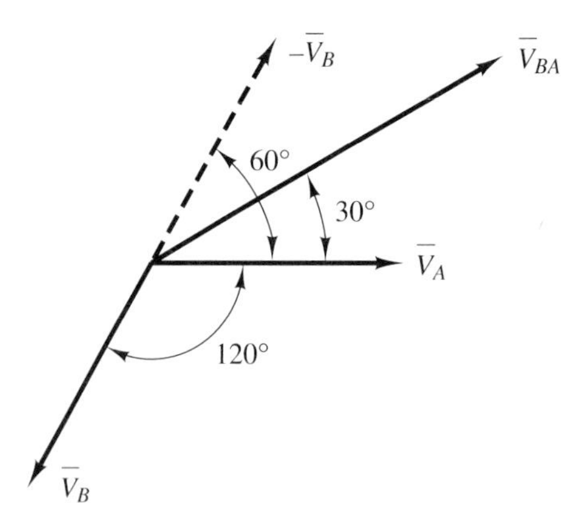


**Figure 6.20** Relation between two phase voltages and the associated line voltage.

The phasor diagram for all three phases of line and phase voltages follows from Fig. 6.20 and is shown in Fig. 6.21.

An alternative to the wye connection is the Δ or *delta connection* (also named for its shape). Instead of a common "minus" terminal, the three phases are connected in a triangle, the corners of which are the terminals to the outside world. A combination of delta-connected source and wye-connected load is shown in Fig. 6.22; all combinations of wye and delta are possible.

The choice between wye and delta in practice is based on consideration of advanced design issues. However, wye generators are more common than delta generators, mainly because of the impossibility of a fourth neutral wire in a delta system. In imperfectly balanced loads, undesirable circulation currents occur. In practice, loads in power distribution systems are unbalanced, so the fourth wire in a wye-connected system is used. They are also used when it is desired to operate single-phase loads (and thus imbalanced) or protect against faults in a balanced system (again an imbalance). Thus Fig. 6.18 represents a common power transmission system.

In the circuit of Fig. 6.22, at the source end (delta connection) the phase voltages are the *same* as the line voltages, while the phase and line currents now differ. By KCL,

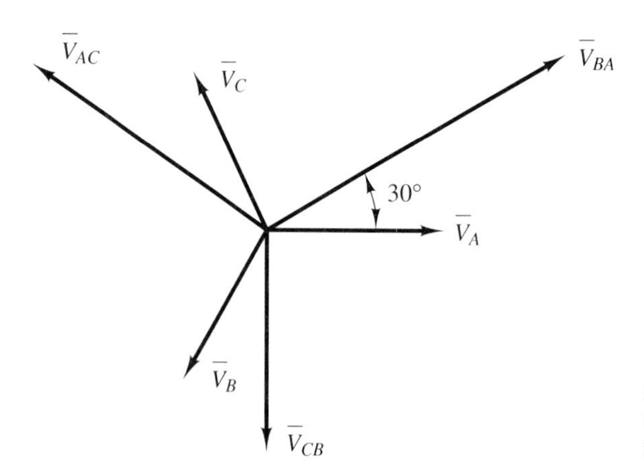


**Figure 6.21** Phasor diagram for all three phases of line and phase voltages for a wye connection.

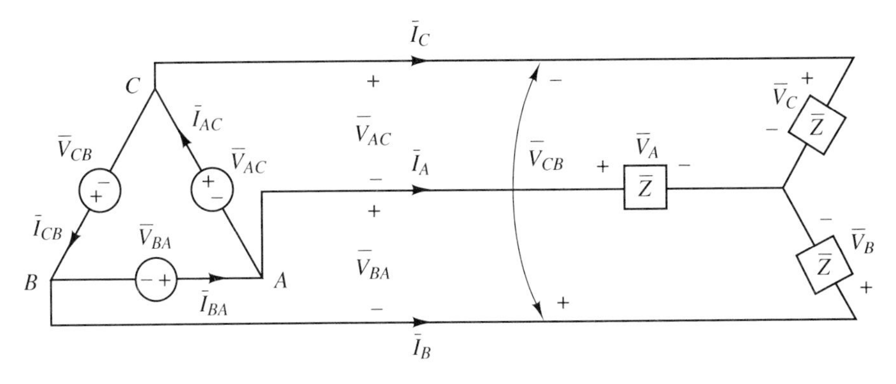


**Figure 6.22** Delta–wye three-phase power transmission system.

$$\bar{I}_A = \bar{I}_{BA} - \bar{I}_{AC} \tag{6.60a}$$

$$\bar{I}_B = \bar{I}_{CB} - \bar{I}_{BA} \tag{6.60b}$$

$$\bar{I}_C = \bar{I}_{AC} - \bar{I}_{CB}, \tag{6.60c}$$

where on the left side are line currents and on the right side are phase currents. The phasor diagram for all three phases of line and phase currents for the delta-connected source is derived analogously to that for the wye-connected source and is shown in Fig. 6.23. The magnitude relations for the currents are also derived analogously:

$$|\bar{I}_{BA}| = |\bar{I}_{AC}| = I_{\text{phase}} \tag{6.61a}$$

$$\begin{aligned} |\bar{I}_A|^2 &= I^2_{\text{phase}} + I^2_{\text{phase}} - 2I^2_{\text{phase}}\cos(120°) \\ &= 3I^2_{\text{phase}}, \end{aligned} \tag{6.61b}$$

so that

$$I_{\text{line}} = \sqrt{3}\, I_{\text{phase}} \tag{6.62}$$

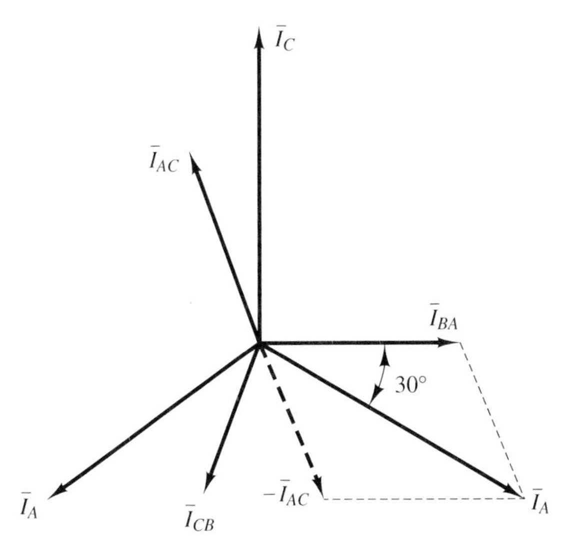


**Figure 6.23** Phasor diagram for all three phases of line and phase current for a delta connection.

for a delta connection. Again the total power is

$$\begin{aligned} p_{\text{tot}}(t) &= 3\, V_{\text{phase}} I_{\text{phase}} \cos(\theta) \\ &= 3\, V_{\text{line}} \frac{I_{\text{line}}}{\sqrt{3}} \cos(\theta) \\ &= \sqrt{3}\, V_{\text{line}} I_{\text{line}} \cos(\theta). \end{aligned} \tag{6.63}$$

**Example 6.9**

Consider a three-phase system with balanced loads $\overline{Z} = 100 + j8\ \Omega$. If the loads are (a) wye- and (b) delta-connected, what is the total power into the loads in (a) and (b) if the effective line voltage is 400 V? What are the line currents in (a) and (b)? What is the power factor?

**Solution** The expression for the total power will be the same for both wye and delta connections, but the line currents will differ. For wye-connected loads (a), we have by Eq. (6.58) $V_p = V_{\text{line}}/\sqrt{3} = 400/\sqrt{3} = 231$ V. The phase current magnitude is $I_p = V_p/Z = 231/(100^2 + 8^2)^{1/2} = 2.3$ A; therefore, for the wye connection $I_{\text{line}} = I_p = 2.3$ A. For delta-connected loads (b), $V_p = V_{\text{line}} = 400$ V, so that the phase current is $I_p = V_p/Z = 400/(100^2 + 8^2)^{1/2} \approx 4$ A. By Eq. (6.62), for delta-connected loads $I_{\text{line}} = \sqrt{3}\, I_p = 6.9$ A. The phase angle is in both cases $\theta = \tan^{-1}(8/100) \approx 4.6°$. By Eqs. (6.59) and (6.63), $p_{\text{tot}}(t) = \sqrt{3}\, V_{\text{line}} I_{\text{line}} \cos(\theta)$ for either case, which for wye (a) comes out to 1.59 kW and for delta (b) is 4.77 kW (a factor of 3 larger than the power into the wye load). The power factor is $\cos(4.6°) = 0.997$ lagging.

### 6.6.3 Wattmeters

For the cases of either balanced or imbalanced loads, the average power (which for balanced loads equals the instantaneous power) can be measured by using two wattmeters. A *wattmeter* is a two-coil measuring device that on one coil detects voltage and on another coil detects current. The output of the device is the average value of the voltage times the current, that is, the average power. A schematic representation is shown in Fig. 6.24. As shown in that diagram, the average power into impedance $\overline{Z}$ is measured.

The ± markings on two of the terminals are polarity markings; these terminals should be connected on the side away from the load. For balanced loads, in theory the total power into the load could be obtained by tripling the value of a single wattmeter connected to one phase impedance. The problem is that the terminals of the impedances are usually not both available. See Fig. 6.25 for examples of this

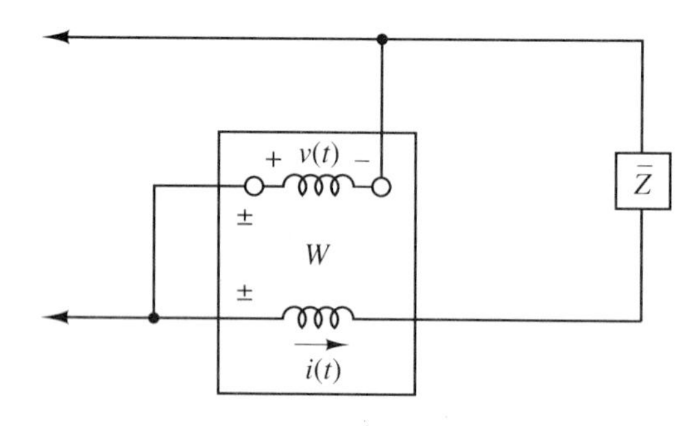


**Figure 6.24** Wattmeter.

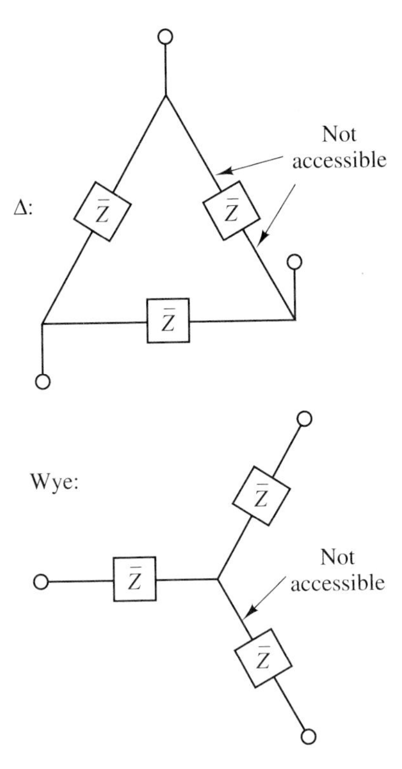


**Figure 6.25** Difficulty of obtaining power into one phase of a three-phase load for delta- and wye-connected loads.

problem for delta- and wye-connected loads. So instead, it will now be shown that two wattmeters connected (for a wye-connected load) as shown in Fig. 6.26 are sufficient to obtain a measurement of total power to the three-phase load. The sum of the two readings is the total average power whether or not the load is balanced.

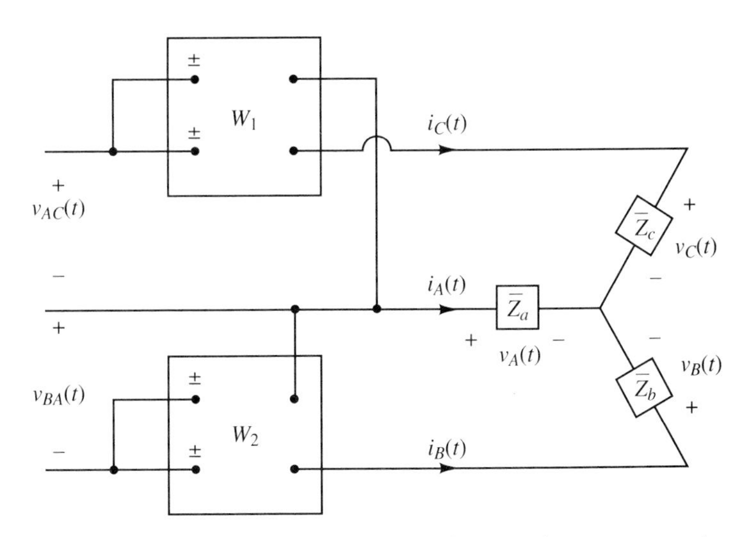


**Figure 6.26** Two-wattmeter connection sufficient for measuring average power into balanced or unbalanced wye-connected loads.

For convenience of notation, designate the average value of a periodic function with angular brackets:

$$\langle p(t)\rangle = \frac{1}{T}\int_0^T p(t)\,dt. \tag{6.64}$$

Using this notation, clearly the desired total average power is

$$\langle p(t)\rangle = \langle v_A(t)\,i_A(t)\rangle + \langle v_B(t)\,i_B(t)\rangle + \langle v_C(t)\,i_C(t)\rangle. \tag{6.65}$$

The reading of wattmeter 1 is

$$\begin{aligned} W_1 &= \langle v_{AC}(t)\,i_C(t)\rangle = \langle v_C(t)\,i_C(t)\rangle - \langle v_A(t)\,i_C(t)\rangle \\ &= \langle v_C(t)\,i_C(t)\rangle + \langle v_A(t)\,i_A(t)\rangle + \langle v_A(t)\,i_B(t)\rangle, \end{aligned} \tag{6.66}$$

where KCL was used $[i_C(t) = -i_A(t) - i_B(t)]$ and the reading of wattmeter 2 is

$$\begin{aligned} W_2 &= -\langle v_{BA}(t)\,i_B(t)\rangle \\ &= -\langle v_A(t)\,i_B(t)\rangle + \langle v_B(t)\,i_B(t)\rangle, \end{aligned} \tag{6.67}$$

so that the sum of the readings is

$$\begin{aligned} W_1 + W_2 &= \langle v_A(t)\,i_A(t)\rangle + \langle v_B(t)\,i_B(t)\rangle + \langle v_C(t)\,i_C(t)\rangle \\ &= \langle p(t)\rangle. \end{aligned} \tag{6.68}$$

(*Note*: The wattmeter has only a positive scale. When the power factor of a balanced load is below $\frac{1}{2}$, one of the readings would be negative and pin the meter the wrong way. In such cases, simply reverse the terminals of that wattmeter and substract its reading from rather than add it to the other reading.)

The correct configuration for a delta-connected load is shown in Fig. 6.27. Similar analysis to that above would again show that $W_1 + W_2 = \langle p(t)\rangle$.

### 6.6.4 Wye–Delta and Delta–Wye Transformations

**Introduction.** At this point it might be asked whether a correspondence can be made between wye and delta connections; that is, as far as the outside world is concerned, is there a delta connection that has the same terminal impedance as a wye-connected load (albeit with different impedances), and vice versa? The answer is yes. And this is a very important set of results, for in practice it is often computationally most convenient to have all three-phase impedances in terms of one kind, particularly wye. In the case of balanced loads, this allows *per phase analysis* to be carried out easily and correctly.

Let the delta phase impedances be called $\overline{Z}_{ba}$, $\overline{Z}_{cb}$, and $\overline{Z}_{ac}$, while the wye phase impedances are $\overline{Z}_a$, $\overline{Z}_b$, and $\overline{Z}_c$ (see Fig. 6.28). To have equivalent networks as far as the outside world is concerned, the relations between the phase impedances of the delta and wye networks must have values such that

$$\overline{Z}_{BA}^{\Delta} = \overline{Z}_{BA}^{Y} \tag{6.69a}$$

$$\overline{Z}_{CB}^{\Delta} = \overline{Z}_{CB}^{Y} \tag{6.69b}$$

$$\overline{Z}_{AC}^{\Delta} = \overline{Z}_{AC}^{Y}. \tag{6.69c}$$

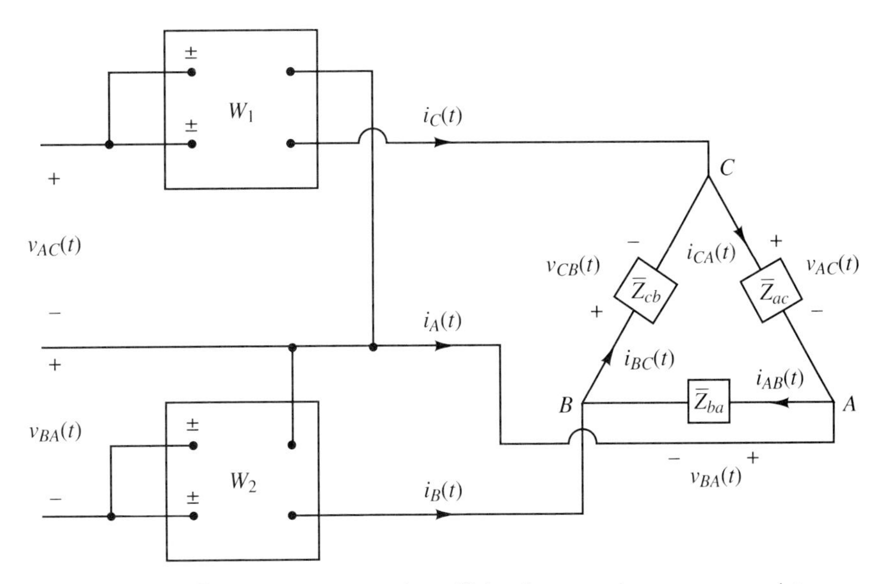


**Figure 6.27** Two-wattmeter connection sufficient for measuring average power into balanced or unbalanced delta-connected loads.

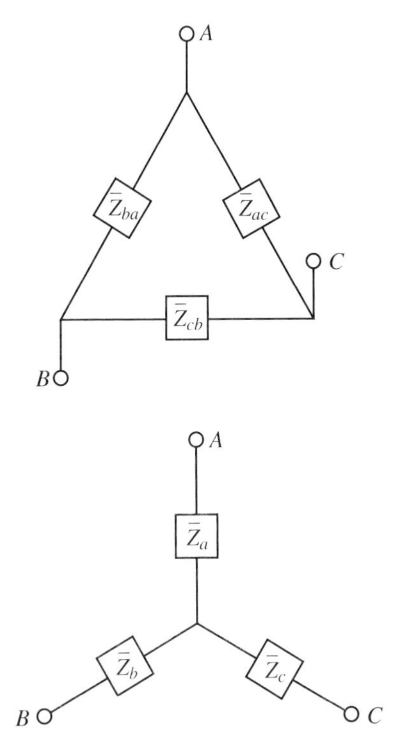


**Figure 6.28** Delta- and wye-connected loads; the relation between them is determined in this section.

**Delta–wye transformation.** Let $\overline{Z}_\Delta = \overline{Z}_{ba} + \overline{Z}_{cb} + \overline{Z}_{ac}$. The three requirements above then translate to

$$\overline{Z}_{ba}||(\overline{Z}_{ac} + \overline{Z}_{cb}) = \overline{Z}_a + \overline{Z}_b \rightarrow \frac{\overline{Z}_{ba}(\overline{Z}_{ac} + \overline{Z}_{cb})}{\overline{Z}_\Delta} = \overline{Z}_a + \overline{Z}_b \tag{6.70a}$$

$$\overline{Z}_a + \overline{Z}_c = \frac{\overline{Z}_{ac}(\overline{Z}_{ba} + \overline{Z}_{cb})}{\overline{Z}_\Delta} \tag{6.70b}$$

$$\overline{Z}_b + \overline{Z}_c = \frac{\overline{Z}_{cb}(\overline{Z}_{ba} + \overline{Z}_{ac})}{\overline{Z}_\Delta}. \tag{6.70c}$$

Performing the operations [all referring to Eqs. (6.70)] (a) + (b) − (c), (a) + (c) − (b), and (b) + (c) − (a) give, respectively,

$$\overline{Z}_a = \frac{\overline{Z}_{ba}\overline{Z}_{ac}}{\overline{Z}_\Delta} \tag{6.71a}$$

$$\overline{Z}_b = \frac{\overline{Z}_{ba}\overline{Z}_{cb}}{\overline{Z}_\Delta} \tag{6.71b}$$

$$\overline{Z}_c = \frac{\overline{Z}_{ac}\overline{Z}_{cb}}{\overline{Z}_\Delta}. \tag{6.71c}$$

So if the phase impedances of a delta connection are known, Eqs. (6.71) give the phase impedances of the wye connection having the same external impedances.

**Wye–delta transformation.** Alternatively, if the phase impedances of a wye connection are known, the phase impedances of an equivalent delta connection may easily be found as follows. Form the quantity

$$\overline{Z}_Y^2 = \overline{Z}_a\overline{Z}_b + \overline{Z}_a\overline{Z}_c + \overline{Z}_b\overline{Z}_c. \tag{6.72}$$

Substituting Eqs. (6.71) into Eq. (6.72) gives

$$\begin{aligned}\overline{Z}_Y^2 &= \frac{\overline{Z}_{ba}^2\overline{Z}_{ac}\overline{Z}_{cb} + \overline{Z}_{ac}^2\overline{Z}_{cb}\overline{Z}_{ba} + \overline{Z}_{ba}\overline{Z}_{cb}^2\overline{Z}_{ac}}{\overline{Z}_\Delta^2} \\ &= \frac{\overline{Z}_{ba}\overline{Z}_{cb}\overline{Z}_{ac}(\overline{Z}_{ba} + \overline{Z}_{ac} + \overline{Z}_{cb})}{\overline{Z}_\Delta^2},\end{aligned} \tag{6.73}$$

or

$$\overline{Z}_Y^2 = \frac{\overline{Z}_{ba}\overline{Z}_{cb}\overline{Z}_{ac}}{\overline{Z}_\Delta}. \tag{6.74}$$

Performing the operations $\overline{Z}_Y^2/\overline{Z}_c$, $\overline{Z}_Y^2/\overline{Z}_b$, and $\overline{Z}_Y^2/\overline{Z}_a$ [using $\overline{Z}_a$, $\overline{Z}_b$, and $\overline{Z}_c$ as given in Eqs. (6.71)] yield, respectively,

$$\overline{Z}_{ba} = \frac{\overline{Z}_Y^2}{\overline{Z}_c} \tag{6.75a}$$

$$\overline{Z}_{ac} = \frac{\overline{Z}_Y^2}{\overline{Z}_b} \tag{6.75b}$$

$$\overline{Z}_{cb} = \frac{\overline{Z}_Y^2}{\overline{Z}_a}. \tag{6.75c}$$

Of special importance is the special case of balanced loads, where $\overline{Z}_{ba} = \overline{Z}_{cb} = \overline{Z}_{ac} = \overline{Z}_{\text{delta}}$ and $\overline{Z}_a = \overline{Z}_b = \overline{Z}_c = \overline{Z}_{\text{wye}}$, for which all of the relations [Eqs. (6.71) and (6.75)] reduce to

$$\overline{Z}_{\text{delta}} = 3\,\overline{Z}_{\text{wye}}. \tag{6.76}$$

All of these relations are useful in three-phase motor and generator and power distribution circuit analyses.

**Example 6.10**

Repeat Example 6.9 using Eq. (6.76).

**Solution** We can just convert the delta impedance to a wye and use the fact that for wye loads, $p_{\text{tot}} = \sqrt{3}\, V_{\text{line}} I_{\text{line}} \cos(\theta) = \sqrt{3}\, V_{\text{line}} [(V_{\text{line}}/\sqrt{3})/Z] \cos(\theta) = V_{\text{line}}^2 \cos(\theta)/Z$. The equivalent $Z_{\text{wye}}$ is just equal to $Z_{\text{delta}}/3$, so that $p_{\text{tot, delta}} = 3p_{\text{tot, wye}}$. From Example 6.9 we found that $p_{\text{tot,wye}} = 1.59$ kW, which gives $p_{\text{tot,delta}} = 3 \cdot 1.59 \text{ kW} = 4.77$ kW. Thus the factor of 3 mentioned in Example 6.9 is again obtained using this approach. Notice that a lot of work was saved using Eq. (6.76).

Let us now review a few of the advantages of three-phase over single-phase ac systems. One was that the instantaneous three-phase power to a balanced load is constant (and equal to $\sqrt{3}$ times the average power of a single phase in terms of line quantities, or 3 times a single-phase power in terms of phase quantities). Not only is this fact computationally convenient, but also it means that in steady state, the developed torque in a three-phase motor is constant.

Physically, when the power in one phase (oscillating at frequency $2\omega$) is at its minimum, the other two phases are at sufficiently high levels to keep the total power exactly constant. And it will be shown when we discuss ac machines in Chapter 15 that a constant-magnitude rotating magnetic field may be set up in a motor by means of three-phase electromagnets. Furthermore, the ubiquitous three-phase induction motor has a starting torque without the special modifications required for a single-phase induction motor (see Chapter 15).

Finally, lower line currents are required for electrical power distribution, which means lower $I^2R$ losses and therefore greater efficiency. Or, for a given efficiency, less copper needs to be used in transmission lines (to obtain sufficiently low resistance).

## 6.7 POWER DISTRIBUTION SYSTEMS

To gain an understanding of the overall structure of the electric energy network requires familiarity with its major components. These include generators, step-up/step-down transformers, power transmission lines, protective circuitry, and typical loads. The theory and construction of generators are covered in Chapter 15. Motors, which are the typical loads dominating the electric power demand, are also covered in the chapters on electromechanical machines (Chapters 13 through 15). Transformers were described in Sec. 6.5. In this section some attention is given to power transmission issues and devices. Typical system structures are also examined.

Like water, food, and other basic necessities, it has generally been found economical to produce electric power centrally in very large quantities. The con-

Edison's 1881 "Jumbo dynamo" on Manhattan Island was by today's standards the genuine "Puny Muni." This first commercial municipal electric power plant was on Pearl Street, and could power up to 1000 lamps at once.

sumer receives electric power not via underground pipes (as for water) or truck or train (as for food) but by wires called transmission lines and by transformers.

In the early 1880s, Thomas Edison opened the first electric power station in New York City, which supplied 30 kW at 110 V dc. Some of today's nuclear power plants can supply 1300 MW—40,000 times as much power as that first station. Consumption has doubled every 10 years since 1900, and ac rather than dc is generated. As noted in Sec. 6.6, today three-phase power transmission systems are used universally. For these, three-phase generators are connected (see Sec. 15.5) as well as three-phase transformers.

Recall that the power dissipated in a resistance R is $I^2R$. If $V$ and $I$ are the voltage across and current delivered to a load via a transmission line having resistance $R_{tr}$, the $I^2R_{tr}$ loss in the line is minimized by minimizing $I$ for a given power $VI$. But note that for a given power $VI$, if $V$ is increased, $I$ will decrease, so the larger $V$ is, the better.

Thus there would be zero transmission loss if $V$ could be made infinite, for then the current would be zero. Naturally, this is impossible even to come close to, for a variety of reasons. But this consideration is why transmission voltages are designed to be so high (typically, hundreds of thousands of volts). Although the resistance of thick transmission cable is quite small (typically, 0.1 Ω per kilometer), the $I^2R_{tr}$ loss is the major expense of power transmission, given that equipment is already in place.

(It may be asked: But isn't the power dissipated in the transmission line also expressible as $V^2/R_{tr}$? The answer is that not $V$ (the line-to-line voltage) but rather $V_{tr}$ (the voltage across $R_{tr}$, a *single* transmission wire from end to end) must be used: $V_{tr}^2/R_{tr}$. The line current $I$ is far easier to measure than $V_{tr}$; hence we always speak of $I^2R_{tr}$.)

Contrary to this requirement that transmission voltages be extremely high is the requirement that user voltages be small (in the hundreds of volts). The latter necessity is due to obvious safety problems of extremely high voltage, as well as impracticality of applying such high voltages to practical loads (electronics, machines, lighting, etc.). For ac power distribution systems, the transformer thus serves a critical purpose: to step up generated voltages for transmission and step down transmitted voltages for consumer usage.

The transformers required to step down voltages to user levels are expensive, so it is not economical to provide every user with such a transformer. More important, it is imperative to avoid having everyone on the same circuit. In the case of a fault (short circuit or other current surge situation), it is desired that as few consumers are adversely affected as possible. Consequently, in the center of the immediate locality of a large number of consumers, a substation is constructed to bring down the voltage to an intermediate level and isolate the cluster of users from other consumers.

Another goal realized at a substation is minimization of the duration of adverse effects of a fault. With relay switches under computer control, appropriate disconnections or connections can often be made to bring power back up to users on a given branch in the event of a fault. Such switching also prevents further destruction of transmission equipment in these situations. It should be noted that the vast majority of faults are momentary; otherwise, equipment usually needs the attention of repairmen.

For these purposes, the substation contains not only power transformers, but also circuit breakers, disconnecting switches, buses and interfaces, coupling capacitors, protective relays, lightning arresters, and so on. The series capacitors, incidentally, can minimize light flicker for users on the same branch as a heavy and fluctuating load such as a factory with motors turning on and off. (Such flicker is noticeable even if the voltage variation is as small as 0.5% if the fluctuations are a few hundred hertz.)

Many of these components are also found in the smaller, cylindrical distribution transformers commonly seen up on power line poles. A typical power distribution system schematic with substations is shown in Fig. 6.29. (A complete and readable discussion of power system components can be found in *The Standard Handbook for Electrical Engineers,* by D. G. Fink and H. W. Beaty [McGraw-Hill, New York, 1987], Sec. 10.)

An important criterion in power transmission is good voltage regulation. If the voltage is greater than about 130 V, light bulbs will burn out faster, and electronic equipment may fail. On the other hand, if the line voltage is less than about 108 V, the brightness of lights is reduced, motors may fail to start or even may burn out, and heating devices will not function. This does not give much room for error, considering that this entire range may be spanned over the extent of the transmission system. Voltages closest to the generators will receive higher voltages, and power line voltage drops will reduce the voltage continuously from there on out.

An interesting question that began with Edison and still is asked today is whether electric power should be transmitted in the form of ac or dc. Edison favored dc, but ac has become almost universal. The reasons are mainly the existence of transformers, which greatly facilitate the voltage step-up/down process described

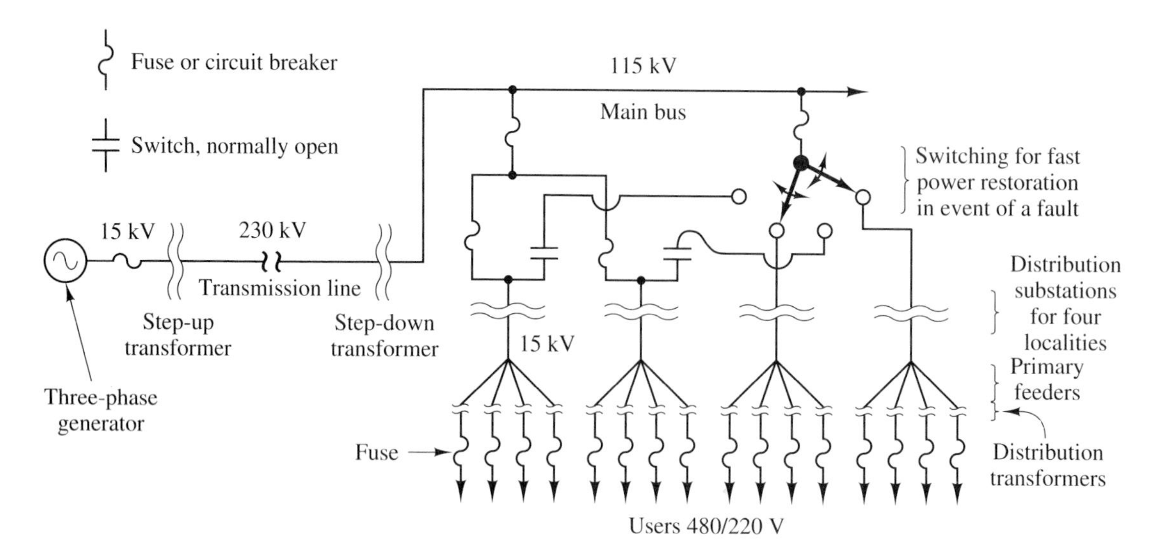


**Figure 6.29** Typical power distribution system.

above, and the relative cheapness and high performance of ac machines (both generators and motors).

However, dc has its advantages: The cost per unit length of a dc transmission line is much cheaper than a corresponding power ac line, and voltage regulation is better because there is zero reactance at zero frequency. It is also a far simpler matter to hook together several generators, because there is no phase matching requirement or requirement that both generators run in phase and at exactly the same frequency.

At sufficient distances (currently over about 400 miles), the cost of ac/dc and dc/ac conversion is outweighed by the savings in transmission line costs. These distances are more common elsewhere in the world, where major power generation facilities may be very far apart. Incidentally, the choice of 60 Hz for ac is somewhat arbitary; in Europe it is 50 Hz. But it is historically tied to the optimized design of machines that run at 1200 rpm = 20 rps = 20 Hz and 3600 rpm = 60 Hz (the speeds of two- and six-pole 60-Hz ac machines; see Chapter 15).

Transmission lines can be classified into three length categories: short (less than 50 miles), medium (50 to 100 miles), and long (over 100 miles). Short lines can be modeled as just a resistance and an inductance in series. Medium lines require in the model a capacitance between lines for accurate representation of circuit behavior. (If the capacitance and frequency are small, $1/(\omega C)$ is huge, so that the capacitance between lines is like an open circuit. There is a fixed capacitance/length, so as the length increases, $C$ increases and makes the parallel $1/(\omega C)$ impedance significant.) Long lines entail a "distributed parameter" model: many medium-length transmission line models in series; each parameter of these portions of the model is a "per length" quantity.

Another reason for these classifications is that the wavelength of an electromagnetic wave at 60 Hz is a little over 3000 miles. So on scales of 100 miles or less, only a tiny fraction of a wavelength exists over the transmission line, so one lumped model suffices. But when there is sufficient variation over the length of the line, a separate circuit model for each constituent "short line" must be used to be able to include that variation in the model.

A phenomenon unique to high-voltage circuits used in power transmission (and sometimes occurring in the high-voltage section of old TV sets) is corona discharge. When exposed to a large electric field, free electrons in the air move under its force. In so doing, collisions with electrons in the molecules of air occur, dislodging those electrons. These new free electrons bombard others still, and soon the air is ionized. The visual effect is a glow, and the aural effect a hissing.

In power distribution systems, energy involved in this process comes from the power being transmitted on the line, so must be viewed as a loss. Large-diameter wires and bundling (several wires wrapped around each other on one cable) minimize the effect. Weather conditions affect the susceptibility of the air to ionization greatly: Losses in fair weather may be 1 kW/mile and in bad weather 20 kW/mile. The latter value would actually dominate over $I^2R$ losses.

A related problem is that of engineering good switching and fuse protection. While for low voltages, switching is straightforward and safe, the possibility for arcing is very real at kilovolt levels. Contributing to the problem is the self-emf (voltage) that develops when we try to interrupt the flow of a large current. By elongating the spark using magnetic fields and/or air blasting, immersing the spark area in oil or vacuum, and other techniques, engineers attempt to minimize both the intensity and duration of arcs. The interruption of circuits may be achieved either by *fuses,* which melt when the current exceeds a specified level for a specified time, or *relays,* which open and close under the action of control circuitry.

The calculations involved in design of major power distribution systems are generally extremely involved and complicated. Thousands of components may be involved, and the theory of fault analysis and stability is quite advanced. Most of the analysis is just tedious applications of complex impedance, phasor diagrams, and other circuit techniques developed previously. Of course, control electronics also complicate analysis.

There is one analysis technique unique to this area of electrical engineering: the *per unit system.* As it is widely used today, any engineer anticipating discussions with power distribution experts must have at least some familiarity with the basic ideas of the per unit system.

In principle, it is a simple idea: Scale all quantities in calculations by "base values" so that those quantities are representable by numbers near unity. Then if some quantity is found to be far different from 1, that is an indicator that something is wrong—either in the calculation or in the actual system. The method was designed in the days before computers in an effort to simplify calculations and because of the advantage just described. And even though computers have made irrelevant some of its advantages, the system persists. The biggest simplification is essentially the removal of transformers from calculations: The per unit voltages on either side of a transformer are the same!

In ac analysis, there are four basic quantities appearing repeatedly in calculation: $\bar{V}$, $\bar{I}$, $\bar{S}$, and $\bar{Z}$. In the per unit system, a variable is represented by the actual value divided by a *base value.* Base values (always real numbers) are chosen for any two of the above four variables. By defining

$$S_b = V_b I_b, \tag{6.77}$$

where the subscript "$b$" denotes "base," the relation $\overline{S} = \overline{V}\overline{I}^*$ translates to

$$\overline{S}_{pu} = \overline{V}_{pu}\overline{I}^*_{pu}, \tag{6.78}$$

where the subscript "pu" denotes "per unit" and where $\overline{V}_{pu} = \overline{V}/V_b$, $\overline{I}_{pu} = \overline{I}/I_b$, $\overline{S}_{pu} = \overline{S}/S_b$. Note that the per unit variables are also complex (phasors) and have the same phase angle as the actual counterpart phasors. Only the magnitude has been scaled, by the base value. Therefore, all per unit variables are dimensionless. Similarly, by defining

$$Z_b = \frac{V_b}{I_b}, \tag{6.79}$$

it follows that

$$\overline{Z}_{pu} = \frac{\overline{V}_{pu}}{\overline{I}_{pu}}. \tag{6.80}$$

Now if any two base values are given numerical values, Eqs. (6.77) and (6.79) fix the other two.

Recall that for a transformer $\overline{V}_1 = (N_1/N_2)\overline{V}_2$. Suppose that we fix $V_{1,b} = V_1$, so that $\overline{V}_{1,pu} = \overline{V}_1/V_{1,b}$ has magnitude 1. By requiring that $V_{1,b} = (N_1/N_2)V_{2,b,}$ we have

$$\overline{V}_{2,pu} = \frac{\overline{V}_2}{V_{2,b}} = \frac{N_2}{N_1}\frac{\overline{V}_1}{V_{2,b}} = \overline{V}_{1,pu}.$$

That is, the per unit voltage on either side of the transformer is the same; the distinction "1" versus "2" is not significant in the "per unit domain."

Of course, once the per unit voltages are determined (by normal circuit analysis applied in the per unit domain), multiplying by the respective base voltages to retrieve the actual voltage phasors will restore the $N_1/N_2$ relation between the magnitudes of $\overline{V}_1$ and $\overline{V}_2$. This characteristic of transformer per unit variables generalizes to any number of coils on a transformer core: All per unit voltages are the same.

For currents, we know that $\overline{I}_1 = (N_2/N_1)\overline{I}_2$. By choosing the relation between base voltages $I_{1,b} = (N_2/N_1)I_{2,b}$, we find that $\overline{I}_{1,pu} = \overline{I}_{2,pu}$. Recall that the relation between primary and secondary currents was derived by noting that for an ideal transformer the total mmf around the magnetic circuit is zero. This principle dictates that the sum of per unit current phasors, each scaled by turns ratios $N_i/N_1$, must be zero. Therefore, the generalization to any number of coils on the transformer core is that the sum of all per unit phasor currents is zero.

We conclude that the equivalent circuit for a transformer with $M$ coils would be as in Fig. 6.30. Note that it is always true that the output kVAs sum to the input kVA. Transformers are rated according to maximum line voltage (typically, in kV) and apparent total (three-phase) power (typically, in kVA), *not* average power because it is the absolute magnitudes that must be met with the physical transformer materials.

Fundamentally, these ratings specify maximum voltage and current allowed, the maximum current inferred from the apparent power rating divided by the maximum voltage rating. Operating the transformer with either too high voltage or

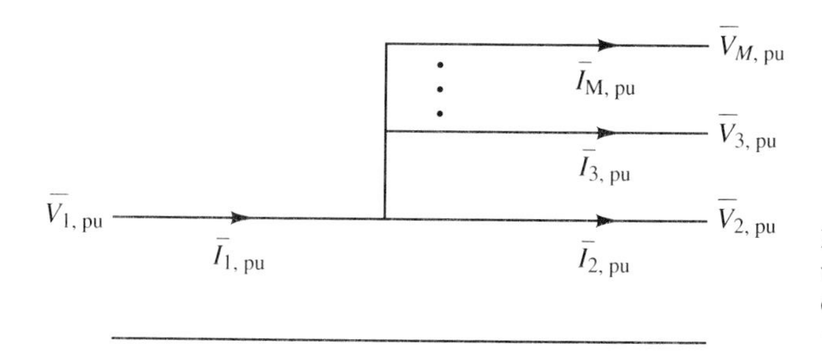


**Figure 6.30** Multitap transformer voltages and currents in the per unit domain; $\bar{V}_{1,\text{pu}} = \bar{V}_{2,\text{pu}} = \bar{V}_{3,\text{pu}} = \cdots = \bar{V}_{M,\text{pu}}$.

current would result in either inefficient operation (overheating) or component failure. Operating too far below rated values merely results in inefficient and ineffective use of the device.

In the rating, the turns ratio is also given implicitly by providing both the input and output line voltages. Thus for a wye-connected step-down power transformer, a typical rating would be "100 MVA at 345/34.5 kV," meaning a maximum allowable apparent power of 100 MVA on the primary winding. For an input voltage of 345 kV, the maximum current is, by Eq. (6.59), $10^8/(\sqrt{3} \cdot 3.45 \cdot 10^5) = 167$ A. There typically will be several branching secondaries at each voltage level, say five at this upper level, each taking 100 MVA/5 = 20 MVA. Consequently, the secondary current is $2 \cdot 10^7/(\sqrt{3} \cdot 3.45 \cdot 10^4) = 335$ A. Keep in mind that the rated voltages and currents are always *line* quantities (easily accessible for measurement); hence the $\sqrt{3}$. The turns ratio $N_1/N_2$ is, by the phasor form of Eq. (6.41), 345/34.5 = 10.

In carrying out power system analysis, we can often reduce the work by a factor of 3 by analyzing "one-line" or "per phase" circuit diagrams. One example was already presented in Fig. 6.30. Considering only one phase as representative of the entire system assumes that the phases are balanced. One must also convert all delta sources and loads to wye equivalents [e.g., see Eq. (6.76)]. The one-line diagram is equivalent to a schematic of a single phase, but it is a sort of shorthand representation.

Thus a simple symbol for a transformer replaces the complete transformer model; similarly, a straight line symbolizes a transmission line that in a calculation would have to be replaced by the appropriate short, medium, or long wire model. The only tricky part is to remember factors of 3 and $\sqrt{3}$. In particular, to write the single-phase equivalent circuit, original delta-connected systems must be converted to wye–wye form by means of the delta–wye transformation derived previously.

**Example 6.11**

Consider the following wye–wye-connected three-phase system, the one-line diagram of which is shown in Fig. 6.31. Each transformer is rated at one-third the total system power, 500 kVA at 13.2/0.48 kV. The per unit impedance of each transformer is $\bar{Z}_{\text{pu}} = 0.01 + j0.05$ pu (looking in from either side). If the load on the transformer is 400 kVA at 490 V and 0.9 power factor, calculate the input voltage and input and output currents to/of the transformer. Also find the actual transformer impedance looking into either end of it, and the load impedance. Draw the per unit circuit diagram. Neglect no-load losses.

**Solution** As this is already a wye–wye configuration, no conversion is necessary. Because it is usually the case that rated apparent power ($S$) and input–output voltages

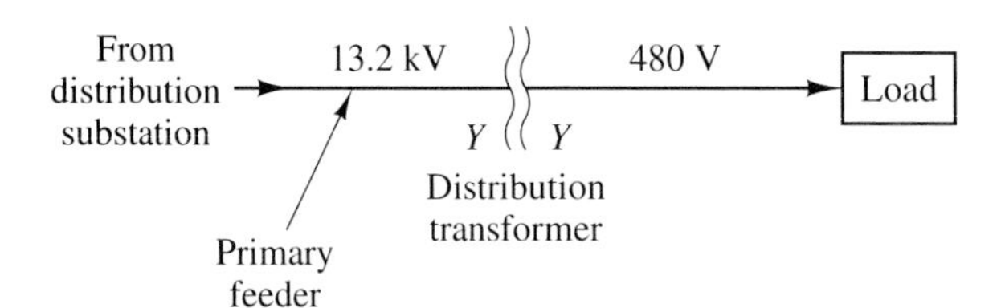


**Figure 6.31** Wye–wye-connected three-phase system for Example 6.11.

of the transformer are given, these are convenient to use as base values. Note that the voltages given are line values; but necessary for calculations with one-line diagrams (equivalently single-*phase* diagrams) such as line current are the corresponding phase voltages.

In a wye-connected system the phase voltages are equal to the line voltages divided by $\sqrt{3}$. Nevertheless, the convention is that the base voltage is always chosen to be the *line* voltage. Also, the base apparent power, $S_b$, is always chosen to be the total apparent power, not the per phase apparent power. Thus $S_b = 500$ kVA, $V_{1,b} = 13.2$ kV, and $V_{2,b} = 480$ V. (The turns ratio is $a = N_1/N_2 = V_{1,b}/V_{2,b} \approx 28$.) This gives $I_{1,b} = (500/3)/(13.2/\sqrt{3})$ A $= 21.9$ A and $I_{2,b} = (13.2/0.480)I_{1,b} = 602$ A. It is desired to find $\bar{V}_{2,\text{pu}}$, for then $\bar{V}_{1,\text{pu}} = \bar{V}_{2,\text{pu}}$ and then $\bar{V}_1 = V_{1,b}\bar{V}_{1,\text{pu}}$.

The per unit diagram along with the various bases and their domains of applicability are shown in Fig. 6.32. The base impedance looking into the input of the transformer is $Z_{1,b} = V_{1,b}^2/S_b = (13{,}200)^2/500{,}000 = 348.5\ \Omega$ and the base impedance looking into the input of the transformer is $Z_{2,b} = V_{2,b}^2/S_b = 480^2/500{,}000 = 0.46\ \Omega$. Therefore, we know that the actual impedances looking into the input and output of the transformer are, respectively, $\bar{Z}_1 = Z_{1,b}\bar{Z}_{\text{pu}} = 348.5(0.01 + j0.05) = 3.48 + j17.4$ and $\bar{Z}_2 = Z_{2,b}\bar{Z}_{\text{pu}} = 0.46(0.01 + j0.05) = 0.0046 + j0.023\ \Omega$ ( $= \bar{Z}_1/a^2$).

The equality of input and output per unit phasors will thus simplify the analysis. Arbitrarily assign zero phase to the load voltage. Then $\bar{V}_L = 490\text{ V} \angle 0°$ and thus $\bar{V}_{L,\text{pu}} = (490/480) \angle 0° = 1.02 \angle 0°$. The per unit load power is $\bar{S}_{L,\text{pu}} = (400/500) \angle \cos^{-1}(0.9) = 0.8 \angle 25.8° = \bar{V}_{L,\text{pu}}\bar{I}_{2,\text{pu}}^*$. Consequently, $\bar{I}_{2,\text{pu}} = (0.8/1.02) \angle -25.8° = 0.78 \angle -25.8°$. The actual output current is $\bar{I}_2 = I_{2,b}\bar{I}_{2,\text{pu}} = 602 \cdot 0.78\text{ A} \angle -25.8° = 472\text{ A} \angle -25.8°$. The input current is $\bar{I}_1 = I_{1,b}\bar{I}_{1,\text{pu}} = 21.9 \cdot 0.78\text{ A} \angle -25.8° = 17.1\text{ A} \angle -25.8°$. The per unit input voltage, by KVL, is $\bar{V}_{1,\text{pu}} = \bar{V}_{2,\text{pu}} = \bar{V}_{L,\text{pu}} + \bar{I}_{2,\text{pu}}\bar{Z}_{\text{pu}} = 1.02 \angle 0° + 0.78 \angle -25.8° \cdot (0.01 + j0.05) = 1.02 + 0.04\cos(-25.8° + 78.7°) + j0.04$

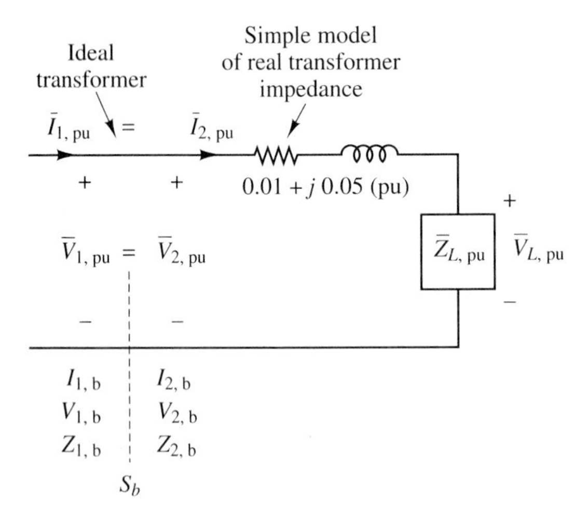


**Figure 6.32** Per unit circuit diagram for Example 6.12.

$\sin(52.9°) = 1.04 + j0.032 = 1.04 \angle 1.76°$. Thus the actual input voltage $= \overline{V}_1 = V_{1,b}\overline{V}_{1,\text{pu}} = 13.7 \text{ kV} \angle 1.76°$.

The per unit load impedance is $\overline{Z}_{L,\text{pu}} = V_{L,\text{pu}}^2/\overline{S}_{L,\text{pu}}^* = (1.01)^2/0.8 \angle -25.8° = 1.3 \angle -25.8°$, so that $\overline{Z}_L = Z_{2,b}\overline{Z}_{L,\text{pu}} \approx 0.6\ \Omega \angle -25.8°$. This, incidentally, agrees, as it must, with the direct calculation $\overline{Z}_L = V_{L,phs}^2/\overline{S}_{L,phs}^* = (490/\sqrt{3})^2/(400{,}000/3) \angle -25.8°$).

In Example 6.11 the conversion to and back from the per unit domain was probably more involved than not using the per unit system. But in realistic problems, the avoidance of transformer manipulations and the presence of near-unity numbers are advantages that make the method desirable. Note, for example, that the per unit voltage magnitude of 1.06 immediately indicates an above-rated quantity being applied to the transformer.

## 6.8 ELECTRICAL SAFETY AND GROUNDING

Electrocution, which is death by electric shock, is one of the leading causes of industrial fatalities (roughly 400 annually in the United States). Furthermore, the percent of accidents that are fatal is very high (10%). Unlike other accidents, one is usually either killed or receives no lasting injury from an electrical accident. An exception to this is contact burns, which may be severe. The worst part of all this is that these accidents often could easily be avoided.

To be shocked, current must enter one part of the body and leave at another part, due to a potential difference between the things the body touches. Hands and feet are the most likely candidates. In this sense the body is like a big resistor. (Recall that this property is used in the body fat measurement technique we discussed in Sec. 2.9.)

When current passes through the body, muscles contract. If the lungs are involved, breathing could stop. If a person is holding onto the culprit conductor, his or her hand will clutch it even more tightly. The current at which a person cannot let go of an object is known as the *let-go* current. Beyond this current, death is likely because the person cannot get away from the shocking conductors. In such cases, try to knock the person away from the hazard with some sort of insulating stick.

If the person has fallen unconscious, two causes are likely. If lungs are involved, breathing could stop; if the heart is involved, the heart could stop. In either case, artificial respiration immediately, and for at least an hour is recommended, as it appears to help even when respiratory failure has not occurred. Naturally, cardiopulmonary resuscitation (CPR) is also recommended if the heart is involved.

Because of the highly varying surface skin resistance, voltage thresholds for dangerous shock are not usually indicated. However, typically let-go may occur at 20 V for 60 Hz and 100 V for dc (death at 40 V ac). Usually, let-go currents are specified. For example, 10 mA for ac (60 Hz) and 75 mA for dc.

From the data above, we see that the household ac can be deadly—a fact the reader hopefully already knew. Shocks can occur when someone is rushed or distracted in working around high voltages. One may think a circuit is disconnected when in fact it is connected. Often, the worker may be ignorant of the safe methods of using equipment. Other times, machinery such as lifting equipment at construction sites may pierce a live power line, resulting in shock. The shock can create an accident worse than the first if the person falls off a ladder, for example, as a result of the shock.

Hints to avoid trouble include: (1) use only equipment in good repair; (2) use protective equipment when required (helmets, gloves, insulated shoes, safety glasses, etc.); (3) double check that switches are disconnected (even locked off) when working around such equipment; (4) have another person present; (5) do not handle electrical equipment when your hands, feet, and so on, are wet—your conductivity rises sharply when wet; (6) if inspecting machinery, use only one hand—and the back of it if possible to avoid hold-on; (7) turn away your face when using switches, circuit breakers, and so on; and (8) be very careful around ladders, cranes, lifters, and so on—find out beforehand where overhead power lines are located.

All equipment used should be certified by a recognized authority as being safe before use. Many design techniques are used to minimize shock hazard, including insulation, fuses and circuit breakers, and grounding.

One of the most frequently confusing issues in wiring regards the third "ground" wire. It is actually very simple to understand in the context of a power distribution system. Instead of the one-line diagram style as shown in Fig. 6.29 with all the substations shown, let us instead consider the simplified three-phase diagram from source to wall outlet shown in Fig. 6.33. The three-phase wye-connected generator is shown on the left. Immediately, a wye–wye three-phase transformer steps the voltages up to transmission levels (e.g., hundreds of kilovolts). The long-distance transmission line is indicated by the four broken lines. The fourth wire is the neutral; on the above-ground wires we see spanning the countryside, it is the small one on top.

Next along the system is the substation three-phase step-down transformer. From there, the three-phase lower voltage is sent to an industrial plant that has its own three-phase transformer. This time one of the secondaries has a center tap that is tied to earth ground. (Recall that, otherwise, the secondary potentials are all floating with respect to the primary.)

Each secondary phase would be single-phase 240 V, in particular, *bc*. Thus *bn* and *cn* are both 120 V single-phase ac usable for normal lighting and office equipment. The single-phase 240 V can be used to operate heavy-duty single-phase machinery on the plant floor. Also, the three phases of the secondary can be used to run any three-phase motors or other equipment on the plant floor. The plant transformer would be replaced by a distribution transformer in home settings.

Finally, we come to the wall outlet. *Two* wires are run from earth ground (*n*) to the wall outlet. One is for the return current from the "hot" wire, *b* or *c*. That is, *the single-phase return path is connected to earth ground.* Thus if we were to touch the "hot" wire with our hand and our feet were on the ground, we would be shocked.

The other wire from earth ground to the wall outlet is for the event of an accidental short circuit within the appliance. It is connected (by the third wire on the appliance plug) to the chassis of the machine (assuming that there is any metal to it, which there usually is for any sizable equipment).

Suppose that the ground were not connected to the chassis and an accidental short circuit occurred, as in Fig. 6.34a. Such short circuits typically happen when, for example, wire insulation within the machine fails. Then the current would pass through the short, through the person, and finally to ground. The person could be

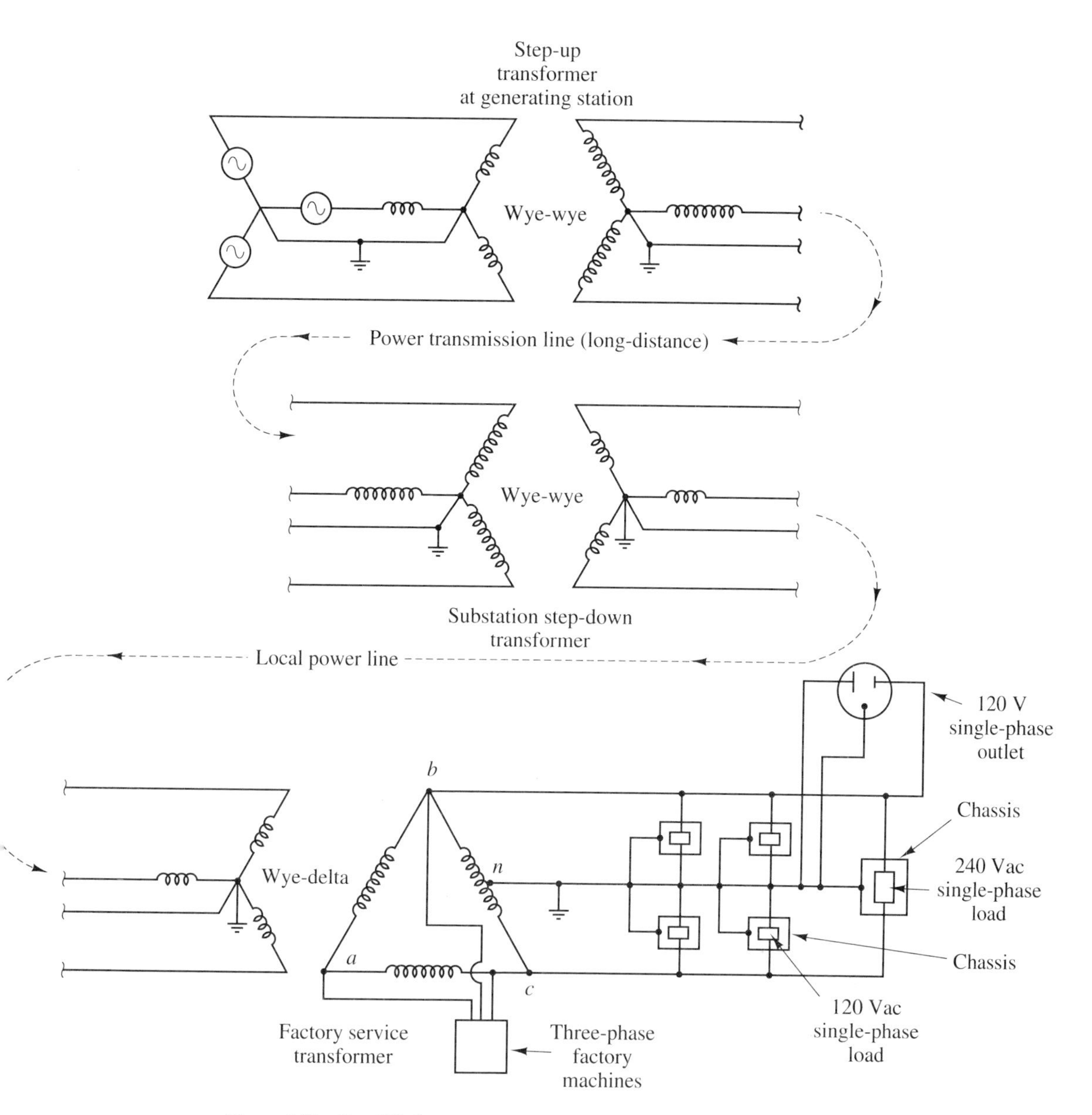


**Figure 6.33** Simplified power transmission system, from generator to wall outlet.

fatally shocked. If, instead, the chassis is grounded, the ground *bypasses the person,* so the person is not shocked (see Fig. 6.34b).

It may be asked: Why not just connect the chassis to the return wire, which itself is grounded? That would save a wire! The problem would be that if we happened to plug the appliance in the wrong way, then *even if no short circuit had occurred*, the chassis would be hot. Of course, even today's two-wire plugs are polarized (shaped as shown in Fig. 6.34c, which also shows a modern three-wire plug), but what if the plug is old and not polarized or the plug or socket wears away

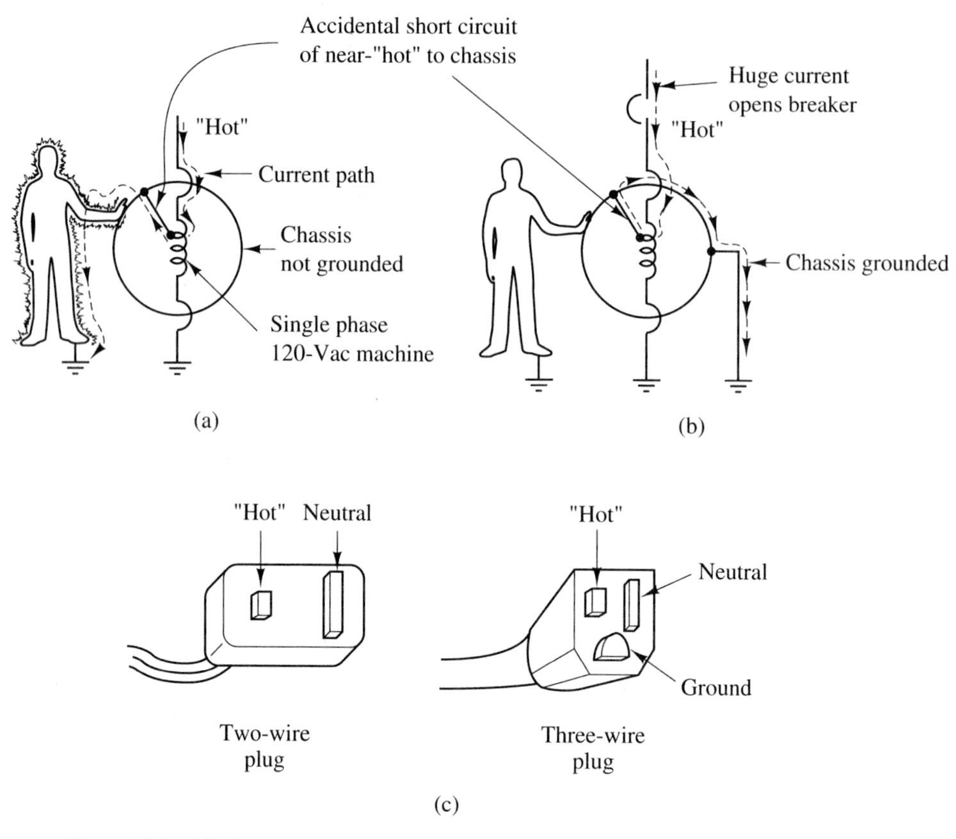


**Figure 6.34** (a) Ungrounded chassis causes shock in event of fault (short from hot wire to chassis); (b) grounded chassis draws fault current away from person, thereby preventing shock; (c) two- and three-wire polarized plugs.

so that both blades are alike? The result would be fatal. Therefore, the chassis is connected by a special third wire which cannot be confused (round instead of a blade) with the hot wire.

Figure 6.33 seems imbalanced; one phase is supplying all single-phase equipment. To minimize imbalances, additional three-phase distribution transformer sets are used to load each phase with single-phase loads of different consumers (home, commercial, etc.).

It also might be asked why the neutral wire of the high-voltage transmission lines is grounded. After all, the fourth neutral wire is there to account for imbalanced loads, and in *any* event power lines in general are dangerous to the touch. One reason is that the NEC (National Electrical Code®) demands it!

Another reason comes from Eq. (6.58), in reference to Fig. 6.20 (which omits the fourth neutral wire). If the neutral node is grounded, none of the three transmission line wires ever has a potential differing from the neutral (ground) potential by more than $1/\sqrt{3} \approx 0.58$ times the line voltage. Therefore, insulation requirements are fixed at minimum levels. For if the lines were floating, the potentials to ground could vary widely among systems and within one system. Many more complex reasons in power system design such as stability considerations and utility interfacing with larger-scale power grids favor grounding of transmission lines. Furthermore, the potential at any point in the system with respect to the earth is known, fixed, and limited. Ground is the potential of any person standing on the earth.

Given that grounding of transmission lines is reasonable, grounding of low-voltage systems is easier to accept. Because the potentials are never more than a few hundred volts from ground, lightning striking the power line near a home has a very convenient path to ground. The tremendous quantities of charge in a lightning bolt are far better off going to ground (a huge capacitor) than anywhere else. Also, if a distribution transformer has a short between primary and secondary windings, a grounded secondary draws the current away from external circuits that could endanger people; see Fig. 6.35.

With no grounding (Fig. 6.35a), when the transformer is shorted, everything in your house would be at 12,000 V! Your feet are on the ground at 0 V, and any connected conductor you touched would kill you. But in Fig. 6.35b, a huge current is drawn because of the effective short to ground caused by the secondary being grounded, and all is safe, especially after the circuit breaker on the transformer blows. (This example should indicate why we always put circuit breakers on the hot wires rather than neutral wires.)

Incidentally, when we recall that in practice secondary and primary transformer windings are wound one directly on the other to minimize flux leakage, a shorted transformer is not far-fetched; they do occur occasionally.

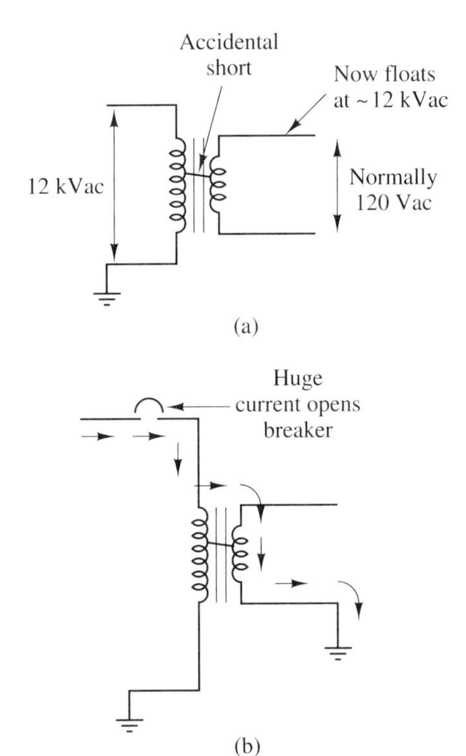


**Figure 6.35** Advantage of grounding. In (a) a transformer short circuits, causing the entire secondary circuit to float at 12 kV. In (b), the short is grounded, causing effectively a short in the *primary,* which opens the primary breaker.

## 6.9 SUMMARY

In this chapter we have considered the application of sinusoidal analysis to the energy conversion/transmission problem. First some essential tools and concepts were introduced. The effective value of a time-varying quantity is the dc voltage or current value that would dissipate the same power in a resistor $R$ as the time-varying voltage or current would dissipate on the time average in $R$. For sinusoids, the effective value is the sinusoidal amplitude divided by the square root of 2.

The average power consumed by a circuit element was derived in terms of the product of the effective values of voltage and current and the power factor, which is the cosine of the phase difference between current and voltage. Although the net power absorbed by reactances is zero, they can contribute to the total line current, the $I^2R_{tl}$ losses for which the user of electricity must pay. So the reactive power, which is the magnitude (including sign of the reactance) of the oscillations of the instantaneous power into the reactance was defined to be the imaginary component of complex power, while the real part is the average consumed power.

To minimize $I^2R_{tl}$ losses and kVA transmission equipment ratings, it is economically desirable for an industrial customer to minimize $I_{eff}$ while maintaining constant $P$. This process is known as power factor correction and involves decreasing the net reactance the power company sees. In most situations in which there is inductive equipment, power factor correction may be accomplished by installing shunt capacitors.

The transformer is an extremely useful device in both power and signal processing circuitry. In power systems, it makes feasible long-distance power transmission by stepping down the current on the lines to minimize $I^2R_{tl}$ losses. In electronics they are used for isolation of circuitry from high-voltage lines, for internal power supplies, for impedance matching, and for oscillators and filters. We studied both ideal and practical transformer models. The ideal model is a fundamental component of all more accurate models.

Three currents or voltages of equal magnitude and frequency but differing from each other in phase by 120° are called balanced voltages or currents. Their sum is zero. A balanced load is a set of three equal impedances connected to a three-phase system. For balanced systems, the instantaneous power is constant and equal to three times the single-phase average power.

Three-phase loads and generators may be connected in either delta or wye connections. For balanced wye connections, the line-to-line voltage is $\sqrt{3}$ times the phase voltage, while the line and phase currents are identical. The converse (switch the words "voltage" and "current") is true for delta connections. If given either a balanced delta or balanced wye load and the other is sought, we merely need to apply the result that the delta impedance is three times the corresponding wye impedance.

Wattmeters were introduced as instruments for measuring average power. It was found that even for unbalanced systems, only two are needed to find the total average power into a three-phase system.

We then considered in some detail the distribution of electric power. These systems are huge, complicated, and interdependent. Major components include generators, transmission lines, transformers, relays, circuit breakers, filter capacitors, and control electronics. An analysis technique used in this area called per unit analysis was described and illustrated by an example.

Our final topic in this chapter was the vital area of electric safety and grounding systems. The physiological facts were reviewed and preventive measures were suggested. A very important design technique of maximizing electrical safety is grounding of systems. We discovered the real reasons for grounding and, in particular, the three-wire plugs that are now the standard in the office and at home.

Excepting sources, so far all of our studies have been restricted to passive devices: those that either consume zero average power or dissipate it. The information revolution would be impossible without (indeed it did not occur until) the development of circuits having the ability to effect power amplification. Power amplification of small signals is one of the chief uses of semiconductor equipment. In Chapter 7 we develop simple circuits that accomplish this.

## PROBLEMS

**6.1.** Find the average value and the effective value of the waveform in Fig. P6.1.

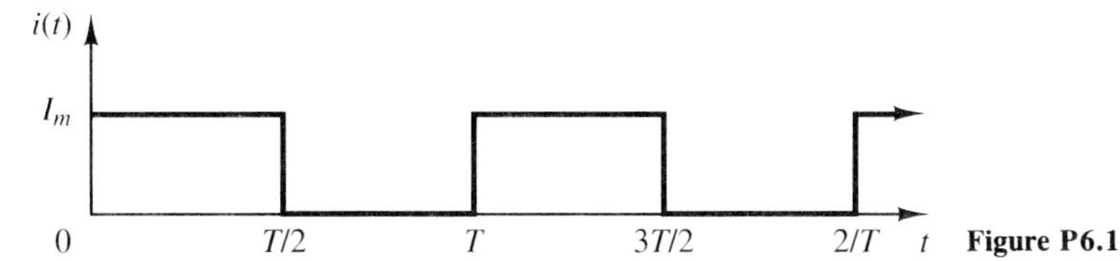


**Figure P6.1**

**6.2.** Find the average value and the effective value of the waveform in Fig. P6.2.

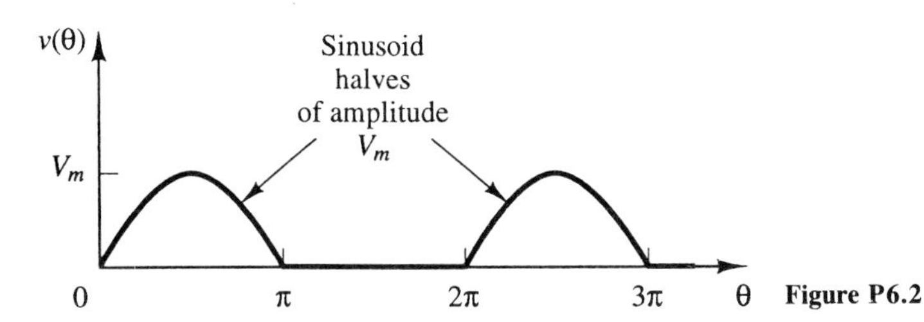


**Figure P6.2**

**6.3.** Find the effective voltage and current phasors for the following voltages and currents.

**(a)** $v_1(t) = 10 \cos(\omega t + \pi/2)$ V

**(b)** $v_2(t) = 20 \sin(\omega t + \pi/2)$ V

**(c)** $i_1(t) = \sin(\omega t - \pi/5)$ A

**(d)** $i_2(t) = \sqrt{2} \cos(\omega t + \pi/4)$ A

**6.4.** For the circuit in Fig. P6.4, $v_s(t) = 40 \cos(1000t)$ V.

**(a)** Find $i(t)$.

**(b)** Find the rms value of $v_s(t)$.

**(c)** Find $\bar{I}_2$ in polar form.

**(d)** Find the average power dissipated in $R_1$.

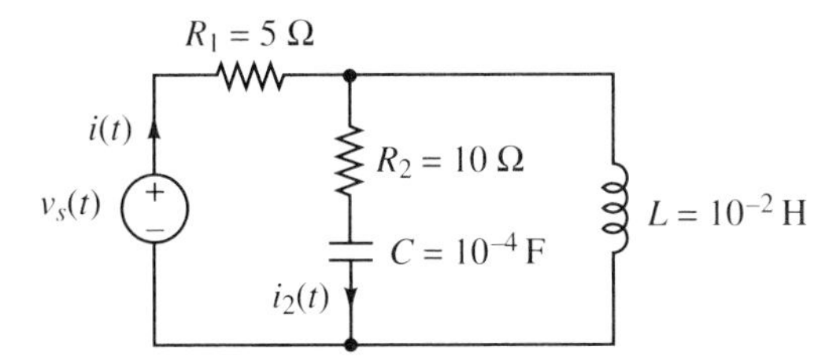


**Figure P6.4**

**6.5.** A sinusoidal voltage $v(t) = 7 \cos(400t + \pi/4)$ V is applied to (i) a resistor $R = 20\ \Omega$, (ii) a capacitor $C = 250\ \mu$F, and (iii) an inductor $L = 100$ mH.

**(a)** Calculate the instantaneous power absorbed by $R$, $C$, and $L$.

**(b)** Using your results from part (a), calculate the average power absorbed by $R$, $C$, and $L$.

**6.6.** A voltage source $v_s(t)$ is connected across a parallel combination of $R$ and $L$, where $v_s(t) = 24 \cos(2\pi \cdot 100 \cdot t)$ V, $L = 160$ mH, and $R = 50\ \Omega$.

**(a)** Draw the instantaneous power waveform $p(t)$ supplied by $v_s(t)$ and absorbed by the parallel $RL$ combination.

**(b)** Find the average power consumed by the $RL$ combination as well as the reactive power that is periodically exchanged between $v_s(t)$ and $L$.

**6.7.** Suppose that in Fig. 6.1, the current $i(t)$ is as shown in Fig. P6.7 and that $R = 10\ \Omega$.

**(a)** Find the effective value of $i(t)$, $I_{\text{eff}}$.

**(b)** Find the average power consumed by $R$ by directly using $I_{\text{eff}}$ from part (a).

**(c)** Find the average power consumed by $R$ by calculating the average of $p(t) = v(t)\,i(t) = i^2(t)\,R$. Do your results agree?

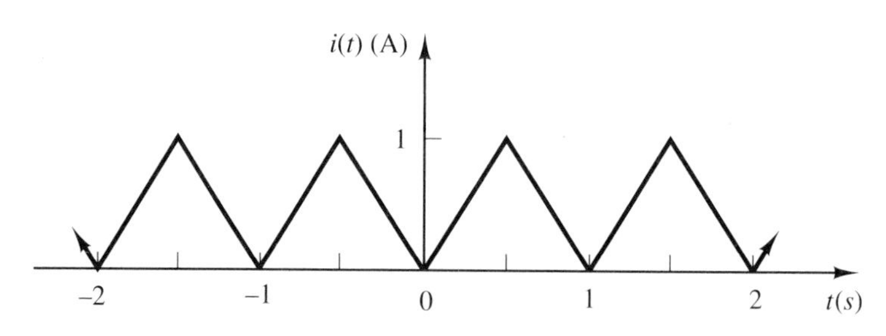


**Figure P6.7**

**6.8.** In Chapters 2 and 4 we studied the resistance of light bulbs assuming dc excitation. Now let the power supply be the standard 110 V ac, 60 Hz; it is applied to a 100-W light bulb. What is the maximum voltage across the bulb at any given time, and what is the resistance of the bulb at illuminating temperature?

**6.9.** In the $RL$ circuit in Fig. 5.19, let $v_s(t) = 12\cos(1500t)$ V, $R = 120\ \Omega$, and $L = 40$ mH.

**(a)** Find the maximum instantaneous power delivered by $v_s(t)$.

**(b)** Find the maximum instantaneous power absorbed by $L$.

**(c)** Find the power factor of the $RL$ combination (leading or lagging?).

**6.10.** For the $RC$ circuit in Fig. 5.31, let $R = 300\ \Omega$, $C = 20\ \mu$F, and $v(t) = 8\cos(500t + 4°)$ V.

**(a)** Find the power factor of the $RC$ combination (leading or lagging?).

**(b)** Calculate the average power consumed by the $RC$ combination.

**6.11.** A series passive circuit fragment may or may not consist of $R$, $R$ in series with $C$, $R$ in series with $L$, or $R$ in series with $L$ in series with $C$. When a voltage source $v_1(t) = 10\cos(200t)$ V is applied, $P_1 = 0$ and $Q_1 = -75$ VAR. When another voltage source $v_2(t) = 8\cos(2000t)$ V is applied, $Q_2 = 30$ VAR.

**(a)** Find $P_2$.

**(b)** Determine the unknown circuit fragment, including parameter values.

**6.12.** For Fig. 6.3, with applied voltage phasor $\overline{V} = V\exp(j\theta_v)$, find $p_x(t)$ and $p_R(t)$ in terms of $V$, $\omega$, $\theta_Z$, $\theta_v$, $\theta_i$, and so on (assuming that $\theta_v > \theta_Z$). Sketch both waveforms, and clearly show the average value of $p_R(t)$ using a horizontal dashed line on your plot. Be sure to indicate explicitly all maximum values and amplitudes. Do your results agree with those stated in Fig. 6.5 about the definitions of $P$ and $Q$?

**6.13.** Suppose a resistor $R$ is connected in parallel with a reactive impedance $jX$. Find the power factor of the equivalent impedance $\overline{Z}_{eq}$ of this parallel combination in terms of $R$ and $X$. Simplify as much as possible.

**6.14.** Show that in a series $RLC$ circuit, the energy dissipated in one cycle of the voltage or current is $I_{max}^2 R\pi/\omega$, where $I_{max}$ is the amplitude of the sinusoidal current into the network at frequency $\omega$. We used this result in a Chapter 5 problem. (*Hint:* The integral of any function over a finite interval is equal to the average value of the function times the interval length. What about the series $RL$ circuit—does the same result hold?)

**6.15.** A 3-Ω resistor is connected in parallel with the series combination of a 2-Ω resistance and a reactive (inductive) impedance equal to $j1$-Ω at the given excitation frequency. The total network (parallel combination) is connected to a sinusoidal power source (of frequency such that the reactance is indeed 1-Ω). Determine the average power dissipated by the 3-Ω and the 2-Ω resistances, and the reactive power absorbed by the inductor, if the total apparent power into the network is 1 kVA.

**6.16.** Given the circuit fragment in Fig. P5.27, with $\overline{Z} = \frac{1}{2} - j\frac{1}{4}\ \Omega$:

(**a**) Give the power factor of $\overline{Z}$. Is it leading or lagging, and why? Is $\overline{Z}$ inductive or capacitive?

(**b**) Draw the power triangle and *fully* label with symbols *and* numerical values.

(**c**) Give the word names of $\overline{S}$, $S$, $P$, and $Q$.

**6.17.** (**a**) In Fig. P5.29, find $\overline{S}$ absorbed by $Z_L$.

(**b**) Find $P$, $Q$, $S$, and the power factor (leading or lagging?).

(**c**) What value of inductance (yes, we are considering an unusual device!) if placed in parallel with $Z_L$ would raise the power factor of the combination to 0.7 leading while maintaining the same value of $P$?

**6.18.** In Fig. P6.18, $M$ designates a motor, consuming average power = 1000 W.

(**a**) Find the capacitance $C$ to achieve a power factor (PF) = 0.85.

(**b**) What is the effective value of the line current before and after addition of the capacitor?

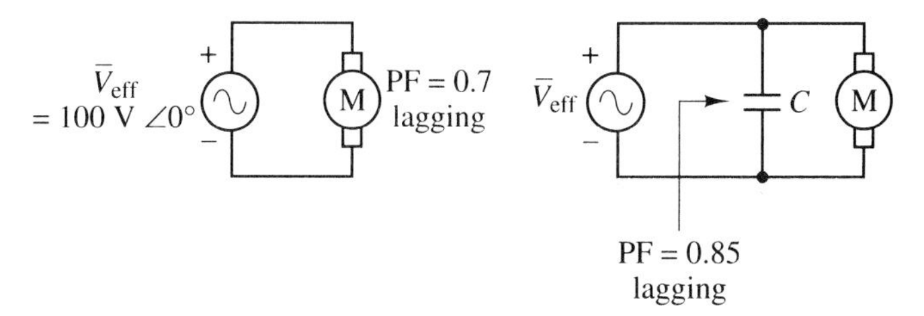


**Figure P6.18**

**6.19.** In the circuit in Fig. P6.19, in which $\overline{V}_s$ is an effective voltage phasor:

(**a**) Find the apparent power drawn by the load.

(**b**) Find the power factor of the load, and whether it is leading or lagging.

(**c**) Find the type of device (inductor or capacitor) and the numerical value (of $L$ or $C$) (i.e., the box with a question mark in it) that should be placed in parallel with the load to change the overall PF seen by $V_s$ to 0.95 leading.

**6.20.** (**a**) In Fig. P5.30, what is the power factor of the load $\overline{Z}_L$? Lagging or leading?

(**b**) What type of device (and the value of $L_a$ or $C_a$) when placed in parallel with the load will change the overall power factor seen by $\overline{V}_s$ to 0.95 leading (same average power to load)? (*Hint:* First find $\overline{S}_L$ consumed by $\overline{Z}_L$.)

(**c**) Find the magnitude of the effective current supplied by the source $\overline{V}_s$ before and after the power factor correction. (After correction, it should be less than before correction.)

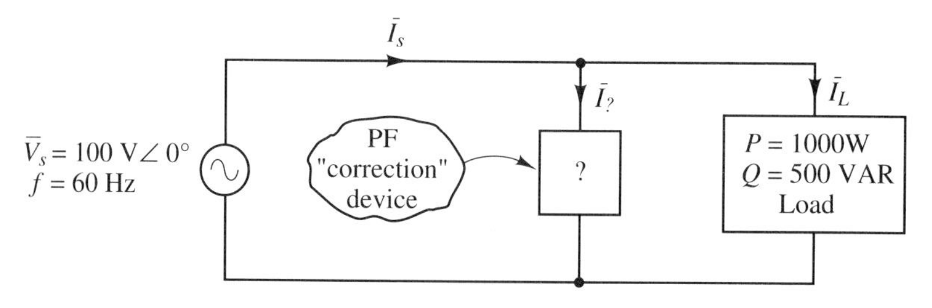


**Figure P6.19**

**6.21.** Suppose we have a 1-Ω resistance in parallel with the series combination of a 1000 μF capacitor and a 1 mH inductor. A 60 Hz sinusoidal excitation is applied to the network. Find the power factor. Is it leading or lagging?

**6.22.** The space heater designed in Problem 4.22 has the following device values: The resistance of the heating element is 10 Ω, and the motor has a resistance of 192 Ω in series with an inductance of 0.486 H. Compute the power factor of this device. Is it efficient in minimizing reactive power?

**6.23.** Suppose that we have an inductor $L = 0.5$ H and a resistance 120 Ω connected in parallel to a power line. Determine the value of parallel capacitance required to correct the power factor all the way to unity for **(a)** $\omega = 60$ Hz, and **(b)** $\omega = 1000$ Hz.

**6.24.** For the circuit shown in Fig. P6.24:

**(a)** Find effective $\bar{I}_1$ and $\bar{I}_2$ in polar form.

**(b)** Find the average power and reactive power consumed by load $A$ and by load $B$.

**(c)** Find the total complex power supplied by the source, in rectangular and polar forms. Identify the physical meanings of each component.

**(d)** Find the source current phasor $\bar{I}$.

**(e)** What value of reactance should be placed in parallel with loads $A$ and $B$ to bring the power factor to unity? Is the required reactance capacitive or inductive?

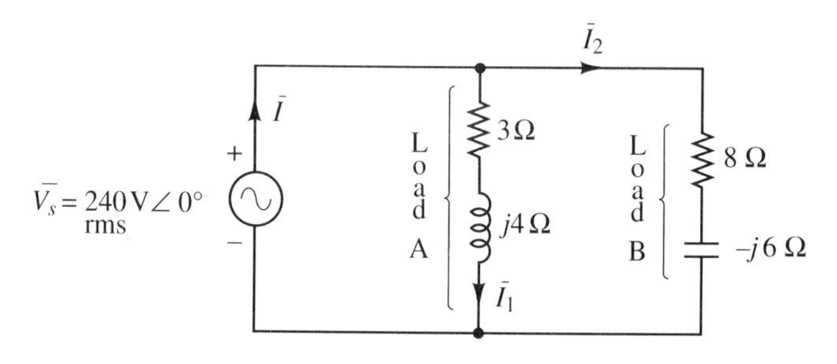


**Figure P6.24**

**6.25.** Loads $A$ and $B$ are connected in parallel with a 240-V rms sinusoidal voltage source; assume zero phase angle for this voltage. Load $A$ is rated at 20 kW, 0.8 PF lagging, and load $B$ is rated at apparent power of 50 kVA, 0.6 PF lagging.

**(a)** Find the complex power supplied by the 240-V ac source, in rectangular and polar forms.

**(b)** Find the effective source current phasor.

**(c)** What value of capacitive reactance, $X_C$, should be connected in parallel with loads $A$ and $B$ to raise the power factor all the way to unity?

**6.26.** Loads $A$ and $B$ are connected in parallel with a 250-V rms $\angle 0°$ sinusoidal voltage source. Load $A$ uses 20 kW at a lagging power factor of 0.65, and load $B$ has impedance $\overline{Z}_B = 1.5\ \Omega + j2\ \Omega$.

**(a)** Find the average and reactive powers absorbed by load $B$.
**(b)** Find the total complex power supplied by the source.
**(c)** Compare the magnitude of the effective source current before and after correction to unity power factor. Do your results make sense?

**6.27.** A small factory has two machine tools $A$ and $B$ which operate at 240 V, 60 Hz. The power consumptions for the two tools are, respectively, 12 kW for $A$ and 5 kW for $B$. We also know that tool $A$ has a lagging power factor of 0.80, and $B$ has a lagging power factor of 0.55.

**(a)** Find the complex powers absorbed by $A$ and $B$.
**(b)** Find the total power factor for the two tools on simultaneously (assume that these are the only machines on at once).
**(c)** Improve the power factor in part (b) to 0.95. What type of component must be used, and what is its value?

**6.28.** For the circuit described in Problem 5.25:

**(a)** Find the average power dissipated by $\overline{Z}_2$ and by $\overline{Z}_{eq}$.
**(b)** Find the complex power into $\overline{Z}_{eq}$ and the power factor of $\overline{Z}_{eq}$. Draw the complete power triangle for $\overline{Z}_{eq}$, fully labeled with your numerical results.

**6.29.** A resistance $R = 10\ \Omega$ and an inductance $L = 5$ mH are in series. The resistor voltage is $v_R(t) = 5.0\cos(377t - 45°)$ (V). Find the complex power, apparent power, reactive power, average power into the network, and the power factor of the network (leading or lagging?). Also draw the fully labeled power triangle.

**6.30.** For the circuit described in Problem 5.31, compute at 50 Hz the average power, complex power, reactive power, apparent power, power factor (including "leading" or "lagging") into the impedance seen by the source, and draw the fully, numerically labeled power triangle.

**6.31.** For the circuit shown in Fig. 5.19, suppose that $v_s(t) = 110\sqrt{2}\cos(2\pi \cdot 60t)$ V, $R = 120\ \Omega$, and $L = 0.5$ H.

**(a)** Find the complex power $S$ applied to the $RL$ circuit.
**(b)** Determine the power factor and the reactive factor.
**(c)** Draw the power triangle.

**6.32.** Suppose that in Fig. 6.9a a load resistance $R = 140\ \Omega$ is connected to the secondary winding, while the voltage source $v_1(t) = 24\cos(2\pi \cdot 60t)$ V (negligible source resistance) is connected to the primary. It is found that the effective value of the load (secondary) current is $I_{2,\text{eff}} = 20$ mA. Assume an ideal transformer.

**(a)** Determine the turns ratio $a$ of the transformer.
**(b)** Find the effective value of the primary current.

**6.33.** For the sinusoidally excited circuit shown in Fig. P6.33 (assuming an ideal transformer):

**(a)** Find the resonant frequency $f_0$ (in hertz).
**(b)** Calculate the quality factor of this circuit. Would it make a good radio tuner circuit?

**6.34.** For the circuit in Fig. P6.34, in which $v_s(t) = 400\cos(2\pi \cdot 200t)$ V (assuming an ideal transformer):

**(a)** Find the average power and the reactive power supplied by the source.
**(b)** Find the average power dissipated in $R_L$.

**6.35.** For the circuit in Fig. P6.35 (assuming an ideal transformer):

**(a)** Find the value of $\overline{Z}_L$ for maximum power delivered to the load ($\overline{Z}_L$).
**(b)** If $\overline{Z}_L = 1 + j2\ \Omega$, find $\overline{V}_1$ and the average power supplied by the source.

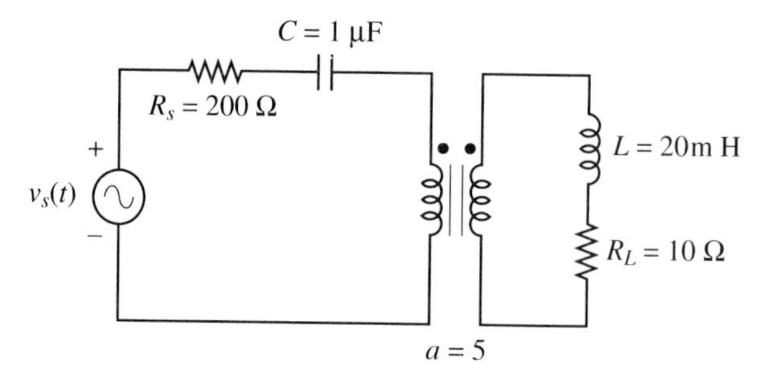


**Figure P6.33**

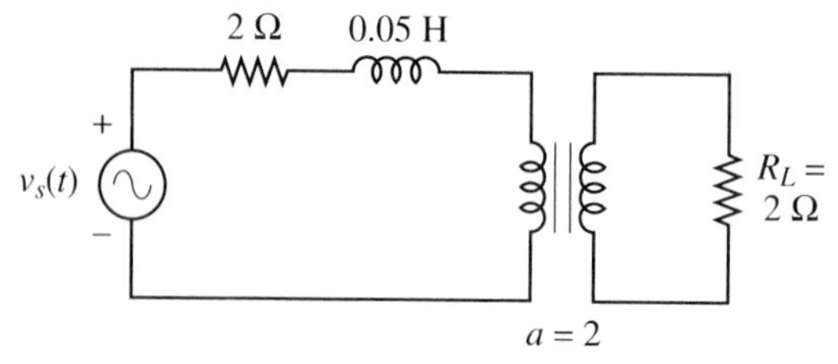


**Figure P6.34**

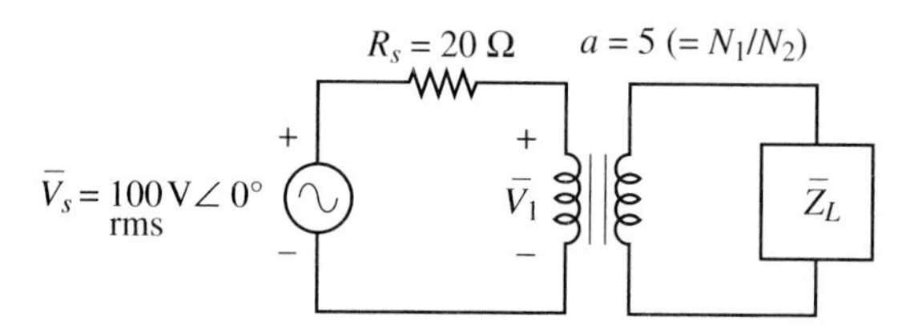


**Figure P6.35**

**6.36.** Suppose that in Fig. P6.35, $\overline{V}_s = 208 \angle 0^\circ$ V rms, $R_s = 20\ \Omega$, and $\overline{Z}_L = R_L = 4\ \Omega$.

(**a**) Find $a = N_1/N_2$ for maximum power transfer to $R_L$.

(**b**) Using $N_1/N_2$ from part (a), find $\overline{V}_1$, $P_{RL}$, $P_s$, and efficiency $(P_{RL}/P_s)$, where $P_{RL}$ is the average power into $R_L$ and $P_s$ is the average power supplied by the source.

**6.37.** Suppose that the power supply of an audio amplifier has the multitap secondary transformer shown in Fig. P6.37. One section of the amplifier needs 80 V rms, another needs 30 V rms, and the lowest-voltage section requires 12 V rms. (Rectifiers will convert these voltages to the required dc values.) Suppose that we are told that the primary has 400 turns. What are the secondary coil turns $N_A$, $N_B$, and $N_C$ as defined in Fig. P6.37? Assume an ideal transformer.

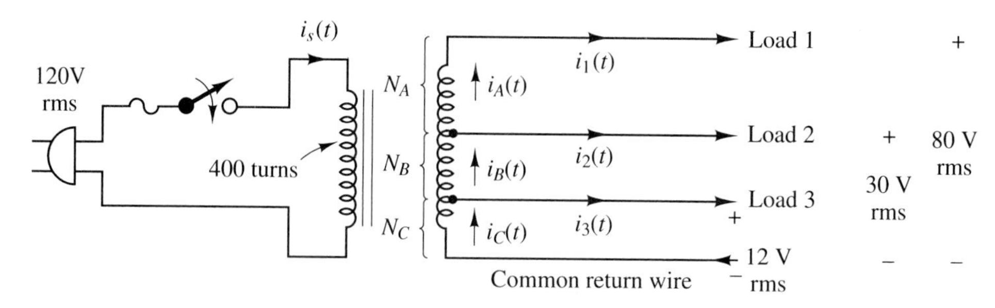


**Figure P6.37**

**6.38.** Suppose that the power supply transformer in Fig. P6.37 drives an audio amplifier having three sections making use of the three available secondary voltages (80, 30, and 12 V rms). Load 1 draws 0.2 A rms, load 2 draws 0.3 A rms, and load 3 draws 0.1 A rms; assume that all loads are resistive. Find the rms value of the source current $i_s(t)$ by using your results in Problem 6.37 and recalling that the sum of mmf values for an ideal transformer is zero. Finally, determine the total power drawn by the amplifier (and compare primary with secondary power).

**6.39.** Suppose that the circuit in Fig. P6.35 is driven by $v_s$, which is a sinusoidal voltage source with frequency $\omega = 1000$ rad/s. (Assume an ideal transformer.) To deliver maximum power to $R_L$, what should be the values of $L$ and $a$? (*Hint:* It will be shown in Problem 6.58 that this occurs if $\overline{Z}_L = \overline{Z}_s^*$.)

**6.40.** Find the impedance $\overline{Z}$ of the circuit in Fig. P6.40 when operating at 60 Hz.

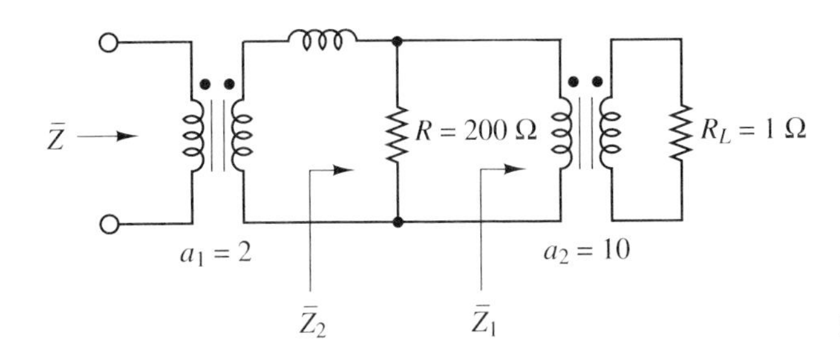


**Figure P6.40**

**6.41.** A 2400 V/240 V transformer is modeled as in Fig. 6.12, where $\overline{Z}_1 = 2.5\ \Omega + j18\ \Omega$, $\overline{Z}_2 = 0.04\ \Omega + j0.20\ \Omega$. Find the effective primary voltage phasor $\overline{V}_s$ when the voltage phasor across the secondary terminals is $\overline{V}_2 = 240\text{ V}\ \angle 10°$. Assume that the load is $\overline{Z}_L = 1.3\ \Omega + j0.5\ \Omega$.

**6.42.** A balanced wye resistive load with each $R = 200\ \Omega$, is connected to a balanced wye-connected three-phase voltage source with $f = 60$ Hz, and each $V_{\text{eff}} = 220$ V.
(**a**) Find the effective line voltage.
(**b**) Find the total average power consumed by the load.

**6.43.** For balanced systems, the fourth "neutral" wire in Fig. 6.18 is unnecessary. In practice, the loads are not balanced. Repeat Problem 6.42 if $Z_A = 100\ \Omega$, $Z_B = 200\ \Omega$, and $Z_C = 300\ \Omega$.

**6.44.** Suppose that in a wye-connected source $\overline{V}_A = 240\text{ V rms}\ \angle 0°$ and $\overline{V}_C = 240\text{ V rms}\ \angle 120°$.
(**a**) Find $\overline{V}_B$.
(**b**) Find $\overline{V}_{BA}$, $\overline{V}_{AC}$, and $\overline{V}_{CB}$.
(**c**) Suppose that in Fig. 6.19 (balanced, wye-connected load), $\overline{Z} = 3\ \Omega\ \angle 10°$. Find the line currents $\overline{I}_A$, $\overline{I}_B$, and $\overline{I}_C$.
(**d**) With the balanced load in part (c), determine the total power supplied by the three-phase source.

**6.45.** Suppose in Problem 6.44 that the transmission lines each have impedance $\overline{Z}_{\text{tr}} = 0.5\ \Omega$. For this situation:
(**a**) What are the line currents?
(**b**) What is the total power absorbed by the load?
(**c**) What is the total power lost in the transmission lines? What is the efficiency in transmission (power supplied by source/load power)?

**6.46.** Suppose that in a 60-Hz delta-connected source $\overline{V}_{CA} = 240\text{ V rms}\ \angle 0°$ and $\overline{V}_{BA} = 240\text{ V rms}\ \angle -120°$. Refer to Fig. 6.23, for which the load is the series combination of 10 Ω and 53 mH.
(**a**) Find $\overline{V}_{CB}$.

**(b)** Assuming cosine-based phasors, determine $v_{CA}(t)$, $v_{BA}(t)$, and $v_{CB}(t)$.
**(c)** Find the load phase voltages $\overline{V}_A$, $\overline{V}_B$, and $\overline{V}_C$. Also give the magnitude of the sinusoidal voltages $v_A(t)$, $v_B(t)$, and $v_C(t)$.
**(d)** Find the source phase currents $\overline{I}_{AC}$, $\overline{I}_{CB}$, and $\overline{I}_{BA}$.
**(e)** Determine the instantaneous load power $p_{\text{tot}}(t)$.

**6.47.** For the circuit in Fig. P6.47, suppose that we wish to use a wattmeter to measure the average power consumed by the lossy inductor in the center branch ($0.2\ \Omega + j10\ \Omega$).
**(a)** Redraw the circuit showing the connections that must be made to the wattmeter; indicate the voltage and current coil connections.
**(b)** Using sinusoidal analysis, predict the value you should read if $\overline{V}_1 = 120$ V rms $\angle 0°$ and $\overline{V}_2 = 10$ V rms $\angle 30°$.

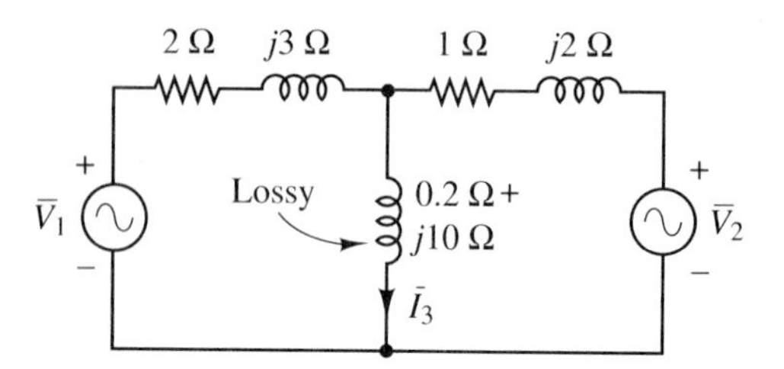


Figure P6.47

**6.48.** Show that in Fig. 6.27, the average power obtained by the delta connected load is $W = W_1 + W_2$, where $W_1 \geq 0$, $W_2 \geq 0$ are the readings of the two wattmeters.

**6.49.** A delta load has impedance 5-Ω + $j$6-Ω between terminals $A$ and $B$ (inductive), 10-Ω between terminals $A$ and $C$ (resistive), and 8-Ω − $j$2-Ω between terminals $B$ and $C$ (capacitive). Convert this delta-connected load to its equivalent wye-connected load.

**6.50.** Figure 6.31 shows a one-line diagram of a wye-wye-connected three-phase system. Draw the equivalent three-phase circuit diagram that the one-line diagram represents. Ignore transmission line impedances.

**6.51.** A 2300 V/230 V 60-kVA wye-wye three-phase transformer is used to drive an industrial motor load. Assume that each phase of the load draws an rms current of 200 A at 0.8 power factor from the secondaries.
**(a)** Find the power consumed by the load.
**(b)** Determine the primary phase and line current rms magnitudes. (*Hint:* For the wye-wye transformer, the per-phase relations are identical to those of the single-phase transformer.)

**6.52.** Figure 6.13 shows the equivalent leakage/loss impedance $\overline{Z}_{\text{tr}}$ which here we will call $Z_{tr,1}$ viewed from ("referred to") the primary side. We can also view this equivalent impedance from the secondary side. It was claimed in the text that in the per unit system these two equivalent impedances are equal. Suppose that a 15-kVA 240 V 110 V 60-Hz single-phase transformer has an equivalent leakage/loss impedance of $Z_{tr,2} = 0.01\ \Omega + j0.05\ \Omega$ when viewed from *the secondary side*. Determine the per unit leakage impedance of the transformer when viewed from the secondary side and when viewed from (referred to) the primary side. They should be equal.

**6.53.** Someone proposes to design a high-voltage dc power transmission line which is to carry 1000 A at 1 million volts from a hydroelectric generator in central Siberia to the industrial Volga River valley 5000 km away (see Fig. P6.53).
**(a)** How much power is sent from the source?
**(b)** If the line is made from two 12-cm-diameter aluminum cables ($\rho = 2.82 \cdot 10^{-8}\ \Omega \cdot \text{m}$), what is the resistance of each cable?
**(c)** How much power in the form of Joule heating is lost in both cables?

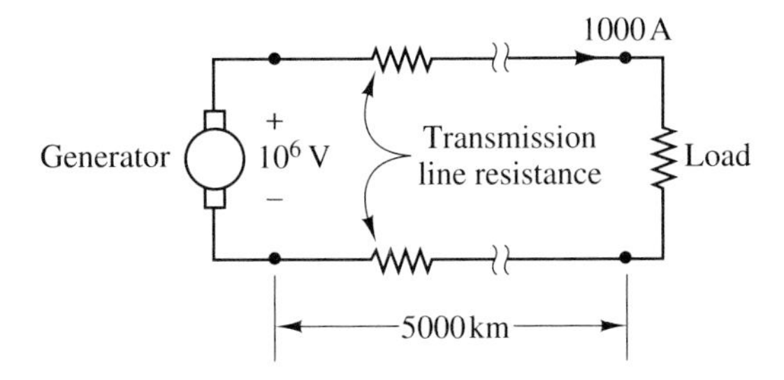


Figure P6.53

**6.54.** A worker servicing a 2200-V power line is suspended in a basket by a long insulating beam, as in Fig. P6.54. A wire from the basket is clamped to the power line. When a nearby tree branch starts to fall onto the line, the worker responds by laying the back of his hand onto the line. Why? Assume that the worker in his perch is very well insulated from the ground.

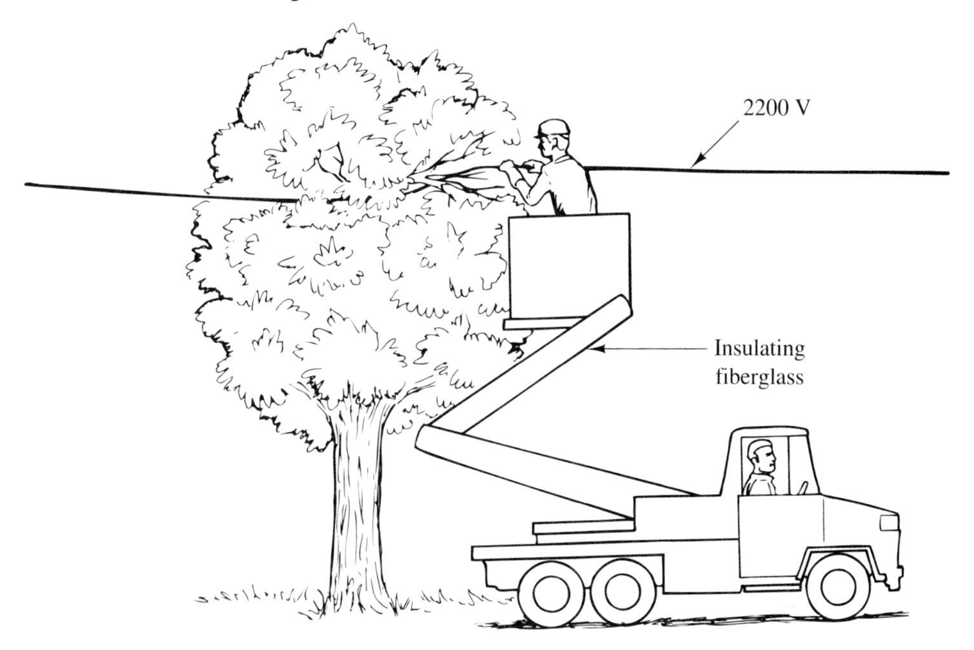


Figure P6.54

**6.55.** One person accidentally touches a 350-V dc conductor on an electric railroad track (subway) and is unable to let go. Another person tries to pull her away, but also becomes paralyzed by the current and is unable to let go from the first victim. Devise some better ways to rescue that person.

**6.56.** Before a lightning stroke occurs, the charged cloud induces a bound charge on the insulators (dielectrics) of a power line. All is fine until the lightning strikes, which suddenly frees the bound charge to possibly cause harmful surge currents in the line. Of course, the effects could be much more disastrous if the lightning actually strikes the line itself. Given the fact that the basic structure of a simplified common lightning arrester involves an air gap between two electrodes, one connected to the protected line, how does it work? To what is the other side connected?

**6.57.** We mentioned that the ground wire in power transmission systems is the one on top. Can you determine why this is? (*Hint:* Think about lightning.)

## ADVANCED PROBLEMS

**6.58.** **(a)** Prove the condition on a load impedance $\overline{Z}_L$ for maximum average power transfer to the load $\overline{Z}_L$ for a sinusoidal source $\overline{V}_s$ with internal impedance $\overline{Z}_s$. (*Hint:* Differentiate $P_L$ with respect to $X_L$ first; then use that value in the derivative of $P_L$ with respect to $R_L$.)

**(b)** Interpret or reconcile your results in part (a) in the context of or with the goals of power factor correction. Bear in mind that when one looks back into one's "wall outlet," there is a multitude of other loads involved in finding "$R_s$" with which to match $R_L$.

**6.59.** Prove that as stated in Sec. 6.5, the actual VA values endured by the coils of an ideal autotransformer are less than the (input VA = output VA) value by the factor $N_s/(N_s + N_c)$. The rating of the transformer is $V_c I_c = V_s I_s$, where $V_c$, $I_c$ and $V_s$, $I_s$ are, respectively, the voltage across and current in the common and secondary windings.

**6.60.** In Chapter 15 we study the three-phase (ac) induction motor. In this problem we look at one phase, which has an applied 60-Hz voltage with amplitude $240\ \text{V}/\sqrt{3} = 139$ V. The model for the induction motor from the point of view of the voltage source is shown in Fig. P6.60. The parameter $s$ represents the rotor speed: If the rotor speed is zero ("locked rotor"/"start"), $s = 1$; at full load, the speed is such that $s = 0.04$. (The precise meaning of the "slip parameter" $s$ will be given in Chapter 15.) For these two conditions, compute the total impedance presented to the voltage source, the current magnitude, and the power factor.

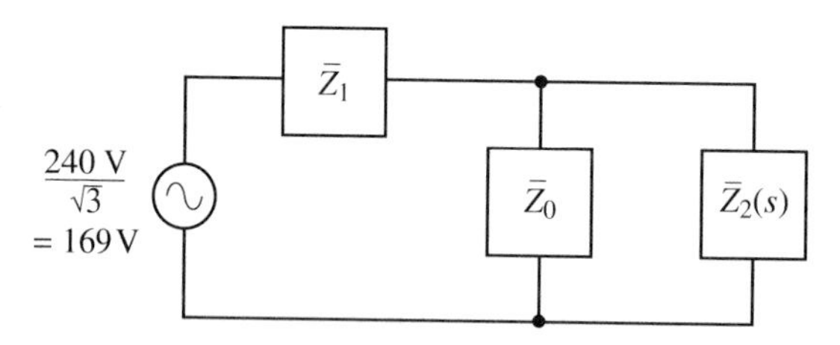


**Figure P6.60**

**6.61.** For the balanced wye-connected load in Fig. 6.26, show that $W_1 = V_{\text{line}} I_{\text{line}} \cos(\theta + 30°)$ and $W_2 = V_{\text{line}} I_{\text{line}} \cos(\theta - 30°)$, where $\theta$ is the impedance angle of the load.

**6.62.** **(a)** Suppose in Problem 6.61 that we measured the average powers $W_1$ and $W_2$. Find an expression for the power factor of the system in terms of $W_1$ and $W_2$. Thus from the wattmeter readings we can determine the power factor of the balanced load.

**(b)** Suppose $W_1 = 219$ kW and $W_2 = 185$ kW. Find the power factor; is it lagging or leading?

**6.63.** In Eqs. (6.54) and (6.55) we found that the instantaneous power into a balanced three-phase load is constant, and thus the instantaneous power is equal to the average power. Does this mean that there is zero reactive power? Some textbooks define a three-phase "total reactive power" $Q_T = 3\, V_{\text{phase}} I_{\text{phase}} \sin(\theta)$. Is there any physical significance to this? It would seem that because there are zero oscillations in total power, such a total "$Q$" would be zero. Resolve the apparent contradiction. How does all this fit into power factor correction?

# Part Three:

# ANALOG ELECTRONICS

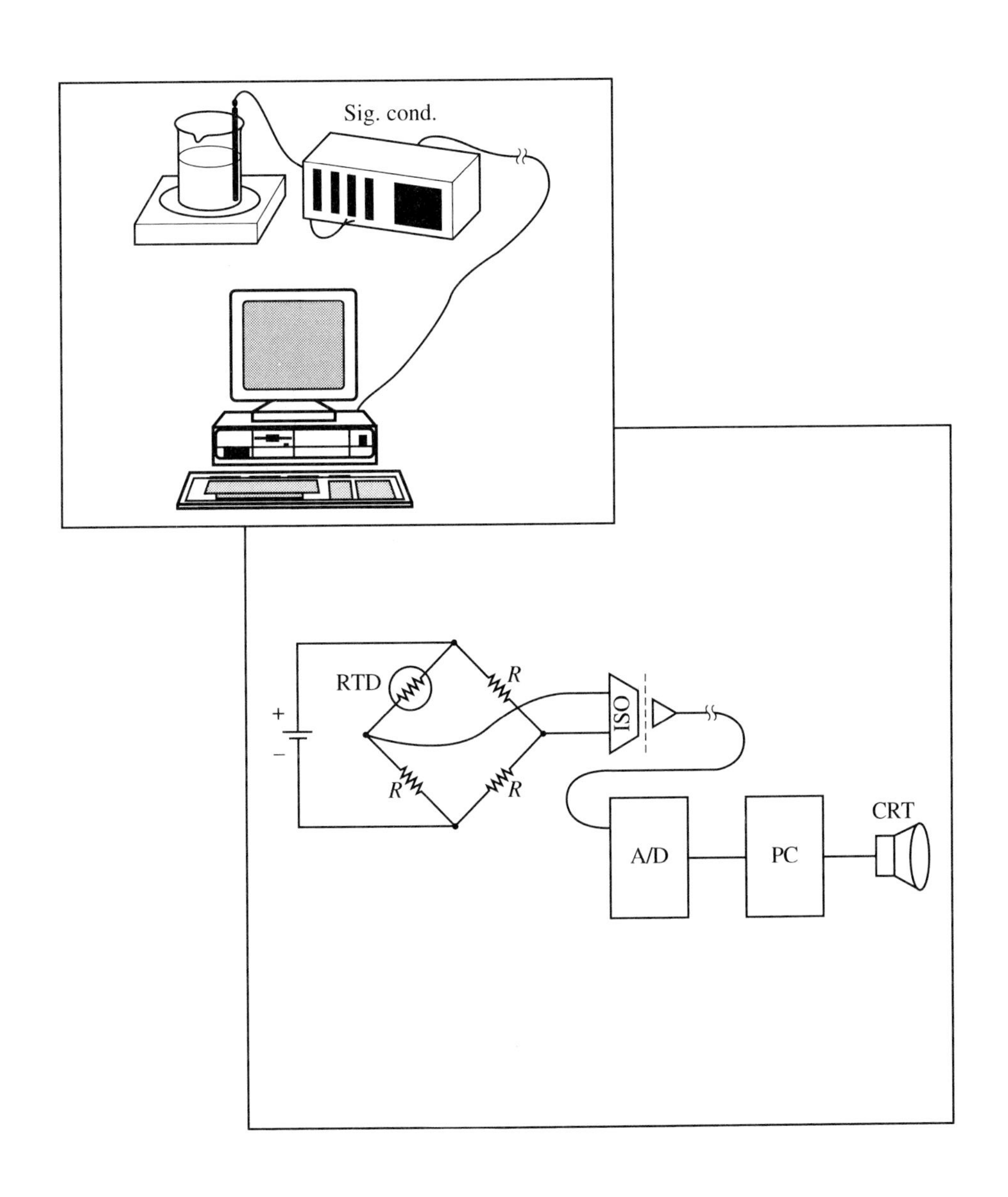

7

# Analog Electronics I: Semiconductor Devices and Circuits

## 7.1 INTRODUCTION

We now have in our possession some very powerful techniques of circuit analysis: voltage and current dividers, KVL, KCL, nodal and mesh analysis, the load line, basic transient analysis, and sinusoidal analysis (phasors, frequency response, complex power, three-phase analysis). Also, we have developed rather substantial knowledge about the physics of circuit devices: electric and magnetic fields, fluxes, and potentials, the mechanism of ohmic electric current, capacitor theory, how batteries work, the decompositions of electric and magnetic fields, electromagnetic induction and its application in inductors and transformers, wattmeters, cathode-ray oscilloscopes, multimeters, and even an introduction to electromagnetic waves.

The question now is: What don't we have? Essentially, *everything* of modern practical interest: circuits based on and/or providing electronic and electromechanic amplification. Today's civilization is based on the manipulation of audio/visual/alphanumeric information and of electrical/mechanical power. The electronics revolution is due to the cheap production of electronic devices crucial to carrying out both of these tasks. Electronic amplification is the foundation of all the chapters on both analog and digital electronics (the next six chapters), and electromechanic amplification is the subject of our final three chapters.

## 7.2 THE CONCEPT OF AMPLIFICATION

Electromechanic amplification is the phenomenon whereby flipping a switch causes a motor to drive a huge load, or a generator to produce gigawatts of electric power.

Electronic amplification is the use of the properties of electrons in solids (or occasionally, gases) to cause a minute input (e.g., measurement) signal to become strong enough to drive a loudspeaker, chart or tape recorder, video screen, or other actuator—in a manner precisely reflecting and preserving the information content in the input signal. It is on electronic amplification that we now focus our attention.

In Sec. 6.5 we found that the transformer amplifies voltage or current, depending on the turns ratio. Why can't we use the transformer to do our signal amplification as described above? The reason is that the power gain of a transformer is always less than 1 (due to losses). Even an ideal transformer (unity power gain) cannot *increase* signal power. To enable a microwatt tape player head to drive speakers requiring a million times that power, we certainly cannot use a device that attenuates signal power. Or, if we wish to send a voice signal over hundreds of miles of telephone wire, we will need something to boost the signal to prevent the lossy wire from obliterating the signal.

How can we achieve such huge power gains? To help answer this question, we consider a mechanical analog of a transformer, the seesaw, shown in Fig. 7.1. Although it takes only a small motion to move the child way up on the other side, this takes a heavy person. In fact, one seesaw takes as much energy to do this way as it would if the support were at the center of the board. A given rate of seesawing requires of the heavy person the same power required to lift the child without the seesaw; the seesaw, like the transformer, is a unity-power-gain device (ignoring friction).

In the transformer, a voltage gain $N_2/N_1$ from primary to secondary is possible only with simultaneous reduction of current by the same factor. Thus the power, the product of voltage and current, is not increased from primary to secondary. Instead, we need the analog of an engine-driven crane (see Fig. 7.2). In this case we wiggle a lift lever with one relaxed hand, and at the output a 10-ton weight follows the same

**Figure 7.1** The seesaw, a unity-power-gain device. The output swing is larger for a given input swing, but the weight of the child that can be lifted is reduced compared with the situation in which the support is at the center of the seesaw board.

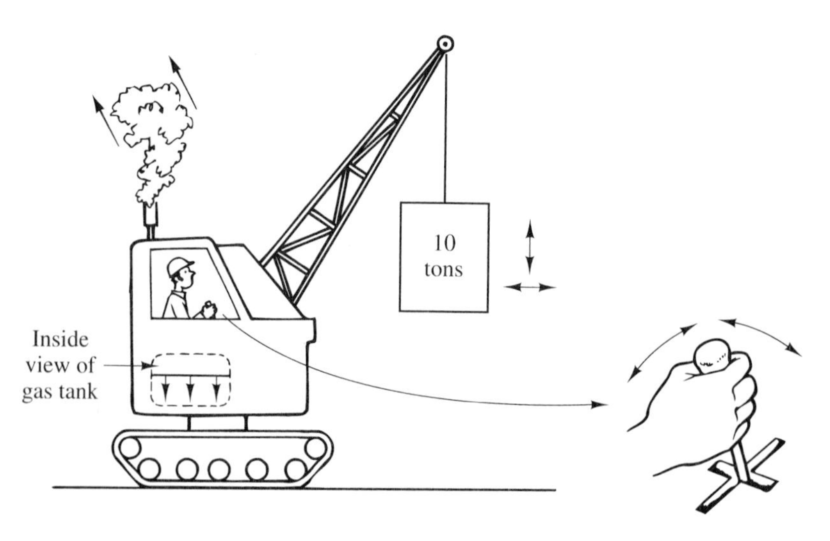


**Figure 7.2** The crane: A greater-than-unity-power-gain device.

wiggle pattern with much wider swing. Both the output swing *and* the weight that can be lifted have dramatically increased. An astronomical power gain is at our fingertips. This is possible only because of the gasoline-guzzling engine under our control (note exhaust and level of gas tank).

To have an analogous circuit element, we need a new device. We need "control of the floodgates," the "flood" being electric current. In fact, the control terminal of one type of these devices is called the gate. In Chapter 5 we noted that information-bearing signals take the form of time-varying voltages and/or currents. Thus it is only reasonable that the control of our floodgates should be done by the signal—by a voltage or current *distinct from* the "flood" current.

What is the source of the flood current? Not a gasoline engine, but rather, a constant voltage source. (As we noted in Sec. 2.10, the voltage source gets *its* energy from chemical energy in the case of batteries, or from the local power generating plant—ultimately, from the sun.) The voltage source must be constant, so that our *signal,* and *only* our signal, controls the time variations in the output. We effectively end up with a time-varying voltage source as powerful as the constant voltage source, but which varies in exact accordance with our weak signal source. This is the essential improvement.

The reason for the term "electronic" rather than just "electric" amplification is that the mechanisms of control in these *power modulators* involve the intimate interactions of electrons on the atomic scale. Not surprisingly, then, to understand such devices fully necessitates knowledge of quantum mechanics.

Because of this, and unlike basic electromagnetism as presented in Chapters 2 and 3, quantitative descriptions of semiconductor devices are not simple. Thus we shall stay away from any detailed quantum mechanical derivations and explanations in this introductory study and focus on elementary presentations of processes and results. Our goal here will be merely to make plausible the simplest models usable for describing circuit behavior of electronic devices.

The device behind electronic amplification is called the transistor, a three-terminal device. While not all transistor types can be said to live up to the name transistor (= *trans*fer res*istor*), the name has stuck. By *transfer resistor* we mean that by some means a signal applied as the input to one terminal of the device is transferred into, or replicated in, the value of the resistance between the other two terminals. To comprehend the basic operation of the transistor requires first a study of a simpler electronic device on which it is based: the diode.

## 7.3 THE DIODE: OPERATION

### 7.3.1 Introduction

Not only is the diode the building block for a transistor, it is also itself an extremely useful device. In this section we investigate its basic behavior; in Sec. 7.4 we consider a few practical uses of it.

The name *diode* stems from its original appearance in vacuum tubes, where electrons traveled one-way between *two* electr*odes* (di-ode). Another name for the modern diode is the solid-state rectifier, or simply the rectifier. *Solid state* contrasts modern electronic devices (made from crystals) with vacuum-tube electronic devices (gaseous state). *Rectifier* indicates that it makes something better (it rectifies). Its most basic way of "making things better" is to transform an ac voltage to a unipolar (single-polarity) waveform. For example, some of the rectifiers we discuss in Sec. 7.4 are typical of those found in "ac adapters" used to operate small electronic equipment (such as radios). This equipment requires dc power, so the available ac wall outlet voltage must be converted.

An input sinusoidal (ac) waveform $v_s$ is shown in Fig. 7.3a, and the rectified output waveform $v_L$ is shown in Fig. 7.3b. The transforming operation is called *half-wave rectification* because only half of the input pulses are retained in the output. This idea of rectification is the most critical step in converting ac to dc.

First let us propose what sort of device would do this, that is, its $i$–$v$ characteristic. The required conditions for an ideal diode in series with a load are that when the voltage across the diode begins to become positive (known as *forward bias*), it behaves as a short circuit, and when the voltage across it goes negative (called *reverse bias*), it is an open circuit. If we had such a device that conducted only one way (see Fig. 7.4a), we would have the desired response (see Fig. 7.5). In Fig. 7.4b is shown the symbol for the diode (rectifier). Its arrow points in the only direction that conventional current can flow in it, and the bar indicates that current cannot go the other way. Note the close analogy with traffic on a one-way street. Also, keep in mind that electrons go the direction opposite that of conventional current.

In Fig. 7.5 we see that this indeed will do the job. For $v_s > 0$, $v_D$ begins to become positive, causing the diode to behave as a short circuit, thereby preventing $v_D$ from becoming any larger. In this case, $v_L \approx v_s$; the positive voltage pulse is preserved in the output. When $v_s < 0$, $v_D$ becomes negative and the diode open circuits. In this case, $v_L = 0$; the negative voltage pulse is absent in the output—the output is zero during the negative input pulses. Thus all we need now is an actual device having the $i$–$v$ characteristic in Fig. 7.4a.

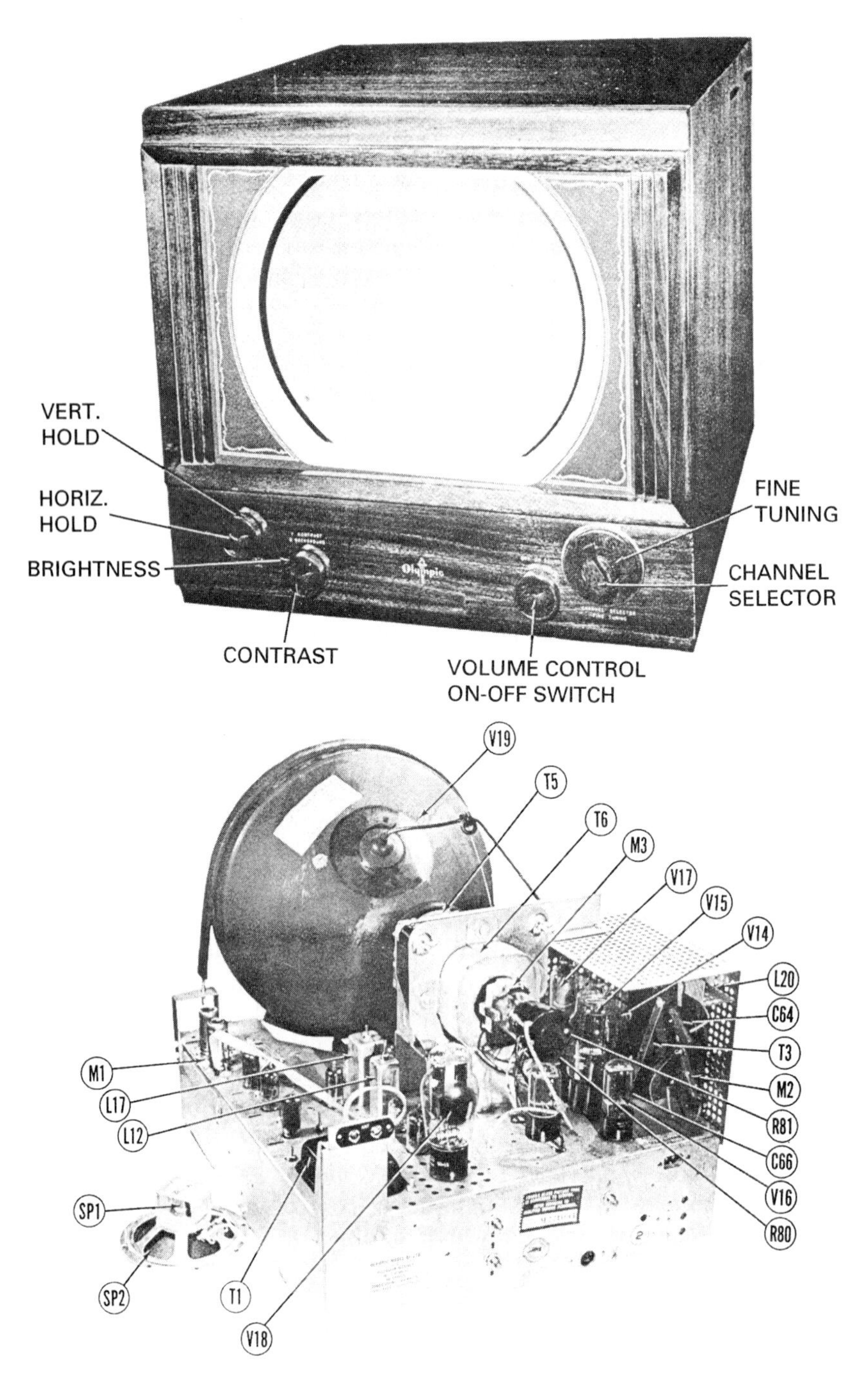


1950 Olympic 12-inch round-screen television; front and inside views. Fixing this old TV in the basement was such fun that the author later took up electrical engineering as a career. Bought new for the author's grandfather by his father, the TV still works, but now takes about 20 minutes to warm up and give a bright picture. Note the large number of vacuum tubes. *Source:* Howard W. Sams & Co.

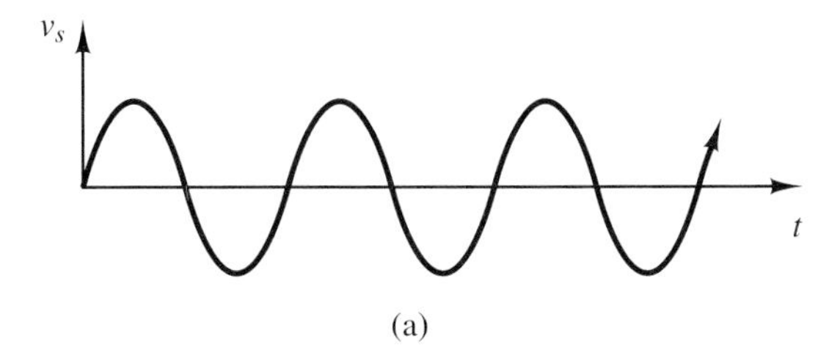


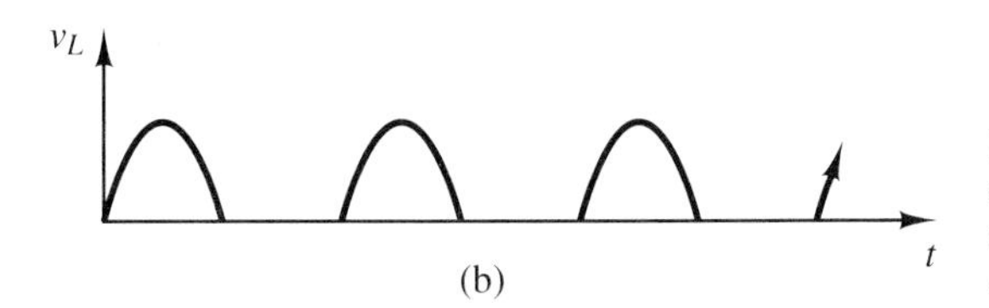


**Figure 7.3** Half-wave rectification of a sinusoidal waveform: (a) original sinusoidal input; (b) half-wave rectified output.

### 7.3.2 Diode Physics

Up to now we have considered only two basic types of material (with respect to electric conduction): one type of charge carrier and one form of electric current. These will not suffice for making a rectifier or for many other electronic devices.

For all of our previous devices, the only two kinds of material were insulators and conductors. The only carrier of electric current was the electron. Finally, in our studies electric current has always been proportional to a driving electric field or voltage; this is ohmic current. Obviously, ohmic current occurs in resistors, where $i = (1/R)v$. Even though $i = C\,dv_C/dt$ for a capacitor, the actual current in the metal plates and wires is proportional to the push of a local electric field. Similarly, the current in the wires of an inductor, $i = (1/L)\int v_L\,dt$, is proportional to the local electric field. The current is not proportional to the *terminal* voltage $v_C$ or $v_L$ in these energy storage devices, however, because of the peculiar electric and electromag-

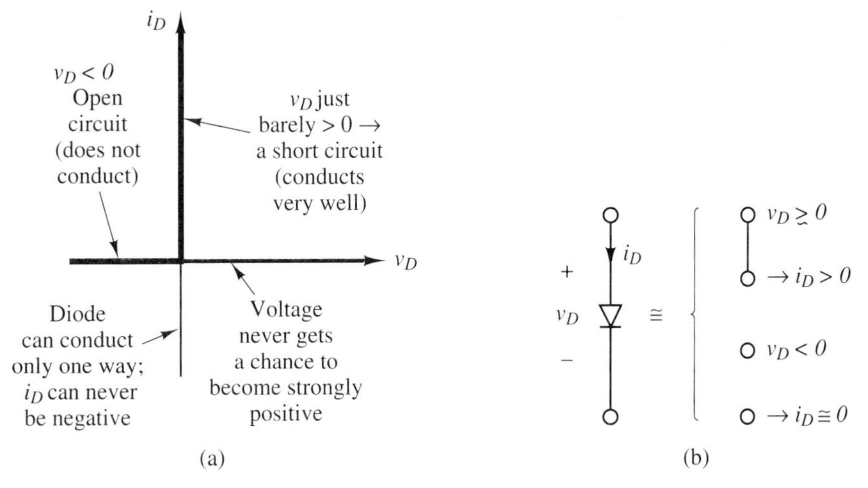


**Figure 7.4** Ideal diode: (a) *i–v* characteristic; (b) circuit models for conducting (forward bias) and nonconducting (reverse bias) modes of diode operation.

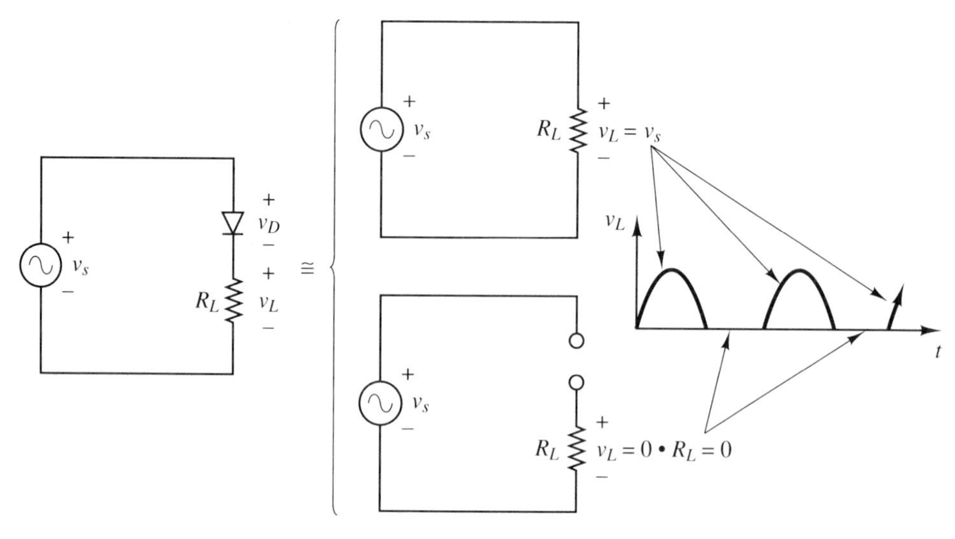


**Figure 7.5** Diode in-circuit as a half-wave rectifier. When $v_s > 0$, $R_L$ is connected to $v_s$, so $v_L = v_s$. When $v_s < 0$, the diode is an open circuit so that $R_L$ is disconnected, so $v_L = 0$.

netic effects in these devices we discussed in Chapters 2 and 3. (Some people reserve the term "ohmic" for terminal voltage–current proportionality, while using "drift" for local electric field–current proportionality. In electronic devices they are not always the same, but for simplicity in this book we use the term "ohmic" to denote both.)

To obtain the benefits of electronic control and amplification, we need a new kind of material called a *semiconductor.* Frequently, two types of charge carrier appear in semiconductors: the usual mobile ("free") electrons and mobile electron deficiencies, called *holes.* Also, it is usually desired that a new type of conduction take place in addition to ohmic conduction: *diffusion.*

A semiconductor is just that—somewhere between an insulator and a conductor. The element most commonly used as a semiconductor today is silicon. Exactly how conductive a semiconductor is can be controlled by its elemental composition (via introduction of impurities into the silicon crystal), by exposure to heat and light, and by the voltages and currents applied to it. The conductivity of a semiconductor can be varied over an extremely wide range: from a resistivity $\rho$ of $10^6\ \Omega \cdot \text{cm}$ down to $10^{-3}\ \Omega \cdot \text{cm}$ or less. Compare these resistivities with those of metals ($\rho = 10^{-6}\ \Omega \cdot \text{cm}$) and good insulators ($\rho = 10^{12}\ \Omega \cdot \text{cm}$) and you see how a semiconductor gets its name.

One way of introducing mobile charge carriers, free electrons and holes, into semiconductors is by a process known as *doping.* Doping is the introduction of impurities into the silicon crystal which, as mentioned above, can greatly increase the conductivity of silicon by increasing the number of mobile charges, either electrons or holes. There are two types of doping called donor and acceptor doping, one to increase conducting electrons and the other to increase conducting holes.

*Donor doping* is the introduction of atoms having an extra electron not required for making its bond with the silicon crystal atoms. The donated electron is free to move throughout the crystal, and the donor atom becomes an immobile positively charged ion; only electrons, not holes are present. Thus as long as the donor-doped region still has all its free electrons in the vicinity, the *net* charge in the region is nevertheless still zero.

*Acceptor doping* is the introduction of atoms deficient by one electron, with respect to making a complete set of bonds within the silicon crystal. This deficiency, called a hole, is free to move throughout the crystal, leaving behind the negatively charged, immobile acceptor ion. In this case there is no free electron. In summary, a conduction electron is made available via donor doping, and a conduction hole is made available via hole doping.

Even without doping, a certain amount of hole conduction can occur in semiconductors due to the inherent (*intrinsic*) presence of holes. These holes are due to thermal motion, which breaks bonding electrons free from the crystal, leaving mobile holes behind. Unlike the situation of doping, in this case each mobile hole has a corresponding mobile electron—both can move around in the crystal. The number of free electrons and holes generated by thermal motion is $n_i = 1.5 \cdot 10^{10}$ holes $cm^3$ in pure silicon at room temperature, out of a total of $5 \cdot 10^{22}$ silicon atoms/$cm^3$ (the subscript $i$ is for "intrinsic"). Often doping can be as high as $10^{19}$ dopant atoms/$cm^3$ with one free hole or electron contributed per dopant atom. Compare this with $10^{23}$ free electrons/$cm^3$ in copper (Sec. 2.9). With heavy enough doping, a semiconductor becomes a good conductor, like metal.

Holes do not appear in significant numbers in insulators because the electrons are bound too tightly. They do not appear in metallic conductors because the binding electrons in metals are already free to move about—so much so that an outer (binding) electron is not even considered to be associated with one particular atom. Moreover, in a metal the positive charge left behind when an electron moves away is an immobile metal ion, not a mobile hole—the same as donor-doped conduction in a semiconductor.

**Example 7.1**

Calculate the resistivity of a slab of silicon having both donor doping concentration $N_d = 10^{15}\ cm^{-3}$ and acceptor doping $N_a = 10^{17}\ cm^{-3}$. Compare this with intrinsic silicon.

**Solution** The total concentration of free electrons is $n = n_i + N_d \approx N_d$ (because $n_i$ is only roughly $10^{10}/cm^3$) and the concentration of holes is $p \approx N_a$. From Sec. 2.9, just after Eq. (2.34), we can straightforwardly extend the relation for resistivity for electron-only conduction to resistivity with both electrons and holes as $\rho = 1/\sigma = 1/[q(q\ \Delta t_e n/m_e + q\ \Delta t_h p/m_h)]$, where $\Delta t_e$ and $\Delta t_h$ are the respective mean times between collisions for electrons and holes, $m_e$ and $m_h$ are the respective effective masses of electrons and holes. In semiconductor studies, the quantities $\mu_e = q\ \Delta t_e/m_e$ and $\mu_h = q\ \Delta t_h/m_h$ are called the *mobilities* of electrons and holes (the $\mu$ is *not* permeability). We can see why $\mu$ is called mobility by equating (for a material with only electrons) the expression for current in Eq. (2.34), $I = \sigma AV/d = \sigma AE = qn\mu AE$, with that for current in Eq. (2.31), $I = Nn_e eAU = nqAU$ (here $Nn_e = n$), to obtain $\mu = U/E$; that is, $\mu$ tells us how fast the charge moves ($U$) for a given applied electric field ($E$). For silicon, the values of mobility are, respectively, $\mu_e = 1350\ cm^2/(V \cdot s)$ and $\mu_h = 480\ cm^2/(V \cdot s)$; thus electrons are nearly three times as mobile as holes in silicon. Therefore, $\rho = 1/[1.6 \cdot 10^{-19}(1350 \cdot 10^{15} + 480 \cdot 10^{17})] = 0.13\ \Omega \cdot cm$. For intrinsic silicon, $n = p = n_i$, so $\rho = 1/[1.6 \cdot 10^{-19}(1350 + 480)10^{10}] = 3.4 \cdot 10^5\ \Omega \cdot cm$. This example should give an idea about the power of doping.

Like holes, the new kind of conduction called diffusion current is essentially unique to semiconductors. Diffusion occurs in certain electronic devices, notably, the *pn* junction diode and the "bipolar" transistor we discuss later in this chapter. It occurs when mobile charge concentration distributions are spatially nonuniform. The mobile charge in a silicon crystal is like ink in a glass of water: It tends to diffuse (i.e., migrate) to a lower, more uniform concentration. Similarly, locally high concentrations of either holes or electrons tend to spread out to uniform, lower concentrations. The greater the steepness of variation in concentration, the faster the diffusion happens, that is, the larger the diffusion current.

The reason electron/hole diffusion occurs only in semiconductors is that in conductors, the charge concentrations never have a chance to build up—they move about and redistribute themselves too easily and quickly for diffusion to be noticeable. In insulators there are very few mobile charges at all, so again local concentrations cannot build up. In semiconductors we have seen that by doping and even just thermal motion, substantial concentrations of free (mobile) charge may be present. And if we have adjacent regions of silicon with different doping levels, there will be *nonuniform* concentration distributions. Therefore, there will be a tendency for the high concentrations of holes and/or electrons to diffuse.

Specifically, if we have a region doped with many acceptors (and thus free holes, called *p* material) beside a region doped with many donors (and thus free electrons, called *n* material), diffusion will take place. This is an extremely steep variation in electron and hole concentrations. Referring to the *pn* junction in Fig. 7.6a, holes will diffuse to the right (into the *n* region) and electrons to the left (into the *p* region). It should be mentioned that the motion of a hole is really the motion of bound electrons, one by one filling the hole at the old hole location, thereby transferring this deficiency to the old bound electron location. A hole may be *created* by the breaking of an electron bond, but its *motion* does not involve bond breaking, only spatial redistribution of bonds.

The electrons in the *p* region (both diffused and intrinsically present) as well as the holes in the *n* region are normally both in the minority and thus are called *minority carriers*. Similarly, electrons in the *n* region and holes in the *p* region are called *majority carriers*.

When an electron deficiency (hole) meets an excess electron (electron), the two annihilate each other. (This process is usually called *recombination,* even though the

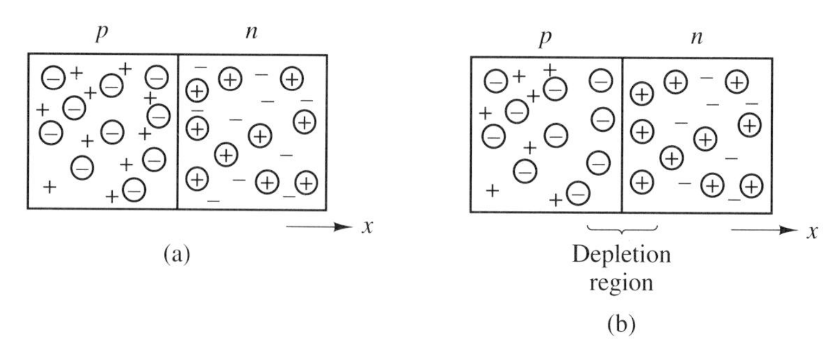


**Figure 7.6** The *pn* junction, showing mobile charges (uncircled) and immobile dopant ions (circled): (a) before equilibration; (b) at equilibrium.

electron and hole may never have previously been "combined.") The result is that the electron deficiency no longer exists, and the once-free electron now takes part in the binding of two silicon atoms. This naturally initially happens most in the vicinity of the junction, where the holes and electrons are encountering each other in large numbers due to diffusion.

Consequently, the region right around the junction becomes depleted of mobile charge and is thus called the *depletion region* (see Fig. 7.6b). For purely illustrative purposes, Fig. 7.6a shows the situation that would occur at the moment when $p$ and $n$ materials are joined together (they are not joined in this manner in practice), and Fig. 7.6b shows the final equilibrium situation.

The dopant ions are left exposed. The depletion region is therefore *not* depleted of immobile charge, but only of mobile charge. It is not neutral: In the $p$ region the net charge is negative, and in the $n$ region there is a net positive charge due to the immobile dopant ions.

Once the depletion region is formed, there are extremely few opportunities for mobile charge annihilation to occur because of the lack of available carriers; now the only significant number of annihilations occurs outside the depletion region. (Note, incidentally, that holes do not "recombine" with acceptor atoms, because as we already know, those bonds are too weak to be maintained at room temperature.) A theorem known as the *principle of detailed balance,* valid for equilibrium, says that there is no net transfer of holes to the $n$ material or electrons to the $p$ material. In fact, there are essentially no minority carriers outside the depletion region in this situation of equilibrium, other than those of the thermally generated electron–hole pairs. Thus the rest of the $p$ and $n$ materials are electrically neutral. This *depletion approximation* is often good enough to describe basic operation of heavily doped junctions and is used extensively in practice.

From Gauss's law [Eq. (2.6)], there is a nonzero electric field in the depletion region due to these immobile charges. (The relative permittivity of silicon is 12.) Assuming that the doping density within each region is uniform, the electric field will increase linearly toward the junction in both halves of the depletion region. This electric field will point from the positive immobile charge ($n$ region) to the negative immobile charge ($p$ region).

Because the potential is the integral of the electric field [Eq. (2.15)], the electric potential will rise in a quadratic way and will be highest at the edge of the depletion region in the $n$ material. The "built-in" rise in potential $V_{\text{bi}}$ in typical silicon $pn$ junctions is around 0.75 V.

**Example 7.2**

Advanced semiconductor physics and statistical mechanics shows that the built-in potential of a $pn$ junction is $V_{\text{bi}} = (kT/q)\ln(N_aN_d/n_i^2)$, where $q$ is the electron charge ($1.602 \cdot 10^{-19}$ C), $k$ is Boltzmann's constant [$1.38 \cdot 10^{-23}$ J/kelvin (K)], $T$ is the temperature (in kelvin). At 300 K, what is $V_{\text{bi}}$ for a $pn$ junction doped with $N_d = 7 \cdot 10^{15}$ donors/cm$^3$ in the $n$ region and $N_a = 8 \cdot 10^{16}$ acceptors/cm$^3$ in the $p$ region?

**Solution** At 300 K, $kT/q = 0.026$ V and $n_i = 1.5 \cdot 10^{10}$ carriers/cm$^3$. Thus $V_{\text{bi}} = 0.026 \ln[7 \cdot 10^5 8 \cdot 10^{16}/(1.5 \cdot 10^{10})^2] = 0.74$ V. Advanced analysis shows that $n_i$ is roughly proportional to $e^{-T_A/T}$, where $T_A$ is a constant. Substitution of this for $n_i$ into the expression for $V_{\text{bi}}$ above shows that $V_{\text{bi}}$ is roughly a linear function of $T$. Also, the higher the doping levels, the larger $V_{\text{bi}}$ will be; that dependence is logarithmic.

Remember that because outside the depletion region there are no uncanceled immobile charges, there is charge neutrality there and thus no electric field or variation in potential. The left-to-right electric field $E_{LR}$ (convince yourself that it should be negative) and built-in potential $V_{bi}$ are shown in Fig. 7.7. The main effect of this electric field with the corresponding potential rise is that it tends to push the diffusing mobile charges back from where they came. For example, the right-to-left field is positive, and thus it pushes the positively charged migrating holes back home ($\mathbf{F} = q\mathbf{E}$). Furthermore, we may view the potential as an energy barrier over which all majority carriers must jump in order to get to the other side (and there become minority carriers).

Naturally, the total current in the diode at equilibrium is zero when its terminals are disconnected from any circuit. Although some current results from charges jumping over the barrier by thermal excitation, an equal and opposite current is due to an equal number of them sent back by the depletion region electric field.

The width of the depletion region and corresponding height of the energy barrier $V_{bi}$ are tied to the doping concentrations in the $n$ and $p$ regions. (*Note*: The fact that mobile charges must pass through the depletion region in jumping over the barrier shows that momentarily, mobile charges *can* be found in the depletion layer, crossing over. They just do not "live" in there very long without being either annihilated by the plentiful opposing mobile carriers from the new side, sent back by the depletion region electric field, or reaching the new side, again to be annihilated.)

One might be concerned that $V_{bi}$ could be used to drive currents through a load resistor—the equivalent of a perpetual motion machine!(?) Can we measure $V_{bi}$ with a voltmeter? The answer to both of these questions is no. For a voltmeter (or the load resistor, or any other circuit element) to make electrical contact with a $pn$ junction, two more junctions must be added: Those between the metal wires going to the voltmeter and the $p$ or $n$ material.

Upon adding these necessary contacts, which have their own built-in potentials, advanced analysis shows that the complete potential diagram would appear as

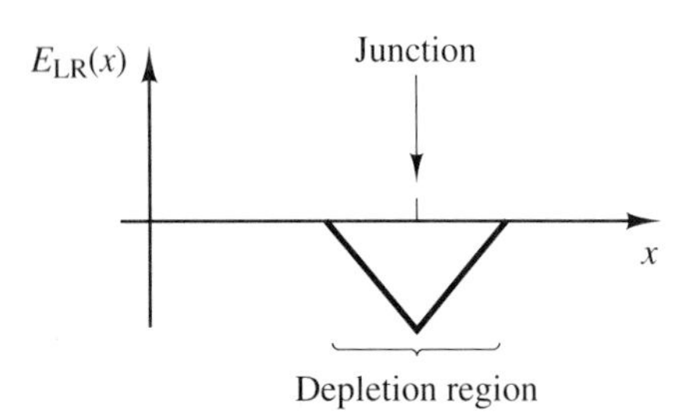


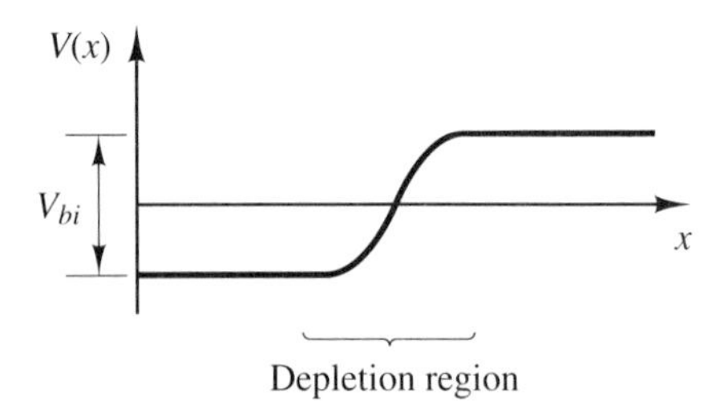


**Figure 7.7** Electric field $E_{LR}$ and built-in potential $V_{bi}$ within diode at equilibrium.

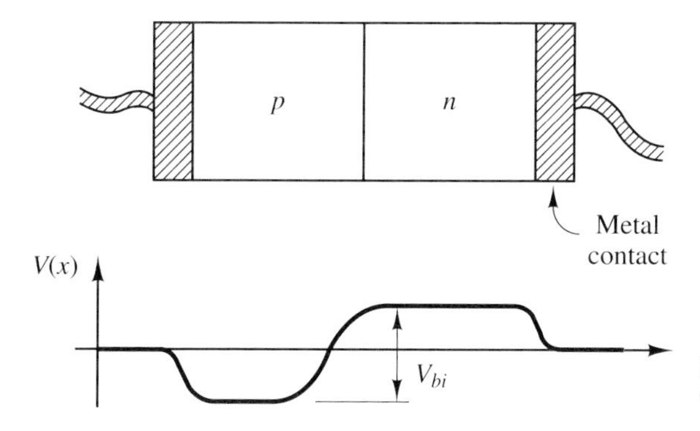


**Figure 7.8** Potential along the diode, with ohmic metal contacts.

in Fig. 7.8. In a manner that is not clearly documented, these metal–semiconductor junctions do simultaneously have very substantial built-in potentials [Gray and Searle],* yet nevertheless, conduct equally well either way (are so-called *ohmic contacts*).

We now come to the climax of this discussion: What happens when we apply a voltage to the *pn* junction? In such a case, the equilibrium situation we have just described is disrupted. Figure 7.9 shows the answer for the case $V_a > 0$. If the applied voltage $V_a$ is positive, the *effective* barrier to electron–hole movement across the junction is reduced. That is, the barrier electric field is partly canceled by the applied electric field, and the depletion layer width is correspondingly reduced.

(Because of the much higher conductivity there than in the depletion region, the electric field due to $V_a$ is very small in the neutral $n$ and $p$ regions and in the low-resistivity ohmic contacts. Thus, nearly all of the electric field due to $V_a$ appears across the depletion region. Recall voltage dividers!)

The reduced number of electric flux lines due to the reduced depletion region electric field requires fewer immobile charges with which to begin and end the flux lines. The fixed (by doping) immobile charge concentration in the depletion region therefore translates into a reduced charged (depletion) region width.

Thus the floodgates of diffusion are opened by $V_a > 0$. For example, electrons from the $n$ region diffuse into the $p$ region, and annihilate holes existing there. Because this diffusion is initiated by us (externally) when we connect the battery $V_a$, this phenomenon is called *electron injection.* Similarly, holes from the $p$ region diffuse into the $n$ region and annihilate the electrons existing there; this process is called hole injection.

Both processes may be referred to by a single term, *minority carrier injection.* Here "minority" refers to the status of the mobile, diffusing charge in its *destination* region, not its source region. Both of the minority carrier injection currents make up the main current through the diode.

It may be asked why the continual annihilation of mobile carriers does not cause the supplies of both injected minority carriers and resident majority carriers to run out. If the majority carriers ran out, the background ions along with the intruding minority carriers would create a large field opposing current even under forward bias. We informally say that this charge would reduce the "forward biased-

*P. E. Gray and C. L. Searle, *Electronic Principles* (John Wiley & Sons, New York, 1969), p. 151.

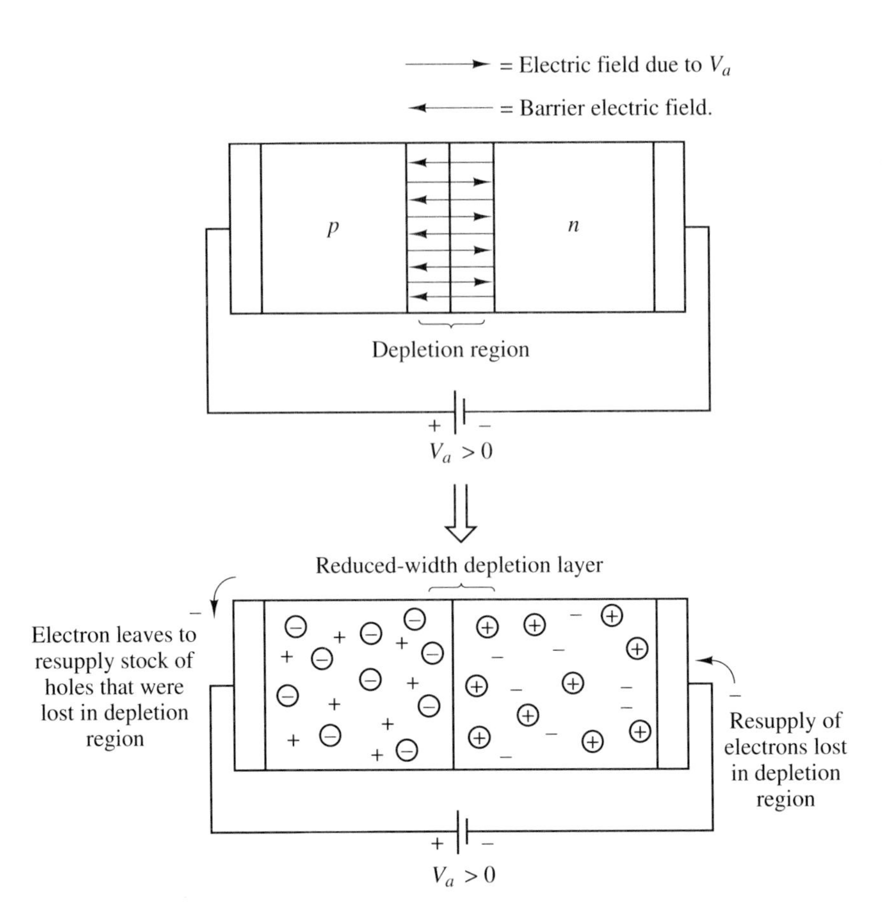


**Figure 7.9** Forward bias of a semiconductor diode.

ness'' of the diode and thus the forward current. If the minority carriers ran out, diffusion would cease and there again would be no forward current.

The reason that the supplies do not run out is essentially that there is a continual resupply from the other side of the junction, through the external circuit. That is, electrons leave the *pn* junction at the *p* side (thereby replenishing the hole population in the *p* region), go through the voltage source (or effectively do so), and return into the diode at the *n* side (thereby replenishing the electron population in the *n* region); see Fig. 7.9. We may conclude that as long as $V_a$ is maintained, a large current will flow through the diode, as there is now only a reduced barrier (and narrowed depletion layer) limiting the current.

On the other hand, if the applied voltage $V_a$ is negative, the situation is as shown in Fig. 7.10. The applied electric field now aids the built-in field in the depletion region. The depletion region widens, and the barrier against current is even higher than it was for the case of no applied voltage. One might conclude that as in the case of no applied voltage, zero current will flow. Actually, there is a small reverse current, due partly to the fact that there are *some* holes in the *n* region and electrons in the *p* region as a result of thermal excitation. Clearly, these minority carriers—unlike majority carriers—are *swept* across the junction by the electric field in the

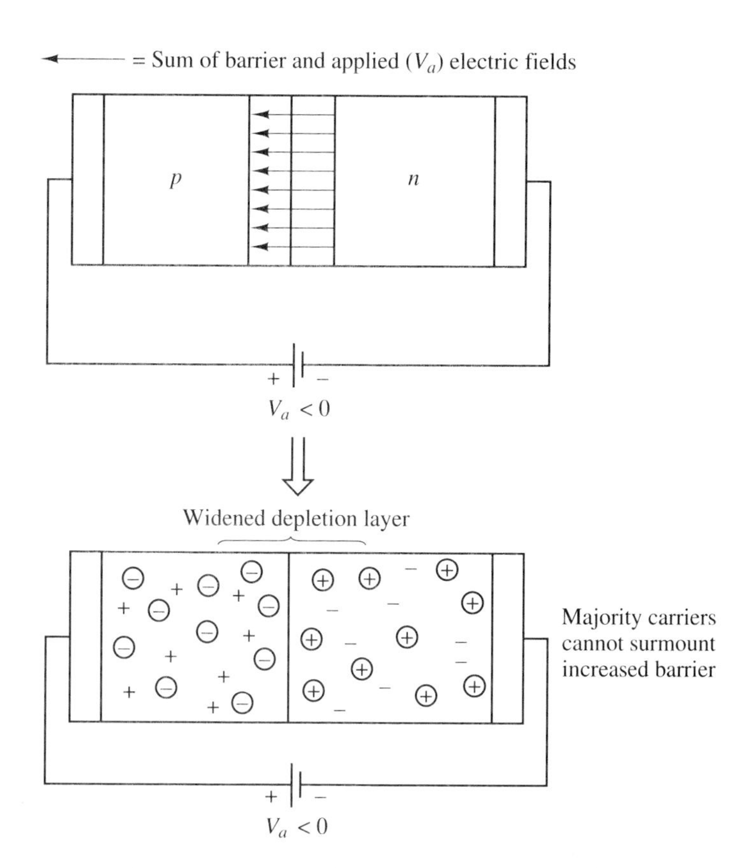


**Figure 7.10** Reverse bias of a semiconductor diode.

depletion region, and become majority carriers. Naturally, this current is called *majority carrier injection.*

In conclusion, we have just what we wanted: very high conductivity for a positive applied voltage (forward bias), and very low conductivity for a negative applied voltage (reverse bias). There are only a few departures from the ideal diode *i–v* characteristic given in Fig. 7.4a, which will now be described one by one. They are indicated graphically in Fig. 7.11, which also shows other practical diode models described below. In Fig. 7.11 and the following, we rename $V_a$, the voltage across the diode, $v$ for genericness.

First, the applied voltage must nearly cancel $V_{\text{bi}}$ for forward conduction to be large. We call this voltage the *contact potential* $V_o$, which is slightly less than $V_{\text{bi}}$; for a silicon diode, it is about 0.7 V. (Often, "built-in potential" and "contact potential" are considered synonyms because they are essentially equal. It should also be noted that, infrequently, germanium is used in place of silicon as the basic semiconductor material; for germanium, $V_o \approx 0.3$ V.) Second, as mentioned above, there is a small reverse current $-I_{\max}$ under reverse bias ($V_a < 0$).

Much more sophisticated treatments of the theory than this show that an excellent model of the actual device forward bias *i–v* characteristic is represented by the formula

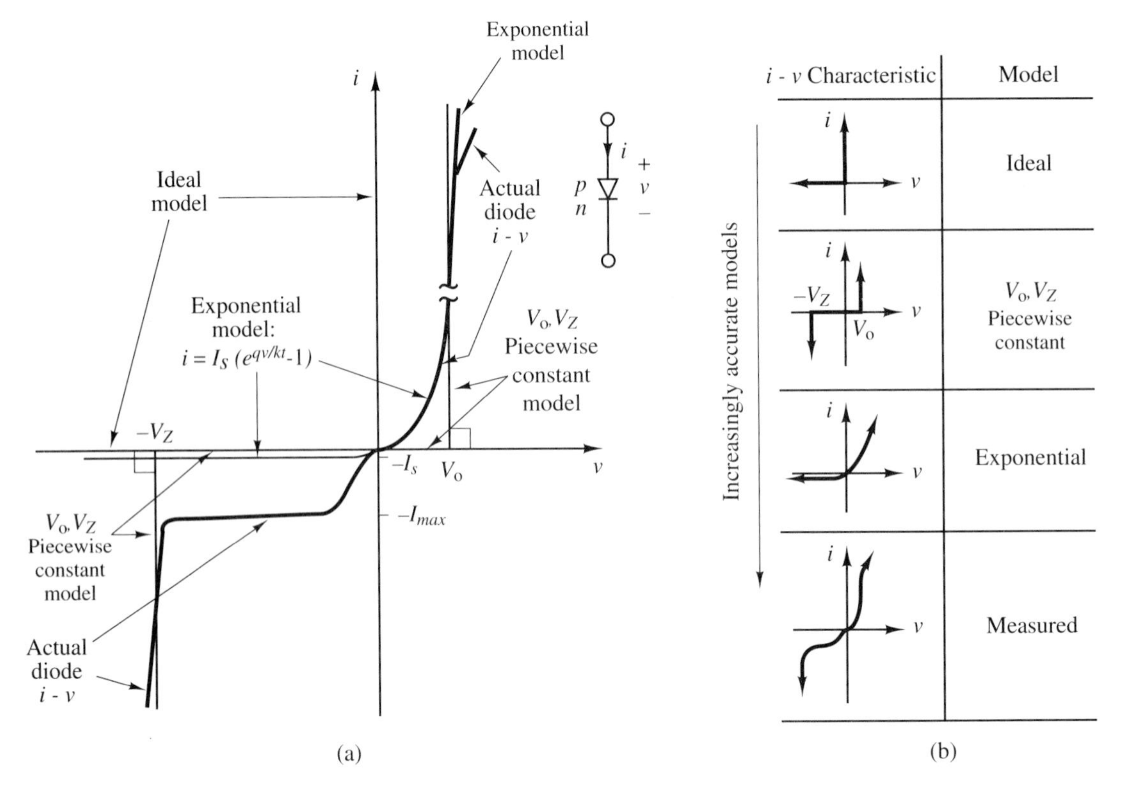


**Figure 7.11** The $i$–$v$ characteristic of an actual diode compared with the ideal diode and other simple models: (a) on same axes; (b) on separate axes, showing increasing accuracy of models.

$$i = I_s\,(e^{qv/kT} - 1), \qquad (7.1)$$

where $v$ is the applied voltage (which we have been calling $V_a$), $q$ is again the electron charge ($1.602 \cdot 10^{-19}$ C), $k$ again is Boltzmann's constant ($1.38 \cdot 10^{-23}$ J/K), $T$ is the temperature (in kelvin), and $I_s$ is called the *saturation current.* [We have changed our symbol for electron charge from $e$ (see Chapter 2) to $q$ to avoid confusion with $e$ of the exponential function.]

The exponential relation essentially reflects the fact that in forward bias the annihilations of electrons injected into the $p$ region and holes injected into the $n$ region are proportional to the injected concentrations, leading to a solution for concentration distributions that is exponential. The $q/kT$ factor is again due to more advanced statistical mechanics considerations. The $-1$ satisfies the physical requirement that $i = 0$ for $V_a = 0$. It also models, to an extent, reverse bias in that for $V_a \ll 0$, $i \approx -I_s$.

**Example 7.3**

Find the current in a diode at 300 K as predicted by Eq. (7.1) when the applied voltage $v$ is (a) 0 V, (b) 0.25 V, (c) 0.5 V, (d) 0.6 V, and (e) 1 V. Assume that $I_s = 0.1$ nA $= 10^{-10}$ A.

**Solution** At 300 K, $kT/q = 0.026$ V, so $i = 10^{-10}(e^{v/0.026} - 1)$. To improve the accuracy of the exponential fit, the exponential in Eq. (7.1) is usually replaced by $qv/(mkT)$, where $m$ is a fitting factor roughly equal to 1.5 for silicon. Thus at (a) $v = 0$ V, $i = 0$ A, (b) $v = 0.25$ V, $i = 0.061$ μA, (c) $v = 0.5$ V, $i = 0.037$ mA, (d) $v = 0.7$ V, $i = 6.2$ mA, and (e) $v = 1$ V, $i = 13$ A. The last value would be enough to destroy some diodes. Momentarily, we shall show how this problem is avoided in practice.

In actual diodes, the reverse current may be much larger in magnitude than $I_s$, but at least Eq. (7.1) is a sufficient model for getting the idea of the shape of the diode $i$–$v$ characteristic. In fact, it agrees extremely well for moderately positive applied voltage $v$, such as the values considered in Example 7.3, parts (a) through (d). On data sheets, the actual maximum reverse mode current $I_{max}$ will be given. The reverse current is not as bad as the exaggerated drawing of Fig. 7.11b suggests: Typically, $I_{max} \approx 10$ μA, while forward currents of 1 A (or even 50 A or more for power rectifiers) are easily obtained in forward bias.

Furthermore, when $v$ reaches a certain negative value denoted $-V_Z$, the actual diode again conducts but now in the direction opposite that for forward bias. The *Zener voltage* $V_Z$ can be anywhere from 1 to 1000 V, depending on the diode construction. Either of two physical processes may be responsible for this conduction mode, depending on the diode construction.

The first process was discovered by Clarence Zener in 1934 in a different context and is due to an effect called quantum mechanical tunneling, in which bound electrons with ostensibly insufficient energy to break free nevertheless break free from the crystal and conduct. It occurs for diodes having small $V_Z$. The other mechanism, descriptively called *avalanche breakdown,* is due to the ionization of resident crystal atoms by minority carriers being driven across the depletion region under strong reverse bias.

At the Zener voltage (which we may informally call the condition of *Zener bias*), the conductivity becomes very large. Thus, whatever the reverse current, the voltage across the diode is $-V_Z$. That is, in the reverse direction (only), the diode acts like a voltage source! Because $V_Z$ is very stable, diodes are often run at Zener bias purposely for the establishing and regulation of voltages in a circuit. Diodes especially designed for this use are called *Zener diodes* and have the symbol given in Fig. 7.12. A typical value of $V_Z$ for a non-Zener diode is 75 V. [Similarly, in forward bias (only), the diode functions identically to a voltage source, now one of value $V_0 \approx 0.7$ V.]

Also, for $|v|$ large, the diode acts like a small resistance $r_d$ in series with $V_0$ or $-V_Z$ for awhile, and for $v$ too large, the diode fries! Normally, a resistor or other device will be placed in series with the diode to limit the voltage across and current in the diode to safe levels.

For rough analysis of circuits that is often sufficient to determine basic be-

**Figure 7.12** Symbol for a Zener diode.

havior, two models simpler than the important exponential model are also commonly used. Both are piecewise constant; the $i$–$v$ characteristic is composed of horizontal ($i = 0$) and vertical ($v = V_Z$ or $V_o$) sections. The first is the ideal diode model where we set $V_o = 0$ and $V_Z = \infty$ (Figs. 7.4a and 7.11b), and the second includes $V_o \approx 0.7$ V and $-V_Z$ in the manner shown in Fig. 7.11b.

Because of their simplicity, yet rough accuracy, these models are both used extensively in practice. The measurable diode $i$–$v$ curve shown in Fig. 7.11a actually looks very similar to the $V_0$, $V_Z$ piecewise-constant model when a large scale is used for the vertical (current) axis.

We may now consider why we had to learn about holes. That is, how have the holes helped out to make this a good rectifying junction? Why do we not just use an "*ni*" junction (*i* being intrinsic—undoped—silicon rather than *p*-type)? We may answer in one way for reverse bias and in another for forward bias. In reverse bias, their parent acceptor atoms contribute to the electric field in the depletion region to help minimize the current. Without them, in reverse bias the depletion region might extend all the way to the metal contacts in the *p* region, and we would not have a nice abrupt junction.

In this situation, known as *punch-through*, the diode would act more like a resistor than a diode. This is because now the metal contact would serve as a huge supply of electrons for reverse current (in the *p* region, there was only a small number of electrons so that the reverse current was very small). Actually, punch-through can happen even in some *pn* junctions when they are strongly reverse biased (recall that reverse biases increase the depletion width). Punch-through is clearly a condition to be avoided.

In forward bias, the holes themselves act to increase the current by annihilating the electrons as soon as they come into the *p* region, making way for yet more electrons to pass through and produce large currents. The same arguments hold for the electrons. Therefore, the *pn* junction is definitely a bipolar (two-carrier) rather than a unipolar (one-carrier) device; both types of charge carrier are vital to achieving high-quality rectification using a *pn* junction.

Although a unipolar junction such as the *ni* junction does rectify, for the reasons above and other design considerations the *pn* junction has better properties in practice. We are now talking about complicated matters, for punch-through may or may not occur in an *ni* junction, depending on the device and complicated physical arguments and concepts (such as the theory of "space-charge-limited currents," which is the phenomenon we mentioned where accumulating charge reduces forward biasedness).

In fact, the *ni* junction is mathematically very difficult to analyze, especially compared with an abrupt *pn* junction. The potential, for example, involves elliptic integrals and Jacobian elliptic functions rather than the simple exponential function for a *pn* junction. Thus even though it is possible to make a good *ni* rectifier, they are not manufactured today, and we are forced to learn about both kinds of charge carriers: holes as well as electrons.

It has also been shown that the diffusion of charge is central to the operation of a *pn* junction in forward bias. Diffusion currents predominantly go one way: holes from *p* to *n* and electrons from *n* to *p* and only minimally the other way, because of the unalterable *p* and *n* doping concentrations. On the other hand, ohmic conduction electric fields reverse with a change in sign of applied voltage $v$. Therefore, we can have cancellation-of-current tendencies (open circuit) for $v < 0$ and addition-of-current tendencies (high conduction) for $v > 0$. In this manner we see that diode operation depends on this new type of conduction called diffusion current.

Now that we have discussed the essential physical processes in a diode, we are ready to apply it to do signal processing operations, such as we discuss in the next section. Furthermore, we are also in a position to understand the basic physical processes in a transistor. With a knowledge of the operation of these two devices, we have the background to appreciate much more complex circuits such as the operational amplifier (see Chapter 8) and the digital electronics ICs we study in Chapters 10 through 12.

## 7.4 THE DIODE: APPLICATIONS

### 7.4.1 Introduction

The applications of diodes are so numerous that it would be impractical to list or discuss them all. Often, a diode will be used for reasons that are clear only to one intimately familiar with the design under consideration. Some of the more common applications are in the conversion/regulation/multiplication of sinusoidal to constant-in-time voltages (known as "dc power supplies"), radio signal detection, voltage spike protectors for integrated circuits, pulse-generating circuits, signal-shaping circuits, LED displays and solar cells, establishing desirable potential differences in transistor amplifiers (via $V_o$ or $V_Z$), logarithmic applications based on the exponential $i$–$v$ relation Eq. (7.1), various operational amplifier circuits (see Sec. 8.8), computer memory devices (see Sec. 11.7.4), and in electromechanical machine-drive systems. Here we investigate only a few of the simplest diode circuits.

### 7.4.2 Half-Wave Rectifier

We have already considered the basic rectifier circuit in Fig. 7.5. If the goal was a dc output when $v_s$ is a sinusoid $v_s = V_m \sin(\omega t)$, we are still not very close. Although the output is unipolar (has only one polarity), it is zero much of the time and greatly varies with time. We can improve the situation by adding a "filter" capacitor across the output, as shown in Fig. 7.13.

For now we assume that $V_Z = \infty$ and $V_o = 0.7$ V. Without the capacitor, the diode remains "off" (nonconducting) until $V_s$ reaches $V_o$. At that point the diode begins to conduct, and the voltage across it remains at $V_o$ volts, even as $v_s$ continues to increase. The balance of the voltage appears across the load resistor as $v_L$. When

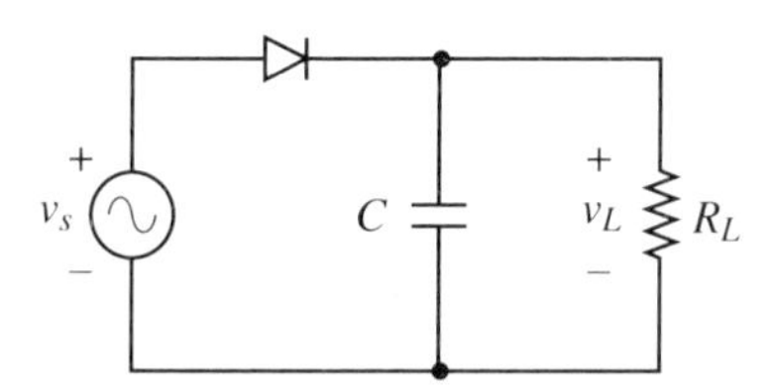


**Figure 7.13** Half-wave rectifier with filter capacitor.

$v_s$ later drops below $V_o$, the diode is again nonconducting ($i = 0$) and the voltage $v_L = i_L R_L = 0$. The cycle repeats indefinitely (see Fig. 7.14a).

To analyze what happens when the capacitor $C$ is placed across $R_L$, we can use our knowledge from Sec. 4.13.1 (see Fig. 7.14b). With the diode conducting, $v_s$ charges $C$ through the extremely small resistance $r_d$ of the diode. Therefore, the time constant $r_d C$ is small and $v_L$ roughly follows $v_s - 0.7$ V. However, when $v_s$ begins to drop, the capacitor has nowhere to discharge except through $R_L$. Recall that current can go only one way through the diode; the capacitor cannot discharge through it. Because $R_L \gg r_d$, the time constant for discharging is much larger. This means that the capacitor voltage, which is $v_L$, tends to remain at its maximum, decaying at the slow rate determined by $R_L C$. When $v_s$ again exceeds $v_L + 0.7$ V, the diode again conducts and the capacitor quickly follows $v_s - 0.7$ V.

The capacitor has reduced the strong ripple in our dc power supply. The larger

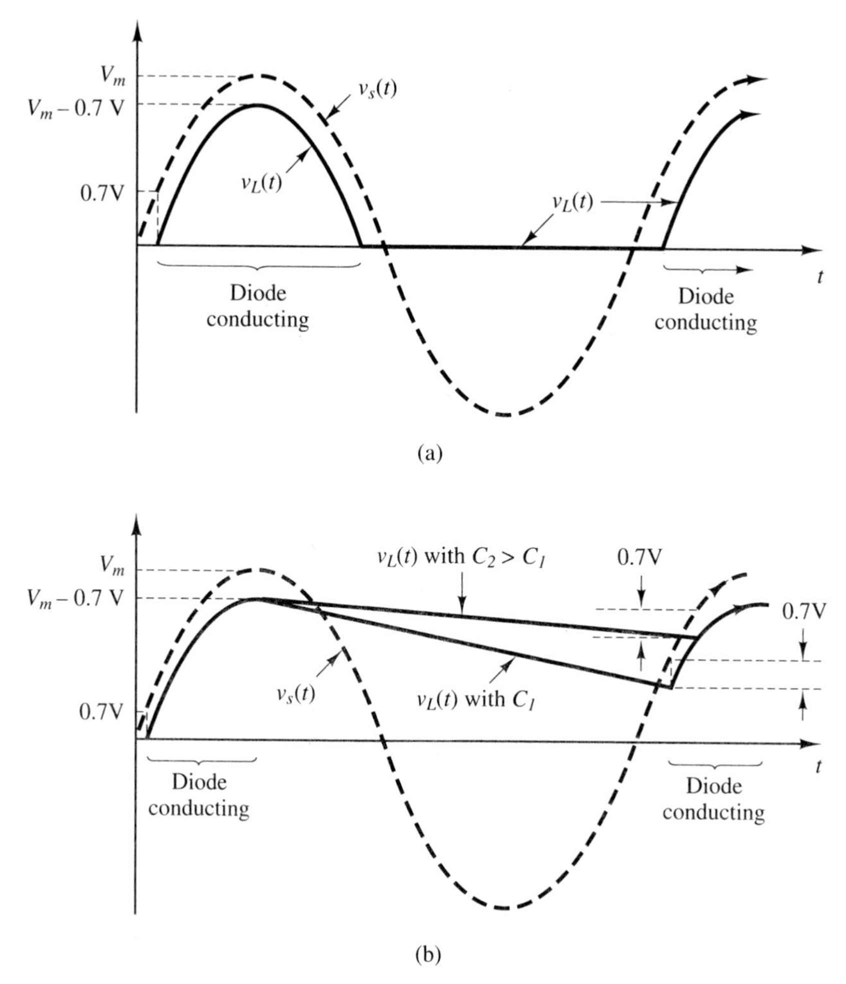


**Figure 7.14** Output voltage $v_L$ of half-wave rectifier assuming that $V_o = 0.7$ V: (a) without capacitor; (b) with capacitor.

the value of $C$, the better the filtering or smoothing effect; the more constant $v_L$ can be made.

### 7.4.3 Full-Wave Bridge Rectifier

A further significant additional improvement is available using the diode bridge circuit in Fig. 7.15a. While the circuit of Fig. 7.13 retains only the positive peaks of the input voltage (*half-wave rectifier*), the bridge circuit in Fig. 7.15a keeps both (*full-wave rectifier*). It does this by reversing the polarity of the negative-going pulse.

Suppose that $v_s > 0$. Then when $v_s$ reaches $2 \cdot 0.7$ V $= 1.4$ V, diodes 2 and 4 conduct and $v_L = v_s - 1.4$ V. This condition is shown in Fig. 7.15b, where the darkened diodes are conducting and the others are not. When $v_s$ drops below 1.4 V, no diodes are conducting until $v_s$ reaches $-2V_o = -1.4$ V, at which time diodes 1 and 3 conduct (see Fig. 7.15c). Now $v_L = -v_s - 1.4$ V. This is because the route of current from the source has been reversed with respect to the load; now the minus terminal of $v_s$ is connected via diode 1 to the plus terminal of $v_L$. The reversal of route cancels the change of sign of $v_s$ so that *again* $v_L > 0$, instead of zero, as would be the case for the half-wave rectifier during this interval. Also, we show in Fig. 7.15d the input–output characteristic, $v_L$ versus $v_s$. We obtain this plot by collecting our results: For $|v_s| < 1.4$ V, $v_o = 0$, while for $|v_s| > 1.4$ V, $v_L = |v_s| - 1.4$ V. Consequently, the waveforms of $v_s$ and $v_L$ are as shown in Fig. 7.16a.

It should be mentioned that the "dead" time as well as the differences in amplitude between $v_L$ and $v_s$ are in Figs. 7.14 and 7.16 exaggerated to help show the phenomena. In practice, $V_m$ is usually much larger than the roughly 5 V used here; that is, $V_m \gg 2V_o$. If it is not, other higher-precision rectifiers can be used (see, e.g., Sec. 8.8.4).

If the same value of capacitance as that used in Fig. 7.14b is added, there is the additional improvement shown in Fig. 7.16b. Because the time before the next conduction mode is roughly cut in half, the ripple (deviation from constant) in $v_L$ is roughly cut in half. (Although the decay is exponential, the time constant $R_LC$ is so large that on this time scale the decay is roughly linear.)

### 7.4.4 Zener-Regulated DC Power Supply

One final major improvement on our dc power supply can be obtained by adding only two components: a resistor $R_A$, in series with a Zener diode across the load. This power supply is shown in Fig. 7.17. Suppose that the output voltage of our capacitor-filtered bridge full-wave rectifier, $v_o(t)$, is always larger than $V_Z$ in the steady state.

Notice that the Zener diode is connected "upside down." Before the Zener diode conducts (before steady state is reached), it is an open circuit. Suppose first that $R_L = \infty$, that is, an open circuit. Then there is no voltage drop across $R_A$ so that the voltage across the Zener diode is $v_o(t)$.

If $v_o(t)$ now exceeds $V_Z$, the voltage across the diode will be $V_Z$ and it will conduct. The voltage across $R_A$ will be $v_o(t) - V_Z$, and its current will by Ohm's law be $i_A(t) = [v_o(t) - V_Z]/R_A$; this is equal to the current in the Zener diode, for $R_L = \infty$. Thus as $v_o(t)$ varies, the operating point on the Zener portion of the $i$–$v$

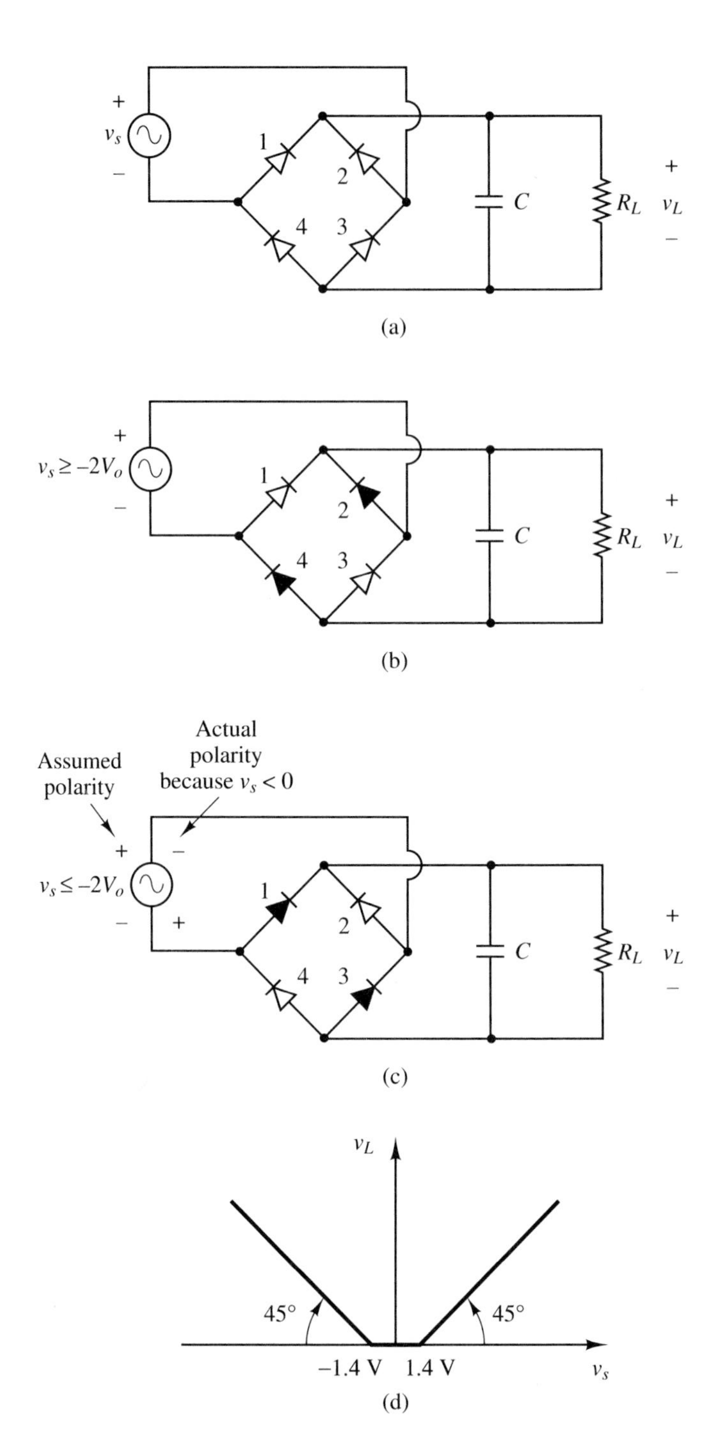


**Figure 7.15** Full-wave bridge rectifier: (a) circuit; (b) operation for $v_s > 2V_o$; (c) operation for $v_s < -2V_o$; at all other times, all diodes are nonconducting; (d) input–output characteristic, $v_L$ versus $v_s$.

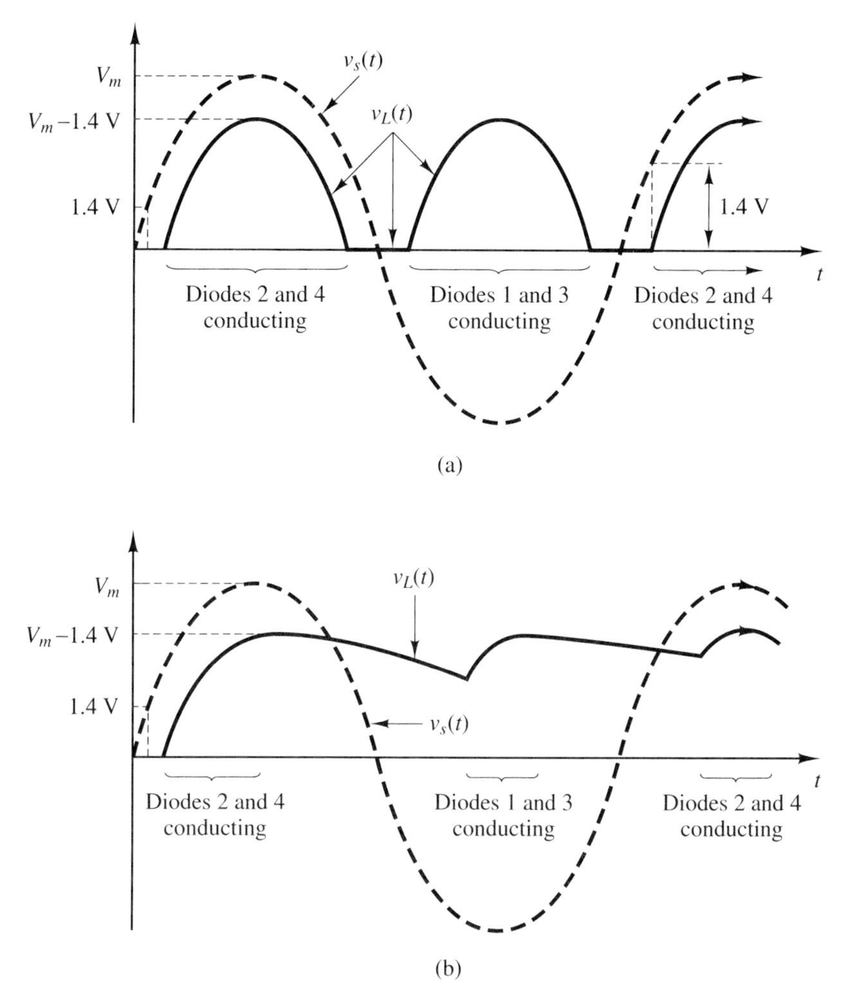


**Figure 7.16** Output voltage $v_L$ of full-wave rectifier assuming that $V_o = 0.7$ V: (a) without capacitor; (b) with capacitor.

characteristic in Fig. 7.11a moves up or down, but the diode voltage remains fixed at $-V_Z$. The voltage across $R_a$ is anything but well regulated—indeed, it contains all the variations. This is fine, because it is $v_L(t)$ that is desired to be and is in fact well regulated: $v_L(t) = V_Z$.

For finite $R_L$, there is a voltage divider when the diode is open, so that $v_L(t) = v_o(t)R_L/(R_L + R_A)$. If $v_o(t)$ now exceeds $V_Z(R_L + R_A)/R_L$, the voltage across the Zener diode will be $V_Z$ and it will conduct. There is no longer a voltage divider relation between $v_o(t)$ and $v_L(t)$. The current in the Zener diode will now be $i_A(t)$ minus the load current: $i_Z = [v_o(t) - V_Z]/R_A - V_Z/R_L = v_0(t)/R_A - V_z/(R_A \| R_L)$. (The latter expression is recognized as the two-source superposition solution.) The load current is $V_Z/R_L$ if the load is a resistor. Thus the current in the Zener diode is a strong function of both the load and variations in $v_s(t)$, *but the voltage across the load is* $V_Z$, independent of the load, and independent of variations in $v_s(t)$! That is, we have a very nearly ideal voltage source for $R_L$.

The combination of $R_A$ and the Zener diode is known as a Zener voltage regulator. If either the input voltage $v_o(t)$ or the load resistance changes with time, the load voltage will remain constant at $V_Z$. The only requirement is that $v_o(t)$ always remains above $V_Z(R_L + R_A)/R_L$; it need not be constant. The "slack," $v_o(t) - V_Z$, is taken up by $R_A$.

The idea of a resistor "taking up the slack" is exceedingly common in electronic circuits. We know that applying substantially more than 0.7 V directly across a forward-biased diode (or $-V_Z$ for a reverse-biased diode) will cause a current so large that the diode will burn out. Thus a series resistor is a typical method of connecting diodes in a manner that will protect them. In Fig. 7.17, for example, the current in the Zener diode is limited to less than $I_A = [v_o(t) - V_Z]/R_A$. Thus in Fig. 7.17 the resistor $R_A$ serves two purposes: It allows the voltage across the Zener diode (and the load) to be fixed at $V_Z$ by taking up the slack, and it limits the current in the diode to safe levels.

### 7.4.5 Maximum-Value Circuit

Another interesting application of diodes is in the determination of the maximum value of several waveforms. The circuit of Fig. 7.18 is most easily analyzed by assuming an ideal diode model ($V_Z = \infty$, $V_o = 0$). Suppose that $v_2(t)$ is the largest of all four input voltages at this particular instant of time, $t = t_A$. Diode 2 will be forward biased because it is assumed always greater than the $-10$-V supply, so that $v_o(t_A) = v_2(t_A)$. Now because $v_1(t_A)$, $v_3(t_A)$, and $v_4(t_A)$ are by assumption all less than $v_2(t_A)$, diodes 1, 3, and 4 must all be reverse-biased [even though $v_1(t_A)$, $v_3(t_A)$, and $v_4(t_A)$ are all assumed to be greater than $-10$-V].

Consequently, whenever another voltage becomes maximum, its diode alone will conduct; $v_o(t)$ will follow the maximum of the four input voltages. It is left as an exercise to show that flipping all the diodes around and changing the $-10$-V battery to $+10$ V will cause $v_o(t)$ to instead be the minimum of the input voltages.

### 7.4.6 Voltage-Limiting Diode Circuit

Another application of diodes uses the same principle as the Zener regulator but is based on $V_o$, not $V_Z$, as the fixed voltage. The problem to be solved is that signals that have been transmitted over some distance may have undesirable noisy, high-amplitude spikes created by other sources and introduced via electromagnetic induc-

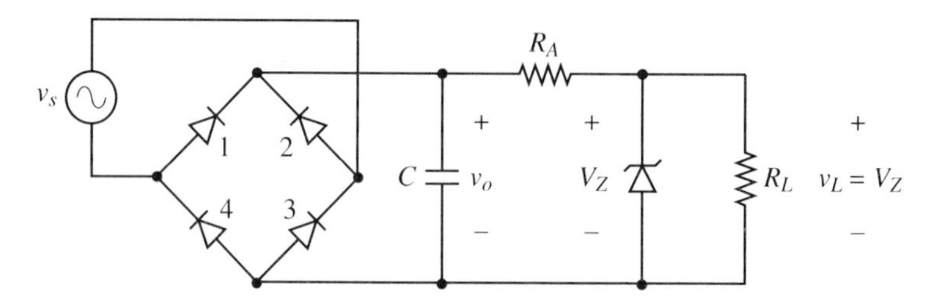


**Figure 7.17** Dc power supply with Zener voltage regulator.

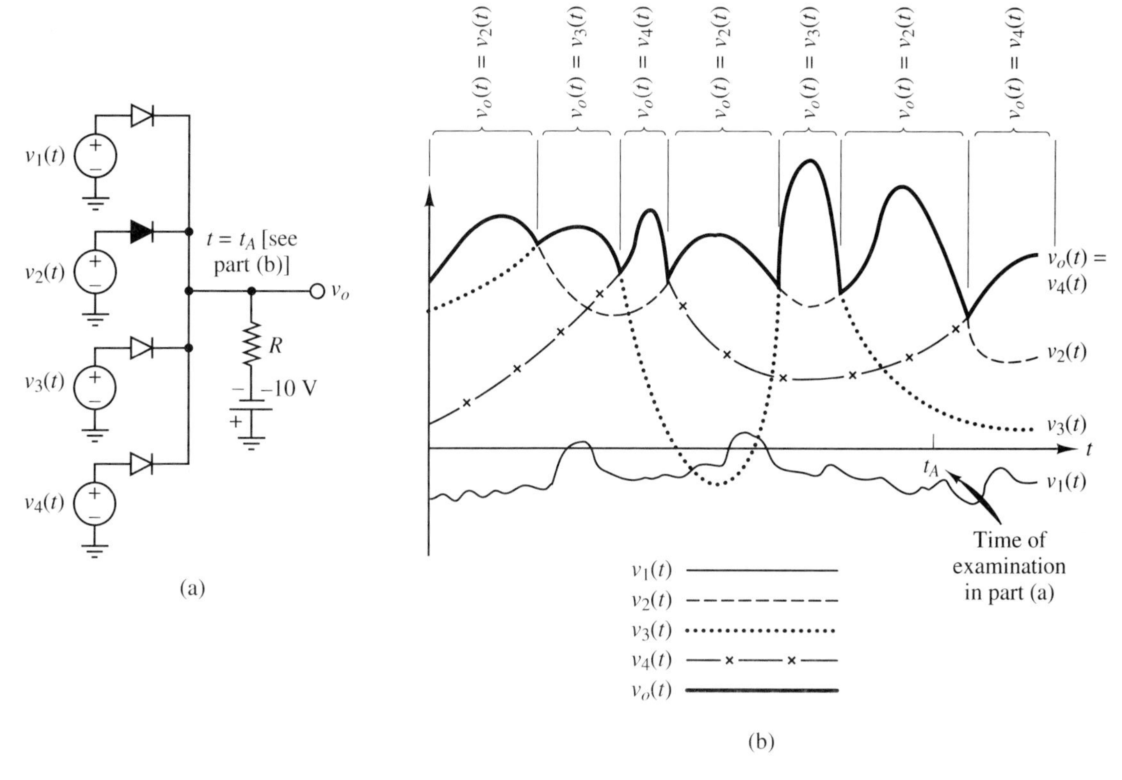


**Figure 7.18** Maximum-value diode circuit: (a) circuit; (b) example waveforms. At this moment $t_A$, $v_2(t_A)$ is the maximum of the four input voltages.

tion. Also, these spikes may be present in the signal itself if the signal undergoes a sudden transition with overshoot. In any case, the spike can be harmful to an integrated-circuit chip subjected to it. Diodes can be used to eliminate the spike as follows.

The circuit of Fig. 7.19a includes two diodes: the upper diode to limit the output voltage $v_o(t)$ to a maximum of $V_B + V_o = 5\ V + 0.7\ \text{V} = 5.7\ \text{V}$ and the lower diode to limit the voltage to a minimum of $-5.7$ V. The portion of the spike above 5.7 V (or below $-5.7$ V) appears harmlessly across $R_s$.

Typical time waveforms for $v_s(t)$, $v_{Rs}(t)$ and $v_o(t)$ appear in Fig. 7.19b. Notice how the spikes are chopped off in $v_o(t)$. The darkened diode in Fig. 7.19a signifies that this diode is conducting at the indicated time $t = t_A$, where, as shown in Fig. 7.19b, $v_s(t)$ exceeds 5.7 V. At $t = t_B$, the upper diode would be nonconducting and the lower diode would be conducting. Note that we must choose diodes for which $V_Z$ exceeds the amplitude of all anticipated spikes for this circuit to function as designed. Finally, in Fig. 7.19c the input–output characteristic is shown, $v_o$ versus $v_s$. Whatever the value of $v_s$, the height of the line at that value of $v_s$ indicates the corresponding output voltage. For $|v_s| \geq 5.7$ V, $|v_o|$ is limited to 5.7 V.

Before moving on to our final diode application in this section, the photodiode, let us briefly review the procedure for solving diode circuit problems. A systematic method is required, especially when several diodes are present. Occa-

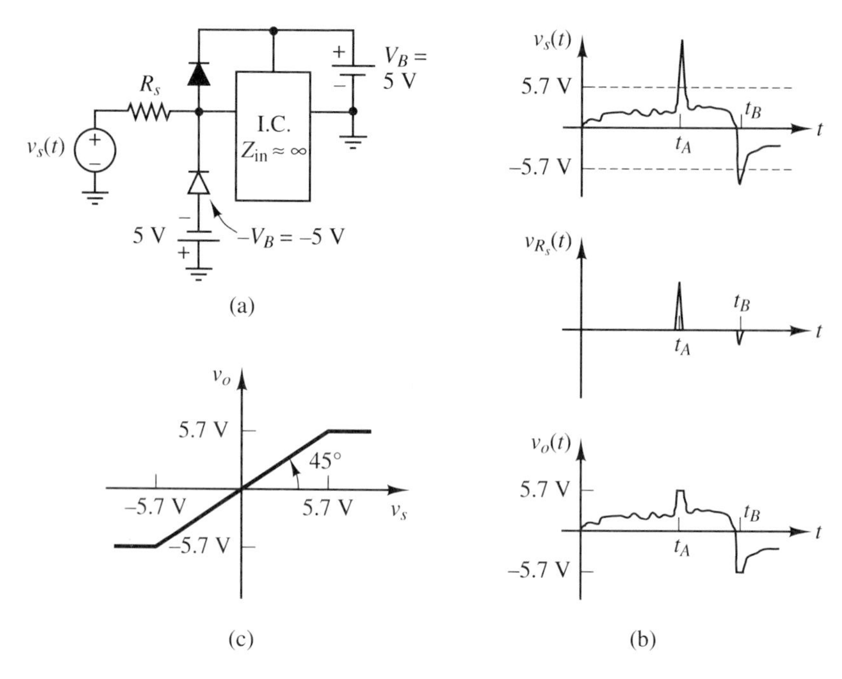


**Figure 7.19** Voltage-limiting circuit: (a) circuit diagram; (b) typical waveforms; (c) input–output characteristic.

sionally, iterative procedures are necessary to end up with a fully self-consistent solution. The immediate goal is a plot of the output voltage $v_o$ as a function of the input voltage $v_s$.

For most simple circuits, it suffices to begin by assuming that all voltages are zero and thus all diodes are open. By mentally increasing the input voltage, determine for each diode conditions on the input voltage that would make that diode forward biased. Equivalently, remove the diode in question and calculate the voltage the diode would see across its open-circuited terminals. Collect the conditions and determine all the distinct intervals, that is, ranges of input voltage where unique combinations of diodes are on. For example, we might have all diodes reverse biased for $v_s < 0$, diodes 1 and 4 forward biased for $0 < v_s < 2$ V, and diodes 2 and 3 forward biased for $v_s > 2$ V. Thus we divide the $v_s$ axis into a number of sections for which the dependencies of $v_o$ on $v_s$ differ.

Next, for each interval draw an equivalent circuit, replacing all diodes by simple equivalent elements. All reverse-biased diodes are replaced by open circuits, unless they are Zener-biased (in which case they are replaced by $-V_Z$-volt batteries). If the ideal diode model is used, all forward-biased diodes are replaced by short circuits. If the $V_o = 0.7$-V model is used, the forward-biased diodes are replaced by 0.7-V batteries. For each such equivalent circuit, valid for only a specified range of input voltage, determine the equation relating the output voltage $v_o$ to the input voltage $v_s$.

Splicing all of these dependences together results in the input–output characteristic curve $v_o$ versus $v_s$ (see, e.g., Fig. 7.15d, in which $v_O$ is called $v_L$). This graph is sufficient to determine $v_o(t)$ for *any* input waveform $v_s(t)$. Merely use the $v_o$ versus $v_s$ graph for each instant in time $t$ to "translate" $v_s(t)$. Doing this for all time yields the output waveform.

### 7.4.7 Photodiodes, Solar Cells, and LEDs

Recall from Sec. 7.3.2 that the reason a diode does not conduct well under reverse bias is that there are few minority carriers. Those that do exist due to thermal excitation are swept across the junction under reverse bias and constitute the reverse current. If somehow we could inject additional minority carriers, the reverse current could be increased and the diode would conduct even under reverse bias.

This is possible if in a *pn* junction one side, made very thin, is exposed to light (see Fig. 7.20a). The exposed side is thin to allow the light to penetrate into the depletion region and other side. The exposed surface area of the junction is made very large to catch as much light as possible. Assume that the diode is reverse-biased by an applied voltage.

When the depletion region is illuminated, light-generated electron–hole pairs will split off due to the depletion-region electric field. The minority electrons in the *p* side will go to the *n* side, attracted by the positive donor ions and ultimately the

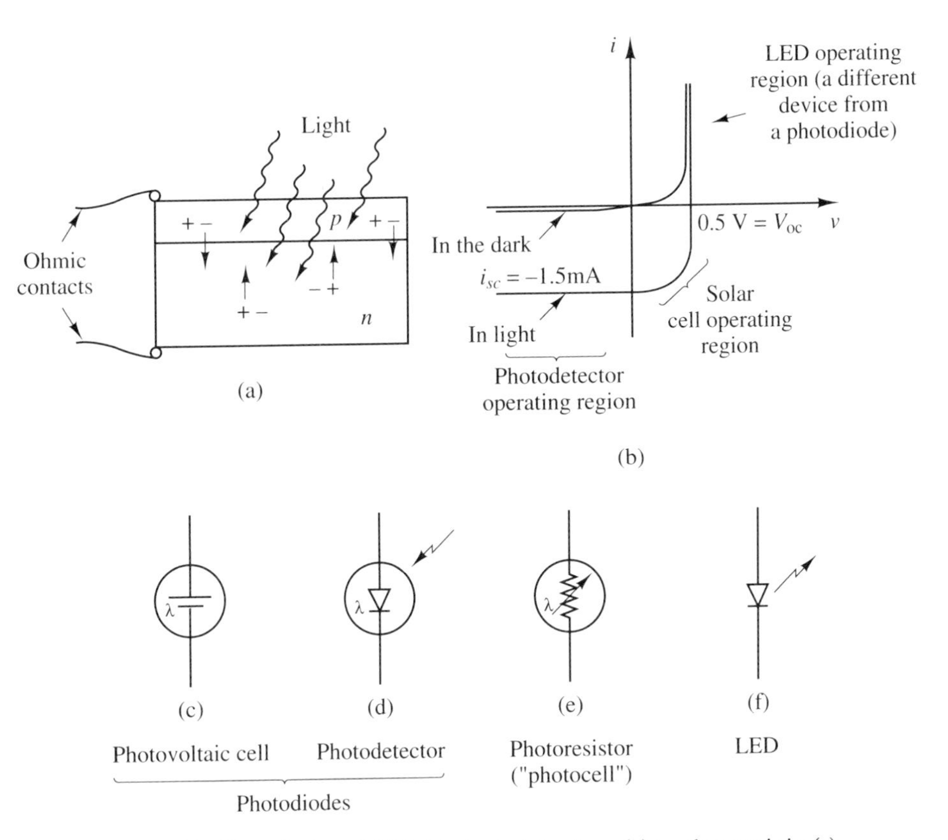


**Figure 7.20** Photodiode and LED: (a) physical structure; (b) $i$–$v$ characteristic; (c) solar cell; (d) photodetector; (e) photocell; (f) LED.

higher-potential $n$ terminal caused by the negative applied voltage. Similarly, minority holes in the $n$ side will drift over to the $p$ side, attracted by the acceptor ions. This movement is called *photocurrent.* The stronger the incident light, the more minority carriers are introduced on both sides, and thus the greater the reverse current.

Different semiconductor materials have different ranges of frequencies over which they are sensitive. For example, cadmium sulfide (CdS) is sensitive over visible light and thus is used in camera equipment (see below). Silicon is plentiful, sturdy, and cheap, and is sensitive to infrared light (invisible light on the low-frequency side); hence it is commonly used for "electric eye" systems, where the light source is designed to be infrared.

The $i$–$v$ characteristic of a photodiode is shown in Fig. 7.20b. Its most conspicuous feature is the fourth quadrant: $v > 0, i < 0$. That is, $p = vi < 0$, so the photodiode is generating electric power! Of course, this energy is merely electromagnetic energy (light) from the sun converted into dc electrical energy. When the photodiode is operated in this quadrant, it is called a *solar cell*; the circuit symbol is shown in Fig. 7.20c. The symbol $\lambda$, by referring to the wavelength of light, indicates that this is a light-controlled device. In this case, the photodiode is connected directly to its load. When it is operated in the third quadrant ($v < 0, i < 0$), it is called a photodetector; its symbol is given in Fig. 7.20d. In this case, the photodiode is reverse biased and the light-dependent reverse current is considered the system output.

The range of reverse current for silicon photodetectors is typically from $-0.01$ $\mu$A in the dark to $-1$ mA or more in bright light. The short-circuit current would then be $-1$ mA in bright light, and the open-circuit voltage is typically about 0.5 V. Unfortunately, solar cells still generally have less than 20% efficiency. Notice that for small positive $v$ and for negative $v$, the photodiode acts like a current source, while for $v$ near the open-circuit voltage it acts like a voltage source. Depending on the load/bias circuit connected, load line analysis will indicate which is the more appropriate model.

A simpler device known as a *photocell* or *photoresistor* is just a slab of semiconductor material whose resistance varies with the level of illumination to which it is exposed. It is not a diode. Cadmium sulfide (CdS) is the most common material for photocells, which are often used as light meters for cameras. The symbol for the photocell is shown in Fig. 7.20e.

To close this section, we briefly consider the reverse process: emission of light from a *pn* junction, the so-called light-emitting diode (LED). LEDs are commonly used as power-on indicators and occasionally, as alphanumeric display units. An increasingly used replacement for the LED in many situations is the liquid crystal display, which we examine briefly in Chapter 9.

Just as light can generate electron–hole pairs, it is possible that annihilations, which are energenically favorable, will also give off light. Recall that only when a diode is forward biased is there much "recombination" (annihilations). Thus it is clear that detectible light will be produced only under forward bias. Special materials such as gallium arsenide and gallium phosphide must be used for light to be emitted; silicon will merely generate heat. Thus although the structure of photodiodes/solar cells and LEDs are similar, their chemical composition is not (silicon does respond to light in photodiodes). The circuit symbol for an LED is shown in Fig. 7.20f.

## 7.5 THE TRANSISTOR: OPERATION

### 7.5.1 Introduction

We began our discussion of diode operation by proposing a desirable operation: rectification. After specifying the required $i$–$v$ characteristic and an example circuit with the proposed device inserted, we went about discussing physically how this behavior can be achieved using semiconductor materials. We were successful, albeit with a few nonidealities to clutter the picture. In this section we proceed in much the same way. This time the desired goal has already been introduced in Sec. 7.2: electronic amplification.

To set the stage, a "black box" approach is proposed (see Fig. 7.21). We need a box with an input port ("door") and an output port, that is, a *two-port network.* The power coming out of the output port $v_L i_L$ ($= i_L^2 R_L$) is to be much larger than the input signal power $v_i i_s$. That is, we seek the greater-than-unity power gain amplifier described in Sec. 7.2.

The contents of the black box were discovered by William Shockley and his co-workers at Bell Laboratories in 1949. If a three-layer *npn* or *pnp* structure is fabricated and connections are made as shown in Sec. 7.5.2, the goal is achieved. We now consider, briefly and qualitatively, the physical processes involved.

### 7.5.2 Transistor Physics

**Bipolar junction transistor (BJT).** To understand the transistor, we may directly apply our knowledge of the *pn* junction. Figure 7.22 shows an *npn* bipolar transistor with some basic connections, which in Sec. 7.6.2 we shall modify to make a practical amplifier. As mentioned at the end of Sec. 7.3.2, *bipolar* indicates that both holes and electrons are critically involved in the device operation. Let us see just how.

In Fig. 7.22 the leftmost region is designated as type $n^+$. The plus means that this region, called the *emitter,* is very heavily doped (with donors). The middle region, called the *base,* is very lightly $p$-doped and is relatively quite narrow. Finally, the rightmost area, called the *collector,* is also lightly doped but is much wider than the base.

The voltage $V_{BE}$[1] is positive, so that the base–emitter *pn* junction is forward biased. Consequently, electrons from the emitter diffuse into the base region and

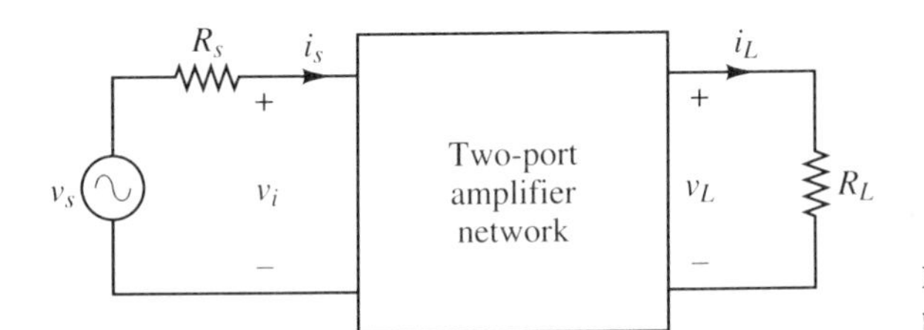


**Figure 7.21** Black box producing signal power gain; $v_L i_L \gg v_i i_s$.

[1] In electronics, the subscript convention for voltage definition opposes that which we have used since Chapter 2: now $V_{ab} = V_a - V_b$. We will now follow this convention.

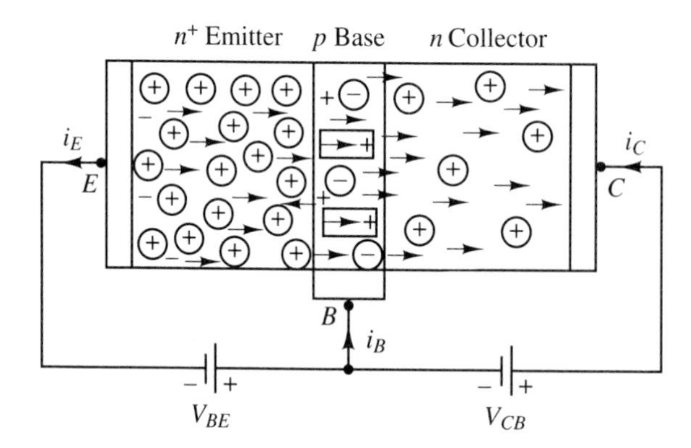


**Figure 7.22** An *npn* bipolar transistor with simplest possible connections.

a few holes from the base diffuse into the emitter. These add up to form the emitter current $I_E$, which is positive because electrons are diffusing to the right (opposite conventional current $i_E$) and holes are diffusing leftward. However, because the base is so lightly doped, the hole diffusion current into the emitter is extremely small relative to the electron diffusion current into the base.

The many arrows in Fig. 7.22 pointing right represent free electrons diffusing toward the collector from the heavily doped emitter, and the + with arrow pointing left is a hole diffusing into the emitter from the lightly doped base. Also, a small fraction of electrons are annihilated by holes in the base; these recombinations are represented by a + and − symbol inside a rectangular box.

If the electrons in the base could reach the metallic base terminal, they would be attracted to it ($V_{BE} > 0$). In that case we would merely have an asymmetrically doped diode operating in forward bias—not particularly interesting. Instead, notice that $V_{CB} > 0$, so that the base–collector *pn* junction is *reverse biased.*

Normally, a reverse-biased junction does not conduct current. But that is only because there are normally very few minority carriers in either the $p$ or $n$ region. Recall from Sec. 7.3.2 that any minority carriers that *do* exist are strongly *swept across* the junction by the depletion region electric field, which in reverse bias is enhanced by the applied voltage, $V_{CB}$.

It is true that there are very few minority carriers (holes) in the collector region, so this component of reverse current is extremely small. However, *there is a large number of minority carriers in the base—the electrons that diffused* (*were "emitted" or injected from the emitter*) *across the forward-biased emitter–base junction.* The pull on these electrons toward the collector terminal far exceeds their attraction to the base electrode because $V_{CB} \gg 0$. It is only because the base is simultaneously half of one forward-biased junction and also half of a different reverse-biased junction that this "emission-collection" behavior is possible.

Now, it will be interesting to see how *control* of this large current is achieved. First, we may view the emitter current as being controlled by the input voltage $V_{BE}$: $i_E = I_S(e^{qV_{BE}/kT} - 1)$. Thus, for even a small variation in the input voltage $V_{BE}$, $i_E$ varies exponentially, and the coefficient of $V_{BE}$ at a warm operating temperature in a room is large and positive: $q/kT = 1/0.026\text{ V} = 38.7\text{ V}^{-1}$ at temperature $T = 300\text{ K} \approx 80°\text{C}$. Thus by varying $V_{BE}$ slightly, large variations in emitter current result; this is signal amplification.

Next note that because of the charge redistributions arising from the attraction between opposite charges, charge neutrality will naturally be (approximately) maintained within the base. Of course, the injection of electrons from the emitter tends to disrupt charge neutrality in the base. Holes in the base slightly diminish this charge imbalance via the process of annihilation of incoming electrons. Note that in the process of annihilation, the original acceptor negative ions become uncanceled, preventing neutrality from resulting. However, the base contact can serve as a source of holes (via repelled electrons exiting the base into the metal contact), thereby restoring the hole supply while some of the injected electrons have been eliminated. Thus the small amount of recombination does tend to favor charge neutrality.

The more common phenomenon is that thermally generated electrons leave the base. With fewer thermally generated electrons left, the corresponding thermally generated holes now act to neutralize the electrons injected from the emitter.

Now, the rate of thermally generated electron–hole pairs $g$ is independent of the number of carriers, while the rate of mutual annihilation $r$ is proportional to the concentrations of mobile charges, $np$, available for annihilation. In the steady state $r = g$ so $np$ = constant; in particular, for intrinsic silicon $np = n_i n_i = n_i^2 =$ constant, so in general (doped or intrinsic), $np = n_i^2$. Thus the stronger the base doping, the faster electrons will be annihilated—leaving newly thermally generated holes "safer" from annihilation.

However, to consider the establishment of base region neutrality, we must consider the transient situation, where electrons are first injected in large numbers into the base. Advanced analysis shows that the ohmic contact of the base region with the base electrode serves as a surface of high generation as well as recombination of holes and electrons. Thus, there being a need for more holes, hole–electron pairs are generated on the metal–semiconductor surface. The electrons created flow into the wire, and the holes created flow into the base region, creating an overall neutral base region. This constitutes a component of positive current into the base.

In any case, this component of the base current is one-time and short-lived, for establishing a large emitter operating current; it need not be considered in steady-state operation. The recombination current, on the other hand is steady, although very small.

Without this source of holes from the base terminal, eventually a negative charge would build up in the base sufficient to halt or significantly diminish the electron injection and thus the emitter and collector currents. In our informal terms, there would be a reduction of the *forward biasedness* of the base–emitter junction. With this source of holes, the injected electron current can be large because the forward bias $V_{BE}$ is not canceled by such a charge imbalance; that is, outside the two depletion regions, charge neutrality in the base is approximately maintained.

The vast majority of electrons injected into the base pass on into the collector, while new electrons enter from the emitter to replace them. The small number that do not are those that are annihilated by base region holes. Although there is a large number of injected electrons in the base, the concentration of injected electrons in the base is still significantly less than the doped hole concentration in the base. (The exceptional circumstance under which this assumption is violated is called *high-level injection,* the complicated explanations of which are outside the scope of this discussion.) This fact makes quantitative prediction of device behavior easier.

Finally, there is one more steady-state current drawn from the base terminal: hole injection into the emitter. Because the base is so lightly doped, the impetus of diffusion of base holes into the emitter is quite small. Nevertheless, it is tied to $V_{BE}$ just as the main emitter current is. Also, the recombination current is tied to $V_{BE}$ because it is tied to the emitter current: A small fraction of the electrons injected into the base recombine, producing the recombination current. The collector current is what is left over after the base current slightly diminishes the injected emitter current.

Consequently, we have the fact that the collector current $i_C$ is *proportional* to the base current, just as we have argued that the base, emitter, and collector currents are all mutually proportional. Thus if we have a very small signal contained in the base current $i_B$, that signal will proportionally be reproduced at a much higher level in the collector current.

It is easier to think of the base current as being the "side effect" than the "controlling force." However, we did mention above how starving the base of holes (via a reduced input base current) will indeed reduce emitter and collector currents. In particular, then, time variations in the base current must be reflected in the large, amplified emitter and collector currents.

Furthermore, it is often the case in physics that what we view as cause and effect may be for our own convenience and in fact physically irrelevant. For example, in a resistor, do we always have to apply a voltage to obtain the current $i = v/R$, or could we apply a current and get $v = iR$? The two are actually simultaneous; but because voltage sources are much more commonly encountered in practice than current sources, we tend to think the former way.

We may therefore conclude that because our signal source is the base current and the collector current follows those variations, the base current controls the emitter and collector currents:

$$i_C = \beta i_B, \tag{7.2}$$

where $\beta$ is a constant strongly dependent on temperature and device specimen.

Moreover, this control of the output (collector) current is proportional, unlike the control via $V_{BE}$, which is exponential. It is far easier to do circuit design using linear relations than using nonlinear relations. Consequently, the bipolar transistor is viewed as a current-controlled amplifier. The field-effect transistor, which we discuss next is a voltage-controlled amplifier.

**Example 7.4**

Suppose that 0.988 of emitter current makes it to the collector: $i_C = 0.988 i_E$. What is $\beta$?

**Solution** To be consistent with our previous physical discussions, we define positive base and collector currents as entering the transistor, and positive emitter current as leaving. The constant 0.988 is known as the $\alpha$ of that transistor, so that $i_C = \alpha i_E$. By KCL, the base current must be what is left over: $i_B = i_E - i_C = (1 - \alpha) i_E = [(1 - \alpha)/\alpha] i_C$, which by Eq. (7.2) is $i_C/\beta$. Thus $\beta = \alpha/(1 - \alpha) = 0.988/(1 - 0.988) \approx 82$.

Another way of viewing the bipolar transistor as a current-sensing device is in terms of its output characteristics. A current-driven device is any device having a small input impedance; a voltage-driven device is any device having a large input impedance. We shall calculate that bipolar amplifiers have relatively small input impedances (typically, a few kilohms), compared with the field-effect transistor (typically, $10^{14}\ \Omega$).

How is it that the bipolar junction transistor, composed of nonlinear diodes, can have a linear gain as in $i_C = \beta i_B$? Note that because the base current is proportional to the collector current, it varies with $v_{BE}$ in precisely the same nonlinear manner. Thus there is no contradiction: If $v_{BE}$ changes, both $i_C$ and $i_B$ will change radically, but in tandem. Of course, we should never forget that it is $v_{BE}$ that is responsible for the electric field that makes large $i_C$ possible.

When there is a small current or voltage signal source connected to a transistor amplifier, $v_{BE}$ *is* altered, albeit minutely. This change in $v_{BE}$ may be reflected in, for example, a change in current flowing through a biasing resistor (see below) but is due to the signal. Now with $v_{BE}$ changed, the amount of base current drawn will be altered a little—drawn from the signal source.

With this understanding of transistor physics, transistor models and circuits are within our grasp. We just need one more concept, that of *biasing*. The term "biasing" is familiar from our discussion of diodes. We said that when the voltage applied to a diode reaches the "turn-on" voltage $V_o \approx 0.7$ V, we have "forward bias" and the diode conducts. Recall that bias means a bent or tendency (in this case, away from $v = 0$); we need such a tendency in the applied voltage to make the diode conduct.

The same thing holds true for transistors. If the base–emitter voltage is not already biased to the turn-on voltage, the base–emitter junction will be reverse biased and there will be no emitter or collector current. The transistor is said to be cut off. A typical input signal has positive and negative voltage and current swings. Yet the requirement is that the base, emitter, and collector currents must all go only one way if cutoff is to be avoided and linear amplification obtained. The way of getting around this is by superimposing the signal onto a fixed, established bias level that guarantees linear operation.

Figure 7.23 shows the idea. A special resistor or set of resistors is used to establish the constant-in-time, forward-bias operating point base current $I_B$. Super-

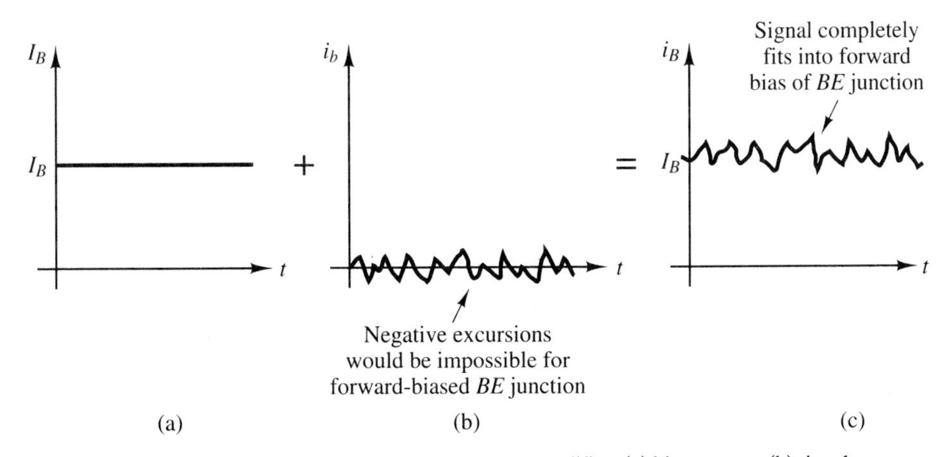


**Figure 7.23** Components of base current in an amplifier: (a) bias current; (b) signal current; (c) total current.

imposed on that is the small signal base current $i_b$, resulting in the total base current $i_B = i_b + I_B$.

This symbology is used for all voltages and currents in transistor circuits. A capital letter $V$ or $I$ indicates the constant-in-time operating point voltage or current. A lowercase $v$ or $i$ with a lowercase subscript indicates a small-signal voltage or current. Finally, a lowercase $v$ or $i$ with a capital letter indicates the total voltage or current. This is all in keeping with our previous use of lowercase for quantities having general time dependence and uppercase for only constant-in-time quantities.

Speaking of symbols, the symbol for the transistor is given in Fig. 7.24a. The arrow in the transistor symbol merely indicates which way conventional current flows: in the *npn* from collector to emitter, and the other way for a *pnp* transistor. The origin of the symbol dates back to the original *point-contact transistor* discovered by John Bardeen and Walter Brattain in 1948. Instead of *pn* junctions, rectifying metal–semiconductor contacts were used. These contacts were just emitter and collector wires touching a base of germanium (hence the term "base"). The physical appearance of the point-contact transistor is shown next to the transistor symbol in Fig. 7.24b; the triangle is just to hold the "cat's whisker" wires in place. Because of a number of inferior performance and analysis characteristics, the point contact transistor was quickly abandoned for the junction transistor (discovered by Shockley in 1949). Bardeen, Brattain, and Shockley won the Nobel Prize in Physics in 1956 for these discoveries.

Consider now the input–output $i$–$v$ characteristics of a transistor. Because it is a three-terminal device, one of the terminals must be shared ("be common") between input and output. The most frequently shared terminal is the emitter—called a *common-emitter* configuration. Figure 7.25 shows the configuration and Fig. 7.26 gives the characteristic curves.

Figure 7.25 just shows the common-emitter configuration for the input–output characteristics in Fig. 7.26 without regard to external circuitry required to produce the voltages and currents shown.

Figure 7.26a shows the base and collector currents as functions of the input voltage $v_{BE}$. Notice the inset, which shows that for small signals, the exponential curve is roughly linear:

$$i_b = \frac{v_{be}}{r_\pi}. \tag{7.3}$$

This resistance $r_\pi$ provides part of the model of the transistor for small variations about the operating point—the *small-signal model.* That is, the $i_B$ versus $v_{BE}$ characteristic is, local to the operating point, essentially a straight line with slope $1/r_\pi$.

We can obtain its value as follows. Recalling the exponential model of the diode

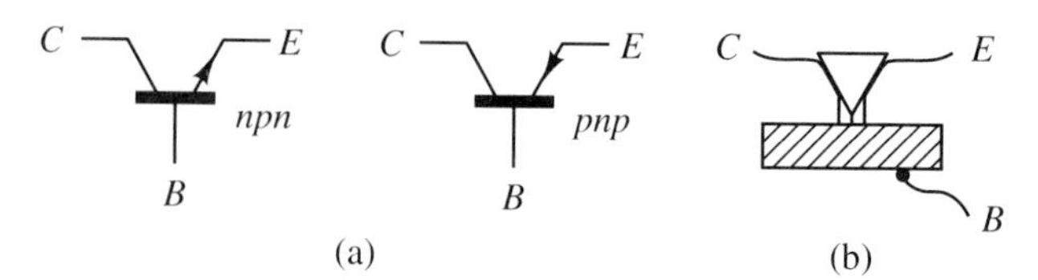


**Figure 7.24** (a) Bipolar transistor symbol; (b) origin of symbol was the point-contact transistor.

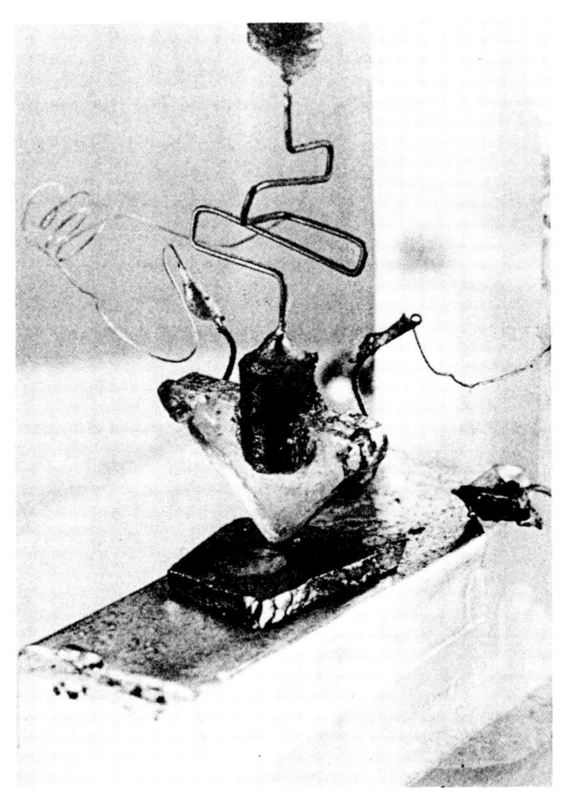

The first transistor (point-contact), which despite its lowly appearance would soon bring on the computer/solid state electronics age. Compare this physical structure with that implied by the common transistor symbols in Fig. 7.24a, and the schematic point-contact transistor in Fig. 7.24b. *Source:* Courtesy of AT&T Archives.

Eq. (7.1), which is accurate in forward bias, we note that in forward bias the "−1" is negligible compared with $e^{qv_{BE}/kT}$. For example, for $v_{be} = 0$, $v_{BE} = V_{BE} \approx 0.7$ V (forward bias), so that $e^{qv_{BE}/kT} \approx 5.10^{11} \ggg 1$. Furthermore, we can use Taylor's expansion of the exponential function $e^x \approx 1 + x$ (valid for $|x| \ll 1$, or in our case $|v_{be}| \ll kT/q \approx 0.026$ V) about the operating point and $v_{BE} = V_{BE} + v_{be}$. This is the same concept as shown in Fig. 7.23, but with the base current replaced by the base–emitter voltage. We may therefore write

$$i_B \approx I_S e^{qv_{BE}/kT} = I_s e^{qV_{BE}/kT} e^{qv_{be}/kT} \tag{7.4a}$$

$$= I_B e^{qv_{be}/kT} \tag{7.4b}$$

$$\approx I_B \left(1 + \frac{qv_{be}}{kT}\right) \tag{7.4c}$$

$$= I_B + i_b, \tag{7.4d}$$

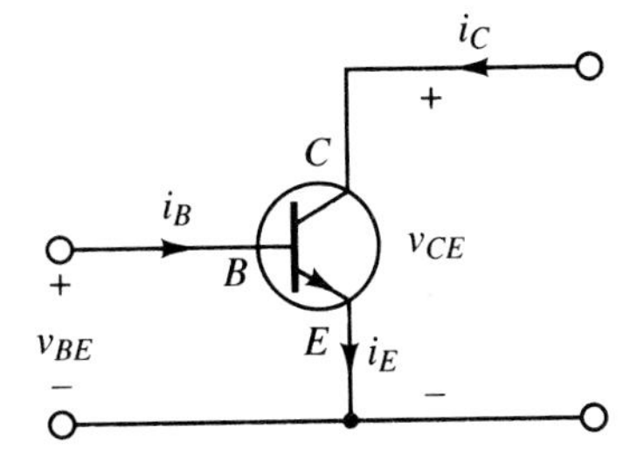


**Figure 7.25** Common-emitter parameters with which to characterize an *npn* transistor.

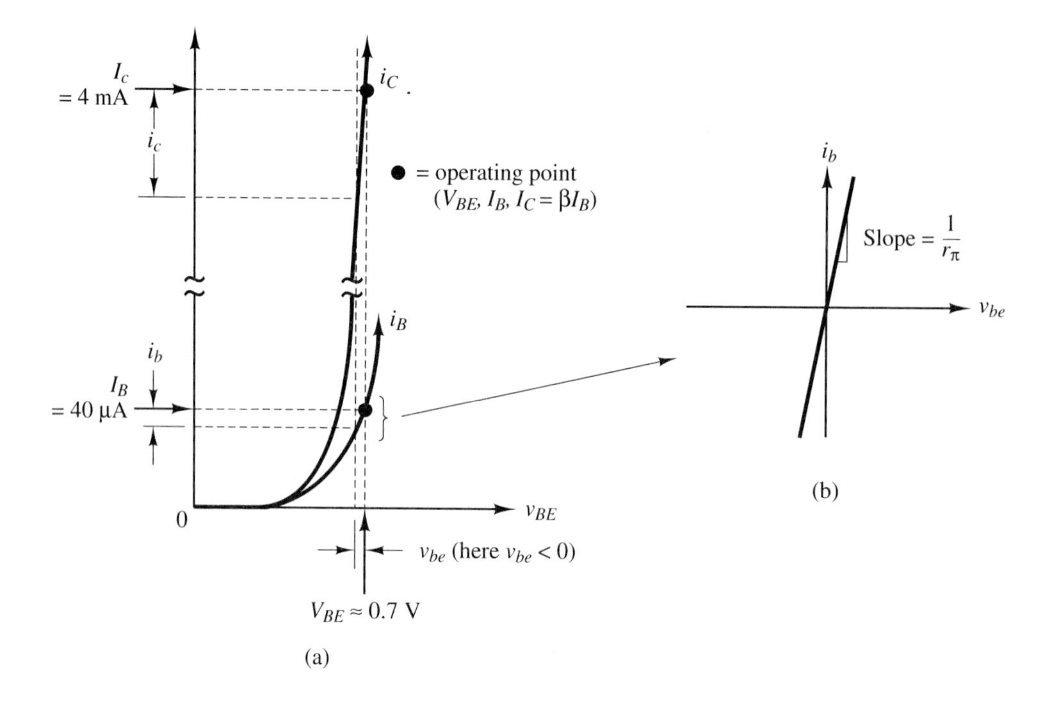


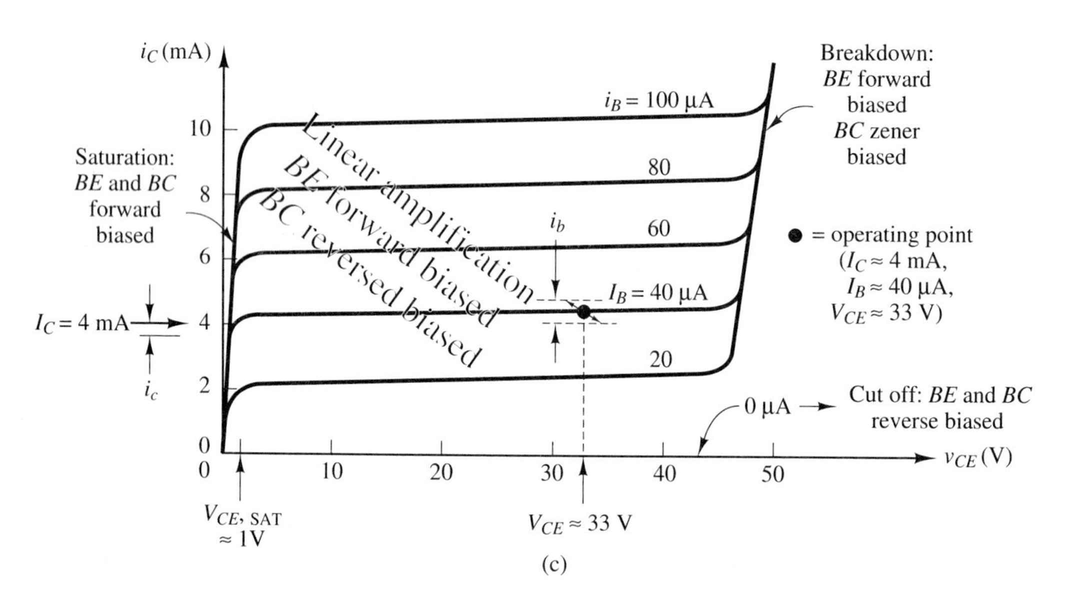


**Figure 7.26** (a) Collector and base currents shown as functions of base–emitter voltage for an *npn* transistor; (b) collector current characteristics for an *npn* transistor.

where

$$I_B = I_s e^{qV_{BE}/kT} \tag{7.5}$$

and

$$i_b = \frac{I_B q v_{be}}{kT}$$

$$= \frac{v_{be}}{r_\pi}, \tag{7.6}$$

where

$$r_\pi = \frac{kT}{qI_B} \tag{7.7a}$$

$$= \frac{\beta kT}{qI_C} \tag{7.7b}$$

All this means that for *small signals* such as $i_b$ and $v_{be}$, the base–emitter junction appears not like a diode (it is already forward biased), but rather, as a resistor. Contrarily, for large signals (such as $I_B$ and $V_{BE}$), it does appear as a diode. Thus our model for the transistor depends on whether we are talking about large signals (such as biasing levels) or small signals (such as audio or other weak measurement signals). In either case, we have both

$$I_C = \beta I_B \tag{7.8a}$$

*and*

$$i_c = \beta i_b \tag{7.8b}$$

because in general, Eq. (7.2) holds. These relations are nothing other than a current-dependent dependent current source. Note also that $r_\pi$ clearly depends on the operating point ($V_{BE}$, $I_C$), as is evident from both Eq. (7.7) and Fig. 7.26a.

The output characteristic appears in Fig. 7.26b. It shows the collector current as a function of the collector–emitter voltage $v_{CE}$ for several equally spaced values of base current.

The *saturation region* for very small $v_{CE}$ (for $v_{CE} < V_{CE,\text{SAT}} \approx 1$ V) occurs because the base–collector ($BC$) junction is still forward-biased. Note that $v_{CE} = v_{CB} + v_{BE}$. For all the curves shown (except the one labeled $i_B = 0\ \mu$A), the base–emitter junction is forward biased, so $V_{BE} \approx 0.7$ V. Thus for $v_{CE} \le V_{CE,\text{SAT}} \approx 0.3\text{ V} < 0.7$ V, it must be true that $v_{CB} < 0$, that is, the $BC$ junction is (to a degree) forward biased. Recall that for linear amplification, it must be reverse biased. Saturation can occur when the input base current is raised so large that $i_C$ can no longer be maintained at $\beta i_B$; further increases in $i_B$ no longer result in increases in $i_C$, hence the term "saturated." Typically, the large $i_C = \beta i_B$ in the collector resistor $R_C$ brings $v_{CE}$ down to $V_{CE,SAT}$ when, e.g., $v_{CE} = V_{CC} - i_C R_C$, where $V_{CC}$ is the power supply voltage.

The *cutoff region* is simply when the base–emitter junction is not forward biased, and hence $i_B \approx 0$ and $i_C \approx 0$. The saturation and cutoff regions are to be avoided when linear amplification is sought. However, we shall see that they both serve as the basis of transistor switching operation.

The *linear amplification region* is the region of horizontal lines; the same, typical operating point is shown in both Fig. 7.26a and b. Here the collector current is proportional to the base current (doubling $i_B$ doubles $i_C$), while it is essentially independent of $v_{CE}$. As long as the *BC* junction is reverse biased, the electrons emitted into the base are all either collected or annihilated regardless of the strength of the collecting field. This is not strictly true, as the slight upward slope of the curves indicates. That is because increasing $v_{CE}$ tends to widen the *BC* depletion region and thus reduce recombination; this is known as the *Early effect* after its discoverer in 1952, J. Early.

Finally, at a certain value of $v_{CE}$, the *BC* junction becomes Zener biased and breakdown occurs, causing the *BC* junction to conduct strongly in the reverse direction. This condition is normally avoided in amplifier design.

From these characteristics, we may consider two circuit models for the bipolar transistor, which are shown in Fig. 7.27. In Fig. 7.27a, the large-signal/dc model is given, which is the model suitable for finding the operating point of the transistor. Figure 7.27b shows the small-signal/ac model, appropriate for determining signal gain for variations about the operating point.

The small-signal model in Fig. 7.27b is referred to as a simplified version of the *hybrid* $\pi$ model. The reason for the term "hybrid" is as follows. The two most important parameters characterizing the small-signal model are $h_{ie} = r_\pi$, the *i*nput resistance for the common-*e*mitter configuration, and $h_{fe} = \beta$, the *f*orward current gain for the common-*e*mitter configuration. Because one parameter, $h_{ie}$, is a resistance, while another, $h_{fe}$ is dimensionless, the parameters are of mixed types (hybrid model). The reason for "$\pi$" is that when a capacitor is introduced between base and collector for modeling high-frequency behavior, the physical appearance of the model has the shape $\pi$.

In transistor circuit analysis, first the operating point must be determined, because knowledge of the operating point is necessary for determining $r_\pi$ of the small-signal model [see Eq. (7.7)]. To find the operating point, we use the large-signal/dc model in Fig. 7.27a, and from that determine the small-signal/ac model in

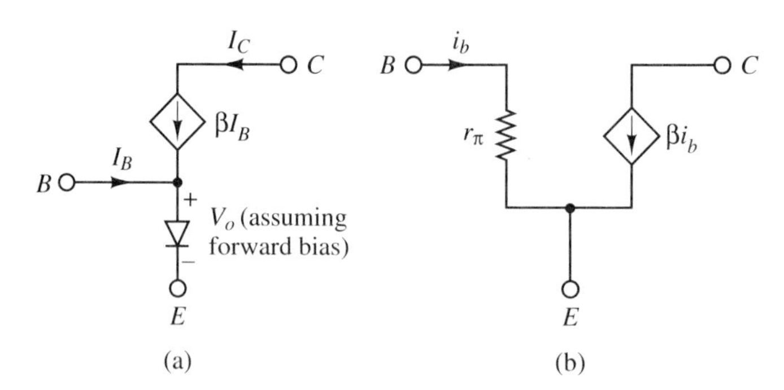


**Figure 7.27** Bipolar *npn* transistor models: (a) large signal/dc; (b) small signal/ac.

Fig. 7.27b. From it we can calculate signal voltage gain, current gain, and power gain. We consider a few examples of transistor analysis in Sec. 7.6.

From our initial studies, we may already see conceptually how power gain is achieved in a bipolar transistor circuit. The input power, $v_{be}\, i_b$, is quite small because $|v_{be}| \ll 0.7$ V and $i_b$ is a tiny recombination/minority diffusion current. The output power is $i_c$ times the load resistance, in a practical amplifier circuit (which we examine in more detail in Sec. 7.6). Now $i_c$ is $\beta i_b$, where $\beta \approx 60$ to 100 for a typical transistor. The load resistance $R_L$ can be whatever value desired (within reason), as long as it does not cause the $BC$ junction to lose its reverse bias. Consequently, $i_c^2 R_L = \beta^2 i_b^2 R_L$ can be far larger than $v_{be} i_b$ and true signal power gain is achieved.

**Field-effect transistor (FET).** The input impedance of the bipolar junction transistor, we noted, was only a few kilohms. As we shall see in Chapter 9, many active sensors have relatively large source resistances. Consequently, a BJT would load down such a sensor and distort the measurement. A better amplifying transistor would be one having very large input impedance, which would minimize loading of the weak source. The *field-effect transistor* (FET) is just such a device. Often, an FET may be the first stage in a multistage amplifier having BJTs for the later stages.

Although the FET has several advantages, particularly for the manufacturing of integrated-circuit chips, the BJT also has competing advantages over the FET. The FET has enormous input impedance, can be used as capacitors and resistors in integrated circuits, is very simple to fabricate in extremely small sizes, and has low power consumption. On the other hand, the BJT has a wider frequency response, better linearity, in some cases is electrically more rugged, and in some cases is more easily designed to handle higher power loads than an FET. For these and other reasons, both the BJT and the FET are still commonly used today.

Therefore, let us see what sort of device an FET is compared with the BJT, and how it works. Whereas the BJT has two *pn* junctions (*bijunction*), the FET has only one main junction (*unijunction*). Whereas the BJT involves both majority and minority carriers in the manner described above (*bipolar*), the FET involves only majority carriers (*unipolar*). Finally, whereas the BJT operated via both diffusion and ohmic currents (bicurrent type), the FET involves only ohmic currents (unicurrent type).

The idea of a field-effect transistor is actually quite simple. An electric field applied from the *gate* terminal is used to modify or modulate the conductivity of the conduction path through a large semiconductor section, called the *channel.* The gate terminal by which the applied electric field is introduced is electrically isolated from the channel. Therefore, negligible input current is required to set up the controlling electric field. Essentially, we end up with a voltage-controlled ohmic-current dependent current source for the channel. This is different from the BJT, where diffusion currents play a major role and thus cannot be characterized by only ohmic currents. Furthermore, our dependent current source is linear with gate voltage rather than base current.

The establishment of a controlling electric field is achieved in one of two ways. One way is by having the gate be heavily doped $p$-type ($p^+$) material and the channel be $n$-type, and applying a voltage that reverse biases that *pn* junction. That is, the gate potential must be negative with respect to the $n$-channel potential. Such a device is called a junction field effect transistor, or JFET.

The other way is by placing an insulator (silicon dioxide) between the metal gate electrode and the semiconductor. In this case the gate potential can be either greater than or less than the channel potential while always maintaining the situation of negligible input current. The effect is very much like a window, which lets light pass through either way (analogous to the electric field) while preventing the flow of cold air (analogous to current) through the window. In fact, silicon dioxide happens to be precisely a window: It is glass! Not surprisingly, this type of FET is called a metal-oxide-semiconductor FET, or MOSFET.

Just as there are both *npn* and *pnp* transistors, there are several types of FET. For an *npn* transistor, the currents are dominated by electrons, not holes. Because of their faster response (electrons are significantly more mobile than holes), *npn* transistors are more commonly used than are *pnp* transistors. Consequently, we concentrated our attention on the *npn* transistor. For the same reason, FETs whose channel current is predominantly electrons are more common than those having hole-type channel current, so we focus on the former type. Doing so also cuts our thinking in half.

For both JFET and MOSFET the channel, of course, has two ends connected to wires via ohmic metal–semiconductor contacts. The end toward which the charge carriers (in our discussion, electrons) normally move is called the *drain,* and the end from which they come is called the *source.* Therefore, conventional current comes out of the drain terminal and flows into the source terminal. (Again, we mourn the fact that electrons are said to have "negative" charge because of this confusion that results.)

Let us now consider briefly how the conductivity of the $n$-type channel of an $n$-channel JFET is controlled by the gate–source potential $V_{GS}$ (see Fig. 7.28). In Fig. 7.28a is shown the basic connection configuration with external voltage sources $V_{GS}$ and $V_{DS}$.

The inset shows a typical operating condition. The depletion region of the reverse-biased gate–channel $p^+n$ junction extends so much farther into the channel than into the $p^+$ gate region that we can ignore the depletion region in the gate. Due to $V_{DS}$, the channel potential increases from left to right. Because $V_{GS} < 0$, the channel is more strongly reverse biased near the drain due to the increased contrast between $V_G$ and the potential of that point within the channel. Therefore, the depletion region is asymmetrical, as shown in the inset. Of course, essentially all the conduction electrons travel through the relatively high-conductivity channel in the gap between the depletion regions.

This gap narrows or widens depending on $V_{GS}$, as shown in Fig. 7.28b. (In the following figures, the acceptor ions and electrons are omitted for simplicity.) Determination of the exact shape of the depletion region requires advanced mathematical analysis. If $V_{GS}$ reaches a critical value called the *pinch-off* (we shall here call it the $V_{GS}$ pinch-off) voltage $-V_P$, the two depletion regions join and the drain current $I_D$ ceases.

Suppose that $V_{GS}$ has an intermediate value, such as that in the inset of Fig. 7.28a. As noted, the depletion region, which depends on both $V_{GS}$ and $V_{DS}$ is asymmetrical. If $V_{DS}$ is very small, the gap between depletion regions decreases as $V_{DS}$ is increased. This tendency decreases $A$ in Ohm's law ($R = \rho d/A$), causing $R$ to

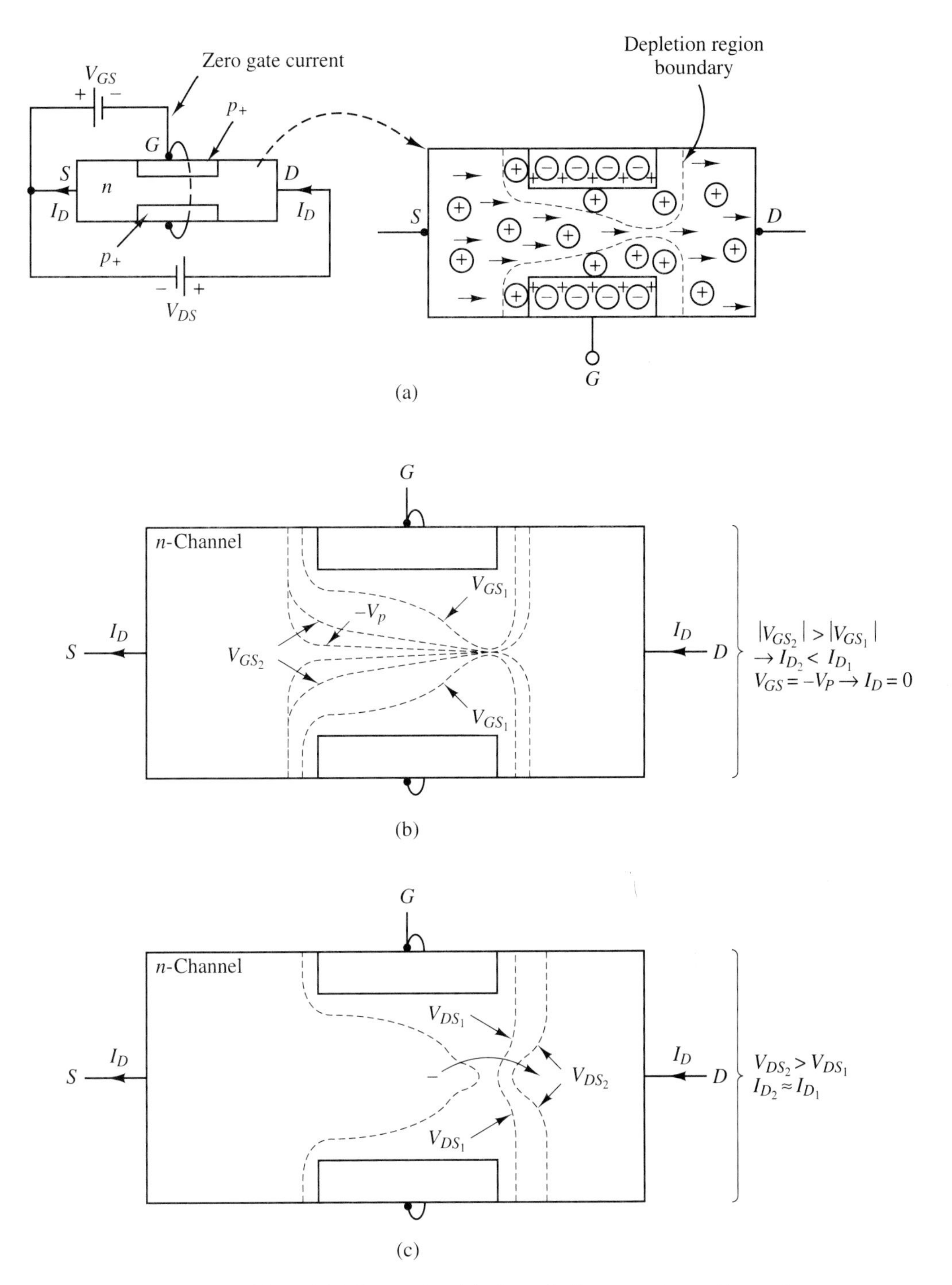


**Figure 7.28** $n$-Channel JFET operation: (a) schematic diagram and normal internal behavior; (b) dependence of depletion-layer boundaries on $V_{GS}$; (c) dependence of depletion-layer boundaries on $V_{DS}$.

increase and $I$ to decrease. However, because $I = V/R$, there is also a tendency for $I$ to increase with $V_{DS}$.

Eventually, again there is pinch-off (we shall call it the $V_{DS}$ pinch-off), but this time the horizontal interruption of the channel caused by the depletion region is far less wide. The number of electrons that can enter the depletion region (under the influence of $V_{DS}$) and make it all the way to the other side (the remainder of the channel) without hitching to a donor (positive) ion depends exponentially on the width of the interruption. Those that get caught repel other electrons from entering, and the current is thus very small. This is the case for $V_{GS}$ pinch-off. Because the interruption is so narrow for $V_{DS}$ pinch-off, current can flow through the depletion interruption because electrons can make it all the way to the other side (see Fig. 7.28c).

It is experimentally verified that the tendencies of drain current decrease (due to increased channel interruption) and drain current increase (due to increased electric field) cancel for a wide range of $V_{DS}$. That is, $I_D$ becomes independent of $V_{DS}$ for $V_{DS}$ beyond the pinch-off value (which depends strongly on the value of $V_{GS}$).

Thus this condition is also called pinch-off even though $I_D$ steadily flows in the channel, with the depletion region interruption. Clearly, this $V_{DS}$ pinch-off ($I_D \neq 0$ and is independent of $V_{DS}$) is behaviorally totally different from the $V_{GS}$ pinch-off (for which $I_D = 0$). It is this $V_{DS}$ pinch-off that is the normal mode of operation of the JFET for linear amplification. This is reminiscent of the independence of the BJT collector current from $V_{CE}$.

The other allowable extreme for $V_{GS}$ other than $V_{GS}$ pinch-off is $V_{GS} = 0$, for which the depletion region is the minimum possible without having the gate–channel junction become forward biased. If for $V_{GS} = 0$ we increase $V_{DS}$ sufficiently to reach the $V_{DS}$-pinch-off mode, we then label $I_D = I_{DSS}$, where $I_{DSS}$ is the *d*rain–*s*ource *s*aturation current. It is called saturation merely because the current can go no higher than this value, with $V_{GS}$ held fixed at a non-forward-biased value. Note that with $V_{GS}$ disconnected, there would be no depletion region, and the channel would be an ordinary resistance.

The symbols for the JFET, both $n$-channel and $p$-channel, are shown in Fig. 7.29. As was the case for the BJT, the symbol harks back to the physical appearance of the original working JFETs. Still, there is similarity with the modern structure in Fig. 7.28a.

Again, the direction of the arrow indicates type: $n$-type if the arrow points toward the channel and $p$-type if it points away from the channel. These arrows conform to the diode arrow convention (recall that the gate–channel interface is a $pn$ junction). The symbol for the gate (not the gate itself!) is usually positioned asymmetrically with respect to the channel, specifically, closer to the *source* terminal, to define the terminals unambiguously.

As noted above, there is a variety of FET devices. The JFET has just been

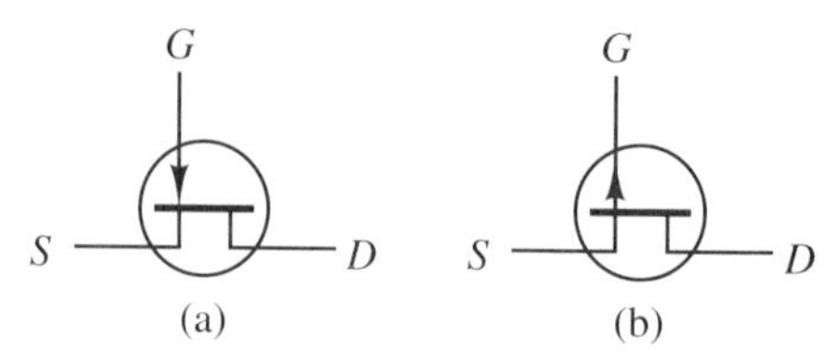


**Figure 7.29** Symbols for the JFET: (a) $n$-channel; (b) $p$-channel.

discussed. We considered briefly the idea behind MOSFETs. There are two types of MOSFET: depletion/enhancement and enhancement-only types. *Depletion* means that the device operates with $V_{GS} < 0$; the greater $|V_{GS}|$ is, the greater the extent of the depletion region (hence the name) and the lower $I_D$. *Enhancement* refers to devices that amplify when $V_{GS} > 0$; the larger $V_{GS}$, the greater the extent of the channel (the channel conduction is enhanced) and the greater $I_D$.

Clearly, the JFET cannot operate in enhancement mode because the gate–channel junction must be reverse biased to avoid undesirable large input currents. But with the "window" in place, the enhancement mode is possible.

Consider first depletion/enhancement FETs (usually abbreviated as "depletion MOSFETs"), shown in Fig. 7.30a. The symbols for the depletion MOSFET are shown in Fig. 7.30b and c. The separation between gate terminal and channel signifies the oxide insulating layer. The way in which the channel can be effectively extended beyond its size for $V_{GS} = 0$ is by having the $n$-channel material on a $p$-type substrate. For sufficiently large and positive $V_{GS}$, a new channel section can be induced in the $p$-type region, enhancing the conduction from source to drain.

This induced channel section is due to the attraction of electrons in the $p$-type material. Neutrality is restored in the $p$ region and electrons for the induced channel are replenished by connection of the substrate terminal $B$ with a low voltage (relative to the gate voltage) such as the source terminal or other available negative voltage. Thus the MOSFET generally has a fourth terminal, unless the connection to the source terminal is internally made, as shown by the source–substrate wire in Fig. 7.30a and dotted line in Fig. 7.30b. Note that for negative $V_{GS}$, electrons are forced out of the $n$-channel, depleting it and thus reducing its conductivity. Eventually, pinch-off occurs, at $V_{GS} = -V_P$.

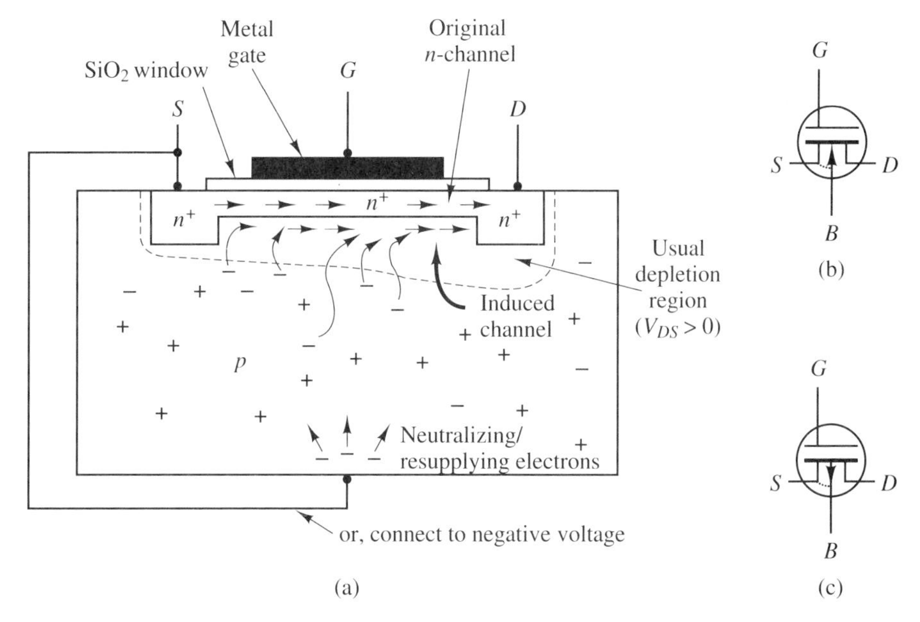


**Figure 7.30** Depletion MOSFET: (a) physical operation schematic; (b) circuit symbol for $n$-carrier device; (c) circuit symbol for $p$-carrier device.

An enhancement-only FET (usually abbreviated "enhancement MOSFET"), shown in Fig. 7.31a, has no $n$-channel at all initially. For a channel to exist, $V_{GS}$ must rise above a threshold value $V_T$. From there, the larger $V_{GS}$, the larger the channel. Its advantage is its extremely simple structure, making high transistor-density fabrication possible. Thus enhancement MOSFETs are used in high-density integrated circuits. Its circuit symbols are shown in Fig. 7.31b and c. The dashed line for a channel signifies that normally no channel is present, while again the dotted line indicates that substrate and source terminals are sometimes internally connected.

The $i$-$v$ characteristics of the JFET, depletion MOSFET, and enhancement MOSFET are shown, respectively, in Fig. 7.32a, b, and c. For $v_{DS}$ sufficiently large, the (nonlinear) amplification region is obtained, wherein $i_D$ depends strongly on the input voltage $v_{GS}$ but is essentially independent of load-driving voltage $v_{DS}$ (note the use of lowercase $v$ and $i$ for the plots in Fig. 7.32 of total voltage and current). It is not shown in Fig. 7.32, but for $v_{DS}$ very large, Zener breakdown for the JFET and dielectric breakdown of the MOSFETs may occur.

Comparing with the analogous characteristics of the BJT (see Fig. 7.26b), we see that the output current $i_D$ depends not on the input current (which is extremely small), but on the input voltage $v_{GS}$. Thus the FET is a voltage-sensing device rather than a current-sensing device such as the BJT. Furthermore, unlike the BJT, for which $I_C = \beta I_B$ (linear input–output relation), the FET is a nonlinear device for large signals/dc. Specifically, complicated mathematical analysis, or simply measurements show that the following quadratic model is quite accurate for most purposes:

$$i_D = I_{DSS}\left(1 + \frac{v_{GS}}{V_P}\right)^2. \tag{7.9}$$

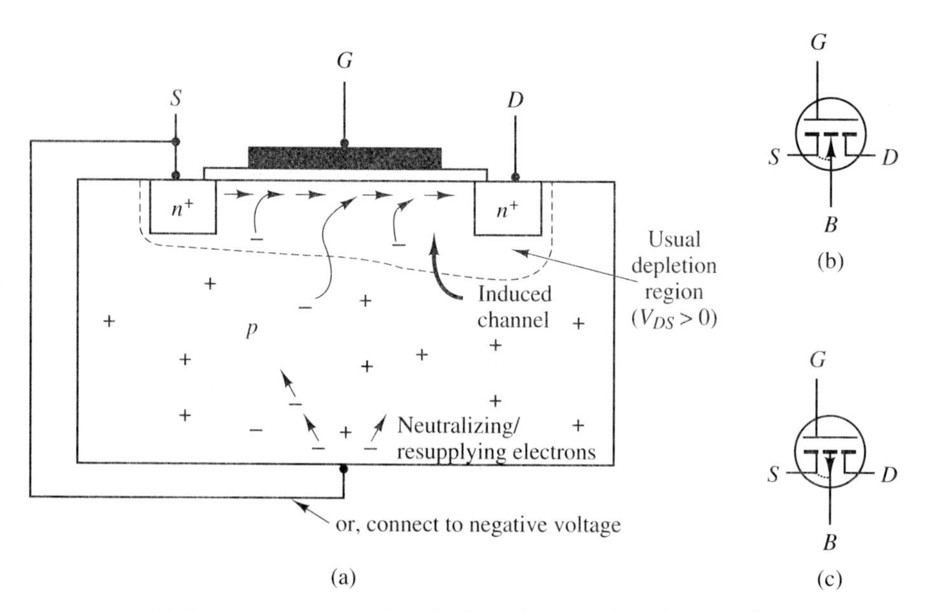


**Figure 7.31** Enhancement MOSFET: (a) physical operation schematic; (b) circuit symbol for $n$-carrier device; (c) circuit symbol for $p$-carrier device.

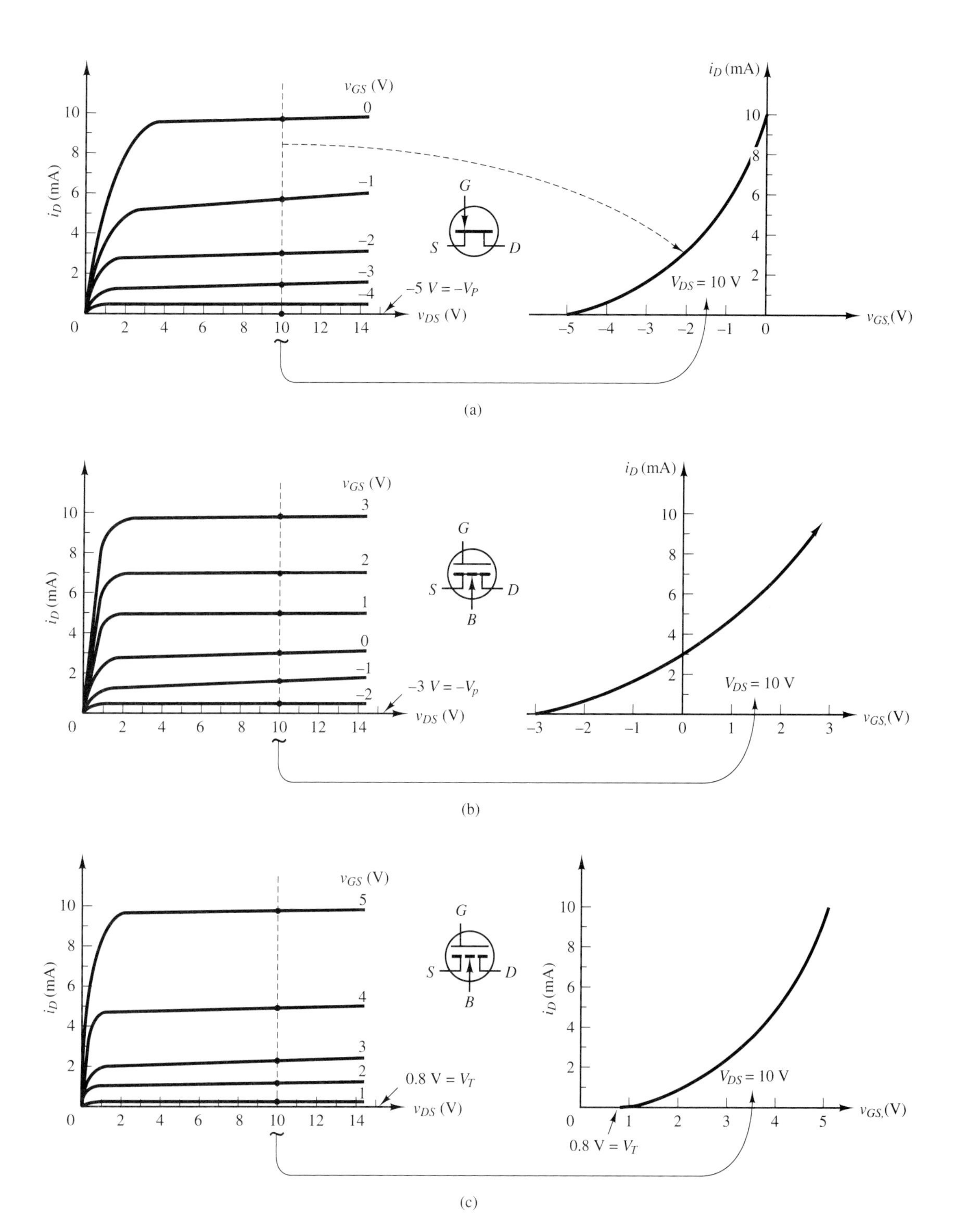


**Figure 7.32** *i-v* characteristics of FETs (common-source configuration): (a) JFET; (b) depletion–enhancement MOSFET; (c) enhancement MOSFET.

Various forms of this equation may be used for the variety of FETs. For example, for the enhancement MOSFET, we use $V_P = V_T$.

We can see this relation qualitatively in the curves, where the spacing of horizontal lines is nonuniform, yet the lines represent uniform increments of $v_{GS}$. These curves can, like Fig. 7.26c, be used to determine the operating point of a transistor circuit. Unlike the BJT, we do not have a simple circuit model for large signals; we must either use graphical techniques or equations such as Eq. (7.9).

Again, variations about the operating point are roughly linear, and there *are* simple circuit models for these. Essentially, we have a dependent current source for $i_d$ which is proportional to the input signal voltage $v_{gs}$. The proportionality constant can easily be determined from Eq. (7.9) as follows. Because Eq. (7.9) is true, then certainly

$$\begin{aligned} I_D &= I_{DSS}\left(1 + \frac{V_{GS}}{V_P}\right)^2 \\ &= I_{DSS}\left(1 + \frac{2V_{GS}}{V_P} + \frac{V_{GS}^2}{V_P^2}\right), \end{aligned} \tag{7.10}$$

and because $i_D = I_D + i_d$ and $v_{GS} = V_{GS} + v_{gs}$, where $I_D$ and $V_{GS}$ are the operating point values, Eq. (7.9) can be rewritten as

$$\begin{aligned} I_D + i_d &= I_{DSS}\left(1 + \frac{V_{GS} + v_{gs}}{V_P}\right)^2 \\ &= I_{DSS}\left[1 + \frac{2(V_{GS} + v_{gs})}{V_P} + \frac{(V_{GS} + v_{gs})^2}{V_P^2}\right] \\ &= I_{DSS}\left(1 + \frac{2V_{GS}}{V_P} + \frac{2v_{gs}}{V_P} + \frac{V_{GS}^2}{V_P^2} + \frac{2V_{GS}v_{gs}}{V_P^2} + \frac{v_{gs}^2}{V_P^2}\right) \\ &= I_D + I_{DSS}\left(\frac{2}{V_P} + \frac{2V_{GS}}{V_P^2} + \frac{v_{gs}}{V_P^2}\right)v_{gs}, \end{aligned} \tag{7.11}$$

or, ignoring the $v_{gs}^2$ term for a Taylor series linearization,

$$i_d = \frac{2I_{DSS}(1 + V_{GS}/V_P)}{V_P}v_{gs}, \tag{7.12}$$

so that we may define the *m*utual- (note the similarity between dependent sources and mutual inductance in Sec. 6.5) or transconductance $g_m$ as

$$g_m = \frac{i_d}{v_{gs}} = \frac{2I_{DSS}(1 + V_{GS}/V_P)}{V_P}. \tag{7.13}$$

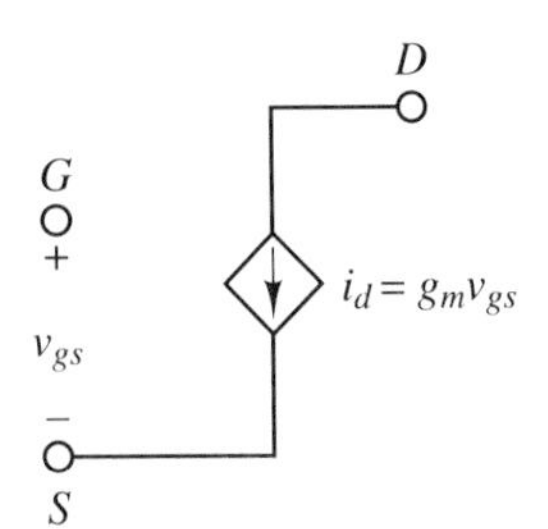


**Figure 7.33** Small signal/ac model of FET.

With $g_m$ so defined, the small signal/ac model of the FET is as shown in Fig. 7.33. [Because $g_m = i_d/v_{gs} = \partial i_D/\partial v_{GS}$, we could also have obtained $g_m$ by the partial derivative of $i_D$ in Eq. (7.9) with respect to $v_{GS}$ (again, to first order).]

For greater accuracy, a drain resistance $r_d$ can be placed in parallel with the dependent current source to model the slight upward tilt of the lines in Fig. 7.32; we do not consider this refinement further here. We consider an example of analysis of a FET amplifier using this model in Sec. 7.6.

## 7.6 THE TRANSISTOR: APPLICATIONS

### 7.6.1 Introduction

We are now in a position to analyze a basic transistor amplifier. That is, we can now fill in the "black box" of Fig. 7.21. This look at transistor circuits will avoid detailed design considerations, which are the domain of advanced electrical engineers; our aim here is to see power or voltage gain in action. We also avoid endless litanies of variations on amplifier circuits and other transistor circuits. One or two representative amplifiers suffice to illustrate the virtues of transistors.

The remarkable property of transistor circuits is that they can have signal power gains of greater than 1. This characteristic was discussed at some length in Sec. 7.2. To run through a list of transistor applications would be pointless, because transistors are in essentially *all* electronic circuits. To begin this brief look into transistor applications, we analyze both a bipolar and a field-effect transistor amplifier.

### 7.6.2 Common-Emitter Amplifier (BJT; CE)

The circuit of Fig. 7.34a is called a common-emitter *npn* transistor amplifier because the emitter is common between the input and the output voltage references. The output voltage is the collector voltage. For dc loads the collector resistor $R_C$ would instead be the load. Otherwise, the black box has been filled in as shown by the dashed line in Fig. 7.34a.

The only basic differences from our original transistor circuit (Fig. 7.22) are that (1) the signal source is now explicitly shown; (2) the load $R_L$ that the collector drives, which could itself be another amplifier in a multistage amplifier, is explicitly shown; (3) the collector voltage is defined with respect to the emitter reference rather than the base ($V_{CE}$ instead of $V_{CB}$); and (4) $V_{BE}$ is derived from $V_{CC}$ so that only one constant voltage source is required. (Note that the circle with $V_{CC}$ beside it is the usual shorthand way of symbolizing a voltage source of voltage $V_{CC}$, with respect to ground.)

In Fig. 7.34a, the signal load $R_L$ is *capacitively coupled* to the collector via $C_3$. Similarly, the input voltage $v_s$ is coupled capacitively to the amplifier by $C_1$. This means that for moderate and high-frequency signals, the coupling capacitor impedance is essentially negligible, while for the constant-in-time operating-point variables it is an open circuit. Recall that the magnitude of capacitive impedance is $1/(\omega C)$, which is small for large $\omega$ and is infinite for $\omega = 0$. Thus for this first-cut analysis we think of capacitors as open circuits for dc biasing and as shorts for incremental signals.

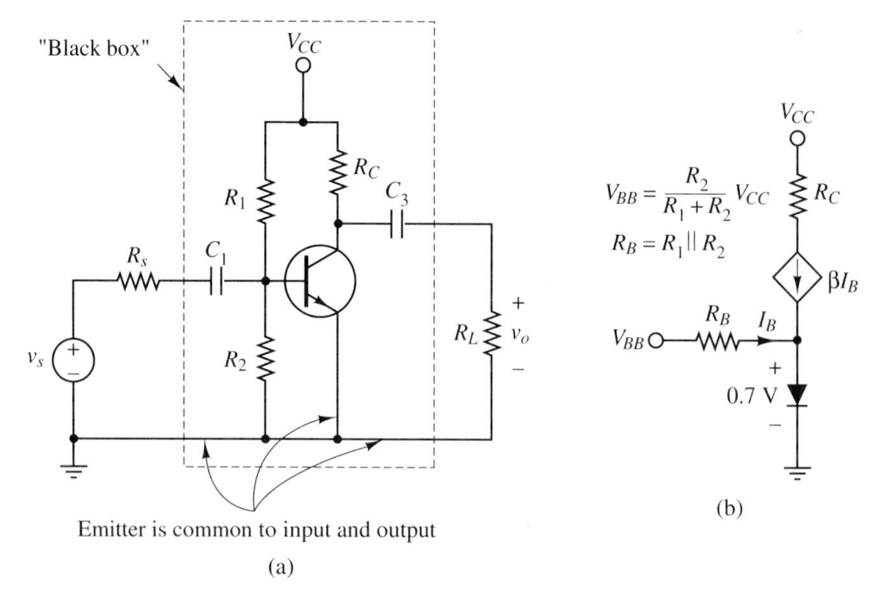


**Figure 7.34** Common-emitter amplifier: (a) component schematic diagram; (b) schematic for bias calculations (constant-in-time voltages and currents).

As noted in Sec. 7.5.2, bipolar junction transistors must be biased in order to be able to amplify bipolar signals (i.e., voltages or currents that go both positive and negative). The type of bias circuit we shall consider here is called the voltage divider configuration: $R_1$ and $R_2$ "divide" $V_{CC}$.

Note that there is pure voltage division *only* if the base terminal is disconnected from the node connecting $R_1$ and $R_2$. Indeed, with the base connected, the potential of that node is $V_o$, the *BE* contact potential. However, we shall now see that the equivalent bias source *is* found by a voltage divider.

To minimize the tedium of calculation, it is convenient immediately to find the Thévenin equivalent of the base-bias source. Recalling that $C_1$ is an open circuit for bias calculations, the equivalent circuit seen by the base looking to the left is as follows. The independent source $V_{CC}$ is turned off to find the equivalent bias resistance $R_B = R_{eq}$, which is easily seen to be

$$R_B = R_1 || R_2. \tag{7.14}$$

The open-circuit bias voltage $V_{BB}$ is easily seen to be the voltage divider $V_{BB} = V_{CC} R_2/(R_1 + R_2)$.

Replacing the bias circuit by its Thévenin equivalent, the capacitors by open circuits, and the transistor with the large-signal model of Fig. 7.27a gives the circuit of Fig. 7.34b. The base current is, by Ohm's law, $I_B = (V_{BB} - 0.7\text{ V})/R_B$, and therefore the operating point is

$$I_C = \beta I_B = \frac{\beta(V_{BB} - 0.7\text{ V})}{R_B}. \tag{7.15}$$

Before even bothering to analyze the small-signal voltage gain by substituting the model of Fig. 7.27b into Fig. 7.34a, notice that $I_C$ is directly proportional to β.

Unfortunately, $\beta$ is highly variable; it varies with temperature, age, and especially individual transistors. If the operating point is strongly dependent on $\beta$, the amplifier performance will be unpredictable. The transistor may even unintentionally go into cutoff or saturation or worse, be destroyed by heating up too much ("thermal runaway," see Sec. 7.7.3). As a matter of fact, thermal destruction is the most common cause of electronic component failure. Furthermore, it is left as an exercise to show that also the small-signal voltage gain is also proportional to $\beta$.

By adding only one resistor, we can largely solve this problem. The amplifier of Fig. 7.35a includes an emitter resistor $R_E$, and the large-signal model is shown in Fig. 7.35b. As with the *coupling* capacitors $C_1$ and $C_3$, the emitter *bypass* capacitor $C_2$ is an open circuit for operating point (bias) variables and a short for moderate- to high-frequency small signals. Its use will be shown when we analyze the small-signal model; for bias analysis it is "not there" (an open circuit). The emitter current in $R_E$ is, by KCL, $I_B(1 + \beta)$. Therefore, again by Ohm's law, the base current satisfies the equation

$$I_B = \frac{V_{BB} - [0.7\text{ V} + I_B(1 + \beta)R_E]}{R_B}, \tag{7.16}$$

or, solving for $I_B$,

$$I_B = \frac{V_{BB} - 0.7\text{ V}}{[1 + (\beta + 1)R_E/R_B]R_B} \tag{7.17a}$$

$$\approx \frac{V_{BB} - 0.7\text{ V}}{R_B + \beta R_E}, \tag{7.17b}$$

where in Eq. (7.17b) we used $\beta + 1 \approx \beta$ because $\beta$ is large and highly variable so that the "+1" is insignificant. The approximation $\beta + 1 \approx \beta$ is omnipresent in bipolar transistor amplifier analysis.

The operating-point collector current is

$$I_C = \beta I_B \tag{7.18a}$$

$$= \frac{\beta(V_{BB} - 0.7\text{ V})}{R_B + \beta R_E}, \tag{7.18b}$$

which for $\beta R_E \gg R_B$ reduces to

$$I_C \approx \frac{V_{BB} - 0.7\text{ V}}{R_E}, \tag{7.18c}$$

which is (roughly) independent of $\beta$.

To determine what this circuit will do to the small input signal $v_s$, we first analyze the small-signal model of the circuit shown in Fig. 7.35c, which is shown without $C_2$ (equivalently, $C_2 = 0$). (Do *not* confuse this case with the case $R_E = 0$ in Fig. 7.34.) Then $C_2$ will be introduced as in Fig. 7.35d. A detailed explanation of how Fig. 7.35c was obtained from the original circuit of Fig. 7.35a is given next.

There are two essential approximation assumptions made in drawing Fig. 7.35c: (1) the frequencies of the input signal $v_s$ and the capacitances are large enough to justify replacing all capacitors by short circuits; that is, we assume that the capacitors were chosen sufficiently large that $1/(\omega C)$ is small relative to other resistances over

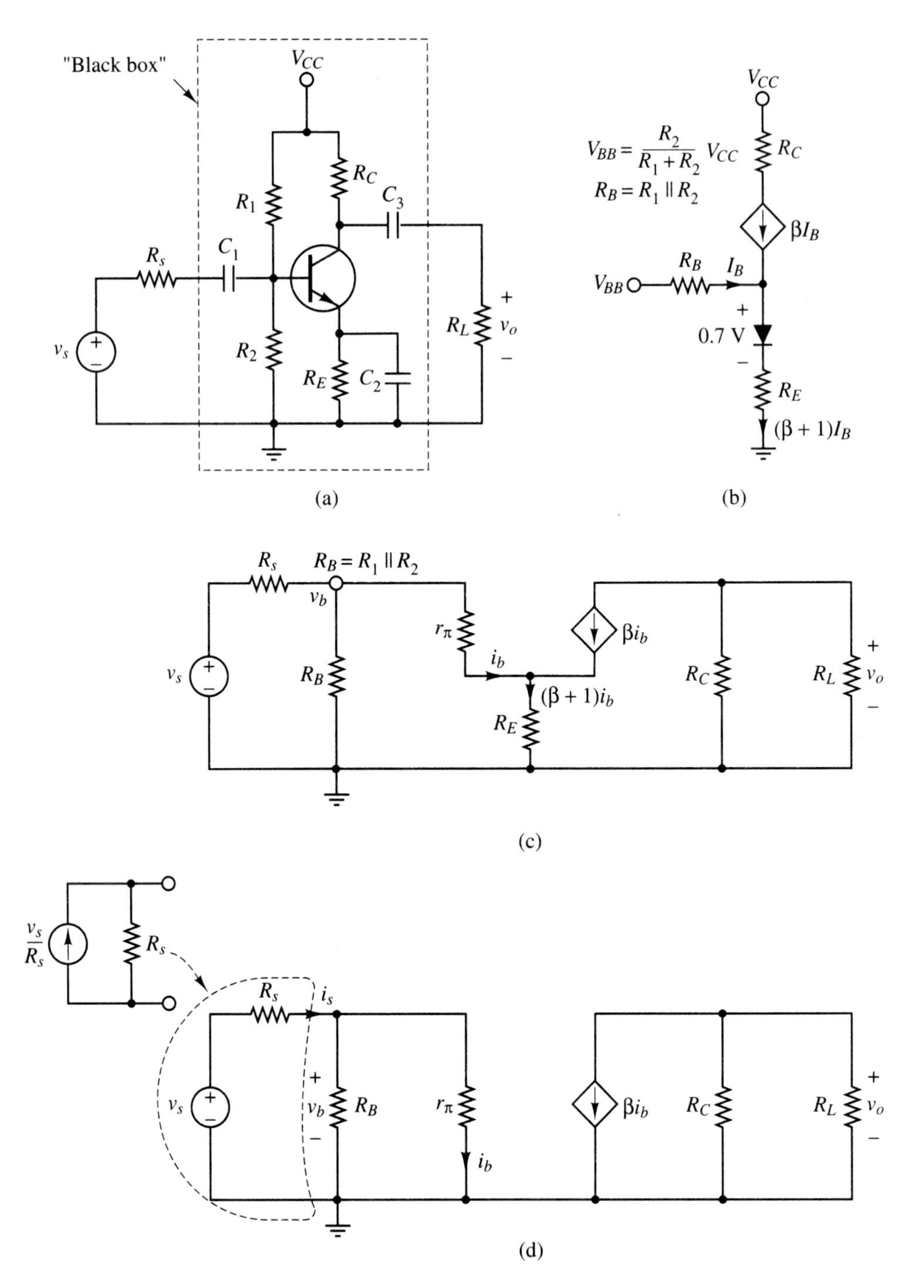


**Figure 7.35** Common-emitter amplifier with emitter resistance $R_E$: (a) component schematic diagram; (b) schematic for bias calculations; (c) schematic for small-signal calculations without $C_2$; (d) schematic for small-signal calculations with $C_2$.

all ω for which $v_s$ has significant energy, and (2) the signals are small enough in magnitude to justify using a linear model ($r_\pi$) for the base–emitter *i-v* characteristic of the transistor, for these variations about the operating point.

One further requirement necessary to obtain Fig. 7.35c from Fig. 7.35a is clear when we again recall that Fig. 7.35c represents the circuit model for how the actual circuit in Fig. 7.35a processes *deviations from the operating point.* For this model, all dc sources must be replaced by sources of zero value because they have zero deviations. Thus the bias voltage $V_{CC}$, having zero signal content or variation about its fixed value, is replaced by a voltage source of zero value—that is, a short circuit. (If there were any independent constant-in-time current sources, they would here be replaced by open circuits.)

Recall that $V_{CC}$ is a node voltage; that is, $V_{CC}$ is the potential of the node so labeled *with respect to the common reference potential* marked with the "ground" symbol. Therefore, the short circuit is between the $V_{CC}$ node and the ground node. Examination of Fig. 7.35c shows that this has been done; recall that $R_B = R_1 || R_2$, and they are indeed in parallel in the small-signal circuit model. With these explanations, the circuit in Fig. 7.35c should now follow directly from the original circuit of Fig. 7.35a (again, for now we take $C_2 = 0$).

The signal *voltage gain* $A_v = v_o/v_s$ can now be calculated from the circuit of Fig. 7.35c. First, KVL and Ohm's law show that

$$v_o = -\beta i_b R_L', \tag{7.19}$$

where

$$R_L' = R_L || R_C. \tag{7.20}$$

In practice, $R_L$ could be either the load resistance or, in the case of a multistage amplifier, the small-signal input resistance of the next amplification stage. Clearly, we must find $i_b$ in terms of $v_s$ in order to obtain $A_v$; we do this now.

Defining the node incremental base voltage $v_b$, KCL at that node takes the form

$$\frac{v_b - v_s}{R_s} + \frac{v_b}{R_B} + i_b = 0, \tag{7.21a}$$

or

$$v_b\left(\frac{1}{R_s} + \frac{1}{R_B}\right) + i_b = \frac{v_s}{R_s}. \tag{7.21b}$$

Note that because of the dependent current source $\beta i_b$, $r_\pi$ and $R_E$ are *not* in series. However, we can say approximately that $r_\pi$ is effectively in series with $\beta R_E$:

$$v_b \approx i_b r_\pi + \beta i_b R_E = i_b(r_\pi + \beta R_E). \tag{7.22}$$

Substituting Eq. (7.22) into Eq. (7.21b) gives

$$i_b\left[1 + (r_\pi + \beta R_E)\left(\frac{1}{R_s} + \frac{1}{R_B}\right)\right] \approx \frac{v_s}{R_s}, \tag{7.23a}$$

or

$$i_b \approx \frac{v_s}{R_s + (r_\pi + \beta R_E)(1 + R_s/R_B)}, \tag{7.23b}$$

which for $R_B \gg R_s$ (true in practice) reduces to

$$i_b \approx \frac{v_s}{R_s + r_\pi + \beta R_E}, \tag{7.23c}$$

which for $\beta R_E \gg R_s + r_\pi$ (again, true in practice) reduces to

$$i_b \approx \frac{v_s}{\beta R_E}. \tag{7.23d}$$

Finally, we substitute Eq. (7.23d) for $i_b$ into our initial expression for $v_o$, Eq. (7.19) to obtain the voltage gain

$$v_o = -\beta i_b R_L' \tag{7.24a}$$

$$\approx -\left(\frac{R_L'}{R_E}\right) v_s, \tag{7.24b}$$

or

$$A_v = \frac{v_o}{v_s} \approx \frac{-R_L'}{R_E}. \tag{7.25}$$

First notice that $A_v < 0$; that is, this amplifier has negative gain. This does not mean attenuation, but amplification with a 180° phase shift between input and output. Second, we see that $A_v$ is independent of $\beta$, which was desired. The resistances $R_L'$ and $R_E$ can be known quite accurately (see Sec. 4.3.1) and are relatively stable over time. Furthermore, for the occasional situation in which $R_L \gg R_C$, $R_L' \approx R_C$ so that $A_v$ is independent of the high-impedance load $R_L$. The main problem now is that the gain $-R_L'/R_E$ may be relatively small (less than 1 to about 10) in order that proper biasing be established. This is a significant reduction in gain relative to the comparable circuit without $R_E$. The reduction is due to a phenomenon called negative feedback, which we explore in detail in Sec. 8.4.

To maintain independence of $\beta$ yet increase the gain, we now include $C_2$, chosen to be large enough so that for the frequency range of our input signal, $1/(\omega C_2) \approx 0$. In this case, for *small signals only,* the emitter resistor $R_E$ is *bypassed*; hence $C_2$ is called a bypass capacitor. For biasing ($\omega = 0$), $R_E$ is still present because $1/(0C_2) = \infty$ (the capacitor is an open circuit). Because $r_\pi$ depends on the operating point and thus on $R_E$, we still expect the gain for this new situation to depend on $R_E$, but we will find that $A_v$ will indeed be larger than for the case $C_2 = 0$.

With $C_2 \neq 0$, the small-signal circuit model of Fig. 7.35d applies. To find the gain in this case, we may proceed as follows. In this case it is convenient to replace the $(v_s, R_s)$ combination by a current source via source transformation, as shown in Fig. 7.35d. With the equivalent current source in place, $R_s$ is now in parallel with $R_B$. Considering this parallel combination as a single resistor, $R_s || R_B$, we see a current divider:

$$i_b = \frac{R_s || R_B}{r_\pi + R_s || R_B} \frac{v_s}{R_s} \tag{7.26a}$$

$$= \frac{R_B}{r_\pi(R_s + R_B) + R_s R_B} v_s, \tag{7.26b}$$

which, if we again assume that $R_s \ll R_B$ and also assume that $r_\pi \gg R_s$, reduces to

$$i_b \approx \frac{v_s}{r_\pi}, \tag{7.26c}$$

so that

$$v_o = -\beta i_b R_L' \tag{7.27a}$$

$$= -\left(\frac{\beta R_L'}{r_\pi}\right) v_s. \tag{7.27b}$$

To understand this expression for $v_o$, recall the expression for $r_\pi$ in Eq. (7.7b) and the operating point collector current for this circuit given in Eq. (7.18) to obtain

$$r_\pi = \frac{\beta kT}{qI_C} \tag{7.28a}$$

$$= \frac{\beta kT}{q} \frac{R_E}{V_{BB} - 0.7 \text{ V}}, \tag{7.28b}$$

with the result that the voltage gain is

$$A_v = \frac{v_o}{v_s} = -\frac{R_L'}{R_E} \frac{V_{BB} - 0.7 \text{ V}}{kT/q}. \tag{7.29}$$

In comparing $A_v$ of Eq. (7.29) with $A_v$ in Eq. (7.25), we see that they differ by only the factor $(V_{BB} - 0.7 \text{ V})/(kT/q)$. Because this factor is much greater than 1 in practice, there has been a rise in $A_v$ without introduction of dependence on $\beta$. One may correctly argue that $\beta$ depends on $T$ in a roughly linear fashion, so the direct dependence of $A_v$ on $T$ is, in that sense, no better than having no emitter resistance at all. However, $\beta$ is an even stronger function of which transistor you pull out of the box ("specimen") than of temperature for a given specimen. Furthermore, do not forget the stabilization of dc behavior achieved by $R_E$. Thus a definite improvement has been made.

**Example 7.5**

To get a feel for what sort of voltage gain is typical for the circuit of Fig. 7.35, let us take some typical parameter values: $V_{CC} = 10$ V, $R_s = 100\ \Omega$, $R_1 = 33$ k$\Omega$, $R_2 = 10$ k$\Omega$, $R_C = 5.6$ k$\Omega$, $R_E = 4.7$ k$\Omega$, $R_L = 2$ k$\Omega$, $\beta = 60$, $C_1 = C_3 = 20\ \mu$F, and $C_2 = 50\ \mu$F. Find the voltage gain.

**Solution** At only 100 Hz, the magnitude of the impedance of $C_1$ and $C_3$ is 80 $\Omega$, and that of $C_2$ only 32 $\Omega$, so at higher audio frequencies they are all indeed effectively short circuits. The base-bias resistor $R_B = R_1 || R_2 = 7.7$ k$\Omega$, so indeed, $R_s \ll R_B$. The effective load resistance $R_L' = R_L || R_C \approx 1.5$ k$\Omega$. From Eq. (7.28b), at 300 K, $r_\pi = 60(0.026)(4700)/(10 - 0.7) \approx 790\ \Omega$, so indeed, $r_\pi \gg R_s$. Therefore, by Eq. (7.29), $A_v \approx -114$. Compare this with the gain without $C_2$, as given in Eq. (7.25): $A_v = -1500/4700 \approx 0.3$—actually, attenuation! This, however, is not to suggest that without $C_2$ the gain will always be less than 1; there is an infinite variety of workable parameter values. The important point is that we have proposed a transistor amplifier having $|A_v| > 1$.

It was promised in Sec. 7.5.1 that the signal power gain $A_p$ of the transistor amplifier could be greater than 1. Let us demonstrate that for the circuit of Fig. 7.35a for $C_2 \neq 0$, by examining Fig. 7.35d. In accordance with Fig. 7.21, the input power is $p_i = v_i i_s$. In this case,

$$v_i = \frac{R_B||r_\pi}{R_B||r_\pi + R_s} v_s \tag{7.30}$$

while

$$i_s = \frac{v_s}{R_B||r_\pi + R_s}, \tag{7.31}$$

so that

$$p_i = v_i i_s \tag{7.32a}$$

$$= \frac{(R_B||r_\pi) v_s^2}{(R_B||r_\pi + R_s)^2}, \tag{7.32b}$$

which for $R_B||r_\pi \gg R_s$ (often true in practice) reduces to

$$p_i \approx \frac{v_s^2}{R_B||r_\pi}. \tag{7.32c}$$

The signal output power is, again in accordance with Fig. 7.21, $p_o = i_L v_L = v_L^2/R_L$. In this case $v_L = v_o$, and

$$p_o = \frac{v_o^2}{R_L} \tag{7.33a}$$

$$= \left(-\frac{R_L'}{R_E}\frac{V_{BB} - 0.7\text{ V}}{kT/q}\right)^2 \frac{v_s^2}{R_L}, \tag{7.33b}$$

so that

$$A_p = \frac{p_o}{p_i} \tag{7.34a}$$

$$= \left(\frac{R_L'}{R_E}\frac{V_{BB} - 0.7\text{ V}}{kT/q}\right)^2 \frac{R_B||r_\pi}{R_L}, \tag{7.34b}$$

For the parameter values in Example 7.5 and with for $C_2 \neq 0$,

$$A_p = \left(\frac{1500}{4700}\frac{10 - 0.7}{0.026}\right)^2 \frac{7700||790}{2000} \approx 4700,$$

a tremendous gain in signal power.

To close this subsection and round out our discussion of the bipolar transistor amplifier, we consider the simpler small-signal model in Fig. 7.36. This is a very useful model for multistage amplifier analysis because it reduces the transistor amplifier to "bare bones." We shall use it again for operational amplifiers in Chapter 8 as a simple model of an enormously complex circuit. All we need to complete it are values for the input and output resistance of our amplifier. For simplicity, we consider here the case $C_2 \neq 0$ (Fig. 7.35d).

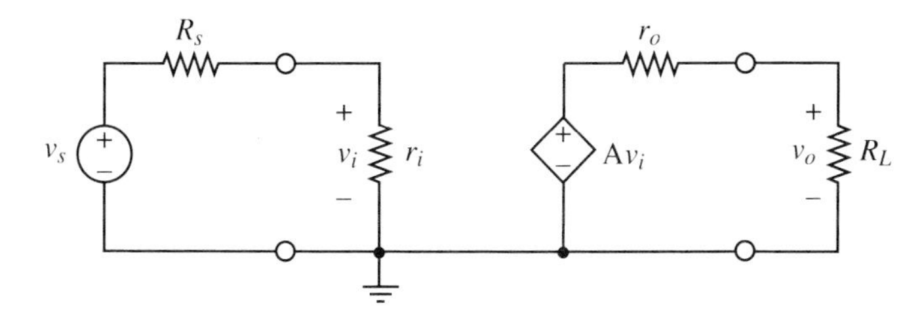


**Figure 7.36** Simple Thévenin-type model for a general transistor amplifier.

Examination of Fig. 7.35d shows, by inspection, that

$$r_i = R_B || r_\pi, \tag{7.35}$$

which for the parameter values chosen in Example 7.5 is $r_i = 7700||790 = 716\ \Omega$. This is a rather low value of input impedance. The FET will give an essentially infinite input impedance and can thus be used as a first stage preceding the amplifier we have just analyzed.

To the load, the entire source and amplifier appear to be a Thévenin equivalent. To find the output resistance $r_o$, which is the Thévenin equivalent resistance seen by $R_L$ looking back into the amplifier, we need $v_{oc}$ and $i_{sc}$. By Eq. (7.27a), $v_{oc} = -\beta i_b R_C$ because on open-circuit conditions $R_L$ is absent, so that $R_L' = R_C$. The short-circuit current $i_{sc}$ is clearly $-\beta i_b$ because zero current flows in $R_C$ when there is a short circuit around $R_C$. Therefore,

$$r_o = \frac{v_{oc}}{i_{sc}} = R_C, \tag{7.36}$$

which for the parameter values chosen above is $r_o = 5.6\ \text{k}\Omega$. This is a rather high value of output resistance. However, there is a bipolar transistor amplifier called an *emitter follower* (see the problems) that has a gain of 1 but transforms the output impedance to a very much lower value. Thus a good amplifier usually has more than one stage, with each stage performing different functions.

### 7.6.3 Common-Source Amplifier (JFET; CS)

We have just studied one practical transistor amplifier, based on the bipolar junction transistor (BJT). Now we similarly examine a practical transistor amplifier based on the junction field-effect transistor (JFET). This amplifier, called the *common-source amplifier* with feedback resistor (see Fig. 7.37a), is quite analogous in structure to the common-emitter amplifier with feedback (compare with Fig. 7.35a). The only difference is that the BJT has been replaced by a JFET.

However, the analysis is somewhat different because of the differences between the JFET and the BJT. For example, the large-signal model for the JFET is not that shown in Fig. 7.27a. Instead, there is no simple circuit diagram for large signals/operating points; Eq. (7.9) with $i_D = I_D$ and $v_{GS} = V_{GS}$ must be used. In fact, determination of the operating point is much more complicated for the JFET than for the BJT, as we shall see below. Also, the small-signal model is not given by Fig. 7.27b, but rather, by Fig. 7.33. In this case the JFET model is slightly simpler than the BJT model of Fig. 7.27b.

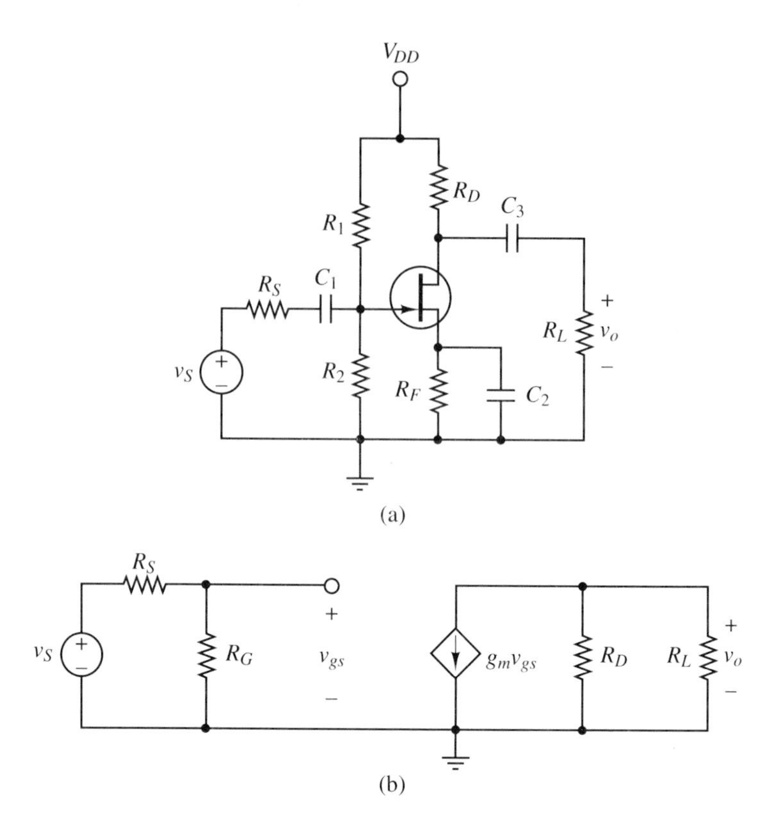


**Figure 7.37** Common-source amplifier with source resistance $R_F$: (a) component schematic diagram; (b) schematic for small-signal calculations with $C_2$.

As before, the purpose of the feedback resistor (here, $R_F$) is to reduce the dependence of the operating point and signal gain on the transistor parameters. In this case the highly variable parameters are $I_{DSS}$, and to a lesser extent, $V_P$.

Also as before, our immediate interest is in computing small-signal gain. This requires use of the small-signal model, which involves $g_m$ as given in Eq. (7.13). Clearly, the operating point gate–source voltage $V_{GS}$ must be determined in order to proceed. To determine $V_{GS}$, we must know $I_D$. This forces us to deal with Eq. (7.9), which will result in a quadratic equation for $I_D$. Let us now determine the operating point $V_{GS}$.

Recall that for determination of the operating (bias) point, the capacitors are all open circuits. Because essentially zero current flows into the gate and zero dc current is in $C_1$, we have a voltage divider of $V_{DD}$ between $R_1$ and $R_2$:

$$V_G = \frac{R_2}{R_1 + R_2} V_{DD}. \tag{7.37}$$

The desired $V_{GS}$ is, by KVL,

$$V_{GS} = V_G - V_F \tag{7.38a}$$

$$= V_G - I_D R_F. \tag{7.38b}$$

(We use a subscript "$F$" for feedback resistor $R_F$ and for the source node voltage $V_F$ instead of "$S$" to avoid confusion with the small-signal source $v_s$.) By Eq. (7.9), we also have for the operating point drain current $I_D$,

$$I_D = I_{DSS}\left(1 + \frac{V_{GS}}{V_P}\right)^2, \tag{7.39a}$$

which by Eq. (7.38b) becomes

$$I_D = I_{DSS}\left(1 + \frac{V_G}{V_P} - \frac{I_D R_F}{V_P}\right)^2 \tag{7.39b}$$

$$= I_{DSS}\left[\left(1 + \frac{V_G}{V_P}\right)^2 - 2\left(1 + \frac{V_G}{V_P}\right)\frac{I_D R_F}{V_P} + \left(\frac{I_D R_F}{V_P}\right)^2\right], \tag{7.39c}$$

which, when terms are collected, forms the following quadratic equation for $I_D$:

$$I_D^2 - 2\frac{V_P}{R_F}\left(1 + \frac{V_G}{V_P} + \frac{V_P}{2 I_{DSS} R_F}\right) I_D + \left[\frac{(1 + V_G/V_P)\, V_P}{R_F}\right]^2 = 0. \tag{7.40}$$

To simplify the results, define $I_P = V_P/R_F$, $\alpha = 1 + V_G/V_P$, and $\beta = V_P/(2I_{DSS} R_F)$ (do not confuse with the BJT $\beta$). Then Eq. (7.40) is simply the quadratic equation

$$I_D^2 - 2 I_P(\alpha + \beta) I_D + (\alpha I_P)^2 = 0, \tag{7.41}$$

the solutions of which are

$$I_D = (\alpha + \beta) I_P \pm \{I_P^2 [(\alpha + \beta)^2 - \alpha^2]\}^{1/2} \tag{7.42a}$$

$$= I_P\{\alpha + \beta \pm [\beta(\beta + 2\alpha)]^{1/2}\} \tag{7.42b}$$

$$= I_P\left\{\alpha + \beta\left[1 \pm \left(1 + \frac{2\alpha}{\beta}\right)^{1/2}\right]\right\}. \tag{7.42c}$$

Finally, we substitute $I_D$ in Eq. (7.42c) and $V_G$ in Eq. (7.37) into Eq. (7.38b) and use the fact that $I_P R_F = V_P$ so that the first term of $I_D R_F$ will be $I_P \alpha R_F = V_P + V_G$, to obtain

$$V_{GS} = V_G - I_D R_F \tag{7.43a}$$

$$= -V_P\left\{1 + \beta\left[1 \pm \left(1 + \frac{2\alpha}{\beta}\right)^{1/2}\right]\right\}. \tag{7.43b}$$

Our mathematical model has produced two solutions, only one of which is physically relevant. We know that for amplification behavior, we must not have gate pinch-off: $V_{GS} > -V_P$. Thus the physically meaningful solution is the one such that the term proportional to $\beta$ in Eq. (7.43b) is negative. Because $2\alpha/\beta > 0$, so that $1 + 2\alpha/\beta > 1$, we must select "$-$" rather than "$+$" in Eq. (7.43b). Therefore, with all device parameters resubstituted [including use of Eq. (7.37)], the operating point gate–source voltage is, finally,

$$V_{GS} = -V_P\left[1 + \frac{V_P}{2 I_{DSS} R_F}\left(1 - \left\{1 + 4\left[1 + \frac{R_2 V_{DD}}{(R_1 + R_2) V_P}\right]\frac{I_{DSS} R_F}{V_P}\right\}^{1/2}\right)\right]. \tag{7.44}$$

A similar expression can be written for the drain current operating point, but it is easier to instead use the value just obtained for $V_{GS}$ in the expression for $I_D$ in Eq. (7.39a). Unfortunately, most of the additive terms in Eq. (7.44) are of nearly the same order of magnitude, so that Eq. (7.44) cannot be greatly simplified. Sometimes people resort to graphical analysis techniques, to avoid working with these quadratic equations.

**Example 7.6**

Let us try out these equations on some typical device parameters for a JFET common source amplifier as shown in Fig. 7.37: $R_1 = 2$ MΩ, $R_2 = 0.5$ MΩ, $R_F = 4.7$ kΩ, $R_D = 4.7$ kΩ, $R_L = 2$ kΩ, $R_s = 100$ Ω, $I_{DSS} = 16$ mA, $V_P = 6$ V, $V_{DD} = 20$ V, $C_1 = C_3 = 20$ μF, and $C_2 = 50$ μF. Find the operating point.

**Solution** Substituting these values into Eq. (7.44) gives $V_{GS} = -4.04$ V and then into Eq. (7.39a) gives $I_D = 1.71$ mA. These are both reasonable values for normal JFET amplification behavior: $V_{GS}$ is negative but greater than $-V_P$, and $I_D$ is positive but less than $I_{DSS}$.

Now we come to the easy part of analysis: substitution of the small-signal model of Fig. 7.33 into the original circuit of Fig. 7.37a. We can obtain the small-signal circuit shown in Fig. 7.37b merely by replacing the hybrid $\pi$ model of the BJT in Fig. 7.35d by the FET small-signal model of Fig. 7.33. The only parameter yet to be obtained is the transconductance parameter $g_m$, which can be calculated from the operating point $V_{GS}$ via Eq. (7.13). For the numerical values of the parameters given in Example 7.6, we thus obtain $g_m = 2 \cdot 16 \cdot 10^{-3}(1 - 4.04/6)/6 = 1.74 \cdot 10^{-3}\ \Omega^{-1}$.

In complete analogy with Eq. (7.27a), we define $R_L' = R_L || R_D$ and $R_G = R_1 || R_2$ and obtain from Fig. 7.37b the output voltage

$$v_o = -g_m v_{gs} R_L' \tag{7.45a}$$

$$= -g_m \left(\frac{R_G}{R_G + R_s}\right) R_L' v_s, \tag{7.45b}$$

so that the voltage gain $A_v$ is

$$A_v = \frac{v_o}{v_s} = \frac{-g_m R_L' R_G}{R_G + R_s} \tag{7.46a}$$

$$\approx -g_m R_L' \tag{7.46b}$$

because typically both $R_1$ and $R_2$ and therefore $R_G$ are many orders of magnitude larger than $R_s$. For our numerical values specified in Example 7.6, $R_G = 2\ \text{M}\Omega || 0.5\ \text{M}\Omega = 0.4\ \text{M}\Omega \gg R_s = 100\ \Omega$, so that Eq. (7.46b) holds and $A_v \approx -1.74 \cdot 10^{-3}\ (4.7\ \text{k}\Omega || 2\ \text{k}\Omega) = 2.4$. Thus this set of parameter values results in a low-voltage-gain amplifier. Actually, the circuit of Fig. 7.37 typically does have substantially lower gain than the BJT equivalent (common emitter).

However, the FET amplifier draws essentially zero current from the signal source; it does not at all load it down. Therefore, a circuit such as that in Fig. 7.37 could be used as an input stage, capable of driving the next bipolar stage without being loaded down. The input impedance of the common-source amplifier is $R_G$; this is why megohm values are chosen for $R_1$ and $R_2$: $r_{\text{in}} = R_G = 0.4$ MΩ; even higher values can be used for $R_1$ and $R_2$. Using the same reasoning as that which led to

Eq. (7.36), the output resistance of the common-source amplifier is $R_D$, in this case 4.7 kΩ. Thus the output impedances of the BJT common-emitter amplifier and the JFET common-source amplifier are comparable.

The power gain of the FET is extremely high because $r_{in}$, being on the order of megohms or more, causes the input power to be extremely small. The output power is comparable to that of the BJT. Because the power gain is so extremely large, FET (and generally input stage) amplifiers are normally characterized by their voltage gain rather than their power gain. Power gain calculations are most relevant at the output stage, where sufficient power must be generated to drive a heavy load such as an audio speaker.

### 7.6.4 The Transistor as a Switch

Previously, we have considered circuits for which the output was linearly proportional to the input voltage. Often, however, one is interested only in on–off behavior. This is the case for alarm circuits or motor controls, for example. Most common of all is the application of switches in computer circuits, which we study in detail in Chapters 10 through 12. Typically in these circuits, one has a low-power signal that is expected to turn on or off the power to a high-power load. The signal source is too weak to drive the load itself; it must control delivery of high power from a source capable of delivering such levels of power.

Both BJTs and FETs are used for switching purposes, and particularly, in computer circuits. For brevity, we consider here only the BJT switch shown in Fig. 7.38a. Now the load is the collector resistance, in this case a mechanical buzzer.

For this application of transistors, we do not want coupling capacitors, because the important content of the input voltage is its steady-state dc value. Typically, the detector will be designed to operate primarily around an "on" voltage or an "off" voltage. The "on" level of input will drive the transistor into saturation, providing large output voltage and current. The "off" level will drive the transistor into cutoff, producing zero output current with which to drive the load.

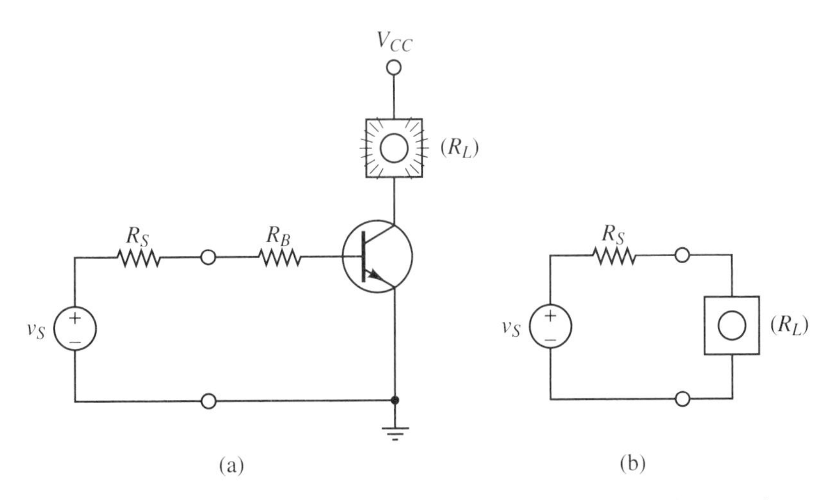


**Figure 7.38** (a) Simple BJT switch applied to the task of signaling a buzzer to be on or off; (b) attempt to run the buzzer directly off the on–off signal source $v_s$, $R_s$.

Neither do we want a feedback resistor, because the dependence of operation on β is no longer a detrimental factor. Indeed, it is the crude but large current gain, β, that makes the circuit useful. Finally, we do not need biasing resistors because there is no reason to establish an operating point in the linear region of Fig. 7.26. The transistor is merely to be in saturation or cutoff, depending on the value of input voltage $v_s$. Only a current-limiting resistor $R_B$ comes between the on–off input signal $v_s$ and the transistor. Finally, there will be no need for small-signal analysis because we have only large deviations in the output, not small perturbations about an operating point.

From these considerations, we will find that analysis of the transistor switch is particularly easy. Suppose that the buzzer is primarily resistive (an efficient buzzer would be) with resistance $R_L = 100\ \Omega$, and it requires at least 30 mA to buzz loudly (and thus have $100 \cdot 0.03 = 3$ V across it). Suppose further that the on–off signal voltage $v_s$ represents "on" and "off" by taking on one of two values: 0 V or 3 V. The internal resistance of the source is $R_s = 1\ \text{k}\Omega$, $R_B = 1\ \text{k}\Omega$, $\beta = 60$, $V_{CE,\text{SAT}} = 1$ V, and $V_{CC} = 5$ V.

Recall from Sec. 4.11.1 that the maximum power extractable from the source if the load were placed directly across it (see Fig. 7.38b) would be $P_{\max} = v_{\text{eq}}^2/(4R_{\text{eq}})$, where here $v_{\text{eq}} = v_{s,\max} = 3$ V and $R_{\text{eq}} = R_s = 1\ \text{k}\Omega$. Thus $P_{\max} = 3^2/(4 \cdot 1000) = 2.25$ mW, yet the buzzer requires that $P_L = I_L^2 R_L = (0.03)^2 100 = 90$ mW, which is much larger than $v_s$ can support. This is true even though the nominal voltage (3 V) is ostensibly sufficient, as the buzzer was shown to have 3 V across it when operating. Because of the severe loading down, the actual voltage across the buzzer would be only $3\text{ V} \cdot R_L/(R_s + R_L) = 0.27$ V. The "3 V" would occur only across an infinite-resistance load (open circuit).

Furthermore, the 2.25 mW calculated above would be true only if the load were matched: $R_L = R_s$, which is not the case here. In reality, the power delivered to the buzzer would be only $v_s^2 R_L/(R_L + R_s)^2 = 0.7\text{ mW} \ll P_L$. Clearly, we need a heftier power source to drive the buzzer—namely, the strong power supply $V_{CC}$, which is capable of delivering many watts of power.

Let us now consider the operation of the circuit of Fig. 7.38a. We need consider only the two values of $v_s$: 0 V and 3 V. Suppose that $v_s = 0$ V. In this case the base–emitter junction of the transistor is reverse biased, so that the transistor is in the cutoff state. This means that $i_C = 0$, so the buzzer does not sound.

If $v_s$ changes to 3 V, the base–emitter junction is strongly forward biased. First we use the large-signal model to analyze the behavior for $v_s > 0.7$ V—in particular, $v_s = 3$ V. Substituting the large-signal model for the BJT from Fig. 7.27a into Fig. 7.38a, we obtain the circuit shown in Fig. 7.39a. Here $R_s + R_B = 1\ \text{k}\Omega + 1\ \text{k}\Omega = 2\ \text{k}\Omega$. The base current is $I_B = (v_s - 0.7\text{ V})/(R_s + R_B) = 1.15$ mA for $v_s = 3$ V. Assuming linear behavior, the collector current would be $I_C = \beta I_B = 60 \cdot 1.15\text{ mA} = 69$ mA.

However, this would give $V_{CE} = V_{CC} - I_C R_L = 5 - 0.069 \cdot 100 = -1.9$ V. A negative collector–emitter voltage is impossible for linear transistor behavior; it implies that the base–collector junction is forward biased. In fact, $I_C \neq \beta I_B$ because the transistor is in saturation. Instead of $I_C = \beta I_B$ as in Fig. 7.39a, we have $V_{CE} = V_{CE,\text{SAT}} \approx 1$ V as in Fig. 7.39b. Referring to the BJT *i-v* characteristics in Fig. 7.26b, we recall that saturation is the situation where further increases in $I_B$ will not yield proportionately larger values of $I_C$. Rather, $V_{CE}$ is fixed at $V_{CE,\text{SAT}}$ and $I_C$ becomes the required value to maintain this, given the values of $V_{CC}$ and $R_L$. In our case, $I_C = (V_{CC} - V_{CE,\text{SAT}})/R_L = (5 - 1)/100 = 40$ mA.

This value of $I_C$, 40 mA, is plenty enough to operate the buzzer. We have achieved our goal of using $v_s$ to control whether or not the buzzer buzzes, even though $v_s$ cannot drive the buzzer itself. If the 40-mA value is considered too high, a small current-limiting resistor can be placed in series with $R_L$ to reduce $I_C$.

### 7.6.5 Silicon-Controlled Rectifier

For the automated control of heavy electric equipment, it can be very convenient to use semiconductor rather than mechanical switches. Furthermore, as we shall see in Sec. 14.4.7, how long a switch is left on can actually end up controlling the speed of a motor. To this end, we consider the very popular electronic switch called the silicon-controlled rectifier (SCR). These SCR switches can be manufactured to handle very large voltages and currents, sufficient to drive heavy electrical machinery.

The SCR is just a *pnpn* structure that behaves like two interconnected transistors. One transistor of the *pnpn* structure is *pnp* and the other transistor of the *pnpn* structure is *npn*. The structure is shown in Fig. 7.40a, the equivalent two-transistor structure is shown in Fig. 7.40b, and its circuit symbol is in Fig. 7.40c. A detailed physical analysis of this device is beyond the scope of this book, but a simplified analysis of Fig. 7.40b should suffice.

Assume that the anode is held at a higher potential than the cathode by an applied voltage. If no voltage is applied to the gate terminal, the base–emitter junction of the *npn* transistor is reverse biased, so that transistor is cut off. Therefore, there is no conduction from anode to cathode. If a positive voltage is now applied to the gate, the *npn* transistor conducts. Also, the *pnp* transistor conducts because, as careful comparison of all voltage levels reveals, its base–emitter junction is forward biased and its collector–base junction is reverse biased (be wary of *pnp* polarities!).

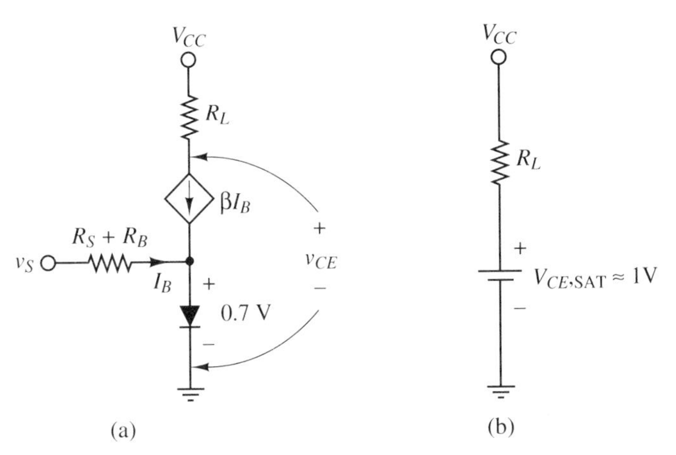


**Figure 7.39** Large-signal analysis of the transistor switch: (a) attempt to use large-signal model of Fig. 7.27a; (b) recognition that for $v_s$ "on" at 3 V, the transistor saturates.

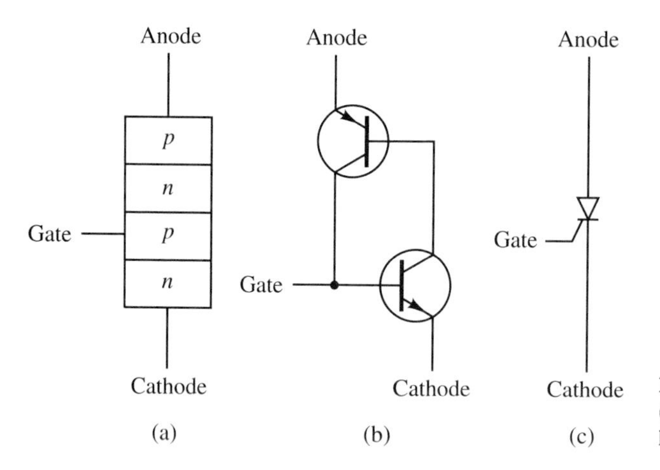


**Figure 7.40** Silicon-controlled rectifier: (a) semiconductor structure; (b) equivalent circuit; (c) SCR symbol.

Now even if, subsequently, the gate voltage is removed, anode–cathode conduction will continue by virtue of the *pnp* collector current supplying the *npn* base. It is not obvious that this would occur, but it does. The only way to stop the anode–cathode conduction is to reverse the polarity of the anode–cathode voltage while the gate voltage is off. Once stopped, the positive gate voltage will again be required to restart conduction.

Because sinusoidal voltages switch polarity every half-cycle, use of the SCR as an on–off switch is easy for sinusoidally excited circuits, which are by far the most common heavy industrial electrical equipment. The "option" of turning off is available every $\frac{1}{60}$ s. A very small, cheap mechanical switch is used to turn off the gate voltage, which is held high as long as the equipment is desired to be on. To turn off, just open the gate switch and as soon as the main ac voltage goes negative, the power will go off and stay off. In Sec. 14.4.7 we shall see an application of the SCR in controlling the speed of a dc motor; SCRs can be used to control either ac or dc equipment.

## 7.7 MORE ADVANCED APPLICATIONS

### 7.7.1 Introduction

The list of applications of transistors and diodes is endless. But to whet your appetite to learn more about electronics, a few additional applications of a more advanced character are described briefly in this section. Complete, quantitative analysis of these circuits is impossible here, as is defining the operation and purpose of each component. However, a partial and qualitative operational description is well within your grasp at this stage and should be gratifying and of interest to all engineers, as well as an indicator of the transition from the "textbook example" to real-world research and development.

### 7.7.2 "Linear" Regulated Power Supply

In Sec. 7.4 we studied a variety of diode-based power supplies. The Zener-regulated power supply offered the best regulation in the face of variations in ac input and load.

In fact, both can be expected to vary considerably in practice. It is common knowledge that ac voltage can be anywhere in the range 110 to 125 V and that a variety of loads should be able to be driven, yet the desired line voltage regulation is often fractions of 1%.

The main limitation of Zener regulators such as that shown in Fig. 7.17 is evident in Fig. 7.11a. For large currents, the Zener conduction region begins to look like a battery in series with a resistor. This Zener-mode resistance causes the voltage across the diode to deviate from $V_Z$ under line or load variations. The variation can be eliminated if the current in the Zener reference diode is kept very small. This can be accomplished by adding a transistor amplifier to the power supply—called the "linear" power supply to distinguish it from "switching" power supplies, which are discussed in Sec. 10.7.5.

Traditional "linear" power supplies such as that shown in Fig. 7.41 work in the following manner. The 120 VAC is stepped down to lower levels by transformer $T_1$. Silicon rectifiers $D_1$ through $D_4$ and a capacitive filter $C_1$ convert the sinusoidal voltage to a roughly constant (ripply dc) voltage $v_o$ somewhat higher than the dc level required at the load ($v_L$). The voltage $v_o$ is fed into the collector of series transistor $Q_1$, which operates as a variable resistor, by its feedback-controlled base current. The power supply output voltage $v_L$ appears at the emitter of $Q_1$. It is sampled via voltage divider ($R_1$, $R_2$) for feedback regulation that ultimately controls the base current into $Q_1$.

A Zener diode $D_Z$ typically is involved in forming a reference voltage with which to compare the output. When the sampled output $v_2$ [roughly $v_L R_2/(R_1 + R_2)$] exceeds the reference voltage $V_Z$ (by the base–emitter contact potential, 0.7 V, of $Q_2$), $Q_2$ is turned on, thereby bypassing some of the base current normally presented to the series transistor $Q_1$. The reduced base current decreases the collector–emitter current of $Q_1$ and therefore reduces the output voltage $v_L$. The bypass current is essentially equal to the Zener diode current, and is in practice quite small, so that $V_Z$ remains very constant.

One major problem with this type of power supply is that the series transistor (serving as a voltage regulator) $Q_1$ eats up a lot of power. This is due to its operation in the linear region. Consider the transistor $i_C$–$v_{CE}$ characteristic in Fig. 7.26b.

For this discussion it may be assumed that the load voltage is constant; that is the purpose of the regulation transistors. Various loads will therefore draw various

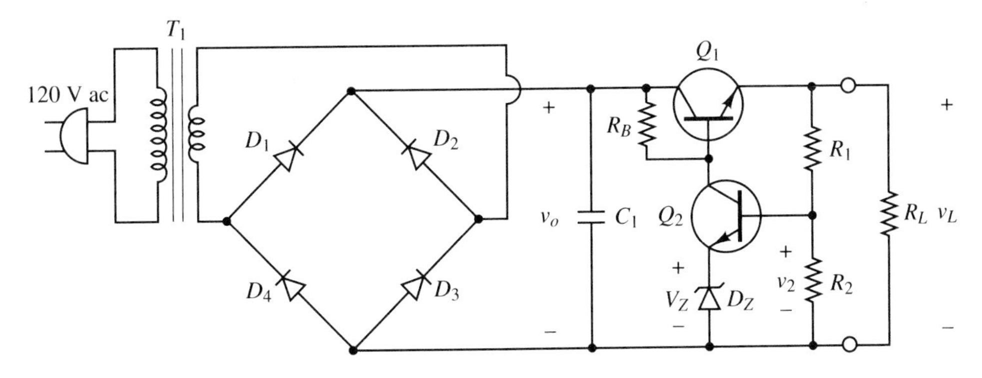


**Figure 7.41** Linear transistorized regulated power supply.

currents, all of which pass through the series transistor. There is nothing we can do about this. But due to the fact that the transistor is operating in the linear region (in the central region of Fig. 7.26b), $v_{CE}$ is fairly large, causing the power dissipated in the series transistor $p_{\text{diss}} = i_C v_{CE}$ to be very large. The fact that both transistors $Q_1$ and $Q_2$ operate in the linear region is why this power supply is called "linear."

If somehow it were permissible to operate the transistor in the saturation region (near the $i_C$ axis in Fig. 7.26b), yet still achieve voltage regulation, $v_{CE}$ would be very small and therefore the dissipated power would be relatively small. As a consequence, the efficiency would be significantly improved. This technique is used in switching power supplies, described in Sec. 10.7.5.

Advantages of the linear power supply are its simple design, good regulation, low cost, low electromagnetic interference, and low output ripple. Disadvantages are its low efficiency and large size and weight/power ratios due to the large transformers, transistor heat sinks, cooling fans, and so on; that is, they are big and heavy. They are used in circuits such as televisions, audio systems, and other systems requiring good regulation and low noise.

### 7.7.3 Transistor AM Radio

The AM radio has long been the most economical form of mass communication; most people in this country can afford one. Amplitude modulation (AM) is also the oldest form of radio audio communication. Today's radio almost always comes with FM also. One reason for the perpetuation of AM, despite its low fidelity compared with FM, is simple: low-cost available space for a would-be broadcasting station. For this reason we should expect AM to be around for quite awhile longer.

Circuits for AM reception have evolved over the years, but the same types of basic stages are used for all radio circuits. For high-quality reception, digital electronics is beginning to take over the jobs of some of these stages. However, linear amplifier circuits still predominate, and it is these that we examine briefly in this subsection.

An overview of the transmission–reception process is shown in Fig. 7.42. Consider first the broadcasting station system, shown in the upper part of Fig. 7.42. For each step, the basic spectrum is shown as well as the time-domain waveform. We begin at the upper left corner of Fig. 7.42. The crooner sings into the microphone to produce the audio signal $a(t)$, which is then amplified by a transistor audio amplifier. To prevent a particular type of distortion that would otherwise occur in the radio receiver (see the problems), the amplifier also adds a dc bias to the signal $a(t)$. The frequencies making up the audio signal $a(t)$ may exist all the way to 10 or 15 kHz. However, due to FCC regulations on AM broadcasting to allow the existence of more radio stations, the amplifier includes a filter that removes all frequencies above 5 kHz. The result is $b(t) = V \cdot \text{low-pass filter}\,[1 + a(t)/a_0]$, where $V$ and $a_0$ are constants. The descriptions above are reflected in the spectra $A(f)$ and $B(f)$.

It should be mentioned that this narrow 5-kHz bandwidth is the reason AM has low fidelity even if received with an expensive stereo system; the high-frequency information just is not there to pick up. The low frequencies are fine, but the high frequencies are missing because they had to be filtered out at the radio station. Contrarily, FM stations are allowed the full audible 15 kHz.

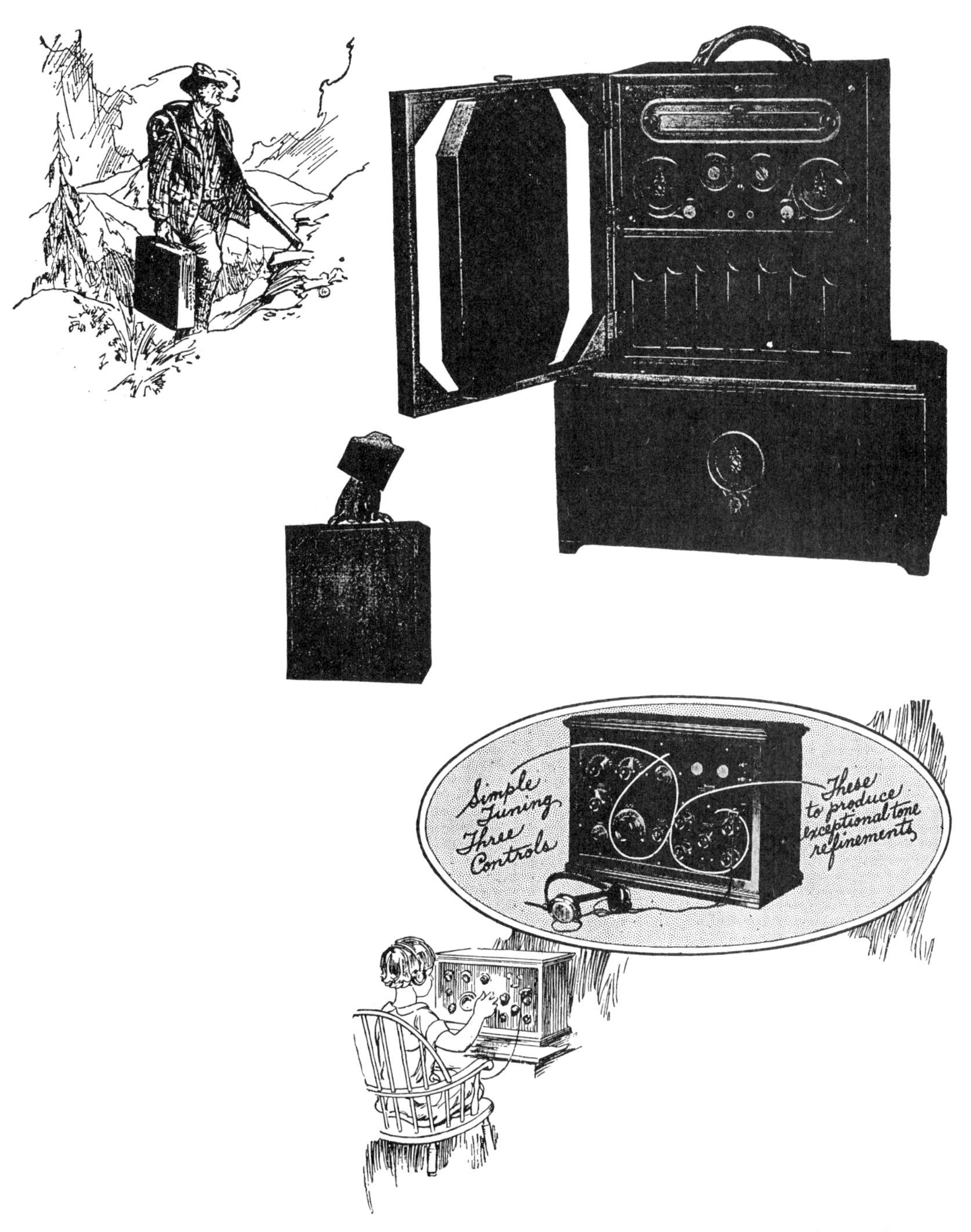


Two major selling points of old AM radios: Above, a "portable" radio, 38 pounds, complete with (marginally) convincing sketch of "walkman" in the wilderness with radio. Below, "only" eight controls are needed to tune in the station and get an acceptable tone—surely easy enough for the child to "master" (rather, "miss")! No explanation in the advertisement on the purposes of the other additional four controls (making a total of twelve). *Source: Scientific American* (June 1925, September 1924).

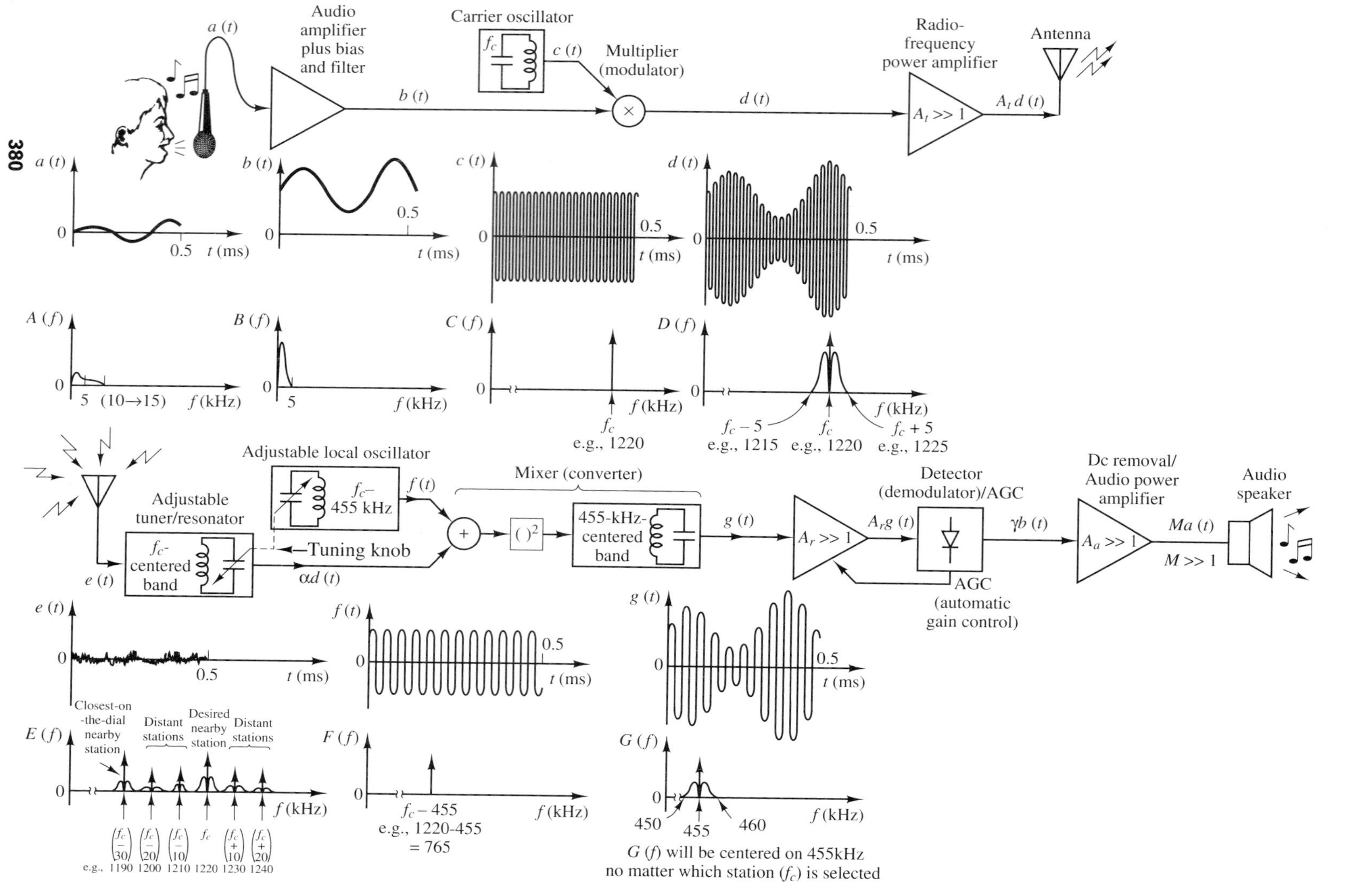


**Figure 7.42** Basic AM broadcast transmission and reception systems.

If all stations tried to broadcast at audio frequencies, they would all mix together in that low-frequency band. Also, the size of the antenna would be impractical. These are two reasons why each station has its own assigned frequency band. The idea of giving each station its own frequency band (10 kHz for each station) over which to broadcast is called *frequency-division multiplexing* of the radio spectrum. Broadcasting circuits are thus designed to multiply the audio signal by a high-frequency *carrier* $c(t)$, at the broadcast center frequency $f_c$ (e.g., 1220 kHz, as found on an AM tuning dial).

Why multiply? Recall that $e^{jf_1t}e^{jf_ct} = e^{j(f_1+f_c)t}$. Therefore, a frequency component $f_1$ in the audio signal, when multiplied by a sinusoid having the carrier frequency $f_c$, now "rides on the carrier"; it now appears as a sinusoid of frequency $f_1 + f_c$. This process, called *modulation,* is at this higher frequency range at which the radio signal is transmitted: the band centered on $f_c$ and containing the audio spectrum. In the receiver, we can demodulate by effectively multiplying $e^{j(f_1+f_c)t}$ by $e^{-jf_ct}$ and thereby retrieve $e^{jf_1t}$ down at the audio-frequency (AF) range.

Because all frequencies in the audio signal are multiplied by the same carrier sinusoid $f_c$, the entire audio spectrum is shifted up to become centered on the carrier frequency. Naturally, our discussion here is simplified to put forth the essential ideas. In practice only real-valued carriers $\cos(f_ct)$ and signals $b(t)$ are used; however, the ideas are the same.

Specifically, recall that by Eq. (6.9), $\cos(f_1t)\cos(f_ct) = \frac{1}{2}\{\cos[(f_1 + f_c)t] + \cos[(f_1 - f_c)t]\}$; thus we can see that the signal component $f_1$ now appears both above and below $f_c$. Consequently, the entire audio spectrum is actually reproduced both above and below $f_c$. This is illustrated in the resulting spectrum, $D(f)$.

In the time domain we see merely that the carrier sinusoid is multiplied by the audio signal, causing the carrier sinusoid amplitude to vary along with the audio-frequency signal $b(t)$. The result is $d(t)$. From here, the radio-frequency (RF) signal $d(t)$ is amplified and applied to the transmitting antenna. This completes the discussion of the radio transmission system.

At the antenna of the receiver, the oscillating electromagnetic field exerts oscillatory forces on free electrons in the antenna wire, causing currents. This current signal results from *all* received electromagnetic waves at *all* frequencies. The time-domain signal $e(t)$ is this jumble of all these signals added together. A part of the AM frequency band (six stations' worth) is shown in the spectrum $E(f)$. Among the stations is our desired station $f_c$. As noted above, each station is given its own 10 kHz band; the stations immediately next to $f_c$ are very unlikely to be local, due to consequent reception selectivity problems, so they are weak. As an example, the next strong station over might be 30 kHz lower than $f_c$ (in practice, hopefully more than this).

The purpose of the tuning knob is to select the station $f_c$ out of the jumble of all stations in $e(t)$. The signal must be filtered and amplified. First the tuner selects the band centered on $f_c$, producing a signal proportional to $d(t)$, denoted $\alpha d(t)$.

It simplifies circuitry to do all the radio-frequency amplifying and other signal improvement operations on a signal whose energy is in only one frequency band—independent of the particular station desired. This is achieved by having a local oscillator in the radio receiver producing a sinusoid whose frequency varies in tandem with the adjustment of the tuning knob: its frequency is $f_{osc} = f_c - 455$ kHz;

this sinusoid is called $f(t)$. That way when the mixer combines $\alpha d(t)$ with $f(t)$, the difference frequency band—the one the filter in the mixer retains—is (the audio spectrum centered on $f_c$) − ($f_c$ − 455 kHz) = (the audio spectrum centered on 455 kHz), no matter what the value of $f_c$ (i.e., no matter what station we select).

This partially demultiplexed signal, $g(t)$, is called an *intermediate-frequency* (IF) *signal,* where $f_{\text{IF}}$ = 455 kHz is the intermediate frequency. Shortly, we shall address the reason for choosing 455 kHz as the IF. The process of normalizing to an intermediate frequency is commonly known as *heterodyning,* and because the frequencies are above the audible range, the term *superheterodyning* is often used.

Now amplification is performed, usually in more than one stage of transistor amplifiers having a combined gain of $A_r$, to produce $A_r g(t)$, which is just an amplified version of $g(t)$. Next, the detector or *demodulator* removes the carrier by another nonlinear process, typically involving a diode and an *RC* low-pass filter. The result of this is a signal proportional to $b(t)$, shown as $\gamma b(t)$. Also, a voltage proportional to the time-average value of $b(t)$, which indicates the signal strength, is fed back to the IF amplifier. If the signal is too strong, $A_r$ is reduced; if the signal is too weak, $A_r$ is increased. This feedback technique, known as automatic gain control (AGC), causes both near and distant stations to have roughly equal loudness.

The last stage of amplification removes the dc bias from $\gamma b(t)$ and amplifies the result, producing a high-power signal proportional to the original audio signal $a(t)$, shown as $Ma(t)$. This signal is now sufficiently strong to drive an audio speaker.

We now consider electronic circuits that implement these functions, from an actual six-transistor AM radio. Each fragment will in turn be considered separately, and finally, the entire circuit will be shown.[1]

First consider the tuning circuit shown in Fig. 7.43. Although no resistance is shown, in practice there is, so we have a parallel *RLC* circuit. The dual of this circuit is the series *RLC* circuit, which we studied in Sec. 5.5.3. Its frequency response is similar to that in Fig. 5.35: It has a peak at the resonant frequency $f_0 = \omega_0/(2\pi)$, which we can vary by turning the tuning knob, which in turn varies the capacitance $C_2$. In particular, we set $f_0 = f_c = 1/[2\pi\{LC_2(\text{tuning knob})\}^{1/2}]$ to obtain our 1220-kHz station.

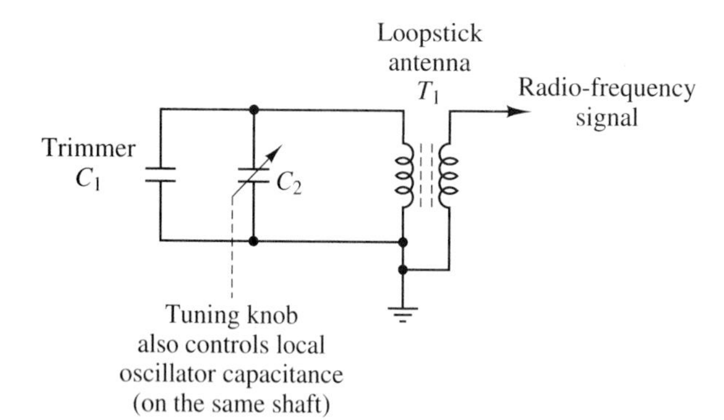


**Figure 7.43** AM radio tuning circuit fragment.

[1](The radio selected is the one analyzed in *Troubleshooting, Servicing, and Theory of AM, FM, and FM Stereo Receivers,* by Clarence Green and Robert Bourque [Prentice Hall, Englewood Cliffs, N.J., 1987]. Also, our analysis follows and adapts the analysis in that book.)

For pocket radios, the transformer $T_1$ is also the antenna itself, called a loopstick antenna because of its physical construction. The bandwidth $[1/(RC)]$ of this $RLC$ combination must be, of course, 10 kHz so that the receiver picks up both 5-kHz-wide audio bands on either side of $f_c$. All other frequencies from other radio stations are rejected by this filter. The trimmer capacitance $C_1$ is for calibration purposes. At this stage, the signal voltage $e(t)$ is on the order of $10^{-6}$ V ($\mu$V), and the signal power is only on the order of $10^{-9}$ W (nW).

The next stage is the mixer/oscillator/filter, shown in Fig. 7.44a. Recalling that capacitors are open circuits and inductors are short circuits for dc, the biasing equivalent circuit is as shown in Fig. 7.44b. Note that $V_{CC}$ = 7.2 V for *all* transistors in this radio. We see that transistor $Q_1$ has essentially the voltage divider bias arrangement considered in Sec. 7.6.2. It results in an operating point and ac gain roughly independent of the $\beta$ of $Q_1$.

For ac analysis of this stage, we argue here only operationally, not quantitatively. When the radio is first turned on, the "tickler" coil (primary of $T_2$) supplies a changing magnetic field which starts the oscillating bank oscillating at its resonant frequency. Recall from our systems discussion that this frequency is $f_{osc} = f_c - 455$ kHz. It would eventually die out, except that a sample of it is fed via $C_4$ to the emitter circuit of $Q_1$.

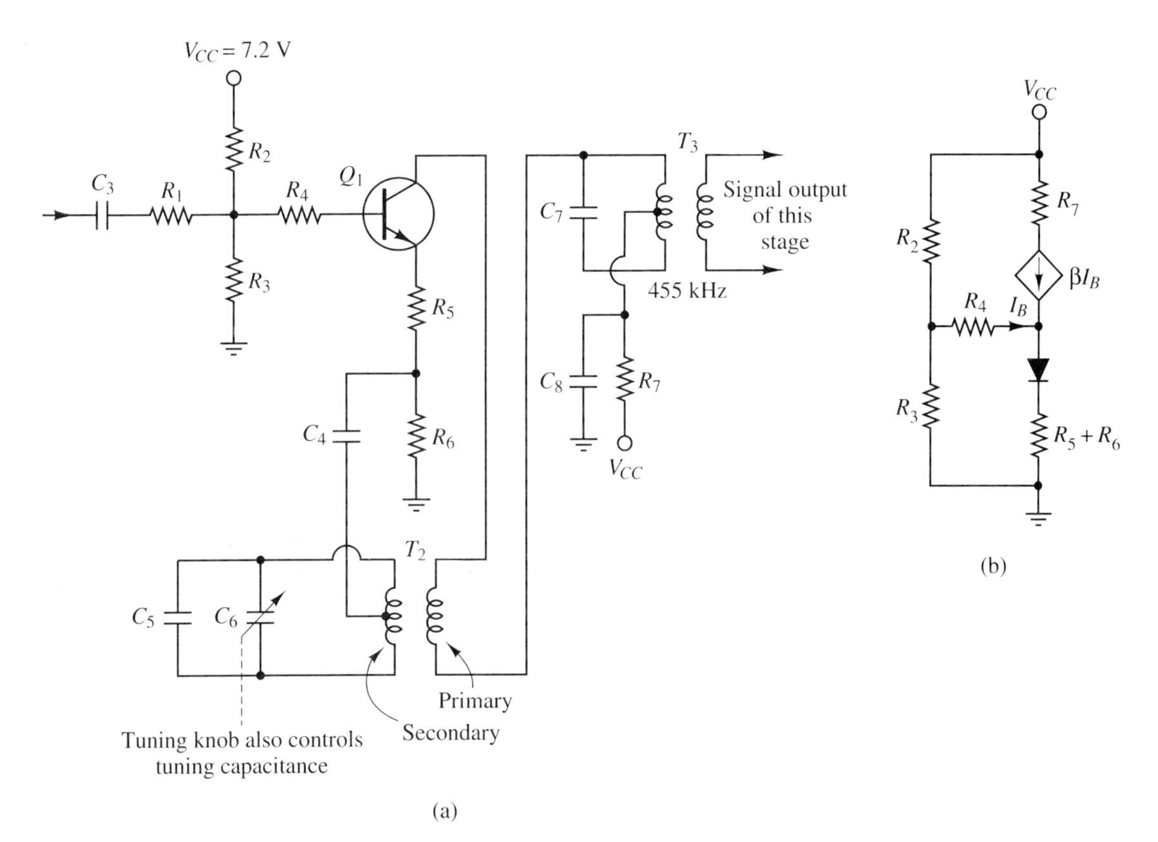


**Figure 7.44** (a) Mixer/oscillator/filter circuit; (b) biasing (dc) equivalent.

We know that this is a common-emitter amplifier with $R_E = R_5 + R_6 \neq 0$, so it has feedback. The way the polarities are arranged, the feedback is in this case positive, as far as the oscillating bank is concerned. Note that the output collector current drives the primary of $T_2$, which keeps the oscillator bank oscillating. As we shall see, one component of the collector current is at $f_{osc}$. In analogy to pushing a swingset at its resonant frequency, the oscillator keeps oscillating with large amplitude.

The 455-kHz filter bank with $T_3$ has its maximum impedance, and thus develops maximum output voltage, at 455 kHz. It has only a 10-kHz bandwidth, so it will actually pass only 455 kHz ± 5 kHz. Identical filter banks $T_4$ and $T_5$ will further sharpen the 455-kHz-centered audio spectrum at later amplification stages. Now, what frequencies are actually contained in the collector current, sent to $T_3$? The answer is evident when we consider that $Q_1$ serves not only as an amplifier for the oscillator, but also as a mixer (converter).

Specifically, $v_{be}$ of $Q_1$ is due partly to the signal voltage applied to the base, and partly to the oscillating voltage from the oscillator $L_2$. By Eq. (7.4b), $i_b = I_B e^{qv_{be}/kT}$. If we let $v_{be}$ due to the oscillator be large, the $v_{be}^2$ term in the Taylor expansion of $e^{qv_{be}/kT}$ will be significant [unlike in Eq. (7.4c), where it was ignored for linear amplification]. This is intentional. We end up with frequency components of the signal multiplying the local oscillator sinusoid. As argued in our systems description, we end up with, among other terms that we reject, a difference frequency signal, which is the audio spectrum centered at 455 kHz.

It may be asked how the value of 455 kHz was chosen as the IF $f_{IF}$. If $f_{IF}$ were too large, the selectivity of the radio would be reduced because the bandwidth would be too high for the given attainable sharpness curves for inexpensive $LC$ filter banks. Similarly, if $f_{IF}$ were too small, the resulting bandwidth of signals the bank would pass would be smaller than 10 kHz, and audio signal would be lost. There are also more advanced issues involved, such as determining how well $f_{osc}$ can track $f_c - f_{IF}$ for the entire AM band for various choices of $f_{IF}$ other than 455 kHz.

The next two stages, shown in Fig. 7.45, are basic voltage amplifiers, but designed specifically for the 455-kHz range. They are essentially common-emitter amplifiers with $R_E \neq 0$, bypassed by bypass capacitors. Because of the high frequencies involved, these bypass capacitors ($C_{10}$ and $C_{14}$) can be as small as 0.01 μF instead of several μF for AF. Rather than capacitor coupling between stages, tuned-to-455 kHz filter coupling is used to further sharpen the cutoff of other stations and emphasize the desired station.

The function of capacitors such as $C_{12}$ and $C_{16}$ is to prevent the signal from going through the power supply and back to a previous stage, causing positive feedback and unwanted oscillations—heard in the output as "whistling" or "howling." $R_{10}$ and $R_{14}$ are collector biasing resistors (recall that $T_4$ and $T_5$ are short circuits for dc biasing). The most notable aspect is $C_9$. It is a large capacitance, and its voltage is thereby designed to represent the average signal power later on in the circuit. This voltage affects the bias point and therefore the signal gain of $Q_2$ and that of the radio overall. This automatic gain control (AGC) is discussed further below.

The demodulator circuit, called an inverting peak detector, is as shown in Fig. 7.46. The signal presented to the diode circuit is $v_1 = A_r g(t)$, from Fig. 7.42. Whenever the signal $v_1$ falls below the capacitor voltage $v_2$, the diode $X_1$ conducts

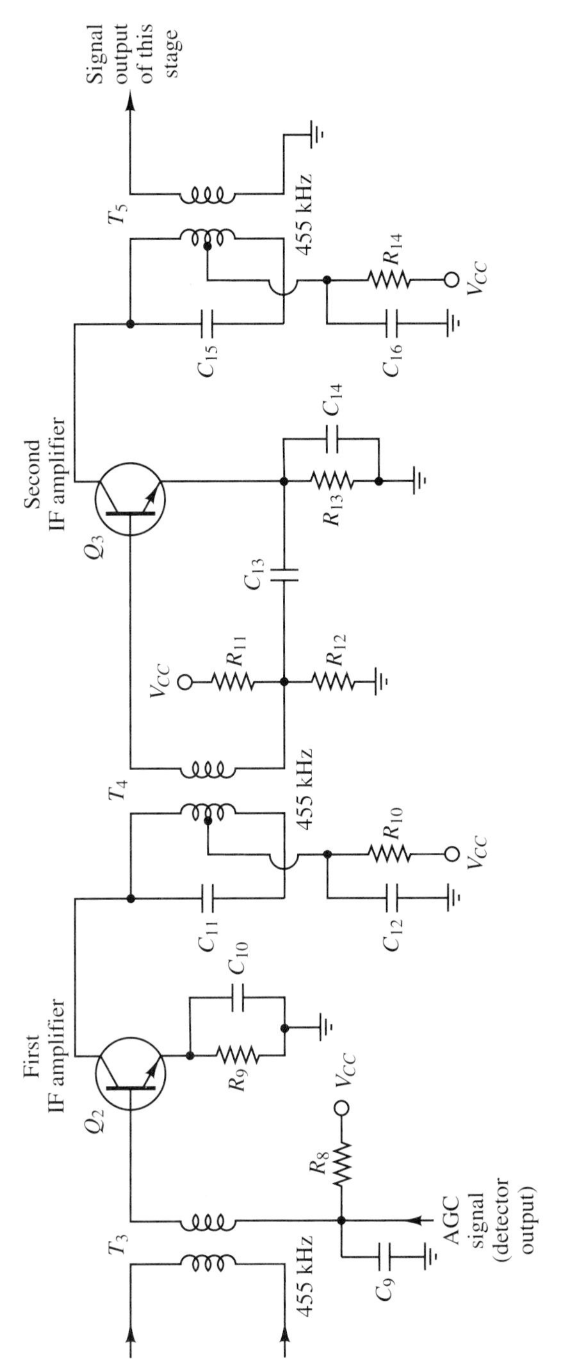


**Figure 7.45** IF amplifier stages of an AM radio.

(is forward biased). In this case, the capacitor charges immediately to follow $v_1$ because the time constant is near zero. That is, the diode is like a short circuit (zero resistance), so that the time constant for the capacitor $C_{17}$ charging toward $v_1$ is $\tau = r_d C_{17} \approx 0$, where $r_d$ is the incremental resistance of the diode under forward bias. Whenever $v_1$ rises above $v_2$, the diode becomes reverse biased, and $C_{17}$ discharges through $R_{15}$, $R_{16}$, and the base circuit of $Q_2$. This time constant is large, so $v_1$ cannot follow the oscillations of $v_1$, only the gradual trend. The result is that $v_2$ is negatively proportional to the *envelope* of $g(t)$, which is $b(t)$—the carrier is removed.

Also, the peak-detected, rectified audio-frequency output voltage $v_3$ can charge or discharge $C_9$ through $R_{16}$. The time constant $R_{16}C_9$ is nearly 50 ms, which is below even the audio frequencies. It therefore reflects the signal level due to station distance, station power, and atmospheric fading. The voltage across $C_9$ is used to control the bias of $Q_2$. Because the peak detector is inverting (output negative), a stronger signal will tend to *reduce* the bias and thus the gain of $Q_2$, and a weaker signal will increase it. By this process, AGC, we have negative feedback and eliminate any deviations from the desired, uniform signal level.

The audio amplifier (see Fig. 7.47) begins by taking a voltage divider portion of the available signal from the IF amplifier output. This voltage divider ratio is selected by the volume control, $R_{17}$ (a potentiometer). The first two audio amplifiers, $Q_4$ and $Q_5$, are standard common-emitter amplifiers. In between these stages is the tone control, potentiometer $R_{23}$. The final transistor amplifier $Q_6$ is the power amplifier to drive the loudspeaker.

Because the speaker is small, it emphasizes high frequencies. To avoid a tinny sound, $C_{24}$ and $C_{26}$ partially "short out" high frequencies, thereby emphasizing low frequencies. The tone control circuit ($R_{22}$, $R_{23}$, $R_{24}$, $C_{21}$, $C_{22}$) works as follows. High frequencies are partially "shorted to ground" via $C_{21}$ and the variable $R_{23\text{A}}$, the latter of which is the tone control. The high frequencies are further attenuated by the low-pass filter $R_{23\text{B}}$ and $C_{22}$. Low-frequency signals traverse primarily through the low resistance $R_{22}$.

Now consider the final stage, the power amplifier $Q_6$. Its bias point is controlled directly by the collector voltage of $Q_5$. Naturally, this is called a direct-coupled power amplifier. Because of the high power levels in $Q_6$, there is the possibility of *thermal runaway*. In this situation, the high power in $Q_6$ causes higher temperature, which increases $\beta$, which increases the emitter–collector current even more, which increases the temperature even more, and so on, until $Q_6$ is destroyed.

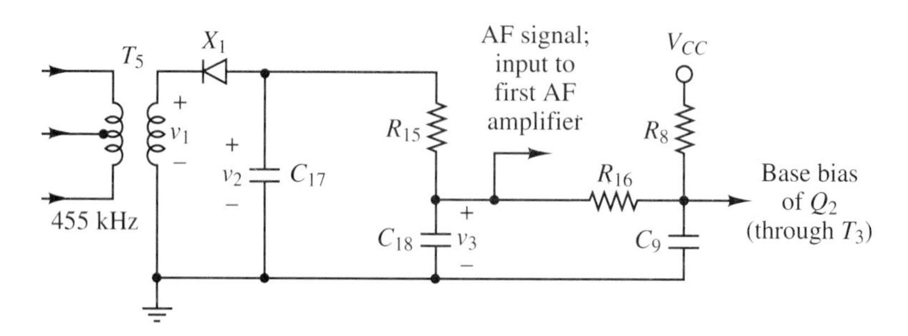


**Figure 7.46** Peak detector and AGC for AM radio.

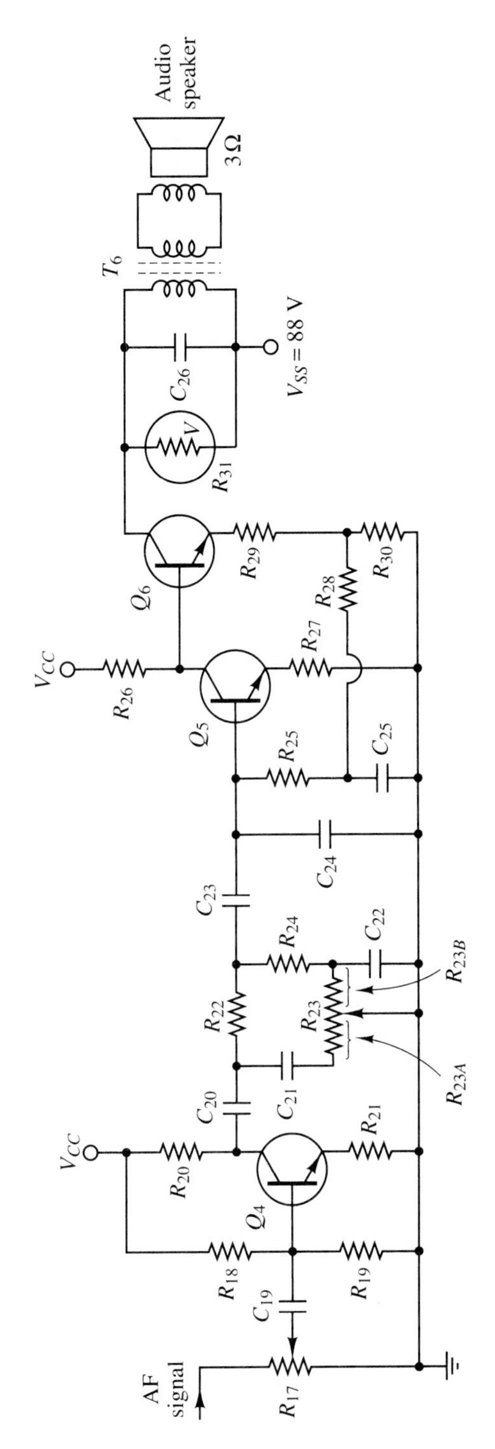


**Figure 7.47** Output AF stages of AM radio.

To prevent this, the *bias* emitter current of $Q_6$ (i.e., its dc component) is used to bias $Q_5$ by a standard voltage divider method (via $R_{29}$ and $R_{30}$). The capacitor $C_{25}$ prevents signal feedback by shorting ac to ground on the way back to $Q_5$. Now if $I_{Q6}$ increases, the emitter voltage of $Q_6$ also increases, which raises the bias point of $Q_5$. However, this increased collector current causes the collector voltage of $Q_5$ to *decrease* due to the increased voltage drop across $R_{26}$ from $V_{CC}$. Therefore, the bias of $Q_6$ is reduced and stabilization is achieved.

Finally, in the output circuit, $R_{31}$ is used to protect $Q_6$ from voltage spikes. Because of the inductance of the audio output transformer $T_6$, a noise current spike could cause a large-voltage spike that could destroy $Q_6$. Because $R_{31}$ is a special kind of resistor whose resistance decreases for very large applied voltages, those spikes would be shorted out and $Q_6$ protected. Of course, the main purpose of $T_6$ is impedance transformation, so that maximum power transfer from $Q_6$ to the 3-Ω speaker is achieved.

The power supply (Fig. 7.48) is a very simple half-wave rectifier. Note the filter capacitance $C_{27}$, which is hundreds of μF (huge). The (fusible resistor $R_{32}$) $C_{27}$ time constant diminishes the 60-Hz ripple to acceptable levels. The speaker circuit is driven by the $V_{SS} = 88$ V source, and $V_{CC}$ is obtained by voltage divider ($R_{33}$, $R_{34}$) and further capacitive filtering ($C_{28}$). The complete circuit is shown in Fig. 7.49.

### 7.7.4 Electronic Camera Flash

Our final circuit for this introduction to electronics is the electronic flash unit of an actual camera. The schematic is shown in Fig. 7.50. An amazing aspect of this circuit is that only two penlight batteries are needed to produce the 370-V charge necessary to flash the xenon flashbulb. This is achieved by charging the capacitor $C_3$ via periodic pulses. The exact shape and frequency of the pulses is not important; only their contribution to the creation of the required large voltage. We consider here only some qualitative descriptions of the pulse circuitry, known as a *blocking oscillator,* and the relatively simple remainder of the circuit.

**Warning:** *This circuit generates huge voltages (400 V); for this reason, actual component values have not been given. The charge and voltage on $C_3$ could be fatal. Our discussion is for educational purposes only, and it is not recommended that the reader construct this circuit.*

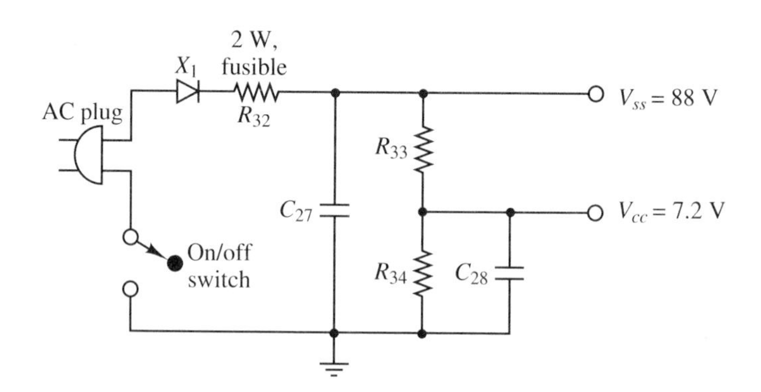


**Figure 7.48** Power supply for AM radio.

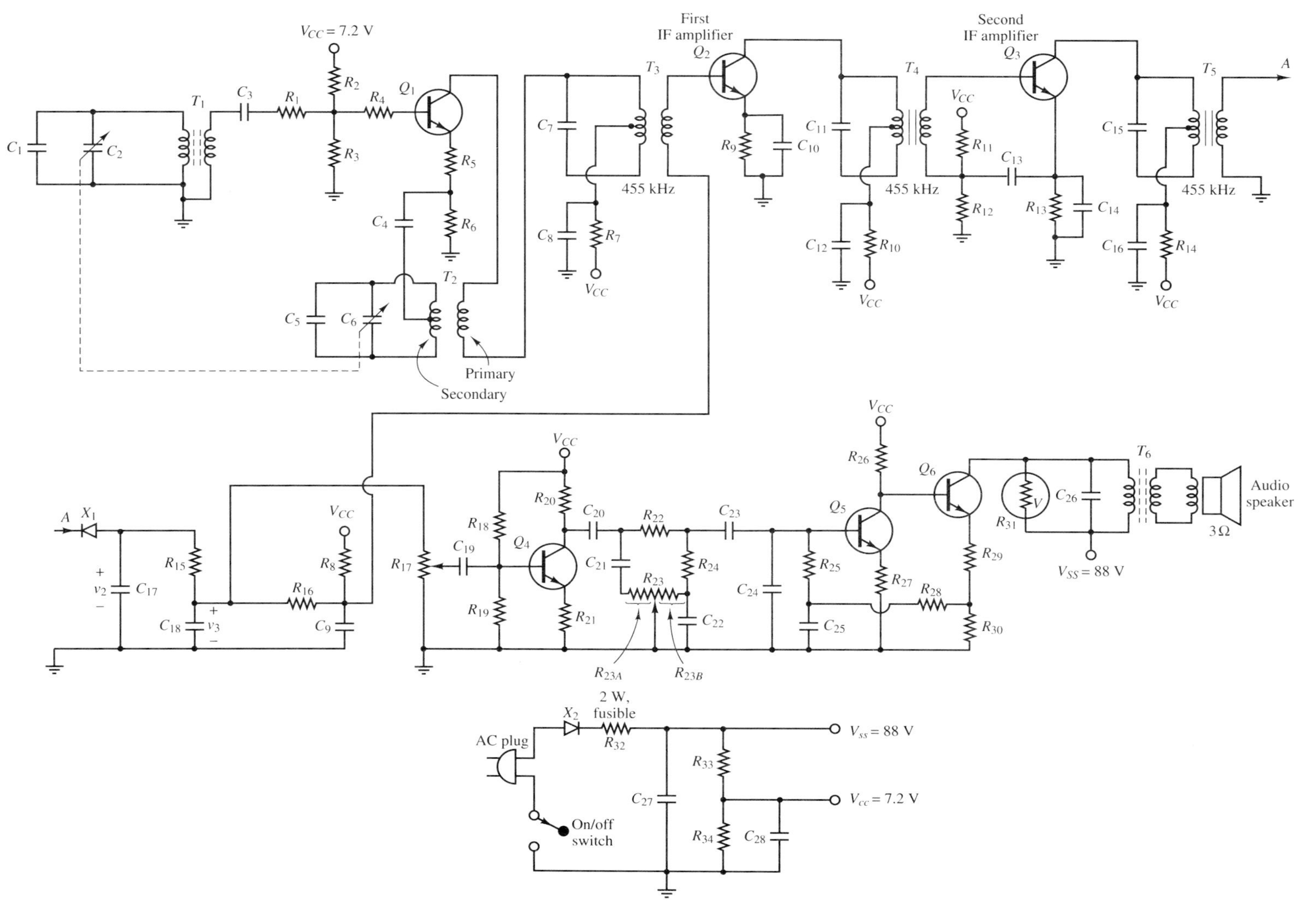


**Figure 7.49** Complete six-transistor AM radio.

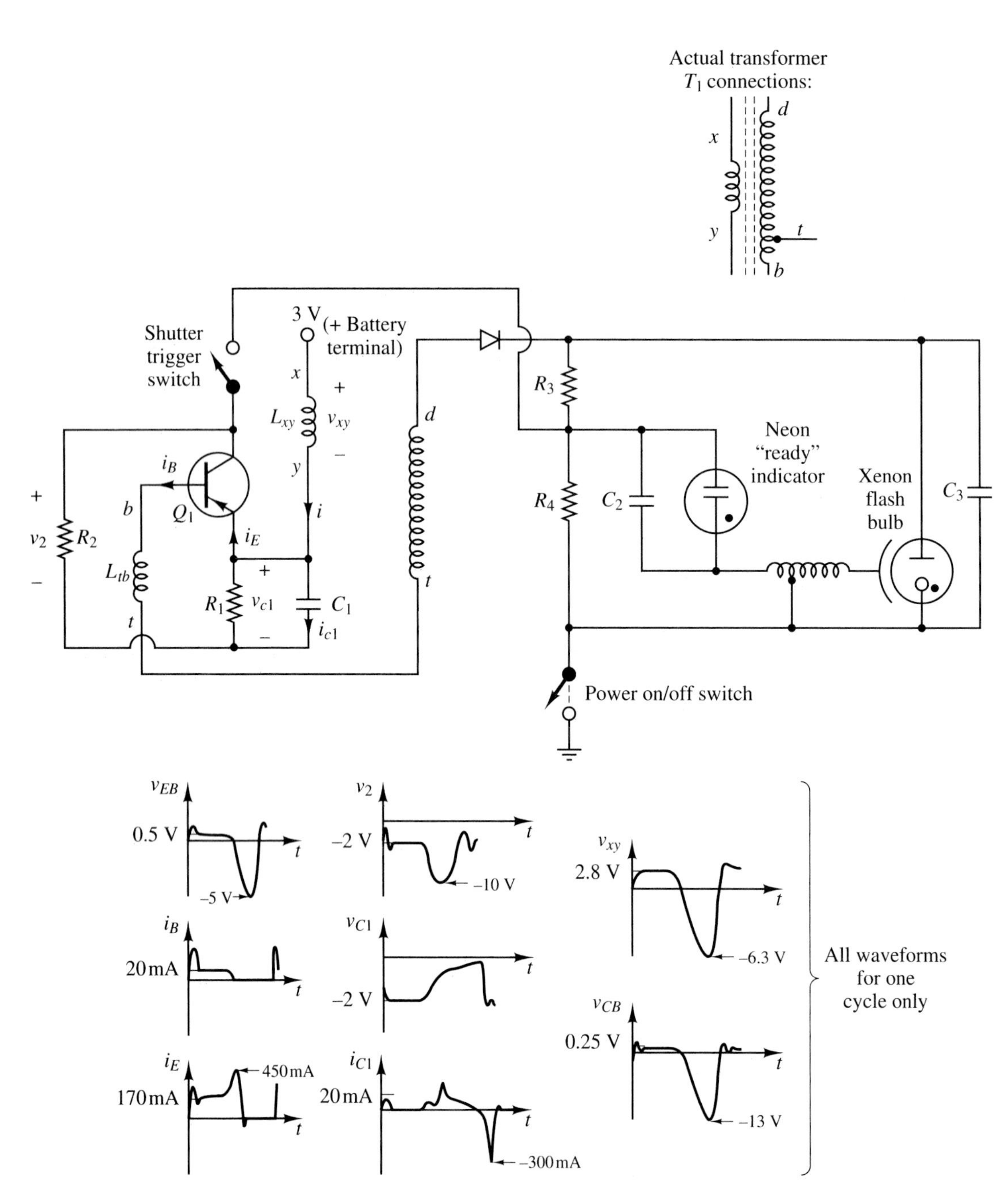


**Figure 7.50** Electronic camera flash circuit.

The steady-state dc bias circuitry may be deciphered by replacing all capacitors by open circuits and all inductors by short circuits. At startup, however, the changes are all sudden, so that capacitors look like short circuits and inductors look like open circuits. Because this analysis is not particularly interesting, we do not describe it here. Suffice it to assume that the bias circuitry is appropriate for desired operation at turn-on time, with all capacitors initially uncharged.

Suppose that initially the bias circuitry causes conduction in transistor $Q_1$. Then collector–emitter current begins to flow from the 3-V battery through $L_{xy}$ (the primary of transformer $T_1$) into the emitter. The increase in the current $i$ in $L_{xy}$ causes it to develop a voltage $L_{xy}\, di/dt$, which forces down the potential of the emitter. Alone, this fact is not so important because it is just a shift in absolute potential level.

However, coil $L_{tb}$ is a secondary winding in which voltage is induced; it is wound so that the base terminal potential is reduced when $i$ is increasing. The coil $L_{tb}$ acts like a voltage source with source resistance $R_1$ applied to the base–emitter junction; the current into the base is limited by $R_1$ and the resistance of $L_{tb}$. The effect is that the base–emitter junction is even more forward biased, which in turn further increases the emitter current. That further increases the voltages across $L_{xy}$ and thus $L_{tb}$, even more forward biasing $Q_1$ and further increasing emitter current.

This positive feedback action continues until the transistor is strongly on. Now the collector–emitter voltage is very small and essentially *constant.* Also, the voltage across $C_1$ is now stabilized; it has charged to its equilibrium value very rapidly through the transistor in its "on" state. However, the emitter current can still continue to increase. The constant voltage across $L_{xy}$ implies that the flux through it increases linearly with time ($v = L\, di/dt$, so $i$ and $\phi$ = integral of constant $v$). However, due to the nonlinearity of flux versus input current (saturation region of the nonlinear $B$–$H$ curve), more-than-linear current is drawn.

The emitter–collector current rises until it just cannot any longer; we now say the transistor is in full saturation. Consequently, $L_{xy}\, di/dt$ and therefore $L_{tb}\, di_b/dt$ decrease. When the voltage across $L_{tb}$ decreases, the base–emitter junction voltage is reduced, which further reduces the emitter current, which further reduces the forward biasedness of the base–emitter junction, and so on.

This collapse of the pulse is all extremely quick. The transistor is cut off, and there is a large voltage pulse due to $d\phi/dt$, during which the base–emitter junction is very strongly *reverse* biased. All the while, the voltage across $C_1$ gradually diminishes by $C_1$ discharging through $R_1$, where $R_1C_1$ is a fairly large time constant, on the order of the duration of the $d\phi/dt$ pulse.

Note the direction of current during the pulse (in the direction of the arrow in the *pnp* transistor); therefore, $v_{C1} < 0$. The transistor thus will stay cut off until $C_1$ discharges through $R_1$ (bringing up the emitter voltage) sufficiently that the base–collector junction is again forward biased, in conjunction with the "background" dc bias circuitry. The amount of time the transistor is cut off ("blocked") is therefore directly influenced by $R_1C_1$.

During these processes, the main secondary, $L_{td}$, develops a huge amplitude pulse. It is huge because the inductance of $L_{td}$ is huge, relative to $L_{xy}$. This huge pulse (over 400 V) is half-wave rectified by the diode, via resistors in the transistor bias circuit. Eventually, a huge dc voltage (380 V) is developed across capacitor $C_3$.

A fraction of this voltage, via the $R_3$–$R_4$ combination, is placed across $C_2$. When that voltage reaches 250 V, the main flash is ready to be set off. This is indicated by the neon indicator flashing, which discharges $C_2$. This capacitor again charges until the indicator blinks again, providing a blinking indication to the photographer that the main capacitor $C_3$ is ready to be discharged through the xenon flash bulb.

(Resistors $R_3$ and $R_4$ are chosen extremely large—in the tens of megohms range; one reason for this is that they form a discharge path for $C_3$. We do not want this type of discharging to occur; hence the large resistances cause that time constant to be thousands of seconds.)

Before triggering the main flash, the shutter trigger switch has 0 V dc on one side and 250 V dc on the other side. Clicking the shutter shorts these, and current is pulled through the trip coil $L_{tr}$. Of course, $L_{tr}\, di_{tr}/dt$ is huge, sufficient to trigger the xenon flash to ionize.

This analysis is grossly simplified. Examination of the voltages and currents on the oscilloscope (see Fig. 7.50) shows that this is indeed a very nonlinear circuit. The frequency of the pulses begins very low and gradually increases to a steady-state value. The dc levels at many points in the circuit also change as $C_3$ charges. Eventually, a steady state is reached, at which point the capacitor is charged and "ready to shoot." So-called "flyback" high-voltage pulse circuits on TV sets are related to this blocking oscillator. These nonlinear pulse circuits are often so difficult to quantitatively analyze that they are designed as much by "tweaking" as by theoretical considerations.

## 7.8 SUMMARY

In this lengthy chapter we have covered a lot of ground. We have introduced the two most important electronic devices: the diode and the transistor. The diode is important as a rectifier of bipolar voltages and currents into unipolar voltages and currents. To power most electronic circuits, dc voltages are required, yet ac is the common voltage available from wall outlets. Therefore, the various power supply circuits we studied are necessary to provide this important power conversion. The diode is also useful for various signal processing applications, a few of which we considered. Also, the diode is often used as a device for protection of other electronic devices, because of its ability to clip off large noise voltage spikes.

For both the diode and the transistor, we went into the physics of these devices in some depth but without getting into a lot of math or quantum mechanics. Our purpose was to motivate diode and transistor behaviors and see how people have achieved those behaviors with physical devices, based on physical laws such as those reviewed in Chapter 2.

The transistor is the foundation of nearly all practical electronic circuits. There are several varieties of device, but two main families: bipolar and unipolar (field-effect). The former necessarily involve both holes and electrons, for reasons we explored in some detail. The latter involve predominantly only one type of charge carrier. Each device has advantages for certain applications. For example, BJTs tend to have better high-frequency performance, while FETs have essentially infinite input impedance and therefore are good input signal detectors.

A theme common to both BJTs and FETs is the necessity of establishing a bias or operating point. A large-signal model of the transistor can be used to design or determine it. Typically, the signal gain depends on the operating point, so the biasing serves a double purpose: Without it, linear operation is unachievable, and with it, the value of the signal gain depends on the particular value of operating point. This gain can be found once the circuit is linearized, which results in a small-signal model. We examined examples of both BJT amplifiers (common emitter) and FET amplifiers (common source).

Just as important as the use of transistors as signal amplifiers is their use as electronic switches. Increasingly, this application is taking over many circuits that used to be linear amplifiers. This is because of the advantages of digital circuits, which we review in the coming chapters. In Sec. 7.6.4 we analyzed a basic BJT switch.

We concluded this chapter with three relatively advanced electronic circuits that have real-world applications: a transistor power supply, an AM radio, and an electronic camera flash circuit. Any time one studies actual commercial circuits, there is much more complexity. Contrary to the simplified textbook cases, these circuits are often the result of weeks or years of research by design teams and are much influenced by empirical information (e.g., what *really* works?) as well as standard analysis. It is hoped that seeing these real-world circuits will interest you more in your daily contact with electronics, if not convert you to an electrical engineer.

## PROBLEMS

**7.1.** What is the most crucial factor in order that an electronic circuit function as a signal amplifier? Why cannot a transformer be used as an amplifier?

**7.2.** Suppose that an amplifier has an input signal power equal to $p_i$ and an output power equal to $p_o = kp_i$, where $k$ is the power gain. If $k > 1$, from where does the extra power come?

**7.3.** One example of a greater-than-unity power gain device is shown in Fig. 7.2. Name two or more other such devices you encounter in your daily routine.

**7.4.** Find the current $i$ in the 1-kΩ resistor for each of the two circuits in Fig. P7.4. Assume ideal diodes ($V_o = 0$, $V_Z = \infty$).

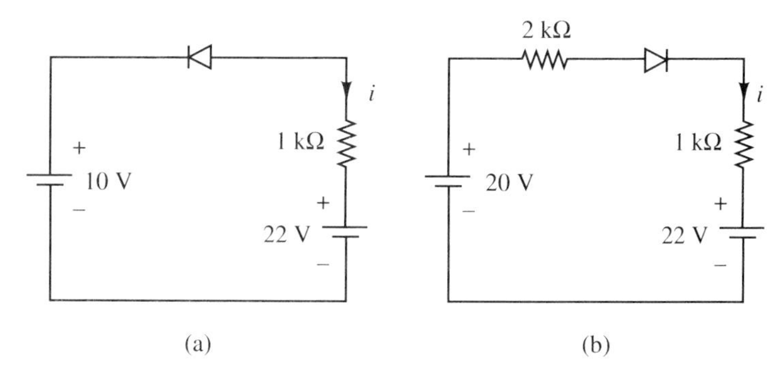


**Figure P7.4**

**7.5.** For the circuit in Fig. P7.5, graph $v_o$ vs. $v_s$. Quantitatively label your axes and all critical points on your axes. Use the following diode model: $V_o = 0.7$ V, $V_Z = 4$ V.

**7.6.** What are three unique features of electronic device materials compared with normal conductors and insulators?

**7.7.** What are the most important factors that control the conductivity of a semiconductor? Which one is most significant?

**7.8.** By what external influences are conduction electrons and conduction holes made available in semiconductors?

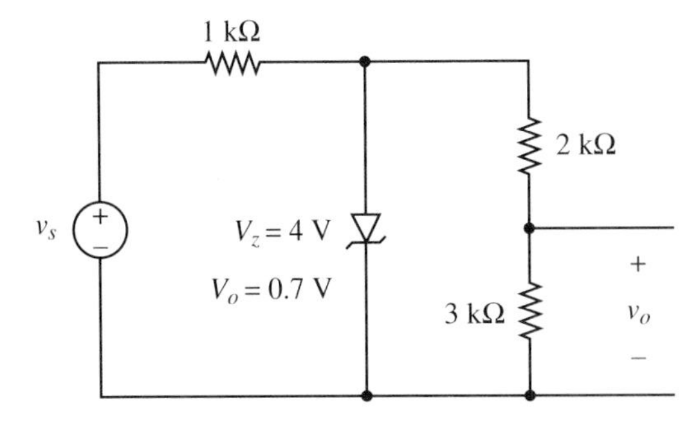

Figure P7.5

**7.9.** Describe the process of hole conduction.

**7.10.** Describe the process of diffusion conduction.

**7.11.** In a $p$-type semiconductor, what are the majority carriers, and what are the minority carriers? Answer the same questions for an $n$-type semiconductor.

**7.12.** What is the unique characteristic of a $pn$ junction, in terms of its conductivity?

**7.13.** In Fig. 7.11, the measured $i$–$v$ characteristic of a real diode is shown. When the bias voltage $v$ reaches $V_Z$, the curve has a very steep slope. Explain this phenomenon.

**7.14.** **(a)** Suppose that a 3-mA current source is placed across a diode such that the positive current out of the current source is directed into the anode of the diode. Does the diode conduct?

**(b)** What if a 3-V battery is connected across the diode, with the higher-potential terminal connected to the cathode of the diode? Does the diode now conduct?

**7.15.** For the half-wave rectifier with filter capacitor in Fig. 7.13, if we model the diode as ideal, compare the output $v_L(t)$ with that shown in Fig. 7.14 if $v_s(t)$ is the same as in that figure.

**7.16.** Repeat Problem 7.15 if $v_s(t)$ is a square wave (5 V for the first half of each period, −5 V for the second half of each period).

**7.17.** **(a)** For the half-wave rectifier shown in Fig. 7.5, draw the waveform $v_L(t)$ if $v_s(t)$ is as shown in Fig. P7.17. Assume an ideal diode.

**(b)** Repeat if $V_o = 0.7$ V.

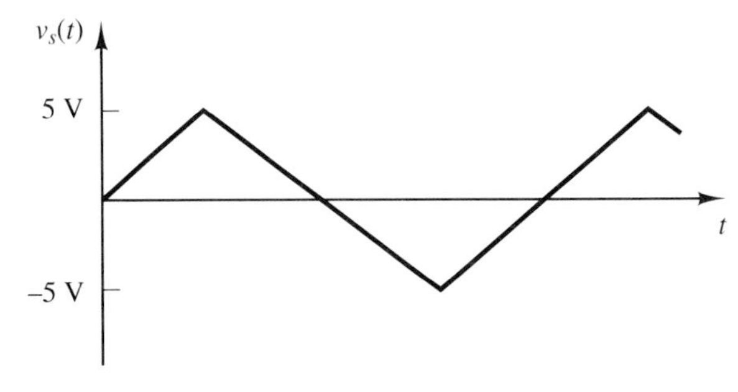

Figure P7.17

**7.18.** For the circuit in Fig. P7.18, determine $v_2$ vs. $v_s$ and graph the input–output characteristic ($v_2$ vs. $v_s$) for the conditions, where always $V_Z = \infty$: **(a)** $V_o$=0 V and **(b)** $V_o$ = 0.7 V. Fully label all critical points and slope(s).

**7.19.** A half-wave rectifier (assume ideal diode) and 1000-μF filter capacitor are used to convert 60 Hz ac to dc for supplying a 100-Ω load (see Fig. P7.19). Due to capacitor discharging, however, $v_1$ will not be exactly dc but will have a ripple. Find an approximate value of the percent of the maximum voltage that the ripple is $100\% \cdot (v_{\max} - v_{\min})/v_{\max}$,

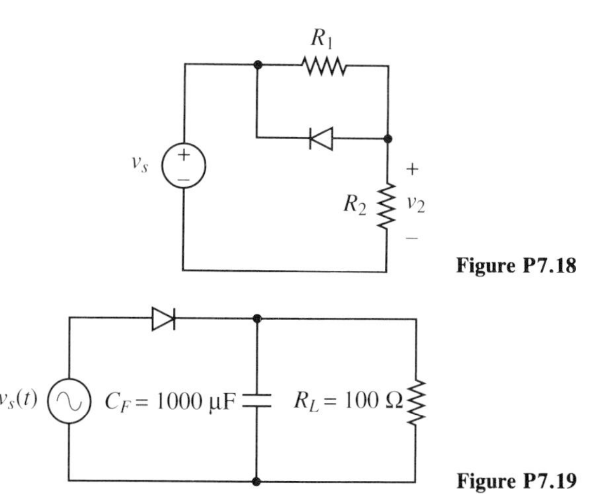


Figure P7.18

Figure P7.19

as well as the approximate duration per cycle that the diode conducts; express as a percent duty cycle ratio (time on/period). Use either a graphical or a rough numerical solution.

**7.20.** Suppose that the diode ($V_o = 0.7$ V, $V_Z = \infty$) in Fig. 7.5 is reversed.
**(a)** Sketch the input-output characteristic ($v_L$ vs. $v_s$).
**(b)** Sketch $v_L(t)$ if $v_s(t)$ is $3 \sin(\omega t)$ V.
**(c)** Place a capacitor $C$ across $R$. Let $RC \approx 2 \cdot 2\pi/\omega$. Sketch $v_L(t)$ if $v_s(t) = 3 \sin(\omega t)$ V.

**7.21.** **(a)** Given the circuit in Fig. P7.21a, draw the $v_L$ vs. $v_s$ graph assuming (i) $V_o = 0$ V, and (ii) $V_o = 0.7$ V. For each case, indicate under what conditions on $v_s$ that each diode conducts. For case (i), give a simple mathematical expression for $v_L$ in terms of $v_s$. For case (ii), repeat by using the function sign $(v) = (1$ for $v \geq 0$; $-1$ for $v < 0)$. It may help you to draw separate equivalent circuit diagrams for different input voltage conditions. Label your graph completely. Give numerical values for all threshold voltages and indicate the functional form of each portion of the graph.
**(b)** For cases (i) and (ii) above, sketch and label the graph of $v_L(t)$ for the given input voltage $v_s(t)$ in Fig. P7.21b. Completely label your graph. Give numerical values for all maximum/minimum/threshold voltages and label important corresponding times on both graphs $v_s(t)$ and $v_L(t)$ (e.g., $t_1$, $t_2$, $t_3$, ...).

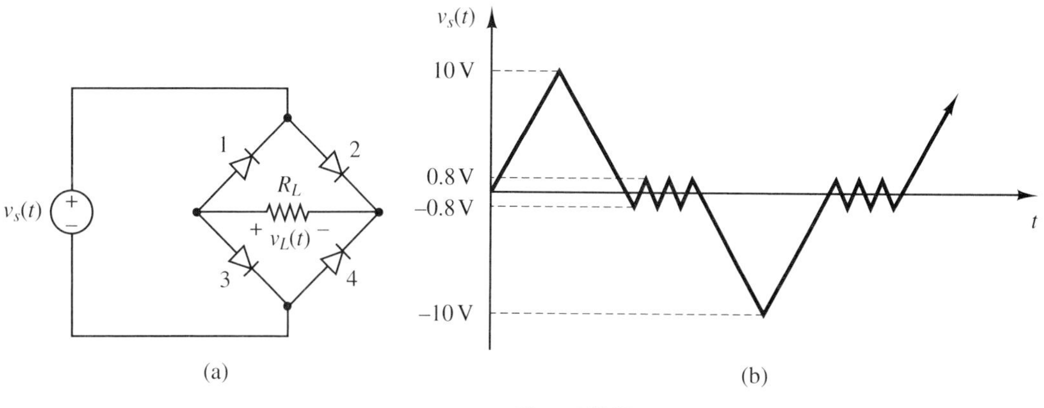


Figure P7.21

**7.22.** **(a)** For the circuit in Fig. P7.22a, graph the output voltage $v_o(t)$ vs. $t$ when the input voltage $v_i(t)$ is as shown in Fig. P7.22b. Assume that the contact potential of the diodes is 0.7 V. Indicate extreme values of the voltage $v_o(t)$ on your graph; qualitatively indicate significant times. [*Hint:* Suppose initially that the current through $R_L$ is zero. Consider ranges of $v_i(t)$ for which distinct combinations of forward and reverse biasing of diodes $D_1$ and $D_2$ occur. If helpful, draw equivalent circuit diagrams for each case. Then use KVL to find $v_o(t)$.]

**(b)** Graph $v_o(t)$ vs. $t$ if $v_i(t)$ is as shown in Fig. P7.22c.

**(c)** Graph the input–output characteristic ($v_o$ vs. $v_i$).

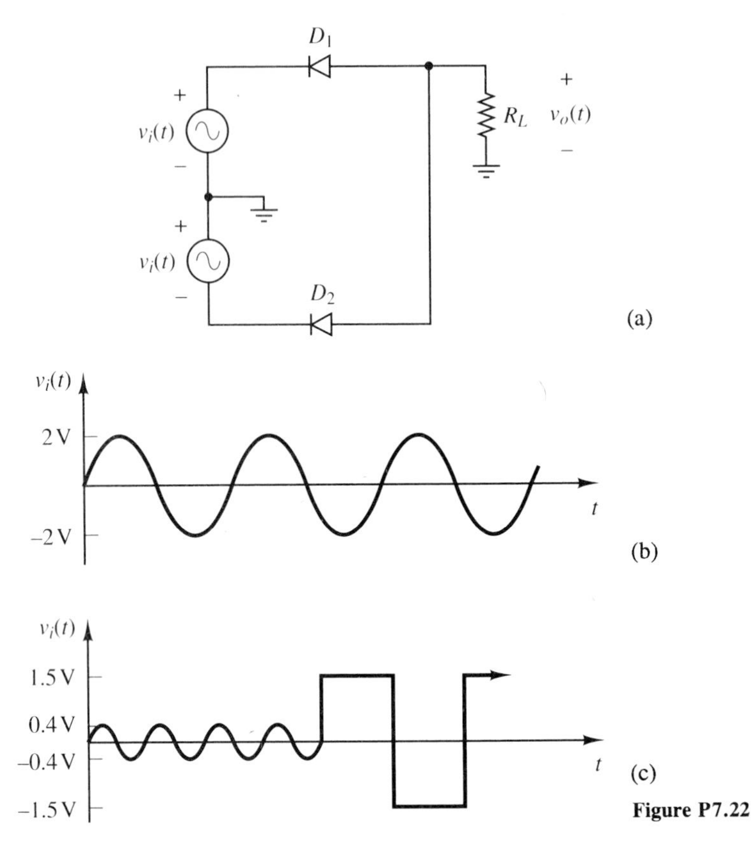


**Figure P7.22**

**7.23.** For the circuit in Fig. P7.23, sketch $i_1(t)$, $i_2(t)$, and $v_o(t)$ if the input voltage is $v_s(t) = 10\cos(\omega t)$ V. Assume that diode contact potentials are both $V_o = 0.7$ V.

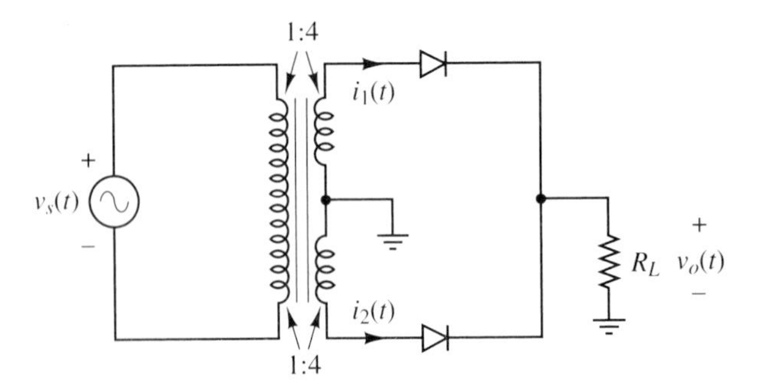


**Figure P7.23**

**7.24.** **(a)** Draw two simplified circuits of the circuit in Fig. P7.24 for the cases $v_s > 0$ and $v_s < 0$. Assume that the contact potentials of both diodes are 0 V (i.e., ideal). (*Hint:* Consider $v_o$ as part of a voltage divider for each circuit.)

**(b)** Based on these circuits, draw a graph of $v_o$ (vertical axis) vs. $v_s$. Label all slopes with correct values.

**(c)** Now sketch $v_o(t)$ given the input waveform $v_s(t)$.

**(d)** What is the average value of $v_o(t)$ in part (c)?

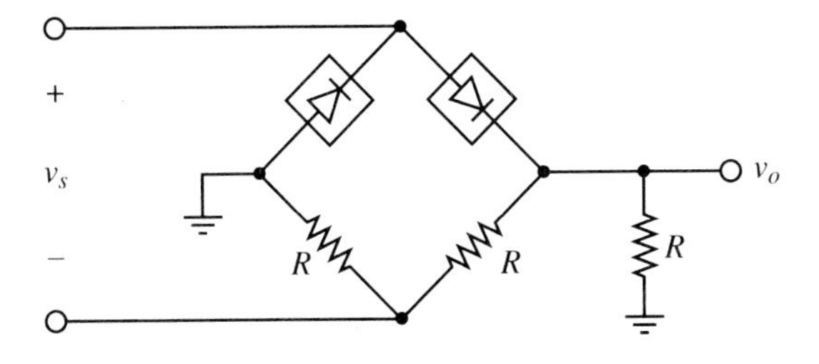


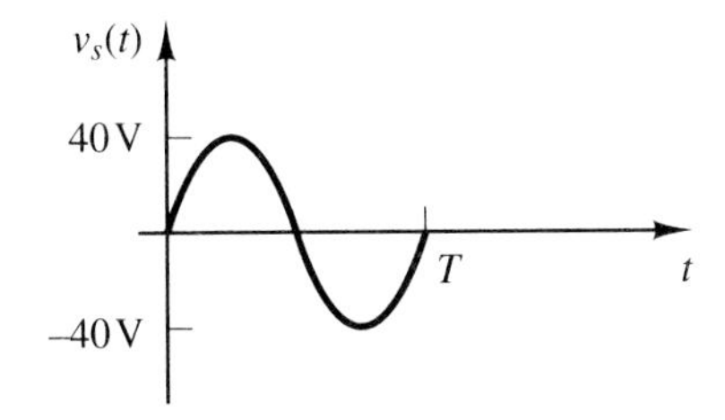


**Figure P7.24**

**7.25.** In the full-wave rectifier shown in Fig. P7.25, which diode is hooked up backwards, and why? Assume that $V_o = 0$ V. Consider the cases $v_s > 0$ and $v_s < 0$.

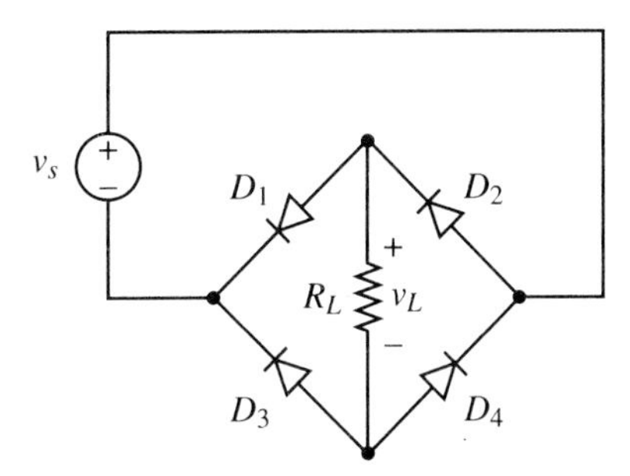


**Figure P7.25**

**7.26.** As discussed in Sec. 7.4.3, an improvement on the half-wave rectifier is obtainable by using three more diodes, as shown in Fig. 7.15. In this problem, a full-wave bridge (assume ideal diodes) with a $C = 1000$-μF filter capacitor supplies a 100-Ω load. Due to capacitor discharging, the load voltage $v_L(t)$ will not be exactly dc, but will have a ripple. If the source voltage $v_s(t)$ is a 60-Hz sinusoid, sketch the output voltage $v_L(t)$. Also, find an approximate value of the percent ripple: $100\% \cdot (v_{\max} - v_{\min})/v_{\max}$.

**7.27.** A simple Zener diode regulator circuit is as shown in Fig. P7.27. Assume that the Zener voltage is $V_Z = 6$ V.

**(a)** Is the diode Zener biased when $v_s = 9$ V?

**(b)** Is the diode Zener biased when $v_s = 24$ V?

**(c)** Calculate $i_R$, $i_D$, and $i_{RL}$ given the conditions in parts (a) and (b).

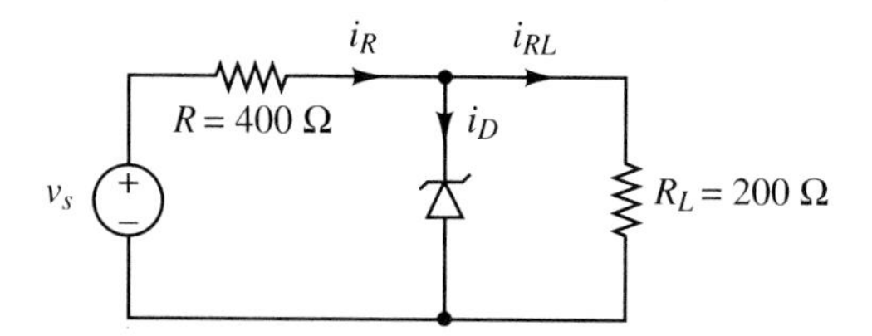


**Figure P7.27**

**7.28.** For the circuit in Fig. P7.28, let $V_B$ range from $-2\,V_Z$ to $2\,V_Z$, and show for three distinctive, representative values of $V_B$ the load line on the double-diode $i$–$v$ curve. Determine the operating point in each case (one operating point on each part of the double-diode $i$–$v$ characteristic). Assume that $V_o = 0$ V.

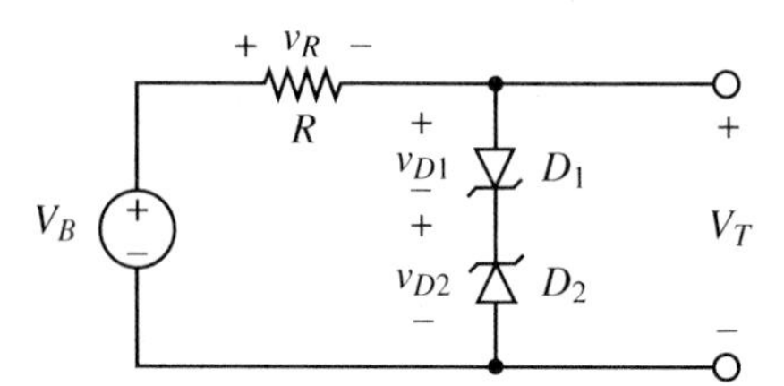


**Figure P7.28**

**7.29.** For the Zener regulator in Fig. 7.17, determine the minimum voltage across and current in $R_A$ and still have regulation occurring (the diode Zener biased). Answer for $R_L$ absent and for $R_L$ present.

**7.30.** For the circuit in Fig. P7.30, sketch $v_o(t)$ given $v_s(t)$ as shown. Assume that the contact potentials of both diodes are 0.7 V. The time axis (abscissa) can be qualitative; the $v_s$ axis (ordinate) must be quantitative.

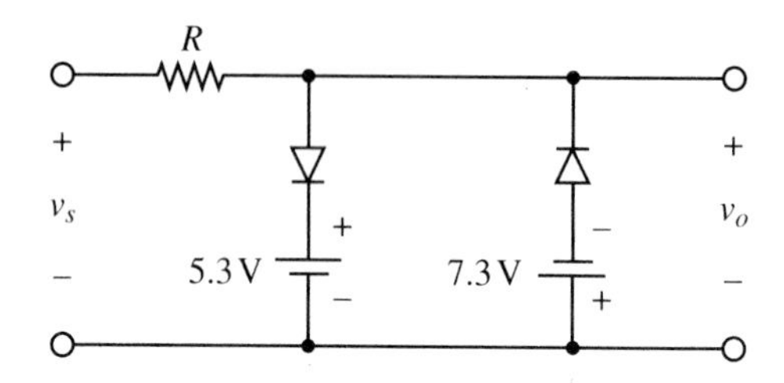


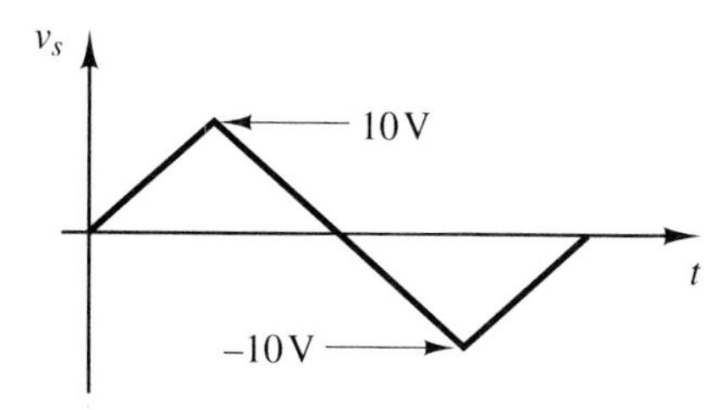


**Figure P7.30**

**7.31.** Suppose that in Fig. P7.28 the voltage source $V_B$ is replaced by $v_s(t)$, a 30-V peak voltage ac voltage source, where $V_Z = 10$ V. Sketch the waveforms of the voltage across each diode and across the resistor. Neglect the 0.7-V drop of forward-biased diodes. (*Hint:* Examine the resistor voltage last.)

**7.32.** A four-input maximum-value diode circuit was shown in Fig. 7.18. Suppose that $v_1(t) = 2\sin(t)$ V, $v_2(t) = 4\sin(2t)$ V, and $v_3(t) = v_4(t) = 0$ V. Furthermore, let $R = 1\ \text{k}\Omega$. Find and plot the current in $R$, $i_R(t)$. Let your time window be large enough to show at least one entire cycle of $v_1(t)$.

**7.33.** **(a)** Suppose that in Fig. P7.33a, $v_1$, $v_2$, and $v_3$ can take on only the values 5 V or 0 V; they are binary signals. This will result in $v_o$ also being only either 5 V or 0 V. Under what conditions is $v_o$ "high" (5 V), and when is it "low" (0 V)? What logical function is performed? Assume that the contact potential of the diodes is 0.7 V.

**(b)** Repeat part (a) for the circuit in Fig. P7:33b. We will study these logical operations in much more detail in Chapters 10 through 12.

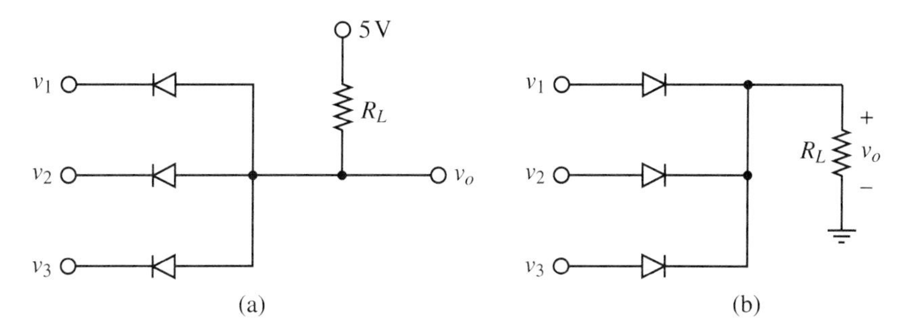


**Figure P7.33**

**7.34.** (**a**) Consider again circuits (a) and (b) in Fig. P7.33. Let $V_o = 0.7$ V, and suppose that again the input voltages $v_1$ and $v_2$ are binary signals; that is, they can take on only one of two values $V_{\text{low}}$ ("low") and $V_{\text{high}}$ ("high"). In Problem 7.33 we assigned these two values: $V_{\text{low}} = 0$ V and $V_{\text{high}} = 5$ V and the power supply voltage $V_B$ we set to $V_B = 5$ V. For what range of $V_{\text{high}}$ will circuit (a) function as it did for $V_{\text{high}} = 5$ V (with the power supply fixed at 5 V)?

(**b**) Now consider the circuit in Fig. P7.34, in which at the first "level" of circuitry we restrict there to be only two inputs $v_1$ and $v_2$. Suppose that $v_{o,a}$ and $v_3$ are given as input to a circuit identical to circuit (a), where $v_3$ is a third binary signal (0 V = "low", $V_{\text{high}}$ = "high"). Again, for what range of values of $V_{\text{high}}$ will the circuit function the same as it would for $V_{\text{high}} = 5$ V (again, $V_B$ is fixed at 5 V)? Does the same restriction on $V_{\text{high}}$ apply for $v_3$ as applies for $v_1$ and $v_2$? Can you think of the implications of this result for more complicated "digital" circuits?

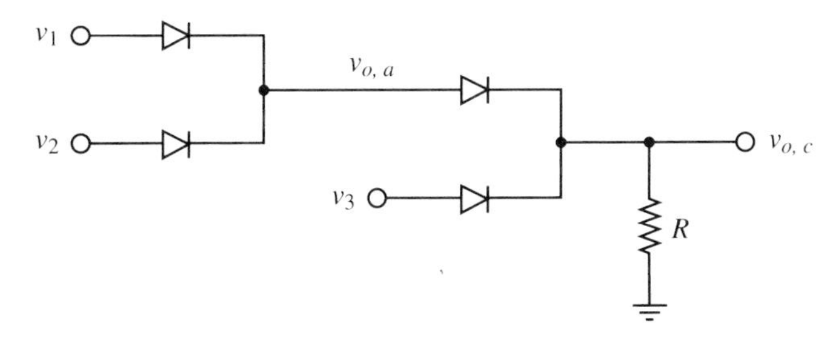


**Figure P7.34**

**7.35.** For the circuit in Fig. P7.35, determine $v_2$ vs. $v_1$ and graph the input–output characteristic ($v_2$ vs. $v_s$) for the conditions $V_o = 0.7$ V, $V_Z = \infty$. Fully label all critical points and slope(s).

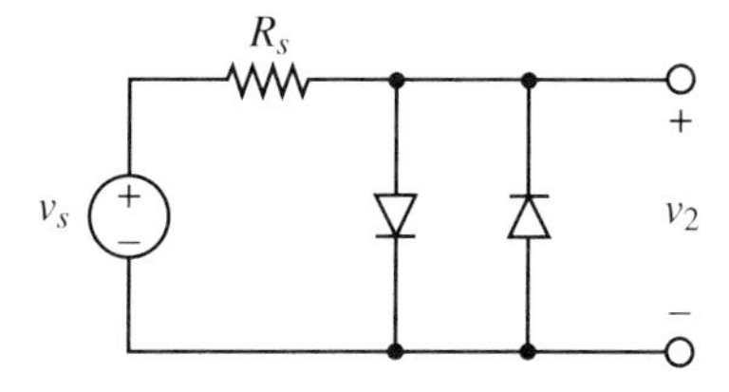


**Figure P7.35**

**7.36.** (**a**) In Fig. P7.36, assume that the contact potentials of both diodes are 0 V. Sketch $v_o$ vs. $v_s$ and fully label your plot quantitatively. Fully indicate your reasoning. Next suppose that $v_s(t) = 5 \sin(\omega t)$. Now sketch $v_s(t)$ vs. $t$, and fully label the vertical axis.

(**b**) Repeat part (a) for the case where both diodes have a contact potential of 0.7 V.

(c) Show changes, if any, to the $v_o$ vs. $v_s$ graph and the waveform $v_o(t)$ if $D_1$ is a Zener diode with $V_Z = 7$ V ($D_2$ is not); still assume that $V_o = 0.7$ V for both diodes.

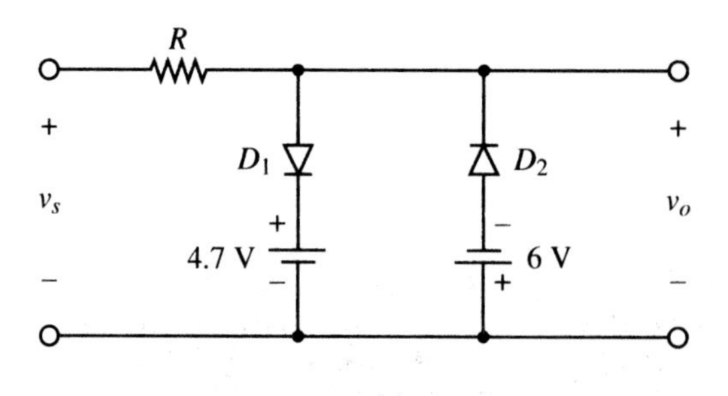


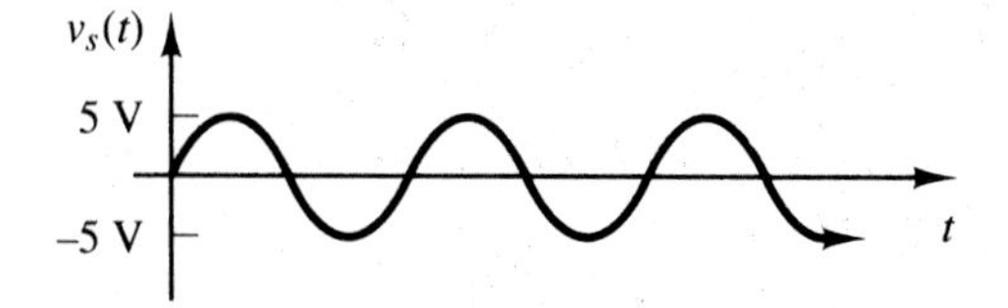


**Figure P7.36**

**7.37.** Draw and fully label the input–output characteristic ($v_o$ on the vertical axis, $v_s$ on the horizontal axis) for the circuit in Fig. P7.37 under the following conditions, where $V_o$ is the contact potential of both diodes (*not* the output voltage $v_o$!):

**(a)** $V_o = 0$ V, $V_Z = \infty$.

**(b)** $V_o = 0.7$ V, $V_Z = \infty$.

**(c)** $V_o = 0.7$ V, and $D_2$ is a Zener diode with Zener voltage $V_Z = 10$ V ($D_1$ is not).

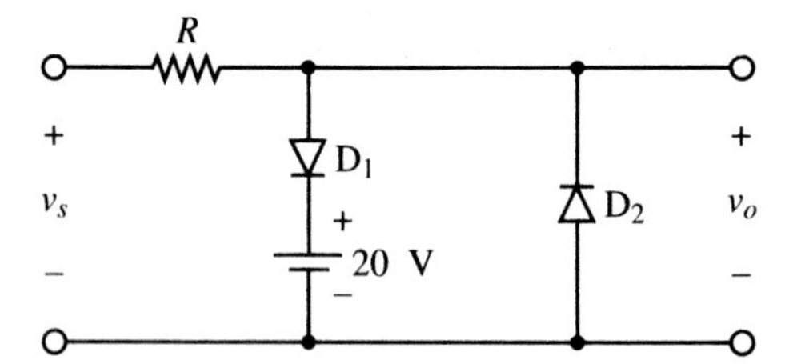


**Figure P7.37**

**7.38.** Figure P7.38a shows a circuit that can be used as a voltage doubler. Assuming ideal diodes, for the given waveform of $v_s(t)$ in Fig. P7.38b, sketch the output voltage $v_o(t)$. If $V_m = 2$ V, what is $v_o(t)$ after transient behavior has ceased? [Assume that $v_{C1}(t) = v_{C2}(t) = 0$ for $t < 0$.]

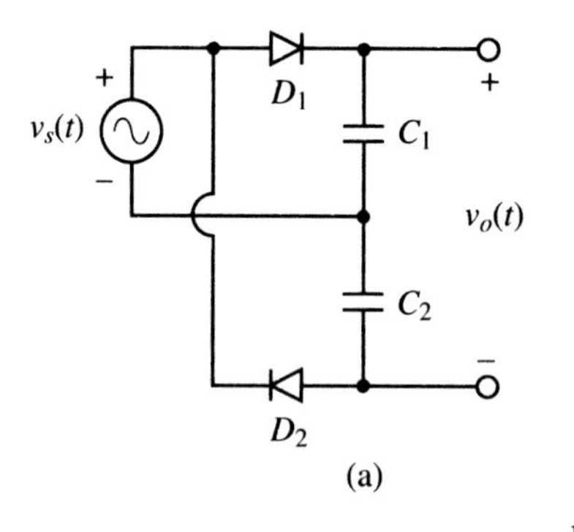


(a)

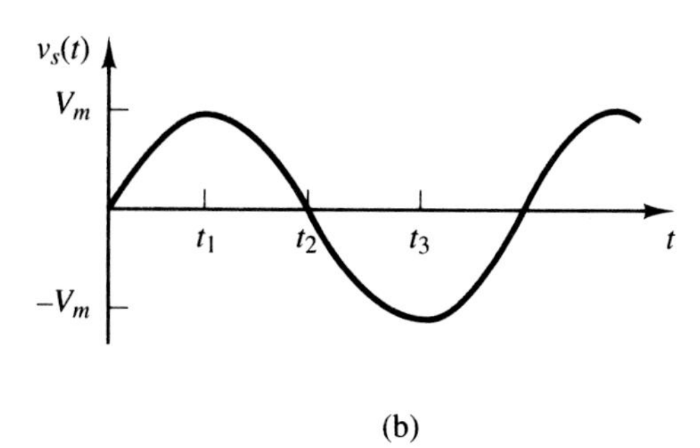


(b)

**Figure P7.38**

**7.39.** Suppose that in the circuit in Fig. 7.5, we consider $v_D(t)$ to be the output voltage, and suppose that the excitation is $v_s(t) = 10 \cos(\omega t)$ V. Plot $v_D(t)$; show all analysis, indicating all the conditions on $v_s$ and resulting expressions (you may draw equivalent circuits if they help you). Additionally, for each case plot $v_D$ vs. $v_s$. Fully label your plots, including any relevant numerical values on the axes. The models to be considered and their $i$–$v$ characteristics are:

(**a**) Ideal diode ($V_o = 0$ V, $V_Z = \infty$).

(**b**) 0.7-V diode ($V_o = 0.7$ V, $V_Z = \infty$).

(**c**) Zener diode ($V_Z = 5$ V, $V_o = 0$ V).

**7.40.** The device shown in the box in Fig. P7.40 is called an optocoupler. If the LED conducts, it shines a light that "turns on" the phototransistor, and the load $R_L$ is supplied by $V_{CC}$. Describe the operation of this circuit connected as shown, and its possible advantages. The switch on the left could be, for example, a manual switch or even another transistor switch. Note that the potential applied to $B$ can be reduced ("reverse biased") to reduce sensitivity, "forward biased" to increase sensitivity, or left disconnected ("floating") if desired. The conductivity in the semiconductor comes from the LED light, not primarily the base wire potential, as is usually the case.

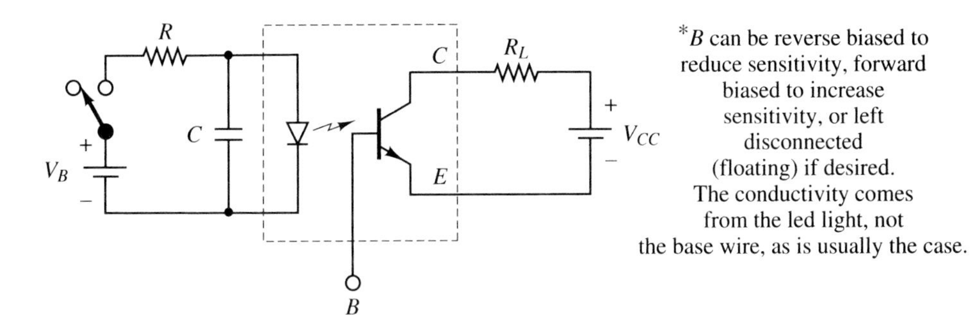
R
C
+
$V_B$
−
C
E
B
$R_L$
+
$V_{CC}$
−
*B can be reverse biased to reduce sensitivity, forward biased to increase sensitivity, or left disconnected (floating) if desired. The conductivity comes from the led light, not the base wire, as is usually the case.

**Figure P7.40**

**7.41.** We determine experimentally that a pocket LCD calculator draws 30 μA except when performing a calculation, in which case it draws 0.3 mA when 3 V is applied. Suppose that we have a box of solar cells, each with the same $i$–$v$ characteristic as that shown in Fig. P7.41, when the lighting is at the usual room level.

(**a**) Draw a circuit using sufficient solar cells to generate the minimum required voltage of 3 V. Assume for simplicity that under each condition the calculator acts as a resistance. Calculate the effective resistance under the two conditions (no calculation vs. calculation).

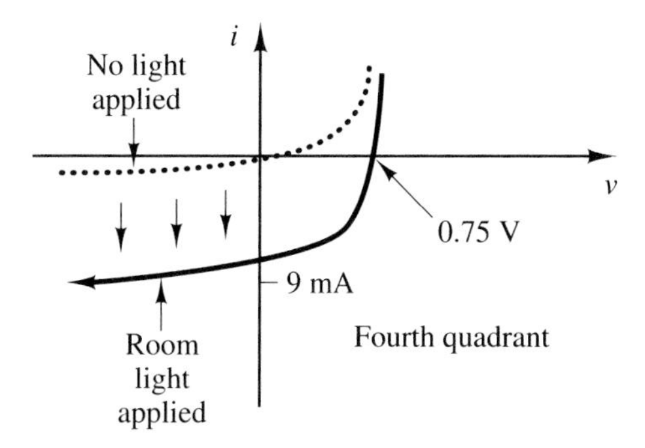
i
No light applied
v
0.75 V
−9 mA
Fourth quadrant
Room light applied

**Figure P7.41**

(b) Graphically show the appropriate load line/operating point graphs relevant for the two situations. Be careful what you use for the source characteristic!

**7.42.** For a BJT, what are the three regions and terminals called? How many *pn* junctions exist; name the regions each separates, and whether each is forward or reverse biased under normal linear transistor operations.

**7.43.** Suppose that the terminal currents of a BJT are defined as follows: $i_B$ is the current into the base, $i_E$ is the current out of the emitter, and $i_C$ is the current into the collector. Assume that the BJT is properly biased.

(a) Using KCL, write down the relations among $i_C$, $i_E$, and $i_B$.

(b) Write down the relations between $i_C$ and $i_B$, between $i_E$ and $i_B$, and finally between $i_E$ and $i_C$.

**7.44.** Figure 7.25 shows the voltage parameter definitions of an *npn* transistor. For the following conditions, determine the region on the collector current characteristic shown in Fig. 7.26(c) in which the transistor operating point is located. Assume that $V_o = 0.7$ V.

(a) $v_{BE} \geq 0.7$ V and $v_{BC} > 0$.

(b) $v_{BE} \leq 0.7$ V.

(c) $v_{BE} \geq 0.7$ V and $v_{BC} < 0$.

(d) $i_C = \beta i_B$.

(e) $v_{CE} = 4$ V and $i_B = 0$.

**7.45.** There are two transistor models used when analyzing transistor amplifiers. Name them, and indicate under what conditions each model is valid. Specifically, for example, how must the two transistor junctions be biased? For linear amplifiers, of what use is the dc/large-signal model given that it does not give the signal voltage gain directly?

**7.46.** What is a FET, and what are its major advantages over the BJT?

**7.47.** (a) How many types of charge carriers exist in a FET?

(b) How many types of charge carrier contribute significantly to the main currents in a FET? What are the charge carriers for a *p*-channel FET and for an *n*-channel FET?

**7.48.** Briefly and qualitatively compare simple FET and BJT small-signal models.

**7.49.** Suppose that we took two diodes and soldered them together as shown in Fig. P7.49. Would the resulting device function as a diode? Why or why not?

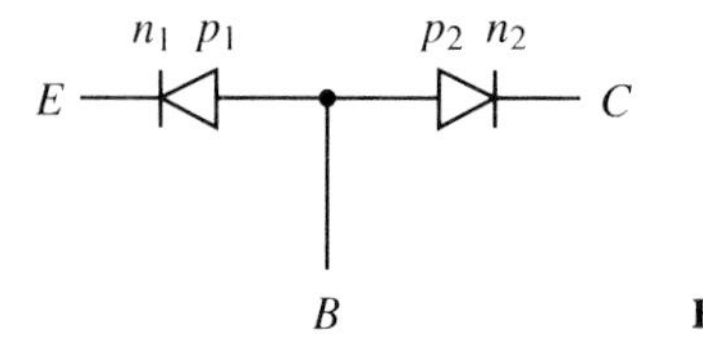


**Figure P7.49**

**7.50.** In the circuit in Fig. P7.50, let $R_E = R_C = 2.5$ kΩ and $R_s = 200$ Ω. Redraw the circuit diagram with the transistor replaced by its large-signal model, assuming a contact potential of $V_o = 0.7$ V. Solve for $V_B$, $V_E$, $I_E$, and $V_C$. Is $i_E > 0$? Is $v_B \leq v_C$? Is the transistor in the linear operating region? Assume that $\beta = 80$.

**7.51.** The voltage across the load resistor $R_D$ in Fig. P7.51a is found to be 6 V at the quiescent operating point. The average drain characteristics are as shown in Fig. P7.51b.

(a) Find the quiescent drain–source current.

(b) Compute the value of the load resistance, $R_D$.

**7.52.** In Sec. 7.6.2 the operating point (dc bias) and amplifying behavior (small-signal

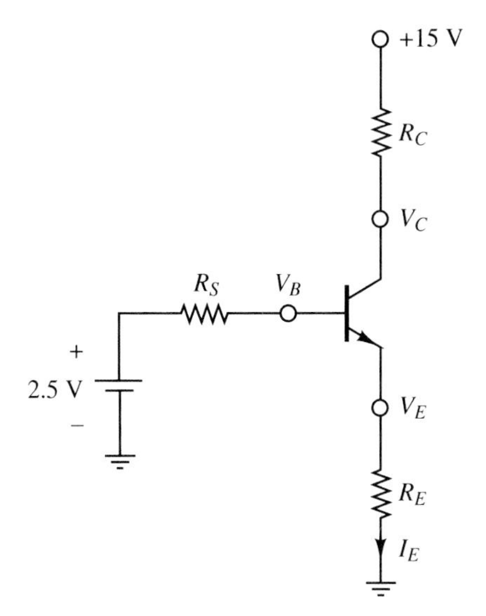


**Figure P7.50**

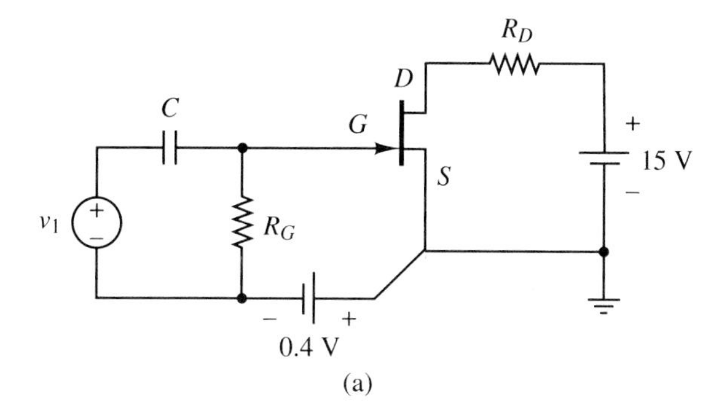


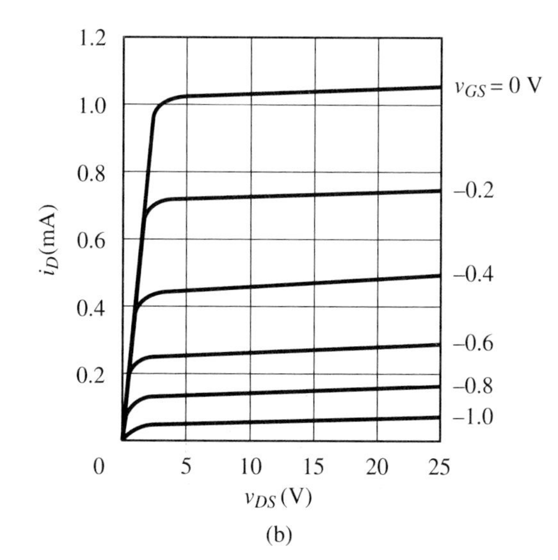


**Figure P7.51**

behavior) are analyzed separately. What is the theoretical basis for doing this (i.e., the assumption responsible)?

**7.53.** For the amplifier in Fig. P7.53, calculate the operating point ($I_C$ and $V_{CE}$) given that $R_1 = 700$ kΩ, $R_2 = 0.5$ kΩ, $R_3 = 5$ kΩ, $\beta = 80$, and $V_{CC} = 10$ V.

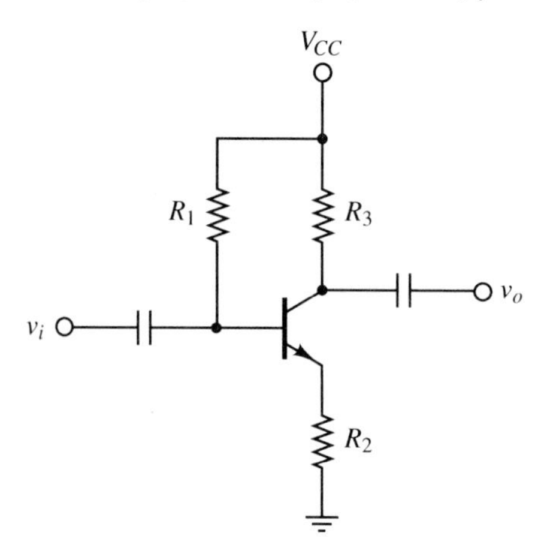


**Figure P7.53**

**7.54.** For the circuit in Fig. P7.54, dc analysis shows that at room temperature, $I_E = 2.6$ mA. Find $r_\pi$ for the small-signal model (assume that $\beta = 30$). Do not assume that $I_C = I_E$. Draw the small-signal circuit model for the entire circuit. What is $v_E/v_1$? What is $v_C/v_1$?

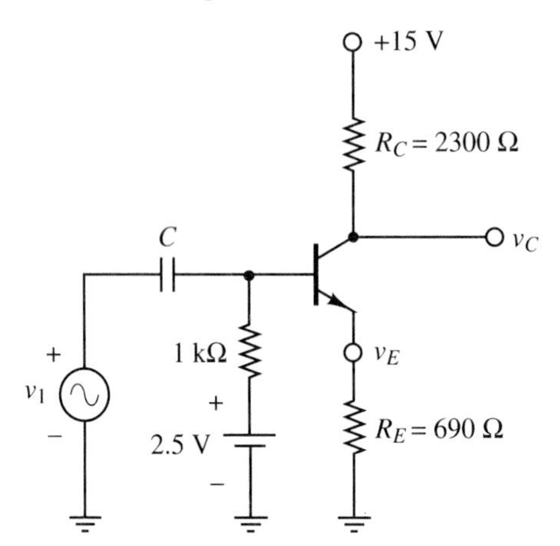


**Figure P7.54**

**7.55.** **(a)** Using the characteristic plot shown in Fig. P7.55a, determine the value of $\beta$ for $V_{CE} = 4$ V.

**(b)** For the circuit in Fig. P7.55b, draw the large-signal (dc) circuit. Use $V_{BE} = 0.7$ V.

**(c)** Find the operating currents $I_{BQ}$ and $I_{CQ}$.

**(d)** Draw the small-signal (incremental) circuit. Label your diagram completely. What is the value of $r_\pi$?

**(e)** Find the small-signal voltage gain $v_o/v_s$.

**(f)** Find the approximate incremental input and output impedances. (Do not include the source resistance $R_s$ in the input impedance calculation!) For this calculation, assume that all resistances are of the same order of magnitude.

**(g)** Suppose that $\beta = 80$, $V_{CC} = 15$ V, and $R_1 = 200$ kΩ. For linear operation, $v_{CB} > 0$. What range of values of $R_c$ will guarantee this?

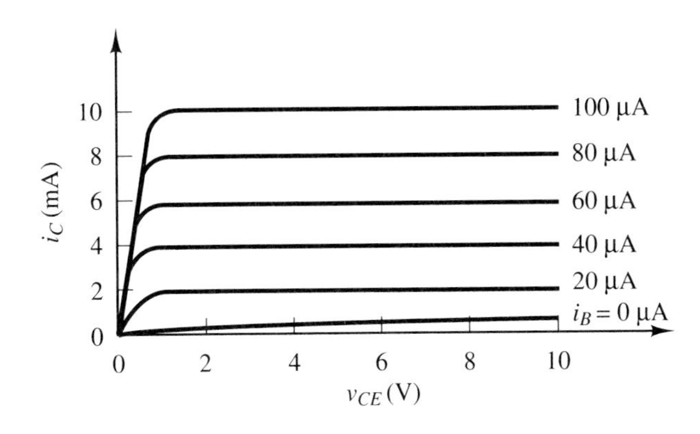


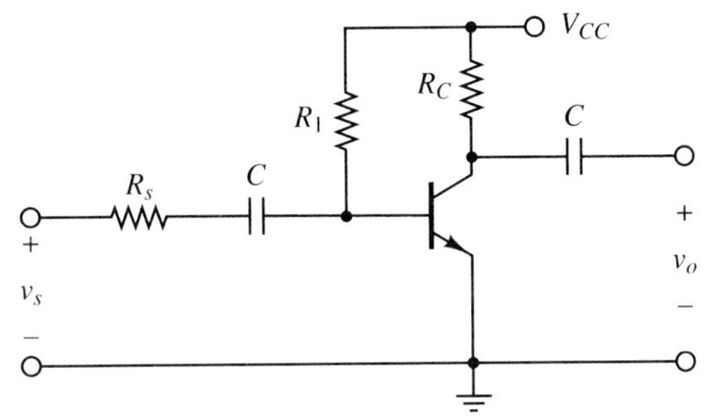


Figure P7.55

**7.56.** The base bias voltage of the circuit in Fig. P7.56 has been supplied by a voltage divider. Redraw this circuit at dc (bias circuit), replacing the voltage divider with its Thévenin equivalent. The capacitor bypasses all ac current. Find the (dc) base bias voltage $V_B$ for $\beta = 100$ and an average value of the current source $i_1$ of $I_1 = 4$ mA. What is the average voltage $V_E$ across the current source? Assume a base–emitter contact potential of 0.7 V.

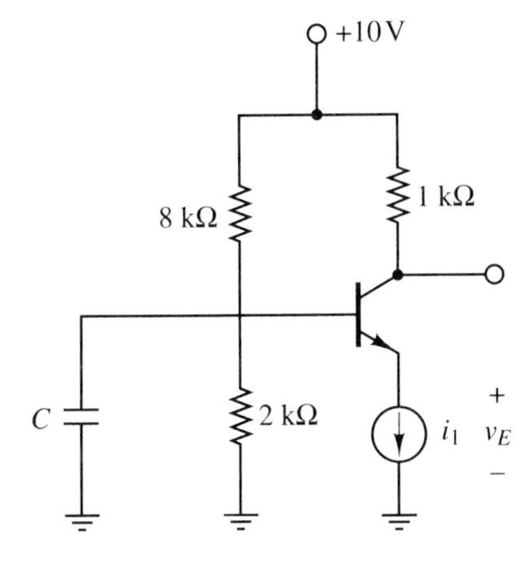


Figure P7.56

**7.57.** For the circuit in Fig. P7.57 ($v_{BE} = 0.7$ V, $\beta = 90$):

**(a)** Draw the dc bias circuit and solve for the base and collector dc currents and the dc voltage from collector to emitter.

**(b)** Draw the small-signal equivalent circuit and solve for the voltage gain of the

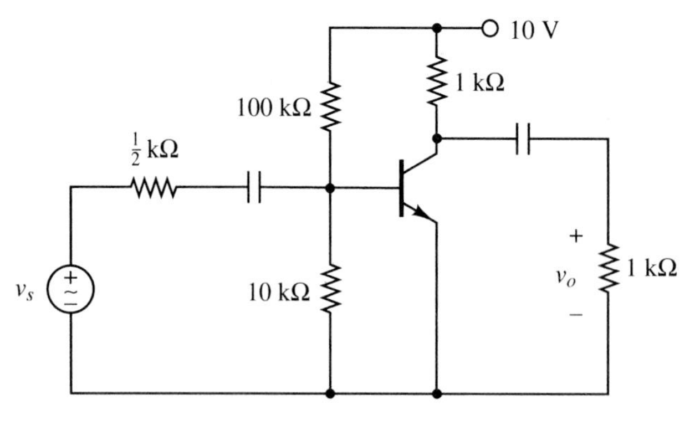


**Figure P7.57**

amplifier, $v_o/v_s$. Consider, as usual, the capacitors as short circuits at the signal frequency.

**(c)** Find the incremental input impedance of the amplifier as seen by the input generator. This does not include the $\frac{1}{2}$-kΩ source resistance.

**7.58.** For the BJT amplifier circuit in Fig. P7.58, assume the following parameter values: BE diode contact potential = 0 V (ideal *pn* junction), $\beta = 100$, $kT/q = 26$ mV. Assume that for dc analysis, capacitors are open circuits, and for ac analysis, capacitors are short circuits.

**(a)** Draw the large-signal (dc) model of this circuit.

**(b)** Using the diagram you drew in part (a), find $I_{BQ}$ and $I_{CQ}$. (*Hint:* To find $I_{BQ}$, use node voltage $V_1$ for a node equation and then use Ohm's law on another branch to eliminate $V_1$. Assume that $\beta + 1 \approx \beta$.)

**(c)** Draw the circuit for ac analysis, substituting the BJT small-signal model for the transistor circuit. What is the value of $r_\pi$? Label your diagram completely!

**(d)** Find the voltage gain $v_o/v_s$ and the output impedance (as seen looking into the $v_o$ terminals). Why might this circuit be useful?

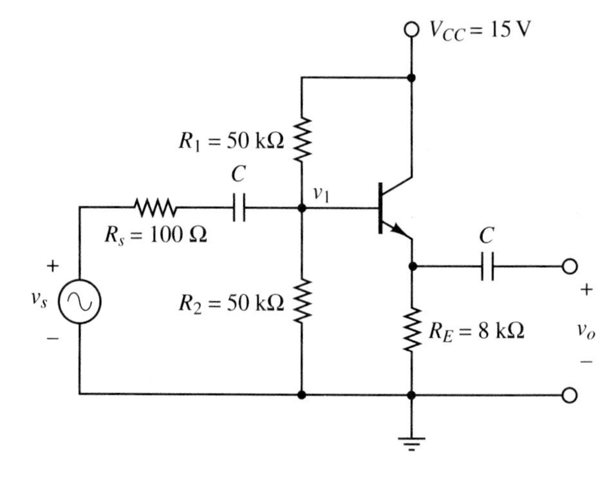


**Figure P7.58**

**7.59.** Figure P7.59 shows a common-collector ("emitter follower") BJT transistor amplifier.

**(a)** Draw the large-signal circuit model.

**(b)** Find the operating point currents $I_B$ and $I_C$.

**(c)** Draw the small-signal circuit model.

**(d)** Considering $R_L$ to be the driven load, find the incremental gains $A_v = v_o/v_i$ and $A_i = i_o/i_i$, as well as the signal power gain $A_p = A_v A_i$.

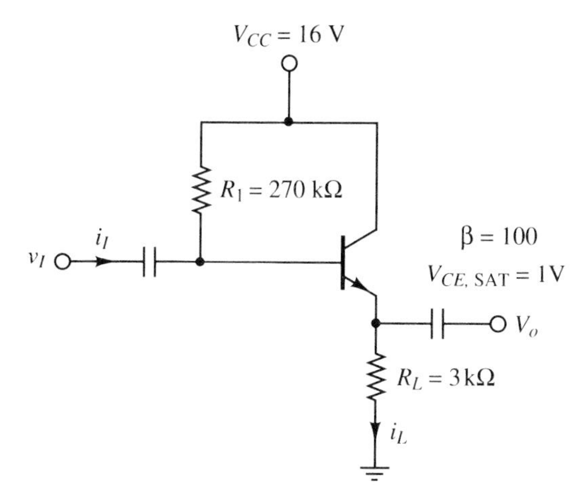


**Figure P7.59**

(**e**) Find the incremental input and output impedances.
(**f**) Characterize this amplifier qualitatively.

**7.60.** The circuit in Fig. P7.60 has a bias configuration known as universal bias. Calculate $I_C$ and $I_B$. Show that the collector current operating point $I_C$ is relatively independent of β, thus making this biasing arrangement very popular in audio-amplifier designs (see, e.g., the AM radio discussed in Sec. 7.7.3).

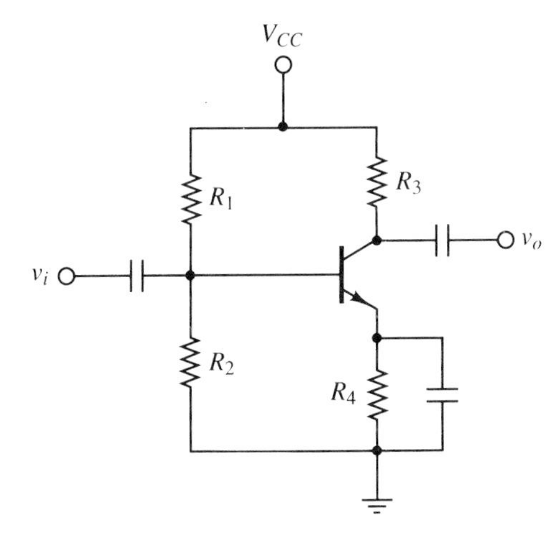


**Figure P7.60**

**7.61.** In order to increase β, the so-called "Darlington pair" is commonly used. In the Darlington pair shown in Fig. P7.61, $Q_1$ and $Q_2$ have the same value of β.
(**a**) Find the relations among the terminal currents $i_B$, $i_C$, and $i_E$ in Fig. P7.61.
(**b**) Find the overall $\beta_D$ for the Darlington pair. First calculate it exactly, and then give a simpler approximate value, assuming a fairly large value of β.

**7.62.** The circuit in Fig. P7.62 is called a common-base amplifier. For numerical answers, let $\beta = 70$, $V_{CE,\text{SAT}} = 1$ V, $R_1 = 5$ kΩ, $R_L = 5$ kΩ, $V_{CC} = 10$ V, and $V_{EE} = -4$ V.
(**a**) Draw the large-signal model of this circuit.
(**b**) Find the operating point currents $I_B$, $I_C$, $I_E$, and $I_L$.
(**c**) Draw the small-signal model, and determine $r_\pi$.
(**d**) Find the small-signal current gain $i_l/i_i$ and voltage gain $v_o/v_i$.
(**e**) Based on your results in part (d), briefly characterize this type of amplifier.

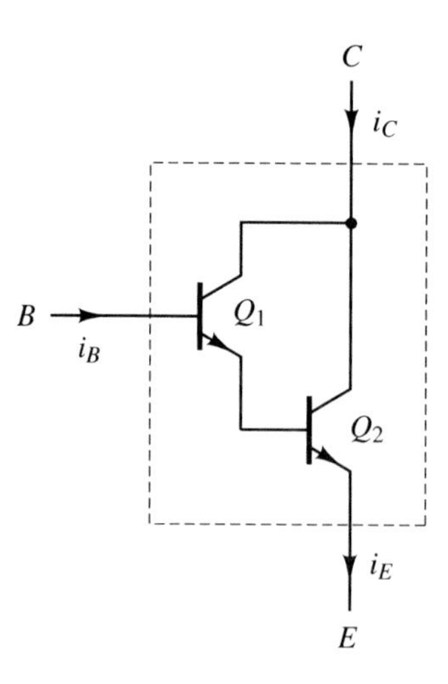


**Figure P7.61**

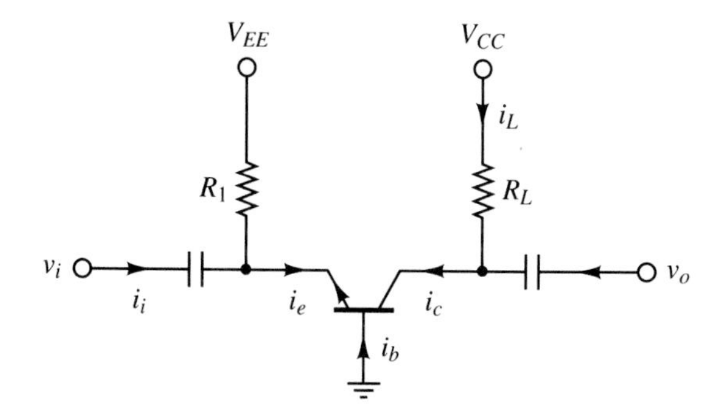


**Figure P7.62**

**(f)** Find the incremental input and output impedances.

**(g)** For what range of $R_L$ will this circuit have a voltage gain of at least 10, yet still operate within the linear region of the transistor?

**7.63.** For the JFET amplifier circuit in Fig. P7.63, assume the following parameters: $I_{DSS} = 10$ mA, $V_P = 5$ V, $V_{DD} = 20$ V. Recall that for dc analysis, capacitors are open circuits, while for ac analysis, capacitors are short circuits.

**(a)** Draw the circuit for dc analysis, leaving the transistor symbol intact as in Fig. P7.62.

**(b)** Using the diagram you drew in part (a) and the $i_D$ vs. $v_{GS}$ curve, find $I_{DQ}$ and $V_{GSQ}$.

**(c)** Using the $i_D$ vs. $v_{GS}$ and $v_{GS}$ curves and the answers to part (b), find $V_{DSQ}$.

**(d)** Draw the circuit for ac analysis, substituting our JFET small-signal model (but simplify by taking $r_d = \infty$) for the transistor symbol. Label your diagram completely.

**(e)** Defining the incremental output voltage as $v_o$, find the voltage gain $v_o/v_s$. [*Hint:* For $g_m$, use the formula $g_m = 2(I_{DSS}I_{DQ})^{1/2}/V_P$.]

**7.64.** For the circuit in Fig. P7.64:

**(a)** Draw the complete incremental circuit.

**(b)** Find $v_s/v_p$ first using symbols and then using the given values $R_P = 1$ MΩ, $R_G = 1$ MΩ, $R_S = 1$ kΩ, and $g_m = 2 \cdot 10^{-3}\ \Omega^{-1}$, $V_{DD} = 20$ V.

**7.65.** Show that the drain current operating point of a JFET [as given in Eq. (7.42c)] can be expressed in terms of only $V_{DD}$, $V_P$, $R_1$, $R_2$, $R_F$, and $I_{DSS}$ as follows:

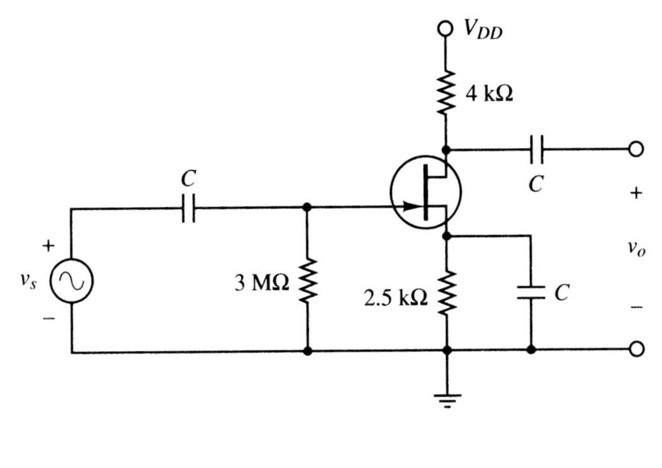


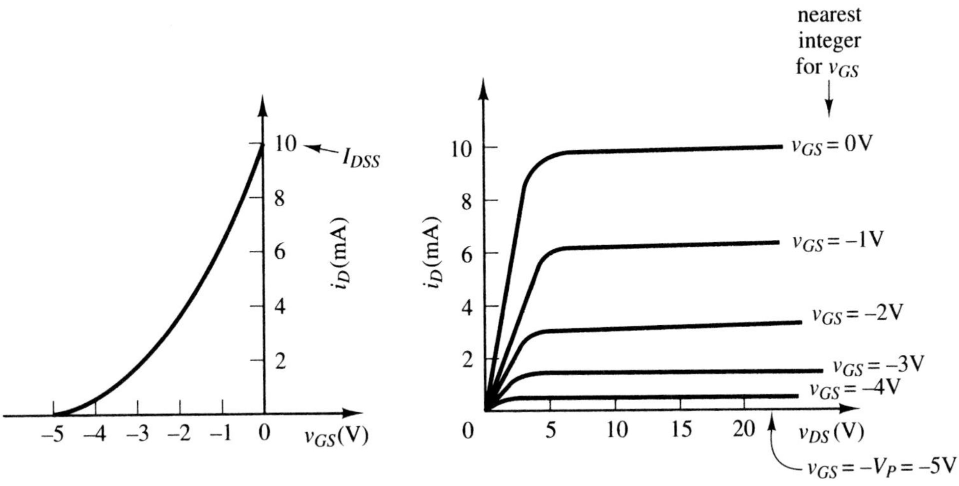


**Figure P7.63**

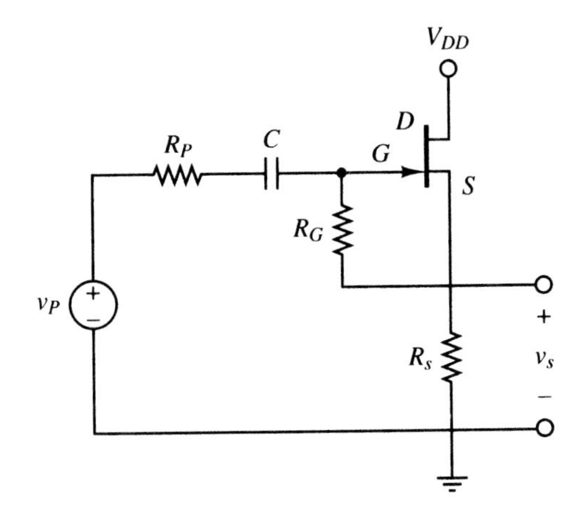


**Figure P7.64**

$$I_D = \frac{V_P}{R_F}\left[1 + \frac{R_2 V_{DD}}{(R_1 + R_2) V_P} + \frac{V_P}{2 I_{DSS} R_F}\right.$$

$$\left. \cdot \left(1 - \left\{1 + 4\left[1 + \frac{R_2 V_{DD}}{(R_1 + R_2) V_P}\right]\frac{I_{DSS} R_F}{V_P}\right\}^{1/2}\right)\right]. \quad \text{(P7.1)}$$

**7.66.** **(a)** Draw the small-signal equivalent circuit of the amplifier shown in Fig. P7.66, and find the gain $v_o/v_i$ if $I_D = 1$ mA. Assume that $I_{DSS} = 3$ mA, $V_P = -2$ V, and that the capacitors are short circuits at operating frequencies.

**(b)** What is the input impedance?

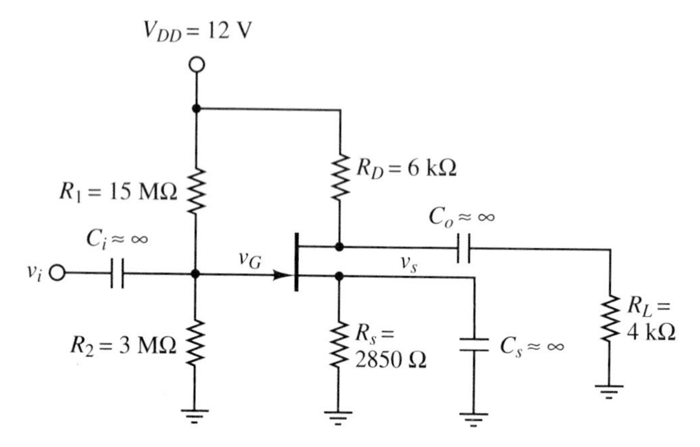


**Figure P7.66**

**7.67.** For the simple transistor switch in Fig. P7.67:

**(a)** Determine the state of the transistor for the cases in which the mechanical switch is off and when it is on.

**(b)** When on, calculate the current that drives the load $R_L$. Assume that $\beta = 100$ and $V_{CE,\text{SAT}} = 0.3$ V.

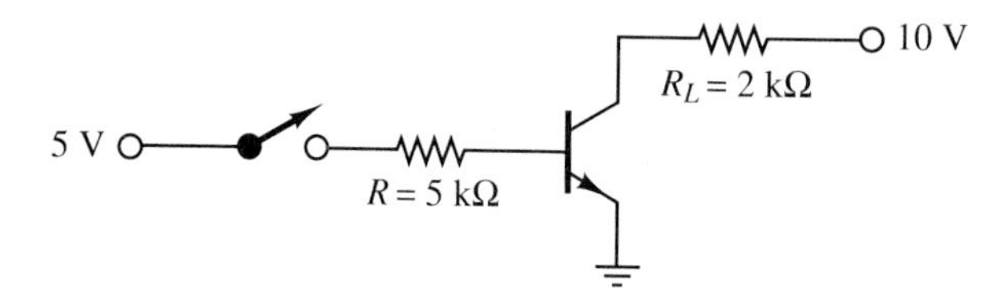


**Figure P7.67**

**7.68.** The circuit in Fig. P7.68 was constructed to convert a 12-V dc power supply to a lower voltage $V_o$, which is independent of the load current $I_o$ drawn by the effective load resistance $R_L$. Is the Zener diode Zener biased? What is $V_o$? What value of $R_L$ will draw a current of $I_o = 100$ mA? If the load this current $I_o = 100$ mA, what is the power dissipated by the transistor when $R_C = 0$, and when $R_C = 66\ \Omega$? From these results, can you determine the function of $R_C$? What is the maximum allowable value of $R_C$? What are the functions of $R_E$ and $C_E$?

**7.69.** Show graphically an example of the kind of distortion that would result were not a suitable dc bias added to the AM radio audio signal before being modulated by the carrier (and later sent to the transmitting antenna for broadcasting).

**7.70.** In Sec. 7.7.3 we studied briefly a peak detector. For the peak detector in Fig. P7.70a, sketch the output voltage $v_L(t)$ if $v_i(t)$ is as given in Fig. P7.70b under the conditions **(a)** $R_L = \infty$, and **(b)** $R_L C \approx 1$ s.

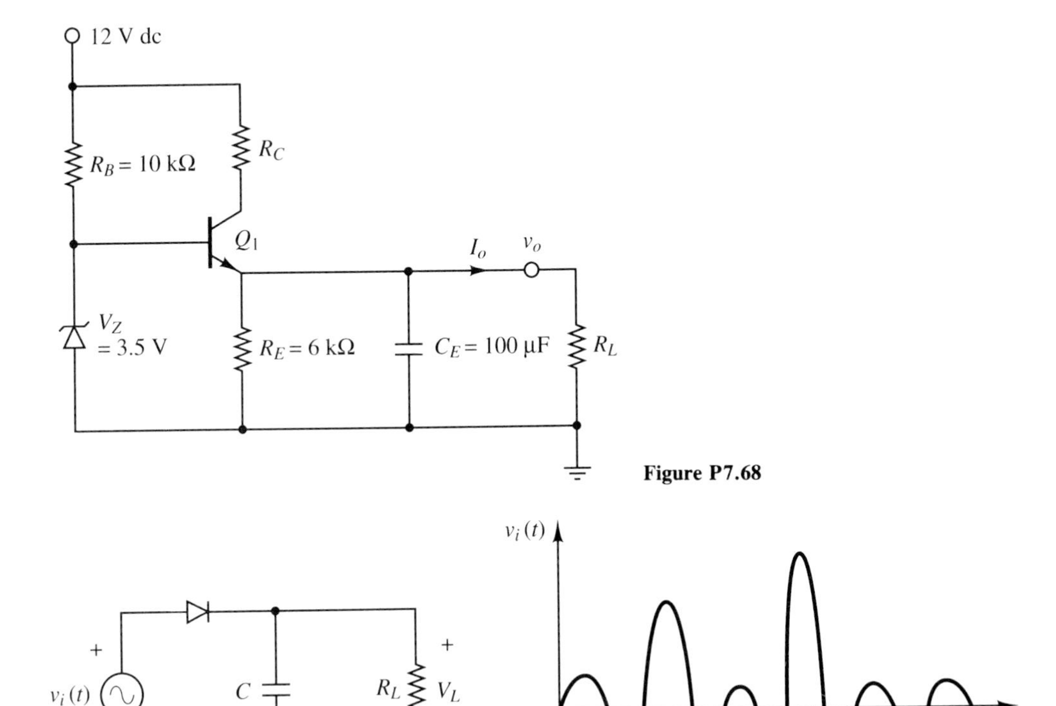


**Figure P7.68**



**Figure P7.70**

## ADVANCED PROBLEMS

**7.71.** A forward-biased diode at room temperature (see Fig. P7.71a) is modeled by the equivalent circuit shown in Fig. P7.71b.

**(a)** Plot the $i_D$–$v_D$ characteristic of the diode using the exponential model and parameter values given in Fig. P7.71a. Plot only up to $i_D = 100$ mA. Make a best

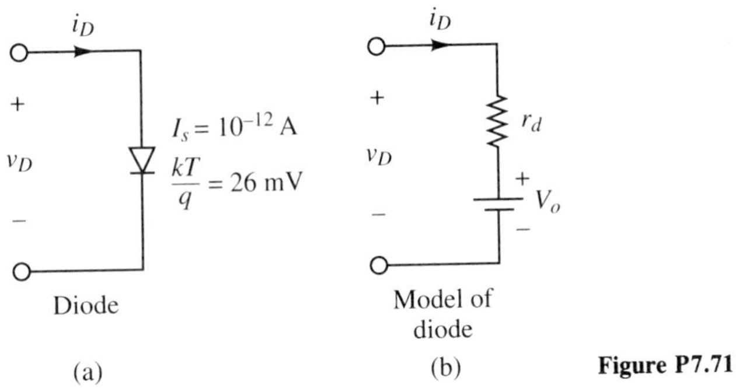


**Figure P7.71**

straight-line fit for strong forward bias, over the range 10 mA $< i_D <$ 100 mA. From this, determine $r_d$ and $V_o$. [*Hint:* It may be convenient to solve the exponential diode equation for $i_D$: $i_D \approx (kT/q)\ln(i_D/I_S)$.]

**(b)** Repeat part (a), making the best fit for 40 mA $< i_D <$ 240 mA. Over which of these two ranges is the linear model much closer to the much more accurate exponential model?

**7.72.** For the circuit in Fig. P7.72, find $v_o(t)$. Assume that the diode has a contact potential of 0 V, and that $R_L C \gg 2\pi/\omega_0$. Assume initially that the capacitor $C$ is uncharged and that $v_i(t \leq 0) = 0$; then at $t = 0$ turn on $v_i(t)$. Also assume that $V_m < V_A < 2V_m$, and that $V_A$ is turned on at a time $t_0$ well in the past: $t_0 \ll -R_L C$, with $v_C(t_0) = 0$.

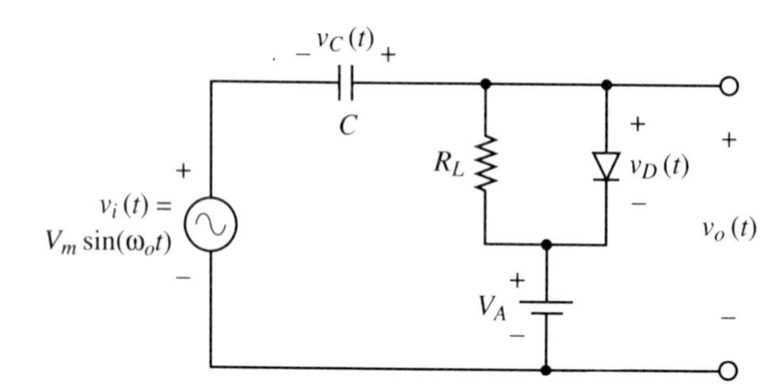


Figure P7.72

**7.73.** In this and the following two problems concerning the audio amplifier in Fig. P7.73, assume that for small-signal analysis (audio frequencies) all capacitors are short circuits and that $\beta$ for both transistors is 100. Find the bias (operating point) values of $V_B$, $I_E$, $I_C$, and $V_C$. Verify that the values of $r_{\pi 1}$ for $Q_1$ and $r_{\pi 2}$ for $Q_2$ shown in the small-signal model are correct. Neglect the current $I_{B2}$ drawn by the emitter follower when calculating the operating point of $Q_1$, but then calculate $I_{B2}$; is it, in fact, negligible compared with other bias currents?

**7.74.** For the audio amplifier in Fig. P7.73:

**(a)** What is the small signal input resistance $R_{\text{in}} = v_c/i_{b2}$ of the emitter follower stage? (*Hint:* Apply KCL to the emitter of $Q_2$; recall that the transformer reflects an impedance $6^2 = 36$ times larger than the 8-$\Omega$ speaker resistance.)

**(b)** What is the small signal gain $v_c/v_i$ of the common emitter stage? [*Hint:* Write KCL on the collector, and use $R_{\text{in}}$ from part (a) as a replacement for the emitter-follower stage.]

**7.75.** For the audio amplifier in Fig. P7.73:

**(a)** What is the emitter follower input resistance $R_{\text{in}} = v_c/i_{b2}$ if the output transformer is removed (and speaker kept)? What, then, will be the resulting gain ($v_e/v_i = v_c/v_i$)?

**(b)** What gain $v_c/v_i$ would this amplifier have without the emitter-follower stage (i.e., if the transformer were connected through $C_0$ to the collector of $Q_1$)?

**7.76.** The circuit in Fig. P7.76 is known as an exclusive or (XOR) gate, which is a logic operator to be described in Chapter 10. We are already in a position to analyze its behavior. Assuming that $\beta = 100$ and $V_{CE,\text{SAT}} = 0.2$ V for each of the three transistors, find $v_o$ when:

**(a)** $v_A = v_B = 5$ V.

**(b)** $v_A = v_B = 0$ V.

**(c)** $v_A = 5$ V and $v_B = 0$ V.

**(d)** $v_A = 0$ V and $v_B = 5$ V.

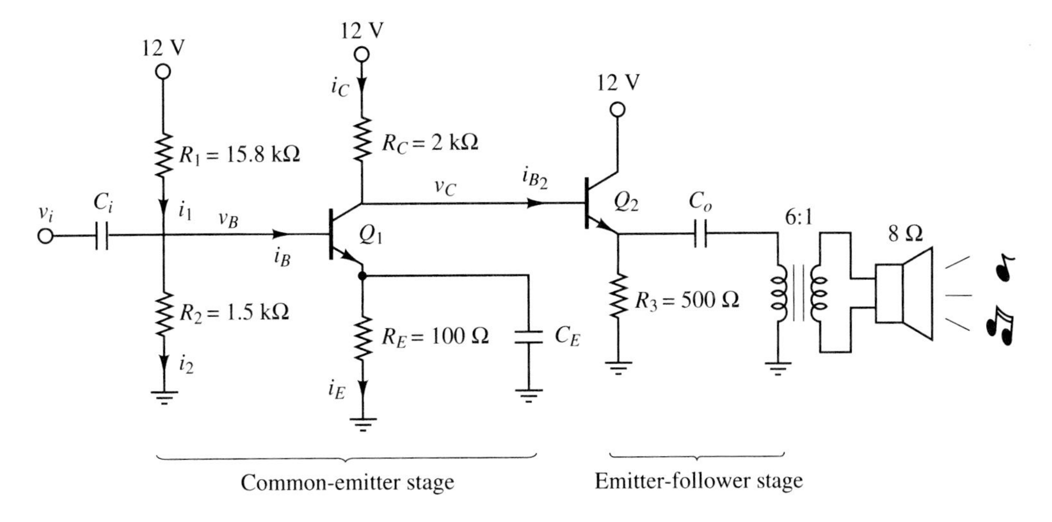


Small-signal model:

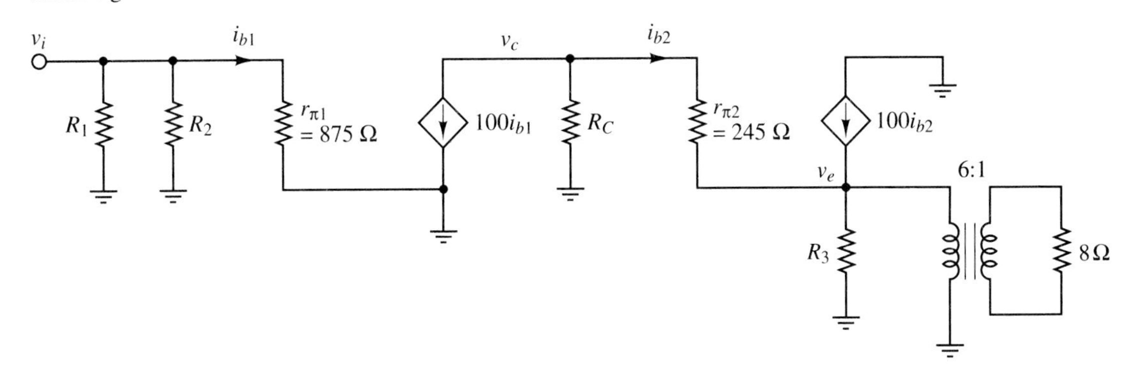


**Figure P7.73**

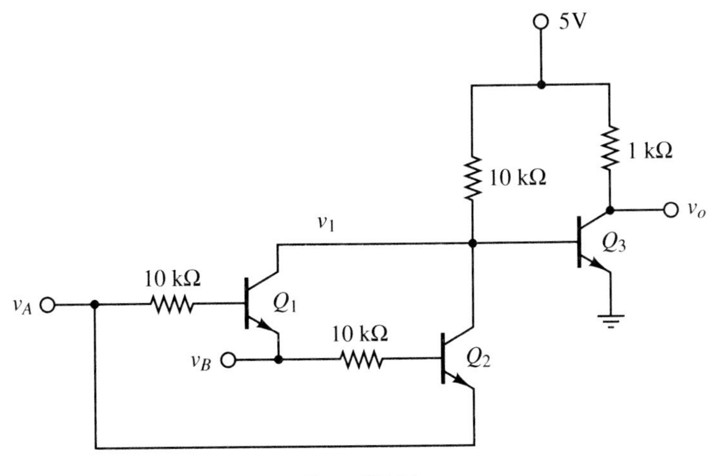


**Figure P7.76**

**7.77.** An automatic switch to turn on a floodlight at night uses a cadmium sulfide (CdS) cell (see Fig. P7.77). This transducer converts luminous intensity to resistance. The CdS cell is connected to a circuit involving two transistors connected in a configuration known as a Darlington pair (see Problem 7.61). The CdS cell has a daylight resistance of 100 Ω and a nighttime resistance of 100 kΩ.

**(a)** What is the daytime value of $v_1$; will the transistors be on or off?

**(b)** At night, what is the maximum current available to the load $i_L$? Assume that $\beta = 30$ for each transistor.

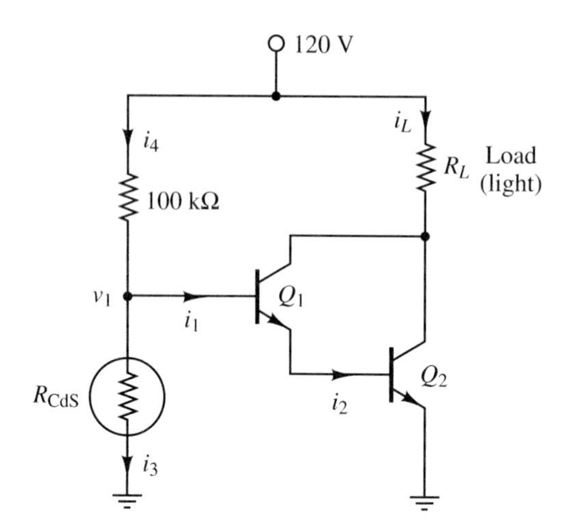


**Figure P7.77**

**7.78.** An alternative type of power supply to those that filter half-sinusoids is the so-called "switching power supply," which we will discuss in Sec. 10.7.5. A simplified schematic is shown in Fig. P7.78. This type of power supply uses a high-speed electronic (transistor) switch to chop a high voltage $V_1$ (about 150 V) so that its average value is $(t_{on}/t_{off})\, V_1$ (typically only 5 V), without significant $p = vi$ losses. These $p = vi$ losses occur in a transistor operating in the linear region, as is the case for half-sinusoidal

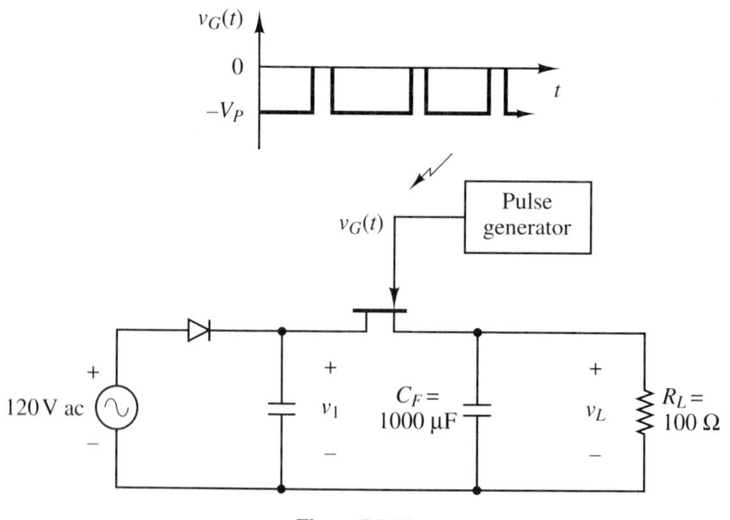


**Figure P7.78**

rectifiers, appropriately called linear power supplies, such as in Sec. 7.7.2. For the switching power supply, at any given instant either $v$ or $i$ of the MOSFET is nearly zero, making $p$ always low. The control circuit adjusts the value of $t_{on}$ to cancel any ripple appearing on $V_1$. If in Fig. P7.78, $t_{off} = 10$ μs, roughly what percent variation, $100(v_{max} - v_{min})/v_{max}$ for $V_L$, will result due to the discharging of $C_F$?

**7.79.** The voltage regulator in Fig. P7.79 consists of a resistor/Zener diode connected to an emitter follower $Q_1$. To protect $Q_1$ from excessive current, $Q_2$ and $R_1$ have been added. Normally, $Q_2$ is cut off, but when the emitter current $i_{E1}$ is large enough, the voltage drop across $R_1$ becomes sufficient for $Q_2$ to conduct and draw off excessive base current from $Q_1$. If the designer wants $Q_2$ to start conducting when $i_{E1} = 200$ mA, what value of $R_1$ is needed? If each transistor has $\beta = 100$, what will be the current through a short-circuit load ($R_L = 0$)? Assume linear operation of both transistors. Comment on the results; is this a good overload current limiter? [*Hint:* The voltage across the Zener diode will be 1.4 V (two base–emitter drops), not $V_Z = 6.8$ V during the short circuit. The Zener diode is then reverse biased and therefore carries zero current.]

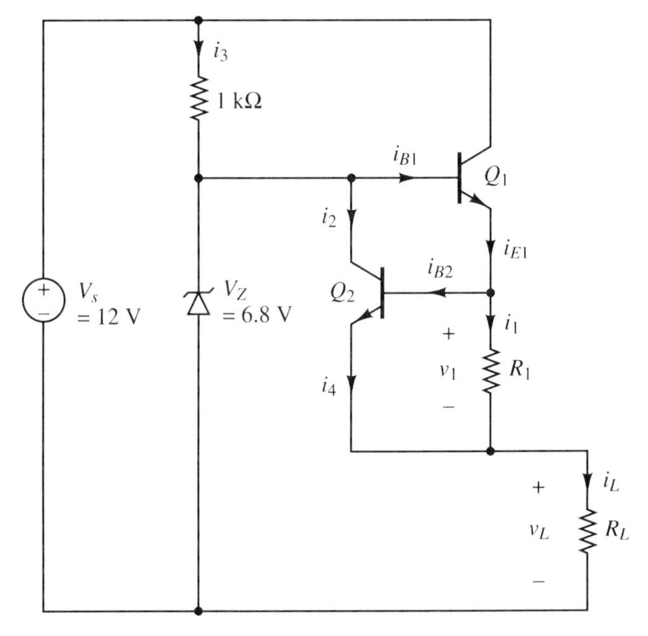


**Figure P7.79**

**7.80.** A persistent problem in the design of computer CMOS circuits is preventing a condition called "latch-up," caused by two unwanted ("parasitic") transistors (modeled as BJTs) caused by the MOSFET geometry and materials (see Fig. P7.80). These transistors are arranged such that if a transient voltage is introduced to the base of $Q_1$, it is possible that the output of $Q_1$ will increase the bias on $Q_2$, which in turn will increase the bias on $Q_1$. This vicious cycle will end up saturating $Q_2$, and thus short-circuiting $V_{DD}$. Use the given small-signal model circuit to compute the small-signal gain; if this gain is >1, latch-up will occur. Should $\beta_1$, $\beta_2$, $R_1$, and $R_2$ be large or small to prevent feedback? Note that the feedback path has been broken and $r_{\pi 1}$ repeated to simulate the feedback.

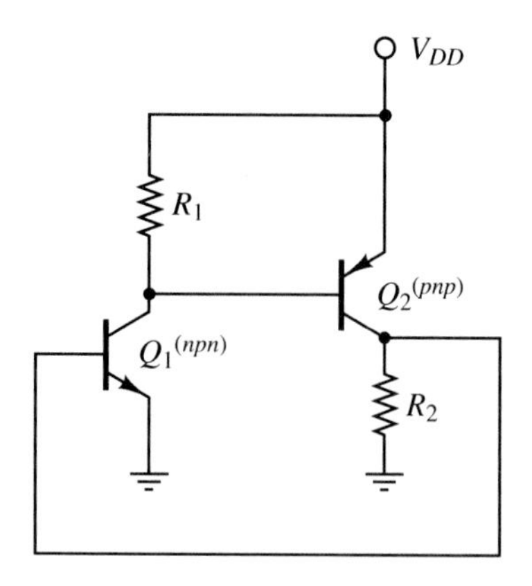


Small-signal model circuit:

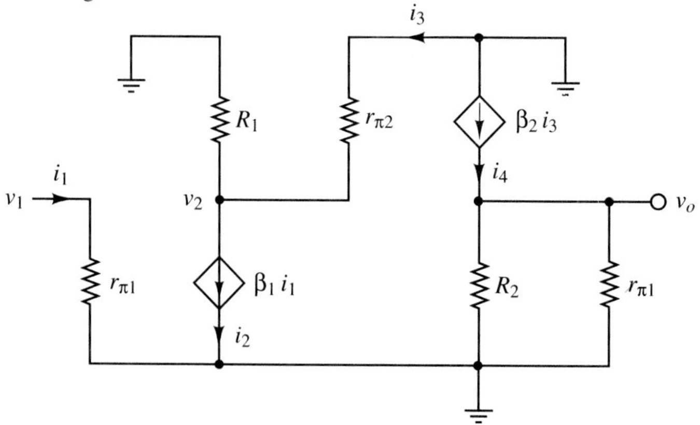


**Figure P7.80**

# 8

# Analog Electronics II: Operational Amplifiers

## 8.1 INTRODUCTION

In our discussions of electronic devices in Chapter 7, a desired behavior was specified and a particular device was shown to approximately provide it. For example, perfect rectification was considered a desired function. Analysis of the *pn* junction showed that it approximately fulfilled this goal. The resulting device was called a diode.

Then signal power amplification was suggested; analysis of the *npn* sandwich subsequently showed that it could be used to accomplish this. That device was called a bipolar junction transistor. Another transistor type, the FET, is useful when it is desired or required that input currents be minimized (input impedance maximized). The ideal diode could be thought of as a voltage-controlled switch, while the transistor could be thought of as a signal-controlled large current source. When combined with other circuit elements, a variety of useful applications was possible. In fact, those two types of devices form the backbone of all signal processing, computational, and communications circuitry.

So what can the operational amplifier offer? A lot: simplified design, low cost (a 741 model costs only about 40 cents), small physical size, nearly ideal behavior, and a huge variety of applications.

Consider a few limitations of the devices so far encountered. The silicon diode, for example, has a potential barrier of about 0.7 V that must be overcome before it will conduct; therefore, it does not behave as a rectifier for small signals.

The situation is far more complicated for transistor amplifiers. A range of criteria must be satisfied simultaneously to ensure acceptable performance: Sufficient signal power gain, high input impedance, low output impedance, sufficient output voltage swing allowance, low dependence on the highly variable gain parameter (such as β for the BJT), minimization of a variety of types of distortion, sufficient bandwidth for the application, and the list goes on. Often, optimization of one criterion is obtained at the expense of some other criterion, so amplifier design is no exception to the law of trade-offs common to all fields of engineering.

However, more complicated design principles and techniques have been developed to meet many of these specifications in the same circuit. Over the years, sophisticated circuits have been developed to the point where the assumption of ideal behavior is in practice a usable approximation of actual behavior. The price of that design has already been paid; the user is now presented with a new "device": A high-performance voltage amplifier known as an *operational amplifier*.

Back in the days of analog computers (computers that manipulate continuous-time rather than discrete-time signals), the concept of a "black box" amplifier was conceived and exploited. It was found that a variety of "operations" (e.g., addition, subtraction, differentiation, integration, multiplication) could be easily implemented using the generic black box and a few additional external components: Hence the name "operational amplifier" (nicknamed *op amp*). No, op amps are not named after André Ampère, so never write "op A" or "op mA."

Although the days of the analog computer have passed, as far as involved computations are concerned, the op amp is here to stay as a simple, accurate, versatile, and cheap module for signal processing. It can even be used in the *interface* circuitry between analog (e.g., measurement) systems and digital (e.g., computation) systems. Applications abound in biomedical signal processing, audio systems, automotive electronics, and anywhere that a transducer signal needs to be amplified or otherwise processed.

In this chapter we study the simplest models of the op amp in detail, including a description of feedback from fundamentals. After studying some of the most basic op amp circuits, a potpourri of op amp applications is presented both to show the breadth of the range of usefulness of this building block and to introduce some practical matters involved in the employment of op amps.

## 8.2 THE COMPARATOR

The physical appearances of some typical op amps are shown in Fig. 8.1; they are quite complicated transistor integrated circuits. For example, even the simplest op amp, the 741, has 24 transistors. Yet for external signal analysis it can be well modeled by the very simple equivalent circuit shown in Fig. 8.2. It is essentially a pair of Thévenin equivalent circuits, one viewed from the input side and the other

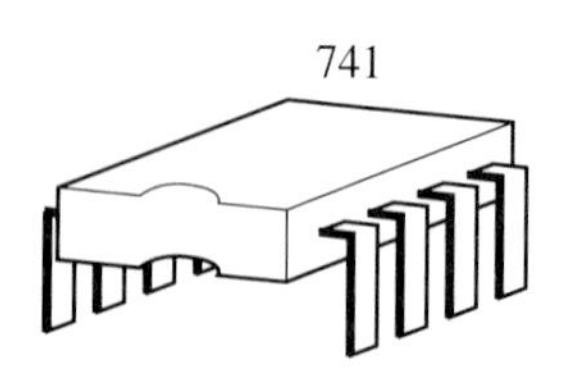

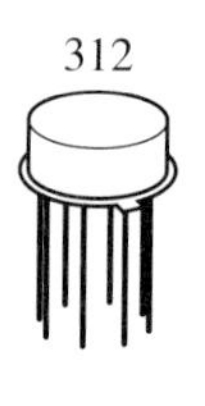

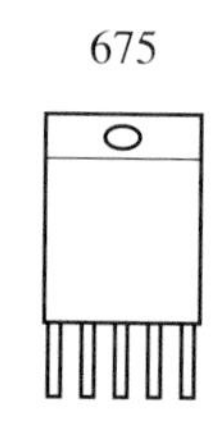


**Figure 8.1** Physical appearances of typical operational amplifiers.

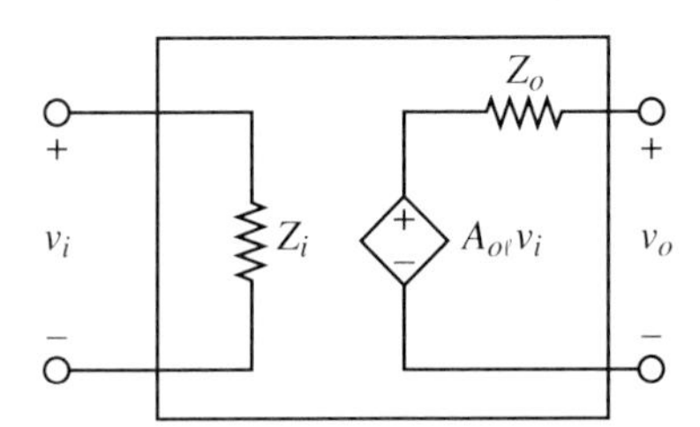


**Figure 8.2** Simplified equivalent circuit for an op amp.

viewed from the output side. No large-signal models and bias calculations are required; that has all been done by the op amp designers.

The input signal $v_i$ is placed across a very large input impedance $Z_i$ (here taken to be purely resistive and typically a few megohms). The output voltage is equal to that across the $v_i$-dependent voltage source, $A_{ol}v_i$, minus the small voltage drop across the quite small output impedance $Z_o$ (typically, 75 Ω or less). Furthermore, the "open-loop" dc voltage gain $A_{ol}$ is on the order of $10^5$ to $10^6$! The meaning of the "open-loop" designation (subscript "ol") will soon be clear, as well as the reasons for wanting such high gain. We shall see that a model even simpler than that in Fig. 8.2 is adequate for most application calculations.

Figure 8.3 shows the usual circuit symbol for an op amp. The output is single-ended; that is, $v_o$ is assumed to be measured with respect to the common (ground) potential. Contrarily, the input voltage is *differential*; that is, $v_i$ is the difference between the potentials $v_-$ and $v_+$ of the two terminals marked, respectively, − and +.

These input terminals are called, respectively, the inverting and noninverting input terminals. *Inverting terminal* means that if the potential of that terminal rises with respect to the noninverting terminal, $v_o$ will *decrease,* and vice versa; the inverse of the behavior at the inverting terminal is reflected in the output. Similarly, the behavior of the noninverting terminal potential is directly reflected in the output.

All this boils down to the fact that if $v_i$ has symbolic polarity as shown in Fig. 8.3, $v_o$ will have (with respect to ground) the same actual polarity as $v_i$. Summarizing, we have

$$v_o = A_{ol}v_i = A_{ol}(v_+ - v_-). \tag{8.1}$$

[Notice that in writing $v_o = A_{ol}v_i$, we are ignoring the small voltage drop across $Z_o$. This is common practice and can be interpreted as assuming that $Z_o$ is very small compared with $R_L$ in Fig. 8.3 (light load).]

Finally, the op amp symbol in Fig. 8.3 shows both a load $R_L$ and the power

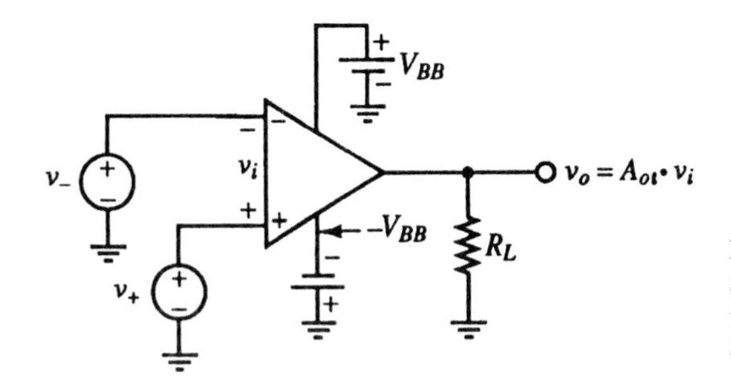


**Figure 8.3** Complete connections of an op amp in open-loop mode. The triangle is the circuit symbol of the op amp.

supply lines. Notice that both positive and negative supply voltages $\pm V_{BB}$ (often $\pm 15$ V or $\pm 5$ V) must be connected, while the ground is not connected to the op amp. These features are characteristic of differential amplifiers and allow the output to swing both negatively and positively, just as a typical signal might.

Usually for brevity the power supply lines are left out of the diagram; in practice, don't ever forget to connect them! It must always be kept in mind that the power supply is the "engine" supplying the raw power that the amplifier manipulates (just as a person operates a crane). In Fig. 8.2 the power supply is buried in the dependent voltage source.

We may well ask: "If $V_{BB}$ is only 15 V, how can the gain be $10^5$? For $v_i = 1$ V, that would mean that $v_o$ would be 10 kV!" The answer is that outside a very narrow linear region in which the gain *is* $10^5$, saturation occurs. A plot of $v_o$ versus $v_i$ is shown in Fig. 8.4. The slope is indeed $A_{\text{ol}} \approx 10^5$ within the linear region; outside the linear region, $v_o$ is limited to near $\pm V_{BB}$. (In practice, the saturation voltages will have magnitude somewhat less than $V_{BB}$, but in this chapter the saturation voltage magnitude will be taken to be $V_{BB}$, called *full $V_{BB}$ saturation*.) Thus the range of $|v_i|$ for which $v_o$ is a linear amplification of $v_i$ is zero to approximately $V_{BB}/A_{\text{ol}} \approx 150\ \mu\text{V}$—*very* small.

A question that must be addressed is: Does the op amp provide power gain as did the transistor amplifiers studied before? Suppose that $v_i = 75\ \mu\text{V}$, so that the linear region applies. Suppose also that a load $R_L = 1\ \text{k}\Omega$ is placed across the output $v_o$ and ground in Fig. 8.3. Using $Z_i = 2\ \text{M}\Omega$, the input power is $v_i^2/Z_i = (75 \cdot 10^{-6})^2/(2 \cdot 10^6) = 2.8 \cdot 10^{-15}$ W. Assuming that $Z_o = 100\ \Omega$ and $A_{\text{ol}} = 10^5$, the output current is $v_o/(Z_o + R_L) = (75 \cdot 10^{-6} \cdot 10^5)/(1100) = 6.82$ mA. So the power to the load is $(6.82\ \text{mA})^2 1000 = 47$ mW, a power gain of $10^{13}$! As before, the additional power comes from the $\pm V_{BB}$ power supply.

Suppose now that $v_-$ is set to 0 V. Then in the linear region, Eq. (8.1) becomes $v_o = A_{\text{ol}} v_+$. But for $|v_+| > V_{BB}/A_{\text{ol}}$,

$$v_o = V_{BB}\,\text{sign}\,(v_+), \tag{8.2}$$

where

$$\text{sign}\,(x) = \begin{cases} 1 & x > 0 \\ -1 & x < 0. \end{cases} \tag{8.3}$$

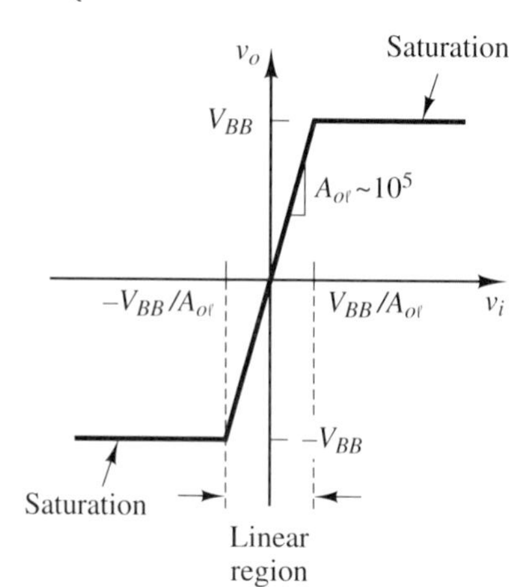


**Figure 8.4** Input–output voltage characteristic of an op amp. Both saturation and linear regions are indicated.

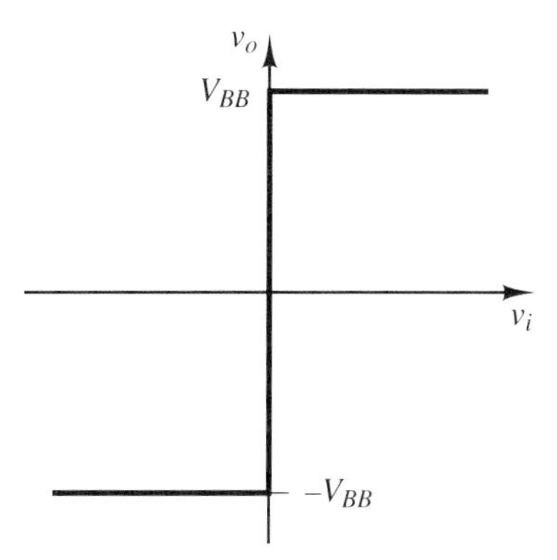


**Figure 8.5** Ideal comparator switch.

Therefore, the op amp output is a noninverting sign detector (positivity detector) of the single-ended voltage $v_+$ for $|v_+| > V_{BB}/A_{ol}$. The ideal noninverting sign detector, often called a *comparator* (comparing $v_+$ with $v_-$), would have an input–output relation as shown in Fig. 8.5. There would be no linear region, so the circuit would function as a perfect switch.

Obvious applications of such a circuit might be the on–off switch for filling a tank when the fluid level reaches a certain threshold, thermostat control of a heater, and so on. Also, it might be useful for determining turn-off switching time during a cycle of an inductive ac machine (motor, etc.). The rationale there would be that sparking due to interruption of currents could be minimized if the switching can happen only at moments when there is zero current flowing.

## 8.3 THE SCHMITT TRIGGER

There can, however, be a problem with the circuit described above. Consider connecting to the noninverting terminal a signal $v_s(t)$ so that $v_+ = v_s(t)$. If the signal voltage $v_s(t)$ contains noise of magnitude larger than $V_{BB}/A_{ol}$, then when the signal is small the noise could cause $v_o(t)$ to flip-flop between $\pm V_{BB}$. This effect is demonstrated in Fig. 8.6. But that is not the desired behavior. In fact, switches or other devices governed by $v_o$ could be worn out prematurely by the rapid, large-magnitude fluctuations in $v_o$.

How can this problem be alleviated? By a simple and old idea: Make the thresholds that must be exceeded in order for $v_o$ to change from one extreme to the other *different* for the two directions of switching. This concept can be found in household spring-snapping light switches, where one must push almost all the way in one direction to turn the switch on, and almost all the other way to turn it off. In this case, the life of both the switch and the light bulb are extended.

In the example involving $v_s$ and $v_o$, the differing thresholds for each direction of change in $v_o$ could be as follows:

$$\begin{aligned} &v_s \text{ must rise above } 0.1 \text{ V for the change } v_o = -V_{BB} \text{ to } v_o = +V_{BB} \\ &v_s \text{ must drop below } -0.1 \text{ V for the change } v_o = V_{BB} \text{ to } v_o = -V_{BB}. \end{aligned} \tag{8.4}$$

Then the behavior would be as shown in Fig. 8.7, which clearly is the desired response to $v_s$.

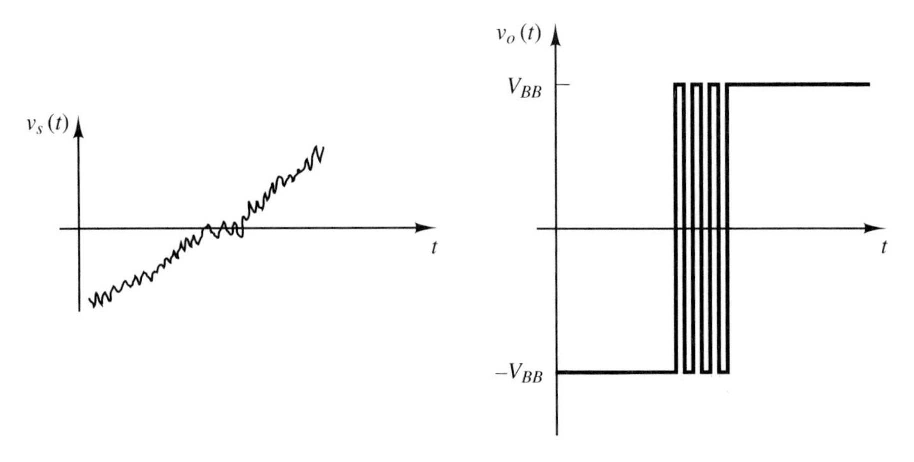


**Figure 8.6** Problem with using ideal comparator as a sign detector of a noisy signal: (a) input signal; (b) thrashing output signal.

Can this effect be implemented electronically using the op amp comparator? It can if a few resistors are added to the circuit. Such a circuit is shown in Fig. 8.8. For this example, suppose that $V_{BB} = 15$ V and assume full $V_{BB}$ saturation. Now $v_s$ is not directly connected to the noninverting terminal, but instead is connected through resistor $R_1$. Furthermore, a sample of the output voltage $v_o$ is fed back through $R_2$ to the noninverting terminal. This is an example of *positive feedback,* discussed in more detail in Sec. 8.4.

In Sec. 8.2 the input and output impedances were given for a 741 op amp. The input impedance was huge and the output impedance was minute. Compared with

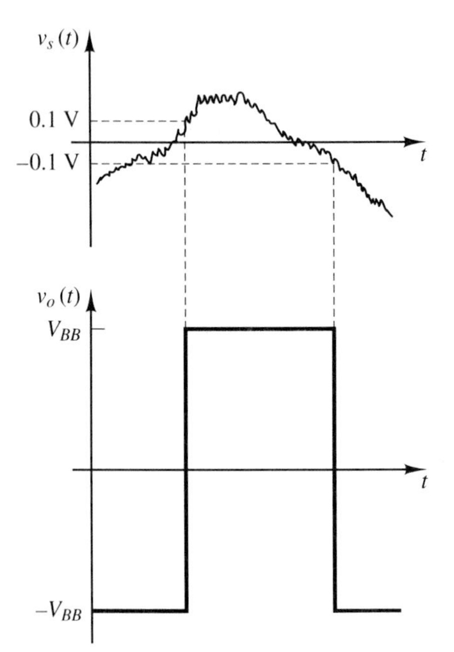


**Figure 8.7** Desired sign detector behavior for noisy signal.

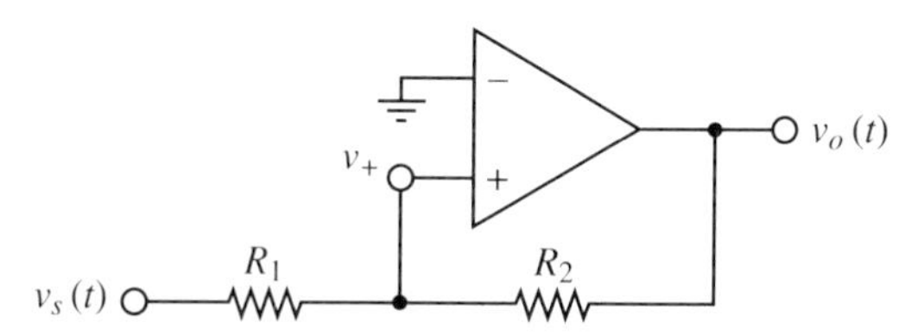


**Figure 8.8** Schmitt trigger circuit.

other circuit parameters, the following ideal op amp assumptions are often valid approximations:

$$Z_i = \infty \quad \text{and} \quad Z_o = 0 \qquad \text{(ideal op amp).} \tag{8.5}$$

Consequently, the current flowing into the noninverting terminal can be neglected compared with the current flowing through resistors $R_1$ and $R_2$. Kirchhoff's current law applied at the noninverting terminal can therefore be written simply as

$$\frac{v_+ - v_s}{R_1} + \frac{v_+ - v_o}{R_2} = 0, \tag{8.6}$$

so that

$$v_+ \cdot \left(\frac{1}{R_1} + \frac{1}{R_2}\right) = \frac{v_s}{R_1} + \frac{v_o}{R_2}, \tag{8.7}$$

leading to

$$v_+ = \frac{R_2 v_s + R_1 v_o}{R_1 + R_2}. \tag{8.8}$$

For use below, the fact that the denominator in Eq. (8.8) is always positive guarantees that

$$v_+ > 0 \quad \text{if} \quad R_2 v_s + R_1 v_o > 0, \tag{8.9}$$

or, rearranging,

$$v_+ > 0 \quad \text{if} \quad v_s > -\left(\frac{R_1}{R_2}\right) v_o. \tag{8.10}$$

Similarly,

$$v_+ < 0 \quad \text{if} \quad v_s < -\left(\frac{R_1}{R_2}\right) v_o. \tag{8.11}$$

Suppose that originally $v_s$ is a large negative number, for example,

$$v_s \ll -V_{\text{th}}, \tag{8.12}$$

where

$$V_{\text{th}} = \frac{R_1}{R_2} V_{BB}. \tag{8.13}$$

Then from Eq. (8.11), $v_+ < 0$, and therefore, by Eq. (8.2), $v_o = -V_{BB}$.

Now suppose that $v_s$ increases. From Eq. (8.10) it can be seen that if $v_s$ increases beyond $-(R_1/R_2)(-V_{BB}) = V_{\text{th}}$, then $v_+$ will become positive. By Eq. (8.2), $v_o$ will

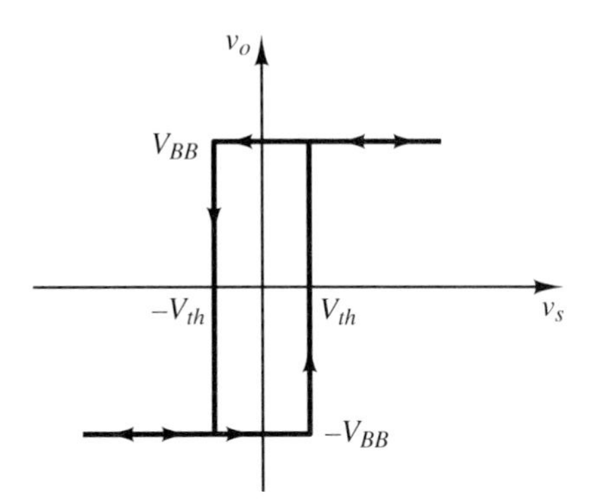


**Figure 8.9** Input–output voltage characteristic of a Schmitt trigger. Note the hysteresis phenomenon, indicated by the differing state-dependent paths.

then swing from $-V_{BB}$ to $+V_{BB}$. Next, suppose that it is desired that the output swing back to $-V_{BB}$. For what $v_s$ will this occur? Equation (8.11) says that $v_+$ will become negative if $v_s < -(R_1/R_2)V_{BB} = -V_{\text{th}}$ (where the present value of $v_o$ is always used). From Eq. (8.2), the fact that $v_+$ is now negative means that $v_o$ will become equal to $-V_{BB}$.

To summarize the behavior:

$$\begin{aligned} &v_s \text{ must rise above } V_{\text{th}} \text{ for the change } v_o = -V_{BB} \text{ to } v_o = +V_{BB} \\ &v_s \text{ must drop below } -V_{\text{th}} \text{ for the change } v_o = +V_{BB} \text{ to } v_o = -V_{BB}. \end{aligned} \tag{8.14}$$

But this is exactly the desired behavior set forth in Eqs. (8.4), where $V_{\text{th}}$ was equal to 0.1 V. Because $V_{BB} = 15$ V, any moderate ($\ll 10^6\ \Omega$) values of $R_1$ and $R_2$ satisfying Eq. (8.13) set equal to 0.1V will work: For example, $R_1 = 500\ \Omega$ and $R_2 = 75$ kΩ.

This circuit is called a no-chatter ("debounced") positivity detector, or *Schmitt trigger,* after O. H. Schmitt, who invented the vacuum-tube version in 1938. The transfer characteristic for an ideal Schmitt trigger is shown in Fig. 8.9. Motion along the curve occurs only for the directions shown.

The fact that the value of the switching threshold depends on the current state of the output ($+V_{BB}$ versus $-V_{BB}$) shows that the circuit has a rudimentary form of "memory." Recall that state-dependent curves such as those in Fig. 8.9 are called hysteresis curves. Another example of hysteresis curves is the *B–H* relation relevant to electromechanical machinery; this was discussed in Sec. 3.8.

## 8.4 FEEDBACK

Feedback is a phenomenon that reaches into most areas of life. Any time that we are rewarded or reprimanded for the quality of our work, words, or behavior, feedback is occurring. A bicycle rider uses feedback to keep balanced on only two wheels. The driver of a car continuously makes little steering adjustments to correct for bumps in the road, poor front-end alignment, and so on. A driver also steps on the brakes or the accelerator to move (hopefully) with the flow of traffic.

There is an automotive example which demonstrates a use of feedback that is exploited in electronic amplifier circuits: cruise control. Cruise control is a very convenient and reliable method of maintaining constant highway speed for the following reason. The output behavior (speed of the car) depends on high-precision, relatively passive components whose characteristics are stable and predictable (and chosen by an electrical engineer(!)). If the throttle were just kept at a constant degree of openness, road surface conditions and variable output power of the engine would eventually lead to loss of control and a car crash. But with cruise control, once the driver selects the desired speed, the feedback circuit will provide whatever variations in the openness of the throttle are necessary to maintain the desired speed.

Figure 8.10 shows the general idea, where the analogy between the engine and the op amp is depicted symbolically. This type of drawing is called a *signal flow diagram.* It is a generic display of the essential "flow of traffic" in a circuit or other control system.

The circle with a plus sign within is called a *summer* because its output, the line with arrow pointing out, is equal to the sum of its two inputs. Note that the two inputs are each multiplied by 1 or $-1$, depending on the $+$ or $-$ sign for each input line indicated outside the summer. The block and triangle here are operators each of whose output is in general equal to the input value multiplied by the number within the box. (Here take the engine to represent $A_{ol}$.) The actual speed of the car is compared with the desired speed. The resulting error signal will be used to determine whether and to what extent the throttle should be opened or closed at this moment.

Another example of feedback was described in Chapter 7. Recall that for the BJT, $\beta$ was highly variable, depending on transistor specimen, temperature, and even operating conditions. It is characteristic of active devices that without some sort of external control, performance can be quite variable or unpredictable. The BJT is in this sense analogous to the automobile engine described above.

Recall that the gain of the common-emitter amplifier was made less dependent on $\beta$ by adding a resistor $R_E$ between the emitter and ground. This was a use of

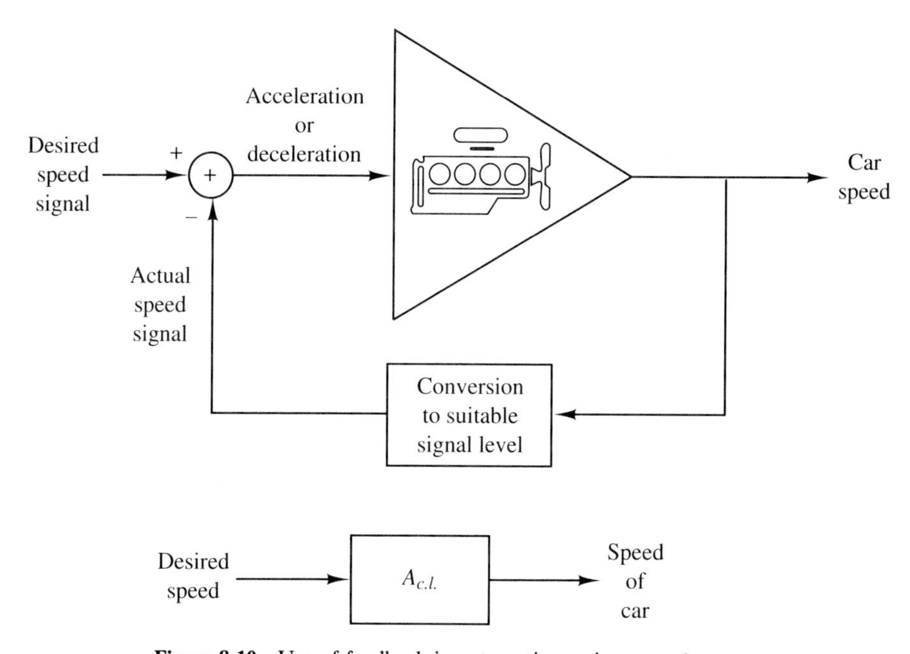


**Figure 8.10** Use of feedback in automotive cruise control system.

feedback. How? First consider the original case without the resistor ($R_E = 0$). The same small-signal model applies to the *bypassed* nonzero emitter resistor, which we analyzed in Sec. 7.6.2 (see Fig. 7.35d); only the operating point, and therefore $r_\pi$, differ. In that case the input equation for the base current $i_b$ was given in Eq. (7.26a) as

$$i_b = \frac{R_B}{r_\pi(R_s + R_B) + R_s R_B} v_s. \tag{8.15}$$

The small-signal voltage gain for $R_E = 0$ can be found by combining Eq. (7.27b) ($A_v = -\beta R_L'/r_\pi$) with the operating point Eq. (7.15) ($I_C = \beta(V_{BB} - 0.7\text{ V})/R_B$) and $r_\pi$ in Eq. (7.7b) [$r_\pi = \beta kT/(qI_C) = R_B(kT/q)/(V_{BB} - 0.7\text{ V})$] to give

$$A_v = \frac{v_o}{v_s} = \frac{-\beta R_L'}{r_\pi} \tag{8.16a}$$

$$= \frac{-\beta R_L'(V_{BB} - 0.7\text{ V})}{R_B(kT/q)}. \tag{8.16b}$$

First notice that in Eq. (8.15), the input (base current) equation, neither $\beta$ nor any other output parameter appears (i.e., there is no feedback), not even via $r_\pi$, which is also independent of $\beta$. Unfortunately, the small-signal gain is *directly proportional* to the highly variable $\beta$. Now consider the case $R_E$ not equal to zero. For simplicity, consider the case where $R_E$ is not bypassed (see Fig. 7.35c). For the case of bypassing, there is clearly still feedback (the presence of the output parameter $\beta$ in the input equation). However, in that case it is harder to isolate a sample of output current in the input equation using our analysis techniques. One must further use the fact that $r_\pi$ is now proportional to $\beta$. In the case of no bypassing, the input equation for $i_b$, Eq. (7.22), is

$$v_b \approx i_b r_\pi + (\beta i_b)R_E, \tag{8.17}$$

and the small-signal voltage gain was given by Eq. (7.24b) as

$$A_v = \frac{v_o}{v_s} \approx -\frac{R_L'}{R_E}, \tag{8.18}$$

which is independent of $\beta$. Now a sample of the output signal, $\beta i_b$, appears in the input equation (8.17). Recall from Eq. (7.24a) that $v_o = -(\beta i_b)R_L'$, so indeed, a scaled version of $v_o$ has been sent back to the input. The resulting small-signal gain [Eq. (8.18)] is approximately independent of $\beta$. A cost has been incurred: $R_E$ must be fairly large for the approximation in Eq. (8.18) to hold, which means that the gain will be small. [How large? For most sources, $R_s \ll R_B$ and $\beta R_E \gg r_\pi + R_s$ (see discussion preceding Eq. (7.23d)), so $R_E \gg (r_\pi + R_s)/\beta > r_\pi/\beta$, so that the gain in Eq. (8.18) is certainly less than that in Eq. (8.16a).]

This compromise is common among electronic amplifiers: Use of feedback inevitably reduces the gain. In an op amp it is $A_{ol}$ that is highly variable (dependent on op amp specimen, temperature, etc.), instead of $\beta$ for the BJT. We shall see momentarily why the extremely high value of $A_{ol}$ of the comparator amplifier is so advantageous: Although the gain is severely reduced in an op amp circuit due to feedback used in the interest of stability, it can still be appreciable.

Feedback in an op amp circuit has already been observed: the Schmitt trigger. The expression for the input voltage $v_+$ in Eq. (8.8) contains a term in the numerator proportional to $v_o$; feedback is present. Notice that this term $+R_1v_o$, contributes positively to $v_+$. This happens because the fed-back signal is connected to the noninverting (+) terminal. Consequently, a rise in output voltage $v_o$ increases $v_+$, which in turn will tend to increase $v_o$. It is by this cyclical process that the saturation voltages are so quickly reached when a threshold is crossed.

Because the feedback contribution is positive, the Schmitt trigger is classified as a positive feedback configuration. Stability is achieved only by way of the saturation voltage barriers; otherwise, voltages would continue to grow in magnitude indefinitely. It is like the case of a student who does well: The teacher compliments the student, who in turn is encouraged to do even better, which in turn makes the teacher even more happy, and so on. "Stability is achieved" only by the student continually receiving 100% on every paper. In positive feedback circuits, minimal time is spent in the linear region of the op amp transfer characteristic.

Operation in the linear region *can* be obtained by using *negative feedback,* where increases in the output tend to *reduce* the magnitude of the input voltage and thereby the output. In fact, because the open-loop gain of an op amp $A_{ol}$ is extremely large, the negative feedback will not only "tend" to reduce the input voltage, but will cause it to be *extremely small,* so small that operation in the linear region is maintained.

How small, and why not zero? The answer has been given by Harold S. Black. In the 1920s he developed the following simple configuration for negative feedback, the flow diagram of which is shown in Fig. 8.11. Control is provided in three steps. First, the output $y$ is fed back through block $F$, where it is scaled so as to produce an estimate of the quantity $x$ to be matched. Second, comparison of $x$ with $Fy$ is made via $\epsilon = x - Fy$. Finally, the error signal $\epsilon$ itself is scaled through $K$ to produce the output. Following the simple operations in Fig. 8.11 gives for the error signal

$$\epsilon = x - Fy \tag{8.19}$$

and for the output,

$$y = K\epsilon. \tag{8.20}$$

The closed-loop gain is just $y/x$, which may easily be found by substituting $\epsilon$ in Eq. (8.19) into Eq. (8.20):

$$y = K(x - Fy). \tag{8.21}$$

Rearranging gives

$$y = \frac{K}{1 + FK}x, \tag{8.22}$$

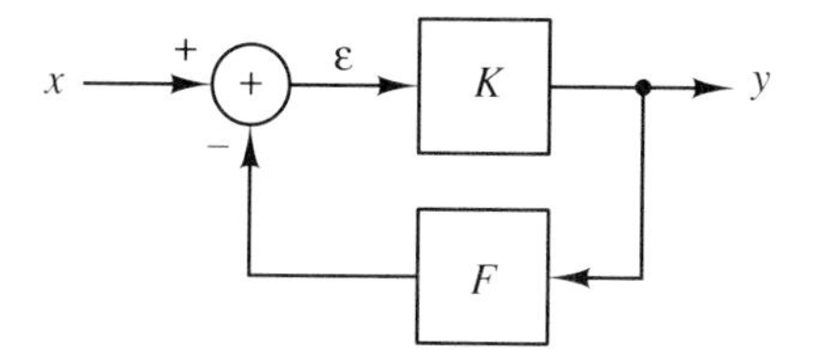


**Figure 8.11** H. S. Black's negative feedback flow diagram.

so that the closed-loop gain $A_{\text{cl}}$ is

$$A_{\text{cl}} = \frac{y}{x} = \frac{K}{1 + FK}. \tag{8.23}$$

Notice that without any feedback (open loop; $F = 0$), Eq. (8.23) would reduce to $y/x = K$. Therefore, $K$ is the open-loop gain $A_{\text{ol}}$. If $K$ is very large, so that $FK \gg 1$, the one in the denominator of $A_{\text{cl}}$ may be neglected, and Eq. (8.23) reduces to

$$A_{\text{cl}} \approx \frac{1}{F}. \tag{8.24}$$

This is just what was sought. The closed-loop gain is independent of the highly variable $K = A_{\text{ol}}$. In practice, $F$ can be made small [but still large enough to maintain the validity of Eq. (8.24)] so that $A_{\text{cl}}$ can be large enough to be useful. Incidentally, by Eq. (8.19), we have for $K \to \infty$, $\epsilon = x - F(x/F) = 0$. So, indeed, the "error" signal $\epsilon$ is driven to zero. It should now be clear why values of $A_{ol}$ on the order of $10^6$ to $10^8$ for op amps are desirable.

## 8.5 A NONINVERTING AMPLIFIER

As a demonstration of negative feedback in an op amp circuit, consider the *noninverting amplifier* shown in Fig. 8.12. If, as before, the input resistance $Z_i$ is taken to be infinite, $v_-$ can be found by the voltage divider

$$v_- = \frac{R_1}{R_1 + R_2} v_o. \tag{8.25}$$

Then, by KVL,

$$v_i = v_s - \frac{R_1}{R_1 + R_2} v_o. \tag{8.26}$$

This is the same as Black's form [Eq. (8.19)] if the following identifications are made:

$$\epsilon = v_i \tag{8.27a}$$

$$x = v_s \tag{8.27b}$$

$$F = \frac{R_1}{R_1 + R_2} \tag{8.27c}$$

$$y = v_o. \tag{8.27d}$$

In addition, the output, as always, is $v_o = A_{\text{ol}} v_i$, so that from Eq. (8.20),

$$K = A_{\text{ol}}. \tag{8.28}$$

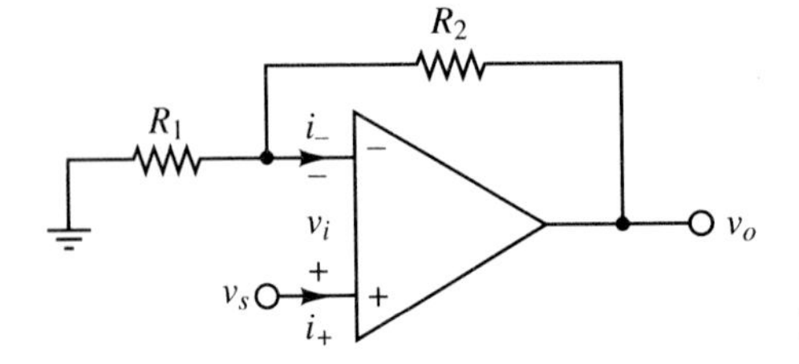


**Figure 8.12** Noninverting amplifier.

Therefore, for very large $A_{ol}$, the closed-loop gain for the noninverting amplifier is, from Eq. (8.24),

$$A_{cl} \approx \frac{1}{F} = \frac{R_1 + R_2}{R_1} = 1 + \frac{R_2}{R_1}, \tag{8.29}$$

which is independent of $A_{ol}$. By choosing $R_2 \gg R_1$, substantial voltage amplification ($A_{cl} \gg 1$) can be achieved.

This result can also be obtained from the circuit diagram directly by making the following assumptions:

$$v_i = 0 \tag{8.30}$$

and

$$i_+ = i_- = 0. \tag{8.31}$$

The first assumption, $v_i = 0$, is equivalent to taking $A_{ol} = \infty$, so that feedback perfectly drives the input voltage to "zero." It is analogous to the perfect car driver, who keeps such persistent and exact control of his car that the motions of the steering wheel appear negligible, even to the point of silencing a back-seat driver.

Of course, the input voltage is not *exactly* zero, or $v_o$ would be zero. But $v_i$ will be of negligible size compared with any other voltages in the circuit. (Recall, for example, that for linear operation $|v_i| < 150\ \mu V$ if $A_{ol} = 10^5$, $V_{BB} = 15$ V.) The value that $v_i$ actually takes will be such that the closed-loop gain has the value in Eq. (8.29).

Do not think of this assumption as meaning that the terminals are shorted together, even though it is commonly referred to as a *virtual short.* On the contrary, the impedance between the two terminals is huge, but the current is negligible [Eq. (8.31)]. We can also use Eqs. (8.19) and (8.27) directly to obtain $v_i = v_s - [R_1/(R_1 + R_2)]A_{ol}v_i$, or, solving for $v_i$, $v_i = v_s/[1 + A_{ol}R_1/(R_1 + R_2)]$. Because $A_{ol}$ is so large and is in the denominator, $v_i$ is very small compared with other voltages (such as $v_s$). Physically, if $v_s$, say, increases, thus increasing $v_i$, then $v_o$ increases dramatically, and by the voltage divider forces $v_-$ up to $v_s$ and $v_i$ to zero: $v_- = \dfrac{R_1}{R_1 + R_2}[A_{ol}(v_s - v_-)] = A_{ol}v_s/(1 + A_{ol} + R_2/R_1) \approx v_s$.

The second assumption [Eq. (8.31)] was of course used earlier (in Sec. 8.3), and again is due to the huge input impedance $Z_i \approx \infty$. In practice, $i_+$ is in the picoampere-to-nanoampere range, whereas signal currents are typically in the microampere range, so Eq. (8.31) is a good assumption. The assumption in Eq. (8.30) implies for the circuit of Fig. 8.12 that $v_- = v_+ = v_s$. Using this fact and Eq. (8.31), KCL at the noninverting terminal reduces to

$$\frac{v_s}{R_1} + \frac{v_s - v_o}{R_2} = 0, \tag{8.32}$$

from which $A_{cl} = v_o/v_s$ in Eq. (8.29) follows directly. The name "noninverting" amplifier stems from the fact that $A_{cl} > 0$, so the sign of the input voltage is kept unchanged in the output voltage. In this chapter, do not take the phrase "inversion of $x$" to mean multiplicative inversion ($1/x$), but rather, polarity inversion ($-x$).

**Example 8.1**

Design a noninverting amplifier having a voltage gain of 10.

**Solution** The circuit diagram appears in Fig. 8.12; all that is necessary is to find $R_1$ and $R_2$. From Eq. (8.29), $1 + R_2/R_1 = 10$ or $R_2/R_1 = 9$. Any reasonable values will work. For example, $R_1 = 5$ kΩ and $R_2 = 45$ kΩ.

## 8.6 AN INVERTING AMPLIFIER

Another example of an op amp circuit with negative feedback is the inverting amplifier, shown in Fig. 8.13. Topologically, the only change from the noninverting amplifier is that the positions of ground and $v_s$ have been reversed. Before applying the simplifying assumption $v_i = 0$ in Eq. (8.30), it is instructive to try to make identifications with Black's feedback form [Eqs. (8.19) and (8.20), Fig. 8.11]. This is done by writing KCL again at the inverting node and making the usual assumption that $i_- = 0$ (see Fig. 8.13):

$$\frac{(-v_i) - v_s}{R_1} + \frac{(-v_i) - v_o}{R_2} = 0, \tag{8.33}$$

so that

$$\begin{aligned} v_i &= \frac{-R_1 R_2}{R_1 + R_2}\left(\frac{v_s}{R_1} + \frac{v_o}{R_2}\right) \\ &= \frac{-v_s R_2}{R_1 + R_2} - \frac{v_o R_1}{R_1 + R_2}. \end{aligned} \tag{8.34}$$

This form does not perfectly match that of Fig. 8.11 and Eq. (8.19) because the input, $v_s$, is not multiplied by 1 as $x$ was in Eq. (8.19). Most important, the coefficient of $v_s$ has sign opposite to that of $x$ in Black's diagram (Fig. 8.11), while the other two terms in Eq. (8.34) have the same signs as the corresponding terms in Eq. (8.19). Replacement of $x$ by $-x$ in Eq. (8.22) instantly shows that the closed-loop gain is negative; hence this circuit is called an *inverting amplifier.*

To make the form as close as possible to Black's form while recognizing that the input is inverted in sign, the coefficient of $v_s$ is made to be $-1$ by multiplying both sides of Eq. (8.34) by $(R_1 + R_2)/R_2$. Then

$$\frac{R_1 + R_2}{R_2} v_i = -v_s - \frac{R_1}{R_2} v_o. \tag{8.35}$$

The left-hand side of Eq. (8.35) is now $\epsilon$, $-v_s$ is $x$, $F$ is $R_1/R_2$, and $K$ is found from

$$v_o = A_{ol} v_i = A_{ol} \frac{R_2}{R_1 + R_2} \epsilon = K\epsilon, \tag{8.36}$$

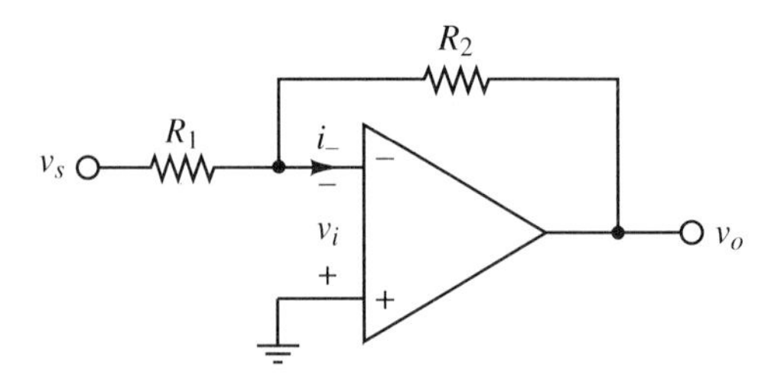


**Figure 8.13** Inverting amplifier.

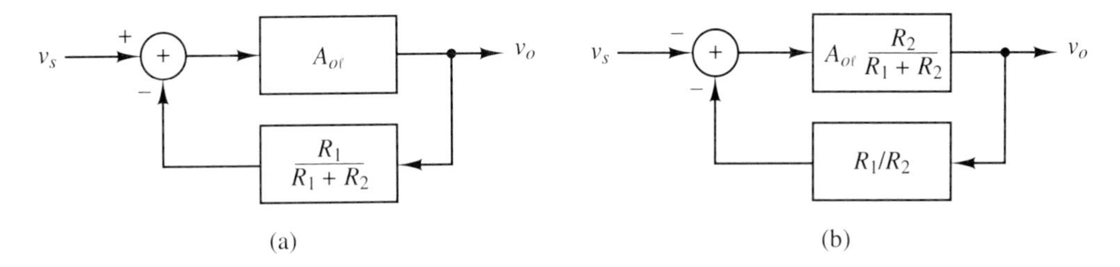


**Figure 8.14** Representation of op amp negative feedback amplifier circuits in Black's form (see Fig. 8.11): (a) noninverting amplifier; (b) inverting amplifier.

from which $K = A_{ol} R_2/(R_1 + R_2)$. If $A_{ol}$ approaches infinity, Eq. (8.22) with $x$ replaced by $-v_s$ (again, because this is an inverting amplifier) gives

$$A_{cl} = \frac{-1}{F} = \frac{-R_2}{R_1}. \tag{8.37}$$

As was true for the noninverting amplifier, $A_{cl}$ is independent of $A_{ol}$ but now is negative, indicating the sign "inversion" effect.

Symbolic block diagrams in Black's form can now be drawn for both the noninverting and inverting amplifiers. The noninverting amplifier fits perfectly and is shown in Fig. 8.14a. The inverting amplifier also fits, except that the sign of $v_s$ is inverted before entering the summer; its block diagram is shown in Fig. 8.14b. Both are examples of negative feedback, and this is indicated in the diagrams by the minus sign "inverting" both feedback lines before entrance into the summers.

As was true for the noninverting amplifier, the assumption $v_i = 0$ [Eq. (8.30)], which is valid for linear operation, allows a very simple analysis of the inverting amplifier. The result will again be Eq. (8.37), which assumes that $A_{ol} \approx \infty$. In this case, $v_- = v_+ = 0$ V, so that KCL at the inverting terminal becomes

$$\frac{-v_s}{R_1} + \frac{-v_o}{R_2} = 0, \tag{8.38a}$$

or

$$\frac{v_o}{v_s} = \frac{-R_2}{R_1}, \tag{8.38b}$$

which is the same result as Eq. (8.37), but far easier to obtain.

Recall that positive feedback led to saturated output most of the time. Has negative feedback resulted in linear operation? The answer is yes, provided that the closed-loop gain $A_{cl}$ [Eqs. (8.29) and (8.37)] and the signal $v_s$ are such that $|v_o| = |A_{cl}v_s| < V_{BB}$. Furthermore, because $A_{cl}$ may be chosen and implemented precisely using resistors $R_1$ and $R_2$, knowledge of the input signal level $|v_s|$ allows designs that guarantee predictable, linear behavior.

**Example 8.2**

Design an inverting amplifier having a voltage gain of −5. What is the maximum input voltage for linear operation?

**Solution** The circuit diagram appears in Fig. 8.13. From Eq. (8.38b), $R_2 = 5R_1$, so if $R_1 = 10$ kΩ, choose $R_2 = 50$ kΩ. Assuming full $V_{BB}$ saturation and $V_{BB} = 15$ V, $v_{s,\max} = 15/5 = 3$ V.

## 8.7 AMPLIFIER INPUT AND OUTPUT IMPEDANCES

The op amp, because of its extremely high open-loop gain, has tremendous advantages over discrete transistor circuits that most people are capable of designing. For example, using positive feedback a polarity detector called the Schmitt trigger can be made whose sensitivity to noise can be reduced to any degree desirable by suitable choice of resistor values. Using negative feedback, amplifiers whose gains can be accurately chosen over a wide range are possible. Yet the formulas for the gain are amazingly simple and involve only two resistor values.

One further advantage occurring for some amplifiers is the following. The input and output impedances of the open-loop op amp are, respectively, huge and minuscule. These are both desirable properties. Recall that the input impedance should be large to minimize loading of the previous stage and the resulting complications that ensue. Similarly, low output impedance makes the amplifier look like a perfect voltage source to any following stages and (within limitations) minimizes the loading effects of them on the present stage.

Under negative feedback, the input and output impedances sometimes become even more extremely favorable. The input and output impedances for the noninverting and inverting amplifiers are straightforward to obtain, but the derivations are omitted for brevity. The results for the noninverting amplifier are

$$Z_{\text{in}} = \frac{A_{\text{ol}} R_1}{R_1 + R_2} Z_i \tag{8.39a}$$

$$Z_{\text{out}} = \frac{R_1 + R_2}{A_{\text{ol}} R_1} Z_o. \tag{8.39b}$$

There are two important points. The first is that the closed-loop input impedance, $Z_{\text{in}}$, is larger by the huge factor of $A_{\text{ol}} R_1/(R_1 + R_2)$ than the open-loop input impedance, $Z_i$. The second point is that the closed-loop output impedance, $Z_{\text{out}}$, is smaller by the factor $(R_1 + R_2)/(A_{\text{ol}} R_1)$ than the open-loop output impedance, $Z_o$. For the inverting amplifier,

$$Z_{\text{in}} \approx R_1 \tag{8.40a}$$

$$Z_{\text{out}} = \frac{R_1 + R_2}{A_{\text{ol}} R_1} Z_o. \tag{8.40b}$$

Thus in this case, the closed-loop output impedance $Z_{\text{out}}$ is again extremely small, but the closed-loop input impedance $Z_{\text{in}}$ is not increased by the large factor over $Z_i$ although it still can be large).

A word of caution: Do *not* try to drive a load $R_L = Z_{\text{out}}$ to maximize power transfer. That would load down the op amp much the same way that the short circuit was said to in the output impedance discussion above. Op amps like to drive networks with large input impedances; an impedance transformer such as the coil–core transformer is often appropriate for maximum power transfer applications.

The op amp is nevertheless often used as an impedance transformer, particularly when large input impedance is desired but not available from an original input stage. In fact, that is one purpose of the so-called voltage follower: a noninverting amplifier having $A_{\text{cl}} = 1$ [$R_2 = 0$ in Eq. (8.29)]. Another is increasing the ability of a previous output stage to drive additional loading output stages; in this case the op amp is called a *buffer* (between signal and load).

All this is not to say that there are no high-current capacity op amps available. Indeed, there even exist 30-A output current op amps whose input current is in the picoamp range. These op amps have to be "hybrid"—the input and output stages are different technologies (e.g., FET input stage, bipolar output stage). There are also high-voltage output op amps, whose outputs (and, of course, whose supply voltage $V_{BB}$) are well over 100 V. These high-capability op amps naturally cost much more than the lowly 741.

## 8.8 POTPOURRI OF OP AMP CIRCUITS

### 8.8.1 Introduction

All of the previous advantages of using op amps are impressive, but what clinches the argument is the vast versatility of the op amp. In this section, several wide-ranging applications are presented that indicate its preeminence in the field of analog signal processing. These examples are only a small sample of what can be and is done with op amps.

To serve as something of a reference section, the circuits described above will be repeated here in their proper order. The progression will be from no-feedback circuits, to positive feedback circuits, and finally negative feedback circuits. *Warning:* There is often more than one configuration that will perform a given function, so variations of these circuits may be encountered in practice. In this section, merely a *representative* implementation of the function using op amps is provided.

### 8.8.2 No-Feedback Circuits

**Simple comparator.** This circuit was discussed in detail in Sec. 8.2. In the linear range, the output voltage $v_o$ in terms of inputs $v_A$ and $v_B$ is

$$v_o = (v_B - v_A)A_{\text{ol}}. \tag{8.41}$$

The circuit diagram is shown in Fig. 8.3, in which $v_- = v_A$ and $v_+ = v_B$. Its transfer characteristic appears in Fig. 8.4, in which $v_i = v_B - v_A$.

**Window comparator.** Suppose that in some industrial process everything is fine as long as the temperature remains between a lower and an upper limit; otherwise, it is desired that an alarm be activated. The circuit for achieving this function is shown in Fig. 8.15a. For convenience, assume that the diodes $D_1$ and $D_2$ are ideal; however, this is not necessary for proper operation.

An explanation of its operation is as follows. If the voltage representing the temperature $v_T$ exceeds the voltage representing the upper limit $V_U$, the input voltage of op amp 1 will be significantly positive. As a result, the output $v_1$ of op amp 1 will be driven to positive saturation $(+V_{BB})$.

Now notice that $v_T$ is connected to the *inverting* terminal of op amp 2. If $v_T$ exceeds $V_U$, it will certainly also exceed $V_L$, causing the input of op amp 2 to be negative. Thus the output $v_2$ of op amp 2 will be driven to negative saturation, $-V_{BB}$.

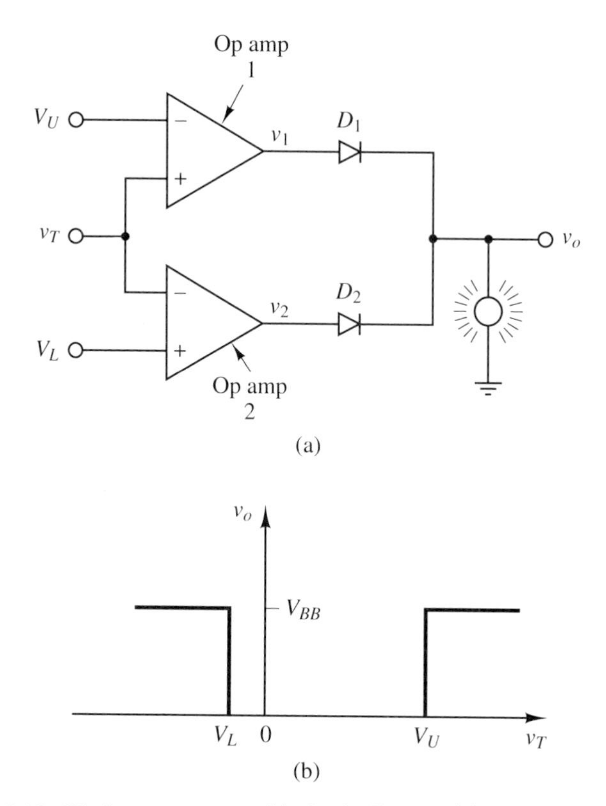


**Figure 8.15** Window comparator: (a) circuit diagram; (b) input–output voltage characteristic.

Because $v_1 > v_2$, $D_2$ will be reversed biased so that $v_o = v_1 = V_{BB}$, which will activate the alarm. In retrospect, $D_1$ is verifiably forward biased, as $v_1 = V_{BB}$ is placed across the alarm impedance and ground (0 V).

If, instead, $v_T < V_L$, the reverse situation occurs. Op amp 2 output voltage $v_2$ goes high, $v_1$ goes low, $D_1$ is reverse biased, and $D_2$ is forward biased. Because $D_2$ is forward biased, $v_o = v_2 = V_{BB}$, and the alarm sounds.

But if $V_L < v_T < V_U$, *both* $v_1$ and $v_2$ are equal to $-V_{BB}$. As they drop below 0 V, both diodes 1 and 2 become reverse biased, leaving the alarm unconnected, so it does not go off. The desired function has been accomplished. Figure 8.15b displays a typical input–output characteristic for an ideal window comparator.

**Simple analog-to-digital converter.** To perform computations on or digitally display measurements efficiently, the analog measurement signal must somehow be converted to a finite number of coded ones and zeros that digital circuits can manipulate. The measurement signal from a transducer typically has a particular value at each instant of time. To obtain a finite set of numbers representing the measured signal, the signal is sampled at evenly spaced, short time intervals. Circuitry to accomplish this sampling is briefly mentioned in Chapter 9.

In addition, the measurement signal can have *any* value describable by an infinite number of "significant digits" (insignificant though most of them may be due

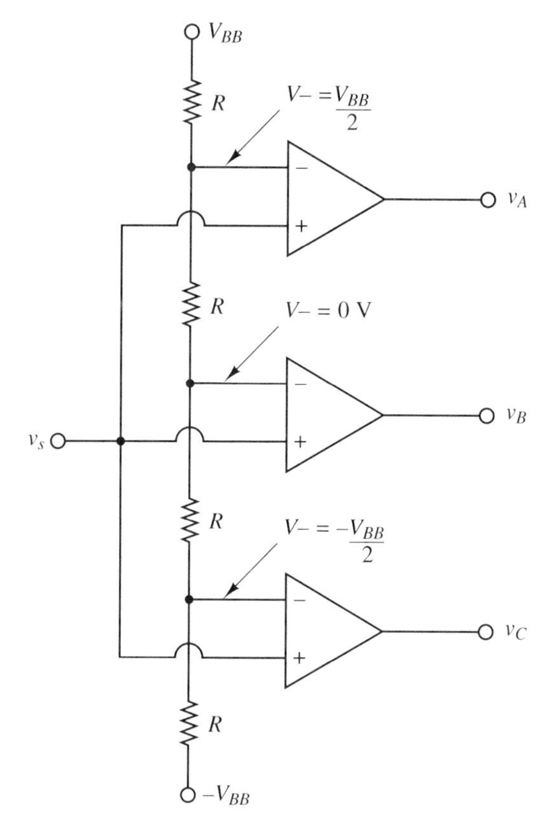


**Figure 8.16** Simplified "flash" analog–digital converter.

to noise and other errors) at a given time within some specifiable range. But the computer can handle only a small number of significant digits, coded in binary form (ones and zeros).

The circuit in Fig. 8.16 can convert a signal $v_s$ taking on a continuous range of values to one of four values. Of course, in a real conversion system there must be many more than four possible values! But the concept could also be used as a range indicator, in which case the distinction among four ranges might be sufficient.

Circuit analysis proceeds as follows. Assume ideal op amps so that zero current flows into the op amp input terminals. Then the voltages of the nodes along the resistor chain that are connected to the inverting terminals of the op amps are evenly distributed between $V_{BB}$ and $-V_{BB}$. Thus for the top op amp $v_- = V_{BB}/2$, for the middle op amp $v_- = 0$ V, and for the bottom op amp $v_- = -V_{BB}/2$. Considering each op amp to be a comparator, the output behavior is summarized in Table 8.1.

**TABLE 8.1** CODING SCHEME FOR SIMPLE A/D CONVERTER

| Range for $v_s$ | $v_A/V_{BB}$ | $v_B/V_{BB}$ | $v_C/V_{BB}$ |
|---|---|---|---|
| $v_s > V_{BB}/2$ | 1 | 1 | 1 |
| $V_{BB}/2 > v_s > 0$ | −1 | 1 | 1 |
| $0 > v_s > -V_{BB}/2$ | −1 | −1 | 1 |
| $-V_{BB}/2 > v_s$ | −1 | −1 | −1 |

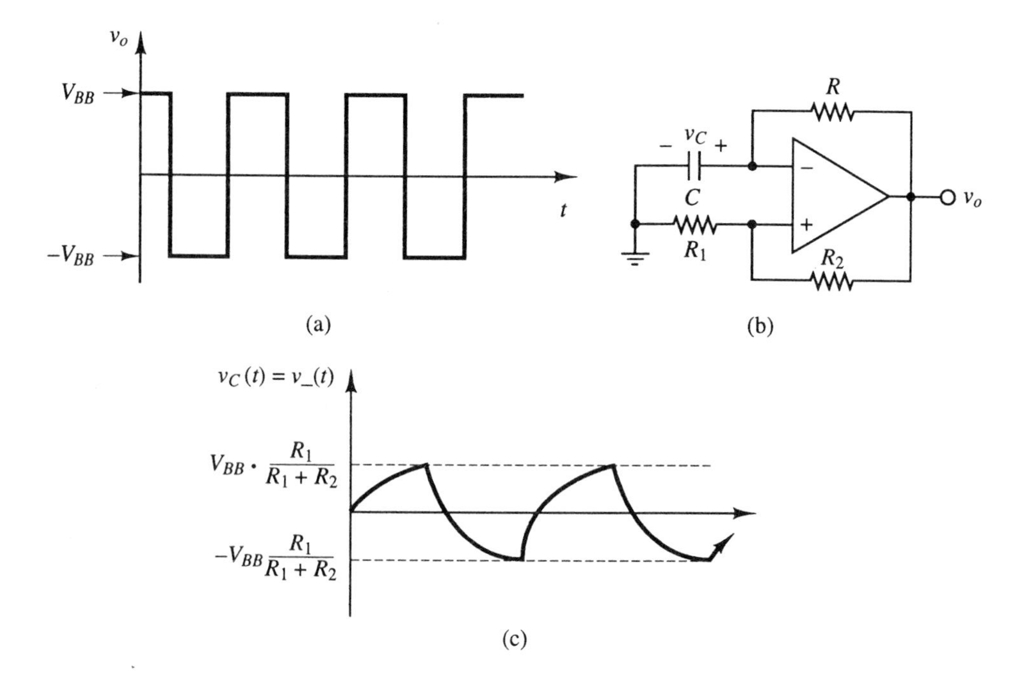


**Figure 8.17** (a) Pulse train; (b) pulse generator circuit; (c) voltage waveform of capacitor in (a).

In a computer, the −1 could be interpreted as zero. In practice, a 0-V and 5-V system is used for digital inputs, and A/D converters are designed to produce these. Because of its high speed relative to other A/D methods, the structure in Fig. 8.16 is known as a flash converter; see Sec. 9.7.6 for more details on A/D conversion.

### 8.8.3 Positive Feedback Circuits

**Schmitt trigger.** This circuit was discussed in detail in Sec. 8.3. The threshold voltage is

$$V_{\text{th}} = \frac{R_1}{R_2} V_{BB}. \tag{8.42}$$

The circuit is shown in Fig. 8.8 and the transfer characteristic appears in Fig. 8.9.

**Pulse generator.** In general, one of the main uses of positive feedback is the generation of many types of waveforms. In particular, availability of a pulse train such as that shown in Fig. 8.17a is necessary for the operation of many digital and analog signal processing systems. An interesting way to obtain one is given by the simple op amp circuit in Fig. 8.17b. Because a sample of the output is fed into the noninverting terminal, we could classify this circuit as having positive feedback. Furthermore, its behavior consists of rapid transitions between, and longer durations

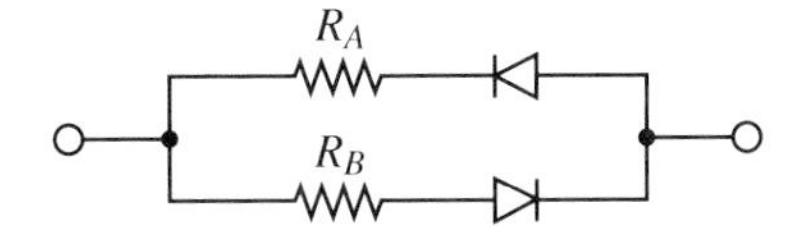


**Figure 8.18** Replacement for $R$ in Fig. 8.17b to achieve asymmetrical pulse train (see Fig. 8.19).

at the saturation values $\pm V_{BB}$. Suppose that $v_o$ is currently equal to $V_{BB}$. Then, using the ideal op amp model,

$$v_+ = \frac{R_1}{R_1 + R_2} V_{BB}. \tag{8.43}$$

Capacitor $C$ charges through $R$ toward $V_{BB}$ until $v_- = v_C$ exceeds $v_+$ as given by Eq. (8.43). After this occurs, $v_o$ swings to $-V_{BB}$, causing $v_+$ to become

$$v_+ = -\frac{R_1}{R_1 + R_2} V_{BB}. \tag{8.44}$$

Again capacitor $C$ charges through $R$, now toward $-V_{BB}$, until $v_- = v_C$ drops below $v_+$ as given in Eq. (8.44). At that point, $v_o$ swings back to $V_{BB}$ and the entire cycle repeats itself. The waveform of the capacitor voltage is shown in Fig. 8.17c.

If $R_1 = R_2$, the period of the pulse waveform produced and depicted in Fig. 8.17 will be approximately $2RC$. If $R$ is adjustable, the oscillator becomes a variable-frequency oscillator. Finally, if $R$ is replaced by the circuit fragment shown in Fig. 8.18, where $R_A$ and $R_B$ have different values, the output pulse train can be made to be asymmetrical, as shown in Fig. 8.19. In this case $R_A$ is used in charging $C$ toward $V_{BB}$, while $R_B$ is used in charging $C$ toward $-V_{BB}$. (At any one time, only one of the two diodes is forward biased.)

**Example 8.3**

Show that the period of the pulse generator in Fig. 8.17 is indeed approximately $2RC$ for $R_1 = R_2$.

**Solution** Consider the output swing from $+V_{BB}$ to $-V_{BB}$. At the beginning of this part of the cycle, the capacitor voltage $v_C = [R_1/(R_1 + R_2)]V_{BB} = V_{BB}/2$. From this value the capacitor voltage swings toward $-V_{BB}$ as the capacitor charges negatively, as discussed above. From our study of transients (Sec. 4.13), the form of $v_C(t)$ is $v_C(t) = Ae^{-t/RC} + v_C(\infty)$. But $v_C(\infty)$ was just stated to be $-V_{BB}$. And at $t = 0$, $v_C$ is $V_{BB}/2$, so that $V_{BB}/2 = A - V_{BB}$ or $A = 1.5V_{BB}$. Therefore, $v_C(t) = V_{BB}(1.5e^{-t/RC} - 1)$. The time $T/2$ at which $v_C(t)$ reaches the other threshold, $-V_{BB}/2$ will be half the period $T$ of $v_C(t)$. After $v_C(t)$ reaches $-V_{BB}/2$ it begins to charge toward $V_{BB}$ until it reaches $V_{BB}/2$, at which point the entire cycle starts again. Therefore, $-V_{BB}/2 = V_{BB}(1.5e^{-T/(2RC)} - 1)$ or $1 = 3e^{-T/(2RC)}$, so that $T = -2RC\ln(\frac{1}{3}) \approx 2.2RC$.

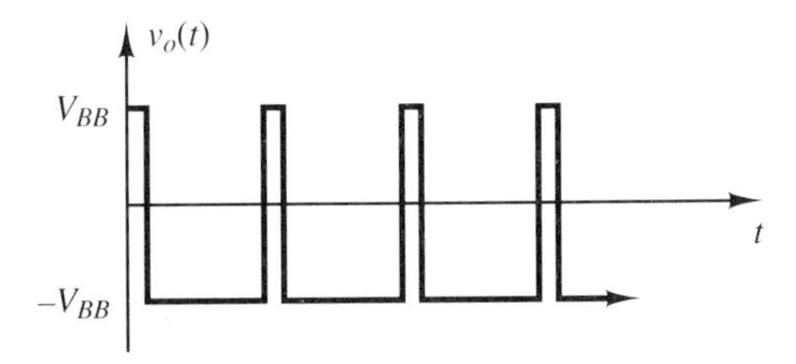


**Figure 8.19** Asymmetrical pulse train using the circuit fragment of Fig. 8.18 for $R$ in Fig. 8.17b, with $R_B > R_A$.

### 8.8.4 Negative Feedback Circuits

**Voltage follower (or buffer).** The input–output relation is trivially $v_o = v_s$. As noted in Sec. 8.7, the purpose of this circuit might be impedance transformation, due to its very high input impedance (Eq. (8.39a) with $R_2 = 0$, or $Z_{in} = A_{ol}Z_i$) and very low output impedance (Eq. (8.39b) with $R_2 = 0$, or $Z_{out} = Z_o/A_{ol}$). It can also be used to increase the number of output circuits $v_s$ can drive.

The circuit is shown in Fig. 8.20a and the input–output characteristic is given in Fig. 8.20b. It is merely a noninverting amplifier with $R_2 = 0$ [see Eq. (8.29)]. The purpose of $R$ is to provide a path for bias currents of the op amp transistors; let it not be forgotten that an op amp is, after all, a (glorified) transistor amplifier. It also minimizes current to protect the op amp in the event that one of the power supplies ($V_{BB}$ or $-V_{BB}$) fails.

**Noninverting amplifier.** This circuit has been discussed in detail. Figure 8.12 shows the circuit configuration and Fig. 8.21 displays the input–output characteristic. The input–output relation in the linear region is

$$v_o = \left(1 + \frac{R_2}{R_1}\right) v_s. \tag{8.45}$$

The input impedance is roughly $Z_i A_{ol} F$, where $F = R_1/(R_1 + R_2)$ and $Z_i$ is the open-loop output impedance of the op amp. The output impedance of the noninverting amplifier is roughly $Z_o/(A_{ol} F)$, where $Z_o$ is the open-loop output impedance of the op amp.

**Inverting amplifier.** Again, this circuit was discussed in detail in Sec. 8.6. The circuit is shown in Fig. 8.13 and the input–output characteristic is graphed in Fig. 8.22. The input–output relation in the linear region is

$$v_o = -\left(\frac{R_2}{R_1}\right) v_s. \tag{8.46}$$

The input impedance of the inverting amplifier is roughly $R_1$ and the output impedance is roughly $Z_o/(A_{ol} F)$, where $F = R_1/(R_1 + R_2)$. Never forget from where the

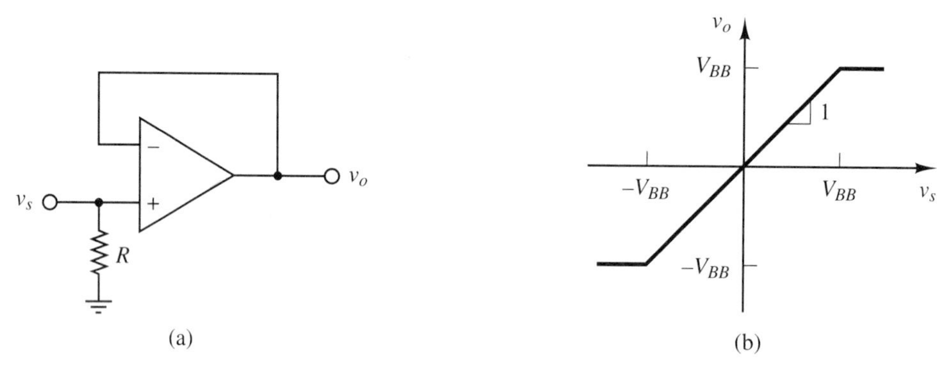


**Figure 8.20** Voltage follower: (a) circuit diagram; (b) input–output voltage characteristic.

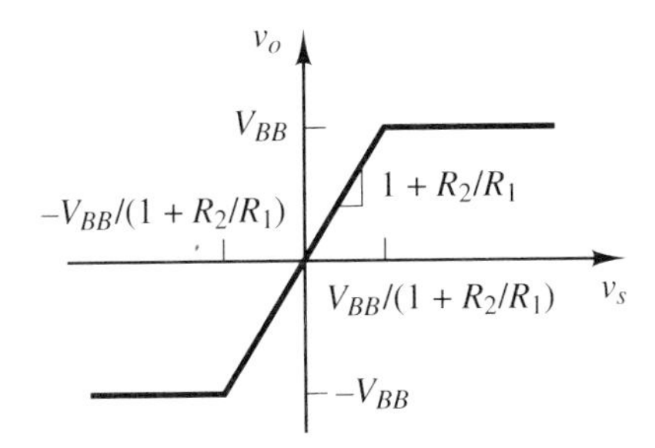


**Figure 8.21** Noninverting amplifier input–output voltage characteristic.

power for amplification comes: from the power supplies $V_{BB}$, $-V_{BB}$, which are often not explicitly shown. The signal $v_s$ is our foot on the gas pedal, and $V_{BB}$ is the engine.

A special case is $R_1 = R_2$, for which $v_o = -v_s$. Such a circuit, called an inverting (voltage) follower, is useful to stick in wherever a given circuit is producing a signal of the undesired polarity.

**Inverting (weighted) summer.** The ability to calculate a weighted sum of analog signals comes in handy frequently. A circuit that performs this operation and incidentally inverts the sign of each summand is shown in Fig. 8.23. Although three input signals are shown, any number could be included by introducing more branches. The circuit is merely a multiple-input inverting amplifier with more inputs tacked on (and thus more terms in the KCL equation). So the input–output relation simply contains $-R_2/R_1$ factors with different values of $R_1$ for each input:

$$v_o = -R_2\left(\frac{v_A}{R_A} + \frac{v_B}{R_B} + \frac{v_C}{R_C}\right). \tag{8.47}$$

One familiar application is the audio mixer. Here the input voltages could be the signals from various microphones picking up sounds from various instruments or voices. To be flexible, $R_A$, $R_B$, and $R_C$ would be variable and set with adjustable knobs by the user. For concreteness, consider the following situation. Suppose that Joe, whose microphone is on input $A$, sings well. On the other hand, Jerry, on input $B$, cannot carry a tune. Then in Fig. 8.23, we would choose $R_B \gg R_A$. And if Jerry is simply intolerable, select $R_B = \infty$ (i.e., cut him off)!

Another simple application would be linear conversion. For example, suppose that we desire to compare the temperatures measured by two transducers but that one transducer is calibrated for Fahrenheit degrees while the other is calibrated for Celsius degrees. The Celsius reading could be converted to Fahrenheit by

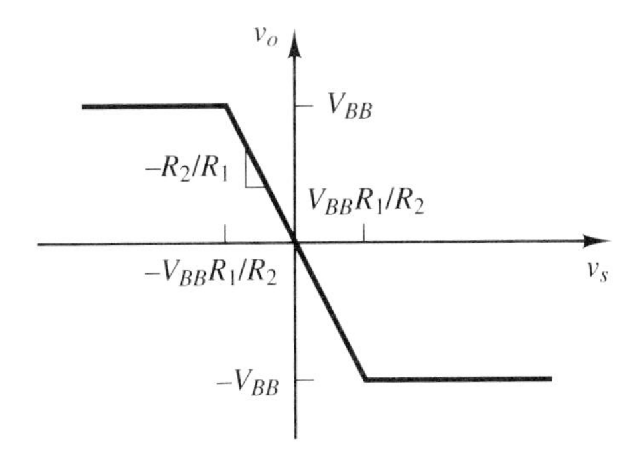


**Figure 8.22** Inverting amplifier input–output voltage characteristic.

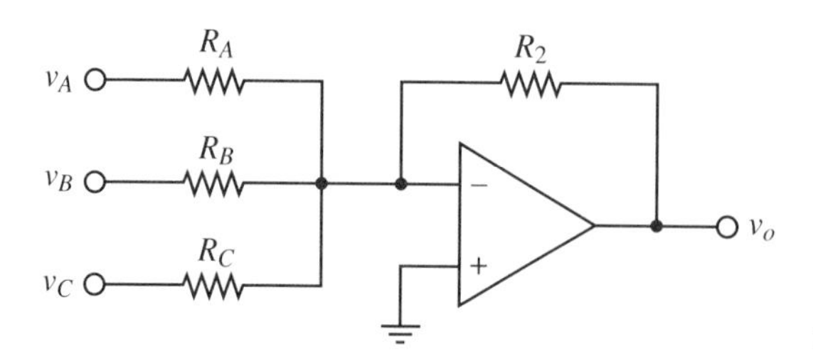


**Figure 8.23** Inverting summer.

$$T_{\text{Far}} = 32° + \tfrac{9}{5} T_{\text{Cel}}. \tag{8.48}$$

The circuit of Fig. 8.23 could accomplish this using only two inputs by letting $v_A$ be a fixed voltage representing 32°, $v_B$ be the measured Celsius temperature reading, and $R_A$ and $R_B$ be chosen to create the coefficients in Eq. (8.48). Actually, it would be helpful to send $v_o$ through an inverting follower to obtain positive coefficients for $v_A$ and $v_B$ in the final output.*

**Differential amplifier; nonideal op amp behaviors.** So far, all the negative feedback amplifiers have used single-ended input signals. This is fine for strong signals originating physically close to the amplifier in a low-noise environment. The "other end" is, of course, implicitly the ground potential.

A circuit is said to be *grounded* if one ("common") node in the circuit is connected to earth. Grounding removes the potential difference between that point of the circuit and earth and provides a path for current to flow should a fault (short circuit) occur. Most high-power circuitry is required to be grounded, as discussed in Sec. 6.8.

But grounding—particularly at more than one physical location—is undesirable for low-level signal circuitry. Often, noise can originate in the ground connection, as the massive ground conductor acts as an antenna to all sorts of noise signals.

Effectively, the situation might appear as shown in Fig. 8.24. A low-level transducer some distance from the amplifier has one end grounded and the other end connected to the amplifier by way of a coaxial cable. The signal wire is encased by a ground-connected sheath to shield the signal electrically from airborne noise signals. The sheath itself has nonzero resistance $R_s$, and the ground acts as a noise voltage source $v_n$, as shown in Fig. 8.24. Circulating currents result which distort the measured signal submitted to the single-ended amplifier.

A way to largely overcome this problem is first to ground the sheath at only the amplifier side. It is still helpful to ground the shield at one end, because this provides a harmless, safe path for all noise-induced currents in the sheath. Now the sheath effectively protects both "ends" of the input voltage source $v_s$, as they are now individual wires running through the sheath to a *differential* (two-"ended") amplifier. Often, the wires are twisted in a helix (*twisted pair*) in the coaxial cable (*coax*; biaxial or *biax* with two signal wires) to further minimize noise. Figure 8.25

* A noninverting summer is also possible, which would obviate the inverting follower required above. However, it has at least two disadvantages: (1) the formula for $v_o$ is more complicated than is that in Eq. (8.47), and (2) the summing point is not at zero volts, so there is *crosstalk*—interference between the different input voltage contributions. So it will not be discussed further here. Of course, for audio applications, the sign inversion is inconsequential.

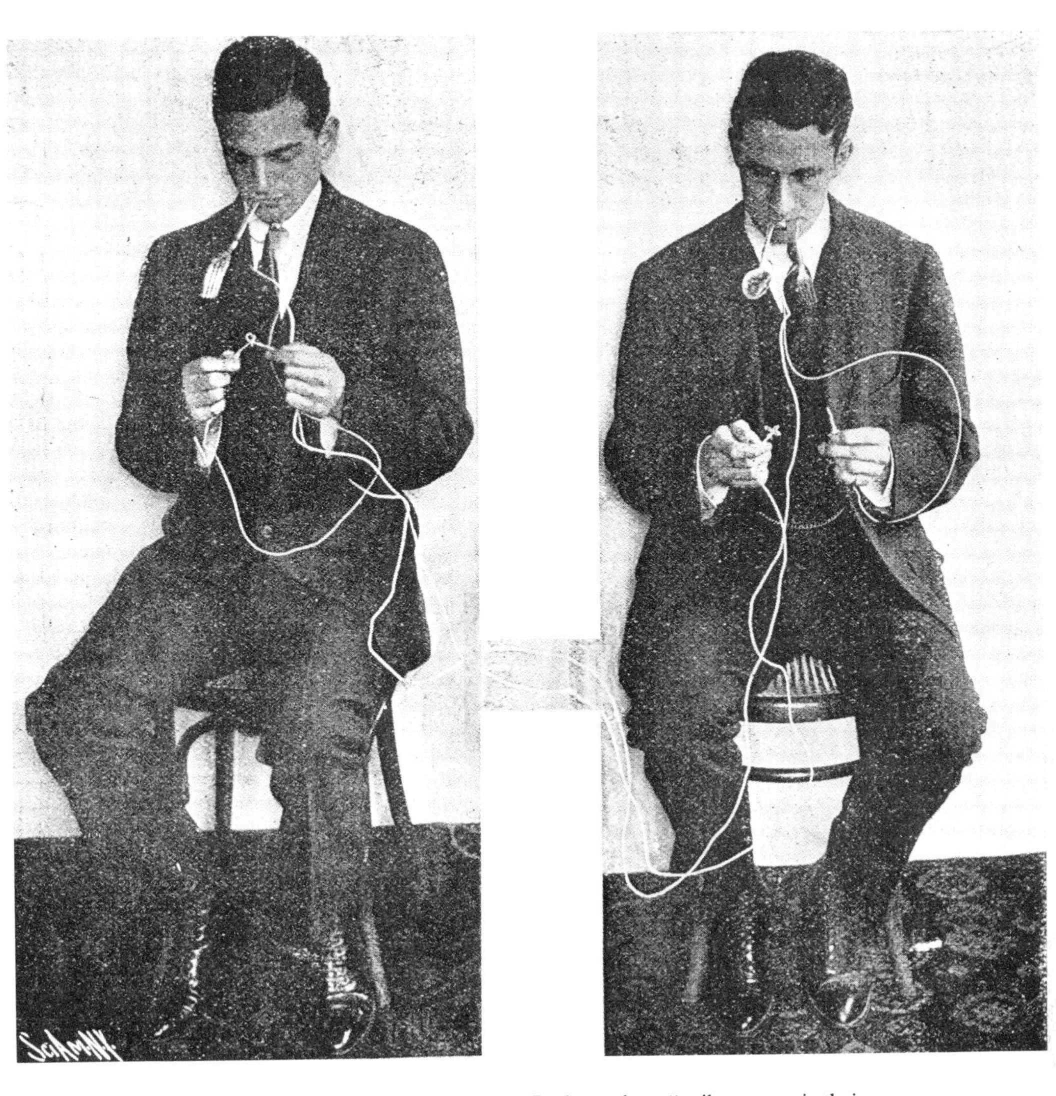

Oral system of telegraphy via human battery. Both men have "a silver spoon in their mouth," as well as an aluminum fork. The receiver holds the wires together, while the Morse code sender connects and disconnects to form the message. The receiver perceives the message as a strong, pungent taste. This is the contact potential due to the alkaline saliva over the aluminum, much like a battery (see Sec. 2.8). Compared to modern analog signal transmission systems using operational amplifiers, this system "would not make it to first base." *Source: Scientific American* (May 30, 1908).

illustrates the arrangement, where the squared-off triangle is the standard symbol for a differential amplifier.

The output of the differential amplifier is $v_o = K(v_B - v_A)$, where $v_A$ and $v_B$ are the potentials of the two wires coming from the transducer and $K$ is the constant gain, designed to be $\gg 1$ for amplification. An op amp circuit to implement this is shown in Fig. 8.26. [This could not be accomplished using either inverting or noninverting

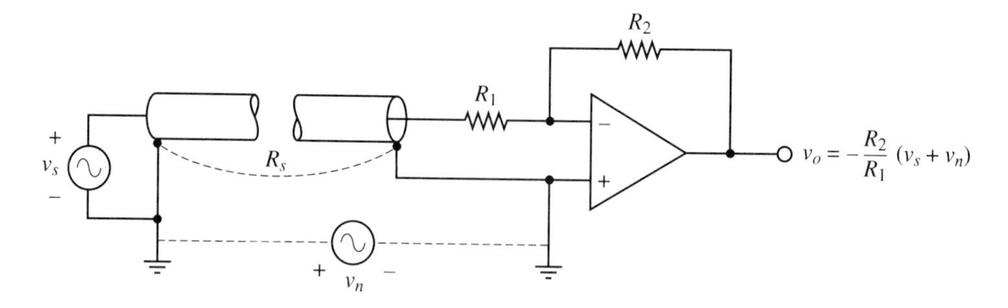


**Figure 8.24** Amplification circuit showing noisy ground loop.

amplifiers because they have single-ended inputs. Also, contrast the differential amplifier, which finds the difference between two signals, with the inverting summer, which finds their sum $\cdot(-1)$.] Using the ideal op amp model, define voltage $v_1$ as $v_1 = v_+ = v_-$. Then KCL at the inverting terminal node gives

$$\frac{v_1 - v_A}{R} + \frac{v_1 - v_o}{KR} = 0, \tag{8.49}$$

and at the noninverting node,

$$\frac{v_1 - v_B}{R} + \frac{v_1}{KR} = 0. \tag{8.50}$$

Solving Eq. (8.50) for $v_1$ gives

$$\begin{aligned} v_1 &= \frac{KR}{R(1+K)} v_B \\ &= \frac{K}{1+K} v_B. \end{aligned} \tag{8.51}$$

Solving Eq. (8.49) for $v_o$ gives

$$\begin{aligned} v_o &= KR\left(\frac{R+KR}{KR^2} v_1 - \frac{1}{R} v_A\right) \\ &= K\left(\frac{1+K}{K} v_1 - v_A\right). \end{aligned} \tag{8.52}$$

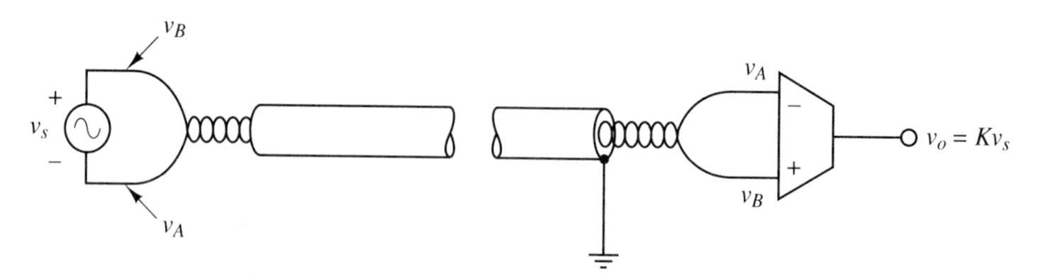


**Figure 8.25** Use of differential input and differential amplifier connected by twisted pair wires to minimize noise.

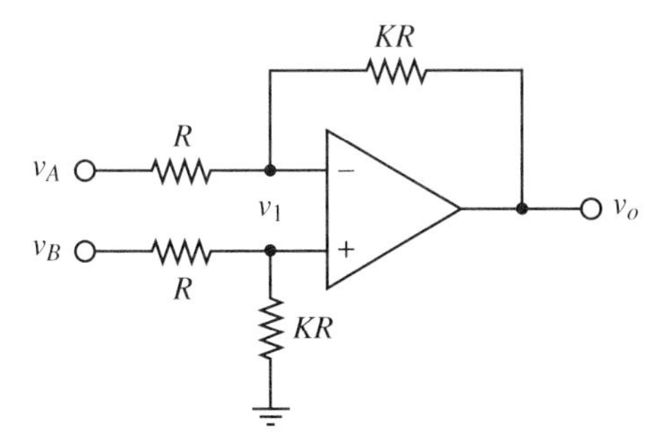


**Figure 8.26** Differential amplifier.

Finally, substitution of $v_1$ from Eq. (8.51) into Eq. (8.52) gives for $v_o$,

$$v_o = K(v_B - v_A), \tag{8.53}$$

the desired result: $v_o$ is just a constant $K$ (called the *differential gain*) times the differential voltage $v_B - v_A$).

The input impedance to ground as seen by $v_A$ is approximately $R$, using the same reasoning as used to find the input impedance of the inverting amplifier. The input impedance to ground seen by $v_B$ is dominated by $R + KR$. So the impedance looking into the terminals $v_A$ and $v_B$ is only on the order of $R$.

Up to this point, various imperfections of the real-world op amp have been ignored. But *anyone* using op amps must be aware of the most significant limitations.

One limitation is the finite bandwidth of the op amp. That is, the open-loop gain depends (and when severely, therefore the closed-loop gain also depends) on the frequency of the input. For example, for dc the gain is on the order of 200,000. For a 1-kHz input, the gain is only 1000, and for 500 kHz it is reduced to unity (doubling the frequency halves the gain). These results are approximate for the 741 IC chip. However, there are plenty of high-bandwidth op amps around, up into the hundreds of megahertz. There are even gallium arsenide amplifiers whose bandwidth is 3 GHz, costing only $8.

Two other important imperfections related to each other, yet distinct, are offset voltage and common-mode gain. Consider Fig. 8.27, where a single voltage $v_c$ is applied to *both* the noninverting and inverting terminals. No matter what the value of $v_c$ may be, the voltage between the two op amp input terminals is exactly zero, so $v_o$ *should* equal zero.

In a real (nonideal) op amp, $v_o$ is not exactly zero. Suppose that $v_c = 0$; equivalently, both input terminals are connected to ground. The output $v_o = v_{\text{off}}$ is what it would be were a nonzero input voltage, $v_{i,\text{off}}$, applied to an ideal opamp. For a 741, $v_{i,\text{off}}$ may reach 1 mV. This voltage when multiplied by $A_{\text{ol}}$ to yield $v_{\text{off}}$, will clearly saturate at $V_{BB}$ for an open-loop op amp. For small input signals, $v_{\text{off}}$ will completely invalidate measurements, assuming the ideal op amp model.

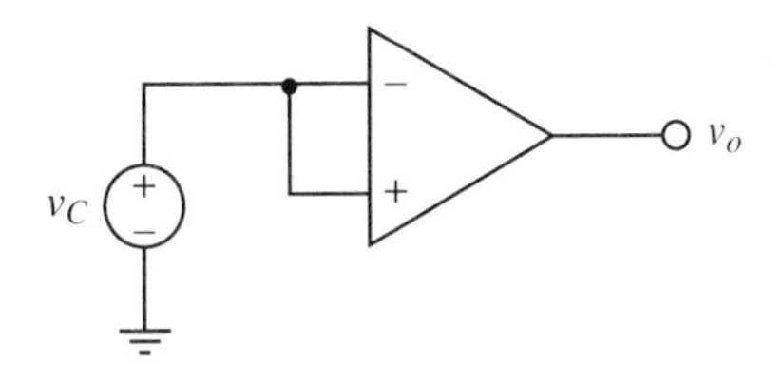


**Figure 8.27** Circuit for determining common-mode gain. Offset voltage may also be obtained by setting $v_c = 0$.

To minimize $v_{\text{off}}$, an adjustment called *offset nulling* can be arranged. For example, a 741 has two pins to which a potentiometer can be connected for this purpose. To perform nulling in a feedback connection (the usual situation), we ground the two input terminals and adjust the "trimming" potentiometer for minimum $v_o$. If no such terminals are provided, nulling can be implemented at a variety of places in the external circuitry.

If $v_c$ in Fig. 8.27 is greater than zero, $v_o$ will, unfortunately, rise in proportion to $v_c$, even though it *should* be zero independent of $v_c$ because the input terminals are connected together. This annoying gain is called the common-mode gain.

To see how this may be relevant to practical amplification of transducer signals, consider the inputs $v_A$ and $v_B$, shown in Fig. 8.28a. They can be decomposed as indicated in Fig. 8.28b; check for yourself to see that the *common-mode signal* plus the difference signal leading to $v_A$ and $v_B$ equal, respectively, $v_A$ and $v_B$. Also notice that the difference-mode component of $v_B$ minus that of $v_A$ comes out to $v_B - v_A$, as it must.

The common-mode signal is just the arithmetic average of $v_A$ and $v_B$. It is called "common" because it contributes equally to both $v_A$ and $v_B$ in this decomposition. Common-mode signals often originate from noise picked up in the wires running from a measurement transducer to a remote amplifier, a frequent situation in a noisy industrial environment. Because both wires are exposed to the same noise, the noise signal will be common to both wires, hence the term "common-mode signal."

The response of the op amp to common-mode and difference-mode signals is summarized in Fig. 8.29. In Fig. 8.29a the offset voltage $v_{\text{off}}$ and common-mode gain $A_{\text{cm}}$ of an op amp (typical value is 6 for a 741 op amp) are shown on a plot of $v_{o,\text{cm}}$ versus the common-mode signal $\frac{1}{2}(v_A + v_B)$. This plot is mainly for conceptual understanding; in practice, as noted above, $v_{\text{off}}$ would actually be saturated at $V_{BB}$. In that case, Fig. 8.29a would refer to an almost perfectly nulled common-mode response. However, in practice one *will* approach approximate nulling by using a nulling potentiometer, so actually Fig. 8.29 is not too far from reality.

The desired difference-mode gain $A_{\text{dm}}$ (which is just $A_{\text{ol}}$), is shown in Fig. 8.29 on a plot of $v_{o,\text{dm}}$ vs. $v_B - v_A$. The total output voltage is $v_o = v_{o,\text{cm}} + v_{o,\text{dm}}$. The ratio $A_{\text{ol}}/A_{\text{cm}}$ is called the *common-mode rejection ratio* (CMRR). Because an ideal op amp would have $A_{\text{cm}} = 0$, it would have an infinite CMRR. A typical op amp has a common-mode rejection ratio on the order of $10^4$.

But what is the CMRR of a differential amplifier with feedback? Assume that the op amp alone (with no feedback connections) has differential gain $A_{\text{ol}}$ and

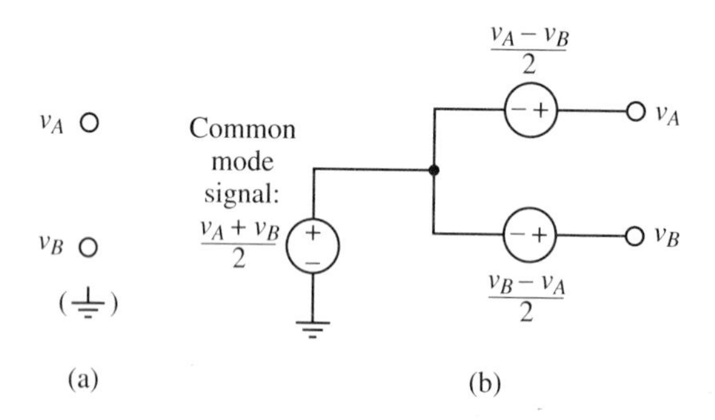


**Figure 8.28** (a) Two signal voltages defined with respect to ground; (b) decomposition of signals in (a) into common-mode and difference-mode signals.

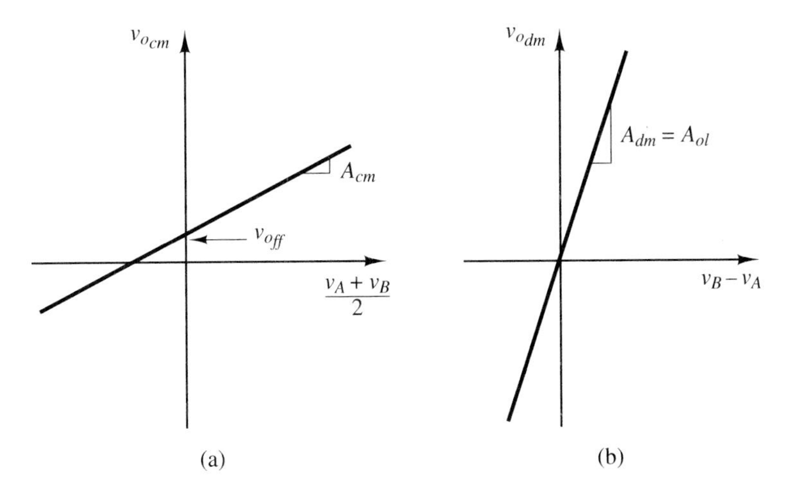


**Figure 8.29** (a) Conceptual diagram showing common-mode output versus common-mode input; (b) difference-mode output versus difference-mode input.

common-mode gain $A_{\text{cm}}$. The differential gain of the differential amplifier was found in Eq. (8.53): it is $A_{\text{dm},f} = K$, where the subscript $f$ indicates that this differential amplifier has feedback.

The common-mode gain $A_{\text{cm},f}$ remains to be determined. Suppose that in Fig. 8.26 we apply a common-mode voltage $v_c$ to both inputs: $v_A = v_B = v_c$. The output $v_o$ for nonzero $A_{\text{cm}}$ is

$$v_o = A_{\text{cm}} \cdot \tfrac{1}{2}(v_+ + v_-) + A_{\text{ol}}(v_+ - v_-), \tag{8.54}$$

assuming zero offset voltage. It should be noted that even though $v_A = v_B = v_c$, the voltages $v_-$ and $v_+$ may not be equal. Voltage $v_+$ may be found by a voltage divider (assuming that $i_+ = i_- = 0$):

$$v_+ = \frac{KR}{R + KR} v_c = \frac{K}{1 + K} v_c. \tag{8.55}$$

Voltage $v_-$ may be found by KCL applied at the inverting input node:

$$\frac{v_- - v_c}{R} + \frac{v_- - v_o}{KR} = 0,$$

or

$$v_- \left( \frac{1}{R} + \frac{1}{KR} \right) = \frac{v_c}{R} + \frac{v_o}{KR},$$

or

$$v_- = \frac{K}{1 + K} v_c + \frac{1}{1 + K} v_o. \tag{8.56}$$

Substituting $v_+$ [Eq. (8.55)] and $v_-$ [Eq. (8.56)] into Eq. (8.54) gives for the output voltage,

$$v_o = \frac{1}{2} A_{\text{cm}} \left( \frac{2K}{1 + K} v_c + \frac{1}{1 + K} v_o \right) - A_{\text{ol}} \frac{1}{1 + K} v_o. \tag{8.57}$$

Because the third of the three coefficients of $v_o$ in Eq. (8.57) dominates ($A_{ol} \gg A_{cm}$), the other two can be ignored. Therefore, the common-mode gain for the feedback differential amplifier, $v_o/v_c$, is approximately

$$A_{cm,f} = \frac{v_o}{v_c} = \frac{A_{cm}K}{A_{ol}}. \tag{8.58}$$

Finally, the CMRR is just

$$\text{CMRR} = \frac{A_{dm,f}}{A_{cm,f}} = \frac{A_{ol}}{A_{cm}}, \tag{8.59}$$

which is the same as the CMRR of the op amp alone. The value, on the order of $10^4$, is quite high, but can be improved significantly by using the instrumentation amplifier, to be discussed next.

**Example 8.4**

An op amp has a CMRR of 70 dB and $A_{ol} = 2 \cdot 10^5$. If it is used as a differential amplifier with gain $K = 100$ (see Fig. 8.26), what is $v_o$ if $v_A = 10$ V and $v_B = 10.05$ V? The situation of a signal only a few millivolts in magnitude (here 50 mV) riding on a common-mode signal of several volts occurs frequently in practice.

**Solution** Often the CMRR is expressed as above in decibels. Converting back from the logarithmic scale gives CMRR $= 10^{70/20} = 3200$. Therefore, $A_{cm} = A_{ol}/\text{CMRR} = 2 \cdot 10^5/3200 \approx 63$. By analogy with Eq. (8.54),

$$v_o = A_{cm,f}\frac{v_A + v_B}{2} + A_{dm,f}(v_B - v_A),$$

assuming zero offset voltage. From Eq. (8.58), $A_{cm,f} = 63(100)/(2 \cdot 10^5) = 0.032$, while as before, $A_{dm,f} = K = 100$. Therefore, $v_o = 0.032(10 + 10.05)/2 + 100(10.05 - 10) = 5.32$ V. So in this case the common-mode signal caused a 6.4% deviation from the expected output of 5.00 V.

**Instrumentation amplifier.** The differential amplifier has two obvious problems: Low input impedance (on the order of $R$) and a gain $K$ that can be adjusted only by simultaneously and accurately adjusting *two* resistors ($KR$). We can tack on two noninverting amplifiers to the inputs of the differential amplifier to eliminate the first problem while making connections to eliminate the second. A further bonus, discussed below, is an improvement in the CMRR.

A schematic diagram for the instrumentation amplifier is shown in Fig. 8.30. The name *instrumentation amplifier* derives from its popularity (due to the advantages noted above) as the input stage of much instrumentation. Its high input impedance [see Eq. (8.39a)] stems from the use of noninverting amplifiers and makes it optimal for amplifying weak signals from transducers. Usually, the entire circuit comes in one package, so the various resistors are well balanced (highly accurate and uniform).

Analysis is easy using the ideal op amp model. First note that the potential of the inverting terminal of the $A$ input op amp is $v_A$ because $v_- = v_+$. Similarly, the potential of the inverting terminal of the $B$ input op amp is $v_B$. Because $i_+ = i_- = 0$ for both input op amps, the current flowing through both feedback resistors $R_2$ has

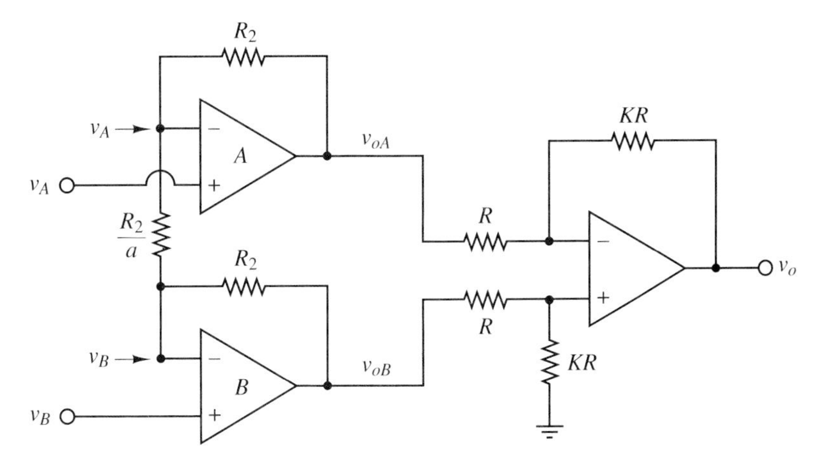


**Figure 8.30** Instrumentation amplifier.

magnitude $(v_A - v_B)/(R_2/a)$. Using this fact and carefully noting the direction of that current, the outputs of the two noninverting amplifiers are found to be

$$\begin{aligned} v_{oA} &= v_A + \frac{v_A - v_B}{R_2/a} R_2 \\ &= v_A + a(v_A - v_B), \end{aligned} \tag{8.60}$$

and similarly,

$$v_{oB} = v_B - a(v_A - v_B). \tag{8.61}$$

From Eq. (8.53), the output of the instrumentation amplifier (which is the output of the differential amplifier stage) is

$$v_o = K(v_{oB} - v_{oA}), \tag{8.62}$$

where from Eqs. (8.60) and (8.61),

$$\begin{aligned} v_{oB} - v_{oA} &= v_B - v_A + 2a(v_B - v_A) \\ &= (1 + 2a)(v_B - v_A). \end{aligned} \tag{8.63}$$

Combining Eqs. (8.62) and (8.63) yields

$$v_o = [K(1 + 2a)](v_B - v_A). \tag{8.64}$$

Thus the instrumentation amplifier acts as a differential amplifier of gain $K(1 + 2a)$. With $K$ fixed by the manufacturer, the resistor $R_2/a$ may be varied by the user, thereby varying the differential gain with a single resistor without upsetting the balance.

It was claimed above that the common-mode rejection of the instrumentation amplifier is superior to that of the differential amplifier. Because the gains of cascaded stages multiply, the CMRR will be the product of the CMRR of the input stage and the CMRR of the differential amplifier (output) stage. Consider the input stage as a black box with two inputs $v_A$ and $v_B$, and outputs $v_{oA}$ and $v_{oB}$, as in Fig. 8.31.

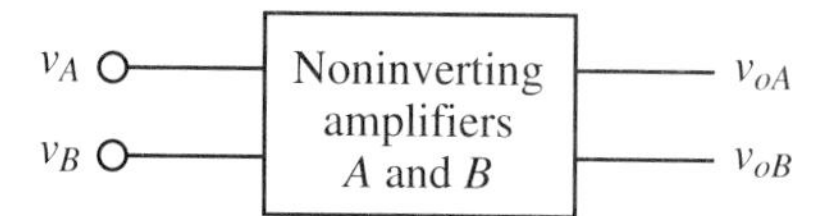


**Figure 8.31** Two-port model of instrumentation amplifier dual-op amp input stage.

The differential-mode gain is just the ratio of the differential output signal component to the differential-mode input signal component:

$$A_{\text{dm, in}} = \frac{v_{oB} - v_{oA}}{v_B - v_A} = 1 + 2a, \tag{8.65}$$

where the second step follows from Eq. (8.63). The common-mode gain of the input stage is analogously defined as

$$A_{\text{cm, in}} = \frac{\frac{1}{2}(v_{oB} + v_{oA})}{\frac{1}{2}(v_B + v_A)} = 1, \tag{8.66}$$

where the second step follows from adding $v_{oA}$ and $v_{oB}$ in Eqs. (8.60) and (8.61). Therefore, the CMRR of the input stage is $(1 + 2a)/1 = 1 + 2a$. Then the overall CMRR of the instrumentation amplifier is

$$\text{CMRR}_{\text{inst}} = (1 + 2a)\frac{A_{\text{ol}}}{A_{\text{cm}}}, \tag{8.67}$$

where the second factor, $A_{\text{ol}}/A_{\text{cm}}$, is just the CMRR of the differential amplifier given in Eq. (8.59). It may now easily be seen that if, for example, $a$ is chosen so that $1 + 2a = 100$, the overall CMRR has been increased to the order of 1 million, a factor of 100 over the CMRR of just a differential amplifier alone.

It should also be noted that new *monolithic* (single-chip) instrumentation amplifiers are now available. Their advantages include highly accurate gain (0.05% error), better CMRR (120 dB at a gain of 1000), low offset voltage (10 μV), and low-temperature drift. All of these are important specifications to keep in mind when selecting an instrumentation amplifier. Yet another is the *slew rate,* the maximum rate of change of the output voltage; representative values are 0.2 to 10 V/μs. Because the maximum rate of change (maximum derivative value) of $1\text{ V} \cdot \cos(2\pi ft)$ is $1 \cdot 2\pi f$ V/s, a slew rate of 1 V/μs could follow signals including a maximum frequency of $f = 1/(2\pi \cdot 10^{-6}) = 160$ kHz.

Usually, two to four gain settings are offered, either at different output pins on the chip or by digital selection. If the desired gain is not offered, we can still use external resistors, a cascaded second lower-grade op amp, or achieve it by computer. In the last option, the instrumentation op amp output is presented to an analog-to-digital converter. The output, in binary number form, can then be scaled by the computer already being used to process, interpret, and display the measurement data. More detail on this procedure is considered in the three chapters on digital electronics, Chapters 10 through 12.

*Sense* and *reference* terminals are also sometimes provided on these monolithic instrumentation amplifiers. The sense terminal is merely the output side of the feedback resistor of the last stage, and the reference terminal sets the dc offset.

Normally, the reference terminal is connected to ground and the sense terminal is connected to the output to complete the feedback loop. However, if it is desired that the output signal ride on a dc offset, the reference terminal can be connected to that offset voltage.

If one has a remote load, running wires from both the output and the sense terminals out to the load will eliminate the signal loss due to transmission wire resistance. This is because making the transmission resistance part of the feedback resistance will increase the gain to compensate for that resistance. Finally, there may be a *guard drive* terminal, which, if connected to signal shields, can improve the CMRR.

One last point on multistage amplifiers. In practice, capacitors of 0.1 to 1 μF are placed locally between each of the power supply pins and ground. This is because the power supply lines act as inductors when a signal spike (e.g., 50 ns, 100 mA) occurs at the output of a later stage. This causes a delayed voltage spike to appear on the power supply pins of previous stages unless a capacitor is there to absorb the sudden delayed current pulse. Such spikes are obviously undesirable.

In addition, because $R_s$ of the power supply is nonzero, the power supply voltage to all preceding chips will be affected by currents drawn from the output of a late stage. For these reasons (undesired feedback), it is always good practice to shunt all IC supply pins with capacitors.

**Precision half-wave rectifier.** In Sec. 8.1, a shortcoming of the diode was pointed out: its nonnegligible contact potential (0.7 V for silicon diodes). This makes precise rectification of small signals difficult or impossible. But the op amp circuit shown in Fig. 8.32 acts as a nearly perfect rectifier (essentially zero "contact potential") and can even provide gain.

Here is how it works. First assume that both diodes are reverse biased and $v_s = 0$. First note that assuming $i_+ = 0$, then $v_+ = 0$ independent of the value of $R_3$.* Now suppose that $v_s$ is positive. Then, because $v_s$ is connected (through $R_1$) to the inverting terminal and $v_+$ is zero, the op amp output voltage $v_{oa}$ tries to go large negative. But as soon as it drops 0.7 V below $v_- \approx 0$ V, diode $D_2$ becomes forward biased and acts like a battery of voltage 0.7 V between $v_- = 0$ V and $v_{oa}$. Thus $v_{oa}$ can go no lower than −0.7 V. Meanwhile diode $D_1$ is open. Therefore, $v_o$, the circuit output voltage, is left hanging at 0 V (no current flows through $R_2$).

If $v_s$ is negative, $v_{oa}$ tries to go large positive. Now $D_2$ is reverse biased and $D_1$ is *immediately* forward biased (for $v_s$ *infinitesimally* positive) because of the huge $A_{ol}$ amplifying $v_-$. Think of $D_1$ with its 0.7 V as now absorbed into the op amp; the circuit now looks just like an inverting amplifier with output voltage $v_o$.

The feedback causes the op amp, irrespective of the 0.7 V, to force $v_-$ to approximately zero. So KCL at the inverting node proceeds exactly as in Eqs. (8.38) (the inverting amplifier) with the result that $v_o = -(R_2/R_1)v_s$ for $-V_{BB}(R_1/R_2) < v_s < 0$. Figure 8.32b summarizes with the input–output characteristic. Notice that

* The presence of $R_3$ does not affect ideal op amp model calculations. Its purpose is to minimize $v_{off}$, which is due to input bias currents. It also limits the current into the op amp if one supply ($V_{BB}$ or $-V_{BB}$) fails. It is often included in practical inverting amplifier circuits. For small source resistances, the value of $R_3$ that cancels the bias currents is approximately $R_1 || R_2$.

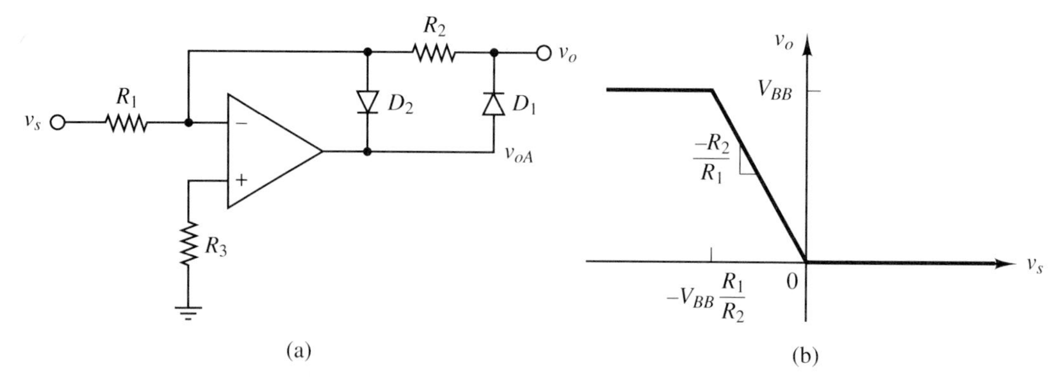


**Figure 8.32** Precision half-wave rectifier: (a) circuit diagram; (b) input–output voltage characteristic.

if $R_2/R_1 > 1$, the circuit not only rectifies but also provides gain; it could therefore be termed an *active rectifier.*

**Example 8.5**

Design an active half-wave rectifier of gain $-5$. What is the maximum amplitude input sinusoid that would not be "clipped" at $+V_{BB} = 15$ V?

**Solution** The gain is $-R_2/R_1 = -5$, so if $R_1 = 10$ kΩ, choose $R_2 = 5$ kΩ. The output will not be clipped if $v_s > -15(\frac{1}{5}) = -3$ V. This means that the maximum amplitude of an input sinusoid would be 3 V (and somewhat less for practical op amps having saturation below $V_{BB}$).

**Integrator.** An op amp circuit was just shown to outperform a simple passive half-wave rectifier. Another op amp circuit will now be shown to provide a better integration operation than that of a simple *RC* passive network. Examine the circuit of Fig. 8.33. It appears to be an inverting amplifier with $R_2$ replaced by a capacitor. What is $v_o$? With the capacitor we should expect time dependence to be crucial. Applying the ideal op amp model, $v_- = 0$ V, so that applying KCL gives

$$-\frac{v_s(t)}{R} + i_C(t) = 0, \tag{8.68}$$

where

$$i_C(t) = -C\frac{dv_o(t)}{dt}. \tag{8.69}$$

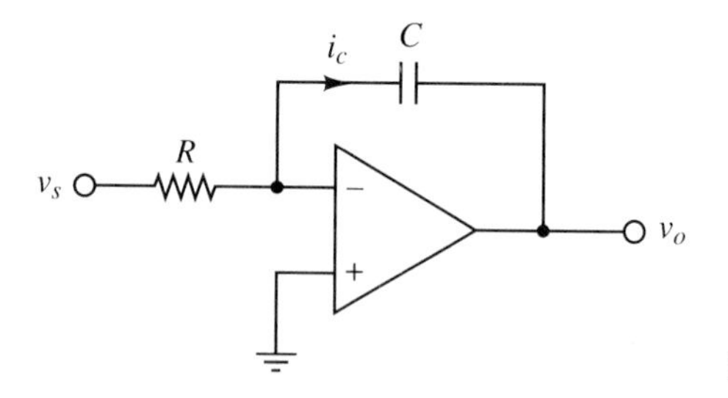


**Figure 8.33** Integrator (low-pass filter).

The solution of the differential equation for $v_o$,

$$\frac{dv_o(t)}{dt} = -\left(\frac{1}{RC}\right)v_s(t), \tag{8.70}$$

is

$$v_o(t) = -\left(\frac{1}{RC}\right)\int_0^t v_s(t')\,dt' + v_o(0), \tag{8.71}$$

where $v_o(0)$ is the value of $v_o(t)$ at $t = 0$.

So, indeed, the op amp is providing time integration of the input signal $v_s(t)$. This circuit may be termed an *active integrator* because it not only integrates but also can amplify if $RC$ is chosen to be less than 1.

One problem is that due to offset and bias currents, $v_o(t)$ tends to "wander off" over long periods of time. So sometimes a large resistor is also connected in parallel with $C$ to prevent too much charge buildup on $C$. Other circuits have a switch placed across $C$ that is closed periodically to "restart" the integration. Because of their extremely small bias currents in the picoamp range (huge input impedance—1/1000 of bipolar op amps), FET-input op amps are recommended for integrators.

How does the behavior of the op amp circuit compare with that of a passive $RC$ network? We recall immediately that for the circuit of Fig. 8.34, the capacitor voltage is the integral of the current:

$$v_C(t) = \frac{1}{C}\int_0^t i_C(t')\,dt' + v_C(0). \tag{8.72}$$

However, $i_C(t)$ is not directly obtainable as an input signal [e.g., $v_s(t)$]. Suppose that the input is a step voltage: $v_s(t) = V_B$ for $t \geq 0$ and zero otherwise. Then in terms of the input signal, $v_C(t)$ was found in Sec. 4.13.1 to be

$$v_C(t) = V_B\,(1 - e^{-t/RC}). \tag{8.73}$$

Yet the true integral of the constant $V_B$ is $(V_B/RC)t$, not the exponential function in Eq. (8.73). Actually, if we examine the plot of $v_C(t)$ in Eq. (8.73) (shown in Fig. 8.35), we find the approximation of $(V_{BB}/RC)t$ to be good for small $t$. [Mathematically, the first term of the Taylor expansion of $v_C(t)$ in Eq. (8.73) is $V_{BB}t/RC$.] Equivalently, for large $RC$ the expression in Eq. (8.73) represents a reasonable integrator. The problem, of course, stems from the differential equation

$$\frac{dv_C(t)}{dt} + \frac{1}{RC}v_C(t) = \frac{v_s(t)}{RC}, \tag{8.74}$$

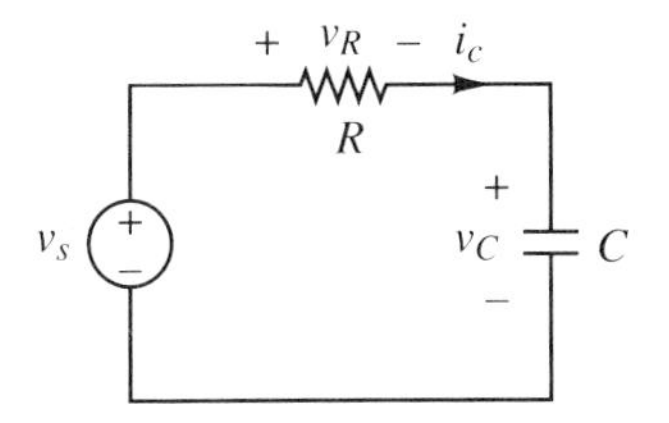


**Figure 8.34** *RC* low-pass filter.

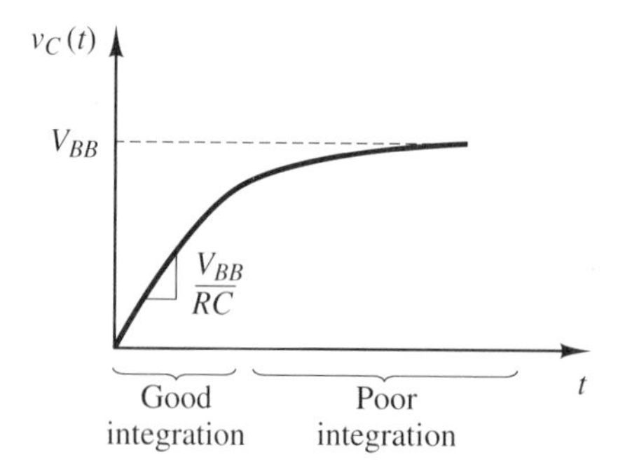


**Figure 8.35** The integral of a constant in time is proportional to time: a ramp. Thus the capacitor voltage of the *RC* filter in Fig. 8.34 approximates an integrator well for small time $t$, but poorly for large $t$.

where it is the presence of the second term that "messes things up." There is no such term in op amp equation (8.70). In fact, the op amp output will follow the $(V_{BB}/RC)t$ dependence for a ramp input until it saturates.

The op amp integrator is used in analog control systems such as PI or PID control (see Sec. 12.6), as well as filtering and signal creation and manipulation applications. As an example of this last application, we merely integrate a square wave (Fig. 8.17) to obtain a triangle wave. Triangle waves are useful, for example, in scanning the electron beam left to right in a TV picture tube.

A further use of op amp integrators is for the improvement of the signal-to-noise ratio (SNR) of dc signals. Recall that averaging is achieved by integration; thus integration performs averaging. Because noise is random, it tends to average to zero over time, while the average of a dc signal is that same dc value. Thus the noise is largely eliminated.

**Differentiator.** The arguments concerning the differentiator are similar to those used above in the discussion of integrators. A circuit diagram for a differentiator is shown in Fig. 8.36. KCL now gives

$$-i_C(t) + \frac{-v_o}{R} = 0, \tag{8.75}$$

where

$$i_C(t) = C\frac{dv_s(t)}{dt}. \tag{8.76}$$

Solving for $v_o(t)$ gives

$$v_o(t) = -RC\frac{dv_s(t)}{dt}. \tag{8.77}$$

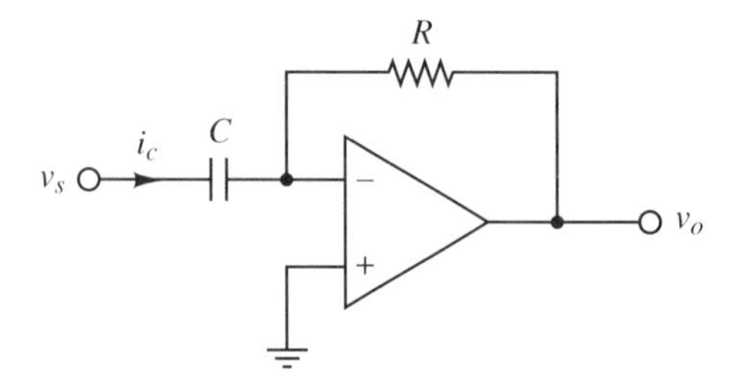


**Figure 8.36** Differentiator (high-pass filter).

Again we search in vain among passive circuits for a true voltage differentiator with strong voltage output (and good isolation characteristics). We can again refer to Fig. 8.34 and now look at $i_C(t)$ (shown in Fig. 8.37) by looking at $v_R(t) = Ri_C(t) = V_{BB}e^{-t/RC}$. If $RC$ is small or we look over a large time window, the pulse appears narrow and looks like the derivative of the step function $v_s(t)$, which is an impulse function $\delta(t)$. But with the op amp circuit, we can have direct differentiation behavior as well as amplification if desired [Eq. (8.77)].

Occasionally, another capacitor is placed across $R$ in Fig. 8.36. If the frequency content of $v_s(t)$ is high enough, some rather complicated arguments show that the phase of the feedback voltage can change 180°, effectively causing *positive* feedback. The result will be the generation of unwanted oscillations and general instability. The additional capacitor effectively eliminates this problem by reducing phase differences.

In any case, it is usually desirable to integrate rather than differentiate if that is possible, because differentiators accentuate high-frequency noise. That is, if we consider the frequency response of a differentiator, the $d/dt$ becomes $j\omega$ in the frequency domain. The magnitude of this frequency response, $\omega$, of course grows with larger frequencies. Often, the higher frequencies are just noise, so the noise is amplified relative to the desired signals. Even small-amplitude high-frequency noise can, under differentiation, drown out the original low-frequency signal.

Nevertheless, there are cases where differentiation is desirable. For example, if one differentiates the pulse waveform in Fig. 8.17a, we obtain a series of extremely short and high pulses that occur at every transition of the pulse train and alternate in sign. Each of these pulses is like the Dirac delta in Figs. A5.1 and A5.2. After rectification, these can be used for displaying the impulse response of a system on an oscilloscope.

Differentiation is also used in image processing (graphics) when *edge detection* is desired. The high frequencies that occur at object boundaries are accentuated. However, usually this type of signal processing is done on a computer, after conversion of the image to digital format, rather than by analog circuitry.

In general, differentiators are useful whenever differentiation provides improved signal shaping. For example, a maximum may be reached gradually, but the first or second derivative may change suddenly at that point. If counting circuits or actuators are to be triggered at the maximum, a differentiator would clearly be an

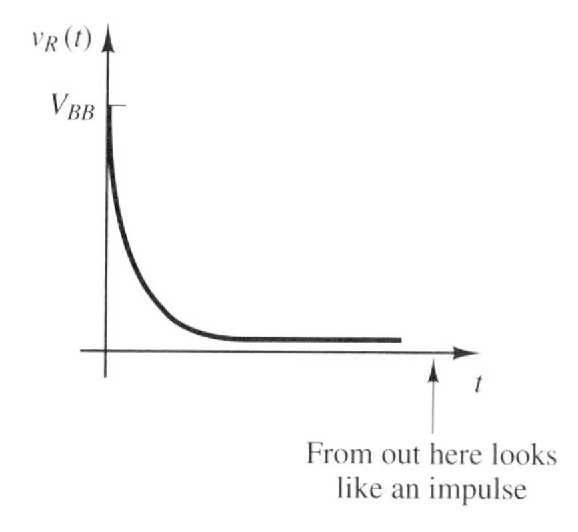


**Figure 8.37** The derivative of a constant in time turned on at $t = 0$ is a delta function (see Fig. A5.1). Thus the resistor voltage of the $RC$ filter in Fig. 8.34 approximates a differentiator well when viewed on a large time scale, but poorly on a short time scale.

improvement over using the original signal. And, of course, in situations such as chemical reaction rate measurements, the time derivative itself is of much more interest than the original signal.

**First-order active low-pass filter.** In Sec. 5.5.3 it was mentioned that we could high-pass filter a signal $v_s(t)$ by placing it across an $RC$ series combination and looking at the resistor voltage $v_R(t)$ [or low-pass filter it by looking at the capacitor voltage $v_C(t)$]. The approximate cutoff frequency was $1/(RC)$, beyond which less than half-power transmission occurred. There are, however, problems with such passive filters. For example, the load impedance $R_L$ that the passive filter drives directly affects the cutoff frequency, as shown in Fig. 8.38. With the load $R_L$ placed in parallel with the filter output, in this case $v_R(t)$ for simplicity, the new cutoff frequency is

$$\omega_o = \frac{1}{(R||R_L)C}, \tag{8.78}$$

which is clearly not equal to the original, presumably designed, $1/(RC)$ cutoff.

The order of a filter refers to the order of the differential equation describing the circuit. It is generally equal to the number of distinct energy storage devices in the circuit. With first-order filters we can only obtain monotonically increasing or decreasing frequency responses (low- or high-pass filters with only a gradual decline at cutoff). Using high-order filters, we can also make bandpass, bandstop, notch, tuning, and all sorts of filters, such as those shown in Fig. 5.25.

However, to achieve higher-order filters, inductors are required. For small $\omega_o = 1/(LC)^{1/2}$ (second-order circuit) (e.g., in the audio range), the required inductor can be large, heavy, lossy, and expensive. Inductance is also a function of temperature for iron-core inductors.

All of these limitations are overcome by using *active filters*—filters with gain, containing op amps. The small output impedance and large input impedance of the op amp greatly reduce the interaction between the filter and surrounding circuitry, eliminating the first problem. The second problem is eliminated in that *any* desired frequency response can be designed with op amps using only capacitors and resistors. Obviously, an op amp has no temperature-dependent iron core; its temperature dependence is minimized by feedback connections. In this section only the simple first-order filter is considered.

The first-order low-pass filter is actually the same circuit as the op amp

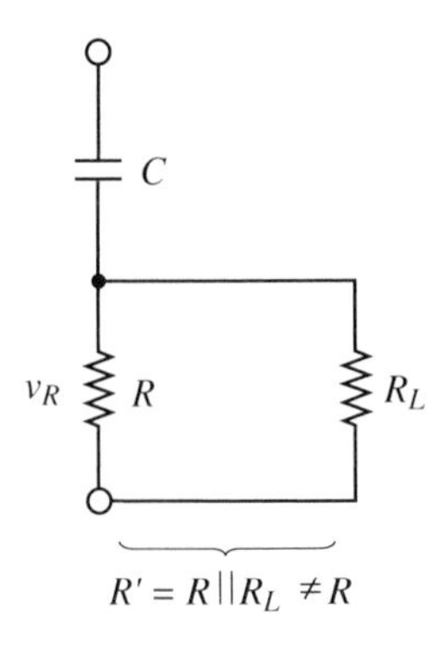


**Figure 8.38** A loaded $RC$ filter will affect the cutoff frequency.

integrator shown in Fig. 8.33! Why? From ac analysis, recall that $d/dt$ becomes multiplication by $j\omega$ in the frequency domain. Therefore, the frequency domain version of Eq. (8.70) is

$$j\omega\overline{V}_o = -\left(\frac{1}{RC}\right)\overline{V}_s, \tag{8.79}$$

so that the magnitude of the frequency response is

$$H(\omega) = \frac{V_o}{V_s} = \frac{1}{\omega RC}. \tag{8.80}$$

For small $\omega$, $H(\omega)$ is large, and vice versa. In fact, Eq. (8.80) implies that $H(\omega)$ approaches infinity for $\omega = 0$. This cannot be, of course; instead, the output saturates well before then. But this is actually a manifestation of the *same* problem encountered in the discussion of the integrator for step inputs ($\omega = 0$), where a large resistor had to be placed across the capacitor $C$ to prevent a huge charge buildup.

Suppose that the resistor placed across $C$ is called $KR$, and typically $K \gg 1$. The formula for the new $H(\omega)$ [no longer that in Eq. (8.80)] is now obtained readily by using complex impedances. The simplification is obtained by recognition of the fact that the differentiator (with or without $KR$) has the structure of an inverting amplifier. Only what originally was $R_2$ has now been replaced by a capacitor (possibly, in parallel with $KR$). So the integrator/low-pass filter circuit in Fig. 8.33 can be considered a complex impedance inverting amplifier (see Fig. 8.39).

In general, the two resistors $R_1$ and $R_2$ of the inverting amplifier now become complex impedances $\overline{Z}_1$ and $\overline{Z}_2$. For this example,

$$\begin{aligned} \overline{Z}_1 &= R \\ \overline{Z}_2 &= KR \,||\, \frac{-j}{\omega C}. \end{aligned} \tag{8.81}$$

By writing KCL for the phasors, a form for the gain identical to that for a resistive inverting amplifier is obtained. Or we can immediately infer that the voltage gain, which was $-R_2/R_1$, is now $-\overline{Z}_2/\overline{Z}_1$. But this complex voltage gain is just the frequency response. Hence the magnitude of the frequency response is, using Eqs. (8.81),

$$\begin{aligned} H(\omega) &= \left|\frac{-\overline{Z}_2}{\overline{Z}_1}\right| \\ &= \left|\frac{jKR/(\omega C)}{KR - j/(\omega C)}\right| \Big/ R \end{aligned}$$

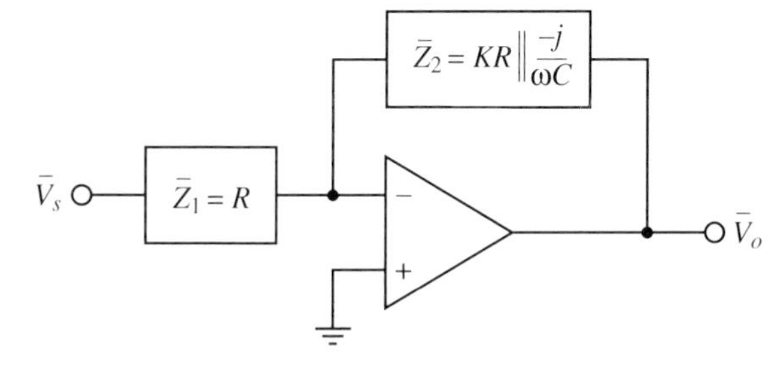


**Figure 8.39** Generalized inverting amplifier. The input and feedback resistances in Fig. 8.13 can be replaced by complex impedances. Then under monofrequency excitation, the gain is just $-\overline{Z}_2/\overline{Z}_1$. The values of $\overline{Z}_1$ and $\overline{Z}_2$ shown represent a practical integrator circuit.

$$= \left| \frac{K}{1 + j\omega KRC} \right| \tag{8.82}$$

$$= \frac{K}{[1 + (\omega KRC)^2]^{1/2}},$$

which is the same form as the passive *RC* low-pass filter but with gain *K*. Note that now the gain at $\omega = 0$ is *K*, not $\infty$. The result of adding *KR* is a less steep, but more stable low-pass filter. It also defines a cutoff frequency $\omega_o$ for which $H(\omega_o) = H(0)/\sqrt{2}$: $\omega_o = 1/(KRC)$, a factor of *K* different from the original passive *RC* filter.

The technique used above of replacing resistors with complex impedances can be used for any linear op amp circuit. The linear resistive inverting and noninverting amplifier structures presented in this chapter were developed with dc or low-frequency signals in mind. But they serve as good examples of resistive frameworks upon which to build more complex circuits. Remember that dc analysis is just the special case of ac analysis, $\omega = 0$.

**Example 8.6**

It is desired to eliminate some 2-kHz noise from a low-frequency (100-Hz) transducer signal. Design an active low-pass filter to accomplish this.

**Solution** Use the circuit of Fig. 8.39. Let $R = 5$ kΩ and $K = 10$. For a margin of safety, choose the cutoff of the filter down at 1 kHz, which is also well above the signal frequency. Then requiring $H(\omega)$ to be equal to $1/\sqrt{2}$ gives $\omega_o = 1/(KRC) = 1/(10 \cdot 5000 \cdot C) = 2\pi(1 \text{ kHz}) = 6280$ rad/s. Solving for *C* gives $C = 3.18$ nF. This specifies the circuit completely, as shown in Fig. 8.39.

A familiar use of the low-pass filter is the Dolby system for tape recorders. The "hiss" on magnetic tape and from extraneous noise sources is relatively constant and distorts the high frequencies in the reproduction of recorded music. If when recording, the high frequencies of the signal to be recorded are boosted (say, by an active high-pass filter—see below), the signal high frequencies will be much stronger than the high-frequency noise components. That is, the signal-to-noise ratio (SNR) is improved. To make the final sound natural, a low-pass filter reduces all high frequencies back to the proper level.

Note that the net gain in SNR is only for noise extraneous to that being recorded. If, for example, an old phonograph recording is being recorded, there will be no gain in its SNR because its noise was boosted along with the music signal.

Finally, it may be pointed out that an op amp is itself a low-pass filter. We pointed this out before by describing the finite bandwidth of the op amp. However, it is not a very good low-pass filter because it has a very gradual decline with frequency (a perfect low-pass filter would have a sudden cutoff at $\omega_o$). Also, its cutoff frequency cannot be controlled precisely.

Thus in practice we start with an op amp with a bandwidth higher than we need and select the desired cutoff by connecting a circuit such as in Fig. 8.39. It should be pointed out that an ideal integrator is *not* an ideal low-pass filter, only a first-order approximation of it. Recall that the ideal filter had unity response in the passband and zero response in the stopband (see Fig. 5.27, dashed curve). A sharper cutoff can be achieved by using higher-order filters (using more reactive elements). Because of its importance in communication systems and signal processing in general, the topic of filter design is a very mature area of research and development.

**First-order active high-pass filter.** Big surprise: The first-order active high-pass filter circuit is the same as that of the active differentiator in Fig. 8.36. The structure is again a complex impedance inverting amplifier. Without the input capacitor added in, $\overline{Z}_1 = -j/(\omega C)$ and $\overline{Z}_2 = R$, so that the magnitude frequency response is

$$H(\omega) = \left| \frac{-R}{-j/(\omega C)} \right| = \omega RC. \tag{8.83}$$

The gain is zero for $\omega = 0$ and rises linearly with $\omega$. But $H(\omega)$ approaches infinity as $\omega$ approaches infinity. This cannot occur; in fact, as was mentioned in the discussion on the differentiator, unwanted oscillations and general instability result for large $\omega$. The effect is somewhat complicated and is due to the finite bandwidth of the op amp (yes, $A_{ol}$ is a decreasing function of $\omega$). The additional capacitance in parallel with $R$ to remedy this problem (call it $C/K$, $K > 1/\sqrt{2}$) changes the frequency response to

$$H(\omega) = \frac{\omega RC}{[1 + (\omega RC/K)^2]^{1/2}} = \frac{RC}{[1/\omega^2 + (RC/K)^2]^{1/2}}, \tag{8.84}$$

which is of a form similar to that of the passive $RC$ high-pass filter, but with the isolation and other advantages of op amps.

Just as the integrator is only a first-order low-pass filter, the differentiator is only a first-order approximation of an ideal high-pass filter. People who are serious about filtering will use high-order low- or high-pass filters based on advanced filter design theory. Various characteristics (minimum ripple in the passband, maximally flat in the passband, etc.) are achieved by certain filter design equations valid for any order. The higher the order, the more sharp the transition from passband to stopband, but the more complex the filter.

**Example 8.7**

Design an active high-pass filter to eliminate some 60-Hz noise from a 500-Hz signal.

**Solution** The circuit of Fig. 8.36 applies, with an additional capacitance $KC$ placed across $R$. From Eq. (8.84), clearly $\omega_o = K/(RC)$, for then $H(\omega_o) = H(0)/\sqrt{2}$. Let $K = 5$, $R = 10\ \text{k}\Omega$, and for a margin of safety let the cutoff frequency be 120 Hz, so that $\omega_o = 2\pi \cdot 120 = 754$ rad/s. Then, solving for the necessary capacitance, $C = K/(R\omega_o) = 5/(10{,}000 \cdot 754) \approx 0.66\ \mu\text{F}$. Now that all of the parameters are known, the circuit may be built.

## 8.9 SUMMARY

The operational amplifier has many advantages over discrete semiconductor amplifiers, including simplified user design, low cost, small physical size, nearly ideal behavior, and a huge variety of applications. Very simple models accurately describe the behavior of an op amp under normal amplification conditions, particularly negative feedback configurations. The simplest is an infinite-gain amplifier that draws zero input current and has zero output impedance.

The concept of feedback, both positive and negative, was examined in significant detail, and the functioning of several example circuits were shown. The stability and isolation properties of these amplifiers far exceed those of simple discrete electronics amplifiers.

As the main idea of op amps is "operations"—applications—in the remainder of the chapter we explored a very wide range of applications of the op amp. Comparators, a D/A converter, the Schmitt trigger, a pulse generator, followers and amplifiers (including differential and instrumentation amplifiers), a rectifier, an integrator, a differentiator, and low- and high-pass filters were analyzed and provide an indication of the scope of applications. Additionally, some of the practical aspects of nonideal op amps were introduced, such as ground loops, common-mode rejection ratio, offset, and extreme frequency behavior.

We could go on and on about the practical op amp applications. Other examples include their use in power supply regulators, waveform generators (one, the pulse generator, was discussed here), digital-to-analog converters (a simple analog-to-digital converter was discussed above), the logarithmic amplifier, the multiplier, voltage-controlled oscillators, higher-order active filters, absolute value and limiting circuits, phase-locked loops, and on and on—the list never ends.

However, from study of this chapter you should be in a position to understand a description of the above circuits and thereby to discuss intelligently some new gizmo that an electrical engineer has added to a non-EE's invention. Furthermore, we now have a veritable arsenal of applications that can be spot-recognized in existing circuits.

## PROBLEMS

**8.1.** Design an op amp circuit with gain of $-50$. What is the gain in dB?

**8.2.** Design an op amp circuit with gain of 20. What is the gain in dB?

**8.3.** An ideal voltage comparator is shown in Fig. 8.3, which has the input–output relation given in Fig. 8.5. Draw the output waveform $v_o(t)$ for the given inputs $v_A(t)$ and $v_B(t)$ in Fig. P8.3, where $v_A$ is connected to the inverting terminal and $v_B$ is connected to the noninverting terminal: $v_- = v_A$ and $v_+ = v_B$.

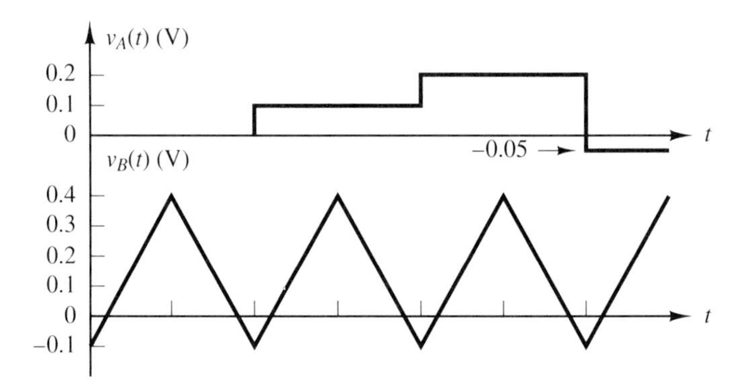


**Figure P8.3**

**8.4.** For the circuit in Fig. P8.4, describe its behavior fully. Then determine the output waveform $v_o(t)$ if the input waveforms $v_A(t)$ and $v_B(t)$ are as shown in Fig. P8.3.

**8.5.** For the circuit in Fig. P8.5a:

(**a**) Determine an expression for $v_o(t)$ as a function of the inputs $v_s(t)$ and $v_x(t)$; assume

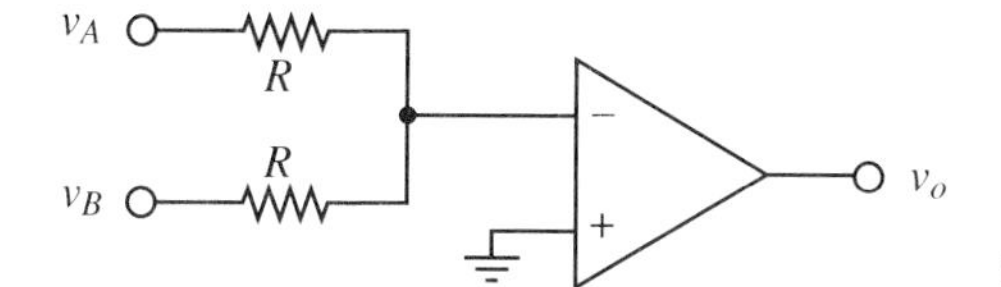


Figure P8.4

that $A_{ol} = \infty$. There may be more than one result, for different conditions on the inputs.

(b) Let $R_1 = R_2$. Plot $v_o(t)$ for $0 \leq t \leq 10$ s given the input waveforms $v_x(t)$ and $v_s(t)$ shown in Fig. P8.5b. Quantitatively label both axes ($v_o$ and $t$).

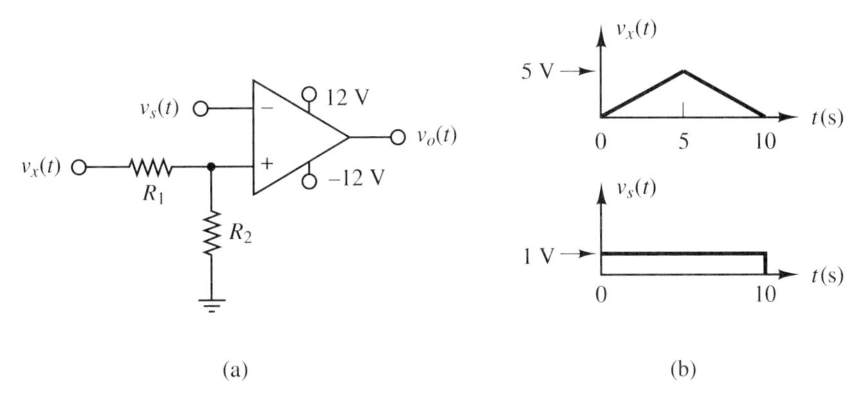


Figure P8.5

**8.6.** Design a Schmitt trigger circuit with the transfer characteristic shown in Fig. P8.6; provide the relation between resistor values. Assume full $V_{BB}$ saturation.

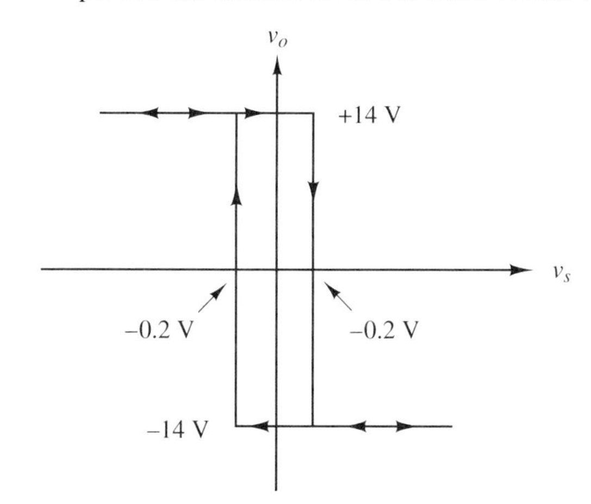


Figure P8.6

**8.7.** Draw and fully label the input–output characteristic ($v_3$ on the vertical axis, $v_s$ on the horizontal axis) of the circuit in Fig. P8.7. Provide all necessary reasoning leading to this graph. Assume that $|v_o| \leq V_{BB}$. Use $V_{BB} = 12$ V, $R_s = 1$ kΩ, $R_1 = 100$ kΩ, $R_2 = 3$ kΩ, $R_3 = 7$ kΩ, $A_{ol} = \infty$, $Z_i = \infty$, $Z_o = 0$.

**8.8.** For the Schmitt trigger circuit shown in Fig. 8.8, find the value of $R_2$ that will just prevent switch chattering if the input $v_s(t)$ is as shown in Fig. P8.8. Assume that $R_1 = 1$ kΩ, op-amp saturation at $V_{BB} = 15$ V, and a peak-to-peak noise voltage of 200 mV

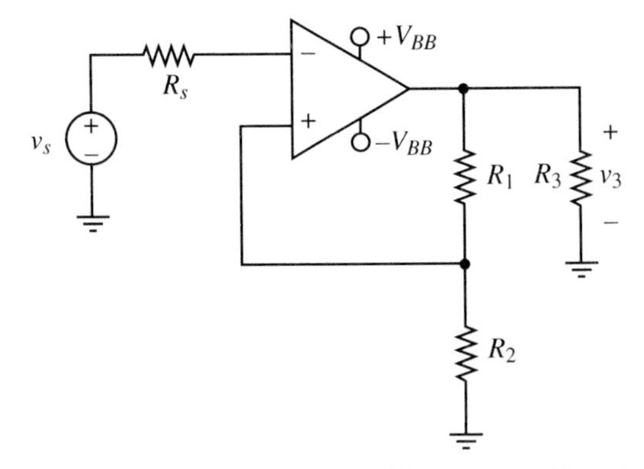


**Figure P8.7**

riding on $v_s(t)$. Sketch $v_o(t)$ on a plot directly below that of $v_s(t)$ for this case where chattering is just barely made impossible. Carefully line up switch times with $v_s(t)$. Assume that initially $v_o = V_{BB}$.

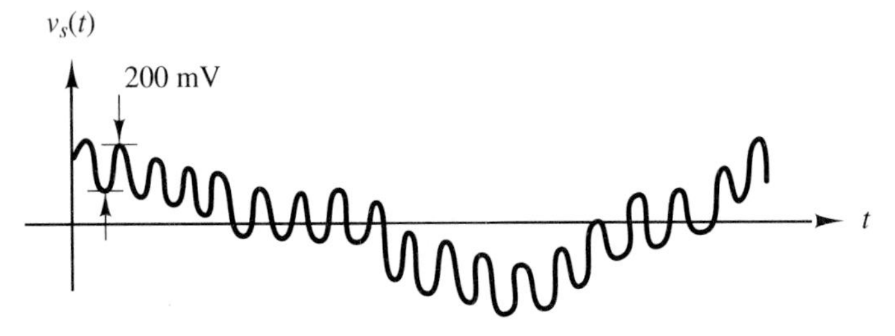


**Figure P8.8**

**8.9.** **(a)** Find the gain $v_o/v_s$ of the circuit in Fig. P8.9. Is the circuit below an inverting or a noninverting amplifier? Assume that $|v_o| \leq V_{BB}$, $A_{ol} = \infty$, $Z_i = \infty$, $Z_o = 0$.
**(b)** What resistance $R_1$ should be selected to achieve a gain having magnitude 6?
**(c)** If the op-amp supply voltage and saturation voltages are $\pm V_{BB} = \pm 12$ V, what restriction must be placed on $v_s$ for linear amplification [with the gain as in part (b)]?

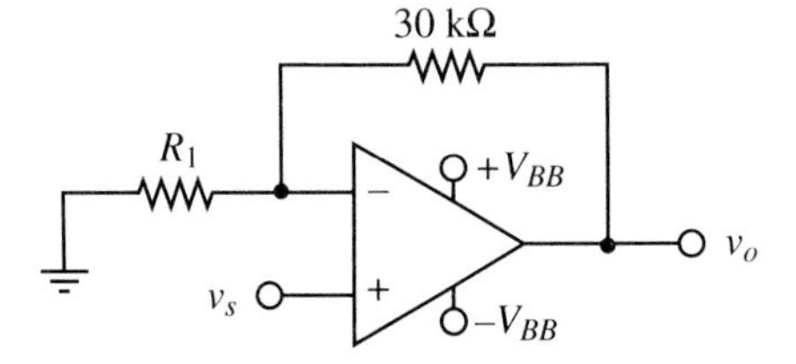


**Figure P8.9**

**8.10.** For the circuit in Fig. P8.10, determine $v_o$ in terms of $v_1$ and $v_2$, assuming linear op amp operation.

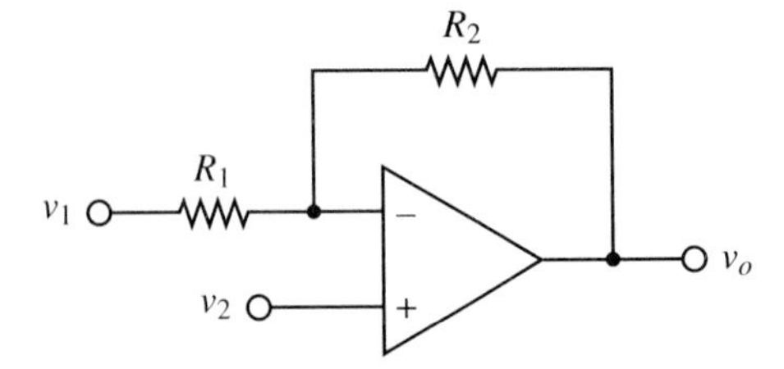


**Figure P8.10**

**8.11.** Find $i_L$ in the circuit in Fig. P8.11 if $R_L = 3\ \text{k}\Omega$.

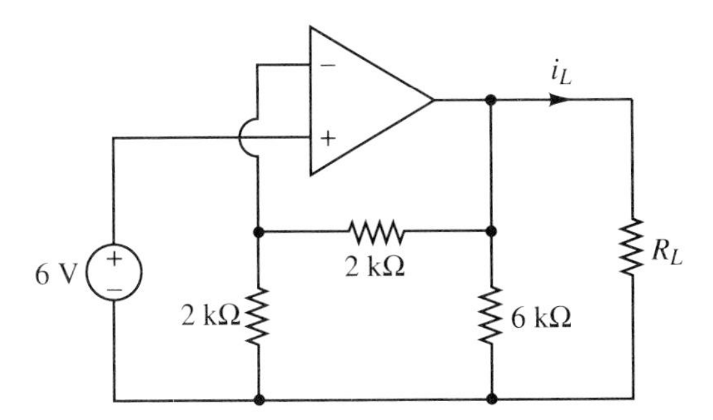


**Figure P8.11**

**8.12.** The window comparator is discussed in Sec. 8.8.2. In Fig. P8.12 is shown a different type of window comparator.

(**a**) Determine the input–output voltage characteristic. When does the alarm go off? Assume zero diode contact potentials.

(**b**) If the alarm $R_L$ is switched with the other $R_L$, what is the behavior?

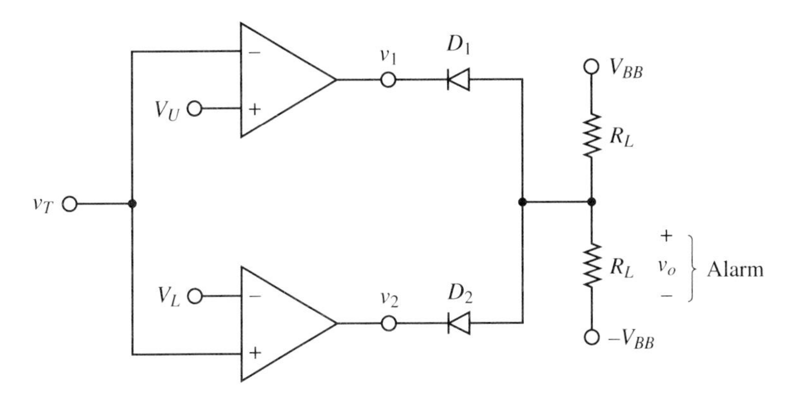


**Figure P8.12**

**8.13.** Assuming that the exponential model for the diode is valid, find $v_o$ in terms of $i_s$ in Fig. P8.13.

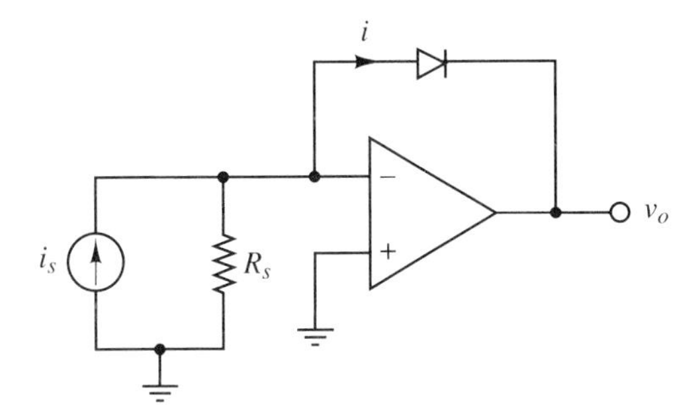


**Figure P8.13**

**8.14.** For the oscillator in Fig. 8.17, suppose that $R_1 = R_2 = 1\ \text{k}\Omega$ and $C = 0.047\ \mu\text{F}$. What should $R$ be to generate a 2-kHz square wave? Repeat for the case $C = 1\ \mu\text{F}$.

**8.15.** For the circuit in Fig. P8.15, find $A$ and $B$ for which $i_L = Ai_s + Bv_1$.

**8.16.** In the circuit in Fig. P8.16, find $R_1$ so that $v_o = i_s \cdot 6\ \text{M}\Omega$.

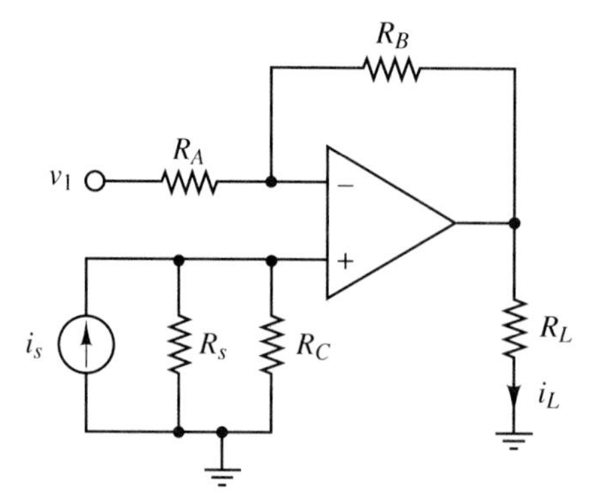


**Figure P8.15**

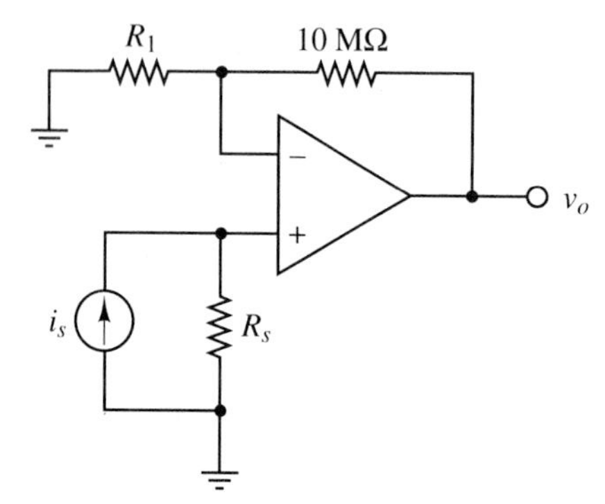


**Figure P8.16**

**8.17.** For the circuit in Fig. P8.17, find the input–output characteristic; that is, determine and graph $v_o$ vs. $v_i$.

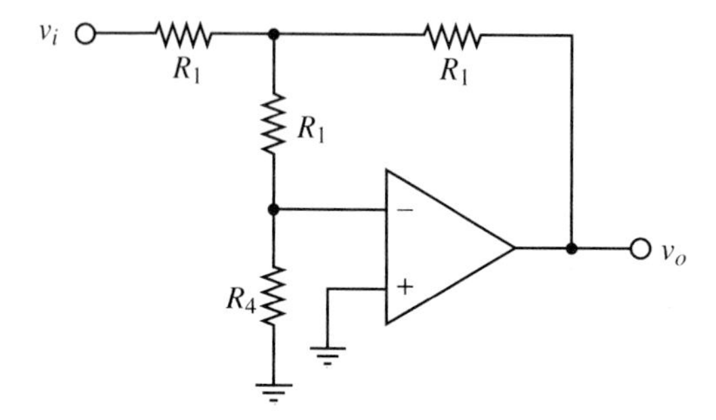


**Figure P8.17**

**8.18.** For the circuit in Fig. P8.18:

**(a)** Find $v_o$ in terms of $v_s$. Simplify as much as possible.

**(b)** Now let $R_1 = 1$ kΩ, $R_2 = 2$ kΩ, $R_3 = 3$ kΩ, and $V_{BB} = 12$ V. Sketch $v_o$ vs. $v_s$; quantitatively label your axes and all critical points on them.

**(c)** Over what range of $v_s$ does your expression in part (b) remain valid (linear operation of op amp)?

**8.19.** For the circuit in Fig. P8.19 within its linear range, find $v_o$ in terms of $v_s$.

**8.20.** **(a)** For the circuit in Fig. P8.20, find $A$ and $B$ such that $v_o = AV_s + BV_A$.

**(b)** For $V_A = 1$ V, $R_1 = 2$ kΩ, $R_2 = 1$ kΩ, and $R_3 = 3$ kΩ, graph $v_o$ vs. $v_s$ showing all important points (giving both $v_o$ and $v_s$ values) and slope(s). Let $V_{BB} = 12$ V, and assume full $V_{BB}$ saturation.

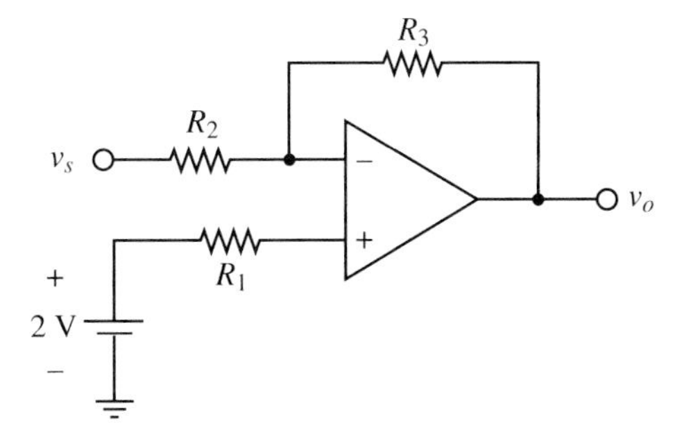


Figure P8.18

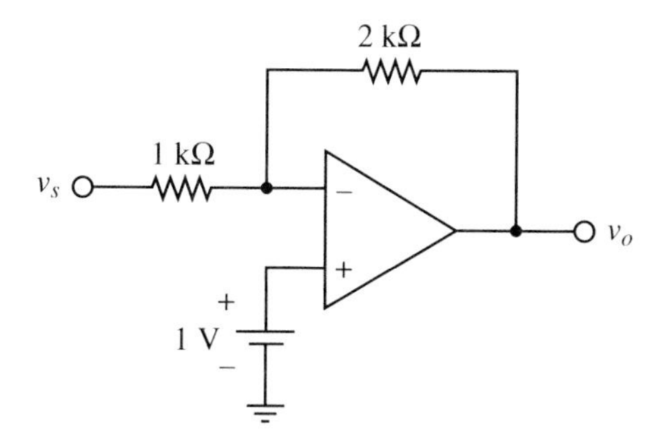


Figure P8.19

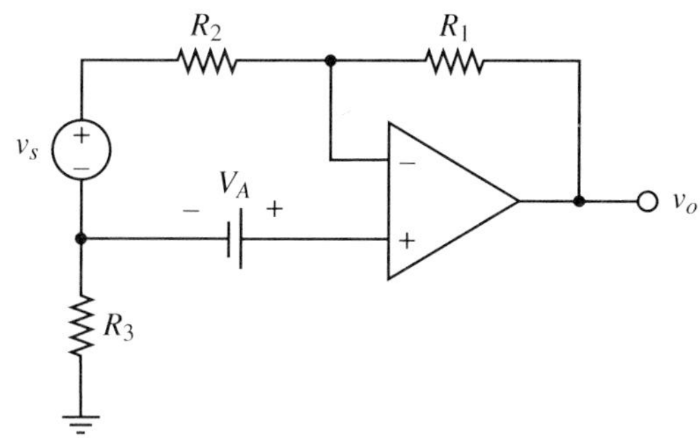


Figure P8.20

**8.21.** An arithmetic multiply/add operation is defined in Fig. P8.21, where $k_1$ and $k_2$ are known constants. Design an op amp circuit with three op amps to implement this operation, and find the relations between the resistances and $k_1$ and $k_2$. (*Hint:* If you use inverting op amps, consider the possibility of cancellation of sign inversions.)

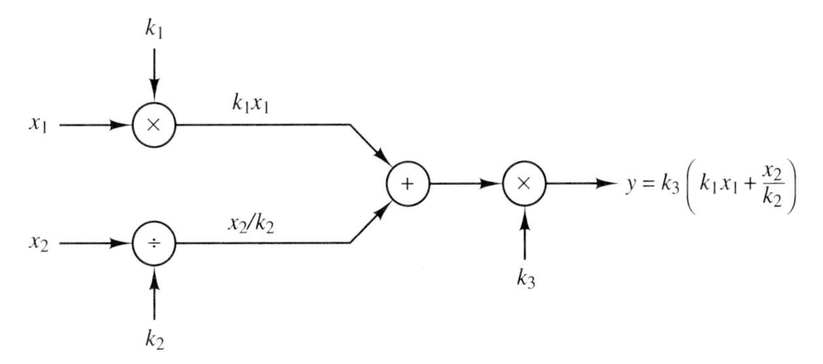


Figure P8.21

**8.22.** For the inverting amplifier shown in Fig. 8.13, if the output impedance of the signal source is very high, or equivalently, if $R_1$ is very large (e.g., 150 kΩ), then $R_2$ must be very large to obtain high gain. This would cause the bias currents to be significant, so that the ideal op amp model would no longer apply, and the behavior would no longer be true inverting amplification. Figure P8.22 shows a circuit that can be used to solve this problem. Let $R_1 = 150$ kΩ, $R_3 = 150$ kΩ, $R_4 = 1$ kΩ, and $R_5 = 99$ kΩ.

**(a)** If we want the voltage gain to be $A_v = v_o/v_s = -100$, find the value of $R_2$ required in Fig. 8.13 if $R_1 = 150$ kΩ.

**(b)** Find the voltage gain of the circuit in Fig. P8.22. (*Hint:* Note that in Fig. P8.22, $R_3 \gg R_4$.)

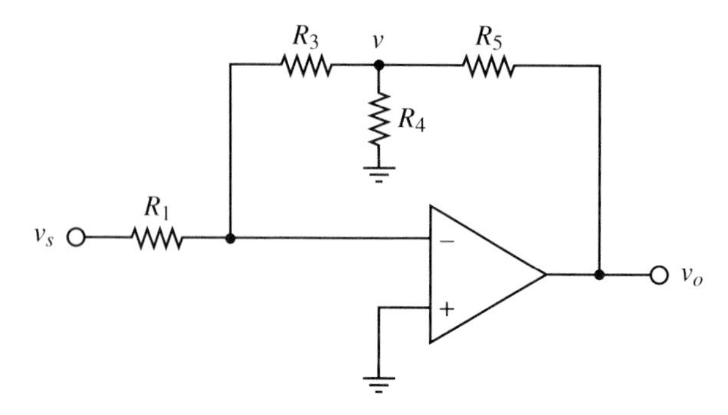


**Figure P8.22**

**8.23.** Voltage references are very important in many circuits and systems applications. Among these are voltage references for A/D converters and signal comparison circuitry. The op amp circuit in Fig. P8.23 (buffer/impedance transformer) provides a method for achieving output gain along with excellent isolation between the load and the reference source. It is useful when the reference voltage source $V_A$ has a large internal impedance $R_s$ (recall that the output impedance of closed-loop op amp circuits is very small; thus impedance transformation is possible). Find the output reference voltage $V_R$ for $V_A = 2$ V, $R_1 = 2$ kΩ, and $R_2 = 6$ kΩ.

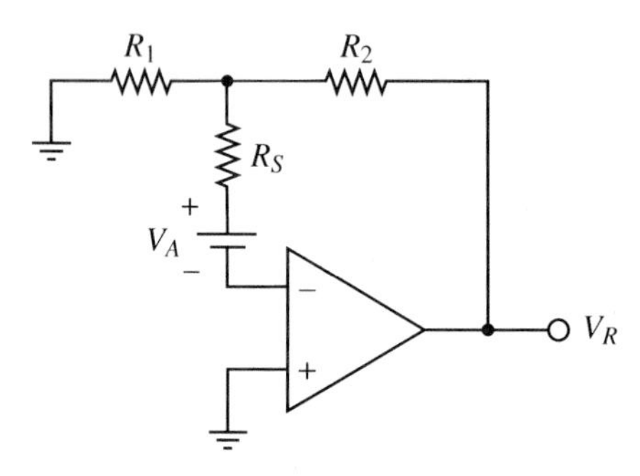


**Figure P8.23**

**8.24.** The circuit in Fig. P8.24 is a current amplifier.

**(a)** Find the current gain $i_o/i_{\text{in}}$ by first finding $v_+$; that is, analyze the circuit one stage at a time. Assume that the values of $R_A$, $R_B$, and $R_L$ are known.

**(b)** Suppose that $R_A = 5$ kΩ, $R_B = 1$ kΩ, and $R_L = 1$ kΩ. Find the magnitude of $i_{\text{in}}$ required to saturate one or both of the op amps, in which case the circuit no longer functions as a linear current amplifier. Find the voltage output of each op amp in terms of $i_{\text{in}}$. Consider separately the conditions for each op amp; then the worst case will be the answer. Which op amp saturates first, or do they both saturate for the same value of $i_{\text{in}}$? Assume 15-V saturation for both op amps. What are $i_o$, $v_+$, and $v_o$ at the saturation threshold (of the system)?

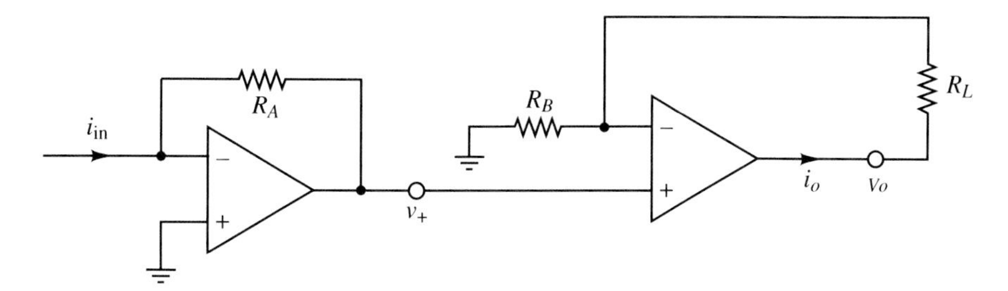


**Figure P8.24**

**8.25.** (**a**) Show that the "subtractor" shown in Fig. P8.25 has output of the form $v_o = Av_2 - Bv_1$, where $A$ and $B$ are constants. What are $A$ and $B$?

(**b**) Suppose that we desired $v_o = 4v_2 - 7v_1$. If $R_1 = 3\ \text{k}\Omega$ and $R_3 = 10\ \text{k}\Omega$, what should $R_2$ and $R_4$ be to perform the desired function? Consider only linear behavior.

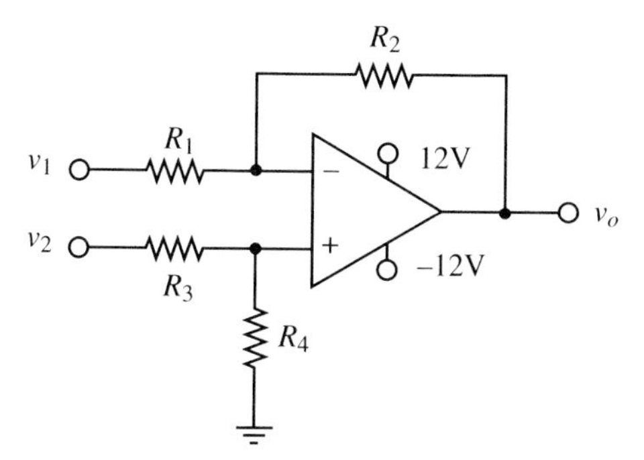


**Figure P8.25**

**8.26.** For the circuit shown in Fig. P8.26, determine $k$, where $k = i_o/v_s$. If $v_s(t) = 30\cos(100t)$ mV, find $i_o(t)$, and the maximum value of the output voltage $v_{o,\max}$.

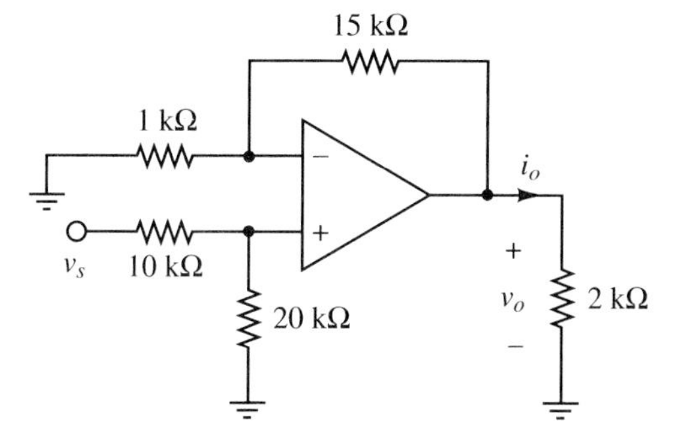


**Figure P8.26**

**8.27.** (**a**) For the op amp circuit in Fig. P8.27, find $v_o$ as a function of $v_1$, $v_2$, and the resistor values. $V_{BB} = 15$ V.

(**b**) What would be the expression for $v_o$ if $v_2$ were instead connected to $v_+$ through a resistor $R_A$, and a resistor $R_B$ were placed between $v_+$ and ground?

(**c**) With the original structure in part (a), suppose that the output voltage were the function $v_o = 5v_2 - 4v_1$ and also that $|v_1| < 1\ v$. What then would be the permissible range of $v_2$ for linear behavior (using the most constricting constraint on $v_2$)? (Use full $V_{BB}$ saturation.)

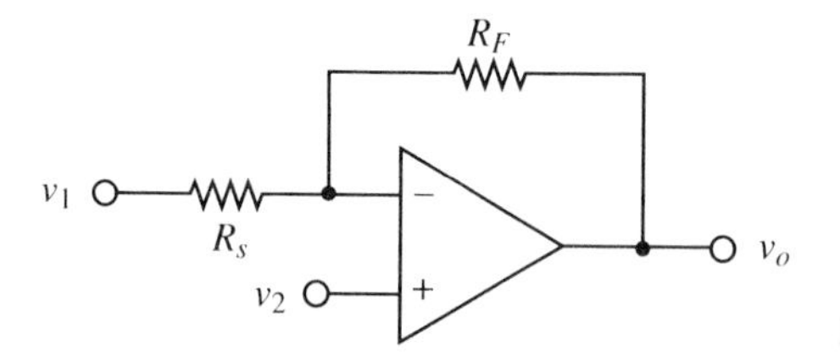


**Figure P8.27**

**8.28.** The circuit shown in Fig. P8.28 has a different gain for $v_i > 0$ from that for $v_i < 0$. Sketch $v_o$ vs. $v_i$. Ignore the 0.7-V diode drop in this problem.

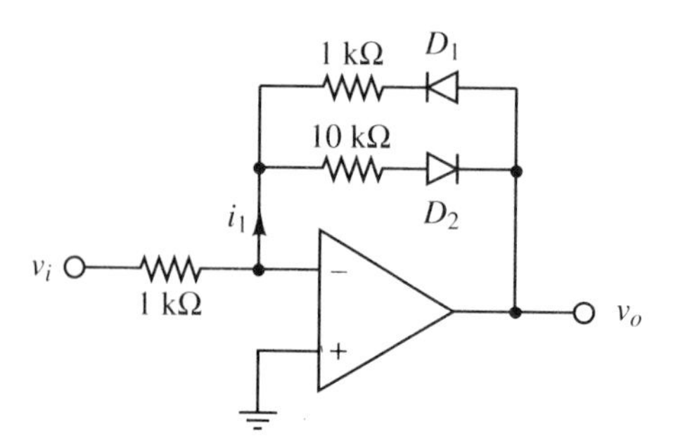


**Figure P8.28**

**8.29.** A "soft limiter" uses two 5.3-V Zener diodes to reduce the gain of an inverting amplifier under certain conditions, as shown in Fig. P8.29. Compute and sketch $v_L$ vs. $v_i$. Assume that the contact potentials of both diodes are 0.7 V.

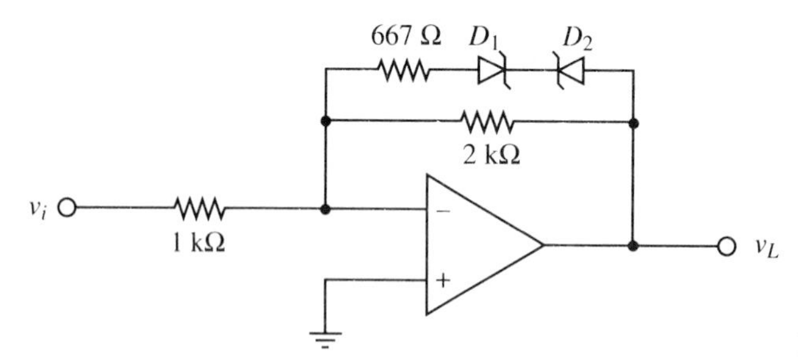


**Figure P8.29**

**8.30.** **(a)** Design an instrumentation amplifier having an overall differential gain of 500 whose differential amplifier has a gain of 50. The circuit is shown in Fig. 8.30, so all you need to do is to select correct resistor values. Choose reasonable values for the resistors where there is flexibility; let $V_{BB} = 14$ V.

**(b)** What magnitude differential input voltage will saturate the output? Use full $V_{BB}$ saturation.

**(c)** What is the overall common-mode rejection ratio (CMRR) for this circuit if the CMRR of the differential amplifier stage is $10^4$?

**8.31.** Consider the circuit in Fig. P8.31. Assume that $|v_o| \leq V_{BB}$, $A_{\text{ol}} = \infty$, $Z_i = \infty$, $Z_o = 0$.

**(a)** Is this circuit an example of negative or positive feedback?

**(b)** What is the value of the potential at the inverting terminal valid for both $v_o$ both positive and negative?

For parts (c) through (f), let the diode contact potential $V_o$ be zero volts.

**(c)** Find $v_o$ in terms of $v_s$ when the diode is reverse biased.

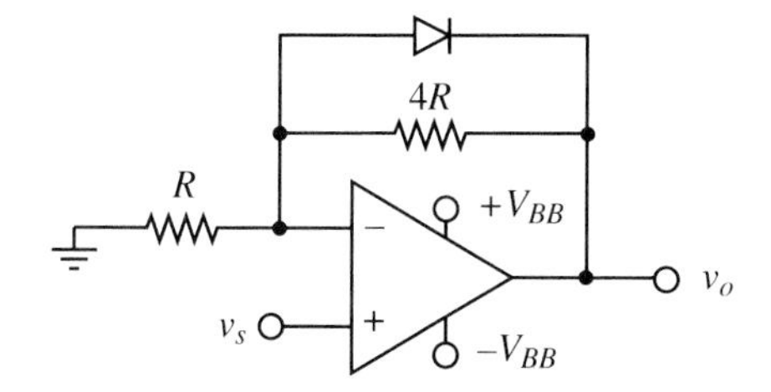


**Figure P8.31**

(**d**) For what condition (on $v_s$) is the diode reverse biased? The voltage $v_o$ should not appear in the condition.
(**e**) If the diode is conducting, what is the output $v_o$ in terms of $v_s$?
(**f**) Draw a fully labeled input–output characteristic ($v_o$ on the vertical axis, $v_s$ on the horizontal axis).
(**g**) Repeat parts (c) through (f) for contact potential $V_o = 0.7$ V.

**8.32.** Find $v_o$ vs. $v_i$ for the circuit in Fig. P8.32 for (**a**) $V_o = 0$ V, and (**b**) $V_o = 0.7$ V.

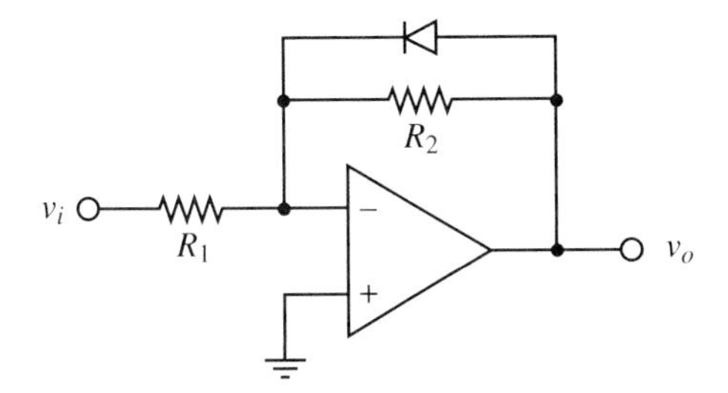


**Figure P8.32**

**8.33.** Suppose that in Fig. P8.29 we short-circuit (essentially remove) the 667-Ω resistor, and we replace the 1 kΩ resistor by a general resistance $R_1$ and replace the 2 kΩ resistor by a general resistance $R_2$. Find $v_L$ vs. $v_i$ for (**a**) $V_o = 0$ V, and (**b**) $V_o = 0.7$ V. Let the Zener voltage of the diodes be $V_Z$.

**8.34.** Find $v_o$ vs. $v_i$ for the circuit in Fig. P8.34 for (**a**) $V_o = 0$ V, and (**b**) $V_o = 0.7$ V.

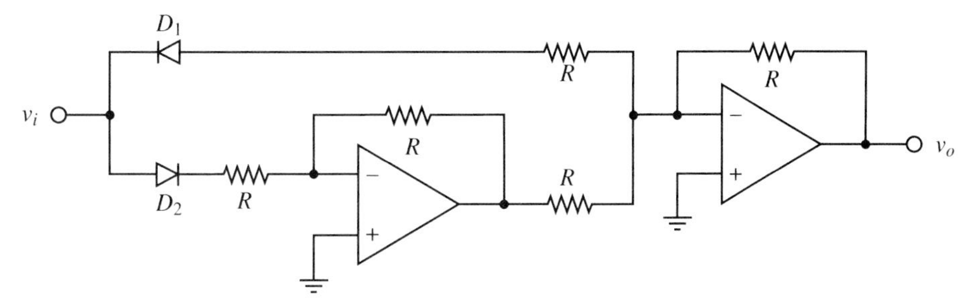


**Figure P8.34**

**8.35.** For the circuit in Fig. P8.35a (in which $V_{BB} = 15$ V):
(**a**) Determine the input–output characteristic $v_o$ vs. $v_i$, assuming that the diodes are ideal and that $R = 1$ kΩ.
(**b**) Draw the output waveform $v_o(t)$ given $v_i(t)$ in Fig. P8.35b, again assuming ideal diodes and $R = 1$ kΩ.
(**c**) Choose the value of $R$ that will give a true full-wave rectifier ($v_o = \alpha|v_i|$), again assuming ideal diodes.
(**d**) Is this a precision rectifier (i.e., if, for example, we assume that the diode contact potentials are 0.7 V, will this affect/degrade the output)?

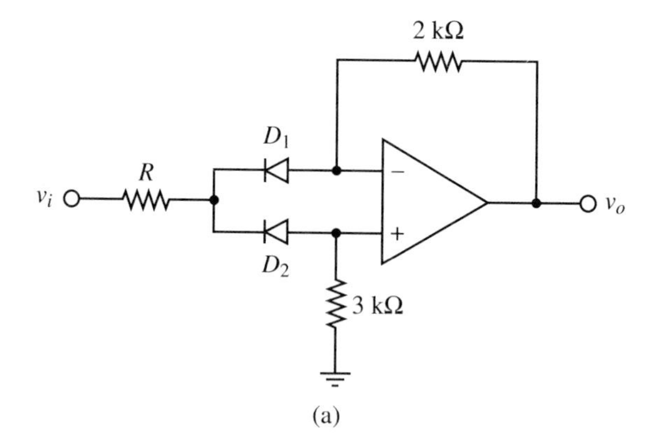


(a)

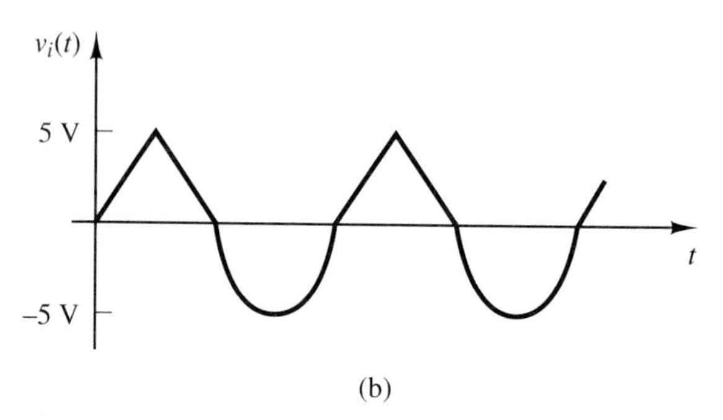


(b)

**Figure P8.35**

**8.36.** For the amplifier circuit in Fig. P8.36, determine $v_o$ in terms of $v_A$ and $v_B$. How would you classify this amplifier?

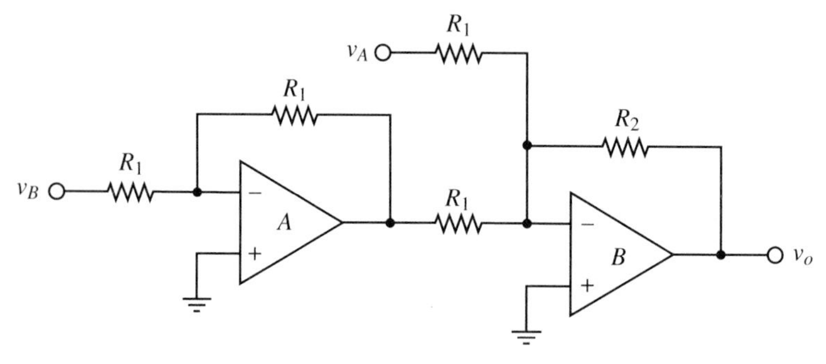


**Figure P8.36**

**8.37.** A three-op amp instrumentation amplifier is discussed in Sec. 8.8.4. In Fig. P8.37 is shown another instrumentation amplifier, which uses only two op amps. Determine the voltage gain of this amplifier. (*Hint:* One easy way in this case is to use the method of superposition, where the output is determined for only one source on at a time, and the results for each input acting alone are added to obtain the total response.)

**8.38.** If $V_{BB}$ = 12 V and we have full $V_{BB}$ saturation, find the value of the input voltage that would cause an op amp-based amplifier to saturate, given that the closed-loop amplifier gain is 80 dB.

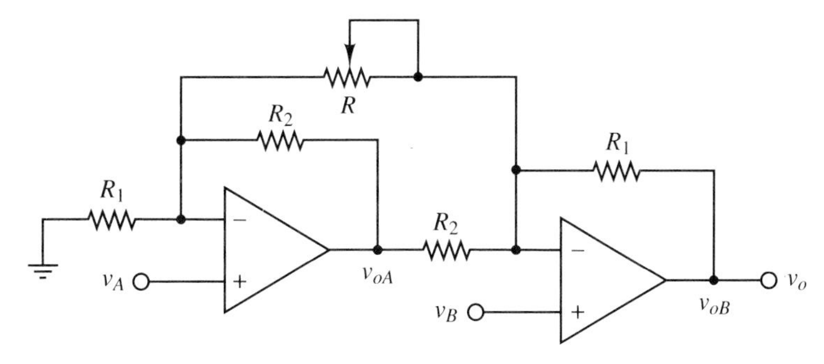


**Figure P8.37**

**8.39.** Suppose we know that for a differential amplifier with a gain of 40 dB, the input common-mode voltages may be as high as twice the input differential-mode voltage levels. If the output common-mode component is to be less than 0.1% of the output, what is the required CMRR in dB? In that case, what is the common-mode gain?

**8.40.** Suppose a 741 op amp has a CMRR of 90 dB and is used in a differential amplifier having gain 30. If the difference mode input voltage is 0.2 V and the common-mode input voltage is 10 V, what components (difference mode and common mode) will appear in the output voltage? What is the total output voltage?

**8.41.** If an op amp has a slew rate of 0.5 V/μs, how long does a Schmitt trigger take to change output state? Let $V_{BB} = 12$ V. Will a 10-kHz "square-wave" output look square? Repeat for a 1-kHz square wave.

**8.42.** If in the low-pass filter shown in Fig. 8.33 we place a 15-kΩ resistor across the capacitor, and if $R = 2$ kΩ and $C = 0.02$ μF, then:

(**a**) Find the frequency response $\overline{H}(\omega)$

(**b**) Determine the −3-dB bandwidth in hertz.

**8.43.** For the integrator in Fig. 8.33 with $V_{BB} = 15$ V, suppose that $v_C(0^-) = 0$ V. Find the output voltage when $R = 500$ Ω, $C = 200$ μF, and:

(**a**) $v_s(t) = 1$ for $t \geq 0$ and zero otherwise.

(**b**) $v_s(t) = A \cos(400t)$ for $t \geq 0$ and zero otherwise.

(**c**) Under what restricting conditions do the results you found in parts (a) and (b) hold true?

**8.44.** Find the frequency response $\overline{H}(\omega)$ of the circuit in Fig. P8.44. Sketch the amplitude of $\overline{H}(\omega)$ [i.e., $H(\omega)$] for the case $R = 200$ Ω, $C = 3$ μF. Indicate the −3-dB cutoff frequency in hertz. Is this a low- or a high-pass filter?

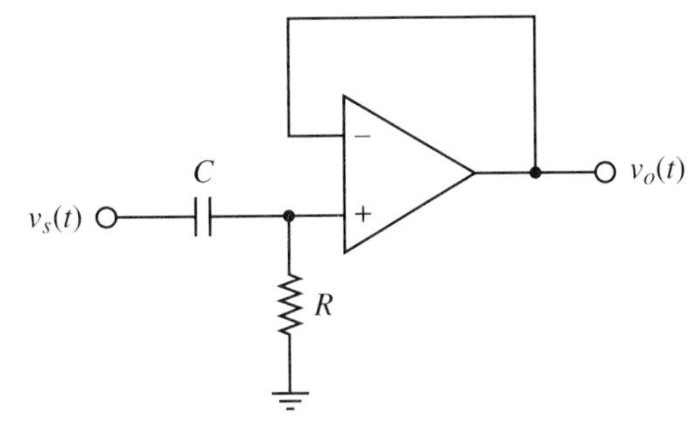


**Figure P8.44**

**8.45.** Suppose that in Fig. P8.44 the resistor and capacitor are swapped. Repeat Problem 8.44 for this case, where again $R = 200$ Ω and $C = 3$ μF.

**8.46.** The circuit in Fig. P8.46 is a bandpass filter.

**(a)** Without making detailed calculations, argue what are the values of the frequency response for $\omega = 0$ and $\omega = \infty$ directly from the circuit.

**(b)** To simplify the analysis, calculate the frequency response for very low and very high frequencies very simply, by ignoring the component(s) having negligible effects over that range of frequency. Suppose that these two models are quite far apart in frequency. What, then, is roughly the maximum value of the frequency response, and in what general frequency range does it occur? Then, using your two models individually, find the approximate lower and upper −3-dB cutoff frequencies of this filter.

**(c)** Suppose that we want a filter having roughly unity gain in the passband and having an upper −3-dB cutoff frequency of 500 kHz and a lower −3-dB cutoff frequency of 50 Hz. Find $R_2$ and $C_1$ if $C_2 = 0.1$ nF.

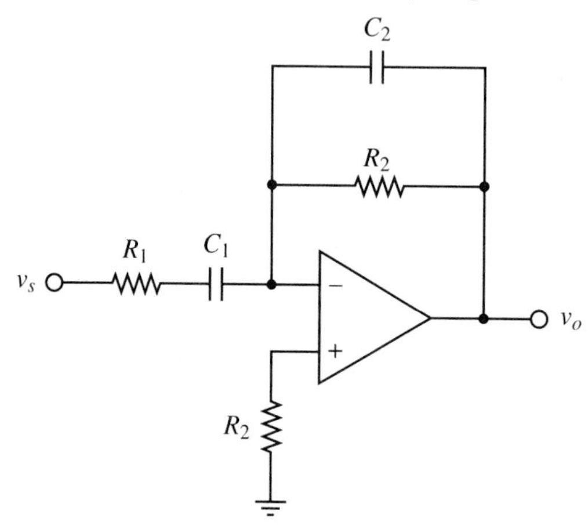


**Figure P8.46**

## ADVANCED PROBLEMS

**8.47.** For the pulse generator circuit in Fig. 8.22, show that if $R_1 = kR_2$, then $T = 2RC\ln(1 + 2/k)$.

**8.48.** Determine the input resistance and voltage gain of the ideal op amp circuit shown in Fig. P8.48. Let $Z_{in}$ of the op amp alone be $\infty$, and $A_{ol} = \infty$.

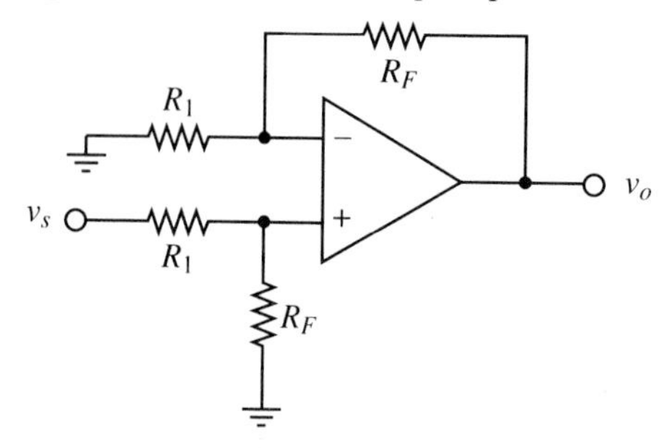


**Figure P8.48**

**8.49.** Show that the closed-loop output impedance of the inverting amplifier shown in Fig. 8.13 is $Z_{out} = Z_o/[A_{ol}F]$, where $F = R_1/(R_1 + R_2)$.

**8.50.** In this problem we design two precision full-wave rectifiers using the precision half-wave rectifier in Fig. 8.38.

**(a)** Using two precision rectifiers from Fig. 8.38 with $R_2 = R_1$, and one noninverting summer (with gain), design a precision full-wave rectifier with unity gain. [*Hints:* (1) You will have to connect your input through a resistor to the *noninverting* terminal for one of the half-wave rectifiers. (2) For this circuit, use resistances equal to $100R$ from the noninverting terminals of all op amps to ground. We use a large value ($100R$) to provide a path for bias currents, yet not significantly reduce voltages by the voltage divider that exists there in two of the three cases.]

**(b)** Using the fact that $-x = x - 2x$, use only one precision half-wave rectifier and a noninverting summer to design a full-wave rectifier with unity gain.

**8.51.** **(a)** A voltage multiplier (see Fig. P8.51) operates by adding logarithmic functions of the inputs, and then taking an exponential function (recall the discussion in Sec. 5.4.2). All the resistors are equal, and the three identical diodes are forward biased so that $i = I_s e^{v/V_T}$. Find $v_o$ in terms of $v_1$, $v_2$, $I_s$, and $R$; you should find that $v_o$ is proportional to the product of $v_1$ and $v_2$. (*Hint:* Consider each stage, one at a time, as building blocks. Then the problem is very easy.)

**(b)** Modify the design in Fig. P8.51 to produce an output proportional to $v_1/v_2$.

**8.52.** It has been found that the forward-biased transistor has a better fit to the exponential model than does the diode [*Operational Amplifiers*, G. B. Clayton, Butterworths, London, 1971, pp. 86–88]. Therefore, the more practical logarithmic amplifier used in practice is that shown in Fig. P8.52. Using the equation $i_C = I_s \exp[qv_{be}/(kT)]$, derive the output voltage $v_o$ in terms of the input voltage $v_s$. What constraints are there on $v_s$ for this circuit to function as intended? Based on this circuit, propose a corresponding exponential circuit (output voltage is exponential of input voltage).

**8.53.** A technician places a sensor over a mother's womb in order to detect the heartbeat of her unborn child. The sensor, however, picks up not only the baby's heartbeat $v_B(t)$, but also interference from the mother's heartbeat $v_M(t)$; in fact, $v_M$ has a magnitude 20 times stronger than $v_B$. To keep the mother's heartbeat from obscuring the desired signal, the technician places a second sensor over the mother's ribs and adjusts $R$ until $v_M(t)$ is subtracted out from the output $v_o(t)$. Refer to the circuit in Fig. P8.53. What value of $R$ will produce this cancellation? If the peak value of $v_B(t)$ is 50 μV, what will be the peak value of $v_o(t)$ when $v_M$ is completely canceled?

**8.54.** The circuit in Fig. P8.54a is a simple "function generator" (with two "functions"). The first stage ("function") is a simple pulse generator. The switchable capacitors select the desired frequency range, while the adjustable resistor $R_v$ fine-tunes the oscillation frequency within the chosen range.

**(a)** Recall that the frequency of oscillation is approximately

$$f = \frac{1}{(R + R_v)C_i}, \qquad i = 1, 2, \text{ or } 3. \tag{P8.1}$$

(i) Suppose that $C_1 = 10$ μF. Choose $R$ and $R_{v,\max}$ so that range 1 is 0.5 Hz $\le f \le$ 50 Hz.

(ii) Now find $C_2$ [with $R$ and $R_{v,\max}$ as in (i)] so that the lower limit of range 2 is 50 Hz. What is the upper limit of range 2?

(iii) Now find $C_3$ so that the lower limit of range 3 equals the upper limit of range 2. What is the upper limit of range 3 in hertz, and why is this an undesirable value?

**(b)** Suppose that the output is sent to the input of the second stage, as in Fig. P8.54b. Sketch the output voltage $v_{o,b}(t)$ as a function of time. If the period of the pulse generator is 0.1 s, what is the maximum value that the output of this second stage will take on? Assume that the initial value of $v_{o,b}$ is zero and that the pulse generator output $v_{o,a}$ is either 15 V or −15 V.

**8.55.** Often it is difficult to measure the impedance of a circuit directly (if it contains, for example, a 20-mile telephone line). In these cases the circuit behavior is described in

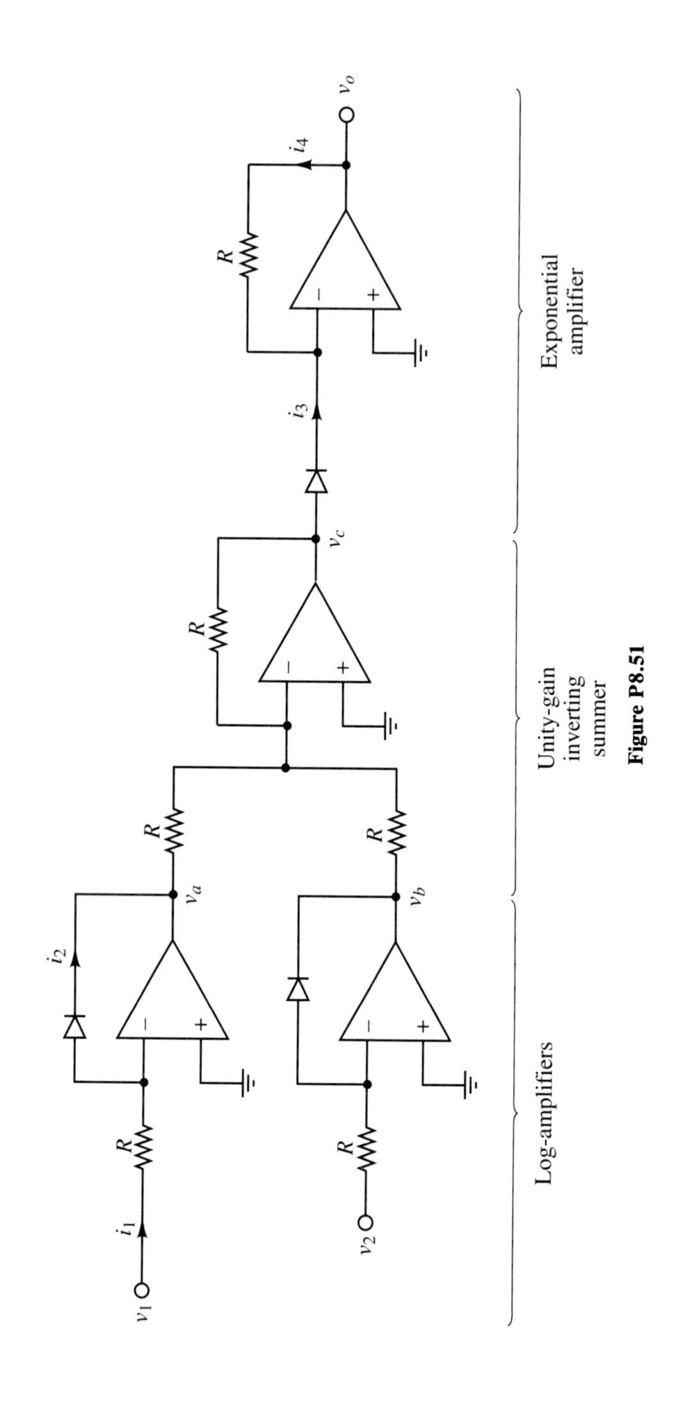

$v_1$
$i_1$
R
$i_2$
$v_a$
R
$v_2$
R
$v_b$
R
R
$v_c$
$i_3$
R
$i_4$
$v_o$
Log-amplifiers
Unity-gain
inverting
summer
Exponential
amplifier


**Figure P8.51**

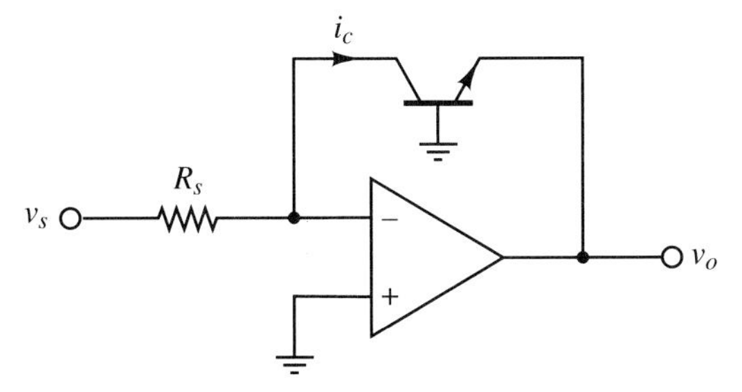


**Figure P8.52**

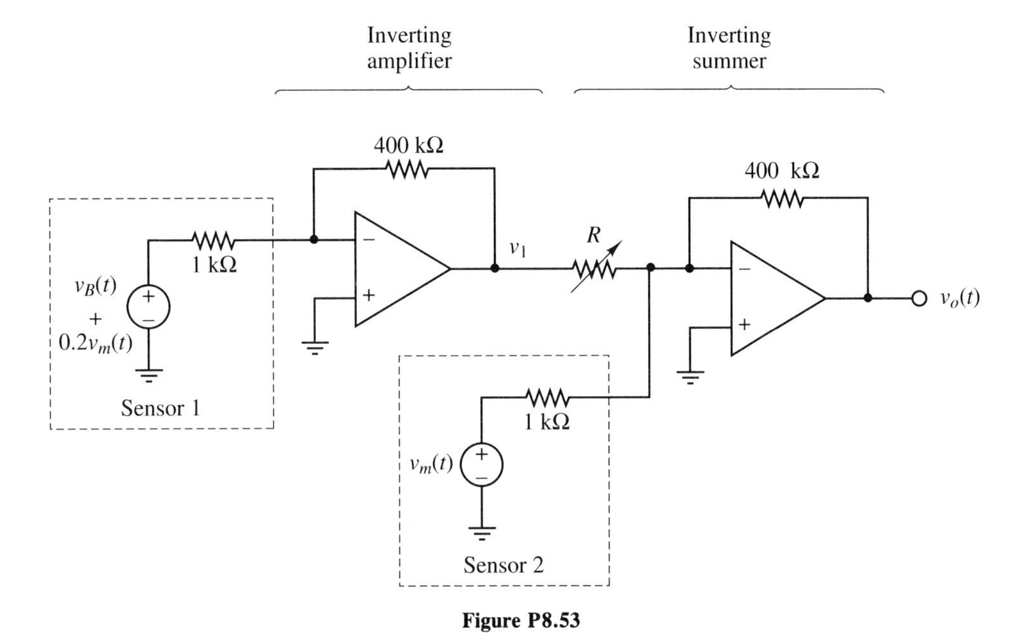


**Figure P8.53**

terms of how it affects sinusoidal voltages of various frequencies by changing their amplitudes and phase angles. Toward this end, the phase meter in Fig. P8.55 will generate an output $v_o$ that is a linear function of the phase differences between the two input voltages. The frequencies of the inputs are the same, of course, but the amplitudes may be different. The clamping diode shunts current away from the inverter whenever $v_c$ is negative so that $v_e$ will have a nonzero average value. (The voltage $v_d$ cannot be less than zero.) The low-pass filter has a time constant $RC$ large enough that only the dc component (average value) of $v_e$ will appear on $v_o$. Sketch $v_a$, $v_b$, $v_c$, $v_d$, $v_e$, and find $v_o$ as a function of $\phi$.

**8.56.** The circuit shown in Fig. P8.56 is called a Wien bridge oscillator. It generates a sinusoidal output on its own (no input), and the frequency may be chosen by the component values. The oscillation is obtained, naturally, by the positive feedback network. A large output occurs at only one frequency—that for which positive feedback actually occurs. This means that the input is exactly in phase with the output, that is, when $\overline{V}_+$ is in phase with $\overline{V}_o$. Normally, it would go into saturation at that frequency. Saturation is prevented by the negative feedback circuit, which partly balances (cancels) the positive feedback input. The diodes further stabilize the output: If it gets too

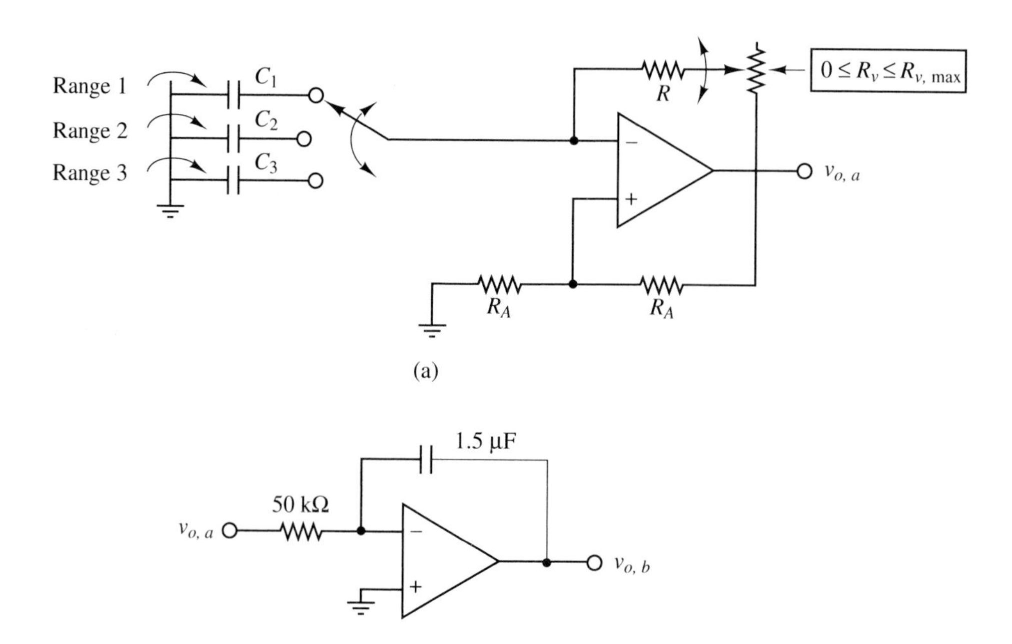


**Figure P8.54**

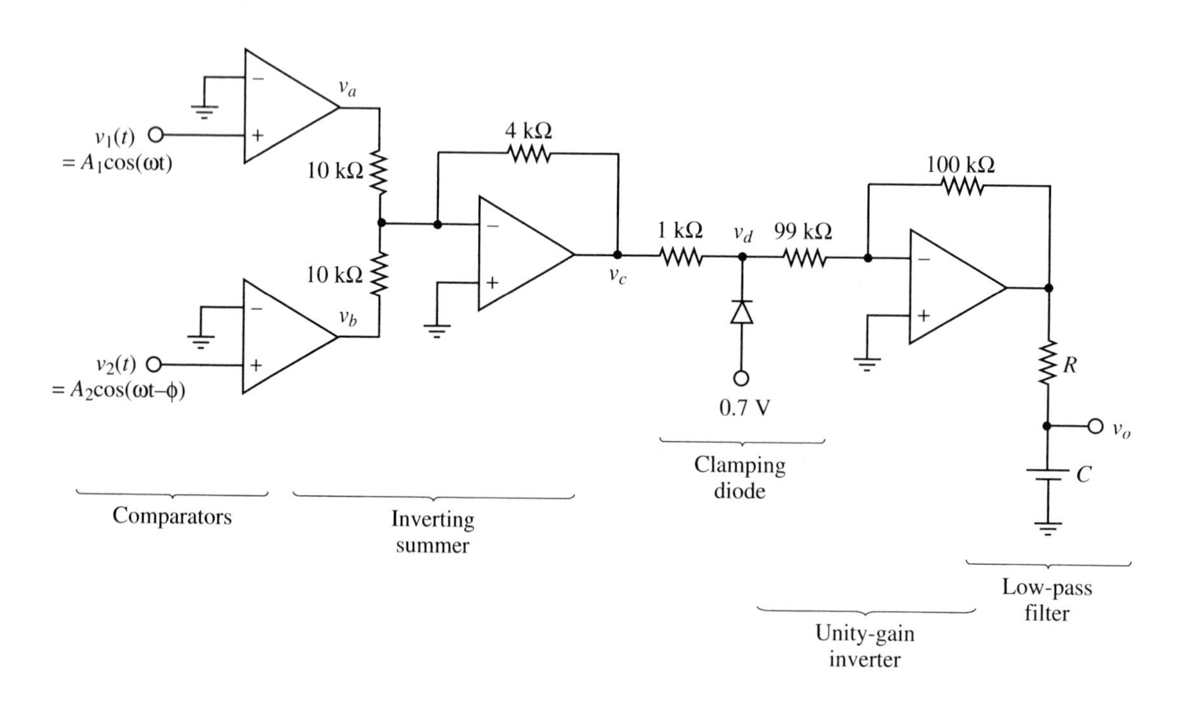


**Figure P8.55**

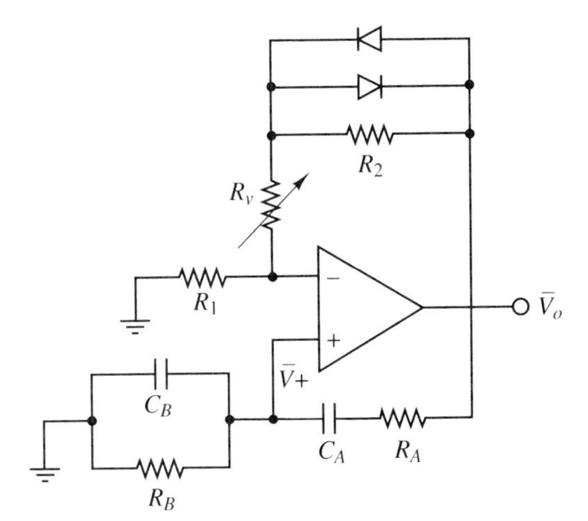


**Figure P8.56**

large (either + or −), one of the diodes conducts, thereby reducing the "$R_2$" in the gain "$-R_2/R_1$." Having $R_v$ variable allows the circuit to be adjusted to achieve desired operation under a variety of operating conditions. Show that the oscillator frequency is $\omega_0 = 1/(R_A R_B C_A C_B)^{1/2}$ by determining when the phase shift between $\overline{V}_+$ and $\overline{V}_o$ is zero. If $R_A = R_B$ and $C_A = C_B$, what is $\omega_0$? If we vary $R_A$, for example, the frequency of the sinusoidal output can thus be adjusted to the value desired.

# 9

# Analog Electronics III: Transducers and Signal Conditioning

## 9.1 INTRODUCTION

In Chapters 7 and 8 we studied the basics of electronics. In Chapter 7 the basic physics of diodes and transistors were described. We then showed how to apply these devices to practical situations such as rectification of ac, amplification of signals, electronic switching, and even some more advanced circuits such as an AM radio and an electronic camera flash unit. In Chapter 8 our transistor amplifier analysis was made easier by considering simple models of the integrated-circuit transistor amplifier known as the operational amplifier. We looked at a wide range of application examples for these versatile chips.

In Chapter 8 we also considered a few of the limitations of actual op amps which require more detailed models and analysis than do the simple ideal dependent voltage source model. Specifically, noninfinite input impedance, nonzero output impedance, finite gain, finite bandwidth, offset voltage, common-mode gain, ground loop distortion, signal feedback through the power supply, and the possibility of stability problems were all considered briefly.

In this chapter we continue our discussion of practical op amp–based circuitry by considering further practical techniques of measurement. For example, we have not said much about $v_s$, the input signal voltage to our op amp. From where does it come? What does it represent physically? We shall answer this by considering a wide range of devices that convert information or energy in a nonelectrical form into information or energy in an electrical form—transducers. Also, we consider what to do with the output voltage, $v_o$. It may drive a display device, actuator, or merely the next stage of signal processing; it may also be converted to digital form for a computer.

A related aspect that we consider is the conditioning of signals. The input signal is in one electrical form, but this is often not the correct form for our ultimate purpose. The terms "signal processing" and "signal conditioning" are essentially synonyms. However, *signal conditioning* emphasizes signal modifications carried out by actual electronic circuits specifically intended to improve the fidelity and format of the signal, and aid in signal transmission over substantial distances. *Signal processing* emphasizes the mathematical operations that one wishes to perform on a signal to extract appropriate information (linear filtering, Fourier analysis, pattern recognition, etc.). Nevertheless, these two terms are commonly used interchangeably.

## 9.2 INTRODUCTION TO TRANSDUCERS

We begin our study of transducers by considering the most basic of all circuit elements: the resistor. However, the essential idea behind transducers is the same for most types. By examining the defining relation between electrical and physical parameters, which includes the parameter to be measured, we usually focus on only the one parameter that we allow to be adjusted independently. The variation of this parameter is considered the input to the transducer. This input necessarily causes (usually only) one of the other parameters to vary significantly; this variation is considered the output of the transducer.

For physical measurements, we depend on the existence of electrical measurement equipment such as the digital voltmeter and/or the impedance bridge. These devices can measure, for example, voltage and current and/or resistance, capacitance, and inductance. Typically, a physical parameter is varied, which also causes the electrical parameter to vary in a known, repeatable manner. The electrical parameter can be measured accurately using the equipment noted above. Using either a simple calculation or another more involved calibration procedure, the change in the physical parameter is inferred; indirect electrical measurement of a physical parameter is thereby achieved. This process is known as *sensing,* and in this case the transducer is called a *sensor.*

The reverse process is actuation, in which case the transducer is called an *actuator.* In this case an electrical parameter such as a voltage or current is varied, which causes in possibly the same device a variation in the desired physical parameter. Unfortunately, this reversibility is not usually possible. For example, a physical motion may cause a resistance and therefore a voltage to vary. It is not possible, however, to cause physical motion by adjusting a voltage across a resistance. However, this fact does not leave us without means of converting electrical energy into mechanical work—that is the subject of Chapters 13 through 15.

When the process is reversible, such as is true for the case of a piezoelectric transducer, the relation between transduction and actuation is similar to that between circuit analysis and circuit design.

For example, by using $I = V/R$, we can infer the current in a given resistor $R$ having a fixed voltage $V$ across it; this is circuit analysis. If, instead, we desire to obtain a current $I$ in that resistive branch having voltage $V$ applied across it, we would select a resistance $R = V/I$; this is circuit design.

Analogously, the voltage across a piezoelectric crystal transducer is $v = K\epsilon$, where $K$ is the piezoelectric constant and $\epsilon$ is the strain on the crystal (not permittivity; the meaning of $\epsilon$ throughout this chapter should be clear from the context). Thus if we apply a strain $\epsilon$, by measuring $v$ we can infer the strain $\epsilon = v/K$, and the crystal is called a sensor. Conversely, by applying a voltage $v$, we can produce a strain $\epsilon = v/K$ by selecting a crystal having piezoelectric constant $K = v/\epsilon$; now the crystal is called an actuator.

## 9.3 RESISTIVE TRANSDUCERS

### 9.3.1 Thickness Measurement

The simplest of all measurements is obtained by solving Eq. (2.36), $R = \rho d/A$, for $d$: $d = AR/\rho$. If the material resistivity $\rho$ and cross-sectional area $A$ of the specimen to be measured are known, measuring $R$ implies the value of its thickness $d$. Tables of resistivity for various materials are available, making this approach practical when one knows the material composition of the specimen to be measured. This method is clearly best for measuring slice thicknesses from a length of material having fixed cross-sectional area $A$.

### 9.3.2 Position Measurement

For measurement of position, the length $d$ of the resistor can again be made to vary, where in this case $d$ signifies position of, for example, a moving shaft. Again by measuring $R$, one can infer position from $d = RA/\rho$. The resistivity $\rho$ and area $A$ of the resistor are held fixed and again are considered known. One might wonder how one can, without cutting, practically vary the length of a solid resistor to correspond to position. Figure 9.1 shows an example that we have already made use of previously: the potentiometer.

In Fig. 9.1a, a linear potentiometer is shown. The coil of noninsulated resistive wire is wrapped around an insulating bar, and the moving shaft contacts it at position $x = X$. The shaft also makes contact with a motionless conducting bar, which serves as the "tap" terminal. In many potentiometers, the resistor element is just a smooth layer of resistive material, so that the wiper makes uniform, continual contact as it slides along. The same principles apply to the transducer in Fig. 9.1b, only now the position variable is the shaft angle rather than distance. Here the shaft itself serves as the "motionless conducting bar" and tap terminal.

Figure 9.1c shows a simple circuit for transforming the position measurement into a voltage output. By knowledge of the voltage divider relation responsible for the measured voltage, we can easily calculate the position. For a more exact measurement that is independent of variations in the reference voltage source $V_r$, we could use the bridge circuit of Example 4.13, with $R_x = R_{ab}$.

It should finally be mentioned that as we have seen in previous chapters, the potentiometer is also useful for the more obvious goal of obtaining human control over a resistance in an electric circuit. An example we studied in Sec. 7.7.3 was the volume control of an AM radio.

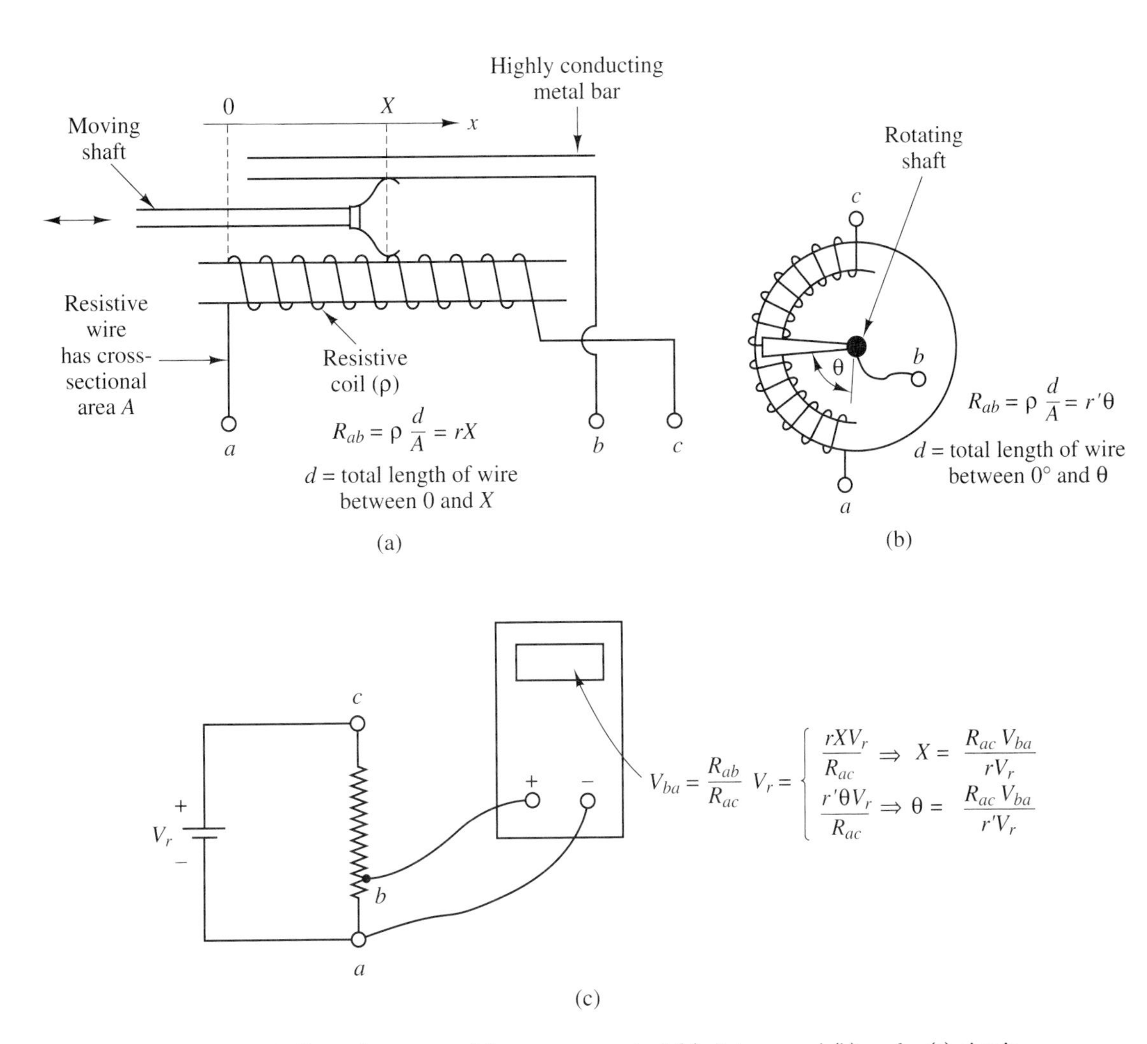


**Figure 9.1** Potentiometer used for measurement of (a) distance and (b) angle; (c) circuit connections for obtaining physical parameters from transducer.

### 9.3.3 Strain Measurement

If instead of taking a tap off a fixed physical resistive structure we squash or stretch a resistor having fixed connections, we get a very important sensor called a *strain gage*. If the resistor is very long and thin (in practice, a long strip of foil glued to the physical object under stress), a small stress will cause a measurable fractional change in resistance. There will be both a change in its length $d$, $\Delta d$, and a change in its area $A$, $\Delta A$. Because $R = \rho d/A$, if resistivity $\rho$ is independent of stress, $\Delta d$ and $\Delta A$ translate into changes in $R$, $\Delta R$. This effect is known as *piezoresistivity*; do not confuse it with the piezoelectric effect, described in Sec. 9.6.2.

Strain gages attached to a stressed object (a *load cell,* or, for example, an airplane wing) are shown in Fig. 9.2a, and the measurement circuit is shown in Fig. 9.2b. The purpose of the accordion-like appearance of the strain gage is to maximize $\Delta d$

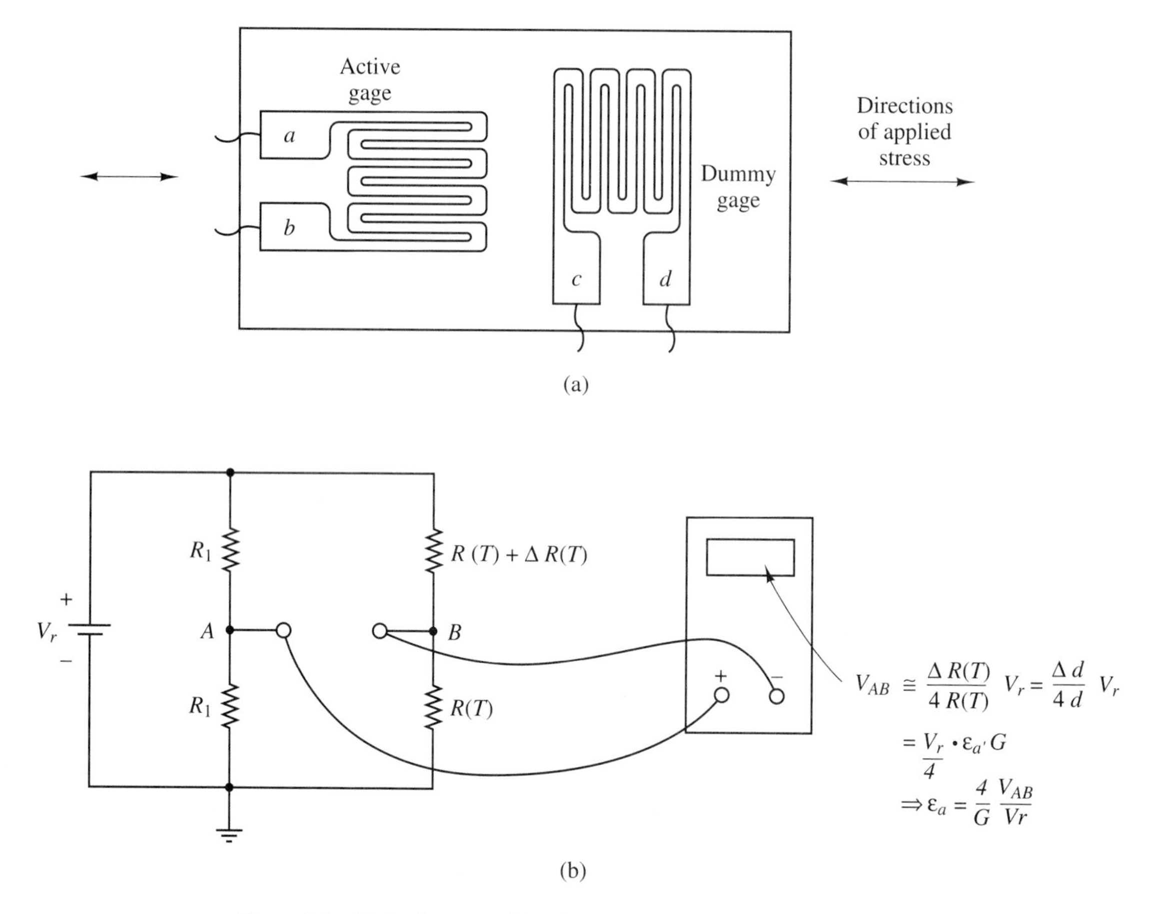


**Figure 9.2** (a) Strain gages; (b) Wheatstone bridge measurement circuit using both gages.

for a given stress in its "long" (axial) direction, and therefore $\Delta R$ and the strain gage sensitivity in that direction.

Two gages are frequently used together: One that is stressed and one that is not. The stressed gage is oriented along the direction in which stress is expected (or is to be applied for testing). The unstressed gage is oriented perpendicular to that direction, and thus exhibits minimal strain. The purpose of the unstressed gage is to provide temperature compensation, as we demonstrate below. Variability of resistance with temperature can far exceed the variation due to stress, so such variations must be accounted for and canceled out.

At a given temperature $T$, let the resistance of each strain gage be $R(T)$ and its resistivity be $\rho(T)$. Thus the resistance of an unstressed gage is $R(T) = \rho(T)d/A$. When tension is applied to the stressed gage, its resistance increases by the increment $\Delta R(T)$. This increment is positive because both its length increases by $\Delta d$ and its area decreases by $\Delta A$ (where $\Delta d > 0$ and we define $\Delta A > 0$). Thus

$$R(T) + \Delta R(T) = \frac{\rho(T)(d + \Delta d)}{A - \Delta A}, \tag{9.1}$$

so that

$$\Delta R(T) = \rho(T)\left(\frac{d + \Delta d}{A - \Delta A} - \frac{d}{A}\right) \tag{9.2a}$$

$$= \rho(T)\left[\frac{Ad + A\,\Delta d - Ad + d\,\Delta A}{A(A - \Delta A)}\right] \tag{9.2b}$$

$$= \rho(T)\left[\frac{A\,\Delta d + d\,\Delta A}{A(A - \Delta A)}\right] \tag{9.2c}$$

$$\approx \rho(T)\left(\frac{A\,\Delta d + d\,\Delta A}{A^2}\right) \tag{9.2d}$$

$$= \rho(T)\left(\frac{\Delta d}{A} + \frac{d\,\Delta A}{A^2}\right), \tag{9.2e}$$

where in Eq. (9.2d) we used the fact that $A \gg \Delta A$. Consequently,

$$\frac{\Delta R}{R} = \frac{\rho(T)(\Delta d/A + d\,\Delta A/A^2)}{\rho(T)d/A} = \frac{\Delta d}{d} + \frac{\Delta A}{A}, \tag{9.3}$$

which we shall use below in calculating the strain. Axial strain $\epsilon_a$ is the fractional change in length divided by the total length, that is, $\epsilon_a = \Delta d/d$, which we note appears in Eq. (9.3). It must be noted that the strain gage is made extremely thin for an additional very important reason: It is easy to stretch. Therefore, it exactly follows the motion of the material to which it is glued; it does not in any way hinder it. This means that $\epsilon_a$ of the strain gage is identical to the strain of the material, which we wish to measure.

To obtain accurate readings, we must compensate for variations of the total resistance due to temperature changes, as noted above. A Wheatstone bridge circuit (see Sec. 4.7) can be used to obtain temperature-independent, sensitive measurements of strain. Instead of adjusting a potentiometer for an ammeter null as in Example 4.13 and Fig. 4.25, one may measure the potential difference between the $A$ and $B$ terminals, labeled $V_{AB}$ in Fig. 9.2b. This configuration also allows for simple temperature compensation by using two strain gages in a way that would not be possible for the voltage divider system in, for example, Fig. 9.1c.

With both strain gages on the right side of the bridge in Fig. 9.2b, we can see how temperature compensation is achieved. If the temperature rises, causing $R(T)$ to increase greatly, the potential at $B$ remains unchanged because *both* resistors (*identical* strain gages) in the voltage divider have increased the same amount. For the potential at $B$ to change, stress must be applied to the upper gage. The resistors on the left voltage divider have high-precision, equal, fixed resistances, so that $V_A = V_r/2$.

In particular, we assume that stress has caused $R(T)$ to increase to $R(T) + \Delta R(T)$. Then the output voltage $V_{AB}$ is

$$V_{AB} = V_A - V_B \tag{9.4a}$$

$$= \left[\frac{1}{2} - \frac{R(T)}{2R(T) + \Delta R(T)}\right]V_r \tag{9.4b}$$

$$= \frac{\Delta R(T)}{2[2R(T) + \Delta R(T)]} V_r \tag{9.4c}$$

$$\approx \frac{\Delta R(T)}{4R(T)} V_r \tag{9.4d}$$

$$= \left(\frac{\Delta d}{d} + \frac{\Delta A}{A}\right)\frac{V_r}{4}, \tag{9.4e}$$

where in Eq. (9.4d) we assume that $2R(T) \gg \Delta R(T)$, and in Eq. (9.4e) we used Eq. (9.3). Dividing both sides of Eq. (9.4e) by $\epsilon_a = \Delta d/d$ gives

$$\frac{V_{AB}}{\epsilon_a} = \left(1 + \frac{\Delta A/A}{\Delta d/d}\right)\frac{V_r}{4} \tag{9.5a}$$

$$= (1 + 2\nu)\frac{V_r}{4}, \tag{9.5b}$$

where $\nu$ is Poisson's ratio, a constant for a particular strain gage metal. Although Poisson's ratio $\nu$ is actually defined as the ratio of lateral strain $\epsilon_l$ to axial strain ($\epsilon_a$), more detailed analysis would show that Poisson's ratio can also be written in this case as $\nu = [\Delta A/(2A)]/(\Delta d/d)$, and hence Eq. (9.5b) is valid. Finally, solving Eq. (9.5b) for $\epsilon_a$, we obtain

$$\epsilon_a = \frac{4}{1 + 2\nu}\frac{V_{AB}}{V_r}. \tag{9.6}$$

In practice, the dimensionless constant *gage factor* $G$ is defined so that

$$\epsilon_a = \frac{\Delta R(T)/R(T)}{G} \tag{9.7a}$$

$$= \frac{4}{G}\frac{V_{AB}}{V_r}, \tag{9.7b}$$

where Eq. (9.4d) was used to obtain Eq. (9.7b) from Eq. (9.7a), so that evidently, within our approximations, $G \approx 1 + 2\nu$. Note, however, that $\nu$ is the fundamental parameter $\epsilon_l/\epsilon_a$, whereas $G$ is purely an experimentally measured parameter determinable if independent measurements of $\epsilon_a$ are possible. This parameter, for example, does not depend on making the only approximately valid assumption that $\rho$ does not change with applied stress; it is an empirical proportionality constant. Therefore, Eq. (9.7b) is the relation commonly used in practice; tables of the value of $G$ exist for various materials.

An additional application of the strain gage idea is the measurement of fluid pressure. The pressure deforms a diaphragm, which strains the resistors (often made of semiconductors). Straightforward calibration converts the strain measurement to a pressure measurement.

**Example 9.1**

A digital multimeter with 0.1-$\mu$V resolution is used to measure $V_{AB}$ in Fig. 9.2b; the reading is 62.5 $\mu$V. Given that the strain gage is constantan ($G = 2$), that the strain gage nominal (unstressed) resistance is 120 $\Omega$ (fixed, room temperature), and $V_r = 3$ V, determine (a) the strain measured and (b) the power consumption of the strained strain gage.

**Solution**

(**a**) From Eq. (9.7b), $\epsilon_a = 4 \cdot 6.25 \cdot 10^{-5}/(2 \cdot 3) = 4.2 \cdot 10^{-5}$.

(**b**) The voltage across the strained gage is roughly $V_r/2$, so that $P_g = (V_r/2)^2/R = 1.5^2/120 = 19$ mW. The value of $R_1$ in Fig. 9.2b would be chosen very large to minimize loading of $V_r$, so that the total power would be roughly $2 \cdot 19 \approx 38$ mW. As a check, note that by equating the right-hand sides of Eqs. (9.7a) and (9.7b), we have $\Delta R(T) = 4 \cdot 6.25 \cdot 10^{-5} \cdot 120/3 = 0.01\ \Omega \ll R(T) = 120\ \Omega$, so using 120 Ω for both gages when computing power consumption is reasonable. This calculation also validates the assumption $2R(T) \gg \Delta R(T)$, made in Eq. (9.4d).

### 9.3.4 Temperature Measurement

In Chapter 2 [see Eqs. (2.33) and (2.34)], we saw that conductivity $\sigma$ was proportional to $\Delta t$, the mean time between collisions of an electron with the atomic lattice of a conductor. Resistivity $\rho$ is the inverse of $\sigma$, so that resistivity increases as $\Delta t$ decreases. As the temperature rises, the likelihood of an electron colliding with the lattice increases because of the increased frequency and excursions of the atomic vibrations. Therefore, in a metal the resistivity increases with temperature.

We saw this phenomenon in Sec. 9.3.3 concerning strain gages. We also saw it in our studies of electric lights in Example 4.19 in Sec. 4.8.3 and in Sec. 4.10 (see Fig. 4.38). There we noted that $R$ was a function of applied voltage $v$; actually, $R$ was a function of temperature, which was increased by more current, which in turn was caused by the increased $v$. In fact, to measure $R$ requires a voltage divider circuit that itself will unavoidably heat the resistor. Hence measuring $R(T)$ to infer $T$ involves significant distortion of the measurement by the measurement apparatus. All resistive thermometers suffer from this type of distortion.

By measuring the resistance $R = \rho d/A$ with only the temperature $T$ changing, we can nevertheless infer the local resistor temperature if $\rho$ of the resistor is calibrated with $T$. Such a resistance thermometer is called a *resistance thermal detector* (RTD). For metals such as nickel, copper, tungsten, and platinum, the relation between $R$ and $T$ (here in degrees Celsius) is linear:

$$R(T) = R(0°)(1 + \alpha T), \tag{9.8}$$

where $R(0°)$ is the resistance at $T = 0$°C and $\alpha$ is the temperature coefficient of the metal. As implied above, $\alpha > 0$ and has units of 1/°C, often written $\Omega/(\Omega \cdot °\text{C})$ because $\alpha$, which is the slope of the $R$ versus $T$ curve, can be found as $\alpha = [R(100°) - R(0°)]/[R(0°) \cdot 100°]$.

The RTD is most often platinum, in the form of a very thin wire coil or film. It has a relatively low sensitivity ($\alpha$ is small) but a very high reliability and linearity. The appearance of RTDs varies with the housing in which these fragile devices are protected; an example is shown in Fig. 9.3a.

Another type of resistive thermometer is called a *thermistor.* In this case the material is a semiconductor (metal oxide or silicon). As we know from Chapter 7, the higher the temperature, the *lower* the resistivity of a semiconductor. This is because more electron–hole pairs are created at higher temperatures, thereby increasing the free charge available for conduction. The physical appearance of a thermistor is often similar to that of a small capacitor, and is shown in Fig. 9.3b.

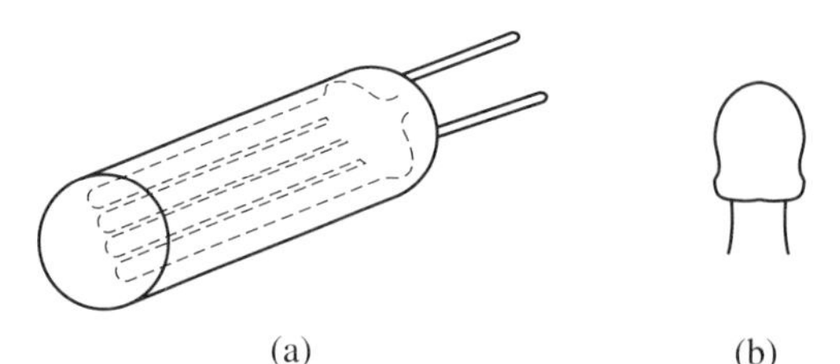

(a) (b) **Figure 9.3** (a) Typical resistance thermal detector (RTD); (b) thermistor.

The temperature dependence of the thermistor is much stronger than that of the RTD—an exponential decrease, as opposed to the linear increase of the RTD. This higher sensitivity is the primary advantage of the thermistor; however, it is less stable and less easily calibrated than is the RTD. The nonlinearity of the thermistor is bothersome and typically, digital computation is required for translation from resistance into temperature.

Physically, the meaning of the exponential dependence is this: Equal increments in temperature cause equal *percent* reductions in resistivity. Consequently, thermistors are specified in terms of the percent reduction per degree Celsius as well as their resistance at a reference temperature (25°C). Typical values are −3% to −5% per °C, from a very broad range of reference resistances (5 Ω to 200 kΩ).

(This logarithmic dependence is also found in hearing, where the volume of the sound must be doubled for the person to sense the changes in volume as being equally incremented. Hence the logarithmic "dB" quantification of sound pressure levels.)

## 9.4 CAPACITIVE TRANSDUCERS

### 9.4.1 Position Measurement

Several types of measurements can be made with a capacitor as well as with a resistor. An advantage of capacitive transducers is that they tend to interfere less with the physical apparatus being measured. For example, consider the capacitor versions of distance (see Fig. 9.4a) and angle (see Fig. 9.4b) transducers. In both cases, there is nothing in the capacitor likely to wear out. Compare these with the wirewound or film wiper pads of potentiometers (Fig. 9.1), which obviously eventually wear out due to continual rubbing of the tap over their surface. However, viewed from another perspective, it must also be pointed out that unwanted capacitance from the physical apparatus can affect the measurements, and this must be taken into account.

As a result of more advanced analysis, the capacitance of concentric cylinders with parameters shown in Fig. 9.4a is

$$C = \frac{\epsilon d}{\log(r_1/r_2)}, \tag{9.9}$$

so that

$$d = \frac{C \log(r_1/r_2)}{\epsilon}, \tag{9.10}$$

where here $\epsilon$ is permittivity. Thus if $C$ can be measured accurately electronically, displacement $d$ can be found from Eq. (9.10). Methods of measuring $C$ include

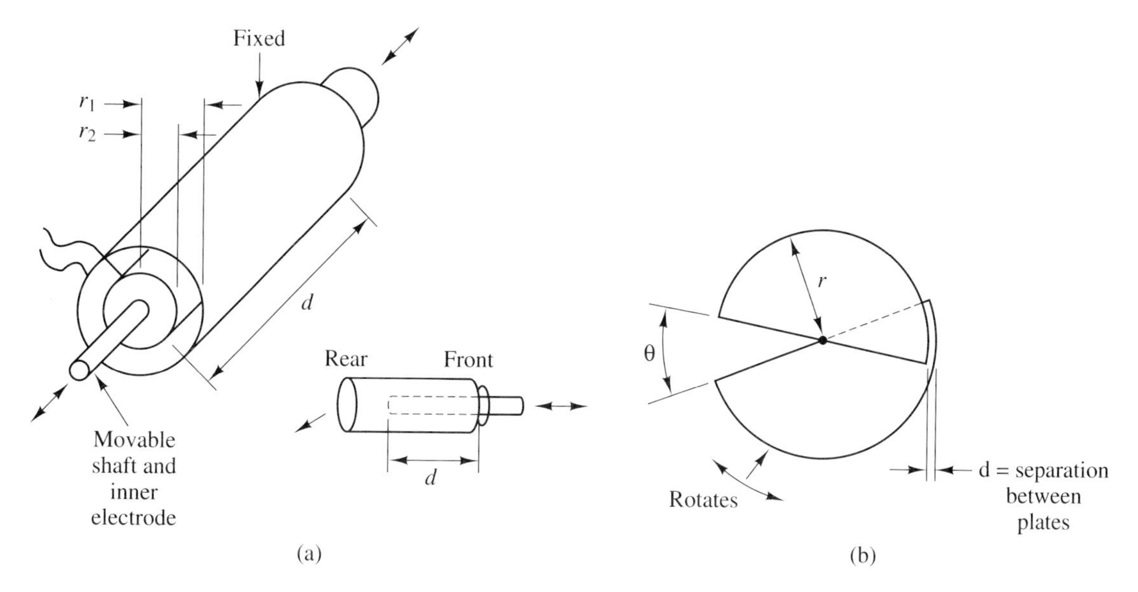


**Figure 9.4** Schematic representations of capacitive transducers for measurement of (a) displacement and (b) angle.

bridge-type nulling circuits as well as resonant frequency measuring circuits that use digital counters (see Chapter 11).

**Example 9.2**

Suppose that with an advanced bridge circuit we have the means of resolving $10^{-15}$ F. If $r_1 = 1$ cm, $r_2 = 0.99$ cm, and initially $d = 2$ cm, what distance can we resolve using the transducer in Fig. 9.4?

**Solution** By Eq. (9.10), $\Delta d = \Delta C \log(r_1/r_2)/\epsilon_0 = 10^{-15}\ \text{F} \cdot \log(1/0.99)/8.85 \cdot 10^{-12}\ \text{F/m} = 4.9 \cdot 10^{-9}\ \text{m} = 4.9\ \text{nm}$. This example shows how accurately displacements can be measured with this type of transducer. While care must be used in setting up the measurement equipment (e.g., minimization of and/or accounting for capacitance of connecting wires), these transducers are very accurate and do not wear out.

For the capacitive structure in Fig. 9.4b, note that the overlapping area of the two semidisks is $A = (\theta/\pi)(\pi r^2/2) = \theta r^2/2$. Substitution of this into $C = \epsilon A/d$ gives

$$C = \frac{\epsilon r^2 \theta}{2d}, \tag{9.11}$$

so that the angular position θ is found to be

$$\theta = \frac{2dC}{\epsilon r^2} \qquad (0 \le \theta \le 180°). \tag{9.12}$$

Just as potentiometers are used not only as position sensors but also as resistance control devices, the same is true of capacitive transducers. The most familiar example is the AM radio tuning knob, which is just a sandwich of structures of the form in Fig. 9.4b. By adjusting θ, we can control the tuning capacitance and therefore the radio station selected; review Sec. 7.7.3 for more details.

### 9.4.2 Liquid Level Measurement

The simplest means of electrically measuring liquid level is by using a potentiometer. The end of one arm is fixed and the other is connected to a float that rises (increases angle) or falls (decreases angle) with the liquid level. Although these are common, a more interesting example, which again cannot wear out, is the capacitive liquid level transducer (see Fig. 9.5).

The liquid is typically dielectric with $\epsilon > 1$. Thus at any given time part of the tank–electrode system has air ($\epsilon = \epsilon_0$) and the rest has $\epsilon > 1$. Depending on the height $d$ of the liquid, the capacitance will change. Exact calculation of $d$ from the measured $C$ is difficult and depends on the geometry of the tank. Usually, one merely calibrates the system to the desired resolution and later looks up the correct height for the measured capacitance at any given time.

### 9.4.3 Pressure Measurement/Microphone

There are two ways to infer liquid and gas pressure from capacitance: The geometrical and dielectric variations with pressure. For the geometrical method, a bendable diaphragm responds to applied pressure by deforming, thus changing the spacing $d$ between plates of a capacitor (see Fig. 9.6a). Roughly, we have pressure $p$ = force/area = $k\ \Delta d$/area, where $k$ is a spring constant. Similarly to the strain gage, $\Delta d$ can be obtained by taking the ratio of capacitance before and after the pressure is applied. Again, because of many complexities not accounted for above, calibration against independently verifiable pressure standards must be done to maximize accuracy. This principle can also be used to make a capacitive strain gage. In response to the strain, the plates bend and $d$ changes.

More directly, we can apply a dc voltage $V$ across the plates through a resistor $R$ to obtain a large charge on the plates (see Fig. 9.6b). Then $v_c = q/C = (q/\epsilon A)d$ is proportional to $d$. In this configuration the charge $q$ is maintained essentially fixed because it is connected to $V$ through a large resistance $R$ ($\tau = RC$ is large). The variations in pressure again vary $d$, which varies $v_c$ in direct proportion. This configuration is used in the capacitor microphone, where the output signal $V - v_c$ is obtained via capacitor coupling (through $C_c$). Because capacitors used to be called condensers, the name *condenser microphone* is still sometimes used. This type of microphone can today be made to have extremely high fidelity, for music recording.

It should also be noted that there is a resistive microphone, called the *carbon microphone*, commonly used in telephones. The density of carbon granules is varied by the sound pressure against a diaphragm; this varies the resistance of the section

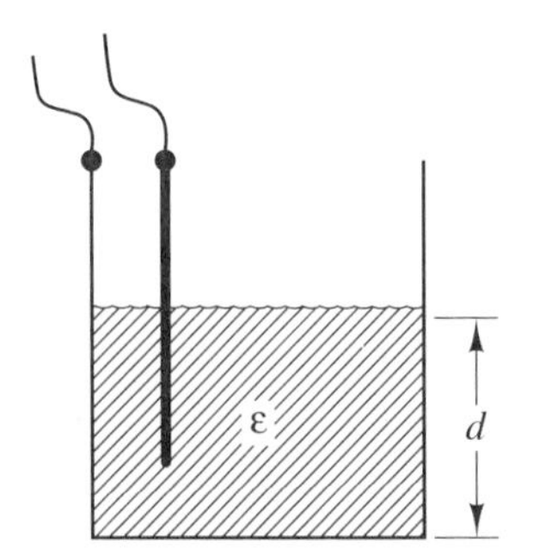


**Figure 9.5** Capacitive liquid level transducer.

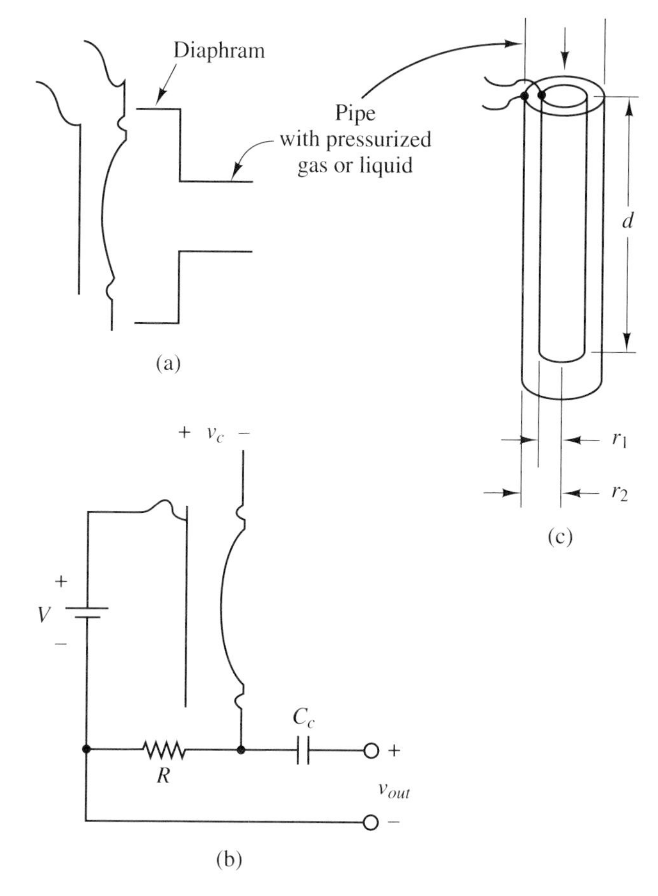


**Figure 9.6** Capacitive pressure transducers for which (a) $d$ varies, (b) $d$ varies (microphone configuration), and (c) $\epsilon$ varies.

of carbon granules. However, it is a low-fidelity microphone and is rapidly being replaced by other inexpensive, smaller transducers. Thus only brief mention is made of it here.

For the dielectric method of pressure measurement, there are no moving parts. Use is made of the fact that $\epsilon$ varies with gas pressure. Therefore, a fixed tubular structure such as that shown in Fig. 9.6c can be placed wherever the value of pressure is desired (e.g., on the end or side of a pipe). Again there are nonlinearities [particularly, $\epsilon$ (pressure)] requiring calibration, but the fundamental capacitance relation, Eq. (9.9), is still used to convert measured capacitance into a value of $\epsilon$.

## 9.5 INDUCTIVE TRANSDUCERS

### 9.5.1 Linear and Rotational Variable Differential Transformers

It may sound like a mouthful, but the essential operation of the LVDT and RVDT is fairly simple to understand. In Sec. 3.10.1 we studied the transformer principle

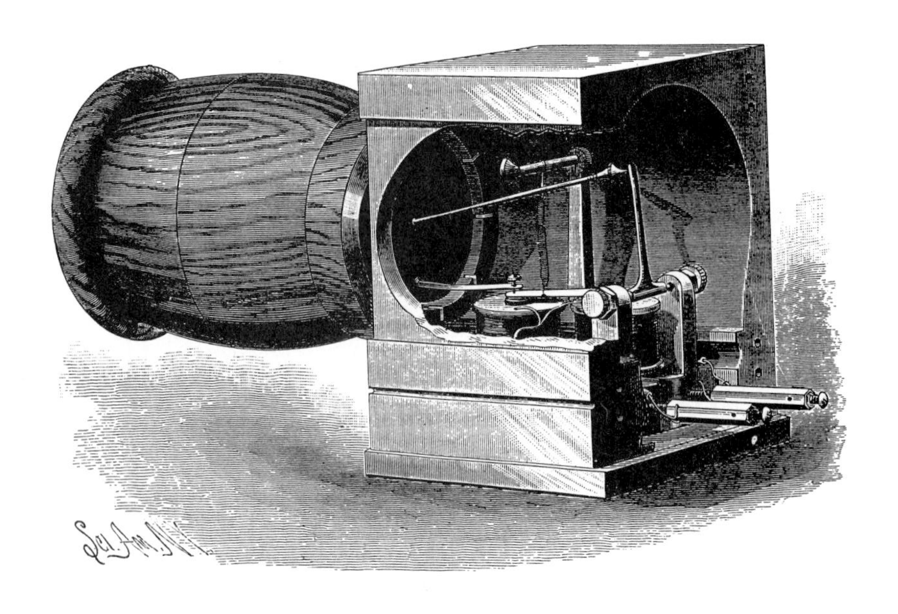

The "house telephone" as it appeared at the end of the nineteenth century (top), and in the more familiar 1920s style. The advertisement (bottom), which shows a lead mine in the background, advertises and proclaims the virtues of lead, not a telephone. *Source: Scientific American* (July 1896, November 1924).

using Faraday's law. Then in Sec. 6.5 we analyzed the transformer in much more detail. In particular, we considered the idea of mutual inductance, in which only part of the flux passing through the primary reaches the secondary coil.

This idea can be carried to an extreme by having a significant fraction of the flux path be through air. The lower the proportion of air path, the more flux reaches the secondary, and vice versa. Furthermore, having two secondaries wound in opposite senses allows for a symmetric, even response that is very close to linear with displacement.

Figure 9.7a shows a practical configuration. When the core is exactly centered on the two secondaries, the two secondary voltages sum to zero. Actually, due to a variety of imperfections, the output voltage never quite goes to zero even when the core is perfectly centered; advanced circuitry can correct for this. When shifted to either side, one of the coils has a higher induced voltage, yielding a net ac output voltage. A phase-sensitive ac voltage detector circuit can distinguish on which side of "zero" the core is located.

The RVDT, a variation on the transformer position-sensing idea that performs measurement of angles, is shown in Fig. 9.7b. The heart-shaped core with off-center shaft causes the flux in the two secondaries to vary when the shaft is rotated. Just as too-large displacements of the core of the LVDT will yield meaningless measurements because of nonlinearities, too-large angles (above about $\pm 40°$) are also unmeasurable with the RVDT.

The electronic circuits for generating ac at the input from a standard dc power supply and for detecting/rectifying the output ac voltage (including "direction") for output electronic devices that require unmodulated signals can be fairly complex. Therefore, integrated packages are available that allow dc input and provide unmodulated (often loosely called "dc") output. These "dc–dc" LVDTs and RVDTs make them much easier to apply in practice.

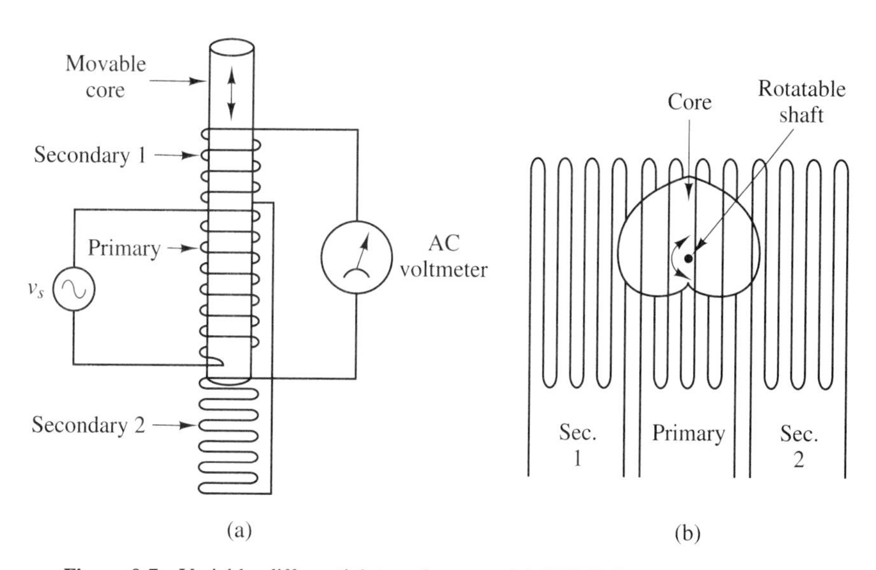


**Figure 9.7** Variable differential transformers: (a) LVDT for measuring linear displacements; (b) RVDT for measuring angular displacements.

### 9.5.2 Microphone/Speaker

The moving-coil microphone is very common as a high-fidelity music transducer. A typical construction is shown in Fig. 9.8a. An E-shaped (in cross section) permanent magnet has a north pole in the center leg and south poles in the outer legs (which actually form one circular leg surrounding the circular inner north pole). A coil is attached to a diaphragm that vibrates in response to sound waves generated by the singer or musical instrument. Because the magnetic flux varies in accordance with the sound information content, a voltage will be induced across the terminals of the moving coil. Many complicated design issues are involved in maximizing the conflicting desirable characteristics of flat frequency response (high fidelity) and high sensitivity.

Applications of both condenser and moving-coil (so-called *dynamic*) microphones are common in today's audio systems. A wide variety of directivity patterns and frequency range emphases are available. For example, a female singer may choose a capacitor microphone that accentuates low frequencies over a moving-coil counterpart. Just as with photography (color emphasis film), the goal is *not always* a perfectly accurate reproduction but a maximally pleasing creation.

The moving-coil transducer is the first of several we shall discuss that are *self-generating*. That is, they need no external source to deliver electric energy containing the measurement information to an amplifier. All the transducers we have discussed before modulated an existing dc or ac electric energy source but could not alone generate an electrical signal.

In all or nearly all cases, these *modulating transducers* are not reversible (e.g., we cannot apply voltages to a potentiometer and expect it to move). Contrarily, self-generating transducers are much more likely to be (but are not always) reversible.

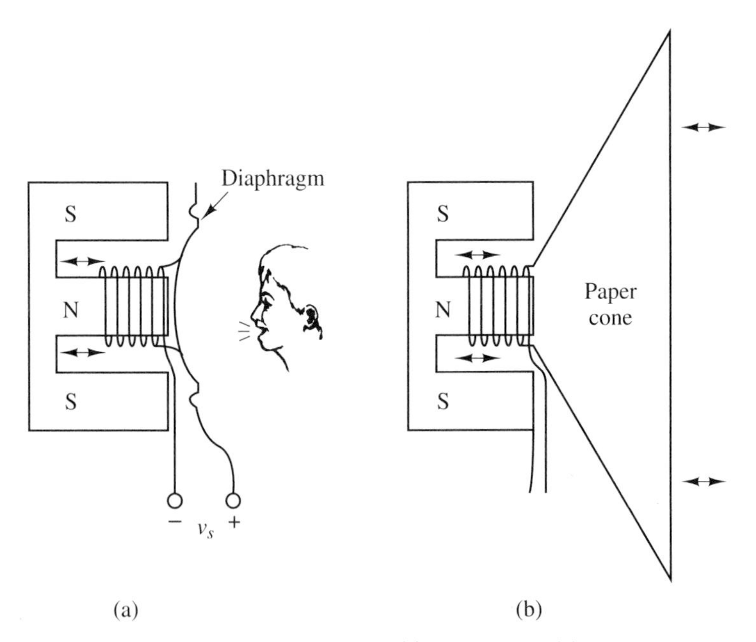


**Figure 9.8** Moving-coil transducer: (a) microphone; (b) speaker.

For example, while the condenser microphone of Sec. 9.4.3 could not be "played" as a speaker by applying voltages, the moving-coil microphone may be modified slightly to function as a speaker. The moving-coil speaker is the most common of all high-fidelity speakers. Referring to Fig. 9.8b, the applied signal voltage causes currents in the coil, and Ampère's force moves the coil in the presence of the permanent magnet. When a paper cone is attached to the coil, audible sound mimicking the signal results.

### 9.5.3 Tape Recorder Head

The tape recorder head is another self-generating, reversible transducer; energy applied in either form is transformed into the other form. In this case the $d\phi/dt$ associated with tiny permanent magnets on the recording tape as it is pulled by the head is one form and the applied/received signal voltage is the other.

The basic principle of magnetic tape recording is the same whether it be a VCR, an audiocassette recorder, or a digital computer data tape storage system. Magnetic recording tape is a layered structure: a polyester base for strength, a back conductive coating to reduce static electricity effects, and, of course, the magnetic coating. This coating is composed of very closely packed magnetic particles, like iron filings, only extremely small.

In the recording mode, the signal to be recorded is amplified and presented to the record head, which converts the time-varying voltage to a time-varying magnetic field. Recall from Chapter 3 that current in a coil produces a magnetic field proportional to the strength of the current.

This magnetic field magnetizes the tape due to the hysteresis/residual magnetization effect. Recall from Sec. 3.8 that for permanent magnets we want a wide hysteresis curve, to maximize the residual field when the excitation is removed. This is true for magnetic tape because we want a permanent record of the excitation (audio signal). The area of the $B$–$H$ curve, which indicates energy lost per hysteresis cycle, is irrelevant here. Contrarily, for machines (Chapters 13 through 15) losses are detrimental, so materials having a narrow $B$–$H$ curve are used in that case.

Clearly, saturation must be avoided for "analog" recordings, because saturation cuts off (distorts) the peaks in the original signal waveform. In digital data and digital audio recordings, only 1s and 0s are stored, so in that case saturation *is* used: one saturation value is 1 and the other is 0.

A plot of particular value for discussing magnetic tape processes is that of *retentivity* (residual $B$) versus applied $H$. Consider first Fig. 9.9a, the $B$–$H$ curve. We discuss the recording process; playback is essentially the reverse. We may assume that we always start from the origin with each new section of tape that goes by because we assume that the tape is initially erased by a previously contacted erase head. We consider $H$ to be due purely to the "magnetizing force" resulting from signal currents in the recording head.

After $H$ is removed (the tape leaves the head), $B$ falls to $B_r$ for the peak $H$ attained, rather than to zero because of hysteresis. The values of retentivity for several signal levels ($H_1$ through $H_4$) are shown in Fig. 9.9a labeled $B_1$ through $B_4$. Using these and many additional values, we can plot the magnetic tape retentivity $B_r$ versus $H$ for each value of $H$ to obtain the plot in Fig. 9.9b. Thus, for a given signal

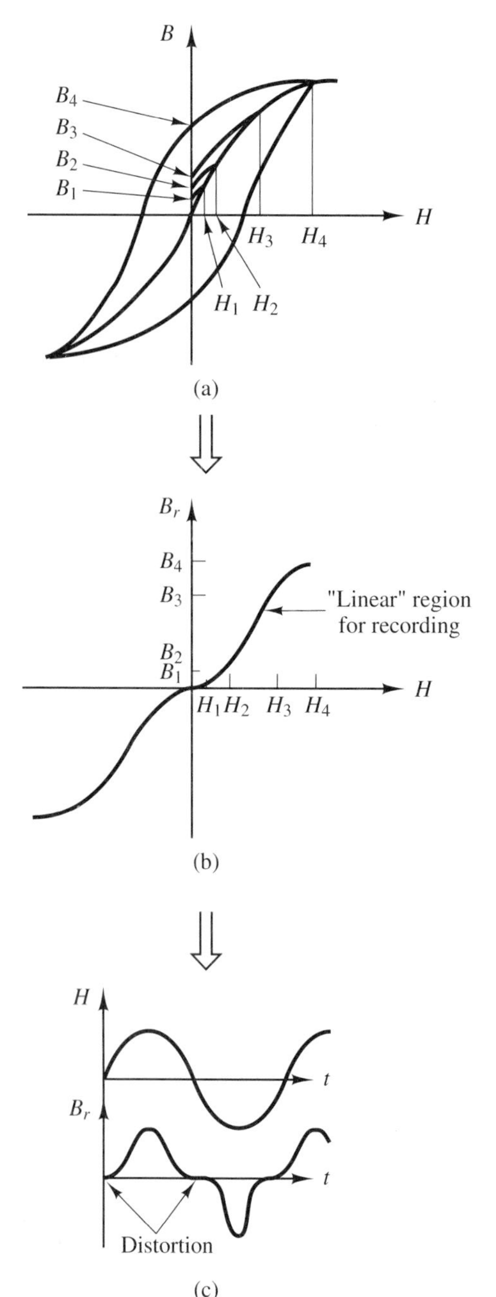


**Figure 9.9** $B$ and $B_r$ for magnetic tape: (a) $B$–$H$ curve; (b) $B_r$–$H$ curve; (c) example $H$ and $B_r$ versus $t$.

($H$) value, such a plot indicates the magnetization level remaining ($B_r$) after the applied signal is removed. After all, the remnant $B$ after $H$ is removed is the key variable; it is what we will have stored on the tape to play back later.

An example input signal (with no bias added) is shown in Fig. 9.9c. We see that distortion results at very low amplitudes due to flattening of the $B_r$–$H$ curve near the origin. To solve this problem, one could take an approach similar to that used in transistors: Add a dc bias so that all signal excursions remain within one of the relatively linear portions of the $B_r$–$H$ curve. However, this limits the amplitude of audio signals unnecessarily because one can use only the linear portion in the first quadrant, instead of the linear portions in both the first and third quadrants.

In the 1940s, some researchers found an improvement "by accident" experimentally, and since then many have proposed theories to explain it. If the audio signal is superimposed on (added to) a very high-frequency signal (150 kHz is typical), the nonlinear portion of the $B_r$–$H$ curve can be avoided, yet both linear portions can be used. This doubles the linear dynamic range attainable compared with using dc bias. Naturally, this is called *ac biasing,* a term probably at least vaguely familiar to most audiophiles.

The concept of ac biasing is illustrated in Fig. 9.10, which shows the high-frequency ac bias added to a low-frequency triangle wave (the signal), and the output. The input (signal applied to the tape head coil, including ac bias) is the sideways signal shown with time progressing downward. The output (what is actually recorded on the tape) is the normally oriented signal shown with time progressing to the right. The output is obtained by finding the $B_r$ value corresponding to the input value at each instant of time. The result is that the *envelope* (dashed line) increases linearly with time! This linear envelope can be recovered merely by low-pass filtering out (see Sec. 5.5; also called demodulating) the bias/nonlinear portions of the

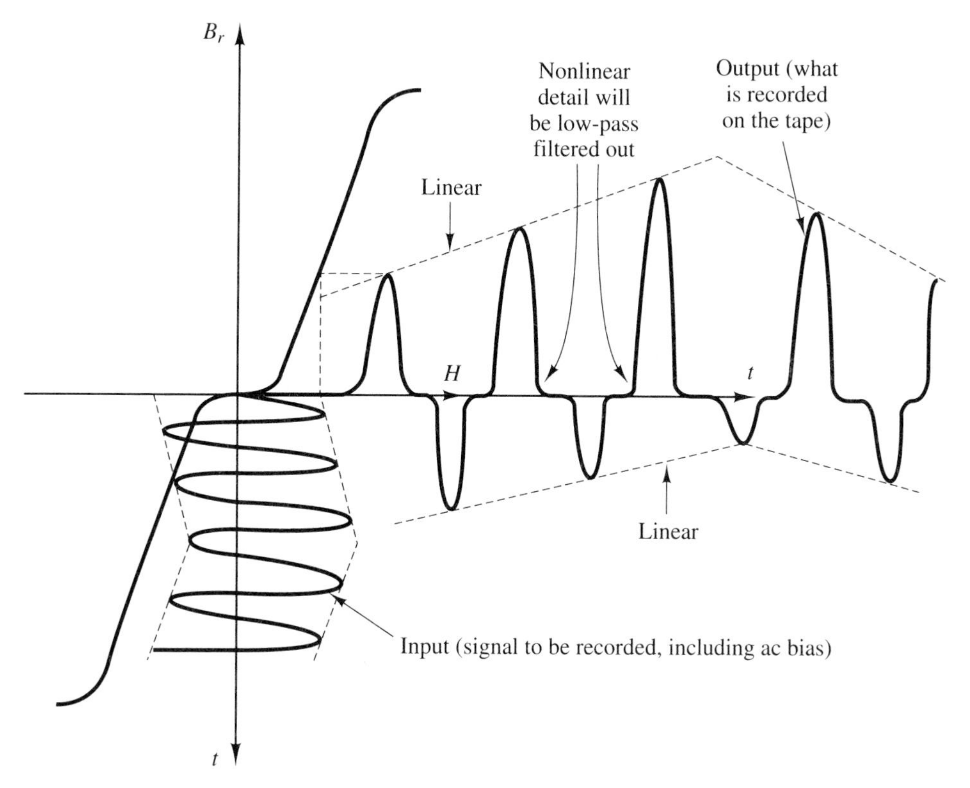


**Figure 9.10** $H$ and $B_r$ versus $t$ with ac bias applied. The result is a linear relation between the original signal and a low-pass version of the output $B_r$ versus $t$.

output. If peak detectors are used to select the maximum magnitude envelope (either positive or negative), the lower magnitude envelope can be nonlinear without distorting the obtained signal.

The ac bias level (amplitude) as well as the signal level must be carefully set for a particular type of recording tape. The bias must be large enough to guarantee avoidance of the nonlinear portion, yet be small enough to maximize the linear portions available to the signal. The signal must be large enough to have a good signal-to-noise ratio but be small enough to preclude saturation distortion.

When *digital tape recording* becomes perfected and made affordable, the analog method just described may eventually be discontinued. That may be many years from now, however. Analog tape recorders and tapes are likely to persist for quite awhile yet.

### 9.5.4 Relay

The electromagnetic relay has always been one of the most important components in industrial control and machinery circuits. A relay is nothing other than a magnetically actuated switch. A typical relay is shown in Fig. 9.11a.

Control signals are often low-voltage lines for safety reasons, yet the machines they control may require very large voltages and currents (e.g, large motors). To isolate the human from the high-power machine circuit, the person can flip a low-voltage switch. This in turn sends current through a relay coil, producing a

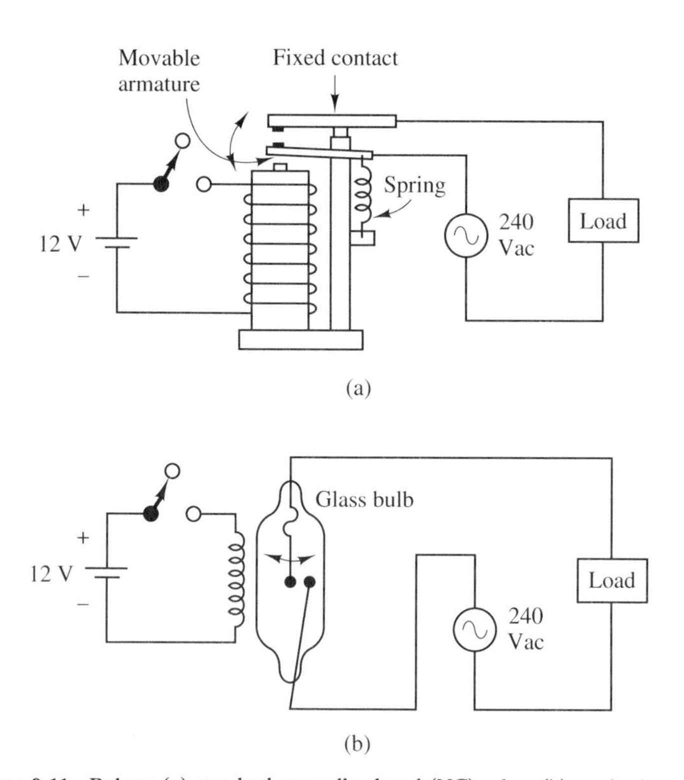


**Figure 9.11** Relays: (a) standard normally closed (NC) relay; (b) reed relay.

magnetic field. The magnetic field in turn closes or opens a high-current/voltage-capability switch to control the machinery. The moving member closing or opening the switch is called the *armature*, a term we shall see much of in other contexts in Chapters 13 through 15.

There are many other purposes for relays. One is the control of numerous high-power machines safely and efficiently from a far-distant control panel. Another is to protect machinery from excessive currents. A sample of the load current can be applied to the relay. Above a certain threshold, the relay can open, thereby turning off that machine. Thus the machine is self-protected.

Yet another common example is the use of microcomputers (see Chapter 12) in modern industrial control systems. The computer can provide only low-power on–off signals, yet it may be required to drive large loads (e.g., a computer-controlled washing machine such as we discuss in Sec. 11.5). The relay provides the required "power buffer" between the computer system and the machinery system.

The relay is an example of an actuator—a transducer that converts electrical information into physical action. It is another instance of a modulating transducer whose transduction is irreversible. Relays are never used as on–off electrical signal generators operated by moving the armature.

One might think that with the computer age, relays must be on the way out. This is true only for relay decision-making (logic) circuits that in the past generated the control signals. These signals are much more efficiently and effectively generated by computers. However, the interface between the low-power computer and the high-power machinery will always remain. Contrarily, often the high-power protective relays discussed above can be effectively replaced by SCRs (see Sec. 7.6.5), so this application may be on the decline. However, because of their long life and low power consumption, even in these situations relays may still be favored over other electronic switches.

Because of the wide range of applications, relays come in many shapes, sizes, and configurations. For example, the controlled switch may be normally open (NO) or normally closed (NC). The voltage required to flip the switch (called the *pickup current*) may be a few volts or hundreds of volts. The turn-on and turn-off times may be just a few milliseconds or may be 25 ms or more. A very popular modern relay called the *reed relay* is shown in Fig. 9.11b. Its contacts are sealed in inert gas and coated with liquid mercury to minimize pitting and extend relay life to billions of on–off switches.

### 9.5.5 Pulse Tachometer

Measurement of the speed of rotation of a shaft has innumerable applications in industry. A simple method of measuring the speed (frequency of rotation) that does not disturb the motion is use of a pulse tachometer. There are two main components (see Fig. 9.12): the *toothed rotor* and the *sensor*. The rotor is made of magnetic material, and the sensor contains a permanent magnet wrapped by a sensing coil.

When the rotor spins, the flux path from the permanent magnet varies because of the nonuniform (toothed) shape of the rotor perimeter. This causes the flux within the coil of the sensor to vary with time. By Faraday's law, $v = d\phi/dt$, a voltage is induced in the coil and is available at the terminals that directly corresponds to the

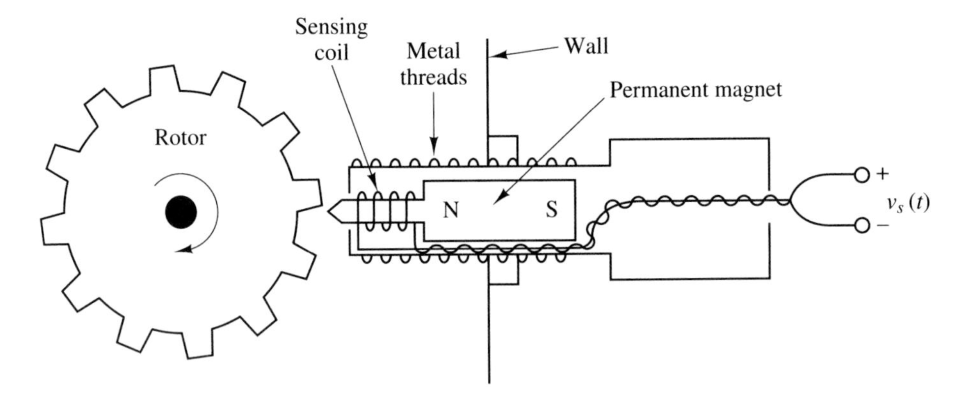


**Figure 9.12** Pulse tachometer with rotor.

motion. Use of gear teeth produces voltage pulses, which may be counted. The faster the rotor spins, the more pulses occur in a given time interval. Electronic circuitry is available for this counting and calculation. Just as the Schmitt trigger uses hysteresis to prevent chatter in switching due to a noisy input signal, counting circuits for the pulse tachometer use hysteresis to prevent false counts due to a noisy pulse waveform.

It would be very difficult to calculate the flux pattern as a function of time, but fortunately this is not necessary. In fact, this comment applies to many magnetic transducers: exact calculation of the effect is usually impossible; empirical design is the rule. Many rotor shapes are possible, each of which produces a particular output pulse waveshape. Also, the probing tip of the permanent magnet can be cylindrical, chisel-shaped, or other shapes. Depending on the application and available conversion circuitry, one particular shape of rotor or probe may be most appropriate.

**Example 9.3**

A pulse tachometer rotor has a tooth every 11.25°.

**(a)** How many pulses should we see on the sensor output for each revolution of the rotor?

**(b)** If we count 10 pulses in a time interval of 50 ms, what is the rotor speed in revolutions per minute (rpm)?

**(c)** Is the pulse tachometer a self-generating transducer? Is it reversible?

**Solution**

**(a)** We see one pulse per tooth and there are $360°/11.25° = 32$ teeth on the rotor, so there are 32 pulses per revolution.

**(b)** Rotor speed in rpm is

$$\text{speed} = (1\ \text{rev}/360°)(11.25°/\text{pulse})(10\ \text{pulses}/50\ \text{ms})(60 \cdot 10^3\ \text{ms/min}) = 375\ \text{rpm}.$$

**(c)** The pulse generator is self-generating; notice that there are no external sources required for its operation. However, the process is not reversible. The forces due to an applied pulse train on the sensing coil would be infinitesimal and nowhere near those required to turn the shaft.

## 9.6 MISCELLANEOUS TRANSDUCERS

### 9.6.1 Thermocouple

Thomas Seebeck in 1821 discovered that a measurable temperature-dependent voltage appears between the ends of two strips of dissimilar metals joined and heated at the other ends, and that current flows if the circuit is completed. This phenomenon is known as the *Seebeck effect.* In Sec. 7.3.2 we noted that there exists a contact potential at the junction between any dissimilar conductors. We did not mention the temperature dependence of this potential. In fact, for junctions between copper and constantan (typical thermocouple metals), the dependence on temperature is reasonably linear up to about 400°C, and between other metals, linearity extends into thousands of °C. However, high-order polynomial approximations are typically used in accurate measurement systems. The greatest advantages of thermocouples are their ruggedness and wide temperature ranges over which they can be used.

Their major disadvantage is that unavoidably there will always be more than one temperature-dependent junction. Figure 9.13 shows a copper–constantan thermocouple with voltmeter. The measured voltage, $V_M$, is, by KVL,

$$V_M = V_1(T_1) - V_2(T_2), \tag{9.13}$$

where $V_1(T_1)$ is the contact potential at the probe junction at temperature $T_1$ to be measured, and $V_2(T_2)$ is the contact potential at the constantan–copper junction at the voltmeter (*cold junction*). (Copper is the usual wiring material at inputs to electronic circuits; at the copper–copper junction there is zero contact potential.)

Unless $T_2$ is fixed at a known reference temperature or unless somehow $V_2(T_2)$ can be canceled, we will not be able to isolate $V_1(T_1)$ and therefore infer $T_1$ from $V_M$ using standard tables or equations. A number of approaches have been taken to solve this problem.

One may establish a known reference temperature at the constantan–copper junction at $T_2$ by means of an ice bath (0°C). Alternatively, the temperature $T_2$ may be measured by an RTD (see Sec. 9.3.4), allowing correction for that junction to be made. Finally, in a third method, an op amp can use an RTD as one of its external

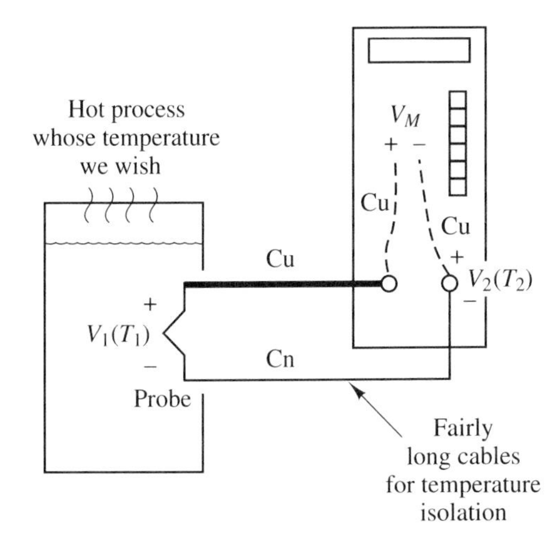


**Figure 9.13** Simple thermocouple application.

resistances. With the op amp input voltage fixed at a well-regulated voltage, the op amp output can be calibrated to cancel $V_2(T_2)$, leaving $V_M = V_1(T_1)$, as desired. Circuits for performing these operations are available in easy-to-use, preassembled card form. Thus it may not be necessary to know in great detail the reference temperature-canceling circuitry. The reason for going to the bother of using an RTD (itself a thermometer) to make a thermocouple thermometer work in such circuits is again the ruggedness and far wider usable temperature range of the thermocouple.

**Example 9.4**

Suppose that the temperature of the room (and therefore that of the cold junction) is known to be a well-regulated 68°F. If we measure 17.21 mV and have no compensating circuitry, what is the temperature of the hot process in Fig. 9.13? The polynomial coefficients given by the National Bureau of Standards (NBS, now NIST) for the copper–constantan junction are: $a_0 = 0.100860910, a_1 = 2.57279 \cdot 10^4, a_2 = -7.67345 \cdot 10^5, a_3 = 7.80256 \cdot 10^7$, $a_4 = -9.24749 \cdot 10^9$, $a_5 = 6.97688 \cdot 10^{11}$, $a_6 = -2.66192 \cdot 10^{13}$, and $a_7 = 3.94078 \cdot 10^{14}$.

**Solution** The measured voltage is $V_M = 17.21$ mV. The subscript $n$ of the coefficient $a_n$ is the power of the contact potential $V_o$ in volts, and the result is the temperature in degrees Celsius:

$$T \approx \sum_{n=0}^{7} a_n V_o^n . \tag{9.14}$$

At the voltmeter side, $V_o = V_2(T_2)$, and at the hot process side, $V_o = V_1(T_1)$. Thus the contact potential $V_2$ at 68°F = 20°C is the real root of Eq. (9.14) with the left-hand side set to 20. The result (by computer) is $V_2(T_2) = 7.91 \cdot 10^{-4}$ V = 0.791 mV. You can check this by plugging 0.791 mV in for $V_o$ on the right-hand side of Eq. (9.14) to obtain $T = 20$°C. In practice, a "lookup table" stored in a microprocessor would be used to invert Eq. (9.14) (see Sec. 11.6.2). Consequently, by Eq. (9.13), the contact potential of the hot process is $V_1(T_1) = V_M + V_2(T_2) = 17.21 + 0.791$ mV $\approx 18$ mV. Now using Eq. (9.14) again, with $V_o = 18 \cdot 10^{-3}$, gives $T_1 = 353$°C = 668°F.

### 9.6.2 Piezoelectric Transducer

Hi-fi enthusiasts may claim that because the phonograph will soon be a technological dinosaur, piezoelectric transducers will go with them. The pickup cartridge on record players uses a piezoelectric crystal to transform needle vibrations from the record grooves into an analogous electrical signal. However, there are many more uses for this crystal, including computer clocks (see Sec. 11.3), accelerometers, pressure sensors, ultrasonic medical imaging transducers (transmit/receive), electronic buzzers, "crystal microphones," and so on.

The piezoelectric effect was discovered by Marie and Pierre Curie (Polish/French) in the 1880s. As described in Sec. 9.2, a force applied to a piezoelectric crystal generates a voltage between the stressed faces. This voltage is due to the rearrangement of atoms caused by the applied force, which results in uncanceled charges of opposite sign being exposed on either side. This would be difficult to visualize accurately without constructing three-dimensional models, because the structure of the most basic piezoelectric material, quartz ($SiO_2$), is very complex (helical, nonlevel hexagonal patterns of silicon and oxygen atoms).

Nevertheless, the result is very tangible. Piezoelectricity has also been obtained

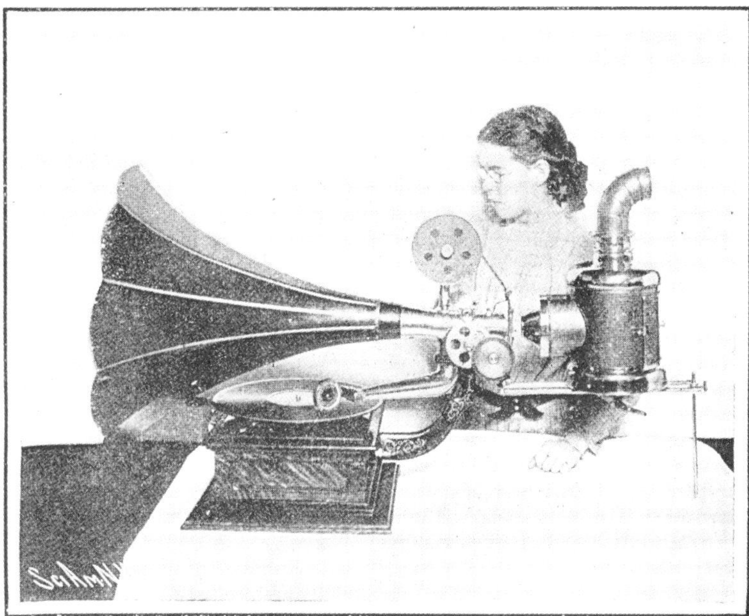

At the top, Edison and his talking machine (using cylinders) in 1888 after 72 hours of continual work on it, at 5:30 AM. "Genius is 1% inspiration and 99% perspiration." Edison's phonograph did not use a piezo-electric crystal, because there was no electronic amplification; the signal was strictly acoustic vibrations. At the bottom, a novel combination of movie projector and phonograph from 1908. The image is projected onto a screen through the same megaphone used to project the phonograph sound. "Either gas or electricity may be used to furnish the light." Note that the audio and visual could be projected simultaneously, though this was well before the first "talkie" (1923). *Source: Scientific American* (April 25, 1908) (bottom photo).

for a variety of materials, including polymers [notably, polyvinylidene fluoride (PVDF)] and ceramics [lead zirconate titanate (PZT)]. For small strains, the developed voltage is proportional to the strain, as noted in Sec. 9.2:

$$v = K\epsilon, \tag{9.15}$$

where $K$ is a piezoelectric constant of the material and $\epsilon$ is the strain on the crystal. Piezoelectric transducers are most responsive to time-varying forces (e.g., oscillations, music, other sounds) rather than to steady forces. The other pressure–force transducers discussed previously are more appropriate for steady forces.

If the exposed sides across which the voltage appears are metallized, the voltage can be measured or used in electronic circuits. In particular, the crystal can respond to sounds by producing a voltage waveform identical to the sound waveform. This application is used in the phonograph pickup cartridges and in crystal microphones.

We also noted in Sec. 9.2 that the piezoelectric process is reversible; if we apply a voltage $v$, there will be a strain $\epsilon = v/K$ on the crystal. In particular, if we apply a rapidly varying voltage (e.g., a high-frequency sinusoid), the piezoelectric transducer will behave as an audio or ultrasonic speaker or "buzzer."

For their use as computer "clocks," a closer study of piezoelectric materials shows that a particular crystal form will have a sharp resonant frequency. This resonance can be so sharp that when excited by a poorly regulated oscillator, the piezoelectric crystal produces an extremely well-regulated and stable oscillation—one that you can set your clock by. Typical computer-application transducers have resonant frequencies ranging from 1 to 50 MHz.

For amplification purposes, the piezoelectric transducer may be modeled as a voltage source placed across a capacitor. As we noted above, applied forces cause charge to be deposited on the crystal surface, which is equivalent to connecting a voltage source across a capacitor. The parallel metallized surfaces with nonconducting quartz in between function exactly like a capacitor, except for the additional piezoelectric behavior. Furthermore, the wires leading from the transducer to the op amp are also essentially a capacitance $C_t$ in parallel with that of the crystal.

We could just amplify the voltage in Eq. (9.15) and obtain an output voltage related to the physical strain or applied pressure. However, the transducer tends to leak charge and also has extremely high output impedance. These factors mean that the transducer is very easily severely loaded by the transmission/input circuitry.

We can use the high gain of an op amp with a large feedback capacitor to minimize this problem (see Fig. 9.14). Regardless of the actual input voltage $v_p$, the *virtual ground* of the high-input-impedance op amp directs all the charge produced by the transducer into the feedback capacitor $C_2$. Note that even though $C_p$ is drawn in series with $v_p$ in Fig. 9.14, it is effectively in parallel with $v_p$, due to the virtual short. However, instead of the charge/current going through the virtual short, it all goes into $C_2$.

The output voltage $v_o$ is $v_o = -q/C_2$, where $q$ is the charge output from the transducer in response to the changing applied force. In this way we obtain a voltage signal $v_o$ either for measuring the force applied to the crystal or for generating a signal analogous to it. Notice that following the suggestions of Sec. 8.8.4, we ground only at the op amp, to minimize ground loop distortion.

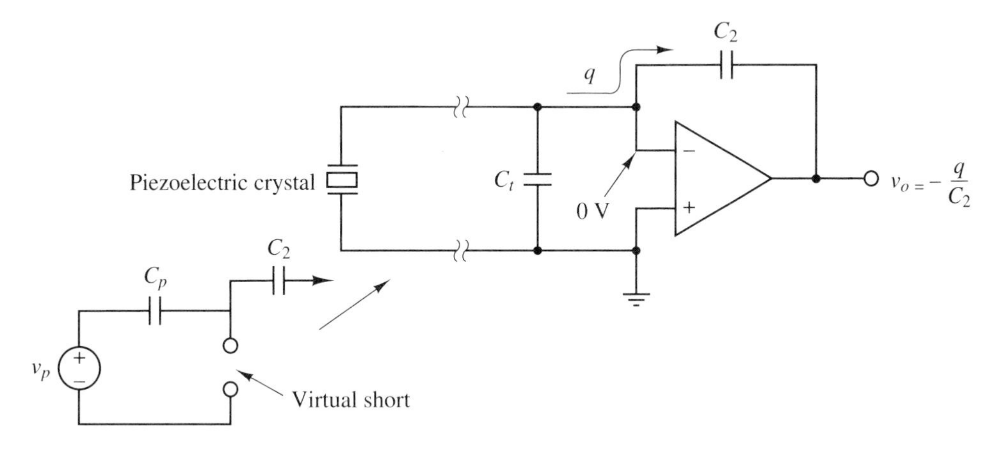


**Figure 9.14** Charge amplifier circuit for piezoelectric transducer.

**Example 9.5**

We noted in Sec. 8.8.4 that for an integrator, we will have to put a resistor across the feedback capacitor $C_2$ to avoid saturation. Must we do the same for the piezoelectric charge amplifier? How does integration relate to charge amplification in circuit analysis terms rather than the physical terms described above?

**Solution** Yes, a resistor $R_2$ placed across $C_2$ is typically used to avoid saturation. Leakage currents could over time charge $C_2$ to have across it a voltage equal to $\pm V_{BB}$, that is, saturate the output. Suppose that the leakage current is 1 nA = $10^{-9}$ A and $C_2$ has been chosen as 100 pF ($10^{-10}$ F); the combined piezoelectric transducer and transmission capacitances sum to a few picofarads. Suppose also that the signal being transmitted has a maximum frequency of 10 kHz. The frequency $1/(2\pi R_2 C_2)$ should be larger than this, say 100 times larger. Thus $R_2 = 1/(2\pi \cdot 10^6 \cdot 10^{-10}) = 160\ \text{k}\Omega$. A steady $I_{\text{leak}} = 1$ nA current into $C_2$ would yield a voltage of $V_{BB} = 15$ V after $C_2 V_{BB}/I_{\text{leak}} = 10^{-10}\ \text{F} \cdot 15\ \text{V}/(10^{-9}\ \text{A}) = 1.5$ s. The $R_2 C_2$ time constant is $1.6 \cdot 10^5\ \Omega \cdot 10^{-10}\ \text{F} = 16\ \mu\text{s} \ll 1.5$ s. Clearly, there will be no drift into saturation.

This "charge amplifier" with $R_2$ is a lossy integrator; without $R_2$ it would be a pure integrator. If we had just an inverting amplifier, the output voltage would be $-R_2 i_{\text{in}}$, where $i_{\text{in}}$ was the current produced by the transducer in-circuit. However, it is the charge that is proportional to the pressure or force applied to the piezoelectric crystal. Charge is the temporal integral of current, so we need some sort of integrator to obtain an output voltage proportional to the applied pressure or force.

### 9.6.3 Hall Effect Transducer

Another transducer whose operating principle was discovered late in the nineteenth century is the *Hall effect,* observed by Edward H. Hall in 1879. The basic idea is shown in Fig. 9.15. If we apply a voltage $V_{ab}$ across terminals $a$ and $b$ of the transducer, current $i$ will flow directly from $a$ to $b$. If, however, a magnetic field **B** is applied as shown, the moving charges will be forced to one side by Ampère's force $\mathbf{F} = q\mathbf{u} \times \mathbf{B}$ [see Eq. (3.11b)]. The resulting charge imbalance creates an electric

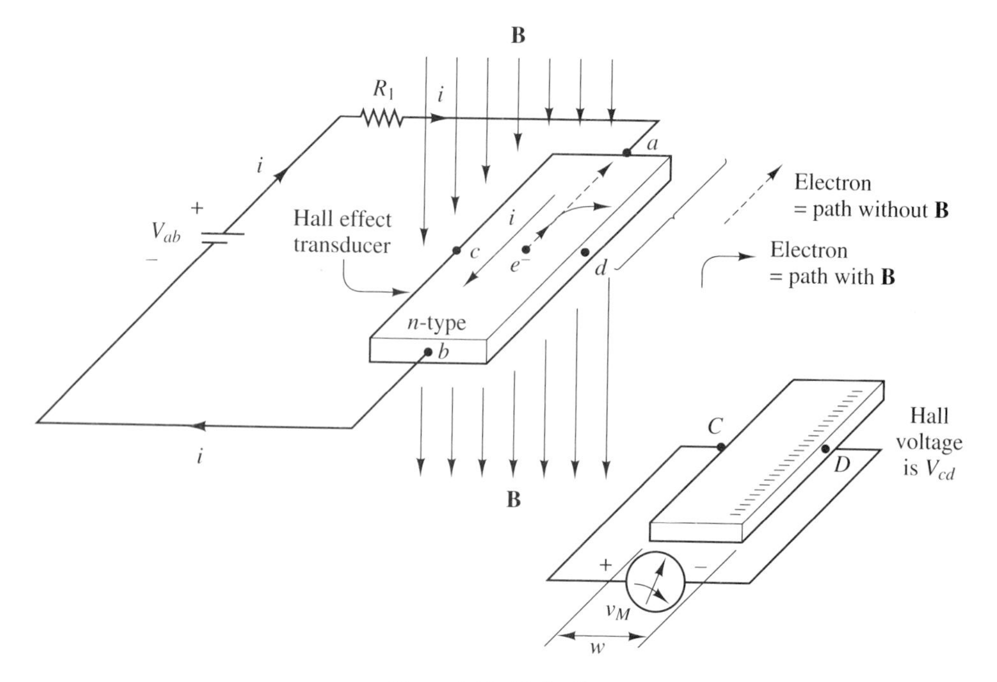


**Figure 9.15** Hall effect.

field that can be measured between points $c$ and $d$ as voltage $v_{\text{Hall}} = v_{cd}$. Quantitatively, the *Hall voltage* is

$$v_{\text{Hall}} = K_H \frac{iB}{w}, \tag{9.16}$$

where $K_H$ is called the Hall constant and $w$ is the width of the slab (see Fig. 9.15). The sign of the mobile charges dictates the sign of $v_{cd}$. Clearly, the Hall sensor is a modulating transducer (modulating the direction of charge travel) which requires the external voltage $V_{ab}$ in order to function.

Although Edward Hall studied the effect using metals, today semiconductors such as indium antimonide are used because in them the Hall constant $K_H$ is much larger than for metals. Furthermore, we note that the Hall voltage generated in a slab of $p$-type material will have opposite sign from that of a slab of $n$-type material. This is because the mobile charges in $p$-type material (holes) have the opposite sign to that of the mobile charges in $n$-type material (electrons). One can therefore measure the Hall voltage of a slab of unknown semiconductor material and determine whether the material is $n$- or $p$-type.

This determination is not, however, the primary use of a Hall sensor. More common is the measurement of current $i_L$ or power $p_L$ in an electrical load. By fixing the applied voltage $V_{ab}$ and placing the transducer in the air gap of a magnetic core (see Fig. 9.16a), then by Eq. (9.16) the Hall voltage will be proportional to the magnetic field $B$. The core concentrates a larger magnetic flux density in its air gap—near the sensor. By wrapping the wire containing the current $i_L$ to be measured

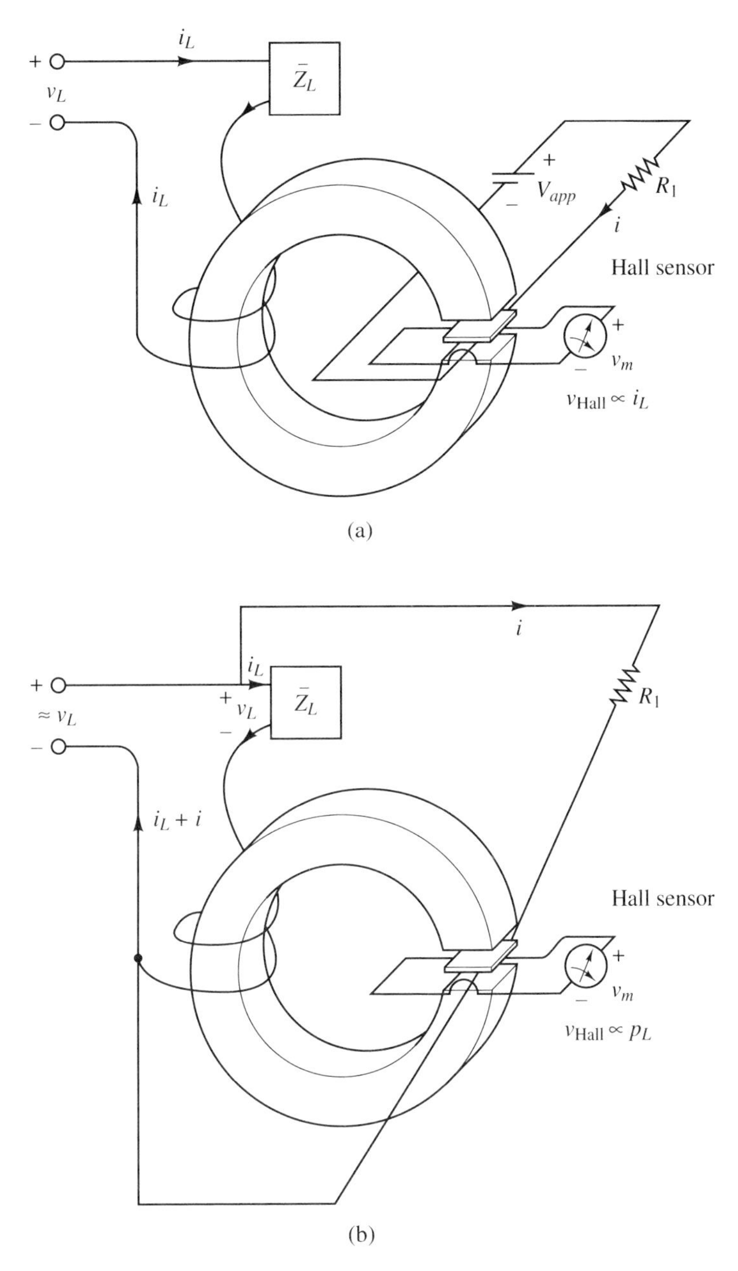


**Figure 9.16** Hall sensor configurations for measuring (a) current and (b) power in a load.

around the core or even by passing it through the core (for large current), a measurement of $i_L$ can be made without breaking the circuit.

If in addition the applied voltage $V_{ab}$ is obtained as a suitable fraction of the voltage across the load $v_L$ (see Fig. 9.16b), then again by Eq. (9.16) the Hall voltage will be proportional to both $v_L$ and $i_L$, and therefore proportional to $p_L$. Because the voltage is multiplied by the current at each instant of time, this application of Hall sensor is known as a *Hall multiplier* or *Hall power transducer.* Note that dc as well as time-varying power and current can be measured with the Hall transducer (which is not true of conventional wattmeters). In more complex setups, one may even obtain power factor measurements from Hall effect transducers.

There are also switching applications of the Hall transducer. When a magnet comes nearby the sensor, a Hall voltage is generated. For example, a Hall effect transducer could produce a pulse once every rotation of a wheel having a small permanent magnet attached (one could make a bicycle speedometer using a Hall sensor). An advantage of this kind of sensing is that it has a large, reliable output, yet does not interfere at all with the rotation. Hall sensors can similarly be used in computer keyboards; such key switches will last a very long time, for there are no switch contacts to wear out. Clearly, the possible applications are very extensive.

### 9.6.4 Liquid Crystal Display

Liquid crystal displays (LCDs) are the prevailing display transducer for portable computers, watches, and many types of modern instrumentation. They have the advantage that they consume extremely low power ($\mu$W), so that portable electronics can function for very long times on one set of batteries. Their disadvantages are that they require external lighting and that they respond rather slowly to changing inputs. The former disadvantage means they cannot be used in the dark, and the latter prevents them from becoming a new form of TV screen. Obviously, they are modulating transducers.

The basic idea used is the same as that behind polarizing sunglasses. Light is usually randomly polarized; that is, the electric field vector (and thus the magnetic field) has a random direction at any point in space and time. (It may be helpful here to review Sec. 3.10.4.) A *polarizer* will allow through it only that light which has the electric field in a particular direction. For the sunglasses application, it has been found that certain directions of polarization produce glare. Thus, if those polarization directions can be "filtered out" by the polarizing sunglass, glare is eliminated without losing the desired light.

Polarizers are also used in LCDs; refer to Fig. 9.17. The incoming light has random polarization. After passing through the polarizer, it has one particular polarization direction. It then passes through transparent glass and a transparent conducting electrode (indium oxide).

Next it passes through a liquid of long molecules such as methoxybenzol-butylaniline (chemists have a way with words). This liquid has an elongated crystalline structure, which can be reoriented by an applied electric field. This reorientation causes incoming light polarized in one direction to be rotated as it passes through the liquid crystal. The result is that if a sufficient external electric field is applied (typically, 3 V across its terminals, designated as "on"), the light passes out the other side of the layer with a polarization rotated away from the direction of the incoming polarized light. The thickness is chosen so that the polarization direction is rotated 45°. If the externally applied electric field is zero (designated as "off"), the polarization is not affected by the liquid crystal layer.

After leaving the liquid crystal, it passes through the other electrode and

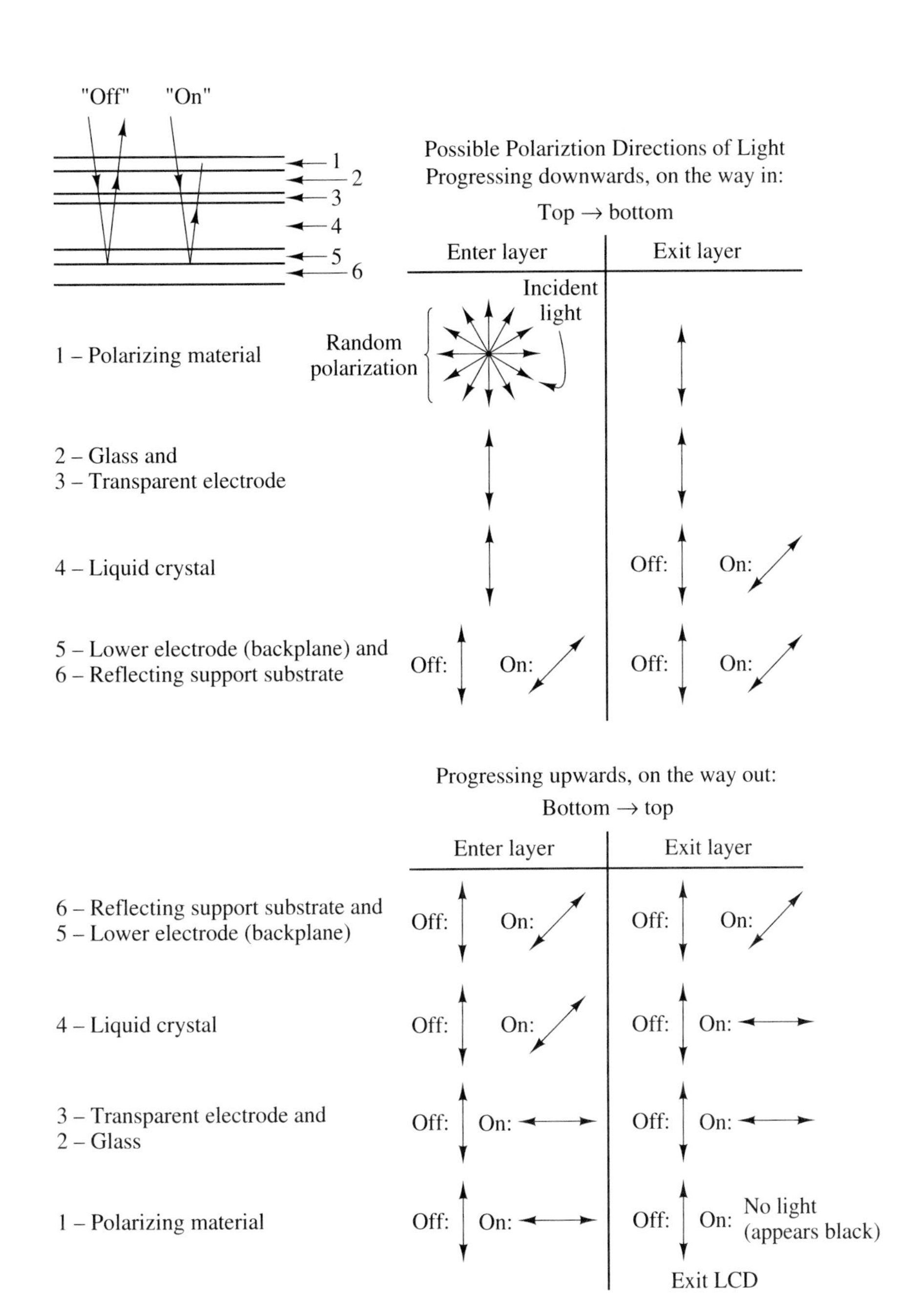


**Figure 9.17** Liquid crystal display structure and operation.

bounces off a reflecting surface, none of which affects the polarization. If the display is "off," the light passes again through the liquid crystal without change. Otherwise, the display is "on" and the polarization direction is changed another 45°, making it *perpendicular* to the original polarization allowed through the polarizer.

Finally, if the LCD is "off," the polarizer allows the light to go through, because its polarization matches that which the polarizer passes. In this case the crystal appears light colored, and indistinguishable from areas of the display unit that have no liquid crystal. Otherwise, if "on," there is *no component of light in the direction the polarizer passes*. Therefore, no light exits the polarizer. In this case the crystal appears black, distinguishing it from areas with no liquid crystal. More details of practical LCD units are given in Secs. 10.6.3 and 10.6.4, including their familiar use in alphanumerical displays.

### 9.6.5 Electro-optical Sensors

In Sec. 7.4.7 we studied solar cells and LEDs. The solar cell is a self-generating sensor; it requires no external excitation. The LED also directly transforms energy—from electric voltage to electromagnetic waves. However, the materials required for an LED differ from those of a solar cell. Neither device is reversible. Nevertheless, the LED and the photodiode (a close relative of the solar cell) are often used as a pair, for a variety of purposes. Usually, infrared light is used in these applications, because the silicon devices are cheaper than visible-light devices (and there is no need for visible light in these applications anyway).

The output signal is always the current that the photodiode allows to pass through. This is determined by the amount of light that the photodiode receives from the LED. There are two basic modulators: electrical and mechanical. In the former case, on–off voltage signals create on–off patterns of light transmitted by a nearby LED and received by the photodiode. In the latter case, the LED is on all the time, but a mechanical system alternately allows and prevents this light from reaching the photodiode.

The purpose of having an electrically modulated LED/photocell pair is to isolate the circuit driving the LED from the load circuit involving the photocell. For example, the pair may serve the same purpose as a relay (Sec. 9.5.4). That is, the source in the LED circuit may be incapable of supplying sufficient power to drive a load connected on the photocell side of the circuit. The "on–off" command is thus transferred from a low-power circuit to a high-power circuit.

Another application is long-distance transmission of digital signals (see Fig. 9.18), which are discussed at length in Chapters 10 through 12. We speak into

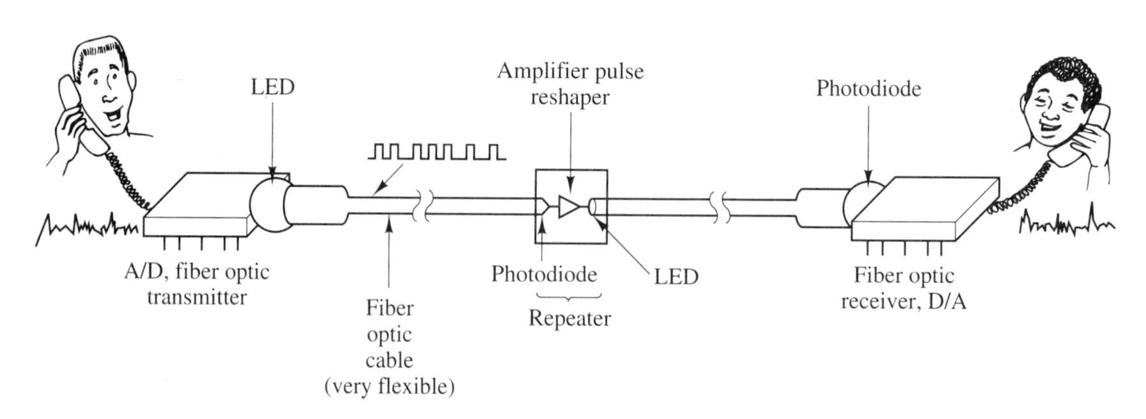


**Figure 9.18** Fiber optic telecommunication transmission system.

the telephone and an analog-to-digital converter (see Sec. 9.7.6) converts the voltage waveform directly corresponding to our voice (analog waveform) into a series of on–off pulses (a digital waveform). An LED or "laser diode" sends digital light signals corresponding to these on–off pulses into one end of a fiber optic cable and a photodiode or phototransistor at the other end receives it. Note that every 40 km or so a "repeater" must be inserted in the data link to reamplify and reshape the waveform because of accumulating attenuation and distortion of the originally square pulse shape. Of course, copper wires must somehow supply power to these repeaters.

Fiber optics cables are gradually replacing copper cable in telecommunications systems because of their low cost and high performance. They are very noise immune and low attenuating and can carry large densities of information per unit time (10 to 100 times as rapid as copper coaxial cable). They are also very small, light, and cheap, being made of silica (from sand).

Yet another application is discussed in Sec. 9.7.4. A special integrated circuit amplifier called an *isolation amplifier* can use optical coupling between the input and output stages to minimize ground loop noise and maximize common mode rejection. (For a review of these concepts, reread Sec. 8.8.4.)

Typical mechanical modulation applications are shown in Fig. 9.19. In Fig. 9.19a, every time an item passes by on the conveyor belt, the light reaching the photocell is interrupted. This causes a counter to increment by one, thereby achieving automatic counting of manufactured items. Notice that the sizes of the objects and their spacing on the conveyor belt need not be uniform if the LED is positioned low. However, in Fig. 9.19a one of the objects will erroneously be counted twice.

Figure 9.19b shows how a digital code (see Sec. 11.6) can be used to represent angular position of the shaft. A stationary row of LEDs illuminates a narrow sector on one side of the disk. The row of photodiodes on the other side of the disk produces a unique on–off pattern for each angular position because of the pattern on the angle-code disk. The more LEDs and photodiodes in the row (i.e., the more digits used to represent the angle), the greater the attainable precision of angular determination. For clarity in the diagram, we have shown a hole in the disk in black and other blocking areas in white.

A huge variety of codes and mechanical configurations is possible. For example, one could have only one hole in the disk. By counting the number of received pulses per unit time, the angular speed (in revolutions per minute) of the shaft could be determined. With more holes in the disk, greater accuracy is achievable.

### 9.6.6 Air Pollution and Smoke Detectors

With the greenhouse effect and ozone depletion, it is essential to be able to monitor air quality accurately. Such information can be used not only for generating the political will to invest in cleaning up the environment, but also in the actual devices for cleaning it up. For example, in an automobile, exhaust gases are analyzed by sensors, and their signals are fed into the automotive microcomputer. The computer can then determine corrective action to minimize further pollution based on these readings; see Sec. 12.7.3 for more details.

Also, the Environmental Protection Agency and other U.S. government

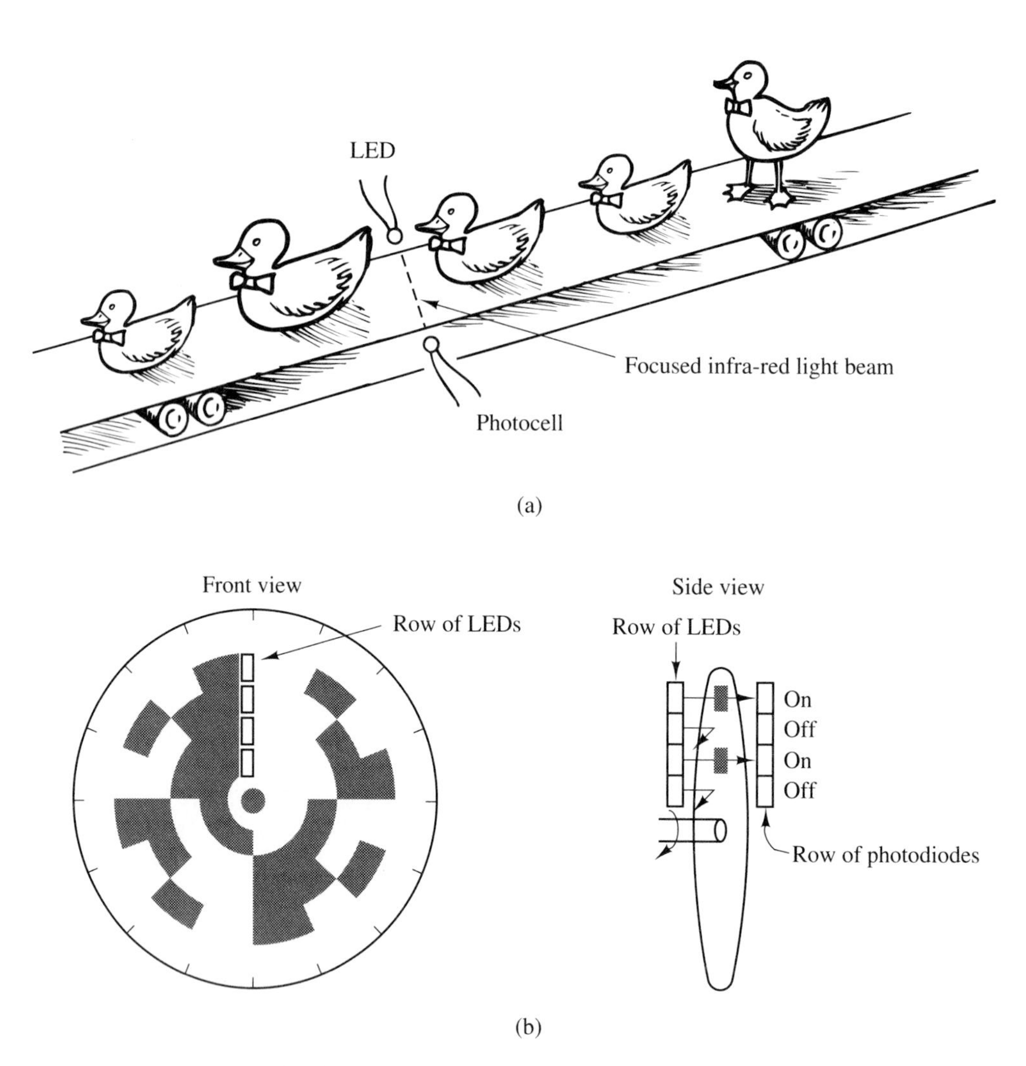


**Figure 9.19** Applications of mechanical modulation in LED–photocell transduction pairs: (a) conveyor belt item counting; (b) angular position encoding and determination.

agencies can use pollution detectors on the road to determine which cars are overpolluting. Systems are being developed that photograph the license plate for police use when a threshold is exceeded. We discuss briefly one method of measuring nitrogen oxides. The intent is not to present the state of the art, but rather, to engender interest in this important area. Clearly, this topic joins with many other disciplines, including chemical engineering, electrical engineering, environmental engineering, civil engineering, and of course, the law.

Figure 9.20a shows one example system that can be used for measuring nitrogen oxides. When nitrogen monoxide is combined with ozone, light is emitted in the visible and near-infrared bands. Such a chemical reaction is called *chemiluminescent*. The emitted light is detected, amplified, and converted to electric current by a *photomultiplier tube*.

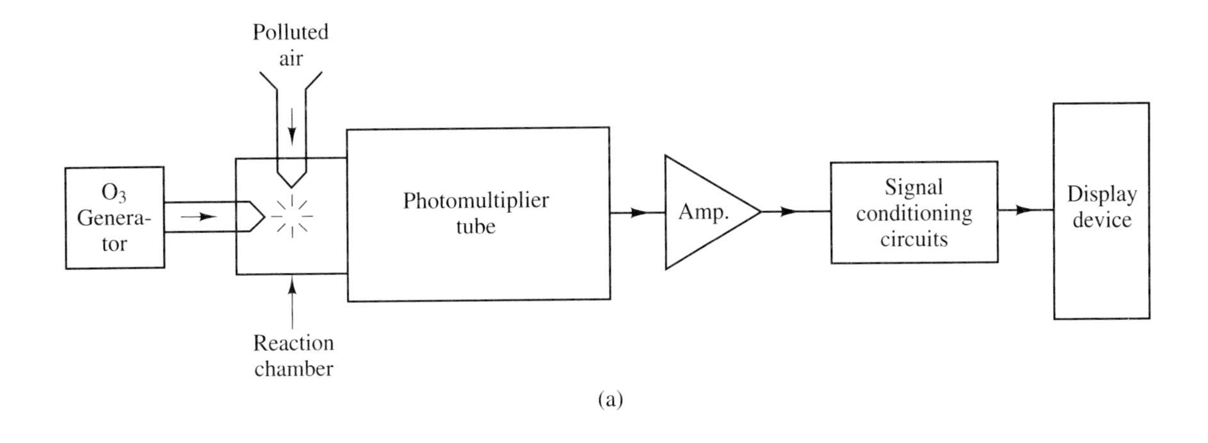


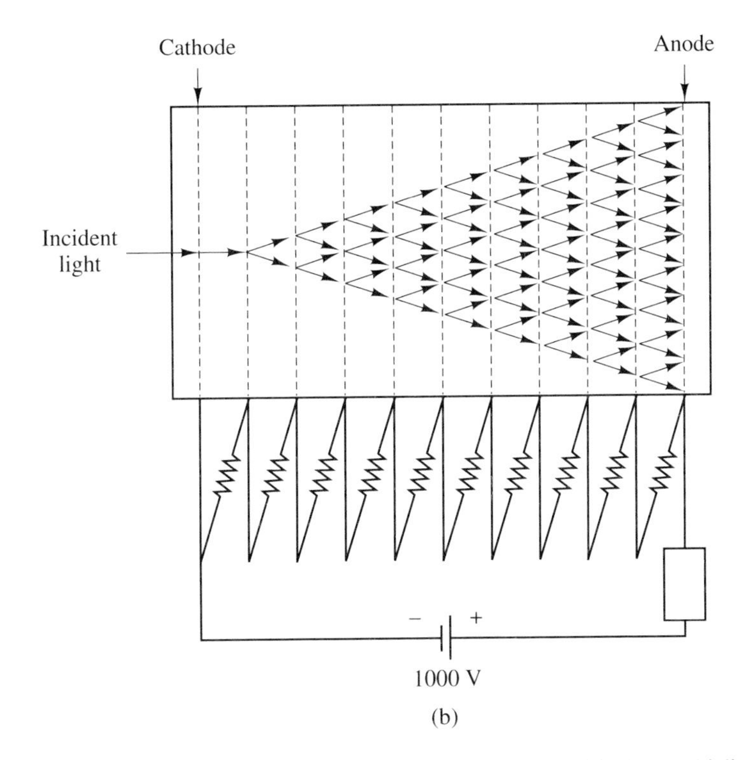


**Figure 9.20** (a) Nitrogen oxide detector system; (b) photomultiplier tube.

A *photomultiplier* (see Fig. 9.20b) is a vacuum tube that has an unheated cathode on one end, followed by a series of intermediate electrodes, and finally, an anode at the other end. When a very large voltage (e.g., 1000 V) is divided over all the electrodes by a voltage divider, there is a significant potential difference between each pair of electrodes.

When light strikes the cathode, some electrons are emitted from the cathode. Because of the large electric field, they are accelerated toward the next electrode. Striking that electrode at high speed, more electrons are emitted from that electrode than struck it. In fact, every electrode liberates more electrons than strike it, and the effect is self-multiplicative: Hence the name "photomultiplier." The light signal is greatly amplified compared with a tube that had only cathode and anode and no intermediate electrodes. Upon leaving the anode, a large current is available for electronic amplification.

At this point (refer again to Fig. 9.20a), the signal is amplified electronically by instrumentation amplifiers and thresholded, filtered, or otherwise processed for human interpretation. It should be noted that experimental investigations are currently exploring the possibilities of using thin films to detect gases. For example, certain oxides deposited on the surface of a semiconductor change the conductivity when exposed to pollutant gases. These detectors promise to be far smaller and cheaper than the ones described above.

While on the subject of gas detectors, consider the household smoke detector. One type of alarm (Fig. 9.21a) uses a radioactive substance that shoots off alpha particles, which are just helium nuclei (charge of +2). The helium ions, in turn, ionize the particles in the air around them, causing the particles to move toward the upper potential plate of the system shown in Fig. 9.21a. Air ions are much lighter than smoke-particle ions; therefore, there will be reduced current if smoke is present. An amplifier amplifies and thresholds this current, and if smoke is detected

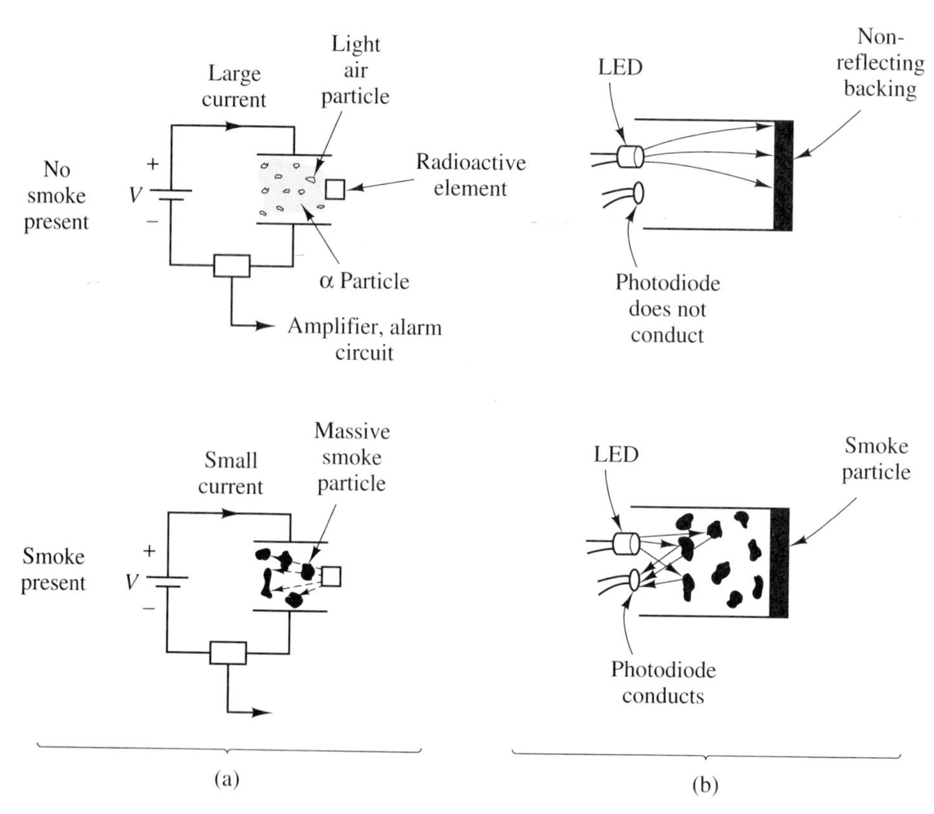


**Figure 9.21** Smoke detectors: (a) radioactive; (b) LED–photodetector.

(reduced current), an alarm goes off. The alarm can be an oscillator circuit driving an audio speaker at an ear-piercing frequency.

The other main type of detector (see Fig. 9.21b) is even simpler. It just uses the infrared LED–photodiode pair we discussed in Sec. 9.6.5. When no smoke is present, the light-absorbing (black) backing reflects no light back to the photodiode, so it does not conduct. When smoke is present, the light is partially reflected to the photodiode because of the large blocking smoke particles, causing the photodiode to conduct. Again, the current in the diode can be amplified and thresholded, setting off an alarm when significant smoke concentration is present.

## 9.7 SIGNAL CONDITIONING ELEMENTS

### 9.7.1 Introduction

Our second and final main topic of this chapter, signal conditioning, should help tie together many previous discussions. Back in the introduction to this chapter we roughly defined signal conditioning as modifications made on signals specifically intended to improve the fidelity and format of the signal and aid in signal transmission over substantial distances. Here we consider some practical issues and devices involved in this process.

In our study of transducers, we have not concentrated on the interfacing between the transducer and the rest of the electronic processing circuitry required to obtain a desirable output. One important problem is the transmission of transducer-based signals over long distances, several meters or more. Without appropriate circuit connections, the typically noisy industrial environment will swamp out the signal and the result will be garbage. Simple methods of minimizing this and related problems are the focus of this section.

We also can make use of circuits developed in Chapter 8 and extensions of them to obtain the kinds of output that are useful to us. The use of instrumentation amplifiers and "isolation" op amps will be considered along with other basic op-amp circuits to accomplish appropriate signal conditioning. For example, transmission of current rather than voltage in order to improve long-distance signal transmission can be accomplished with specialized circuitry.

Fortunately, most of these complex devices necessary to improve performance come in integrated forms, so they can be hooked together much like building blocks. That is, detailed design calculations can often be avoided with these easy-to-use circuits. The trade-off is that they tend to be much more expensive than most of the other electronic equipment we have discussed previously.

Finally, everyone knows that today digital computers are universally used for data acquisition, storage, and processing. A brief discussion of the terms "analog" and "digital" as well as methods for conversion back and forth between the two signal representations concludes the new material for this chapter.

We close the chapter with a typical signal conditioning circuit, showing several of the elements discussed earlier in the section. The interface between digital and analog worlds will bring us to the next major area of study in Chapters 10 through 12: digital electronics.

### 9.7.2 Current Loops

In industrial applications it is common that the signal from a transducer sensing a physical process must be transmitted over a long distance to the electronic control/processing/output panel. In such cases, the *iR* voltage drop due to the transmission wire resistance can be serious. Because the signal is typically presented to the transmission line as a very small voltage, the voltage received over at the panel will be very weak.

Another problem with distant transducers is getting power to the initial amplifier, which must usually be located at the transducer site (and/or to the transducer if it is a modulating transducer). Separate power wires would have to be run over to that amplifier/transducer, causing additional cost as well as the possibility of additional noise on the signal wires.

Finally, electromagnetically induced voltages due to noise sources such as high-power electromechanical machines (motors/generators) can destroy the signal quality. This problem is particularly severe for voltage transmission.

A method of overcoming these problems is to transmit the signal in the form of current rather than voltage. Figure 9.22 shows a simple op-amp circuit that will convert a voltage to a current as well as simultaneously achieving signal power gain.

The output current $i$ can easily be determined by noting that this is a negative feedback circuit. Consequently, $v_- = v_+ = v_s$ and hence $i = v_s/R$. The current $i$ in $R$ is the same as that in the transmission wires $R_{\text{tw}}$ and the driven load $R_L$. Therefore, the transmitted current ($i$) is directly proportional to the transducer voltage $v_s$ *independent of the resistances of the transmission wire and load resistance!* This is true, of course, only as long as $2R_{\text{tw}} + R_L$ is small enough that the op amp does not saturate.

Unfortunately, the voltage-to-current converter in Fig. 9.22 does not solve the problem of having to run power wires out to the transducer. There are electronic circuits that will accomplish this and they come in single-chip form (e.g., Analog Devices AD693). However, these are very advanced circuits, and worse, their detailed schematic diagrams are confidential. As long as one need not understand in detail how they work, they come in very handy in practice. All that is required is the series circuit in Fig. 9.23: the transducer/voltage-to-current converter and the remote power supply (e.g., 24 VDC) and signal-receiving electronics.

After initial amplification, the signal typically enters the voltage-to-current

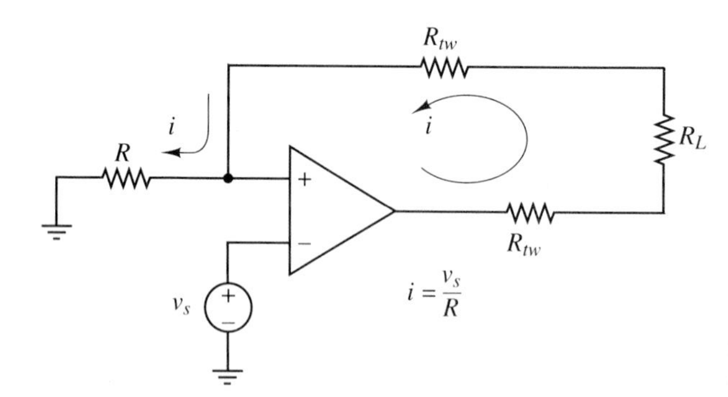


**Figure 9.22** Simple voltage-to-current converter.

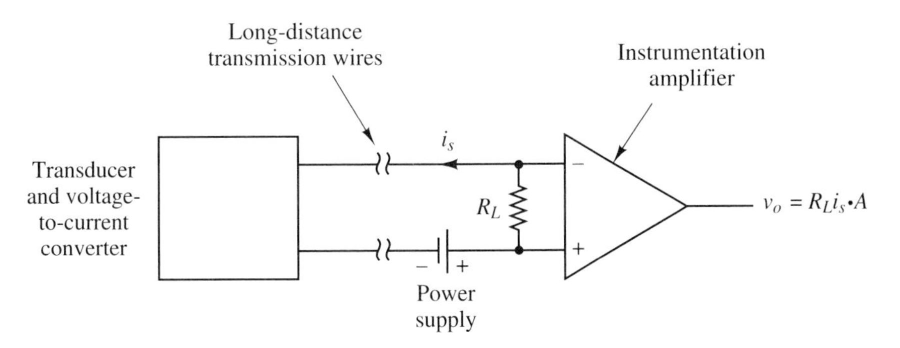


**Figure 9.23** Block diagram of practical voltage-to-current converter in-circuit.

converter as a 0- to 10-V voltage signal. For transmission, the signal is converted to a variable current. The standard range of current is from 4 to 20 mA. The reasons for having 4 mA instead of 0 mA for the no-signal current are that (1) we can distinguish between zero signal (4 mA) and a power failure (0 mA), and (2) the 4 mA is used as the basis for powering the voltage-to-current converter and transducer assembly.

For example, 2 mA may be selected from the 4 mA and regulated (perhaps by Zener diodes) to provide transistor biasing. Another 2 mA may be used for supplying power to the transducer assembly, assuming that it is a modulating transducer such as a strain gage (which requires external electric power).

In the Burr–Brown model XTR101 voltage-to-current converter, two input op-amps serve as an instrumentation amplifier input stage used to control an output op-amp current source. Two 1-mA current sources are available for driving an RTD, strain gage, and so on. It is prudent to put a diode in series in the current loop to prevent accidental reverse current from destroying the voltage-to-current converter IC chip.

Whenever the power supply is derived from the 4- to 20-mA signal lines, exacting regulation circuits must be used. That is, the transmitted current, which varies with the signal over its entire range of 4 to 20 mA, must be sampled to obtain *constant,* well-defined voltages or currents to bias the internal op-amp transistors.

What do we do with the current on the receiving end to convert the signal back into a voltage? The answer is extremely simple: Just connect a resistor across the differential inputs to an instrumentation amplifier and run the transmitted current through it. In Fig. 9.23 this simple arrangement is shown, where for simplicity we use the basic op-amp symbol to represent the complete instrumentation amplifier (with gain $A$).

For example, suppose that the instrumentation amplifier accepts differential input voltages up to 5 V. Then $R_L$ in Fig. 9.23 should be chosen to be the maximum signal current, 20 mA, divided into 5 V: $R_L = 5\text{ V}/0.02 = 250\ \Omega$. Thus an adequate current-to-voltage converter is just a resistor.

While on the subject of level conversions (consider, for example, conversion from 0 to 1 V to 4 to 20 mA), some related terminology in the area of signal conditioning should be defined. The *zero* of a signal under a particular representation is the voltage or current representing zero signal. Thus 0 V would be the zero before conversion to current and 4 mA would be the zero after conversion. The *span* of a signal under a particular representation is the range of voltage or current covered by maximum allowable deflections of the signal. Thus before conversion the span would be $1 - 0 = 1$ V, and after conversion the span would be $20 - 4 = 16$ mA.

As another example, the output current for the Burr–Brown XTR101 voltage-to-current converter chip is $i_o = 4\ \text{mA} + (0.016 + 40/R_s)v_s$, where $R_s$ determines the span. Thus if the maximum input voltage were 100 mV, which should correspond to $i_o = 20$ mA ($i_o - 4\ \text{mA} = 16$ mA), we should choose $R_s = 40/(0.016/0.1 - 0.016) \approx 280\ \Omega$.

Frequently, conversion within the voltage representation is desired. For this we can use the summing amplifier described in Sec. 8.8.4 and shown in Fig. 8.23. When we introduced this amplifier, recall that we considered a temperature conversion from Celsius to Fahrenheit. This *zero span* operation is precisely the type of linear modification common in signal conditioning. One instrument produces a signal with a given zero and span, while the circuit it drives requires a different zero and span. Often, a simple op-amp circuit such as that shown in Fig. 8.23 is sufficient to achieve such conversions.

One final advantage of current loops is that transmission of current signals dramatically reduces electromagnetic interference. Consider the circuit of Fig. 9.24a. The input current source $i_s$ represents the output of our 4- to 20-mA voltage-to-current converter. The resistances $R_{\text{tw}}$ and $R_L$ are again the transmission wire and current-to-voltage resistances, as described above.

The voltage source $v_n$ represents the *electromagnetic interference* (EMI). It has the form of a voltage because $v_n \approx d\phi/dt$, where $\phi$ is the net magnetic flux passing through the current loop. Typical high-powered industrial equipment (e.g., motors) produces strong stray magnetic fields. In practice, *twisted pairs* are used to reduce $v_n$, but it will always be present to some extent. Another source of erroneous voltages is unwanted thermocouples (see Sec. 9.6.1). Unavoidable connections between dissimilar metals in the transmission circuit can generate significant voltages in the loop, especially in harsh industrial environments with large temperature gradients.

If the current source $i_s$ is of fairly high quality, $v_n$ will not affect $i_s$. This is true in practice, using integrated voltage-to-current converters, with $R_L$ and $R_{\text{tw}}$ ensured

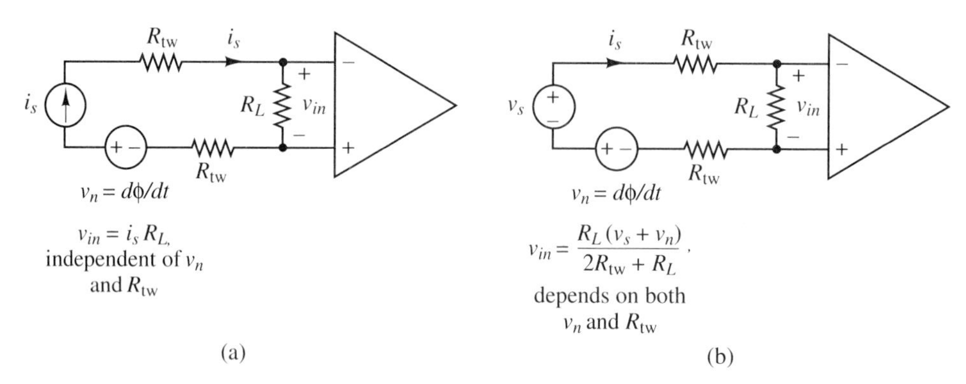


**Figure 9.24** Effects of electromagnetic interference (EMI) on (a) current transmission and (b) voltage transmission.

to be within a stated range. (The allowable range for the BB-XTR101 is merely that for which the voltage across $i_s$ remains between 11.6 and 40 V, a generous permissible variation.)

Because in Fig. 9.24a there is only one path for current to flow, *the current in* $R_L$ *is guaranteed to be* $i_s$, so that the received signal voltage is $v_{in} = i_s R_L$, *independent of* $v_n$. Thus the effects of $v_n$ are truly minimal; all $v_n$ does is cause the voltage across the current source $i_s$ to change slightly. As long as that voltage stays within the allowable range (e.g., 11.6 to 40 V), that is the only effect that $v_n$ will have; $v_n$ will not affect $v_{in}$.

Compare Fig. 9.24a with Fig. 9.24b. In the latter we are attempting voltage rather than current transmission. Now $i = (v_s + v_n)/(2R_{tw} + R_L)$ is the current, *which depends on* $v_n$ and therefore corrupts the received signal voltage $v_{in} = iR_L$. These figures dramatically illustrate the advantage of the current loop with respect to noise immunity for dc to moderate-frequency transducer signals. They also show explicitly that current transmission results in $v_{in}$ being independent of the transmission wire resistances $R_{tw}$, while for the voltage transmission, $v_{in}$ depends strongly on $R_{tw}$.

### 9.7.3 V/F and F/V Converters

We have just seen that current loop transmission has definite advantages over voltage transmission. It is much less susceptible to electromagnetic interference, is independent of transmission line voltage drops, and obviates the need for a voltage supply at the transducer or an extra transmission cable to run power out to the transducer site. In extremely noisy environments, all this may not be enough.

We have already seen the essential ideas now to be applied back in Sec. 9.6.5, where we considered the fiber optics communication system of Fig. 9.18. The same idea can be implemented inexpensively for transducer signal transmission by using two IC chips known as *voltage-to-frequency* (V/F) and *frequency-to-voltage* (F/V) *converters*. The V/F chip converts the moderately varying input voltage from the transducer (after initial amplification) into a series of pulses. The repetition frequency of the pulses is designed to be directly proportional to the input voltage: hence the name "V/F converter."

These pulses can either be transmitted directly over wires or converted to an optical signal by an LED. In the latter case, fiber optic cable and a photodetector at the other end will be required to transmit and convert the pulse train optical signal back into an electrical signal. In either case there is the possibility of using special counting circuits to convert the result into binary numbers for computers (see Sec. 9.7.6 and Chapters 10 through 12), or of converting the pulse train back into the original "analog" signal format. The F/V converter is required for transforming back into the original analog signal format.

There is tremendous noise immunity with V/F and F/V systems. The reason is shown in Fig. 9.25 for the case in which the signal is transmitted in electrical form. In Fig. 9.25a we show an "analog" signal such as would be produced by a transducer, and its conversion into a "frequency" pulse-train signal for a V/F converter. Clearly, the frequency of pulses in the converted signal is tied directly to the voltage level of the original signal.

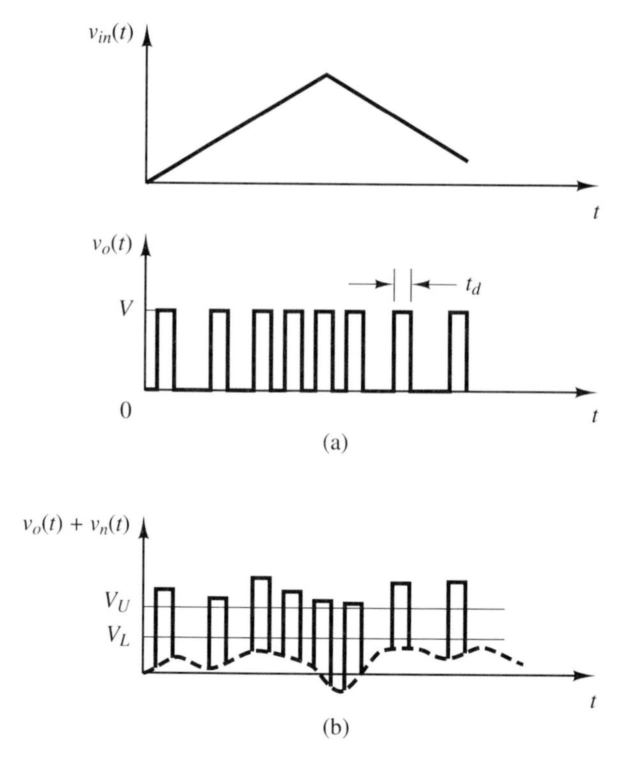


**Figure 9.25** (a) "Analog" signal; (b) after conversion into a pulse train whose frequency is proportional to the value of the original "analog" signal.

In Fig. 9.25b, a noise voltage due to EMI is superimposed on the pulse train waveform. However, as long as a "high" or "on" can still be correctly interpreted as a "high" at the receiving end, *absolutely no distortion results.* For example, in Fig. 9.25b suppose that anything above $V_U$ is called "high" and anything below $V_L$ is called "low." Then despite the extremely low pulse amplitude/noise amplitude ratio (high noise level), the detection decisions will be made perfectly. Pulse detection electronics accomplish this "on–off" decision at the receiving end. The information in the original signal, encoded into frequency, is uncontaminated. Such is one of the tremendous advantages of digital representations, as we discuss further in Sec. 9.7.5.

In practice, the V/F and F/V operations are usually available from the same IC chip merely by changing connections. These operations can be described basically as shown in Fig. 9.26. In Fig. 9.26a we introduce the *one-shot.* The one-shot produces a single pulse of a well-defined duration $t_d$ every time its input voltage falls to zero. It does not produce another pulse unless the input voltage again falls from positive to zero; it produces only one "shot," hence its name.

Figure 9.26b presents a typical V/F converter block diagram, along with illustrative waveforms in Fig. 9.26c for the input voltage $v_s$, the capacitor voltage $v_c$, and the output voltage $v_o$. Suppose that the input signal $v_s$ is a triangle wave as shown, and suppose that $v_c$ is initially nonzero. Then at first the constant "reference" current

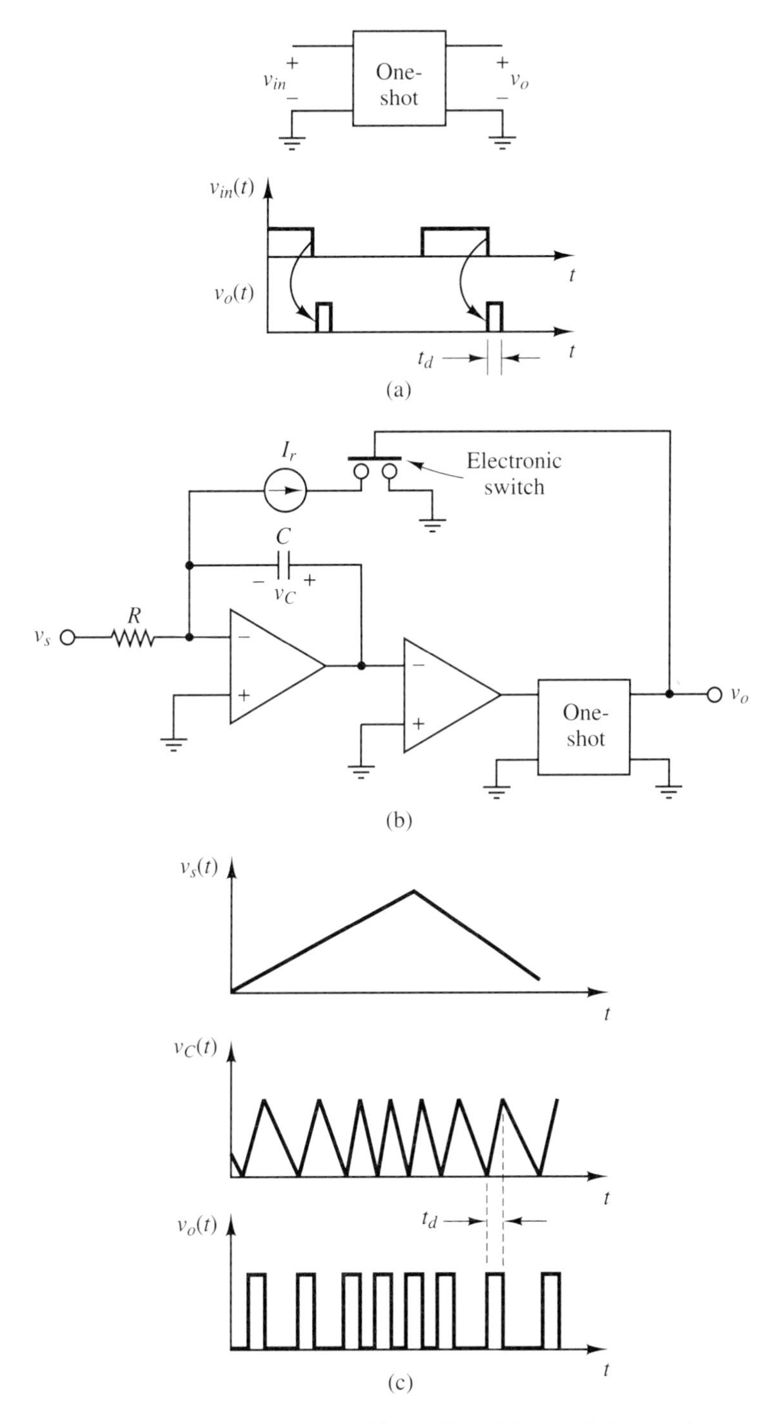


**Figure 9.26** V/F and F/V converters: (a) one-shot, with example input and output waveforms; (b) voltage-to-frequency (V/F) converter block diagram; (c) representative V/F waveforms; (d) frequency-to-voltage (F/V) converter; (e) representative F/V waveforms.

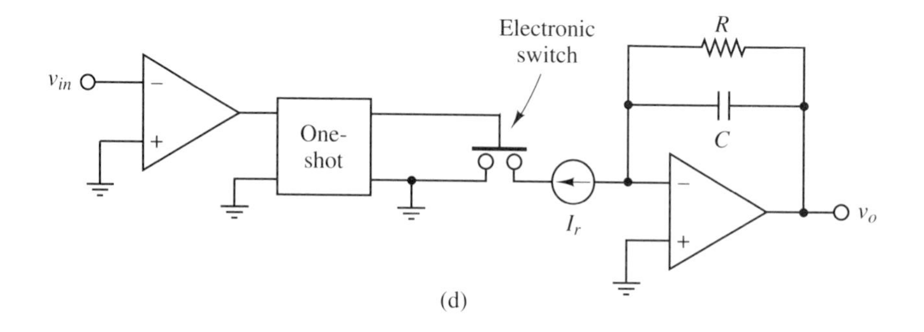


(d)

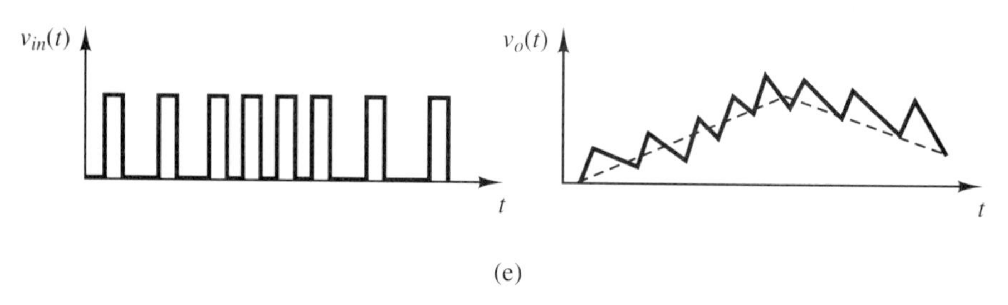


(e)

**Figure 9.26** (Continued).

source $I_r$ is disconnected and the left op amp functions as a negative-slope integrator. Therefore, $v_c$ decreases; that is, it discharges at a rate proportional to $v_s$.

When $v_c$ falls below zero, the comparator (right op amp) triggers the one-shot. The one-shot outputs a pulse of fixed duration $t_d$ and also electronically connects the current source $I_r$. This current source steadily charges the capacitor for the duration $t_d$.

When the one-shot "times out" after time $t_d$, $I_r$ is again electronically disconnected. Now the capacitor discharges again at a rate dependent on $v_s$. Because this rate depends on $v_s$, the time at which $v_c$ falls below zero again depends on $v_s$; this is the coding of the signal into frequency. The one-shot must wait on this $v_s$-dependent discharge time to fire again.

Each of the one-shot pulses is uniform in length, but the frequency of occurrence depends on $v_s$. Thus the output voltage $v_o$ is a train of on–off pulses, the instantaneous frequency of which is directly proportional to the input signal $v_s$, just as in Fig. 9.25. As discussed above, this pulse train can efficiently be transmitted to a remote location without being significantly corrupted by EMI.

Figure 9.26d shows how the same IC that was used for V/F can be used to invert the process (F/V) at the receiving end. Now the terminal previously connected to $v_s$ in Fig. 9.26b is connected to the output of the integrating op amp, forming a lossy integrator because of $R$.

The input voltage, the frequency-encoded signal pulse train, is now presented to the comparator input (now the op amp on the left). The comparator output again triggers the one-shot when its output falls to zero, at the end of an input pulse. This connects the reference current source $I_r$, charging the capacitor for the one-shot duration $t_d$. After $t_d$, $I_r$ is disconnected and the capacitor begins to discharge through $R$ for the time duration that had encoded the original signal $v_s$. This causes the output voltage, $v_o$ (which is equal to the capacitor voltage), to fall for that duration. A new cycle begins when the one-shot is again triggered by the termination of a new input pulse.

Clearly, the output voltage has a close resemblance to the original $v_s$ way back at the transducer (see Fig. 9.26b). In practice, the match is far better than shown here because in Fig. 9.26 we have exaggerated the processes of operation. Also, the input signal $v_s$ can be anything from dc to an oscillating signal with frequency up to about 10 kHz, and having values from 0 V up to 10 V or more.

### 9.7.4 Grounding and Shielding Revisited, and Isolation

In Sec. 8.8.4, Figs. 8.24 and 8.25, we introduced the need for differential amplifiers. With single-ended inputs from transducers far from the amplifier (see Fig. 8.29), a ground loop could occur. This ground loop is due to the fact that earth ground potential varies significantly, even dramatically, in a noisy, industrial environment. By using ground as one of the transmission wires, a series noise voltage $v_n$ is added to the signal, distorting the results. Note that for small signals (mV or μV), a $v_n$ of only μV could significantly distort or even drown out the desired signal.

To avoid this problem, two transmission wires configured in a twisted pair—neither connected to earth ground—serve to transmit the signal. These wires could either transmit a signal voltage or be the current loop of Sec. 9.7.2. This pair of wires can be encased in an electromagnetic shield (see Fig. 8.25), grounded at one point to minimize electromagnetic interference. The shield is grounded at only one point because this prevents large currents due to $v_n$ from flowing in it, which in turn could be coupled to the signal transmission wires. These are the most basic principles of grounding in signal transmission.

Incidentally, recall from the Chapter 2 problem involving a current-carrying "cylindrical pipe" that the magnetic field inside it was zero. Thus one might expect that it is OK to have current flowing in the shield. However, often the shield is braided wire, with small holes in it. Also, the ends produce fringing fields that access the signal wires at their exit from the shield. Finally, the calculation of zero $B$ field was for a constant-in-time, straight wire. The influence of the fact that neither of these assumptions is true in practice is a question for further study. In sum, it is better to be safe than sorry: Ground the shield at only one place.

Another aspect of good versus bad grounding practice is shown in Fig. 9.27. When we draw a circuit such as that shown in Fig. 9.27a, we really mean that all the "ground" wires are to be connected together as in Fig. 9.27b. In reality, the "ground" in Fig. 9.27 may not be an actual connection to earth, merely a *common* connection. Unfortunately, "common" and "ground" are used interchangeably in practice.

If we consider *real* wire with finite resistance, the circuit of Fig. 9.27b should be drawn more accurately as in Fig. 9.27c. We clearly see that what we thought was 0 V at the transmission cable shield is instead $V_1 \gg 0$. In fact, Fig. 9.27c illustrates a worst-case scenario because contributing to the error voltage are the much higher currents driving the chart recorder and computer. The farther from the true 0-V ground, the more currents accumulate errors from each connected device.

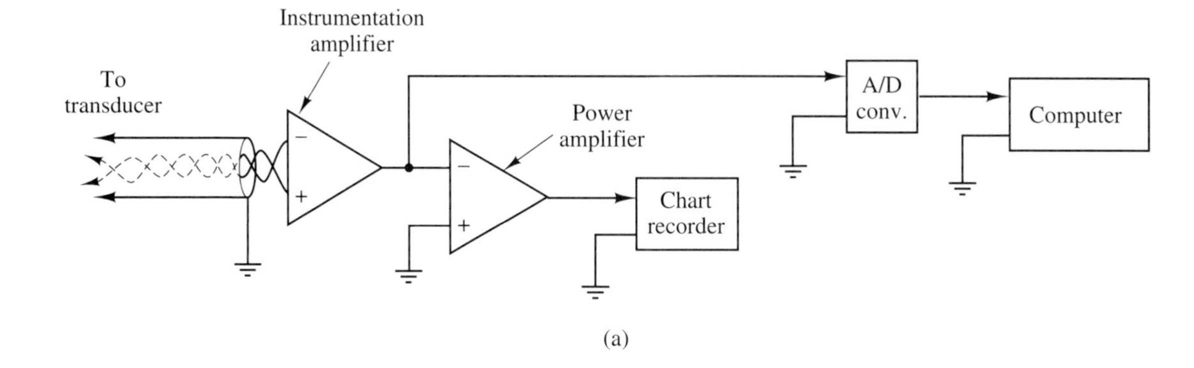


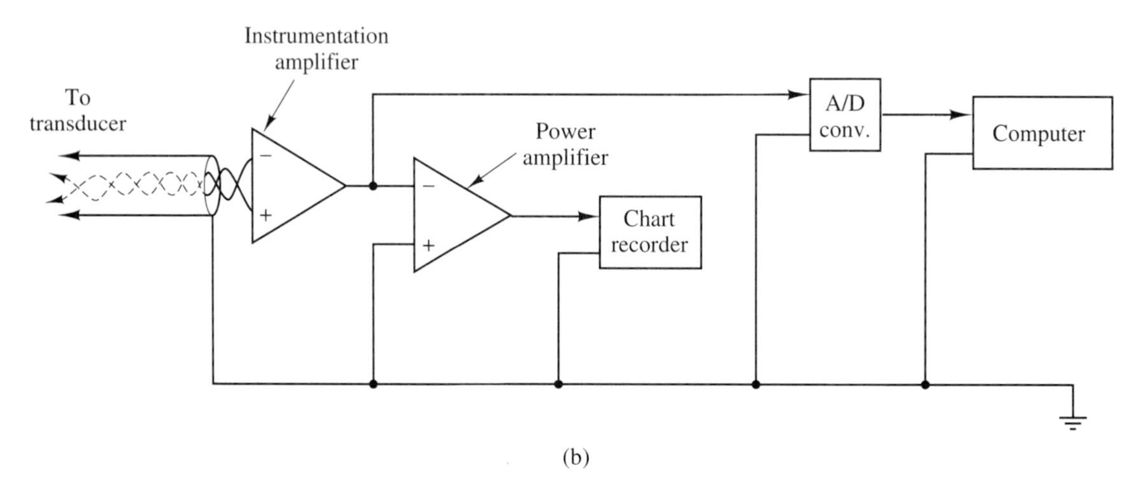


**Figure 9.27** Virtue of running separate ground wires: (a) original schematic diagram of measurement system; (b) equivalent schematic, showing wires as having zero resistance; (c) same schematic showing actual wire resistances; (d) improved circuit.

A much improved circuit is shown in Fig. 9.27d. Because each circuit has its own path to ground, there is no cumulative error from all devices in the circuit. Now the voltage at the transmission cable shield is $V_1' = i_1 R_1'$, which (noting that $i_1$ is very small) is probably orders of magnitude less than $V_1$ in Fig. 9.27c: $V_1 = i_1 R_1 + (i_1 + i_2)R_2 + (i_1 + i_2 + i_3)R_3 + (i_1 + i_2 + i_3 + i_4)R_4 + (i_1 + i_2 + i_3 + i_4 + i_5)R_5$. Furthermore, lots of noise originates from the digital circuitry in the A/D converter and the computer. These lines should be kept away from the "analog" ground for that reason alone.

There are many methods of protecting nonsignal voltages and currents from entering the signal path. The most obvious is to keep power supply wires and any wires containing high-frequency content away from signal wires. Another tip is to minimize wire lengths whenever possible; this will reduce signal losses and capacitive noise coupling and distortion. Shields around transmission cables have already been discussed as ways of minimizing EMI. Guard wires/boxes are also used occasionally. These guards are placed where leakage currents are suspected to be flowing in a

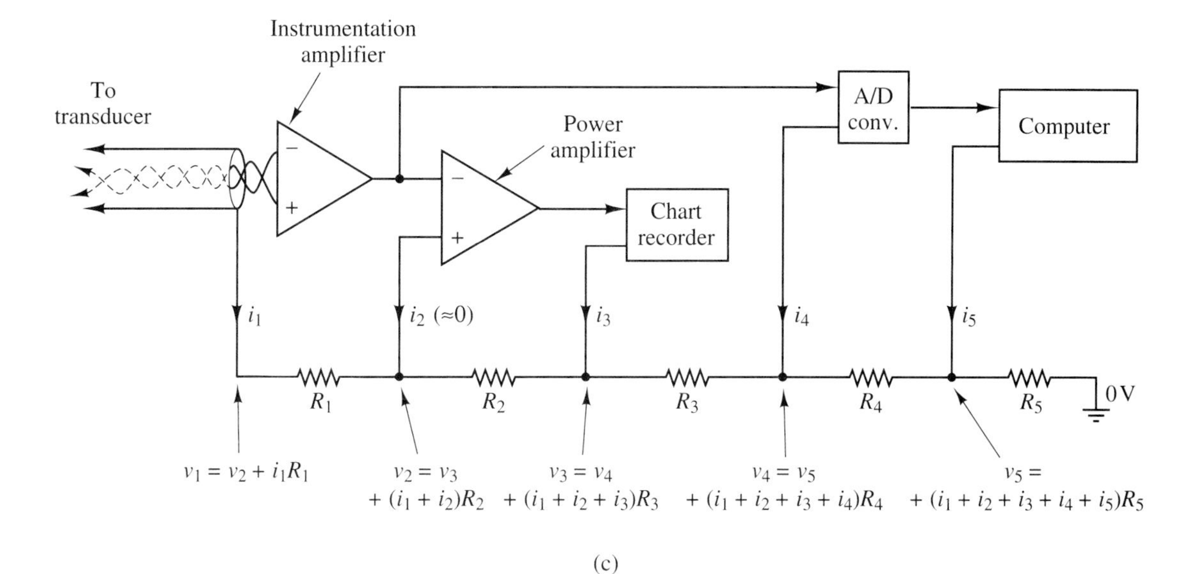


(c)

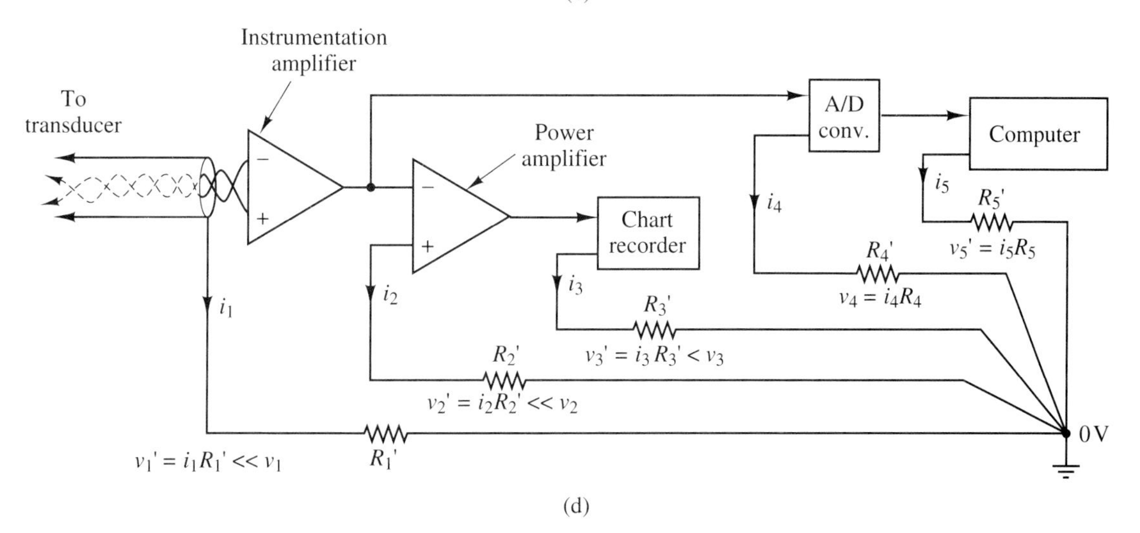


(d)

**Figure 9.27** (Continued).

circuit. Wherever that is, the guard is attached and connected with a separate wire to ground. The guard intercepts the unwanted current and diverts it to ground by a very low impedance path. It is really only a variation on shielding as discussed above.

More than just a variation is the concept of an integrated circuit isolation amplifier. An isolation amplifier is one that completely breaks direct connection between the ground on its input side and on its output side. The benefits are that ground loops are completely broken and much higher CMRR is attainable.

Furthermore, most instrumentation amplifiers can handle only about 10 V of common-mode voltage input before being damaged. Because of the complete isolation of potential levels on either side, isolation amplifiers typically can take 2 kV or more of common-mode input voltage. Such high levels are actually attainable in certain harsh industrial environments using high-voltage equipment.

Another situation when use of isolation amplifiers is mandatory is when the transducer must be connected to the remote ground. Such is the case for thermocouples and other transducers that have conductive connections to the object being monitored. The isolation amplifier allows this, yet still breaks the ground loop.

Yet another circumstance where isolation is required is in the use of electrodes connected to patients for medical monitoring or treatment. Clearly, there must be no possibility of large unwanted currents entering the body; electrocution would result. It should be noted that some disadvantages of isolation amplifiers compared with instrumentation amplifiers are that the latter are more linear, more stable with respect to temperature, and have far larger bandwidths (1000 times larger).

There are currently three methods of isolation: Transformer coupled, optically coupled, and capacitor coupled. Transformer coupling was the original way of isolating circuits. In fact, power supplies with transformer inputs provide a degree of isolation protection from ground faults because the secondaries are floating. They protect against dc faults, not ac (ac would be passed on to the secondary). There are also "dc–dc converters" that use transformer coupling to provide isolated dc power supplies to drive isolated circuits (including isolation amplifiers), given only one input dc power supply. Amazingly, the required miniature torroid transformer for the transformer-coupled isolation amplifier actually fits on the same chip!

An example of this type of isolation amplifier (BB 3656) is shown in Fig. 9.28a. We see that not only are the power supplies for the input and output amplifiers transformer isolated, but the signal is transformer coupled. The only external power supply required is a 15-V dc power supply. The pulse generator converts the dc to a carrier of 750 kHz, and feeds this to the transformer primary (upper right). Secondaries (shown at the bottom) provide ac voltages that are rectified and filtered to provide isolated power supplies for the input and output amplifiers.

The input amplifier modulates its output (which could be dc, which would be incompatible with transformer coupling) at the same carrier frequency, making it suitable to place on the transformer primary. Because this process is sometimes known as *chopping,* the transformer-coupled isolation amplifier is sometimes called a chopping amplifier.

Note the demodulator in the feedback path; this must be here to cancel the effects of modulation. Here essentially the same idea is used as was used with the sense terminal of monolithic instrumentation amplifiers discussed in Sec. 8.8.4. When there is an unwanted effect in the output line, it can often be canceled in the negative feedback path because of automatic gain compensation to drive the input voltage of that op amp to zero.

Finally, a demodulator on the secondary turns the modulated signal back into its original form (again, this might be dc). The second amplifier is just a buffer to drive loads (note that it has unity gain). Notice that because the two sides are isolated and have different grounds, different ground symbols are used. This is required practice for drawing isolated systems.

Figure 9.28b shows an optically coupled isolation amplifier. The input and

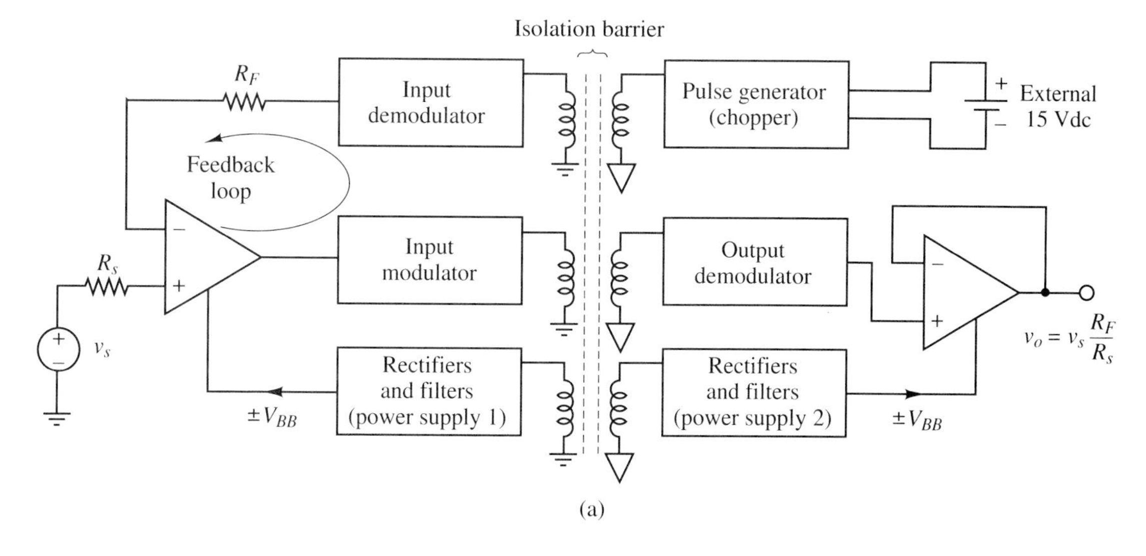


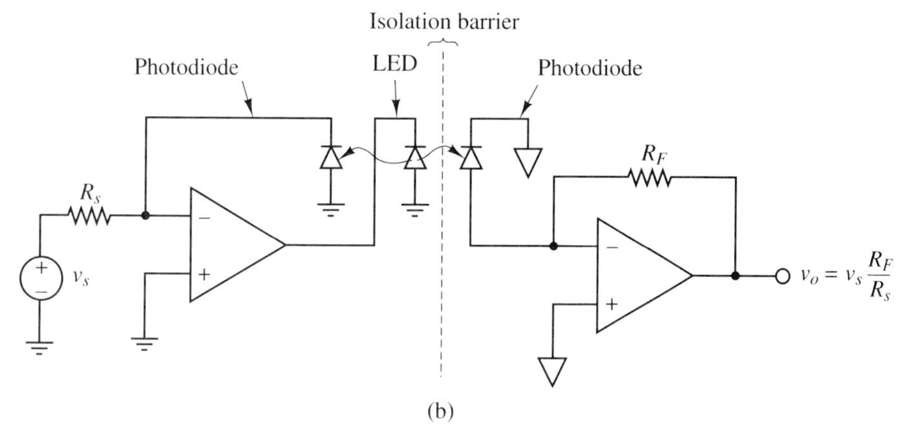


**Figure 9.28** Integrated-circuit isolation amplifiers: (a) transformer coupled; (b) optically coupled.

output amplifiers must be supplied by isolated dc supplies, such as are available from the "dc–dc converter" mentioned above. The LED–photodiode pairs use the same concept as is used in fiber optics. However, here the signal is not chopped or turned into on–off signals as was true in the transformer-coupled amplifier and fiber optic system, respectively. Instead, the original signal is linearly, optically coupled to the isolated system.

One might argue that the nonlinearity of the LED–photodiode pair would make linear amplification impossible. To correct for this, the old idea of replicating the nonlinearity in the negative feedback to cancel out the nonlinearity is used, as shown in Fig. 9.28b.

Optical isolation is superior to other isolation methods when safety is the dominating issue, because the only connection from input to output is light. Optical coupling is thus the method of choice for the patient monitoring equipment considered above. For brevity we merely state here that capacitor coupling uses the same ideas as transformer coupling, but with somewhat better results. The capacitor is a more nearly perfect device than is a transformer. However, transformer-coupled isolation amplifiers are still common, and perhaps are still more common than the relatively recently introduced capacitor amplifiers.

In all these amplifiers, the signal typically can be anywhere from dc up to 50 to 70 kHz. The current path from input to output is broken, yet there is signal transmission without significant distortion or attenuation.

### 9.7.5 Analog versus Digital Signals

In both Chapter 8 and this chapter, we have informally made the distinction between analog and digital signals. Let us briefly restate these definitions.

An *analog* signal is one that is truly analogous to the original process creating the signal. It is one that has a definite and unique value for every moment in time. Its value can be quantified by any real number, even one that "never ends," such as $\pi$. The pressure in a partially opening valve is shown in Fig. 9.29a, and an analog voltage $v(t)$ representing it is shown in Fig. 9.29b. They are "the same" except that one is a pressure (measured in gage pressure, pounds per square inch relative to ambient air pressure, or psig) and the other is a voltage. The voltage representation has the advantage that it can be amplified and processed using electronics.

As we have noted previously, a computer requires input signals to be discretized in both value and time. We will informally define a *digital* signal as one that has been sampled in time and discretized in value. A more complete discussion of the word "digital" is presented in Sec. 11.6.1.

For now, we simply present the basic idea in Fig. 9.30. Shown is the analog voltage waveform of Fig. 9.29b (solid line) and the resulting *sequence* $v(n)$ (dots) after sampling and value discretization. The sampling interval is 0.1 s, and the voltage resolution is only 0.5 V.

Notice that at each sampling instant, the value may have to be rounded up or down, because only a finite number of values are permissible (on this scale, only 10, covering the range 0 to 5 V). This is true in general because a computer can only

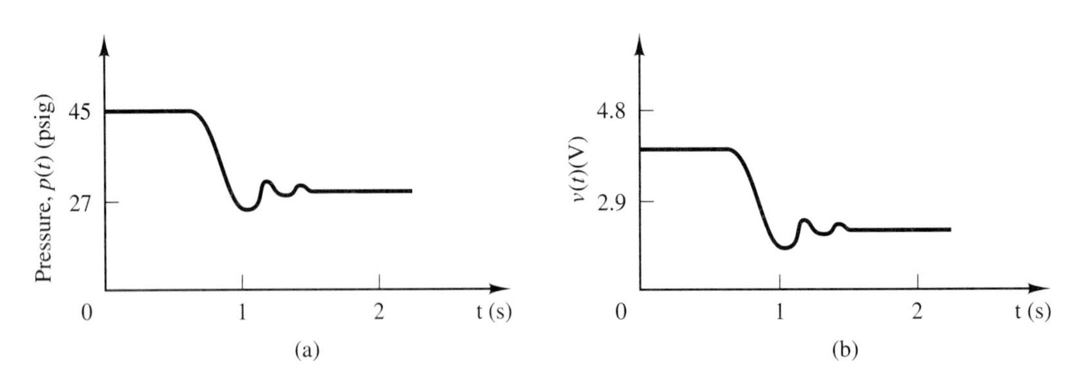


**Figure 9.29** (a) Pressure and (b) analog voltage waveforms.

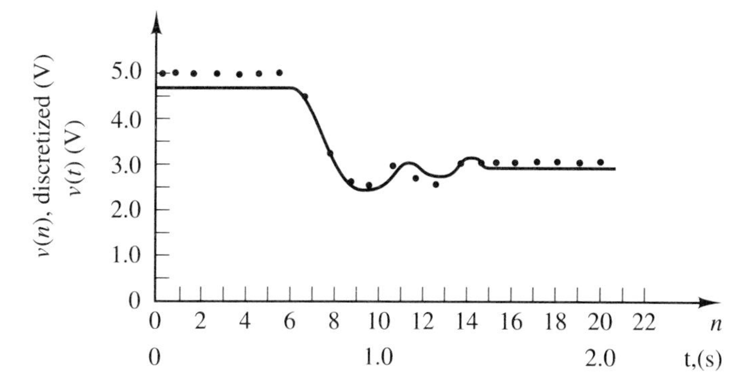


**Figure 9.30** Digital representation of the analog voltage waveform.

process numbers with a small, fixed number of digits. (The same is true of us when we work with pencil and paper; after three or four digits of long division, we are usually ready to call it quits.) The result is that if you look at the sequence $v(n)$, we cannot distinguish the last peak in $v(t)$; it is completely flattened out. Both the sampling interval (or sampling rate) and the discretization resolution are crucial parameters in the sampling of analog signals.

Once sampled and converted to binary form as discussed in Sec. 11.6, the data are in an extremely convenient form: a finite array of numbers having a fixed number of digits. This array can be stored on hard disks or diskettes, magnetic tape, or semiconductor memory chips. All of the advantages of digital processing are now available.

Some of these advantages are that analog processors drift with time and temperature, whereas on–off decision making is completely immune to such distortions. We noted in Sec. 9.7.2 the tremendous noise immunity possible in data transmission using digital rather than analog representation. Only a 1/0 decision must be made correctly for digital, as opposed to the need to exactly reproduce an easily distorted waveshape for analog transmission.

Analog electronics tends to be physically large and expensive compared with equivalent digital electronics. Digital processing is as flexible as your ability to program a computer. Analog processing is only as flexible as the particular hardware you have soldered into place (not very). Also, the sky is the limit, with potential complexity available using digital signal processing, whereas practical analog processing is relatively very limited in this regard.

Analog data cannot be stored without degradation over time or during copying as can digital data. For example, whenever a copy is made of a cassette tape of music, the fidelity is reduced in the copy. Not so of a copy of digital data; there is zero degradation. Thus the digital information can be perfectly stored "forever." About the only advantage of analog processing is that it can be faster. However, even this advantage is now dubious with the cheap, incredibly fast processors such as those we study in Chapter 12. In the following section we study more specifically how to transform between analog and digital representations using electronic devices called A/D and D/A converters.

### 9.7.6 A/D and D/A Converters

A significant fraction of the industrial uses of digital circuitry (see Chapters 10 through 12) involves measurements and control. Essentially, observation of and response to physical systems are involved. Physical systems are usually characterized and affected by continuous-time and continuous-valued signals. However, as we noted in Sec. 9.7.5, computers require both discrete time and discrete value because they have only a finite amount of storage space and a finite amount of time they can give to numerical calculations. Furthermore, computers are designed to process numbers in base 2 (1/0) form. Thus the systems for conversion to and from the computer signal format are extremely important signal conditioning elements.

For example, recall the pressure signal discussed in Sec. 9.7.5. If digital circuitry is to be used to process such a signal, the signal must first be sampled and represented in discrete-valued binary [base 2 (1/0)] form. See Sec. 11.6 for details on the binary numbering system. Briefly, recall that in base 2 an integer is represented as

$$d_{N-1} \cdots d_1 d_0 = \sum_{n=0}^{N-1} d_n 2^n, \tag{9.17}$$

where each of the $N$ binary digits (*bits*) $d_n$ is either a 1 or a 0. The bit pattern $d_{N-1} \cdots d_1 d_0$ in Eq. (9.17) can represent integers 0 through $2^N - 1$, depending on the values of each of the bits $d_n$. The sampling and conversion of the continuous-valued pressure value to an $N$-bit binary form such as in Eq. (9.17) is called *analog-to-digital* (A/D) conversion (also abbreviated ADC).

In a digital audio system, for example, the audio signal is represented by binary numbers such as the one in Eq. (9.17). This allows computer processing, convenient storage (e.g., CD form), and vastly improved fidelity. However, ultimately the binary numbers representing the audio signal must be converted back to continuous voltages that can drive an audio loudspeaker. This process is called *digital-to-analog* (D/A) *conversion* (also abbreviated DAC).

Happily, both of these operations can now easily and efficiently be performed using reasonably priced integrated circuits (IC chips). This was not the case until the mid-1980s. We may therefore satisfy our curiosity with only descriptions or block diagrams when that is convenient, as we will do for many other digital ICs in Chapters 10 through 12. Our discussion is simplified because a full treatment is well beyond the scope of this book. We must leave out many important but complex issues. Furthermore, even this treatment will have to make occasional reference to later chapters on digital electronics (Chapters 10 through 12). Although the material here is mostly self-contained, it will be beneficial to reread this section after reading Chapters 10 through 12.

First, a brief look at the A/D converter. In Sec. 8.8.2, a simple A/D converter was presented. However, to be usable as a base 2 equivalent digital signal, the op-amp outputs would have to be encoded using combinational logic, a type of digital circuitry we discuss in detail in Secs. 10.2 and 10.3. Also, digital circuitry represents "on" by 5 V or 3.5 V and "off" by 0 V, not $\pm V_{BB} \approx \pm 18$ V as for the linear op amps in Sec. 8.8.2. These modifications pose no serious problem.

The main problem with the circuit in Sec. 8.8.2 is that an $N$-bit representation requires $2^N$ op amps, one for each possible combination of bit values [see Eq. (9.17)]; thus an 8-bit converter would require 256 op amps! The advantage of such an A/D converter would be its high conversion speed, a big issue in digital circuitry/computers. The only delays would be the sum of the signal delay through one op amp plus the delays through the encoder. An encoder is a digital electronics code-generating module, discussed in Sec. 10.4, which in this instance produces the required base 2 representation ($N$-bit output) from the op-amp "on–off" patterns (the $2^N$ 1/0 inputs). The type of A/D converter we have just described does exist in IC form. It is called a *flash converter* because it is so fast.

A less expensive, slower, but still efficient alternative is the *successive approximation method.* This works by the familiar method of number estimation in which the next estimate is always the midpoint of the current possible interval in which it is assumed that the actual number lies. The current possible interval is thereby successively divided by 2. In base 10, if the number were 6 and the range 1 to 10, the first estimate would be 5 and successive estimates would be 7.5 (midway between 5 and 10), 6.25 (midway between 5 and 7.5), 5.625, 5.8125, and so on. This technique is particularly convenient for base 2 representations. This is because the successively halved possible intervals are merely defined by groups of successively fewer *least significant bits* (LSBs). Here the LSBs are $d_n \cdots d_0$ in Eq. (9.17), where $n$ begins at $N$ and is decremented by one at each step.

For comparison with the analog (continuous-valued) input in order to determine the new "possible interval," the digital approximation is converted to "analog" using a D/A converter, which is discussed below. The allowed range of input analog voltage is typically 0 to 10 V.

Consider an eight-bit binary representation. If a 1 on the currently *most significant bit* (MSB) causes the approximation to be higher than the analog input, it is reset to 0; otherwise, it is left at 1. All lower significant bits are initially set to 0 because 10000000 is the midpoint between 0 and 11111111; the higher significant bits, already determined, define the level on which the range currently under consideration rests. Each of the eight bits in turn is determined similarly, from the MSB down to the LSB.

An internal "clock" which initiates each of the foregoing steps in the conversion in sequence is derived from the main system clock (see Sec. 11.3). It allows fast conversions, in keeping with the speed of other digital components. In particular, it controls the sample-and-hold circuit that holds the inputs to the A/D converters at stable values during the conversion process (see the problems). See Fig. 9.31 for the circuit symbol for the A/D converter showing connections. These ICs cost about $10 apiece.

A digital-to-analog converter is a simpler device than the A/D (and thus about half the price); as noted above, the D/A converter is itself a building block in the successive approximation A/D converter. For this conversion back to a voltage, each higher significant bit (1 or 0) of the binary number to be converted is weighted with a higher power of two in accordance with Eq. (9.17). These weights are implemented by resistor values in the D/A op-amp circuit. Referring to the four-bit D/A converter in Fig. 9.32, KCL gives the following set of equations:

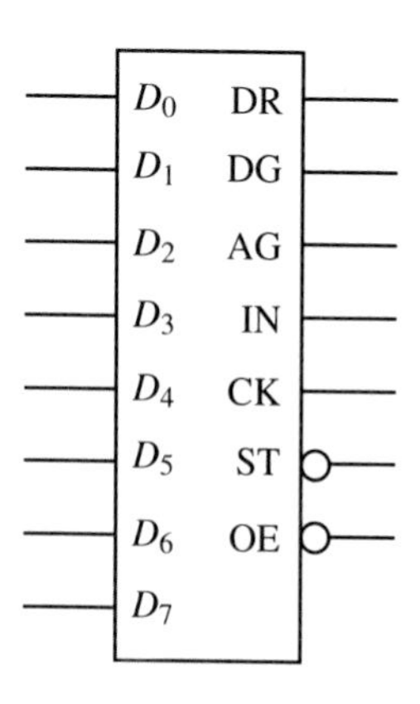


DR = data ready
OE = output enable (high impedance if disabled – not set to 0).
ST = start approximation
CK = clock
IN = data input
AG = analog ground
DG = digital ground

**Figure 9.31** A/D converter using the successive approximation method.

$$
\begin{aligned}
\frac{V_a - 0}{2R} + \frac{V_a - D_0}{2R} + \frac{V_a - V_b}{R} &= 0 \\
\frac{V_b - V_a}{R} + \frac{V_b - D_1}{2R} + \frac{V_b - V_c}{R} &= 0 \\
\frac{V_c - V_b}{R} + \frac{V_c - D_2}{2R} + \frac{V_c - 0}{R} &= 0 \\
\frac{0 - V_c}{R} + \frac{0 - D_3}{2R} + \frac{0 - V_o}{2R} &= 0.
\end{aligned}
\tag{9.18}
$$

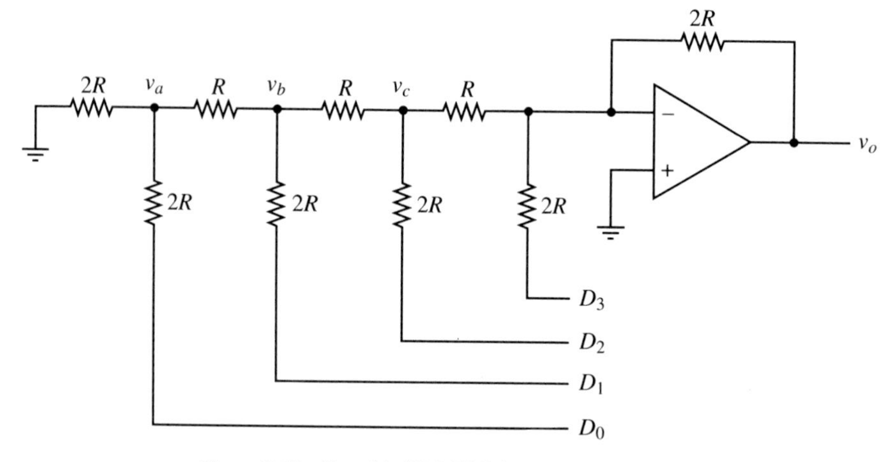


**Figure 9.32** Four-bit "*R*-2*R*" D/A converter circuit.

Standard substitution circuit analysis techniques are easily applied in this case, beginning with the top equation and working down to the bottom, finally yielding

$$V_o = -\left(D_3 + \frac{D_2}{2} + \frac{D_1}{4} + \frac{D_0}{8}\right). \tag{9.19}$$

This is clearly the required operation for conversion from binary voltages ($D_i$) to the corresponding output voltage $V_o$; the value coded in the binary digits has been decoded using an op-amp implementation of Eq. (9.17). Note that $V_o$ is baseless; it is not base 2, base 10, or any other base. Only when coded in digits is it represented in a particular base. This is why it is called an "analog" representation; it is just a voltage analogous to the physical quantity in question and is not in digit format. Note, however, that it is discrete in value, because it is a direct conversion from the discrete, digital signal $D_3 D_2 D_1 D_0$, which has only $2^4 = 16$ possible values.

There are other possible arrangements, but the advantage of this "*R*-2*R*" circuit is that only two values of resistance, one double the value of the other, are required for any number-of-bit conversion. In Fig. 9.33 is shown an eight-bit integrated circuit D/A converter, with the various peripheral connections. It is also based on the *R*-2*R* ladder network. The symbol $I_o$ stands for "output current" here, not "input number zero." An externally supplied op amp is required to convert $I_o$ to a good voltage source. The D/A converter chip has internal high-speed conversion switches to maximize conversion rate; it also has many features and alternative modes too numerous to discuss here.

**Example 9.6**

Suppose that a binary 1 at any preceding digital circuit outputs used to form the D/A conversion inputs ($D_0$, $D_1$, $D_2$, and $D_3$) appears as 3.5 V.

(**a**) What is the smallest spacing between possible output voltages of the D/A converter?

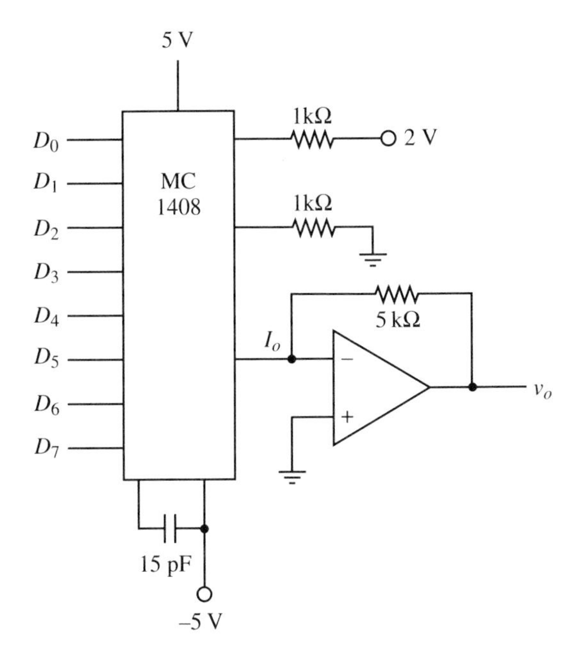


**Figure 9.33** Eight-bit IC D/A converter with eight-bit binary input.

(**b**) What is the output voltage corresponding to an input bit pattern 1010?

(**c**) Draw the input/output relation for this D/A converter.

**Solution**

(**a**) From Eq. (9.19), this voltage is 3.5 V/8 = 0.4375 V.

(**b**) The corresponding output is $-3.5 \cdot (1 + 0/2 + 1/4 + 0/8) = -4.375$ V.

(**c**) See Fig. 9.34.

**Example 9.7**

Suppose that a 4-bit A/D converter takes an "analog" input from 0 V (binary representation: 0000) to 10 V (binary representation: 1111).

(**a**) What is the resolution of this A/D converter?

(**b**) Show the steps in converting 4.6 V to a 4-bit binary representation by successive approximation.

(**c**) Draw the input/output relation for this A/D converter.

**Solution**

(**a**) The resolution is the smallest difference between two input voltages that would produce different binary outputs. Because of uniform level spacing, this would be 10 V/(number of levels) = 10 V/$2^4$ = 0.625 V.

(**b**) The first bit (MSB) by itself (with all others 0) contributes 10 V/2 = 5 V. This is too high, so turn it back off. Simultaneously, turn on the next MSB, which contributes 10 V/$2^2$ = 2.5 V. This is too small, so leave it on and turn on the next MSB, which can contribute 10 V/$2^3$ = 1.25 V. The sum, 2.5 V + 1.25 V = 3.75 V, is still too small, so leave it on. Finally, the LSB contributes 10 V/$2^4$ = 0.625 V. When added to 3.75 V, this gives 4.375 V, the best approximation of 4.6 V that this A/D converter can produce. The approximation as a function of time is summarized in Fig. 9.35a.

(**c**) See Fig. 9.35b.

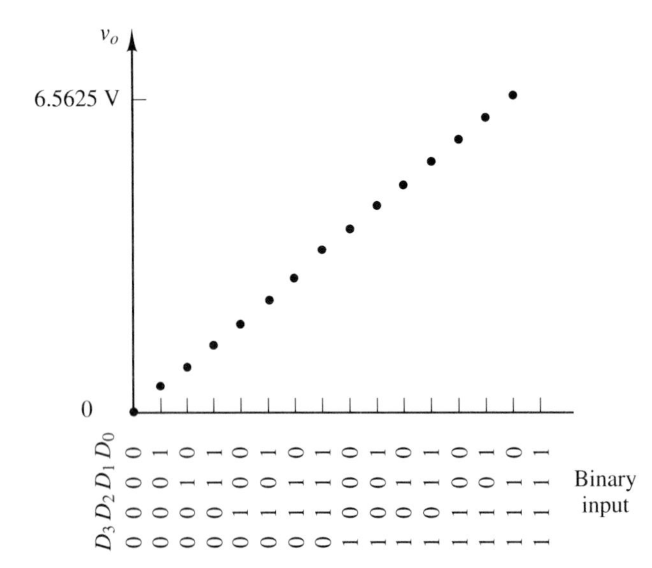


**Figure 9.34** Input–output relation for D/A converter of Example 9.6.

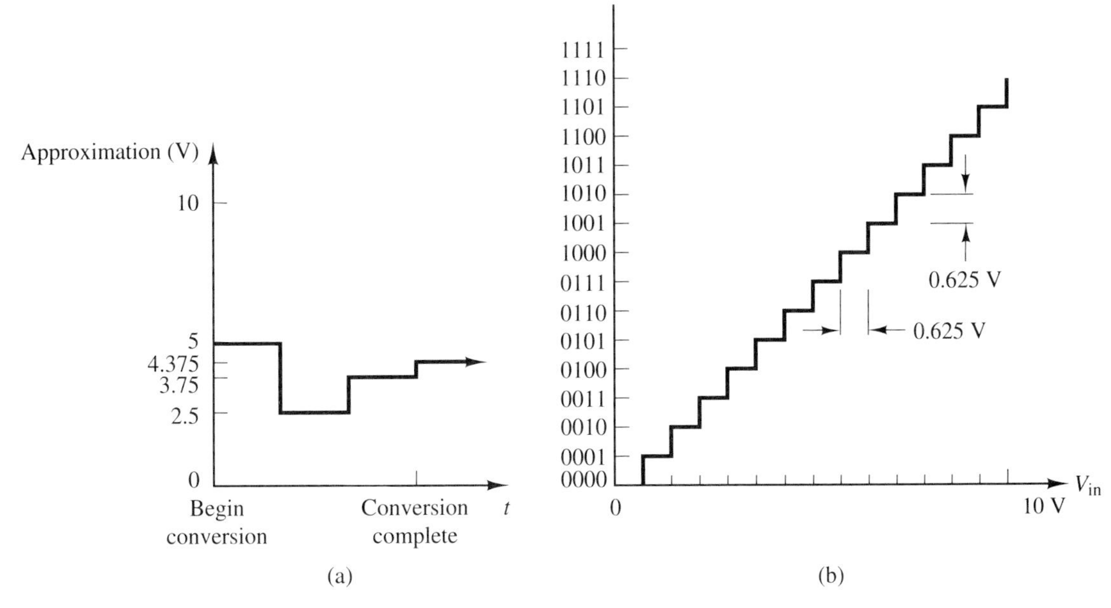


**Figure 9.35** (a) Result of successive approximation method A/D as a function of time; (b) A/D input–output characteristic (digital output versus analog input).

### 9.7.7 Typical Signal Conditioning System

(This subsection greatly benefited from a discussion with a Burr–Brown technical support representative; Fig. 9.36 is constructed from applications suggestions in the 1989 Burr–Brown catalog.)

As a way of briefly putting together some of the signal conditioning elements studied in this chapter, we will here look at one application example. In Fig. 9.36 is shown a temperature monitoring system; we look now at a few of its details.

This temperature system has a cold-junction compensating RTD in addition to the main thermocouple sensor. Its current comes from one of the 1-mA sources in the XTR101 voltage-to-current ("current transmitter") chip. Proper voltage bias levels are established by running 1 mA in addition to the other 1 mA through the 2.5-kΩ resistor. Thus the voltage across that resistor is 2 mA · 2.5 kΩ = 5 V. The resistor $R_s$ is selected to provide the appropriate span. The voltages across the RTD and the thermocouple oppose; the RTD compensates for the cold-junction temperature as described briefly in Sec. 9.6.1. The 15-Ω resistor is used to help calibrate these voltages.

The 0.01-μF capacitor attempts to shunt common-mode noise, and the capacitor on the XTR101 output filters out noise on the power supply lines. The diode protects the XTR101 from unexpected polarity reversal. The squiggly lines represent the twisted pair current loop.

On the receiving end is an ISO100 optically coupled isolation amplifier. Note that it requires dual isolated power supplies ($\pm V_{BB}$) for its input and output stages. The input $V_{BB}$ also drives the current loop. The 250-Ω resistor converts the 4- to

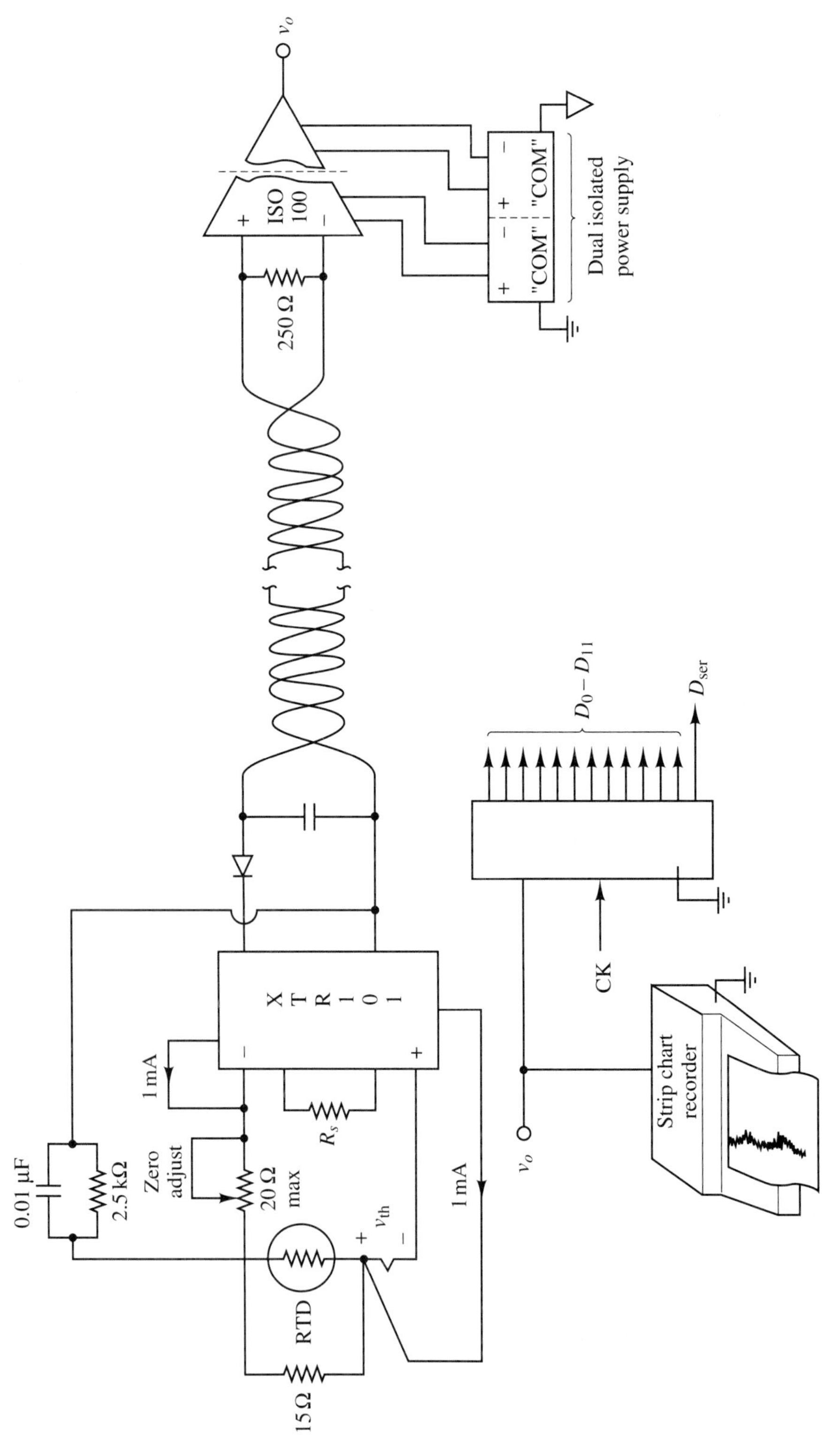


**Figure 9.36** Temperature monitoring system.

20-mA current to a 1- to 5-V input voltage to the isolation amplifier. The output voltage, $v_o$, is shown connected to an A/D converter. This A/D converter provides a 12-binary-digit representation of $v_o$, suitable for input to a computer. It also has a serial output (discussed in Sec. 11.4) and a clock input (see Sec. 11.3). Finally, $v_o$ is also connected to a strip chart recorder for an analog hardcopy in graphical form.

## 9.8 SUMMARY

In this chapter we studied a variety of practical devices and issues relevant to real-world electronic measurement systems. We began by considering transducers made from the basic circuit elements: resistors, capacitors, and inductors. By varying one physical parameter of the element, its electrical parameter would change in a predictable way. The resistor is very versatile; we found that it is used in strain and thickness gages, fluid level sensors, and temperature sensors. The capacitor and inductor are also valuable as transducers, allowing us to measure other quantities, such as pressure and angular velocity.

There are two main categories of transducers: modulating and self-generating. Modulating transducers need an external electrical source of energy to function, while self-generating transducers convert directly from the sensed energy form into the desired electrical form. Examples of modulating transducers were all those in the preceding paragraph, as well as Hall effect transducers, liquid crystal displays, and a host of others. Solar cells, thermocouples, the moving-coil microphone, and the tape recorder head are examples we studied of self-generating transducers.

It should be mentioned in passing that many transducer sensors today are being made on a chip. This makes the sensors easy to mass produce, extremely small, low in power consumption, and inexpensive. Furthermore, there is now the possibility of sensor arrays—for example, with a two-dimensional array of thermocouples, one can obtain a temperature map. These predominantly silicon-based sensors, appropriately called microsensors, are the sensor of the future in many applications.

The concept of optical coupling is useful not only for a variety of measurement transducers (we studied the counter, angular position, and smoke detector), but also as an optimal means of signal transmission, via the fiber optic cable. Advantages such as data handling capability, low error rate (or distortion), and low cost are why many telecommunications systems are becoming optically linked. To accomplish this, either full-blown A/D, D/A converters or the cheaper V/F, F/V converters will perform the necessary signal transformations for driving diodes digitally. The choice of module depends on the application.

Another use for optical coupling is the isolation between an input stage and an output stage of an instrumentation amplifier. The output stage may have a very high, undesirable common-mode voltage. Using optical or other isolation techniques, these high common-mode voltages can be removed from the signal successfully and safely. Isolation amplifiers are also helpful for breaking ground loops and for establishing near-perfect separation of driving and driven circuits. The latter factor is essential for biomedical instrumentation.

Transmission of current instead of voltage is another way of minimizing noise and maximizing signal strength over long transmission distances. The building blocks are current converters, "transmitters," and current-to-voltage converters (usually, resistors). These circuits are then known as current loops. In addition to the minimization of noise and elimination of voltage drop problems, they also make possible two-wire transmission. Power is sent out to the transducer site, and the signal is returned—all on the same two wires.

We also reviewed in Sec. 9.7.4 grounding and shielding practices. One could devote months or years to learning all the fine points of this "art." Instead, we looked at some of the basic rules of wiring up measurement systems. Using separate wires to ground for different parts of the system, grounding at only one point, encasing low-power signals with a grounded shield, and separating signal cables from harmful higher-power lines will minimize noise.

The last major topic we covered provides a smooth transition between this trio of chapters on analog electronics and the next trio on digital electronics. This topic is the conversion of analog signals into digital signals, and vice versa. Two methods have been studied in some detail: The flash converter (Sec. 8.8.2) and the successive approximation converter (Sec. 9.7.6). The V/F converter (Sec. 9.7.3) can be considered a rudimentary A/D converter, except that it requires a special counter and other electronics to interface correctly with a computer. The standard *R*-2*R* D/A converter served to illustrate the reverse process, from a digital signal to an analog one.

We now embark on a major study of digital electronics. Without question, op-amp circuits, computer circuits, and electromechanical machines are the three most important types of circuits that you will encounter in both your professional and personal lives. We have finished looking at op-amp circuits; let us now consider the others. With the entire nine chapters on fields, circuits, and electronics behind us, we are fully prepared for these remaining topics.

## PROBLEMS

**9.1.** Figure P9.1 shows a light control system. Identify the transducers in this system, and explain the function of each. What would be a practical use for such a system? Suppose that the amplifier is constructed so that when light shines on the photocell, current is fed into the relay. For your practical application, should the relay be normally closed or normally open? Why? Comment on the physical placement of the CdS photocell and the light bulb.

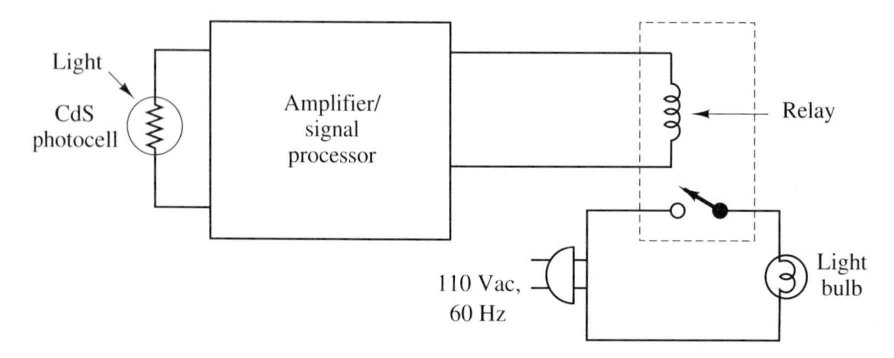


**Figure P9.1**

**9.2.** Decide whether the following transducers are modulating or self-generating: (**a**) thermocouple, (**b**) piezoelectric, (**c**) piezoresistive, (**d**) LED/photodiode, (**e**) radioactive smoke detector in Fig. 9.21a.

**9.3.** When we rotate the volume control on a stereo, we seem to get the most "leverage" (increase in sound per angular increment) at low volumes. Yet from Fig. 9.1b we know that equal increments in angle yield equal increments in resistance. Resolve the apparent discrepancy.

**9.4.** We have a tiny cube of carbon ($\rho = 1.375 \cdot 10^{-5}\ \Omega \cdot \text{m}$). We measure the resistance across two opposite faces to be 4 m$\Omega$. What is the length of a side of the cube?

**9.5.** In a laboratory class, the teacher brought out a coil of annealed standard copper wire and asked a student to determine the diameter of the wire accurately. The student measured the length of the wire to be 8.2 m and the resistance to be 0.6 Ω. The temperature in the room was 20°C, for which the resistivity of annealed standard copper is 1.724 μΩ · cm. What should the student find to be the diameter of the wire? (Express your answer, if you have a table, in AWG gage.)

**9.6.** (**a**) In Fig. 9.1c for the case of a linear potentiometer (see Fig. 9.1a), $v_{ba} = 1$ V when $X = 2$ cm:

(i) Find $v_{ba}$ when $X = 1$ cm.

(ii) Find $X$ when $v_{ba} = 2$ V.

(**b**) In Fig. 9.1c for the case of a rotary potentiometer (see Fig. 9.1b), $v_{ba} = 1$ V when $\theta = 15°$:

(i) Find $v_{ba}$ when $\theta = 30°$.

(ii) Find $\theta$ when $v_{ba} = 3$ V.

**9.7.** Copper telephone wires are used to connect a home user to an exchange module. Suppose that we are at the exchange module and wish to know the length of wire. The wire is inaccessible and makes all kinds of twists and turns. Propose a method by which we can determine the length by a simple electrical measurement.

**9.8.** In an industrial plant we are considering running some signal wires from the process location to a distant control room. Our wires must present a total resistance not exceeding $R$ ohms; otherwise, the signal will be attenuated too much. Suppose that we have gate No. 39 copper wire (diameter of 10 mils). If the maximum tolerable resistance is $R = 7\ \Omega$, how far can we run the wires?

**9.9.** In an ultrasonic medical imaging system, we need to have automatic coordinates determination of the ultrasonic transducer (see Fig. P9.9). This is so that when the medical ultrasonographer moves the transducer over the patient's body in an irregular fashion, the signals are correctly oriented to produce an image without geometrical distortions. Suppose that the potentiometers are identical and that we have measured that when their angle is 45° each resistance is 2 kΩ.

(**a**) In terms of $L_1$, $L_2$, $\theta_1$, and $\theta_2$, determine the $x$, $y$ coordinates of the transducer.

(**b**) Suppose we measure that at a given transducer position, $R_1 = 1.4$ kΩ and $R_2 = 0.3$ kΩ. What is the transducer position?

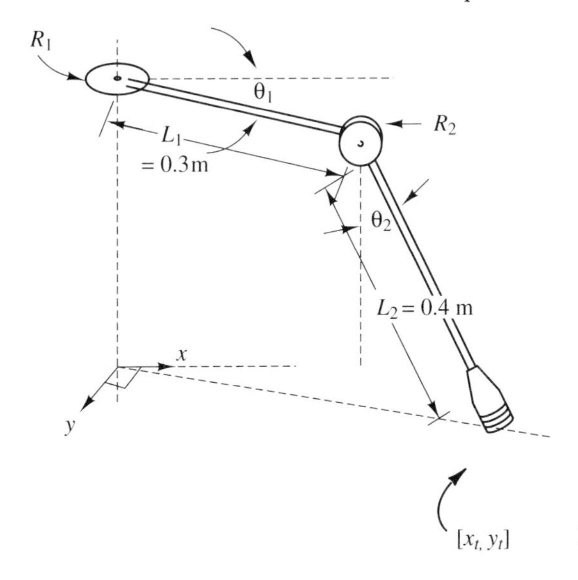


**Figure P9.9**

**9.10.** Figure 9.2 shows a typical usage of strain gages. Suppose now that the two gages are isoelastic wire gages with a nominal resistance of 500 Ω and a nominal gage factor of 3.5. Also assume that the directions of the applied stress are as shown in the figure and that the dummy gage is again not affected by the strain.

**(a)** If $V_r = 6$ V and we measure $V_{AB} = 0.3$ mV, find the strain applied to the active gage.

**(b)** In part (a), find the change of the resistance of the active gage when the strain is applied.

**(c)** Strain is commonly expressed as microstrain. Based on your answer in part (a), explain why. Express your answer in part (a) in microstrain.

**9.11.** **(a)** Calculate the change of resistance $\Delta R$ of a strain gage when the strain applied is $\epsilon_a = 8 \cdot 10^{-5}$, $G = 2.1$, and $R = 100\ \Omega$.

**(b)** For the strain gage described in part (a), suppose that we had measured $\Delta R = 0.05\ \Omega$. What, then, was the strain applied to the gage?

**9.12.** When we write $\epsilon_a$, are we meaning the strain of the gage or of the base material? In Eq. (9.5b), we introduce $\nu$, Poisson's ratio, for the particular strain gage metal: $\nu = \epsilon_l/\epsilon_a$. If the whole point of using a strain gage is to investigate the material on which the strain gage is mounted, why do we use $\nu$ of the strain gage material rather than $\nu$ of the base material?

**9.13.** Recall that the resistance-temperature relation of certain metals used in resistance thermal detectors (RTDs) is linear [see Eq. (9.8)]. Figure P9.13 shows an example application of temperature measurement using an RTD.

**(a)** Determine the relation between $T$ and $V_{AB}$. Assume that $R(0°\text{C})$ of the RTD is $R$. Let $V_r$, $R_1$, and $R$ be known constants.

**(b)** In part (a), suppose platinum is used for the RTD. Calculate the temperature $T$ given that $\alpha = 0.00392$, $R = 200\ \Omega$, $V_r = 3$ V, and $V_{AB} = -0.125$ V.

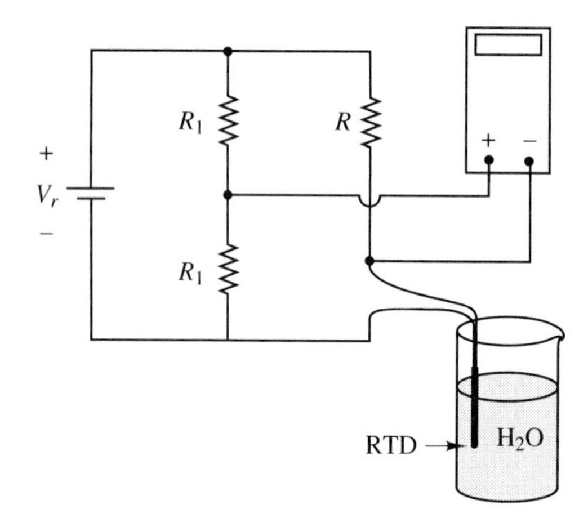


**Figure P9.13**

**9.14.** A chemical process requires minimal variations in temperature from the optimal value. Figure P9.14 shows the circuit used to control the temperature. The RTD is made of platinum: $R_T(0°\text{C}) = 200\ \Omega$, $\alpha = 0.00392$. The op amp is used as a comparator whose output is high or low, causing the heating system to turn on or off, respectively. What should the value of $R$ be if we desire the temperature to be about 85°C?

**9.15.** Suppose that we wish to use the sensitive thermistor thermometer, but desire an output voltage that varies linearly with temperature, instead of the exponential decrease of the thermistor. Show how the circuit in Fig. P9.15 achieves this (at least in principle).

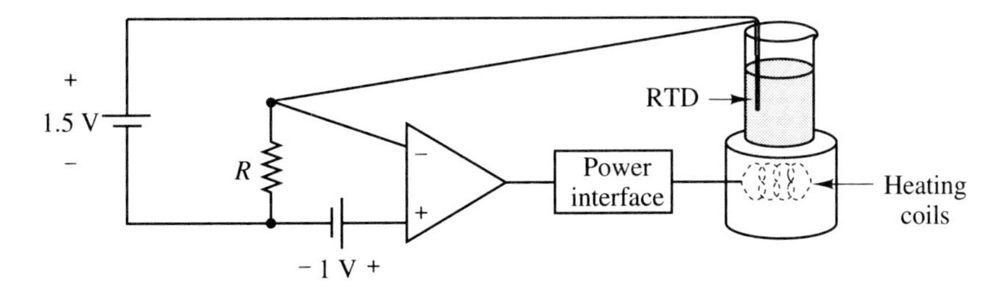


**Figure P9.14**

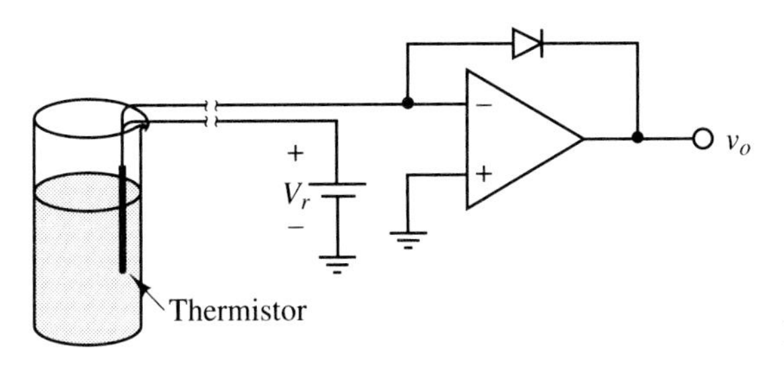


**Figure P9.15**

Assume that the op amp and associated components are at the fixed room temperature $T_o$. Suggest an improved circuit for more accurate or dependable performance.

**9.16.** A platinum RTD has a resistance of 300 Ω at 50°C. Find its resistance at 170°C given that $\alpha = 0.00392$.

**9.17.** Thermistors have been used in TV sets as major component protectors. Explain this use of a thermistor.

**9.18.** A refrigerator thermostat uses a thermistor sensor and is connected in a Schmitt trigger-type circuit, as shown in Fig. P9.18. The thermistor has resistance linearly modeled as $R_T = 1287\ \Omega - 88.7T$, where $T$ is in degrees Celsius.

(**a**) Find $v_1$ in terms of $v_o$, $R_1$, $R_2$, and $R_T$.

(**b**) Find the positive- and negative-going transition temperatures.

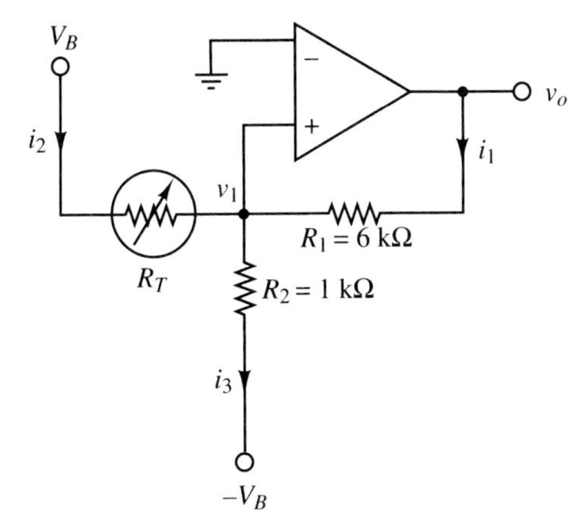


**Figure P9.18**

**9.19.** The capacitive angle transducer shown in Fig. P9.19 is of the general variety found in AM radio tuners (see also Problems 9.20 and 9.54). It consists of two half-circle plates with spacing $d$ between them.

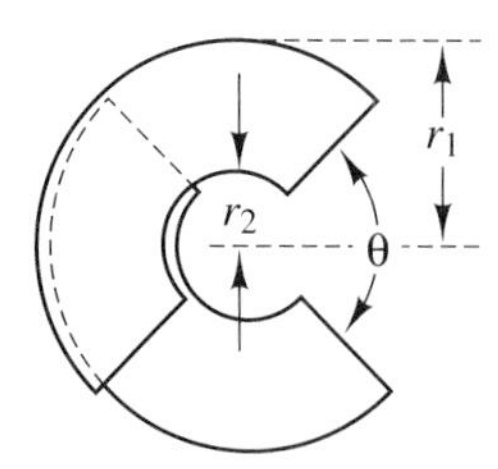


Figure P9.19

(a) Find the capacitance $C$ of the transducer when the angular position is $\theta$. Assume that the capacitor is in air, so that $\epsilon = \epsilon_0$.
(b) Suppose that $r_1 = 2$ cm, $r_2 = 1$ mm, $d = 1$ mm, and $\theta = 45°$. Find $C$.
(c) Determine the expression for $\theta$ in terms of the capacitance $C$ (assuming that $r_1$, $r_2$, and $d$ are known constants).

**9.20.** A closer model of the AM radio tuning capacitor is that shown in Fig. P9.20. This stacked type is used because it performs the same function as the usual rotary capacitor, but with much increased capacitance for a given angle, without requiring impractical transducer radius. Repeat Problem 9.19 for this capacitor.

Stack of five parallel disk pairs

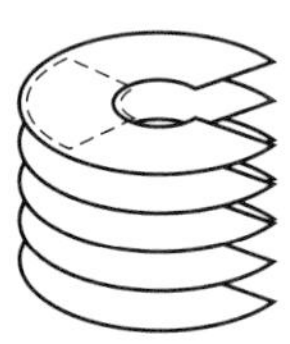

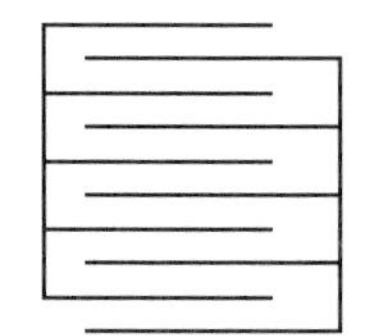

Figure P9.20

**9.21.** Refer to Example 9.2.
(a) If we remeasured the capacitance and it had changed by 20 pF, determine the change in displacement that had occurred.
(b) If $r_2$ were changed from 0.99 cm to 0.9 cm, what would be the change in capacitance in the new ($r_2 = 0.9$ cm) capacitor for the same change of displacement in the old ($r_2 = 0.99$ cm) capacitor?
(c) Which transducer in part (b) is more sensitive to changes in displacement, the old capacitor or the new one?

**9.22.** Not all devices serve well as transducers. Suppose that someone proposes to use the capacitor in Problem 2.44 as a transducer for measuring the distance between two objects. One connects one ball to each object and measures the capacitance, which thereby implies $R$, the distance between the objects. Why is this an inconvenient transducer?

**9.23.** A capacitive thickness transducer measures the thickness of a plastic sheet as it passes between the plates, as shown in Fig. P9.23. The plates are spring-loaded, so that they remain in contact with the plastic as its thickness varies between 1 and 2.5 mm. The dielectric constant of the plastic is $\epsilon_r = 15$. Each capacitor plate is a square 1 cm by 1 cm. Find the maximum and minimum readings of the ammeter A as the sheet moves through. When will the ammeter be highest—at the thickest or thinnest point in the plastic sheet?

**9.24.** In Sec. 9.4.2 a capacitive liquid level transducer was introduced. In Fig. P9.24 we have another example of a capacitive liquid level transducer.
(a) For $L$, $W$, and $d$ given, find the capacitance $C$ and the expression for $d$ in terms of $C$.
(b) If $W$ is decreased to $W/2$, what is the change in capacitance? Is the transducer more or less sensitive to changes in $d$ when $W$ has been decreased?

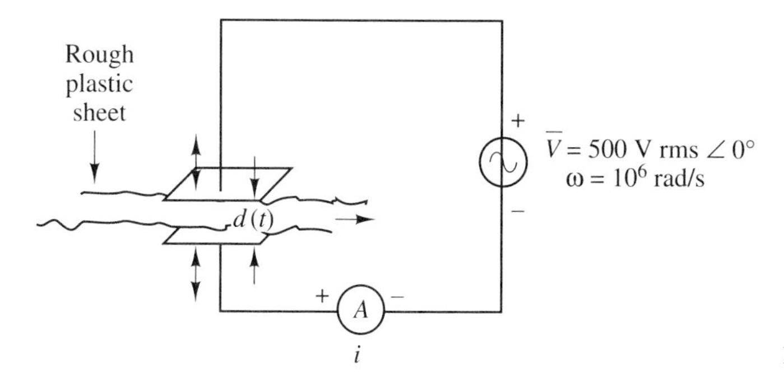


**Figure P9.23**

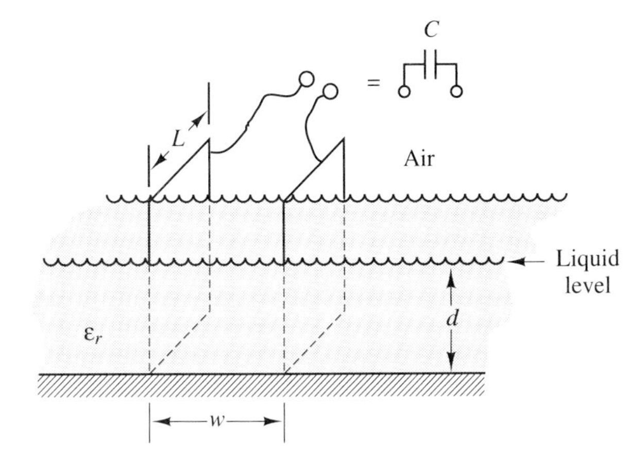


**Figure P9.24**

**9.25.** A real LVDT might have the voltage-displacement characteristic shown in Fig. P9.25. Comment on the use of this device for measuring large displacements.

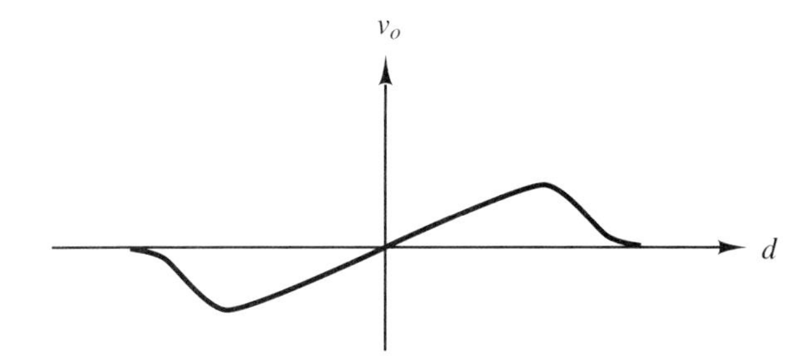


**Figure P9.25**

**9.26.** One may view an electromagnet as a transducer—from electric into mechanical energy via magnetic fields. Suppose that an electromagnet used for arranging scrap iron in an auto junkyard has to be able to pick up pieces weighing as much as 2 tons. The electromagnet has a coil of 1000 turns wound on a cast steel cylinder ($\mu_r = 1300$). More advanced analysis (see also Problem 9.55) shows that the force in $N$ is $F = B^2A/(2\mu_0)$. Assuming that the effective area through which flux passes is 1 $\text{ft}^2$, that the core has a mean flux path length of 1.2 m, that there is a 2-mm average air gap between the scrap metal at contact regions and the magnet (due to irregular shape of the scrap), find the required current to be fed to the electromagnet to pick up the 2-ton heap. (*Hint:* Use for Ampère's circuital law [Eq. (3.10)] a path including the air gap and the core.)

**9.27.** Suppose that a pulse tachometer is mounted on the axle of the front wheel of a bicycle to serve as a speedometer. The rotor has 10 teeth (see Fig. P9.27).

**(a)** If the diameter of the front wheel is $D = 27$ in. and the rider rides at 15 miles/hour, how many pulses does the tachometer generate each second?

**(b)** If $n$ pulses are measured per second, what is the speed of the bicycle in miles per hour?

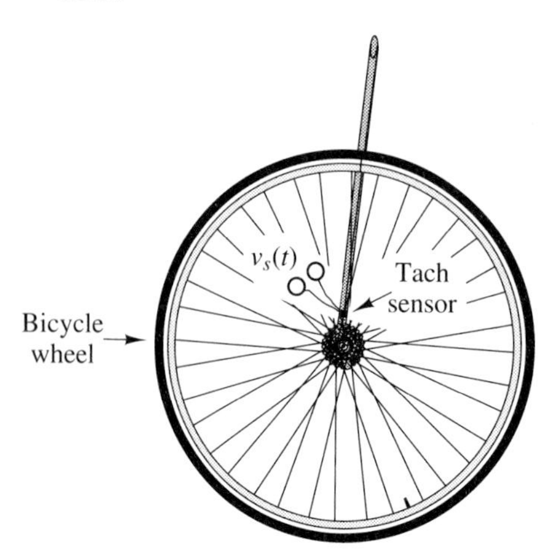


**Figure P9.27**

**9.28.** An analog tachometer has an output voltage proportional to the rotor speed. To achieve constant rotor speed by continuously and automatically correcting for speed variations due to load variation and mechanical vibrations, the following design is proposed. This design is also to have adjustable speed, controllable by the setting of a potentiometer. Connect the inverting terminal of an op amp differential amplifier to the potentiometer, which forms a voltage divider of a reference voltage. To the noninverting terminal, connect the tach output. Use the op amp output signal to control a power boosting amplifier, which supplies the motor with electric power. Sketch the overall circuit diagram and explain its function. Exactly how is speed adjusted?

**9.29.** Copper–constantan thermocouples are widely used in temperature measurement systems. For example, extremely small probes may be used in a grid for measuring localized internal body temperature profiles for ultrasonic hyperthermia cancer treatment. Suppose that in Fig. 9.13, a single probe is surgically implanted in the human body, near the skin surface. The reference junction is kept at the constant temperature of 20°C. We know from solving Eq. (9.14) at $T = 20°C$ that $V_2(20°C) = 0.791$ mV. Suppose that at time $t_1$ we measure $V_{M1} = 0.709$ mV, and at time $t_2$ we measure $V_{M2} = 0.859$ mV. Give the body temperature at the probe location in the body at times $t_1$ and $t_2$. Was treatment started before $t_1$, between $t_1$ and $t_2$, or not at all yet?

**9.30.** Suppose that in a phonograph pickup system, the actual piezoelectric transducer voltage vs. strain characteristic were a quadratic dependence on strain rather than a linear one as in Eq. (9.15). How might such a crystal still be used for sensitive, linear reproduction of music from the grooves on the phonograph record?

**9.31.** A charge amplifier is required for the piezoelectric transducer because of its limited signal power output, its high output impedance, and its special form of output (charge). Figure 9.14 shows a charge amplifier circuit. Suppose that the piezoelectric constant is $k$, and the strain applied to the crystal is $\epsilon(t)$. Find the output voltage waveform $v_o(t)$.

**9.32.** Suppose that a king commissioned his jeweler to build a new silver crown, but suspects the jeweler used silver-plated tin and pilfered the silver. The king's scientist states that electrons are the main charge carriers in silver, while holes are the main carriers in tin. Using the Hall effect, he uses the setup in Fig. P9.32 to determine the composition of the tin. Assuming that the scientist's claim is right and the measured polarity is as shown, is the jeweler guilty or not guilty?

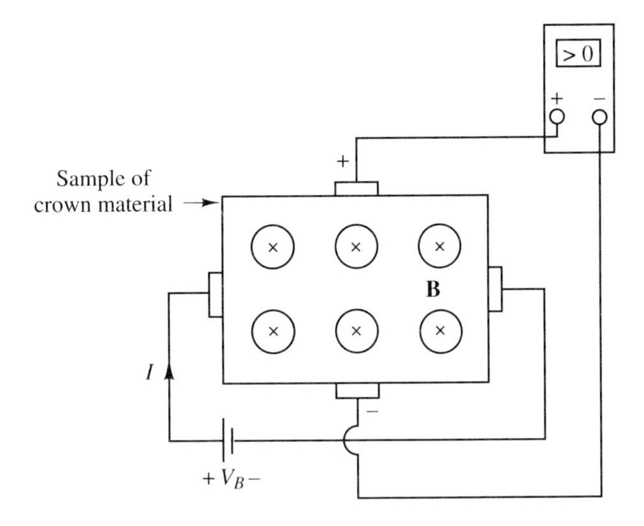


**Figure P9.32**

**9.33.** Electrooptical sensors can be used to detect the level of a liquid by threshold. Figure P9.33 shows such a system; when the water level reaches the height of the fiber optic connections to the prism, the switch to the pump-in drive is shut off. Given the fact that there is near-total reflection from a glass–air interface and near-total transmission for a glass–water interface, explain the operation of this system. If there is a large reverse current detected in the receiving photodiode, should the pump be turned off or left on?

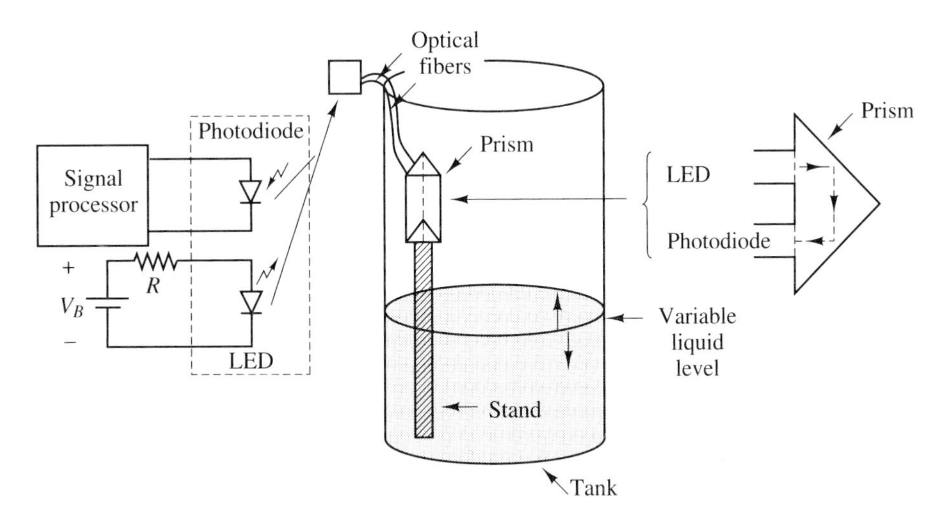


**Figure P9.33**

**9.34.** If we merely wish to measure the ionic content of a solution rather than distinguishing specifically between different types of ions, conductivity measurement offers a fast, reliable, inexpensive approach. To measure the conductivity of a given solution, we define a cell probe such as that shown in Fig. P9.34. The insulating parts of the cell restrict the cross-sectional area to be $A$ between the electrodes. The cell is thus constructed of an insulating material, with metallic pieces at both ends and a hole in the bottom to let the liquid in. When measuring the conductivity of a liquid, the sample is let into the cell from below. Suppose that the cell has the following dimensions: $A = 1\ \text{cm}^2$ and $l = 2$ cm. If the measured resistance between the plates is 40 kΩ:

**(a)** Find the conductivity of the liquid.

**(b)** Suppose that we are given the following list of material conductivities, and are told it is one of these. Which is it?

| | |
|---|---|
| Distilled water | $0.5\ \mu\Omega^{-1}/\text{cm}$ |
| Drinking water | $50\ \mu\Omega^{-1}/\text{cm}$ |
| Ethyl alcohol | $3\ \mu\Omega^{-1}/\text{cm}$ |
| Mercury | $10^{10}\ \mu\Omega^{-1}/\text{cm}$ |
| Seawater | $4 \cdot 10^{4}\ \mu\Omega^{-1}/\text{cm}$ |

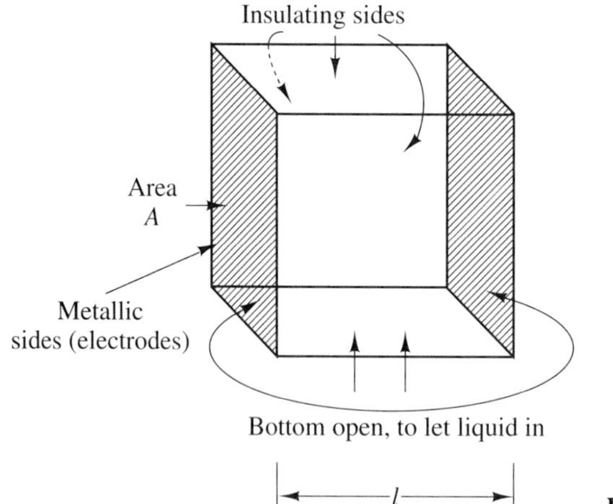


**Figure P9.34**

**9.35.** The parameter pH (p for natural logarithm and H for hydrogen ion concentration) is a logarithmic unit of measure that describes the degree of acidity or alkalinity of a solution (0 for an extreme acid, 7 for neutral, and 14 for an extreme base): $\text{pH} = -\ln[\text{H}^+]$. To measure pH, a pH sensing electrode, a reference electrode, and a high–input impedance instrumentation amplifier that translates the signal into something the user can read are needed. The relation between the pH value, absolute temperature $T$ (°K), and the voltage $v$ generated by the sensing electrode is given by

$$v = E_1 - 0.198T[\text{pH } 7] \qquad \text{mV} \tag{P9.1}$$

where $E_1$ is a constant voltage, dependent on the reference electrode, which serves as the reference potential to which to compare the pH electrode potential because its value is independent of pH. Refer to Fig. P9.35, where we assume that $K = 10$.

**(a)** Suppose that first we use a calibration pH 10 solution and measure $v_o = -100$ mV at $T = 0$°C. Then we measure the temperature to be 20°C, and later measure $v_o$ of an unknown-pH solution to be 4 V. What is the pH value?

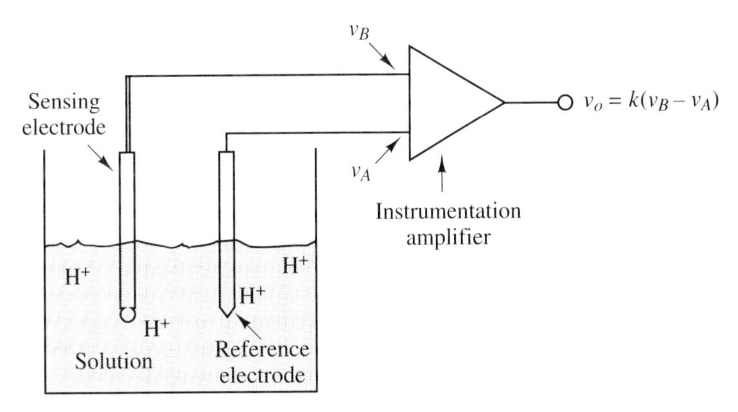


**Figure P9.35**

(b) If we have an error in our temperature measurement, will this be more severe, less severe, or as severe as a degradation in pH measurements that are near 7 vs. far from 7? Why?

**9.36.** Level conversion is quite an important type of signal conditioning. Suppose that the output voltage of a transducer is from zero to 100 mV. By using a voltage-to-current converter, we desire a current ranging from 4 to 20 mA. Find the linear relation between the voltage and the current.

**9.37.** Figure 9.22 shows a simple voltage-to-current converter. If the output voltage of a transducer is 6.317 mV, $R = 10\ \Omega$, $R_L = 100\ \Omega$, and $R_{tw} = 0.1\ \Omega$, find the current in and voltage across the load resistor $R_L$. If we wish to end up with a voltage gain of 5, what must be the new value of $R$?

**9.38.** If we wish to implement the relation obtained in Problem 9.37 using a circuit similar to that in Fig. 9.22, that circuit will have to be modified. Show the necessary modification, and provide numerical values for $R$ and the component you added, for the case $R_L = 100\ \Omega$.

**9.39.** In Fig. 9.24a, if $i_s(t) = 10 + 5\cos(2\pi \cdot 100t)$ mA, $v_n(t) = 30\cos(2\pi 100t)$ mV, and $R_L = 100\ \Omega$, calculate $v_{in}(t)$. Did you find that the noise $v_n(t)$ significantly affect the voltage $v_{in}(t)$?

**9.40.** In Fig. 9.23, suppose that the amplifier accepts differential input voltages of up to 0.3 V, and the maximum signal current is 20 mA.

(a) What must $R_L$ be to make full use of the dynamic range of the instrumentation amplifier?

(b) What voltage is applied to the amplifier when the signal current is 4 mA (the minimum value possible)?

(c) Add a second amplifier stage that converts the range to 0 to 10 V for the signal current ranging from 4 to 20 mA. Assume that the instrumentation amplifier has a voltage gain of 30.

**9.41.** For the one-shot system shown in Fig. 9.26a:

(a) Sketch the output waveform for the given input waveform in Fig. P9.41a. As shown, assume that $t_d$ is smaller than any of the on/off intervals in $v_{in}(t)$.

(b) If $v_{in}(t)$ has very short-duration pulses (shorter than $t_d$), what will happen? Demonstrate this effect by finding $v_o(t)$ for $v_{in}(t)$ in Fig. P9.41b.

**9.42.** The basic requirement for the V/F converter is that the repetition frequency of the pulses is proportional to the input voltage; that is, $f = kv$, where $f$ is the frequency of the pulses generated and $v$ is the voltage to be converted ($k$ is the proportionality

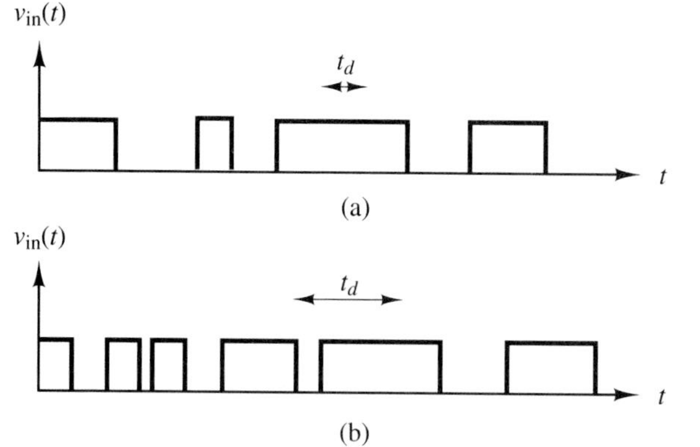


**Figure P9.41**

constant). Let $k = 0.1$ Hz/mV. Suppose that we measured 38 pulses in 0.20 s. Estimate what must have been the value of that input voltage. Is this an instantaneous voltage or an average voltage? How can we obtain greater temporal resolution, and how can we obtain greater resolution of the value of $v$?

**9.43.** Figure 9.27c and d show two methods of grounding. Suppose that $R_1 = R_2 = R_3 = R_4 = R_5 = 4$ MΩ and $R_1' = R_1 + R_2 + R_3 + R_4 + R_5 = 5 \cdot 4 = 20$ MΩ, $R_2' = R_2 + R_3 + R_4 + R_5 = 4 \cdot 4 = 16$ MΩ, $R_3' = R_3 + R_4 + R_5 = 12$ MΩ, $R_4' = R_4 + R_5 = 8$ MΩ, and $R_5' = R_5 = 4$ MΩ. This means that for Fig. 9.27c and d, the same size of wire and the same distance from the ground are assumed. We also assume that $i_1 = 1$ mA, $i_2 = 10$ μA, $i_3 = 100$ mA, $i_4 = 50$ mA, and $i_5 = 20$ mA. Calculate the actual voltages of the terminals that have been supposedly grounded for Fig. 9.27c and d, and compare the voltages at each terminal for the two methods of grounding. Which of the methods has lower interference?

**9.44.** A data acquisition system is shown in Fig. P9.44, in which the shield is attached to earth at both ends. Unfortunately, there exists a potential between the two earth points: $\Delta V = V_1 - V_2 \neq 0$, where $V_1$ is the earth potential at the transducer side and $V_2$ is the earth potential at the display/controller side. The resistance of the shield is 0.3 Ω.

**(a)** If $\Delta V = 1$ V, find the current in the shield in amperes.

**(b)** Discuss ways of reducing this current and the interference it produces with the desired transducer signal.

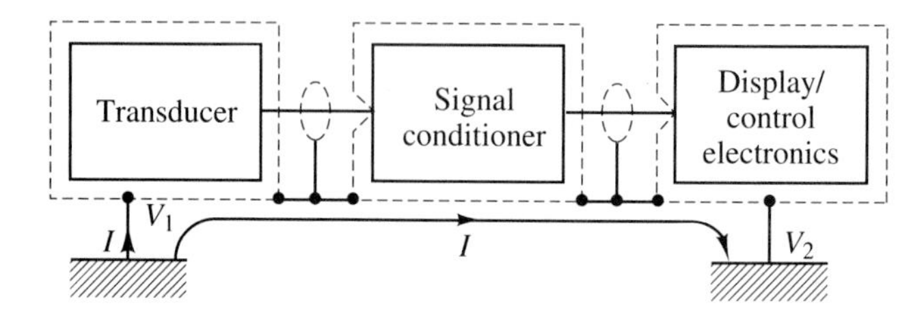


**Figure P9.44**

**9.45.** The reverse current $i_R$ of a photodiode is proportional to the light intensity applied to it, In; that is, $i_R = k \cdot \text{In}$, where $k$ is the sensitivity coefficient. The light intensity In generated by an LED is a function $f(\cdot)$ of the current $i$ in it: In $= f(i)$, where $f(\cdot)$ is an increasing function of $i$. Suppose that in Fig. 9.28b, the two photodiodes are identical and are placed symmetrically with respect to the LED. Find the output of the amplifier, $v_o$, in terms of resistor values and $v_s$.

**9.46.** Determine which of the following are analog, which are digital, or which are a hybrid of analog/digital. State for each whether they are discrete or continuous in time and whether they are discrete or continuous in value.

(**a**) The speed of a car.

(**b**) The number of electrons completely passing through the cross section of a copper wire during 1 s.

(**c**) The instantaneous rate of electrons passing through the cross section of a copper wire, rounded to the nearest integer.

(**d**) The Dow Jones industrial average.

(**e**) The time readout for a three-hand watch.

(**f**) The time readout for a "digital" watch.

**9.47.** Figure 9.29b shows a voltage $v(t)$ that represents pressure. For a digital computer to process the signal, we sample the signal $v(t)$ periodically every $T = 0.1$ s.

(**a**) Draw the resulting sequence $v_d(n) = v(nT)$, where $n$ is the sample index. Do not discretize in value (as was done in Fig. 9.30), only in time.

(**b**) If we use a 12-bit binary code to represent the signal $v_d(n)$, what is the resolution in mV/value level? Assume that 4.8 V is the maximum voltage to be represented.

**9.48.** Figure 9.32 shows a four-bit D/A converter. Let 1 be represented by 5 V and 0 be represented by 0 V, for the digital input signals $D_0$ through $D_3$.

(**a**) If the input digital signal is $D_3 D_2 D_1 D_0 = 1011$, find the output voltage $v_0$.

(**b**) Repeat part (a) for the case $D_3 D_2 D_1 D_0 = 1001$.

**9.49.** Suppose that we have two four-bit D/A converters, as shown in Fig. 9.32. Using one op amp, some resistors, and the two four-bit D/A converters, design an eight-bit D/A converter. Be sure to specify values or ratios of resistors to obtain the correct output. Do not worry about sign changes or other global scaling factors.

**9.50.** In Example 9.7, suppose that an eight-bit A/D converter is used in place of the four-bit A/D converter.

(**a**) Determine the new resolution.

(**b**) Find the factor by which the resolution has improved.

(**c**) What is the resolution of an $N$-bit A/D converter?

**9.51.** A simplified portion of a computer tomography system is shown in Fig. P9.51. The x-rays pass through the person's head and, in proportion to the amount not absorbed by the body, hit the scintillation crystal. This crystal sparkles (emits light) when hit with x-rays. A photodiode right behind the crystal then detects the light. It produces a small voltage and current, the latter of which is in this case 5 nA. Model the photodiode as a current source with $R_s = \infty$ and $I = 5$ nA. (This is appropriate if we recall the $i$–$v$ characteristic of a photodiode, and recall that the voltage $v_-$ is very close to zero.) What is the voltage measured at the output of the preamplifier op amp?

**9.52.** On the deflection yoke of a TV, there are 15,750 horizontal traces of the electron beam per second and 60 fields per second. Each field is a half-line-density picture, and the lines of succeeding fields are interleaved; thus there are 30 complete frames (pictures) per second. To move the electron beam horizontally across the screen, a current ramp function as shown in Fig. P9.52 is required. Sketch the required form of the voltage applied to the yoke coil. Is the function performed by the yoke differentiation or integration?

**9.53.** The $i$–$v$ characteristic of a liquid electrolyte can be used for identification of substances (traces of metals in alloys, ores, organic materials, drugs and vitamins, etc.) and in the determination of chemical reaction rates. Such studies are called polarography (because of the resulting polarization of the liquid). Clearly, a stiff voltage source (independent of the load) is required because the $i$–$v$ curve can be very nonlinear. Argue that for the circuit in Fig. P9.53, $v_o$ is independent of the load current $i_o$—even if the reference voltage source internal resistance $R_s$ is substantial.

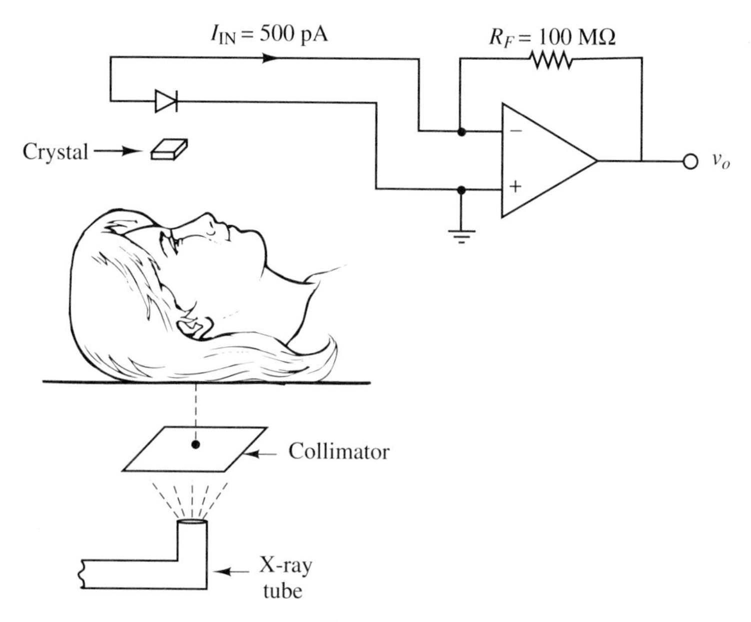


**Figure P9.51**

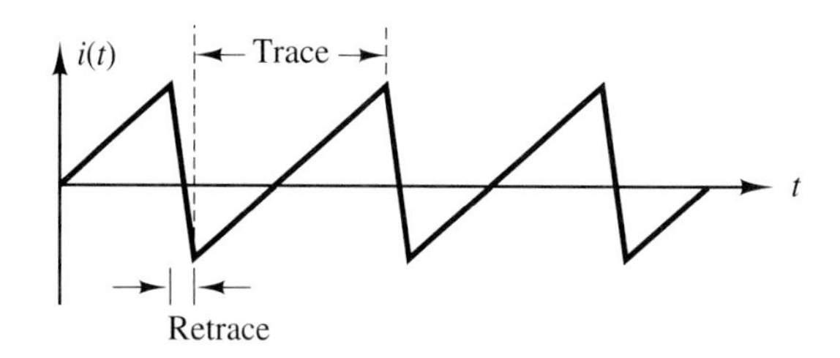


**Figure P9.52**

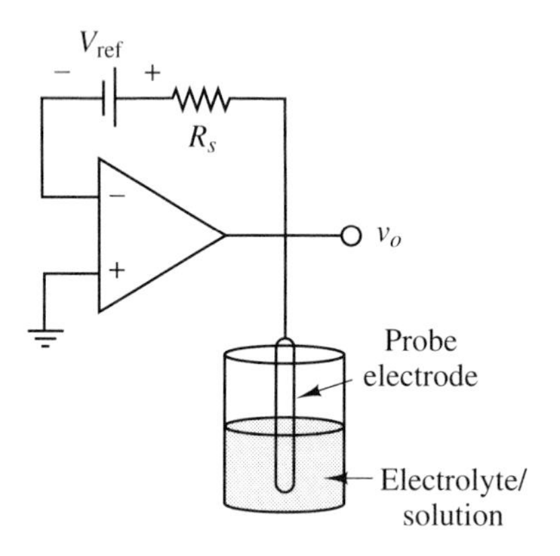


**Figure P9.53**

## ADVANCED PROBLEMS

**9.54.** The AM radio dial is nonlinear: the lower frequencies (e.g., 550 to 650 kHz) are spread out much more than are the higher frequencies (e.g., 1300 to 1600 kHz). As we noted in Problems 9.19 and 9.20, AM radios use the angular capacitor transducer for tuning.

What is the reason for the nonlinearity? Does it have to do with the rotary capacitor transducer?

**9.55.** A relay is pictured in Fig. P9.55.

(**a**) By using the energy balance upon movement of the armature an amount $dx$ in time $dt$: electrical energy input ($dW_e$) + mechanical energy input ($dW_m = F\,dx$) = increase in stored magnetic energy ($dW_s$), show that the force on the armature is $F = \frac{1}{2}\phi^2 d\,\mathcal{R}/dx$, where $\phi$ is the flux in the core and $\mathcal{R}$ is the total reluctance of the core (permeability $\mu_{\text{core}}$) and air gap. (*Hint:* By Ampère's circuital law [Eq. (3.10)], we have $Ni = H_{\text{core}}\,d_{\text{core}} + H_{\text{gap}}\,x$, where $d_{\text{core}}$ is the mean length around the magnetic core and $x$ is the air gap width. Also, electric power is $p = vi$, and energy is $p \cdot dt$. Assume that the flux in the gap is equal to the flux in the core, and that the effective areas of the core and gap for the flux are equal. Finally, the magnetic field energy density is $BH/2$.)

(**b**) Show that this agrees with the result for the force used in Problem 9.26 for the electromagnet, namely that the attractive force is $F = B^2A/(2\mu_0)$.

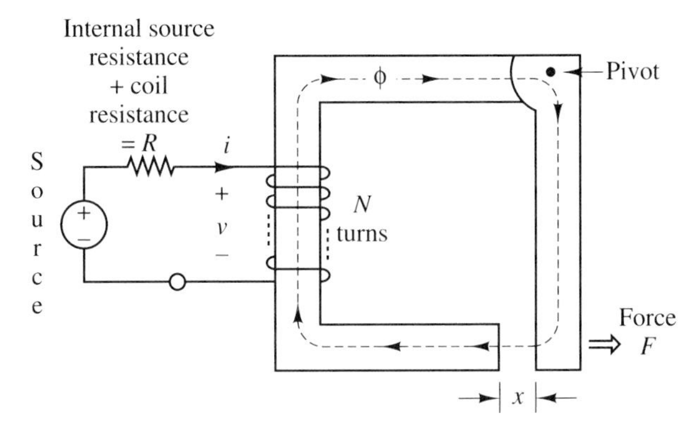


**Figure P9.55**

**9.56.** The Doppler flowmeter can be used to noninvasively measure the speed of blood in human blood vessels. It relies on the change in frequency of reflected ultrasound from moving particles in a liquid. (The principle is the same as the familiar change in pitch of a train whistle as the train passes by, and in the case of electromagnetic radiation, the "red shift" used to determine the trajectories of stars.) The Doppler frequency shift $\Delta f$ can be derived to be $\Delta f/f_0 = 2u/c$, where $f_0$ is the transmitted frequency, $u$ is the "line-of-sight" velocity of the liquid relative to the receiver, and $c$ is the speed of sound. Because we cannot have a line of sight in the blood vessel (we cannot have the transducer inside the vessel!), we must take our measurements at an angle $\theta$, as shown in Fig. P9.56a.

(**a**) For the practical measurement configuration shown in Fig. P9.56a, determine the received $\Delta f$ in terms of $u$.

(**b**) Suppose we assume that the transmitted sound signal is $s(t) = A_m \cos(2\pi f_o t)$ and the received signal is $r(t) = A_m' \cos[2\pi(f_0 + \Delta f)t + \phi]$. Figure P9.56b shows a schematic "flowgraph" representation of a system that could be used to determine $\Delta f$. If we require the output at label "5" to be exactly $\Delta f$, how should parameter $k$ be chosen?

**9.57.** The circuit in Fig. P9.57 may be viewed as a voltage-to-current converter with both positive and negative feedback. Find $i_L$ as a function of $v_s$ under the conditions that this circuit behaves as a linear amplifier. Simplify your results as much as possible. What is the condition for which linear amplification occurs? What is the behavior of $v_o$ if this condition is violated?

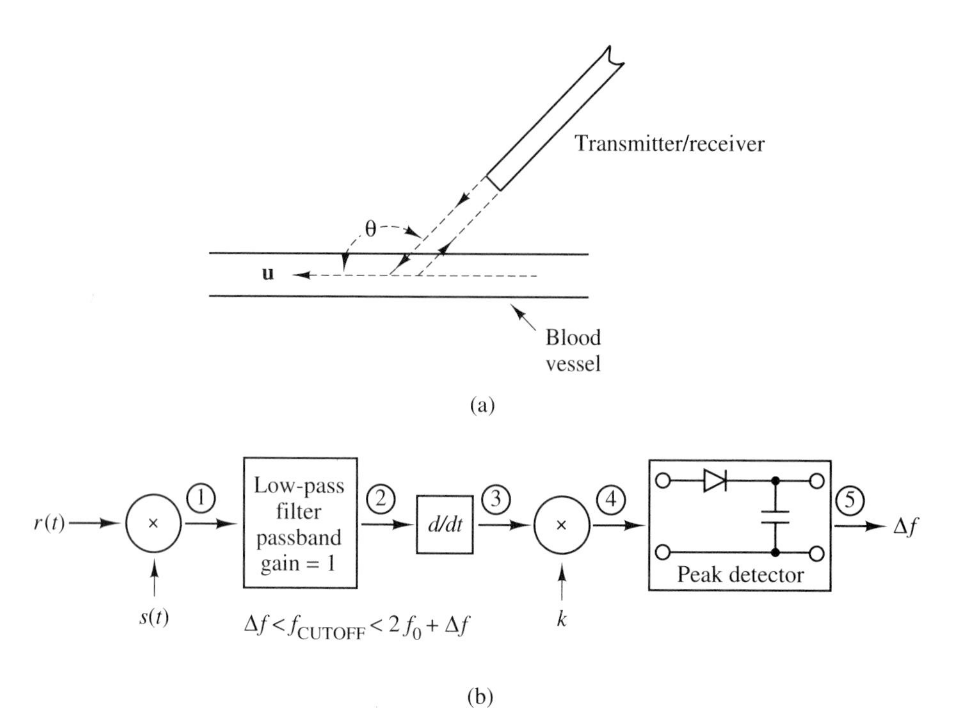


**Figure P9.56**

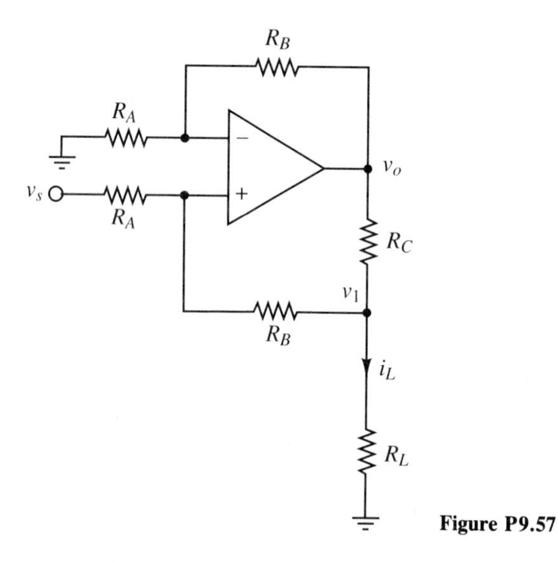


**Figure P9.57**

**9.58.** A sample-and-hold circuit (see Fig. P9.58a) is constructed to aid in analog-to-digital conversion by temporarily holding $v_o$ constant while A/D circuits calculate the computer representation of $v_o$ (the number suitable for storage and use in a computer). A digitally generated square wave is run through a differentiator ($\tau = R_3 C_3 = 500$ ns), whose output will turn on the MOSFET when $v_G > 1$ V. Sketch $v_G$ vs. time. (*Hint:* Remember that if you think that a derivative is infinite, it will only saturate the op amp, at which pont $C_3$ may discharge through $R_3$.) When the MOSFET is on, $C_2$ quickly charges toward $v_1$ because of the very small drain–source resistance (1 to 2 Ω) when

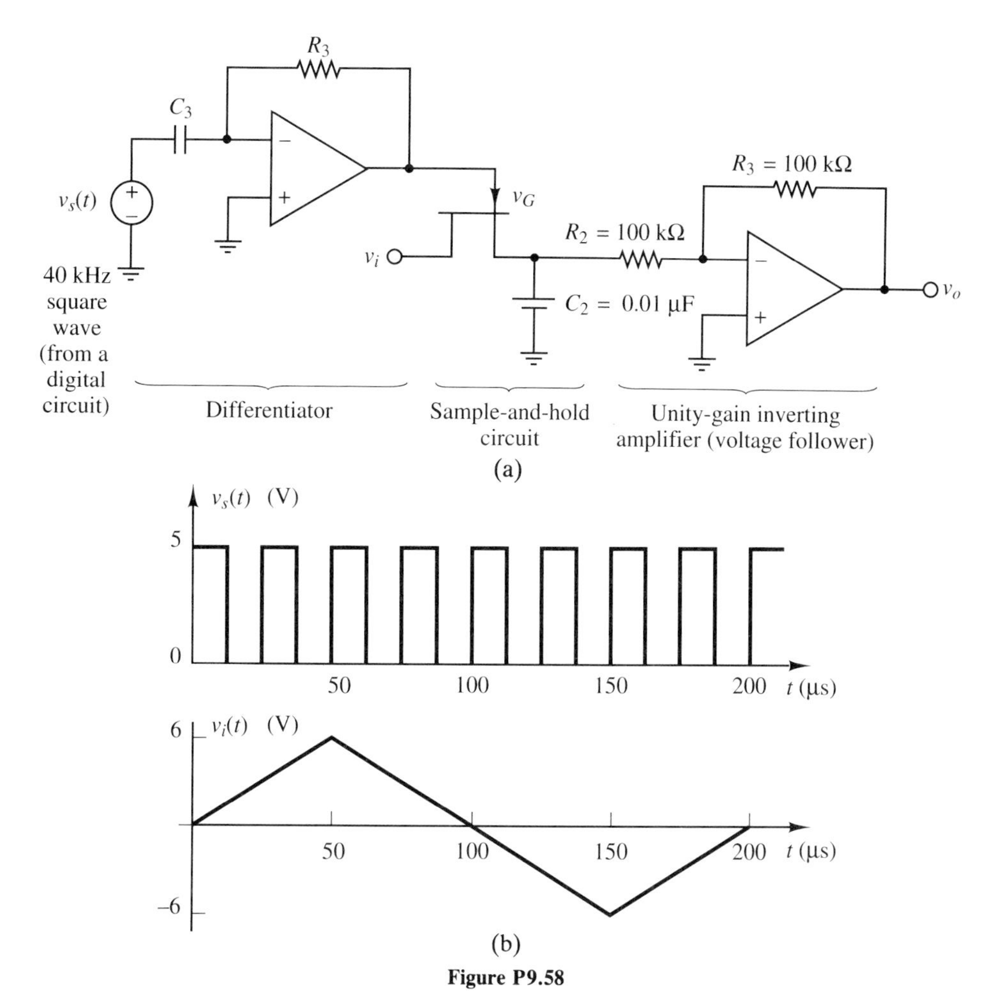


**Figure P9.58**

conducting. What is the time constant of its discharge (when the MOSFET is off)? How does this compare with the period of the 40-kHz square wave? If $v_i(t)$ is a 5-kHz triangle wave [see Fig. P9.58b, which also shows the 40-kHz square wave $v_s(t)$], sketch $v_o$. Sketch $v_o$ again if $R_2 = R_4 = 2.5\ \text{k}\Omega$.

**9.59.** When we sample continuous-time signals using an A/D converter, there can be a problem if we do not sample frequently enough. A high-frequency component of a signal can, under such sampling, result in a sampled sequence identical to a lower-frequency signal component due to the periodicity of sinusoids. This distorts the sequence and is known as aliasing.

**(a)** Consider the two sinusoids $v_1(t) = \sin(200t)$ and $v_2(t) = \sin(550t)$. If these sinusoids are sampled, respectively, every $\Delta t_1 = 0.2$ s and $\Delta t_2 = 0.02703$ s, find the sequences $v_{1,d}(n) = v_1(n\,\Delta t_1)$ and $v_{2,d}(n) = v_2(n\,\Delta t_2)$ (where the subscript "$d$" stands for "discrete"). Show explicitly that aliasing occurs, by closely comparing the arguments of the sinusoids.

**(b)** Suppose now that we have two sinusoids $v_1(t) = \sin(\omega_1 t)$ and $v_2(t) = \sin(\omega_2 t)$, and that we sample these waveforms as follows: $v_{1,d}(n) = v_1(n\,\Delta t_1)$ and $v_{2,d}(n) = v_2(n\,\Delta t_2)$. If $\omega_1$, $\omega_2$, and $\Delta t_1$ are known and fixed, what values of $\Delta t_2$ will cause $v_{2,d}(n)$ to be equal to $v_{1,d}(n)$ for all values of $n$?

# Part Four: DIGITAL ELECTRONICS

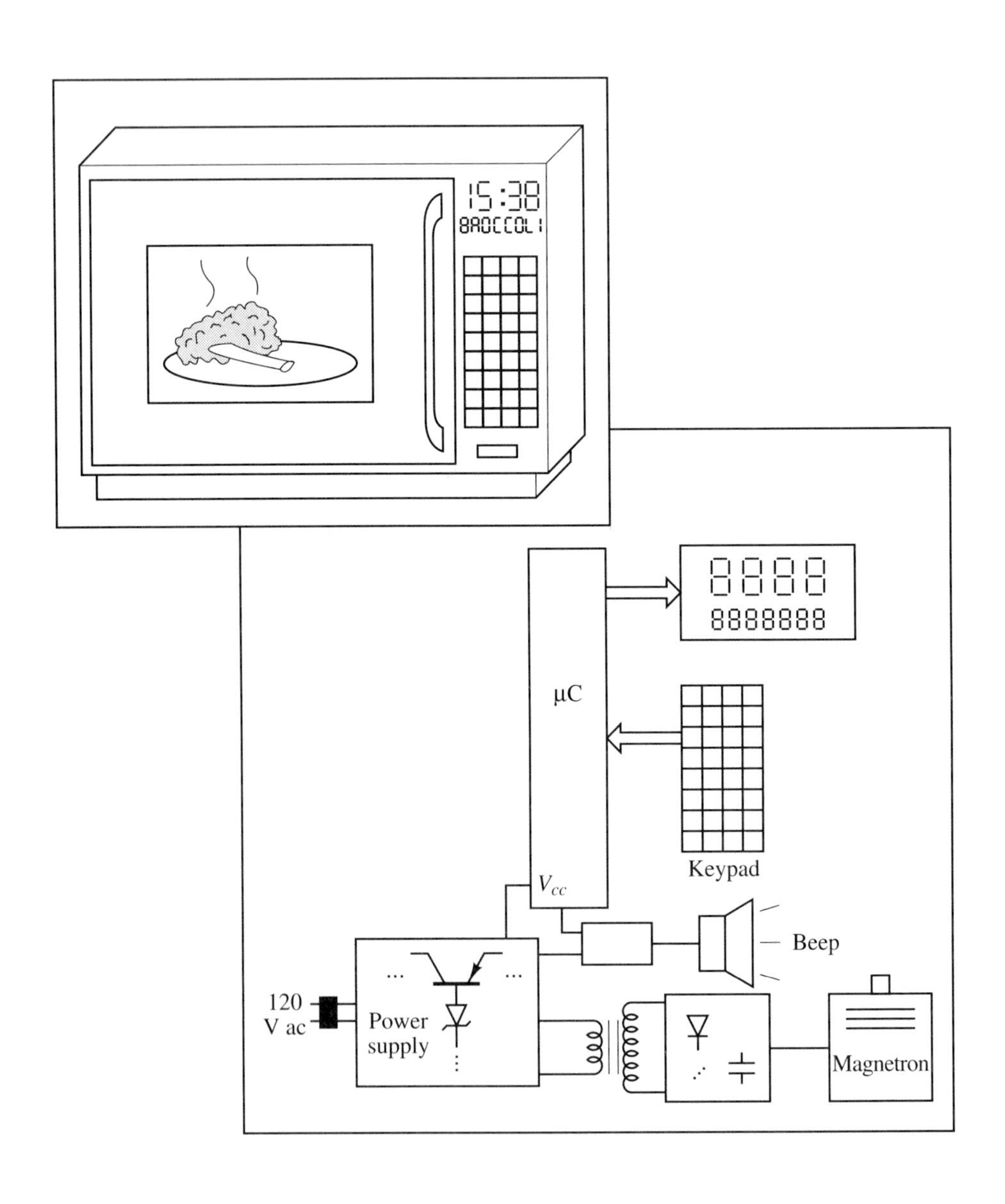

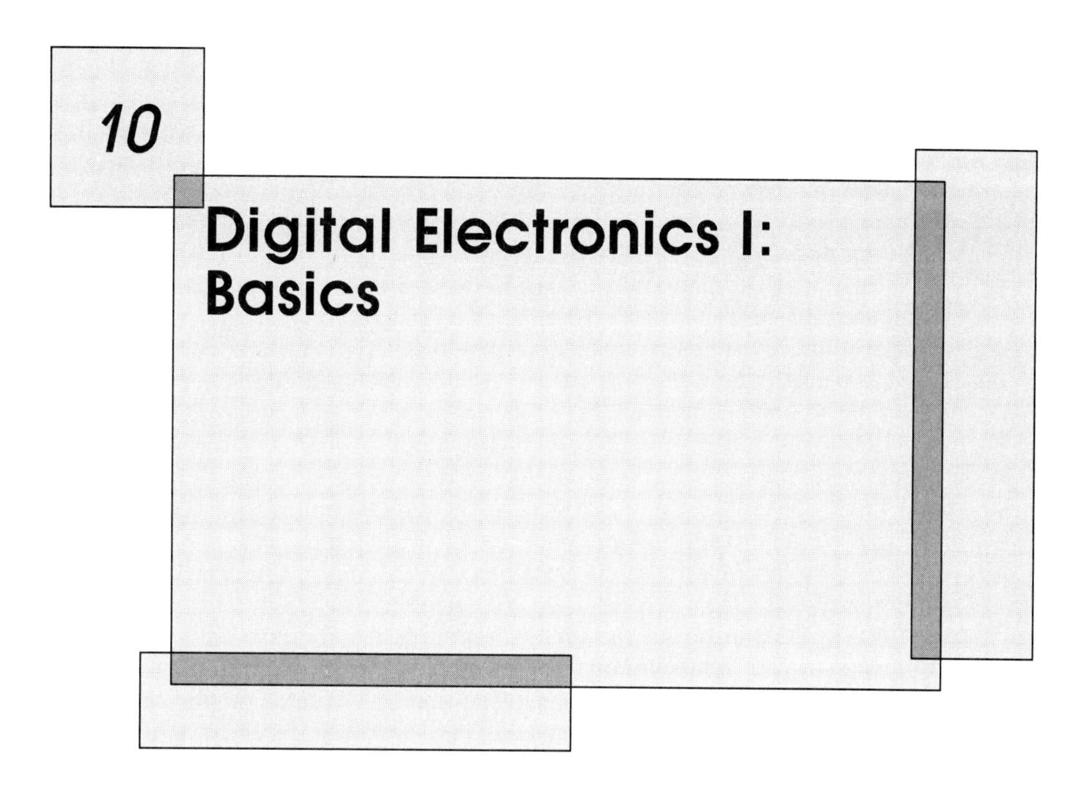


## 10.1 INTRODUCTION

The extent to which digital electronics pervades American life today is clear. At the store, we used to show the price on an item to the clerk, who would then "ring it up" on the register. Now not only does the digital bar code eliminate the need for the clerk to read each price, but the clerk need not even tell the customer the total—a digitally synthesized voice does that.

In the past, a watch would have to be wound every day or so, and reading the time involved interpreting the position of two "hands." Now we replace the battery once a year or so, and read the time to within a second accuracy on a digital display. We used to have to spend hours or days completely retyping a paper when producing a revision; now we merely modify a file of digital information and ask for a new hardcopy.

We used to have to turn an oven dial to a specific temperature and "watch the clock" to cook. Now we hold down a button until "corn on the cob" appears on a vacuum flourescent readout on a microwave oven. The scratchy sound of the phonograph record has been replaced by the clean sound of the digital compact disk. Even the traditional method of information storage on cassette tapes will soon be outmoded by digital tape recording, an entirely different (digital) format of storage of the sound information.

The list could go on indefinitely. The important advantages that the digital perspective to signal processing offers include noise immunity, ease in storage and transfer of data with no degradations, flexibility in processing the data of interest, and the extremely low cost, high speed, small size, and light weight of digital hardware. Finally, when the signals of interest are expressed in digital format [a pattern of ones (1s) and zeros (0s)], they may be processed by the digital computer. This opens up a whole world of signal processing possibilities that would otherwise be impractical because of complexity, noise, and size of components. Therefore, in consumer products, industrial plants, and research laboratories, the irreversible trend is toward complete digitization.

Nearly all of our previous studies have been with continuous-variable devices: the voltage, current, and time variables all could take on continuous ranges of values and were defined over continuous time. These systems and variables are referred to as *analog* because the behaviors of the variables involved directly correspond to those of the natural phenomena that produced them or that they represent. Because most physical phenomena are of a continuous nature, continuous variables varying in tandem with them are said to be analogous to those phenomena, or "analog" for short. The two-hands clock, the oven dial, the phonograph record, and the traditional cassette tape recording are all examples of analog devices.

In our discussion on discrete semiconductor devices in Chapter 7, however, a simple *electronic switch* was considered. The electronic switch forms the basis of the logical gate, which in turn is the foundation of all digital electronics. The voltage output of a switch can take on only two possible values, not a continuous range. It would be termed a binary variable, not an analog variable. In this chapter we examine binary logic more systematically and show how practical hardware can implement it. We also introduced the differences between analog and digital signals in Sec. 9.7.5, and hardware to convert from one format to the other in Sec. 9.7.6.

Another crucial new concept will be that of sequential circuits: Systems that change state in accordance with digital data and the rules of binary logic. In that case, the value of the electronic switch is "read" only at discrete instants of time, not continuously. Digital computers can deal only with this sort of discrete variable: Discretized in both value and time.

An important goal of these chapters on digital electronics, Chapters 10 through 12, is gaining familiarity with typical digital devices likely to be encountered on the job. If an electrical engineer is going to succeed in explaining how a digital circuit will be introduced into a previously mechanical and/or analog system, it is essential that the non-EE have familiarity with both the digital concepts and the digital system building blocks to be used.

## 10.2 FUNDAMENTAL GATES

Everyone knows that a gate is an obstruction that can be removed under only certain circumstances. If the person guarding the gate outside an industrial plant says that it is OK to proceed, the gate will be opened and the visitor may pass through. The key word is "if." Following the "if" is the condition under which the answer to the question of whether we may proceed is "yes": The person guarding the gate says that it is OK.

It is now appropriate to pause and ask: Under what conditions will the guard say that it is OK to proceed? Suppose in this case it is OK if the visitor is on the list of guests and if the visitor is carrying no cameras or sound recording equipment. It is also OK if the "visitor" happens to be an employee.

This example is a simple instance of *binary logic*. Logic is the use of reason to describe cause and effect. The logic associated with this situation may be systematically described by any of a number of forms. All of them involve the definition of a sufficient number of binary variables—variables that have one of two values: yes or no.

For the purpose of brevity as well as the management of more complicated situations, we define the value 1 to mean "yes" and 0 to mean "no." We may further express the situation that a binary variable has the value 1 as that variable being "high," while that for it being 0, "low." This last interpretation points toward eventual circuit implementation of logic, where "high" might be 3 V and "low" might be 0.1 V.

Let $D_0$ be the binary variable signifying whether the visitor is on the list of guests, $D_1$ signify whether the visitor has cameras or sound recording equipment, $D_2$ represent whether the visitor is actually an employee, and $D_3$ be whether the guard says "OK." We can then make up an exhaustive list of all possibilities, and the respective outcomes, as shown in Table 10.1. Notice that $D_3$ is the *output*, or dependent variable, while the others are *input* or independent binary variables.

Such a table is called a *truth table*. The input variables appear on the left, and the output for a given input combination appears separately on the right. The order of input variables from left to right is of course arbitary, but once fixed, the order of the pattern of input variable 1s and 0s is chosen to follow binary counting. The leftmost input variable is considered the *most significant bit* and the rightmost the *least significant bit*. Recall from Eq. (9.1) that in base 2, an integer is represented as

$$d_{N-1} \cdots d_1 d_0 = \sum_{n=0}^{N-1} d_n 2^n, \tag{10.1}$$

**TABLE 10.1** TRUTH TABLE[a]

| $D_2$ | $D_1$ | $D_0$ | $D_3$ |
|---|---|---|---|
| 0 | 0 | 0 | 0 |
| 0 | 0 | 1 | 1 |
| 0 | 1 | 0 | 0 |
| 0 | 1 | 1 | 0 |
| 1 | 0 | 0 | 1 |
| 1 | 0 | 1 | 1 |
| 1 | 1 | 0 | 1 |
| 1 | 1 | 1 | 1 |

[a] $D_3$ = whether guard opens gate, where $D_0$ = visitor on list of guests, $D_1$ = visitor has cameras or sound recording equipment, $D_2$ = visitor is an employee.

where each of the $N$ binary digits (bits) $d_n$ is either a 1 or a 0. The bit pattern $d_{N-1} \cdots d_1 d_0$ in Eq. (10.1) can represent integers 0 through $2^N - 1$, depending on the values of the bits $d_n$. In Table 10.1, the bit pattern $d_2 d_1 d_0$ when substituted into Eq. (10.1) gives the row number as well as specifies the unique combinations of the values of the logic variables $D_2$, $D_1$, and $D_0$.

**Example 10.1**

Find the value (in base 10) of $10111_2$.

**Solution**

Solution

$$10111_2 = 1 \cdot 2^0 + 1 \cdot 2^1 + 1 \cdot 2^2 + 0 \cdot 2^3 + 1 \cdot 2^4 = 1 + 2 + 4 + 16 = 23.$$

More will be said about binary representation in Sec. 11.6. For now, convince yourself that the row order in Table 10.1 is indeed binary counting, by substituting the digits of successive rows into the right-hand side of Eq. (10.1) to obtain 0, 1, 2, 3, 4, 5, 6, 7. For many cases, the main reason for ordering the rows such that the input value combinations when viewed as a multidigit binary number are ordered according to binary counting is merely convenience. That is, once the pattern of 1s and 0s for binary counting is familiar, we may by habit set up the truth table without forgetting or duplicating any rows.

There is also another reason, however: ease in determining input–output relations. From the table it is easy to organize our thoughts and arrive at the precise condition that $D_3$ is "high" (the guard says "OK"): $D_3$ is "high" if ($D_0$ is "high" AND $D_1$ is NOT "high"), OR if ($D_2$ is "high"). (See also Sec. 10.5).

Notice that AND, OR, and NOT are written in capital letters; these are the primary logical operators. The AND operator has at least two inputs. Its output is high only when all its inputs are high. The OR operator also has at least two inputs. Its output is high only when at least one of its inputs is high. The NOT operator has only one input; its output is the logical opposite of its input. That is, its output is high if its input is low, and its output is low if its input is high.

The relation between $D_3$ and its inputs can be summarized in equation form if the above definitions are introduced. First, however, we look more closely at these operators, including examples of their effects on binary waveforms. Suppose that $D_{\text{AND}}$ is high only when $D_a$ and $D_b$ are high. The truth table for the two-input AND operator is as shown in Table 10.2. Conventionally, the output of the AND operator, $D_{\text{AND}} = \text{AND}(D_a, D_b)$, is written as the logical product of $D_a$ and $D_b$: $D_{\text{AND}} = D_a D_b$.

**TABLE 10.2** TRUTH TABLE FOR *AND* OPERATOR

| $D_a$ | $D_b$ | $D_{\text{AND}} = D_a D_b$ |
|---|---|---|
| 0 | 0 | 0 |
| 0 | 1 | 0 |
| 1 | 0 | 0 |
| 1 | 1 | 1 |

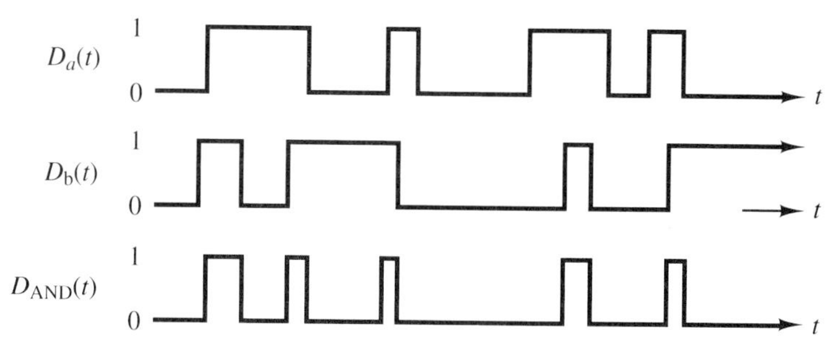


**Figure 10.1** Binary waveforms for Example 10.2 (AND operator).

The logical product is identical to the arithmetic product when, respectively, 1 and 0 are used to signify "high" and "low."

**Example 10.2**

The two binary waveforms $D_a(t)$ and $D_b(t)$ shown in Fig. 10.1 are inputs to an AND operator. Find the output waveform $D_{AND}(t) = D_a(t)D_b(t)$.

**Solution** Look for times when both $D_a(t)$ and $D_b(t)$ are high. These will be the only times when $D_{AND}(t)$ is high. [*Note:* Usually the notation "$(t)$" accompanying binary signals is omitted for brevity.]

Next, suppose that $D_{OR}$ is high when either $D_a$ or $D_b$ is high or both are high. The truth table for the two-input OR operator is as shown in Table 10.3. Conventionally, the output of the OR operator, $D_{OR} = OR(D_a, D_b)$, is written as the logical sum of $D_a$ and $D_b$: $D_{OR} = D_a + D_b$. The logical sum is, for three of the four entries, identical to the arithmetic sum when, respectively, 1 and 0 are used to signify "high" and "low." But for the fourth entry $D_a = D_b = 1$, the logical sum is not 2 (the arithmetic sum), but rather, 1. This is in agreement with our intuitive notion of "or"; "either input high" and "both inputs high" have the same effect—the output is high. Although there is a difference in meaning between the logical and arithmetic sums, there is no chance of confusion, given the context.

In this book we use "$D$" exclusively for binary variables ("$D$" for "digital" or "discrete") both for binary logic and binary arithmetic expressions. Incidentally, the manipulation of binary variables according to the rules of logic is called *Boolean algebra* after George Boole, a pioneer in this type of mathematics working in the 1850s.

**Example 10.3**

Given the same inputs $D_a(t)$ and $D_b(t)$ as in Example 10.2, find $D_{OR}(t) = D_a(t) + D_b(t)$.

**TABLE 10.3** TRUTH TABLE FOR *OR* OPERATOR

| $D_a$ | $D_b$ | $D_{OR} = D_a + D_b$ |
|---|---|---|
| 0 | 0 | 0 |
| 0 | 1 | 1 |
| 1 | 0 | 1 |
| 1 | 1 | 1 |

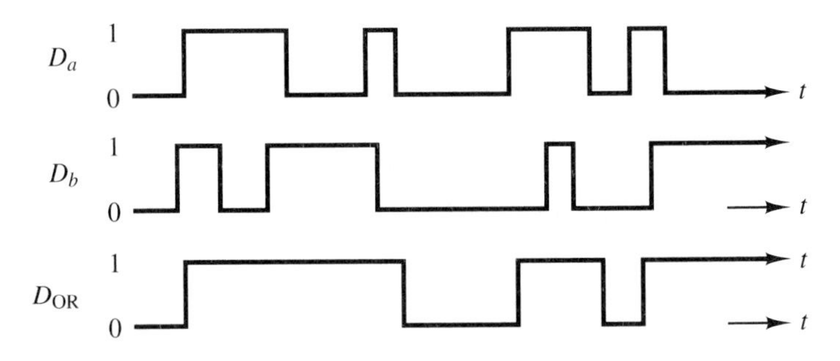


**Figure 10.2** Binary waveforms for Example 10.3 (OR operator).

**Solution** Look for times when either $D_a$ or $D_b$ or both are 1. These will be the times when $D_{OR} = 1$ (see Fig. 10.2). [Henceforth, the "$(t)$" notation will be omitted.]

The truth table for the NOT operator, also known as an inverter, is shown in Table 10.4, where the NOT operation on $D_a$ is written as $\overline{D}_a$: $D_{NOT} = \overline{D}_a$. The overbar indicates logical negation, often simply called the *complement.* Again there is a difference from the arithmetic counterpart: The arithmetic negative of 1 is $-1$, while the logical negative (complement) is 0. Also, the arithmetic negative of 0 is 0, while the logical negative (complement) of 0 is 1. Finally, note that $D_a\overline{D}_a = 0$ because either $D_a$ or $\overline{D}_a$ is always 0.

**Example 10.4**

For $D_a$ as given in Example 10.3, sketch the waveform of $\overline{D}_a$.

**Solution** Just change all 1s to 0s and all 0s to 1s. The waveforms $D_a$ and $\overline{D}_a$ are both shown in Fig. 10.3.

Using these simple definitions, it is now straightforward to express the output binary variable of our gate problem, $D_3$, in terms of $D_0$, $D_1$, and $D_2$:

$$D_3 = D_0\overline{D}_1 + D_2. \tag{10.2}$$

Clearly, the equation form is the most concise representation of the problem stated earlier in words: It is OK for the visitor to pass through the gate ($D_3$) if the visitor is on the list of guests ($D_0$) AND is carrying no cameras or sound recording equipment ($\overline{D}_1$), OR the "visitor" happens to be an employee ($D_2$).

A circuit that implements the NOT operator is called a logical inverter and has the symbol shown in Fig. 10.4. The triangle represents a gate that passes its input to its output unchanged, like the op amp voltage follower. The bubble on its output is the symbol for logical negation, *wherever* it may appear in a digital circuit diagram.

**TABLE 10.4** TRUTH TABLE FOR *NOT* OPERATOR

| $D_a$ | $D_{NOT} = \overline{D}_a$ |
|---|---|
| 0 | 1 |
| 1 | 0 |

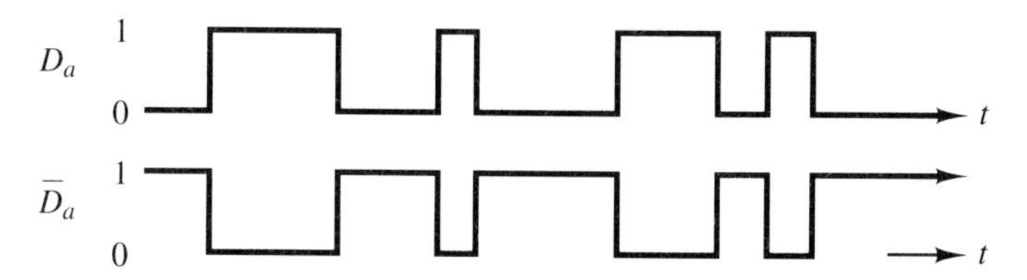


**Figure 10.3** Binary waveforms for Example 10.4 (NOT operator).

Similar types of circuitry to those used to implement the NOT operator can be used to implement the AND operator. The symbol for the AND operator is shown in Fig. 10.5. Also, the OR operator has a symbol, shown in Fig. 10.6. These circuits are called logic *gates*. In light of our example about the visitor to an industrial plant, it should be evident why the name "gate" is used. A circuit implementation of the expression for $D_3$ is shown in Fig. 10.7; this concludes the industrial gate design example.

**Example 10.5**

Often in computer systems, we wish to ignore some inputs and pay attention to others. This is done by means of a *mask*. If an element of the mask is 1, the input to that mask element—be it 0 or 1—is allowed to pass on through to the output. If the mask element is 0, the input data are blocked; the output is 0 regardless of the input. A four-line mask is shown in Fig. 10.8. How could we implement this mask using logic gates?

**Solution** The mask can be implemented using AND gates, as shown in Fig. 10.9. [In practical computers, the 1s and 0s of the masks (the fixed inputs to the AND gates) would instead be binary contents of memory elements that could be changed to achieve desired behavior for different situations.]

**Example 10.6**

The switch to connect the motor of a paint-stirring system is driven by a relay which in turn is directed by a binary on–off signal. Suppose it is desired that the switch can be on only if the main power switch is turned on and if it is not true either that there is paint in the stirring tank (which would be old paint) or that the motor has been left in gear, in which case the initial load on the motor is too great.

Let $D_0$ signify that there is old paint left in the tank, $D_1$ indicate that the clutch is in gear, $D_2$ signify that the main power switch is turned on, and $D_3$ be the go-ahead signal to start the motor. Find the truth table, a binary logic equation for $D_3$ in terms of $D_0$, $D_1$, and $D_2$, and a logic circuit implementing the system.

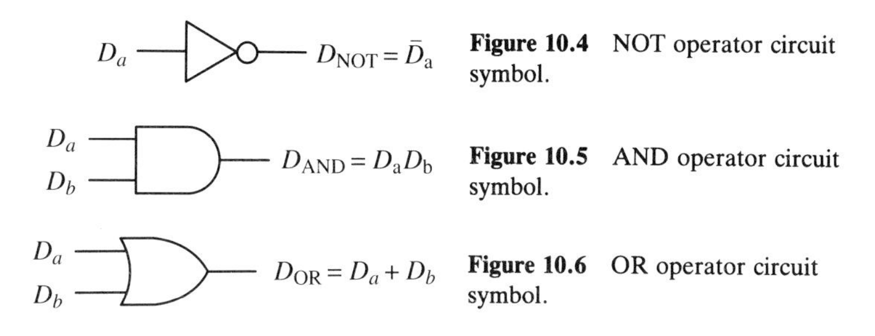


**Figure 10.4** NOT operator circuit symbol.

**Figure 10.5** AND operator circuit symbol.

**Figure 10.6** OR operator circuit symbol.

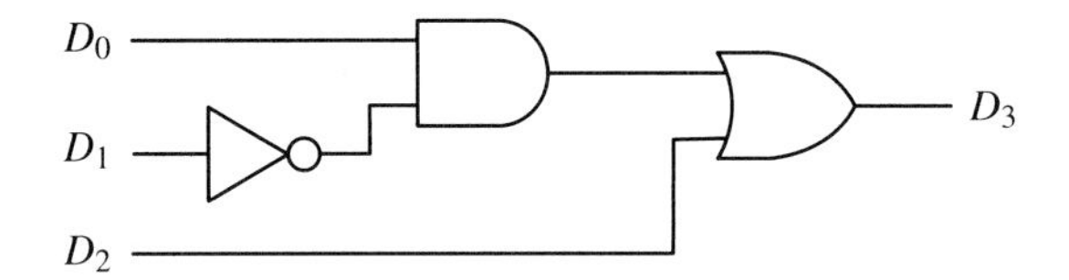


**Figure 10.7** Logic circuit implementation of Table 10.1 (visitor at the gate).

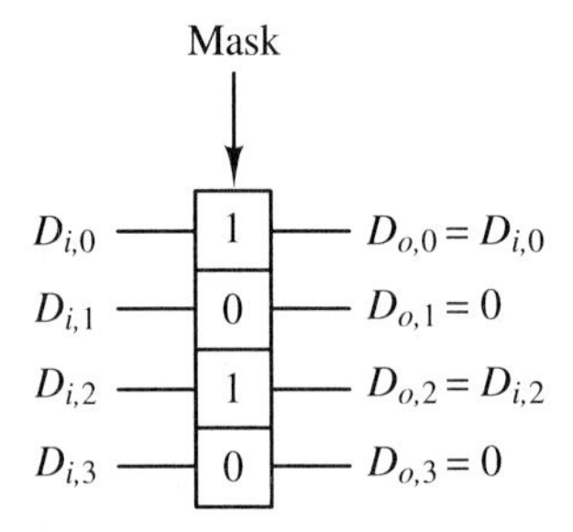


**Figure 10.8** Binary variable mask for Example 10.5.

**Solution** All of the results follow intuitively. The truth table is given by Table 10.5. From the word description of this problem, the logic equation is simply $D_3 = D_2[\overline{D_1 + D_0}]$. The most straightforward circuit implementation is the direct translation of the equation into gates. Thus $D_3$ is the output of an AND gate whose inputs are $D_2$ and the output of the logical complement of an OR gate whose inputs are $D_1$ and $D_0$. The complement of the OR gate is represented by a single gate called a NOR gate (short for NOT OR). The two-input NOR gate $D_{\text{NOR}} = \text{NOR}(D_a, D_b) = \overline{D_a + D_b}$ has the truth table shown in Table 10.6.

The NOR gate has the circuit symbol that is just an OR gate with a logical negation bubble at the output (see Fig. 10.10). As noted above, a bubble anywhere in a logic circuit indicates logical negation. With this symbol for the NOR gate, a circuit implementing the stirring system start signal is given in Fig. 10.11.

**Example 10.7**

For the same $D_a$ and $D_b$ input waveforms as in the previous timing examples, find $D_{\text{NOR}} = \overline{D_a + D_b}$.

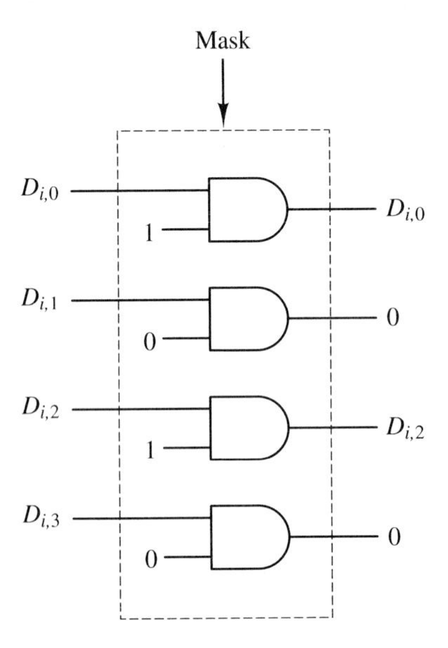


**Figure 10.9** Logic circuit implementation of mask in Fig. 10.8.

**TABLE 10.5** TRUTH TABLE FOR EXAMPLE 10.6[a]

| $D_2$ | $D_1$ | $D_0$ | $D_3$ |
|---|---|---|---|
| 0 | 0 | 0 | 0 |
| 0 | 0 | 1 | 0 |
| 0 | 1 | 0 | 0 |
| 0 | 1 | 1 | 0 |
| 1 | 0 | 0 | 1 |
| 1 | 0 | 1 | 0 |
| 1 | 1 | 0 | 0 |
| 1 | 1 | 1 | 0 |

[a] $D_3$ = start motor, where $D_0$ = paint left in tank, $D_1$ = clutch in gear, $D_2$ = main power switch on.

**TABLE 10.6** TRUTH TABLE FOR *NOR* OPERATOR

| $D_a$ | $D_b$ | $D_{\text{NOR}}$ |
|---|---|---|
| 0 | 0 | 1 |
| 0 | 1 | 0 |
| 1 | 0 | 0 |
| 1 | 1 | 0 |

**Solution** Just invert the $D_{\text{OR}}$ waveform that was found in Example 10.2. Alternatively, look for times when both $D_a$ and $D_b$ are low—these will be the only times when $D_{\text{NOR}}$ is high. The waveform $D_{\text{NOR}}$ appears in Fig. 10.12; $D_a$ and $D_b$ are repeated for reference.

As we might suspect, just as there is the NOR operator, there is also a NAND operator (NOT AND), which is the complement of AND: $D_{\text{NAND}} = \text{NAND}(D_a, D_b) = \overline{D_a D_b}$, which has the truth table appearing in Table 10.7. The circuit symbol for the NAND gate appears in Fig. 10.13.

**Example 10.8**

For the same $D_a$ and $D_b$ waveforms as before, find $D_{\text{NAND}} = \overline{D_a D_b}$.

**Solution** Look for times when both inputs are high. These are the only times when $D_{\text{NAND}}$ is low. The waveforms $D_a$, $D_b$, and $D_{\text{NAND}}$ are all given in Fig. 10.14.

Incidentally, another simple implementation of the logical function $D_3$ for the paint stirring example results from writing $D_3$ from its truth table as $D_3 = D_2 \overline{D}_1 \overline{D}_0$, which can be implemented with a three-input AND gate and NOT gates as shown in Fig. 10.15. We shall soon clarify the logical rules involved in ensuring that such different looking circuits (those in Figs. 10.11 and 10.15) have the same output.

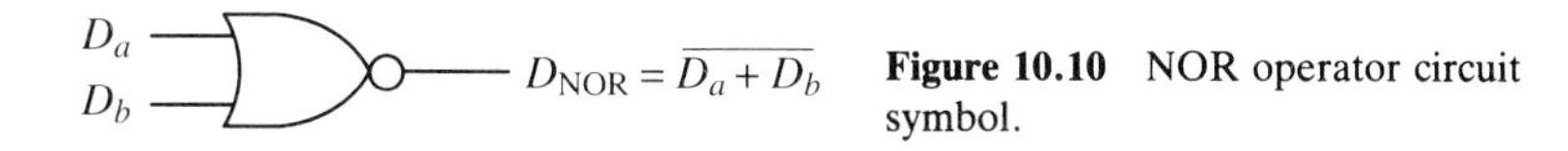


**Figure 10.10** NOR operator circuit symbol.

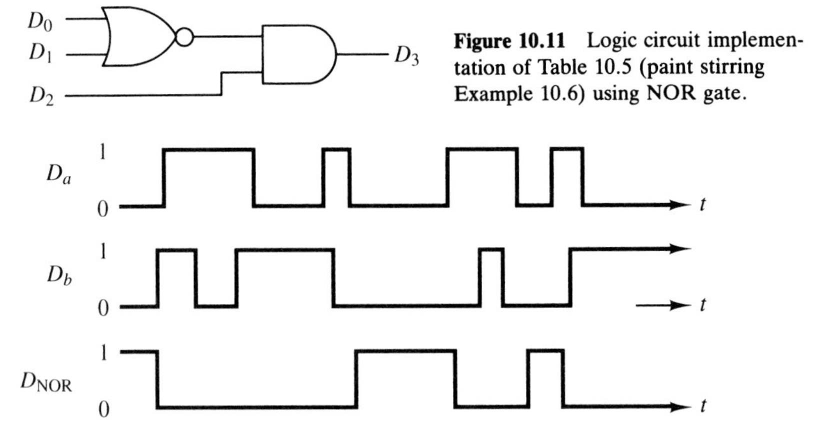


**Figure 10.11** Logic circuit implementation of Table 10.5 (paint stirring Example 10.6) using NOR gate.

**Figure 10.12** Binary waveforms for Example 10.7 (NOR operator).

## 10.3 THE RULES OF LOGIC AND FOOTBALL

Another example will help indicate the kinds of thinking involved in digital design. Every autumn, the NFL professional football league has 16 regular season games followed by two playoffs. The winner of the second playoff goes on to the Super Bowl to contest for the national championship.

In this example we consider the prospects for the Cleveland Browns to enter the playoffs, just before the last game of the regular season. Thus all teams have played 16 games. A win is considered 1 point, a loss 0, and a tie $\frac{1}{2}$ point (these are game result points and do not refer to the scores in individual games). Suppose that the situation is as follows. The Browns' win–loss–tie record is 9–5–1, so their points total is 9.5. The Patriots' record is 9–4–2 or 10.0 points. These two teams are closest in records, so assume that it will be one of them that becomes a wild card team in the playoffs.

(In each of the two major conferences, the AFC and the NFC, there are three divisions: east, central, and west, each composed of four or five teams. The wild card teams in each conference are the division nonchampions within the given conference who have the three best records. So even though they are not division champs, the wild card teams still get to enter the playoffs.)

**TABLE 10.7** TRUTH TABLE FOR *NAND* OPERATOR

| $D_a$ | $D_b$ | $D_{NAND}$ |
|---|---|---|
| 0 | 0 | 1 |
| 0 | 1 | 1 |
| 1 | 0 | 1 |
| 1 | 1 | 0 |

$D_a$, $D_b$ → $D_{NAND} = \overline{D_a D_b}$

**Figure 10.13** NAND operator circuit symbol.

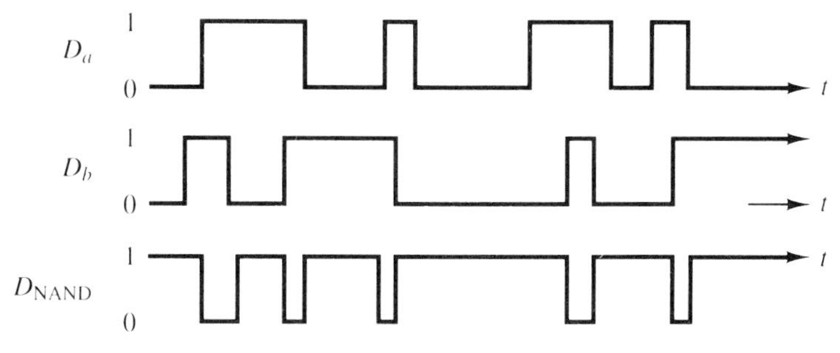


**Figure 10.14** Binary waveforms for Example 10.8 (NAND operator).

Our interest is only in the fate of the Browns. If the Browns lose their last game, it is over for them no matter what the Patriots do. If the Browns tie and the Patriots lose, then tie-breaking conditions would determine who goes to the playoffs. However, for simplicity we will omit this possibility, and assume that only if the Browns win their last game is there hope for them. If they win (and therefore finish with 10.5), the Patriots must either lose (so that their total is 10.0) or tie *and* lose the tiebreaking conditions compared with the Browns, now to be given.

In real pro football, there are eight levels of tiebreaking conditions. For our purposes, we consider only three. That is, if two teams, such as the Browns and Patriots, have the same final score, the decision will be made as follows. Whoever has the better won–lost–tied percentage record in the head-to-head games (Browns versus Patriots) will enter the playoffs. But if that record is a tie, whoever has the better won–lost–tied percentage record in the games played within the conference will enter the playoffs. If the Browns and Patriots are also equal in this sense, the winner of a coin toss will enter the playoffs (we have omitted the third level of comparison through the seventh).

We begin to solve this problem by breaking the problem into subproblems and defining the binary variables. Let $D_{Bp}$ be 1 if the Browns will go to the playoffs—it is the output binary variable. Let $D_{Bw}$ be 1 if the Browns win their last game and 0 if they either tie or lose, $D_{Bt}$ be 1 if the Browns tie their last game and 0 otherwise (so, obviously, $D_{Bw}$ and $D_{Bt}$ cannot both be 1, and if they lose—no win and no tie—then $\overline{D}_{Bt}\overline{D}_{Bw} = 1$).

Similarly, let $D_{Pw}$ and $D_{Pt}$ be 1 if the Patriots win or tie, respectively, while $D_{Pl} = \overline{D}_{Pt}\overline{D}_{Pw}$ is 1 if they lose. Let $D_t$ be the result of the tiebreaker determination should it be necessary: If the Browns win the tiebreaker, it is 1; 0 otherwise. Below, $D_t$ will be dissected according to its definition above. Finally, let $D_{Pt,t} = D_{Pt}D_t$, which is 1 if the Patriots tie in their last game and they lose the tiebreaker conditions.

All of the conditions can be enumerated in tabular form via the truth table, Table 10.8. Notice that because a team cannot both tie and lose a game, lines 4 and 8 of the truth table never occur. The output function can then be chosen to form

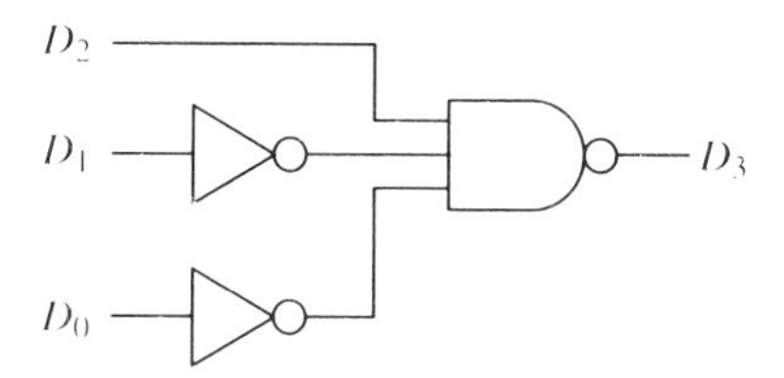


**Figure 10.15** Alternative logic circuit implementation of Table 10.5 (paint stirring Example 10.6) using NAND gate.

**TABLE 10.8** TRUTH TABLE[a]

| $D_{Bw}$ | $D_{Pl}$ | $D_{Pt,t}$ | $D_{Bp}$ | |
|---|---|---|---|---|
| 0 | 0 | 0 | 0 | |
| 0 | 0 | 1 | 0 | |
| 0 | 1 | 0 | 0 | |
| 0 | 1 | 1 | × | |
| 1 | 0 | 0 | 0 | |
| 1 | 0 | 1 | 1 | Browns win and Patriots tie but Patriots lose tiebreaker conditions |
| 1 | 1 | 0 | 1 | Browns win and Patriots lose |
| 1 | 1 | 1 | × | |

[a] $D_{Bp}$ = Browns go to the playoffs, where $D_{Bw}$ = Browns win their last game; $D_{Pl}$ = Patriots lose their last game; $D_{Pt,t}$ = Patriots tie in their last game and lose the tiebreaker conditions involving the Browns.

either 0 or 1 for those conditions—whichever is easiest for us to do. Such circumstances are referred to as *don't care* conditions and are marked $X$ in Table 10.8.

The fact that there are three possible outcomes (win, lose, or tie) requires two binary lines for every event, which makes this example challenging. Also, we have "unused" possibilities because there are four ($2^2$) possible sets of values for the two binary variables available to represent only three possibilities. This is a common situation in digital design.

From the truth table, the 1s can be read off to form an equation for the output. The output, $D_{Bp}$, must be 1 if the inputs are either 101 or 110:

$$D_{Bp} = D_{Bw}(\overline{D}_{Pl}\overline{D}_{Pt,t} + D_{Pl}\overline{D}_{Pt,t}). \tag{10.3}$$

The term in parentheses is an example of the "exclusive or" operator $D_{\text{XOR}} = \text{XOR}(D_a, D_b) = D_a\overline{D}_b + \overline{D}_a D_b$, which is 1 if precisely one (not both) of its inputs is 1, and which has the truth table given in Table 10.9. The circuit symbol for the XOR operator is given in Fig. 10.16.

**Example 10.9**

For the inputs $D_a$ and $D_b$ used in previous examples, find $D_{\text{XOR}} = D_a\overline{D}_b + \overline{D}_a D_b$.

**Solution** Simply look for all times when one, and only one, of the inputs is 1. These will be the times when $D_{\text{XOR}} = 1$. All relevant waveforms are given in Fig. 10.17.

Although the XOR function can be used to implement Eq. (10.3), there exists a simpler alternative in this case. Because of the *don't care* input 111, Eq. (10.3) can

**TABLE 10.9** TRUTH TABLE FOR *XOR* OPERATOR

| $D_a$ | $D_b$ | $D_{\text{XOR}}$ |
|---|---|---|
| 0 | 0 | 0 |
| 0 | 1 | 1 |
| 1 | 0 | 1 |
| 1 | 1 | 0 |

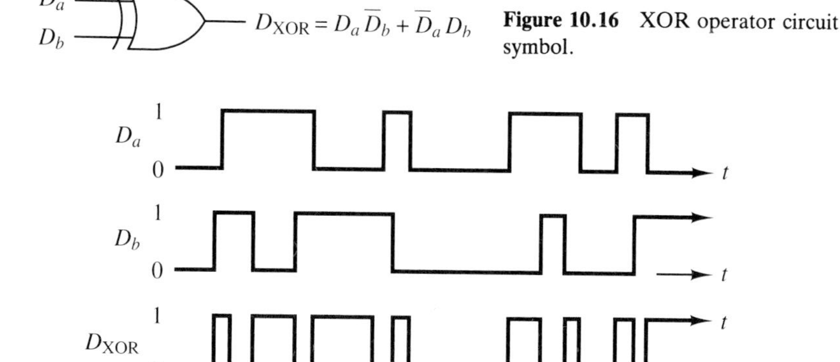


**Figure 10.16** XOR operator circuit symbol.

**Figure 10.17** Binary waveforms for Example 10.9 (XOR operator).

be simplified. If the output for the impossible 111 condition is for convenience set to 1, Eq. (10.3) becomes

$$D_{Bp} = D_{Bw}(D_{Pl} + D_{Pt,t}). \tag{10.4}$$

Another simplification can be made regarding $D_{Pl} = \overline{D}_{Pt}\overline{D}_{Pw}$. The truth table appearing in Table 10.10 is helpful. From our previous discussion, Table 10.10 indicates that $D_{Pl} = \text{NOR}(D_{Pt}, D_{Pw}) = \overline{D}_{Pt}\overline{D}_{Pw}$. We thus have the useful general relation

$$\overline{D_a + D_b} = \overline{D}_a\overline{D}_b, \tag{10.5}$$

which we tacitly made use of in the second realization of the paint stirring problem (Fig. 10.15). Thus the circuit diagram can at this point be drawn as shown in Fig. 10.18.

Equation (10.5) is a particular case of one of *DeMorgan's laws* of binary logic, named after the mathematician Augustus DeMorgan. The general form is

$$\overline{D_1 + D_2 + D_3 + \cdots + D_N} = \overline{D}_1\overline{D}_2\overline{D}_3 \cdot \ldots\ldots \cdot \overline{D}_N, \tag{10.6a}$$

or symbolically,

$$\overline{\sum_{n=1}^{N} D_n} = \prod_{n=1}^{N} \overline{D}_n, \tag{10.6b}$$

**TABLE 10.10** TRUTH TABLE[a]

| $D_{Pt}$ | $D_{Pw}$ | $D_{Pl}$ |
|---|---|---|
| 0 | 0 | 1 |
| 0 | 1 | 0 |
| 1 | 0 | 0 |
| 1 | 1 | 0 |

[a] $D_{Pl}$ = Patriots lose their last game, where $D_{Pt}$ = Patriots tie that game and $D_{Pw}$ = Patriots win that game.

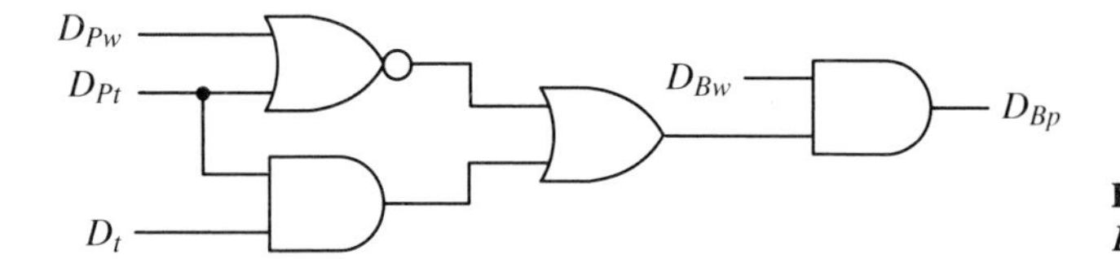


**Figure 10.18** Logic circuit diagram for $D_{Bp}$ (Browns go to playoffs).

where Π is the symbol for product. This equation merely states that NORing $N$ signals ($N \geq 2$) will produce 1 only if each of the inputs is 0, that is, if the AND of all of the logically negated inputs is 1.

Just as KVL and KCL are dual electric circuit laws, there is a law of binary logic that is dual to Eq. (10.6), except that in this case the two laws are just different expressions for the same law. This dual form, DeMorgan's other law, can be obtained from Eq. (10.6) by (1) replacing all variables $D_i$ by $\overline{D}_i$ (and therefore also all $\overline{D}_m$ by $D_m$), and (2) taking the logical complement of both sides of the equation. The result is

$$\overline{D_1 D_2 D_3 \cdot \ldots \cdot D_N} = \overline{D}_1 + \overline{D}_2 + \overline{D}_3 + \cdots + \overline{D}_N, \tag{10.7a}$$

written symbolically as

$$\overline{\prod_{n=1}^{N} D_n} = \sum_{n=1}^{N} \overline{D}_n. \tag{10.7b}$$

Equation (10.7) says that if you NAND together $N$ signals and get 1, at least one of the inputs had to be 0; that is, ORing the logically negated inputs must give 1. Both of these simple and intuitive laws are extremely useful in digital design for simplifying logical equations and reducing the number of gates required in an implementation of a logical function.

**Example 10.10**

Express $\overline{D}_a + \overline{D}_b + D_c$ in terms of the NAND function.

**Solution** By DeMorgan's law, Eq. (10.7), $\overline{D}_a + \overline{D}_b + D_c = \overline{D_a D_b \overline{D}_c} =$ NAND $(D_a, D_b, \overline{D}_c)$.

Returning to our football example, all that still needs to be determined is $D_t$, which is 1 if the Browns win the tiebreaking conditions and is 0 otherwise. Let $D_{B\mathrm{HH}w}$ be 1 if the Browns have the better record in head-to-head games with the Patriots, and $D_{\mathrm{HH}t}$ be 1 if they are tied on that count. Similarly define $D_{B\mathrm{CG}w}$ to be 1 if the Browns have a better in-conference games record than the Patriots and $D_{\mathrm{CG}t}$ be 1 if their in-conference records are identical. Finally, let $D_{B\mathrm{CT}}$ be 1 if the Browns win the coin toss that would occur in the event of head-to-head and conference game tied records.

For simplicity, suppose for the moment that there is only the head-to-head condition, and if that is a tie, the coin is tossed. The truth table for $D_t$ would then be Table 10.11. Again the last two entries are don't cares because no one can both tie and win any contest. Letting the don't cares be 1 is again convenient, for it then gives the tiebreaker $D_t$ as

$$\begin{aligned} D_t &= D_{B\mathrm{HH}w} + \overline{D}_{B\mathrm{HH}w} D_{\mathrm{HH}t} D_{B\mathrm{CT}} \\ &= D_{B\mathrm{HH}w} + D_{\mathrm{HH}t} D_{B\mathrm{CT}}. \end{aligned} \tag{10.8}$$

**TABLE 10.11** TRUTH TABLE[a]

| $D_{BHHw}$ | $D_{HHt}$ | $D_{BCT}$ | $D_t$ |
|---|---|---|---|
| 0 | 0 | 0 | 0 |
| 0 | 0 | 1 | 0 |
| 0 | 1 | 0 | 0 |
| 0 | 1 | 1 | 1 |
| 1 | 0 | 0 | 1 |
| 1 | 0 | 1 | 1 |
| 1 | 1 | 0 | × |
| 1 | 1 | 1 | × |

[a] $D_t$ = Browns win tiebreaking conditions, where $D_{BHHw}$ = Browns have better record in head-to-head games with Patriots; $D_{HHt}$ = Browns tied in head-to-head games with Patriots; $D_{BCT}$ = Browns win coin toss if coin toss occurs.

In Eq. (10.8), the general relation

$$D_a + \overline{D}_a D_b = D_a + D_b \tag{10.9}$$

has been used. This *absorption law* merely uses the fact that requiring $D_a$ to be 0 in the second term is superfluous because the entire expression is *always* 1 if $D_a$ is 1 (whether or not $D_b$ is 1). Other absorption laws, which can easily be proven with truth tables, are

$$D_a D_a = D_a \tag{10.10}$$

$$D_a(D_a + D_b) = D_a \tag{10.11}$$

$$D_a(\overline{D}_a + D_b) = D_a D_b \tag{10.12}$$

$$D_a D_b + D_a \overline{D}_b = D_a, \tag{10.13}$$

as well as obvious identities such as $D_a + 0 = D_a$, $D_a + 1 = 1$, $D_a 0 = 0$, and $D_a 1 = D_a$.

## 10.4 MULTIPLEXERS/DEMULTIPLEXERS, ENCODERS/DECODERS

Now suppose that if head-to-head records are tied, the conference game records are checked before resorting to the coin toss. This time the number of binary variables is 5, so the truth table would be 32 lines long—a bit tedious. Instead, notice that $D_t$ can be determined from the shortened truth table, Table 10.12, with variables instead of 1s and 0s for output variable entries.

From Table 10.12, the expression for $D_t$ is readily found to be

$$D_t = \overline{D_{CGt}}\,\overline{D_{HHt}}\,D_{BHHw} + \overline{D_{CGt}}\,D_{HHt}\,D_{BCGw} + D_{CGt}\,\overline{D_{HHt}}\,D_{BHHw} + D_{CGt}\,D_{HHt}\,D_{BCT}. \tag{10.14}$$

The straightforward circuit implementation of the *sum of products* form in Eq. (10.14) is shown in Fig. 10.19. With $D_t$ fully determined, it may be spliced into

**TABLE 10.12** TRUTH TABLE[a]

| $D_{CGt}$ | $D_{HHt}$ | $D_t$ |
|---|---|---|
| 0 | 0 | $D_{BHHw}$[b] |
| 0 | 1 | $D_{BCGw}$[c] |
| 1 | 0 | $D_{BHHw}$[b] |
| 1 | 1 | $D_{BCT}$[d] |

[a] $D_t$ = Browns win tiebreaker conditions in terms of $D_{CGt}$ = Browns and Patriots tied in their conference game records, $D_{HHt}$ = Browns and Patriots tied in head-to-head games.

[b] If HH (head-to-head) records are not tied, the only way the Browns can win the tiebreaker is if the Browns have a better HH record. If the Browns lose on HH, conference game records are not considered—the Patriots go to the playoffs.

[c] If HH records are tied and CG (conference game) records are not, the only way the Browns can win the tiebreaker is if the Browns have a better CG record.

[d] If CG and HH records are both tied, the tiebreaker is determined by coin toss.

Fig. 10.18, thus completing our football playoffs example. However, we still have a bit more to say about the circuit in Fig. 10.19.

The circuit in Fig. 10.19 is known as a 4 : 1 *multiplexer* or MUX. The 4 : 1 characterization indicates that from four inputs come one output; note the four AND gates. A multiplexer is a logical device that on the basis of a small number of *control*

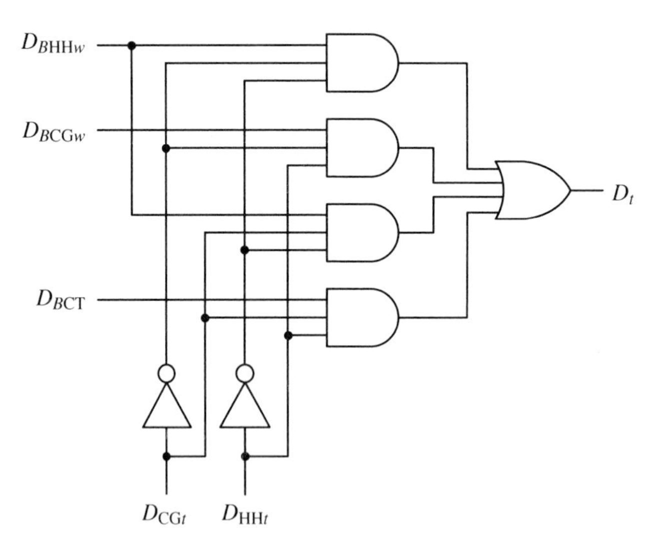


**Figure 10.19** Logic circuit implementation of Table 10.12 (Browns win tiebreaker conditions).

lines, selects one of several possible *data* inputs to be transferred to the output. This is precisely what is done in Table 10.12.

Here the *control lines* are $D_{\mathrm{CG}t}$ and $D_{\mathrm{HH}t}$, and the *data lines* are $D_{\mathrm{HH}w}$, $D_{B\mathrm{CG}w}$, and $D_{B\mathrm{CT}}$. On the basis of the values of $D_{\mathrm{CG}t}$ and $D_{\mathrm{HH}t}$, one of $D_{\mathrm{HH}w}$, $D_{B\mathrm{CG}w}$, and $D_{B\mathrm{CT}}$ is selected to be the output, $D_t$. In this case, note that two of the four data lines are identical, namely those with $D_{B\mathrm{HH}w}$. Thus in this case only one of three lines is selected by $D_{\mathrm{CG}t}$ and $D_{\mathrm{HH}t}$.

Just as each gate has a symbol, the multiplexer has a standard symbol, as shown in Fig. 10.20. The generic symbols for the input data lines are $D_0$, $D_1$, $D_2$, and $D_3$, while the control lines are labeled $D_{S0}$ and $D_{S1}$ and the output is $D_Y$. The letters inside correspond to the binary variable symbols typically used by the manufacturers of the given chip and found on their data sheets.

One extra feature in the standard multiplexer not shown in Fig. 10.19 is an *enable line,* $D_E$. In certain digital circuits, particularly "sequential" or dynamic digital systems, it may be appropriate to produce the logic combination making up $D_t$ at only certain times. For example, the data lines or control lines may have valid inputs only after a certain time has elapsed since a change in the system. If the output were continuously available, including times when the inputs were not valid, all sorts of problems or *glitches* could adversely affect the rest of the circuit. Or, there may be several MUXs, only one of whose outputs is desired at any given time. The enable provides this selection.

In practical MUXs, the complement of the enable is presented as a fourth input to the internal AND gates, which are now four-input AND gates. When $D_E = 0$, the enable line presents a 1 to each of the AND gates, and therefore does not affect the output. But if the enable line is 1, the outputs of all the AND gates will be 0 regardless of the other control and data signals. This is exactly the same idea as used in the "mask" Example 10.5. In some realizations, the output when disabled is held at a high-impedance state rather than at 0.

Because the circuit is considered functional when $D_E$ is *low,* the enable input to the MUX is called *asserted low* and labeled $\overline{D}_E$ rather than $D_E$. The enable line of a given device is not always asserted low, but for the purpose of illustration was chosen to be for this example. When $\overline{D}_E$ is high, the MUX is enabled.

Note also that because the bubble is placed on the outside of the device, the letter "$E$" rather than "$\overline{E}$" is placed on the inside. We can think that once the

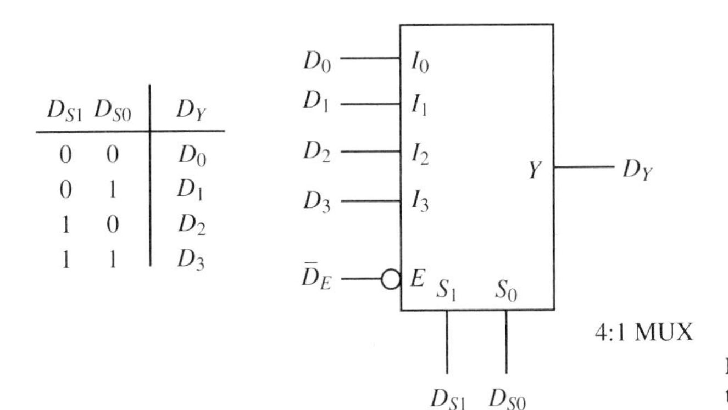


| $D_{S1}$ | $D_{S0}$ | $D_Y$ |
|---|---|---|
| 0 | 0 | $D_0$ |
| 0 | 1 | $D_1$ |
| 1 | 0 | $D_2$ |
| 1 | 1 | $D_3$ |

**Figure 10.20** 4 : 1 Multiplexer truth table and circuit symbol.

negation is done, a normal enable function exists within the MUX. This convention for denoting asserted low is pretty much standard today. Throughout digital circuitry, asserted low signals are used for convenience and/or gate minimization. Although it sometimes complicates matters, we must get used to it because it is reality.

However, the important concept to be grasped here is that a slight bit of extra complexity has resulted in the introduction of a new tool, in this case the multiplexer. What began as some *combinational logic*—just implementation of a logical expression, Eq. (10.14)—ended up as a control/data structure. This is typical of digital circuitry: Certain combinations of the basic gates bring forth entirely new capabilities. Other examples we shall soon encounter are memory and arithmetic computation.

*Note:* Surprisingly, in data sheets specifying logic symbols and pin assignments, there is no universally held convention on numbering of bit significance. That is, sometimes the order of binary digits agrees in meaning with Eq. (10.1), and other times it is the reverse!

Forest of telegraph wires on Broadway and John Streets, New York, 1890. To dramatically reduce the required number of lines, the multiplexing concept has been applied. For analog systems, in frequency division multiplexing each caller modulates a different frequency, all on the same line. In today's digital systems, predominantly time division multiplexing (implemented by devices such as the multiplexer in Fig. 10.20) is used for toll calls, where one sample from each caller is sent at a time, in sequence, allowing many simultaneous calls per line. For analog local calls, however, the one-message-per-line idea is still commonly used. *Source:* Courtesy of AT&T Archives.

**TABLE 10.13** TRUTH TABLE FOR EXAMPLE 10.11 (MULTIPLEXER EXAMPLE)

| $D_c$ | $D_b$ | $D_a$ | $D_o$ |
|---|---|---|---|
| 0 | 0 | 0 | 0 |
| 0 | 0 | 1 | 1 |
| 0 | 1 | 0 | 0 |
| 0 | 1 | 1 | 0 |
| 1 | 0 | 0 | 0 |
| 1 | 0 | 1 | 1 |
| 1 | 1 | 0 | 1 |
| 1 | 1 | 1 | 1 |

**Example 10.11**

Implement $D_o = D_a\overline{D}_b + D_cD_b$ using an eight-input multiplexer.

**Solution** The easiest way is to fill in a truth table as given in Table 10.13. Then the inputs to the eight data lines $I_m$ in Fig. 10.21 are the values of $D_o$ for each particular combination of values of the actual inputs $D_a$, $D_b$, and $D_c$ for a given row of the truth table. The multiplexer output will therefore be $D_o$.

It should come as no surprise that there is a dual to the multiplexer: The demultiplexer. Instead of selecting one signal from several input signals and sending that to the output, the demultiplexer sends a single input signal to one of several output lines. An obvious application would be the selection of one of several output devices to which to send the input signal. In general, the number of possible input (output) lines for the multiplexer (demultiplexer) is $2^m$ for $m$ control signals. This is because $2^m$ is the number of possible combinations of 1s and 0s for a set of $m$ binary variables; this also determines the number of rows in a truth table.

If the control lines are again $D_{S0}$ and $D_{S1}$ and the output lines are $D_0$, $D_1$, $D_2$, and $D_3$, the input data $D_{\text{in}}$ is sent to the appropriate output line as indicated in Table

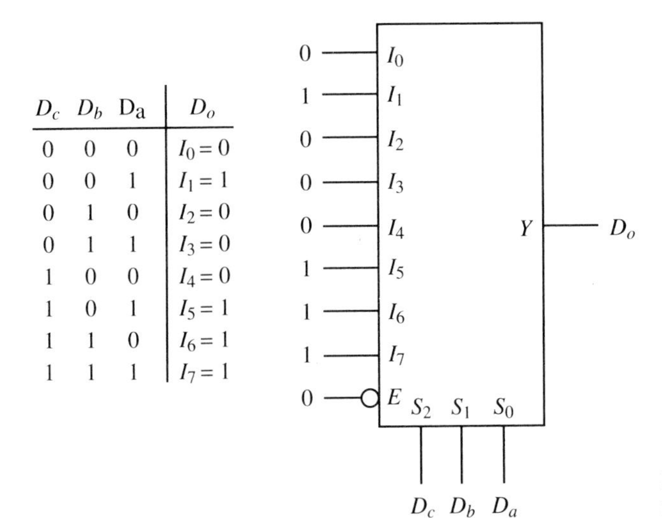


**Figure 10.21** Connections of multiplexer for Example 10.11.

**TABLE 10.14** TRUTH TABLE FOR DEMULTIPLEXER[a]

| $D_{S0}$ | $D_{S1}$ | $D_{in}$ appears on: |
|---|---|---|
| 0 | 0 | $D_0$ |
| 0 | 1 | $D_1$ |
| 1 | 0 | $D_2$ |
| 1 | 1 | $D_3$ |

[a] $D_0$ through $D_3$ are output lines, $D_{S0}$ and $D_{S1}$ are control lines, and $D_{in}$ is the input binary data.

10.14. A simple implementation of this behavior requires four 3-input AND gates and two inverters. A 1 : 4 demultiplexer is shown in Fig. 10.22. The simplified circuit symbol for a demultiplexer appears in Fig. 10.23. Note that for this application the "enable" input is actually used as the data input.

The concept here is again the same as that of the "mask" Example 10.5 in our initial study of the AND operator, but here there is only one input data line. Also, now only one of the outputs is enabled at any one time, and its output will be equal to the single data input. Contrarily, the mask allowed several outputs to be enabled simultaneously, each with independent input data lines.

An extension of the multiplexer/demultiplexer idea is that of the *encoder/decoder*. Consider the $N:1$ multiplexer. Instead of choosing one of $N$ data lines to form a single output, from the $N$ inputs form $M$ outputs, where $1 < M \leq N$. The $M$ lines are considered "encoded" because the data have taken a different form from that of the inputs, and often the number of lines has been reduced. If it is reduced, and if the mapping is to be reversible, not all possibilities in the input data lines must be used or be significant. That is, there are only $2^M$ possibilities for the coded data. To reverse the mapping, the decoder is used, which has $M$ inputs and $N$ outputs, where again $1 < M \leq N$.

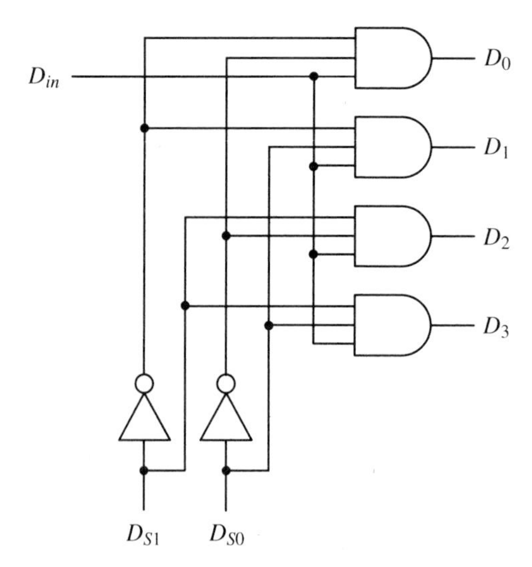


**Figure 10.22** Circuit implementation of demultiplexer.

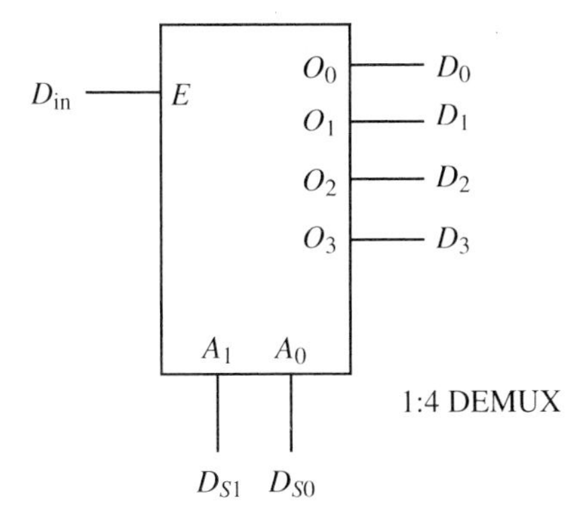


**Figure 10.23** 1 : 4 Demultiplexer logic circuit symbol.

**Example 10.12**

In truth tables, the rows are always ordered according to binary counting: The binary combination in the next row is that of the current row plus 1. When binary counting is to be transmitted, it is often desirable that only one binary "digit" change at a time. Discuss a system that accomplishes this.

**Solution** Suppose that there are three binary variables. Clearly, when adding 1 to 001 to obtain 010, two variables have changed value simultaneously. If this is undesirable, the *Gray code* can be used instead of the binary code to detect data transmission errors easily. The Gray code is obtainable from the binary code (*encoding*) by reproducing the leftmost input variable in the leftmost output variable and then, two by two, XORing the rest of the way from left to right. The XOR outputs are then the Gray code. The operation is shown in Table 10.15. Thus a basic three-variable binary-to-Gray encoder is shown in Fig. 10.24.

To go back from Gray to binary (i.e., *decode*), reproduce the leftmost variable of the input Gray code in the output binary code. Then again going from left to right, XOR successive resulting binary digits with the current Gray code digit to get the next binary digit. Satisfy yourself that this scheme works. This circuit is given in Fig. 10.25.

Unlike the universal applicability of multiplexers and demultiplexers, the structures of encoders and decoders depend entirely upon the application. They often are manufactured to perform a variety of conversions between number representations. The most common and important of these conversions is the binary-

**TABLE 10.15** TRUTH TABLE FOR ENCODING OF DATA $D_2D_1D_0$ TO THE GRAY CODE $D_cD_bD_a$

| Binary code $D_2\,D_1\,D_0$ | Gray code $D_c\,D_b\,D_a$ |
|---|---|
| 000 | 000 |
| 001 | 001 |
| 010 | 011 |
| 011 | 010 |
| 100 | 110 |
| 101 | 111 |
| 110 | 101 |
| 111 | 100 |

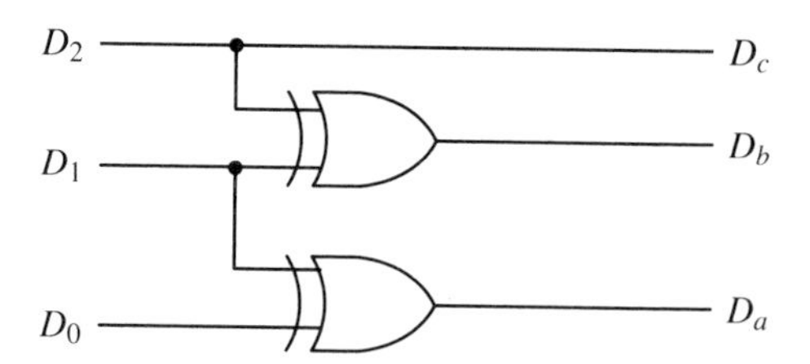


**Figure 10.24** Gray code encoder logic circuit diagram.

coded decimal (BCD, discussed in Sec. 11.6) or hexadecimal-to-seven segment display/decimal selection (see the problems) used in driving alphanumeric displays.

A decoder can also be used to select one of many output lines just as a demultiplexer does, but there is no data line that is routed; the selected line addressed by the inputs is held fixed at 1 while all the others are held at 0 or are disconnected. Thus the decoder is often the source of the "chip enable" inputs of other devices.

**Example 10.13**

Suppose that each memory location for number storage in a computer is given an address number. How might we gain access to the memory cell?

**Solution** A decoder is the solution, as indicated in Fig. 10.26. Suppose that there are 16 memory elements. This is, of course, for illustration; in practice, there may be thousands or millions. The necessary decoder in this example is called either a 4 : 16 line decoder [$4 = \log_2(16)$ input lines control the 16 output lines] or a 1-of-16 decoder (indicating that a 1 appears on only one of the 16 outputs at any given time). The 1 that appears at a specific memory location determined by $D_3 D_2 D_1 D_0$ will be used as an enable for a "read" or "write" operation. We shall study digital memory systems further in Sec. 11.7. This example should help show the practical necessity of decoders.

One interesting and useful variation on the encoder is the priority encoder. A typical 8-to-3 priority encoder is shown in Fig. 10.27. Suppose the encoder was to be used in a digital system as a device request-for-services handler. To each of eight devices is assigned one of the eight "request" lines; when a device wants services (e.g., use of a data transmission line), it "raises its hand" and sets its line high. On the output of the encoder is the binary code corresponding to that device. The device having this address will obtain access to the main data line.

A problem can arise when two or more devices call for services at the same time. Which code will appear on the output? The priority encoder uses the following rule: The line of the highest index ($D_0$ through $D_7$) gets its "address" on the output. Such a rule at first seems arbitrary. But such a rule allows the digital designer to set up priorities.

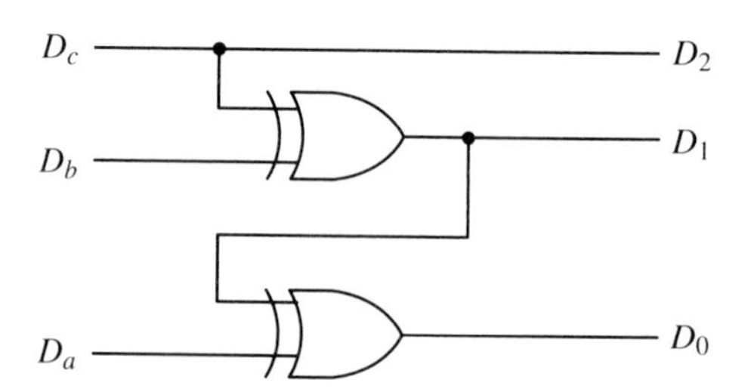


**Figure 10.25** Gray code decoder logic circuit diagram.

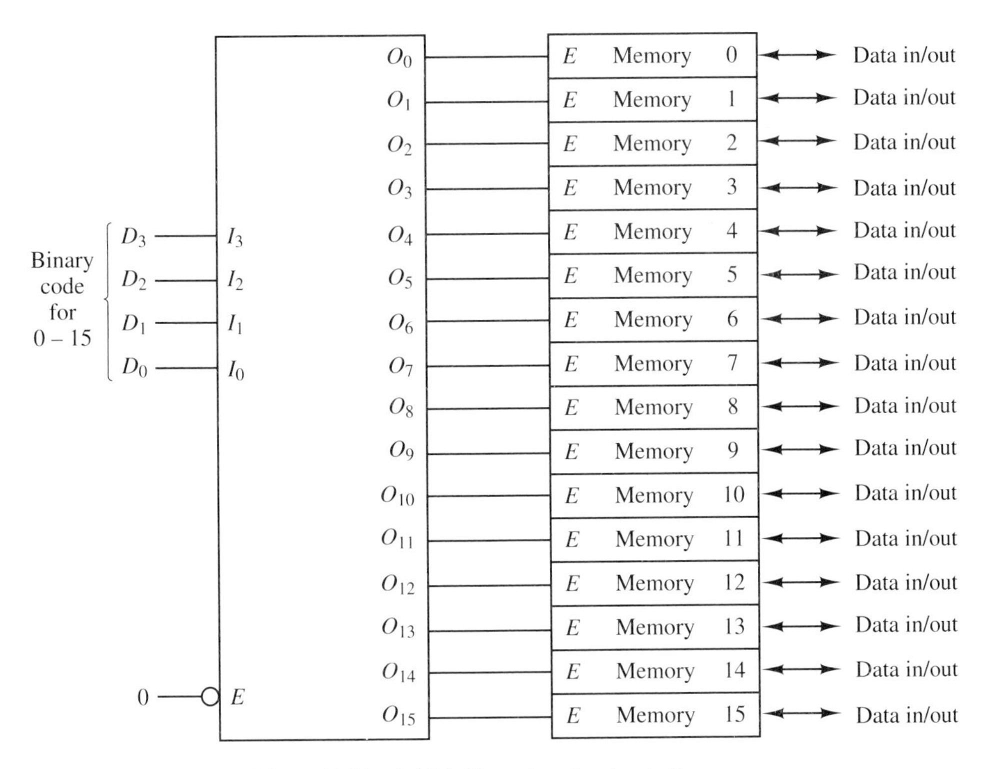


**Figure 10.26** 4 : 16 Address decoder circuit diagram.

For example, suppose that the main data line is connected to some memory systems, some line printers, and some high-quality plotters. Most likely, the memory systems would have higher priority to use the main data line than would the line printer, and the line printer would have higher priority than the plotter. The plotter can wait. But if some data is not written to memory, it could be lost if a power emergency should occur. So, for example, the designer might assign the memory system to $D_7$, the line printer to $D_4$, the plotter to $D_2$, and other devices with other priorities to the other lines. If $D_7$ is high, its address will appear on the output regardless of whatever $D_i$ may be on. If $D_7$ through $D_5$ are low and $D_4$ is high, then whatever the values of $D_0$ through $D_3$, the address of $D_4$ appears on the output.

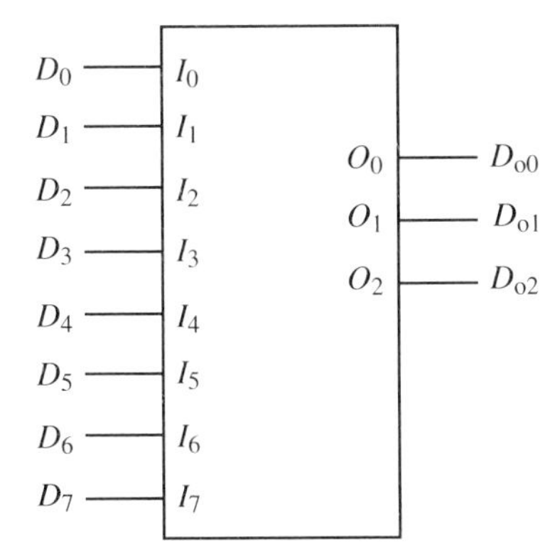


**Figure 10.27** 8 : 3 Priority encoder logic circuit symbol.

## 10.5 ANALYSIS AND REDUCTION PROCEDURES

It is worthwhile at this point to pause for a moment to consider a few design principles and methods. To begin, there are two main forms by which to express conveniently any relation involving binary variables: the *sum of products* (SOP) and the *product of sums* (POS). Equation (10.14) is an example of SOP and Eq. (10.4) is a simplified example of POS [where in Eq. (10.4) one of the multiplicands is just a variable, not a sum].

From the previous studies, it should be evident that there are many ways to express a given function of binary variables, just as is true for continuous variable algebra. And just as the laws of algebra often help simplify continuous variable expressions, DeMorgan's laws, absorption laws, and others such as commutative and distributive laws can simplify binary variable logical expressions (Boolean algebra). The sum of products and product of sums forms are related by DeMorgan's laws. SOP expressions have the form

$$\sum_{m=1}^{M} \prod_{n=1}^{N} d_{m,n}, \tag{10.15}$$

where $d_{m,n}$ is either $D_n$, $\overline{D}_n$, or 1 (in which case the variable is not included in that product term). The corresponding POS expression has the form

$$\overline{\prod_{m=1}^{M} \sum_{n=1}^{N} \overline{d}_{m,n}}. \tag{10.16}$$

Thus, equating of Eqs. (10.15) and (10.16), which makes use of both of DeMorgan's laws, gives

$$\sum_{m=1}^{M} \prod_{n=1}^{N} d_{m,n} = \overline{\prod_{m=1}^{M} \sum_{n=1}^{N} \overline{d}_{m,n}} \tag{10.17}$$

$$\text{SOP} = \overline{\text{corresponding POS}},$$

where the "corresponding" expression involves the logical complements of each of the variables in the original expression. Incidentally, each product term in the SOP expression is called a *minterm,* and each sum term in the POS expression is called a *maxterm.*

**Example 10.14**

Find the product-of-sums expression for $D_3 = D_0 D_1 \overline{D}_2 + \overline{D}_0 \overline{D}_1$. Draw both the SOP and POS circuit implementations of $D_3$.

**Solution** We could step through using DeMorgan's first law, Eq. (10.6), by noting that $\overline{\overline{D}}_a = D_a$. Double-negating the given SOP expression gives

$$\begin{aligned} D_0 D_1 \overline{D}_2 + \overline{D}_0 \overline{D}_1 &= \overline{\overline{D_0 D_1 \overline{D}_2 + \overline{D}_0 \overline{D}_1}} \\ &= \overline{(\overline{D_0 D_1 \overline{D}_2})(\overline{\overline{D}_0 \overline{D}_1})} \\ &= \overline{(\overline{D}_0 + \overline{D}_1 + D_2)(D_0 + D_1)}, \end{aligned}$$

where in the second step Eq. (10.6a) is used with $D_0 D_1 \overline{D}_2$ and $\overline{D}_0 \overline{D}_1$ considered as single binary variables, and in the last step where DeMorgan's other law, Eq. (10.7),

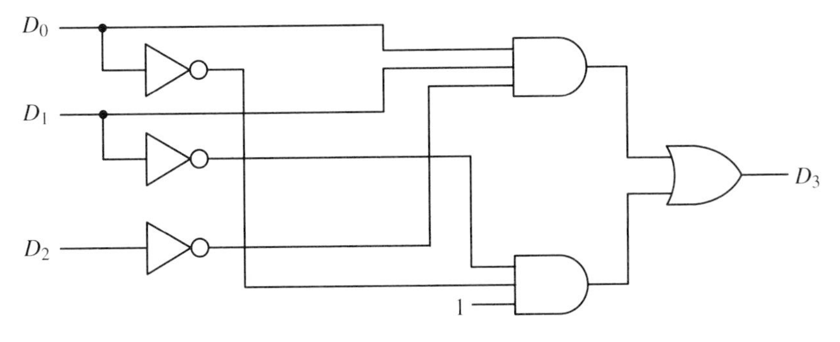


**Figure 10.28** SOP logic circuit implementation of $D_3$ in Example 10.14.

is used. More directly, in Eq. (10.17) set $M = 2$, $N = 3$, $d_{1,1} = D_0$, $d_{1,2} = D_1$, $d_{1,3} = \overline{D}_2$, $d_{2,1} = \overline{D}_0$, $d_{2,2} = \overline{D}_1$, and $d_{2,3} = 1$; the result immediately follows from the right-hand side of Eq. (10.17). The circuit directly formed from the given SOP representation of $D_3$ is given in Fig. 10.28, while that of the POS found above appears in Fig. 10.29.

Note that in the SOP diagram (Fig. 10.28), for the purposes of illustration a three-input AND gate was used for only two inputs: $\overline{D}_0$ and $\overline{D}_1$. The third input is tied high to 1 ($d_{2,3} = 1$). In a hardware situation where available gates have more than the desired number of inputs, the others must be tied to appropriate levels. Convince yourself that the choices indicated in Table 10.16 are appropriate in that they do not affect the intended function.

Example 10.14 has shown how to convert an equation from SOP form to POS form; the same procedure could be used to do the reverse. In many circumstances,

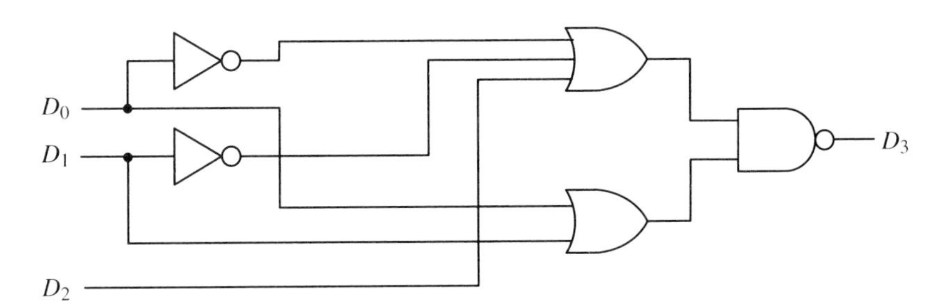


**Figure 10.29** POS logic circuit implementation of $D_3$ in Example 10.14.

**TABLE 10.16**[a]

| Function | Tie unused inputs |
|---|---|
| AND | 1 |
| OR | 0 |
| NAND | 1 |
| NOR | 0 |
| XOR | 0 |

[a] For gates with three or more inputs, this shows what to do with any unused inputs so as not to affect the output function.

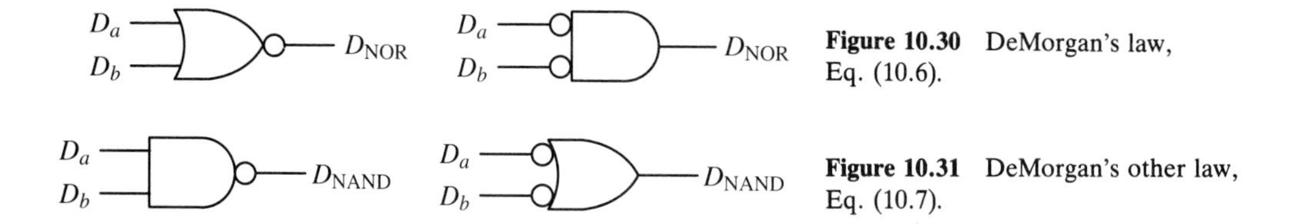


**Figure 10.30** DeMorgan's law, Eq. (10.6).

**Figure 10.31** DeMorgan's other law, Eq. (10.7).

we have a circuit diagram that is neither purely SOP or POS. We may apply a simple procedure to convert to, for example, SOP form. From DeMorgan's law, Eq. (10.6), we know that NOR$(D_a, D_b) = \overline{D}_a\overline{D}_b$, which in circuit form appears as shown in Fig. 10.30. From DeMorgan's law, Eq. (10.7), NAND$(D_a, D_b) = \overline{D}_a + \overline{D}_b$, which in circuit form is given in Fig. 10.31.

Another tool used above and frequently in practice is the double-negation identity

$$\overline{\overline{D}}_a = \text{NOT}(\text{NOT}(D_a)) = D_a. \tag{10.18}$$

In a circuit, this seemingly unlikely combination often appears as a negation bubble at the output of one gate (such as a NOR or NAND gate) sent as input to another gate having a negation bubble at its input. Figure 10.32 presents an example of this, where the circuit of Fig. 10.32a is redrawn in Fig. 10.32b using the result in Fig. 10.31. Because of Eq. (10.18), Fig. 10.32b can be equivalently drawn as shown in Fig. 10.33. Together, these ideas can be used to convert combinational logic as appearing in a given circuit diagram into a desired form.

Incidentally, in Fig. 10.32b, why was the uncomplemented $D_c$ input into a bubble here, while for the 4 : 1 MUX example studied in Sec. 10.4 (Fig. 10.20), the *complement* of the enable signal, $\overline{D}_E$, was shown connected to the enable bubble at the entrance to the chip? The reason is that in the case of the MUX, the enable line is predetermined to be asserted when low. We must arrange the input to satisfy this. Contrastingly, the bubble at the lower OR gate input in Fig. 10.32b is merely a consequence of the use of a Boolean algebra transformation (DeMorgan's law) applied to the original circuit, which used the uncomplemented $D_c$. If the same function is to be performed after the transformation as was before, the same inputs must be applied after the transformation as were before.

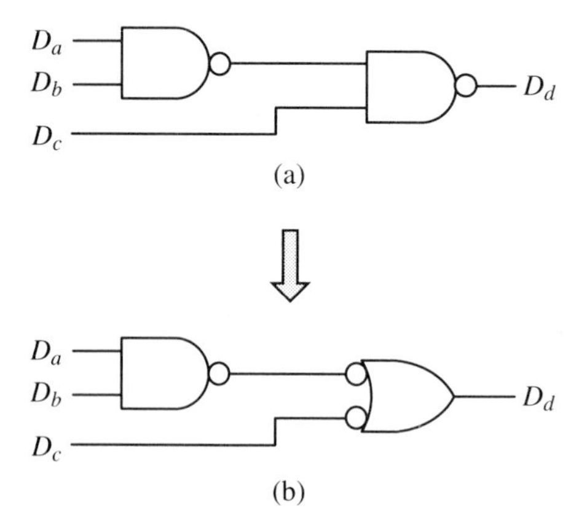


**Figure 10.32** Double-negation principle (see also Fig. 10.33): (a) original circuit; (b) use of DeMorgan's law, Eq. (10.7) (see Fig. 10.31).

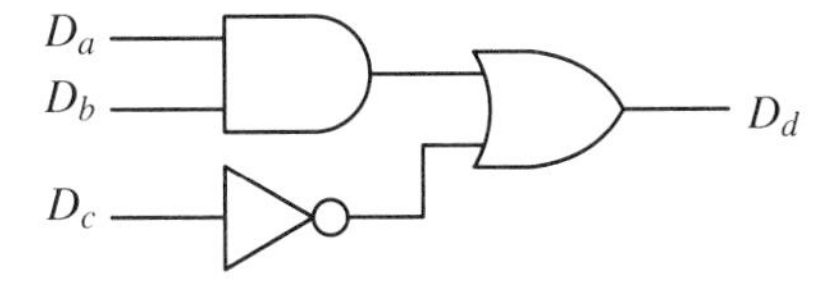


**Figure 10.33** Simplification of circuit of Fig. 10.32b using double-negation principle (and inverter).

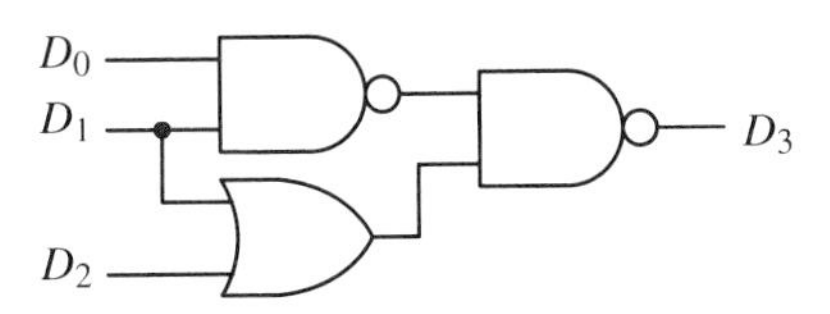


**Figure 10.34** Two-level logic circuit for Example 10.15 (to be converted to SOP form).

**Example 10.15**

Convert the *two-level* digital circuit of Fig. 10.34 to an equivalent one in SOP form.

**Solution** Notice that because of the mixed form, the expression for $D_3$ is not immediately evident. To obtain SOP form, recall that the "sum" is of "products," so the products are to be formed first. Thus in the first level, convert any (N)OR gates to (N)AND gates to result in the products. If the output gate is an (N)AND gate, it must be converted to an (N)OR to yield the sum (see Fig. 10.35a).

When completed, remove any double negatives (as we should always do in English). Also, convert the remaining bubbles at the inputs of gates to inverters, or push the bubbles back to the previous output and make that gate NOR or NAND. The completed circuit is shown in Fig. 10.35b. The final SOP expression can now be read directly from this diagram; it is $D_3 = D_0 D_1 + \overline{D}_1 \overline{D}_2$.

If there are more than two "levels" of gates, we may decompose the network into sections like the above and do the same kind of analysis in two-level stages.

Of considerable interest and convenience is the fact that NOR gates alone and NAND gates alone are capable of implementing *any* combinational logic functions. It was shown above that AND and OR gates are interchangeable provided the appropriate negation bubbles are also present. The only other function necessary is NOT. Suppose that the two inputs of the NAND gate are tied together, to $D_a$. Then if $D_a$ is 1, the output will be zero, and if $D_a$ is 0, the output will be 1. Thus

$$\text{NAND}(D_a, D_a) = \overline{D}_a, \tag{10.19}$$

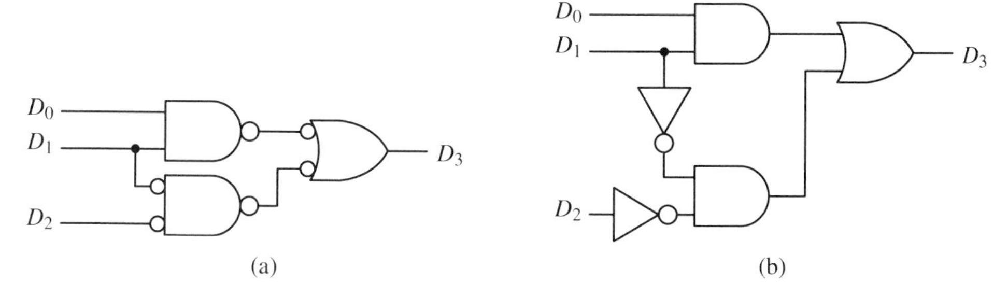


**Figure 10.35** (a) Conversion of sum at first level of Fig. 10.34 to a product using Fig. 10.30 and negation. Also, the product at the second level of Fig. 10.34 is converted to a sum using Fig. 10.31. (b) Simplification of (a) using double negation and inverters.

which may also be seen from examination of the first and fourth rows of the NAND truth table, the rows of identical inputs. The output is the complement of the identical inputs. By similar reasoning,

$$\text{NOR}(D_a, D_a) = \overline{D}_a, \tag{10.20}$$

which can also be seen by noting that the first and fourth rows of the NAND and NOR truth tables are identical.

Because of this versatility of NAND and NOR gates, it is often convenient to convert to "all-NAND" or "all-NOR" representations. Often, there are four NAND gates per chip ("quad NAND"). If all gates are the same, the required number of chips can be as low as one-fourth the number of gates. In fact, it is desirable to have minimal variety of gate types to avoid connection errors, and this may take precedence over the simplicity of the expression. Generally, what happens to be available in the component layout locally often dictates the final form of expression. Although theoretically any digital system can be implemented using only NAND gates, for a variety of reasons that is often not practical.

Which is more convenient to use: NAND or NOR? Often, NAND gates are used because they arise naturally in the implementation of SOP expressions, and SOPs are often more intuitive to form and understand than are POSs. Consider the following example.

**Example 10.16**

Implement $D_o = D_a D_b + \overline{D}_a \overline{D}_c + D_a \overline{D}_b D_d + D_c \overline{D}_d$ using NAND gates.

**Solution** First draw the AND/OR (direct) logic diagram as indicated in Fig. 10.36. Now negate *both* ends of each line between the AND gates and the OR gate. This does not change the value of the output; this circuit is given in Fig. 10.37a.

Next, recognize that the OR gate with inverted inputs is equivalent to a NAND gate. The NAND implementation shown in Fig. 10.37b follows naturally. As you might guess, the POS expression leads naturally to an all-NOR implementation.

When space and component count minimization are important, as is usually the case, the simplification of logical expressions can be beneficial. Although in the above examples several simplifications were pointed out as they arose, no systematic,

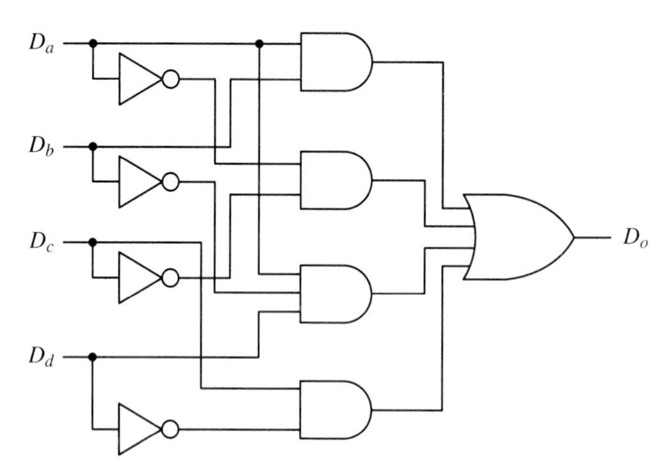


**Figure 10.36** Straightforward SOP implementation of $D_o$ in Example 10.16.

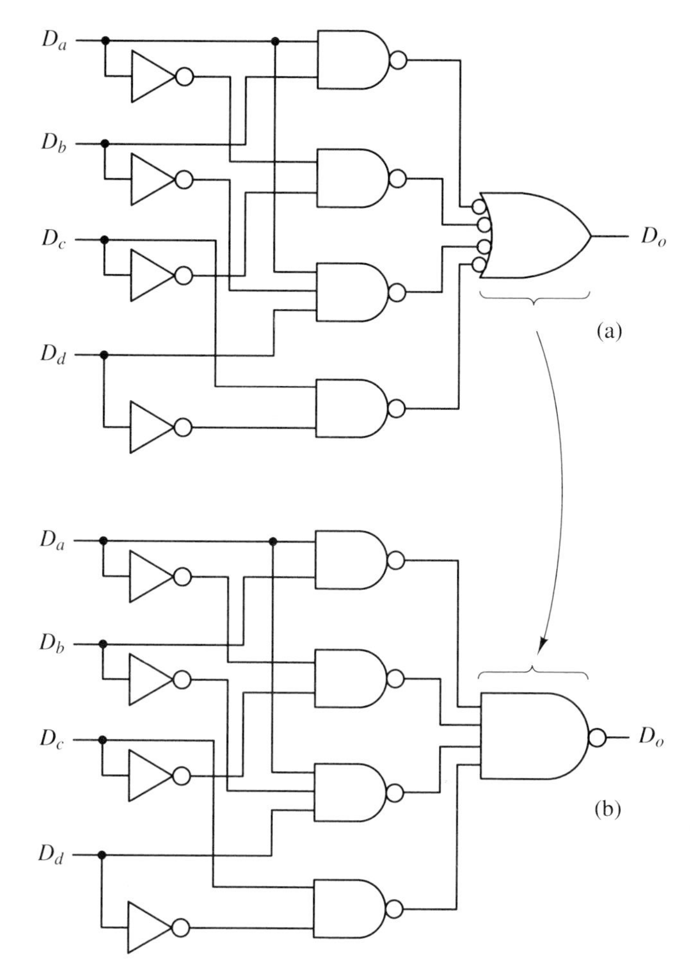


**Figure 10.37** Conversion of the circuit of Fig. 10.36 to an all-NAND implementation: (a) first the inputs of the final stage are negated, which requires also negation of NAND outputs; (b) Fig. 10.31 is used to convert the output NOR of (a) to a NAND gate.

general procedure was given. One such procedure involves the *Karnaugh map,* after M. Karnaugh, who introduced the map method in 1953. First the construction of the Karnaugh map is discussed. Then its use in logical expression minimization is considered.

The Karnaugh map is really nothing more than a particularly illuminating form of a truth table. Most uses of Karnaugh maps are for two to four input variables. Given in Fig. 10.38 are the Karnaugh maps for one, two, three, and four input variables from the corresponding truth table outputs.

The horizontal or vertical "width" indicated for each of the input variables indicates the slice of the Karnaugh map throughout which that input variable is 1; elsewhere it is 0. In the map element location having a particular unique combination of inputs is placed the corresponding output from the truth table. For example, in the four-variable truth table, the map element location having $D_a = 1$, $D_b = 0$, $D_c = 1$, and $D_d = 1$ is located in the third row and fourth column of the Karnaugh map. The value placed in this map element location (either 1 or 0) will be taken as the output variable value from the corresponding row of the truth table: row 13. The thirteenth row is selected because that is the row having $D_a = 1$, $D_b = 0$, $D_c = 1$,

Truth Table

| $D_a$ | $D_{out}$ |
|---|---|
| 0 | $D_0$ |
| 1 | $D_1$ |

Karnaugh Map

| | $D_a$ |
|---|---|
| $D_0$ | $D_1$ |

(a)

Truth Table

| $D_b$ | $D_a$ | $D_{out}$ |
|---|---|---|
| 0 | 0 | $D_0$ |
| 0 | 1 | $D_1$ |
| 1 | 0 | $D_2$ |
| 1 | 1 | $D_3$ |

Karnaugh Map

| | | $D_a$ |
|---|---|---|
| | $D_0$ | $D_1$ |
| $D_b$ | $D_2$ | $D_3$ |

(b)

Truth Table

| $D_c$ | $D_b$ | $D_a$ | $D_{out}$ |
|---|---|---|---|
| 0 | 0 | 0 | $D_0$ |
| 0 | 0 | 1 | $D_1$ |
| 0 | 1 | 0 | $D_2$ |
| 0 | 1 | 1 | $D_3$ |
| 1 | 0 | 0 | $D_4$ |
| 1 | 0 | 1 | $D_5$ |
| 1 | 1 | 0 | $D_6$ |
| 1 | 1 | 1 | $D_7$ |

Karnaugh Map

| | | | $D_a$ | |
|---|---|---|---|---|
| | $D_0$ | $D_2$ | $D_3$ | $D_1$ |
| $D_c$ | $D_4$ | $D_6$ | $D_7$ | $D_5$ |
| | | $D_b$ | | |

(c)

Truth Table

| $D_c$ | $D_c$ | $D_b$ | $D_a$ | $D_{out}$ |
|---|---|---|---|---|
| 0 | 0 | 0 | 0 | $D_0$ |
| 0 | 0 | 0 | 1 | $D_1$ |
| 0 | 0 | 1 | 0 | $D_2$ |
| 0 | 0 | 1 | 1 | $D_3$ |
| 0 | 1 | 0 | 0 | $D_4$ |
| 0 | 1 | 0 | 1 | $D_5$ |
| 0 | 1 | 1 | 0 | $D_6$ |
| 0 | 1 | 1 | 1 | $D_7$ |
| 1 | 0 | 0 | 0 | $D_8$ |
| 1 | 0 | 0 | 1 | $D_9$ |
| 1 | 0 | 1 | 0 | $D_{10}$ |
| 1 | 0 | 1 | 1 | $D_{11}$ |
| 1 | 1 | 0 | 0 | $D_{12}$ |
| 1 | 1 | 0 | 1 | $D_{13}$ |
| 1 | 1 | 1 | 0 | $D_{14}$ |
| 1 | 1 | 1 | 1 | $D_{15}$ |

Karnaugh Map

| | | | $D_a$ | | |
|---|---|---|---|---|---|
| | $D_0$ | $D_2$ | $D_3$ | $D_1$ | |
| | $D_4$ | $D_6$ | $D_7$ | $D_5$ | $D_c$ |
| $D_d$ | $D_{12}$ | $D_{14}$ | $D_{15}$ | $D_{13}$ | $D_c$ |
| $D_d$ | $D_8$ | $D_{10}$ | $D_{11}$ | $D_9$ | |
| | | $D_b$ | | | |

(d)

**Figure 10.38** General truth tables and corresponding Karnaugh maps: (a) one input variable; (b) two input variables; (c) three input variables; (d) four input variables.

and $D_d = 1$. We may think of the input data values as coordinates for the Karnaugh map cells.

Notice the (unavoidably) irregular pattern of map element location numbering in the three- and four-variable cases. The reason for the strange ordering is that when moving from cell to cell, only one variable changes value (check this), similar to the Gray code property. This characteristic of Karnaugh maps is known as *logic adjacency*. Mapping theory shows that when the groupings using the Karnaugh map are made, redundancy is eliminated and we end up with a minimal SOP representation of the original logical expression. In fact, the Karnaugh map method automatically uses Eqs. (10.9) through (10.13) to best advantage, so that when done, the result is as simple as possible.

**Example 10.17**

Construct the Karnaugh map for the truth tables given in Fig. 10.39. Using the given tables as a model and using the outer "band marks" to find the right map element locations, the Karnaugh maps are as shown.

**Solution** Solution:
The solutions are given beside the truth tables.

Alternatively, the information may have been given as an SOP formula that has not yet been minimized (i.e., simplified). For example, the following SOP expressions may have been given instead of (but equivalent to) the truth tables:

Example 10.17a:

$$D_{\text{out}} = \overline{D}_c D_b \overline{D}_a + D_c \overline{D}_b \overline{D}_a + D_c \overline{D}_b D_a + D_c D_b D_a \tag{10.21a}$$

Example 10.17b:

$$\begin{aligned} D_{\text{out}} = {} & D_b D_c D_d + \overline{D}_a \overline{D}_b \overline{D}_d + \overline{D}_a D_b \overline{D}_c \overline{D}_d + D_a \overline{D}_b \overline{D}_c \overline{D}_d \\ & + \overline{D}_a D_b D_c \overline{D}_d + D_a D_b D_c \overline{D}_d + \overline{D}_a \overline{D}_b \overline{D}_c D_d \\ & + D_a \overline{D}_c D_d + \mathrm{D}_a \overline{D}_b D_c D_d. \end{aligned} \tag{10.21b}$$

It should be emphasized that the minterms specify where on both the truth table and Karnaugh map the output is 1. For all uncomplemented input variables specified in the minterm assign 1s and assign 0s for complemented input variables appearing in the minterm. The simultaneous specification of the values of all the input binary variables locates a 1 for the output variable for the corresponding row of the truth table or cell of the Karnaugh map.

What about when one or more of the input variables is missing from a given minterm? For simplicity, denote the minterm by $D_x$ and the missing variable in question by $D_y$. Then because $D_x = D_x D_y + D_x \overline{D}_y$, 1s are entered on all truth table lines for which $D_x = 1$ for both $D_y = 1$ and $D_y = 0$. Thus a minterm missing any variables indicates that the expression has been at least partially simplified. In general, 1s are entered for all truth table rows specified by inputs satisfying $D_x$—regardless of the values of input variables not specifically included in that minterm.

Notice that the SOP expression in Eq. (10.21b) has already been partly simplified, and the fully minimized expression is sought. The more variables in each minterm, and the more minterms, the more likely it is that the expression can be simplified using Karnaugh maps.

Now that the Karnaugh map has been constructed, use can be made of its

Truth Table

| $D_c$ | $D_b$ | $D_a$ | $D_{out}$ |
|---|---|---|---|
| 0 | 0 | 0 | 0 |
| 0 | 0 | 1 | 1 |
| 0 | 1 | 0 | 0 |
| 0 | 1 | 1 | 0 |
| 1 | 0 | 0 | 1 |
| 1 | 0 | 1 | 1 |
| 1 | 1 | 0 | 0 |
| 1 | 1 | 1 | 1 |

Karnaugh Map

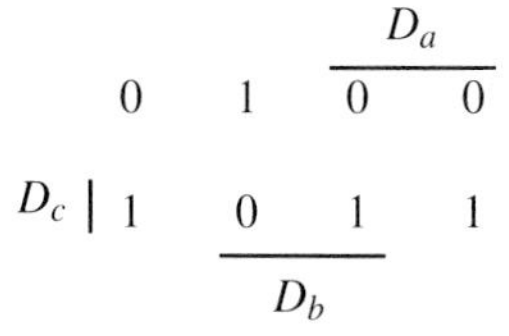

| | | | $D_a$ | $D_a$ |
|---|---|---|---|---|
| | 0 | 1 | 0 | 0 |
| $D_c$ | 1 | 0 | 1 | 1 |
| | | $D_b$ | $D_b$ | |

(a)

Truth Table

| $D_c$ | $D_c$ | $D_b$ | $D_a$ | $D_{out}$ |
|---|---|---|---|---|
| 0 | 0 | 0 | 0 | 1 |
| 0 | 0 | 0 | 1 | 1 |
| 0 | 0 | 1 | 0 | 1 |
| 0 | 0 | 1 | 1 | 0 |
| 0 | 1 | 0 | 0 | 1 |
| 0 | 1 | 0 | 1 | 0 |
| 0 | 1 | 1 | 0 | 1 |
| 0 | 1 | 1 | 1 | 1 |
| 1 | 0 | 0 | 0 | 1 |
| 1 | 0 | 0 | 1 | 1 |
| 1 | 0 | 1 | 0 | 1 |
| 1 | 0 | 1 | 1 | 1 |
| 1 | 1 | 0 | 0 | 0 |
| 1 | 1 | 0 | 1 | 1 |
| 1 | 1 | 1 | 0 | 0 |
| 1 | 1 | 1 | 1 | 1 |

Karnaugh Map

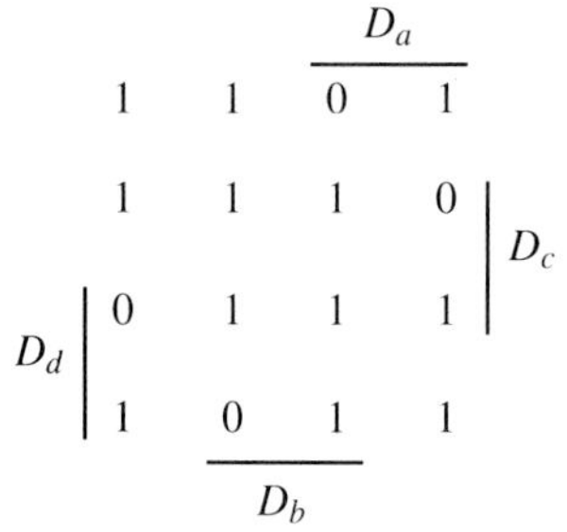

| | | | $D_a$ | $D_a$ | |
|---|---|---|---|---|---|
| | 1 | 1 | 0 | 1 | |
| | 1 | 1 | 1 | 0 | $D_c$ |
| $D_d$ | 0 | 1 | 1 | 1 | $D_c$ |
| $D_d$ | 1 | 0 | 1 | 1 | |
| | | $D_b$ | $D_b$ | | |

(b)

**Figure 10.39** Truth tables and corresponding Karnaugh maps for (a) Example 10.17a and (b) Example 10.17b.

reduction capability. In Example 10.17a, we might be able to arrive at the minimal sum of products by using only the relatively simple truth table. But the Karnaugh map transforms complicated considerations of combinations of input variables that produce a 1 in the output into simple pattern recognitions.

The procedure is to draw loops enclosing as many 1s as possible that fall into a square or rectangular shape consisting of two, four, or eight cells. Always draw loops around the largest such blocks. Also, the Karnaugh map actually is horizontally and vertically cyclical, so the top "wraps around" to the bottom and the left edge "wraps around" to the right edge. Simple checks will show that corresponding elements on those edges are indeed different in only one input variable. In addition, cells common to more than one block may be included in both loops. Thus in Example 10.17a, we may draw the loops in Fig. 10.40 to identify the minterms. Similarly, in Example 10.17b we may draw loops on the Karnaugh maps as shown in Fig. 10.41.

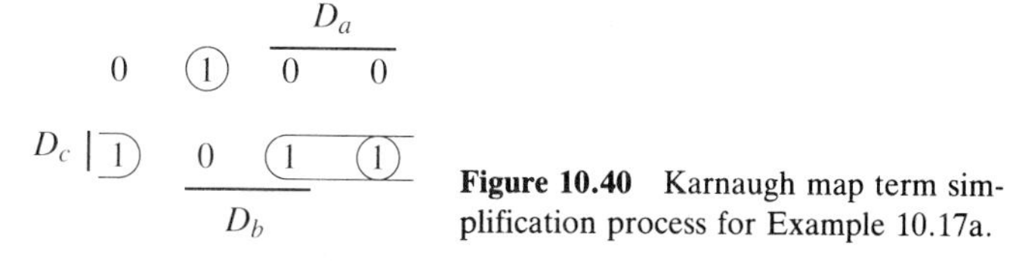


**Figure 10.40** Karnaugh map term simplification process for Example 10.17a.

By examining the conditions that may or may not exist on each input variable common to all cells in a loop, the minterm is read directly from the map. The reader should verify that the final expressions obtained are:

Example 10.17a:

$$D_{\text{out}} = \overline{D}_a D_b \overline{D}_c + D_a D_c + \overline{D}_b D_c. \tag{10.22a}$$

Example 10.17b:

$$D_{\text{out}} = \overline{D}_c \overline{D}_b + D_b D_c + \overline{D}_a \overline{D}_d + D_a D_d \tag{10.22b}$$

It should now be clear why we choose the largest possible blocks: That minimizes the number of variables in the corresponding minterm and hence simplifies the final expression. Comparison of Eq. (10.21b) with Eq. (10.22b) especially shows the power of the Karnaugh map method.

Despite the value of the Karnaugh map, there are limitations on its effectiveness and appropriateness. When the number of variables is greater than four or five, the maps become unwieldy and impractical. The rules of the method given above were a simplified version of a more complicated set of rules that is guaranteed to yield the minimal SOP expression. Because of this, just looking closely at a truth table or considering the word specification of the task to be done can actually sometimes yield better and faster results than will Karnaugh map construction. Nevertheless, the Karnaugh map is an often-used simplification method beyond head-scratching and short of computer-aided design programs, which themselves use the math behind Karnaugh maps (e.g., absorption laws).

## 10.6 SIMPLE INPUT–OUTPUT DEVICES

### 10.6.1 Introduction

In this chapter we have been developing a set of fundamental concepts and building blocks used in digital circuitry. These are used both for basic logic circuits and in the design of sophisticated systems. Our study would not be complete without at least

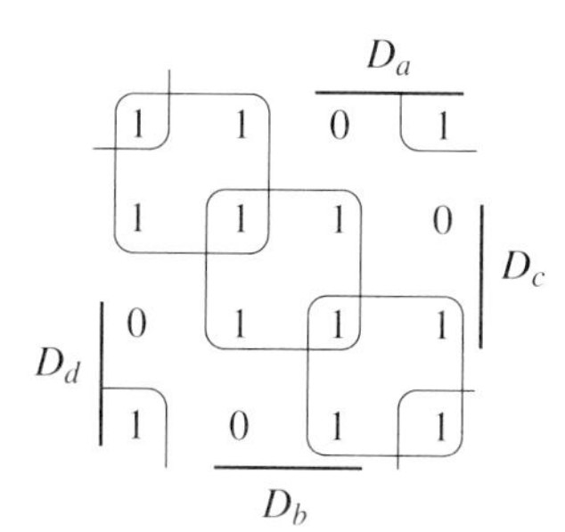


**Figure 10.41** Karnaugh map term simplification process for Example 10.17b.

a brief discussion of devices that allow communication between the circuit and the human. The complexity of these devices ranges from the simplicity of a slide or lever switch all the way to the extreme sophistication of three-dimensional movable viewpoint video displays. The former is an input device and the latter an output device. Only a few of the simplest input–output devices are discussed here; our aim is merely to provide rudimentary communication capabilities.

### 10.6.2 Switches

The switch is the most direct means of setting a 1 or a 0 on a binary line. A variety of switches exists to maximize convenience for particular situations. For example, a good switch for conveying a momentary alert, sequential data entry, or for initiation of a process is the momentary connection pushbutton switch. As long as it is held down, connection is maintained. A familiar example of this switch is the keyboard character key.

If the switch is to be set temporarily but frequently switched, a toggle (push on, push again off) pushbutton switch may be appropriate. A common application of this switch is the power on–off switch. Switches involving less frequent switching include low-cost slide switches and high-reliability locking lever and rocker switches.

Finally, for digital circuitry a very convenient option is the group of eight or more switches on a package the size and shape of an integrated-circuit (IC) chip. These minute switches are either rocker or slide switches, intended for only occasional switching (such as mode/configuration selection) or for circuit testing/development. They are known as DIP switches because they come in a *dual in-line package*, as do most chips. Dual in-line packages, which are small, rectangular, and have two rows of pins along the long sides, are described further in Secs. 10.7.2 and 10.7.6.

Probably the most important issue in the use of switches for digital systems is the "bouncing" problem. We normally think that once a switch is switched on, a connection is made and there is no more to say about it. This is true on the human time scale but not on the time scale of digital systems, where switching is occurring millions of times per second. The mechanical switch actually makes a series of momentary connections and disconnections over a period of about 50 ms or so before finally settling to a stable connection. This is unavoidable because of the imperfect smoothness of switching materials.

Consequently, mechanical switches determining 1s and 0s of binary signals are usually coupled to the electronics by a "debouncing" circuit. The problem is similar to that of the positivity detector with noisy input studied in Sec. 8.3. A Schmitt trigger was the solution of that problem, and it can also solve the switch debouncing problem for *single-throw switches* such as DIP and pushbutton switches. "Single-throw" means that the movable lever or "pole" is "thrown" to connect and disconnect it with a single stationary terminal. A single-pole, single-throw switch is shown in Fig. 10.42.

The Schmitt trigger debouncing circuit appears in Fig. 10.43. Note that it is an inverting Schmitt trigger, hence the inverter symbol (bubble). These *Schmitt trigger inverters* come in integrated-circuit form specially designed for digital circuits, obviating the need for external feedback resistors. (Do not be alarmed that there is only one input to the Schmitt inverter: Recall that $v_s$ was the only input to the op amp

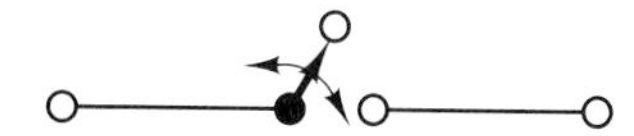

**Figure 10.42** Single-pole, single-throw switch.

inverter. As usual, power supply connections are omitted.) Operation of the debouncing circuit is as follows.

If the switch is left in the open position, $V_C$ is high—the capacitor $C$ has had a chance to charge to 5 V through $R_2$ after a time on the order of $R_2C$ = 100 ms. Upon the first connection of the switch contacts, $C$ discharges rapidly to 0 V through $R_1$ over a time on the order of $R_1C$ = 10 μs—far shorter than the bouncing intervals. When momentary loss of contact occurs, $C$ charges very slowly through $R_2$ toward 5 V. Before it gets very far in charging, contact is reestablished, which instantaneously brings $V_C$ back to 0 V.

The Schmitt upper and lower thresholds are well separated, so that the momentary $R_2C$ charging does not *at all* alter the output voltage $V_o$ from its new "saturation" level caused by switching. Rather than being $\pm V_{BB}$, the saturation levels of the digital Schmitt trigger are the same as the logic voltage levels of standard digital gates. These voltage levels are discussed in Sec. 10.7.5.

Also common is the (single-pole) double-throw switch (see Fig. 10.44). Instead of just a connection between two wires, a double-throw switch connects the movable lever to one of two stationary terminals. In between connections it is connected to neither. For this type of switch, the common method of debouncing uses an *RS flip-flop*. A flip-flop is a circuit with memory, just as the Schmitt trigger has memory, as noted in Chapter 8. It is discussed in detail in Sec. 11.2.2. We mention it now because it is good practice to use debounced switches for most digital circuits, and you are likely to find them on many existing circuit diagrams. (However, the effects of bouncing can be far more disastrous in sequential circuits than in the simple combinational circuits of this chapter.) Suffice it to say that just as the Schmitt trigger detects the first time the upper threshold is crossed, the *RS* flip-flop detects the first time the switch makes contact (or loses contact, in the case of turning the switch off).

In the circuit diagram of Fig. 10.45, the inputs are $S$ and $R$ and the output is $Q$. Notice that $\overline{Q}$ is also available as an output, a feature that can come in handy. When the switch first contacts the upper line $S$, the output is "set" to 1. It remains at 1 even through there may be momentary loss of contact at the mechanical switch.

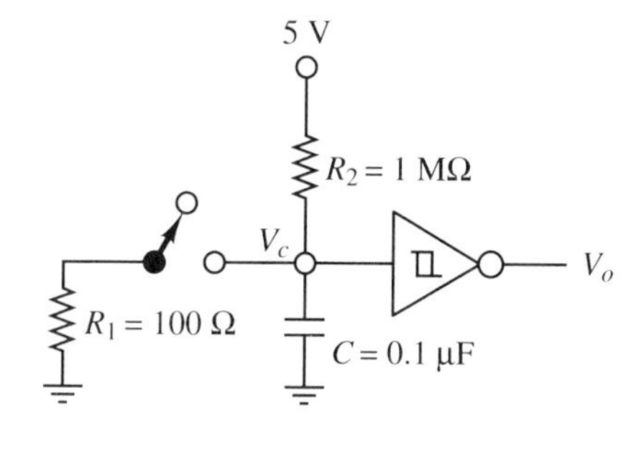


**Figure 10.43** Schmitt trigger switch debouncing circuit.

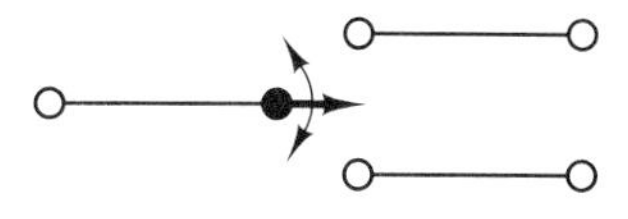

**Figure 10.44** Single-pole, double-throw switch.

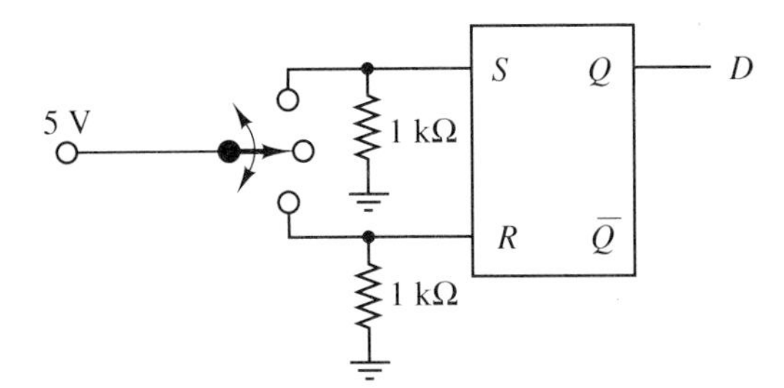


**Figure 10.45** *RS* flip-flop debouncing circuit for single-pole, double-throw switch (see Fig. 10.44; for discussion of *RS* flip-flop, see Sec. 11.2.2).

When the switch first contacts the lower line *R*, the output is "reset" to 0 and remains there even in the presence of "bounce" (now, loss of contact with the lower terminal). The two terminals are separated sufficiently that a bounce does not involve contact with the other terminal, only loss of contact with the desired terminal.

### 10.6.3 LED and LCD Readout

We reconsider here briefly the *light-emitting diode* (LED), discussed in Sec. 7.4.7. Another type of readout, the *liquid crystal display* (LCD), was discussed in Sec. 9.6.4. The LCD is currently the favored alphanumeric character display device because of its low power consumption and high readability in daylight. Although the LED is being replaced by liquid crystal displays (LCDs) in commercial equipment, the LED is still a useful tool for experimentation and digital circuit development.

A 1 on a binary line is evident when an LED hung from it glows. Figure 10.46a shows a gate driving an LED. There is not much to say here, except to review how the current-limiting resistor is chosen. Suppose that the LED operates with a current of 10 mA and the maximum voltage across it and the current-limiting resistor is 5 V. Then the series resistor must have resistance (5 − 1.7)/0.01 = 330 Ω, where 1.7 V is the voltage across the LED when conducting (i.e., when lighted). (Actual gate output voltage for a 1 may be somewhat less than 5 V; see Sec. 10.7.5, Fig. 10.50. The LED will, however, still light using this design.)

A modification must be made to get the longest life possible out of a liquid crystal display. Instead of applying a dc voltage to turn on the display as discussed in Sec. 9.6.4, a 20 to 300 Hz pulse train is applied to the backplane so that the time-average electric field applied to the crystal is zero. Doing so has been found to extend the life of the display greatly.

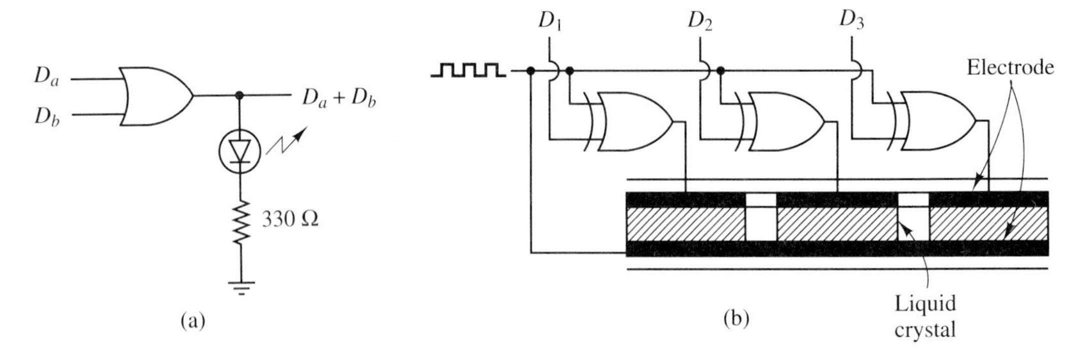


**Figure 10.46** Circuit for (a) LED and (b) LCD readout of the output of a logic gate.

Now instead of applying 0 V to have the display off, we just apply the same pulse train to the other electrode. Instead of applying a constant, different voltage to the other electrode to turn "on" the display, we apply the logical NOT of the pulse train. If we refer to Table 10.9, we see that if we hold $D_a = 1$, then $D_{\text{XOR}} = D_b$, and if we hold $D_a = 1$, then $D_{\text{XOR}} = \overline{D}_b$. Therefore, we just apply our on–off signal to $D_a$ of an XOR gate, and apply the pulse train to both the backplane and the other input of the XOR gate, $D_b$. The output of the XOR gate, $D_{\text{XOR}}$, is applied to the upper terminal of the LCD. Figure 10.46b shows the connections for several on–off LCD indicators as might be used in an alphanumeric display (see the next section).

### 10.6.4 Alphanumeric Displays

A variety of devices exists for display of digital system outputs in forms easily understood by humans—much more easily than the 1, 0 patterns of individual LEDs. For example, the *seven-segment display* is used to display numbers in base 10. Each digit is composed of seven segments, as shown in Fig. 10.47. Each segment is an LED or LCD. When any of $D_0$ through $D_7$ are low (0 V), the corresponding $D_a$ through $D_f$ segment(s) are on.

Much more will be said about conversion from base 2 to base 10, base 16, and the *binary-coded decimal* (BCD) representation of a number in Sec. 11.6.2. The last representation, BCD, is just a four-line binary representation of *a single decimal digit*; each digit is given its own set of four binary lines. These four lines are decoded to seven binary lines, one for each display segment. Every character has been assigned a distinct on–off code for each of the seven segments. Possible display characters are 0, 1, 2, 3, 4, 5, 6, 7, 8, 9, A, B, C, D, E, F—the 16 characters of base 16. Naturally, base 10 numerals 0 through 9 are of the greatest immediate interest.

By placing several single-digit displays side by side, multidigit numbers can be displayed. There are even extra lights for displaying decimal points on seven-segment displays. Operations involved in producing such numbers are discussed in Sec. 11.6.2.

Necessary decoders and drivers to supply required currents are available in chip form that take standard digital gate outputs (in BCD) as inputs to be displayed. A driver is a circuit providing all interfacing electronics required between usual digital devices (e.g., gates) and a special device, here a seven-segment display. They make digital circuit design much easier than it otherwise would be.

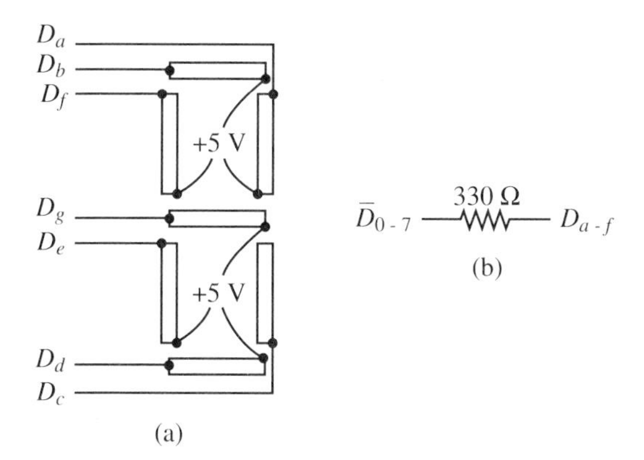


**Figure 10.47** (a) Seven-segment display; (b) method for connecting display inputs to desired binary data lines.

The actual segments may be either LED or LCD. Typically, LCD displays come only in multidigit packages. As noted in the preceding section, unlike LEDs, the LCD requires a special 20- to 300-Hz square wave to operate. This square wave is usually supplied by LCD driver chips.

A more advanced alphanumeric display is capable of producing any of 128 characters, including the entire alphabet with upper- and lowercase. A four-character package is shown in Fig. 10.48a. These characters, along with the input codes necessary to produce them, are shown in Fig. 10.48b. Instead of a seven-segment display, a dot matrix display is used. That is, each character is made up of a rectangular pattern of $7 \times 5 = 35$ dots, each of which is either on or off, depending on the shape of the written character. Naturally, there are $\log_2(128) = 7$ input binary data lines.

Internal circuitry decodes the seven lines into 35 on–off lines for the dots, drives the display, and even has memory for continued display as well as many other features. Examples of these appear in new medical imaging equipment, control panels of many types, computer peripheral equipment, and "smart" oscilloscopes.

### 10.6.5 Other Input–Output Devices

In a computer or other digital system, there are many other forms of input and output. Laser printers, line printers, plotters, and video displays, modems (modulator/demodulators used for computer data communication via telephone), and even memory systems are all considered forms of output. Similarly, modems, memory systems, keyboards, and a variety of transducers may be considered forms of input.

Discussing these in detail would take us far afield from this introduction to digital systems. But being aware that, for example, the two main things you stare at when sitting in front of a personal computer (PC) are its primary input device (keyboard) and its primary output device (video screen) is a good first step to learning about digital systems.

## 10.7 PRACTICAL ISSUES

### 10.7.1 Introduction

Up to now, all of our digital devices have been ideal, even abstract. The way in which a "1" or a "0" is implemented in practice has not been addressed. Logical symbols have been used for NOT, AND, OR, and so on, without regard to how they are realized with electronic circuits. It should come as no surprise that the reason for doing this has been to focus on the new basic concepts, and there were a great deal of them. In Chapter 11 we continue this idealistic approach for the most part. But there does come a time to touch bases with reality, and now is as good a time as any.

### 10.7.2 Speed–Power Trade-off

There are many characteristics of digital circuitry that determine under what circumstances a circuit behaves as designed. Three dominating characteristics, which are interrelated, are speed, size, and power requirements. Speed would not seem to be a big issue—after all, a transistor can switch pretty quickly. For example, an early

(a)

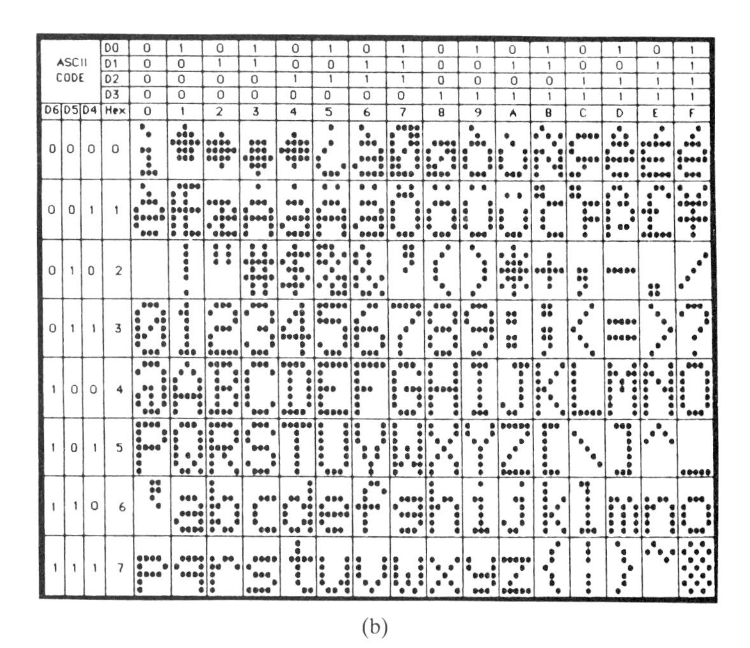


(b)

**Figure 10.48** A 128-Character-type, four-"digit" display panel: (a) physical appearance; (b) character codes and appearance. (Courtesy of Siemens Components, Inc., Optoelectronics Division.)

type of digital circuitry called *resistor–transistor logic* (RTL) had a propagation time through a typical gate of only 50 ns.

For the types of circuitry discussed in this chapter, that would seem to be fine. But the industry is, of course, driven by survival of the fittest. In particular, the capability of rapid computation is of paramount importance. Also, the longer it takes a gate to change state (1 to 0 or 0 to 1), the more difficult design becomes because everything can be held up by one slow gate. So when *transistor–transistor logic* (TTL) and other technologies came along with 5- to 10-ns propagation times or better, the game was up for RTL.

And it was not only speed that did RTL in. Other issues, such as susceptibility to noise and so-called *fanout* (see Sec. 10.7.3) were also unfavorable to RTL. One measure of noise susceptibility is the voltage difference between a 1 and a 0. If that difference is small, a small level of noise can erroneously cause a 1 to look like a 0, or vice versa. At the output of a gate, the difference between a 1 and a 0 for RTL was only 0.6 V. For TTL, it can be up to nearly 2 V.

On the issue of power consumption per gate, RTL and TTL are comparable: about 5 to 10 mW/gate (even here TTL wins out). But because of all the other disadvantages, RTL is not even manufactured anymore. RTL has no important redeeming qualities.

Here lies the key to understanding why a type of gate implementation called complementary symmetry metal-oxide semiconductor (*CMOS*, a particular form of MOS we studied in Chapter 7) is extremely popular even though as originally produced it is no faster than RTL. It consumes 0.5 to 1 mW/gate under active use. Furthermore, when no switching is occurring, the power drain is essentially zero. This greatly reduced power consumption makes it ideal for battery-powered systems and for being able to cram huge numbers on a chip (i.e., achieve high "gate density") without the chip burning out due to overheating. The names RTL and TTL are descriptive of the components linking digital signals in the circuits composing the gates. The name CMOS describes the physical construction of the transistors as well as the transistor configurations used in the circuit. Both RTL and TTL use bipolar transistors, while CMOS uses field-effect transistors. (It is the extremely high input impedance of FETs that makes CMOS such a low-power device.) Because within each technology similar varieties of gates are available, technologies such as TTL and CMOS are called *logic families.*

Usually, speed increases in digital circuitry are attained at the cost of higher power consumption. Because of the differing requirements for various digital systems, in some circumstances power minimization is most important, while in others speed takes top priority, and in still others a compromise is best.

Consequently, no single family or subcategory of a family has taken universal preeminence. For example, those who want higher speed than TTL and do not mind the higher power dissipation and initial cost can use a family called *emitter-coupled logic* (ECL), which has propagation delays of only the order of 1 ns. For those who want maximum speed at any cost (including a sizable one to their bank account), the gallium arsenide (GaAs) family offers propagation times on the 0.25-ns range. And for higher-than-TTL gate density using bipolar transistors, *integrated injection logic* ($I^2L$) has been developed. But for high availability at low cost with high speeds,

TTL has been the most popular family. With CMOS speeds approaching those of TTL, many designers are recommending that all new designs use the high-speed version of CMOS because of its noise immunity, high gate density, and low power consumption.

A practical aspect of specifying and selecting integrated circuits (ICs) is the package type. By far the most common type for general-purpose digital electronics is the *dual in-line package* (DIP) device. These are the familiar plug-in units (recall the 741 op amp, for example) that have two rows of pins on opposite sides of the chip. They typically plug into a socket that is soldered onto a circuit board, or they may themselves be soldered onto the circuit board. In either case, the chip pins must be strong enough to withstand insertion into the socket or circuit board holes. The new, narrow versions of this type of package are called "skinny-DIPs."

Another increasingly common package type is the *surface mount* (SM) device. By no longer requiring the pins to be rigid enough to plug into a socket or withstand insertion through holes in the circuit board, the pins can be made shorter and narrower than those of DIPs. The pins are directly soldered to the face of the circuit board—hence the name "surface mount."

Advantages of surface mount devices are smaller size (0.2 to 0.9 times the size of the corresponding DIP), ease in making chips with large numbers of pins (e.g., 44 pins—all the way around the chip instead of on two sides only), and automation of assembly of digital circuits (a robot selects the required part and automatically puts it in the correct spot on the circuit board). They are, however, currently more expensive than DIPs. Yet another type, called *pin grid array*, has huge numbers of pins in concentric squares coming through the bottom of the chip. They are currently 10 times as expensive as their DIP counterparts.

One precautionary note on using CMOS. CMOS chips should be handled with care, as they are prone to destruction from static charges. For example, any unused pins (including, for example, those of unused gates in a multigate package) must be attached to 0 V or the supply voltage, whichever is appropriate under the circumstances. Otherwise, the voltages on those pins could wander off to high voltages that would destroy the chip. Similarly, things such as inserting a chip with the power on or leaving unused chips outside their protective foam or tube casing are ill advised. These warnings are not nearly as severe as was true in the past, because now CMOS chips have protective diodes at their inputs to provide a discharge path for offending charge; nevertheless, it is prudent to be on the safe side.

The sensitivity is especially keen because of the highly insulated MOSFET gates. (Do not confuse an FET gate with a logical gate.) The FETs can behave like very small capacitors so that the voltage $V = q/C$ across the insulation can be very large even for a minute charge $q$ on the transistor gate. The result can be a destroyed chip.

There are many variations on both TTL and CMOS. For example, to speed up TTL, a device called a *Schottky diode* is used. A Schottky diode is a slab of $n$-type material joined to a metal slab rather than a slab of $p$-type material. The observed behavior is similar to that of a diode, except that the Schottky diode changes from conducting to nonconducting states much more quickly than does a normal diode. This is because of its low contact potential. It can be used to speed up a transistor switch by placing it across the base and collector terminals of a normal bipolar junction transistor. Fortunately, this speed advantage is not accompanied by greatly increased power consumption.

It is, however, accompanied by a much smaller reverse voltage ($V_z$) than normal *pn* junction diodes have. Therefore, it cannot be used in typical analog electronics applications, where significant reverse voltages can be expected. We find it in digital electronics because the low reverse voltage is not a factor, whereas speed is critical. Note that this is another instance where a rectifying junction is possible without holes, as discussed in some detail in Sec. 7.3.2.

### 10.7.3 Fanout

Fanout is one measure of the convenience of using a particular type of digital logic circuitry. It indicates the number of circuits that a single gate output can drive and still maintain the intended logic levels. The idea is similar to that of power supply loading: If you hang enough circuits off a battery, eventually none of them will function because of the too-severe load on the battery. The output of a logic gate is in a real way a supply of energy to those gates to which it is connected, and under overload the gate can be destroyed.

The fanout for RTL was only about 4, whereas that for TTL is roughly 10. Because its input current drain is extremely small, one might expect CMOS to have a huge fanout capability. However, at high speeds, capacitive effects limit CMOS fanout to roughly that of TTL, or a small amount higher.

The simplest fanout calculations occur for TTL. Each gate to be driven requires a certain amount of current for each state. These values are given on the data specification sheets for the device. Their meanings are illustrated in Fig. 10.49. Typical values would be $I_{IL}$ = input low-state current requirement = 40 $\mu$A and $I_{IH}$ = input high-state current requirement = (−)1.6 mA. The output current that a preceding gate is capable of supplying might be $I_{OL}$ = output current available at low-state = (−)0.4 mA, $I_{OH}$ = output current available at high-state = 16 mA. Thus we find that the fanout for driving low states is $I_{OL}/I_{IL}$ = 10 and also that for driving high states is $I_{OH}/I_{IH}$ = 10. Obviously, in general, the two ratios may not be

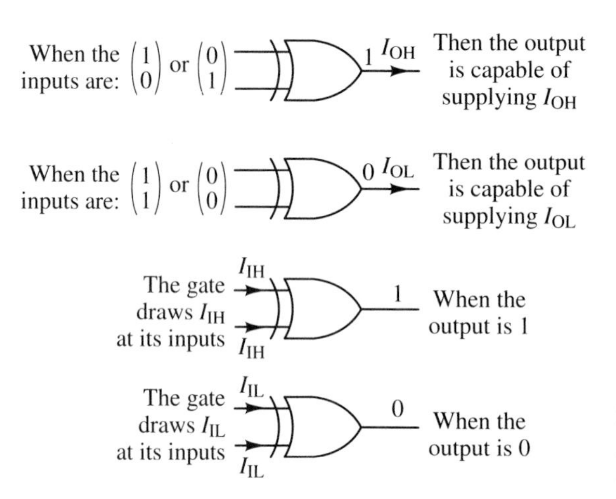


**Figure 10.49** Meanings of gate currents given on specification sheets: $I_{OH}$, $I_{OL}$, $I_{IH}$, and $I_{IL}$.

equal, and the minimum must be considered the usable fanout estimate. The numbers that the sheets give you are usually worst-case, so the fanout you compute should be usable.

**Example 10.18**

Refer to the circuit of Fig. 10.50. Also, the following data are obtained from the IC data sheets:

NOR, NAND, and INVERTER gates:

$I_{IH} = 20\ \mu A,\ I_{IL} = -0.4\ mA.$

XOR gate:

$I_{OH} = -0.4\ mA,\ I_{OL} = 8\ mA;\ I_{IH} = 40\ \mu A,\ I_{IL} = -0.8\ mA,$

where $I_{IH}$ and $I_{IL}$ are the maximum currents required to hold a gate input line at high and low, respectively, and where $I_{OH}$ and $I_{OL}$ are the maximum currents available at the output when it is high and low, respectively.

**(a)** What is $V_2$?

**(b)** Find $I$ with the switch in the upper position (upper input of the XOR gate is 1).

**(c)** Find $I$ with the switch in the lower position (upper input of the XOR gate is 0).

**Solution**

**(a)** Clearly, $V_2$ is at 0 state. From TTL data sheets, $I_{IL} = -0.8$ mA flows in the 1-kΩ resistor between $V_2$ and ground. Therefore, $V_2 = -1000(-0.0008) = 0.08$ V. (Read Sec. 10.7.5 to see why this is a reasonable value.)

**(b)** With the switch in the upper position, the two inputs are 1 (flip-flop output) and 0 (lower input). Therefore, the output of the XOR is 1. (Again, see Sec. 10.7.5 to verify that 4.6 V is considered a 1.) The current required of the XOR output, $I$, is equal to $I_{IH,INV} + I_{IH,NAND} + 20I_{IH,NOR} = 22 \cdot 20\ \mu A = 0.44$ mA, which exceeds $I_{OH}$ of the XOR gate. Therefore, the circuit will not function correctly. A *buffer* (see Sec. 10.8) is needed to drive all these gates.

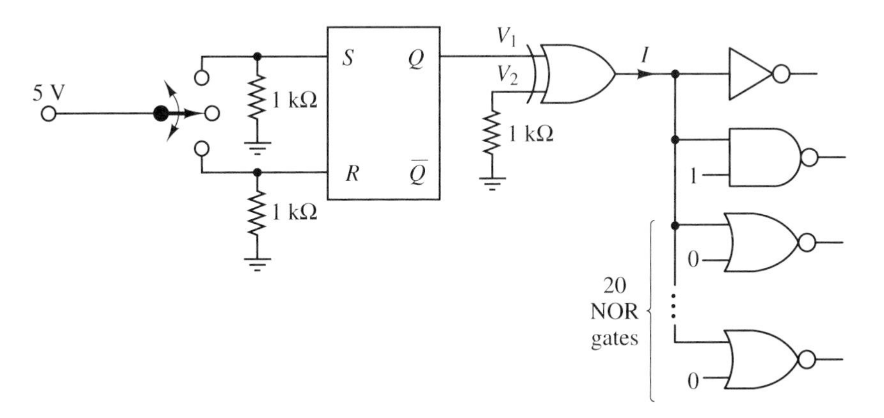


**Figure 10.50** Circuit for Example 10.18 involving a fanout calculation.

(c) With the switch in the lower position, the two inputs to the XOR gate are both 0, so the XOR output is 0. Therefore, the current $I = I_{IL,INV} + I_{IL,NAND} + 20I_{IL,NOR} = 22(-0.4 \text{ mA}) = -8.8 \text{ mA}$, which exceeds the magnitude of the allowable current $I_{OL}$ of the XOR gate. Of course, the failure in (b) obviated this calculation—a failure of either requires use of buffers.

### 10.7.4 Integration Scale

Another advantage of CMOS over TTL is its small size per gate. This allows huge numbers of gates per chip. There are several categories for gate density, as shown in Table 10.17. SSI is used for op amps, gate packages, and other relatively simple circuits. MSI is appropriate for shift registers, adders, comparators, and other moderately complicated digital devices. LSI is the density required for microprocessors, A/D converters, and other sophisticated circuitry. Finally, VLSI is necessary for the immense requirements of computers and memory systems. Notice that only the various MOS types, including CMOS, are available for this category; TTL is not. As intuition would suggest, the higher the density and complexity of the chip, the more expensive it is. The amazing fact is that even VLSI is affordable, often only $10 to $30, depending on the device.

### 10.7.5 Voltage Requirements

Although much effort has been directed toward making pin-for-pin compatibility between the major families, we must be aware of the requirements of each family before just plugging them in. Most important are the power supply and input–output voltage specifications. For TTL, the supply voltage must be near 5 V. CMOS can be effectively run on supply voltages ranging from 3 to 15 V (or 2 to 6 V for the high-speed version). The noise immunity voltage difference defined above (voltage difference between a 1 and a 0) is proportional to the supply voltage. Therefore, for noisy industrial environments, CMOS using a higher voltage supply is most appropriate.

**Power supply.** Often overlooked in presentations of digital circuitry is an essential building block: the power supply. The first thing to check in a "dead" system is whether the power supply is connected. Fortunately, power supplies are

**TABLE 10.17** CATEGORIZATION OF DIGITAL TECHNOLOGY TYPES AND CORRESPONDING COMPONENT DENSITY SCALES, DEFINED IN TERMS OF MAXIMUM GATES PER CHIP

| Density category | Gates/chip (or × 5 to 10 for transistor count) | Digital families available |
|---|---|---|
| Small-scale integration (SSI) | 1–10 | TTL, CMOS |
| Medium-scale integration (MSI) | 10–100 | TTL, CMOS |
| Large-scale integration (LSI) | 100–10,000 | TTL, CMOS |
| Very-large-scale integration (VLSI) | 10,000+ (current limit: 200,000) | MOS only |

available as predesigned units that need only be plugged into a given system. Still, a few basic principles of operation help in the selection of the appropriate type of supply and in obtaining the comfortable feeling that all basic elements have been considered.

To make all our digital circuitry go, it must be supplied with energy. We always begin with 110 to 120 V sinusoidal voltage from the wall outlet. Two options are possible: linear and switching power supplies. The linear power supply was discussed briefly in Chapter 7. It is suitable for circuits such as linear amplifiers requiring low noise distortion, so that the small signals involved are not overwhelmed by noise.

For digital systems, the noise immunity is greater than that of linear circuitry. All that is needed is lots of dc power to run all those ICs. Therefore, *switching power supplies* are ideally suited to digital electronics. Switching power supplies use entirely different methods for producing the desired dc level that, while relatively noisy, are relatively very efficient. In a linear power supply, large amounts of energy are expended just running the series power transistor. In a switching power supply, that energy can instead be used to operate the astronomical numbers of microscopic transistors occurring in computer systems.

As hinted in Chapter 7, the series transistor of a switching power supply does not operate in the linear region. Instead, it switches back and forth between cutoff ($i_C \approx i_E$ small) and saturation ($v_{CE}$ small) regions, thus always maintaining low dissipated power $i_C v_{CE}$. The efficiency of this type of power supply is typically 70% rather than the 50% or less of linear power supplies.

An ac voltage is first rectified and filtered in a switching power supply. It is then chopped into a square wave of frequency 100 kHz to 1 MHz. Voltage step-down transformation is done *after* the switching (at 100 kHz to 1 MHz) rather than at the 60-Hz ac input. This is because for a given voltage, flux density, and number of turns, the required core area of a transformer is inversely proportional to the frequency. Therefore, the higher the frequency, the smaller the core area, which allows use of a smaller core and therefore a physically smaller transformer and overall power supply.

Although the switching is a square wave rather than a sine wave, a sinusoid at the switching frequency dominates the behavior. However, the presence of high-frequency signals makes these supplies inappropriate for such systems as radios and tape recorders because of their susceptibility to noise. Conversely, digital circuits are largely immune to this noise; waveforms are not processed, only strings of binary information.

The switching circuitry creates pulses at 120 V, which are then stepped down to the desired range. The exact desired output dc level is achieved by controlling the pulse width or the pulse frequency of the square-wave oscillator. The smaller the fraction of time that the voltage is "high," the smaller will be the output dc voltage once filtered.

As with the linear power supply, feedback circuitry is used to regulate the output voltage in the face of load variations. Detailed study of the design of these complicated power supplies is well beyond the scope of this book; they are available in ready-made units for about $100. They typically have 5-V and ±12-V dc outlets that are adjustable for calibration.

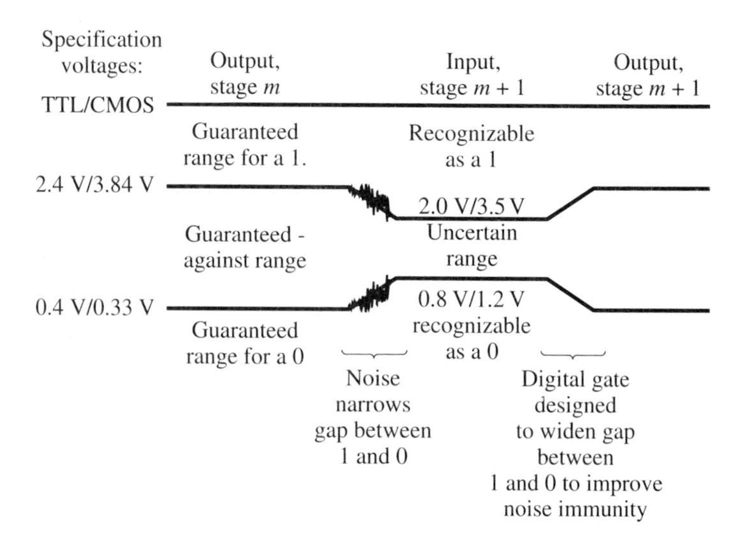


**Figure 10.51** Voltage level requirements for TTL (bipolar) and CMOS (FET) logic families. The upper levels define "1" and the lower levels define "0." The numerical voltages are thresholds between a 1/0 and forbidden zone. A wide separation between 1 and 0 is desirable. The output of stage $m$ has wide separation; this is reduced by noise by the time the input to stage $m + 1$ is reached. Stage $m + 1$ reestablishes the wide separation between 1 and 0.

**Gate input–output voltages.** The basic idea of input–output voltage level specifications is as shown in Fig. 10.51. A gate is designed so that relatively wide input voltage ranges are recognizable as a 1 or a 0, and so that the output produces 1 and 0 levels that are more widely separated. This increase in the gap between the 1 and 0 voltages from input to output minimizes the probability that noise will cause an error (a 1 changing to a 0, or vice versa).

Typical threshold values for both TTL and CMOS (based on a 5-V power supply) for inputs and outputs are also given in Fig. 10.51. As long as the input voltage requirements are satisfied, the outputs are guaranteed to be at the well-separated levels indicated in Fig. 10.51. This separation is known as the noise margin.

### 10.7.6 Chip and Pin Identification

Because all chips of the same size appear identical, an alphanumeric code is always printed on the chip for identification. Furthermore, a number of manufacturers offer the same list of digital ICs. As we might guess, the prefix to the part number begins with a code indicating the manufacturer. For example, see Table 10.18. Rest assured, however, that the company name or symbol can also be unambiguously found somewhere on the chip!

Many design techniques in both TTL and CMOS have led to improved speed, usually at the cost of higher power consumption. Each of these subfamilies is identifiable from the coded part number. Presumably to reinforce the idea that the new technologies are pin-for-pin compatible with the old, the prefixes of the codes for a wide-ranging series of chips include the arbitrarily chosen number 74, a number apparently originally used by Texas Instruments. (Its main function now is to

**TABLE 10.18** PREFIX CODE FOR CHIP MANUFACTURER

| Prefix | Manufacturer |
|---|---|
| SN | Texas Instruments, Motorola |
| DN (TTL), MN (CMOS) | Panasonic |
| MC | Motorola |
| DM (TTL), MM (CMOS) | National Semiconductor |
| None | Signetics |

distinguish between the standard line and the more stringent specification "military" line, designated with 54 rather than 74. Primarily, the '54 series has a wider temperature range over which it will function properly.)

Other letters of the prefix identify the technology type, or *series* (see Table 10.19). In integrated circuits, the word "technology" refers to the device fabrication type. After all of this, we finally get to the two- or three-digit device number. In the following discussions, we refer to parts and information appearing on digital IC parts lists such as are published in IC Master, Hearst Business Communications, Garden City, NJ (published yearly).

**Example 10.19**

A loose IC is found that has the label MM74HC08N. Identify it.

**Solution** The MM indicates the device is by National Semiconductor and is CMOS. That this is a standard line (normal temperature) part is clear from the 74. It is a high-speed CMOS, as indicated by HC. (Its speed is similar to the LS TTL subfamily, but it has the low power consumption of CMOS ICs). The part itself ('08) is a quad, two-input NAND gate. The N at the end is a package designation. N is the code for dual in-line packages (DIPs).

Once the required part number has been determined, a specification sheet is essential to obtain. Such sheets can be found in, for example, the *TTL Databook* by Texas Instruments, Inc. On the sheet, a diagram appears that shows the connections that must be made to each pin.

The pins are numbered as follows. First orient the chip pins pointing down and

**TABLE 10.19** PREFIX CODE FOR CHIP TECHNOLOGY

| Prefix | Technology |
|---|---|
| 74 | Original TTL |
| 74S | Schottky TTL |
| 74LS | Low-power Schottky TTL |
| 74ALS | Advanced low-power Schottky TTL |
| 74F | Fast (oxide isolation) TTL |
| 4000 | Original CMOS |
| 4000B | Improved CMOS |
| 74HC | High-speed CMOS |
| 74HCT | High-speed TTL-compatible CMOS |
| 10K, 100K | ECL |

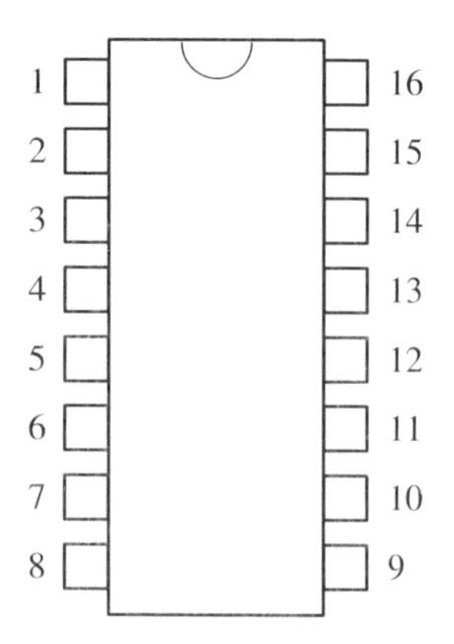


**Figure 10.52** Top view of IC chip; pin numbers.

the notch end away from you. Pin 1 is the pin farthest away from you on the left side. The numbers increase coming toward you on the left side. They continue on the right side increasing away from you. Thus pin 1 is opposite the highest pin number on the right side, and we might describe the counting as increasing in the counterclockwise direction. See Fig. 10.52 for the example of a 16-pin chip.

It should be mentioned that on digital IC parts lists, the words "quad," "hex," "triple," and "dual" refer to the number of duplicated items that are found on one chip. So if your design requires four inverters, a hex inverter ('04) will take care of the design with two unused inverters left over.

Also, although most of the gates in this chapter have had only two inputs, often more are necessary. Most common are two-, three-, and four-input gates. However, the NAND gate number '30 has eight inputs. There are times when a large number of inputs must all be high in order that something occurs. Any unused inputs can be tied high so that they do not interfere with the important inputs. As will be discussed in Sec. 11.6, groups of eight binary variables (called *bytes*) have great significance in computational structures. This device could look at a byte of memory and declare whether it is filled with all 1s.

The abbreviation "OC" in devices '01, '06, and others on digital IC parts lists stands for *open-collector output.* This means that a low is still near 0 V, but a high is no longer fixed at a high voltage level. Instead, the output is left floating. Otherwise, the device functions as a normal logic gate. To cause the output to be at a high voltage (e.g., 3 to 5 V) when the gate output is a 1, a *pull-up* resistor $R$ is connected between the output and the power supply, as shown in Fig. 10.53. (*Note:* "OC" if used for MOS chips means *open drain.* Note also that "$V_{CC}$" is often used when "$V_{DD}$" is meant. These remnants from BJT technology die hard and can be confusing.)

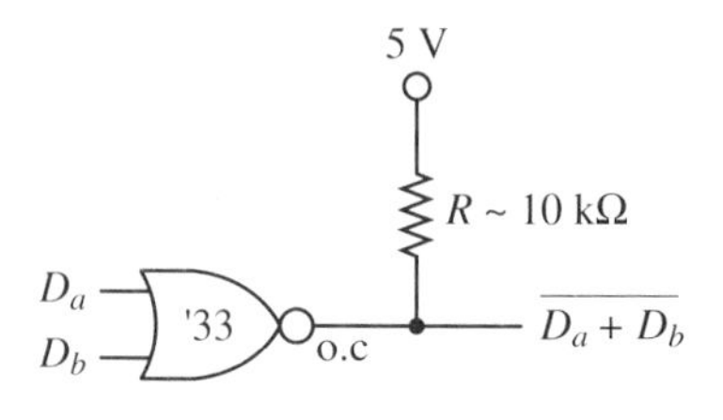


**Figure 10.53** Pull-up resistor connection used for open-collector gate outputs.

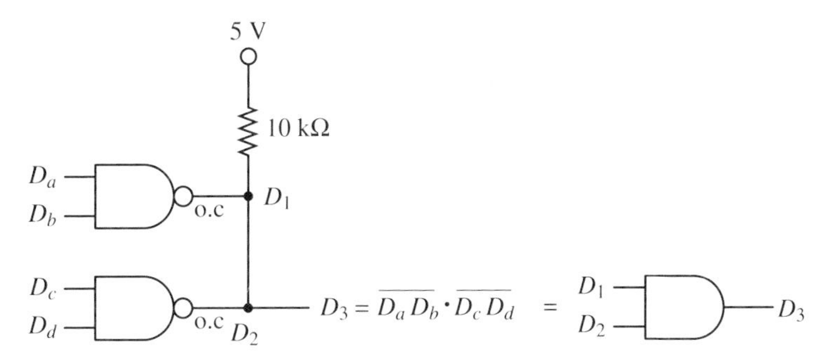


**Figure 10.54** Wired-AND using open-collector gates.

One reason for the open-collector option is that we may reduce gate count in some cases by using *wired-AND* logic. An example appears in Fig. 10.54. If either $D_1$ or $D_2$ is low, the output $D_3$ will be low. If both are high, the pull-up resistor brings $D_3$ high. The logical AND of $D_1$ and $D_2$ has been accomplished without an electronic gate.

Another reason for using open-collector gates occurs when, for example, we are using the results of TTL logic as inputs to CMOS logic running at a higher power supply. By leaving the TTL output floating, the high voltage can be defined to match that of the CMOS to follow by attaching a pull-up resistor to that high voltage. This would not be possible if the high were internally derived from the 5-V power supply of TTL, as is otherwise the case.

### 10.7.7 When Things Go Wrong

As is true of all circuitry, a design usually does not work the first time you try it out. Typically, a prototype design is developed on a *breadboard*—a board with arrays of holes into which ICs, wires, resistors, and so on, may be inserted. It often comes with a power supply that can be connected to desired rows of pins.

When the circuit does not work, first check that power is not only plugged in and turned on, but also whether the row you have connected to power supply pins on the chips is, in fact, the row you have elsewhere connected to the power supply. It is also possible that the power supply is faulty. Check it out with a digital multimeter (voltmeter).

Second, make sure that no chips were plugged in backward. Third, check to see whether a wire has popped out somewhere (or in printed circuit boards, whether all soldering connections are intact). Fourth, check every connection to every pin and make sure that they match the order on the specification sheets. Often, if some pins are unused it is easy to lose pin count, or the wires you think are lined up correctly are all one pin off.

Fifth, get out your logic probe. The logic probe is a hand-held device that is connected to the 5-V power supply. It has an LED that lights when the sharp metal tip on the end of the probe touches a "high" voltage level. Knowing the circuit you are trying to build and the pin layout for each chip, trace through the logic until you find a 1 that should be a 0, or a 0 that should be a 1. Pull out the associated chip if necessary to test it in isolation with the logic probe (elsewhere on the breadboard).

The chip could be faulty; for example, it could have been destroyed by one of the errors described above.

For chip testing, there is another device called a *pulser* that spits a train of alternating 1s and 0s out from its metal tip. To test an AND gate, tie one input high and touch the pulser tip to the other input pin. The logic probe should blink rapidly when its metal tip touches that AND output pin.

The overriding essential is that you know beforehand what your circuit should be doing, and that you can predict at each point in the circuit whether it should be a 1 or a 0 for a given set of inputs. Without this knowledge, the frustration of a sea of meaningless 1s and 0s can be overwhelming. But with this knowledge, troubleshooting simple logic circuits can be a straightforward procedure as dependable as the rules of logic themselves.

## 10.8 MISCELLANEOUS DEVICES

There is a seemingly endless variety of available digital ICs. Clearly, in just a few chapters on digital electronics there is no way each of the devices in such a list can be discussed in detail. Let us now mention just a few of the less self-explanatory items that are commonly used.

One unfamiliar item is the buffer/driver, '06 (which also inverts) and '07 (pictured in Fig. 10.55). These gates pass anything (or their complement if the buffer is inverting). Their purpose is to increase fanout over the usual limit of 10, or drive higher-current loads such as lamps or relays. The outputs can support 40 mA—two to three times what a standard gate can drive. We must be aware that they also introduce delay, just as any gate does. When timing is important, it can be critical to account for all delays in a circuit. The buffer can also be attached to NAND (device '37) or NOR (device '33) gates.

An output variation reminiscent of the open-collector idea that appears on, for example, buffer '125 is the *tri-state buffer*. In this case there are three possibilities: a 1, a 0, or a high-impedance (disconnected) state. A special enable input disables the output by putting it in the high-impedance state rather than 0 state. When enabled, the operation is that of a normal buffer, transferring input binary information to the output. Figure 10.56 illustrates the operation.

Of what use is the third state? It facilitates putting several devices capable of sending messages on the same line. Only one device is allowed to "talk" on the line at once even though output signals may be continuously generated at every connected device (recall the priority encoder). The third state is like disconnecting someone's telephone line. The one whose line is disconnected may talk all he or she wants on the disconnected phone, but his or her words will not be transmitted. The single device allowed to use the line is enabled, bringing the whole line up and down (1 and 0) in accordance with the message produced by that device. This line called a *bus,* is used extensively in all computer systems. We shall return to the topic of the bus in Secs. 12.3 and 12.4.2.

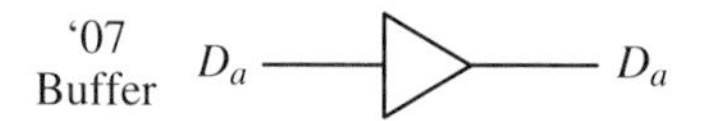


**Figure 10.55** Buffer for driving heavy loads and/or increasing fanout.

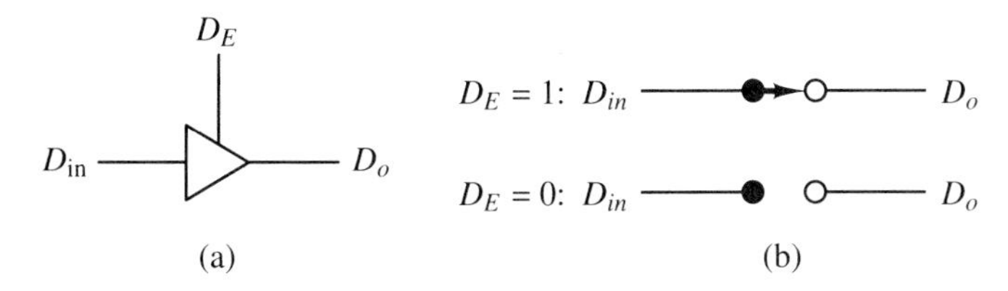


**Figure 10.56** Tri-state buffer: (a) circuit symbol; (b) operation.

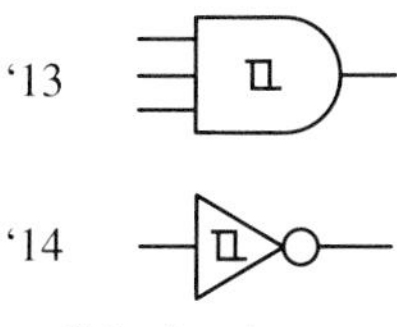


**Figure 10.57** Schmitt trigger AND and NOT circuit symbols.

Devices '13 and '14 (see Fig. 10.57) have Schmitt trigger inputs. As discussed in Sec. 10.6.2, the Schmitt trigger provides chatter-free low-to-high transitions (debounced switching). In device '13, they precede an otherwise normal four-input NAND gate and in device '14, an inverter. Note that the same Schmitt trigger symbol depicting an inverting hysteresis loop is used whether or not inversion actually takes place.

A device using both Schmitt trigger inputs and tri-state gates is the '242 transceiver. A transceiver is a digital circuit that controls flow of information: in which direction a digital signal can go. Consider the logic diagram in Fig. 10.58. The inverter without a bubble is just a noninverting buffer. If GBA is high, the tri-state inverter it controls (the upper one) will let $B$ be transferred to $A$ (left to right); then $A$ will equal $B$. Otherwise, that tri-state inverter will be an open circuit (high

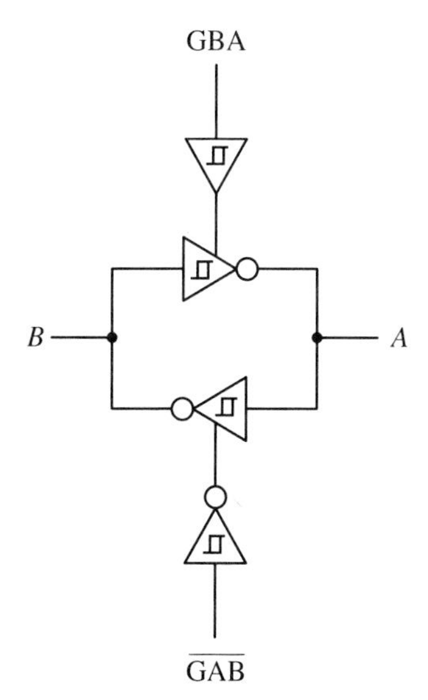


**Figure 10.58** Bidirectional transceiver used in the '242 IC.

impedance state). Similarly, if $\overline{\text{GAB}}$ is low, the control on the tri-state converter it controls (the lower one) will let $A$ be transferred to $B$ (right to left); otherwise, that tri-state will be open. This type of circuit is used in digital communication "buses," where sometimes we want digital information to go one direction, and other times the opposite direction. A prime example is reading and writing to digital memory. The '242 includes transceivers for four lines simultaneously controlled, hence the name *quad transceiver.* We shall note uses of the transceiver in Sec. 12.4.3.

Another important digital device is the parity generator/checker (e.g., the '280). This device takes nine digital (1/0) inputs and determines whether there is an odd or even number of inputs, referred to as the *parity* of that set of lines. If the number of inputs equal to 1 is even (0, 2, 4, 6, or 8), the even output is high (1) and the odd output is low (0) (even parity). Otherwise, the even output is low and the odd output is high (odd parity). A main use for this device is in error detection for digital data transmission. A *parity bit* is appended to a set of data so that it guarantees that the data sent have, for example, even parity. If the data on the receiving end are found to have odd parity, an error has been detected. We discuss this device again in Secs. 11.6.2 and 12.4.6.

The AOI (AND-OR-INVERT) is a handy logic fragment performing a frequently needed logical operation, a SOP (see device '50 in Fig. 10.59). Note that both $D_x$ and $\overline{D}_x$ must be provided as inputs. It is the extra $D_x$ that makes this chip "expandable." Often, this chip is used in conjunction with the expander chip '60 that is a four-input AND ("$D_x$") function that also outputs the NAND ("$\overline{D}_x$").

As noted previously, it would serve little purpose to define each and every available device in isolation. Instead, in Chapters 11 and 12 we introduce several of the others in a more organized fashion. In this chapter we have dealt strictly with devices whose output depends only on the current inputs. In Chapter 11 we look into the far more powerful sequential circuits, which form the basis of the computer. An industrially significant computer is the microprocessor, introduced in Chapter 12.

## 10.9 SUMMARY

Our purposes in embarking on the complicated world of digital electronics are threefold. First, the industrial world is increasingly becoming dominated by digital circuitry of all kinds. Therefore, the engineer who is unaware of this field is increasingly viewed as outdated, while the engineer who is knowledgeable is valued.

A suggested study parallel to these chapters (Chapters 10 through 12) would involve use of a personal computer (PC). Such a study would include the use of word processors,

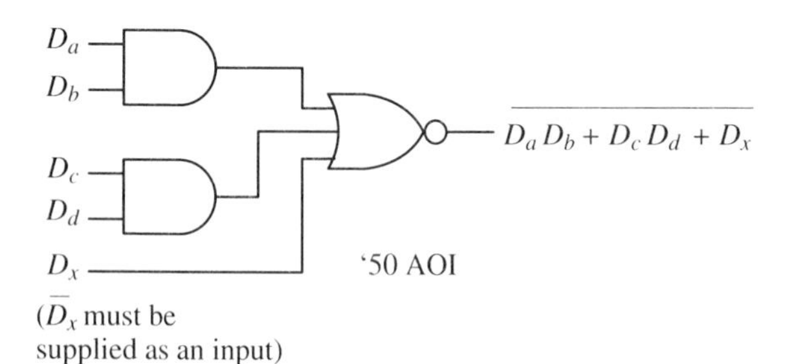


**Figure 10.59** Expandable AND-OR-INVERT single-chip combination.

spreadsheets, operating systems, high-level programming languages, computer networking such as local area networks (LAN), and a host of supporting digital gadgetry.

Although outside the scope of a typical one-semester basic electrical engineering course, these topics are essential to provide a complete view of the field of digital electronics and to maximize its usefulness to the nonelectrical engineer. Such knowledge tends to be very specific to particular systems and is very subject to change; it is therefore best left in the hands of manufacturers' operating manuals. After studying these chapters (Chapters 10 through 12), a perusal of a digital electronics catalog will be both fun and thought provoking. All the animals in our menagerie and many others will be there, along with their prices!

Second, a nonelectrical engineer must be conversant with a digital design engineer because together they must adapt digital circuitry to new situations that have traditionally used analog electronics or even no electronics. If the electrical engineer can effectively communicate with the non-EE, the project is more likely to succeed.

Finally, the nonelectrical engineer has a unique perspective from which to view digital electronics. True creativity occurs when previously disconnected bodies of knowledge are profitably brought together. With the extremely high versatility and low cost of digital circuitry (most chips we discuss in the SSI and MSI density categories cost less than $1), this emerging technology has endless possibilities for profitable application in other fields of engineering. However, this can occur only if the non-EE has the necessary background.

In this chapter we have sought to introduce some of the most important issues and circuits of basic logic design. The connection between the familiar situation of "the man at the gate" and binary logic was made, and the transition to a circuit implementation was fairly straightforward. This transition was facilitated by introduction of the truth table, an enumeration of all possible outcomes in a given logical system. Any digital logic system can be constructed using a few simple gate types, such as the NAND and NOR operators.

To whet our interest in digital electronics (and perhaps our TV appetites, too), the example of professional football playoff conditions was considered in some detail. Out of this case study came several new concepts and devices: Don't-care entries in a truth table, the XOR operation, DeMorgan's laws of logic, multiplexers and demultiplexers, and the general idea of data control logic. A sequel to these ideas was the encoder/decoder. The Gray code and BCD code are used for various purposes and must be converted to and from binary or base 10 representations. Priority encoders are ingeniously constructed to allow the digital designer to create a 10-level prioritizing hierarchy for competing devices in a very flexible manner.

Next, the sum-of-products and product-of-sums expressions provided a more formal, systematic way of expressing logical relations. The relation [Eq. (10.17)] between these two forms is quite simple. To reduce chip count and the possibility for human error or circuit malfunction, the Karnaugh map method of logical expression reduction was presented. It can greatly reduce the tedium of unsystematic application of DeMorgan's laws while simultaneously producing an optimum logic design. Its practical applicability is usually limited to four input variables.

A brief study of input–output devices is essential if a circuit is ever going to be built. For without these transducers we cannot control and observe what a digital circuit is doing; such a circuit is useless. We studied the most basic input device: the switch. The main consideration unique to digital circuitry regarding switches is the bouncing problem. The Schmitt trigger can be used to debounce a switch, as can the RS flip-flop, studied in detail in Sec. 11.2.2. Basic output devices are LEDs and alphanumeric displays. The host of other "peripherals" (input–output devices) was mentioned, but detailed studies of many of these are beyond the scope and level of this book.

The last main area of study was consideration of a variety of practical issues involved in making a circuit function. While the goal was not to attain the level of a technician—

impossible without laboratory sessions—at least some of the mystique surrounding digital design was removed: for example, which end of the chip is up; how to count the pins; what kind of power supply to use; an overview of some of the main technologies used in digital design; the "fanout" limitation; scales of integration in existence and their uses; how to read a part number; what an open-collector version of a device means.

Finally, a few miscellaneous digital devices were considered briefly. The buffer, as far as ideal logic goes, has no purpose. Its value is in its ability to drive large electrical loads—either specialized high-current loads or just more of the same chips (to increase fanout). A special kind of buffer, called the tri-state buffer, is useful in digital data communication systems. The third state, other than 1 and 0, is a disconnected state that gives the line to which that output is connected over to some other device that needs it.

The Schmitt trigger makes a guest appearance on the list as an input tack-on to gates for improved switching behavior in cases where that is critical. That type of gate is useful for connecting mechanically switched inputs to the digital system, as it debounces. The transceiver, a device that controls the direction of information flow in computer systems, is based on tri-state logic elements. The parity checker can be used in such data transmission for error detection (and ultimate elimination by, for example, retransmission of erroneously received data). Finally, the AND-OR-INVERT chip was thrown in for its practical utility and interest.

## PROBLEMS

**10.1** What is a truth table? Determine the truth table given the following conditions: Task $T$ can be completed only if both tasks $A$ and $B$ are completed, or if both tasks $B$ and $C$ are completed. First assume that $A = 1$, $B = 1$, $C = 1$, and $T = 1$ signify that the respective task is completed. Then repeat, assuming now that each logic variable is 0 if the respective task is completed.

**10.2.** For the given binary waveforms and gates shown in Fig. P10.2, sketch the output waveforms $F_1(t)$ and $F_2(t)$.

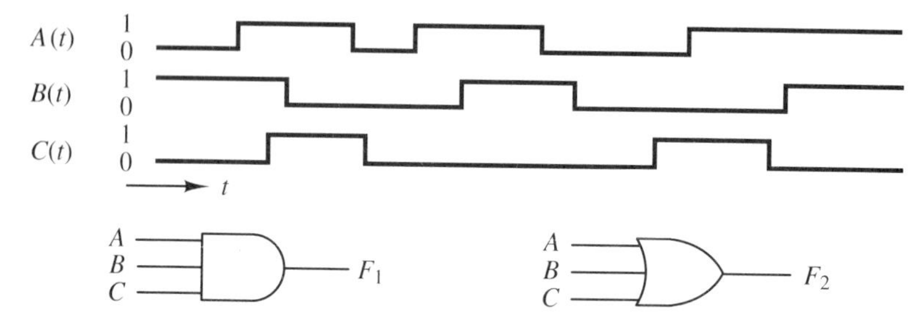


**Figure P10.2**

**10.3.** For the circuit shown in Fig. P10.3:
(**a**) Write down the logic expression for $F$.
(**b**) Use a truth table to check your answer.

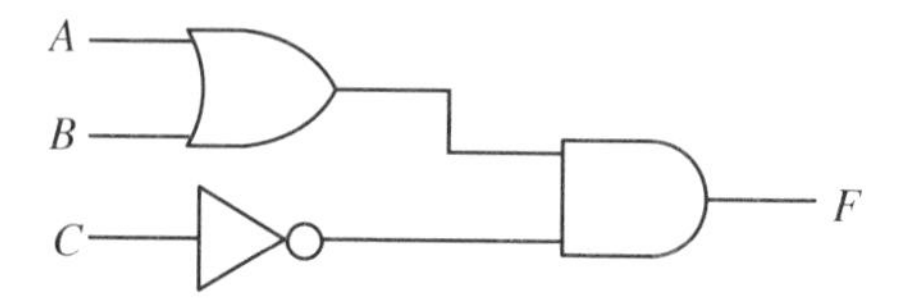


**Figure P10.3**

**10.4.** The truth table of a logic circuit is shown in Table P10.4.
(a) Write down a three-term SOP logical expression for $F$.
(b) Simplify that logical expression to one involving only two terms, each involving only one logical variable.
(c) Draw the logic circuit of the simplified expression.

**TABLE P10.4**

| $A$ | $B$ | $F$ |
|---|---|---|
| 0 | 0 | 1 |
| 0 | 1 | 1 |
| 1 | 0 | 0 |
| 1 | 1 | 1 |

**10.5.** For the circuit shown in Fig. P10.5, find the logic expression for the output $F$ and draw the output waveform $F(t)$ for the given input waveforms $A(t)$, $B(t)$, and $C(t)$.

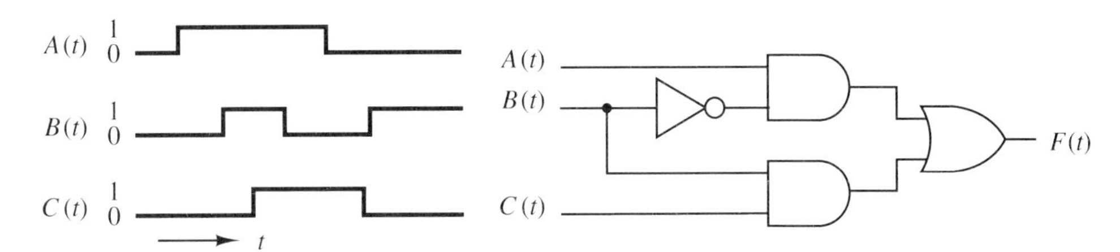


**Figure P10.5**

**10.6.** Based on the logic values shown in Fig. P10.6 which were determined by measurements using a logic probe, determine which (if any) of the gates is/are malfunctioning.

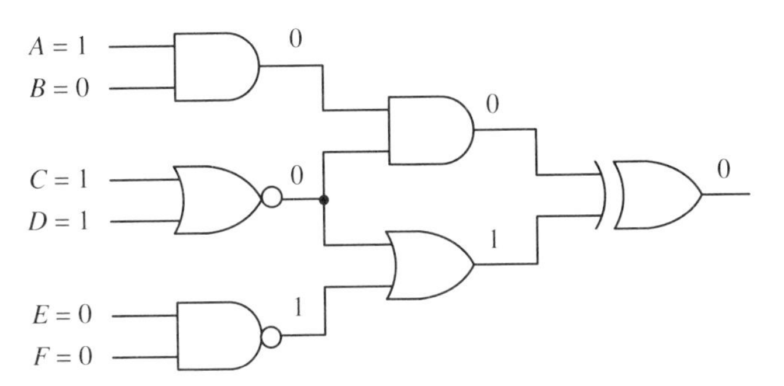


**Figure P10.6**

**10.7.** If one input to an XNOR gate (i.e., an inverted XOR gate), $D_A(t)$, is as shown in Fig. P10.7 and the output $D_o(t)$ is also as shown, what must the other input, $D_B(t)$ be?

**10.8.** Fig. 10.22 shows a 1 : 4 demultiplexer. If we want to assign $D_{in}$ to $D_2$, to what should the control lines be set? What are the outputs $D_0$, $D_1$, $D_2$, and $D_3$, assuming that $D_{s1} = 1$ and $D_{s0} = 1$?

**10.9.** What is a multiplexer? What is a demultiplexer? Give examples of their uses.

**10.10.** What are the differences between a multiplexer/demultiplexer system and an encoder/decoder system?

**10.11.** Suppose that in Fig. 10.26, $D_3 D_2 D_1 D_0 = 1001$, and that the data stored in the

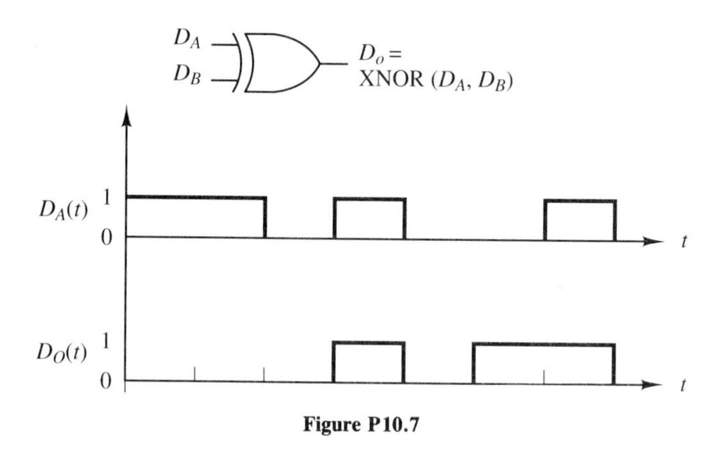


**Figure P10.7**

memory are as shown in Fig. P10.11, with the numerical order of memory elements arranged in increasing order from top to bottom, as shown in Fig. 10.26. Specify the retrieved eight-bit set (byte) of data.

| |
|---|
| 01111001 |
| 11000110 |
| 10011101 |
| 10000001 |
| 00111010 |
| 11000010 |
| 01101111 |
| 10010010 |
| 11011111 |
| 00101101 |
| 11111100 |
| 00011000 |
| 01000011 |
| 11100000 |
| 10011001 |
| 00011010 |

**Figure P10.11**

**10.12.** Four people have access to a computer for processing their data. Each person has his or her own data line, but there is only one input to the computer. Assume that only one person tries to access the computer at a time. Design a circuit using a 4 : 2 encoder and a 2 : 4 multiplexer that will select the correct data line $D_0$ through $D_3$ of person 0 through 3 when one of those people sets his or her request line ($R_0$ through $R_3$) high. Draw a circuit diagram using only block modules for the encoder and multiplexer. Specify using logic tables what the encoder and decoder will do, but do not design their internal gate structure. Assume that if no one wants the data line that it is by default assigned to person 0.

**10.13.** Regarding Problem 10.12, suppose that person 2 is now to have top priority for gaining access to the computer, person 0 is next on the totem pole, then person 1, then person 3. Using a block diagram priority encoder, show a circuit diagram that will connect the proper line to the computer.

**10.14.** A logic circuit consisting of only NOR gates is shown in Fig. P10.14.

**(a)** Determine a logic expression for $F$.

**(b)** Construct the truth table for $F$.

**(c)** Simplify the logic function to the form $(w + x)y$, where $w$, $x$, and $y$ are one of $A$, $B$, or $C$.

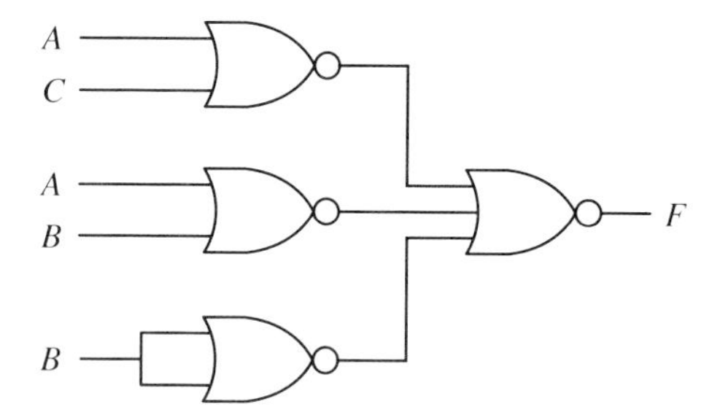


**Figure P10.14**

**10.15.** Using truth tables, prove Eqs. (10.10) through (10.13).

**10.16.** Simplify the following using DeMorgan's laws, absorption laws, and others, such as commutative and distributive laws.

**(a)** $F = ABC + AB\overline{C}$

**(b)** $F = \overline{A}\,\overline{B}\,\overline{C} + \overline{A}\,\overline{B}C + \overline{A}B\overline{C} + AB\overline{C}$

**(c)** $F = \overline{A}\,\overline{B} + \overline{A}\,\overline{B}\,\overline{C} + A\overline{B} + AB\overline{C}$

**(d)** $F = (A + B)\overline{C}(A + C)$

**(e)** $F = A\overline{B} + (\overline{A} + BC)(B + \overline{C})$

**10.17.** K-maps are very useful in simplifying logic expressions. Try to give a theoretical basis of K-maps. Show a three-variable example of how this simplification is achieved using a K-map.

**10.18.** **(a)** In Fig. P10.18, put a square around the K-map element for which all the inputs are 1, and put a triangle around the element for which all the inputs are 0. Provide the value of the logic function $F$ in each case.

**(b)** Provide the simplified SOP expression for $F$ using the K-map.

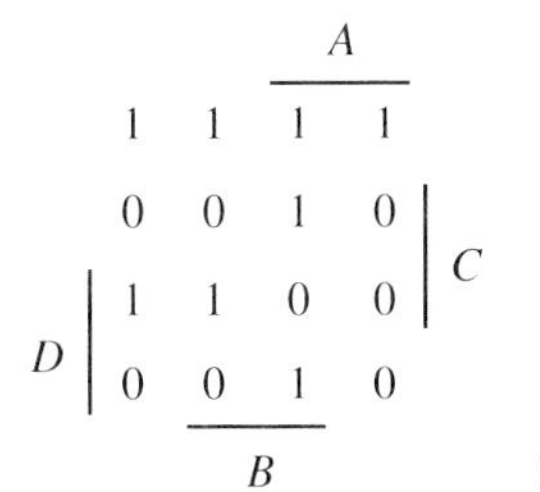


**Figure P10.18**

**10.19.** Write down the logic expressions according to the K-maps given in Fig. P10.19.

**10.20.** A meeting of the chairs of the board is called. Jim, one of the nonchair board members, thinks the meeting will be a total failure if any of the following happen:

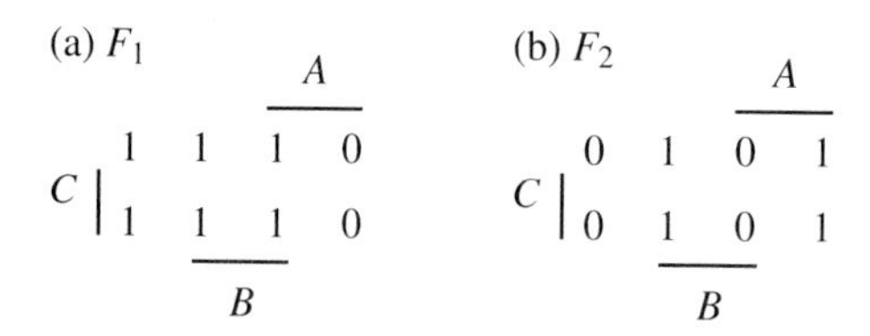


(c) $F_3$

A

0 1 1 0

0 1 1 0 | C

D | 1 0 0 1

1 0 0 1

B

(d) $F_4$

A

0 1 0 1

C | 1 0 1 0

B

**Figure P10.19**

(i) None of them shows up.
(ii) Alice is absent but Chris shows up (Chris creates trouble).
(iii) Alice is present but so is Bill (they argue).
(iv) Clarence is absent but Bill shows up (Bill is incompetent to lead).

Define the logic variables: $A = 1$ if Alice comes, $B = 1$ if Bill comes, and $C = 1$ if Chris comes. How might Jim simplify his statement of conditions that the meeting is a failure? Write the logic expression for $F$ (the logic variable signifying that the meeting is a failure), and simplify it using a K-map. Then state the simplified result in words.

**10.21.** Suppose that we are given the following logic expression:

$$F = \overline{A}\,\overline{B}\,\overline{C}\,\overline{D} + AB\overline{D} + A\overline{C}\,\overline{D} + A\overline{B}\,\overline{D} + ABCD + A\overline{B}\,\overline{C} + \overline{A}\,\overline{B}\,\overline{C}D.$$

**(a)** Draw a circuit diagram implementing this unsimplified expression for $F$. How many gates are required? (Assume that the available gates can have up to only four inputs each.)

**(b)** Write the truth table and draw the K-map. Use it to find a simplified SOP expression for $F$, and draw a circuit implementation of this simplified expression for $F$. Now how many gates are required?

**10.22.** The logical expression for a two-input XNOR gate is $F = \overline{A}\,\overline{B} + AB$.

**(a)** Write down the truth table of the XNOR gate.

**(b)** Using the four-input multiplexer shown in Fig. 10.20, implement the logic expression for $F$ by setting $D_{S1} = A$, $D_{S0} = B$, and $y = F$. Give appropriate values to the inputs $D_0$ through $D_3$.

**10.23.** Use only NAND gates to implement the following logic functions.

**(a)** $F = AB + BC$

**(b)** $F = A + C + BD + \overline{B}\,\overline{D}$

**(c)** $F = (A + \overline{C})(B + C)$

**10.24.** Employ only NOR gates to implement the following logic functions.

**(a)** $F = (A + \overline{B})(B + \overline{C})$

**(b)** $F = (\overline{A} + \overline{B} + \overline{C})(A + B + C)$

**10.25.** For the circuit shown in Fig. P10.25 to implement the desired logic manipulation, that is,

$$B_1 = A_1 + A_2 + A_3 + A_4$$

$$B_2 = \overline{A}_1(A_2 + \overline{A}_3A_4),$$

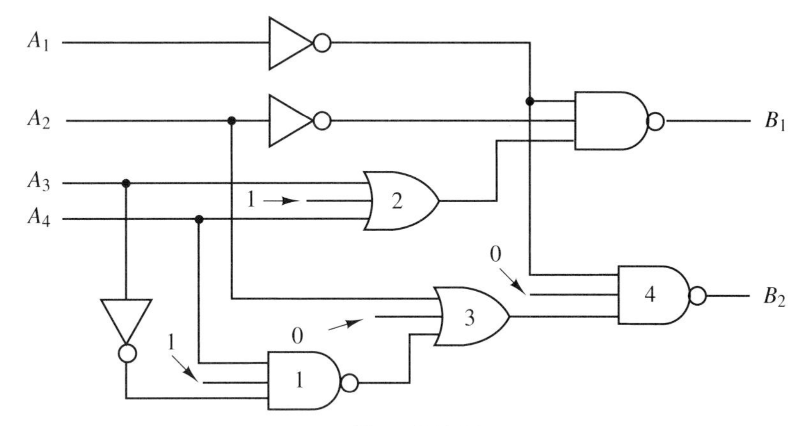


**Figure P10.25**

the unused inputs of the gates must be wired properly. The given expressions for $B_1$ and $B_2$ are the functions that would result were gates 1 to 4 two-input instead of three-input, with the 1/0 input eliminated. Only three-input gates are available, so the four unused inputs have been tied to 0 or 1. Determine which have been connected incorrectly.

**10.26.** Simplify the following logic functions by using Karnaugh maps.

(**a**) $F = AC\overline{D} + AB + BC\overline{D} + \overline{A}BCD$

(**b**) $F = AB + DC + \overline{A}DC$

(**c**) $F = ABC + A\overline{B}C + BD$

(**d**) $F = A\overline{B} + AB + \overline{C}D + CD$

**10.27.** Give the truth table in Table P10.27:

(**a**) Find a logic expression for $F$.

(**b**) Simplify your expression using a Karnaugh map.

(**c**) Draw a logic circuit that implements the simplified expression for $F$.

**TABLE P10.27**

| $A$ | $B$ | $C$ | $F$ |
|---|---|---|---|
| 0 | 0 | 0 | 1 |
| 0 | 0 | 1 | 1 |
| 0 | 1 | 0 | 0 |
| 0 | 1 | 1 | 1 |
| 1 | 0 | 0 | 0 |
| 1 | 0 | 1 | 0 |
| 1 | 1 | 0 | 0 |
| 1 | 1 | 1 | 1 |

**10.28.** What is a bouncing problem? How can it be solved?

**10.29.** Seven-segment displays are still widely used. Another method is a matrix display. Figure P10.29a is a 5 × 7 matrix that can be used to display the 10 Arabic numbers (0, 1, 2, . . . , 9). An example (the numeral 0) is given in Fig. P10.29b. Symbols "."s in the figures indicate the matrix points and "*"s indicate the points that are on. If we want to display 1, 2, 7, and 9, which of the points in the matrix should be on? Answer using figures such as that in Fig. P10.29b.

**10.30.** In Example 10.18 we determined that a NOT, a NAND, and 20 NOR gates could not be driven by the XOR gate. Suppose that the XOR gate is driving only other XOR

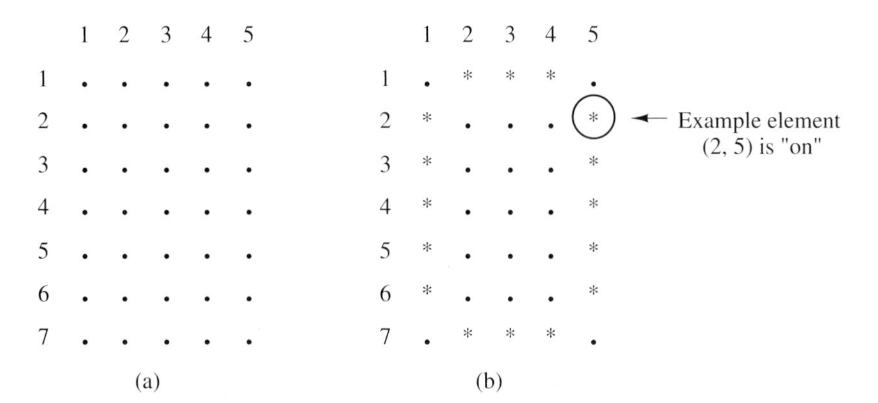


**Figure P10.29**

gates. Calculate the maximum number of XOR gates it can drive. (Use data from the text.)

**10.31.** In Fig. 10.54, two NAND OC gates are wired together to implement AND-OR-NOT logic. If the NAND OC gates are replaced by NOR OC gates, what logic function is implemented? (Assume that each NOR gate also has two inputs.)

**10.32.** Why must we use buffers in digital systems? What is the main advantage of tri-state buffers?

**10.33.** What sort of device package is recommended when chip pin count and chip density on a circuit board are to be maximized but cost is to be held down?

**10.34.** Classify the following integrated circuits according to the following categories: SSI, MSI, LSI, or VLSI: (**a**) a 4 : 16 address decoder, (**b**) a hex inverter, and (**c**) an Intel 27011 EPROM 1-Mbit (1 million bit storage) memory chip. (We study computer memory systems in detail in Chapter 11.)

**10.35.** Identify the following integrated circuit chip: MC74LS251N. Refer to Sec. 10.7.6 and to any digital IC parts list.

**10.36.** For the Gray code encoder circuit of Fig. 10.24, give the output waveforms for the case in which the input waveforms are as given in Fig. P10.36.

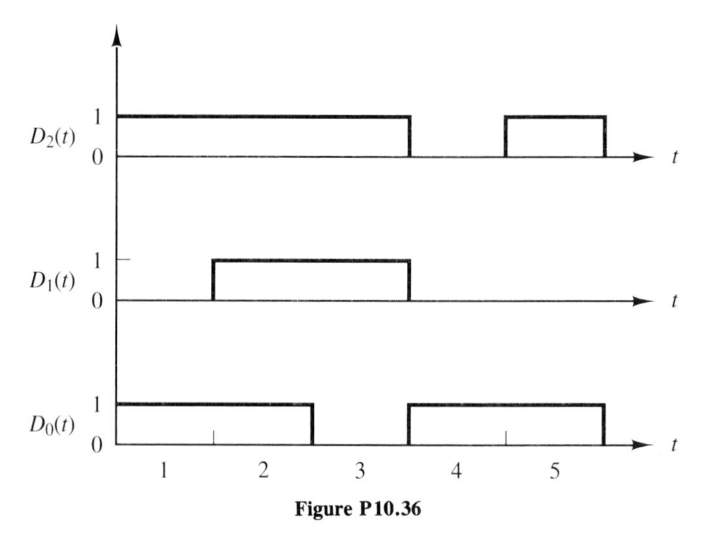


**Figure P10.36**

**10.37.** Repeat Problem 10.36 for the case in which $D_2(t)$, $D_1(t)$, and $D_0(t)$ as given in Fig. P10.36 are actually presented instead, to the inputs $D_c$, $D_b$, and $D_a$, respectively, of the Gray code decoder in Fig. 10.25.

**10.38.** Suppose that in the circuit of Fig 10.43, the switch is switched from open to closed and then back open. Plot $v_c(t)$ and $v_o(t)$ versus time, showing time intervals on the order of 50 ms. Do so for the following conditions.

(**a**) Capacitor $C$ is removed from the circuit.

(**b**) Capacitor $C$ is back in place.

## ADVANCED PROBLEMS

**10.39.** Consider the following alarm system. If the temperature of the environment is not higher than 90°F, or if the moisture is over 95% and the dust density is over 0.00001%, the buzzer will sound to alert monitoring personnel.

(**a**) Write down the truth table.

(**b**) Write an expression for the logical function dictating when alarm is to sound.

(**c**) Use primary logic gates to implement the logic system.

**10.40.** In a stairway, there is a light ($L$) with two switches (switch $A$ and switch $B$) at the top and bottom ($L = 1$ means that the light is on, $A = 1$ means that switch $A$ is on, etc.). When going upstrairs and throwing the associated switch, the light is to turn on. Upon leaving the stairs and throwing the associated switch, the light is to turn off. The same kind of behavior must also hold for going downstairs. Devise such a digital switch system.

(**a**) Determine the corresponding truth table.

(**b**) Show that $L = \overline{A}B + A\overline{B}$.

(**c**) Sketch the circuit diagram.

**10.41.** The two-input XOR gate is explained in Section 10.3. Three such gates are connected together as shown in Fig. P10.41.

(**a**) Write down the output logic function $F$ in terms of $D_1$, $D_2$, $D_3$, and $D_4$.

(**b**) If only one of the inputs is one and all the rest are zero, what is the output $F$? If any two of the inputs are 1 and the other two are zero, what is $F$? What if three inputs are 1, or all four inputs are 1?

(**c**) What can you conclude from parts (a) and (b)?

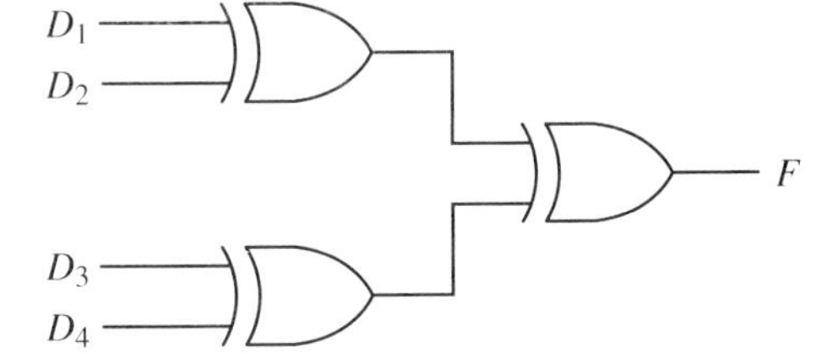


**Figure P10.41**

**10.42.** In this problem you will design an encoder that performs the coding as shown in Fig. P10.42.

(**a**) Write down the truth tables for the two output binary variables $C_1$ and $C_2$. Designate "don't care"s by "x."

(**b**) Draw the Karnaugh maps, drawing the largest loops possible.

(**c**) From your Karnaugh maps, write down the simplified expressions for the two outputs and implement them with simple gates.

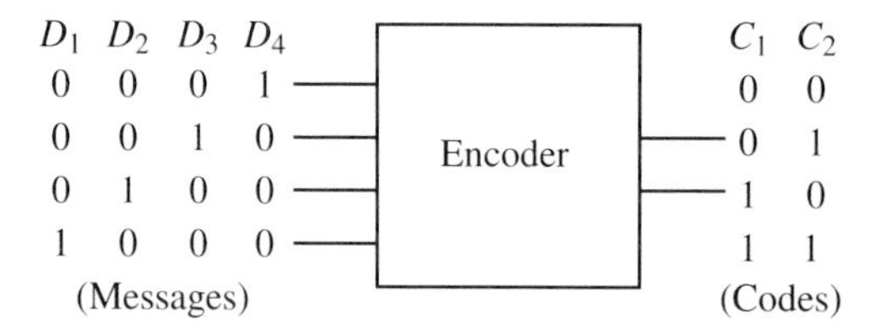


**Figure P10.42**

**10.43.** Use a Karnaugh map to simplify the logic circuit shown in Fig. P10.43.

(**a**) Find the logic expression for the circuit output $F$.

(**b**) Simplify that expression using DeMorgan's laws (etc.) to get a SOP expression. Then draw the Karnaugh map and use it to simplify the SOP expression.

(**c**) Sketch the simplified circuit.

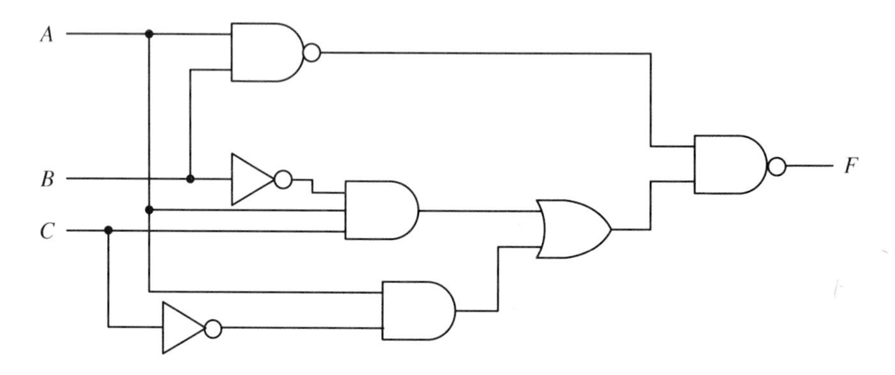


**Figure P10.43**

**10.44.** Gray code has the characteristic that adjacent words differ by only one bit (i.e., 000, 001, 011, 010, 110, 111, 101, 100 is the three-bit Gray code sequence). Devise a binary code–Gray code converter by using NAND gates.

**10.45.** A simplified voting machine is designed such that its output is one when more than half of its inputs are ones. We assume that there is only one voting location. Table P10.45 shows the truth table for the voting machine with three inputs, that is, three voters. Design the voting machine using only NOR gates (recall that any logic function can be implemented with only NAND or with only NOR gates). First draw the Karnaugh map and write the simplified SOP expression. Then rewrite the simplified SOP expression using DeMorgan's laws to express it in terms of NOR operations.

**TABLE P10.45**

| $A$ | $B$ | $C$ | $F$ |
|---|---|---|---|
| 0 | 0 | 0 | 0 |
| 0 | 0 | 1 | 0 |
| 0 | 1 | 0 | 0 |
| 0 | 1 | 1 | 1 |
| 1 | 0 | 0 | 0 |
| 1 | 0 | 1 | 1 |
| 1 | 1 | 0 | 1 |
| 1 | 1 | 1 | 1 |

**10.46.** Siemens DL-3416 16-segment display, shown schematically in Fig. P10.46, is a more advanced alphanumeric display than the seven-segment display. It is capable of producing any of 64 characters, including the entire alphabet. Depending on the shape of the character, each of the 16 segments is either on or off. Determine which segments should be on in order to produce the following characters: "5", "7", "%", "&", "*", "M", "[", "\", and ":". (1 represents on, 0 represents off.) Make a table whose columns are 1 or 0 depending on whether or not a particular segment (a through p) is on, and whose rows are the characters above desired to be represented.

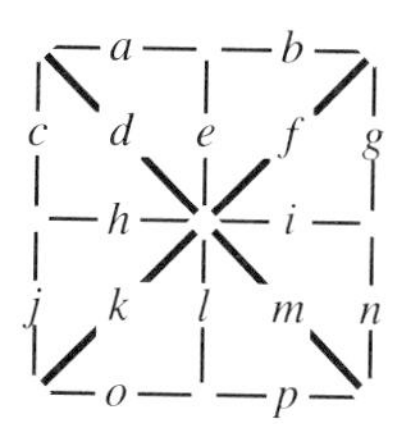


**Figure P10.46**

**10.47.** The time delay of a signal passing through a gate sometimes has a significant effect on the behavior of the circuit. Consider the logic circuit shown in Fig. P10.47a.

**(a)** Write down the logic expressions of $F_1$, $F_2$, and the output $F$.

**(b)** Now take the delay times of the gates into consideration; letting the delay times of the AND, OR, and NOT gates be the same value $t_o$, draw the output waveforms $F_1(t)$, $F_2(t)$, and $F(t)$ for the given input waveforms $A(t)$, $B(t)$, and $C(t)$, which are shown in Fig. P10.47b. What unexpected feature did you see on the output waveform $F(t)$? (This phenomenon is called *static glitch.*)

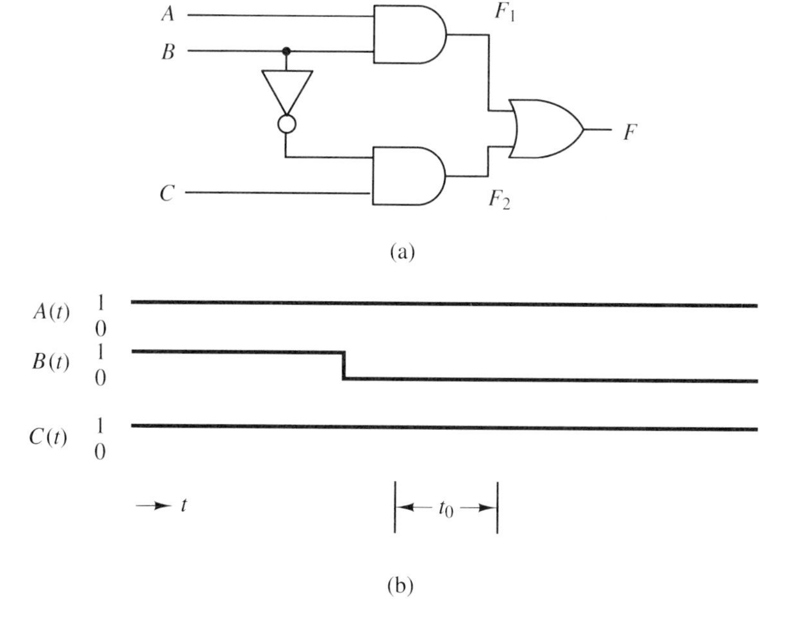


**Figure P10.47**

# 11

# Digital Electronics II: Sequential and Arithmetic Systems

## 11.1 INTRODUCTION

In the first part of our study of digital electronics, we concentrated on the basic ideas surrounding binary logic and binary variables. Although many powerful techniques and devices are now in our repertoire, some essential ones are yet missing. All of the devices considered in Chapter 10 were memoryless: The output was a function of current inputs only, aside from propagation delays.

But as is true for all industrial processes, automatic repetition is where the real power lies. For example, if there are three stages in mixing paint, we could have three containers, each with the appropriate stirrer and mixing duration. It would be far less costly to have a procedure whereby the same mixing tank could be used for all three.

This would require some new elements, however. The first is memory. We must keep track of what stages have already been completed. Next, counters would be handy to step through the three stages and time each stage in progress. Most important, there must be a set of instructions directing when to begin counting and stirring, which stirrer is to be connected when, and in emergency conditions what action to take.

In this chapter we consider some elements and ideas that are crucial to automatic computational and control systems. A *sequential system* is a system like the automated paint stirrer in that there are several states or stages. Depending on the state, certain actions will take place. And transitions from state to state take place in a well-defined, orderly and sequential fashion. In order to recall in what state the machine is, memory systems are necessary. The most basic of these is the *flip-flop*, which stores only one piece of binary information (on or off).

Arrangements of these flip-flops can be configured to be able to count and store numbers. Counting must be done using binary variables (base 2) in digital systems, yet humans think in base 10. Therefore, a study of number systems is necessary to convert what we want done into something a digital system can do. Counting is the simplest arithmetic procedure, but everyone knows that in any practical situation, addition, subtraction, multiplication, and division may be required. These operations are introduced in the context of describing what a personal computer (PC) does when we type a number into it.

The chapter concludes with a discussion of the types of memory structures that are most used in digital systems. This last topic in particular, and the A/D and D/A converters presented in Sec. 9.7.6, prepare the way for our final digital study of the next chapter, the microprocessor.

## 11.2 THE FLIP-FLOP

### 11.2.1 Introduction

The concept of *state* in any system is applicable whenever the system output depends not only on present inputs, but also on past values of the output and/or inputs. When there is a dependence upon signals from the past, there must be storage of that information.

We have already considered one example of state: the Schmitt trigger. The response of the Schmitt trigger to its present input depends on the past values of the input. For the noninverting Schmitt trigger, the output will stay high when the input is set to 0 V or even slightly negative if the output is currently high. But the output is currently high only if the most recently exceeded threshold is $+V_{thr}$ (not $-V_{thr}$). Therefore, the Schmitt trigger is a kind of memory device, a fact that makes possible its main application, a no-jitter switch. It is used in digital circuitry to reduce the rise time (sharpen the edge) of "clock" waveforms and debounce mechanical switches. Its memory behavior is a direct consequence of positive feedback.

We have already seen in Chapter 3 that a magnetic core is another memory device, although again the idea of memory is not the major application in mind. In that case the magnetic flux density $B$ as a function of applied magnetic field intensity $H$ depends on which saturation region was most recently attained. For the magnetic core, the reason for the *memory* is the magnetic inertia of the multitude of little dipoles making up the iron. They tend to remain in the polarity to which they were most recently forced by the applied $H$. This concept is used in the context of computers as the basis for design of magnetic memory devices such as diskettes, tapes, and in the past, rotating drum massive magnetic memories.

Anyone who has used a pocket calculator will substantiate that one of the most helpful keys is the memory button. As soon as we attempt something more than $A + B = C$, it becomes convenient to store intermediate results in memory. Otherwise, we are forced to write it down on paper or memorize it—using two of the oldest memory devices.

What is the way in which the pocket calculator stores the number? Or how does the microwave remember that you have selected "broccoli"? There are a number of methods of implementing electronic memory, but one of the most pervasive is the flip-flop. The flip-flop, as we shall see, is just a couple of logic gates connected together with (positive) feedback. Based on the current value of the output (which depends on the past input), the input will cause the output either to stay where it is, flip to a 1, or flop to a 0—hence the name.

Like the *B–H* and Schmitt trigger examples, but unlike the calculator (which stores an entire number), the flip-flop stores only a 1 or 0. Of course, we shall see that combinations of flip-flops are in practice used to store entire numbers in binary form. For now, though, the task at hand is to store a single binary digit, or *bit* of information.

### 11.2.2 The *RS* Flip-Flop

Why not use a Schmitt trigger as a memory element? The reason is as follows. It is true that the output of a Schmitt trigger is a binary variable and that the Schmitt trigger has memory as stated above. The problem is how the change is implemented. A *set* operation to set the output to 1 requires the input to exceed one threshold, while the *reset* operation to change the output to 0 requires the input to drop below a different threshold. The region in between thresholds serves as the situation in which neither set nor reset is requested.

Consequently, there are three distinct input ranges; the input must be a trinary rather than binary variable. These are required to implement the three distinct operations involved with a memory device: set, reset, do nothing. Consequently, there must be two binary signals over which to distribute these commands because a single binary variable provides only two possibilities. Thus in practice two binary inputs are used rather than the single trinary input of the Schmitt trigger. In the *RS* (reset/set) flip-flop now to be discussed, these are the set ($D_S$) and reset ($D_R$) lines. In Sec. 11.2.4 we introduce the "*D*" flip-flop, in which case the set and reset operations are contained within one binary (*data*) variable, but another enable line is required to provide for the "do nothing" command.

It is thus perhaps most natural to implement binary memory using binary logic gates with positive feedback and two binary inputs, as opposed to trying to use the Schmitt trigger. The most simple flip-flop to understand is the *RS* flip-flop, shown in Fig. 11.1. It consists of two NOR gates, each of whose outputs is sent to one of the inputs of the other NOR gate. For reference, the truth table for a NOR gate is reproduced in Fig. 11.1. To analyze the circuit behavior, the most important thing to keep in mind is that both of the NOR gates continue to function as NOR gates

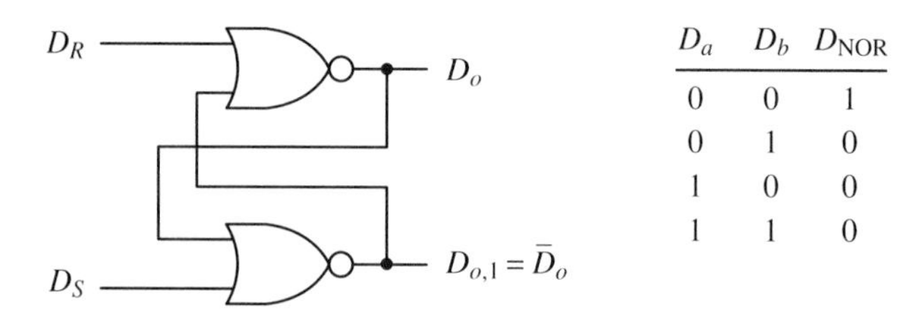


| $D_a$ | $D_b$ | $D_{\text{NOR}}$ |
|---|---|---|
| 0 | 0 | 1 |
| 0 | 1 | 0 |
| 1 | 0 | 0 |
| 1 | 1 | 0 |

**Figure 11.1** *RS* flip-flop with NOR truth table for reference.

even with the positive feedback. Also, recall that there are only two possible values for the outputs: 1 and 0.

The output of the upper NOR gate, $D_o$, is designated as being the state of the flip-flop. The output of the lower NOR is the logical complement of the state ($D_{o,1} = \overline{D}_o$), as will now be shown. Suppose that the output of the upper NOR is 1: $D_o = 1$. Right off the bat we know that both of the inputs ($D_R$ and $D_{o,1}$) of the upper NOR must be 0, or $D_o$ would have to be 0.

The fact that the output of the lower NOR, $D_{o,1}$, must be zero is consistent with the requirements on its inputs. Specifically, because at least one of its inputs ($D_o$) is 1, $D_{o,1}$ must be zero regardless of the value of $D_S$. Now suppose that $D_o = 0$. If $D_S = 0$, then $D_{o,1}$ will be 1 because both of its inputs are zero. With $D_{o,1} = 1$ sent to become one of the inputs of the upper NOR gate, $D_o$ will remain at 0 regardless of the value of $D_R$ because one of the inputs of that NOR ($D_{o,1}$) is 1. The consequences of this discussion are (1) that if $D_o = 0$ and $D_S = 0$, $D_o$ will remain at 0 independent of the value of $D_R$ and (2) that if $D_o = 1$ and $D_R = 0$, $D_o$ will remain at 1 independent of the value of $D_S$.

But what if $D_o = 0$ and $D_S$ is changed to 1? We might guess that $D_o$ will change from 0 to 1 ("set"). Similarly, if $D_o = 1$ and $D_R$ is changed to 1, $D_o$ will "reset" to 0. Let us show that these statements are indeed true.

Suppose that $D_o = 0$ and $D_S$ becomes 1, and assume that $D_R$ is 0. Then because one of the inputs of the lower NOR is 1 ($D_S$), $D_{o,1}$ will become 0. This will cause both of the inputs of the upper NOR to become 0 (recall that we assumed that $D_R$ is now 0), and hence $D_o$ becomes 1. Indeed, the flip-flop has been set. Similarly, suppose that $D_o = 1$ and $D_R$ becomes 1, and assume that $D_S = 0$. Then because one of the inputs of the upper NOR is now 1 ($D_R$), $D_o$ will become 0. This will cause both of the inputs of the lower NOR to become 0 (recall that we assumed that $D_S$ is now 0), and hence $D_{o,1}$ becomes 1. Indeed, the flip-flop has been reset. Once the set or reset has occurred, the situation is that of the preceding paragraph: The output will remain fixed, whether or not that set or reset line stays equal to 1, until a reset or set is issued to change state again.

It has now been shown that $D_S$ will set the flip-flop if it is not set, and $D_R$ will reset the flip-flop if it is currently set; otherwise, the flip-flop will remain in its present state. Thus the output is fixed by whichever was raised high last: 1 if $D_S$ was, and 0 if $D_R$ was. Also, examination of both of the steady-state situations shows that in each, $D_{o,1} = \overline{D}_o$, as claimed. The combination $D_S = D_R = 1$ must be avoided; because of slight delay differences between gates, one or the other gate would determine the output in an unpredictable fashion.

The analysis above has proved that the desired operation of an *RS* flip-flop has been achieved. It may be summarized in a *state transition table,* as shown in Table 11.1. The flip-flop state $D_{o,b}$ is the previously assumed steady-state value of $D_o$ (*before* the $D_S$ and $D_R$ inputs became as shown in that row of Table 11.1). Similarly, $D_{o,a}$ is the new steady-state value of $D_o$ based on the inputs $D_S$ and $D_R$ for that row of Table 11.1 (*after* $D_S$ and $D_R$ have become as shown in that row of Table 11.1). Referring to the cross-coupled NOR diagram, the state $D_o$ may be expressed in equation form as

$$D_{o,a} = \overline{D_R + \overline{(D_S + D_{o,b})}} = \overline{D}_R(D_S + D_{o,b}), \tag{11.1}$$

**TABLE 11.1** STATE TRANSITION TABLE FOR *RS* FLIP-FLOP

| Inputs | | Flip-flop state |
|---|---|---|
| $D_S$ | $D_R$ | $D_{o,a}$ |
| 0 | 0 | $D_{o,b}$ |
| 0 | 1 | 0 (whatever the value of $D_{o,b}$) |
| 1 | 0 | 1 (whatever the value of $D_{o,b}$) |
| 1 | 1 | Avoid $D_S = D_R = 1$. |

where the second expression results from use of DeMorgan's law, and similarly,

$$D_{o,1,a} = \overline{D}_{o,a} = \overline{D}_S(D_R + \overline{D}_{o,b}), \tag{11.2}$$

where in using $D_{o,1} = \overline{D}_o$ it is assumed that $D_R = D_S = 1$ is avoided.

We may present the same information in the form of a *state transition diagram*, as in Fig. 11.2a. Each bubble represents a distinct state, labeled by the number in the bubble. Each arrow indicates a possible means by which the state may change. The conditions for that change are written beside the arrow; the form is "input values/resulting output value."

In the case of the *RS* flip-flop, the state variable within the bubble, $D_o$, is also the output variable. (In more complicated circuits there will be more than one state variable, and the output(s) will be some specified function(s) of the state variables. Thus although specification of the output as well as the state may seem redundant for the *RS* flip-flop, in general it is not.) The input values are, of course, $D_S$, $D_R$. Figure 11.2b shows the standard circuit symbol for the *RS* flip-flop. Note that the state variable output is conventionally labeled "$Q$."

**Example 11.1**

For the $D_S$ and $D_R$ binary waveforms shown in Fig. 11.3, predict the output binary waveform $D_o$. Assume that the output $D_o$ is initially 0.

**Solution** Whenever $D_S$ goes high, $D_o$ becomes 1 whether or not it was previously. Whenever $D_R$ goes high, $D_o$ becomes 0 whether or not it was previously. The resulting waveform is shown in Fig. 11.3. From this example the ability of the *RS* flip-flop to debounce switches, as discussed in Chapter 10, should be clear. For instance, the second pulse of $D_S$ in the Fig. 11.3 is only a momentary contact, yet the output "catches" this set and remains 1 until $D_R$ pulses.

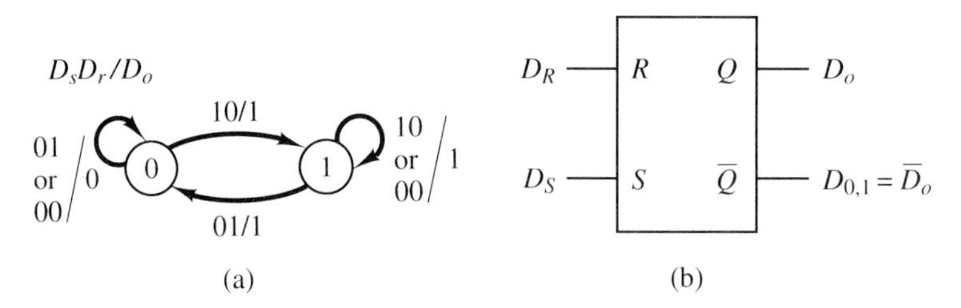


**Figure 11.2** (a) State transition diagram for *RS* flip-flop; (b) *RS* flip-flop circuit symbol.

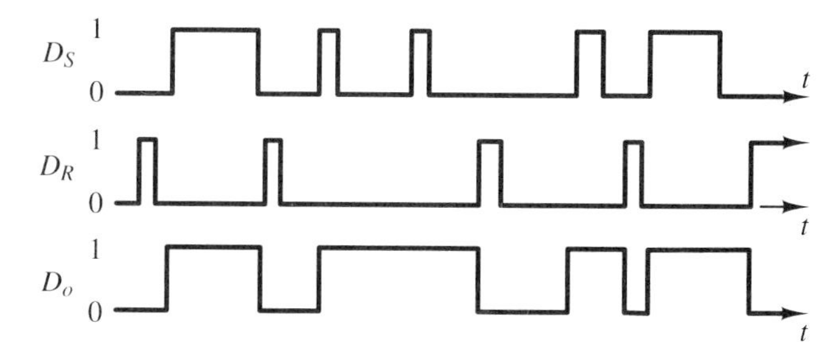


**Figure 11.3** Binary waveforms for Example 11.1 (*RS* flip-flop).

### 11.2.3 The *JK* Flip-Flop

Just as in Chapter 10 it was found convenient to have enable/disable control over the multiplexer, it is usually helpful to be able to cut off the flip-flop inputs from the set/reset lines. Often, for the purpose of reading data off lines in an organized fashion, the enable should take place at regular, well-defined instants in time rather than having the inputs continually connected.

For example, in the toggle flip-flop to be introduced in Sec. 11.2.4, the state will change if the input is 1. Presumably, only one toggle (switching) is intended at a time, but if the toggle line were continuously enabled, the state would oscillate rapidly in an uncontrolled fashion.

But if, instead, the enable line is a pulse train, the *rising* (*or falling*) *edge* of a pulse can be used to enable (i.e., sample) the inputs *momentarily*. Such a pulse train is typically generated in computer circuits of all kinds and is, not surprisingly, called the *clock* signal (see Sec. 11.3). In this book the clock signal is denoted by $D_C$. The method of effecting *edge triggering* in a typical flip-flop is discussed below. Thus, as desired, the enable (sampling of input signals for setting the memory state) is done once per clock cycle, on the rising (or falling) edge only.

As just noted, the command "toggle" is often a helpful feature of a memory cell. In electric circuits, a toggle switch is one for which the same action is used in successive repetition to turn the switch on, off, on, and so on; a simple example is a lamp with a pull chain. In digital electronic circuits, the toggle function is analogously one in which the state can be made to change from 1 to 0 to 1, and so on, by repeatedly presenting the same inputs.

It occurred to a bright engineer to modify the *RS* flip-flop circuit to make use of the forbidden (and therefore unused) input combination $D_R = D_S = 1$ for this toggle feature. However, the previous discussion has indicated that any toggle function must involve only momentary sampling—edge triggering—to avoid multiple uncontrolled toggles. The following description will describe how both the toggle and edge triggering features can be incorporated into the flip-flop.

First, the basic *RS* configuration will have to be modified if the edge triggering is to be achieved. The essential operational sequence is that while the clock is high, the set and reset inputs settle to the intended values they will have at the sampling time. Sampling occurs at the falling edge of the clock in the particular flip-flop to be discussed, an actual logic design for the manufactured chip 74LS112. The main circuit alteration required for edge triggering involves two additional "stages" of gates preceding the NOR gates of the *RS* flip-flop.

It should be mentioned that in some devices, the data are sampled on the rising edge. The choice for rising versus falling edge triggering depends on the particular location and function of the device within a digital system. Often within one circuit, some data transfers are to occur on the rising edge of the clock, while others occur on the falling edge.

We may think of the edge triggering concept as analogous to a water lock system in a canal on a slope with a boat traveling downhill. The water level within the lock is first brought to the upper level by closing the lower gate and opening the upper gate. Then the boat goes in. These two steps are like the enabling of the input data to enter the first stage of gates.

Next, the water level within the lock is brought down to the lower level on the exit side of the lock, which necessitates closing the entrance door (upper gate) and opening the exit door (lower gate). This is analogous to the delay time when the first stage of gates is disabled and the second set of gates is enabled, a period that begins at the falling edge of the clock and ends when the flip-flop is set (or reset). The only difference is that while canal locks are slow to fill or empty, this particular stage is *very* short in a flip-flop. Finally, as soon as the boat leaves, the lock must be set up for the next boat, during which time no boats can enter or leave the lock. This is analogous to the interval after the clock falls and before it rises again.

It is clear from the analogy above that two levels of gates are needed in front of the main *RS* cross-coupled NOR configuration: one to let in (i.e., enable) inputs, and another to let those inputs set/reset the flip-flop. These gates will be opened and closed by the clock and the state of the flip-flop.

To implement edge triggering we first consider a "gated" *RS* flip-flop, where transitions can occur only if the clock is low. Therefore, in the expression for $D_o = \overline{D}_R(D_S + D_{o,b})$ in the *RS* flip-flop [Eq. (11.1)], $D_S$ is replaced by $\overline{D}_C D_S$. To allow for the eventual toggle function, $D_o$ will no longer be conditioned on $\overline{D}_R$; hence the starting point for development is $D_o = \overline{D}_C D_S + D_{o,b}$. A two-level sum of products expression for $D_o$ is determined by using the simple absorption and DeMorgan's rules:

$$\begin{aligned} D_{o,a} &= \overline{D}_C D_S + D_{o,b} = \overline{D}_C D_S + \overline{D}_C D_{o,b} + D_S D_{o,b} + D_{o,b} \\ &= (\overline{D}_C + D_{o,b})(D_S + D_{o,b}) = \overline{(D_C \overline{D}_{o,b})}\;\overline{(\overline{D}_S \overline{D}_{o,b})} \\ &= \overline{D_C \overline{D}_{o,b} + \overline{D}_S \overline{D}_{o,b}}. \end{aligned} \tag{11.3}$$

Similarly, $\overline{D}_{o,a} = \overline{D}_S(D_R + \overline{D}_{o,b})$ [Eq. (11.2)] is modified to remove $\overline{D}_S$ to allow toggling and substitution of $\overline{D}_C D_R$ for $D_R$ to allow gating, and becomes

$$\begin{aligned} \overline{D}_{o,a} &= \overline{D}_C D_R + \overline{D}_{o,b} = \overline{D}_C D_R + \overline{D}_C \overline{D}_{o,b} + D_R \overline{D}_{o,b} + \overline{D}_{o,b} \\ &= (\overline{D}_C + \overline{D}_{o,b})(D_R + \overline{D}_{o,b}) = \overline{(D_C D_{o,b})}\;\overline{(\overline{D}_R D_{o,b})} \\ &= \overline{D_C D_{o,b} + \overline{D}_R D_{o,b}}, \end{aligned} \tag{11.4}$$

so that the circuit appears as in Fig. 11.4. The reason that the inverters for $\overline{D}_S$ and $\overline{D}_R$ were done with NAND gates is that to implement the edge-triggering and toggle features, other inputs will be presented to those NAND gates, as will now be discussed.

The circuit in Fig. 11.4 is a gated *RS* flip-flop. Notice that the set line is now

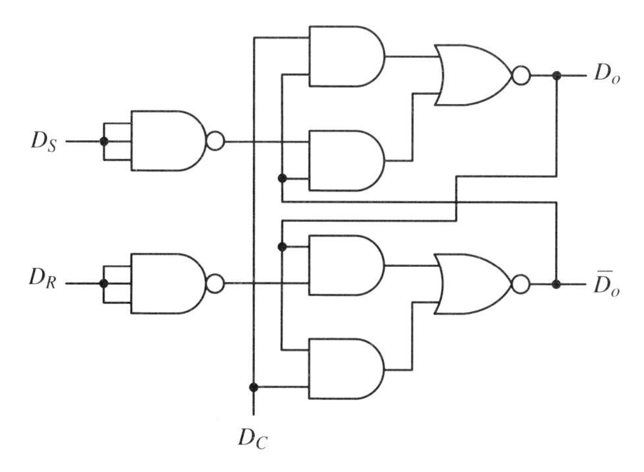


**Figure 11.4** Gated *RS* flip-flop helpful for understanding edge triggering and toggling.

on the top of the circuit diagram, even with $D_o$, rather than on the bottom with $\overline{D}_o$, as was the case with the original *RS* flip-flop. Similarly, the reset line is now on the bottom.

To give the flip-flop a toggle capability for $D_R = D_S = 1$, the additional lines marked ** are shown in Fig. 11.5. Because with the additional toggle feature for $D_S = D_R = 1$ it is no longer completely appropriate to call $D_S$ and $D_R$ strictly set/reset lines, new symbols $D_J$ and $D_K$ replace $D_S$ and $D_R$, respectively. When completed, the new flip-flop will be called an edge-triggered *JK* flip-flop.

For the moment, ignore the clock inputs to the NAND gates. A moment's thought will bring forth the motivation for NANDing $D_J$ with $\overline{D}_o$ and $D_K$ with $D_o$. Consider first the NAND truth table for $D_J$ with $\overline{D}_o$, shown in Table 11.2. This table can also be written in the form of Table 11.3. Thus if $\overline{D}_o = 1$, the NAND($\overline{D}_o$,$D_J$) "passes" $D_J$ as though $\overline{D}_o$ were not there; otherwise, the output of the NAND gate is what it would be if $D_J$ were 0—$D_J$ is disabled. Consequently, only if the flip-flop is currently reset (0 state) is $D_J$ enabled to set it. This view of a NAND gate as an enabler/disabler is very important in digital design. Similarly, the output of the other input NAND gate can be expressed by the truth table, Table 11.4. Clearly, only if

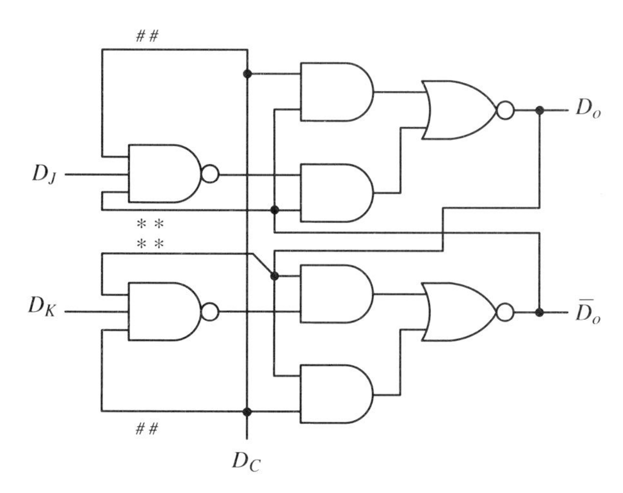


**Figure 11.5** Edge-triggered *JK* flip-flop.

**TABLE 11.2** TRUTH TABLE FOR NAND $(\overline{D}_o, D_J)$ IN FIG. 11.5

| $\overline{D}_o$ | $D_J$ | NAND $(\overline{D}_o, D_J)$ |
|---|---|---|
| 0 | 0 | 1 |
| 0 | 1 | 1 |
| 1 | 0 | 1 |
| 1 | 1 | 0 |

**TABLE 11.3** FUNCTIONAL AND CONCISE WAY OF REWRITING TABLE 11.2

| $\overline{D}_o$ | NAND $(\overline{D}_o, D_J)$ |
|---|---|
| 0 | 1 |
| 1 | $\overline{D}_J$ |

the flip-flop is set is $D_K$ enabled to reset it. Together, these enable/disable functions allow $D_J = D_K = 1$. If $D_J$ and $D_K$ are both 1, then for either state of the flip-flop, only the 1 from the input line ($D_J$ or $D_K$) that can change the state will be enabled. When the clock changes, as discussed next, the flip-flop will "toggle."

What is the reason for connecting the clock inputs to the NAND gates (shown with ##) given that the clock already gates the flow at the second stage? A very interesting idea is used here that implements the "canal" analogy described above to achieve edge triggering.

The NAND gates are *designed* to have a relatively long delay between the time when a change occurs in their inputs and the time that the corresponding change appears in their outputs. "Long" is in comparison with all other gates (AND and NOR) in the flip-flop. Suppose that $\overline{D}_o = 1$, so that the flip-flop output does not disable passage of $D_J$ at either the NAND or the AND gates. Then essentially we need to study the circuit fragment in Fig. 11.6.

The timing waveforms for $D_J$, $D_C$, $D_a$ = NAND $(D_C, D_J)$, and $D_o$ are shown in Fig. 11.7. A similar diagram would apply to a reset operation via $D_K$. The shaded regions of the input and output waveforms are used to label them as unspecified. When an input is shaded, that means that whatever the input was for this time interval, 1 or 0, will not affect the rest of the waveforms. A shaded output indicates that the information necessary to determine that output for the shaded time interval lies outside the time interval displayed on the timing diagram. Shaded regions occur in most sequential digital circuit timing diagrams.

**TABLE 11.4** OUTPUT OF OTHER NAND GATE IN FIG. 11.5

| $D_o$ | NAND $(D_o, D_K)$ |
|---|---|
| 0 | 1 |
| 1 | $\overline{D}_K$ |

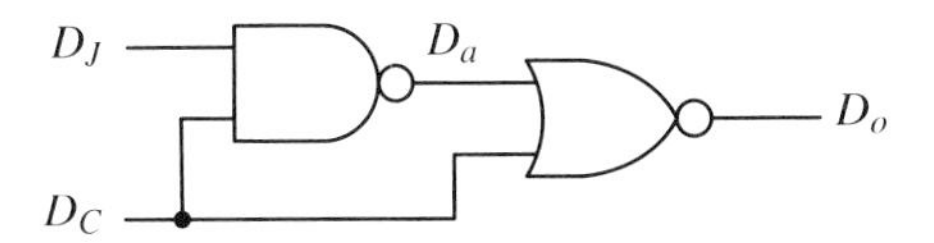


**Figure 11.6** Circuit fragment providing a window of enabling of $D_J$ for setting of flip-flop on the falling of the clock, due to the time delay through the NAND gate.

In this discussion, we concentrate on Fig. 11.6, but also make reference to the full circuit diagram in Fig. 11.5. Before the clock falls ($D_C = 1$), $\overline{D}_J$ is enabled through the NAND in the same way that $D_J$ is enabled for $\overline{D}_o = 1$, as discussed above. However, $D_J$ is simultaneously disabled at the NOR. The latter fact is due to the truth table for the NOR gate given in Table 11.5. It is seen that this NOR gate enables $\overline{D}_a$ only after the clock falls. When the clock falls, $\overline{D}_a$, containing the latest value of $D_J$, is enabled at the NOR (via the preceding AND gate stage in the complete circuit) but simultaneously $\overline{D}_J$ is disabled back at the NAND.

However, because of the long delay through the NAND, the "latest value of $\overline{D}_J$" is available for awhile longer—long enough to pass through the NOR and set the flip-flop if it was not set (or toggle it). These values of $\overline{D}_J$ originate from $D_J$ before the clock fell and become available at the output of the NAND gate as they propagate through it. Once $D_o = 1$, $\overline{D}_o$ also becomes 0. This very quickly disables further changes from $D_J$, via the AND gate in the main circuit diagram. It also maintains the current value of $D_o$ until a clocked reset occurs. Hence the enabling of $D_J$ to set $D_o$ high lasts for a very short time: From the fall of the clock until the flip-flop is set. True edge triggering has been achieved.

The circuit symbol for the *JK* flip-flop appears in Fig. 11.8, and the state

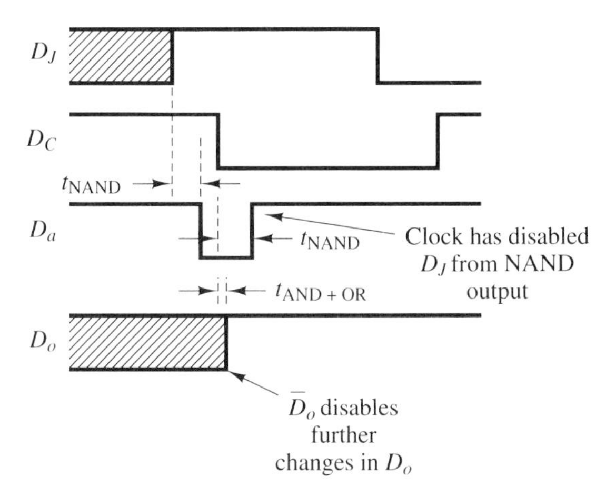


**Figure 11.7** Timing diagram for *JK* flip-flop.

**TABLE 11.5** TRUTH TABLE FOR NOR GATE IN FIG. 11.6

| $D_C$ | $D_o$ |
|---|---|
| 0 | $\overline{D}_a$ |
| 1 | 0 |

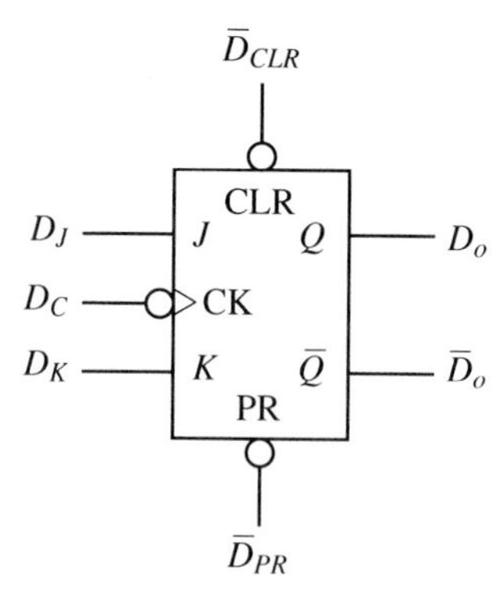


**Figure 11.8** *JK* flip-flop circuit symbol.

transition diagram in Fig. 11.9. The state transition table is given by Table 11.6. In the circuit symbol (Fig. 11.8), note the small triangle at the clock (CK) input. It indicates that this input is edge triggered. The bubble outside the flip-flop at the clock input indicates that this is a falling edge-triggered device; no bubble would infer a rising edge-triggered device. (A falling edge trigger is an "asserted low" version of a rising edge trigger.)

The *RS* flip-flop is not a clocked flip-flop, while this falling edge-triggered clocked *JK* flip-flop is. To reflect this fact, the value of $D_o$ before the falling edge of the clock will now be labeled $D_{o,bc}$ for "before the presently occurring falling edge of the clock pulse." Similarly, the value of $D_o$ after the falling edge will be labeled $D_{o,ac}$ for "after the falling edge of the presently occurring clock pulse."

Finally, notice the "PR" and "CLR" inputs. When asserted, the PR will "preset" the flip-flop independent of what the clock is doing, while the "CLR" line will "clear" or reset the flip-flop independent of the clock. Such input lines that act independent of the clock are called *asynchronous inputs.* Notice that for the typical flip-flop, the PR and CLR lines happen to be asserted low (negation bubble outside device). This is because in Fig. 11.5, they are fed directly to the NAND and AND gates as third inputs, where as we have seen, the logic is all asserted low. See data sheets for this final addition to the circuit diagram.

**Example 11.2**

Determine the output waveform $D_o$ if the $D_J$ and $D_K$ inputs are as shown in (a) Fig. 11.10a and (b) Fig. 11.10b.

**Solution** In part (a) (see Fig. 11.10a), they are respectively equal to $D_S$ and $D_R$ of the *RS* flip-flop Example 11.1, and in (b) (see Fig. 11.10b) $D_K$ is modified as shown. The division lines on the time axis represent falling transitions of the clock. Use of the state transition table is sufficient to determine the $D_o$ waveform for both cases. Notice that in part (b) there are instances where $D_J = D_K = 1$ at a clock transition; the result is a toggle. Also, note that on the ninth clock fall there is no toggle even though during

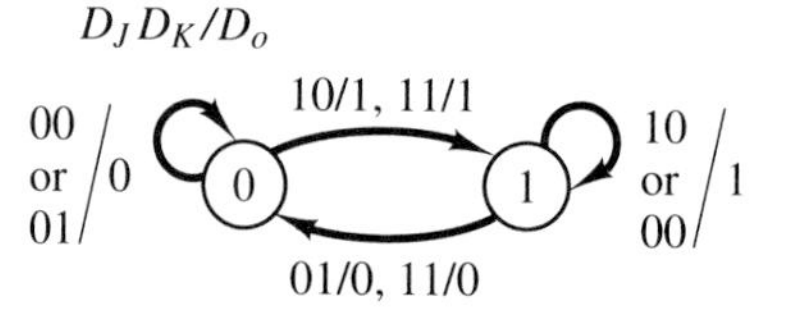


**Figure 11.9** State transition diagram of *JK* flip-flop.

**TABLE 11.6** STATE TRANSITION TABLE OF *JK* FLIP-FLOP

| Inputs at falling edge of clock | | Flip-flop state $D_{o,ac}$ |
|---|---|---|
| $D_J$ | $D_K$ | |
| 0 | 0 | $D_{o,bc}$ |
| 0 | 1 | 0 (whatever the value of $D_{o,bc}$) |
| 1 | 0 | 1 (whatever the value of $D_{o,bc}$) |
| 1 | 1 | $\overline{D}_{o,bc}$ |

that pulse $D_J = D_K = 1$. The reason is that $D_J$ was not still equal to 1 on the *fall* of the clock (hash mark).

### 11.2.4 The *D* and *T* Flip-Flops

A couple of variations on the *JK* flip-flop are used widely: the *D* and the *T* flip-flops. The *D* (for *data* or *delay*) *flip-flop* is just a *JK* flip-flop whose *J* and *K* inputs are designed always to be mutual logical complements: $D_J = \overline{D}_K = D_D$.

The name "data" is used because once each clock cycle, a sample of the single data line $D_D$ is stored in the flip-flop, as opposed to the *RS* or *JK* flip-flops, where a sample of one of two input control lines is stored. The other name, "delay," is used because transitions of $D_D$ occurring before the falling edge of the clock appear only *later*, at the falling edge of the clock. Thus the waveform of the state variable $D_o$

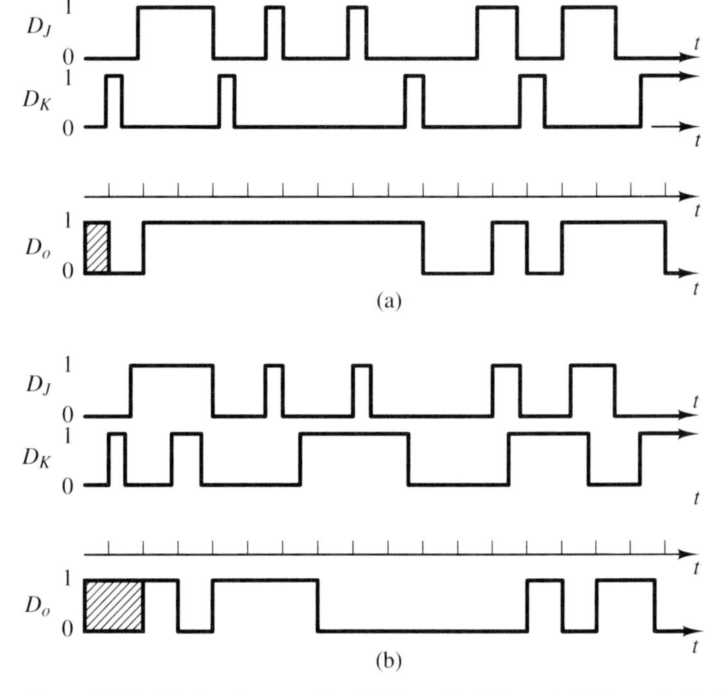


**Figure 11.10** Timing diagrams for (a) Example 11.2a and (b) Example 11.2b.

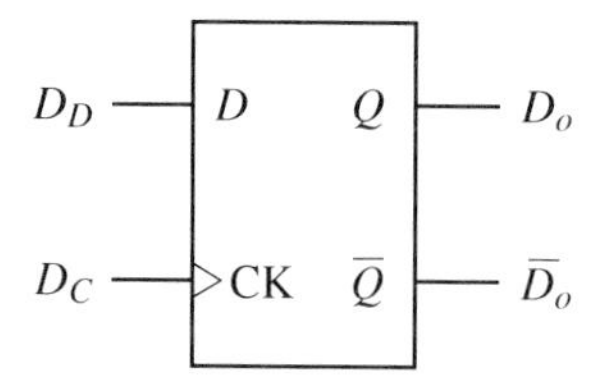


**Figure 11.11** $D$ flip-flop circuit symbol.

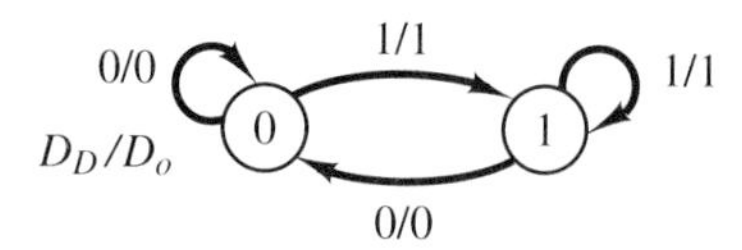


**Figure 11.12** State transition diagram for $D$ flip-flop.

appears as a delayed version of the data line $D_D$. The circuit symbol for the $D$ flip-flop is given in Fig. 11.11, while the state transition diagram is given in Fig. 11.12. The state transition table appears in Table 11.7. Notice that Table 11.7 consists of only the second and third rows of the state transition table for the $JK$ flip-flop (Table 11.6) because for those rows only $D_J = \overline{D}_K$.

The $T$ (*toggle*) *flip-flop* is simpler yet. This time the $J$ and $K$ inputs are held equal: $D_J = D_K = D_T$. Now the state transition table is made up of the first and fourth rows of the $JK$ state transition table (Table 11.6) (not used by the $D$ flip-flop) because those are the rows for which $D_J = D_K$. Clearly, if $D_T = 1$ at the falling edge of the clock, the flip-flop toggles, and otherwise it remains as it was before the clock pulse. The circuit symbol for the $T$ flip-flop is shown in Fig. 11.13, its state transition diagram in Fig. 11.14, and its state transition table in Table 11.8.

**Example 11.3**

A circuit designer is faced with coming up with a convenient on–off switch for the hand-held remote control "zapper" of a new TV set. How could a flip-flop be used, and which kind would be most suitable?

**Solution** When the on–off button on the zapper is pressed, a unique five-binary-variable code word modulates a carrier at roughly 40 kHz, which in turn drives an LED that emits infrared light when on. The 40-kHz carrier separates the desired code from infrared noise such as turning on incandescent light bulbs, sunlight variations, and so on. This modulated carrier, in turn riding on the infrared electromagnetic wave, is transmitted from the zapper to the TV set. Thus there is a sort of double modulation: one for the mechanics of transmission (radio-wave production) and one for noise avoidance.

The TV set receives the infrared signal via an LED infrared detector; the on–off

**TABLE 11.7** STATE TRANSITION TABLE FOR $D$ FLIP-FLOP

| Input at falling edge of clock | Flip-flop state |
|---|---|
| $D_D$ | $D_{o,ac}$ |
| 0 | 0 (whatever the value of $D_{o,bc}$) |
| 1 | 1 (whatever the value of $D_{o,bc}$) |

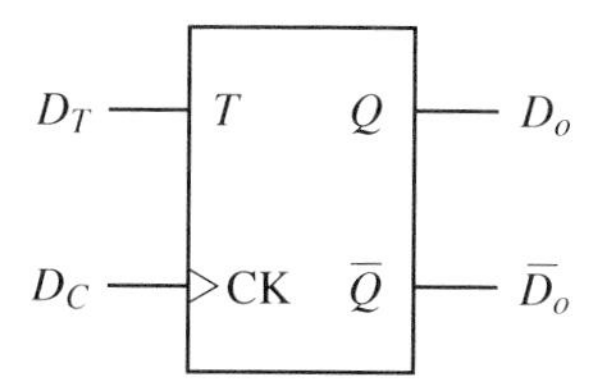


**Figure 11.13** $T$ flip-flop circuit symbol.

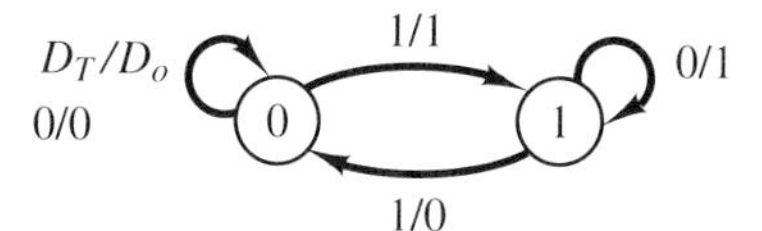


**Figure 11.14** State transition diagram for $T$ flip-flop.

pattern of the resulting signal is, of course, at 40 kHz. The 40 kHz is claimed by a tuned filter and then demodulated, leaving the originally intended code. A decoder translates this code to indicate that in this case the on–off flip-flop (within the TV set) is to be toggled.

For zapper (remote) control, clearly a toggle is most efficient; only one button on the zapper need be spent for the on–off switch. Thus the appropriate flip-flop is a $T$ flip-flop. The voltage and power associated with the TV on–off switch far exceed allowable levels for digital circuitry, so isolation will have to be arranged. This is usually done by means of a relay. The $T$ flip-flop could energize a relay, which in turn would mechanically close a switch of suitable power capability, turning the TV set on or off.

### 11.2.5 Timing Parameters

As noted near the end of Chapter 10, it takes time for a binary signal to propagate through a digital gate. As flip-flops are combinations of gates, the same delays occur. However, timing can be more critical here: Timing may be the deciding factor determining whether or not a flip-flop is set, which can have drastic consequences.

Consider a $D$ flip-flop that changes its output on low-to-high clock transitions, as does the 74LS74. Specification of the practical operation of this flip-flop is summarized by the parameters now to be defined and shown in Fig. 11.15. All of the specified values of these parameters are required in order for the output to be predictable and recognizable as $D$ flip-flop behavior. That is, they are *worst-case* values for the 74LS74.

$t_S(L)$ = *setup time* ($S$ subscript) for a 0 (low—$L$ argument of $t_S$) on the input. This is the minimum time that the input has to have been stable at the intended value before the clock transition.

**TABLE 11.8** STATE TRANSITION TABLE FOR $T$ FLIP-FLOP

| Input at falling edge of clock $D_T$ | Flip-flop state $D_{o,ac}$ |
|---|---|
| 0 | $D_{o,bc}$ |
| 1 | $\overline{D}_{o,bc}$ |

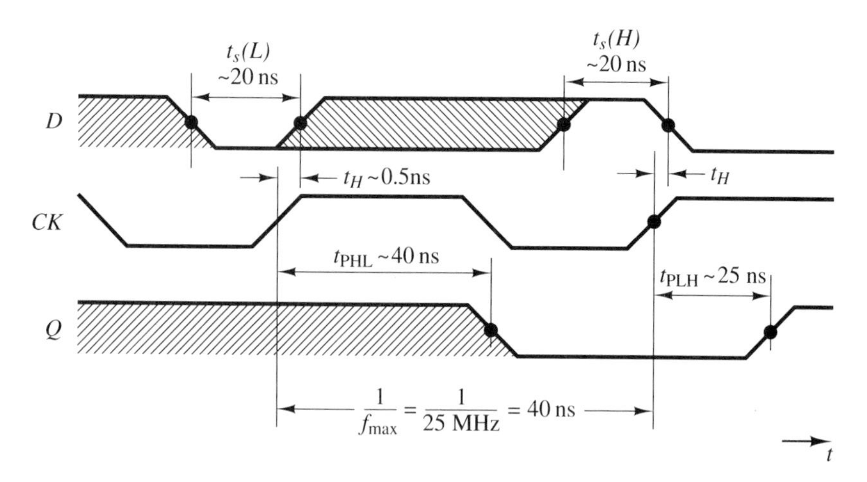


**Figure 11.15** Practical timing diagram for $D$ flip-flop.

$t_S(H)$ = setup time for a 1 (high—$H$ argument of $t_S$) on the input. Otherwise, it has the same meaning as $t_S(L)$.

$t_H$ = *hold time* ($H$ subscript). This is the minimum time past the clock transition that the input must remain at its intended value. Usually, $t_H$ is zero or very small relative to other timing parameters.

$f_{MAX}$ = maximum clock frequency allowed.

In addition, the propagation times from the clock transition to a set/reset change in the output are $t_{PHL}$ and $t_{PLH}$ for high–low and low–high transitions, respectively. These two parameters are fixed; their values must be known to integrate this flip-flop successfully into a larger digital system.

Note that state transitions (transitions of $Q$) are triggered by the rising edge of the clock. The diagonal lines of the signals in Fig. 11.15 roughly represent the actual appearance of high-to-low transitions of the signal shown. The timing parameter values are, of course, derivable from considerations such as were discussed in our study of the $JK$ flip-flop timing diagram (Fig. 11.7).

If the timing rules are not followed, the behavior of the flip-flop will be determined by which 1 or 0 within the flip-flop circuit reaches its destination first. That depends on the particular gate structure of the flip-flop circuit. This situation, known as a *race condition*, obviously must be avoided in all designs. As by now you might have guessed, timing issues are among the biggest headaches in sequential digital design.

At a functional level, the flip-flop as described here is a good indication of memory operation in large and very-large-scale integrated circuits (ICs) as well as in the discrete digital circuits studied here so far. However, the logistics of transferring information to and from the cells and cramming as many memory cells as possible onto a chip for large memory systems necessitates details specific to the manufacturing technology which may not be recognizable, as, for example, cross-coupled NOR gates. But these changes are merely technicalities required to implement the familiar memory functions based on the principles discussed here.

## 11.3 THE DIGITAL CLOCK

The digital clock is an essential element for sequential circuit operation. Like the power supply, the clock is often taken for granted in presentations of digital electronics. Perhaps in both cases this is because they may be purchased as integrated units. However, it is again true that there are options. We consider here two possibilities: The astable multivibrator and the voltage-controlled oscillator/crystal clock.

An outline description is now presented for a simple clock circuit known as an *astable multivibrator.* It is called "astable" because it never stops switching. The term "multivibrator" refers to the fact that there are two (multiple) states between which the output may switch (or vibrate): high (1) or low (0). Incidentally, the flip-flop (Sec. 11.2) is sometimes called a *bistable multivibrator*, and the one-shot (see Sec. 9.7.3) is called a *monostable multivibrator.* For this discussion, refer to Fig. 11.16.

Let $V_{DD}$ be the power supply voltage. In this discussion it is important to remember that the outputs of digital gates are held strongly at 0 or 1 (i.e., 0 V or $V_{DD}$) whenever the input exceeds/drops below the threshold voltage. Therefore, we can consider that $v_1$ will vary in a continuous way above and below thresholds, while $v_2$ and $v_3$ will slam back and forth between 0 V and $V_{DD}$.

Suppose that the input $v_1$ is at this moment somewhere above the 1/0 threshold voltage $V_T$ so that $v_2 = 0$ V, and therefore $v_3 = V_{DD}$. Notice that $v_1$ is the potential on one side of capacitor $C$, while $v_3$ is the potential of the other side of $C$. Also recall that for practical logic gates such as CMOS, the input current to each gate is essentially zero. Therefore, we see that if $v_2 = 0$ V and $v_3 = V_{DD}$, $C$ will charge exponentially through $R$ to a voltage of $v_C = V_{DD}$. This will drive the potential $v_1$ to 0 V—$V_{DD}$ volts below $V_{DD}$ (i.e., $v_3$).

When $v_1$ falls below $V_T$, $v_2$ will suddenly switch to $V_{DD}$ and $v_3$ will suddenly switch to 0 V. Now $C$ discharges the opposite direction through $R$ so that $v_C$ approaches $-V_{DD}$. Because $v_3$ is held by the gate at 0 V and $v_C$ is becoming negative, it must be true that $v_1$ is rising. When $v_1$ reaches $V_T$, $v_2$ will switch to 0 V and $v_3$ will switch to $V_{DD}$ and one cycle has been completed.

Naturally, this analysis is very reminiscent of that of the pulse generator discussed in Sec. 8.8.3; it has merely been specialized for digital circuitry. The period of the clock circuit in Fig. 11.16 is roughly $3RC$, which to derive requires a more detailed analysis than that given here and which for accuracy in any case would have to be calibrated using an oscilloscope or other measurement equipment.

The clock described above is the simplest possible clock, and naturally is the cheapest. For most digital circuitry, a more rigid standard of accuracy of frequency and wave shape is desirable. The $3RC$ period can wander with time and temperature.

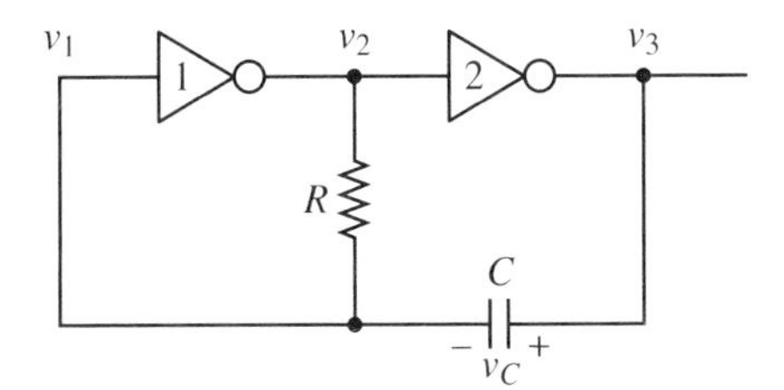


**Figure 11.16** Simple clock circuit: The astable multivibrator.

Instead, for most applications quartz crystal oscillator circuits are used. The crystal is often externally connected to the oscillator chip, although there are also chips that include the quartz crystal. The crystal works on the piezoelectric principle (see Sec. 9.6.2). A driving oscillation is generated by, for example, a *voltage-controlled oscillator* (VCO). A VCO is an oscillator whose output frequency is proportional to an input DC voltage (similar to the V/F converter of Sec. 9.7.3 except that the VCO output is sinusoidal). The driving oscillation excites the resonant frequency of the crystal, which produces a voltage at that very constant and precise resonant frequency. Of course, the VCO will still function if the crystal is replaced by a capacitor, but with decreased stability and accuracy.

Typical frequencies are 12 MHz for most applications and 20 MHz for higher-speed requirements (corresponding roughly to a clock period of 50 to 100 ns). The particular clock frequency used is usually not critical but must be especially well regulated when deriving "wall clock" time measurements such as in timers. We can always derive a reduced desired clock pulse frequency by the frequency-dividing techniques described in Sec. 11.4. For most applications the 12 MHz is overkill, but as long as the supported circuitry has small-enough setup times, this does not matter. A clock chip costs only a few dollars anyway; some of these require only power supply and crystal connections, which makes their use extremely simple.

## 11.4 COUNTERS, FREQUENCY DIVIDERS, AND REGISTERS

Now that the flip-flop has been fully discussed, its use in the construction of a number of handy digital devices should be painless. Consider, for example, the circuit in Fig. 11.17. For the moment, focus on the rightmost flip-flop, FF0. Notice that both of the inputs of this clocked *JK* flip-flop are tied to 1. This means that the state of FF0 will toggle every falling edge of the clock. Consequently, the timing waveforms

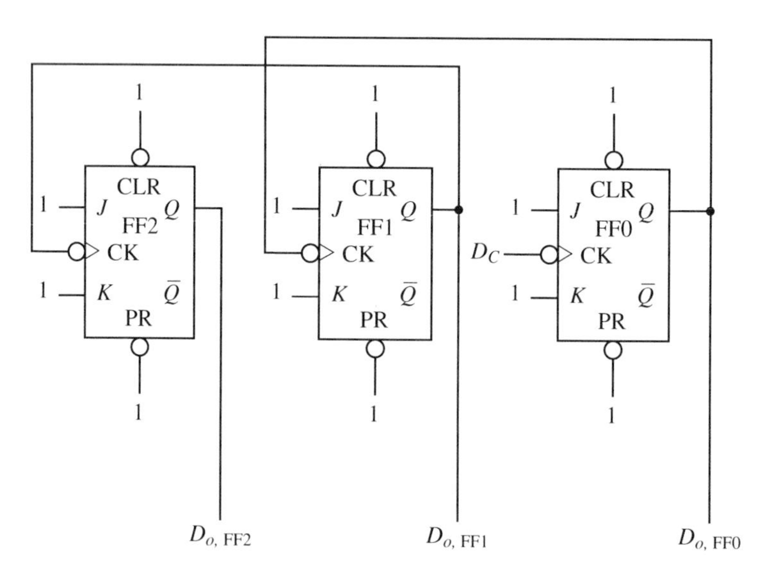


**Figure 11.17** Asynchronous three-bit counter.

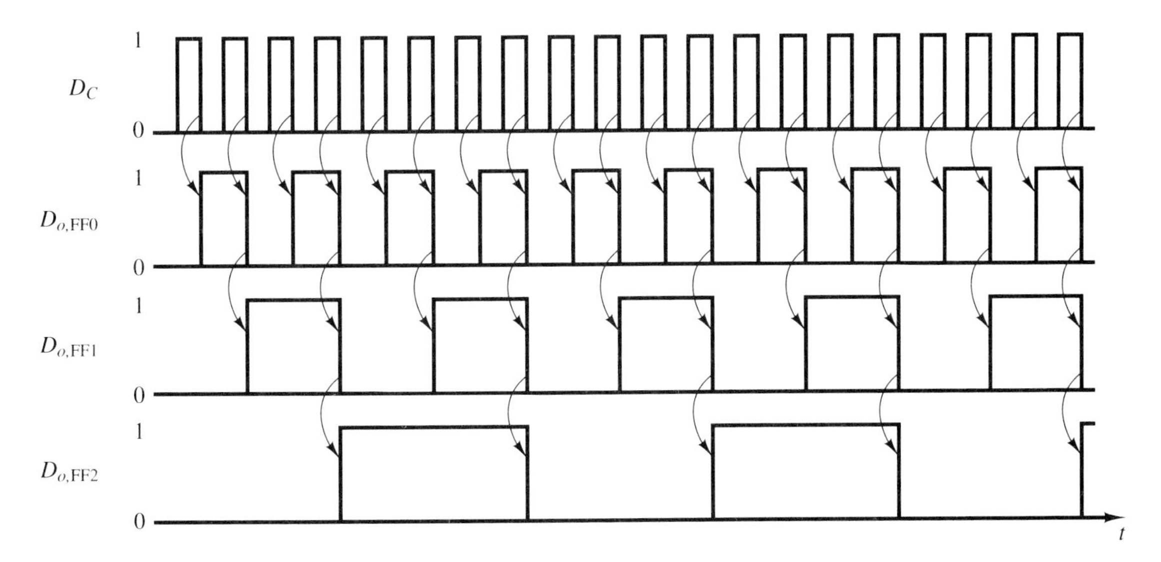


**Figure 11.18** Timing diagram of the three-bit counter in Fig. 11.17.

for the clock $D_c$ and the flip-flop output $D_{o,\text{FF0}}$ are the uppermost two waveforms in Fig. 11.18.

The arrows in the timing diagram indicate that the transition made at the arrowhead is due to the transition made previously at the back end (tail) of the arrow. Although the resulting transition is shown directly below the causing transition, it actually occurs slightly later by a time equal to the propagation delay through the flip-flop. The state of flip-flop FF0, denoted $D_{o,\text{FF0}}$, evidently completes one up-down period every two clock cycles (one clock cycle to bring the state up, another to bring it back down).

Effectively, a half-clock-frequency binary signal has been generated: the state of the flip-flop FF0. Now consider the remainder of the circuit of Fig. 11.17. Notice that the half-clock-frequency waveform $D_{o,\text{FF0}}$ is connected to the clock input of another JK flip-flop having both $J$ and $K$ inputs tied to 1, FF1. The state of this second flip-flop, $D_{o,\text{FF1}}$, will clearly be a pulse train having half the frequency of the state of the first flip-flop, or *one-fourth* the frequency of the original clock. Similarly, the third flip-flop, FF2, has state $D_{o,\text{FF2}}$, which oscillates at one-eighth of the clock frequency. We may obtain pulse trains with lower and lower frequency by adding more flip-flops connected this way.

Of great interest and utility is the fact that this frequency-division circuit can also be viewed as a binary counter. For if the states of the flip-flops farther along the chain are considered higher significant "digits," the pattern formed is identified as the binary count. Thus, for example, $D_{o,\text{FF2}}\, D_{o,\text{FF1}}\, D_{o,\text{FF0}}$ will sequentially generate the successive input variable patterns of a three-input truth table.

Also, a quick look at any three-input truth table shows that moving from right to left, each input binary variable oscillates down the table between 0 and 1 at half the rate of the one just to its right, just as in decimal counting each digit cycles through at one-tenth the rate of the digit immediately to the right. It is left as an exercise to show that if the flip-flops were rising edge instead of falling edge, the counter would count in binary *reverse* order.

We may begin and end counting where desired by using gates to detect the upper number. When detected, an asynchronous signal is sent to the PR and CLR lines appropriate for each flip-flop to form the desired starting number. Because the PR and CLR lines are asserted low, we must use the logical negations of the 1s and 0s making up the starting number. Thus if for a particular digit of the starting number a 1 is desired, the PR line must be made low when the count is to restart, while CLR is always held high. Such a three-digit ("three-bit") counter circuit is shown in Fig. 11.19, where $D_{s,2}D_{s,1}D_{s,0} = 010$ (two) and $D_{e,2}D_{e,1}D_{e,0} = 110$ (six) are, respectively, the binary representations of the starting and ending numbers.

Notice that only $D_{o,\text{FF2}}$ and $D_{o,\text{FF1}}$ had to be checked for 1s while $D_{o,\text{FF0}}$ was not checked for 0. This is because in binary counting, six will be reached first ($D_{o,\text{FF0}} = 0$, before $7 = 111$, for which $D_{o,\text{FF0}} = 1$) anyway when $D_{o,\text{FF2}} = D_{o,\text{FF1}} = 1$. Also, a NAND gate was used instead of an AND gate, so that all negation bubbles for PR and CLR cancel; the output of the NAND is used to clear $D_{o,\text{FF2}}$ and $D_{o,\text{FF0}}$, and to preset $D_{o,\text{FF1}}$ to restart at two. If the initial states were 000, the count would be 0,1,2,3,4,5,6,2,3,4,5,6,2,3,4,5,6,2, . . . .

If the starting point is 0 and the ending point is $N$, the most significant bit serves as a divide-by-$N$ counter. We may cascade divide-by counters having different moduli to obtain ever-lower frequency counters with unusual factors. For example, there are 86,400 seconds in a day. A 1-Hz pulse train could be the clock of a divide-by-10 followed by a divide-by-6 (modulo 6) counter, producing minutes. Further dividing by 60 to produce hours and then again by 24 provides a day count. Of course, such a circuit could also be called a "count-to-86,400" circuit except that the digits are not the normal base ten decimal digits. This procedure is called cascading of counters. See the problems for analysis of a digital hours/minutes/seconds clock.

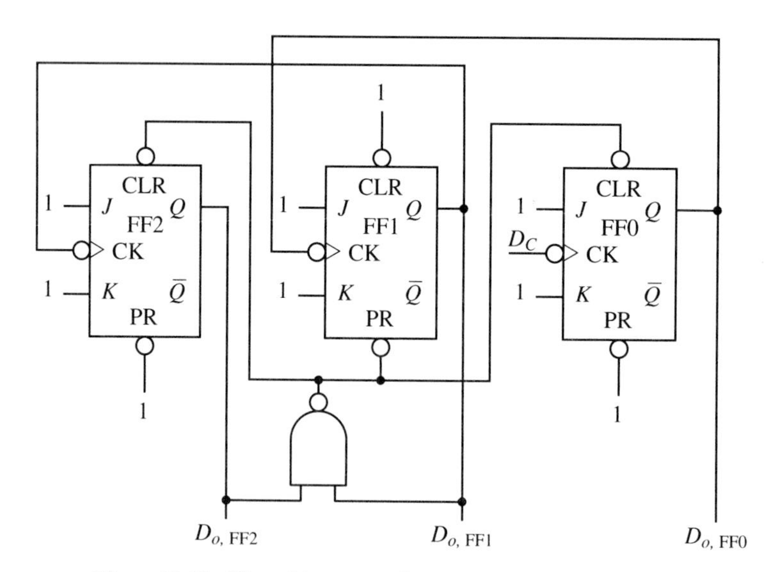


**Figure 11.19** Three-bit counter that starts at two and ends at six.

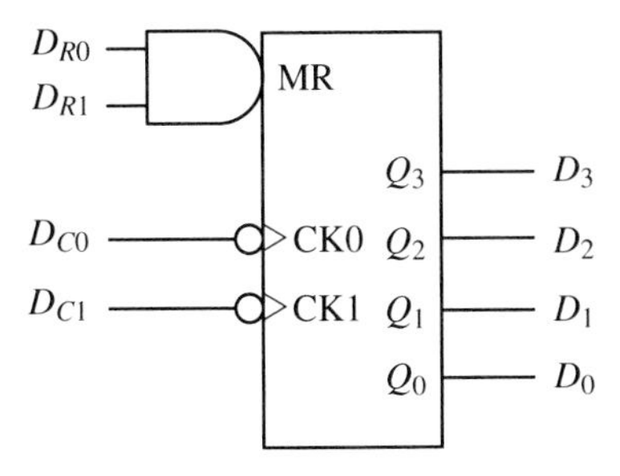


**Figure 11.20** Asynchronous ripple counter circuit symbol.

Counters such as that in Fig. 11.17 are called "*asynchronous*" or "*ripple*" *counters* because the timing of the transitions in all but the first flip-flop are governed by propagation delays through flip-flops rather than all occurring on the clock transition. This can be a limitation on how fast the input clock can be run. For proper operation, the input clock period must exceed the propagation delay through *all* the flip-flops, not just one; otherwise, the outputs will be unpredictable. But it is a simple, convenient counter for low-frequency work. Naturally, it is available in integrated form; a typical logic symbol appears in Fig. 11.20.

This circuit consists of two independent counters: one that just counts from 0 to 1 ("divide-by-2," driven by CK0) and one that counts from 0 to 7 (divide-by-8," driven by CK1). The two counters can be cascaded to make a four-bit (0 to 15, or divide-by-16) counter by using $Q_0$ to drive CK1 and clocking CK0. Having them independent merely allows for added applications flexibility. The reason for the two resets ANDed together to make MR (master reset) will be made more clear in the following example.

**Example 11.4**

Make connections using the IC asynchronous ("ripple") counter to make a *decade counter*, one that counts from 0 to 9 and then resets.

**Solution** A three-bit counter counts only from 0 to 7, so a four-bit counter is needed. To make the count 0,1,2,3,4,5,6,7,8,9,0,. . . the counter should be reset at 10 to zero. But if we count to 10 in binary, we obtain the pattern $D_3 D_2 D_1 D_0 = 1010$ (more detail on converting base 10 to binary is given in Sec. 11.6.2). The first occurrence of $D_3 = D_1 = 1$ in binary counting happens to be 1010, so the reset should be done when $D_3 = 1$ AND $D_1 = 1$. Consequently, just connect $D_3$ to $D_{R0}$ and $D_1$ to $D_{R1}$; the internal AND gate does the rest for us.

Because it is common that at least two things must be true for a reset to occur, an AND gate is provided internally on the counter chip for convenience. If only one line need be 1 for a reset to occur in some other application, just tie both inputs $R_0$ and $R_1$ to that line. The decade counter circuit is given in Fig. 11.21. Several of these connected together could make an electronic odometer for a car (see the problems) or a digital radio station readout on a modern radio receiver.

If we connect all CK inputs to the clock $D_C$ and tie the output of the AND of all lower significant binary digits (henceforth called "bits") to both the $J$ and $K$ inputs, the same binary counting can be done in a synchronous manner. This allows higher rates and more predictable behavior because each stage is valid after only one stage of propagation delay.

Examination of binary counting shows that each time all lower significant bits

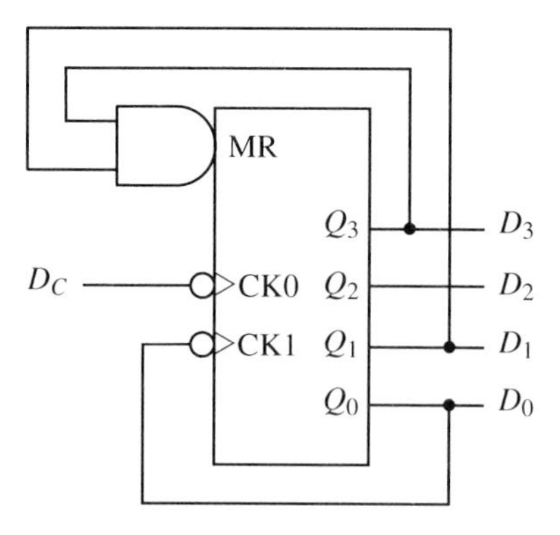


**Figure 11.21** Asynchronous decade counter circuit.

are 1, the next count causes the next higher significant bit to change state from 0 to 1 or from 1 to 0. Of course, the least significant bit toggles every clock cycle, so its $J$ and $K$ inputs are both tied to 1. The circuit of Fig. 11.22 is a three-bit synchronous counter.

**Example 11.5**

Determine the three-line output sequence produced by the circuit of Fig. 11.23. Assume that the circuit is initially cleared and that at $t = 0$ (beginning of cycle), flip-flop 2 is preset.

**Solution** Initially (at $t = 0$), $D_{FF2}\,D_{FF1}\,D_{FF0} = 100$. At the next clock pulse, the old 1 of FF2 is transferred to FF1, the old 0 of FF1 is transferred to FF0, and the old 0 of FF0 is transferred back to FF2, yielding 010. Similarly, at the next clock pulse the pattern is 001, and then the cycle begins again with 100. This type of counter is called a *ring counter* because the sole 1 "travels" "around" a ring. It is *not* a base 2 counter.

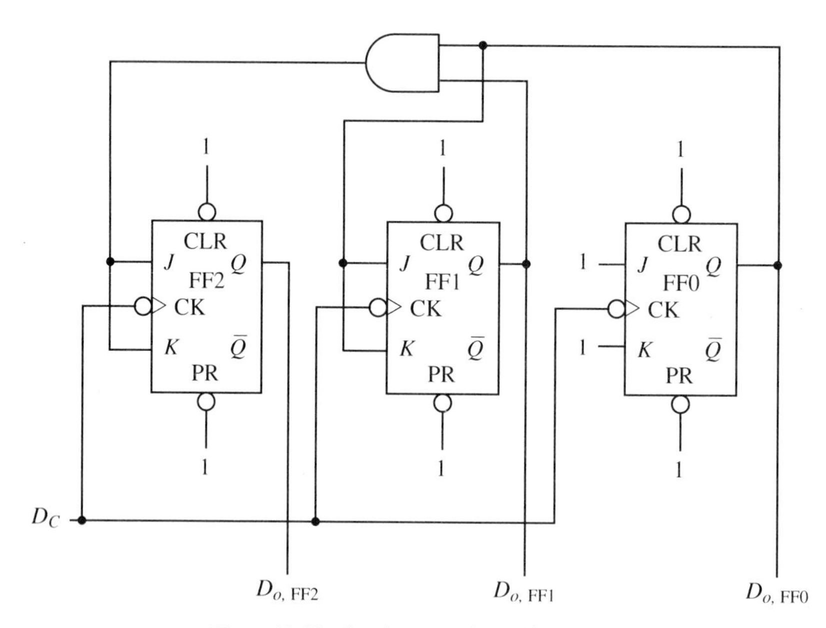


**Figure 11.22** Synchronous three-bit counter.

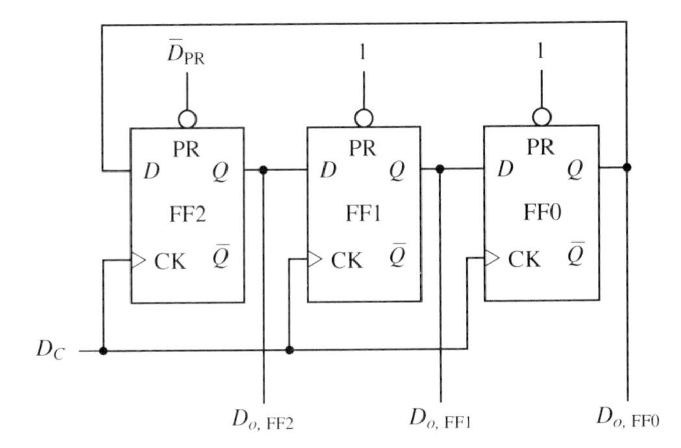


**Figure 11.23** Circuit for Example 11.5.

**Example 11.6**

Suppose that the feedback line to $D$ of FF2 in Fig. 11.23 is taken from $\overline{Q}$ of FF0 rather than from $Q$, as indicated originally in the figure. Find the output pattern sequence $D_{FF2}\,D_{FF1}\,D_{FF0}$ produced, where again $D_{FFi}$ is $Q$ of FF$i$, $i = 0,1,2$. Initially, the counter is cleared.

**Solution** At the first clock transition, the old 0 of FF2 goes to FF1 and the old 0 of FF1 goes to FF0. However, the *complement* of the old 0 of FF0, a 1, is fed back to FF2. Convince yourself that the pattern sequence is 000, 100, 110, 111, 011, 001, 000, 100, . . . . This circuit is called a *Johnson counter* even though it, like the ring counter, does not perform counting. These circuits should be called ring and Johnson sequence generators, not counters.

The ripple and synchronous circuits above have both satisfied the requirement that the memory elements count in base 2. The Johnson and ring counters step repetitively through other sequences. Another application of a string of flip-flops is the *data register.*

Back at the beginning of Chapter 10, a register well known to each of us was mentioned: the cash register. In many ways the binary register is similar. The cash register can display (price) data directly, or the results of an addition, subtraction, or other arithmetic operation. It has a small number of available digits. Once displayed, the numbers remain until a *clear* command is issued by the clerk to get ready for the next customer.

It is particularly this last feature that distinguishes a data register from just a set of wires containing a set of digital signals. The signals on a set of wires may change at random in response to either outside influence (pushing of keys, measurement signals, etc.) or the asynchronous outputs from some previous circuit stage. With a data register, the signal contents of the set of wires feeding the register are frozen in time within the register at the fall (or rise) of the clock pulse until the next clock pulse, unless an asynchronous clear command is issued.

When circuit design ensures that the signals in the wires are "read" only when they represent valid data and only when it is otherwise appropriate to destroy current register contents, a highly controlled computational or logical system is possible. The register "remembers" the data as long as convenient, and no longer. Just how long

This "player violin" is powered by a small but powerful electric motor and electromagnets. It exceeds the capabilities of the human player: Duets and quartets are given as easily as solos. "The sweetness is remarkable, as are also the harmony and volume of tone." *Source: Scientific American* (May 30, 1908).

is appropriate depends on the application, but in computers it could be fractions of a microsecond and in monitoring systems it could be milliseconds.

For example, a logical system in a power plant could check the status of an entire range of processes periodically. If any were finished, certain action would be taken; if any were failing, other action would be taken. Many times during the process, various calculations used to optimize production could be performed. All of this work would be done in a systematic, clocked fashion using registers to hold the currently important data "at the fore."

A variety of possibilities exists for register configurations. *Parallel-in, parallel-out* is the most straightforward: It is just several flip-flops arranged side by side. The only tie between them is that the same clock signal is presented to each flip-flop. Therefore, we may think of this type of register as a multibit clocked flip-flop set.

A common task is the transmission of multibit binary numbers over a transmission line (e.g., a telephone line) one bit at a time. Usually, the receiver requires the received data to be converted back to parallel (multibit) form for storage. For the receiver, a *serial-in*, *parallel-out* register fits the bill. After 8 clock cycles, an eight-bit number is ready to load in parallel to a memory device or is ready for computations. To reverse the process, *parallel-in*, *serial-out* registers will clock-out a multibit number to a single transmission line.

Finally, *shift register* is often the term given to serial-in, serial-out. These registers are helpful either for data delays or for calculations. One piece of binary data may be ready before other data with which it is to "join forces." While the other "slow" data is getting ready, the faster bit is slowed down by the shift register: Kind of like loading down the fastest hiker on the trail with everybody's food supplies or sending him or her off to climb three extra mountains on the side before the next stop for a break.

For calculations, a shift of a binary number to the right implements division by 2, while a shift left implements multiplication by 2. It should not be surprising that shift registers are very common in the arithmetic units of computers!

**Example 11.7**

Connect two 4-bit shift registers *A* and *B* so that the contents of register *A* are copied to *B* without erasure or alteration of the original contents of *A* at the conclusion of the operation.

**Solution** The trick is to feed the serial output of *A* to its serial input as well as to the serial input of register *B*. The required circuit is given in Fig. 11.24. Suppose that *B* was initially clear and that *A* had 1011. Then the *A* and *B* registers would become, respectively, 1101 and 1000 after one clock pulse, 1110 and 1100 after two clock pulses, 0111 and 0110 after three, and finally, 1011 and 1011 after four. Thus after four clock pulses, the copying is complete and A again has its original contents. (The four inputs P1, P2, P3, and P4 allow presetting of the internal *RS* flip-flops, provided that the preset enable PE is high.) Notice that register A is functioning as a ring counter.

While special-purpose registers are available to perform each of these types of function, a general-purpose *universal shift register* can be controlled to do whichever is desired at any particular moment. Parallel input signals are $D_0$ through $D_3$, parallel output signals are $Q_0$ through $Q_3$, $D_{SR}$ is serial input to be shifted to the right, $D_{SL}$ is an input line for the "shift left" operation, $D_{SR}$ is an input line for the "shift right" operation, $Q_0$ can be taken as the serial output for "shift left," and $Q_3$ can be used as the serial output for "shift right." An asynchronous master reset is labeled MR and the clock is CK. The function carried out by the universal register is controlled by the lines $S_0$ and $S_1$.

Use of the universal register in the four situations is illustrated in Fig. 11.25. The labeling acronyms are formed from $S$ = "serial," $P$ = "parallel," $I$ = "input," and $O$ = "output." The use of "×" indicates "don't care" because those lines are

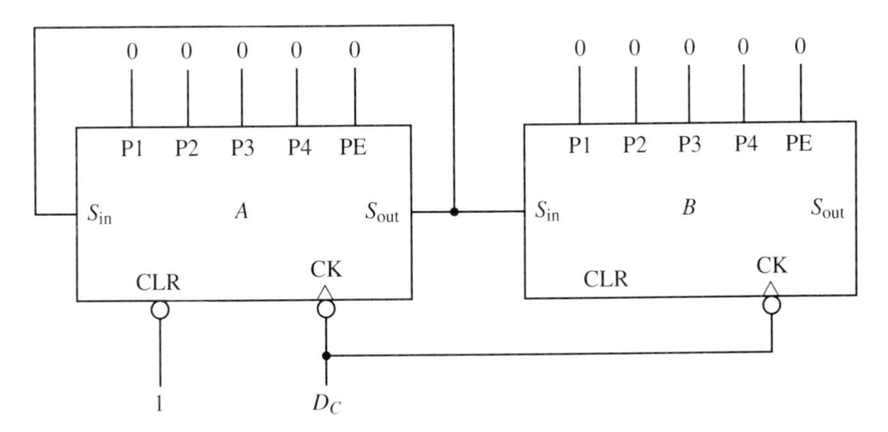


**Figure 11.24** Two shift registers connected so that the contents of one can be copied into the other.

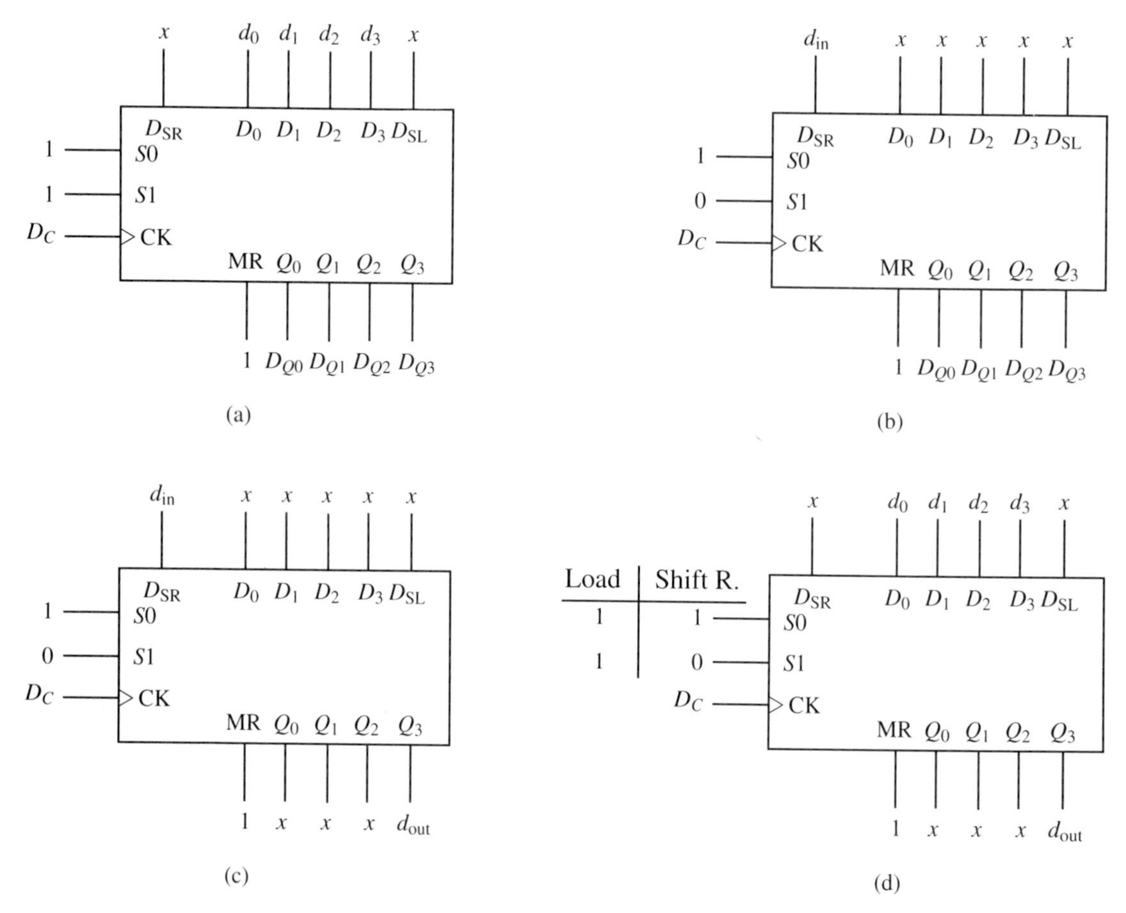


**Figure 11.25** Universal shift register configurations: (a) PIPO, parallel input/parallel output; (b) SIPO, serial input/parallel output; (c) SISO, serial input/serial output; (d) PISO, parallel input/serial output.

disconnected or unimportant under that configuration. All "shift right" examples have analogous "shift left" counterparts not shown; data sheets have information for all possible situations. In this diagram, the labeling convention for one particular manufacturer is given.

Notice that in Fig. 11.25, $D_3$ is the *least* significant bit. Unfortunately, as noted in Chapter 10, there is no consensus on this, which makes digital design even more challenging for an unnecessary reason.

## 11.5 STATE MACHINE DESIGN EXAMPLE

The design of a state machine (i.e., a sequential digital circuit) requires several steps. Each step will be stated and illustrated by means of a familiar example: the cycle of a microprocessor-controlled washing machine. (To date, microprocessor-controlled washing machines are still uncommon—only top-of-the-line models are. Apparently, this is because of the difficulty of interfacing with the 110-VAC side.)

An early "state machine" of sorts—a 1906 electric dishwasher. From its looks, it would seem to be more at home in a torture chamber than in a kitchen. No microprocessors in this system. *Source: Scientific American* (April 28, 1906).

We must begin by enumerating the distinct states possible for the machine and the associated outputs. For the washing machine, the list in Table 11.9 describes basic operation for the wash and rinse cycles. It should be mentioned from the outset that practical washers may have features different from and/or additional to those of the simplified example to be described here for illustrative purposes.

To simplify notation, only those outputs that are 1 ("on") are shown. The binary variables are: $D_w$ = fill with water, as specified by the user (hot wash, warm rinse/warm wash, warm rinse/cold wash, cold rinse); $D_t$ = timer on; $D_a$ = mechanically set up the motor to produce agitate motion (engage agitate belt, disengage spin tub belt); $D_s$ = mechanically set up the motor to spin the outer tub in order to drive the water from the tub and clothes (engage spin tub belt, disengage agitate belt), simultaneously pumping out water with drain open. (When $D_s = 1$, initially the starting winding is connected; at high speed, centrifugal forces flip a switch connecting the steady-state run winding and disconnecting the starting winding.) The 1-minute pauses between certain states that occur in actual machines have been omitted for simplicity.

Notice that in both of the safety wait states, the timer is off. The only way to restart the machine where it left off is to close the lid. The timer is also off during fill; filling stops when and only when the tank is full of water. Fullness of the tank

**TABLE 11.9** DEFINITION OF SUCCESSIVE STATES OF A WASHING MACHINE CYCLE AND THE REQUIRED HIGH OUTPUTS

| State | Description | High outputs |
|---|---|---|
| 0 | Stop | None |
| 1 | First fill | $D_w$ |
| 2 | First agitate | $D_t$, $D_a$ |
| 3 | First spin/pumpout | $D_t$, $D_s$ |
| 4 | First safety wait: lid open | None |
| 5 | Second fill ("rinse") | $D_w$ |
| 6 | Second agitate | $D_t$, $D_a$ |
| 7 | Second spin/pumpout | $D_t$, $D_s$ |
| 8 | Second safety wait: lid open | None |

is detected by a liquid level (pressure) sensor. (Thus if it took an hour to fill, that would not otherwise disrupt the cycle.)

Next, the conditions under which transitions from one state to another occur are itemized. New "input" variables needed here are $D_{\text{on}}$ = washer turned on; $D_l$ = lid open; $D_f$ = tub full (automatically detected), $D_{\text{wl}}$ = "water level" button, which cuts short the filling if desired by the user; $D_{14}$ = time in current state has reached 14 minutes; $D_8$ = time in current state has reached 8 minutes. These last two signals are obtained from the timer.

It is perhaps easiest to draw a state diagram first (see Fig. 11.26), and from that write out the state transition table, Table 11.10. (In each state bubble, the binary codes used below to signify the state are also shown, as well as the output variables that are 1 for that state.)

It is understood that for *any* transition to occur, the clock must fall (or rise, depending on the hardware) and the input data lines must be 1 as shown. For this example, the required input combinations for transitions are all very simple (single- or double-variable expressions). In general, they could be any combinational logic expressions.

We may now code the states with binary state variables. The number of state variables required is the smallest integer $N$ for which $N \geq \log_2$ (number of states). So in this case, $N \geq \log_2(10) = \log_{10}(9)/\log_{10}(2) \approx 3.17$, so $N = 4$. For this application, seven combinations of state variables are not used (an inefficient design). We had to choose $N$ to be the lowest integer larger than the nonintegral $\log_2$ (number of states). Thus the number of possible states, $2^N$, exceeds the number of required states.

The unused states (here states 9 through 15) in theory are never accessed. Because in practice they might be by accident (e.g., when the system is turned on), no-condition transitions from disallowed states to allowed states can be arranged to prevent "lock-up" in a disallowed state. In this example, all of the disallowed states will be sent to state 0 (stop).

If we use base 2 counting from 0 to 7 for assignment of the binary state variables, the state transition table, Table 11.11, may be constructed. The ",*o*" in the subscript indicates old values of the state variable, and ",*n*" indicates new values.

From Table 11.11 we can determine individually the new state variables in

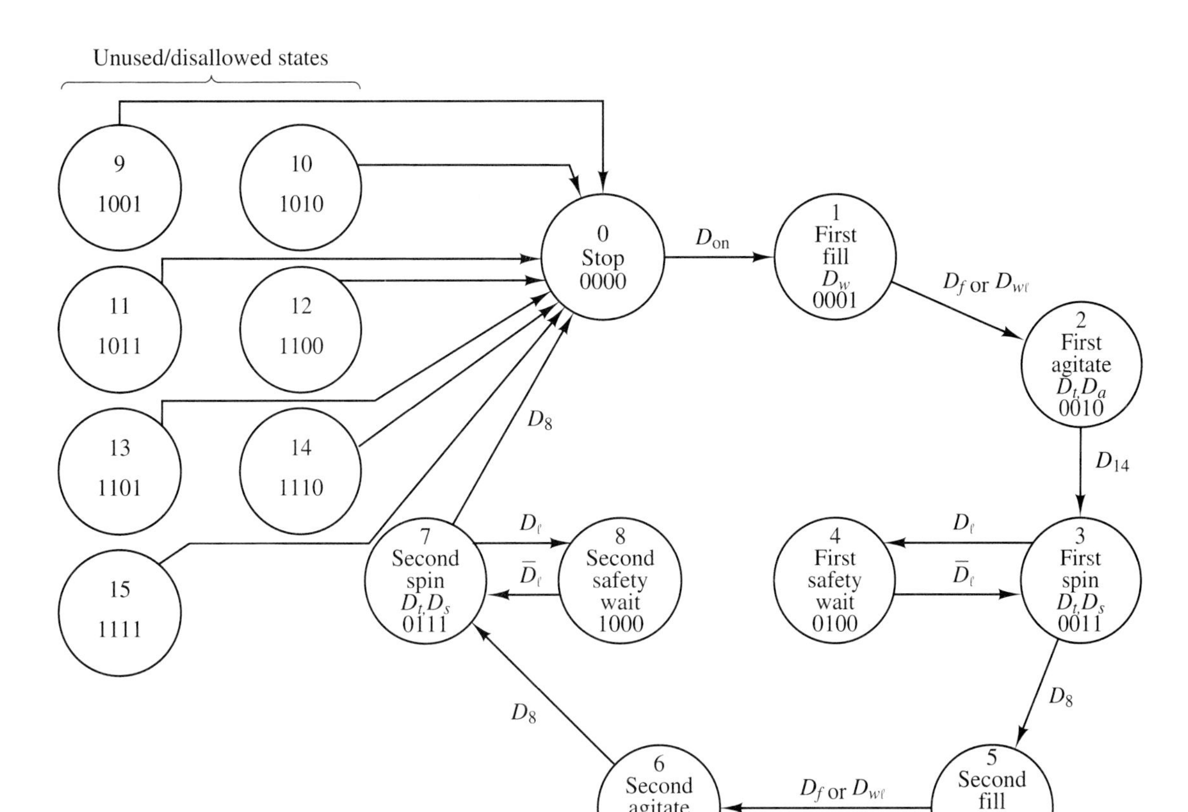


**Figure 11.26** State transition diagram of washing machine cycle.

**TABLE 11.10** STATE TRANSITION TABLE FOR WASHING MACHINE CYCLE

| State transition Old→New | High input required for transition |
|---|---|
| 0→1 | $D_{on}$ |
| 1→2 | $D_f + D_{wl}$ |
| 2→3 | $D_{14}$ |
| 3→4 | $D_l$ |
| 4→3 | $\overline{D}_l$ |
| 3→5 | $D_8$ |
| 5→6 | $D_f + D_{wl}$ |
| 6→7 | $D_8$ |
| 7→8 | $D_l$ |
| 8→7 | $\overline{D}_l$ |
| 7→0 | $D_8$ |

**TABLE 11.11** DETAILED STATE TRANSITION TABLE FOR WASHING MACHINE CYCLE INCLUDING VALUES OF STATE VARIABLES

| Transition number | State transition: Old state $D_{3,o}D_{2,o}D_{1,o}D_{0,o}$ | State transition: New state $D_{3,n}D_{2,n}D_{1,n}D_{0,n}$ | High input required for transition |
|---|---|---|---|
| t1 | 0000 | 0001 | $D_{\text{on}}$ |
| t2 | 0001 | 0010 | $D_f + D_{\text{wl}}$ |
| t3 | 0010 | 0011 | $D_{14}$ |
| t4 | 0011 | 0100 | $D_l$ |
| t5 | 0100 | 0011 | $\overline{D}_l$ |
| t6 | 0011 | 0101 | $D_8$ |
| t7 | 0101 | 0110 | $D_f + D_{\text{wl}}$ |
| t8 | 0110 | 0111 | $D_8$ |
| t9 | 0111 | 1000 | $D_l$ |
| t10 | 1000 | 0111 | $\overline{D}_l$ |
| t11 | 0111 | 0000 | $D_8$ |
| 1001, 1010, 1011, 1100, 1101, 1110, 1111→0000 | | | None |

terms of the old. Thus $D_{0,n}$ will be 1 if the old state was 0000 and the input $D_{\text{on}}$ is 1 (call this logical combination $D_{t1}$), or if the old state was 0010 and the input $D_{14}$ was 1 (call this $D_{t3}$), and so on. Consequently, the expressions for $D_{0,n}$ through $D_{3,n}$ are (where the subscript "$t$" denotes "transition")

$$
\begin{aligned}
D_{0,n} &= D_{t1} + D_{t3} + D_{t5} + D_{t6} + D_{t8} + D_{t10} \\
D_{1,n} &= D_{t2} + D_{t3} + D_{t5} + D_{t7} + D_{t8} + D_{t10} \\
D_{2,n} &= D_{t4} + D_{t6} + D_{t7} + D_{t8} + D_{t10} \\
D_{3,n} &= D_{t9},
\end{aligned}
\tag{11.5}
$$

where

$$
\begin{aligned}
D_{t1} &= \overline{D}_3\overline{D}_2\overline{D}_1\overline{D}_0 D_{\text{on}} & D_{t7} &= \overline{D}_3 D_2\overline{D}_1 D_0(D_f + D_{\text{wl}}) \\
D_{t2} &= \overline{D}_3\overline{D}_2\overline{D}_1 D_0(D_f + D_{\text{wl}}) & D_{t8} &= \overline{D}_3 D_2 D_1\overline{D}_0 D_8 \\
D_{t3} &= \overline{D}_3\overline{D}_2 D_1\overline{D}_0 D_{14} & D_{t9} &= \overline{D}_3 D_2 D_1 D_0 D_l \\
D_{t4} &= \overline{D}_3\overline{D}_2 D_1 D_0 D_l & D_{t10} &= D_3\overline{D}_2\overline{D}_1\overline{D}_0\overline{D}_l \\
D_{t5} &= \overline{D}_3 D_2\overline{D}_1\overline{D}_0\overline{D}_l & D_{t11} &= \overline{D}_3 D_2 D_1 D_0 D_8, \\
D_{t6} &= \overline{D}_3\overline{D}_2 D_1 D_0 D_8
\end{aligned}
\tag{11.6}
$$

in which for brevity $D_0$ through $D_3$ represent the old values of the state variables.

Note that the disallowed states 1001 through 1111 automatically go to 0000 (stop state) by default because none of those combinations of state variables are included in the minterms for the new state variable values to become 1. In turn, they cannot be accessed from any allowed state.

Because of the abundance of variables (four state variables and five input variables), Karnaugh maps would be of little help here. In practice, a direct implementation using a *programmable logic array* (PLA; see Sec. 11.7.4) may be most convenient. Or we may use individual logic gates, factoring out common terms where possible to minimize the required number of gates.

Once these next-state signals are available, they can be inputs to four $D$ flip-flops, one for each state variable. The outputs of these flip-flops (having common clock inputs) are the old values of the state variables. An overall circuit diagram would appear as in Fig. 11.27. All that is necessary is a four-bit register and sufficient combinational logic to form $D_{i,n}$ and the outputs $D_w$, $D_t$, $D_a$, $D_s$ from the inputs and $D_{i,o}$, $i = 0$ to 3. The details of these combinational logic blocks are for simplicity not shown explicitly. After one clock cycle, the "new" values of the state variables become "old" and cause the new outputs, while the "old" values of the state variables and outputs are erased. The "old" values of the state variables are used along with the inputs shown in Fig. 11.27 to determine their "new" values.

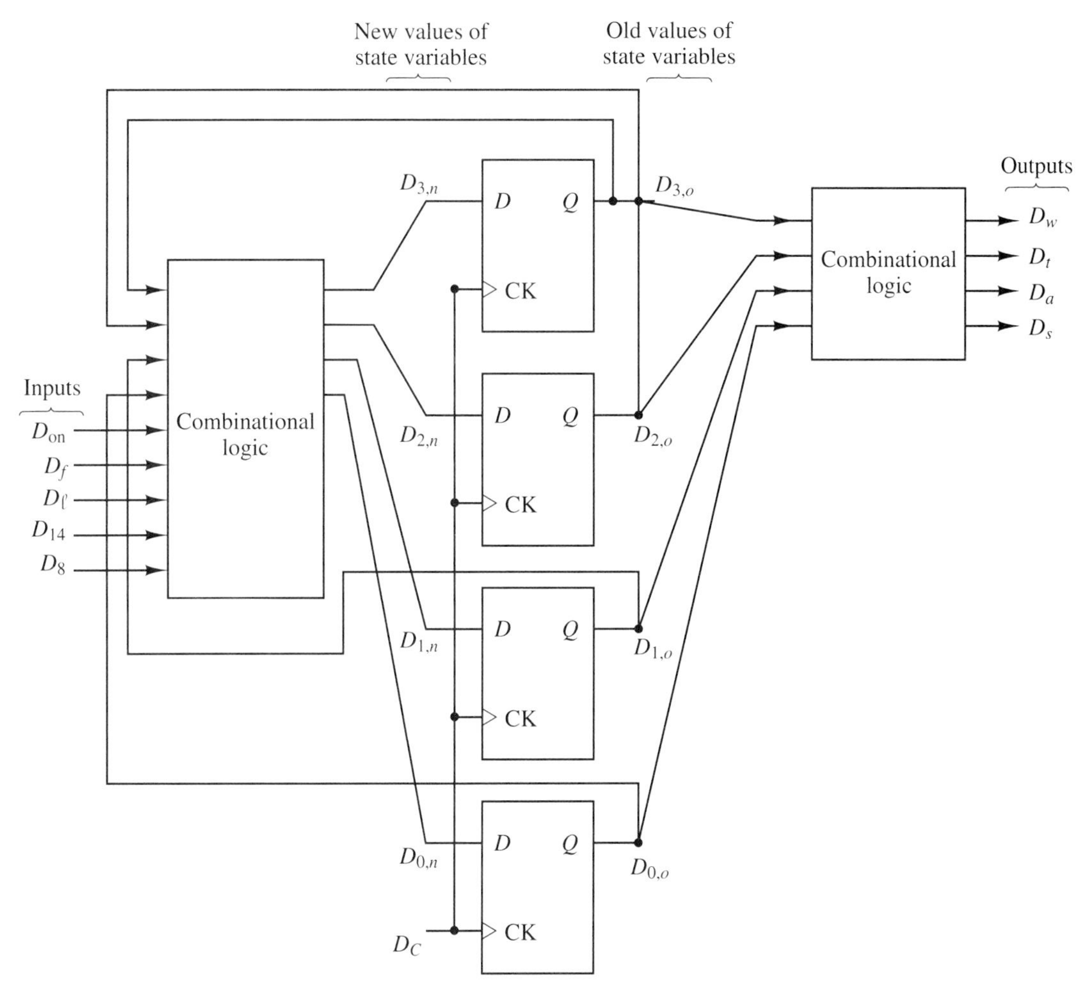


**Figure 11.27** Basic circuit diagram for washing machine cycle digital circuit showing the four state-variable flip-flops, inputs, outputs, and their connections.

## 11.6 NUMBER REPRESENTATIONS AND BINARY ARITHMETIC

### 11.6.1 Introduction

Up to now, we have made only a loose connection between logic circuits and the binary number system. The connection has mainly been that just as there are two logical states (e.g., high and low), there are two unique numerals in the binary number system (1 and 0). A little thought will reveal why binary arithmetic is used in computers.

The correspondence just mentioned is crucial. Also, pragmatic electronics issues favor binary. It is much simpler to distinguish between two voltage levels than among several, especially in the presence of noise. And the use of magnetic materials for storage of numerical data must be binary, for the magnetic orientation of a memory cell can practically be only in one of just two directions.

People, however, think in base 10. If we are to use binary arithmetic systems to do number crunching, conversions between binary and base 10 are required. In fact, another number system, base 16, also comes in handy when using digital electronics. In Sec. 11.6.2 we shall consider the mechanics involved in making our base 2 equipment usable by base 10 people.

Before diving in, however, a word on "digital." It is no mere coincidence that the word "digit" means both "an integer from 0 through 9" and a "finger" or "toe." The reason for both the choice of the set of 10 basic integers (as opposed to some other-sized set of distinct integers) and the use of "digit" to mean such an integer and also "finger" is that people have 10 fingers and originally counted with their fingers. The 10 fingers constituted the only handy (pun not intentional) distinct physical symbols for number representation; the Arabic numerals 0 through 9 later became associated assigned written symbols. To represent numbers beyond these ranges, multiple digits were required. The reality of 10 fingers on humans is the only conceivable reason that base 10 took such prominence; it has no other special distinguishing qualities.

It is interesting that one of the unique digits should be zero rather than the show of no fingers. That is, 0123456789 were the assigned unique 10 numerals as opposed to the more intuitive (10 things, show all 10 fingers) assignment 123456789A, where A would be a single-digit representation of 10. In the latter case we would have 11, not 10, distinct symbols for base 10, the symbol 0 being used *only* for the representation of zero and not, for example, in the representation of 10, 20, and so on.

Because of the existing convention, to represent decimal (base 10, "deci" always referring to 10) numbers, we have to use powers of a *two*-digit number, 10. Contrastingly, for base 2, we use various powers of the *single*-digit number 2—merely because we have that higher single-digit symbol available from our base 10 set. For calculations in base 2, however, only 0 and 1 are used, just as 0 through 9 are used in base 10. The reason for representing the base by "10" (in any base) and allowing 0 as a digit value is that it greatly facilitates multiplication and sum-of-powers representations.

Integers are different from numbers defined on a continuous range (*reals*) in that each integer is separated by 1 from the two nearest other integers; "real" numbers can be infinitesimally close to other "real" numbers. More generally, though, we may define sets of numbers such that each is separated from the nearest others in the same set not by 1, but by some fraction of 1.

It is these discretized number sets that are used in computers. This is because a computer can store only a finite accuracy representation of a number. The smaller that the *discretization* fraction is, the closer that calculations within that number set simulate real number calculations; that is, the more accurate the computer calculations will be.

Although "digit" originally meant only integers (and then only 0 through 9), *digital* has lately come to mean the more generalized discretization just mentioned. This is because in the finite accuracy representation of a number, a small number of digits are used; for example, a six-digit representation of $\pi$ is 3.14159 ("digit" originally implying base 10), the various digits being 3, 1, 4, 1, 5, and 9.

The terminology is further confused by the use of "digital" to mean *finite binary representation*. That is, "digital" refers to representing a number by a finite-length combination of 1s and 0s; for example, an eight-binary-digit ("bit") representation of $\pi$ is 11.001001. Furthermore, "digital" is even used to refer to binary logic circuitry in which only logical inference and no numerical computation at all is involved! Finally, the term *digital signal* often implies both discretization of the value of the signal and *sampling* of the continuous-time signal, that is, discretization of time as well as signal value. Our only hope out of this confusing wilderness of terminology is context—within a given situation the meaning of the word "digital" is usually apparent.

### 11.6.2 Binary Representation and BCD

**Introduction.** As noted in Eq. (10.1), an integer is represented in base 2 as

$$d_{N-1}\cdots d_1 d_0 = \sum_{n=0}^{N-1} d_n 2^n, \tag{11.7}$$

where each binary digit (bit) $d_n$ is either a 1 or a 0. As is the case for base 10, the physical position of $d_n$ within the number $d_{N-1}\cdots d_n\cdots d_0$ signifies the power of 2 it multiplies (its "significance"). The leftmost bit is the *most significant bit* (MSB) and the rightmost bit is the *least significant bit* (LSB). In binary arithmetic using digital electronics, it is often convenient to speak of groups of digits:

1 binary digit = 1 bit.
4 bits = 1 nibble.
8 bits = 2 nibbles = 1 byte.
? bytes = 1 word.

*Byte* was apparently a modification of "bite," meaning a morsel not of food but of bits; nibble was an amusing name for half a byte. A *word* is taken to mean the size of the digital representation of a number on a particular computer. Large, relatively high precision computers have large words, and small computers have small words. Hence the number of bytes per word varies and therefore is unspecified above.

Typical sizes of words are 1 word = $\frac{1}{2}$, 1, 2, 4, or 6 bytes (i.e., 4, 8, 16, 32, or 48 bits). It is the words that are added together, multiplied, stored, and retrieved as individual numbers, and so on. A larger number of bits per word naturally allows both greater precision and range of numbers that can be represented.

The *nibble* size is useful both for representing decimal numbers in binary (binary-coded decimal, BCD) and for representing binary numbers in base 16 (*hexadecimal*). This is because for both BCD and hexadecimal, one digit can be represented in binary by a nibble. Bytes are useful in that they are the length of the binary codes for a single keyboard character in the ASCII (American Standard Code for Information Interchange) system, with an extra bit left over. Nibbles and bytes are also often the "denominations" of various hardware devices, or even microprocessors (in which case the nibble or byte constitutes a word).

To keep track of all the 1s and 0s in long binary representations, it has been found useful to think and write the equivalent hexadecimal (base 16) number. This conversion is extremely easy: Each nibble starting from the LSB and going left is represented by a *hex* digit. They naturally run from 0 to 15, as detailed in Table 11.12. One byte is therefore just two hex numbers instead of eight 1s and 0s: 10100001 = A1. Going back and forth is as easy as using Table 11.12, which one quickly memorizes. Which would you prefer to remember: AF3B or 1010111100111011? So when base 2 binary codes are being notated, hex is used exclusively in practice; get used to it!

Number representations are rather lifeless unless they are actually used for something specific. A question much more interesting is: What happens when we type a number into a personal computer? We have the vague notion that computers work with only 1s and 0s; yet here we are typing in 5s, 8s, and $10^{-4}$. The question of computer data entry should be vital to anyone who spends lots of time sitting at a terminal, which is most of us these days.

This would not only be interesting to find out as a way of demystifying the

**TABLE 11.12** BINARY, HEX, AND DECIMAL COUNTING FROM 0 TO $15_{10}$

| Base 2 (binary) | Base 16 (hex) | Base 10 (decimal) |
|---|---|---|
| 0000 | 0 | 0 |
| 0001 | 1 | 1 |
| 0010 | 2 | 2 |
| 0011 | 3 | 3 |
| 0100 | 4 | 4 |
| 0101 | 5 | 5 |
| 0110 | 6 | 6 |
| 0111 | 7 | 7 |
| 1000 | 8 | 8 |
| 1001 | 9 | 9 |
| 1010 | A | 10 |
| 1011 | B | 11 |
| 1100 | C | 12 |
| 1101 | D | 13 |
| 1110 | E | 14 |
| 1111 | F | 15 |

computer, but it is also important to look into. For the base 10 keyboard is the lifeline between people and computers. Furthermore, in learning about the processes involved, several otherwise dry topics will come to life.

For example, there will come a point when we simply *must* do some binary addition and binary division to get the job done. To prepare for this, in this section we systematically consider the four arithmetic operations: addition, subtraction, multiplication, and division. Then when required in the subsequent discussion, we can conveniently call upon these methods as the need arises.

We will also be forced to deal with the conversion between base 10 and base 2, and the involvement of the BCD representation of numbers in that conversion. Finally, the hardware required to implement these operations will be addressed explicitly: gates and registers can be used to realize all of them. Thus we shall simultaneously have fun and learn a lot of important digital concepts.

**Binary arithmetic: Addition and subtraction.** Digital computers work with binary signals. All of the complicated operations a computer performs can be accomplished by combinations of add, subtract, multiply, and divide. In this and the next subsection we consider each of these operations in turn. By the end of this section we shall be using them to demonstrate how a computer is able to do simple arithmetic given a sequence of keystrokes in base 10.

For starters, suppose that we have two numbers in base 2 that we would like to add, each one byte long. Just as decimal numbers are added, we put one number below the other and add corresponding digits, beginning with the LSB. From simple counting rules, the single-bit possibilities are $0 + 0 = 0$, $0 + 1 = 1$, $1 + 0 = 1$, and $1 + 1 = 2_{10} = 10_2$. In the last case, the 1 of 10 serves as a "carry" if several digits are to be added and the 0 as the sum for that bit. An example should clarify the procedure.

**Example 11.8**

Add 01100101 to 00011100.

**Solution** Substitution of these numbers into Eq. (11.7) yields, respectively, 101 and 28, so we know that the result should be 129. Following the procedure above,

$$\begin{array}{ll} 11111\phantom{000} & \leftarrow \text{carries} \\ 01100101 & \leftarrow \text{augend} \\ \underline{00011100} & \leftarrow \text{addend} \\ 10000001 & \leftarrow \text{sum,} \end{array}$$

which is $2^7 + 2^0 = 129$, as required.

**Example 11.9**

Add 11000000 to 01000000.

**Solution** Substitution of these numbers into Eq. (11.7) yields, respectively, 192 and 64, so we know that the result should be 256. Following the procedure above,

$$\begin{array}{r} 11\phantom{0000000} \\ 11000000 \\ \underline{01000000} \\ 100000000, \end{array}$$

**TABLE 11.13** TRUTH TABLE FOR SUM AND CARRY FOR A FULL ADDER

| $d_{c,i-1}$ | $d_{2,i}$ | $d_{1,i}$ | $d_{s,i}$ | $d_{c,i}$ |
|---|---|---|---|---|
| 0 | 0 | 0 | 0 | 0 |
| 0 | 0 | 1 | 1 | 0 |
| 0 | 1 | 0 | 1 | 0 |
| 0 | 1 | 1 | 0 | 1 |
| 1 | 0 | 0 | 1 | 0 |
| 1 | 0 | 1 | 0 | 1 |
| 1 | 1 | 0 | 0 | 1 |
| 1 | 1 | 1 | 1 | 1 |

which is $2^8 = 256$, as required. However, the number is no longer only one byte long. To represent 256 we need either more bits, a floating-point representation of the form $A \cdot 2^m$, where $0 \leq A < 1$ and $m$ is a binary integer, or a means of indicating *overflow*.

The close tie between binary logic and binary arithmetic is made explicit when a circuit implementation for binary addition is sought. A basic element for single-bit addition can be obtained by examination of the truth table. There are three inputs for the $i$th bit sum: the bits $d_{1,i}$ from the augend and $d_{2,i}$ from the addend, and the carry bit $d_{c,i-1}$. Also, there are two outputs: the sum $d_{s,i}$ and the carry $d_{c,i}$. The truth table is given by Table 11.13.

The Karnaugh map for $d_{s,i}$ is shown in Fig. 11.28, which does not help, because there are no groups to circle—only "islands." But if we write out the terms as $d_{s,i} = \overline{d}_{c,i-1}(\overline{d}_{2,i}d_{1,i} + d_{2,i}\overline{d}_{1,i}) + d_{c,i-1}(\overline{d}_{2,i}\overline{d}_{1,i} + d_{2,i}d_{1,i})$, we see that $d_{s,i}$ is 1 if $\{d_{c,i-1} = 0\}$ AND $\{\text{XOR}(d_{2,i},d_{1,i}) = 1\}$, OR $\{d_{c,i-1} = 1\}$ AND $\{\text{XOR}(d_{2,i},d_{1,i}) = 0\}$. But this is just $\text{XOR}(d_{c,i-1}, \text{XOR}(d_{2,i},d_{1,i}))$.

The Karnaugh map for $d_{c,i}$ is shown in Fig. 11.29. Drawing the three obvious loops gives $d_{c,i} = (d_{1,i} + d_{2,i})d_{c,i-1} + d_{1,i}d_{2,i}$. Recognizing that the option $d_{1,i} = d_{2,i} = 1$ is already included in the second term and applying absorption, $d_{c,i}$ can be written as $d_{c,i} = d_{c,i-1} \cdot \text{XOR}(d_{1,i}, d_{2,i}) + d_{1,i}d_{2,i}$. This form allows use to be made of $\text{XOR}(d_{1,i}, d_{2,i})$, which was already generated for $d_{s,i}$. The complete *full-adder* circuit is given in Fig. 11.30. (There are also *half-adders* that do not take a carry input.) When the carry out $d_{c,i}$ is fed into the carry-in of the next stage, multibit binary addition is possible. By systematizing the conditions associated with known outcomes of binary addition, we have "fooled" the logic gates into doing arithmetic. This idea is fundamental to all computers.

What about negative binary numbers? There are two common approaches: sign-magnitude and 2's complement. *Sign-magnitude* is exactly what humans do in base 10: Just stick a minus sign out front. In digital circuits, the leftmost bit is used not for the MSB but for a sign flag: 1 if negative, 0 if positive. Thus a byte has only seven significant bits. Naturally, if more bits are needed, additional bytes may be

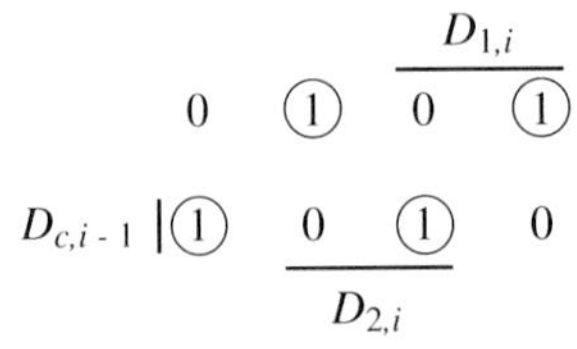

| | | | $D_{1,i}$ | |
|---|---|---|---|---|
| | 0 | (1) | 0 | (1) |
| $D_{c,i-1}$ | (1) | 0 | (1) | 0 |
| | | $D_{2,i}$ | | |

**Figure 11.28** Karnaugh map for $d_{s,i}$ (sum output of $i$th adder stage).

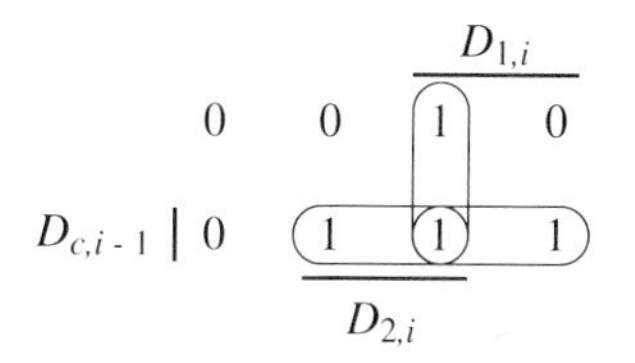


**Figure 11.29** Karnaugh map for $d_{c,i}$ (carry output of $i$th adder stage).

used for number representation and storage; for those bytes all eight of the bits are available for significant digits of the number. Because of the easy detection of sign, this form is commonly used in the representation of floating point numbers.

But once addition of negative numbers or subtraction is encountered, a complication arises. It is common knowledge that when using pencil and paper to subtract two numbers, we subtract the number with the smaller magnitude from the other and give the result the sign of the larger magnitude number. For example, to compute $55 - 120 = -(120 - 55)$, we write

$$\begin{array}{r} 120 \\ -\underline{\phantom{0}55} \\ 65 \end{array} \rightarrow -65.$$

We could implement this digitally with the binary *comparator*, the four-bit version of which is shown in Fig. 11.31.

Because of pin limitations, the four-bit magnitude comparator is the type found in IC form. The design of a comparator is easily accomplished using combinational logic and the techniques of Chapter 10: Truth tables and Karnaugh maps for a three-output system. Its use is also completely straightforward. Just connect the four bits of number $A$ to $A_0$–$A_3$ and those of number $B$ to $B_0$–$B_3$. Set the $I_{A>B}$ and $I_{A<B}$ inputs to 0 and the $I_{A=B}$ to 1. If $A > B$, the $O_{A>B}$ line will be high. If $A < B$, $O_{A<B}$ will be high, and if $A = B$, then $O_{A=B}$ will be high.

The four-bit comparator is easily extended to byte comparison by using two four-bit ICs (see Fig. 11.32). Just connect the $O_{A<B}$, $O_{A=B}$, and $O_{A>B}$ of the lower order bits to $I_{A<B}$, $I_{A=B}$, $I_{A>B}$ of the higher-order chip as shown. This is why such strange names as $I_{A<B}$ are used for an input; they refer to the digital signals connected to them when more than four bits are being compared. The lowest four bits, however, always must have $I_{A>B} = I_{A<B} = 0$, $I_{A=B} = 1$.

Returning to the task at hand, the requirement of a comparison operation for subtraction is inconvenient, as is the need for different methods for addition and subtraction. In *2's complement,* the following trick is used. In this discussion, let $D_m$ represent a single byte representation of a number, except for the sign bit such as would appear in a data register; that is, $D_m = d_6 d_5 d_4 d_3 d_2 d_1 d_0$. If we add $2^7$ to the byte $D_m$, the original bits of the byte remain the same:

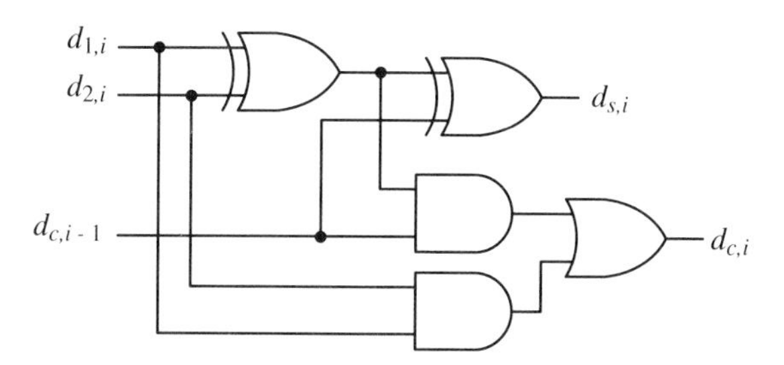


**Figure 11.30** Full-adder logic circuit.

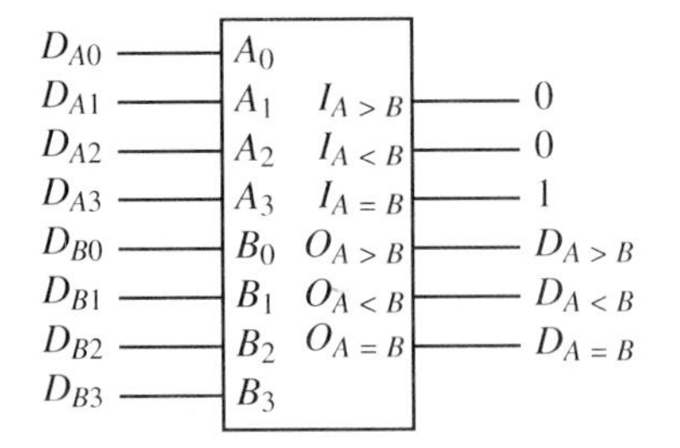


**Figure 11.31** Four-bit magnitude comparator circuit symbol.

$$\begin{array}{r} 1\ 0\ 0\ 0\ 0\ 0\ 0\ 0 \\ d_6\, d_5\, d_4\, d_3\, d_2\, d_1\, d_0 \\ \hline (1)\ d_6\, d_5\, d_4\, d_3\, d_2\, d_1\, d_0. \end{array}$$

Therefore, for all of the bits within the byte-sized representation, $D_m + 2^7$ "=" $D_m$, where the quotation marks around equals mean equality only for all seven bits in the representation. Substituting $D_1 - D_2$ for $D_m$ gives

$$D_1 - D_2 \text{ "=" } D_1 + (2^7 - D_2) = D_1 + D_3, \tag{11.8}$$

where $D_3 = 2^7 - D_2$ is called the *2's complement* of $D_2$. The reason for writing the subtraction as in Eq. (11.8) is that $D_3$ can be obtained very easily from $D_2$, and thus turns subtraction into addition. We know that $2^7 - 1 = 1111111$. But if each bit of $D_2$ is logically complemented and this complemented byte is added to $D_2$, the result is also 1111111. For example, if $D_2 = 0101110$, then $\overline{D}_2 = 1010001$ and $D_2 + \overline{D}_2$ is

$$\begin{array}{l} 0101110 \\ \underline{1010001} \\ 1111111. \end{array}$$

Consequently, $2^7 - 1 = D_2 + \overline{D}_2$, and therefore $D_3 = 2^7 - D_2 = D_2 + \overline{D}_2 + 1 - D_2 = \overline{D}_2 + 1$. Thus, as far as the bits being kept track of are concerned, we can use addition [note the + sign in Eq. (11.8)] to perform subtraction, as long as the number being subtracted is first converted to 2's complement.

Of course, if $|D_1| \geq |D_2|$, then $D_1 - D_2 \geq 0$. How can the sign of the difference be detected using 2's complement arithmetic? Well, if $|D_1| \geq |D_2|$, then $D_1 - D_2 \geq 0$ or $D_1 + 2^7 - D_2 \geq 2^7$, which is 1 larger than one byte can hold. That is, adding the 2's complement of $D_2$, $2^7 - D_2$, to $D_1$ causes a MSB carry. Thus a carry indicates a nonnegative difference. Conversely, there is no MSB carry in the addition of $D_1$ and the 2's complement of $D_2$ if $|D_1| < |D_2|$. Therefore, no MSB carry

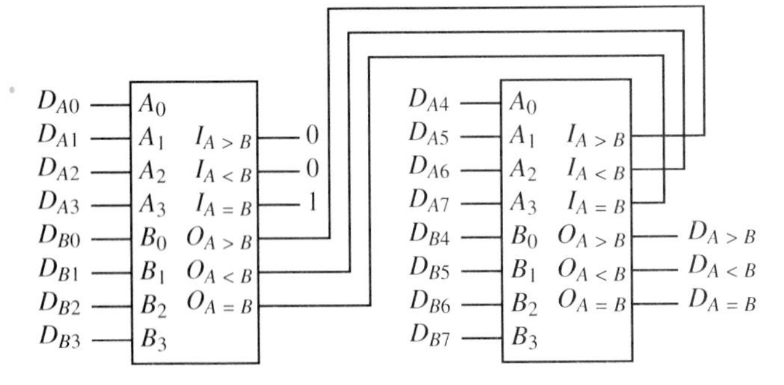


**Figure 11.32** Eight-bit (byte) comparator circuit.

resulting from performing the right-hand side of Eq. (11.8) indicates a negative difference.

We see that once again the connection between binary logic and binary arithmetic has come to the rescue. In Sec. 10.2 we saw how arithmetic and logical products and sums were closely related (e.g., AND versus multiply and OR versus add). Previously in this section, we saw how binary logic could produce the same results as and thereby simulate arithmetic addition. Now we see that for base 2, the arithmetic ("2's") complement is closely related to the logical complement of the logic variables contained in the byte-sized representation. A few examples of the 2's complement method will indicate how convenient it is. No comparisons of binary numbers will have to be done, and no *borrowing*. Once the 2's complement is found, it is as easy as binary addition.

**Example 11.10**

Find 1000110 − 0101011.

**Solution** First, complement the second number: 1010100 and add 1: 1010101. Now just add it to the first number:

$$\begin{array}{rrl} & 1 & \\ & 1000110 & \leftarrow \text{minuend} \\ & \underline{+\,1010101} & \leftarrow \text{2's complement of subtrahend} \\ 1 & \leftarrow 0011011 & \leftarrow \text{difference.} \end{array}$$

Because of the final carry, we know the result is positive. We note that in base ten, the problem is 70 − 43 = 27, as conversion of all three binary numbers to base 10 shows.

**Example 11.11**

Find 0011100 − 1111000.

**Solution** First, complement the second number: 0000111 and add 1: 0001000. Now just add it to the first number:

$$\begin{array}{rr} & 11 \\ & 0011100 \\ & \underline{+\,0001000} \\ 0 & \leftarrow 0100100. \end{array}$$

Because of no carry, we know that the result is negative. We note that in base 10, the problem is 28 − 120 = −92. The negative sum, 0100100, may be converted to the more recognizable (sign-magnitude) binary form −1011100 by subtracting 1 and complementing: Subtracting 1 from 0100100 gives 0100011 and complementing that gives the desired result after tacking on the minus sign bit. In the register the eighth bit (the sign bit) is thus set to 1 if 2's complement subtraction is used in conjunction with sign-magnitude representations, as is often the case.

One last use for the carry is to detect overflow. Once a subtraction problem has been converted into a sum, the same following rule holds for either addition or subtraction. If the sign bits of the two operands differ, overflow is impossible. If they agree, then overflow occurs if the MSB carry differs from those two identical sign bits. In practice, one equivalently adds the sign bits: If the carrys into and out of the sign bit differ, overflow has occurred.

For simplicity in Examples 11.10 and 11.11, we did not bother to add the sign bits because we knew that we were adding numbers having differing sign bits (so there was no possibility of overflow). In practice, the sign bits are always added for overflow checking.

**Example 11.12**

Find the following sums and differences using binary arithmetic: (a) $7 - 1$, (b) $1 - 7$, (c) $7 + 7$, (d) $-7 + -7$, using three data bits and one sign bit.

**Solution**

**(a)**

```
                111
  0111  \       0111
- 0001  /  →  + 1111
              ------
               10110 = 6 (MSB carry means positive result of subtraction;
                       carries into and out of sign bit agree, so no overflow
                       has occurred.)
```

**(b)**

```
                  1
  0001  \       0001
- 0111  /  →  + 1001
              ------
               01010 = −0110 (No MSB carry means negative subtraction
                       result; no carry into and out of sign bit, so no
                       overflow has occurred.)
```

**(c)**

```
   111
   0111
 + 0111
 ------
  01110   Carry into sign bit differs from carry out of sign bit, while sign
          bits of operands agree; overflow has occurred.
```

**(d)**

```
                  1
- 0111  \       1001
- 0111  /  →  + 1001
              ------
               10010   Carry into sign bit (0) differs from carry out of sign
                       bit, while sign bits of operands agree; overflow has
                       occurred.
```

**Binary arithmetic: Multiplication and division.** Multiplication is a sequence of adds and shifts, as we all know from pencil-and-paper multiplication in base 10. In base 2 the identical technique is used. Two single-byte integers (excluding sign bit) will now be multiplied to illustrate; notice that the result is two bytes long.

**Example 11.13**

Multiply 1001011 by 0100101.

**Solution**

```
        1001011   ←multiplicand
      × 0100101   ←multiplier
      ---------
        1001011
       0000000
      1001011
     0000000
    0000000
   1001011
 + 0000000
 -------------
  0101011010111   ←product,
```

which is the binary form of the product 75 × 37 = 2775. As opposed to hand multiplication in decimal, there are only two alternatives for each addition step: add 0 or add the multiplicand.

We could perform the same multiplication operation using shift registers. Each clock pulse, a shift or add would be done according to a microprogram (see, e.g., Chapter 12). To implement multiplication, two byte-sized registers would be hooked together side by side so that the serial output of one is the serial input of the other.

In the less significant byte, the multiplier (in this case 0100101) is initially placed, in the more significant byte 0000000 is initially placed, and in the end the two bytes together will contain the product 0101011010111. The multiplicand 1001011 is stored in another register. There will be 7 (add)-shift steps, one for each bit in the byte (excluding sign bit). The word "add" is in parentheses because if the current multiplier bit is 0, we just shift; there is no point in adding 0. The shifts will be to the right.

The LSB of the multiplier contributes to the LSBs of the result. So if the result of the first (add)-shift step is placed in the MSBs, when all seven steps have been done it will be where it should be [including all added accumulations from subsequent (add)-shifts]. Hence the example above would appear as follows (compare with the pencil-and-paper method shown above):

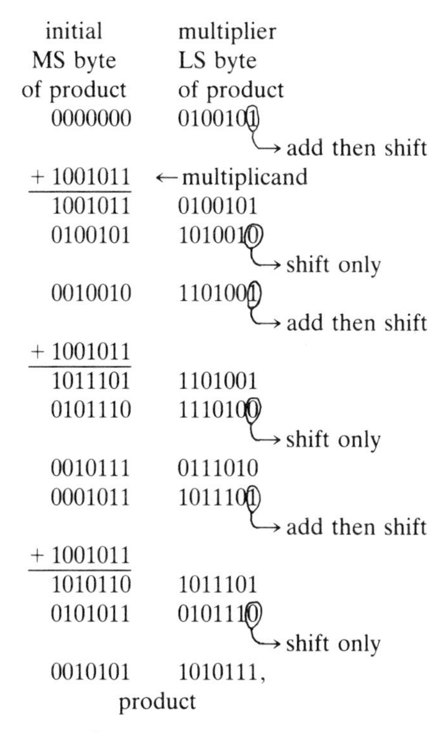


that is, 101011010111, which agrees with the result of the pencil-and-paper method.

Division also proceeds in much the same way we are familiar with "on paper." The only difference is that in the trial-and-error method, there are only two possibilities: Either the divisor fits into the dividend or partial remainder (so put a 1 in this quotient bit) or it does not (so put in a 0). Instead of doing the comparison in our heads, we shall always first try out 1. If the difference is positive, put a 1 for that quotient digit and retain the difference as a partial remainder. If the difference is negative, that means it did not fit. In that case the dividend or partial remainder is restored to its value before the subtraction and a 0 is the quotient bit. An example should illustrate further.

**Example 11.14**

Find 1000101/0101011 to eight-bit accuracy.

**Solution** Remember that for subtraction we shall add the 2's complement of the divisor, 0101011, which is 1010101. If the carry of that addition is 1, the result is positive and no restoration is necessary; otherwise, it is negative and the dividend or partial remainder is restored to its value before the subtraction.

```
                        1.1001101    ←quotient
   0101011          |1000101.0000000  ←dividend
      ↑            + 1010101
   divisor     1←    0011010
                                      (no restore)
                      011010 0
                    + 101010 1
               1←     000100 1
                                      (no restore)
                       00100 10
                     + 10101 01
               0←      11001 11
                                      (restore)
                        0100 100
                      + 1010 101
               0←       1111 001
                                      (restore)
                         100 1000
                       + 101 0101
               1←        001 1101
                                      (no restore)
                          01 11010
                        + 10 10101
               1←         00 01111
                                      (no restore)
                           0 011110
                         + 1 010101
               0←          1 110011
                                      (restore)
                             0111100
                          +  1010101
               1←            0010001
                                      (no restore).
```

The equivalent base 10 problem to that just solved is 69/43 ≈ 1.6046. The answer obtained above, 1.1001101, converted to base 10 is ≈ 1.6016. A more accurate answer naturally could be gained by using more bits; the finite bit answer obtained by this procedure is always a little too small.

Using our results for this example as a guide, it is straightforward to implement binary division with registers. The dividend is placed as a two-byte number (to allow for the possibility that the result is an integer a full byte long). Notice how the carry bit is fed directly into the right side of the two-byte register as a quotient bit.

```
        dividend
        1000101 0000000
      + 1010101 ←2's complement of divisor
   1←   0011010     (no restore; shift)
        0110100 0000001
      + 1010101
   1←   0001001     (no restore; shift)
        0010010 0000011
      + 1010101
   0←   1100111     (restore and shift)
        0100100 0000110
      + 1010101
   0←   1111001     (restore and shift)
        1001000 0001100
      + 1010101
   1←   0011101     (no restore; shift)
        0111010 0011001
      + 1010101
   1←   0001111     (no restore; shift)
        0011110 0110011
      + 1010101
   0←   1110011     (restore and shift)
        0111100 1100110
      + 1010101
   1←   0010001     (no restore; shift)
        0111101 1001101
               \_______/
             8-bit quotient.
```

The eight LSBs now contain the quotient to eight-bit accuracy. The location of the binary point is easily kept track of. In practice, there would be seven preceding steps where a leading byte initially containing 0s would be placed to the left of the dividend. Each of those would shift in 0s because of there would be no carries, and the final result would be 0000001.1001101, as required.

It should also be mentioned that many minicomputer systems such as personal computers (PCs) use iterative techniques such as the Newton–Raphson method to perform division. Also, intrinsic functions such as sine, cosine, inverse tangent, exponential, and natural logarithms are often performed by series solutions (Taylor or Chebyshev expansions). In such cases, the necessary coefficients are retrieved from *look-up tables* stored in the computer (see Secs. 11.7 and 12.7.3).

**Keyboards, ASCII, and conversions.** The goal of the immediate discussion will be to gain a degree of understanding about what happens when we type numbers into a typical computer. How does the computer handle scientific notation? If the keys are in base 10 and digital circuitry uses only 1s and 0s, what sort of conversion is made, and how? Exactly what kind of representation is used? It clearly must involve only a small number of digits, due to hardware constraints. Removal of the mystery behind the keyboard is not only satisfying but also involves manipulations typical of human–computer interfacing, formats typical of many computational circuits, and basic binary arithmetic operations.

For the purposes of illustration, consider use of the computer to determine a pressure differential and then express the value in a different unit system. Specifically, let the pressure $P$ in a tank be 10.3 atm, and the ambient pressure $P_0$ be 1.042 atm. It is desired to convert the pressure difference $P - P_0$ into $N/m^2$, where 1 atm $= 1.013 \cdot 10^5$ $N/m^2$. We all know how to do this in decimal: $(10.3 - 1.042) \cdot 1.013 \cdot 10^5 = 9.38 \cdot 10^5$, to three significant places.

A computer, however, must work in binary. Depending on the computer and the application, the computation may be done either in purely base 2 or in *binary-coded decimal* (BCD). In either case, a conversion from the keystroke to a binary code must first be done. And naturally in the end, we shall want the output to be in readable decimal form, not binary or BCD or ASCII.

It is of interest that BCD was developed mainly because business calculations demanded no roundoff error as occurs in base 10-to-base 2 conversions. Using specialized algorithms developed for BCD, base 10 calculations could be performed in base 10 using the binary signals of BCD representations.

When a key is pressed on a modern keyboard, a digital scanner discovers that a switch has been connected on the given row and column of keys (including numbers and letters, and also possibly qualified by a control character enabling a special character set). Typically, an ASCII code (seven bits long) is generated for the particular key. In data transmission situations, an eighth bit, called a *parity bit*, is included that often is used to check for transmission errors due to noise or other reasons (recall Sec. 10.8 for details). The ASCII code for numbers is 011 followed by the binary-coded decimal (BCD) form. The BCD and ASCII codes for numerals 0 through 9 are shown in the Table 11.14.

It would seem that BCD does not differ from usual base 2, until we consider *multiple*-decimal-digit numbers. For example, 21 is coded as 0010 0001, whereas

**TABLE 11.14** ASCII AND BCD CODES FOR THE DECIMAL DIGITS 0 THROUGH 9, AND ASCII CODE FOR THE BINARY POINT

| | ASCII | |
|---|---|---|
| Decimal | | BCD |
| 0 | 011 | 0000 |
| 1 | 011 | 0001 |
| 2 | 011 | 0010 |
| 3 | 011 | 0011 |
| 4 | 011 | 0100 |
| 5 | 011 | 0101 |
| 6 | 011 | 0110 |
| 7 | 011 | 0111 |
| 8 | 011 | 1000 |
| 9 | 011 | 1001 |
| . | 010 | 1110 |

the base 2 representation of 21 is 10101. In BCD, each decimal digit is represented by an individual four-bit base 2 binary code (a nibble). Because each nibble represents one *decimal* digit, and not successively more significant nibbles in the base 2 representation, different procedures from those used in binary arithmetic must be used in performing arithmetic with numbers expressed in BCD representations.

One reason for using BCD is that it matches the decimal-digit-at-a-time process of data entry using keyboards. It provides a hybrid decimal digit/binary digit form that can be manipulated with digital circuitry either "as is" (BCD arithmetic) or after further conversion to base 2. Also, when it comes time to display results in decimal form for people, each BCD nibble can drive a separate single-decimal-digit display pattern such as on an LCD or a CRT display.

Suppose our computer is one that performs calculations in base 2. How can the original BCD be converted to base 2? The number is still fundamentally in decimal, even though each of the individual BCD digits has a four-bit base 2 representation. Let the number in base 10 be 10.3. The four ASCII binary codes generated will be 011 0001 for 1, 011 0000 for 0, 010 1110 for ".", and 011 0011 for 3. Notice that the first three digits of the code are 011 for numbers, as stated previously, but 010 is used for the decimal point. The BCD digits are thus 0001 for 1, 0000 for 0, and 0011 for 3.

Incidentally, just because in base 10 there is only the single digit, 3, to the right of the decimal point, there may be an unlimited number of bits to the right of the binary point in the base 2 representation of the same number, 10.3. Therefore, error will necessarily be incurred as a result of conversion. This is generally true for all finite-digit base conversions.

The first thing to do is to express the number in *scientific notation* (*floating-point* form); we assume that our computer uses floating-point arithmetic. In a normalized floating-point form, the decimal point is moved just to the left of the leftmost nonzero digit, and the corresponding power of 10 is appropriately incremented. Thus $10.3\ (\cdot 10^0) = 0.103 \cdot 10^2$.

Now the decimal floating-point form must be converted to binary floating-point form. The conversion is $W \cdot 10^n = W \cdot 2^{m+x}$, where $n$ and $m$ are integers. In our example, $W = 0.103_{10}$ and $n = 2_{10}$ ($W$ and $n$ are now in BCD). Taking the $\log_{10}$ of $10^n = 2^{m+x}$ yields $n \cdot \log_{10}(10) = n = (m + x) \cdot \log_{10}(2)$, so that $m + x = n/\log_{10}(2)$. In the current example, $m + x = 2/\log_{10}(2) \approx 2_{10}/0.30103_{10}$. Both $n$ ($= 2_{10}$ in this example) and $\log_{10}(2) \approx 0.30103$ are converted to binary as follows.

**Base 10-to-base 2 conversion and BCD.** Both a BCD integer ($n$) and a BCD fraction (0.30103) need to be converted to base 2. Remember that BCD is still base 10. Thus the procedure for conversion of integers and fractions from base 10 to base 2 must be obtained. First consider integers.

It is commonly known that when a integer $M_{10}$ is divided by 2 it is odd if the remainder is 1 and even if the remainder is 0. Because in base 2 all the bits except the LSB represent whole multiples of 2, the LSB is precisely the remainder of the number divided by 2. That is, the LSB is zero if the integer is even and the LSB is 1 if it is odd. Thus we set LSB $= d_0 =$ remainder of $M_{10}/2$. Define $[M_{10}/2]$ as the integer portion of the quotient of $M_{10}/2$; then

$$M_{10} = 2 \cdot [M_{10}/2] + d_0. \tag{11.9}$$

The quantity $[M_{10}/2]$ can serve as a "new" $M'_{10}$, which itself can be broken down in the same way as

$$M'_{10} = 2 \cdot [M'_{10}/2] + d_1. \tag{11.10}$$

Substitution of $M'_{10}$ for $[M_{10}/2]$ in Eq. (11.9) gives

$$M_{10} = 2 \cdot (2 \cdot [[M_{10}/2]/2] + d_1) + d_0, \tag{11.11}$$

where $d_1$ (a 1 or a 0) is recognized as both the remainder of $[M_{10}/2]/2$ and the second LSB in the base 2 representation of $M_{10}$.

The latter fact is true for the following reason. Just as the even- or oddness of $M_{10}$ indicated how many times $2^0 = 1$ fits into $M_{10}$ without involving $2^{m>0}$ (i.e., $2^m$ where $m$ is any integer greater than zero), the even- or oddness of $[M_{10}/2]$ indicates how many times $2^1$ fits into $M_{10}$ without involving $2^{m>1}$ [notice how $d_1$ is multiplied by $2^1$ in Eq. (11.11)]. Continuing this process shows that the remainders of successive division by 2 therefore yield the binary digits in the base 2 representation.

**Example 11.15**

Convert 50 to base 2.

**Solution** Read the following diagram from bottom to top.

$$\begin{array}{rl} & 0 \rightarrow 1 = d_5 \\ 2 & \overline{|1} \rightarrow 1 = d_4 \\ 2 & \overline{|3} \rightarrow 0 = d_3 \\ 2 & \overline{|6} \rightarrow 0 = d_2 \\ 2 & \overline{|12} \rightarrow 1 = d_1 \\ 2 & \overline{|25} \rightarrow 0 = d_0 \\ 2 & \overline{|50} \end{array}$$

Taking the top bit as the MSB, the binary number is 110010. Check for yourself that substitution of 110010 into Eq. (11.7) gives 50.

**Example 11.16**

The integer $n$ required for the conversion of the pressure $P_1 = 10.3$ was 2. Following the steps gives us

$$\begin{array}{rl} & 0 \rightarrow 1 = d_1 \\ 2 & \overline{|1} \rightarrow 0 = d_0 \\ 2 & \overline{|2} \end{array}$$

and thus $2_{10} = 10_2$, as you might have guessed.

How can this general conversion process be implemented using the given BCD, which is made up of nibbles of 1s and 0s? Consider first only the least significant nibble $d_{B,3}d_{B,2}d_{B,1}d_{B,0} = d_{B,3} \cdot 2^3 + d_{B,2} \cdot 2^2 + d_{B,1} \cdot 2^1 + d_{B,0} \cdot 2^0$, where the "$B$" stands for BCD and where we recall that a BCD digit is a base 2 representation of a single decimal digit. Division by 2 gives

$$\frac{d_{B,3}\,d_{B,2}\,d_{B,1}\,d_{B,0}}{2} = 0 \cdot 2^3 + d_{B,3} \cdot 2^2 + d_{B,2} \cdot 2^1 + d_{B,1} \cdot 2^0 + d_{B,0} \cdot 2^{-1}. \qquad (11.12)$$

Just as division of a base 10 number shifts all digits to the right one place with respect to the decimal point, division of a base 2 number shifts all digits to the right one place with respect to the binary point.

The digit "shifted out" normally just goes on to lower significant digits. But with BCD, a correction must be made. It is easiest to explain by example. Suppose that we have the BCD number 13, which has BCD representation 0001 0011. It is desired to convert this number to base 2. If it is divided by 2, all digits are shifted right.

Normally, the 1 shifted out of the tens digit (originally representing 10) should contribute 10/2 = 5 to the ones digit. But instead, the shifted pattern is 0000 1001—the shifted-in 1 has contributed 8 = 1000, not 5. Consequently, if the MSB of a BCD digit is 1 after shifting into it, 3 must be subtracted from the result.

Also, the 1 shifted out of the ones digit is naturally the desired remainder and is saved as the LSB in the true base 2 representation being calculated. It can be fed into the serial input of a shift register that will eventually contain the base 2 representation, shifting right one place at each successive division by 2. When completed, the block of all the shifted-in remainders will be the desired base 2 representation. Explicitly, the steps are:

```
13:             Base 2 conversion
0001  0011      000000
divide by 2:
0000  (1)001    100000
       └→ subtract 3 (add 2's complement of 3 = 1101)
0000  1101
      ----
0000  0110
divide by 2:
0000  0011      010000
divide by 2:
0000  0001      101000
divide by 2:
0000  0000      110100
divide by 2:
0000  0000      011010
divide by 2:
0000  0000      001101 = 13₁₀.
```

Notice that we had to divide by 2 six times—the number of binary digits in the base 2 register; the last few times 0s were shifted out.

It is also required to convert a fractional number from BCD to binary. Procedures similar to those for integers are used except that they are reversed. For integers, the number is successively divided by 2 and each remainder serves as a bit in the base 2 representation. For fractions, the number is successively multiplied by 2 and the carry-outs or overflows become bits in the base 2 representation. Specifically, if the result of a multiplication by 2 is greater than 1, the binary digit is a 1; otherwise, it is a 0. The fractional portion is then remultiplied by 2.

**Example 11.17**

Convert 0.423 to a six-bit base 2 representation.

**Solution**

$$
\begin{array}{rr}
 & 0.423 \\
 & \times \quad 2 \\ \hline
d_{-1} = 0 \leftarrow & 0.846 \\
 & \times \quad 2 \\ \hline
d_{-2} = 1 \leftarrow & 1.692 \\
 & 0.692 \\
 & \times \quad 2 \\ \hline
d_{-3} = 1 \leftarrow & 1.384 \\
 & 0.384 \\
 & \times \quad 2 \\ \hline
d_{-4} = 0 \leftarrow & 0.768 \\
 & \times \quad 2 \\ \hline
d_{-5} = 1 \leftarrow & 1.536 \\
 & 0.536 \\
 & \times \quad 2 \\ \hline
d_{-6} = 1 \leftarrow & 1.072,
\end{array}
$$

and thus the result is $.011011 = 2^{-2} + 2^{-3} + 2^{-5} + 2^{-6} \approx 0.4219 \approx 0.423$.

**Example 11.18**

The fraction required for the conversion of the pressure $P_1 = 10.3$ was 0.30103. Following the steps,

$$
\begin{array}{rr}
 & 0.30103 \\
 & \times \quad 2 \\ \hline
d_{-1} = 0 \leftarrow & 0.60206 \\
 & \times \quad 2 \\ \hline
d_{-2} = 1 \leftarrow & 1.20412 \\
 & 0.20412 \\
 & \times \quad 2 \\ \hline
d_{-3} = 0 \leftarrow & 0.40824 \\
 & \times \quad 2 \\ \hline
d_{-4} = 0 \leftarrow & 0.81648 \\
 & \times \quad 2 \\ \hline
d_{-5} = 1 \leftarrow & 1.63296 \\
 & 0.63296 \\
 & \times \quad 2 \\ \hline
d_{-6} = 1 \leftarrow & 1.26592 \\
 & 0.26592 \\
 & \times \quad 2 \\ \hline
d_{-7} = 0 \leftarrow & 0.53184 \\
 & \times \quad 2 \\ \hline
d_{-8} = 1 \leftarrow & 1.06368,
\end{array}
$$

and thus $0.30103_{10} \approx .01001101_2$. Note that this number is $\log_{10}(2)$, which will always be the same for all base 10-to-base 2 conversions. Therefore, it can be stored as a constant instead of being recalculated repetitively.

When the base 10 fraction is in BCD, a method analogous to that used for integer conversion is appropriate. Now instead of dividing by 2 (shifting right), the number is multiplied by 2 (shifted left). The binary digits appear in an initially blank register to the left, rather than on the right as was the case for integers.

For integer conversion, when a 1 was shifted out from one BCD digit and into another on the right, 3 had to be subtracted because $10/2 = 5$ was intended but 8 was contributed. For fraction conversion, before a 1 is shifted into a BCD digit to the left, 3 is *added* to reverse the effects of that same problem, now encountered in reverse. Incidentally, essentially the same process is used to convert from base 2 to BCD. Again repeated multiplication by 2 is necessary, but now the base 2 number is on the right, and empty BCD nibbles await shift-left outputs from the base 2 register.

**Completion of floating-point conversion example.** The division $n/\log_{10}(2)$ is carried out in base 2 in the manner described earlier in this section to yield $0000110.10100100_2$. The integer portion of the result is given to $m$ and the rest is assigned to $x$. For our example, $m = 0000110_2$ and $x = .10100100_2$. We may check this by noting that $2_{10}/0.30103_{10} \approx 6.644_{10}$, so that $m = 6_{10} = 110_2$ and $x = 0.644_{10} \approx .10100100_2$.

Next, $W = 0.103_{10}$ is converted from the original BCD to binary to yield $W = .00011010_2$, and then the product $A = W \cdot 2^x$ $(0 \leq x < 1)$ is computed in base 2, the result being $A = .00101001_2$. The binary fractional number $A$ is everything in the original number $10.3_{10}$ except the $2^m$ factor. The value of $2^x$ in base 2 used in calculating $A$ is found by a means of look-up table. Essentially, for each possible value of $x$, a corresponding value of $2^x$ in base 2 is stored in the table.

While the use here of 7 bits for $m$ is typical, $A$ may be given 24 bits for high accuracy; the calculations for $A$ will be correspondingly much more accurate than those given here. The twenty-fourth bit has a weight of $5.9 \cdot 10^{-8}$, which is more accurate than any pressure measurements we have are likely to take. The extra accuracy helps reduce errors in roundoff that accumulate in computations.

By comparison, the eight-bit representations used in our example can at best be accurate to only $2^{-8} \approx 0.004$. With binary, even eight places is not very accurate and roundoff can become a problem more quickly than in decimal. A decimal integer having $M = 8$ digits would require $L = 27$ binary digits to be fully represented; this is because 27 is the smallest value of $L$ such that $2^L - 1 \geq 10^M - 1$. Solving this inequality for $L$ gives $L \geq [M/\log_{10}(2)]$, where here $[x]$ means take the least integer greater than $x$. A rough estimate can be obtained as follows:

$$\text{number of binary digits required } L \geq 3.32 \text{ (number of decimal digits } M). \tag{11.13}$$

The final result of the conversion to binary floating point is put in a normalized form such that the leftmost digit is always a 1. This requires, in our example, shifting the binary point of $A$ two places to the right, and necessarily decreasing $m$ by 2 so that the binary representation is a fraction, or *mantissa,* of $.10100100_2$ with exponent $0000100_2$. That is, $P = 10.3_{10} \approx .10100100_2 \cdot 2^{0000100_2} = A' \cdot 2^{m'}$.

It is the pair $A'$, $m'$ that is the string of 1s and 0s which the computer actually uses to perform floating-point arithmetic operations. The pair constitutes a word. (Not shown is the sign bit for the number, nor other modifications that may be made to the exponent for convenience. For example, to avoid wasting a bit to indicate negative exponents, the exponents may be "biased" so that the largest negative exponent is represented by $0000000_2$, while the largest positive exponent is $1111111_2$.)

Using the unbiased exponents, the value of ambient pressure in our numerical example, $P_0$, is $0.10000111_2 \cdot 2^{00000001_2}$. Also, we may check using only eight-bit conversions and arithmetic that the conversion factor $1.013 \cdot 10^5$ from *atm* to *N/m*$^2$ has approximate base 2 representation $.110001_2 \cdot 2^{0010011_2}$.

From this point on, the standard binary arithmetic procedures for subtraction of $P_0$ from $P$ and subsequent multiplication by the factor $1.013 \cdot 10^5$ in base 2 as presented earlier in this subsection can be used to compute the final result, the pressure difference in N/m$^2$. But remember that these operations will be used in combinations appropriate for floating-point operations.

For example, addition and subtraction can be performed only after the mantissa and exponent of the smaller exponent number are shifted to match the exponent of the other number. Doing so allows mere addition or subtraction of the two mantissas; the result will have the common exponent, and it may have to be renormalized to maintain the 0.1xxx form.

For multiplication, the two mantissas are multiplied using standard base 2 multiplication, while the two exponents are added. For division, the dividend mantissa is divided by the divisor mantissa, and the divisor exponent is subtracted from that of the dividend. Again, the results will have to be renormalized.

After obtaining the final base 2 result of any calculations that may be desired, essentially all of the conversion steps above can be carried out in reverse to translate the final answer from base 2 floating point to base 10 floating point.

We should be aware that while the base 2 floating-point representation described here is typical, there is no universal standard. For example, some computer manufacturers use a normalized form 1.xxxxx instead of .1xxxxx. Also, some manufacturers use more or fewer bits for either the mantissa or the exponent.

We have laboriously worked through many of the details involved in a simple calculation for several reasons. First, the tedium itself indicates how important automation is even for what at first seems a simple calculation. This characteristic is typical of digital computing. We assume that the arithmetic is simple because we never see all the details; only the digital designers are aware of it. Even they have automated procedures such as CAD for dealing with all of the above and other circuit implementation details.

It is hard to imagine how monotonous, intricate, and extensive the details are until we are actually forced to carry out a complete procedure. With all of the work above, we only did about half of the required steps. Left out were many details that make heavyweight computer design the domain of specialists. For example, when a key is pressed, the computer encounters an "interrupt" command that grabs its attention, indicating that there has been new input. This sets into sequence various acknowledgments and other preparations. Various gimmicks, such as the character-repeating feature, a 20-character "buffer" for those who type fast, and alternate character sets make the keyboard more attractive to the user without much extra hardware.

## 11.7 MEMORY DEVICES

### 11.7.1 Introduction

The flip-flop is a basic memory device that stores one bit of information. Several flip-flops arranged together can form a register, suitable for storing a number (e.g., a byte). The term "memory" usually refers to devices that can store many numbers or arrays of bits, accessible by addressing mechanisms.

The purposes of memory are many. The instructions a computer uses to function are stored in various types of memory. All data bases (personnel, payroll, customer information) reside in memory. Look-up tables for special functions (e.g., trigonometric) are retrievable from memory. Temporary *scratch pads* for intermediate results of computation are implemented using memory devices. If your computer can tell you what day it is, that data is hiding away in a memory register somewhere. Your CD, VCR, and cassette players read their entertainment from memory devices.

We may think of memory as a book. Several aspects of computer memory will be introduced using this analogy. All books can be read. Textbooks cannot be written into (*read-only memory*), whereas composition books can (*writable* or *programmable memory*). If we write with pencil, the writing can be erased (*erasable* or *alterable memory*); if we use ink, it cannot be erased. If the pages are made of newspaper, in a few years the information is lost due to disintegration of the paper. This would be called *volatile storage* (in a computer, information would be lost as soon as the power is turned off). If they are made of cloth paper, the storage of information is much more permanent (*nonvolatile*).

If the book is on your desk and is only two pages long, it will not take you long to locate and read a certain word in the book (*fast memory*). If, however, it is in the stacks of the downtown library and the book is 1000 pages long, it will take a long time to access the desired word (*slow memory*). If the book has an index, it is easy to locate the desired information wherever it is in the book (*random access memory*) using the page number (*address*). Without an index, we may have to read from page 1 all the way to where the desired information is on page 992 (*sequential memory*).

If the desired word is the 251,422nd word in the book, most people searching for that word would prefer to be told that it is the 23rd word on page 751. Use of a shortened address within a small area of computer memory (*page*) is called *paged addressing*.

Suppose that you are writing a paper on some topic. You are likely to have at home several books from the library that are highly relevant to that topic. This is far more efficient than running back and forth between your home word processor and the main library stacks. Your small, temporary home library, which changes when you have a new paper to write on a different topic, is equivalent to a computer *cache memory*. Most computers have relatively small amounts of maximally fast memory because of its high expense. The computer makes use of this memory for rapid data processing by copying into it the most recently accessed data from slow memory. This small, fast memory is called a *cache*.

There are many types of memory, each applicable to or economical for certain situations. For example, the *hard disk* of a personal computer must retain its memory when power is removed and be capable of both storing huge amounts of information and accessing it at fairly high speed. Main processor memory may be erased when the power is turned off (volatile), but must require the absolute minimum time possible to store and retrieve information to maximize computing/processing speed.

Floppy disks or diskettes represent an economical and portable alternative to the hard disk, but each disk can store only relatively small amounts of data and is much slower than the hard disk. If the information is stored on magnetic tape, typically we must begin looking for the desired location of information at the beginning of the tape (sequential memory). Contrarily, if electronic (semiconductor) memory is used, every location can be reached in the same time (random access).

Implementation of logic functions can be achieved by storing the information contained in the minterm expression: the 1s and 0s of the truth table. In this case, once the minterm expression is stored, it need only be "read" and not "written." Thus *programmable logic arrays* (PLAs), discussed below, are appropriate for implementing combinational logic using memory.

Techniques similar to those required to fabricate and to operate the PLA are used in *read-only memory* (ROM). ROMs are used for storing specialized programming instructions used repeatedly in small computers, computer startup instructions, look-up tables, and anything else that we wish to have available on a permanent basis. And then there are various hybrid memories, having qualities of more than one standard type of memory. An entire chapter could easily be devoted to memories alone; we shall merely view a passing parade of a few of the most commonly used types.

### 11.7.2 Read-Only Memory

Read-only memory (ROM) is just that: once initially written, its information cannot be changed; only read. Consider again for a moment the book analogy. If we have a word processor and desire only a few copies of a book, it may be written "at home" and printed on a home laser printer. A *programmable ROM* (PROM) is made "at home" using a special writing device. But if we expect many copies to be sold, that is best done by a publisher. Similarly, a *masked ROM* is made in large quantities at a factory per instructions sent by the customer ("author").

Neither ROM nor PROM can be reprogrammed or erased; changing your mind about their contents means throwing the memory chip away. As the price of making a masked ROM is in the $1000 range, clearly erasable memory should be used until we are certain that the specified contents are correct. In mass production, the masked ROM is actually cheaper per chip than other ROMs.

ROMs have many uses, including the look-up table alluded to briefly above. Look-up tables contain coefficients or function values that would take too long to compute or for any other reason would be impossible for a computer to calculate at the time the values are required by the computer. The independent variable is converted to an address using the same code as was used to order the table. The output word of the memory contains the value of the dependent variable. In Sec. 11.6.2 we discussed one example of the look-up table used in the process of the conversion of base 2 floating point to base 10 floating-point representation: $2^x$ with $0 \leq x < 1$.

Another common application of ROM is program storage for small computers.

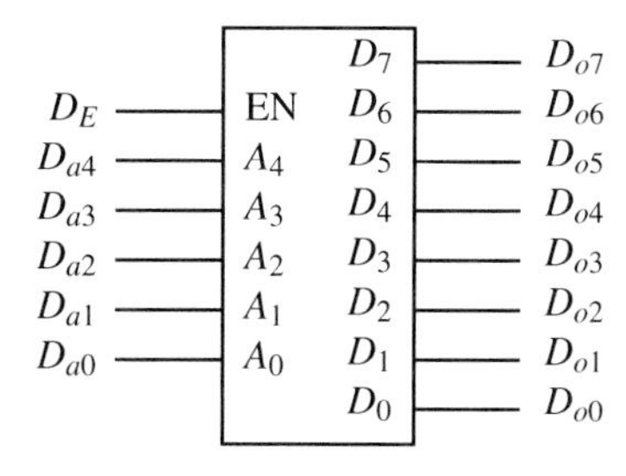


**Figure 11.33** A 32-byte ROM/PROM circuit symbol.

These programs must not be lost when the power is turned off; the nonvolatility of ROM guarantees this.

A typical ROM chip appears in Fig. 11.33. Operation is quite simple: Put the address on the address lines $D_{a4} \cdots D_{a0}$ and soon the values of the complete byte having that address appear on the lines $D_{o7} \cdots D_{o0}$. Because there are five address lines, there are $2^5 = 32$ bytes of storage in this chip. The enable line on the chip makes building up larger memories very easy: Put the same address on $D_{a4} \cdots D_{a0}$ of all the chips, but enable only the chip containing the byte you want. For this reason, the *enable* (EN) line is also called the *chip select* (CS) line.

The PROM is essentially the same except that it must first be written to. This is done by the user by plugging the PROM into a special writing device and programming it. Originally, the PROM has a 1 stored at every memory location (cell). The writing device essentially "blows fuses" to designate selected cells as 0. Clearly, once it is written no changes can be made. (If you want to be sure to remember to go to the dance on time, "take your date to the PROM"; even if in the meantime the power goes out, the PROM will not forget that stored time and day.)

The erasable PROM, EPROM, uses isolated transistor gates rather than fusible links to store 1s and 0s. Thus it can be erased and rewritten. However, it cannot be erased selectively. Through a quartz window on top of the chip, ultraviolet light can be used to erase the entire chip "at once" (actually, after a half hour of exposure).

An EPROM chip diagram appears in Fig. 11.34. For both reading and programming, the desired address is put on lines $D_{a10} \cdots D_{a0}$. For reading, bring the chip

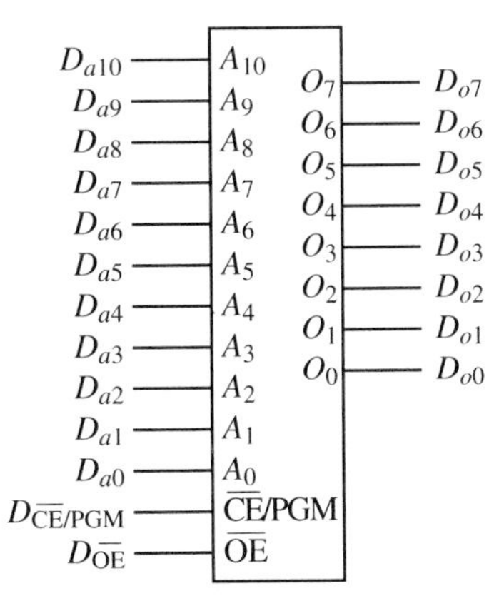


**Figure 11.34** A 2K-byte EPROM circuit symbol.

enable/program ($\overline{CE}$/PGM) line low and the output enable low. Soon afterward the desired memory contents will appear on output lines $D_{o7} \cdots D_{o0}$. For programming, bring the output enable line high. Put the desired data on the output lines (now functioning as inputs). When a large pulse is put on the $\overline{CE}$/PGM line, the data will be written at the specified address. Note that this memory holds $2^{11} = 2048$ bytes or "2K bytes." In memory terminology, the K signifies *roughly* 1000, and implies the nearest larger integral power of 2; hence $2^{10} = 1024$ is called "1K" and $8192 = 2^{13} = 8$K.

Finally, the electrically erasable PROM, EEPROM, allows individual bytes to be modified selectively. It can both be modified and yet remain unchanged after the power has been turned off and on again. Because of this convenience, it is used for TV channel, volume, and picture settings, so that when you hit the "zapper" (remote control) ON button, you begin right where you left off. From here, use of the zapper buttons will change these settings according to your whim.

At this point, it is tempting to ask: If it can be written to, why is it called "read-only"? The reason is that it is too slow to be used in most writing situations; it takes about a millisecond, whereas other fast RAM memory may take only 50 ns to write (i.e., RAM memory writes in 1/20,000th the time required by the EEPROM). The EEPROM is consequently viewed as a "mostly read" device (the read operation is less than 1 μs). If it were not for the slowness and the low density (bytes/chip possible), this type of memory would be ideal: Writable and nonvolatile. Anyone who has ever lost hours of typing on a word processor because of a power outage understands the value of nonvolatility.

An example EEPROM chip is shown in Fig. 11.35. Unlike most memory packages, this one takes all its information serially. The *serial data input* takes in both the address and the data, one bit at a time. The reason that serial operations are more common for this type of memory is that it is used for less frequent accesses. Therefore, the advantage of a low pin count (number of pins on the chip) is not outweighed by the longer access time for serial devices. Now that the data is entered serially, a clock signal must also be present to distinguish "00" from "0." EEPROMs are used not only in TV sets but also in electronic security systems, robotics, computer terminals, and other devices.

Incidentally, a compact disk (CD) is a special form of read-only memory. Examples are audio and video disks, and data base storage disks such as indexes of journal publications. To produce the CD, the manufacturer digs out the 1s and 0s pits and you are stuck with them, no matter how bad the music or movie is! The consumer can only read the disk, using a laser beam.

The CD is classified as read-only optical memory. However, it is actually called CD-ROM rather than just ROM because the term ROM implies electronic, random access memory where any ("random") byte can be accessed in the same period of

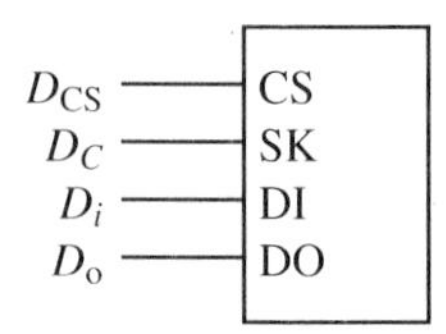


**Figure 11.35** EEPROM circuit symbol.

time. The CD-ROM may have 20,000 tracks, each containing vast amounts of binary data that must be accessed sequentially.

To give an idea of how slow a CD-ROM can be, consider the fact that a search for a particular address in a CD-ROM can take 3 million times as long as a similar search in a RAM. However, subsequent serial data collection is about as fast as that of magnetic disks (150K bytes/s). Furthermore, a CD-ROM can hold the equivalent of 200,000 typed pages of text.

### 11.7.3 Random Access Memory

As noted above, an EEPROM has the very important write and nonvolatility features, but it is too slow and has too low density to be practical for most situations. If we are willing to give up the nonvolatility, the write option is available economically at high density and high speed in the form of random access memories (RAMs). Unfortunately, the memory referred to by the acronym RAM really means "read and write memory" and should be called RWM. After all, semiconductor ROM is also random access (randomly selected bytes equally quickly accessible). The confusing terminology is, alas, written in stone (i.e., not erasable).

Typically, the RAM is used in conjunction with more permanent memory such as magnetic media. In such cases, the information likely to be needed at high speed (such as a computer program) is loaded into RAM. In this way, nonvolatility and high speed are combined.

There are two main categories of RAM: static and dynamic. *Static RAM* (SRAM) is made up of flip-flop-like elements; as long as power is on, the stored data remains. *Dynamic RAM* (DRAM) is made up of tiny capacitors; in 1 ms the data will be lost if it is not "refreshed" by some external circuitry.

Clearly, dynamic RAM is an inconvenience. Its only reason for existing is that because of its extremely simple storage method, very high densities and low cost are possible. The memory cells are stored in huge rectangular arrays. Because the huge number of memory cells necessitates very long addresses, the address must be given in two parts: first the row address, then the column address. This procedure is known as *address multiplexing*. For large memory requirements, dynamic RAM is the cheap way to go.

It should also be mentioned that RAMs sometimes come with a lithium "backup" battery (with expected lifetimes of 10 years) to keep it going when the main power is turned off. This way, a RAM can be made effectively nonvolatile. However, if the battery fails, everything is lost.

An example static RAM (SRAM) chip is shown in Fig. 11.36. Notice that

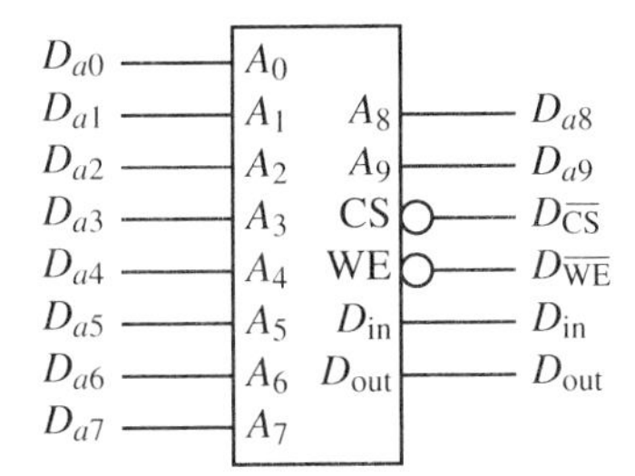


**Figure 11.36** SRAM circuit symbol.

unlike ROM, there is a single bit output. Thus the address lines specify only a single bit. To access a byte at a time, eight chips would be tied together. The reason for single-bit access is pin count minimization. The more pins, the more expensive the chip. RAMs are typically used for very large memory arrays; hence most of the pins are used up on the address. Of course, in theory any memory size can be constructed using either technique.

Operation of the RAM is simple. When the WE line (write enable) is low, a write of the bit on the *data in* line ($D_{in}$) will take place at the designated address. When the WE line is high, a read will place the contents of the bit at the designated address on the *data out* ($D_{out}$) line. For flexibility, a separate output enable (OE) input line connects or disconnects outputs from the bidirectional I/O lines; tri-state logic makes this possible. When the outputs are not enabled, they are put in the high-impedance state—essentially disconnected from the rest of the world. As usual, the CS is the chip select or enable line.

In an expanded (multichip) memory, the same address bits and common OE and WR lines are sent to all chips; additional address bits select the chip.

**Example 11.19**

Design a 64K RAM using eight 8K RAM chips.

**Solution** There are 16 address lines required because $2^{16} = 65{,}536 = 64K$. Three will be used to select which RAM is enabled ($2^3 = 8$); the other 13 lines specify which address within one RAM chip is requested ($2^{13} = 8192 = 8K$). The three lines are decoded by a 3:8 decoder such as the 74HC138. The circuit diagram appears in Fig. 11.37a.

The 8K static RAM chip selected for this example is the NMC6164. Notice the two chip select pins on the RAM. They aid in interfacing the memory with a microprocessor, a concept discussed (for a different device) in Sec. 12.4.3. Both must be enabled to enable the chip. Here they are tied together. There are also two enables pins on the decoder. These make cascading decoders easy for large decoder implementation. In Fig. 11.37b is shown a typical timing diagram for a RAM, which should now be self-explanatory. On the left is a read operation, and on the right is a write operation.

### 11.7.4 Programmable Logic Array

In the introduction to this section, mention was made of the ability to use memory to implement combinational logic. Essentially, each row of the truth table that has an output of 1 has an associated pattern of 1s and 0s for the input variables. This pattern can be stored using the same technology used by the PROM or the ROM.

The AND of that pattern with the actual input pattern yields one of possibly several minterms in the complete expression for the output logical variable. The *programmable logic array* (PLA) allows for several output variables. The required minterms for each output variable are ORed together to form the outputs. An example best illustrates the use of a PLA.

**Example 11.20**

Implement the expressions

$$D_{oa} = D_1 D_2 \overline{D}_3 + \overline{D}_2 \overline{D}_3$$

$$D_{ob} = D_1 \overline{D}_2 + \overline{D}_1 \overline{D}_2 D_3 + D_2 D_3.$$

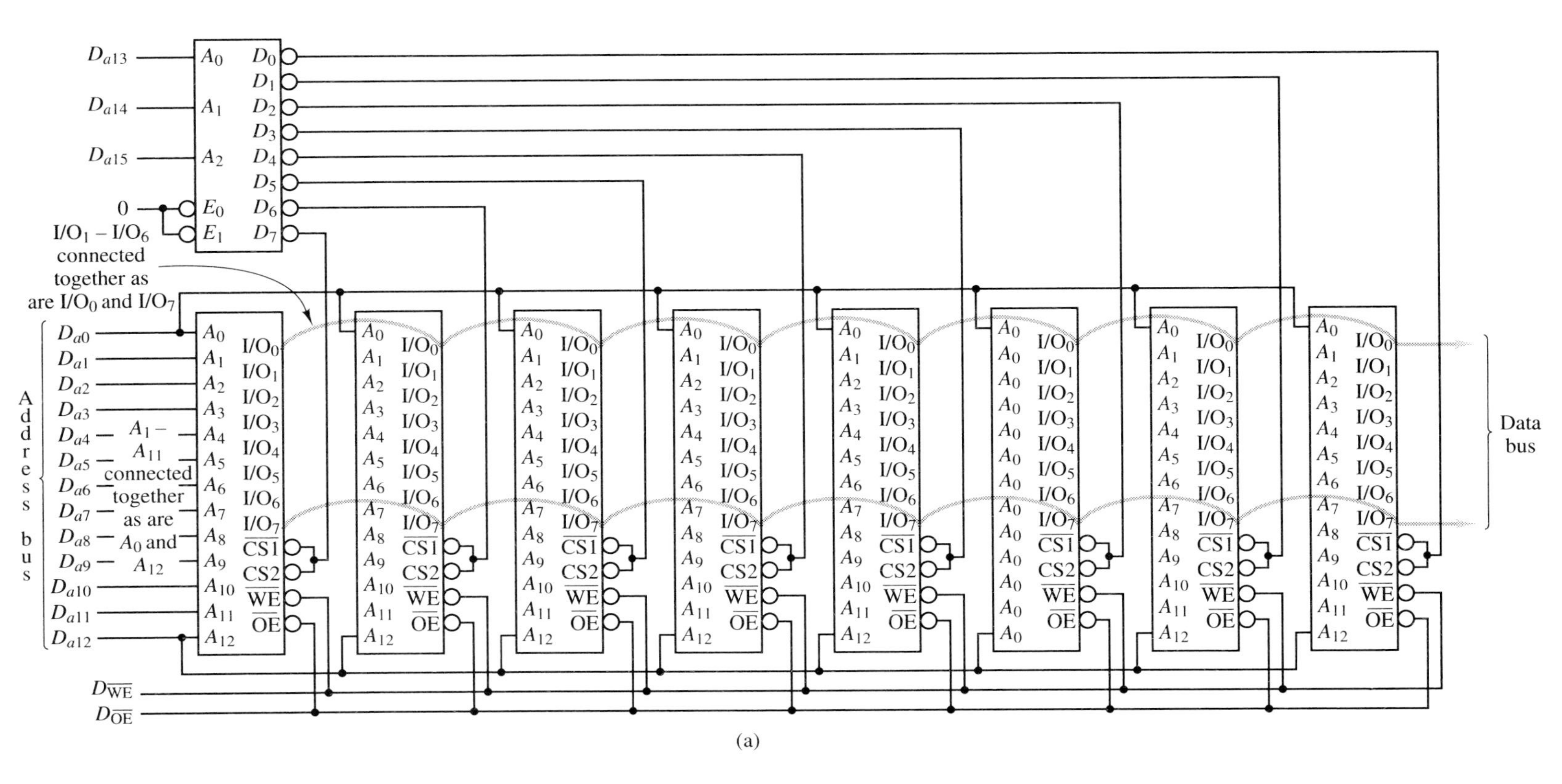


**Figure 11.37** A 64K-byte RAM: (a) circuit; (b) timing diagram.

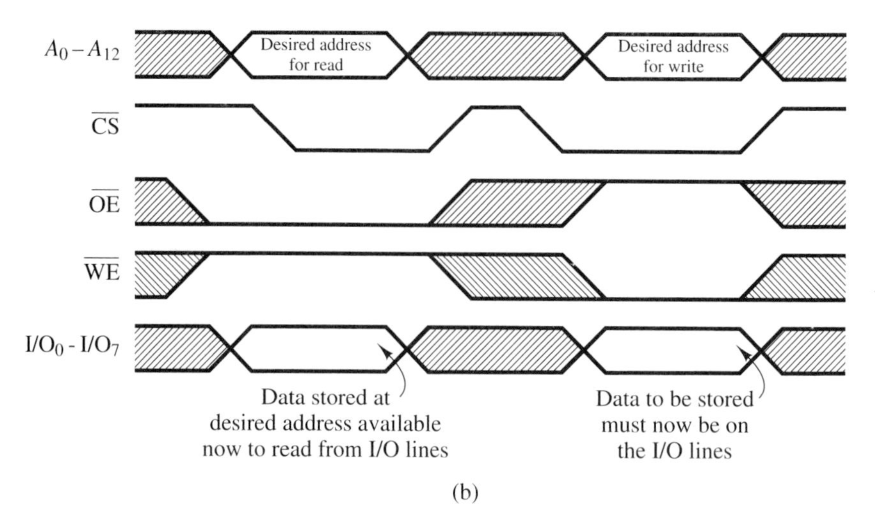


**Figure 11.37** (Continued).

**Solution** A portion of a PLA is shown in Fig. 11.38. A dot at an intersection implies connection. (Note the diode between a fuse and a connection. This diode is needed to prevent a short circuit. If two lines are connected, one high and one low, current would flow from the high to the low as a short circuit. Diodes prevent current from completing such a short circuit loop. This idea is also used in PROMs.)

For a minterm, all the dotted variables along the vertical line leading to a given AND gate are ANDed together. Hence the (merely symbolic) AND symbols at the lower region of the array. To make up the output variables, certain of these minterms are ORed together: those dotted minterms along the horizontal line leading to that OR gate. Hence the (merely symbolic) OR symbols at the lower right-hand region of the array. There are also single-output versions called PALs (programmable array logic, for want of imagination).

(Alternatively, we could consider the input variable bit pattern to be a ROM address. The value of the output variable for that row of the truth table (1 or 0) would be stored at that address. This, however, would be less efficient than the PLA because the PLA stores only the terms producing 1, not those producing 0.)

### 11.7.5 Magnetic Memory

For mass storage that is nonvolatile yet can be both read from and written to, magnetic memory devices are the most economical. However, data read/write times for magnetic memory can be orders of magnitude slower than those of semiconductor memory. For example, access time to find the beginning of data can take $\frac{1}{2}$ s on a floppy disk. As noted previously in the discussion about finding a word in a book, sequential memory is inherently slow if random access is desired. Magnetic memory is often sequential, or a hybrid of sequential and random access.

Another example of the difference between serial and random access is the setting of digital time clocks. For example, because there is no keypad on a digital alarm clock, the time must be set by holding down a time-advance button until the desired time shows on the display. But the clock on a microwave oven can be set to any time equally quickly just by entering the time on the keypad. The former method

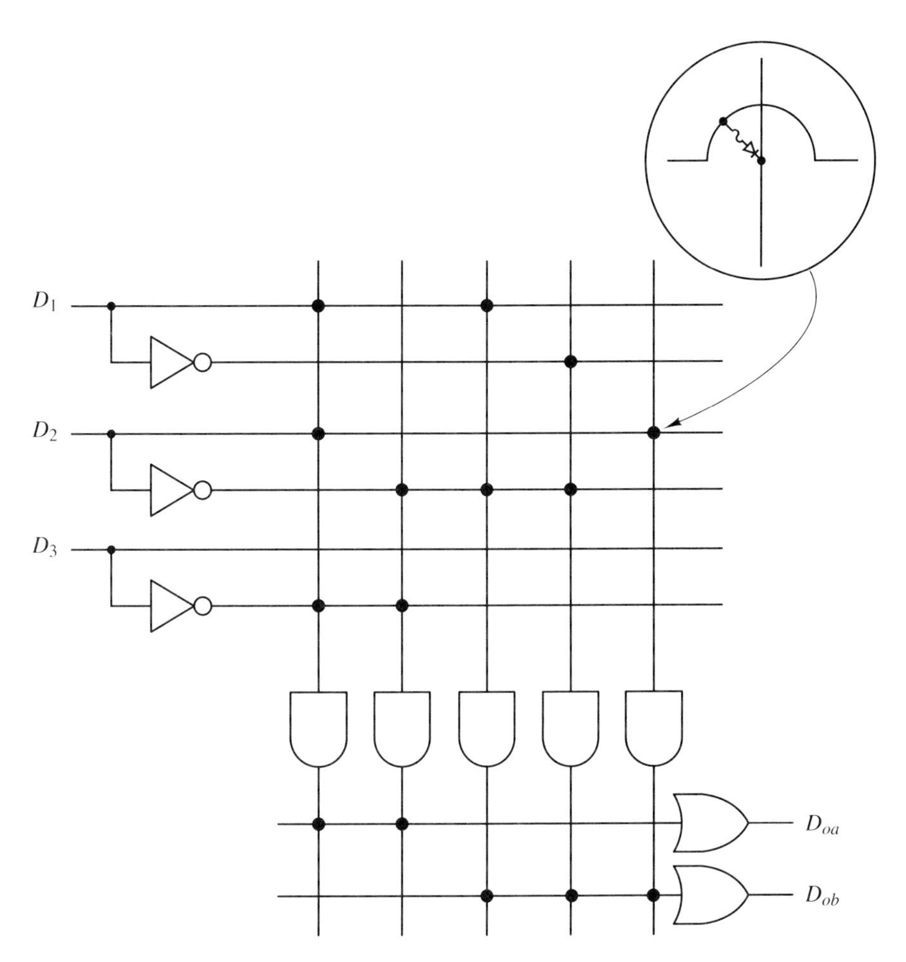


**Figure 11.38** Portion of a PLA logic circuit. Inset shows fusible link with protective diode, which is found at each junction with large dot; the fuses of other junctions were selectively blown.

is serial, which is slower and more tedious than the random-access method for the microwave oven clock.

Of course, if the data is stored sequentially, once the beginning is found the other data is recoverable extremely quickly. One convenient aspect of magnetic memory is that it is usually very simple to use. In personal computers, the hard disk is already connected, and the floppy disk is simply stuffed into the drive.

The hard disk can store millions of bytes (megabytes) of data, whereas the floppy disk or diskette can store less than 1 megabyte. A hard disk is expensive but relatively fast and large. A floppy disk or diskette is very portable and cheap but less convenient for access. A hard disk has additional capacity due (among other factors) to several disks spinning on the same axle, each with its own read/write head.

Both floppy and hard disks use a magnetic film covering a material with greater strength. The floppy disk (and also magnetic tapes) use Mylar plastic as the substrate, while the hard disk uses solid aluminum (hence the name "hard").

In both floppy and hard disks, the data is arranged and retrieved in concentric tracks much like those on a phonograph LP. However, the disk is further separated into many sectors in order to speed up access to the beginning of a desired data block. That is, it is a hybrid of serial and random access.

A magnetic tape is also divided into tracks: Each track is a fraction of the width of the tape and runs along the entire tape. Some of the tracks carry the address and the others carry the data. Magnetic tapes are notoriously slow, but can be essentially as large as we want. They are used for "backing up" computers—retaining a copy of all important data, programs, and text files in each user's account.

## 11.8 SUMMARY

In this chapter we have entered the world of sequential systems—systems that move in an orderly fashion from state to state in order to carry out an automated process.

This was made possible by the memory element known as the flip-flop. Although the *RS* flip-flop is commonly used in IC chips, the *JK* edge-triggered flip-flop is the workhorse of sequential systems. This is because its state can change only when the clock makes its periodic low-to-high or high-to-low transition. If the input lines are designed to be "ready" for this transition, there will be no ambiguity about whether a flip-flop will be set or reset. The use of clocked flip-flops brings order out of chaos in number-crunching systems. Successful use of *JK*, *T*, or *D* flip-flops requires proper timing. The timing diagram was introduced as a way of visualizing the essential parameters.

The ability to read a timing diagram, combined with the material in Sec. 10.7 is sufficient for being able to extract vital information from data sheets for many digital devices. Previously, a data sheet might have seemed impossible to understand.

Two common digital clocks were examined briefly: The astable multivibrator and the quartz crystal resonator. The latter is more popular because of its high performance; the former is simpler and cheaper. If a pulse train having a frequency lower than that of the clock is desired, counters may be used to generate it. If a higher frequency is required, use a faster clock, but be sure that the new frequency does not exceed the maximum allowable for the other digital circuitry being used in the system.

Not only were base 2 counters such as the ripple and synchronous counters considered, but also circuits such as the Johnson and ring counters, which produce other sequences. The variety of sequences that may be generated using groups of flip-flops is limited only by our imagination. Usage of counters in industrial control systems using digital IC technology is pervasive. In particular, the microprocessor, investigated in Chapter 12, uses counters. Closer to home, the alarm clock has a binary counter.

If a set of flip-flops is used for data storage rather than fixed sequence generation, the device is called a register. For different applications, the appropriate access format for reading data from and writing data to the register may be either serial or parallel or both. The universal shift register provides all the possibilities on one chip; the desired mode is selected by control lines.

An example of sequential machine design was presented all the way from desired behavior to flip-flop connections. A nonelectrical engineer can make the connection between the various stages in a wash cycle and those of an industrial process. In either case the design procedure is the same.

First enumerate the distinct desired states in the automated cycle. Then determine what control signals must be asserted to move from state to state. Draw a state transition diagram and table. From these, determine the combinational logic involving the present state and input variables that will set each of the "next state" variables to 1. Finally, the flip-flop outputs are considered the present state, while their inputs are the next state.

Digital electronics is concerned with the manipulation of binary data, predominantly representing numbers. The human thinks in base 10, but the computer must process binary data. Any involvement with computational electronics requires that the user be familiar with the number representations used in the electronic system. Otherwise, the myriad of 1s and 0s will be completely meaningless.

We studied the most common digital number representations: base 2, base 16, and BCD. The basic operations of addition, subtraction, multiplication, and division were introduced for a definite purpose: to learn what happens when we type a number into a personal computer. Otherwise lifeless subjects such as ASCII code, look-up tables, BCD operations, and conversions between number systems became important when actually used to answer this question. Essentially, binary logic simulates arithmetic operations, and digital circuitry simulates binary logic (and thus binary arithmetic).

We shall see in Chapter 12 that arithmetic in a computer is carried out by a combination of hardware and software. An example of a hardware implementation of arithmetic was the full-adder designed in this chapter. Multiplication and division required the repetition of a sequence of commands (add/shift and negativity decisions for division) taking place in just two physical data registers. This is an example of software-repetitive instructions that take the place of replicated hardware. Software is the key to the power of microprocessors and computers in general. In this chapter we have gently introduced it without getting into detailed computer programming codes.

The instructions to carry out multiplication (for example) must be stored somewhere. We investigated a fair variety of hardware that stores binary data, both electronic and magnetic (and even a note on optical). Volatility refers to the loss of data when the power is turned off. ROM is nonvolatile memory such that once defined, its contents are not alterable. Variations on ROM technology have allowed limited writing capability (PROM, EPROM, EEPROM). RAM is high-speed read/write memory; unfortunately, it is volatile. PLA is a ROM specifically designed for implementation of combinational logic. Magnetic memory is nonvolatile, slow read/write memory that is well suited for long-term mass storage.

With this and Chapter 10 as well as Sec. 9.7.6 on A/D and D/A converters in mind, we can now proceed to discuss microprocessors. Microprocessors are the main contact that many nonelectrical engineers have with digital electronics. Therefore, a study of them should be the most rewarding of all digital electronics discussed in this book. However, that material would be entirely incomprehensible without knowing these two preparatory chapters.

## PROBLEMS

**11.1.** What characteristics do sequential systems have?

**11.2.** What does "memory" mean in an electronic circuit?

**11.3.** Why is the Schmitt trigger called a kind of memory device?

**11.4.** What are the memory elements of a sequential circuit called? How many bits can one memory element store?

**11.5.** By what mechanism is memory achieved in an electronic memory cell? A memory device that can store one bit is shown in Fig. P11.5. Analyze the behavior of the circuit and determine which input "sets" and which input "resets."

**11.6.** Give an advantage and a disadvantage of the basic *RS* flip-flop compared with the *JK* flip-flop.

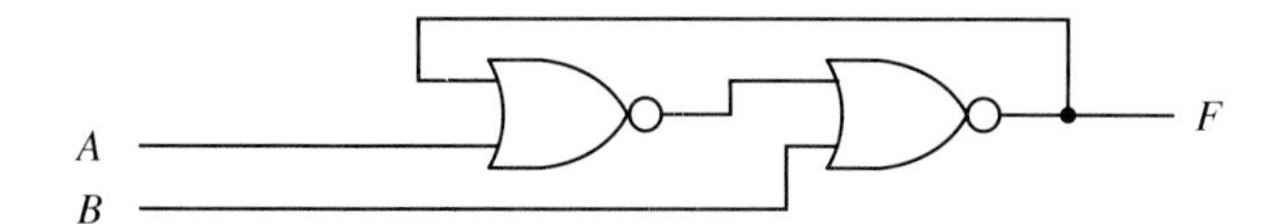


**Figure P11.5**

**11.7.** In Fig. 11.2, the *RS* flip-flop is implemented using NOR gates. We can also structure an *RS* flip-flop using NAND gates. Figure P11.7a shows how an *RS* flip-flop is formed using NAND gates. Determine the state transition table and state transition diagram. Given the waveforms $D_s(t)$ and $D_R(t)$ in Fig. P11.7b, draw the waveform $D_o(t)$.

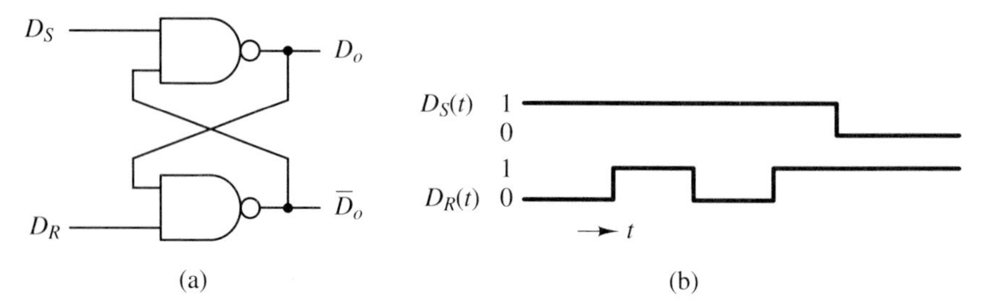


**Figure P11.7**

**11.8.** The clock, as explained in Sec. 11.3, is used to sample the input signals for setting the memory states. A clocked (or gated) *RS* flip-flop is shown in Fig. P11.8a. Explain the effect of the clock signal $D_C$ and determine the output waveform of $D_o$ given the inputs $D_C$, $D_R$, and $D_S$ (see Fig. P11.8b). Note that until the clock is high and a set or reset occurs, the output cannot be determined from the timing diagram unless the initial state of the flip-flop is given (or known).

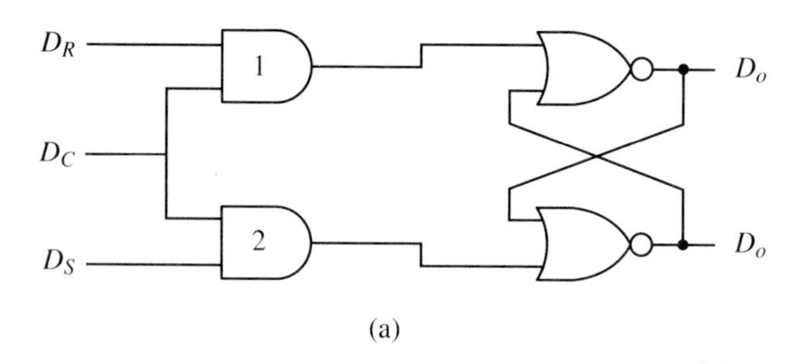


(a)

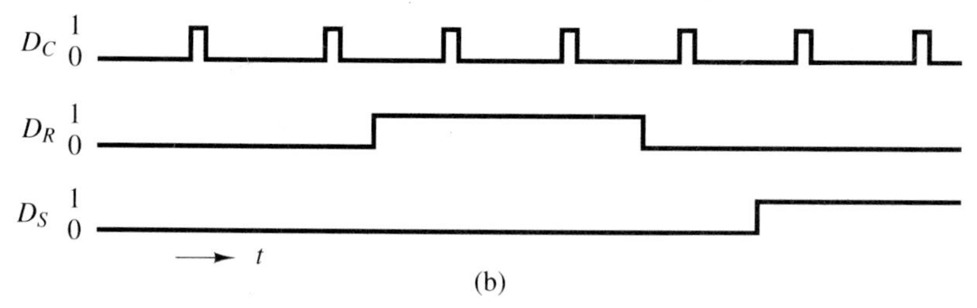


(b)

**Figure P11.8**

**11.9.** In Sec. 11.2 the behavior of the edge-triggered flip-flop is analyzed in detail, and a water lock system is used to explain the concept of edge triggering. In your own words, explain edge triggering. What are a few advantages of an edge-triggered flip-flop compared with an *RS* flip-flop? The symbol of a positive (rising-edge) triggered *JK* flip-flop is shown in Fig. P11.9a. Determine the output waveform $D_o$ given the input waveforms in Fig. P11.9b.

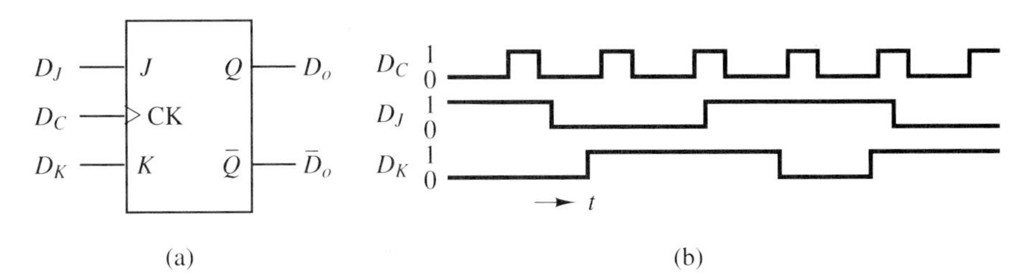


**Figure P11.9**

**11.10.** *D* flip-flops and *T* flip-flops are also commonly used in digital systems. If we have only *JK* flip-flops available (see Fig. P11.9a), how can we convert them to *D* and *T* flip-flops? Draw the circuit diagrams and state transition diagrams.

**11.11.** In Sec. 10.6.2 we qualitatively discussed two debouncing circuits (see Figs. 10.43 and 10.45). A simple debouncing circuit with the "innards" of the *RS* flip-flop is shown in Fig. P11.11. Suppose that initially the switch is in contact with the upper terminal. Next, we move the switch to make contact with the lower terminal. Unfortunately, multiple contact and noncontact events occur (bouncing), which can not be avoided. Explain explicitly how the bouncing does not, however, affect the output *Y* in Fig. P11.11.

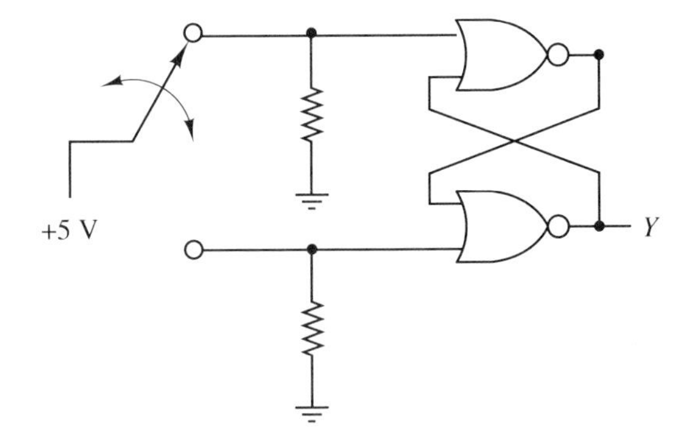


**Figure P11.11**

**11.12.** The clock plays a very important role in digital circuits. A simple clock circuit, shown in Fig. 11.16, is analyzed in Sec. 11.3. Given that $R = 10\ \text{k}\Omega$ and $C = 0.001\ \mu\text{F}$, calculate the period.

**11.13.** How many flip-flops are required to implement a modulo-10 counter?

**11.14.** A three-bit shift register is shown in Fig. P11.14a. For the given specified input and clock waveforms in Fig. P11.14b:

(**a**) Assuming that the initial states of the flip-flops are 101, draw the output waveform.

(**b**) Determine the output sequence.

**11.15.** For the shift register working in SISO mode with output "shift right" (Fig. 11.25), determine the output waveform at point $D_{out}$ for the waveforms of the input data $D_{in}$ and clock $D_C$ in Fig. P11.15. Assume the initial flip-flop states to be $D_0 D_1 D_2 D_3 = 1011$ and the flip-flops to be positive edge triggered.

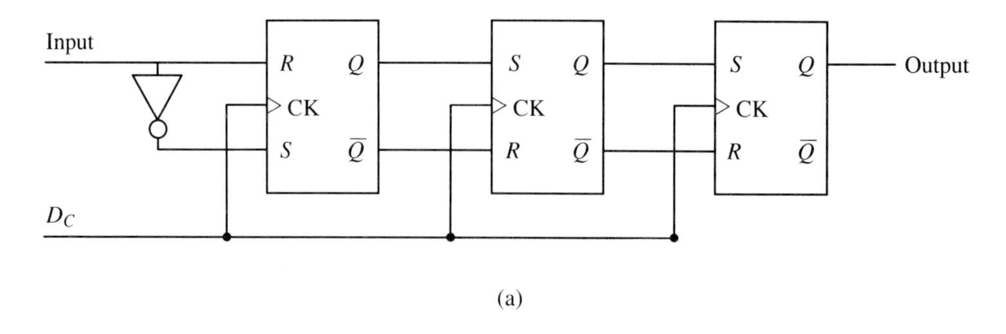


(a)

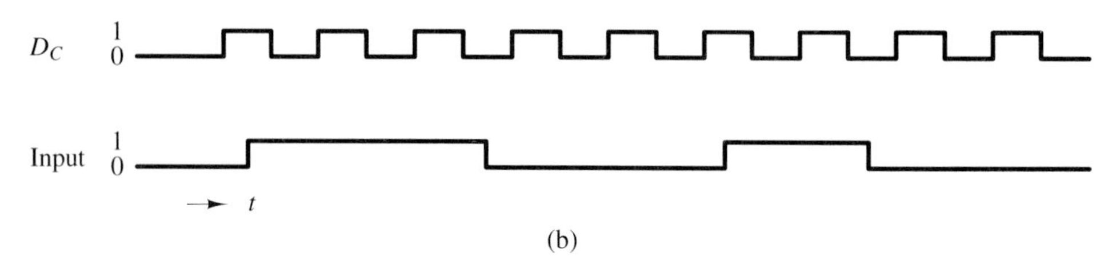


(b)

**Figure P11.14**

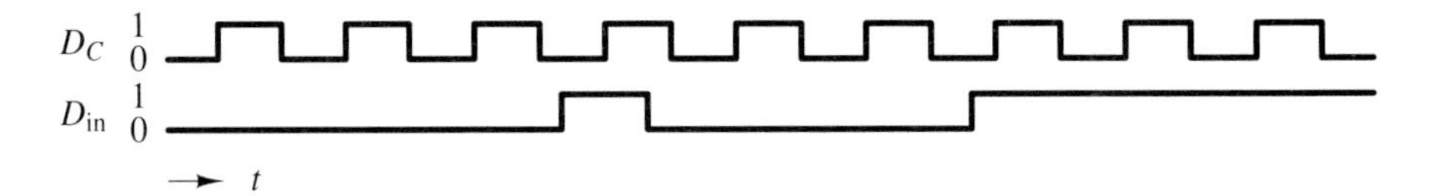


**Figure P11.15**

**11.16.** (**a**) Find the base 10 representations of 1101, 1001, and 111 (given in base 2).
(**b**) Find the base 2 representations of 16 and 13 (given in base 10).

**11.17.** Complete the following conversions.
(**a**) $101011001111_2 = (?)_{16}$ (**b**) $1000110111_2 = (?)_{16}$
(**c**) $\text{A7FE9}_{16} = (?)_2$ (**d**) $\text{7CA2}_{16} = (?)_2$

**11.18.** Complete the following binary arithmetic.

(**a**)
$$\begin{array}{r} 10010110 \\ +\,01001111 \\ \hline \end{array}$$
(**b**)
$$\begin{array}{r} 10110110 \\ -\,00111001 \\ \hline \end{array}$$
(**c**)
$$\begin{array}{r} 00011011 \\ -\,10110001 \\ \hline \end{array}$$

**11.19.** The arithmetic in Problem 11.18 can also be done in hex. Perform the same arithmetic in hex and check your answers against those in Problem 11.18. Use the same methods that you use for base 10 "pencil-and-paper" arithmetic, but applied to base 16 (e.g., carrying and borrowing).

(**a**)
$$\begin{array}{r} 96 \\ +\,4\text{F} \\ \hline \end{array}$$
(**b**)
$$\begin{array}{r} \text{B}6 \\ -\,39 \\ \hline \end{array}$$
(**c**)
$$\begin{array}{r} 1\text{B} \\ -\,\text{B}1 \\ \hline \end{array}$$

**11.20.** What are two commonly used ways of representing negative numbers? Find both representations of the following negative binary numbers, using 1 for the negative sign bit in both cases.
(**a**) −0100101
(**b**) −1011011
(**c**) −0011011

**11.21.** Using 2's complement, complete the following binary arithmetic (which does not involve sign bits). Remember that for positive numbers, 2's complement is identical to sign-magnitude representation.

(**a**) 1100011 − 1011010

(**b**) 1010111 + 1100001

**11.22.** Why do we use BCD? Determine the BCD versions of decimal numbers 276, 1206, and 6409.

**11.23.** Express the following BCD numbers in decimal form.

(**a**) 01110101

(**b**) 01001001

(**c**) 011101100101

(**c**) 0001001000111000

**11.24.** Using the "full adder" shown in Fig. 11.30, how can we form a four-bit full adder? Sketch the logic circuit in block diagram form. The block diagram of a full adder is given in Fig. P11.24.

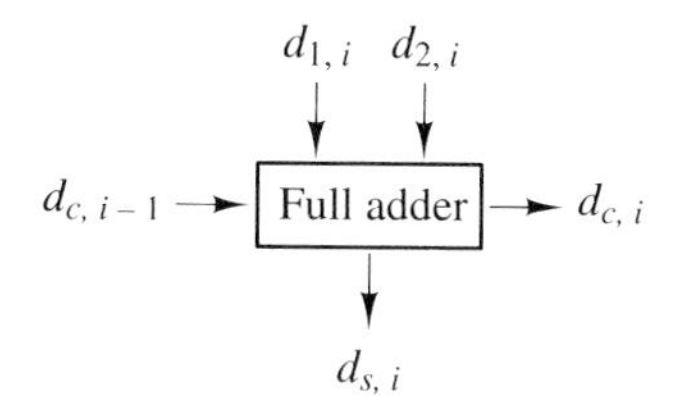


**Figure P11.24**

**11.25.** We studied the four-bit comparator in Sec. 11.6.2. The truth table of a one-bit magnitude comparator is shown in Table P11.25.

(**a**) Determine the logic circuit.

(**b**) How can this circuit be used to compare two binary numbers, each having more than one bit?

**TABLE P11.25**

| one-bit binary number $A$ | one-bit binary number $B$ | $I_{A<B}$ | $I_{A=B}$ | $I_{A>B}$ |
|---|---|---|---|---|
| 0 | 0 | 0 | 1 | 0 |
| 0 | 1 | 1 | 0 | 0 |
| 1 | 0 | 0 | 0 | 1 |
| 1 | 1 | 0 | 1 | 0 |

**11.26.** Complete the following multiplications of eight-bit binary numbers.

(**a**) 10101011 × 00010010 = ?

(**b**) 00111100 × 10001010 = ?

**11.27.** Find to eight-bit accuracy the following quotient of eight-bit binary numbers: 11100101/10001110.

**11.28.** Convert the following numbers to sign-magnitude base 2 representations to an accuracy of five bits to the right of the binary point and 8 bits to the left of it.

(**a**) 32.412

(**b**) −8.75

(**c**) −7.25

**11.29.** What does "ROM" mean? Why is it called nonvolatile? How can you obtain data from ROM?

**11.30.** What does "nonvolatile" mean? Which of the following memory devices are nonvolatile: RAM, ROM, PROM, EPROM, EEPROM, tape drive, and hard disk?

**11.31.** What do the acronyms PROM, EPROM, and EEPROM mean?

**11.32.** What does RAM really mean? Distinguish between static RAM and dynamic RAM.

**11.33.** A PLA is a combinational two-level AND-OR device that can be programmed to realize any sum-of-products logic function, as shown in Sec. 11.7.4. Figure P11.33 shows a portion of a PLA logic circuit. Find the logic functions the circuit implements: namely, $F_1$ and $F_2$ in terms of the inputs $A$, $B$, and $C$.

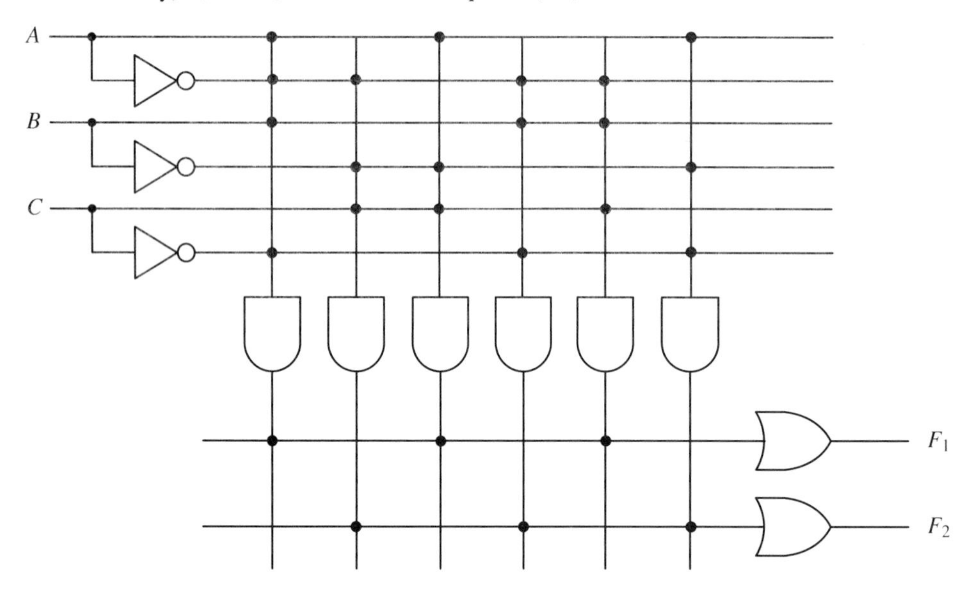


**Figure P11.33**

## ADVANCED PROBLEMS

**11.34.** An interesting application of multivibrators is in controlling the speed of a small electric motor. The instantaneous motor speed is proportional to the duty cycle (ratio of pulse "on" time to "off" time) of the pulse train, which is set by the user. Figure P11.34a is a simple motor controller, and Fig. P11.34b shows the details of the 7404 integrated circuit. Analyze the circuit and determine how the motor speed is controlled (e.g., which circuit element controls the speed).

**11.35.** The operation of *RS* flip-flops can be improved with a master–slave connection. Analyze the master–slave flip-flop in Fig. P11.35, first at the block diagram level.

**(a)** Determine the state transition diagram.

**(b)** What kind of flip-flop is it functionally (*RS*, *JK*, *D*, or *T* flip-flop)?

**(c)** Do the changes of the inputs at times other than the edge of the clock affect the output state?

**(d)** Is it an edge-triggered flip-flop? If it is, which edge? Note that the logic diagram of a clocked *RS* flip-flop is shown in Fig. P11.8a.

**(e)** A full understanding of the master–slave configuration requires analysis at the gate level. Perform this analysis using the detailed analysis of the *JK* flip-flop in Sec. 11.2.3 as a guide.

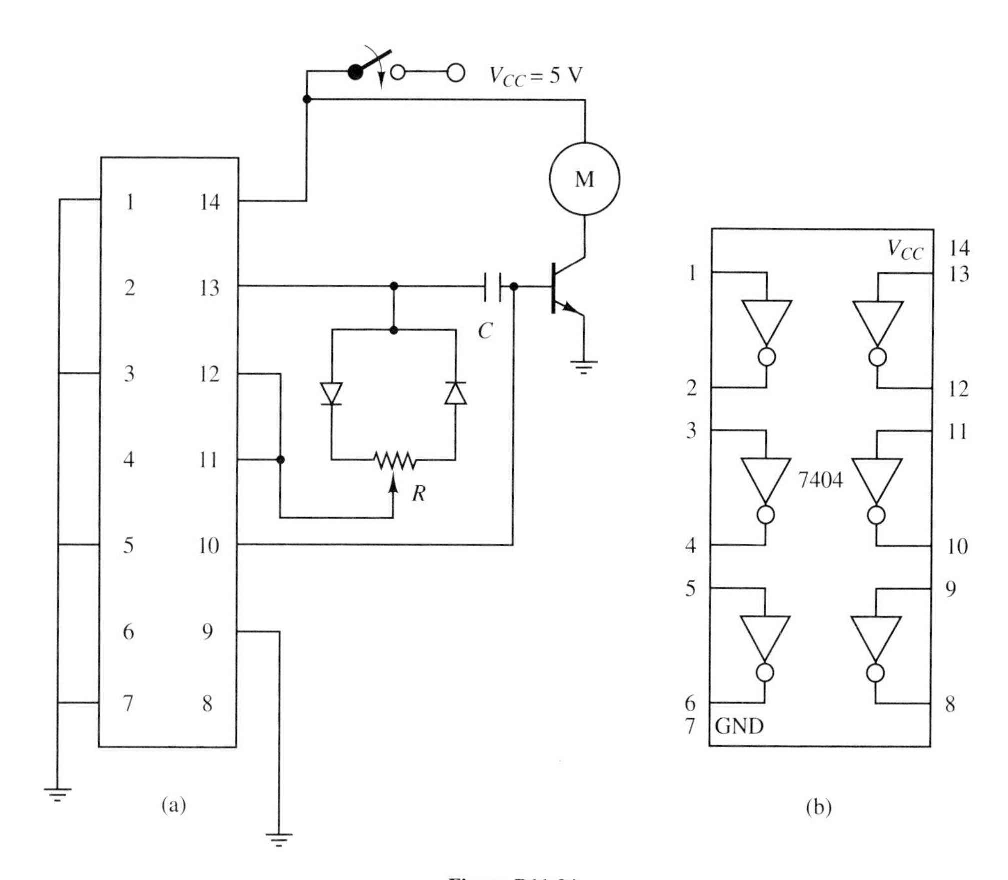


**Figure P11.34**

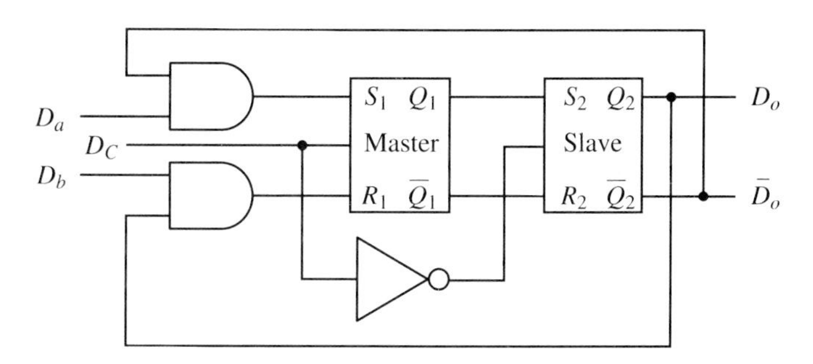


**Figure P11.35**

**11.36.** The block diagram of an alarm clock is shown in Fig. P11.36.
(**a**) Determine how many flip-flops are needed for each counter.
(**b**) Explain how the alarm clock works.

**11.37.** Another simple clock generator distinct from that in Fig. 11.16 and having an enable line is shown in Fig. P11.37a. Analyze the behavior of the circuit and sketch the output waveform for the given enable signal in Fig. P11.37b. Let the power supply voltage $V_{DD}$ represent logic "1" and assume that the output of the XOR gate, $V_1$, is "0" before "enable" goes high, that is, $V_C = V_1 = 0$ V at $t = 0$.

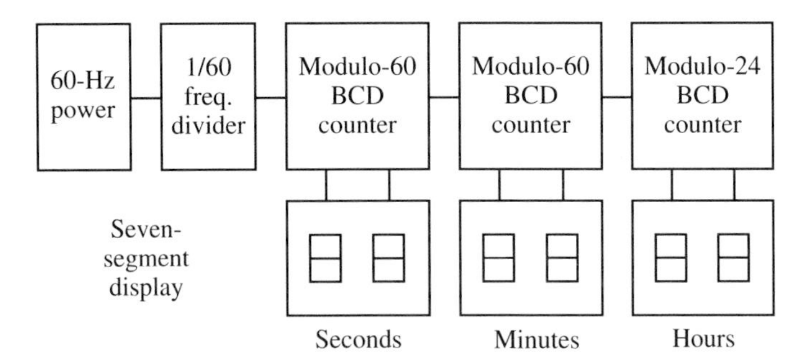


**Figure P11.36**

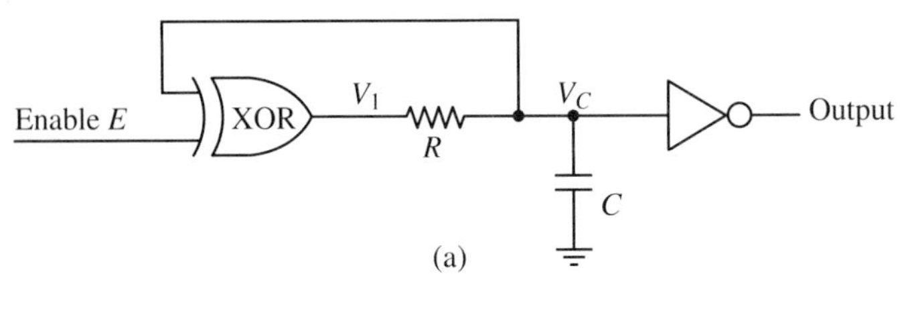


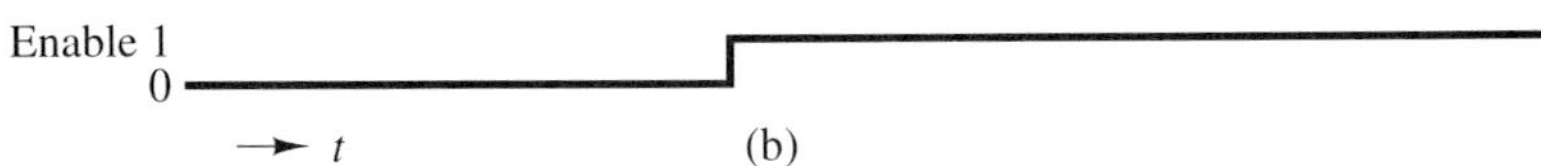


**Figure P11.37**

**11.38.** An asynchronous three-bit counter is shown in Fig. 11.17.

**(a)** What is the maximum number to which the counter can count?

**(b)** If the flip-flops were rising edge triggered instead of falling edge triggered, the counter would count in binary reverse order, as stated in Sec. 11.4. Show how it functions as a binary reverse order counter using a timing diagram. (*Hint:* Let the initial states of the flip-flops be 111.)

**11.39.** In an electric odometer for a car, the analog value of speed could be converted to the number of binary pulses per second. For example, 45 miles per hour could be represented by 45 binary pulses per second. Draw a block diagram of this kind of circuit which could be used to count the number of binary pulses in a given period and display the counted number in decimal. (*Hint:* You may use a tachometer asynchronous BCD counters, a timer, and seven-segment display devices and their decoders as building blocks.)

**11.40.** Figure P11.40 shows a synchronous counter circuit.

**(a)** Write the logic expressions for $T_0$, $T_1$, $T_2$, and $T_3$.

**(b)** Draw the output waveforms $D_{o,\text{ff0}}$, $D_{o,\text{ff1}}$, $D_{o,\text{ff2}}$, and $D_{o,\text{ff3}}$, assuming that the initial state of each flip-flop is zero. Draw an arrow at the time when the counter reaches $8 = 1000_2$. Keep in mind that because of AND gate delays, the output of one flip-flop cannot alter any other flip-flop state until the next trigger edge.

**(c)** Draw the state transition diagram; what happens if one of the unused states occurs accidentally?

**(d)** What is the maximum binary number to which the circuit will count?

**11.41.** Figure P11.41 shows a circuit that compares two binary numbers given as input in serial format.

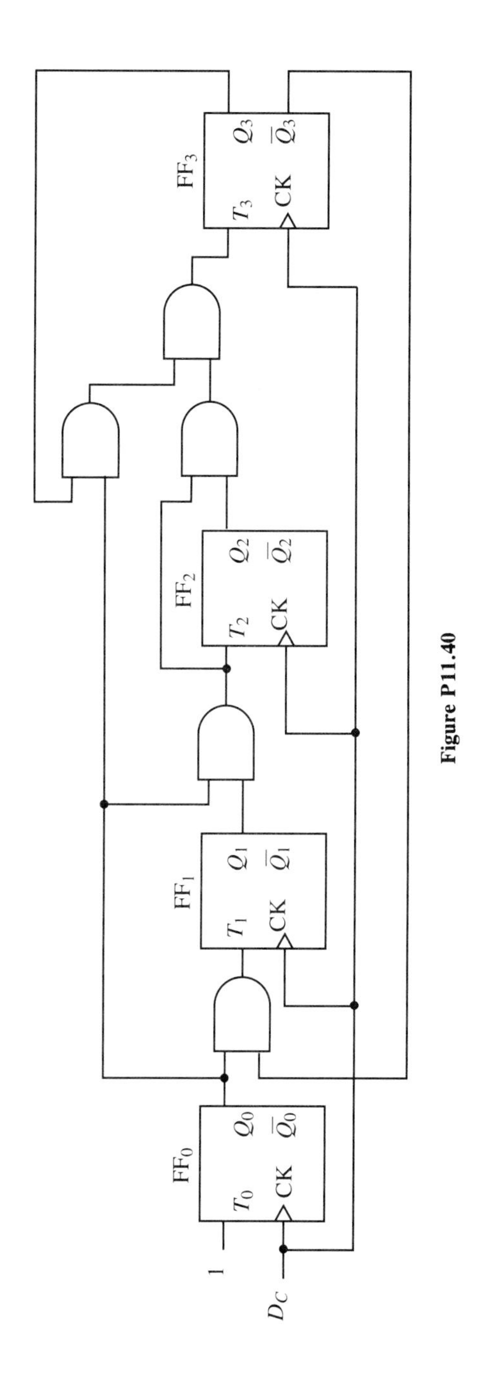


**Figure P11.40**

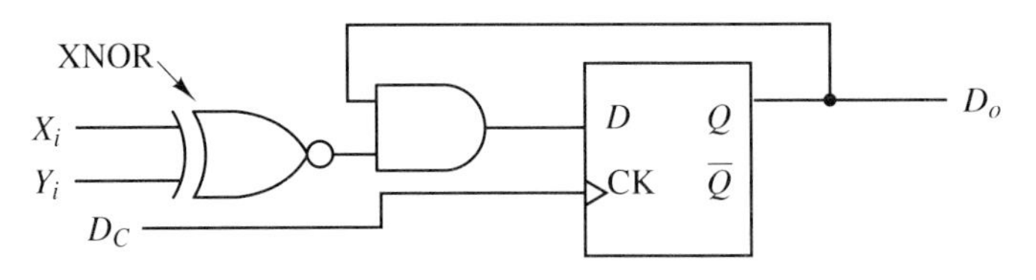


**Figure P11.41**

(**a**) Draw the state transition diagram.
(**b**) Analyze how it completes comparison of two arbitrary length binary numbers.
(**c**) In what sense are the two input numbers compared? (*Note:* Before comparison, the initial state of the flip-flop must be $D_o = 1$.)

**11.42.** Suppose that the values of the four-bit binary number in the upper significant nibble of a register are to be compared continually with that in the lower significant nibble; see Fig. P11.42a. For the bit waveforms shown in Fig. P10.42b, sketch the output waveforms $I_{A<B}(t)$, $I_{A=B}(t)$, and $I_{A>B}(t)$.

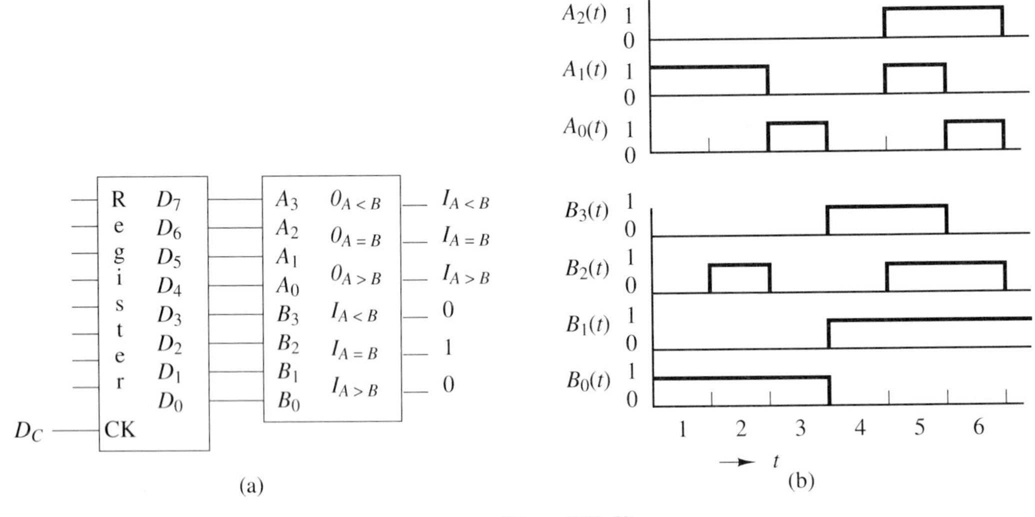


**Figure P11.42**

**11.43.** Using a truth table and K-map, design a two-bit comparator. Draw the logic diagram using gates, and then redraw using a PLA instead.

**11.44.** Design a three-bit comparator using *only* a truth table (do not attempt a six-variable K-map), and draw the PLA circuit diagram. The key to making this problem easy is to spot patterns rather than figuring out all 64 cases one at a time. For example, only 32 lines of the truth table need be written out; make two columns in the output: one for $A_2 = 1$ and one for $A_2 = 0$. Use the patterns of dots on the PLA from the truth table and the solution should take only a few minutes. It will, however, take a few large sheets of paper.

**11.45.** The pattern-generating circuit shown in Fig. P11.45 can generate two patterns controlled by the input X. This idea was used in early digital communication systems as a means of generating "flag" sequences that signify the start of a message.

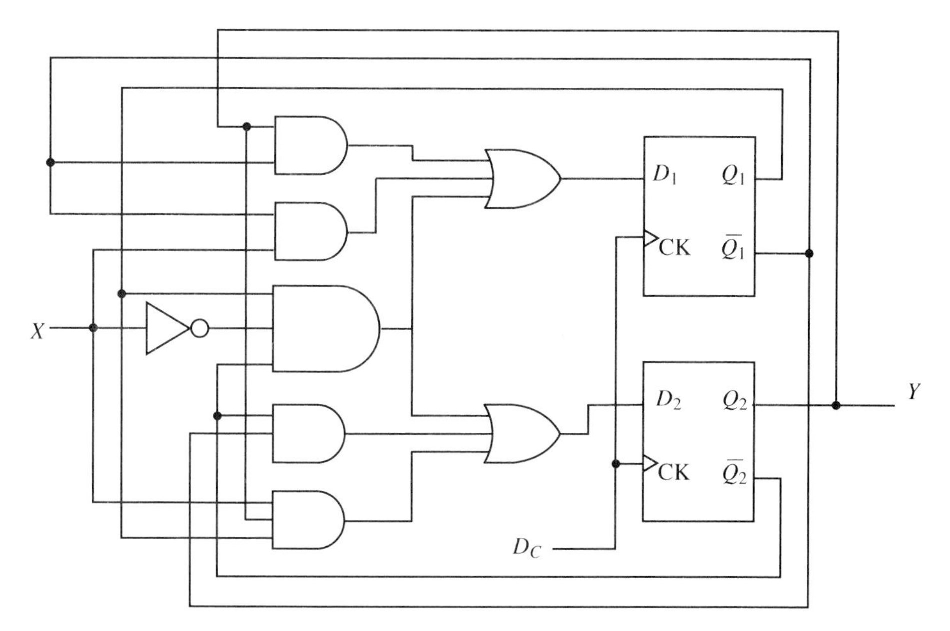


**Figure P11.45**

(**a**) Write down the input expressions for $D_1$ and $D_2$.
(**b**) Determine the state transition table and diagram.
(**c**) Determine what patterns it can generate when the input $X$ is held fixed at 1 and at 0.

**11.46.** Figure P11.46 shows the state transition diagram of a sequence detector that detects the presence of sequence "101." Design the detector, such that when a sequence "101" is received, the output is 1, and otherwise the output is 0.

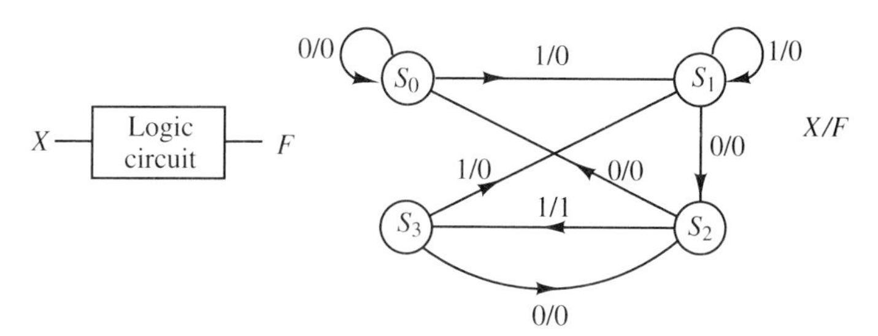


**Figure P11.46**

**11.47.** As explained in Sec. 11.6.2, when we press a key in a modern keyboard (such as PC keyboard), the computer receives the ASCII code of the key pressed. Determine what the computer receives when we type the following numbers on the keyboard.
(**a**) 32.5
(**b**) 42.7
(**c**) 0.0098

**11.48.** The representations of floating-point numbers using exponents and mantissas have been discussed in Sec. 11.6.2. Figure P11.48 shows a standard single-precision storage format which is used in the common personal computer. $S$ is the sign bit, $m'$ is the true exponent (eight bits long), and $A'$ is the mantissa (23 bits). The value of a number with this representation is given by $(+/-)(1.A') \times 2^{m''-B}$, where $B = 1111111_2 = 127_{10}$ is the bias of the exponent.

| Bit no: 31 | 30        23 | 22        0 |
|---|---|---|
| $S$ | $m'' =$ biased $m' = m' + B$ | $A'$ is the mantissa |

**Figure P11.48**

Change the following binary numbers to single-precision binary floating-point format as presented in Fig. P11.48.
(**a**) −11.101011
(**b**) 101.11000001
(**c**) 11010.0101
(**d**) −0.0001110111

**11.49.** For the following single-precision numbers expressed in the format discussed in Problem 11.48 and shown in Fig. P11.48, find the value in decimal.
(**a**) 10001011110000000000000000000000
(**b**) 00100000010100000000000000000000

12

# Digital Electronics III: Microprocessors

## 12.1 INTRODUCTION

Our final chapter on digital electronics is devoted to its major industrial and consumer applications: microprocessor automation, monitoring, and control. The word *microprocessor* essentially means "small computer on a single chip." Familiar examples of devices containing microprocessors in the home are microwave ovens, "smart" washing machines and dishwashers, hand calculators, stereo radio receivers, TVs, VCRs, CDs, automobiles, personal computers, answering machines, modern typewriters, 35mm cameras, toys, modern home heating and lawn sprinkler systems, and photocopiers. Also, the microprocessor is making its way into music synthesizers, animation sequences in films, and virtually every modern entertainment medium.

In industry, a large fraction of chemical, mechanical, and electrical processes and measurement equipment are controlled or are controllable by the microprocessor. Increasingly, old industrial control and monitoring systems are being replaced by microprocessor-driven digital systems. Finally, the microprocessor is becoming a vital tool in spacecraft such as satellites, where an on-board computer obviates much of the need for computations traditionally done at "ground control."

The microprocessor represents the pinnacle of low-cost computation and control, and its capabilities will continue to improve. Because it is such a sophisticated device, there are many levels of study of microprocessor-based circuits. Of course, this is nothing new to us. For example, digital ICs may be studied at either the logic gate level or at the transistor level. Analog ICs such as the op-amp can be studied as either a high-performance voltage-dependent voltage source or as an amplifier with 25 or more transistors. In both cases we have briefly studied lower-level rudimentary circuit behavior, not for the purpose of *designing* ICs, but to gain an appreciation of their basis.

Our main interest has been in the upper (gate or amplifier) level, which led to multitudes of interesting and practical applications circuits. We were generally able to ignore the lower (transistor) level, retaining only a few essential parameters that quantify significant departure from ideal behavior. This is possible because of advanced design techniques used by specialists at the lower level to produce an IC that is simple to use.

What will be our level of study of the microprocessor? In this case there are even more possible levels. There is, of course, the individual transistor level, which no one except experts having access to computer-aided design systems will attempt. Another area of study is that of the fabrication techniques used to make the chip. Fabrication techniques are again highly involved, and their study would serve little purpose here other than to broaden the reader's knowledge.

At a higher level, we begin to reach details needed for hooking up and programming a microprocessor in a way that will perform the desired task. Here there are two broad areas, hardware and software, each of which may again be studied at various levels. These areas were alluded to briefly in the summary of Chapter 11.

One of the definitions of "ware" is "manufactured articles or goods—often used in combinations such as 'tinware'." Well, computers are not made of tin; rather, they are made of digital devices operating under the control of a set of detailed instructions. The devices are called *hardware* and the instructions are called *software.* Permanent instruction sets/lists burned into ROMs are called *firmware* because they are a hybrid of hardware and software (nonvolatile programs).

Hardware may be studied at the gate level, the chip level, or the system level. Chapters 10 and 11 concentrated on the gate and chip levels of abstraction. With the microprocessor, it is also possible to take a brief look at the system level. The system level involves physical systems, transducers, interface electronics (e.g., A/D and D/A converters), other peripherals (input/output devices and memory systems), and of course the microprocessor itself, all under the direction of the applications program.

One example of *hardwired* (hardware-based) control is the priority encoder, which we studied in Chapter 10. The hardware designer connects binary lines to the encoder in accordance with his or her preconceived priority scheme. Once connected, the selection strategy is fixed by the hardware.

Software may be studied at the 1/0 pattern level, assembly language level, higher-level language level, and package level. The 1/0 level is the only level we examined earlier, in Chapters 10 and 11. Recall that the multiplexer in the football playoffs example sent one of its inputs to its output under the direction of the pattern of 1s and 0s found on its select lines. This pattern of 1s and 0s could be selected and modified at will to change the function of the multiplexer. We also considered selection of a memory location by means of a binary address bus. Once enabled, the data stored at that address would be available to be sent to another desired location for processing. The idea of using control lines to dictate the flow of information is central to programming any computer.

*Assembly languages* are discussed further in Sec. 12.5. They are sets of commands with abbreviated names, each of which implies a certain pattern of 1s and 0s required to select the given command. An example is: "ADD r" meaning "add the contents of register r to the accumulator register." Only relatively simple commands are available at this level; more complicated operations must be constructed from the basic set of commands.

At the level of languages such as C, BASIC, and FORTRAN, more complicated operations can be performed by single commands: For example, sin (5*t) in FORTRAN. These languages are far easier to use but require much more memory and execution time than do assembly language programs.

Finally, there is the level of package, which may be most familiar to us. Word processors, spreadsheets, graphics packages, and data base manipulation programs are prewritten and have many helpful features that make them *user friendly*. All you have to do is select the desired operations from a *menu* of possibilities. But to maximize speed and available memory, many microprocessor system programs are written at the assembly level.

The main goal of this chapter is to provide an overall picture of the microprocessor-based system. Reaching this level of understanding will involve introductory studies of many of the foregoing aspects of these systems, with particular attention to the microprocessor itself. It is quite likely that on your job, the call from "on high" will be "stick a microprocessor here" or "let's digitize this control system." It has even been said that any consumer product costing over $100 is a good candidate for microprocessor monitoring and control. Having read this chapter, you should be able to work more successfully with the design team, composed of digital designers, programmers, and others, in what could be a very rewarding and exciting enterprise. So let's get started!

## 12.2 WHAT IS A COMPUTER?

Everyone knows the common purposes of computers: automated computation, system control, and storage, display, and communication of information. The digital computer offers exact, convenient, and fast realization of all of these. Less familiar is the overall structure of a computer and the reasons for the large variety of computers in existence today.

Let us first consider briefly the available range of computer power. Computers range in size from massive supercomputers dominating entire rooms down to the microprocessor computer-on-a-chip. If we are working in the highly complicated fields of weather prediction, state-of-the-art medical/biological imaging, fluid mechanics computations, astronomical calculations, geophysical characterizations, and so on, supercomputers are required. A few statistics should indicate the power of a modern supercomputer.

With all eight processors working, the Cray Y-MP8/684 at the Ohio Supercomputer Center has a calculation speed of 200 *megaflops* (i.e., $2 \cdot 10^8$ floating-point

The first major computer, the ENIAC from 1946: 18,000 vacuum tubes consuming 175 kW to perform 5,000 additions per second, and twenty 10-digit registers each two feet long. A typical comparable machine today would be on the order of 1,000 times faster and 1,000 times smaller. *Source:* Smithsonian Institution photo No. 53192.

operations per second—*flops*). That is on the order of 200,000,000 floating-point multiplications each second! Note that multiplication on the Y-MP takes about 17 instructions per floating-point operation, addition and subtraction take about 4, and division takes about 60 to 100 instructions.

The Cray has 64 megawords of main (fastest) RAM memory. An intermediate semiconductor memory system called the Solid State Device has 128 megawords and communicates with main memory at 1250 megabytes/s. Its conventional magnetic disk memory has a 30-gigabyte ($3 \cdot 10^9$ bytes) capacity, and it communicates with main memory at "only" 10 megabytes/s. Finally, it has a new tape system called the Silo: the Silo houses 6000 magnetic tapes storing a total of 1.2 terabytes (i.e., $1{,}200{,}000{,}000{,}000 = 1.2 \cdot 10^{12}$ bytes), weighs 8400 pounds, takes 25 s to load or access a tape, and having located data, moves it in and out at a rate of 4 megabytes/s.

There are computer systems at all levels between that of this massive system and that of the microprocessor. The microprocessor is far slower than the Y-MP, by a factor on the order of 10,000 plus or minus an order of magnitude. Variations in speed depend greatly on the operating programs used, the application, and the particular hardware structures involved (known as *computer architectures*). An example of a computer in between is the Hewlett–Packard HP 835 minisupercomputer, which runs at about 1 megaflop (15 million instructions per second, i.e., 15 Mips), and has 16 megabytes of memory.

A recent *benchmark study* compared the speed of the Y-MP with that of a microprocessor when both were solving the same computational problem. A benchmark study is one that uses a program designed to be typical of those commonly used in practice. Concerning the microprocessor, an average instruction takes on the order of 10 to 20 clock cycles, and a floating-point operation may take on the order of 20 or more instructions. So if the clock frequency is 1 MHz, this would give on the order of 0.003 megaflop. If a *math coprocessor* such as is found in IBM PC ATs is used, the speed can be several times greater. The result was thus that the Y-MP is about 60,000 times as fast as this microprocessor. Such is the wide range of computing power available in modern digital computers.

While we naturally would like maximum speed, economic and physical limitations determine the actual machine chosen for a given application. A Cray Y-MP does not quite fit into a VCR! The untold millions that the absurd Cray-based VCR would cost clearly would make it unaffordable and unappealing to all but the junk bond kings. But an 8051 microprocessor costing less than $10 will serve the purpose, and is a single 40-pin DIP.

For applications in between, such as computer-aided design, medium-sized scientific computations, and many advanced control systems, powerful workstations or mainframe systems are applicable. A mainframe is typically a fast computer with large memory resources which is made available to many users simultaneously (time sharing/multiuser). The workstation typically incorporates extensive graphics capabilities available to an individual user, as well as fast data processors. It often shares memory resources with other workstations.

Having discussed the overall range of existing computers, let us consider the basic structure of all computers. The washing machine example of Sec. 11.5 involved several of the essential elements of a computer system. Referring to Fig. 12.1, the washing machine circuit diagram can be drawn in block form: Several inputs are fed into a logic/memory module, which determines the values of several output variables.

The several steps of the wash cycle are similar to lines or blocks of code in a computer program. At each step, various commands (output variable values) are given and decisions are made, based on both the inputs and the current contents of memory. One essential aspect missing from the washing machine system but present in all computers is an arithmetic capability. Suppose that instead of a washer it were a chemical mixer, and that certain components were to be added in until desired concentrations were obtained. That would require some simple arithmetic on measurement data, performed on a continual basis. Some sort of *arithmetic logic unit* (ALU) is needed, which can be found in all computers.

When the *program* becomes longer and more complicated, a more structured command sequencing is required. Moreover, a small set of commands tends to be

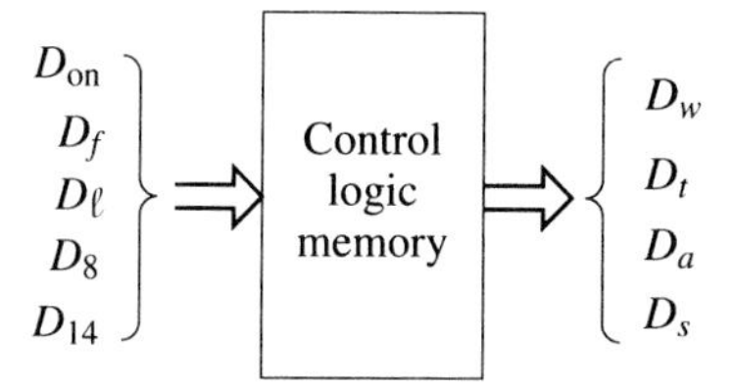


**Figure 12.1** Washing machine binary inputs (on left), digital processing circuitry (block), and binary outputs.

used repetitively. It would be nice to be able to call on these *routines* sequentially from one hardware location rather than replicate the *instruction* sequencing with many identical pieces of hardware wherever required in the complete system. Finally, if plans are changed, it is clearly far easier to modify a list of instructions than it is to replace the hardware (digital logic) carrying out the old function with that implementing the new function. Of course, we are describing the computer program constructed from a programming language.

These changes from the simple-minded washer circuit are what make computer systems such highly efficient and flexible electronic control machines. A digital computer has most or all of the elements shown in Fig. 12.2; however, this diagram is especially representative of microprocessor systems. The computer programmer needs to know the instruction set (programming language) and how to enter the program into the appropriate memory. Once that is stored and the input–output connections to the computer are made, the computer can perform the intended functions.

Within the computer is memory sufficient to store (in binary form) the application program (in RAM), the system management program (*operating system*) (in ROM), look-up tables (in ROM), numerical results of calculations (if applicable) or other data relevant to the process (in RAM), and constant parameter values used in those calculations (in ROM). As are most microprocessor system design issues, the type of memory hardware most appropriate for these various kinds of information is application specific.

Also, several registers (which can be found in Fig. 12.2) are required to keep track of the following pieces of information: where we are in the program (program counter), what type of instruction we are doing (instruction register), the result of the current calculation or data manipulation (accumulator), status of the accumulator (flags register), values of the operands of whatever operation is currently under process (temporary registers), *scratch pad* for intermediate calculation results values (B, C, . . . registers), address of the memory location currently of most interest (memory address register), and the starting address of the block of information necessary to return and restore the program to where it was before a *subroutine* was called or other interruption occurred (stack pointer).

Translation of the binary *commands* into specific sequences of action is accomplished by an *instruction decoder.* The collection of registers, instruction decoder, ALU, and input/output control logic is known as the *central processing unit* (CPU).

Electronics that match particular input and output devices to the input–output pins of the microprocessor chip are called *interfaces.* Examples include A/D and D/A converters, buffers, and drivers. Recall that in Chap. 10 many typical input and output devices were discussed. Input–output devices, interfaces, and supporting memory systems dominate the cost of small computer systems, such as microprocessor-based systems.

Data, address, and control multibit transmission lines (called *buses*) connect the computer with its input–output devices (peripherals) through the interfaces. And of course, a power supply must be connected as well as a clock. (The clock is usually, but not always, generated within the microprocessor chip. Usually, an external quartz crystal must be supplied.)

The above-mentioned components are the basic elements of a typical com-

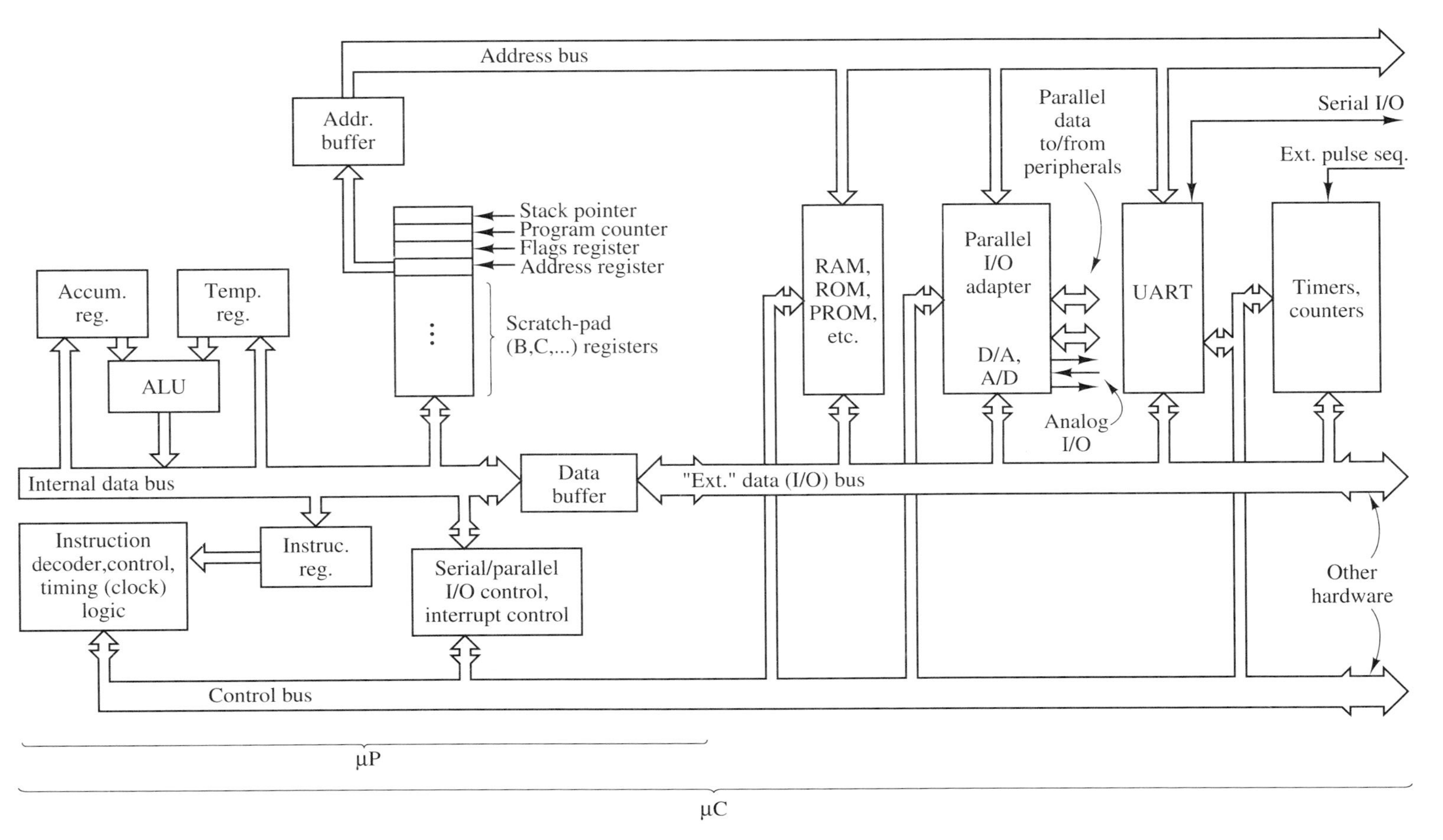


**Figure 12.2** Fundamental elements of microprocessor and microcontroller systems.

puter system. A microprocessor system is just a small-scale computer system with the CPU on one chip. We look at these components in more detail as we study microprocessors.

## 12.3 MICROPROCESSOR TYPES

It would be simpler if there were only one kind of microprocessor; in reality, there are thousands. When we just want to automate a system with a microprocessor, the range of available ICs can be bewildering. Furthermore, the differences between microprocessor ICs can be subtle—far more subtle than can be clarified in one chapter on the entire subject of microprocessors.

However, there are a few major categorizations. When we read that a chip is a "μP," or microprocessor, this implies that the chip does not include all of the electronics necessary for a complete computational system. Several supporting ICs must be obtained and connected to create a working digital system. The μP is designed to move lots of data into the ALU, do arithmetic calculations, and move it out to mass storage or display for human consumption.

A μP typically contains little or no user-accessible on-chip memory; external memory systems must be supplied. Counters and timers commonly needed for creation of pulse signals of lower frequency than the clock must be externally supplied. And input–output circuitry (*ports*) must be provided by specialized support chips. The personal computer processing chip is always a μP.

When we read that a chip is a "μC", a *microcomputer,* this implies that the chip includes all of the essentials for a small computer system. All of the resources above [memory, timers, input–output (I/O) support, etc.] are included on the processor chip. It should be pointed out that μCs are also amenable to expansion using external support chips (e.g., extra RAM and ROM). This flexibility is very advantageous when improvements on an existing design are proposed and where more sophisticated operations are desired while still minimizing chip count.

Obviously, having all essentials on one chip means that the computing power of the system will be less than that of the μP-based system, where the functions are spread over several chips. But because the cost of microprocessor-based systems is proportional to the number of individual chips required, a single-chip system is usually by far the cheapest.

The latter category of processor, the μC, is most common in consumer products, where large numbers of identical systems are manufactured and sold. ("Large numbers" in the auto industry means hundreds of thousands or even millions. At these levels, custom-made chips may be cost-effective.) Here cost minimization is critical. Included in this category are most *dedicated control systems*, in which the microcomputer is "dedicated" to controlling a single process or group of processes. This is to be contrasted with the μP, which tends to be either a general-purpose computational resource constantly being reprogrammed or a part of a very large control system requiring more processing power than the μC can provide. The μC is usually programmed once or infrequently.

The instruction set of the μC is geared more toward bit manipulations than toward moving data in and out of memory in huge chunks. The bit-at-a-time manipulations are required for process control applications involving many individual I/O lines. They are also required to make maximal use of the limited number of pins on the μC IC chip, which must serve the entire computer system as opposed to only a portion of it as is the case with the μP.

In fact, many microprocessor pins are *programmable:* In different situations their lines have entirely different functions. An IC DIP chip is usually limited to 42 pins; even surface mount chips cannot go much higher than 64. However, the pin grid array allows 168 pins on the Intel 860 64-bit μP family, with goals of 5 million transistors and 40 MHz in the near future for personal computer/workstation graphics systems. But these run into the price range of hundreds or thousands of dollars.

A word on terminology. The microcomputer (μC) is often also called a *microcontroller* because of its predominant use in controls applications. Also, it should be noted that in common usage, the word "microprocessor" refers generically to both μPs and μCs.

Within these two categories (μP and μC), perhaps the most basic distinguishing trait of a microprocessor chip is the number of bits on the *data bus*. The available bus widths are 4, 8, 16, and 32 bits (and a few are beginning to be 64, such as the Intel i860). Four bits means that only $2^4 = 16$ levels are possible within one word. Double or higher precision extends this range to 256 or more, but this requires many instructions per arithmetic operation. That is, the operation must be performed in "chunks," which makes computation relatively slow. If lots of rapid calculation is required, only an eight-or-more-bit machine is practical. On the other hand, if control of only a small number of lines such as exists in a microwave oven is desired, the four-bit microcomputer has often been the most economical approach. New designs increasingly use higher-than-four-bit machines.

When many calculations are required on a continual basis, and there are large numbers of inputs and outputs, an eight- or 16-bit μC or μP is best. An example we consider in Sec. 12.7.3 is the microprocessor control of an automotive engine. Hybrid 8/16-bit μPs are used in some personal computers. That is, the data bus has eight bits, but arithmetic operations in the CPU are performed on 16-bit numbers. The reason for not "going all the way" to a full 16-bit machine is that the 8/16 can be used in place of the equivalent 8/8 predecessor; the 8/16 is fully compatible with the old 8/8, but the new machine is faster. This obviates the costs of system redesign.

The larger the number of bits (the wider the bus), the faster the computer can become because fewer calls to memory are required for each operation. Also, higher accuracy becomes feasible. Of course, as noted in Chapter 11, when floating-point representations are used, extremely large ranges of numbers can be represented even using moderate numbers of bits. Also, special floating-point processor ICs are available for such number crunching to free the main microprocessor for other required tasks.

Very wide buses such as 32- or 64-bit buses are currently used only for the most advanced *workstation*, memory control, or communications control systems. As the technology continues its progression toward more on a chip for less money, the wider data buses will be used more frequently.

In tandem with but distinct from the data bus width is the *address bus* width. An eight-bit address bus can select from $2^8 = 256$ memory locations (words, which for an eight-bit data bus are bytes). A 32-bit address bus can select from 4 gigabytes.

This is the tremendous range of *address space* sizes occurring on available microprocessors. Of course, even the largest μCs have only about 32K or so of on-chip memory; extensions of memory size must be made externally. But for dedicated systems, the on-chip 32K or less is usually sufficient. It is only when large amounts of data are to be processed or accumulated that such large memories are required, a common example being the personal computer.

Then there is, of course, the set of functions available on a given chip. For instance, several processors offer multiplication and division, whereas others offer only addition and subtraction. But if no multiply or divide operations are required in a given control example, there is no point in spending more money for them or having precious chip and memory space devoted to them. This philosophy clearly applies to other expensive microprocessor features.

Processing speed is another important factor distinguishing microprocessors. Data sheets specify it in various forms: maximum clock frequency, duration of shortest instruction (or its inverse: Millions of instructions per second, or Mips), duration of long instructions such as division, duration of register transfers, duration of a *machine cycle* (time for a standard read or write operations, which serve as building blocks of instructions), or megaflops.

The myriad of parameters makes comparison very difficult. Furthermore, programming skill and application details influence the obtained performance at least as much as the hardware specifications. Machine cycle/shortest instruction (typically one to four machine cycles) times range from about 1 μs (1 Mip) on several 8-bit machines to 0.1 μs (10 Mips) or better on 32-bit machines. Do not confuse these Mips with megaflops! Floating-point operations take many instructions. Instructions take from one to several machine cycles. Machine cycles take roughly three to six clock cycles.

Ease of programming is another factor. Some processors may easily be programmed using high-level languages, whereas others must be programmed in their own particular *assembly language*. This issue of programming can be a factor overriding all others. If the programming team is experienced in working with the assembly language of one family of chips, they will be inclined to push for use of that same family when a new project is begun.

One simple specification of a microprocessor is the number of input–output lines. Clearly, this must be sufficient for the physical system to be controlled/monitored. However, we can multiplex one line to carry input–output from more than one physical transducer or other I/O device.

Not to be overlooked are more technical differences between processors. The number of registers, the methods of addressing memory, the bus structures, features such as *direct memory access* (DMA, the capability of circumventing usual data paths during large data transfers to maximize speed) interrupt capabilities, available data manipulation/transfer instructions, and many other factors contribute to an informed opinion on the applicability of a given processor to the application at hand. Decisions on these issues require far more knowledge and experience than can be imparted in a single chapter on microprocessors, and must be made by specialists.

Table 12.1 lists several of today's most popular μPs and μCs, together with some of the more basic parameters discussed above. Where no values are given, the data was not available or was not applicable to the particular processor. The "x" in

**TABLE 12.1** CURRENT MICROPROCESSORS AND MICROCONTROLLERS AND THEIR FEATURES

| Manufacturer | Model | Machine type | Data bus width (bytes) | CPU width (bytes) | On-chip (EP)ROM (bits) | On-chip RAM (bits) | Maximum off-chip memory expansion (bits) | Cycle/instruction time (μs) or maximum clock time (shown in MHz) | Number of interrupts | Number of I/O lines | Multiply and divide? | Comments and applications |
|---|---|---|---|---|---|---|---|---|---|---|---|---|
| Texas Instruments | TMS1000 | μC | 4 | 4 | 2K | 128 | Not expandable | 15 | 0 | 19–24 | No | <$1, microwave oven, washing machine, dishwashers, etc.; first single-chip μC |
| National Semiconductor | COP800 | μC | 8 | 8 | 1K–4K | 64–256 | 32K | 1 | 8 | 20–40 | No | 75 million sold; automotive, other dedicated applications |
| Microchip Technology | PIC16C5x | μC | 8 | 8 | 512–2K | 32 | Not expandable | 20 MHz | | 12–20 | No | |
| Intel Corp. | 8048 | μC | 8 | 8 | 0–2K | 64–128 | 4K | 1.4 | | 27 | No | Was the leading 8-bit μC chip UART |
| Intel Corp. | 8051 | μC | 8 | 8 | 4K–8K | 128–256 | 256K | 1 | 5 | 1–4 | Yes | Newer, leading 8-bit μC (on-chip EEPROM, A/D converter, 2 timers, serial peripheral interface) |
| Texas Instruments | TMS370 | μC | 8 | 8 | 4K–16K | 129–512 | | | | 22–55 | Yes | Features of 8051 plus D/A converter, LCD drivers |

(Continued)

**TABLE 12.1 (Continued)**

| Manufacturer | Model | Machine type | Data bus width (bytes) | CPU width (bytes) | On-chip (EP)ROM (bits) | On-chip RAM (bits) | Maximum off-chip memory expansion (bits) | Cycle/instruction time (μs) or maximum clock time (shown in MHz) | Number of interrupts | Number of I/O lines | Multiply and divide? | Comments and applications |
|---|---|---|---|---|---|---|---|---|---|---|---|---|
| Motorola Microprocessor Group | 6804/5 | μC | 8 | 8 | 1K–16K | 64–304 | 1 | 2–5 | 16–32 | | Mult. only | <$3, A/D, EEPROM |
| Motorola Microprocessor Group | 6801/6301/68HC11 | μC | 8 | 8 | 2K–24K | 128–768 | 64K–2M | | | 30 | Div. only | Used in Japanese consumer products |
| Rockwell International Mitsubishi Electric, Western Design | 6500/6501/65C256/50740/37700 | μC | 8 | 16 | 2K–32K | 64–512 | Only 37,700, Expandable to 16M | | >2 | 52 | No | 8 timers, 2 UARTs, 8-channel A/D |
| Zilog Inc. | Z8/Super8 | μC | 8 | 8 | 0–8K | 256 | 128K | 0.6–3 | 37 | 40 | Yes, on Super8 | DMA, timers, UART; not compatible with Z80 |
| Zilog Inc. | Z80A | μP | 8 | 8 | 0 | 0 | 64K | 1–6 | 3 | 256 | No | Leading 8-bit μP; 8080/8085 (old standard μP) instruction set |
| Texas Instruments | 7000 | μC | 8 | 8 | 0–4K | 128–256 | 64K | 1.25 | 4–6 | 20–32 | Mult. only | A/D, UART |
| Intel Corp. | 8085AH/80C85 | μP | 8 | 8 | 0 | 0 | 64K | 0.67 | 5 | 256 | No | General-purpose 8-bit μP |
| Zilog Inc. | Z180/HD-64180 | μP | 8 | 8 | 0 | 0 | 512K–1M | 4 MHz | | | Yes | DMA, UART, 2 timers, cache, same |

| | | | | | | | | | | | | |
|---|---|---|---|---|---|---|---|---|---|---|---|---|
| | | | | | | | | | | | | instruction set as Z80 |
| Motorola Microprocessor Group | 6800/6802/6809/6309 | μP | 8 | 8 | 0 | 0–128 | 64K | 1 | | | Mult. only | Very popular μP |
| Western Design | 650x/65C0x | μP | 8 | 8 | 0 | 0 | 64K | 0.1 | | | No | |
| Intel Corp. | 8088 | μP | 8 | 16 | 0 | 0 | 1M | 0.6 | | | Yes | <$10, leading 8/16 μP used in IBM PCs |
| Zilog Inc. | Z280 | μP | 8 | 16 | 0 | 0 | 16M | 10 MHz | | | Yes | 256 cache; for multiuser system, upgrade of Z80 |
| NEC | 78K | μC | 8 | 16 | 16K | 384 | | 0.125 | | | Yes | Hard disk drive audio control etc. |
| Western Design | 65C816/65C802 | μP | 8 | 16 | 0 | 0 | 16M | 12 MHz | | 8 | No | Used in Apple computers |
| Intel Corp. | 80186/80188 | μP | 8 | 16 | 0 | 0 | 1M, 64K I/O | 0.125 | | | Yes | DMA, interrupt controller, clock, 3 timers; software compatible with 8088 |
| Intel Corp. | 8086 | μP | 16 | 16 | 0 | 0 | 1M | 0.5 | | | Yes | Leading 16-bit μP; software compatible with 8088 |
| Intel Corp. | 80286 | μP | 16 | 16 | 0 | 0 | 1G, virt. 16M, phys. | 0.3 | | | Yes | <$30. Used in IBM PC/AT; can use high-level language |
| Intel Corp. | 8096/8097 | μC | 16 | 16 | 8K | 500 | 64K | 1; 3 for div. | 8 | 40 | Yes | A/D, UART, 4 timers, high-speed I/O; developed for Ford Motor Co. automotive |

(Continued)

**TABLE 12.1** (Continued)

| Manufacturer | Model | Machine type | Data bus width (bytes) | CPU width (bytes) | On-chip (EP)ROM (bits) | On-chip RAM (bits) | Maximum off-chip memory expansion (bits) | Cycle/instruction time (μs) or maximum clock time (shown in MHz) | Number of interrupts | Number of I/O lines | Multiply and divide? | Comments and applications |
|---|---|---|---|---|---|---|---|---|---|---|---|---|
| | | | | | | | | | | | | electronic control |
| National Semiconductor | HPC16000 | μC | 16 | 16 | 0–16K | 256–512 | 16M | 0.13 | | 52 | Yes | <$10. A/D, DMA, 8 timers |
| Siemens | 80C166 | μC | 16 | 16 | 32K | 1K | | 20 MHz | 32 | 76 | Yes | Industrial control, A/D, 2 timers |
| Zilog Inc. | Z8000/ Z16C00 | μP | 16 | 16 | 0 | 0 | 8M–48M | 0.3 | | | Yes | Good for real-time control; DMA, counters; power is in between 8086 and 68000 |
| Allied Signal Microelectronics, LSI Logic, United Technologies Microprocessor Group | 1750A | μP | 16 | 16, 32 | 0 | 0 | 64K–1M | 4.1 Mips | | | Yes | Airborne for floating point, $760–$2000 |
| Motorola Microprocessor Group | 68000/ 68020 | μP | 8, 16, 32 | 16, 32 | 0 | 0 | 4G virt. | 0.4 | | | Yes | Used in Apple Macintosh II, Sun workstation ($5–$7); 512-byte cache, floating-point |

| | | | | | | | | | | | |
|---|---|---|---|---|---|---|---|---|---|---|---|
| | | | | | | | | | | | unit; not compatible with 6800 |
| National Semi-conductor | 32000 | μP | 8, 16, 32 | 16, 32 | 0 | 0 | 4G virt. | 1–10 Mips | | Yes | DSP accelera-tion, DMA, multiplier, interrupt unit $11–$600; used in FAX, office machines |
| Intel Corp. | 80386 | μP | 32 | 32 | 0 | 0 | 64T virt. 4G phys. | 33 MHz | | Yes | Used in UNIX systems, DOS, windows, $185, trans-cendental functions with math processor; runs on 8086/8088 code |

a model name, for example 650x, indicates that there may be several models, such as 6500, 6501, . . . . When several *families* are given on one line, all processors on that line have comparable characteristics despite certain unlisted differences. Some of these differences *are* implied in entries with, for example, 64K to 1M.

In practice (and in Table 12.1), letters are often interspersed within the model number. These letters indicate variants such as technology type (e.g., CMOS), maximum allowable clock frequency, and precisely which features are available on that particular chip. When a number is similar to that of another model by the same manufacturer, this may *or may not* indicate compatibility with software or with peripheral hardware of the other chip.

Memory expansion capability external to the chip is usually limited by the number of address lines on the address bus. Therefore, the address bus width is roughly the base 2 log of the number in this column of Table 12.1. For example, an off-chip memory expandable to 64K implies a 16-bit address bus because $2^{16} = 64K$. These addresses are known as *physical memory,* abbreviated "phys." in Table 12.1.

The abbreviation "virt." in Table 12.1 for expandable memory means *virtual memory.* When an address is presented by the microprocessor, it may be accompanied by other signals that tell whether this address refers to main RAM memory or some mass storage device such as a hard disk. If it is from a hard disk, the associated *page* of hard disk memory beginning at the desired address is loaded into RAM from the hard disk. This event is known as a page fault, and the retrieval process is called *swapping.* This procedure allows much larger memory to be accessed than the address bus would imply, yet still maintain fairly high access speed on the average. This *virtual memory* technique is also used on computer systems much larger than those based on microprocessors.

## 12.4 MICROPROCESSOR HARDWARE

### 12.4.1 Introduction

As mentioned above, any computer system is composed of hardware operating under the direction of software. Our goal in studying microprocessors is to gain a familiarity with the types of components required for these systems as well as some basic principles and terminology. With this knowledge, a nonelectrical engineer can intelligently converse with a microprocessor design team. Such discussions between experts on the nonelectrical aspects and experts on the electrical aspects are sure to maximize the chances for successful application of microprocessor-based systems in physical control situations.

True working knowledge of microprocessor systems can be obtained only by extended study and practical experience. This chapter should serve as a qualitative introduction. Chances are that nonelectricals are more likely to service the system hardware than they are to debug the software. Furthermore, software is highly microprocessor specific. Therefore, we concentrate most of our attention on hardware, and then consider software briefly.

A word on where this is all going. In this section and Sec. 12.5 we provide some basic hardware and software information typical of microprocessor systems. Not forgetting the ultimate aim of applications in control, the short study of simple process controls that follows will help put the preceding discussions in perspective. Finally, we conclude the chapter with a consideration of two familiar real-world applications: microwave ovens and automobile engine control.

### 12.4.2 The Three Buses

Within and extending out from a microprocessor computer/controller system are three groups of binary lines, or buses. These buses connect together the various components of the microprocessor-based system. The three buses are the address bus, the control bus, and the data bus. As mentioned in Sec. 12.3, depending on the microprocessor model, these buses have various widths.

The address bus carries the location of a desired memory cell or input–output device. It is a unidirectional line: from the microprocessor to the memory or I/O peripherals. (Incidentally, in most discussions the designations "input" and "output" are defined with respect to the IC under discussion, in this case the microprocessor.)

The *control bus* transmits binary signals determining various actions such as read, write, enable, request for services (interrupts), acknowledge reception of services, and bus arbitration (who gets control of the data bus in situations such as DMA). Although individual lines may be unidirectional, some lines go each way. Finally, the *data bus is* used for transferring data between memory, I/O devices, and the microprocessor. The same lines are used for sending and receiving data; hence the data bus must be a bidirectional bus.

The address and control buses are driven by open-collector drivers. Here the purpose for using open collectors is impedance matching. Associated with any transmission line is a characteristic impedance determined by the geometrical and material construction of the line. This is the ratio of the forward-traveling voltage divided by the forward-traveling current of a transmitted signal anywhere along the line. A detailed study of signal propagation over wires shows that signal reflections travel throughout the line if the end of the line is not terminated with an impedance matching that of the line.

Impedance matching can be achieved by tying a resistance of that value between the receiver input and the power source. (Remember, the FET receiver input impedance is huge; it alone cannot match that of the line.) For the same reason, the open-collector driver is also connected to a resistor equal to the transmission line characteristic impedance.

Usually, the data-carrying bus is called the data bus internally to the microprocessor and is called the I/O bus externally. To minimize pin count, some microprocessors time-multiplex data and addresses on the I/O bus. That is, first an address is put on the line and sent to the appropriate I/O or memory chips, and then the data to or from that chip is transferred on the I/O bus. Control signals may also be time-multiplexed on the data bus.

### 12.4.3 I/O Ports

Between the ends of the I/O bus and the chips it supports are bidirectional buffer/drivers known as *transceivers*, which we introduced in Sec. 10.8. They transmit information in one direction or the other, depending on whether a read or a write is taking place. To share this bus with possible contenders such as DMA, the output buffers must be tri-state. Also, peripheral devices hold onto the data bus only while they are putting data on it during a "read" from them; otherwise, they are in the third high-impedance state.

For signals going out to peripherals, the buffers are connected to registers known as *latches* (see below). They are necessary because the data from the microprocessor may be valid or available for only a short time, while use of the data by the peripheral may occur somewhat later. These registers are said to latch the data. For signals coming in from the peripherals, either a register within the microprocessor or one at the buffer site will take in the data. The tri-state buffers and latches are known as *I/O ports*.

For added flexibility there are I/O ports in IC form that have many programmable features, as well as multiple, multiplexible I/O output ports that minimize chip count and maximize the use of available lines. The programming instructions to operate these complicated chips come from control lines sent by the microprocessor.

**Example 12.1**

Describe and give a block diagram showing the interface between a 6800 series μP, an MC6821 *peripheral interface adapter* (PIA), and two peripheral devices.

**Solution** As with all our examples in this chapter, we cannot be as complete in the descriptions as would be necessary to really get the system to work. That is the job for digital design teams. However, a flavor of the kinds of signals that are required is obtained by considering the "pinout" of the PIA. Thus we shall consider each of the lines into and out of the PIA and show a block diagram. Refer to Fig. 12.3.

*D0–D7 = data lines.* These are bidirectional data lines. Output drivers are tri-state and remain in the high-impedance state unless a "read" from the peripheral is ordered by the microprocessor unit (MPU). (*Note:* Here "output" is defined with respect to the 6821, not the μP.)

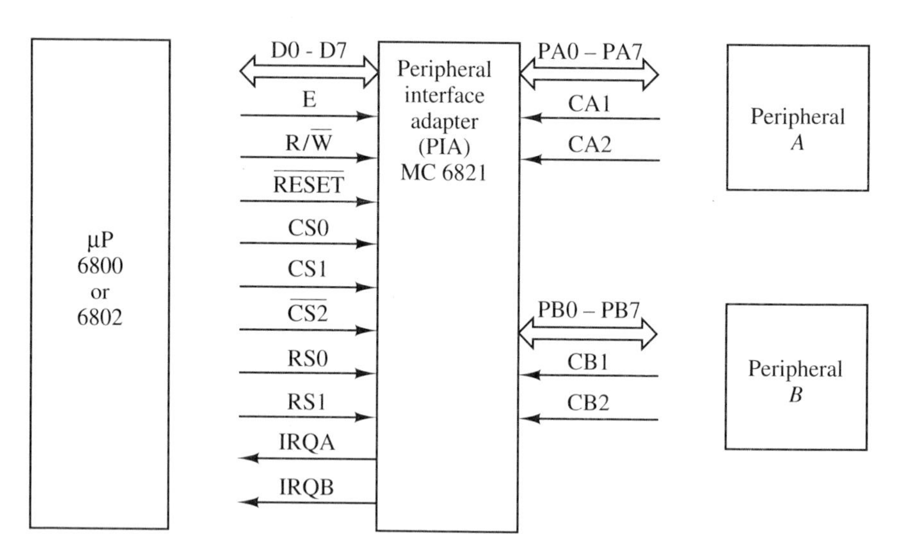


**Figure 12.3** Peripheral interface adapter connections for Example 12.1.

$E$ = *enable*. The clock for this device is connected here.

$R/\overline{W}$ = *read/write*. This signal is generated by the MPU, and controls the direction of D0–D7.

$\overline{RESET}$. This resets all PIA register bits to 0.

*CS0, CS1, CS2* = *chip select*. All three of these signals must be asserted for the chip to be enabled. Reason for having three CS lines: This is good for address demultiplexing. For example, the three CS lines can be connected to address lines $A_{13}$, $A_{14}$, and $A_{15}$. Then the pattern $A_{13}A_{14}A_{15}$ must be 110 for this chip to be addressed.

*RS0, RS1* = *register select*. In conjunction with CRA-2 and CRB-2 (the second bits of the two internal control registers CRA and CRB), these lines select which particular register of the six within the PIA is to be read from or written to. The six registers are data direction registers, peripheral data registers, and control registers for the two peripherals *A* and *B*.

$\overline{IRQA}$, $\overline{IRQB}$ = *interrupt request*. These interrupt the MPU either directly or via priority interrupt circuitry for service to the respective peripheral device. These are "OC" (actually, open drain), so they can be connected in a wired-OR configuration. The values of these interrupt request lines are stored in bits 7 and 8 of the control registers.

*PA0–PA7, PB0–PB7* = *peripheral data lines*. Input or output data from peripheral device *A* or *B*. Input versus output mode for each bit of these lines is determined individually by the binary values stored in the corresponding bits of the data direction registers *A* and *B*.

*CA1, CB1* = *interrupt input*. These set interrupt flags in the control registers mentioned above; these interrupt lines are set by the peripheral devices *A* and *B*.

*CA2, CB2* = *peripheral control lines*. These can be either interrupt control lines or a peripheral control output, as determined by the *A* or *B* control register. For more details, see data sheets.

### 12.4.4 Latches

In Fig. 11.4 we studied the gated *RS* flip-flop. For this device, the output could be set or reset by the *S* and *R* inputs, respectively, only when the enable line was asserted. At all other times, the *S* and *R* inputs were blocked and the output remained at the value it had just when the enable line became not asserted. A gated *D* flip-flop is a gated *RS* flip-flop with an inverter between the *R* and *S* inputs. Thus, whenever the data is low, that is equivalent to a reset in an *RS* flip-flop, and when the data is high, that is equivalent to a set.

A gated *D* flip-flop is also known as a transparent *D* latch, abbreviated *D latch* or simply *latch*. The word "transparent" refers to the fact that when enabled, the device is as transparent as a wire to changes in the input; that is, aside from the propagation delay, the output follows the input continuously. It is to be contrasted with the *D* flip-flop, which can change value only on the falling (or rising) edge of the clock.

Why would anyone want to use a latch? After all, presumably in sequential machine design all signals are designed to be valid on the clock edge. Why allow the latch output to change before then? Consider the addressing sequence used in microprocessor input–output systems. As noted previously, the bus output pins of a microprocessor are often multiplexed between data and address. First the address enable line is asserted, signifying that the coming signal is to be an address (rather than data). Shortly afterward, the valid address appears on the bus pins.

If a transparent latch is used, this address is ready to be sent to the appropriate memory device after only the propagation delay through the latch. If a *D* flip-flop is used, its output is not defined until the address enable line makes the transition to not asserted. There is no point in waiting this extra time; it just slows everything else down. Having the address at the memory chip early allows the memory chip to prepare for the required I/O.

However, much of the time the word "latch" is used to mean an edge-triggered *D* flip-flop. This is probably because of the convenience of the word, based on its analogous common usage as a fastener that may be unfastened. In fact, any data register used in setting up data for transfer is commonly referred to as a latch. The meaning should be apparent from the context, but if not, the data sheets for the flip-flops at hand should resolve the question.

**Example 12.2**

In Example 12.1 we studied a PIA for a Motorola 6800 μP. In this example we will show latch connections to an Intel μP, the 8085. Specifically, show a transparent latch used with the 8085 μP and the EPROM model 2716.

**Solution** The circuit diagram for these connections is shown in Fig. 12.4. The 8085 time-multiplexes the address/data lines as described above for lines AD0–AD7 ("AD" for address/data). ALE = address latch enable. When ALE goes from 1 to 0, the address is latched (i.e., frozen in the 74LS373). This transition is made when a valid address is on address/data lines AD0–AD7. The $\overline{\text{output control}}$, shown grounded, puts the 373 in a high-impedance state when high. The $\overline{OE}/V_{pp}$ line of the EPROM is low for a read and is pulsed at 21 V for programming, which can be done out-of-circuit. Therefore, in-circuit it is fixed at 0 V. Recall that the higher voltage of 21 V is required for writing to PROMs. The MEMR line is a memory read signal from the 8085 μP.

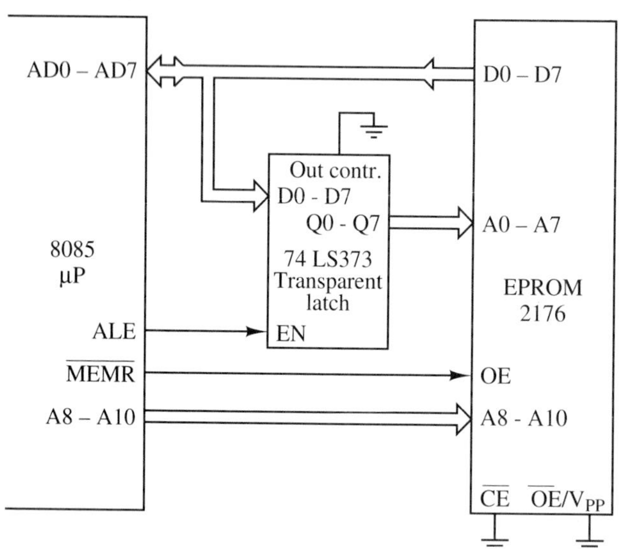


**Figure 12.4** Transparent latch connections for Example 12.2.

### 12.4.5 System Buses

Depending on the complexity of a particular application, a μC or μP may be more suitable, as discussed in Sec. 12.3. If a μC is used, all of the essentials are on one chip except for possible expansions such as external memory. These systems are likely to fit on one circuit board. For μP applications, the external circuitry may be extensive and require several boards.

Fortunately, hardware design is facilitated today with ready-made boards containing the μP, the memory systems, the I/O support circuitry, and so on. These boards have *edge connectors*—plug-in connectors with metal slabs for every binary line needing external connections or power supply line. Complementary boards for each resource are selected which are designed to operate on the desired microprocessor chip. To compose a complete system, they are all plugged into a standardized *backplane* or *motherboard* consisting of sockets for several boards and the connections between sockets.

This set of connection lines between boards is called the *system bus*. There are several *standards* for system buses. A standard in this context is just a complete set of specifications for both the hardware constituting the bus and the protocol for using it.

The protocol involves *handshaking* procedures in which request and acknowledge sequences are defined. As usual, the digital design community has a very colorful and amusing set of terminology—one reason people find computer work so much fun! A typical handshaking scenario goes like this: (1) the sender says the data is ready, (2) the receiver acknowledges data reception, (3) the sender acknowledges the acknowledgment, (4) the receiver acknowledges the acknowledgment of the acknowledgment, (5) now ready for new transmission. A very cordial couple of handshakers—imagine if this were the standard etiquette for party invitations! Actually, parts 3 and 4 are not as unreasonable as they first appear. Part 3 is implemented by a turning off of the *data valid* line, and part 4 is implemented by a "ready for new data" assertion.

Example names of bus standards currently in use, with their data/address maximum bit widths, are IEEE standards: P961 [STD (for "Standard"), 8/24], 796 (Multibus, 16/24), P1296 (Multibus II, 32/32), 1014 (VMEbus, 32/32), and P896 (Futurebus, 32/32).

In addition, there is the IEEE 488 general-purpose interface bus for connecting systems involving more than one microprocessor, and for linking a microprocessor with high-speed peripherals on a parallel basis. It is frequently used in control applications.

Finally, there are serial buses known as *local area networks* (LANs) that connect multiple computer systems and all kinds of peripherals over moderate distances such as between nearby rooms and buildings. Bus standards for these arrangements are known by names such as Ethernet and Localtalk. And there are serial computer links such as Arpanet that span the entire United States. Multiprocessor communications systems are complex and well beyond the scope of this introductory chapter.

### 12.4.6 Serial Transmission

Parallel communication is economical using buses such as the IEEE 488 up to only on the order of 15 feet. Note that an eight-bit parallel bus cable is several times as expensive as a serial bus. For these reasons, many peripheral devices use serial communications. Examples include laser printers when operated at long distances from their computers, and terminal keyboards. One example where serial communication is mandatory is the use of telephone lines for digital data transmission. Serial communication takes place one byte at a time, and each byte is sent a bit at a time (LSB first). Alphanumeric data may be coded in ASCII bytes for transmission.

In the case of serial communication, there are also standard buses. The most popular general-purpose serial communications standard is the RS-232 (RS for "recommended standard"), which is a single-wire system referenced to common ground. It is limited to a range of 50 feet when operated at 9600 bits/s. Of course, with additional drivers at the end of the bus, distances can be extended because it is equivalent to "starting over."

Incidentally, the term for bits per second of serial transmission is *baud*, named after Emile Baudot, a pioneer of telegraphy. Typical baud rates are 300, 1200, and 9600, with the highest naturally being most desirable. Today, 9600 is becoming very widespread. To minimize the possibility of error due to noise, the voltage representing 1 is often stepped up to 12 V rather than 5 V.

Better noise immunity than that of the RS-232 is available using the RS-422 line because it is a differential bus: Instead of common ground reference, a two-wire system is used. The two wires are often arranged as a twisted pair to minimize noise penetration. Of course, the price paid is double the wire. The advantage is that it can transmit as far as 4000 feet at 100K baud.

At the sending end of the serial bus is a driver to handle line losses and the possibility of serving multiple receivers on the other end. Incidentally, current loops could be used instead (and occasionally are, at 20 mA), but they are unnecessarily inefficient, cause interference in neighboring data lines, and are not standardized to the digital circuitry to which they connect. Having high and low voltages rather than current is convenient for digital circuitry. The noise immunity issues discussed in Chapter 9 are not as crucial for digital as for analog systems.

Also required for serial transmission is a method of determining when a new transmitted word begins. Clearly, the clocks of the two devices will not match in phase or in frequency—that is, the transmission is asynchronous. Preceding the word to be transmitted is a *start bit.* The receiver runs a counter a fixed factor higher in frequency than the baud rate. It counts to the estimated middle of the start bit and samples at each estimated bit time from there to the end of the word. This way, if the receiver clock is a little too fast or slow, by the end of the word the sampling may have wandered off the center of the correct bit, but not into an erroneous bit.

The larger the count rate/baud rate ratio, the better the estimation of the center of the start bit; a typical value is 16. This time recalibration procedure begins anew with each subsequent byte, so errors do not accumulate. There is also a *stop bit*, for the purpose of error detection; if the expected stop bit is not received, something went wrong and the word should be resent.

Conversion from the parallel output of the transmitting device to serial transmission is accomplished in the usual method described in Chapter 11: Parallel-in-serial-out (PISO) shift registers for transmission and SIPO registers for reception. These shift registers are connected to a buffer PIPO register for data holding, error handling, and so on.

Together with control logic and a register for status information, the entire transmit–receive system is called a *universal asynchronous receiver/transmitter* (UART). Status information includes the result of parity error checking, determination of an undetected stop bit, register ready signal to receive or transmit (handshaking signals), and so on. As we may note from Table 12.1, many μCs contain UARTs; otherwise, separate UART ICs are available. A UART should be used on both ends of the transmission line.

**Example 12.3**

Show the interface between a μP, a UART (8250), and the RS-232 serial bus.

**Solution** Again, the various signals will only be defined. Precise timing, handshake, and other control arrangements must be determined using the data sheets with the particular μP and application in mind. The basic diagram is shown in Fig. 12.5.

*ALE = address latch enable.* This drives the address strobe, ADS, to select the appropriate register in the UART.

*RTS, CTS = request to send, clear to send.* These are handshaking signals for signal transmission.

*DSR, RLSD, DTR = data set ready, receive line signal detect, data terminal ready.* More handshaking/verification lines.

*SOUT = TXD = serial out/transmitted data.* The serial data to be transmitted is put on this line.

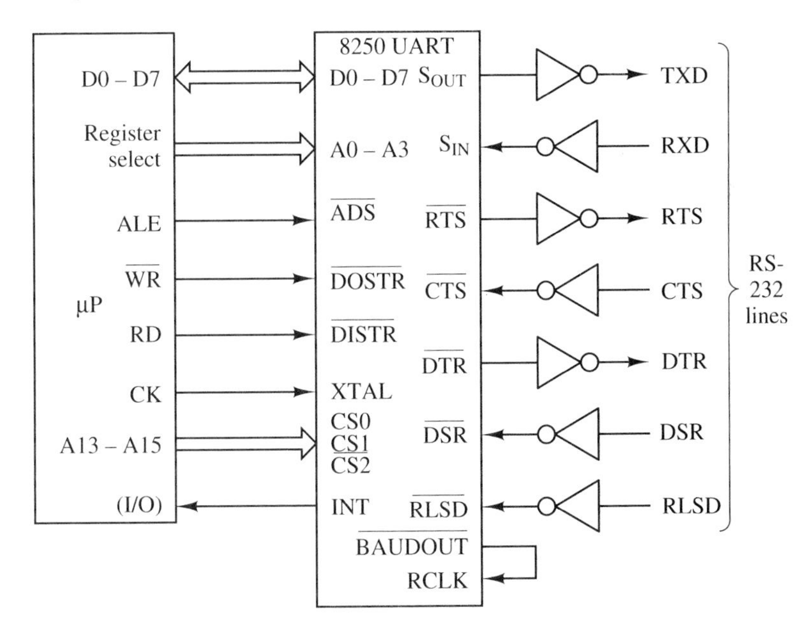


**Figure 12.5** UART connections for Example 12.3.

*SIN* = *RXD* = *serial in/received data.* The incoming serial data will appear on this line.

*BAUDOUT, RCLK* = *transmit clock, receive clock.* The transmit clock is internally generated, and typically used also for the receive clock.

*INT* = *interrupt.* Usual interrupt signal (see Sec. 12.4.7).

### 12.4.7 Interrupts

Everyone has probably had the experience of a co-worker who cannot be bothered or interrupted. While that person may make maximum headway on his or her task, other important tasks are unduly delayed. It is better for the overall project when the worker is open for questions.

The same philosophy is used in task management of computers. To take an extreme example, without interrupts a major component failure endangering human lives in a human transportation system could be ignored by the controlling microprocessor. With interrupts, appropriate responses to external events can be made expeditiously. For example, in the case of a component failure, a backup system could be connected, the faulty system removed, and a warning signal issued to the operator that the system is in need of repair. An interrupt can also serve as innocuous a purpose as indicating that an A/D conversion is completed.

Each input–output device requiring attention of the microprocessor has an interrupt request line. The interrupt request and related control lines form part of the control bus. The complete set of interrupts is prioritized, so that the computer knows which request to handle first in the event of two simultaneous requests. This conflict is not as unlikely as it might seem, for "simultaneous" only means occurring during the same machine or instruction cycle. One of several approaches to setting priorities uses the priority encoder discussed in Sec. 10.4.

The alternative to prioritized interruptions is *polling*, where periodically the processor examines each request line in turn to see whether it is high. Naturally, this is simpler but less efficient. Imagine that every two minutes your teacher had to ask every student individually whether he or she had a question; you would never get past Ohm's law! Classrooms use the *interrupt* method; the request line is a raised hand or voice. Alternatively, in some systems the polling loop can be initiated only when an interrupt has been detected.

Obviously, interrupts are asynchronous—they are not constrained to occur on the clock. A latch will catch the request and hold it until the microprocessor completes the instruction it is currently executing.

Each interrupting device will require specialized service. In ROM resides a section of code that was designed to fulfill such requests. This code is called an *interrupt service routine* (ISR). A common function of the ISR is to make the data bus available to the requesting (I/O) device for data transfers.

There will be an ISR for each device; hence the interrupt request must be accompanied by the starting address of the ISR for the interrupting device. Because the ISR starting address "points" to the appropriate starting point in ROM, this type of interrupt is called *vectored interrupt.* Vectored interrupt structures are particularly helpful if there is a large number of devices frequently requesting interrupts.

When the microprocessor responds to an interrupt, it stores the contents of all registers on the stack, including of course the *next instruction address*. This way, when the interrupt servicing is completed, the microprocessor can resume whatever it was doing before it was interrupted as though nothing had happened. It also orders the requesting device to turn off its interrupt line; this prevents the processor from becoming stuck on that device.

### 12.4.8 Programmable Timers

Any operation depending on the measurement of time can make use of *programmable timers*. For example, we may need the time it takes for 25 pulses to occur to detect the speed of a rotating axle. In another situation, the frequency of a pulse train or count of a finite series of pulses may be required. In these cases, the input to the timer is a pulse train. Alternatively, perhaps an output device cannot handle data as fast as the microprocessor tends to produce it. Then a timer can wait a suitable time for the output device before sending another piece of data. In this case, the pulse train is derived internally from the μC clock.

It is possible to program the microprocessor to "twiddle its thumbs" for the required time interval, or to do the counting itself. However, that prevents the μC from doing other more useful things. Having timers "on board" the μC is clearly an attractive feature for most control applications, freeing the CPU for other tasks.

Common uses for the programmable timer include square-wave generation, frequency measurement, interval measurement, generation of single pulses of controlled duration, pulse-width comparison, and frequency comparison.

**Example 12.4**

Show the MC6840 programmable timer IC with the lines it would run to a μP such as a 6800 μP.

**Solution** This programmable timer has three 16-bit binary counters and latches that are independently clocked and controlled. It contains three control registers (one per counter), and status and buffer registers. It can also initiate interrupts and/or single-output signals. There are eight modes determined independently for each of the three timers via the control registers. The MC6840 is shown in Fig. 12.6.

*D0–D7* = *bidirectional data lines.* The output lines are tri-state.

*CS* = *chip select.* This enables the data lines.

$R/\overline{W}$ = *read/write.* This line controls the direction of the data lines. If high, a read (data moves from timer to MPU), and if low, a write (data moves from MPU to timer for programming it).

*E* = *clock.*

*IRQ* = *interrupt request.*

$\overline{RESET}$. If low, it initializes all registers and counters. The counters count down, so the initial value of the timer registers is the maximum count value.

*G1–G3* = *gate inputs.* These function as triggers or as clock gates to timers 1, 2, and 3, respectively.

$\overline{C1}$–$\overline{C3}$ = *clock inputs.* These clocks are what cause decrements of the individual counters.

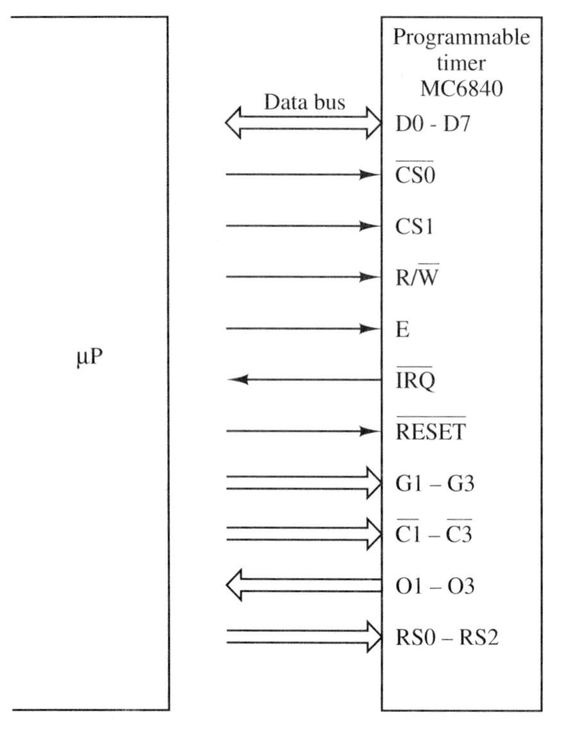


**Figure 12.6** Programmable timer connections for Example 12.4.

*O1–O3 = timer outputs.* These single-line timer outputs are determined by control register bits. They can be a controllable duty-cycle square wave or a single pulse.

*RS0–RS2 = register select.* These select the register accessed by the MPU. The data in the registers appears on the eight data lines.

### 12.4.9 Connecting a System

We have now discussed all the major elements of a microprocessor-based system. These elements include memory systems (covered in Chapter 11), I/O devices (transducers, printers, keyboards, etc.), I/O ports and other interface equipment, and the microprocessor itself. Assembling this hardware into a complete system is often straightforward, although doing so requires "glue logic." The most difficult aspects are what to do with timing lines (e.g., handshaking) and making chip select lines do what they are supposed to do, at the correct times. In short, interfacing is the challenge.

Before moving on, let us briefly discuss an actual microprocessor chip. A popular µC, the 8051, is shown in Fig. 12.7. Its various lines are described as follows.

*P0.0–P0.7 = Port 0.* During operation, Port 0 serves as either a bidirectional I/O port or as the low-order address/data multiplexed bus during read/writes from/to external memory. During programming of the EPROM, this port contains the code bytes either being written to the EPROM or being read from it for verification of the code. These lines are "open drain" and are therefore in a high impedance state when held high.

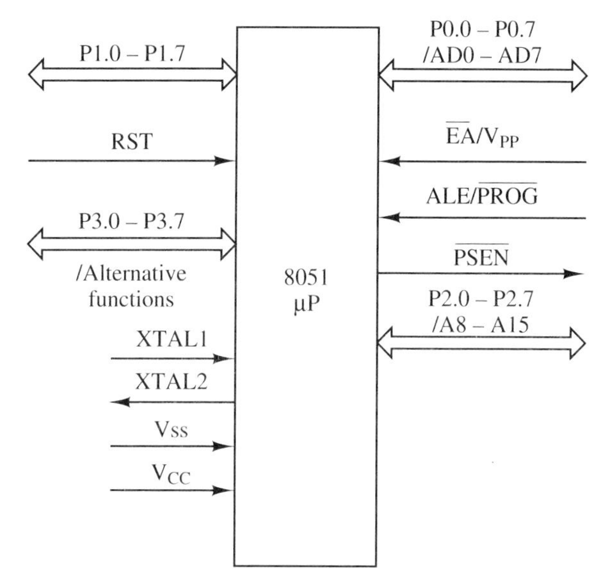


**Figure 12.7** Connections of 8051 microcontroller.

*P1.0–P1.7 = Port 1.* A bidirectional I/O port during normal operation, it has internal pull-up resistors (not open drain). During programming and verification of the EPROM, this port receives the low-order address byte.

*P2.0–P2.7 = Port 2.* A bidirectional I/O port during normal operation, it has internal pull-ups. During fetches to/from external memory, it emits the high-order address byte for 16-bit addresses; for 8-bit addresses, it contains the value stored in a special function register.

*P3.0–P3.7 = Port 3.* A bidirectional I/O port during normal operation, it has internal pull-ups. Its various bits are also used as the serial I/O port, the two external interrupts, the two timer inputs when they are being used as counters of externally generated pulses, and external data memory read/write lines (*strobes*).

*RST = reset.* A high on this pin for two machine cycles while the oscillator is running resets the μC.

$ALE/\overline{PROG}$ = *address latch enable/program pulse input.* During normal operation, this line determines when to latch the low-order address on Port 0, as opposed to its multiplexed use as the data bus. When programming the EPROM, the write pulses are given to this pin (program pulse input).

$\overline{PSEN}$ = *program store enable.* This is the read strobe line to external program memory.

$\overline{EA}/V_{pp}$ = *external access enable/programming supply voltage.* During normal operation, this line must be tied low if external program code is to be accessed; it is tied high for internal program execution. During EPROM writing, the 21-V supply voltage is input here.

*XTAL 1,2 = connect lines for the external quartz oscillator crystal.*

These are all the lines required for basic microcontroller execution. Notice how many pins are multipurpose (multiplexed). Everything is orchestrated by the program stored in ROM/EPROM. As mentioned repeatedly, interfacing the μC or μP with peripheral devices is an advanced subject, beyond the scope of an introductory course on electrical engineering.

Unfortunately, gathering and gluing together the necessary components is less than half the battle. The real killer is program writing, debugging, and documentation. We now consider a few aspects of programming.

## 12.5 MICROPROCESSOR SOFTWARE

The set of instructions required to operate a μP or μC is called the *programming language*. Today, programming a microprocessor can be done by using a high-level language such as C, a microprocessor-level language called an assembly language, or a combination of both. At the high level, programming is relatively easy, but the resulting code is long and runs relatively slowly compared with the equivalent assembly-level code. Therefore, high-level programming is fine for processing that can be done at suboptimum speed, while assembly-level programming is used to create code for time-intensive or time-critical work or when available memory is scarce.

Every microprocessor has its own assembly language, whereas C is a more general language that may be compiled to run on many microprocessors. At the assembly level, within the group of microprocessors made by a single manufacturer, software compatibility means that upgraded models should work using assembly code written for the previous models.

The steps to programming a microprocessor in *assembly language* are straightforward. First, the physical problem to be solved or system to be controlled must be defined from general goals down to specific loops and calculations. A flowchart may be valuable as a programming and documentation aid. Then the programmer must combine his or her knowledge of the microprocessor hardware and assembly language to create from these specifications an assembly language program. Typically, this is entered into a file on a personal computer (PC) having a filename ending in ".asm," for example, to indicate that it is assembly code: prog.asm would be a filename for the program "prog."

Once entered, the program is *assembled* by an assembler program on the PC. The assembler converts the programming statements into words of binary code that the microprocessor can interpret. The microprocessor could be that of the PC, or for product design/industrial control problems it might be another microprocessor-based system. In the latter case, if the PC and applications microprocessors are different, the assembler is called a *cross-assembler*. The resulting binary code is then burned into PROM and the system is ready to go.

That is, it is ready to go if there are no errors. During assembly, the assembler detects syntactical errors, but not errors of logic, which are harder to find and correct. It would be foolhardy to burn a PROM on version 1 of a program. At the very least, start with an EPROM or EEPROM so that you do not have to throw the chip away when an error is found. Incidentally, cards can be purchased for EPROM programming that fit into personal computers.

As a tool in *debugging* assembly language programs (i.e., finding and eliminating errors), an *in-circuit emulator* is often available for the particular microprocessor that will be using the program. These are actual hardware printed circuit boards that often can be plugged into a PC or into the microprocessor socket of an existing microprocessor-based system which simulate and report everything that would be going on in the actual microprocessor hardware, were it connected. During this simulation, contents of registers, memory, and the step-by-step sequence of instructions can be examined at any point in the program. The only hitch is that an in-circuit emulator costs thousands of dollars and is valid for only a single microprocessor.

A particularly popular high-level programming language for microprocessor applications is the "C" language. One reason for this is that it is easy to "get at" specific I/O and memory locations just as is required in assembly language programming for microprocessor systems. For example, the C language has addressing modes much like those of assembly language. Therefore, C is, in the programmer's eyes, a shortcut.

Being a high-level language, it has many advanced functions available from *function libraries* as well as FORTRAN-like constructs such as IF/ELSE, FOR, and DO WHILE. Also, compared with other high-level languages, C code is relatively short, highly modular (making it easy to understand), and *portable* (the same code can be made to run on other computer systems with little additional work). Thus the current trend is toward writing most of the code for the microprocessor the easy way—in C—and spending the more laborious time involved in generating assembly language on only the innermost loops and speed-critical code.

The steps involved in creating a C program are similar to those required for assembly programming. Using the editor on the PC, create a file containing the C program; the filename must end with ".c". Within the PC is the code necessary to compile the program and to link it with prewritten code for functions or routines called by your program that exists in a section of PC memory called a *library*. During compilation, the code is internally converted to assembly code, and then assembled. The compiled and linked code is executable, meaning that it is in binary form suitable for running on the microprocessor. Again, the microprocessor could be that of the PC, or for product design/industrial control problems, it might be another microprocessor-based system.

Mixing C and assembly code is as easy as writing the main C code and replacing the lines or sections to be written in assembly language with assembly code. The assembly code can (on some compilers) be inserted "in line"—right in the C code, with a label indicating that it is assembly code. In this case the C compiler will hop right over this code as "already compiled." Alternatively, the assembly code can take the form of external routines that are linked up with the main program after compilation of the C program.

Because assembly languages are microprocessor specific, the C compiler/assembler will also have to be microprocessor specific. This, combined with different assembly languages and in-circuit emulators for each microprocessor, should make clear why code developers like to stay with the same microprocessor for different applications. Although compilation is a two-step process in which C is translated to assembler and then the assembler is translated to binary, often this is all done via one compilation command issued by the programmer.

Finally, packaged software is developed and marketed for particular microprocessors (usually, those in personal computers), which implements commonly used tasks such as word processing, spreadsheets, and data base management. Often, these programs are written in C and then translated to *machine code* for the specific μP. Such programs are usually very complicated, yet they are very easy to use. The ease of use is achieved by presenting menus that list on a video screen the most commonly desired possible actions (edit a file, print it out, etc.). We are reminded of the op amp: Because of a very sophisticated design (which would be difficult for anyone other than circuit designers to comprehend), it is extremely easy to use.

A non-EE is not at all likely to have to deal with the details of personal computer operating systems or packaged software at any level below the user level. More likely, he or she will encounter *embedded* or dedicated controller hardware on the job, particularly as it pertains to control of mechanical or chemical systems. It is therefore helpful to have at least a rudimentary idea of how a program may be used to direct the actions a microprocessor takes. As discussed above, we may write programs that direct microprocessor control in either assembly language or an upper-level language such as C. For brevity, however, we must end our discussion of software here.

## 12.6 THE CONTROL CONCEPT

The primary use of μPs and μCs for nonelectrical engineers is in process control. The regulated parameters under control could be the temperature of the process, the flow rate of a fluid, the time the process is allowed to continue, the position and movement of robotic "arms," the position or speed of a motor, or the concentrations of chemical components in a mixture; the list could go on indefinitely.

We have already studied one electronic control system: The operational amplifier with negative feedback connections. In that case, the regulated parameter was the output voltage and the system under control was the open-loop op amp. The negative-feedback structure served to guarantee that the output voltage remained equal to the input, scaled by the closed-loop gain. The error signal was $v_+ - v_-$.

In a typical microprocessor-based control system, a physical system rather than an electronic amplifier is under control. Figure 12.8 shows a block diagram of a general control system configuration. The physical process to be controlled and its outputs are monitored by a collection of sensors. Measurement signals from sensors that are inherently digital are sent directly into the microprocessor controller. Analog sensor signals are sampled and converted to digital representations by an A/D converter. A single A/D converter can be multiplexed to handle several inputs; it is cyclicly connected to each input in turn for conversion.

Based on the measured data, a model of the physical system, and any user inputs, the microprocessor determines the appropriate response to send to the physical system to achieve the desired behavior. This collection of responses is sent to the actuators, which again may be a mixture of digital and analog devices (relays, solenoid valves, motors, voltage levels, etc.). In the case of analog actuators, the digital microprocessor output must be passed through a D/A converter. Also, the microprocessor may present the (human) user with a display of important process

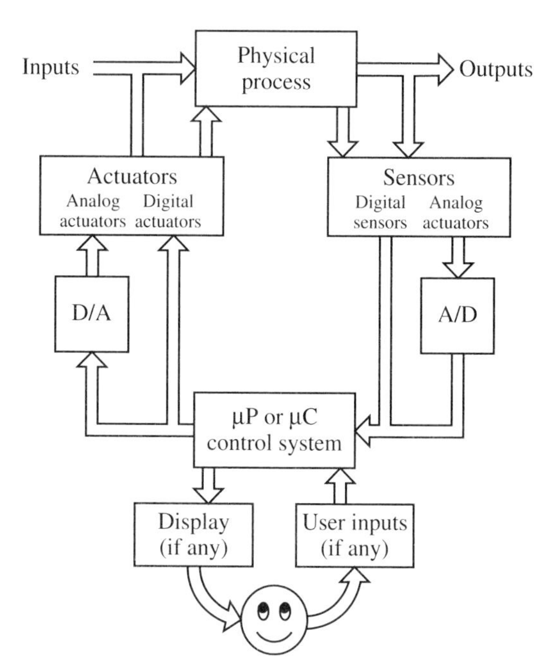


**Figure 12.8** Elements of a control system.

data in alphanumeric or graphical form using alphanumeric displays, video screens, or printers.

Negative feedback is used to reduce the error between the actual physical and desired behaviors. The desired behaviors may either be specified at the time of operation by the user, or they may be stored in ROM for fully automated systems. Also, the values of the model parameters are stored in ROM. Adaptive systems such as those performing automotive engine control can actually change the model parameter values. This is desirable when experience indicates that due to wear, age, or other conditions the old parameter values are no longer accurate.

Computations involved in providing the required feedback response are carried out by the microprocessor. Thus the calculations usually must be fairly simple to guarantee real-time control. That is, the response must be available quickly enough in relation to the dynamics of the physical system that it is effectual and that the system remains stable. Of course, the sensors and actuators also contribute to this delay.

What sorts of control responses are feasible? There is clearly a trade-off between speed and accuracy. The fastest and simplest computationally is the on–off response, or "bang-bang" control. Combined with hysteresis (recall the Schmitt trigger, for example), this type of control is very common in practice. When a threshold is reached, the corrective input (such as the flame in a gas heating system) is turned completely on. When a lower threshold is crossed, the input is turned completely off. In a gas heating system, the conventional thermostat relies on the curling and uncurling of a metallic spiral in response to temperature changes to flip a mechanical (mercury) switch on and off. The switch then powers a relay which opens/closes the main gas valve.

Unfortunately, the on–off control system has limited ability to maintain precise control, precisely because of its all-or-nothing approach. In fact, overshoot is a very likely consequence. There is also a delay in reaction; correction takes place only after the error grows to a sizable amount. For home heating, these imperfections are usually tolerable. For industrial process control, they frequently are not.

If the control reaction is made proportional to the error rather than fixed, then even very small errors will cause a corrective reaction, and the correction will be more immediate. This type of control is known as *proportional control.* Of course, there still will exist delays between the controlled variable and the control system corrective response.

Presumably, the reason a feedback control system is necessary is the presence of unavoidable disturbances or other variability in the physical system. Without any such disturbance, an open-loop approach would suffice. The disturbance may be slowly or rapidly varying compared with desired dynamic behavior, or it may be a combination of rapid and slow variations.

If the variation of the disturbance is slow or the disturbance is constant, it will tend to cause the output variable to drift. The proportionality factor required to overcome this type of disturbance may be very large—so large that the control system may not be able to produce it. Even if it could, the inevitable delays combined with the large control responses would cause instability.

If, instead, the *rate* of correction is made proportional to the error, the constant disturbance can eventually be canceled, without the large unstable proportional gain. This type of control is called *integral control.* The longer an error has been present, the larger will be the required correction. Of course, if the disturbance has a more rapidly varying component, the integral control cannot respond rapidly because it is based on an average of past errors rather than the instantaneous error. Thus a combination of proportional and integral control (PI, or *proportional–integral control*) is commonly used. In fact, automobile cruise control systems use PI control.

Actually, even proportional control may not be sufficient to respond to rapid disturbances. For these, an anticipatory response is required. That is, if the rate of change of the error is high, an extra "oomph" can be generated to cancel the predicted (or projected) larger error that would be impossible for the proportional control to anticipate or follow. But this is nothing other than including in the control response a term proportional to the derivative of the error.

A system combining all of these components is called a PID (*proportional–integral–derivative*) *controller.* The proportional control responds to dynamic disturbances, the integral control corrects for offsets to achieve steady-state accuracy, and the derivative control damps out high-frequency noise and improves stability.

When implemented using microprocessors, all the operations must be carried out in discrete time. Consequently, integration becomes a sum and differentiation becomes a difference. In equation form, the corrective response $u(n\,\Delta t)$ to the error $e(n\,\Delta t)$ at discrete time $n\,\Delta t$, where $\Delta t$ is the sampling interval, is

$$u(n\,\Delta t) = K_1\left\{e(n\,\Delta t) + K_2 \sum_{m=0}^{n} e(m\,\Delta t) + K_3[e(n\,\Delta t) - e((n-1)\,\Delta t)]\right\}. \qquad (12.1)$$

The values of $K_1$, $K_2$, and $K_3$ are determined in practice by various modeling strategies outside the scope of this course. To prevent wandering, $K_2$ may be reset to zero if $u(n\,\Delta t)$ saturates. This equation can also be written in a recursive form in which the summation is replaced by adding on the new error $e(n\,\Delta t)$; the time-consuming sum need not be repetitively recomputed.

More advanced control strategies are possible and are in fact used in demanding controls applications. The state variable formulation, in which a cost function is minimized, results in a matrix equation for the proper response that may be solved by computer. Of course, this is usually beyond the capability of a small, low-cost system. Nevertheless, automotive emission/fuel injection control systems use results of optimal control theory.

## 12.7 APPLICATIONS EXAMPLES

### 12.7.1 Introduction

Applications of the microprocessor are limited only by the ability of the imagination to recognize a complex system in need of control. Transportation engineers would immediately point to the possible introduction of microprocessors in complex passenger conveyor systems (lifts, escalators, elevators, moving sidewalks, etc.) in situations such as airports. The microprocessor can be used in advanced systems to respond to a wide variety of traffic, load, machine failure, and positional situations. Traffic stoplight systems on roads present extremely complex variations on this theme. All sites can be scanned and contribute to overall system behavior many times each second.

To "bring home" at least a few of the topics covered in this chapter, we consider briefly two applications of microprocessors: the microwave oven and automotive engine control. In both cases, acquiring detailed information about the electronic circuits is not possible, due to the proprietary nature of today's consumer electronics. Even small portions of circuits are kept confidential by all sorts of legal protection. Furthermore, we will *never* gain access to the assembly code directing the operations, so only educated guesses can be ventured about actual control sequences. Nevertheless, many of the issues raised previously in this chapter come into play in these examples.

Incidentally, the presence of a keypad on any consumer product indicates that the product contains a microprocessor. Examples are the modern microwave oven and remote control televisions. Other microprocessor-based instruments, such as automotive engine control units, do not require human numeric input and therefore do not have a keyboard.

### 12.7.2 Microwave Oven Control

The method of cooking used in microwave ovens is the rapid pushing of food molecules back and forth using a high-frequency electromagnetic field. Fundamentally, energy is transmitted from a magnetron to the food via electromagnetic waves traveling through air (see Sec. 3.10.4).

A *magnetron* is a device that combines large-magnitude electric and magnetic fields to drive electrons into a resonance chamber. The walls of the resonance chamber act as an anode of a high dc voltage that attracts electrons emitted from a nearby heated filament within the magnetron. They also act as a parallel rc network that responds to oscillations at its geometrically determined resonant frequency, 2.45 GHz.

The designed choice of resonant frequency forms a compromise between the absorption coefficient of microwave energy for foods and therefore cooking speed (which rises with frequency) and penetration depth into the food (which decreases with frequency). At 2.45 GHz, a typical effective penetration depth is about 2 inches; from there to the rest of the food, heat conduction from the outer layer cooks the inside.

The high-frequency oscillations induced in the resonant cavity are created by the combined forces of the electric and magnetic fields on the emitted electrons, which causes them to swirl rapidly in an outward spiral. With induced currents at 2.45 GHz in the anode, there must be an accompanying electromagnetic field at the same frequency. This field is received by an antenna within the cavity designed to respond maximally to that frequency. From there, the 2.45-GHz signal is directed by a waveguide to an inlet to the cooking chamber, distributed evenly in the cooking chamber by a fan, and into the food directly and via single and multiple reflections from the cooking chamber walls.

Where does the microprocessor come into play? A modern microwave oven has many features controlled by a microprocessor. These include multiple "power levels" that we may select to control the rate of cooking (including defrosting); preprogrammed "menus" that automatically select power and cooking time for common foods according to the type, size, and desired doneness of the food; temperature control that heats up food to a desired temperature and keeps it there by automatically turning the microwave power on and off; and various other timing modes, such as turning the oven on at a particular time of day. These are all selected by means of a keypad, typically having 32 keys.

The keypad, which is arranged in a rectangular array, is scanned by the microprocessor. Thousands of times per second, a cycling through all rows is performed, sending a high pulse to all buttons in each row simultaneously. All columns are monitored continuously. If a column goes high, which can happen only if its button on the currently selected row is high, the microprocessor can determine which button has been pressed.

The effect of a pressed button is to aid in the programming of the oven for the next cooking cycle. Also, a button such as "cancel" can interrupt the cooking cycle. This sort of programming changes with each use of the oven, so it is stored in read/write memory (RAM) that resides on the μC chip. On-chip ROM contains the basic program for carrying out the functions selected by the user. It also contains data such as the time required to cook two ears of corn-on-the-cob.

A common display system for microwave ovens is the *vacuum fluorescent display*, which is very readable in daylight. It operates in much the same way that a TV screen does: Electrons are emitted from a heated cathode and accelerated to the anode, where they strike a fluorescent material to produce a glow.

One vacuum fluorescent unit for time display uses seven-segment displays, and another unit uses 14-segment displays for alphabetic display. Because it requires a higher voltage (e.g., 28 V or more) than the CMOS μC (5 V), conversion circuitry between logic voltage levels may be required. Such circuitry is provided by a combination of on-chip resources called *vacuum fluorescent display drivers* and off-chip voltage divider-type configurations that convert the output from the display drivers back down to usual CMOS levels. The vacuum fluorescent display also requires a low-amplitude sinusoidal voltage generated at the low-voltage transformer to drive its filament.

The same lines that drive the display at high voltage levels may also be used to sense the keypad (via time multiplexing of the lines), which must be fed back to the microprocessor at 5 V. This procedure makes best use of the limited number of pins on the μC chip.

Depending on the model, there may be one or two 42-pin μC chips (one may be dedicated to driving the displays), or one larger chip (e.g., 64 pins). Typical specifications and features are a four-bit data word, eight-bit instruction words, 100 basic instructions, 200 nibbles of RAM, and 4K bytes of ROM, 0.3-MHz clock, 3 to 10 μs/instruction, four interrupts, 32 I/O lines, and on-chip eight-bit timers and display drivers. Example μC numbers are NEC's 7508 and 7537, and Motorola's 6805.

One of the interesting portions of the electronic control unit is the temperature control circuit. Using a thermistor probe (see Sec. 9.3.4) inserted into the food being cooked, the temperature of the food can be ascertained. The resistance of the thermistor decreases with temperature (e.g., 7 to 1 kΩ over a 100°F range), resulting in a voltage representative of food temperature. This voltage is compared with one representing the desired temperature. To form the voltage representing the user-specified desired temperature (stored in binary form in RAM), a simple D/A converter involving a resistor network is used. If the actual temperature is lower than the desired temperature, the magnetron will be turned on; if it is higher, the magnetron will be left off.

The magnetron is either on full power or completely off. Controlling the duty cycle is the only method of controlling the average heating power. If the temperature probe signal climbs slowly and smoothly above the desired temperature reference level (set point) and mere thresholding is used, overshoot will occur as well as delay in the desired response. Now the desired behavior is the opposite of that achieved using hysteresis. To maximize efficiency, the magnetron is continually turned on and off for very short periods when the temperature is close to correct, and left on or off for longer intervals when the current temperature is far off. This does not harm the magnetron.

This is accomplished by adding to the temperature probe signal a small magnitude oscillation (see Fig. 12.9) before thresholding. The effect, called *dithering*, is that if the temperature is centered on the set point, the duty cycle will be 50% because the oscillation spends half the time positive and half the time negative.

Other sections of circuitry attend to basic tasks such as turning the magnetron off when the door is opened, sounding a crystal-driven beeper oscillator when the cooking cycle is over or (for example) when the user must turn food over, high-voltage generation for the magnetron, initializing the digital circuitry, and browning the food (if this feature is included).

Another interesting fact is that the oven uses the line frequency (60 Hz) to

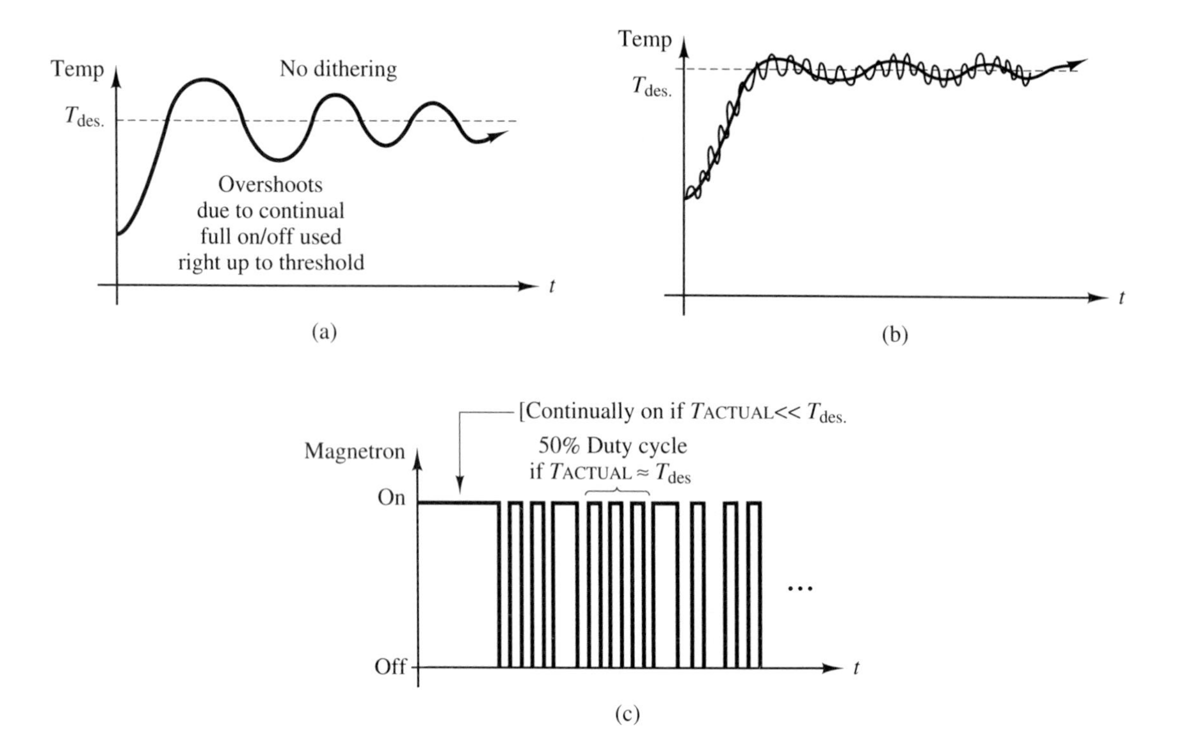


**Figure 12.9** (a) Transient temperature response of microwave oven without dithering; (b) with dithering; (c) on–off time dependence of magnetron.

generate "wall clock" time for its clock and cooking timing. (This is also true for other ac-powered digital time clocks.) The clock signal frequency is derived directly from the ac wall voltage frequency because it is very stable over long periods of time. From the sinusoidal 60-Hz voltage, a transistor or op amp is saturated to form a square wave suitable for the μC.

The circuit generating the supply voltages to drive all circuitry is a linear single-transistor power supply rather than a switching power supply because that is the cheapest solution in this case.

### 12.7.3 Automotive Engine Control

Although microprocessors were introduced as early as 1971, the first microprocessor to appear in an automotive application was not until 1977. This was an electronic ignition system for Oldsmobile. By 1980, Ford was using the 6800 μP, a microprocessor appearing in Examples 12.1 and 12.4. As early as 1981, Delco/Motorola was supplying two-board, 22-chip electronic control unit systems for engine control in the Cutlass model.

Today, 32-bit μCs are being used in these systems. An average car now includes half a dozen or more microprocessors, which govern braking, suspension, transmission, steering, and voice informational systems. In the future, microprocessors will be used for such advanced capabilities as navigational aids and collision avoidance.

At the time microprocessors were introduced, mechanical engine control sys-

Spark plugs have not changed dramatically in appearance since 1909, but the advertising images have.

tems were highly developed. However, pollution and fuel economy standards set by the American people through their government demanded better controls than these systems could produce. Initially reluctant to change, manufacturers of automobiles are now very excited about these applications of microprocessors. This is true to the point that a major concern is whether sufficient wiring space and materials is practical for all the new things they want to do with microprocessors.

The major impetus has, however, been pollution. Although toxic pollutants make up only 1% of the volume of exhaust gases from cars, that 1% is deadly. There are three main toxins; they have the following effects on people exposed to significant doses. Carbon monoxide (CO) is attracted to blood cells more than is oxygen; hence the person is deprived of oxygen. Hydrocarbons (HC), major contributors to smog and low-altitude ozone, are carcinogenic, are nerve toxins, and irritate mucous membranes. Nitrogen oxides and dioxides (generically labeled $NO_x$) cause rapid central paralysis, are blood poisons, and damage lung tissues as well as contribute to smog.

The more people drive, the more they expose themselves to these poisons. To maintain the lifestyle made possible by cars and stay in business under the new regulations, the automobile manufacturers began to use microprocessors as a way of optimizing engine performance.

As mentioned above, the first critical control to be implemented with microprocessors was in electronic ignition. The first electronic ignition systems used only power transistors to realize the on–off switching of the ignition coil. However, the precise timing (optimal "advance angle") of this switching significantly affects emissions and efficiency. Furthermore, the optimum timing depends on both the speed and the load in a very nonuniform manner. Fortunately, both the load and speed can at any instant be estimated.

Figure 12.10 dramatically shows the difference between the old mechanical advance and the microprocessor-determined advance angle. The height on the three-dimensional map is the advance angle, which is plotted against load and speed. While the mechanical system was able to crudely estimate the advance angle, the simple surface shown does not represent the optimal surface.

Contrarily, the microprocessor-based system is capable of using the true complicated relationship for all speed–load conditions. All that is required is to have the map stored in sequential memory as a look-up table, and then to decode the instantaneous speed–load measurement pair to the corresponding address of the look-up table. The speed and load measurements can be obtained by associated transducers placed within the engine.

The optimal timing values are determined in the factory by a process known as *mapping*. Mapping is merely subjecting a car to all possible load/speed/advance conditions and measuring such relevant quantities as pollutants and efficiency. The optimal advance would be that for which a chosen combination of dependent variables is considered "best." Of course, there are complications, such as the

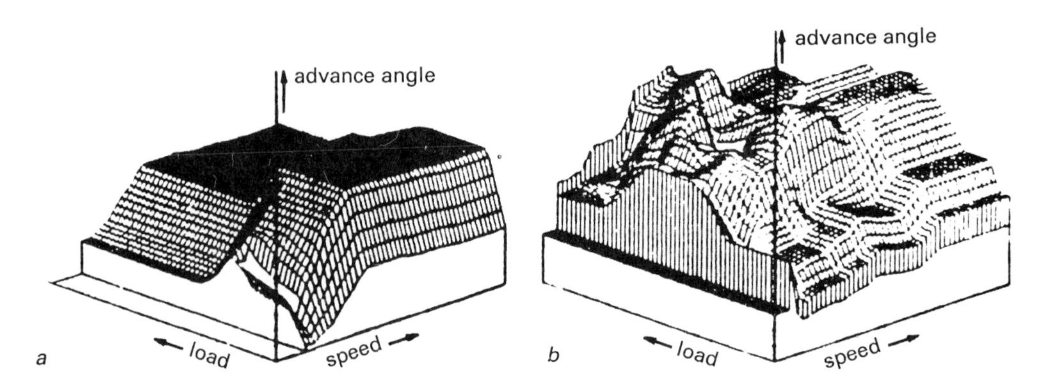


**Figure 12.10** (a) Three-dimensional surfaces relating best obtainable ignition advance to engine load and speed for ignition system without microprocessor control; (b) much more detailed relation achievable with a microprocessor system; this truly optimal dependence is stored in the microprocessor memory. *Source:* IEE Proceedings, Vol. 133, Pt. A, No. 4, June, 1986 from "Automobile Electronics" by M. H. Westbrook.

phenomenon called *knocking*, that can occur under an otherwise optimal advance. Knock sensors are employed to detect the onset of knocking so that corrective and preventative action can then override the look-up table values.

The other major engine control directly affecting emissions is that of the air/fuel mixture. In the automotive internal combustion engine, the optimal ratio of air to fuel is 10,000 : 1 for volume (e.g., gallons) and 14.7 : 1 for mass or weight (e.g., pounds). Traditional carburetor systems do not meet this ratio in practice because of a number of mechanical limitations.

A microprocessor-controlled fuel injection system can make use of closed-loop control of the air/fuel mixture to achieve the optimal ratio. The feedback is provided by a sensor measuring the concentration of oxygen in the exhaust. Using this information, the microprocessor computes the necessary time within a fixed cycle period to hold open the fuel solenoid valve(s). It then holds high for this time duration a digital line that actuates the valve. The duty cycle (on-time/off-time) translates to a known corresponding air/fuel ratio. Hundreds of calculation cycles can be done in 1 second. The volume of this controlled air/fuel mixture reaching the combustion region (determining how fast the engine runs) is, of course, determined by the accelerator pedal via the throttle plate.

Naturally, much more goes into the calculation of the fuel injector duty cycle than just the exhaust oxygen concentration. Other factors include the loading as indicated by the intake air flow sensor, the temperature both within and outside the engine, the engine speed, idle/severe acceleration demands detected by the throttle plate position sensor, and possibly other input measurements. In a mapping procedure similar to that described for ignition timing, appropriate formulas and/or look-up tables for the required "metering" (controlling of fuel injection) are derived for the particular engine design in question.

Further complicating the situation is that special operating conditions such as startup, warm-up, severe loading, breaking/downhill operation, and idling all have particular engine requirements. These requirements sometimes preclude closed-loop operation and often necessitate use of air/fuel ratios other than 14.7 : 1.

For example, the exhaust oxygen concentration sensor does not function properly when cold; during startup and warm-up open-loop control is mandatory. When starting the engine, much richer mixtures are needed (lower air/fuel ratio) because of problems making the fine gasoline spray suitable for burning when the temperature is cold and due to the slow cranking speed.

During severe loading, again richer mixtures are needed to meet the performance demanded by the driver; this temporarily overrides the goal of maintaining minimal pollution. During idle, smooth engine operation and avoidance of stalling are the primary concerns. During braking or downhill, *overrun cutoff* interrupts the supply of fuel to the engine to reduce fuel consumption and help brake.

The sensors provide all the information needed for the microprocessor to determine in which condition the engine is running. Because of the rapid computations, the responses to new situations are smooth and quick.

One other engine control function other than ignition timing and fuel injection is exhaust gas recirculation. If part of the exhaust gas is fed back to the air intake, the cylinder temperature can be reduced because the exhaust does not burn; this reduces $NO_x$ emissions. Once again several factors enter into the precise amount of exhaust to recirculate. The microprocessor uses both sensor data and stored characteristics to determine the setting of the recirculation valve.

As noted previously, both μCs and μPs are used in automotive electronics. Engine control is the most complex and requires a multichip system. The microcomputer system consists of the μP, RAM, ROM, PROM, A/D and D/A converters, timers, I/O bus control systems, pulse shapers, op amps, and so on. It also often has a form of memory called *keep-alive memory* (KAM). This is merely RAM that remains energized via the car battery after the engine is turned off. With this read/write, mostly nonvolatile memory (nonvolatile until the car battery is disconnected), the microprocessor system can actually adapt its controls to changes in the engine due to age and wear. It merely detects nonoptimal behavior and changes stored engine parameters accordingly.

## 12.8 SUMMARY

This introductory look at microprocessors began with a summary of several of their applications areas. Two major components make up a microcomputer system: hardware and software. We noted that hardware can be studied at many levels of abstraction: physical construction (the chemical layers on a chip forming the transistors and other components on a chip), transistor behavior, gate connections, IC structure and interconnection, system level, and even multiprocessor interaction. Similarly, software can be studied and programmed at the levels of machine code (1/0 combinations), assembly code, high-level language such as C, and the user level of software packages.

The levels most commonly of interest to microcontroller systems are the IC structure and interconnection levels for hardware and the assembler or C language levels of software. Software for microprocessors tends to become complicated very quickly when interaction with peripherals is involved—as it invariably is. Therefore, its development is best left to specialists. Nevertheless, a brief look at the programming process was presented to give a flavor of what is involved in directing a multipurpose microprocessor to perform a specific task.

A microprocessor-based system is nothing other than a small computer system interfaced with the transducers, actuators, and other input–output devices that allow the system to detect and respond to a dynamic system it is observing or controlling. To get an idea of the range of computing power available from today's computers as well as to put the microprocessor in proper perspective, a few statistics were reviewed. While the computing power of a microprocessor is very low compared with monstrous supercomputers, it is small, cheap, and sufficient for many industrial needs.

The main components of the microprocessor itself are its arithmetic and logic unit, specialized registers and control logic, clock, RAM and ROM (if on-chip), and its three communication buses. A microcontroller (μC) has all of the above plus extras such as timers, serial I/O controllers, and other interface equipment such as D/A and A/D converters. It is essentially a complete computer system (excepting peripherals), with capabilities for off-chip expansion in many cases.

A microprocessor (μP) does not have these extras and may not even have any on-chip RAM or ROM. But it has faster computational potential, and much greater memory expansion capability than the μC. The μC is used for embedded controllers typically appearing in consumer appliances; the μP is for general-purpose high-speed computational systems such as personal computers and workstations or more advanced controlling systems.

Other than speed and memory capacities, the main distinguishing features among microprocessors are the widths of its data and address buses and its instruction set. The choice of a particular microprocessor can be a complicated process if we try to optimize the selection on many fronts. The decision is sometimes simplified by overriding concerns such as previous acquisition of development systems for and experience with a particular microprocessor family.

From these general concerns, we then focused on particular hardware devices and software languages. Buses, programmable I/O ports, transceivers, and transparent latches were all described and illustrated with specific examples. Not only are there the three buses internal to the microprocessor, but in extended systems there may be I/O buses connecting peripherals to the data bus, system buses connecting the μP board to memory, I/O, and other boards, and local/national area networks that hook together computer and microcomputer systems.

Multicomputer network systems and many peripheral devices communicate using serial transmission lines. The speed of data transfer is measured in bits/second, or baud. Standard protocols and hardware configurations such as RS-232 are generally followed, which makes serial communication hardware design much easier than it otherwise would be. The standard hardware interface device is called a UART.

When a peripheral device needs access to the data bus to transmit some data to the microprocessor, it must grab the attention of the microprocessor so that it will open the data bus to the device and properly direct the flow of data. Interrupts are the procedure for doing this. For each different type of device, the microprocessor has a special interrupt service routine to handle the request. As it does for all routines, the microprocessor saves the contents of all its registers on the stack so that when the routine is finished, it can begin where it left off before the interrupt.

Programmable timers are handy devices for measuring time intervals and implementing desired control sequences. Direct memory access allows rapid transfer of data between peripherals. This is accomplished by circumventing the microprocessor; the microprocessor relinquishes the data and address buses to the peripherals and DMA controller (DMAC). DMAs are used more in μPs than in μCs, although some μCs do have an on-chip DMAC.

The process of writing, testing, and implementing computer instruction code was then considered briefly. For readers who have already programmed in upper-level languages such as FORTRAN, assembly language has some unfamiliar aspects. Specification of particular registers and memory locations is not necessary in high-level languages, but it is in assembly. Consequently, assembly language has a flexible set of addressing modes: immediate, direct, register, indirect, relative, and so on. Because one line of assembly code translates directly to one binary instruction, even simple operations may require many lines of code.

By comparison, an upper-level language such as C is easier to read and write but is translated to much longer assembly code (by the compiler) than would be necessary in direct assembly coding. These languages have DO loops, IF/THEN structures, and many math functions available to expedite the writing of code. Today, a mixture of C and assembly code is typical, with only the most computation-intensive sections written in assembly.

To place this short study of microprocessors into perspective again, the basic ideas of simple control were reviewed. On–off control and proportional/integral/derivative control methods are the most common control techniques currently used in industrial control. Optimal control techniques are used in critical situations such as aircraft/spacecraft, engine controls, and nuclear reactors.

Finally, two examples of the use of microprocessor in familiar consumer products were considered briefly: the microwave oven and automotive engine control. In the microwave oven example, on–off control (cooking time and duty cycle) can be selected numerically using a keypad or chosen automatically from a menu of foods. Here the translation from user input to output digital control waveform is fairly straightforward.

In the automotive engine control application, the controlled variables are connected in more indirect and complex ways to the user input (gas pedal). In both cases the microprocessor uses sensors, be they a keypad or an exhaust gas concentration transducer, to determine instantaneous information required to infer the proper control output.

Anyone working today in consumer product research and development is likely to come across digital systems. For the more laboratory-oriented engineer, the microprocessor system used will probably be that in a personal computer. In this context the microprocessor can be used to automate experiments, record reams of data, and perform control functions analogous to those described above.

Finally, in manufacturing, robotic operations are increasingly commonplace. Here the actuators may be stepper motors and the sensors might be position, angle, and speed transducers. It is hoped that this three-chapter introduction to digital electronics has been interesting and will later become practically useful.

## PROBLEMS

**12.1.** At what distinct levels may hardware be studied?

**12.2.** At what distinct levels may software be studied?

**12.3.** What does "nonvolatile" mean? Which of the following memory devices are nonvolatile: RAM, ROM, PROM, EPROM, EEPROM, magnetic tape, and hard disk?

**12.4.** Name at least four examples of computational jobs that computers can perform efficiently that would be impractical for manual human computation.

**12.5.** Name three computer systems and qualitatively compare their speeds. If you own a programmable calculator, try to find out the number of operations it performs per second. If you work on a PC, ask an expert to provide some statistics on how fast that PC is relative to other available computers. What are your computational requirements? How fast a computer and what memory capacity are necessary for you to do your work efficiently?

**12.6.** Give a typical, rough conversion between $N_{\text{flop}}$ = number of floating-point operations per second (flops) and $N_{\text{inst}}$ = number of instructions per second.

**12.7.** In Section 12.2 we discussed the calculation speed of the Cray Y-MP8/684. If a scientific computation requires 3000 additions, 4000 subtractions, 500 multiplications, and 600 divisions, estimate how long it will take the supercomputer to perform the calculation. Assume 60 instructions per floating-point division.

**12.8.** *Gone With the Wind*, a famous novel by Margaret Mitchell, has about 1300 pages in its popular paperback version. Assume that each page has 50 lines of text, and each line contains, on the average, 60 characters. If it takes 1 byte to store 1 character, how many bytes of storage would be necessary to store the book electrically? How does this figure compare with typical RAM space on a modern PC?

**12.9.** A paperback novel contains 0.8 megabyte of information. How many such books would fit in a Silo magnetic memory storage system?

**12.10.** Review the meanings of the terms *RAM* and *ROM*. What are their uses in a computer system?

**12.11.** On Planet Blarg, computer technology has developed that is remarkably similar to ours, except for one crucial factor: A bit in a Blarg computer can have one of three states (low, medium, or high) rather than only our two (low or high). A certain Blarg PC is advertised (on Blarg) as having 1 Mbyte of RAM. How much RAM would an Earth PC need to have a similar memory capability?

**12.12.** What does the CPU represent, and what systems does it contain?

**12.13.** What are the differences between a μP and a μC? List three models of μP and μC from Table 12.1. Why and when would we use a single-chip computer? What systems does a single-chip computer have on chip?

**12.14.** What is an "8-bit machine"? Does this refer to the data bus or the address bus or both?

**12.15.** How many distinct buses does a μP or a μC have? What are they called, and how are they used?

**12.16.** What is the difference between a data bus and an I/O bus?

**12.17.** How does increasing the data bus width also increase the speed of the microprocessor?

**12.18.** How does increasing the address bus width also increase the speed of the microprocessor?

**12.19.** **(a)** How many memory locations can be selected using a 16-bit address bus?
**(b)** If we have 64 memory locations, how many address bus lines are needed? In both parts, generalize your answers to the case in which the specified parameter is $n$.

**12.20.** What is the purpose of virtual memory?

**12.21.** Table 12.1 lists several μPs and μCs. From the table, determine the machine types and data bus widths of the Intel 80386, Motorola 68000, and Texas Instruments 7000. Which would be most appropriate as a simple dedicated controller?

**12.22.** Must bidirectional transceivers be used on the address bus? How about on the data bus? In each case, state why. What if the data and address lines are time-multiplexed on the same bus?

**12.23.** What is the use of a driver? Why do we use open-collector drivers?

**12.24.** What is a transceiver? What is a latch?

**12.25.** What is a transparent latch, and when would it be useful?

**12.26.** Using two AND gates, an inverter, and an RS flip-flop, construct a D latch and briefly explain its operation.

**12.27.** What is a system bus? What is the commonly used protocol for communication through system buses between connected devices?

**12.28.** What is the largest integer number that can be stored in the EPROM 2716 in Fig. 12.4? (Express your number in base 10.) How much memory (number of bits) can be stored in that memory chip?

**12.29.** What is a programmable pin on an IC chip? Name an example from the chapter.

**12.30.** Give an advantage and a disadvantage of time-multiplexed lines on a microprocessor or peripheral IC.

**12.31.** List some names of bus standards currently in use (both parallel and serial).

**12.32.** What is a main use for handshaking other than establishing connections between the I/O device and μP?

**12.33.** When and why is serial transmission used instead of parallel transmission?

**12.34.** A serial port is examined and found to have three wires. What is the purpose of the third wire?

**12.35.** Compare the two serial data transmission standards: RS-232 and RS-422. Why is RS-232 more susceptible to noise?

**12.36.** If we wish to use a 9600-baud serial bus between devices that are 500 ft apart (without using repeaters), should we use the RS-232 or the RS-422? Why?

**12.37.** Why are interrupts so important in computer systems? Describe the interrupt process.

**12.38.** What are the differences between "polling" and "interrupt" for requesting services of the CPU?

**12.39.** What are the meaning and usage of a vectored interrupt, and when is it efficient to have this feature?

**12.40.** What is "glue logic"?

**12.41.** Programmable timers such as the MC6840 explained in Sec. 12.4.8 are very flexible. From where does the flexibility come? What are some operations that a programmable timer can perform?

**12.42.** How might timers fit into a handshaking system?

**12.43.** What is "DMA," and for what is it used?

**12.44.** By consulting a microprocessor data manual, answer the following questions. When a DMA controller is working, can the CPU access memory? Why or why not? What does the microprocessor do during a DMA data transmission? Must it write its register contents onto the stack?

**12.45.** List all the types of data that may be found on port 0 of an 8051 μC, and in which direction(s) the data may go.

**12.46.** When EPROM is mentioned in the control line descriptions in Sec. 12.4.10, is this an external EPROM or one on the 8051 itself?

**12.47.** What is the main reason that EPROM rather than PROM is preferred in μP system design?

**12.48.** What does an assembler do? Is it machine specific?

**12.49.** When a subroutine is called, all information necessary to restore the microprocessor to where it was before the call is dumped onto the stack. The stack is a section of RAM similar to a deck of playing cards: If you put some cards on the top, they will be the first ones to be drawn (provided that no one shuffles the stack). This organization of information storage and retrieval is known as "last in, first out," or LIFO. The address of the top of the stack is kept track of by the stack pointer register. When information is dumped on the stack ("pushed"), the stack pointer is incremented to the correct amount. When information is retrieved from it ("popped"), it is decremented. Suggest another use of pushing and popping that could be handy when programming.

For Problems 12.50 through 12.53, use the following 8051 μC assembly code statement definitions and/or address mode conventions as defined:

MOV A, #E4H stores the number E4H into accumulator A.

[# designates that the number afterward is the actual hexadecimal number (thus the H), not an address location—immediate addressing.]

MOV A, R3 copies into the accumulator the contents of register R4 (register addressing).

MOV A, 54H copies into the accumulator the contents of memory location 54 (hexadecimal) (direct addressing).

MOV A, @R3 copies into the accumulator the contents of the memory element whose location is specified in register R3 (indirect addressing).

ADD A, R1 adds the contents of the accumulator and register 1, and stores the result in the accumulator.

SUBB A, R1 subtracts the contents of register 1 from those of the accumulator and stores the result in the accumulator.

**12.50.** Read the following assembly code. Explain the meaning of each statement. Given that $(20)_{10}$ and $(15)_{10}$ are stored, respectively, in memory locations 44H and 48H, what will be the result in R3 after the completion of the code? Assume that initially the accumulator has $7_{10}$ stored in it.

```
MOV R1 44H
MOV R2 48H
ADD A R1
SUBB A R2
MOV R3 A
```

**12.51.** Identify the addressing mode for the source operand and the destination operand type of the following instructions.

(**a**) MOV R1 A

(**b**) MOV A #05H

(**c**) MOV A @R2

(**d**) MOV R3 56H

**12.52.** Write three lines of code using the 8051 μC assembly language commands above to compute Z = X + Y, where X is stored in the memory location referred to in R3 and Y is stored in R1. Store the result Z in memory location F8H.

**12.53.** Trace the content changes of the stack and the memory locations 30H, 34H, and 38H in Fig. P12.53b for the assembly language code in Fig. P12.53a.

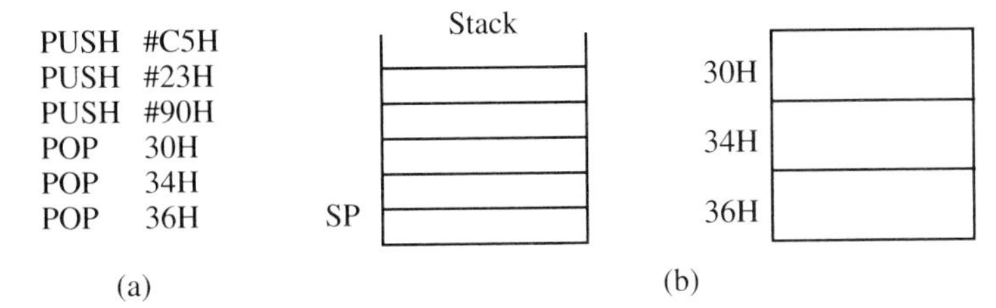


**Figure P12.53**

**12.54.** What is a cross-assembler?

**12.55.** What is the name of the program that converts high-level language code into executable machine code?

**12.56.** If you have available a manual on the C language, read the following C code and determine the operation it completes.

```
main()
{
        int i,k;
        int x,y,z;
        k=100;
        x=0.0;
        y=0.0;
        z=0.0;
        for(i=1;i<=k;i++) {
           x=x+i;
           y=y+i*i;
           z=z+i*i*i;
        }
        printf("x=%d  y=%d  z=%d",x,y,z);
}
```

**12.57.** It was mentioned in Sec. 12.6 that Eq. (12.1) could be written in a recursive form that did not involve recalculating the sum at every time step. Derive the recursive form of Eq. (12.1) for $u(n\,\Delta t)$.

**12.58.** Two separate rooms in a house are heated by a microprocessor-controlled heating system. The room temperature is held constant in one room by an integral-control algorithm, and in the other by a proportional-control algorithm. Which algorithm will respond better to a sudden change in room temperature (e.g., a door to the outside is opened)? Which will respond better to a gradual change in temperature (e.g., slow, overnight cooling)? Why?

**12.59.** Virtually all high-level computer languages (C, FORTRAN, Pascal, etc.) allow the use of floating-point variables and numbers. An example is the assignment statement in C: `pi = 3.14159;`

However, many microprocessors can manipulate only integers, not scientific notation floating-point numbers. Postulate why many microprocessors do not allow for floating-point numbers and calculations, but only integer arithmetic.

**12.60.** In high-level language programs where speed is critical, portions of the program are sometimes written in assembly language. This is done to speed up the time-critical sections of the program. Why is a compiler not able to generate machine language that is as fast as carefully human-written assembly code?

**12.61.** A microcomputer chip executes a MOV instruction from RAM to a register. What is the likely sequence of events for the three buses that will perform this operation?

**12.62.** Propose a simple block diagram for a microcontroller-based microwave oven on/off switching system. Use a flip-flop, switches, and any other combinational logic required to state your ideas clearly.

**12.63.** Suppose that the 2-in. penetration depth of microwaves into food refers to the $-10$-dB depth. What is the power absorption coefficient $\alpha$ of the food, assuming exponential decay $\exp\{-\alpha x\}$ where $x$ is the depth in inches? Also express the relation in decibels per inch.

**12.64.** Identify a problem with "keep-alive memory" described in Sec. 12.7 for automobiles.

**12.65.** A car has a mechanical automatic transmission system. Transducers are already implanted in the car that produce digital outputs for engine rpm and the actual car speed. A third actuator/mechanical link is available to shift the gears up or down, according to electrical signals. Upgrade the transmission to an electronic automatic transmission system, with cruise control. Specifically, come up with a simple block diagram, and describe its operation at the system level.

**12.66.** Suppose that you are the systems manager of a small newspaper. The newspaper gets Associated Press stories directly from Washington through a serial teletype line. These stories currently have to be manually retyped into the system before they can be typeset. You are asked to design an addition to the system such that this process is automated and the retyping is eliminated. Draw a simple block diagram showing main data and address lines with direction arrows. (*Hint:* Use an 8051 to collect data and when the size becomes acceptable, say 1 K, use a DMA controller to transfer it directly to the disk.)

**12.67.** The production of the local soft-drink bottling plant has increased, and management wants to automate the packaging line. Soft drinks are packaged in 6-, 12-, and 24-packs. The cans come by at a fast rate through the assembly line. Currently, a mechanically operated system produces the 6-, 12-, and 24-packs. There are 10 packaging lines. Electromechanical devices already exist to switch between packaging lines. You are to design a control system that will direct the first six cans to the first line, the second six cans to the second line, and so on. After 10 lines, it comes back to the first line and counts out 12-packs, and so on. Draw a very simple block diagram showing the needed components of the system.

# Part Five:
# ELECTROMECHANICAL MACHINES

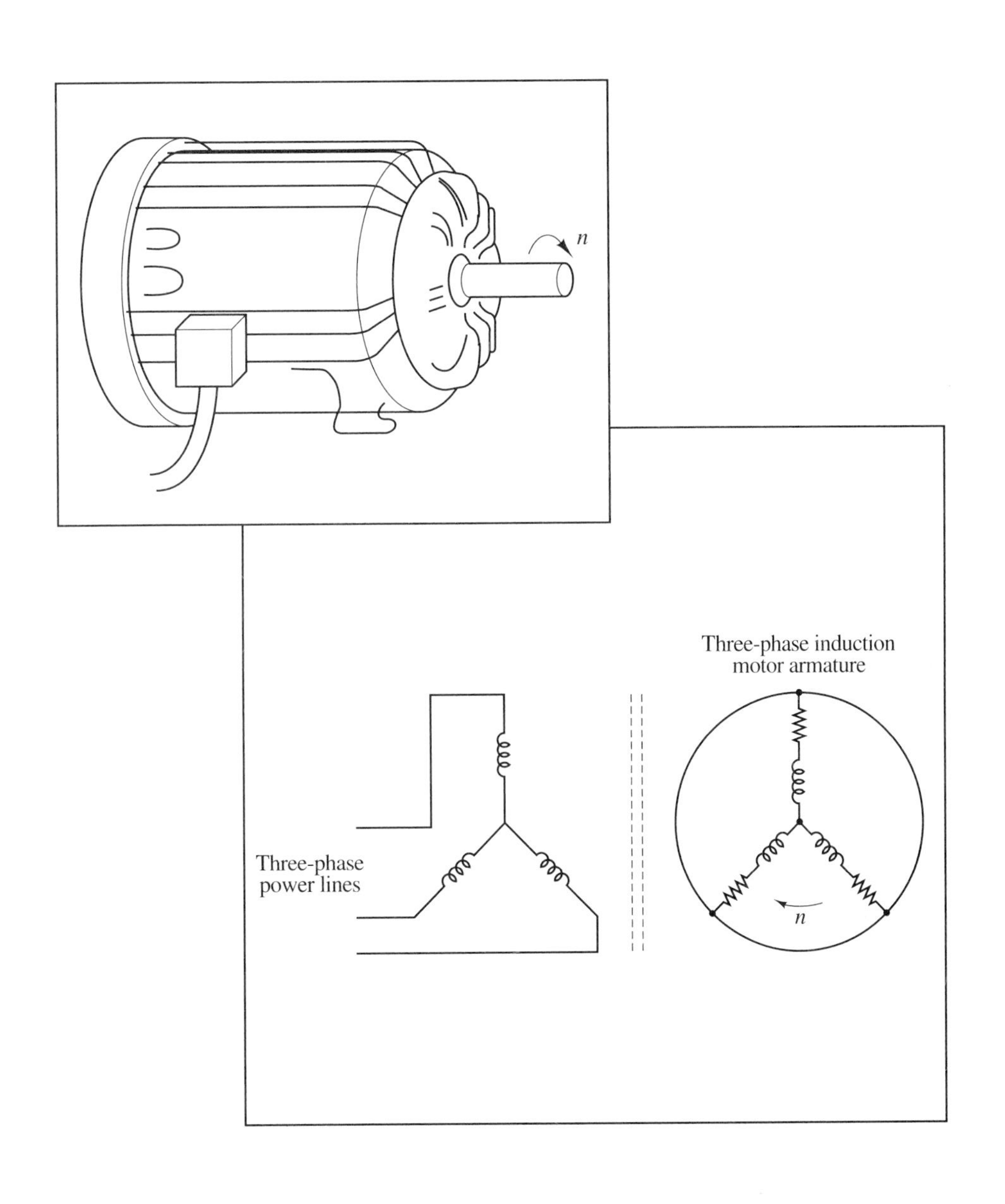

# 13

# Electromechanical Machines I: Basics

## 13.1 INTRODUCTION

It has been said that the quality of our electromechanical machinery directly determines our standard of living. Perhaps this is an overstatement; however, it certainly is true that without such machinery the current way of life for Americans would collapse. Two-thirds of all electricity in the United States goes into running motors, and of course essentially all our electricity comes from rotating electric generators.

The first dc motor was constructed by Michael Faraday around 1820. The first practical dc motor was produced by Joseph Henry in 1829, and the first commercially successful motor came out in 1837. The three-phase ac induction motor was invented by Nikola Tesla around 1887. Today, motors and generators of all types have entered most phases of American life.

Examples of motors in the home include fans, vacuum cleaners, washing machines and driers, garage door and can openers, dishwashers, refrigerators, microwave ovens, hair driers, electric shavers, tape and compact disk players, VCRs, some lawn mowers, and so on. In industry, there are pumps, stirrers, cranes and other lifting equipment, conveyors, compressors, crushers, fans, stokers, endless machine tools of all sizes, railroad train propulsion, and so on.

There is thus little question that the non-EE will at some point use or otherwise deal with motors. Depending on one's particular line of work, either dc or ac generators may also be present, particularly at remote sites where connection to power lines is not feasible (e.g., construction sites or certain vehicles) or at electric power plants.

"In our cities and large villages, everyone knows how vexatious travel is made by a little snow. When horses are used as the motive power the extra resistance offered by a few inches of snow on the track necessitates the use of one or more additional pairs of horses to each car; and when, as in the case of a heavier fall of snow, it becomes necessary to bring out the snow plow, it is not uncommon to see eight or ten pairs of horses working hard to clear the track. Under conditions like these, the electric railway has peculiar advantages in having almost unlimited power for direct application to the work of clearing away the snow." The sweeper motor ("of the usual waterproof type") rotates at 1200 rpm and the carriage motor at 620 rpm. "The flier motors are provided with rheostats by means of which the speed of the [snow] brushes is controlled." *Source: Scientific American* (November 12, 1892).

The goals of this introductory study of electromechanical machinery contained in our final three chapters, Chapters 13 through 15, are essentially (1) to study the construction of motors and generators and the dynamics of their operation, and (2) to understand some of the factors and calculations relevant to the intelligent selection of a machine for a given application. Of course, the former of these must first be addressed, and is the subject of this chapter.

## 13.2 MAGNETIC CIRCUITS

Most of the theory necessary for understanding electromechanical machinery was developed in the brief study of electromagnetic theory in Chapters 2 and 3. Those laws are sufficient to describe motor and generator phenomena, when augmented

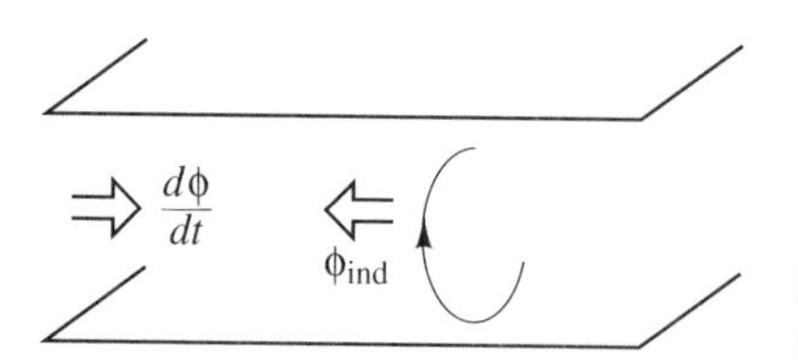


**Figure 13.1** Eddy current generation in an iron core.

with a few additional practicalities. One such item we shall look at briefly here is eddy currents. From there we concentrate on the magnetic circuit concept, and its importance in the study of machines.

First recall that hysteresis was a phenomenon in magnetic material that causes loss of energy in the form of heat. Recall also that for large fields, the *B–H* curve levels off into saturation; some machines depend on this for their operation, and all designs must account for it. Usually, a machine should be operated in the vicinity of the knee of the *B–H* curve (see Fig. 3.17a). If operated in the linear region, then just a bit more current would yield much more flux; if operated in the saturation region, the law of diminishing returns dominates. Thus to get the most out of the machine, the knee is often optimal.

Another loss factor is eddy currents. By Faraday's law, a time-varying magnetic field applied to magnetic material will induce electric fields within the material. By Lenz's law, these electric fields produce circulating currents flowing in such a direction as to oppose $d\phi/dt$, as shown in Fig. 13.1. The magnetic material is resistive to these currents, so thermal losses result.

A common way of reducing this type of loss is to laminate the magnetic material—that is, construct it as a stack of thin layers of magnetic material separated by insulating glue (Fig. 13.2). The induced eddy currents are then largely restricted to one layer and are smaller, resulting in lower power loss. Also, if the metal alloy itself is not highly conducting, eddy currents will be minimized. Eddy currents are actually another name for expressing (or a particular case of) the phenomenon of *inductive heating,* which is the occurrence of heat-producing currents in any conductive material exposed to time-varying magnetic fields.

As with hysteresis and $I^2R$ losses, eddy currents can contribute to motor heating, which can in some cases result in overheating. When overheated, a major danger is that the winding insulation can fail and a short circuit (*fault*) can occur. This causes expensive repair costs and possible interruption of all operations on the circuit to which the failed machine is connected.

Many magnetic devices are composed of objects having simple shapes in which the cross sections of material containing the magnetic flux are constant throughout (or at least along one dimension). Two examples are relays and transformers. Under these conditions, electric circuit concepts can be used in the analysis of the device. The devices are then called *magnetic circuits.*

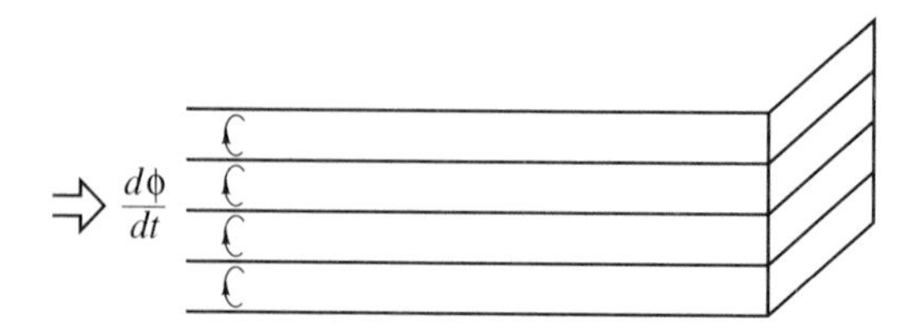


**Figure 13.2** Minimization of eddy currents can be effected by laminating the core to restrict eddy currents.

The reasons for briefly studying magnetic circuits in this chapter are (1) to increase experience in using simplified magnetic equations, (2) to provide a background for further study of magnetic devices (relays, transient behavior of motors, etc.), and (3) most important, to gain an appreciation of the great significance of air gaps in magnetic systems. We shall see that a simple magnetic circuit is the means of creating a constant background magnetic field required for operation of electromechanical machines. In our example we also consider the effects of an air gap; any rotating machine will inevitably have an air gap between the rotating member and the stationary member.

In Chapter 3, analogies were drawn between magnetic and electric quantities and equations. Under the simple geometries just described and assuming no magnetic saturation, the following simple relations hold:

$$\phi = BA \tag{13.1a}$$

$$B = \mu H \tag{13.1b}$$

$$\mathcal{R} = \frac{d}{\mu A} \tag{13.1c}$$

$$\mu_r = \frac{\mu}{\mu_0} \tag{13.1d}$$

$$\begin{aligned} \mathcal{F} &= Hd \\ &= NI = \phi\mathcal{R}, \end{aligned} \tag{13.1e}$$

where $\mu_0 = 4\pi \cdot 10^{-7}$ N/A$^2$. (In these final chapters, $\phi$ is taken to mean magnetic flux $\phi_m$.) These, together with a "$K\mathcal{F}L$" law (the sum of mmfs and magnetic potential drops around a magnetic circuit is zero), are sufficient for simple magnetic circuit calculations.

In particular, we may write for a series magnetic circuit having $M$ windings and $L$ differing material/mechanical sections

$$\mathcal{F}_{\text{net}} = \sum_{i=1}^{M} N_i I_i = \sum_{i=1}^{L} H_i d_i = \sum_{i=1}^{L} \phi_i \mathcal{R}_i, \tag{13.2}$$

where $N_i$ and $I_i$ are the turns and current in the $i$th coil and, referring to the $i$th section of material in the Amperian ($K\mathcal{F}L$) loop, $H_i$ is the magnetic field intensity, $d_i$ the length, $\phi_i$ the flux, and $\mathcal{R}_i$ the reluctance. This is just a discretized statement of the closed-contour integral magnetic potential equation (3.10) (Ampère's circuital law) presented in Sec. 3.3. The process is entirely analogous to how Eq. (2.30) in Sec. 2.8 was discretized to become KVL via Eq. (4.4) in Sec. 4.5.2.

A word of caution: Although the sums in Eq. (13.2) are equal, *individual terms between the first sum and the other two are not, in general.* Thus

$$H_i d_i = \phi_i \mathcal{R}_i \neq N_i I_i. \tag{13.3}$$

This is because equality of the first sum in Eq. (13.2) with the other two depends on a *closed-loop* integral. It would be the same error as equating an $IR$ drop to a battery voltage in a series electric circuit in which many resistors are in series with many batteries. Terms in the second and third sums in Eq. (13.2) *are* identical; they are merely statements of the definition of reluctance: $\mathcal{R} = d/(\mu A)$.

Another warning: Magnetic circuit analysis assumes operation in the linear region of magnetic material. This assumption is clearly made by the dependence of reluctance on permeability, which is meaningful only for linear operation ($B = \mu H$). Unfortunately, in practical machines the assumption of linearity is often not valid.

**Example 13.1**

A cast steel rectangular core (see Fig. 13.3a) has a 200-turn coil wrapped around one leg. The coil carries a current of 2 A. The dimensions of the core are as shown. Find the flux $\phi$ and the reluctance $\mathcal{R}$ of the magnetic circuit.

**Solution** The mean length around the core is shown as the dashed line, and for this approximate calculation is the line midway between the inner and outer perimeters of the core:

$$d_{\text{mean}} = 2(20 - 2\cdot 2) + 2(18 - 2\cdot 2)$$
$$= 60 \text{ cm} = 0.6 \text{ m}$$
$$A = 4\cdot 3 \text{ cm}^2 = 0.0012 \text{ m}^2.$$

The **B** lines in Fig. 13.3 go clockwise when viewed from the front, as can be verified by the right-hand rule. Recall that if $B < 1$ T (approximately), the linear portion of the cast steel $B$–$H$ magnetization curve applies, so that $\mu_r = 1300$ is the relative permeability. Assuming that this is true, the magnetic flux density is

$$B = \mu H$$
$$= \frac{\mu NI}{d_{\text{mean}}} \tag{13.4}$$
$$= \frac{1300\cdot(4\pi\cdot 10^{-7})\cdot 200\cdot(2)}{0.6} = 1.04 \text{ T},$$

which is close to 1 T. For this approximate calculation, then, use of the linear region is satisfactory. If $B$ had been too large (much greater than 1 T), the $B$–$H$ curve would have to be used to obtain $B$ from $H = NI/d_{\text{mean}}$. Now the flux is easily obtained as

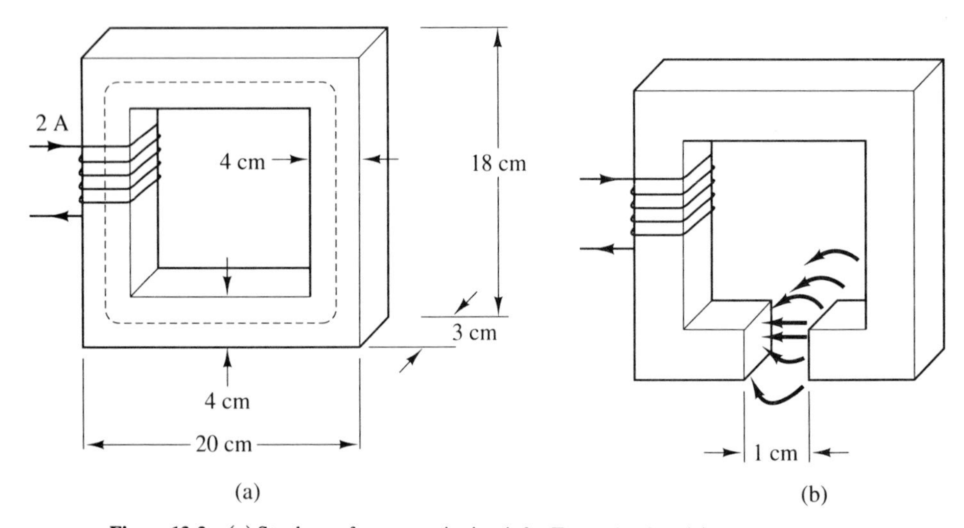


**Figure 13.3** (a) Steel core for magnetic circuit for Example 13.1; (b) same steel core as in (a) but with an air gap cut into it (see Example 13.2).

$$\phi = BA = (1.04)0.0012 = 1.25 \text{ mWb},$$

and the reluctance is

$$\mathcal{R} = \frac{d_{\text{mean}}}{\mu A} = \frac{0.6}{1300 \cdot 4\pi \cdot 10^{-7} \cdot 0.0012} = 3.05 \cdot 10^5 \text{ A} \cdot \text{t/Wb}.$$

**Example 13.2**

Now suppose that a 1-cm *air gap* is cut into the core (see Fig. 13.3b). What is the value of reluctance of the gap compared with that of the steel, and what current would now be required to obtain the same flux as in Example 13.1?

**Solution** Now the mean length in the core is 1 cm less than before:

$$d_{\text{mean, steel}} = 60 - 1 = 59 \text{ cm} = 0.59 \text{ m}.$$

The reluctance of the core is then 59/60 times its previous value:

$$\mathcal{R}_{\text{steel}} = \frac{59}{60} \cdot 3.05 \cdot 10^5 = 3.00 \cdot 10^5 \text{ A} \cdot \text{t/Wb}.$$

If the cross-sectional area of the core is $xy$, the equivalent area of the gap is, experimentally, approximately $(x + d_{\text{gap}})(y + d_{\text{gap}})$ due to field fringing effects. For example, if you placed a paper with iron filings around the gap, the field pattern would spread out a little beyond the main cross section. A lower density of field lines due to the spreading out is by definition a reduction in $B$ (flux density). Thus

$$A_{\text{gap}} \approx 5 \cdot 4 = 20 \text{ cm}^2 = 0.002 \text{ m}^2.$$

The gap length is 0.01 m, so the reluctance of the gap is

$$\mathcal{R}_{\text{gap}} = \frac{0.01}{4\pi \cdot 10^{-7}(0.002)} = 3.98 \cdot 10^6 \text{ A} \cdot \text{t/Wb}.$$

Note that $\mathcal{R}_{\text{gap}} \approx 13 \cdot \mathcal{R}_{\text{steel}}$. To obtain the same flux as in Example 13.1, a larger current will be required because the total reluctance has increased. Using $K\mathcal{F}L$,

$$\begin{aligned} \mathcal{F}_{\text{net}} &= \mathcal{F}_{\text{steel}} + \mathcal{F}_{\text{gap}} \\ &= (\mathcal{R}_{\text{steel}} + \mathcal{R}_{\text{gap}})\,\phi = NI, \end{aligned} \tag{13.5}$$

so that

$$I = \frac{1.25 \cdot 10^{-3}(3.98 \cdot 10^6 + 3 \cdot 10^5)}{200} \approx 27 \text{ A}.$$

In Example 13.1, only $\frac{1}{13}$ as much current was required to obtain the same flux.

It is worthwhile to note a few of the consequences of this example. Recall that energy density equals $H\,dB$, or rather the integral of $H\,dB$ from the minimum value of $B$ (say, the residual $B$) to the operating $B$. Because with the air gap the required current is much larger, and because $H$ is proportional to that current, more energy is required to build up the same flux.

We have also seen that because $B$ is proportional to flux (for a fixed cross-sectional area) and because the mmf is proportional to $H$, the energy density is proportional to the product of the flux and the mmf. But the mmf is, in turn, equal to the product of flux and reluctance. The flux in the circuit above is everywhere approximately uniform, including within the air gap. The total energy input will therefore be equal to the sum of the energies in each section of the circuit, each section containing energy proportional to the reluctance of that section times the square of the flux.

As calculated, the air gap will consequently contain the largest concentration of energy, because it has by far the greatest reluctance (due to its low $\mu$), while containing the same flux as other sections of the magnetic circuit. (Again, recall from Chapter 3 that energy density goes as $\phi^2\mathcal{R}$.)

The main lesson here is that air gaps are extremely significant in the design of magnetic devices. Often the effect is undesirable, as it was in the above example. But in machines, they can be "just what the doctor ordered." For it is the air gap with which the conductors of a rotor are in contact. It makes sense to direct all the energy there, to best make use of it. The iron core of the electromagnets does this directing.

At the same time, too much of a good thing is bad: Notice that the air gap reluctance is also proportional to the gap length. So the longer the gap, the more current is required to set up the desired flux. Thus we want the air gap, but it should be as narrow as feasible. In practice, other nonidealities, both mechanical and electromagnetic, limit the narrowness of the gap.

Note that it is essential to have the iron core. Without it, the required current would be $I = Bl/(\mu_o N) = 2480$ A! So while the air gap is mechanically unavoidable in a rotating machine, it is helpful in concentrating the energy where we want it, but it does greatly increase the current required to achieve a given flux (and therefore $B$, which is responsible for the magnetic forces causing motor motion).

## 13.3 THE SIMPLEST ELECTROMECHANICAL CONVERTERS

### 13.3.1 Introduction

By their very nature, electromechanical machines (often here abbreviated "machines") are dynamic; there are moving parts and time-varying electromagnetic quantities. Recall that for constant-in-time electric flux (i.e., charge), a nonzero electric field exists, but the magnetic field is zero. Also, for constant-in-time magnetic flux, there is a magnetic field but no electric field.

From Faraday's law, if $d\phi/dt \neq 0$, there is an induced voltage and therefore an electric field (and, of course, a magnetic field associated with $\phi$). From Ampère's circuital law, if $I = dq/dt \neq 0$, there is a magnetic field (and, of course, an electric field associated with and driving $q$). So if there is a changing-in-time flux (either magnetic or electric), there will be *both* nonzero electric and magnetic fields. Recall that together, they are called the *electromagnetic field.* These electromagnetic fields will always be present during normal operation of machines. Note, however, that at the frequencies of practical machines, electromagnetic *waves* (Sec. 3.10.4) are negligible.

### 13.3.2 Linear Generator

The simplest possible machine is the linear generator, shown in Fig. 13.4. It consists of two conducting rails a distance $l$ apart having a spatially uniform, constant-in-time

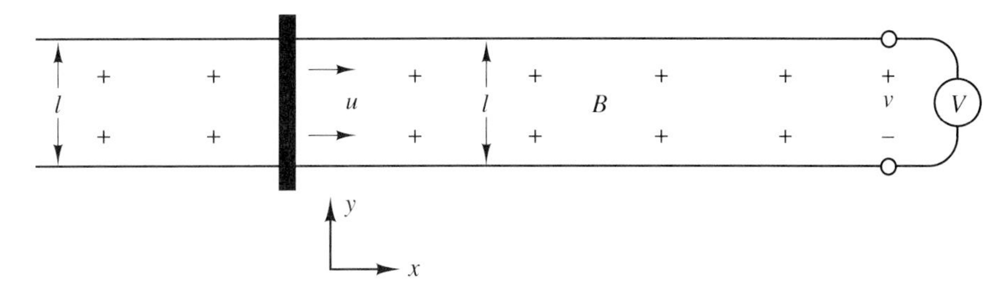


**Figure 13.4** Simple linear electric generator.

magnetic field **B** passing between them. A conducting bar lies across them that makes electric contact with both rails and is shown moving to the right at speed $u$. A voltmeter $V$ displays the potential difference between the rails. From Faraday's law, this nonzero voltage is induced because the magnetic flux enclosed by the voltmeter-rail-bar circuit is decreasing with time:

$$\begin{aligned} v &= \frac{d\phi}{dt} \\ &= \frac{d(BA)}{dt} = \frac{B\,dA}{dt} \\ &= Bl\,\frac{dx}{dt} \\ &= Blu. \end{aligned} \tag{13.6}$$

As an exercise, apply Lenz's law from Chapter 3 to verify that the polarity of $v$ in Fig. 13.4 is correct. Equation (13.6) indicates that if $u \neq 0$, there is both an electric field (which would be $v/l$ if the rails were parallel plates) and a magnetic field $B$—an electromagnetic field. Incidentally, if $B$ was changing-in-time, then even if $u = 0$ (no motion) there would be a nonzero voltage $v$ (which would be $v/l$ if the rails were parallel plates) and thus an electromagnetic field. And, of course, if $u \neq 0$ and $B$ spatially varies (with respect to $x$ or $y$), there will again be an electromagnetic field. There are many ways to obtain an electromagnetic field, and many of them occur in machines.

Suppose that **B** remains as shown in Fig. 13.4, but the track is tilted "into the paper" by an angle $\theta$ (see Fig. 13.5). For $\theta = 0$, the motion of the moving bar is perpendicular to the magnetic field lines, but no longer for $\theta \neq 0$. The important issue in the determination of $v$ is the time rate at which the number of flux lines passing through the circuit is changing (here, decreasing). Clearly, the number is not

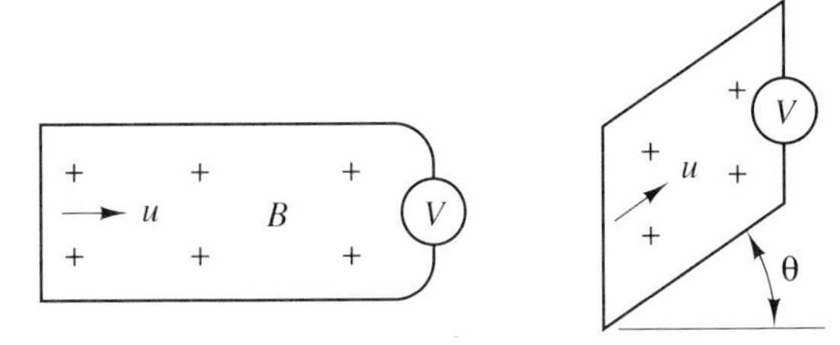


**Figure 13.5** Effect on the enclosed flux of tilting the linear generator with respect to the background $B$ field.

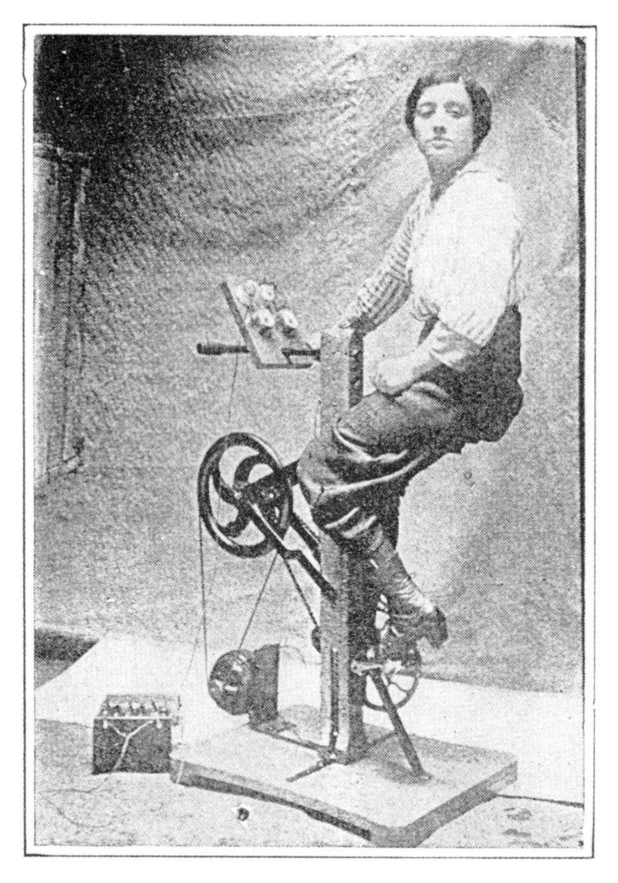

A 1913 exercycle. The flywheel drives a generator: "The energy which the human motor produces, instead of going to waste, is stored up in the accumulator (this model, which charges a 12 V, 25A battery is for "persons with strong muscles"). . . . In this way it is possible to use one or several small lamps for domestic lighting, so that the work done with the exerciser during the daytime can be taken advantage of in the evening. . . . The fact of using electric lamps gives a stimulus to the use of the machine for the purpose of exercise and is apt to prevent the daily exercise from being neglected." *Source: Scientific American* (November 22, 1913).

decreasing as fast as before; before in proportion to $u$, now in proportion to $u \cos(\theta)$. Therefore, we now have

$$v = Blu \cos(\theta). \tag{13.7}$$

Note that if $\theta = \omega t$, which could be accomplished by rotating the circuit at angular velocity $\omega$ in a *uniform* magnetic field, a sinusoidal-in-time voltage would be developed: $v = Blu \cos(\omega t)$. However, neither the linear motion-single loop circuit nor the uniform magnetic field are valid assumptions in practical machines. Nevertheless, the operation of both ac and dc generators is based on Faraday's law.

### 13.3.3 Single-Loop Rotating Generator

Everyone knows that essentially all practical machines have rotational motion, as opposed to the linear generator discussed above. The reason is mainly that rotational motion is inherently efficiently *cyclical.* The linear motion described above is either one-shot (nonrepeating) or reciprocating, which are impractical and/or inefficient types of motion compared with rotation. So attention is now given to the operation of a simple rotating generator. In passing, we refer to one notable exception in transportation: linear induction motors, which may be used in future personal rapid transit systems. These operate on rails and function (e.g., individual destination selection) like a horizontal elevator—fully automated under computer control, but at 70 mph.

Suppose that (e.g., by hand) the single-loop coil in Fig. 13.6a is rotated clockwise. Because the flux passing through the loop changes with time as the loop is cranked, there is an induced voltage observable at the terminals marked with $v$. Again, check the polarity with Lenz's law from Chapter 3. Note that in the second half of the rotation the polarity will reverse.

As shown in Fig. 13.6a, the coil is horizontal, so the left conductor is moving up while the right conductor is moving down, as indicated by the arrows. This case is analogous to the $\theta = 0$ case of Fig. 13.5 because the conductors are moving perpendicular to the flux lines. Consequently, in this position the induced voltage $v$ is maximum. Even though for this position $\phi = 0$, $d\phi/dt$ is at its maximum.

In Fig. 13.6b, the loop has a maximum flux passing through it, but $d\phi/dt = 0$, so there is no induced voltage. This is because turning the coil a tiny bit either way does not significantly change the number of flux lines passing through the loop; compare with Fig. 13.6a. In general, at an angle $\theta$ from the horizontal, the area exposed to flux is $2RL \sin\theta$ where $L$ is the depth of the loop and $R$ is the radius of the loop. Thus $\phi = 2RLB \sin\theta$ so that $v = d\phi/dt = 2RLB \cos\theta$ so that $v$ is maximum at $\theta = 0$ and $v = 0$ at $\theta = 90°$.

In practice, the external magnets are curved as shown to minimize the air gap effects described in Sec. 13.2, when the loop is replaced by an iron core. This curvature causes the magnetic field to be significantly radial rather than purely left to right. The result is that for moderate loop angles, $d\phi/dt$ and therefore the induced voltage remain relatively constant. Some of the radial lines that passed through one side of the loop when it was horizontal now change to pass through the other side as the loop is rotated, which results in a constant $d\phi/dt$ (see Fig. 13.7a). So instead of a sinusoidal-induced voltage, an alternating flat-topped induced voltage is in fact measured. Figure 13.7b summarizes the behavior for one cycle (rotation).

It is often written that a single conductor cutting across magnetic flux experiences an induced voltage. The statement is true, as it derives from Ampère's law ($\mathbf{F} = q\mathbf{u} \times \mathbf{B}$, where $\mathbf{F}$ is the force on charge $q$ moving at velocity $\mathbf{u}$ with respect to magnetic field $\mathbf{B}$, and the electric field causing the voltage is $\mathbf{E} = \mathbf{F}/q$).

However, it may be confusing in that in practice there are no "single conductors," only circuits. Furthermore, by Faraday's law as studied here the induced emf

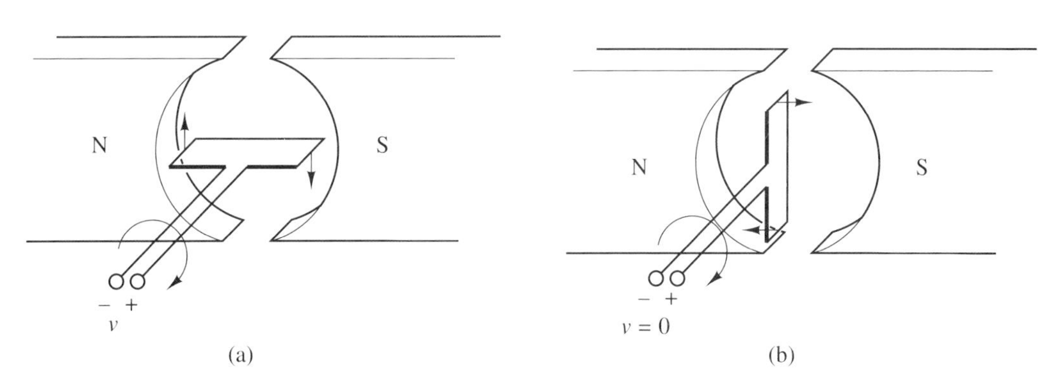


**Figure 13.6** Single-loop rotating generator at two angular positions.

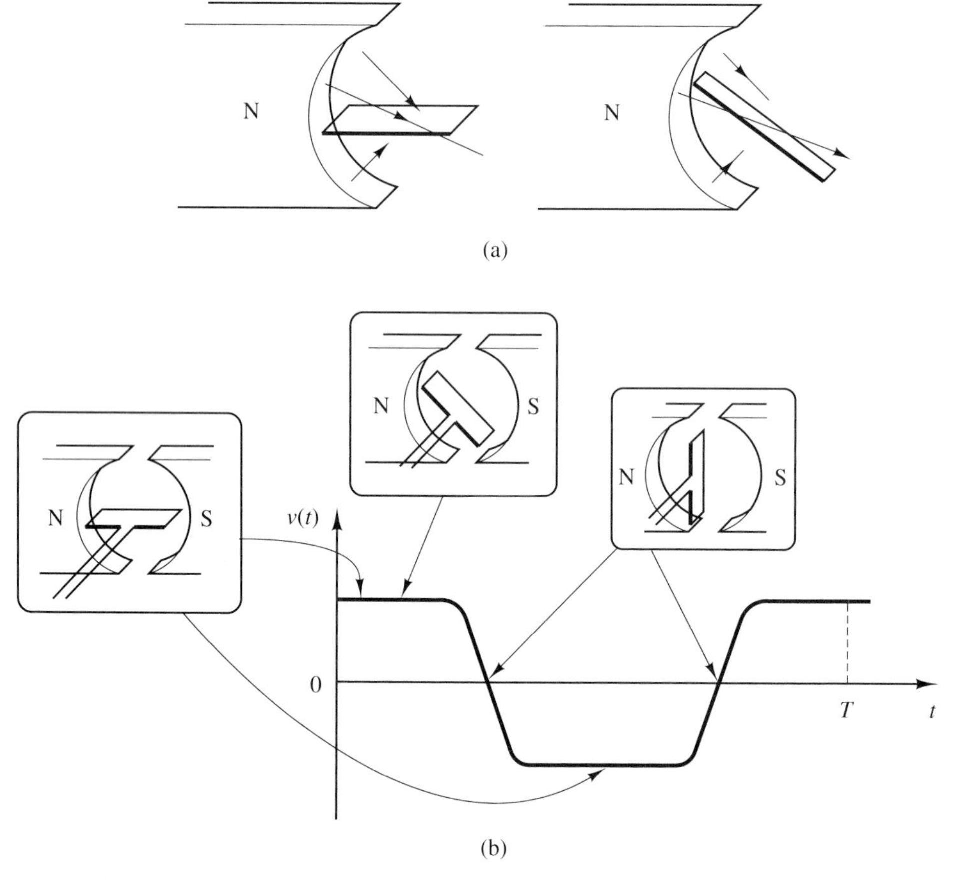


**Figure 13.7** (a) Demonstration of how the generated voltage levels off around its peak; (b) generated voltage waveform with angular positions of the loop indicated.

equals the time rate of change of flux *enclosed by the circuit or loop* (*or coil*). So in this book a $d\phi/dt$-type explanation will always be given instead of "conductors cutting flux." This point is addressed to clarify the intentions of other presentations. The "single-conductor" explanation is often used because it provides a uniformity of interpretation of the equations for induced voltages and torques [after all, *it is* clearly individual current-carrying conductors (sides of a coil) that experience torque forces in machines].

Another point of clarification regards pole flux. Up to this point, only two-pole magnetic fields have been considered; the flux leaving the north pole nearly equals that entering the south pole. For reasons described in Sec. 13.4.2, more than two poles are often used. As the fluxes from all these poles interact, it is impossible to identify *pole pairs* (see Fig. 13.8). Therefore, the term "flux per pole" has been introduced, meaning the total flux leaving or entering any particular pole. Usually, the poles are constructed such that the fluxes of all the poles are equal, yielding one flux per pole value for the entire field system.

In the analysis and design of machines, time averages are often used to describe

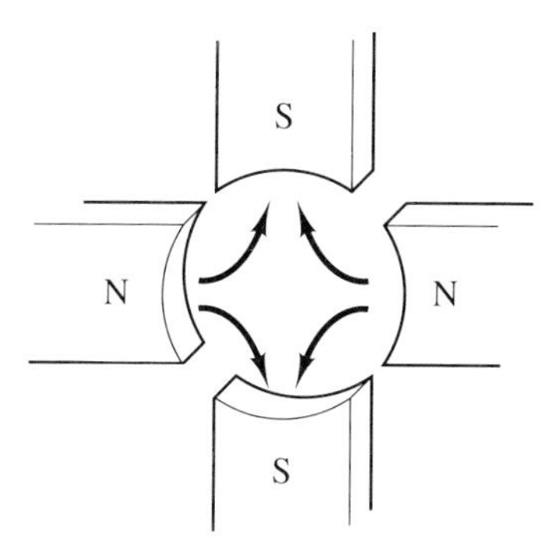


**Figure 13.8** Four-pole magnetic field.

steady-state behavior. This simplifies the analysis as well as recognizes that only approximate values of certain relevant quantities (such as $\phi$) are ever known.

Consider the single-loop generator. Although the average value of $v(t)$ in Fig. 13.7b is zero, that of $|v(t)|$ is not. In fact, the absolute value of $v(t)$ is taken in dc machines, soon to be discussed. For now, then, let $v_{avg}$ represent the average value of $|v(t)|$. Thus, for the single-loop generator described above, the average induced voltage can be estimated by noting that in a quarter-turn, the flux passing through the loop changes from zero to $\phi$:

$$v(t) = \frac{d\phi}{dt}$$
$$v_{avg} \approx \frac{\Delta\phi}{\Delta t} = \frac{\phi}{1/4\,T} = \frac{4\phi}{T}, \tag{13.8}$$

where $T$ is the time in seconds for a complete rotation. If the loop is cranked at $n$ revolutions per minute, $1/T = n/60$, so that

$$v_{avg} = \frac{4n\phi}{60}. \tag{13.9}$$

**Example 13.3**

A single-loop generator spins in a flux of 0.1 Wb at a speed of 78 rpm. What is the average of the absolute value of the voltage across the loop terminals?

**Solution** From Eq. (13.9), $v_{avg} = 4(78)(0.1)/60 = 0.52$ V.

Naturally, for this generator to be at all useful, it will have to be connected to some load $R_L$ (see Fig. 13.9). To verify the direction of the induced current, note that (in accordance with Faraday's law) the generator loop acts like a "battery" in

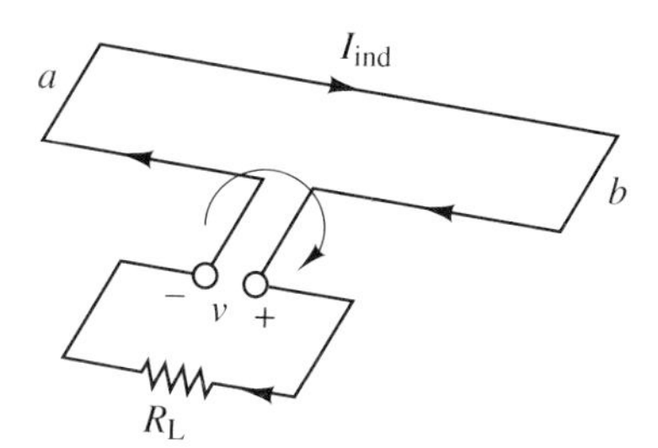


**Figure 13.9** Load resistor $R_L$ connected to terminals of loop in which voltage is generated during rotation.

that positive current flows out of the positive terminal. (As noted earlier in this section, the direction of this current will reverse with the induced voltage during the second half of a rotation.)

Once the load is connected, a current $I_{ind}$ will flow in the loop. As hinted at above, there will be torque-producing forces on certain conductors of the loop due to the induced current they carry while immersed in the background magnetic field causing that current. In Fig. 13.10, such forces will exist on conductors $a$ and $b$ and are determined from Ampère's force $\mathbf{F} = I_{ind}\mathbf{l} \times \mathbf{B}$.

The total force on the conductors $a$ and $b$, each of length $L$, has magnitude $I_{ind}LB$ because $\mathbf{B}$ is uniform along the conductor and is always perpendicular to it. (The other sides experience inward, non-torque-producing forces also dictated by Ampère's force. Be careful when checking this with the right-hand rule: Point your fingers in the direction of $I_{ind}\mathbf{l}$ and then in the direction of $\mathbf{B}$ *using an angle less than 180°*.) The direction of the forces on conductors $a$ and $b$, shown in Fig. 13.10, can also be verified by using the right-hand rule (Chapter 3). Although the $\mathbf{B}$ field is shown in Fig. 13.10 as left to right, in practice it is directed more radially, as described above. The resulting forces will be better tangents to the circular path of the loop and thus increase the induced torque.

These forces, acting on the lever arm of length $R$, are noticed as a torque. Because the torques acting on conductors $a$ and $b$ both act in the counterclockwise direction and are of equal magnitude, the total torque magnitude is

$$\text{torque} = 2FR = 2I_{ind}LBR. \tag{13.10}$$

The fact that the counterclockwise-induced torque opposes the clockwise cranking means that we have to crank harder to maintain the clockwise rotation. Of course, were cranking ceased, the induced current would become zero (because $d\phi/dt$ would become zero) as would the opposing torque; the loop would *not* begin rotating counterclockwise.

This opposing torque is necessary for the conservation of energy; "you can't get something for nothing." For if there were no opposing torque, then once the rotor was brought to a given speed, no further energy would be required to keep it

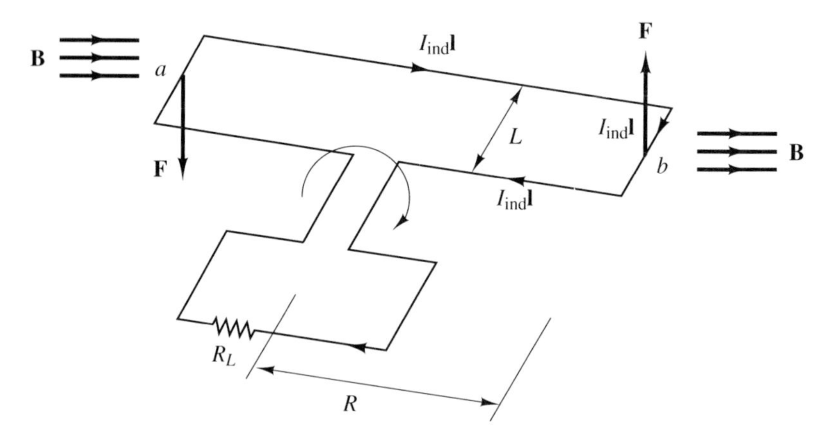


**Figure 13.10** Forces on the loop conductors caused by the induced current in the loop, in the presence of the background $B$ field.

going. Yet there would be a continual output power $I_{ind}^2 R_L$ which could be made as large as desired by setting up the rotation at a sufficient speed!

**Example 13.4**

Suppose that the induced current at a given time in the rotating loop is 10 mA. Also suppose that $B = 0.1$ T and $L = R = 0.1$ m. What torque does the loop experience?

**Solution** By Eq. (13.10), the torque is simply $2(0.01)(0.1)^3 = 2 \cdot 10^{-5}$ N · m.

Another effect of the induced current is the creation of an accompanying magnetic field called the *armature reaction field.* Using the right-hand rule gives the direction of this magnetic field (see Fig. 13.11). The total magnetic field around the loop will be the vectorial sum of this field and the background field; therefore, the original magnetic field will become distorted by the armature reaction. More will be said about this in Sec. 13.4.6.

Incidentally, this field does *not* produce forces on the coil conductors. At the conductors, the field is circumferential, as shown in Fig. 13.11. Although the field is everywhere perpendicular to $I_{ind}$, the forces on all sides of the conductor cancel. The current in a wire cannot produce a force on itself.

What has been accomplished? By expending mechanical energy to overcome the opposing induced torque and frictional losses, voltage and current have been induced in the rotating loop that are capable of supplying power to an electrical load $R_L$ such as a light bulb or radio. That is, this device has converted mechanical energy into electrical energy.

### 13.3.4 Single-Loop Rotating Motor

Now that the basic ideas concerning mechanical-to-electrical energy conversion have been addressed, attention will be given to the reverse process: electrical-to-mechanical energy conversion. This process is accomplished by what is known as a motor. However, the apparatus (for this simple study) is *identical* to that for a generator!

Actually, all the relevant analysis has already been done. But instead of cranking the loop, the loop is now supplied with (externally generated) current. Suppose that current is sent into the loop in the same direction as the induced current flowed in Fig. 13.9. As shown in Fig. 13.10, there is a magnetic force acting with an associated torque to rotate the loop counterclockwise. (As before, the **B** in $\mathbf{F} = I\mathbf{l} \times \mathbf{B}$ is supplied by the external, curved magnets.)

No longer is a torque being supplied (by cranking) in the clockwise direction,

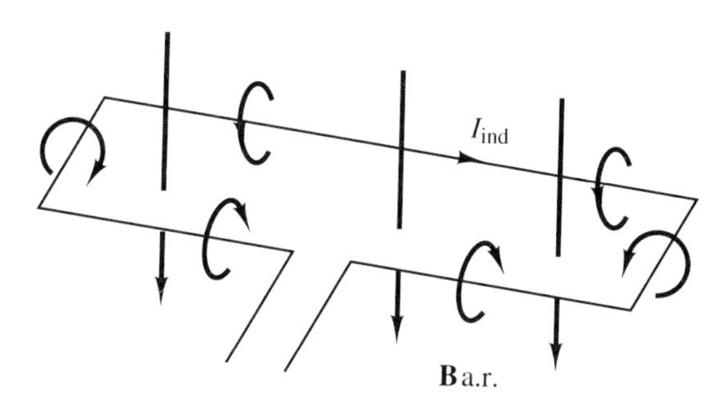


**Figure 13.11** Armature reaction magnetic field.

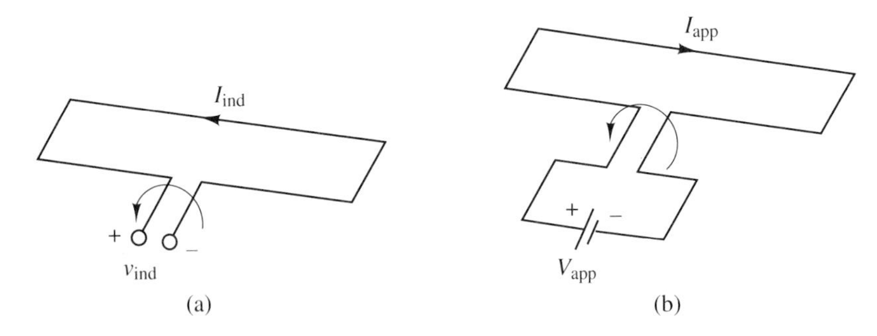


**Figure 13.12** Comparison of induced (a) and applied (b) voltages and currents in a single-turn rotating loop motor.

so the loop will rotate counterclockwise—the motor works. Note, however, that during the second half of rotation, the direction of the applied current will have to be reversed; otherwise, the torque direction will also reverse and the motor will not spin.

But the story is not quite complete; recall that a rotating loop in a background magnetic field acts as a generator. The induced voltage will, however, have polarity opposite that in the previous discussion of the simple generator because now the loop is rotating counterclockwise instead of clockwise. Thus the direction of induced current will be as shown in Fig. 13.12a.

Now recall that the direction of applied current is that in Fig. 13.9 or 13.12b. Clearly, the induced current opposes the applied current. This is analogous to the fact that in a generator, the induced torque opposes the applied torque. Here, the (opposing) induced current will require a larger applied current to keep the motor spinning a given speed; the force on the conductors is due to the *net current* $I_{app} - I_{ind}$. Of course, if the applied current is removed, the motor will eventually stop, and simultaneously the induced current will vanish.

### 13.3.5 Review and Some Terminology

It is fitting at this point to pause briefly for review and a little terminology. Table 13.1 summarizes the behavior of generators and motors. A few relevant definitions and clarifications are now stated.

The electric potential or voltage (applied or induced) is often called an *electromotive force* (emf), even though it is *not* a force. (A better term would be "electromotive impetus.") Therefore, the voltage induced in a generator is often called the *generated emf*, while that induced in a motor is called the *counter* or *back emf* to emphasize its oppositional polarity. Also, the average generated voltage is

**TABLE 13.1** SUMMARY OF MOTOR AND GENERATOR EFFECTS

| Generator | Motor |
|---|---|
| $v_{induced}$ *produces* loop current | $v_{induced}$ *opposes* applied loop current |
| $\text{Torque}_{induced}$ *opposes* applied torque | $\text{Torque}_{induced}$ *produces* loop rotation |

usually given the symbol $E$ (do not confuse with an electric field), while applied voltages are usually given the symbol $V$. The magnetic potential $\mathcal{F}$ is often called a *magnetomotive force* (mmf), even though it is not a force.

The apparatus of background field magnets, the rotating loop, and all accompanying parts is often called a *dynamo,* an abbreviation of the term "dynamo-electric" machine, coined by Werner and William Siemens in 1867. In a dynamo, the loop and all parts that rotate are called the *rotor,* while the stationary magnets and all parts that do not rotate are called the *stator.* Note that to increase the induced voltage in a generator or induced torque in a motor, not just one but many turns comprise the rotor winding.

The background field is in practice a permanent magnet only for low-power applications (toys, speakers, etc.). Instead, it is usually an electromagnet—that is, a coil having an impressed current and surrounding a core of magnetic material the purpose of which is to increase and direct magnetic flux (see Fig. 13.13). The first use of electromagnets for this purpose was in 1845 by Charles Wheatstone and others.

This background field system is, of course, just a magnetic circuit with an air gap, and is equivalent to a primary-only "transformer" run at dc. All we want is a constant-in-time magnetic field in the loop region, which is the "air gap" we studied in our magnetic circuit example. The coil (or magnet) that creates the background field is called the *field winding* or magnet; it is often abbreviated as just "the field." Note that in the present discussion the field is on the stator, but as discussed in Chapter 15, on some machines it is on the rotor.

The winding that is not the field is called the *armature.* Although in this case it is placed on the rotor, on some machines it appears on the stator. What purposes, then, define the field and armature? The field is the coil creating the background field that causes the main induction behavior (voltage in a generator, torque in a motor). The armature is the coil in which the main induction takes place. The word "main" here means "effectual in producing the desired behavior"; there are other induced quantities that may appear in the field winding.

In discussions of mechanical gears (sprocket wheels), the distance from one point on a gear tooth to the corresponding point on the next tooth is called *pitch* (see Fig. 13.14a). As noted before, the field in a motor or generator may have more than

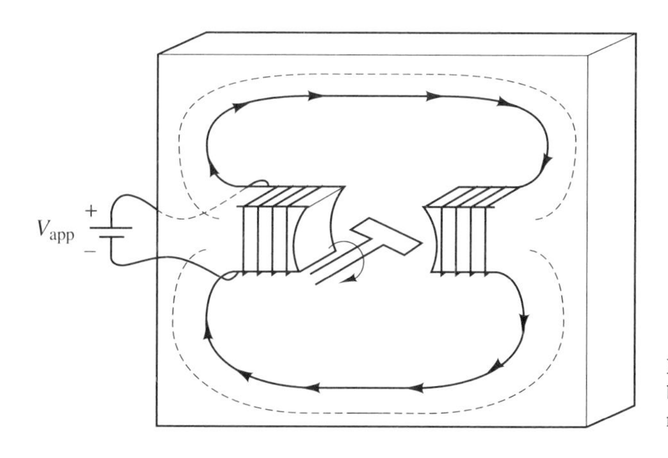


**Figure 13.13** Electromagnet background field for rotating loop motor or generator.

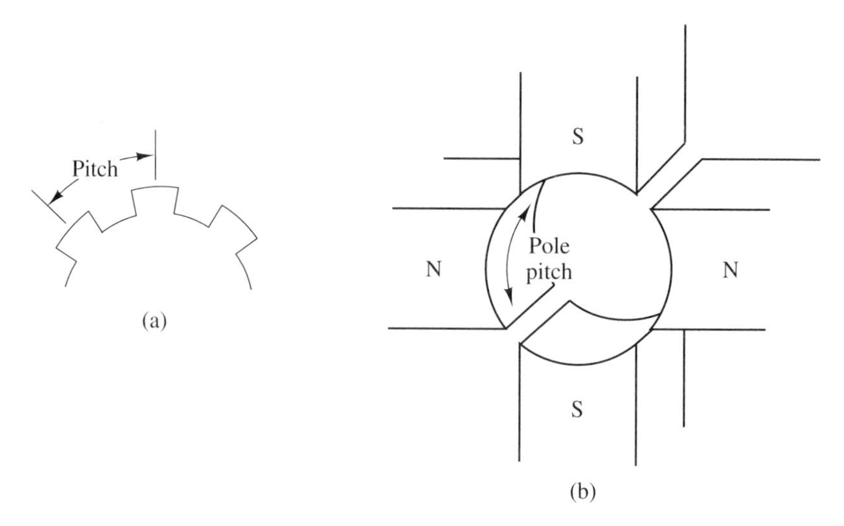


**Figure 13.14** Pitch defined for (a) gears and (b) magnetic field poles.

two poles. The distance from a point on one pole to the corresponding point on the next pole is analogously called the *pole pitch* (see Fig. 13.14b).

## 13.4 COMMUTATION AND ARMATURE WINDINGS

### 13.4.1 Commutation

At this point, all of the physical principles necessary for generator and motor action have been described. A remaining task is to develop *practical* machines. For example, in Fig. 13.12b the power supply for the armature (on the rotor) must spin. It is hardly practical to have to spin the city power plant. Two simple ways to keep the power plant from becoming dizzy are now discussed briefly.

The easiest way is to connect some *slip rings* to the ends of the loop, as shown in Fig. 13.15a. Electrical contact to the power supply is then accomplished by *brushes* that rub against the metal rings. The resulting waveform associated with continual rotation is shown in Fig. 13.15b, where the voltage is that induced for generators or that applied for motors. Note that the voltage alternates in polarity (although it is not sinusoidal). So this machine, which except for the slip rings is the same as that of the previous discussions, is considered an ac machine (although not a very good one).

Figure 13.16a shows a rectifying dynamo that could be classified as a dc machine. The rectification is achieved by what is known as a *commutator.* One meaning of the word "commute" is "exchange." This is precisely what the commutator does; every half rotation the ends of the loop connected with the brushes are swapped or exchanged. It is the mechanical equivalent of a full-wave rectifier and was invented in 1832 by Hippolyte Pixii of France (see Figure 13.16b).

Notice that the brushes are oriented midway between the field poles. This fact, when combined with the fact that the junction line of the U-shaped commutator segments is parallel to the loop orientation, means that the exchange will take place

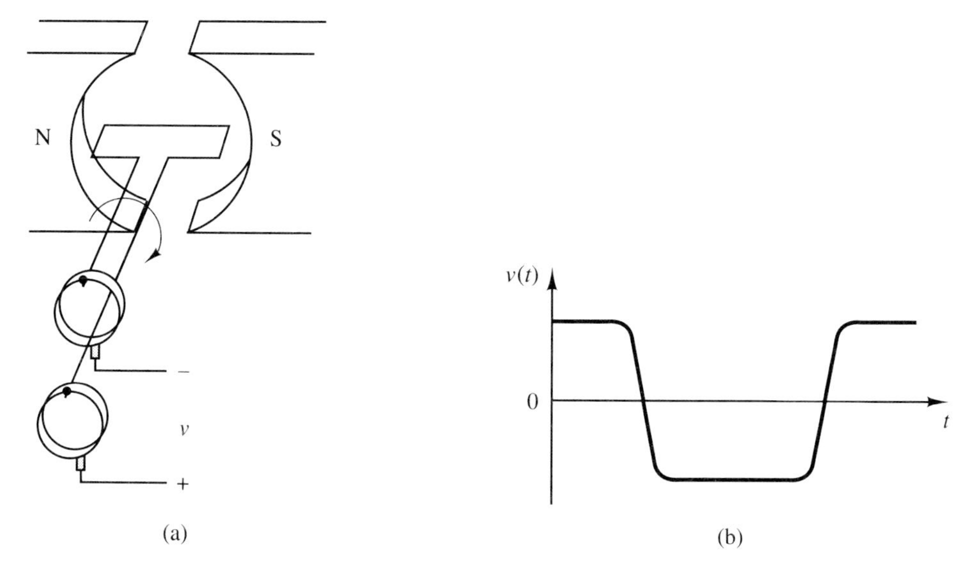


**Figure 13.15** (a) Slip rings with which to connect the rotating loop to external source or load; (b) generated voltage waveform alternates in sign.

at points where the voltage induced in the loop (and thus the terminal voltage $v$) is zero. This is because at the time of exchange, the coil will be vertical, at which point $d\phi/dt = 0$, as noted in Sec. 13.3.3.

If the brushes were aligned with the poles, the second half of the alternating voltage would be reversed and Fig. 13.16c would result. The sudden voltage changes

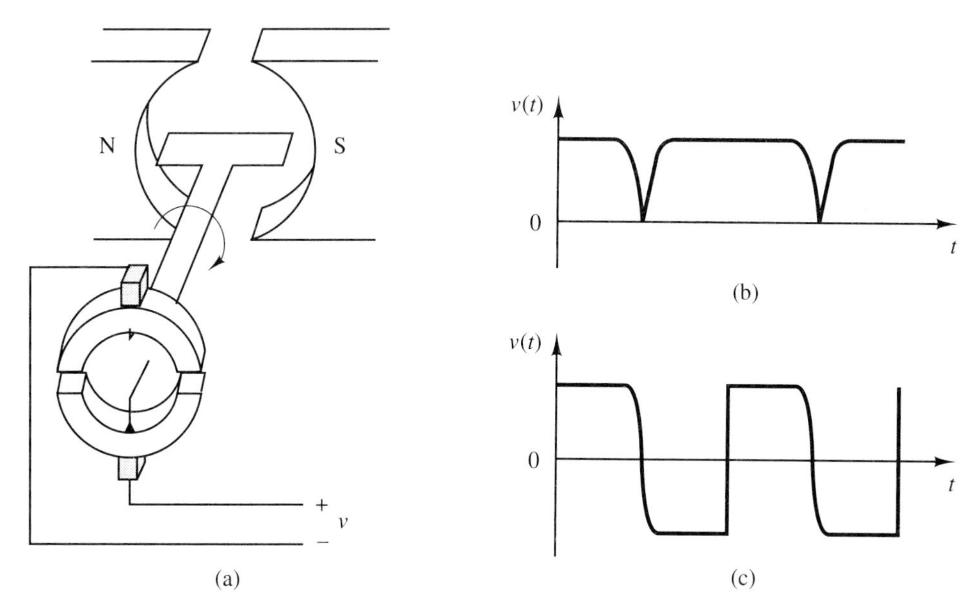


**Figure 13.16** (a) Commutator used for connecting the rotating loop to external source or load; (b) generated voltage waveform does not change in sign; (c) rectification lost if commutator switching done at the wrong time/place.

might be large enough to cause sparking; in addition, the resulting voltage is no longer rectified (single polarity). But if the brushes are midway between the poles, for there to be continuous rotation of the loop the voltage induced (in a generator) or applied (for a motor) must be single polarity, as shown in Fig. 13.16b.

It is interesting to note that whereas the output voltage of the generator was a particular flat-topped waveform as shown in Fig. 13.15, the commutator with dc applied will present a far steeper voltage waveform to the rotor coils. If the commutator were removed and a perfect sinusoidal waveform applied, what would be the effect? That is, in these two cases we do not have precisely the reverse of the generator situation because applied voltage is not identical to that produced by a rotating generator. The answer is that the rotor would spin, but not with precisely uniform-in-time torque. Because of the moment of inertia of the rotor, the effect will not be noticed for any reasonable approximation of the waveform in Fig. 13.15.

### 13.4.2 Windings: General Considerations

Up to now, the discussion of the armature has been confined to a single-turn loop. But as noted above, practical machines have many turns per coil, and in fact many coils comprise the armature winding. Furthermore, the winding is typically distributed over the circumference of the rotor. Such a multiturn winding is thus called a *distributed winding*.

To understand the rationale for distributed windings, consider the typical rotor core shown in Fig. 13.17. The slots in the core contain the conductors of a wound rotor, as shown. The slotted rotor was proposed by Antonio Pacinotti of Italy in 1860. For large voltages, the four coils shown in Fig. 13.17 would be connected in series, while for large currents they would be connected in parallel.

But why should the coils be "distributed"? The benefit is that when one coil is in a *dead zone* (region where its induced voltage is low or zero), another coil at a different location on the rotor is in a *live zone* (with large induced voltage). Furthermore, there is only a small amount of room in a slot, and wires have a finite thickness. Therefore, if many turns are desired, many slots distributed over the rotor surface will be required to hold them all.

One of the main reasons for multiple poles complements that for distributed windings: With many poles, a greater percentage of the distributed windings can be under $d\phi/dt \gg 0$ or $IlB \gg 0$ conditions at any particular time. A four-pole machine would appear schematically as in Fig. 13.18. Note that now instead of only two commutator segments, there are several, shown near the center of the rotor. Of course, highly insulating material must be placed in between the segments.

Notice that in Fig. 13.18, geometrically opposite poles have the same magnetic polarity. Why? The answer involves another aspect of multiturn distributed wind-

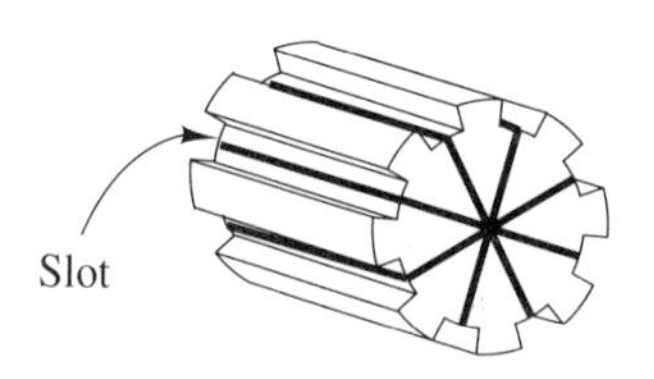


**Figure 13.17** Rotor construction.

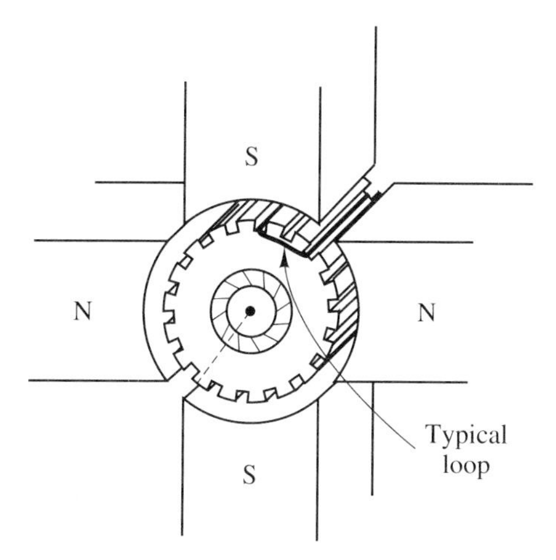


**Figure 13.18** Four-pole field with rotor.

ings. Now, instead of each loop being centered on the rotor axis, with a north pole on one side and a south pole on the other (see Fig. 13.6), the loop face is placed on the rotor *surface*, passing by one pole at a time (see Fig. 13.18).

For this configuration, which is convenient for multipole fields and distributed windings, it makes sense to have successive poles that a given coil passes by be alternating. This type of winding was introduced by Friedrich von Hefner-Alteneck of Germany in 1872. The earlier setup in Fig. 13.6 (in which the loop rotates about its own central axis) was simpler to describe and visualize, while the more complex arrangement in Fig. 13.18 (in which the loop rotates around the rotor perimeter) is more practical.

To visualize the distributed windings and determine the net effect of various types of winding connections, it is helpful to "split open" and "unroll" the rotor and poles at, for example, the dashed radial segment in Fig. 13.18 in the manner shown in Fig. 13.19. Such a visualization is called a *developed diagram.*

Also, it will be very convenient to have available a simple electrical equivalent for a winding loop. For convenience, the present discussion is restricted to generators. Eventually, however, it will be clear how particular winding connections affect torque in a motor. If the small resistance (source resistance $R_s$) of the loop is ignored, the loop can be well modeled as a time-varying voltage source (see Fig. 13.20).

What number of armature windings coils (which equals the number of commutator segments) is appropriate? That depends on the total generated voltage desired. Possibly, large differences in induced potential may exist between two commutator segments, which could cause sparking at switching time. Sparking across segments can typically occur at about 15 V. Thus the greater the total voltage desired, the more commutator segments there must be to spread the voltage out and eliminate or reduce sparking. (The absence of the possibility of sparking is an advantage of commutatorless ac machines, discussed in Chapter 15.)

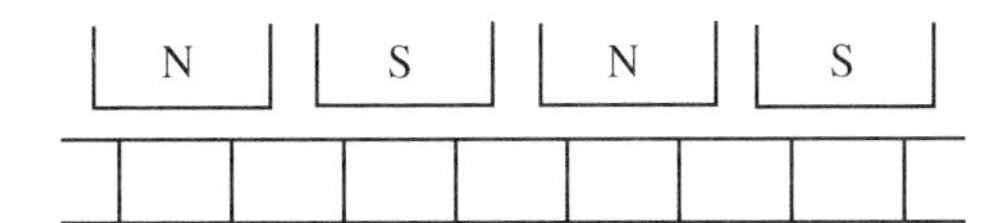


**Figure 13.19** Developed diagram or "unrolling" of field and rotor to facilitate analysis.

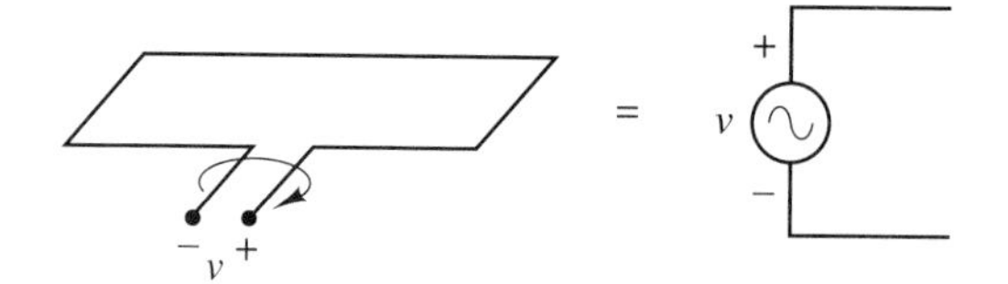


**Figure 13.20** Simplest possible model of a loop in which a voltage is being generated.

How should the armature windings be connected and distributed? There are many possibilities, some of which are more complicated than can be covered in this brief treatment. A few complexities that will be largely ignored in this book are multielement loops and multiplex windings. Multielement loops would result if in the winding diagrams each loop of the winding is replaced by a multiturn coil. *Multiplex windings* would result if the winding winds around the rotor (and returns to the starting point) more than once before electrically closing on itself; for example, twice would be called a *duplex winding.*

Another detail of construction is the fact that sides of two adjacent coils are often physically close together, and in fact are placed in the same rotor slot; this arrangement is therefore called *double entry.* The two most common basic winding arrangements are called *lap* and *wave windings.* Here only *simplex* (once around before closing) lap and wave windings are considered.

In Secs. 13.4.3 and 13.4.4, the characteristics of lap and wave windings are

"Ploughing by electricity." In the early 1880s, two so-called Gramme motors would drag the bidirectional plow back and forth at 266 feet per minute (full speed) by 1/2-inch cables, creating a double furrow. The side wagons themselves were propelled by the motors. Power for the machines was generated by an 8-hp engine 1300 feet from the field. Note the large crowds of interested people on the sidelines.

derived from basic principles. The results are that the generated voltage is larger for wave windings than for lap windings because more coils are in series in wave windings. Conversely, lap windings are appropriate for higher current applications, as they have more coils in parallel.

### ☆ 13.4.3 Lap Windings

One of the meanings of the word "lap" is "to fold over in layers," an apt description of this type of winding. Figure 13.21 shows an eight-turn simplex lap winding for a four-pole field. The commutator-and-winding assembly is assumed rotating such that in the "unrolled" diagram it is traveling to the right. Coils moving off the right edge, of course, actually reenter the left edge. The poles in reality would be located above the page and "look down on" the coils passing by, but this cannot be represented as such in the figure.

To make full use of available current, the number of brushes is equal to the number of poles. If there were fewer, many of the coils would simply be unconnected because *net* voltages over several poles will be zero; this point will be more clear later. Notice that the brushes are opposite the poles instead of between them, as was the case in Fig. 13.16a; the reason for this too will soon be clear.

There is a commutator segment for each of the loops, which are labeled 1 through 8. The long wire shown above the poles is actually no longer than any other loop top when the rotor is "rolled back up." The times zero, $t_1, \ldots, t_7$ are the times at which commutator segment switching takes place. On the diagram they are located so as to indicate the position at which the right edge of coil 1 will be at those times. Therefore, the snapshot shown in Fig. 13.21 is $t = 0$.

The goal of this investigation is to determine the terminal output voltage waveform $v_T(t)$. To accomplish this, in Fig. 13.22 is shown the equivalent electrical circuit of the machine of Fig. 13.21; it, too, is a snapshot for $t = 0$. Notice the absence of poles. Instead, the phases of the equivalent voltage sources for the loops (as shown in Fig. 13.20) differ, depending on the position of the corresponding loop relative to the poles.

To determine the output voltage it is helpful to consider the voltage waveform of an

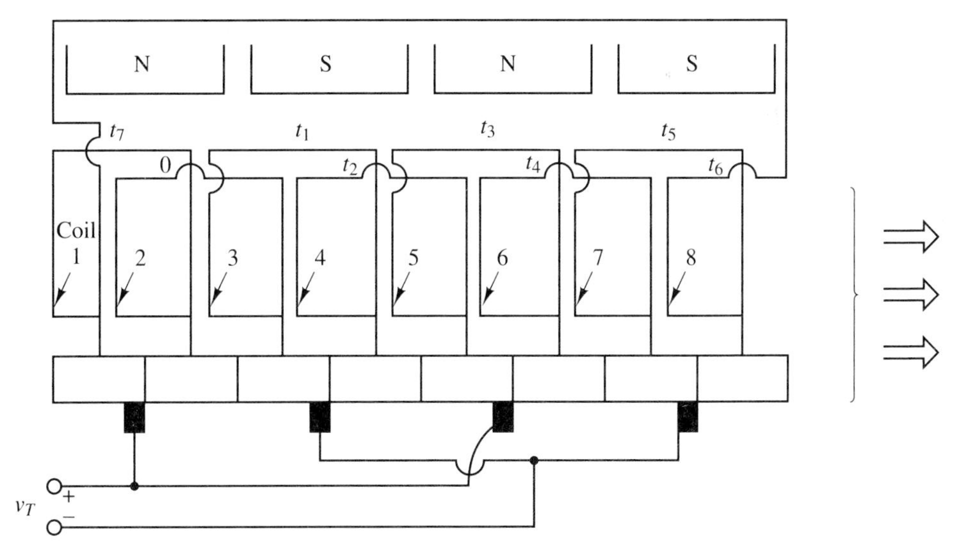


**Figure 13.21** Eight-turn lap winding with commutator and field.

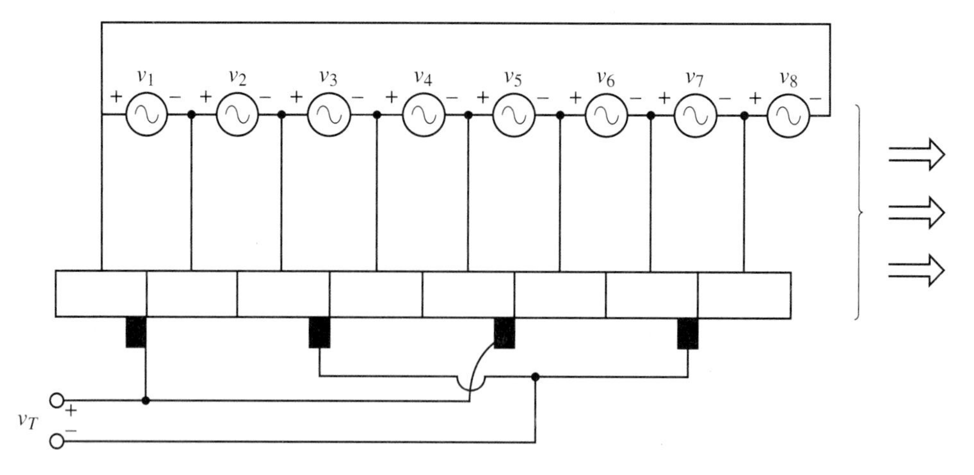


**Figure 13.22** Simplified circuit model of eight-turn lap winding generator.

individual loop as it passes by the "row" of alternating poles (see Fig. 13.23). Recall that the induced voltage is $d\phi/dt$. As the coil moves to the right in Fig. 13.23a, many flux lines toward the south pole are passing through the coil that were not before. At the same time, many flux lines from the north pole that formerly passed through the loop no longer do. Therefore, $d\phi/dt$ and thus the generated voltage are large (and negative).

But as the coil moves a bit to the right in Fig. 13.23b, very little change in flux takes place; the vast majority of flux lines through the loop is the group toward the south pole. Those flux lines are in the center of the loop and are not leaving the coil, and nothing is changing much at the edges; thus in this case, the induced voltage is essentially zero. The conclusion is that when *the right edge* of a coil passes under a north/south pole, the induced voltage in the loop will be large and positive/negative. More generally, we may say that the induced voltage is in phase with the flux *at the coil side "cutting" flux lines* (in this case, the right coil).

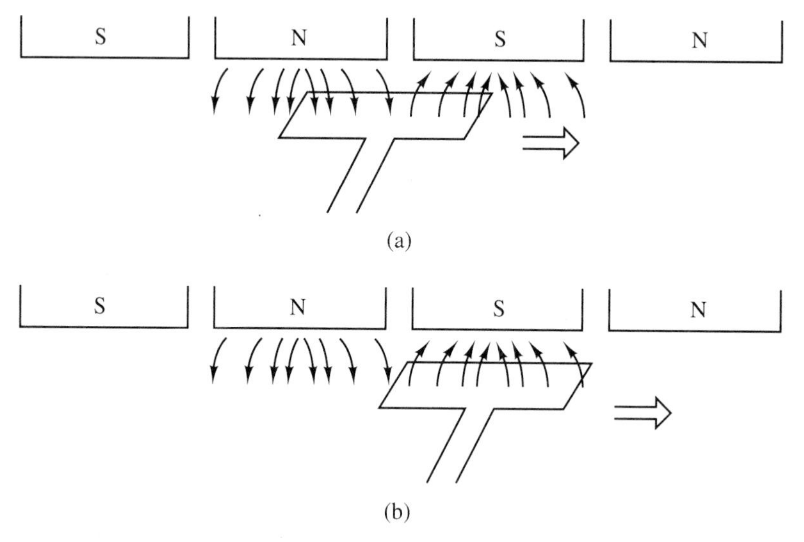


**Figure 13.23** Demonstration that the maximum $d\phi/dt$ occurs in (a) when the right conductor falls under a pole; the minimum occurs in (b) where the right conductor is between poles.

One more relevant fact is that the phases of the induced voltages of the following pairs of loops are equal, due to the fact that the loops are exactly two pole pitches apart: $v_1$ and $v_5$, $v_2$ and $v_6$, $v_3$ and $v_7$, $v_4$ and $v_8$. Careful study of Fig. 13.21, using the facts given above, will verify that the generated voltage waveforms are as shown in Fig. 13.24. Those waveforms were computer generated, using the assumption that the maximum magnitude of the generated voltage in one loop is 1 V. A flat-topped alternating polarity waveform (see Fig. 13.7) was the model for the induced voltage, with brief, linear transitions between the extrema.

Keep in mind that as $t$ increases, the coil and commutator rotate (in Fig. 13.21, move to the right), but the brushes do not. So as the motion proceeds, the segments of the commutator to which the brushes are connected switch repeatedly. Each such switch is indicated in the waveform graphs of Fig. 13.24 by a small arrow. Note that at $t = t_8$, one entire rotation has been completed. Also, each waveform is one-fourth of a north–south pole cycle ahead of the previous waveform.

The terminal voltage for all time can now be found. Recall that Fig. 13.22 was drawn for $t = 0$. In fact, for $0 \le t \le t_1$, using KVL in the electrical diagram (Fig. 13.22) shows that the following equalities are valid:

$$v_T = \begin{cases} v_1 + v_2 \\ -(v_3 + v_4) \\ v_5 + v_6 \\ -(v_7 + v_8), \end{cases} \tag{13.11}$$

which illustrates the fact that there are four parallel paths. It can also be seen from the waveform plots in Fig. 13.24 that

$$(v_1 = v_5) = -(v_3 = v_7) \tag{13.12a}$$

and

$$(v_2 = v_6) = -(v_4 = v_8), \tag{13.12b}$$

which are seen to satisfy the connection requirements of Eq. (13.11).

The questions remain: Is $v_T$ single-polarity, and does it have huge sudden changes in value? For $t < t_1$, it was shown above that, for example, $v_T = v_1 + v_2$. At $t = t_1$, the brushes make contact with the "new" commutator segments formerly to the left. From the electrical diagram in Fig. 13.23, it is found that

$$v_T = \begin{cases} v_8 + v_1 & t_1 < t < t_2 \\ v_7 + v_8 & t_2 < t < t_3 \\ v_6 + v_7 & t_3 < t < t_4 \\ v_5 + v_6 & t_4 < t < t_5, \end{cases} \tag{13.13}$$

and so on. Visually, we can use these formulas in Fig. 13.25 to obtain $v_T(t)$ for a particular time interval by adding the circled voltages for any interval ("column").

The result of adding together the appropriate waveforms for the sequential time intervals (commutator positions) is shown in the computer-generated plot shown in Fig. 13.25b. As before, a piecewise linear form of induced voltage waveform was assumed, as shown in Fig. 13.25a. On the left are plots for a moderately sharp ramp, while on the right a nearly perfect pulse was used. The voltage $v_T$ shown in Fig. 13.25b is always positive: The commutator is functioning.

Notice that the voltage is 2 V most of the time and dips down only 1 V. The maximum is twice that for a single-loop generator rotating at the same speed and exposed to the same flux (rectified version of Fig. 13.25a). The average value of $v_T$ (usually written $E_G$) is somewhat less than 2 V, due to switching-waveform interaction factors. The fraction of time $v_T(t)$ is away from 2 V is reduced for the more perfect pulse shape on the right, thereby increasing $E_G$.

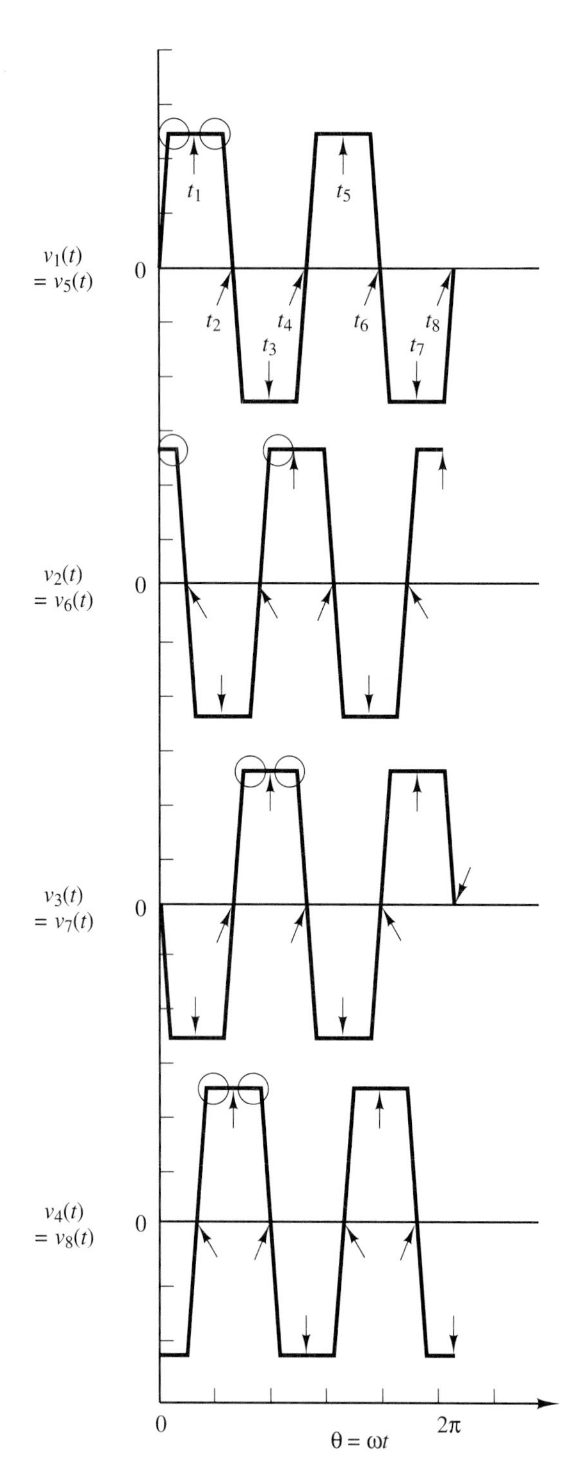


**Figure 13.24** Generated voltage waveforms for all eight lap winding loops. Arrows indicated times of commutator switching and circles indicate windings connected for that time interval, whose sum is the terminal voltage.

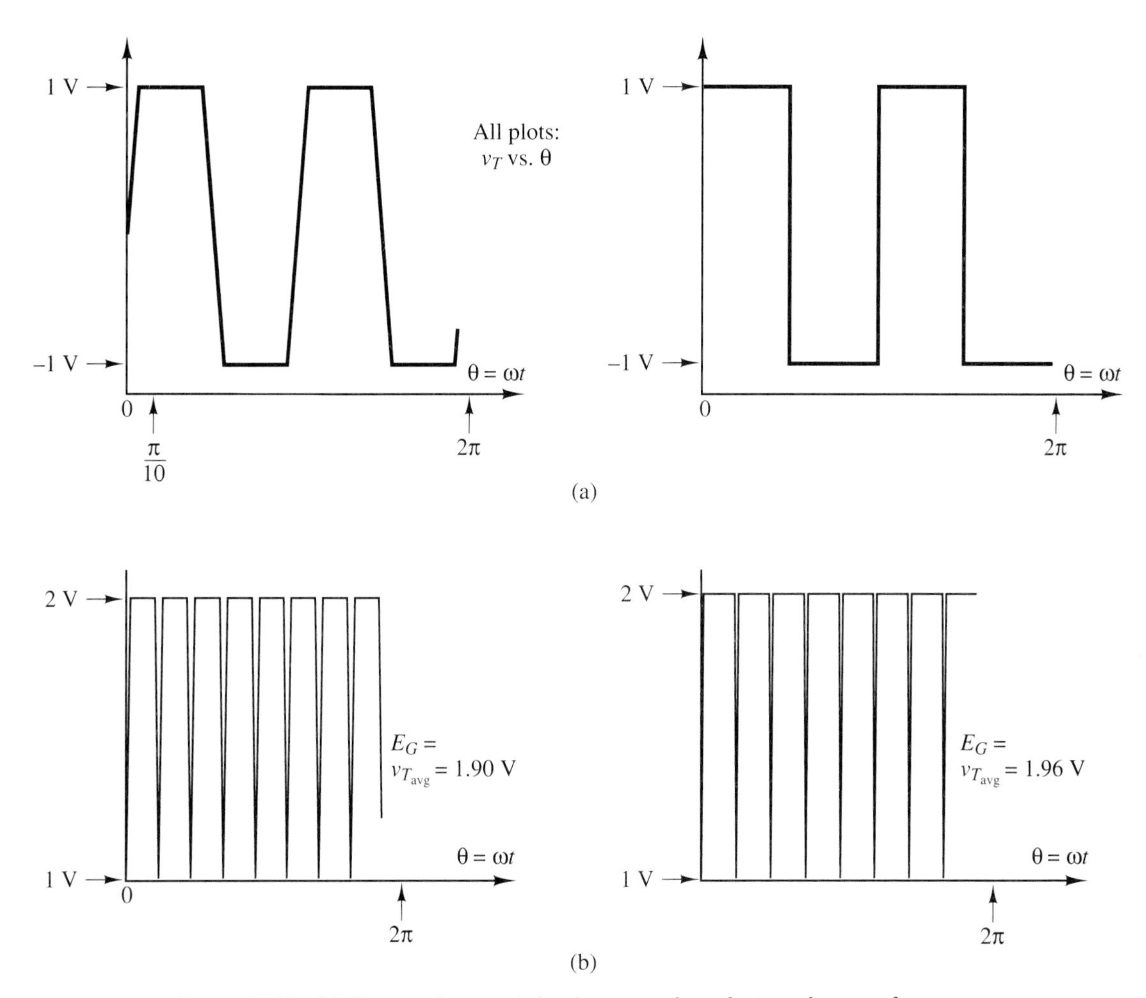


**Figure 13.25** (a) Single-coil generated voltage waveform for two degrees of squareness; (b) corresponding terminal voltage waveforms of eight-turn lap winding generator.

Similar style of analysis could be applied to determine the output current of the generator when connected to a given load, but the analysis is more complicated, so will not be attempted here. The important point is that there are four (the number of poles) parallel branches, each with two (8/4) coils in series. Because of the large number of parallel branches, lap windings are good for low-voltage, high-current applications.

### ☆ 13.4.4 Wave Windings

The *wave winding* is an alternative method of arranging the coils about the rotor. Instead of sequential coils being nearly adjacent as for the lap windings, in wave windings sequential coils each traverse approximately halfway around the rotor, as shown in Fig. 13.26. A study of the diagram will show that the name "wave" is self-explanatory. For this arrangement, each time the winding goes around the rotor, it "gains one" commutator segment on the previous time around ("wave") until it closes on itself (i.e., returns to segment 1—in this case after three

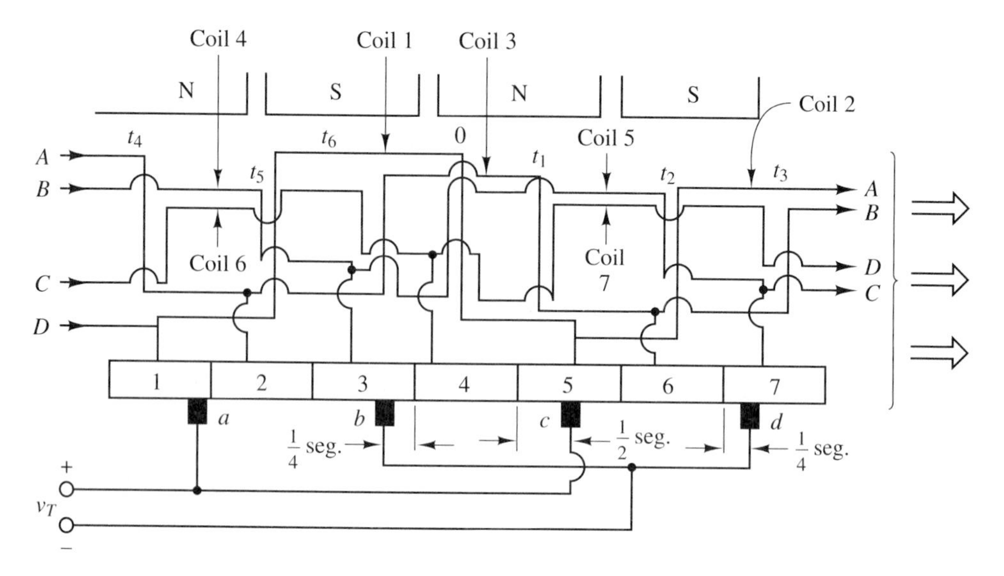


**Figure 13.26** Seven-turn wave winding with commutator and field.

times around). Therefore, this winding is called *progressive wave*; there are also analogously defined digressive wave windings.

While lap windings can have either an even or odd number of commutator segments (although typically even), the number of commutator segments in a wave winding must be odd. Otherwise, the coil would not end up at the starting point after completing the entire winding circuit. So in Fig. 13.26 are shown seven coils instead of eight. Also note that the brushes are not separated by an integral number of commutator segment pitches. This is because while now, as before, the brushes must be evenly spaced around the commutator circumference, the number of segments is now odd.

It would seem reasonable to have a brush located at exactly each quarter of the way around the rotor. However, note that there is *always* a coil connected across brushes $a$ and $c$ that is approximately directly under the south pole (at the time shown, $t = 0$, this is coil 1). It remains so for essentially the entire time it is connected across brushes $a$ and $c$. From the previous discussions, a coil directly under a pole has its sides midway between poles; hence the induced voltage is nearly zero: $v_{ac} = 0$. For large numbers of coils and commutator segments, there is not much motion of a conductor over one commutator interval, so $v_{ac}$ will always remain zero. Therefore, brush $c$ is actually not needed and often will not be included. The same reasoning shows that brush $d$ is unnecessary.

Close examination of Fig. 13.26 shows that for lap windings all the loop voltages have differing phases. In fact, each coil is 1/(number of coils) of a single north–south period ahead of the previous coil. The waveforms are readily obtained from Fig. 13.26 and are shown in Fig. 13.27.

The electric circuit equivalent, shown with only the two necessary brushes, is shown in Fig. 13.28. This time the form for $v_T(t)$ is more involved. Let $\delta = \frac{3}{4}t_1$, where $t_1$ is the time for one commutator segment to pass by. Then $v_T(t)$ has the following form:

$$v_T(t) = \begin{cases} v_1 + v_2 + v_3 + v_4 \quad [= -(v_5 + v_6 + v_7)] & 0 < t < \delta \\ v_1 + v_2 & \delta < t < t_1 \\ v_6 + v_7 + v_1 + v_2 \quad [= -(v_5 + v_4 + v_3)] & t_1 < t < t_1 + \delta \\ v_6 + v_7 & t_1 + \delta < t < t_2, \end{cases} \tag{13.14}$$

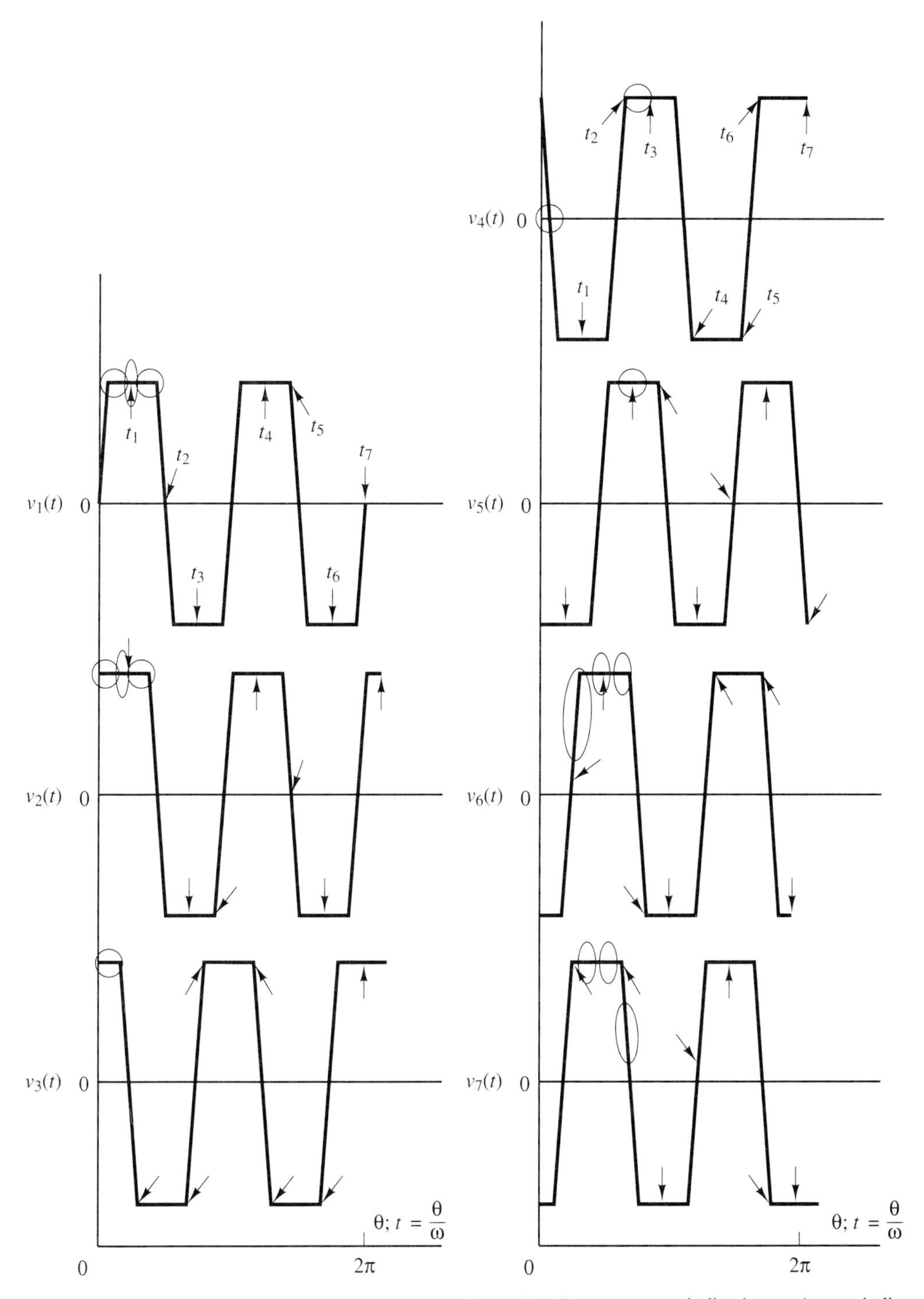


**Figure 13.27** Generated voltage waveforms for all seven wave winding loops. Arrows indicated times of commutator switching and circles indicate windings connected for that time interval, whose sum is the terminal voltage.

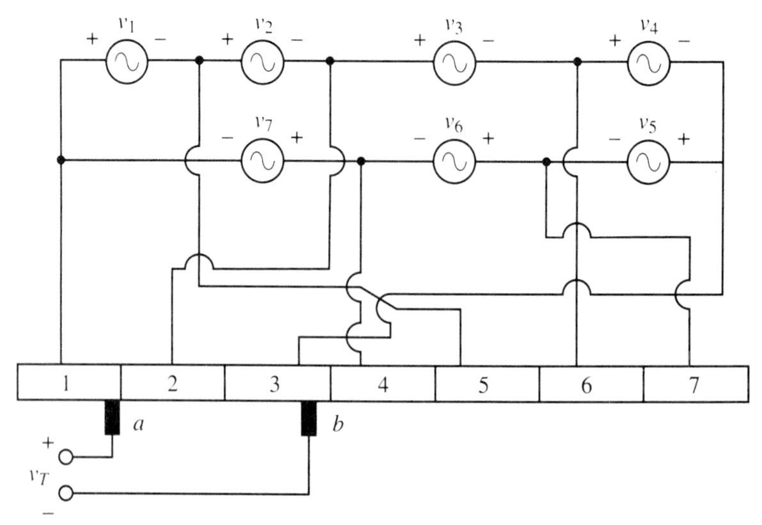


**Figure 13.28** Simplified circuit model of seven-turn wave winding generator.

and so on. The quantity $t_1 - \delta = t_1/(\text{number of poles})$ is usually very small for large numbers of coils and poles. So for most of the time, $v_T(t)$ is made up of four (in this case) series coils, which tends to make the average value of $v_T(t)$ higher than for a lap winding. Also, there are only two parallel paths in wave windings, so the current capacity will be less than that for a lap winding.

Computer-generated plots of $v_T(t)$ are shown in Fig. 13.29a for the seven-coil wave winding, and in Fig. 13.29b for a 21-coil wave winding. The basic single-turn waveforms in Fig. 13.25a were used to generate the left and right plots, respectively (as in Fig. 13.25b). Notice that the average generated voltage for the seven-coil winding is larger than that of the eight-coil lap winding in Fig. 13.25b even though it has one *less* coil. This is of course due to the series characteristic of wave windings.

The behavior for the seven-coil winding is rather poor; its time variation is very large. For the larger number of turns in Fig. 13.29b, things look much better; the percent ripple is much reduced, and $v_T$ spends a much greater percentage of time at its maximum value.

### 13.4.5 Adjacent Coil Separation

A significant discrepancy between the seven- or eight-coil windings discussed in Secs. 13.4.3 and 13.4.4 and windings actually found in motors is the spacing between adjacent coils. It can be seen from Fig. 13.21 that for lap windings, adjacent coils are exactly one-half a pole pitch apart. For wave windings, Fig. 13.26 shows that adjacent coils are approximately one-half a pole pitch apart.

Were this true in actual motors, it could be one of several possible causes for sparking because the maximum potential of a single loop could appear between adjacent commutator segments at the time of brush contact. The mechanism for such sparking is discussed in Sec. 13.4.6. This large discrepancy in potential between adjacent commutator segments is also partially responsible for the output voltage (particularly that for the wave winding) having a very large ripple.

In actual dynamos, having hundreds of windings, the coil sides of adjacent coils are physically very close together (see Fig. 13.30). The potentials of each of the coil ends are about the same and nearly zero at the point of brush contact. (This is true because of the brush–poles orientation if interpoles, which are discussed next, are

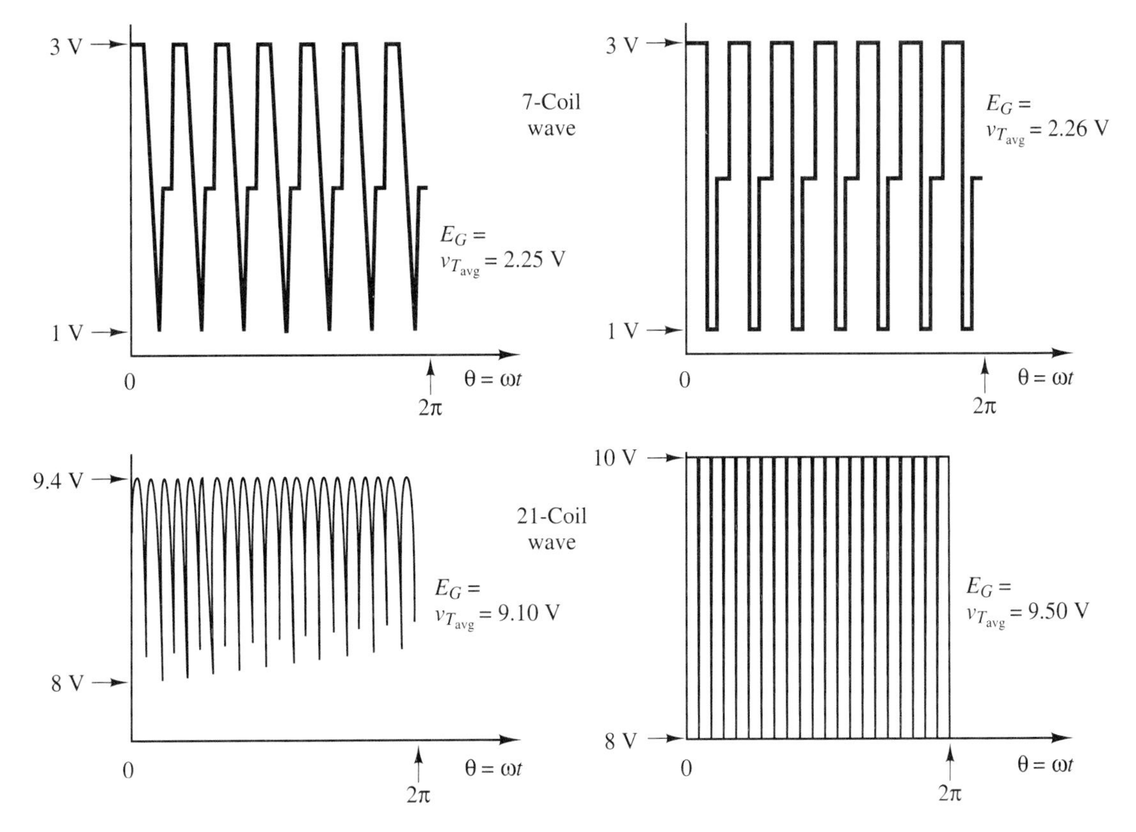


**Figure 13.29** Terminal voltage waveforms of seven-turn wave winding generator for two degrees of squareness of a single-turn waveform.

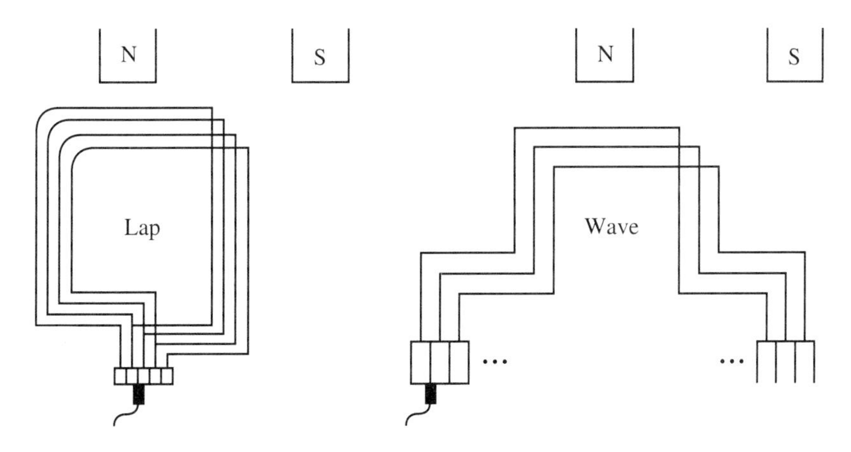


**Figure 13.30** Successively connected practical many-loop windings are close together with respect to the field poles.

used.) This situation, due to the large number of loops, results in a decreased amount of ripple in the output voltage waveform. Of course the reason for analyzing seven- or eight-coil windings was the relative simplicity of the winding and circuit diagrams, and of the construction of the total voltage waveform.

### 13.4.6 Armature Reaction

As mentioned briefly in Sec. 13.3.3, an effect of armature current is the creation of a distorting magnetic field called the *armature reaction*. For the two-pole, single-loop case it was found that the armature reaction field was perpendicular to the rotating loop face and so was time varying. For a wound-rotor two-pole machine, many components of armature reaction from all the loops sum to a relatively constant armature reaction field, as shown in Fig. 13.31. There it is assumed that the rotor is turning counterclockwise, so the applied current (which exceeds the induced countercurrent) flows clockwise when viewed from the top. Of course, it is the total (net) current that forms the armature reaction field, so the net current dictates its direction. Note that the armature reaction is the $B$ resulting from the TOTAL current. In a motor, this total current is the difference between the applied and induced currents, while in a generator it is the induced current.

Figure 13.31 shows (a) the rotating rotor with total armature current directions, (b) the induced current and its associated field (which is opposite in direction to but is overcome by the field due to the applied current), (c) the field with no rotating armature, and (d) the armature reaction field alone (which is valid for a motor; for a generator, it would have opposite direction because then $I_{APP} = 0$, so $I_{IND} = I_{TOT}$, so $I_{TOT}$ has the same direction as $I_{APP}$). Figure 13.31e shows the appropriate flux density vector sum diagram and also the total field (for clarity, omitting the rotor). In Fig. 13.31e, the vertical line is the *mechanical neutral*, where the brushes should be placed in the absence of armature reaction to minimize voltages on coils being switched.

The line at an angle $\theta$ from the vertical is the equivalent such line when the armature reaction field is included, called the *magnetic neutral*. A problem with the remedy of physically shifting the brushes is that $\theta$ depends on the load current $I_A$ (for a generator). The better solution is to leave the brushes at the mechanical neutral and introduce *interpole* windings physically in between the poles and electrically in series with the armature winding, as shown in Fig. 13.32. (Actually, in practice the poles may again be slightly shifted to minimize sparking for a typical load.) The magnetic field resulting from the armature current passing through these interpole windings is designed to cancel the armature reaction field. Because this field is proportional to the armature current, the cancellation can be made fairly complete whatever the load.

Certain electric toy race cars have an adjustment to maximize speed. The adjustment is alignment of the brushes with the magnetic neutral. In the case of such small permanent-magnet motors, there are no interpole windings.

For more than two poles, the situation is largely the same. Examining either lap (Fig. 13.21) or wave (Fig. 13.26) windings shows that, ignoring armature reaction, there is nearly zero $d\phi/dt$ at switching time and therefore optimal voltage rectification in any particular loop.

Of course, there may be currents introduced from other in-series coils that are

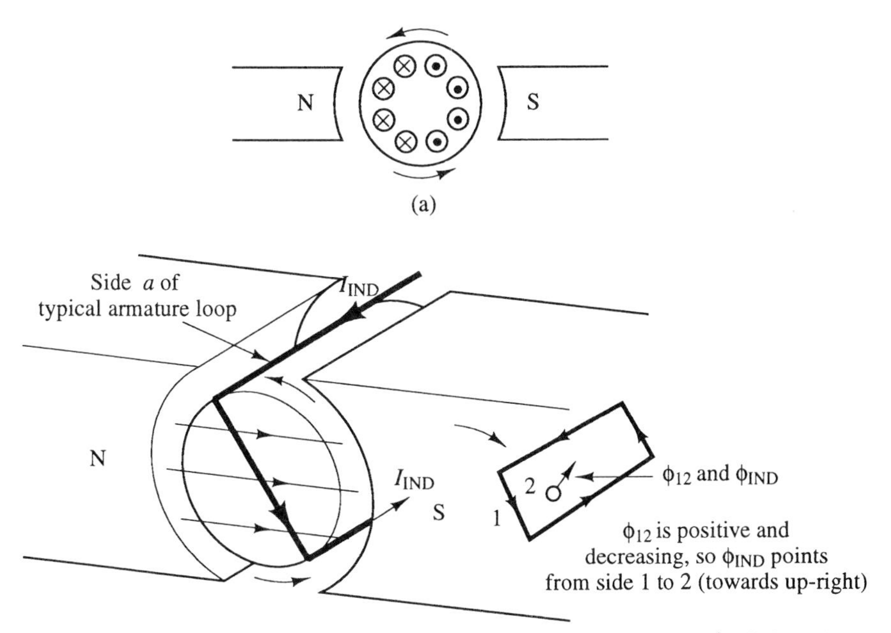


$I_{APP}$ is in *opposite* direction from $I_{IND}$ for CCW rotor rotation (check $I_{TOT}\ell \times \mathbf{B}$ for side $a$ where $I_{TOT}$ and $I_{APP}$ are in the same direction because $|I_{APP}| > |I_{IND}|$ assuming a load is connected, so that the input electrical power $I_A V_{APP} > 0$, where $I_A$ is the total armature current. Note that $I_A = 0$ for frictionless motor operation ($P_{in} = 0$). Note: For all currents and discussion in this figure, we are assuming operation as a motor (not a generator).

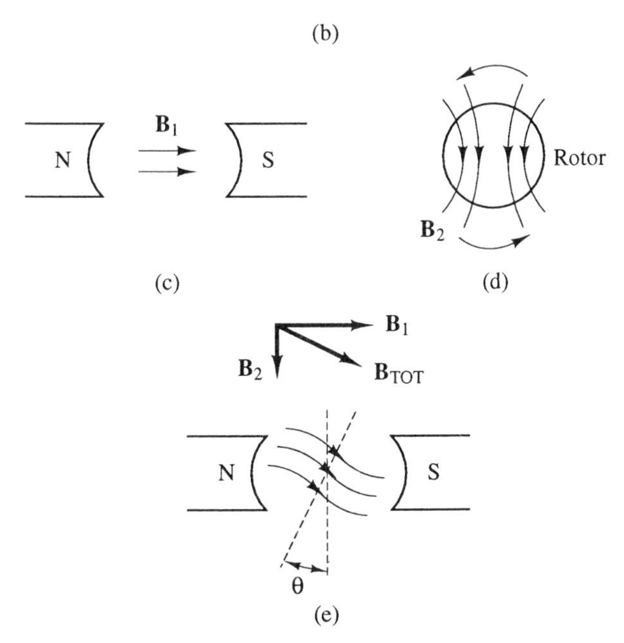


**Figure 13.31** (a) Directions of total current in a rotating rotor; (b) direction of induced current in the rotor; (c) background field; (d) rotor armature reaction field; (e) resultant field. The magnetic neutral angle is shifted an amount depending on the rotor current.

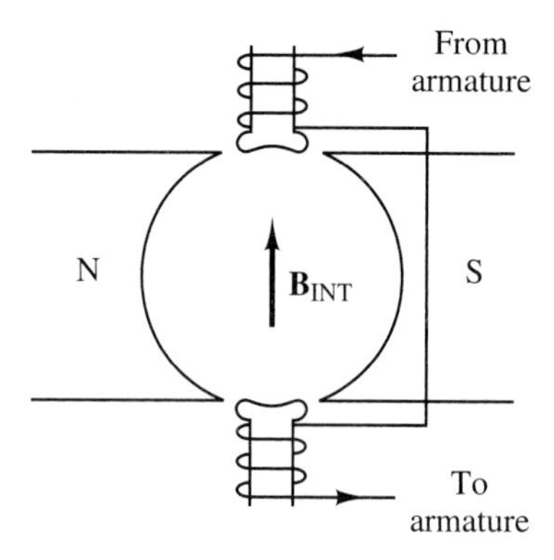


**Figure 13.32** Interpole and its corrective magnetic field.

experiencing a nonzero $d\phi/dt$. After all, the brushes *must* carry the load current, which presumably is large and nonzero. However, in this *one* coil at the brush, $di/dt$ is *large*; the loop current is in the process of rapidly changing from one extremum to the other. Consequently, the voltage of *self-induction* $L\ di/dt$ (as opposed to the field-induced voltage $d\phi/dt$) is quite large. The distinction between $d\phi/dt$ and $L\ di/dt$ is the same as that between mutual and self-induction in a transformer (see Sec. 6.5).

The current associated with the self-induction voltage may circulate within the individual coil during the time when the brush short-circuits its ends, as shown in Fig. 13.33. But when the brush no longer completes the circuit, the short-circuit current suddenly stops, causing $v = L\ di/dt$ to become *very* large. This voltage can be responsible for sparking.

The interpoles therefore must supply flux sufficient to (1) make $d\phi/dt$ nearly zero for the connected coils (for good output voltage behavior), and (2) provide an emf within the connected coil to cancel (at least partially) the *reaction voltage* $v = L\ di/dt$ (and thereby minimize sparking). (*Note:* This discussion on commutation and sparking is greatly simplified; see Del Toro and Cook for more details.* However, the discussion above includes some of the fundamental mechanisms involved.)

## 13.5 FUNDAMENTAL DYNAMO EQUATIONS

### 13.5.1 Average Induced Voltage

At this point the reader should find very easy to comprehend the general, approximate expression for the average value of generated voltage, $E_G$ (which, ignoring armature impedance is the average value of $v_T$):

$$E_G = \frac{ZPn\phi}{60a} = k_e n\phi \approx v_{T,\text{avg}} \tag{13.15}$$

where

$$k_e = \frac{ZP}{60a}, \tag{13.16}$$

*A. L. Cook, C. C. Carr, *Elements of Electrical Engineering* (John Wiley & Sons, NY, 1954), pp. 147–150.

V. Del Toro, *Basic Electric Machines* (Prentice Hall, Englewood Cliffs, NJ, 1990), pp. 327–331.

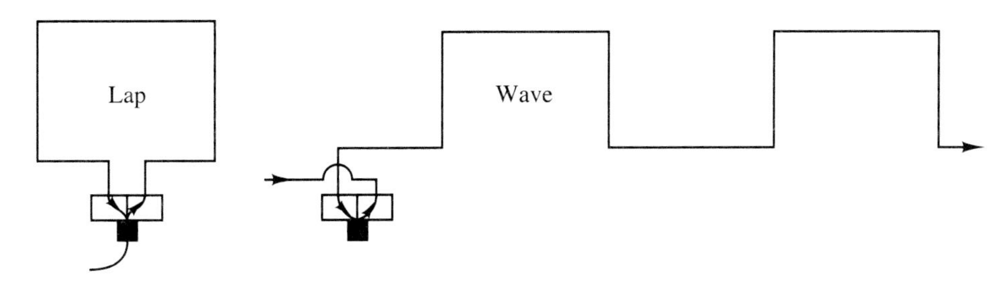


**Figure 13.33** Shorting and breaking as one cause of sparking in lap and wave windings.

in which $P$ is the number of poles, $Z$ the number of conductors in the armature slots (*not* an impedance), $a$ the number of parallel paths on the armature, and $n/60$ the number of revolutions per second ($n$ being the number of revolutions per minute). The number of parallel paths $a$ is equal to $P$ for lap windings and 2 for wave windings.

It makes sense that $E_G$ is proportional to $P$, because for larger $P$, more coils (whose voltages are added together) are under $d\phi/dt \gg 0$ conditions at any one time. There are two "conductors" per loop in the armature slots. Therefore, it is also reasonable that $E_G$ should be proportional to $Z/a$ because $Z/a$ is (twice) the number of coils connected in series. Note that $k_e$ has been defined as a machine constant that is independent of particular operating and load conditions.

The expression $E_G = k_e n\phi$ thus clarifies and emphasizes the fact that $E_G$ is proportional to both the speed and the flux. If we recall that induced voltage in a single loop equals $d\phi/dt$, and remember that it is rectified, averaged, and summed with voltages from other loops to form $E_G$, $E_G = k_e n\phi$ makes sense. For increasing either $n$ or $\phi$ will increase $d\phi/dt$ and therefore the induced voltage.

If the parameters of the two-pole one-turn generator are substituted into Eq. (13.15) ($Z = 2$, $a = 1$, $P = 2$) then $E_G = 4n\phi/60$, just as derived back in Eq. (13.9). Suppose that $P = 4$, $\phi = \frac{1}{8}$ Wb, and $n = 60$ rpm. Then $E_G = Z/(2a)$. For a one-turn coil (Fig. 13.26a), $E_G$ is then $2/2 = 1$ V. For the eight-turn lap winding considered previously, $E_G = 16/8 = 2$ V, which agrees well with the computer (waveform-based) calculation. For the seven-turn wave winding, $E_G = 14/4 = 3.5$ V, which exceeds the computer estimation by a large percentage. But for the 21-turn wave winding, $E_G = 42/4 = 10.5$ V, which is closer to the waveform-based estimate.

In real machines, $Z$ is on the order of hundreds, not tens, and for $Z$ this large, $E_G$ in Eq. (13.15) is sufficiently accurate. Anyway, $\phi$ is typically not known accurately; it is the *proportionalities* (which *are* approximately valid for $Z$ large) that are most important. Also important, in summary, are the relations $E_{G,\text{wave}} > E_{G,\text{lap}}$ and $I_{\text{lap}} > I_{\text{wave}}$, where the currents are those that can be delivered without severe loading effects.

**Example 13.5**

Consider a four-pole dc generator with an armature composed of 100 loops. The armature spins in a flux of 0.1 Wb at a speed of 600 rpm. What is $E_G$ if the generator is (a) lap wound versus (b) wave wound?

**Solution** For 100 loops, $Z = 200$, so in Eq. (13.15), $E_G = 200(4)(600)(0.1)/(60a) = 800/a$. For lap windings, $a = P = 4$, while for wave windings $a = 2$. So the respective average terminal voltages $E_G$ are (a) 200 V and (b) 400 V.

### 13.5.2 Average Induced Torque

Just as $E_G = ZPn\phi/(60a)$ was a multiloop generalization of $v_{T,\text{avg}} = 4n\phi/60$ for a single loop, two-pole generator, there is a similar multiloop generalization of the torque $T = 2ILBR$ for a single loop motor. The total average torque developed is

$$T = Z'I_{\text{loop}}LBR, \tag{13.17}$$

where $Z'$ is the number of conductors "directly under" poles. Obviously, $Z'$ is an approximation, as there are many conductors only partly under the influence of a pole. As before, $R$ is the radius of the armature, $L$ the length of a side of a current-carrying coil "cutting the flux lines," and $B$ the per-pole magnetic flux density. So the only difference from the single-loop case is that now there are many more conductors experiencing torque, thereby contributing to the total average induced torque.

It is helpful to express $Z'$ in terms of the total $Z$ and $I_{\text{loop}}$ in terms of the total armature current $I_A$. Figure 13.34 shows that the number of conductors "directly under" a pole will be the total $Z$ times the fraction of circumference covered by poles, namely $(Px)/(2\pi R)$, where $x$ is the width of a pole, which is usually roughly equal to the width of an armature coil loop. Thus $x/(2\pi R)$ is roughly the fraction of the circumference occupied by poles. Note that $L$ is the depth of the rotor, and $\phi = BLx$ because $LX$ is the loop area. The loop current $I_{\text{loop}}$ is simply the total armature current $I_A$ divided by the number of parallel paths $a$. Substituting these relations into Eq. (13.17) gives

$$\begin{aligned} T &= \frac{ZPxI_A}{2\pi Ra}(LBR) \\ &= \frac{ZPI_A\phi}{2\pi a} \\ &= k_t I_A \phi, \end{aligned} \tag{13.18}$$

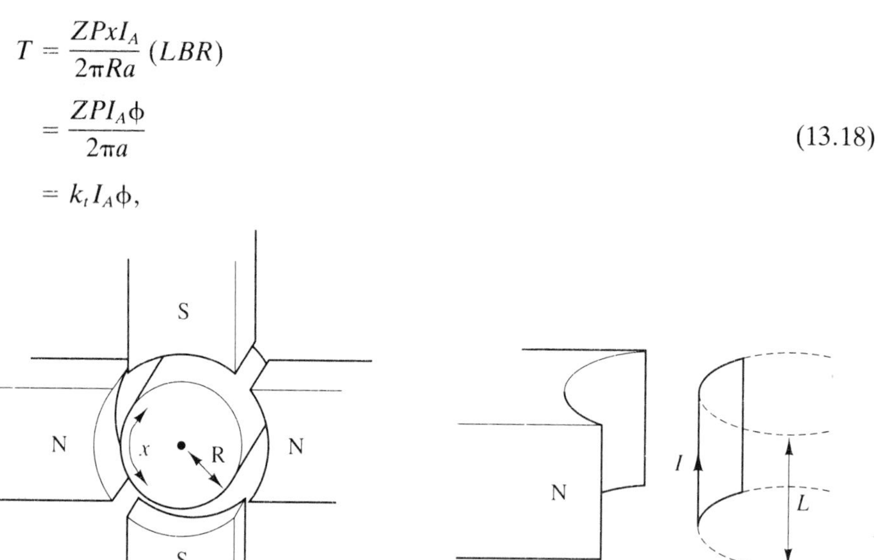


Top view
(a)

Side view
(b)

**Figure 13.34** (a) Top and (b) side views of a rotor, with parameters required for calculation of the approximate average torque on the rotor.

where

$$k_t = \frac{60k_e}{2\pi} \tag{13.19}$$

and where $B = \phi/(Lx)$ was used, $Lx$ being the area of a loop. The constant $k_t$ emphasizes the fact that the torque is proportional to both the armature current and the flux. If it is remembered that magnetic force equals $ILB = L(IB)$, and that flux is proportional to $B$, we see that Eq. (13.18) matches exactly the dependence we would expect. Incidentally, the loop in Fig. 13.34 in the position shown experiences zero torque because the torques on the near and far vertical conductors cancel. Maximum torque occurs halfway between poles, similar to the phenomenon in Fig. 13.24 for generated voltage.

**Example 13.6**

Suppose that the machine of Example 13.5 carries an armature current of 5 A. What torque opposes the motion of the rotor?

**Solution** By Eq. (13.18), the torque is $T = 200(4)(5)(0.1)/(2\pi a) = 63.7/a$ N · m. Therefore, the torque is 16 N · m for a lap-wound rotor and 32 N · m for a wave-wound rotor.

It should be remembered that the same torque expression applies for motor action as for the opposing torque in a generator. One torque merely aids rotation, while the other opposes it. Similarly, the general expression for the average generated voltage (often called "generated emf") discussed in Sec. 13.5.1, is the correct one for the *counter-* or *back-emf:*

$$E_B = \frac{ZPn\phi}{60a} = k_e n\phi. \tag{13.20}$$

Equation (13.20) looks suspiciously close to the torque expression in Eq. (13.18). Recall that electrical power is voltage times current and that rotational power is angular velocity times torque. Here $\omega = n(2\pi/60)$, so

$$P_{\text{electrical}} = E_B I_A = \frac{ZPn\phi I_A}{60a} \tag{13.21}$$

$$P_{\text{mechanical}} = \omega T = \frac{ZPI_A\phi}{2\pi a} \cdot \frac{2\pi n}{60} = E_B I_A = P_{\text{electrical}}. \tag{13.22}$$

This is merely a conservation principle; mechanical power delivered equals electrical power developed (ignoring losses).

### 13.5.3 Terminal versus Induced Voltage

It is of interest to quantify the relation between the *actual* terminal voltage of a dynamo and its (average) induced voltage. They are not the same because of losses in the resistance of the armature coil, $R_A$. (In the steady state, the terminal voltage and current are nearly constant in time for dc machines, on which we focus in this introduction. Consequently, the impedance of the armature coil can be well modeled as purely resistive.)

Very simple circuit equivalents are given in Fig. 13.35 for a generator

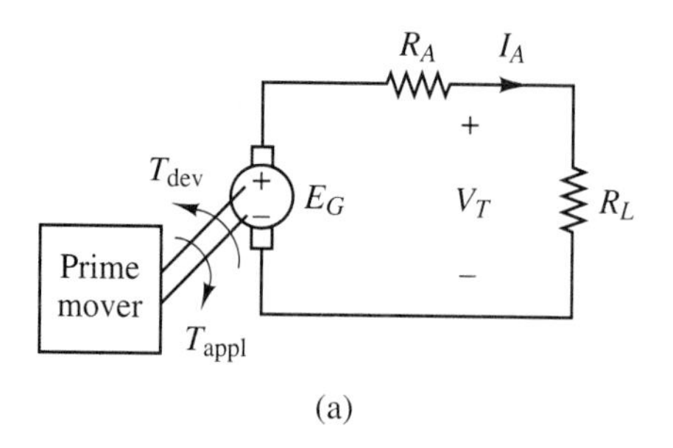


(a)

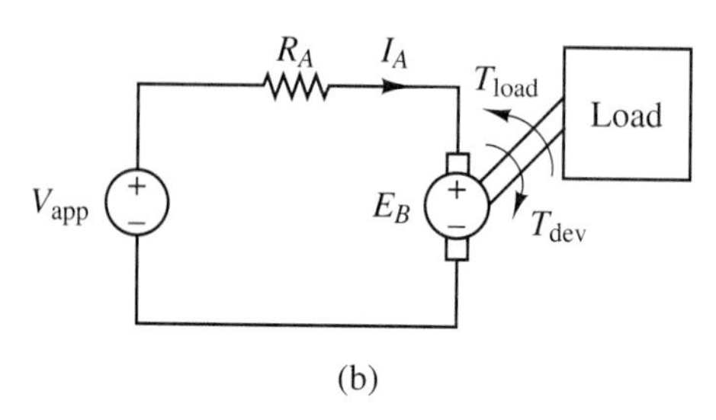


(b)

**Figure 13.35** Simple circuit diagrams for a dc machine: (a) generator; (b) motor.

(Fig. 13.35a) and a motor (Fig. 13.35b). Note that $E_G$ and $E_B$ are, of course, constant voltage sources given by Eq. (13.15); the new symbol refers to the fact that they are produced by a dynamo with rotation. Also shown are the prime mover (e.g., turbine) that turns the generator, the load driven by the motor, and the directions of applied and developed (opposing) torques. Similarly, Fig. 13.35b shows the load and the opposing developed and load torques. In Fig. 13.35a the actual rotation direction is that of the applied torque, and in Fig. 13.35b it is that of the developed torque, assuming that each machine is operating as intended.

In both cases, it is tacitly assumed that the background field is generated by some external circuit; this is *separate excitation*. Also, in circuit diagrams, the symbols for the prime mover/load and opposing torques are usually omitted for simplicity.

Because electrical power flows *from* a generator and *into* a motor, the polarity conventions for the armature current $I_A$ differ for the two dynamo uses. The induced voltage for a generator is labeled $E_G$ for the generator and $E_B$ for the motor. The terminal voltage is labeled $V_T$ for the generator and $V_{APP}$ (for applied) for the motor. By KVL, the relations are then

$$V_T = E_G - I_A R_A \qquad \text{(generator)} \tag{13.23a}$$

$$V_{APP} = E_B + I_A R_A \qquad \text{(motor).} \tag{13.23b}$$

In a typical motor operating at rated conditions, $E_B$ is about 0.8 or 0.9 times $V_{APP}$; similar relations between $E_G$ and $V_T$ apply for a generator. There are many examples using Eqs. (13.23) in Chapter 14.

If Eqs. (13.23) are multiplied by $I_A$, simple power conservation equations result:

$$V_T I_A = E_G I_A - I_A^2 R_A \qquad \text{(generator)} \tag{13.24a}$$

$$V_{APP} I_A = E_B I_A + I_A^2 R_A \qquad \text{(motor).} \tag{13.24b}$$

For the generator/motor, the first term is the output/input electrical power, the second term is equal to the input/output mechanical power $\omega T$, and the third term represents ohmic heat loss in the armature windings.

### 13.5.4 Fundamental Speed Equation

In a generator, the speed is presumably fixed by the *prime mover* that cranks the rotor; solving for the speed is therefore not particularly helpful. In the case of the generator, the dependent variable is the generated voltage (and possibly the flux). But for a motor it is often desired to calculate the speed for a given operating condition; now the speed is the dependent variable. From the equation for the back emf it is easy to find:

$$n = \frac{E_B}{k_e \phi} = \frac{V_{APP} - I_A R_A}{k_e \phi}. \tag{13.25}$$

This equation will prove very helpful in the discussion of motors in Chapter 14.

### 13.5.5 Conservation of Angular Momentum

Newton's second law is a statement of conservation of linear momentum; the net force on an object is equal to the rate of change of its linear momentum $m\mathbf{u}$: $\mathbf{F} = m \cdot d\mathbf{u}/dt$. For rotational dynamics, force is replaced by net torque $T_{net}$ on the rotor, mass by moment of inertia $J$, and velocity by radial velocity $\omega$:

$$T_{net} = \frac{J d\omega}{dt}. \tag{13.26}$$

In electromechanical machines,

$$T_{net} = T_{app} - T_{dev} - T_{rot} \quad \text{(generator)} \tag{13.27a}$$

$$T_{net} = T_{dev} - T_{load} - T_{rot} \quad \text{(motor)}, \tag{13.27b}$$

where $T_{app}$ is the cranking torque, $T_{dev}$ the developed (induced) torque, $T_{rot}$ the frictional torque in the absence of any other physical loading, and $T_{load}$ the load torque that the motor is driving.

The important point here is that when there is an imbalance of torques (i.e., $T_{net} \neq 0$), $d\omega/dt \neq 0$ and the motor will change speed [recall that $n = (\pi/30)\omega$]. This equation is very helpful in describing how a motor will start from rest and build up to a steady-state speed, as well as describing behavior under a change of load.

## 13.6 SUMMARY

Basic concepts involved in electromechanical energy conversion have been the subject of this chapter. First we reviewed a few practical facts concerning magnetic materials, such as the saturation of the flux density $B$, the associated hysteresis phenomenon between $B$ and $H$, and eddy current loss mechanisms.

Each of these properties has ramifications in machine operation and design. For example, to minimize eddy current losses, iron cores are often made in layers (laminated). And both types of losses, as well as $I^2R$ losses, limit the power rating of the machine because of possible overheating. The saturation property is shown in Chapter 14, to allow a dc generator to function despite an increased load. Finally, many small machines are based on permanent magnets, which exist only because of the residual $B$ in the hysteresis loop. Indeed, the starting of some generators depends on residual magnetism even though once started, that region of the hysteresis loop is no longer critical.

For simplicity of analysis, design, and construction, many magnetic devices are designed to have high degrees of symmetry. With such symmetry, the equations of electromagnetism take on very simple forms. The magnetic analysis of such devices is called magnetic circuit analysis.

One result of particular importance was that air gaps "soak up energy." That is, to set up a certain flux in a magnetic device requires much larger current if an air gap is present than if not. And when it is there, the highest energy concentrations may be found in the gap. This is actually a useful property of air gaps in machines, for it is in the air gap that the forces on/induction in conductors occurs.

The simplest machines, both generators and motors, were considered next. Faraday's law and Ampère's force are both crucial to the operation of any electromechanical machine. First to be considered was the linear generator. As the conducting crossbar moved along the rails, the flux passing through the loop consisting of part of [the rails, the crossbar, and the voltmeter circuit] changed with time. Consequently, a voltage was developed that could in principle be used to perform electrical work.

A more practical machine that is hardly any more complicated is the rotating loop generator. When spun in a uniform magnetic field, the generated voltage is an oscillating flat-topped waveform. Although the average value of this voltage is zero, the average of its absolute value is not. So if a commutator is used to connect the loop alternately in reversing polarity, a rectified voltage with a nonzero average value results. When a load is connected, the current that flows interacts with the magnetic field to produce forces on the loop conductors tending to slow the generator down. These currents also produce a magnetic field that distorts the background field; the phenomenon is called armature reaction.

When the prime mover is removed and in its place a mechanical load is connected, and when the electrical load is replaced by an electrical power source, the generator becomes a motor. Now there is still generator action, but it produces currents that oppose the applied currents and act as a drain on the applied electrical power source. Otherwise, the principles of operation are similar to the effects occurring in a generator.

Subsequently, some terminology used repeatedly in the study of machines was defined. The distinctions between terms such as armature, field, stator, rotor, back emf, and pole pitch should now all be clear. Then a more detailed discussion on the commutator and windings was presented.

The armature windings are distributed over the surface of the rotor in patterns such as lap or wave windings. In detailed discussions on these windings, it was shown both with circuit analysis and computer simulation that the generated voltage is larger for wave windings than for lap windings because more coils are in series in wave windings. Conversely, lap windings are appropriate for higher current applications, as they have more coils in parallel. Distinction in qualitative behavior between simple eight-loop windings and windings found in actual machines showed that in the latter case the performance is more ideal. A further look at armature reaction suggested the use of interpoles to minimize the deleterious effects it brings about (in particular, sparking and consequent destruction of brushes and commutator segments).

At this point the reader is in a position to comprehend the basic approximate equations of dynamos: Average generated voltage, average induced torque, terminal/applied versus generated/back emf equations, the fundamental speed equation, and conservation of angular momentum and power. These last relations form the basis of the theory of dc and ac machines considered in Chapters 14 and 15. At last, practical machines will be analyzed with simple terminal relations, and application examples and issues will be discussed.

The studies in this chapter should provide a solid basis from which to proceed to the real-world analyses and issues now to be presented.

## PROBLEMS

**13.1.** What factors cause the loss of energy (magnetic heating) in magnetic material?

**13.2.** Explain how eddy currents are produced and how they can be minimized.

**13.3.** Review Chapter 3. Compare magnetic circuits with electric circuits. State the analogies that apply.

**13.4.** What are the relationships between $\phi$, $B$, $\mathcal{R}$, $\mu_r$, and $\mathcal{F}$, and what are the equations for a magnetic circuit?

**13.5.** What is flux fringing? Explain why the maximum magnetic potential drop occurs in the air gap.

**13.6.** A ferromagnetic ring with a mean diameter of 20 cm is wound with 300 turns of wire. The current in the wire is 2.0 A and produces a magnetic flux density $B$ of 0.9 T. Assuming a linear $B$–$H$ relation and a cross-sectional area of 4 cm$^2$ for the ring, calculate the flux $\phi$, the reluctance $\mathcal{R}$, and the relative permeability $\mu_r$ of the ring.

**13.7.** The toroidal ring with rectangular cross section shown in Fig. P13.7 has a current of 1.5 A, $r_i = 12$ cm, $r_o = 15$ cm, $h = 1.5$ cm, and $\mu_r = 4000$.

**(a)** If the flux $\phi$ in the ring is 0.2 mWb, determine the number of required turns, N.

**(b)** Find the magnitude of $B$; is it proper to assume that $B = \mu H$?

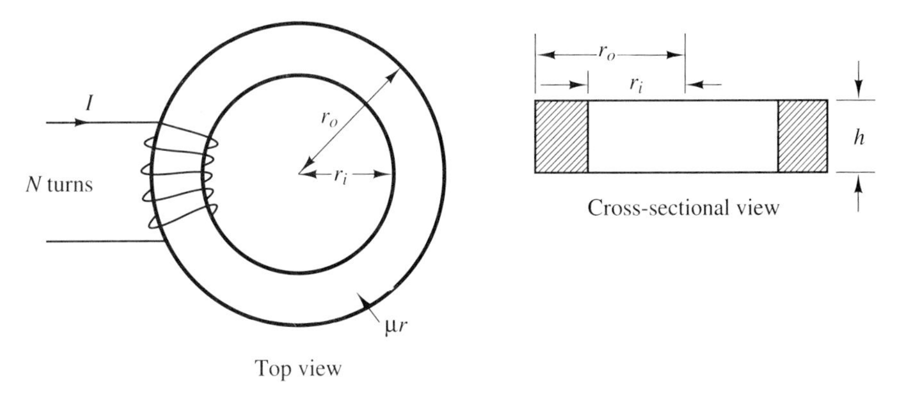


**Figure P13.7**

**13.8.** If an air gap of 0.8 mm in thickness is cut in the ring of Problem 13.7, what current is required to maintain the same flux in the ring? What is the percent change in required current? Neglect flux leakage.

**13.9.** A toroidal core with mean radius $r_m = 20$ cm is wound with two coils, as shown in Fig. P13.9a. The cross-sectional area of the coils is 5 cm$^2$. The current is 2.5 A in coil 1 and 1.2 A in coil 2, in the directions shown. The core has $\mu_r = 2500$ for $B < 1$ T, and otherwise is roughly proportional to $H$ with a smaller proportionality constant, as shown in Fig. P13.9b. Find the clockwise flux in the core using the mean path length.

**13.10.** Repeat Problem 13.9 if the direction of the current in coil 2 is reversed. Be careful that the answer you obtain is realistic.

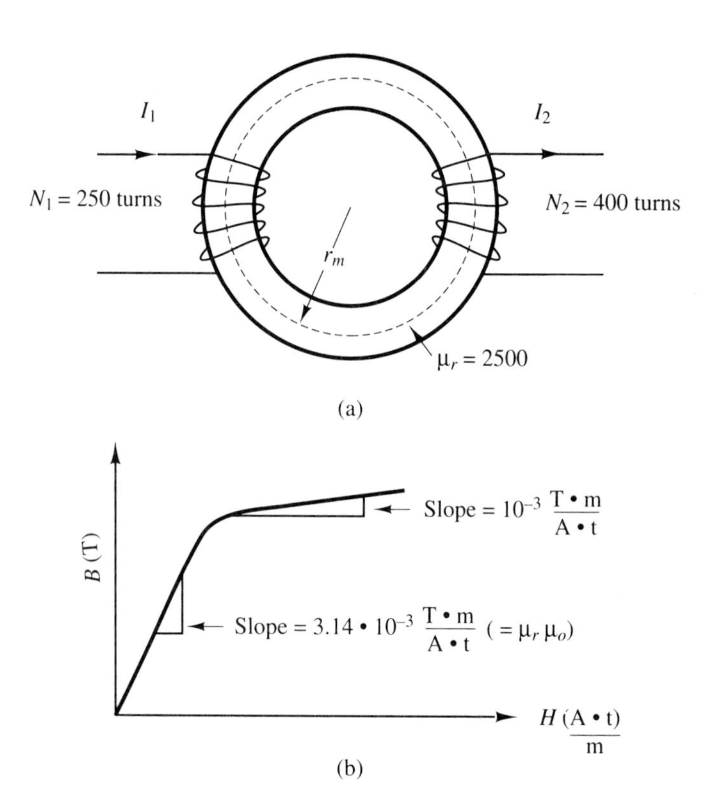


**Figure P13.9**

**13.11.** If the net flux in the core of Fig. P13.9 is zero and $I_1$ is held fixed at 1.5 A, determine the current in coil 2 ($I_2$).

**13.12.** Suppose in the core of magnetic material in Fig. 13.3a that all legs have width 1 cm, and also that the core is built up to a depth of 1 cm. Also, the outer width of the core is 12 cm and so is the outer height. Finally, suppose that a 440-turn coil carrying 3 A produces flux in the square-shaped cast steel core.

**(a)** What is the magnetic flux density within the core, assuming operation in the linear region of the *B–H* curve (see Fig. P13.12)?

**(b)** What is the magnetic flux density within the core, using the *B–H* curve shown in Fig. P13.12?

**(c)** Which of these two values (or both) is (are) the correct value? Are we operating in the linear region of the *B–H* curve?

**(d)** Calculate the flux in the core in mWb.

**13.13.** Find the current in the coil in Fig. P13.13 such that the magnetic flux density within each (identical) outer leg is 0.4 T. What is $B$ in the center leg?

**13.14.** Suppose that for the magnetic core shown in Fig. 13.3b, all legs have width $t$, and also that the core is built up to a depth $t$, where $t = 10$ cm. Suppose that the width of the gap is not 1 cm (as in Fig. 13.3b), but rather $d_g = 0.2$ mm. A coil of 500 turns wound on the left leg carries a current $I = 1.5$ A. Determine the magnetic flux density in the air gap. Assume that the mean path length of the magnetic circuit is $d_{\text{mean}} \approx 1.6$ m; neglect flux leakage and assume a linear *B–H* curve.

**13.15.** If a magnetic flux density of 1 T is required in the air gap of the core in Fig. P13.14, what is the required current in the coil?

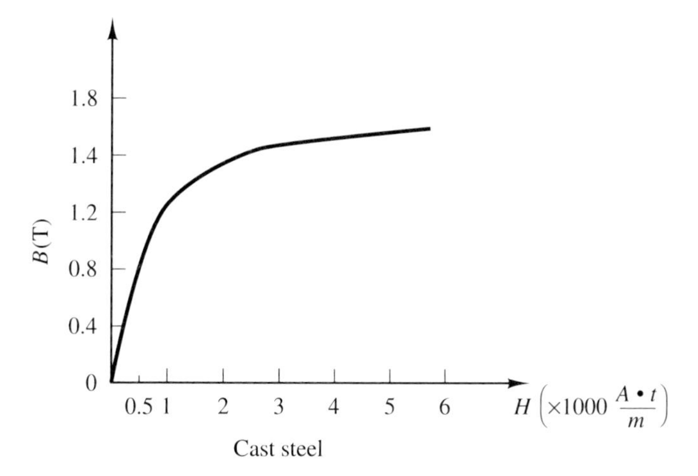


**Figure P13.12**

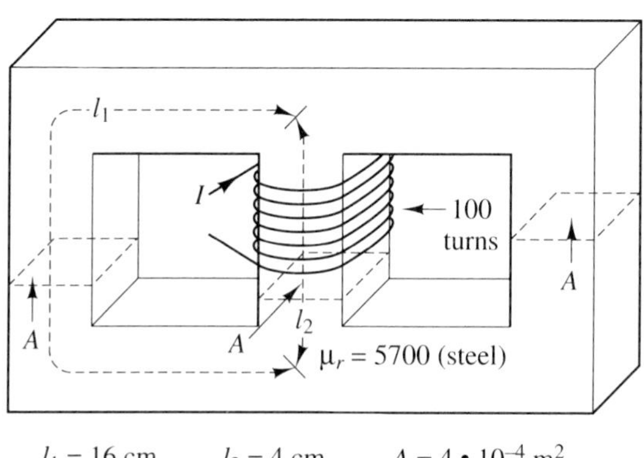


**Figure P13.13**

**13.16.** Consider the motor–generator core shown in Fig. P13.16. The barrel-shaped magnetic material in the center is the rotating armature. Above and below it are the poles (this is a two-pole machine). Two coils of $N/2$ turns each create the background $B$ field of 0.9 T in the armature, when driven by the current source $I_o = 20$ A. Assume that the effective area at the armature/pole interface is 5 cm × 2 cm = $10^{-3}$ m$^2$ and that the various reluctances are $\mathcal{R}_{\text{gap}} = 2 \cdot 10^6$ A · t/Wb (each gap), $\mathcal{R}_{\text{arm}} = 2 \cdot 10^5$ A · t/Wb, $\mathcal{R}_{\text{right leg}} = \mathcal{R}_{\text{left leg}} = \mathcal{R}_{\text{leg}} = 8 \cdot 10^5$ A · t/Wb, and $\mathcal{R}_{\text{poles}} = 2 \cdot 10^5$ A · t/Wb (each pole). Find the required number of turns on the pole pieces. Use the concepts from electric circuit analysis to help you (draw an electric/magnetic circuit hybrid).

**13.17.** **(a)** Develop the idea of a magnetic flux divider and an mmf divider. Show magnetic cores exemplifying these ideas and provide the defining relations.
**(b)** Can we also develop "nodal analysis" and "loop analysis" for magnetic circuits? What about Thévenin equivalents?

**13.18.** **(a)** From the $B$–$H$ curve in Fig. P13.3b, estimate $\mu_r$ of cast steel.
**(b)** In the toroid shown in Fig. P13.18, there is a flux of $\phi = 0.56$ mWb. How many turns must there be in the coil if the core is cast steel? Could you use the result in part (a) to help you?

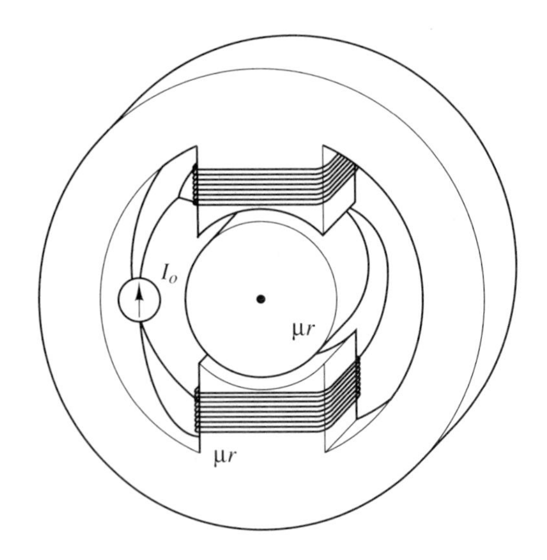


**Figure P13.16**

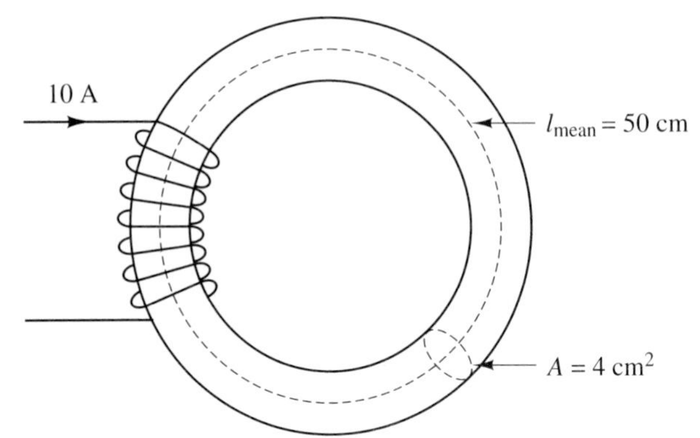


**Figure P13.18**

**13.19.** For the transformer shown in Fig. P13.19, here operated at dc, find the flux $\phi$ and magnetic flux density $B$ in each leg of the core. What is the total reluctance of the core? The depth of the core is $w$ meters. The left coil has $N$ turns, while the right coil has $2N$ turns. Current $I$ (amperes) flows into both coils in the directions shown. Specify the direction of flux lines (i.e., of **B**): clockwise or counterclockwise.

Mean length, permeability, and width of left leg: $L$, $\mu_A$, $2d$.
Mean length, permeability, and width of right leg: $L$, $2\mu_A$, $d$.
Mean length, permeability, and width of top leg: $L$, $\mu_A$, $d$.
Mean length, permeability, and width of bottom leg: $L$, $\mu_A$, $d$.
($L$ and $d$ are in meters and $\mu_A$ is in N/A$^2$.)

**13.20.** We wish to establish a magnetic flux $\phi$ of $6 \cdot 10^{-4}$ Wb everywhere in the toroid shown in Fig. P13.20. Relevant parameters are $\mu_0 = 4\pi \cdot 10^{-7}$ N/A$^2$, $\mu_{ra} = 5700$, $\mu_{rb} = 1300$, $A = 10^{-3}$ m$^2$, $L_a = 0.5$ m, $L_b = 0.3$ m, $N = 100$ turns.
(**a**) What is the required magnetic flux density $B$ everywhere in the toroid?
(**b**) What will be the required mmf across section $a$ to obtain $\phi$ in section $a$?
(**c**) What will be the required mmf across section $b$ to obtain $\phi$ in section $b$?

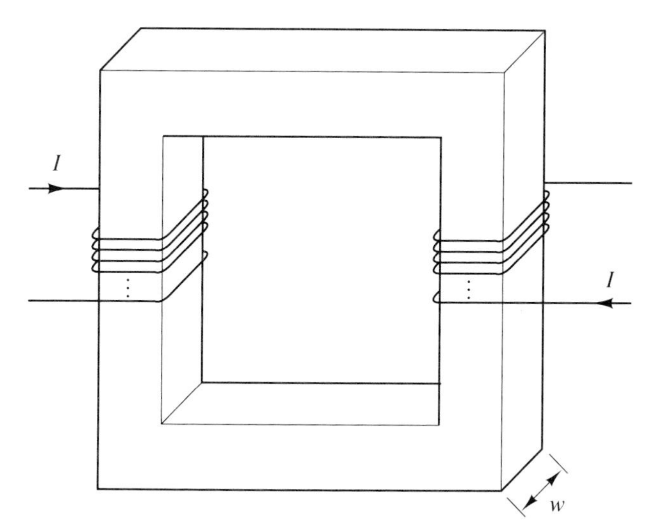


**Figure P13.19**

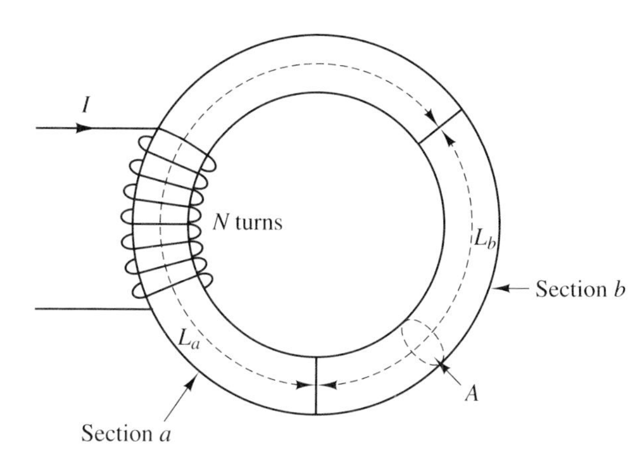


**Figure P13.20**

(**d**) What is the total required mmf?
(**e**) What is the required current $I$?

**13.21.** A magnetic circuit with cross-sectional area 10 cm$^2$ with a coil having $N = 100$ turns is shown in Fig. P13.21. A flux of $\phi_L = 0.15$ mWb is required in the air gap. Assume that $\mu_r = 3000$ and neglect flux leakage.

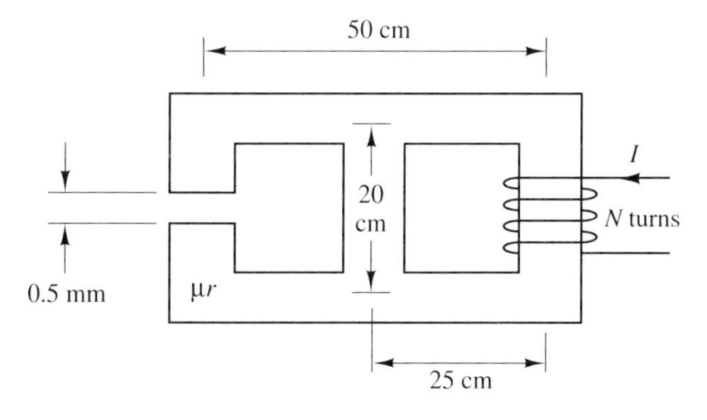


**Figure P13.21**

(**a**) Find the total reluctance seen by the coil. (*Hint:* Find the equivalent reluctance seen by the source using a series–parallel combination of reluctances.)

(**b**) Determine the flux in the center leg, $\phi_C$.

(**c**) Find the required current $I$ and the flux in the right leg, $\phi_R$. Note that you can use nodal analysis with magnetic potential nodes, and also use "mmf dividers" to simplify your analysis (see Problem 13.17).

**13.22.** Consider a core having the same general shape as that in Fig. P13.21, but not necessarily the same dimensions or number of coil turns as in Problem 13.21. If the current in the coil on the right leg is 1.5 A, the flux in the gap is 2 mWb, and the flux in the center leg is 1 mWb, determine the number of turns on the coil. Assume for this problem that the core reluctance is $1 \cdot 10^5$ H$^{-1}$.

**13.23.** Consider the magnetic circuit shown in Fig. P13.23. The two halves are made of different magnetic materials, having relative permeabilities $\mu_{r1}$ and $\mu_{r2}$. Because of fringing effects, the effective area of the air gap is assumed to be 5% larger than the corresponding (adjacent) cross-sectional area.

(**a**) Determine the total reluctance of the circuit, seen by either coil acting alone.

(**b**) If $N_1$, $I_1$, and $I_2$ are known, calculate $N_2$ such that the total magnetic flux in the core and air gaps is zero.

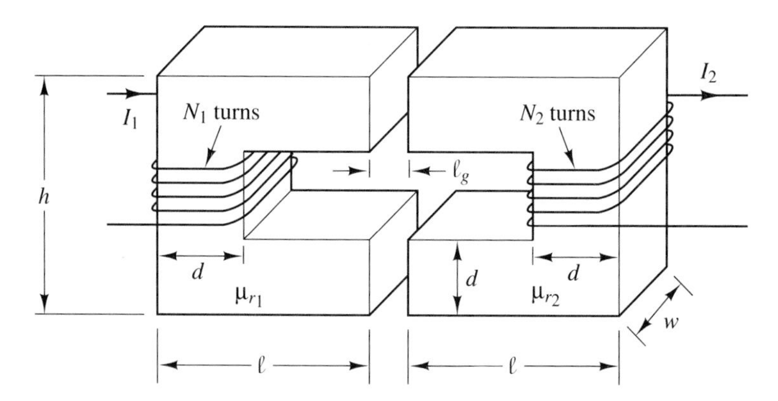


**Figure P13.23**

**13.24.** A 400-turn coil is wound on the center leg of the sheet steel core ($\mu_r = 5700$) shown in Fig. P13.24. A flux of 0.1 mWb is required in the air gap. Ignoring fringing effects, calculate the coil current $I$. Are we correct in assuming linearity ($B = \mu H$)?

**13.25.** The linear generator shown in Fig. P13.25a is used to measure the magnetic flux density and power a load $R_L$.

(**a**) If the conducting bar is moving at a speed of 10 cm/s and the measured voltage is 10 mV, what is the magnetic flux density? What is the flux passing through the generator at the time shown? What is the magnetic force opposing our movement of the bar?

(**b**) Now suppose that we use a single-loop rotating generator as shown in Fig. P13.25b. If the flux is equal to that calculated in part (a), how many revolutions per minute must the rotor be cranked to measure 10 mV (as an average voltage)? What is the average developed torque?

**13.26.** Review Faraday's and Lenz's laws. How can they be applied to a linear electric generator?

**13.27.** A rectangular loop of length $b$ and width $c$ as shown in Fig. P13.27 is pulled with

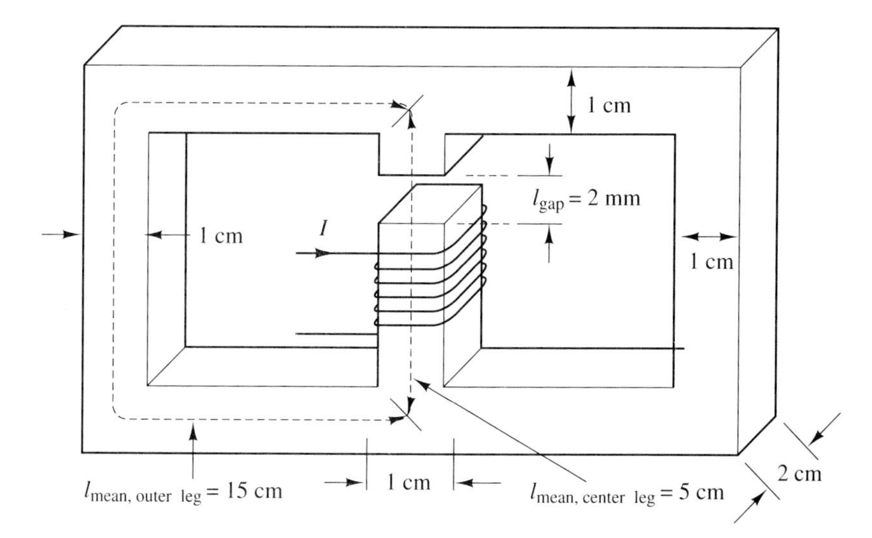


**Figure P13.24**

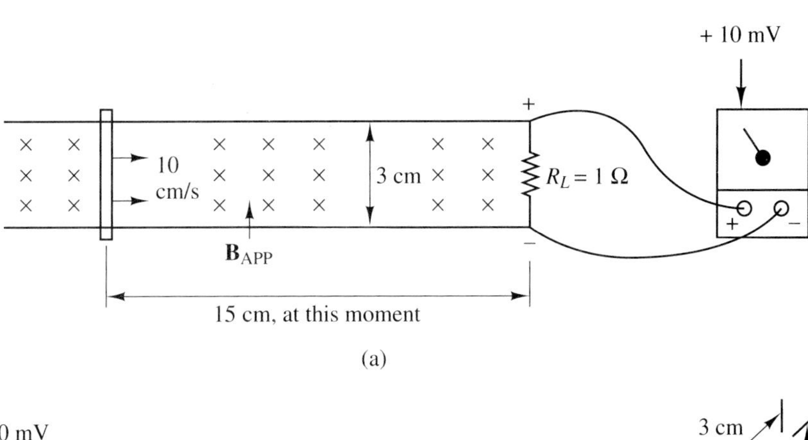


(a)

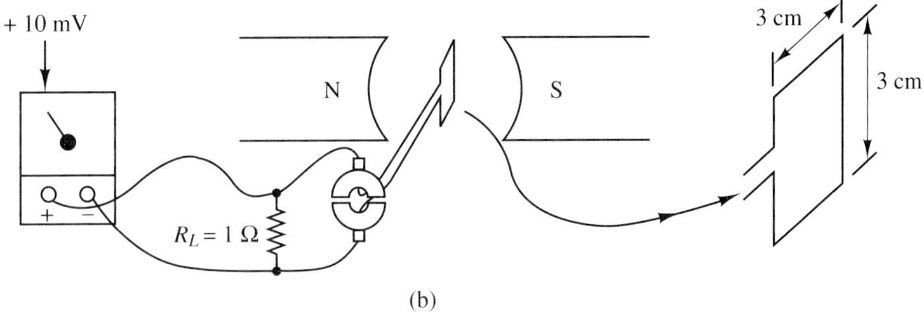


(b)

**Figure P13.25**

uniform speed $u$ through a region of length $t$, where a uniform $B$ field has been set up. Plot the flux $\phi$ through the loop and the induced emf V as functions of $x$. Assume that $t = 20$ cm, $B = 1$ T, $b = 12$ cm, $c = 4$ cm, and $u = 2$ m/s.

**13.28.** If the rectangular loop of Fig. P13.27 has resistance $R = 2\,\Omega$, plot the power dissipated in the loop as a function of $x$.

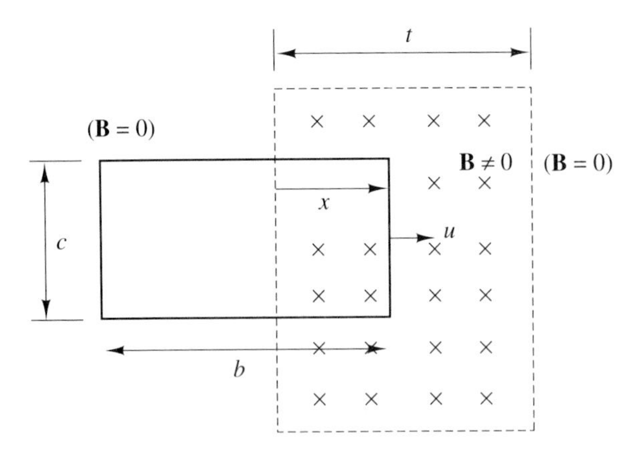


Figure P13.27

**13.29.** Figure P13.29 shows a conductor of length $l$ caused to move at a speed $u$ along a set of parallel rails. A resistor $R$ is connected to the terminals $xy$. Assume that $B = 0.5$ T, $u = 5$ m/s, $l = 50$ cm, and $R = 100\ \Omega$. Find the direction and magnitude of the current in the loop.

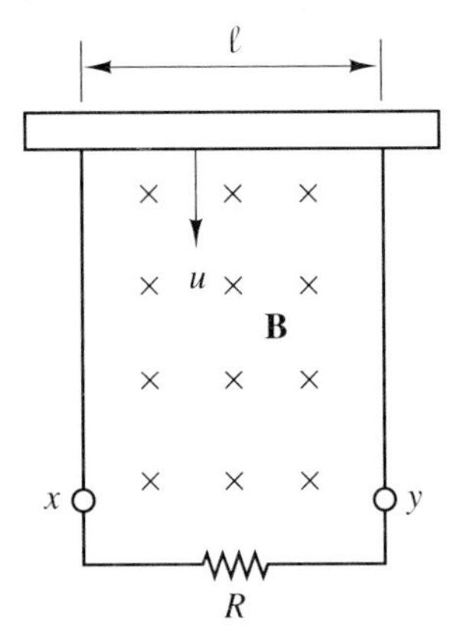


Figure P13.29

**13.30.** If a force of 0.1 N in the same direction as $u$ is applied to the conductor in Fig. P13.29, determine the direction and magnitude of the current through $R$. Note that the conductor accelerates to the value at which the magnetic and applied forces on the conductor balance, and from there it remains moving at that constant speed (steady state). Assume that steady state has already been reached.

**13.31.** Explain how the voltage is generated in a single-loop rotating generator. At what angular position of the loop will the induced voltage be maximum? Sketch the waveform of the induced voltage.

**13.32.** With a load resistor connected to a single-loop generator, indicate the direction of current, the component magnetic fields, the resultant field, and the direction of the mechanical force on each conductor.

**13.33.** What is the armature reaction magnetic field? How can it affect the total magnetic field?

**13.34.** Compare the rotating generator with the rotating motor. Describe the physical meanings of Eqs. (13.23a) and (13.23b), and Eqs. (13.27a) and (13.27b).

**13.35.** How do the curved pole faces minimize the air effects and contribute to a smoother dc output voltage waveform?

**13.36.** What is the function of the commutator, and how does it accomplish its intended effect?

**13.37.** Compare lap and wave windings. How do they differ?

**13.38.** What are the fundamental dynamo equations for average induced voltage and average induced torque in a dc machine?

**13.39.** A six-pole generator has an armature that is wound with a coil having 360 induction conductors. If the voltage and current generated in each conductor for a particular connected load and prime mover are, respectively, 2.0 V and 4 A, find the total voltage and current generated if the armature is (**a**) lap wound, and (**b**) wave wound. Also, find the power generated in each case.

**13.40.** A six-pole generator has an armature that is wound with $Z$ conductors. The rotational speed and flux are such that 3 V is generated in each conductor. The total generated voltage is 720 V. Determine the value of $Z$ if the armature is (**a**) lap-wound, or (**b**) wave-wound.

**13.41.** A two-pole dc generator has a wave-wound armature containing 360 conductors. The field is 1 T around each pole face, and 75% of the armature conductors are effectively under the pole faces. The armature has length 15 cm and radius 10 cm, and the armature current is 50 A. If the generator is turning at a speed of 3600 rpm, find the torque and mechanical power induced in the armature which oppose the prime mover.

**13.42.** Find the generated emf in the armature described in Problem 13.41.

**13.43.** A lap-wound armature is used in a 12-pole dc generator. There are 432 conductors in the armature. The flux per pole is 40 mWb, and the generator spins at 600 rpm. Find the average induced voltage.

**13.44.** Repeat Problem 13.43 for the case in which the armature is wave-wound.

**13.45.** Suppose that a 500-Ω load resistor is connected to the generator described in Problem 13.43. Neglecting the internal armature resistance, find the resulting countertorque on the shaft of this machine. Assume for simplicity that the armature current equals the load current.

**13.46.** Repeat Problem 13.45 for the case in which the armature is wave-wound.

**13.47.** Consider the realistic situation of a loop on an armature moving past alternate polarity poles as the armature rotates. Suppose that the loop shown in Fig. P13.47 is moving "up" from N to S (armature rotating clockwise). When the loop is squarely in front of the N pole, determine the magnitude and direction of the net force on the loop in terms of $I$, $B$, and $d$. Repeat for the case in which the loop is halfway from the N pole to the next (S) pole.

## ADVANCED PROBLEMS

**13.48.** (**a**) Write $K\mathcal{F}L$ for the magnetic circuit shown in Fig. P13.48 and determine the reluctance of the entire structure. Simplify as much as possible by factoring out common terms.

(**b**) Suppose that $\mu_r$ is so large that the core term in $K\mathcal{F}L$ and $\mathcal{R}$ can be ignored. Using these simplifications, reexpress $\mathcal{R}$, and solve for the flux assuming all other parameters are given.

(**c**) This device is a relay if the bottom right side can slide. It is known that the magnetic energy density equals $\frac{1}{2}BH$ so that the total stored magnetic energy is $W = \frac{1}{2}BH(AL) = \frac{1}{2}(BA)(HL) = \frac{1}{2}\phi\mathcal{F} = \frac{1}{2}\phi^2\mathcal{R}$. Using your answers from part

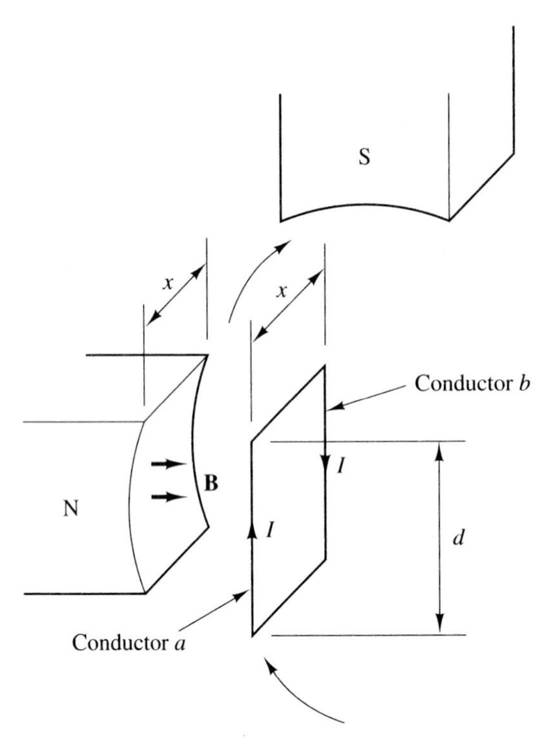


**Figure P13.47**

(b), find $W$ as a function of all parameters, including the gap length $g$. From an energy balance, one may show that the attractive force pulling the relay closed is $f = \frac{1}{2}\phi^2 \, d\mathcal{R}/dg$. In terms of $N$, $I$, $A$, $\mu_0$, and $g$ find the force. As far as the dependence of force on gap length $g$, does your answer agree with common sense?

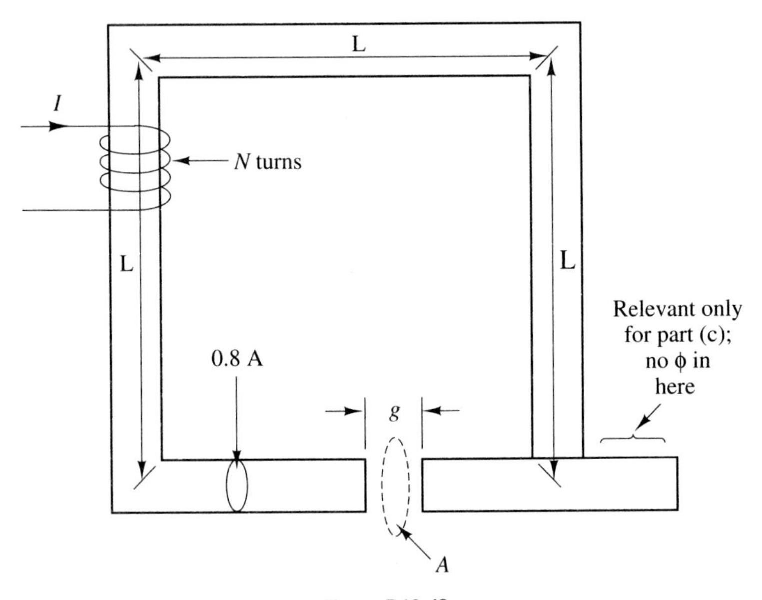


**Figure P13.48**

**13.49.** In Fig. P13.49, the magnitude of the $B$ field is increasing with time $t$ as follows:

$$|\mathbf{B}|(t) = 2t \text{ tesla.}$$

**(a)** Is the induced flux from the coil pointing into or out of the page? Why?
**(b)** With the coil wound as shown, what will be the direction of current flow when viewed from the top? Does this depend on which way the coil is wound?
**(c)** Based on this current direction, determine the direction of the force on the wires on each side of the coil.
**(d)** Find the numerical value of the voltage $v$.
**(e)** Find the numerical value of the load power.
**(f)** Find the numerical value of the magnitude of the current in the loop and load.
**(g)** Find the magnitude of the force on each of the wires at $t = 1$ s.
**(h)** If we wait long enough, what will happen to the coil?
**(i)** If instead, $|\mathbf{B}|$ was initially constant and very large and then was rapidly decreased, what would happen to the coil?

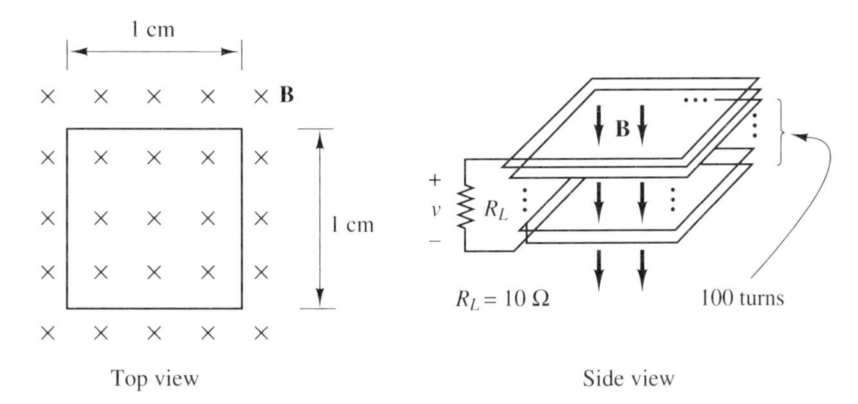


**Figure P13.49**

**13.50.** A characteristic of lap and wave windings was that lap windings have $a = P$ parallel current paths while wave windings have $a = 2$ parallel paths. Suppose instead that we have a six-pole machine. This will require $2P - 1 = 11$ commutator segments for a wave winding and 12 for a lap winding. Draw the simplex lap and wave windings for $P = 6$ analogous to Figs. 13.22 and 13.27. Verify that the number of parallel paths agrees with the general formulas. How many loops are in series at any given time for lap and wave windings for this six-pole machine?

**13.51.** The number of current paths in a machine having $m$ "plex" (number of times winding encircles armature before closing on itself) is $a = 2Pm$ for lap windings and $a = 2m$ for wave windings ($P$ is the number of poles). Using this result, find the number of parallel paths in a 12-pole machine which is **(a)** simplex wave wound, **(b)** duplex lap wound, **(c)** triplex lap wound, and **(d)** quadruplex wave wound.

**13.52.** The linear dc machine shown in Fig. P13.52 is connected to a battery voltage of 100 V, through a variable resistance $R$. Neglect the friction between the conductor and the rails.

**(a)** Assume that $R = 5\ \Omega$ and $B = 1$ T. What is the starting current of this machine?
**(b)** What is the steady-state speed of the conductor?
**(c)** If the resistance is changed to $3\ \Omega$ after the steady state is reached, will the speed of the conductor change?

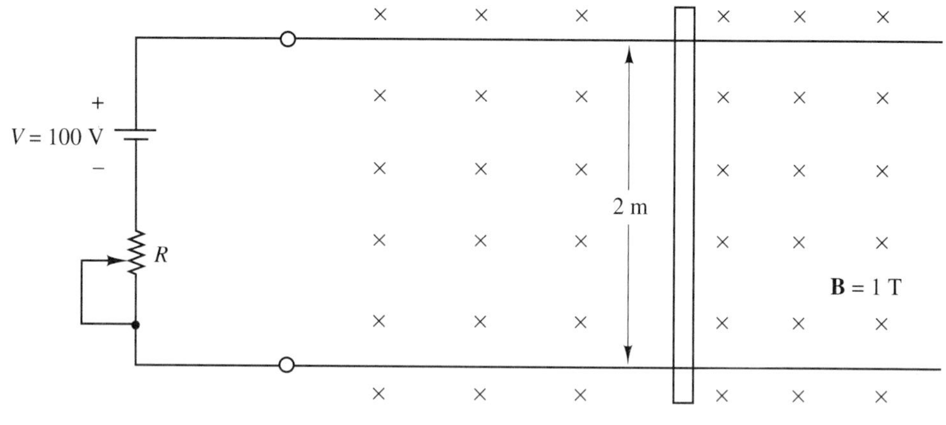


Figure P13.52

**13.53.** If the generator described in Problem 13.52 suddenly runs into a region where the magnetic flux density is uniform but reduced to 0.8 T, determine the new steady-state speed of the moving conductor.

**13.54.** Assume that the variable resistance $R$ of the machine described in Problem 13.52 and shown in Fig. P13.52 is now set to 5 Ω, and $B = 1$ T. Suppose that after steady-state speed is reached, a force of $F_{app} = 5 \text{ N } \hat{\mathbf{x}}$ (to the right) is applied to the movable conductor in Fig. P13.52.

**(a)** Find the new steady-state speed of the conductor.

**(b)** Find the power absorbed by the conductor. How much power is the battery producing or consuming? Calculate and explain the difference between these two power values. Is this machine acting as a generator or as a motor?

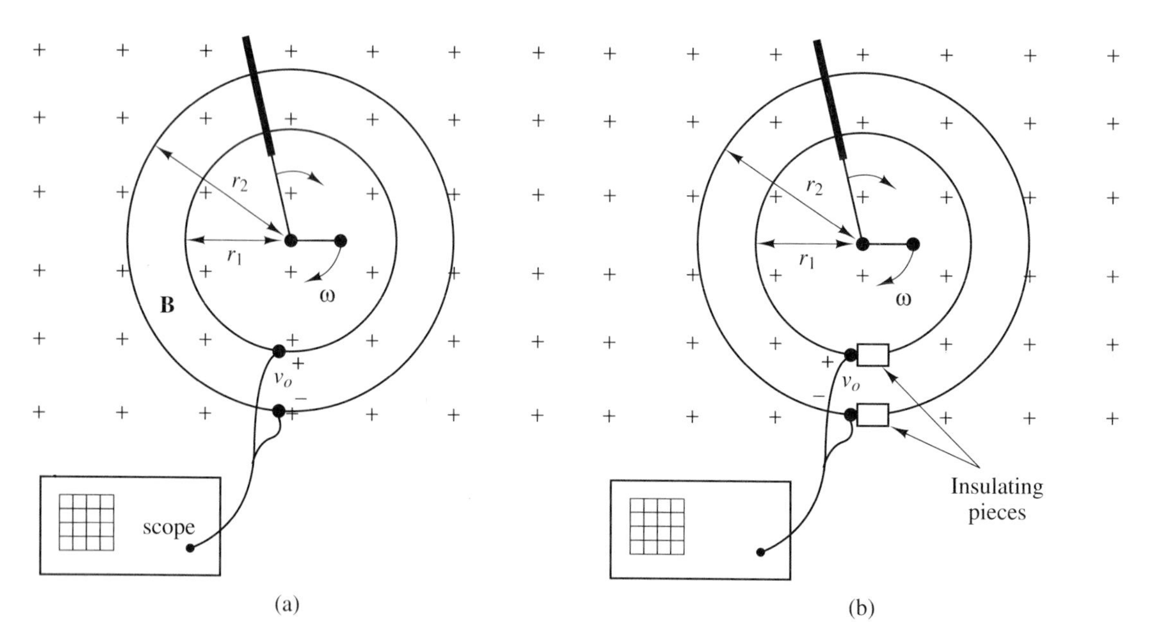


Figure P13.56

**13.55.** Concerning the situation described in Problem 13.54, suppose that the applied force is now directed to the left of the conductor rather than the right. What would the new steady-state speed be? Is this machine acting as a generator or as a motor?

**13.56.** Two concentric circular rings of metal are arranged in a uniform $B$ field, as shown in Fig. P13.56a. A radial conductor is connected to a crank as shown. On the right, the two wire probes from an oscilloscope are connected, one to each ring.

(**a**) As the conducting bar is cranked, what, if any, is the voltage $v_o(t)$ that would be seen on the screen of the oscilloscope? If nonzero, plot $v_o(t)$ as a function of time, and provide a quantitative formula for $v_o(t)$, assuming that the crank is turned at $n$ revolutions per minute.

(**b**) If insulating pieces are now inserted as shown in Fig. P13.56b, repeat part (a).

14

# Electromechanical Machines II: DC (Commutator) Machines

## 14.1 INTRODUCTION

Having had a fairly thorough introduction to the basic physics of dynamos and elements of their construction, it is now appropriate to begin to consider the realities of electrically hooking one up. In this chapter, machines with commutators are examined. They are called dc machines because the terminal voltage and current are usually nearly constant in time.

Dc machines have an important place both in the history of motors and in current practice. Dc motors in particular have in no way been made obsolete by the proliferation of ac power. We shall find that there are applications where both dc generators and motors are still the machine of choice. And even in cases where other machines are now favored, the dc machine may still be found in use and so must be understood by the user.

First, dc generator and then dc motor connections and resulting performance will be investigated. As always, the most convenient way of describing particular hookups is by way of circuit diagrams. Working with circuit diagrams also conveniently limits attention to the few most important quantities, which simplifies analysis.

There are four basic configurations: Separately excited, shunt, series, and compound connections. Each type is discussed in turn; first the generators and then the motors. In each case some applications examples and issues are presented at the end of the discussion. Finally, two recent additions to the repertoire of "dc" motors are introduced: The stepper motor (actually, a sort of hybrid which does *not* have a commutator) and the micromachine.

## 14.2 DC GENERATOR CONFIGURATIONS AND OPERATING CHARACTERISTICS

### 14.2.1 Separately Excited Generator

As the name implies, the excitation for the voltage-inducing magnetic field (called "the field") is separate in origin from the armature circuit. Some external power source provides field current $I_F$ to the field winding, which is the source of flux $\phi$. Keep in mind that this excitation is constant-in-time (dc, $\omega = 0$), so that the impedance of the field winding is $R_F + j0 \cdot L_F = R_F$—purely resistive.

The circuit diagram is shown with no load connected (and therefore $I_A = 0$) in Fig. 14.1. (This figure is just Fig. 13.35a with the field added to the diagram.) Note that in Fig. 14.1, $R_A$ is implicitly included in the dynamo symbol, a common practice. If $I_F = 0$, then $\phi = 0$ and therefore $E_G = 0$, according to previous analysis. However, there is a small, nonzero residual $B_r$ that causes a nonzero flux. If the rotor is spinning, this nonzero flux causes a nonzero $E_G$, called $E_r$ ($r$ for residual). Noting that $E_G/n$ is proportional to $\phi$ and therefore $B$, and that $I_F$ is proportional to $H$, a plot of $E_G/n$ versus $I_F$ should be and is identical to the corresponding $B$–$H$ saturation curve (see Fig. 14.2). For a fixed cranking speed $n$, we can merely plot $E_G$ versus $I_F$. In steady-state operation, a value $I_{F,\text{op}}$ is used to create a strong flux $\phi$, resulting in a particular value of $E_G$. But it should be clear that in order to double $E_G$ for a fixed $I_F$, all we must do is double $n$. It is assumed that no load is connected, so $V_T = V_{T,\text{nl}} = E_G$ in Fig. 14.2.

Because $I_L$ (load current) $= I_A$ for this circuit, the connection of a load will reduce the terminal voltage $V_T$ from $E_G$ to $E_G - I_L R_A$. A plot of $V_T$ versus $I_L$ is actually the generator output characteristic, also called the *external characteristic*. An example of such a plot is shown in Fig. 14.3. For $I_A = I_L = 0$, as mentioned, $V_T = E_G$, and there is a roughly linear decrease in $V_T$ for increasing $I_L$. This is precisely the "practical voltage source" discussed in Chapter 4: An ideal source ($E_G$) in series with a source resistance ($R_A$). As can be seen, $R_A$ is small; the negative slope is quite small.

Sometimes an additional droop is plotted (shown here as a dotted curve) that represents the effect of armature reaction. With the use of interpoles this droop can be minimized. Also shown with a dashed line is the behavior that would occur if $R_A$ were zero and the voltage source were perfect—no matter what the value of $I_L$, $V_T$ would equal $E_G$. Note that in circuit analysis terms, $R_A$ is just the internal resistance of the voltage source (generator). Finally, a typical resistive load line is superimposed, and the intersection between it and the generator output characteristic is labeled as $Q$, the operating point.

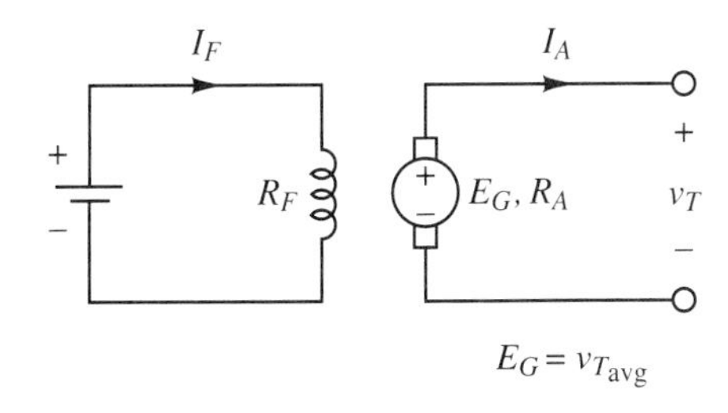


**Figure 14.1** Circuit diagram of a separately excited dc generator.

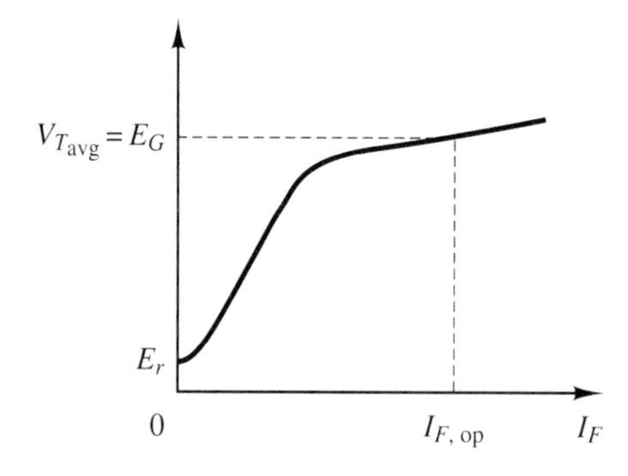


**Figure 14.2** Buildup curve for separately excited and shunt dc generators.

As with all electromechanical machines, there is a particular operating point called the *rated* condition. Rated or *full-load* values are those for which the machine was designed to operate most efficiently and are specified on the metal *nameplate* on the machine. In the example of Fig. 14.3, the generator and load are well matched in that the operating point is near the rated condition.

The full-load value of $V_T$ is somewhat less than $E_G$, although not excessively lower. A means of quantifying this is the parameter called *voltage regulation,* which is the percent change in $V_T$ due to the $I_A R_A$ drop. If $I_{A,\text{fl}}$ is the rated (full-load) current and $V_{T,\text{nl}}$ and $V_{T,\text{fl}}$ are the no-load and full-load terminal voltages, respectively, then

$$\begin{aligned} \text{voltage regulation} &= \frac{V_{T,\text{nl}} - V_{T,\text{fl}}}{V_{T,\text{fl}}} \cdot 100\% \\ &\approx \frac{I_{A,\text{fl}} R_A}{E_G - I_{A,\text{fl}} R_A} \cdot 100\% \\ &= \frac{100\%}{(E_G / I_{A,\text{fl}} R_A) - 1}, \end{aligned} \tag{14.1}$$

where the second equality is approximately valid if interpoles are used. Note that the smaller the voltage regulation, the closer the generator approximates the desired ideal voltage source. The reason $V_{T,\text{fl}}$ is used for the denominator is that its use

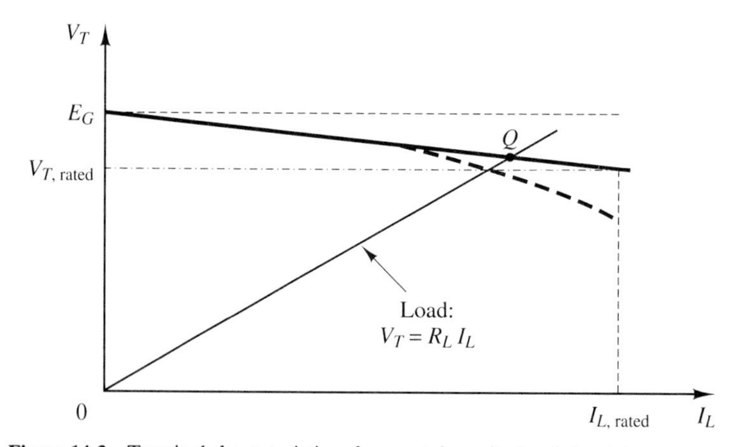


**Figure 14.3** Terminal characteristics of separately excited and shunt dc generators.

will give a worst-case (larger) value for voltage regulation than if the larger $V_{T,\mathrm{nl}}$ were used.

### 14.2.2 Shunt Generator

As is evident from Fig. 14.4, the shunt generator differs from the separately excited generator only in that it is *self-excited:* The field winding is connected in parallel with (shunted to) the generator terminals (and load). By KCL, $I_A = I_L + I_F$.

It might at first appear that the generator could not build up voltage. That is, the field current $I_F$ is nonzero only if $E_G$ is nonzero, yet $E_G$ is nonzero only if $I_F$ is! However, there is a small nonzero residual magnetic flux density and therefore a nonzero generated voltage $E_r$ when the rotor is cranked, so there is a "spark to start the fire." Again, for steady-state calculations the field winding can be considered a pure resistance $R_F$.

The same $V_T$ versus $I_F$ curve as for the separately excited generator (Fig. 14.2) applies for the shunt generator, except in one important way: The field current is no longer independently controllable, but depends on $E_G$. At steady state, the operating point is found by determination of the intersection of the *saturation curve* ($V_T$ versus $I_F$) with the field *load line* $V_T = I_F R_F$. Keep in mind that during buildup, the field inductance is significant. Thus the building-up operating point of the resistive generator model will not match that of the resistive load line.

A startup scenario with no load connected could proceed as follows (see Fig. 14.5). (The staircase shape of the saturation building curve is in reality continuous; it has been so drawn to help show the individual steps listed below toward buildup. You will notice that it is an example of positive feedback.

1. The rotor is cranked. Recall that $E_G(I_F = 0) = E_r \neq 0$, assuming that the generator has been used before. (If not, we can start by momentarily giving the field a boost from another dc source.)
2. $E_r$ causes a (small) field current; the operating point therefore moves to the right in Fig. 14.5.
3. Because of the increased field current and therefore flux, $E_G$ and consequently, $V_T = E_G - I_A R_A$ also, increase. This moves the operating point up.
4. The larger terminal voltage in turn increases the field current.
5. Steps 3 and 4 continue to cycle until steady state is reached (and thus intersection of the generator curve with the resistive field load line stabilizes).

*A word of caution:* If the field load line lies above the buildup portion of the saturation curve ($E_G$ versus $I_F$), $E_G$ will never build up. But recall that the saturation

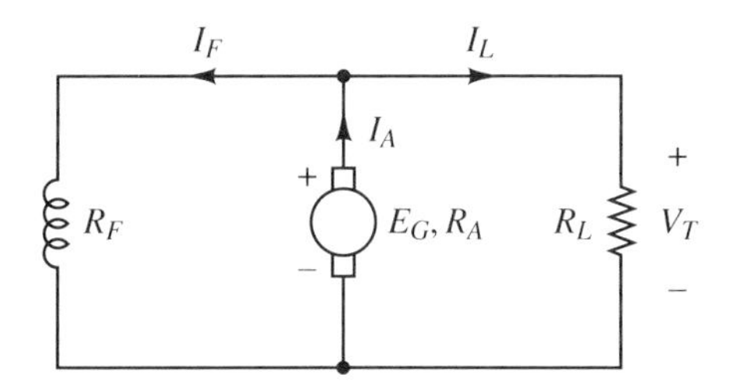


**Figure 14.4** Circuit diagram of a shunt dc generator.

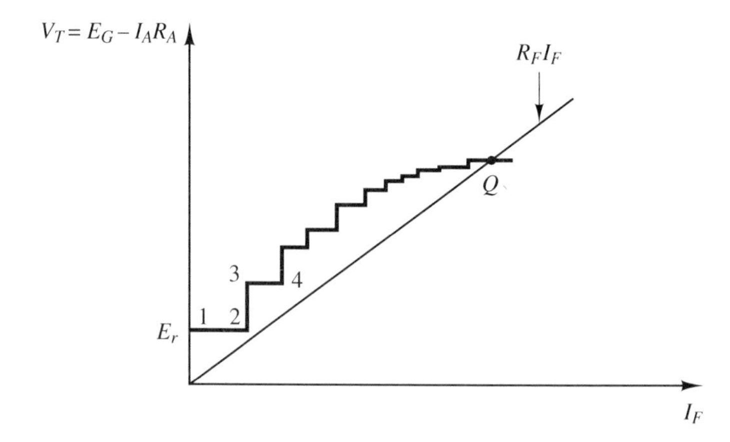


**Figure 14.5** Buildup curve for shunt dc generators, with steps for self-buildup indicated schematically. The field load line is also shown.

curve can be scaled up and down by changing the cranking speed, $n$. For a larger $R_F$ (steeper load line), merely crank faster and $E_G$ will build up.

Once the steady state has been reached, what happens when a load resistor $R_L$ is connected? First note that due to a current divider, $I_L = R_F I_F / R_L$, so that now $V_T = E_G - I_A R_A = E_G - (1 + R_F/R_L) R_A I_F$ because $I_A = I_F + I_L$. It might initially appear that

1. $V_T$ would decrease according to a voltage divider (but not yet $E_G$).
2. $I_F$ would therefore decrease, causing reduced flux.
3. $E_G$ would decrease due to reduced flux, causing $V_T$ to decrease further. (The cranking speed $n$ is assumed constant throughout this scenario.)
4. Steps 2 and 3 continue to cycle until $E_G = 0$.

Such a generator would not be very effective!

Fortunately, one fact has been left out: It can be seen in Fig. 14.5 that the operating point has been designed to lie in the saturation region. Therefore, decreasing $I_F$ in step 2 even fairly substantially would not cause a significant reduction in $E_G$ because of the flatness of the curve. The result is a measurable but not devastating drop in $E_G$.

The external (output) characteristic is similar to that of the separately excited generator (Fig. 14.3). Again there is a designated full-load or rated load current for normal (most efficient) operation. For example, it is most efficient to operate near the knee of the buildup curve, yet still within saturation. To predict the resulting steady-state rated terminal voltage explicitly in terms of the load current $I_L$, $I_A = I_L + I_F$ and $V_T = I_F R_F$ must be substituted into $V_T = E_G - I_A R_A$:

$$
\begin{aligned}
V_T &= E_G - \left(I_L + \frac{V_T}{R_F}\right) R_A \\
&= \frac{1}{1 + R_A/R_F}(E_G - I_L R_A).
\end{aligned}
\tag{14.2}
$$

There is also additional drooping due to the weakened flux (and therefore $E_G$) and any armature reaction not canceled by interpoles. In conclusion, the voltage regulation is comparable with that of the separately excited generator (see Fig. 14.3). Notice that now instead of $V_{T,\text{nl}} = E_G$, we have in Eq. (14.2) for $I_L = 0$, $V_{T,\text{nl}} = E_G/(1 + R_A/R_F)$; however, in practice this difference is only a few percent.

**Example 14.1**

Consider a dc shunt generator. Given that $R_A = 1\ \Omega$ and $R_F = 100\ \Omega$, and the no-load terminal voltage is 100 V, what is the terminal voltage for a load of $R_L = 2\ \Omega$? What is the load power?

**Solution** $V_{T,\text{nl}} = 100\text{ V} = E_G - I_A R_A = E_G - (V_{T,\text{nl}}/R_F)R_A = E_G - (100/100)\cdot 1$. Therefore, $E_G = 101$ V. (Often, $E_G$ is simply taken to be $V_{T,\text{nl}}$ because the resulting error is so small.) With $R_L = 2\ \Omega$, the terminal voltage is

$$V_{T,\text{fl}} = \frac{R_L || R_F}{R_L || R_F + R_A} E_G$$

$$= \frac{1.96(101)}{1.96 + 1} = 67\text{ V}.$$

The load power is $V_{T,\text{fl}}^2/R_L = (67)^2/2 = 2.2$ kW. Notice that in this example it has been assumed that $E_{G,\text{nl}} = E_{G,\text{fl}} = E_G$. This is valid if the generator is operated in the saturation region, as is usually the case, for then $E_G$ is relatively independent of $I_F$ and therefore the load. Note that we still have $E_G = k_e n\phi$, but $\phi$ is saturated and relatively independent of variations in $I_F$.

### 14.2.3 Series Generator

The series generator (Fig. 14.6) is also self-excited because the field current is due to the armature current. Now, however, the field winding is connected in series with the load instead of in parallel with it; hence $I_F = I_L = I_A$. The terminal voltage is now

$$V_T = E_G - I_A(R_A + R_S). \tag{14.3}$$

For no load, $I_F = I_L = I_A = 0$, so $E_G = E_r$ (very small). But when a load is connected, $I_L = I_A = I_F$ increases, so $\phi$ also increases, which in turn increases $E_G = k_e n\phi$. Thus the remarkable fact that to build up voltage, a load must be applied! The external characteristic (for fixed $n$) is given in Fig. 14.7. Because $I_F = I_L$, in this case the buildup curve is identical to the external (output) characteristic.

For reference, the saturation curve has been included (the dashed curve). It can be obtained by measuring $V_T = E_G$ under no load and separate excitation. The

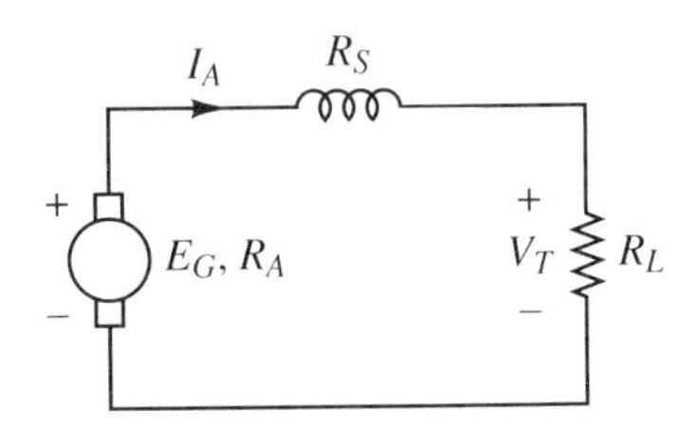


**Figure 14.6** Circuit diagram of a series dc generator.

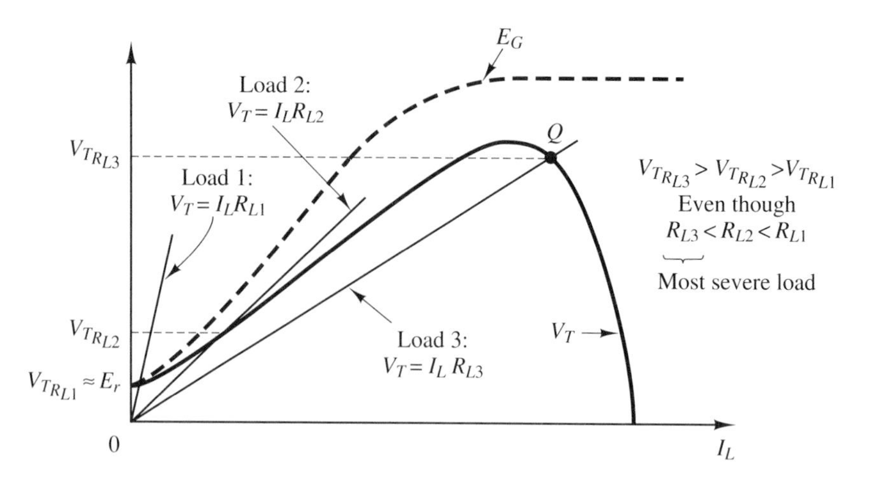


**Figure 14.7** Buildup curve for series dc generators, also indicating the generated voltage characteristic.

difference between the $V_T$ curve and the saturation curve is the $I_A(R_A + R_S)$ drop and any leftover armature reaction.

Two important characteristics of Fig. 14.7 are (1) $V_T$ varies greatly (poor voltage regulation) for low $I_L$, and (2) eventually armature reaction and $I_A(R_A + R_S)$ take over, causing $V_T$ to drop to zero. The former aspect is helpful in compound generators (discussed next), while the latter can be thought of both as a constant current source and/or current-limiting behavior under short-circuit conditions (a "fault"). Simplistically, we can think of a shunt generator as a good voltage source and a series generator as a good current source.

### 14.2.4 Compound Generator

Although one student replaced "mp" by "nf" in the word "compound," the compound generator is actually the most versatile of the dc generators and is the type most often used. An advantage of the shunt generator was the relative independence of the terminal voltage on the load current. A disadvantage was that when $I_L$ became large enough, there was nonetheless a drooping of $V_T$. An advantage of the series generator was that $V_T$ actually increased with the load current (for low to moderate values of $I_L$). However, it had poor voltage regulation.

If a little of the series generator behavior is combined with that of the shunt generator, advantages of both result. This is accomplished by placing a few series windings between the shunt field winding and either the armature (*long shunt*) or the load (*short shunt*). Physically, the series windings are wrapped around the field poles over the shunt windings. The behaviors of the long and short shunt generators are comparable.

By choosing the number of series windings, the rated terminal voltage can be either below $V_{T,\text{nl}}$ (*undercompounding*), equal to $V_{T,\text{nl}}$ (*flat compounding*), or above $V_{T,\text{nl}}$ (*overcompounding*). In all of the cases above, the series flux aids the shunt field flux; such compounding is therefore called *cumulative compounding*. It is also possible to reverse the series coil terminals and have the series flux oppose the shunt field; such compounding is called *differential compounding.*

The reason for flat compounding is not only that the rated-load terminal voltage can be equal to the no-load voltage, but also that for a variety of loads, the terminal voltage will be roughly constant.

(It should now be clear why $R_F$ for "field" is used for the shunt field winding while $R_S$ for "series" is used for the series field: Both "shunt" and "series" begin with *s*, yet both may appear in the same machine.)

The external characteristics for series only, shunt only, and all types of compounding generators are shown together in Fig. 14.8. For this set of plots, each generator was spun at such a speed and supplied with such flux that with no load connected, all the generators (except the series generator) had the same terminal voltage $V_T$. Then increasing loads $R_L$ were applied and the resulting terminal voltage recorded and plotted. In this way the effect of loading on each type of generator can be compared with that on the other types.

Note that the severe droop of the differential compounding generator, like that of the series generator, is useful in some cases for current limiting. If by mishap $V_T$ becomes zero (short circuit), $I_L$ will not become dangerously large but will hold at $I_{L,\text{sc}}$.

The compounding effect can be summarized as

$$\phi_{\text{total}} = \phi_{\text{shunt}} + \phi_{\text{series}}, \tag{14.4a}$$

where

$$\begin{aligned} \phi_{\text{series}} &> 0 \quad \text{cumulative compounding} \\ \phi_{\text{series}} &< 0 \quad \text{differential compounding,} \end{aligned} \tag{14.4b}$$

where the sign is controlled by the hookup polarity of the series winding. Flat compounding is most desirable, except when the load is considered remote, and the *IR* transmission line drops are considered extra, requiring overcompounding.

An alternative characterization of the same machines as shown in Fig. 14.8 is given in Fig. 14.9. Here the defined reference point is the full-load point rather than the no-load point. That is, each generator was spun at sufficient speed and given sufficient flux to have a common terminal voltage for a specified full load ($R_L$). Then

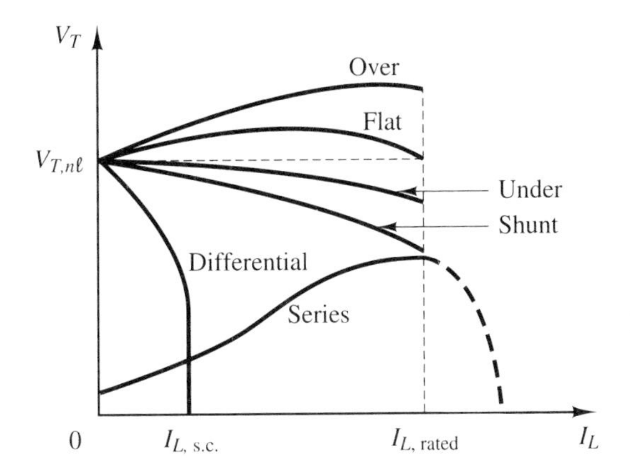


**Figure 14.8** Terminal characteristics of shunt, series, and compound generators on the same axes for comparison. They all have a common no-load terminal voltage.

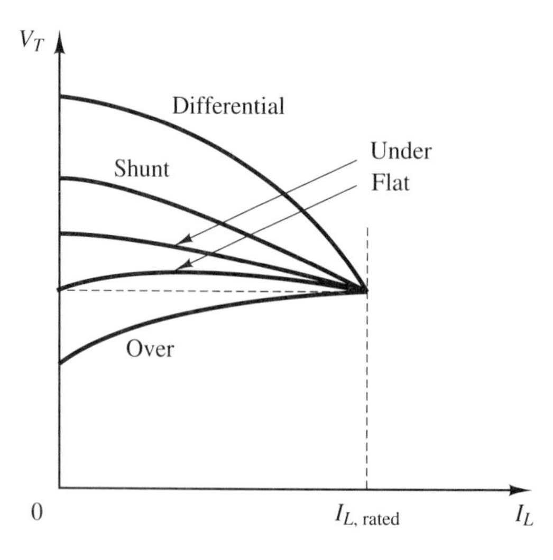


**Figure 14.9** Terminal characteristics of shunt, series, and compound generators on the same axes for comparison. They all have a common rated terminal voltage.

the load was decreased and $V_T$ recorded. Now at no load the machines have differing terminal voltages.

Notice that the curves are positionally reversed compared with those in Fig. 14.8. The reason for this can be seen by considering, for example, the differential compounding generator. To obtain a specified $V_T$ at full load, it would develop a huge $V_T$ at no load because of the tremendous drooping characteristic of differential compound generators, now functioning in reverse.

The circuit diagram for a long shunt compound generator is given in Fig. 14.10. Clearly, the following relations hold:

$$I_A = I_S = I_F + I_L \tag{14.5a}$$

$$I_F = \frac{V_T}{R_F} \tag{14.5b}$$

$$V_T = E_G - I_A(R_A + R_S). \tag{14.5c}$$

Notice that the equation for terminal voltage, Eq. (14.5c), is identical to that for a series generator, Eq. (14.3). In practice, for flat compounding $R_F$ is the total resistance of the shunt field winding plus the resistance of a potentiometer that can control the level of the no-load $\approx$ full-load voltage.

**Example 14.2**

Determine the number of series turns required on each pole of a compound generator to enable it to maintain the voltage at 240 V between no load and a full load of 20 kW.

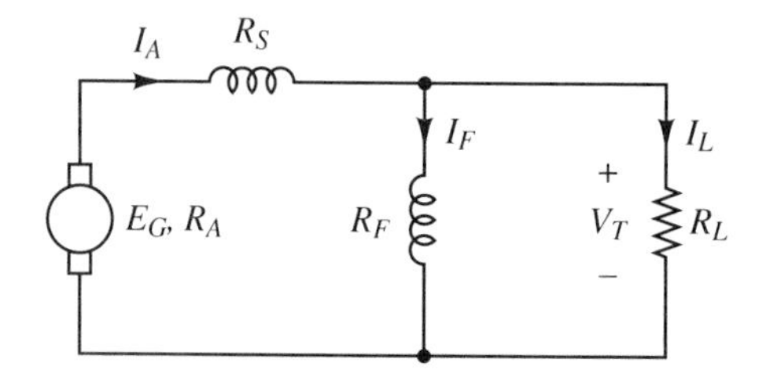


**Figure 14.10** Circuit diagram of a compound dc generator.

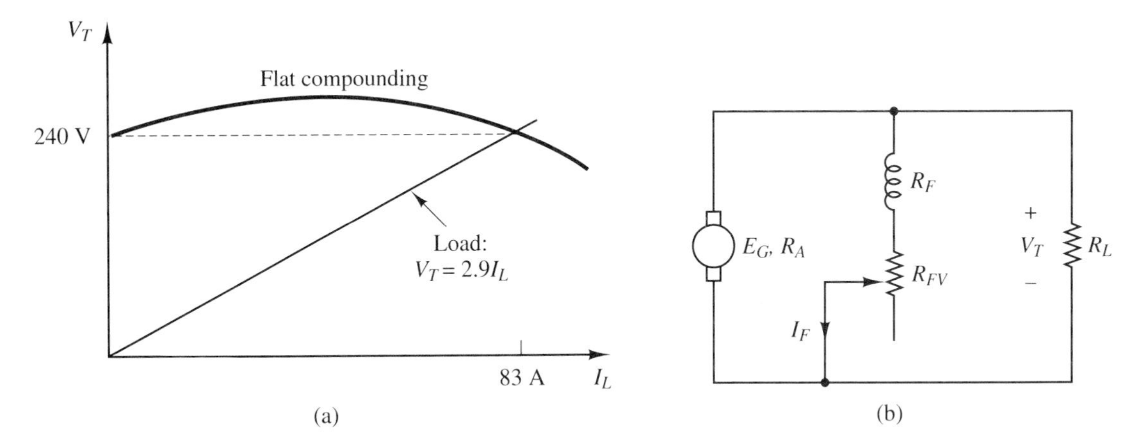


**Figure 14.11** (a) Desired output characteristic for the compound generator of Example 14.2; (b) shunt generator circuit diagram for Example 14.2.

By installation of a variable resistance $R_{FV}$ in series with the shunt field winding, the field current could be varied (without the series winding) so as to maintain the no-load voltage at full load as follows: At no load, the field current was 4 A and at full load, 5 A. The number of turns per pole on the shunt field winding is 600.

**Solution** Clearly, flat compounding is sought (see Fig. 14.11a). The variable resistor in series with the field winding (see Fig. 14.11b) accomplished this without a series winding. The object of flat compounding is that $R_{FV}$ will not have to be readjusted every time a new load is connected. From Fig. 14.8 we see that although the terminal voltage varies somewhat with the load, it is much more constant than with no compounding. With the series winding connected, the only reason for readjusting $R_{FV}$ would be for the infrequent case where the terminal voltage is desired to be changed. (In that case, the flat compounded terminal characteristic will merely be shifted up or down as a whole.) Now the additional required flux to account for the $I_A R_A$ drop (and armature reaction) is accomplished not by having to adjust $R_{FV}$, but rather by introducing the series winding with a number of turns to be determined (see Fig. 14.12). The purpose of $R_D$ is discussed later; it has only a minute effect. Keep in mind that throughout this example, the rotor speed is maintained constant.

First note that $R_L = V_T^2/P_L = 240^2/20{,}000 = 2.88\ \Omega$, and $I_{LQ} = V/R_L = 240/2.88$ $(= P_L/V_T = 20{,}000/240) = 83.3$ A, where $Q$ means *operating point.* In the solution of this problem, we ignore the $I_A R_S$ drop. This is valid because the series winding will be

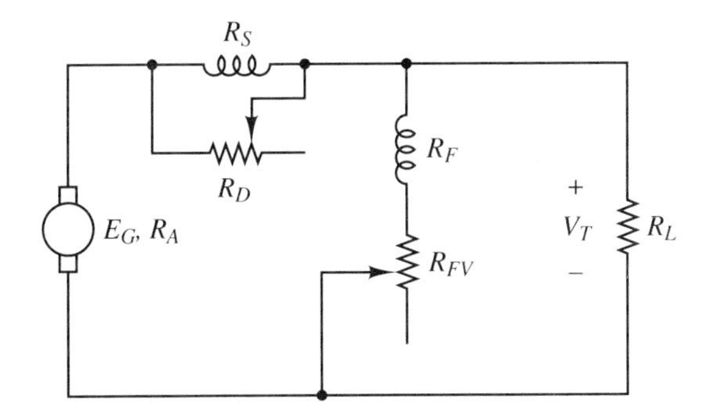


**Figure 14.12** Compound generator circuit diagram for Example 14.2, with diverting resistor shown across the series winding.

only a few turns of thick wire having very low resistance (to minimize heat loss). The number of turns can be found from

$$\mathcal{F}_{\text{series/pole}} = N_{\text{series/pole}} I_{\text{series}}. \tag{14.6}$$

Therefore, we need $\mathcal{F}_{\text{series/pole}}$ and $I_{\text{series}}$. Again, Fig. 14.11b depicts the situation before addition of the series coil.

With $R_{\text{FV}}$ initially set at a relatively large value, $I_F = 4$ A at *no load* ($R_L = \infty$). When the load *is* applied, $V_T$ decreases due to increased $I_A R_A$ and any leftover armature reaction not corrected for by interpoles. Because $V_T$ decreases, so does $I_F$. The only way to increase $V_T$ is to decrease the setting of $R_{\text{FV}}$. If that is done, $I_F$ will increase and therefore the flux and consequently $E_G$ and $V_T$. The value of $R_{\text{FV}}$ is decreased until $V_T$ is again equal to its no-load value.

Call the field current at no load with $R_{\text{FV}}$ at its initial setting $I_{F1}$ (= 4 A) and with $R_{\text{FV}}$ at its final setting (full load) $I_{F2}$ (= 5 A). The goal is to accomplish the increase in flux due to the decrease in $R_{\text{FV}}$ with the series coil and without changing $R_{\text{FV}}$. Because $R_{\text{FV}}$ is held fixed at the value it had at no load without the series winding for the situation of full load with the series winding, $R_F + R_{\text{FV}}$ is also fixed. Therefore, the field current with the series winding at full load will be 4 A because the same resistance has the same voltage (240 V) across it. That is, because the series winding is meant to take the place of the variation in $R_{\text{FV}}$, the value of $I_F$ used is the original value for no load, namely 4 A.

Notice that the armature current for the full-load case for the circuit of Fig. 14.11b is $I_A = I_{F2} + I_L = 83.3 + 5 = 88.3$ A, while the armature current for the circuit of Fig. 14.12 (again at full load) is $I'_A = I_{F1} + I_L = 87.3$ A. Because these two values are so close, we will assume that $I'_A = I_A$. Therefore, the required generated voltage and thus the total required flux for the two cases is the same.

We now make another simplifying assumption—we assume that we are in the linear region of operation even though we know that actually we are in saturation. This way we can obtain a simple solution that is nevertheless in the right ballpark. Consequently, we can speak of mmf as being flux times reluctance.

Because the reluctance of the dynamo is not significantly affected by the introduction of the series winding, the equality of change in flux from no load to full load translates to equality of change in mmf ($\Delta\mathcal{F} = \mathcal{R}\Delta\phi$), which is

$$\begin{aligned}
\Delta\mathcal{F}_{\text{series/pole}} &= \mathcal{F}_{\text{series/pole,now}} - \mathcal{F}_{\text{series/pole,before}} \\
&= \mathcal{F}_{\text{series/pole,now}} - 0 \\
&= N_{\text{series/pole}} I_{\text{series}} - 0 \\
&= \Delta\mathcal{F}_{\text{shunt/pole,without series}} \\
&= N_{\text{shunt/pole}} \Delta I_F \\
&= N_{\text{shunt/pole}} (I_{F2} - I_{F1}) \\
&= 600 \cdot (5\text{ A} - 4\text{ A}) \\
&= 600\text{ A} \cdot \text{turns/pole}.
\end{aligned} \tag{14.7}$$

The series current is

$$I_{\text{series}} = I_F + I_L. \tag{14.8}$$

The load current is, from above, 83.3 A. Thus

$$I_{\text{series}} = 83.3\text{ A} + 4\text{ A} = 87.3\text{ A}. \tag{14.9}$$

Finally,

$$N_{\text{series/pole}} = \frac{\mathcal{F}_{\text{series/pole}}}{I_{\text{series}}} \tag{14.10}$$

$$= \frac{600}{87.3} = 6.9 \text{ turns/pole.}$$

Although it is possible to make fractional turns, in practice the next higher integer is used, which will overcompound the generator. To exactly flat compound, a potentiometer could be placed across the series winding. This variable *diverter* resistor $R_D$ will reduce $I_{\text{series}}$, and therefore the flux, sufficiently to flat compound. This adjustment is very slight compared with the adjustment required for the circuit of Fig. 14.11b, and for most purposes, $R_D$ would not be readjusted. That is, for a wide range of values of $R_L$, the terminal voltage will be 240 V. Again, the final circuit would appear as shown in Fig. 14.12.

### 14.2.5 Efficiency

As in any mechanical or electrical system, energy conservation must be satisfied. A simple formula for efficiency of a dc generator can now be understood readily as it pertains to each of the machine types above:

$$\frac{\eta}{100\%} = \frac{\text{output power}}{\text{input power}}$$

$$= \frac{I_L V_T}{I_L V_T + I_A^2 R_A + (I_A^2 R_S) + [I_F^2 (R_F + R_{\text{FV}})] + P_{\text{core}} + P_{\text{rot}}} \tag{14.11a}$$

$$\approx \frac{I_L V_T}{E_G I_A + P_{\text{rot}}}, \tag{14.11b}$$

where in Eq. (14.11b) $E_G I_A = \omega T_{\text{dev}}$ is the developed power and is equal to all terms in the denominator of Eq. (14.11a) (the total power) except for $P_{\text{rot}}$. Also, the first term in parentheses in Eq. (14.11a) would not be present for the shunt generator, and the second term in parentheses in Eq. (14.11a) would not be present for a series generator.

Rotational losses $P_{\text{rot}}$ are due to bearing friction, windage (air viscosity/turbulence), and brush friction. Other losses, such as hysteresis and/or eddy current losses $P_{\text{core}}$, may also be included in the denominator of Eq. (14.11a) for greater accuracy. We review this topic after studying dc motors; see the end of Sec. 14.4.6.

## 14.3 APPLICATIONS OF DC GENERATORS

As noted in Sec. 6.7, electric power generation was originally dc. Thus dc generators were extremely important long ago. But when the advantages of ac caused the changeover from dc to ac, the importance of dc generators declined. However, dc generators can still be found, particularly in situations where line ac may not be available and the load to be run is dc; that is, in cases where the power generation is physically close to the dc load. Also, they are occasionally driven by an ac motor to provide ripple-free, precisely controlled dc voltage, a condition sometimes stringent for dc motors.

In the following brief review, each of the four generator configurations is considered in turn. It should be remembered that this and the other applications sections are only representative guides. For example, there is behavioral overlap, so more than one type of machine may be used in any given application. Naturally, this fact complicates the process of choosing the appropriate machine.

**1.** *Separately excited.* There are very few practical applications; the description was merely an instructional vehicle for the understanding of the other configurations. However, occasional permanent magnet applications do occur, such as tachometers based on the fact that $V_T$ under light load is nearly equal to $E_G$, which is proportional to shaft speed.

**2.** *Shunt.* Because of the $V_T$ droop, again there are not many applications of this type. However, it too is valuable for gaining understanding of the compound generator.

**3.** *Series and differential compound.* There are welding and electroplating applications where the constant current and current limiting behaviors are desired. Otherwise, it too is little used.

**4.** *Cumulative compound.* This is the main configuration for practical dc generators. Often, they may be connected to a diesel engine for applications where remote dc is needed. An example would be diesel locomotives, in which a diesel engine drives a dc or rectified ac generator that in turn powers the driving dc motors. (Dc motors are chosen here for their high torque, a characteristic discussed further in the Sec. 14.4.4.) As another example, electric cars will probably use the driving dc motor as a dc generator to help charge the battery and slow the vehicle simultaneously during braking. Cumulative compound generators are also used as dynamometers (torque-measuring instruments) for the testing of turbine engines and other vehicles. Finally, the current-limiting characteristic of the shunt component of the compound generator makes the compound generator appropriate for applications such as electric shovels, which are frequently subjected to loads that produce stalling.

An apparently obvious application of dc generators would be the conversion of ubiquitous ac voltages to dc for dc motors. However, this would require an ac motor whose shaft drives a dc generator which could thereby provide dc voltages. It is usually cheaper to convert ac voltages to dc electronically by transformers and rectifiers. But the ripple produced by rectifiers may in some cases be sufficiently objectionable that a generator/motor pair may be worth the cost.

## 14.4 DC MOTOR CONFIGURATIONS AND OPERATING CHARACTERISTICS

### 14.4.1 Introduction

Recall from Fig. 13.35b that all that needs to be done to obtain a motor from a generator is to switch the source and load. Specifically, (1) do not crank, but instead apply a voltage $V_{APP}$, and (2) remove $R_L$ and instead apply a mechanical load to the

His and hers electric cars. On the left, a 1900 "New Form of Electric Automobile" built by Mr. D. L. Davis, superintendent of the Salem Electric Light Company of Salem, Ohio. Two one-horsepower electric motors drive each of the front wheels independently. "The general appearance of the vehicle is handsome, owing to its compact construction." On the right, "A powerful Electric Touring Runabout. This (1907) runabout is capable of covering about 80 miles across country at an average speed of between 15 and 20 miles per hour. It has twenty-four 150-ampere-hour cells, a 3-horse-power motor, and 5-inch pneumatics." It was a simpler and less demanding era. *Source*: *Scientific American* (November 9, 1907).

rotor shaft. In addition, the conventional direction of armature current is shown flowing *into* the armature rather than out, as was the case for the generator. Similarly, the line current $I_L$ flows from the power source to the motor, instead of from the generator to the electrical load $R_L$. These conventions guarantee positive-valued current variables.

Just as was true for generators, "full-load," "load," and "rated" are all synonyms unless otherwise stated. Again, there are four configurations: separately excited, shunt, series, and compound; each is discussed in turn.

### 14.4.2 Separately Excited Motor

The construction of the separately excited motor (Fig. 14.13) is identical to that of the separately excited dc generator except for the input–output reversal as discussed above. The two most important facts for simple calculations are (1) $E_B = V_{APP} - I_A R_A$, and (2) $\phi$ is independent of the mechanical load.

If there are no rotational losses and no other loads ($T_{load} = 0$), then $T_{net} = T_{dev} = k_t I_A \phi$. In steady state, $dn/dt = 0$, so $T_{net} = 0$, and thus $T_{dev} = 0$. Because $\phi$ is constant, $I_A$ must be zero. From the KVL equation, it follows that for $I_A = 0$, $E_B = V_{APP}$. Of course, in any real motor there will be loads and losses, in which case $I_A$ is not zero and $E_B < V_{APP}$. A simple calculation might be to determine the no-load speed given the parameters at full load.

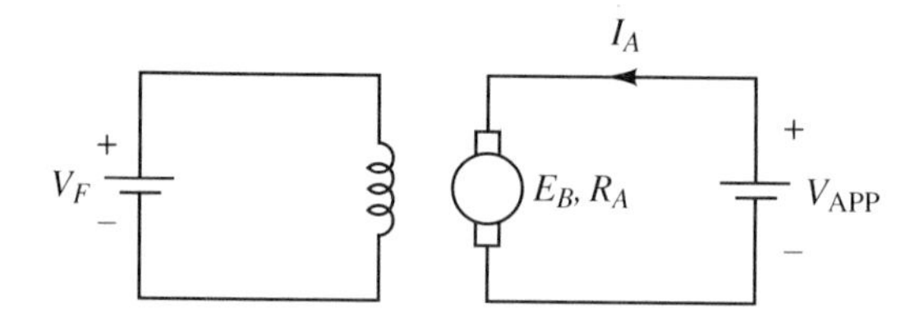


**Figure 14.13** Circuit diagram of separately excited dc motor.

**Example 14.3**

A separately excited motor operating at full load spins at $n_{fl} = 1600$ rpm with $I_{A,fl} = 40$ A (the subscript "fl" means "full load"). The applied voltage is $V_{APP} = 240$ V and the armature resistance is $R_A = 0.5\ \Omega$. Assuming no rotational losses: (a) What is the no-load speed $n_{nl}$? (b) What is $T_{dev}$? (c) What is the speed regulation?

**Solution**

(**a**) At full load,

$$\begin{aligned} E_{B,fl} &= V_{APP} - I_{A,fl}R_A \\ &= 240 - 40 \cdot 0.5 = 220 \text{ V} \\ &= (k_e\phi)n_{fl} = 1600 \cdot (k_e\phi). \end{aligned} \tag{14.12}$$

This yields

$$k_e\phi = \frac{220}{1600} = 0.138 \text{ V/rpm}$$

Recall that at no load, $\phi$ and therefore $k_e\phi$ are unchanged because of the separate excitation. In addition, because it is assumed there are no losses, $I_{A,nl} = 0$. Hence

$$\begin{aligned} E_{B,nl} &= V_{APP} - I_{A,nl}R_A \\ &= V_{APP} = 240 \text{ V} \\ &= (k_e\phi)n_{nl}, \end{aligned} \tag{14.13}$$

so that

$$n_{nl} = \frac{240}{0.138} = 1739 \text{ rpm}.$$

Of course, $n_{nl} > n_{fl}$, as intuition would predict.

(**b**) There is zero developed torque at no load; the requested torque is therefore the developed torque at full load.

$$\begin{aligned} T_{dev} &= \frac{P_{dev}}{\omega_{fl}} \\ &= \frac{E_B I_A}{n_{fl}\pi/30} \\ &= \frac{220 \cdot 40}{1600\pi/30} = \frac{8.8 \text{ kW}}{168 \text{ rad/s}} \\ &= 52.5 \text{ N} \cdot \text{m}. \end{aligned} \tag{14.14}$$

(**c**) *Speed regulation* is defined analogously to voltage regulation [Eq. (14.1)] for a generator, and is again desired to be as small as possible:

$$\text{speed regulation} = \frac{n_{nl} - n_{fl}}{n_{fl}} \cdot 100\%$$
$$= \frac{1739 - 1600}{1600} \cdot 100\% = 8.7\%. \qquad (14.15)$$

Suppose that the separately excited field is reduced (by reducing the field current). What happens? The speed cannot change instantaneously, so $n$ initially remains at its previous value. That fact combined with the decreased flux causes $E_B = k_e n\phi$ to decrease. In turn, $I_A = (V_{APP} - E_B)/R_A$ increases, thereby increasing the torque, which is proportional to $I_A$:

$$T = k_t I_A \phi. \qquad (14.16)$$

The increased $I_A$ tends to increase $T$ while the decreased flux tends to reduce it. But the net effect is that $T$ is increased. We can isolate the dependence of $T$ on $\phi$ by writing

$$T = k_t \frac{V_{APP} - k_e n\phi}{R_A} \phi, \qquad (14.17)$$

which, at normal operating speeds, increases numerically with decreasing $\phi$. The increased torque will result in $dn/dt > 0$, so the motor will speed up. The result may seem counterintuitive: if the field flux is reduced, the motor speeds up. Where does the power to do this originate? It comes from the power source $V_{APP}$ via the increased armature current.

### 14.4.3 Shunt Motor

The shunt motor, again analogous to its generator counterpart, is shown in Fig. 14.14. The relevant relations are now

$$I_L = I_F + I_A \qquad (14.18a)$$

$$I_F = \frac{V_{APP}}{R_F + R_{FV}} \qquad (14.18b)$$

$$V_{APP} = E_B + I_A R_A, \qquad (14.18c)$$

where again $I_L$ represents line current, not load current (there is now no electrical load). (There may or may not be a variable resistor $R_{FV}$ placed in series with the shunt field winding. If it exists, it may be used to control the motor speed.)

It is interesting to observe that while the speed was typically fixed for a

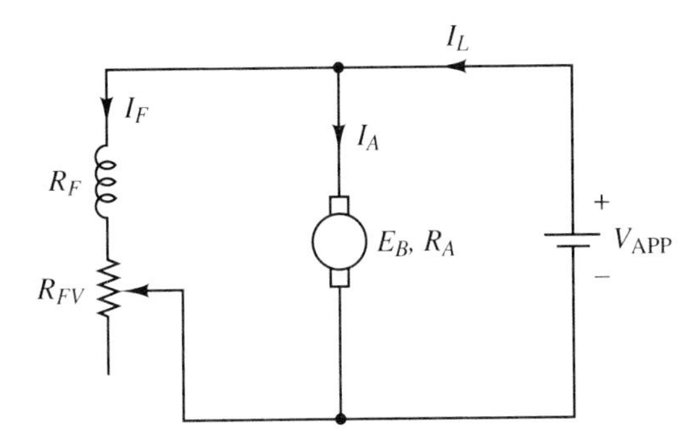


**Figure 14.14** Circuit diagram of shunt dc motor, showing variable field resistance for simple speed control.

generator (by the prime mover cranker), now it is the applied voltage $V_{\text{APP}}$ that is assumed constant, independent of the load. Often $V_{\text{APP}}$ is thought of as an *infinite bus* (ideal voltage source), capable of delivering any needed current while maintaining constant voltage.

Furthermore, for the case of motors, the distinction between separately excited and shunt motors is *artificial.* Unlike the shunt generator, the voltage across the field winding is *fixed* by $V_{\text{APP}}$—just as fixed as it is for the separately excited motor. We merely use the same "infinite bus" to supply both the armature and field windings.

In this machine, $\phi$ is proportional to $I_F$. Because $I_F$ is proportional to $V_{\text{APP}}$, $\phi$ is also proportional to $V_{\text{APP}}$ and is therefore independent of the (mechanical) load. [This is reminiscent of the shunt generator $V_T$ being independent of the electrical load $R_L$; parallel fields give constant outputs (speed or terminal voltage).] What *does* vary under loading is $I_A$. The developed torque for the shunt motor is

$$T_{\text{dev}} = (k_t\phi)I_A, \tag{14.19}$$

so the torque is in this case proportional to the armature current.

The speed is found from

$$n = \frac{E_B}{k_e\phi}. \tag{14.20}$$

The flux is independent of the loading, so $n$ is proportional to only the back emf. If an additional load is placed on a shunt motor spinning in steady state, of course the motor will slow down some, but it is found that the reduction is less than about 10% for rated load compared with no load. Consequently, the shunt motor is often characterized as a relatively constant speed machine, just as the shunt generator can be thought of as a relatively constant terminal voltage machine. In both cases the "constant" quantity is usually easily adjustable—an advantage of dc shunt machines.

A further informative relation easily obtained is that between the torque and the speed. It is found by substituting the equation for $I_A$ in terms of the speed into the expression for the developed torque (which is typically expressed in terms of $I_A$). The expression for the speed in terms of the armature current is

$$n = \frac{V_{\text{APP}} - I_A R_A}{k_e\phi}. \tag{14.21}$$

Solving for $I_A$, we obtain

$$I_A = \frac{V_{\text{APP}} - k_e\phi n}{R_A}, \tag{14.22}$$

which when substituted into $T_{\text{dev}} = (k_t\phi)I_A$ gives

$$T_{\text{dev}} = \frac{k_t(V_{\text{APP}} - k_e\phi n)\phi}{R_A}. \tag{14.23}$$

For a fixed shunt field flux $\phi$, the developed torque [Eq. (14.23)] can be written

$$T_{\text{dev}} = T' - Kn, \tag{14.24}$$

where $T' = k_t V_{\text{APP}}$ and $K = k_t k_e \phi^2/R_A$ are machine constants for fixed $V_{\text{APP}}$, the former having units of N · m. Thus if armature reaction is neglected, the torque for a shunt motor decreases in a linear way with increasing speed. Also of interest is the fact that for a particular value of $n$ [in Eq. (14.24), $T'/K$], there is zero developed torque.

**Example 14.4**

At a speed of 500 rpm, a 10-hp load is driven by a dc shunt motor with an applied voltage of 300 V. If the motor has an efficiency of 80% and $R_F = 100\ \Omega$, $R_A = 1\ \Omega$, (a) what is the power lost in rotation? (b) What is the speed regulation? Ignore core losses.

**Solution**

(**a**) Summarizing the given information, $V_{\text{APP}} = 300$ V, $P_{\text{out}} = 10\ \text{hp} \cdot 746\ \text{W/hp} = 7.46$ kW, $n_{\text{fl}} = 500$ rpm so that $\omega_{\text{fl}} = (2\pi/60)n_{\text{fl}} = 52.4$ rad/s, $R_F = 100\ \Omega$, so that by Eq. (14.18b) $I_F = V_{\text{APP}}/R_F = 300/100 = 3$ A, $R_A = 1\ \Omega$, and $\eta = 0.8$. From the efficiency, the load current may be found as

$$\eta = \frac{P_{\text{out}}}{P_{\text{in}}} = \frac{P_{\text{out}}}{V_{\text{APP}} I_L},$$

or

$$0.8 = \frac{7.46 \cdot 10^3}{300 I_L},$$

or $I_L = 31.1$ A. Therefore, by Eq. (14.18a) $I_A = I_L - I_F = 31.1 - 3 = 28.1$ A. The back emf is by Eq. (14.18c) equal to $E_B = V_{\text{APP}} - I_A R_A = 300 - 28.1(1) = 271.9$ V. Consequently, the developed power is $P_{\text{dev}} = E_B I_A = 271.9(28.1) = 7.64$ kW. Thus the rotational power loss is $P_{\text{rot}} = 7640 - 7460 = 180$ W. For a check on conservation of power, note that

$$P_F + P_A + P_{\text{dev}} = I_F^2 R_F + I_A^2 R_A + P_{\text{dev}} = 900 + 789.6 + 7640 = 9330\ \text{W}$$

$$= P_{\text{in}} = V_{\text{APP}} I_L = 300 \cdot 31.1 = 9330\ \text{W}.$$

So, yes, power is conserved.

(**b**) At full load, using Eq. (14.20), $E_B = 271.9\ \text{V} = k_e \phi n_{\text{fl}} = (k_e \phi)500$, so that $k_e \phi = 0.544$ V/rpm. But for a shunt motor $\phi$ is constant, independent of the load, so $k_e \phi_{\text{nl}} = k_e \phi_{\text{fl}}$. Also, at no load $I_A = 0$ or is negligible, so that $E_B \approx V_{\text{APP}} = 300\ \text{V} = 0.544 n_{\text{nl}}$. Solving for $n_{\text{nl}}$ gives $n_{\text{nl}} = 551$ rpm. By Eq. (14.15), the speed regulation is equal to $(551 - 500)/500 \cdot 100\% = 10.2\%$.

As an example of an explanation of motor dynamics, the process of starting a shunt motor will now be described. Before starting the motor, it is assumed that $V_{\text{APP}}$ has not yet been connected. Therefore, $I_A = 0$, $\phi = 0$, $E_B = 0$, and $n = 0$. When $V_{\text{APP}}$ is connected, then (after inductive transients) $I_F = V_{\text{APP}}/R_F$, so $\phi$ becomes its independent-of-load value.

Due to inertia, $n$ does not instantaneously change, so $n$ is still about zero. Therefore, $E_B = k_e n \phi$ is about zero (or $E_r$). But because $E_B + I_A R_A \approx I_A R_A = V_{\text{APP}}$, $I_A = V_{\text{APP}}/R_A \gg 0$. Therefore, $T_{\text{dev}} = k_t \phi I_A \gg 0$. If it so happens that $T_{\text{dev}} > T_{\text{load}}$, then $T_{\text{net}}$ will be positive and consequently $dn/dt > 0$ and the motor begins to spin; if not, the motor will stand still.

In the former case, as $n$ increases, $E_B = k_e n \phi$ increases, causing $I_A = (V_{\text{APP}} - E_B)/R_A$ to decrease. This in turn causes $T_{\text{dev}} = k_t \phi I_A$ to decrease (recall that $\phi$ is constant), although $T_{\text{net}}$ and therefore $dn/dt$ may still be positive. Eventually,

$T_{dev}$ decreases to the point at which $T_{net} = T_{dev} - T_{load} = 0$ (i.e., $T_{dev} = T_{load}$). Of course, at this time $dn/dt$ becomes zero. Also at this point, $I_A = T_{load}/(k_t\phi)$ (constant for a constant $T_{load}$) so that $E_B = V_{APP} - T_{load}R_A/(k_t\phi)$ is constant for a constant load. (Note also that $dn/dt = 0$ implies that $dE_B/dt = 0$.) This, then, is the steady-state operating condition. That is, because the load and therefore $I_A$ is constant, $T_{net}$ and consequently $dn/dt$ will stay zero. Note that

$$n_{\text{steady state, load}} = \frac{E_B}{k_e\phi} = \frac{V_{APP} - T_{load}R_A/(k_t\phi)}{k_e\phi}. \tag{14.25}$$

What happens if the load is removed so that $T_{load} = 0$? First, $I_A$ cannot change instantaneously (because of the inductive armature coil), so temporarily $I_A$ remains at $T_{\text{old load}}/(k_t\phi)$, where $T_{\text{old load}}$ is just the value of $T_{load}$ in the previous discussion. Therefore, $T_{dev} = T_{\text{old load}}$, so that $T_{net} = T_{\text{old load}} - 0 \gg 0$. Consequently, $dn/dt > 0$ and the motor speeds up, as intuition would suggest. But this means that $E_B = k_e n\phi$ will also increase, causing $I_A = (V_{APP} - E_B)/R_A$ to decrease. When $I_A$ decreases, so does $T_{dev} = k_t\phi I_A$. The speed continues to increase until $E_B = k_e n\phi$ becomes equal to $V_{APP}$, at which point $I_A = (V_{APP} - E_B)/R_A$ becomes zero, in turn causing $T_{dev}$ to drop to zero. Now $T_{net} = 0 - 0 = 0$, so $dn/dt = 0$. Therefore, a new steady state has been reached at no load. The steady state speed is

$$n_{\text{steady state,nl}} = \frac{E_B}{k_e\phi} = \frac{V_{APP}}{k_e\phi}. \tag{14.26}$$

Incidentally, this expression agrees with the evaluation of Eq. (14.25) at $T_{load} = 0$. Similar reasoning would indicate what new steady state would be reached if the load were increased rather than removed.

Figure 14.15 presents typical time dependencies of all the quantities described above. Startup is shown from $t = 0$ to the settling time, and subsequently the load is removed and the transition to the new steady state is shown. Note the very large starting armature current, which could be dangerous. It can be avoided by putting a rheostat in series with the armature; more will be said about this below.

### 14.4.4 Series Motor

The series motor equivalent circuit is shown in Fig. 14.16. Now the line current is the same as the armature current:

$$I_L = I_A. \tag{14.27}$$

As always, $\phi$ is proportional to $B_{field}$. If the motor is operated "below the knee" of the saturation curve (below saturation) as is often the case (especially at startup), $B$ is linearly proportional to $H$, which in turn is proportional to $I_F$. But $I_F = I_A$, so the flux is proportional to $I_A$. This fact is responsible for the remarkable torque relation

$$T_{dev} = k_t I_A \phi = k' I_A^2, \tag{14.28}$$

where $k' = k_t\phi/I_A$ is a machine constant, again due to the fact that $\phi$ is proportional to $I_A$. This dependence on $I_A^2$ rather than $I_A$ is responsible for the very large starting torque of the series dc motor. It is this large starting torque that makes the series motor the motor of choice for certain applications.

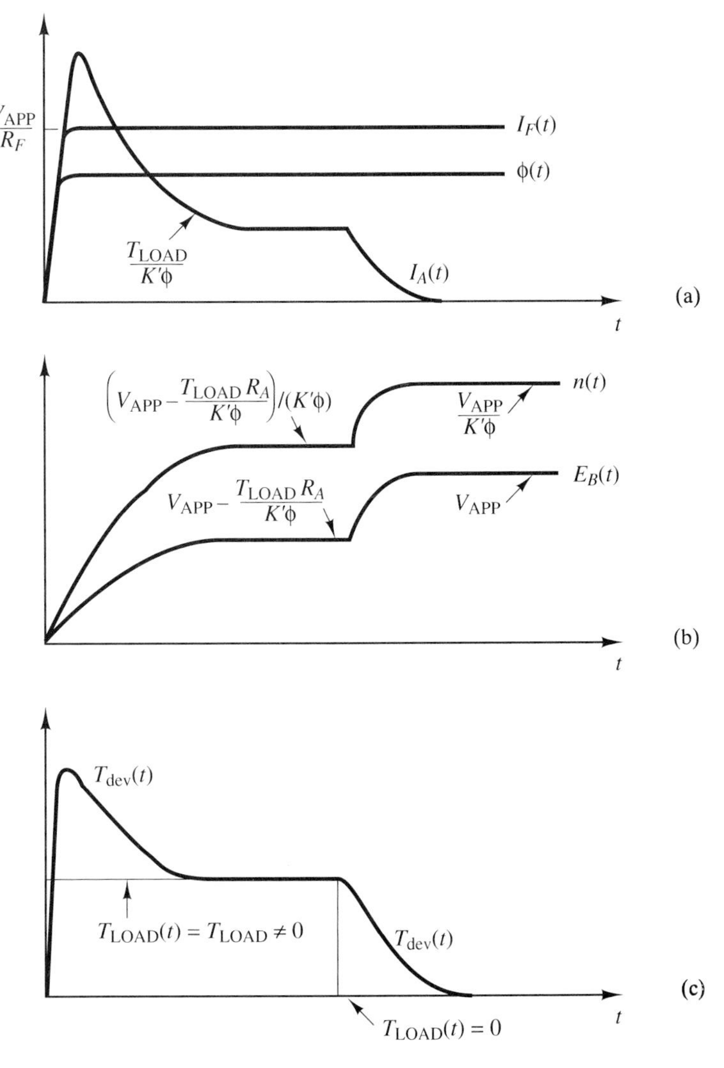


**Figure 14.15** (a) Field and armature currents and flux from startup of a shunt dc motor to steady state, and after removal of mechanical load; (b) speed and back-emf waveforms; (c) load torque and developed torque waveforms.

The equation of the speed for a series motor is different from that for a shunt motor because $\phi$ is proportional to $I_A$ (let the constant $K$ be defined so that $k_e\phi = I_A/K$):

$$n = \frac{V_{APP} - I_A(R_A + R_S)}{I_A/K} = \frac{KV_{APP}}{I_A} - n', \tag{14.29}$$

where $n'$ is a constant having units of rpm.

The most interesting aspect of the speed relation (14.29) is the $1/I_A$ dependence. It means that $n$ can become very large if $I_A$ becomes very small, a situation that could occur if all loading is removed. This condition, which will be examined

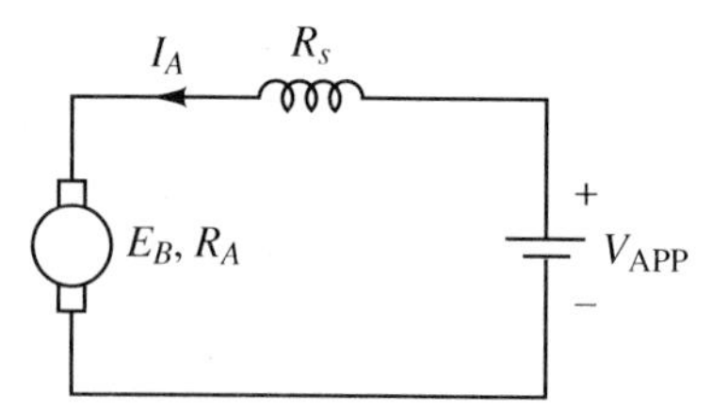


**Figure 14.16** Circuit diagram for series dc motor.

again shortly, is known as *runaway,* and can lead to destruction of the motor. A circuit breaker would be of no use here because runaway is associated with a small, not a large current. Often, normally closed centrifugal switches are installed which open if, for example, $n$ exceeds $1.5n_{fl}$. Examination of Eq. (14.29) shows that if $V_{app}$ is removed by such switches at this speed, so will be the source of the problem.

Again, an expression can be easily found for the torque in terms of the speed. Solving Eq. (14.29) for $I_A$ gives

$$I_A = \frac{KV_{APP}}{n + n'}, \tag{14.30}$$

which when substituted into Eq. (14.28) gives

$$T_{dev} = \frac{\boldsymbol{K}V_{APP}^2}{(n + n')^2}. \tag{14.31}$$

where the constant $\boldsymbol{K}$ is defined as $\boldsymbol{K} = k'K^2$. Thus the torque decreases with the speed according to an inverse square law and increases with the square of applied voltage. [*Note:* If $n$ is replaced by $\omega_{rotor} = n \cdot 2\pi/60$ in Eq. (14.31), then that equation takes the form $T_{dev} = K_a[V_{APP}/(R_A + R_s + K_a\omega_{rotor})]^2$, so that there need be only one machine constant $K_a$ in the torque equation.

*Proof*. Let $k$ be defined so that $\phi = kI_F = kI_A$. Then $V_{APP} = E_B + I_A(R_A + R_s) = k_e n\phi + I_A(R_A + R_s) = k_e nkI_A + I_A(R_A + R_s) = I_A(R_A + R_s + k_e kn)$, so that the developed torque by Eqs. (13.18) and (13.19) is $T_{dev} = 60k_e I_A\phi/(2\pi) = 60k_e kI_A^2/(2\pi) = 60k_e k[V_{APP}/(R_A + R_s + k_e kn)]^2/(2\pi)$. Letting $K_a = 60\,k_e k/(2\pi)$ gives the expression for $T_{dev}$ that was to be shown: $T_{dev} = K_a[V_{APP}/(R_A + R_s + K_a\omega_{rotor})]^2$.

**Example 14.5**

A dc series motor has a rated power of 10 hp and $R_A = R_S = 1\ \Omega$. It is connected to a 220-V bus. For a given load the armature current is measured to be 30 A, the speed is 800 rpm, and the developed torque is 40 N · m. If the load is increased so that $I_A$ increases to 50 A, what is the new speed and torque? Assume that the new load is still in the linear range of the magnetization curve.

**Solution** By Eq. (14.29), $1/K = [V_{APP} - I_A(R_A + R_S)]/(nI_A) = [220 - 30(2)]/(800 \cdot 30) = 6.67 \cdot 10^{-3}\ \Omega \cdot$ min/rev. Again by using Eq. (14.29) at the new load, $n_{new} = [220 - 50(2)]/(6.67 \cdot 10^{-3} \cdot 50) = 360$ rpm. Finally, by Eq. (14.28), $T_{new} = T_{old}(I_{A,new}/I_{A,old})^2 = 40 \cdot (50/40)^2 \approx 63$ N · m.

Because of the interesting differences in dynamics from those of a shunt motor, starting and no-load scenarios will now be presented for the series motor. After all, what good are steady-state equations if the process that leads to them is not at all understood?

Before starting the motor, it is assumed that $V_{APP}$ has not yet been connected. Therefore, $I_A = I_S = 0$, $\phi$ (which is now proportional to $I_S = I_A$) is zero, $E_B = 0$, and $n = 0$. When $V_{APP}$ is connected, $n$ initially remains zero because of inertia and therefore $E_B = k_e n\phi = 0$. Consequently, $I_A = V_{APP}/(R_A + R_S) \gg 0$. This causes a very large $T_{dev} = k' I_A^2$. If it happens that $T_{dev} > T_{load}$, then $T_{net}$ will be positive so that $dn/dt > 0$ and the motor begins to spin. As $n$ increases, what happens to $E_B$? Surely, $n$ tends to increase $E_B$, but what about the flux $\phi$? The flux changes because of the series field connection: $\phi = kI_A$, where $k$ is a constant. But $I_A = (V_{APP} - E_B)/(R_A + R_S)$, so the back emf can be expressed in terms of the speed, rather than in terms of $\phi$:

$$E_B = \frac{k_e n k(V_{APP} - E_B)}{R_A + R_S}. \tag{14.32}$$

Notice that $E_B$ appears on both sides of the equation. Solving for $E_B$ and defining $k_1 = k_e k$ yields

$$\begin{aligned} E_B &= \frac{1}{1 + k_1 n/(R_A + R_S)} \frac{k_1 n V_{APP}}{R_A + R_S} \\ &= \frac{V_{APP}}{\left(\dfrac{R_A + R_S}{k_1 n}\right) + 1}. \end{aligned} \tag{14.33}$$

From this equation it can be seen that as $n$ increases, the *denominator* of $E_B$ decreases, so that $E_B$ *increases*. (Note that all other terms on the right-hand side are constants.)

As a result of the increase in $E_B$, $I_A = (V_{APP} - E_B)/(R_A + R_S)$ decreases, so that $T_{dev} = k_t\phi I_A = k_t k I_A^2 = k_2 I_A^2$ decreases ($k_2 = k_t k$). Eventually, $T_{dev}$ decreases to the point where $T_{dev} = T_{load}$, so that $T_{net} = 0$ and $dn/dt = 0$. At this point, $I_A = (T_{dev}/k_2)^{1/2} = (T_{load}/k_2)^{1/2}$. Therefore,

$$\begin{aligned} E_B &= V_{APP} - I_A(R_A + R_S) \\ &= V_{APP} - \left(\frac{T_{load}}{k_2}\right)^{1/2}(R_A + R_S), \end{aligned} \tag{14.34}$$

which is a constant for constant load. Consequently, $T_{dev}$ will remain equal to $T_{load}$ so that $T_{net}$, and therefore $dn/dt$, stay zero. The steady-state speed is

$$\begin{aligned} n_{\text{steady state,load}} &= \frac{E_B}{k_e\phi} \\ &= \frac{V_{APP} - (T_{load}/k_2)^{1/2}(R_A + R_S)}{k_1(T_{load}/k_2)^{1/2}} \\ &= \frac{V_{APP}(k_2/T_{load})^{1/2} - (R_A + R_S)}{k_1}. \end{aligned} \tag{14.35}$$

Finally, suppose that the load is removed; that is, $T_{load}$ is changed to zero. The armature current $I_A$ cannot instantaneously change, so $I_A$ temporarily remains at $(T_{old\,load}/k_2)^{1/2}$, where $T_{old\,load}$ is the value of $T_{load}$ in the preceding paragraph. This means that $T_{dev}$ will temporarily stay at $T_{old\,load}$, making $T_{net} = T_{old\,load} - 0 \gg 0$, causing a very rapid increase in speed ($dn/dt \gg 0$).

This increase in speed causes $E_B = k_e n\phi$ to increase (as it did for the starting scenario in the preceding discussion). Therefore, $I_A = (V_{APP} - E_B)/(R_A + R_S)$ decreases, and consequently, $T_{dev} = k' I_A^2$ decreases. Eventually, $T_{dev}$ decreases to the point where $T_{dev} = T_{load} = 0$, so that $T_{net} = 0$ and $dn/dt = 0$. But then $I_A = (T_{load}/k_2)^{1/2} = 0$, which means that $\phi = 0$. With $I_A = 0$, $E_B = V_{APP} - 0 = V_{APP}$.

What is the steady-state speed? It is

$$n_{\text{steady state,nl}} = \frac{E_B}{k_e\phi} = \frac{V_{APP}}{0} = \infty. \tag{14.36}$$

This is actually the "runaway" condition mentioned briefly earlier in the section. The infinite speed can also be obtained by substituting $T_{load} = 0$ into Eq. (14.35). It can now be seen what happens when a centrifugal switch opens under excessive speed conditions. With the armature disconnected, $E_B$ never reaches $V_{APP}$ because $V_{APP}$ is no longer connected, the flux is immediately cut off, and $I_A = 0$, so $T_{dev} = 0$. Electrical power delivered to the motor ceases, and along with it the developed torque.

Figure 14.17 presents typical time dependencies of all the quantities described above. Startup is shown from $t = 0$ to the settling time, and subsequently the load is removed and the transition to the new steady state is shown.

### 14.4.5 Cumulative Compound Motor

Just as the compound generator combined the desirable features of the series and shunt generators, the (cumulative) compound motor combines the high starting torque of the series motor with part of the good speed regulation of the shunt motor. Its circuit diagram (long shunt) appears in Fig. 14.18. Now the relevant equations are

$$I_L = I_A + I_F \tag{14.37a}$$

$$\phi = \phi_{shunt} + \phi_{series} \tag{14.37b}$$

$$T_{dev} = k_t(\phi_{shunt} + \phi_{series})\, I_A \tag{14.37c}$$

$$n = \frac{E_B}{k_e(\phi_{shunt} + \phi_{series})}. \tag{14.37d}$$

Of note is the fact that for the same armature current, the torque for the compound motor exceeds that of the shunt motor. As is true for the shunt motor, under increased loading the speed drops, thereby decreasing $E_B$ and therefore increasing $I_A$. But this increases $\phi_{series}$, which is proportional to $I_A$. The denominator of $n$ therefore increases and $n$ becomes smaller than $n$ for the shunt motor under otherwise identical conditions. In fact, the speed regulation of the compound motor is worse than that of the shunt motor; that is the price paid for the improved starting torque.

This time the torque has two terms:

$$T_{dev} = k_t(\phi_{shunt} + \phi_{series})I_A, \tag{14.38}$$

the first being linear in $I_A$ and the second being proportional to $I_A^2$.

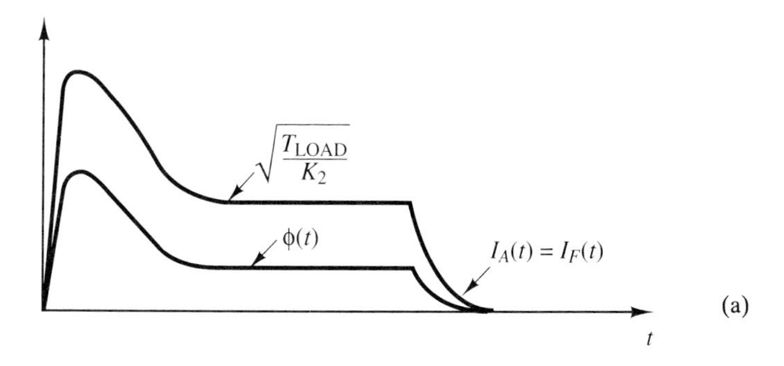


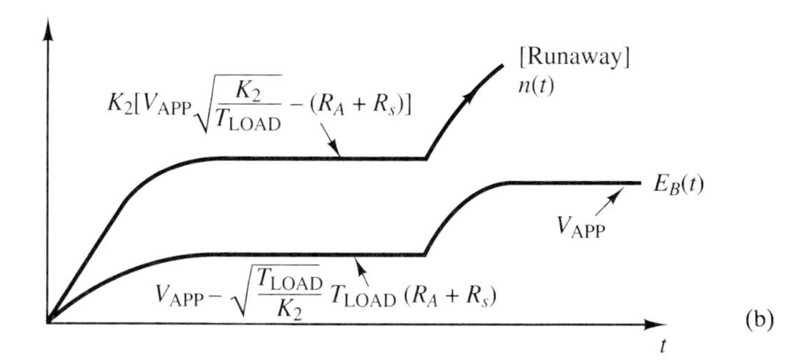


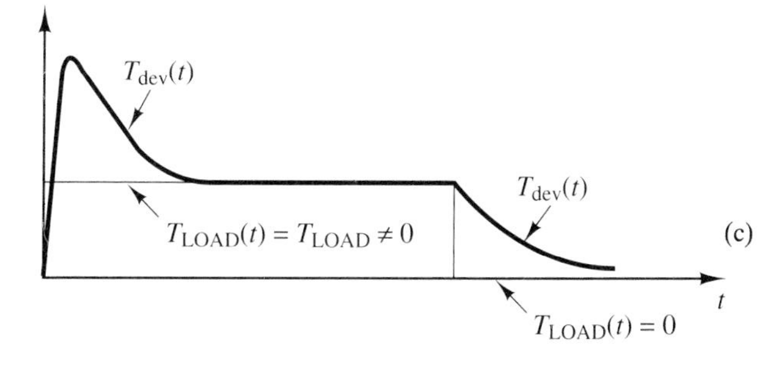


**Figure 14.17** (a) Field and armature currents and flux from startup of a series dc motor to steady state, and after removal of mechanical load; (b) speed and back-emf waveforms; (c) load torque and developed torque waveforms.

As before, an expression for $T_{dev}$ can be found in terms of the speed $n$. Explicitly writing out the speed equation (14.37d) in terms of $I_A$ gives

$$
\begin{aligned}
n &= \frac{E_B}{k_e(\phi_{shunt} + \phi_{series})} \\
&= \frac{V_{APP} - I_A(R_A + R_S)}{k_e(\phi_{shunt} + kI_A)},
\end{aligned}
\tag{14.39}
$$

where $k = \phi_{series}/I_A$ is a constant for the machine, so that

$$
n(\phi_{shunt} + kI_A) = \frac{V_{APP} - I_A(R_A + R_S)}{k_e}. \tag{14.40}
$$

Solving for the armature current yields

$$
I_A\left(kn + \frac{R_A + R_S}{k_e}\right) = \frac{V_{APP}}{k_e} - n\phi_{shunt}, \tag{14.41}
$$

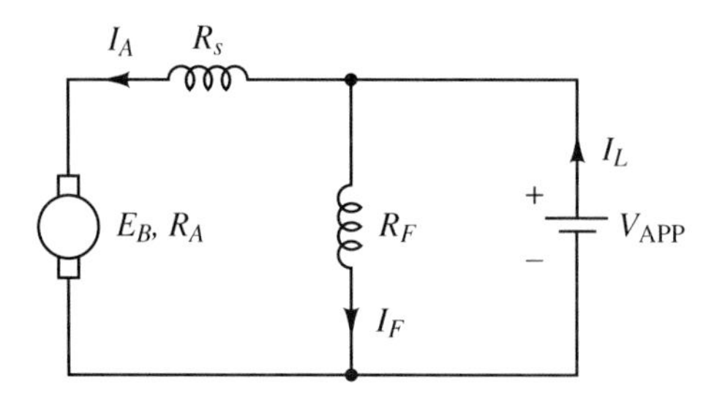


**Figure 14.18** Circuit diagram of the compound dc motor.

which gives

$$I_A = \frac{V_{APP} - k_e n \phi_{shunt}}{k_1 n + R_A} = \frac{K'(n' - n)}{n + n''}, \tag{14.42}$$

where $k_1 = k_e k$, $K' = \phi_{shunt}/k$, $n' = V_{APP}/(k_e \phi_{shunt})$, and $n'' = (R_A + R_S)/(k_e k)$ are all machine constants for fixed $V_{APP}$. Substituting Eq. (14.42) into Eq. (14.38) gives

$$T_{dev} = \frac{K''(n' - n)}{n + n''} + K'''\left(\frac{n' - n}{n + n''}\right)^2, \tag{14.43}$$

where $K'' = k_t K' \phi_{shunt}/I_A$ and $K''' = k_t k K'^2$ are machine constants for fixed $V_{APP}$. This expression shows that there is a rather complicated dependence on $n$. The torque is clearly maximum for $n = 0$, and there is a nonlinear decrease in torque with increasing $n$ until $n = n'$, at which point $T_{dev} = 0$.

### 14.4.6 Summary of Characteristics

Figures 14.19 through 14.21 show typical performance curves derived from the dynamic equations derived in previous sections. Any two curve sets are sufficient to

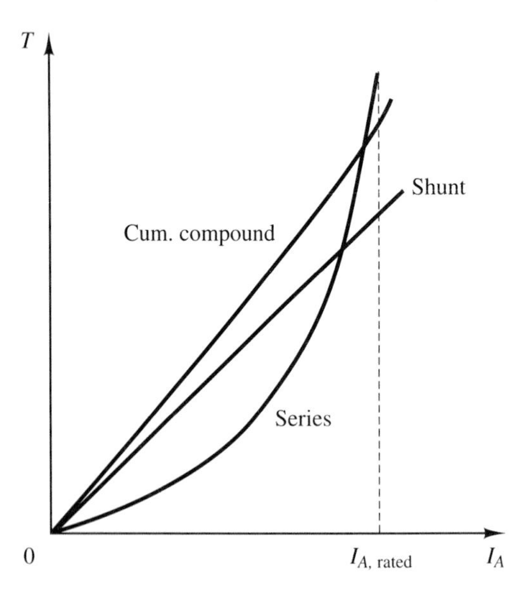


**Figure 14.19** Torque versus armature current for shunt, series, and compound motors.

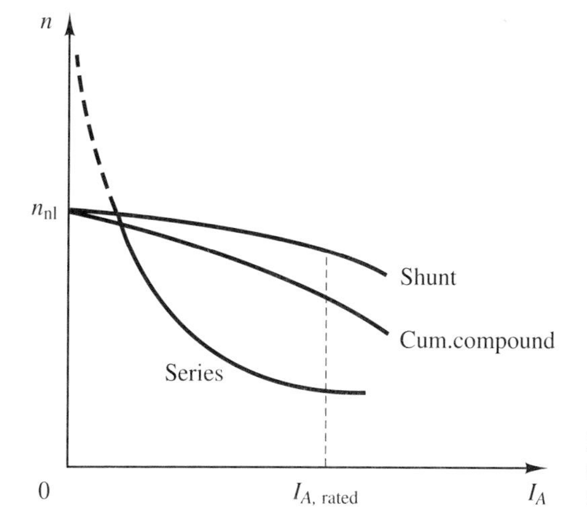


**Figure 14.20** Speed versus armature current for shunt, series, and compound motors.

describe the motor behavior. However, in particular circumstances one curve may be more convenient than another. No curve is shown for the separately excited motor because for comparable (and large) power applications, such motors are infrequently encountered in practice.

The first set of curves (Fig. 14.19) is the torque versus armature current. As shown in Sec. 14.4.3, the shunt motor torque is linear in $I_A$, whereas the series motor torque is quadratic in $I_A$. Intuition verifies that in all cases the torque increases with armature current. The cumulative compound is a mixture, although as a result of design much closer to shunt motor behavior.

At startup, the back emf will be zero so that the armature current will be quite large. The armature current will gradually decrease as $n$ and therefore $E_B$ will

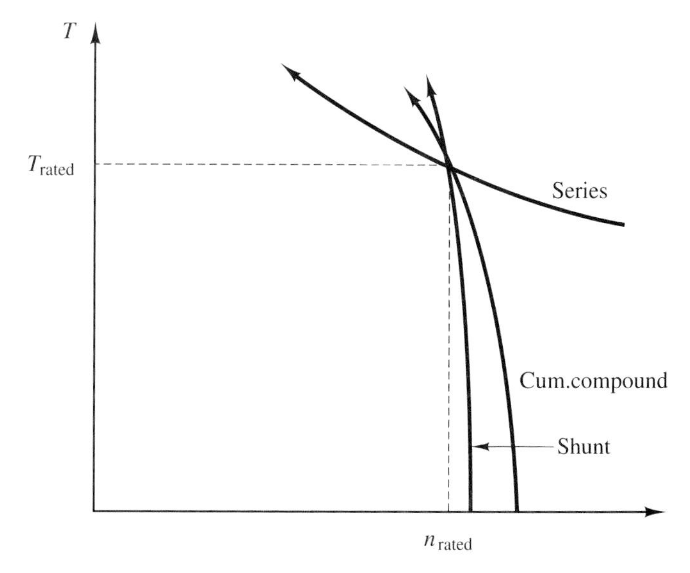


**Figure 14.21** Torque versus speed curves for shunt, series, and compound motors.

increase. Thus operation begins at the right end of those curves and proceeds left during startup. In fact, unless special precaution is taken, $I_A = (V_{APP} - \approx 0)/R_A$ ($R_A$ small and $V_{APP}$ large) could become dangerously large. For this reason, often a motor is started in several stages, each of which (by way of a selected series resistor) limits the current to a safe value (e.g., $1.5I_{A,\text{fl}}$). If the switching can be made speed dependent, the possibility exists for automatic starting.

The second plot (Fig. 14.20) is the speed versus armature current. The linear decrease of $n$ with $I_A$ for the shunt motor is evident, and the cumulative compound behavior is similar. Armature reaction effects are included, so the curves exhibit nonlinearities. The dependence of the speed on the inverse of $I_A$ is clearly seen in the series motor curve. The dashed portion of the curve corresponds to runaway.

Finally, Fig. 14.21 is a composite of torque versus speed plots for the three motor types. Both the shunt and compound motor torques depend on $n$ in a near-linear fashion, with steep, negative slopes. The steep slope indicates that the speed is relatively independent of the (load) torque. The series motor curve shows the $1/(n + n')^2$ behavior. Note the extremely high series motor torque for zero or low speed, as described above. Of course, in all cases the torque decreases with increasing speed.

Finally, before closing this section, it may help to summarize some distinctions between the types of variables and their effects on generator and motor performance. These distinctions are tabulated in Table 14.1.

It may also be helpful to contrast the expressions for conservation of power of dc generators and motors as follows. In all cases, efficiency is the output power divided by the input power. The output power of a generator is $I_L V_T$, and that of a motor is $\omega T_L$. The input powers are:

***Generator***

$$P_{\text{in(mech.)}} = \omega T_{\text{in}} = \underset{(\text{elect. } E_G I_A = \text{mech. } \omega T_{\text{dev}})}{P_{\text{dev}}} + \underset{(\text{mech. losses: windage, friction})}{P_{\text{rot}}}$$

where also

$$P_{\text{dev}} = \underset{(\text{elect. load: } V_T I_{\text{load}})}{P_{\text{out}}} + \underset{(\text{elect. losses: all } I^2R \text{ terms})}{P_{\text{copper}}} + \underset{(\text{magnetic losses: hysteresis heating, eddy currents})}{P_{\text{core}}}$$

***Motor***

$$P_{\text{in(elect.)}} = V_{\text{APP}} I_{\text{line}} = \underset{(\text{elect. } E_B I_A = \text{mech. } \omega T_{\text{dev}})}{P_{\text{dev}}} + \underset{(\text{elect. losses: all } I^2R \text{ terms})}{P_{\text{copper}}} + \underset{(\text{magnetic losses: hysteresis heating, eddy currents})}{P_{\text{core}}}$$

where also

$$P_{\text{dev}} = \underset{(\text{mech. load: } \omega T_{\text{load}})}{P_{\text{out}}} + \underset{(\text{mech. losses: windage, friction})}{P_{\text{rot}}}$$

**TABLE 14.1** VARIABLE AND LOSS CLASSIFICATIONS FOR GENERATORS AND MOTORS

| Variable class | Generator variable or effect | Motor variable or effect |
|---|---|---|
| Independent variables (inputs) | $n$ (with required $T$ externally supplied); $\phi_{\text{field}}$, if separately excited | $V_{\text{APP}}$ (with required $I_L$ supplied); $\phi_{\text{field}}$, if separately excited |
| Dependent variables (outputs) | $V_T$ (with required $I_{\text{load}}$ drawn by load); $\phi_{\text{field}}$, if self-excited; $I_A$ | $n$ (with required $T_{\text{load}}$ overcome); $\phi_{\text{field}}$, if self-excited; $I_A$ |
| Input and output powers | $P_{\text{in(mech.)}} = \omega T_{\text{in}}$; $P_{\text{out(elect.)}} = V_T I_{\text{load}}$ | $P_{\text{in(elect.)}} = V_{\text{APP}} I_{\text{line}}$; $P_{\text{out(mech.)}} = \omega T_{\text{load}}$ |
| Effect of electric and magnetic losses, $P_{I^2R}$, $P_{\text{core}}$, where $R = R_A, R_F, R_S, \ldots$, $P_{\text{core}}$ = eddy current, hysteresis heating | Diminish $P_{\text{out(elect.)}}$ from $P_{\text{dev}} = E_G I_A$ $(= \omega T_{\text{opposing}})$ | Diminishes effect of a given $P_{\text{in(elect.)}}$ |
| Effect of mechanical losses $P_{\text{rot}}$ (windage, friction, etc.) | Diminishes effect of a given $P_{\text{in(mech.)}}$ | Diminishes $P_{\text{out(mech.)}}$ from $P_{\text{dev}} = \omega T_{\text{dev}}$ $(= E_B I_A)$ |

### 14.4.7 Motor Control

To stop a motor, the simplest thing to do is to turn it off. However, for large motors it may not be safe to leave the rotor spinning for the long periods of time before the rotor stops by friction. Therefore, many methods of braking have been used. Obviously, the simplest method is mechanical braking. An ingenious method, known as *dynamic braking,* consists in having the back emf drive a resistive load to dissipate the energy in heat.

This approach, mentioned in Sec. 14.3 in the context of electric cars, is a principal method of slowing down locomotives. In that case, the energy is dissipated in grids of resistors on the roof of the locomotive. We can also throw the motor into reverse (called *plugging*); the results are not as devastating as when the same technique is applied to a car! Of course, current-limiting resistors are used to accomplish the stopping in a safe and controlled fashion.

To reverse motor direction either for stopping or for desired opposite spin, either the shunt (and/or series) field winding or the armature winding (but not both) may be reversed. The armature is usually reversed for one or more of several reasons: (1) disconnecting the field could lead to runaway (series motor), (2) the highly inductive field (carrying large currents) could spark terribly when disconnected, and (3) the armature circuit may be already designed to be "disconnectable" for other reasons.

The complicated topic of speed control is well beyond the scope of this book and is thoroughly treated in advanced texts exclusively on machinery. Suffice it to say that the speed can be controlled by way of variation of $V_{\text{APP}}$, $R_A$, or $\phi$. It was mentioned in passing that a potentiometer in series with the field winding of a shunt motor could vary $\phi$ and therefore $n$.

However, there are far more versatile and efficient ways to control the speed.

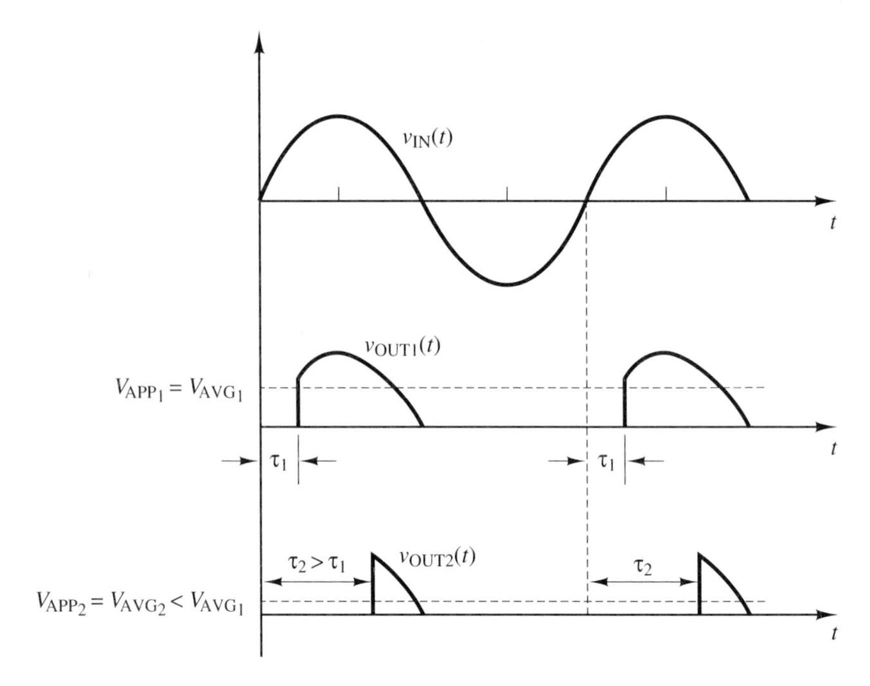


**Figure 14.22** Basic idea of pulse-width modulation for speed control of a dc motor for two values of turn-on delay $\tau$; average voltage ($V_{APP}$) decreases as $\tau$ decreases.

For example, with silicon-controlled rectifiers (SCRs; see Sec. 7.6.5), and ac voltage can be rectified and chopped (see Fig. 14.22). The delay parameter $\tau$ is controllable, and the average value of the chopped waveform depends on $\tau$; hence the (average) value $V_{APP}$ may be so controlled. Increasingly, computers are used for optimal control of electromechanical machinery.

## 14.5 APPLICATIONS OF "DC" (COMMUTATOR) MOTORS

Before turning to the specifics of dc motor applications, a word concerning motor application in general may be helpful. Of the factors influencing motor selection, some of the more important include available power source type and voltage, speed and speed regulation requirements (e.g., whether adjustable speed is a requirement), and of course power and steady-state torque requirements.

Also important to specify are the required starting torque, maximum allowable starting current, acceptable startup time, and the type of load (continuous, pulsing, periodic, etc.). Yet another consideration is efficiency. For example, a high-speed induction motor (see Chapter 15) with reduction gears may end up being cheaper than a low-speed induction motor. Finally, the price tag may override many other considerations. Familiarization with available products, consultation with motors specialists, and as complete knowledge as possible of the characteristics of the load to be driven are the main ingredients to successful motor selection.

A plot of the torque–speed relation of the load can be of great help in determining the class of motors most appropriate. By plotting the load and motor

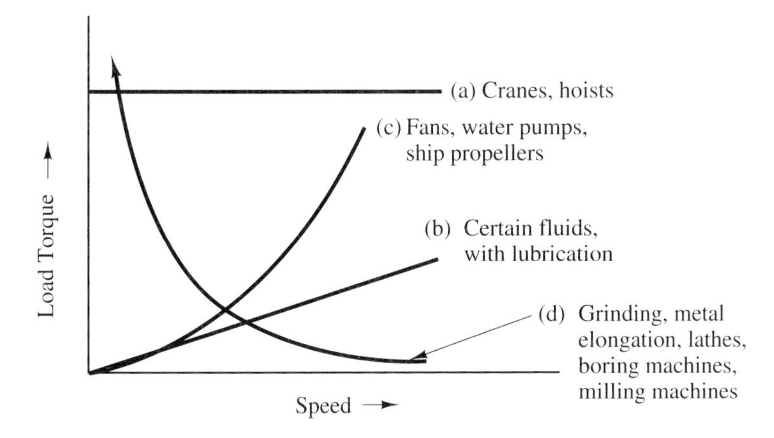


**Figure 14.23** Torque versus speed characteristics of several representative loads. (From C. Wadhwa, *Generation, Distribution, and Utilization of Electrical Energy*, Wiley, New York, 1989.)

torque–speed curves on the same set of axes, the intersection point will indicate the (approximate) operating point. If the load line is above the motor curve, the motor will not rotate. In particular, if the load line is above the motor curve at $n = 0$, the motor will not start, even if the motor curve is above the load line at rated load.

Some examples of torque–speed relations for various types of loads are given in Fig. 14.23. The intercept with the vertical axis gives the starting torque, which may be greater or less than that at rated speed. Cranes and hoists have constant torque with increasing speed. The torque increases with the square of the speed for viscous-medium applications such as fans, water pumps, ship propellers, and so on. For other less viscous media, the dependence may be only linear. In grinding, metal elongation, lathes, boring machines, milling machines, and so on, the power (which equals the product of torque and speed) is roughly independent of speed; thus the torque is inversely proportional to the speed.

In Fig. 14.24, two load line curves are plotted on the same axes as the torque–speed curve for a series motor. The higher load curve indicates that this load is way too heavy for the machine under consideration, while the other load can be handled well by the given motor.

Other load considerations vary with the application. For example, the variation of the load with time is an important factor that to address properly goes beyond the scope of this book. Load variation is important because peculiarities in the transient behavior of machines will determine which machine satisfies the load variation requirements.

Some loads may be continuous and constant (fans, pumps, etc.), some may have pulsating loads (reciprocating pumps, textile looms, etc.), some may be intermittent (cranes, hoists, etc.), and others are impact loads (rolling mills, forging hammers, etc.). However, the very problematic loads such as pulsing and impact loads may be smoothed by the use of flywheels. It should also be noted that in general, the use of control electronics (another advanced subject beyond the scope of this book) is broadening the application scope of many machines by improving the overall characteristics via electronic means.

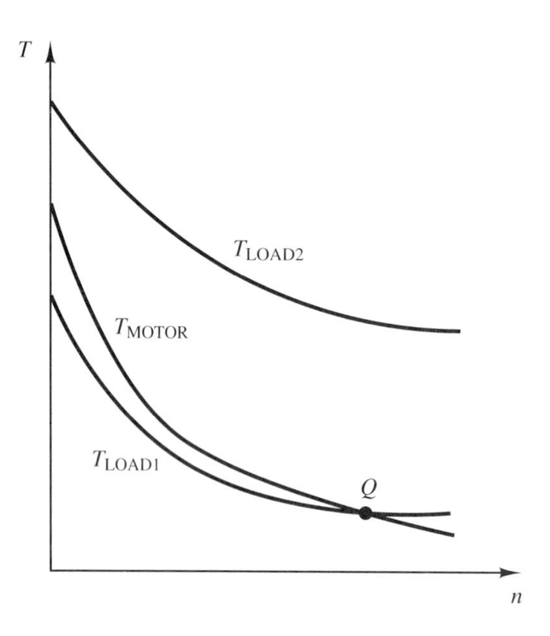


**Figure 14.24** Example of plot of load torque on same axes as developed torque. Because the motor torque exceeds $T_{\text{load }1}$ it will accelerate the load; the other load will not accelerate because $T_{\text{load }2} > T_{\text{motor}}$.

Note that although worst-case scenarios should be considered it is not wise to grossly overestimate the typical load, for that will result in inefficient motor operation as well as unnecessarily high capital cost. In high-power applications, issues such as peak power, power factor, frequency of starts, and transient behavior complicate the decision-making process. But time and money invested in the motor selection process may be well worth the expense; installation of the wrong type of motor could be far more costly.

Dc motors are used in a variety of circumstances where large-range speed controllability or high low-speed/starting torque is required. Consider in turn each of the four configurations:

**1.** *Separately excited motors.* Often, the separate excitation is provided by permanent magnets in very low power applications. Examples of permanent-magnet motors are toys and aircraft control systems. Permanent-magnet separate excitation is also being used in ac synchronous motors (see Chapter 15) for prototype electric cars. As noted previously, the distinction between separately excited and shunt motors is conceptually somewhat artificial.

Advantages of permanent magnet dc motors are steady (low) power and fast response. These are sometimes used as *servomotors* in computers, numerically controlled apparatus in which starts and stops must be made swiftly and accurately. (Servomotors are just motors having the same characteristics for forward and reverse rotation, which are applied in position/rotation control situations; the term does not refer to a particular structure. In particular, the magnitude and direction of their torque is directly determined by the magnitude and sign, respectively, of their input control voltage.) For applications requiring position control, stepper motors are optimal (see Sec. 14.6).

**2.** *Shunt motors.* These are used in applications requiring wide ranges of speed and speed independence from load. The speed may best be controlled by varying the field strength. A few examples would be pump drives, fans, mixers, machine tools, printing presses, agitators, blowers, and compressors.

**3.** *Series motors.* The high torque at startup and low speeds is the main advantage. Also, the ability to perform high-torque reversal makes dc series motors desirable in some situations. Some applications are cranes, locomotives, vacuum cleaners, hand tools, electric cars, coke-iron machinery, and drawbridges. Some speed control may be realized by altering connections of a tapped field coil or by using resistors. In the electric car application, the controllability of speed makes dc motors in this respect advantageous over ac motors.

Use of hydraulic series motors in movable bridges (drawbridges, retractable bridges, etc.) should interest civil engineers. Wide-ranging, robust speed control is essential in order to avoid violent impact upon closing of the bridge.

Another interesting application of series motors is the propulsion of modern ships. (In some cases, synchronous generator–motor pairs, covered in Chapter 15, are used.) Rather than direct coupling of the diesel engine or turbine to the propeller through reduction gears, the prime mover drives a generator, which in turn powers a motor.

In this way speed and direction of the propeller are more easily controlled, many prime movers can easily be used together (past the generators) without complicated mechanical coupling, and resulting electrical power is available for other needs on the ship, to mention a few advantages. Tugboats, icebreakers, fireboats, and ferries are some of the types of ships that can make profitable use of electric propulsion. Obviously, there are no gains in efficiency, only small losses, in introducing electric machines. The benefits above must outweigh the higher initial cost and loss in efficiency for electric machines to be used.

Also, the starting motor of an automobile is a series dc motor. The auto starting motor (typically four-pole, four-brush) requires surprisingly high power (a few kilowatts), which means a very large starting current (hundreds of amperes). Both the drain on the battery and the large amount of heating can be endured by the motor and battery for only short times. Protracted starting times (over 30 s) can drain the battery and ruin the motor by overheating it.

The *universal motor,* so named because it can be run from either ac or dc, is a "dc" commutator series motor and can be found in electric typewriters (soon to become museum pieces), food mixers, drills and other hand tools. Under ac input, the function of the commutator is to keep the armature and the field in phase. That is, not only is ac applied to the armature, but because of the series field connection, ac is also applied to the *field* winding. So those two alternations cancel, while the coil positional alternation due to rotation is also canceled via the commutator. In this way, the torque direction is always the same, and therefore the commutator machine can be run on ac. Actually, the same could be done with a shunt motor, but the starting torque would not be sufficient to be practical.

**4.** *Compound motors.* High starting torque and slow speed are combined with fairly uniform speed over greatly varying load conditions (although not as uniform as for shunt motors). They are good for situations where there may be intermittently high and low loads. Applications include conveyors, elevators, crushers, bulldozers, and balers.

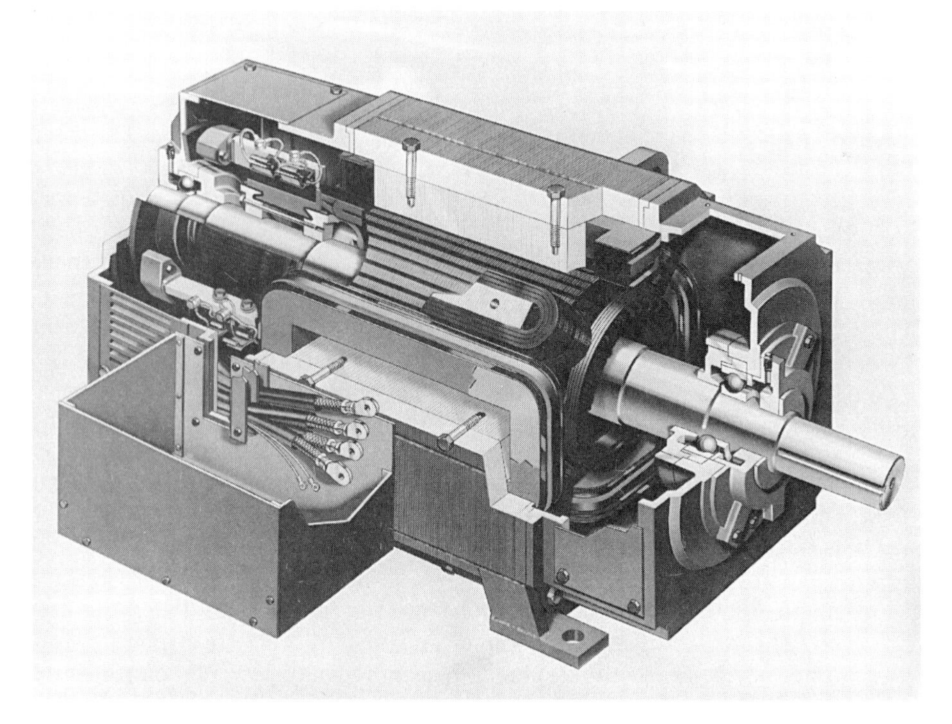

"Drip-proof guarded lightly stabilized" compound long shunt motor. 500 to 600 hp, 4-pole, 1750 rpm, 500 V armature and 300 V field. Used for printing presses, extruders, paper-making/boxboard, metals industry. Must be used indoors to protect commutator. *Source:* Reliance Electric.

As mentioned in the short discussion on speed control, "dc" motors are often powered by ac sources but with electronic rectification in between. This allows for flexible waveshaping, which can ultimately control the speed. In this way, control of speed and high starting torque—two advantages of commutator machines—are made efficiently available using ac power supplies.

The concept of the universal motor shows that, in fact, the dc operation of commutator machines is merely a special case of ac operation: $\omega = 0$. This is the reason for the quotes on "dc" in this section title. In fact, M. R. Lloyd and K. K. Schwarz remark (in *Electric Motor Handbook*)* that "there is no basic difference between ac and dc machines." And there are "two groupings: (1) stationary and (2) rotating field machines," with respect to a stationary observer. (The meaning of the rotating field is discussed in Chapter 15.) However, this statement is controversial and maybe a bit of an overstatement. For example, how could we operate a rotating magnetic field machine with dc?

## 14.6 STEPPER MOTORS

In robotics and precise experimental data measurement systems, it is often convenient to have control over shaft position as well as speed. Moreover, the ability to

* B. J. Chalmers, ed., *Electric Motor Handbook* (Butterworths, Boston, MA, 1988), p. 133.

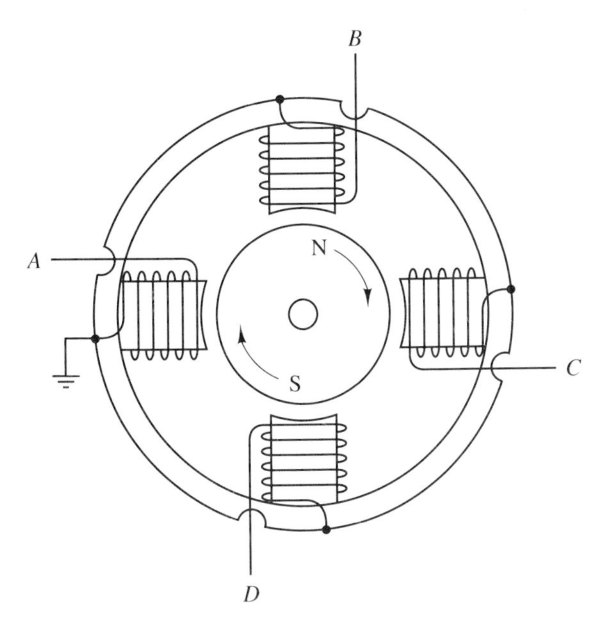


**Figure 14.25** Stepper motor.

exert such control by means of digital signals allows the direct involvement of computers (e.g., microprocessor control). Precision, automation, and repeatability result. Due to the discrete steps in which it moves, such a motor is called a *stepper motor*.

The stepper motor has as inputs a set of on–off signals connected to a set of stator coils. For certain sequences of these inputs, the soft-iron rotor (often a permanent magnet) moves in a controlled fashion to a desired location. We may think of the operation as the conversion of numerical data or sequences into angular positions. It is interesting that no digital-to-analog converter is required; the motor itself is in one sense a sort of D/A converter. Although the same results can be accomplished using automatic feedback control units and conventional motors, the stepper motor is a simple and inexpensive alternative.

Figure 14.25 shows a simplified stepper motor wiring diagram. Under the most simple operating conditions, successive stator electromagnets are turned on and off so that with each change of connections, the rotor is attracted to a position 90° clockwise from its current position.

The more poles on both stator and rotor, the smaller can be the step size. Also, we may decrease the step size by *half-stepping*, which involves the increase in possibilities associated with having more than one stator coil energized at once. Using 50 poles ("teeth" on the rotor) we can specify and obtain an angular resolution of about 2°. Each stopping point is an unchanging angle, and the rotor is held there by magnetic forces. Therefore, errors are not cumulative and drift is not a problem.

If the rotor is not a permanent magnet and consists merely of steel laminations, it is called a *variable reluctance stepper motor*. It then operates on the same principle as does a relay: It responds to forces in the direction of the gradient of the inverse of the reluctance. When this gradient is moved from position to position, the rotor follows under action of that force.

Applications include robotics, computer disk drives, line printers, $x$-$y$ plotters, precise holding and cutting of machine tools, many situations involving computer control, and measurement-taking lab equipment. Because of the discontinuous nature of operation, the stepper motor is limited to applications having power requirements of less than a few horsepower. If overloaded, they will merely slip and eventually overheat.

## 14.7 MICROMOTORS AND MICROMACHINES

With the maturation of processes for fabrication of microchips, it was inevitable that such technologies would be applied to machines. Extremely small motors, called *micromotors*, are possible using these methods. The uses for machines-on-a-chip abound, from robotics applications to microsurgery. Operation is based on electric forces rather than magnetic, as coils at the micro scale are impractical.

One simple example is the *electrostatic top motor.* By specialized template and etching techniques, a set of electrode stripes (the stator), a ground electrode, and a rotor having a conducting surface can be fabricated on a substrate. The operation is similar in principle to the stepper motor: The applied voltage is sequentially connected (via external electronics) to successive electrode stripes. The stator is only 2 mm in diameter.

More elaborate structures are also possible, having, for example, 12 rotor and 18 stator poles. The poles can be oriented either above the rotor (*top-drive*) or radially outside (*side-drive*). A typical rotor radius is only 0.1 mm.

So far, performance of these machines is rather poor. In one report, the input voltage frequency was decreased from 100 Hz to 20 Hz. At 60 Hz, the motor began to spin, and slowed down with the decreasing input frequency. But at 20 Hz it stopped rotating. Another report discussed very high rotational frequencies: 2000 to 6000 rev/s. But the devices self-destroyed after about 3 minutes. Continued research in this area appears promising and is not limited to electromechanical machines. For example, micro flow sensors and micro mechanical joints, gears, and turbines are already being made.

## 14.8 SUMMARY

The machines under consideration in this chapter, all having commutators (with the exception of stepper motors and micromachines), are called dc machines because the terminal voltage and current are usually nearly constant in time. Analysis procedures were reduced from the previous detailed field relations to simple circuit diagrams.

There are four basic types of generators: separately excited, shunt, series, and compound generators. The first has an externally supplied background field. As a result, the terminal voltage droops with increased load only with the $I_A R_A$ drop (ignoring armature reaction). This agreeable output characteristic is obtained at the price of the need of an external electrical energy source. The other three types of generators are all self-excited; the voltage-generating field is developed and maintained by the dynamo itself. This very important improvement in generators was developed by Wheatstone, the Siemens brothers, and others in the 1860s.

The field voltage for a shunt generator is the terminal voltage. Use of the residual field or an external source can be used to start the buildup of generated voltage upon turning the rotor. Normally operated in the saturation region, the output voltage is again relatively constant for a wide range of load currents, although it does droop.

In the case of a series generator, the field is in series with the load, so a load must be connected in order for the generated voltage to build up. The series generator has poor voltage regulation but has the advantage that the generated voltage *increases* with the load current (to a point). By combining the series and shunt field windings, we obtain a compound generator. The series winding compensates for the drooping with load that occurs in the shunt generator. We have the possibility of under-, flat, or overcompounding. A short discussion about efficiency and applications concluded Secs. 14.2 and 14.3 on generators.

Because all that needs to be done to convert a generator to a motor is reverse the roles of source and load, it is a straightforward matter to consider the corresponding four types of dc motors. But now instead of the generated or terminal voltage and voltage regulation being the parameters of interest, it is the developed torque and speed regulation that fundamentally characterizes the motor. In all cases the developed torque is proportional to the product of the flux and armature current.

For the separately excited motor, we have independent control over the flux (via the field current) and the armature current (via the applied voltage) and therefore the torque. The only real difference for the shunt motor is that the flux and armature current are both dependent on $V_{APP}$ and so are no longer independently controllable. The armature current is approximately linearly related to the speed, so the torque–speed curve is roughly linear.

The same is typically true of the compound generator, it being a perturbation of the shunt generator. But in the series motor, the flux is now proportional to the armature current, so the torque is proportional to the square of the armature current. And the mathematics shows that the torque–speed relation for the series motor is roughly an inverse relation. This means that the series motor is well suited to constant-power applications, where the required starting torque is very high and the high-speed torque requirement is low.

After a brief look at motor control, application examples and issues concerning dc motors were addressed at some length. The general procedures and elements involved in motor selection were reviewed, many of which also apply to selection of ac motors.

The chapter concluded with some brief descriptions of some "dc" motors that are of recent interest: stepper motors and micromotors. In the former case, the asset is the ability to numerically specify the angular position of the rotor and the sequence of motions to arrive there. In the latter case, the incredibly small size is the main advantage; much more research is needed in making them practical. Our final topic will be ac machines, primarily motors, as they are the ac machines one is most likely to encounter on the job.

## PROBLEMS

**14.1.** For review of Chapter 13 materials, consider the following situation. A four-pole motor has a total of 360 conductors on its armature; refer back to Fig. 13.34. The armature is $L = 20$ cm long and has a radius of $R = 8$ cm. Each pole takes up $x =$ 9 cm of the perimeter of the air gap. The field is 1.2 T in the air gap. The armature is lap wound and is rotating at 1200 rpm. The current in each armature loop is 20 A. Find:
- **(a)** The percent of the conductors that are active (directly under a pole).
- **(b)** The torque developed on the armature.
- **(c)** The counter (back) emf $E_B$.
- **(d)** The mechanical power developed.

**14.2.** What is the most convenient form of separate excitation for generators, and why?

***14.3.** If the no-load voltage of a separately excited generator is 125 V at 900 rpm, what will be the voltage if the speed is increased to 1200 rpm? Assume constant field excitation.

**14.4.** For a dc generator, which *I–V* characteristic is used to determine the operating point under full-load conditions?

**14.5.** Which is better: a high or a low value of voltage regulation, and why?

**14.6.** If the shunt field requires $E_G \neq 0$ to make the nonzero $\phi$ and therefore nonzero $d\phi/dt$ necessary to have $E_G = d\phi/dt$ be nonzero in the first place, how can the generator ever build up voltage?

**14.7.** Name three reasons why the voltage at the terminals of a shunt generator could fail to build up, and present a method of correcting each problem.

**14.8.** Why is a shunt generator operated in saturation? Is this necessary for a series generator?

**14.9.** Using Eq. (14.11a), write an expression for the efficiency of a shunt generator in terms of only resistor values, $P_{core}$, $P_{rot}$, and the terminal voltage and current.

**14.10.** For Example 14.1, write the power balance equations for no-load and full-load conditions. Show that within rough calculation differences, power is conserved.

**14.11.** A shunt generator has an open-circuit terminal voltage of 144 V. When loaded, the voltage across the load is 120 V. Determine the load current when the field circuit resistance is 100 Ω and the armature resistance is 0.52 Ω. In the process, find $R_L$. (*Source:* Ryff)

**14.12.** The no-load terminal voltage of a shunt generator is 100 V, and the full-load terminal voltage is 77 V. $R_A = 0.4\ \Omega$, $R_F = 80\ \Omega$. Find $I_L$, $I_A$, the load resistance $R_L$, and the generated and output powers. Ignoring $P_{rot}$, calculate the efficiency.

**14.13.** **(a)** Find the load current drawn from a shunt generator for which $R_F = 100\ \Omega$, $R_A = 0.5\ \Omega$, given that the open-circuit and full-load terminal voltages are, respectively, 115 V and 100 V. In the process, find the load resistance $R_L$.

**(b)** Suppose that the load is replaced by a new load of resistance $R_L = 1\ \Omega$. Find the new values of $V_T$, $I_L$, $I_A$, and $I_F$.

**14.14.** Suppose that when a shunt generator produces 15 kW of load power, we determine that $V_T = 90$ V and $E_G = 100$ V. The field resistance is 80 Ω.

**(a)** Find the value of $E_G$ that must exist if the load and cranking speed are changed such that when on the same machine $V_T = 100$ V, 25 kW is produced. (This calculation will require the determination of the value of $R_A$.)

**(b)** For both the originally stated conditions and those in part (a), find $V_{T,\text{nl}}$. You should find that each is almost identical to the associated generated voltage.

**(c)** Considering the two values of $V_T$ to be $V_{T,\text{fl}}$ for the two situations described above, find the two corresponding values of voltage regulation.

**(d)** Find the load resistance $R_L$ for each of the two stated operating conditions.

**14.15.** Which type of generator can operate as a current source: a compound generator, a shunt generator, or a series generator? Why?

**14.16.** Why might someone undercompound or differentially compound a compound generator?

**14.17.** Why is overcompounding done?

**14.18.** Why is there still a potentiometer $R_{FV}$ in series with the field winding in a flat-compounded generator?

* *Source:* Peter F. Ryff, *Electric Machinery,* Prentice Hall, © 1988, pp. 75, 76, 104, 105. Reprinted by permission of Prentice Hall, Englewood Cliffs, New Jersey.

**14.19.** What is the purpose of a series diverter resistor in a compound generator?

**14.20.** A long-shunt compound generator has the following parameter values: $V_T = 120$ V, $R_F = 100\ \Omega$, $R_A = 0.5\ \Omega$, $R_S = 0.1\ \Omega$, and $R_L = 20\ \Omega$. Calculate:
(**a**) The rated line (i.e., in this case, load) current.
(**b**) The generated voltage.
(**c**) The power to the load.
(**d**) The power generated by the armature.
(**e**) The efficiency, neglecting rotational losses.
(**f**) If the developed torque is 15 N · m, what is the speed in rpm?

**14.21.** A short-shunt compound generator delivers 4 kW to a load. The terminal voltage is 100 V. The armature, field, and series resistances are, respectively, 0.6 Ω, 100 Ω, and 0.05 Ω. Find $E_G$ and the series, load, field, and armature currents. Draw a schematic diagram showing all current directions.

**14.22.** A load draws 70 A from a 400-V short-shunt compound generator for which $R_S = 0.06\ \Omega$, $R_F = 150\ \Omega$, and $R_A = 0.1\ \Omega$. Find $E_G$, $I_A$, and the efficiency if $P_{rot} = 300$ W.

**14.23.** A short-shunt compound generator delivers 50 A at 500 V to a resistive load. The armature, series field, and shunt field resistances are 0.16, 0.08, and 200 Ω, respectively. Calculate the generated emf and armature current. If the rotational losses are 520 W, determine the efficiency of the generator. (*Source:* Ryff)

**14.24.** A short-shunt compound generator has 1000 turns per pole on the shunt field and 4.5 turns per pole on the series field. If the shunt and series field ampere-turns per pole are 1400 and 180, respectively, calculate the power delivered to the load when the terminal voltage is 220 V. Repeat for a long-shunt connection. (*Source:* Ryff)

**14.25.** For the generator in Problem 14.23, calculate the voltage regulation. (*Source:* Ryff)

**14.26.** Is the field flux of a shunt motor independent of the load? What about the field flux of a shunt generator? Why?

**14.27.** Suppose that a dc shunt motor draws 30 A when 150 V is applied. If $P_{rot} = 400$ W, $R_F = 140\ \Omega$, and $R_A = 0.6\ \Omega$, what are $P_{in}$ and $P_{out}$ and the various $I^2R$ losses? From these, estimate the efficiency of the machine.

**14.28.** A 12-hp 150-V shunt motor spinning at 850 rpm draws a line current of 90 A; $R_A = 0.3\ \Omega$, $R_F = 120\ \Omega$. Suppose that we increase the mechanical load and also drop $V_{APP}$ down to 100 V, with the result that now the line current is 75 A. Make the fairly realistic assumption that $P_{rot}$ is proportional to the speed $n$.

**14.29.** Suppose that we are told that at a speed of 1000 rpm, a shunt motor has a back emf of 105 V. The applied voltage is 200 V, $R_F = 100\ \Omega$, and $R_A = 0.3\ \Omega$. First draw from memory the shunt motor schematic diagram, with all current directions properly indicated. Then find the torque and speed at no-load and at full-load conditions ($P_L = 10$ kW). Let $P_{rot} = 100$ W for both conditions.

**14.30.** In Example 14.4, solve again for the no-load speed, but this time do not neglect the no-load armature current. You should find that the result is nearly identical with that in Example 14.4. Assume that rotational losses are proportional to speed.

**14.31.** A 120-V shunt motor draws 23 A and spins at 1200 rpm; $R_A = 0.20\ \Omega$, $R_F = 200\ \Omega$, and $P_{rot} = 350$ W. Calculate:
(**a**) The developed power.
(**b**) The output power.
(**c**) The output torque.
(**d**) The efficiency at full load. (*Source:* Adapted from Ryff)

**14.32.** A 220-V shunt motor draws 40 A. $R_A = 0.40\ \Omega$, $R_F = 140\ \Omega$, and $P_{rot} = 600$ W. Calculate:

(a) The developed power.
(b) The output power.
(c) The input power.
(d) The field and armature copper losses.
(e) The efficiency at full load.
(f) Demonstrate approximately conservation of power.
(g) If $\omega_{fl} = 104.72$ rad/s and $I_{A,nl} = 0$ A, what is $n_{nl}$? Also give the speed regulation.

**14.33.** (a) A 120-V shunt motor draws 10 A and spins at a speed of 1500 rpm ($R_A = 0.6\ \Omega$, $R_F + R_{FV} = 100\ \Omega$). Suppose that the load is increased to the point where now the motor draws 25 A. What is the motor speed at the second load?
(b) Recall that $R_{FV}$ can be used in a shunt motor to adjust the speed. Suppose that now the field potentiometer is adjusted so that $R_F + R_{FV}$ is reduced to 60 Ω. Find the new speed if the new line current is 40 A.

**14.34.** A 240-V shunt motor takes 20 A when running at 960 rpm. The armature resistance is 0.2 Ω. Determine the no-load speed, assuming negligible losses. (*Source:* Ryff)

**14.35.** A 120-V shunt motor has the following parameters: $R_A = 0.40\ \Omega$, $R_F = 120\ \Omega$, and rotational loss 240 W. On full load, the line current is 19.5 A and the motor runs at 1200 rpm. Determine:
(a) The developed power.
(b) The output power.
(c) The output torque.
(d) The efficiency at full load. (*Source:* Ryff)

**14.36.** A 460-V shunt motor drives a 50-hp load at 900 rpm. If the shunt field resistance is 57.5 Ω, the armature resistance is 0.24 Ω, the full-load speed is 900 rpm, and the efficiency is 82%, determine:
(a) The rotational losses. (b) The speed regulation. (*Source:* Ryff)

**14.37.** A 220-V dc shunt motor has an armature resistance of 0.3 Ω. Calculate:
(a) The resistance required in series with the armature to limit the armature current to 80 A at starting.
(b) The value of the counter emf when the armature current has decreased to 30 A with the resistor still in the circuit. (*Source:* Ryff)

**14.38.** A 10-hp 200-V shunt motor has a full-load efficiency of 85%. The armature resistance is 0.25 Ω. Determine the value of the starting resistor placed in series with the armature to limit $I_{start}$ to $1.5\,I_{fl}$. Neglect the shunt current, thus $I_{start}$ and $I_{fl}$ are armature current values. What is the value of the counter emf when the current drops to full-load value, assuming that the starting resistor is still in the circuit? (*Source:* Ryff)

**14.39.** A 100-kW belt-driven shunt generator runs at 600 rpm on a 250-V bus. When the belt breaks, it continues to run as a motor (supplied by the bus), then taking 8 kW. What will be its speed? The armature resistance is 0.02 Ω, and the field resistance is 125 Ω. (*Source:* Ryff)

**14.40.** A 240-V shunt motor has a field resistance of 100 Ω, an armature resistance of 0.5 Ω, and has rotational losses of 300 W at full load. When operating at full load and rotating at 2400 rpm, the line current is 30 A. Find:
(a) The developed torque, the back emf, and the developed power.
(b) The output power.
(c) The output torque.
(d) The efficiency at full load. (*Source:* Adapted from Ryff)

**14.41.** Repeat Problem 14.40 for a series motor having all the same parameters as the shunt motor in that problem except that the (series) field resistance is 0.5 Ω (and there is no shunt field winding).

**14.42.** A 100-V series motor has an armature resistance of 0.2 Ω and a series resistance of 0.05 Ω. If the output power is 6 hp at 600 rpm, what is the speed at 2 hp? Assume that $P_{rot} = 300$ W for both conditions.

**14.43.** A 120-V series motor draws 40 A while spinning at 800 rpm and generating a torque of 50 Nm. $R_A = 0.8\,\Omega$, $R_S = 0.1\,\Omega$. Suppose that the mechanical load is increased such that the line current is now measured to be 70 A. What are the new torque and speed? Assume that the magnetization curve is linear over this range of operation and ignore mechanical losses.

**14.44.** A 100-V series motor has $R_S = 0.08\ \Omega$ and $R_A = 0.10\ \Omega$. When supplying a 4-hp (2984-W) load, the rotor spins at 600 rpm. Suppose that the load is now changed such that the new rotor speed is 130 rpm. Find the new output power and, for both conditions, $E_B$ and the efficiency. [*Hint:* Express the new output power in terms of the new armature current, the old power (2984 W), and the new and old speeds using $P = KnI_A^2$. Substitute this into the power balance equation to find $I_A$ and therefore $P$.]

**14.45.** A series motor takes 40 A at 460 V while hoisting a load at 6 m/s. The armature plus field resistance is 0.48 Ω. Determine the resistance to be placed in series with the motor to slow the hoisting speed to 4 m/s. Assume linear operation on the magnetization curve. (*Source:* Ryff)

**14.46.** Consider the following 200-V long-shunt compound motor for which $R_S = 0.01\ \Omega$, $R_A = 0.4\ \Omega$, and $R_F = 120\ \Omega$. At no load the total flux is $\phi_{nl} = 45$ mWb, $I_{L,nl} = 12$ A, and $n_{nl} = 1200$ rpm. At full load, $\phi_{fl} = 50$ mWb and $I_{L,fl} = 80$ A. Find $n_{fl}$ and the torque $T_{fl}$ as well as the speed regulation. (*Hint:* You cannot use $k_e\phi$ for both conditions because $\phi$ changes, but you can solve for $k_e$.)

**14.47.** Cite the dependence of developed torque on armature current for the four basic types of dc motor.

**14.48.** When load-independent speed is sought, should the speed regulation be low or high? Why?

**14.49.** Is it always true that speed increases with increasing field flux in a dc motor? If not, cite a counterexample.

**14.50.** Which of the following two motors, in other respects comparable, will have the higher torque, especially at start: series or shunt? Why?

**14.51.** Give a physical explanation for the runaway condition of a series motor from the viewpoint of Eq. (14.29).

**14.52.** If one desires a relatively constant-speed dc motor near rated speed, should the series motor be used? If not, suggest a better alternative.

**14.53.** Without looking at the tabular summary in the chapter, write the power balance equations for dc generators and motors.

**14.54.** In Fig. 14.22, should $\tau$ be large or small to obtain high speed using an SCR-controlled dc motor?

**14.55.** How can the universal motor work on ac and yet have a commutator?

**14.56.** The stepper motor is not a particularly high torque motor. Wherein lies its value?

# 15

# Electromechanical Machines III: AC (Commutatorless) Machines

## 15.1 INTRODUCTION

Although there are many uses for dc machines, some of which have just been described in Chapter 14, there are also serious disadvantages that in some applications make them unattractive.

Dc machines tend to be expensive, owing in part to the relatively complicated commutator system. In fact, the commutator limits the power and speed capabilities of the dc motor. Furthermore, for a given power rating, dc motors are generally far larger (physically) than their ac equivalents. Also, the commutator usually must be exposed to the environment for cooling, which hastens its deterioration. Carbon buildup and sparking problems also plague dc motors, as well as the necessity for ac-to-dc conversion (for some motors). Of course, today the clumsy mechanical commutator can often be replaced by high-performance, cheap electronic rectification circuitry.

In addition, the largest currents flow in the rotating component (rotor), making external electrical connections more difficult and expensive. Also, it is easier to insulate a stationary coil (for high-power applications) than a moving coil, because there are no weight restrictions on the former. In an ac motor the large voltages and currents appear on the stationary member. Furthermore, the extensive, delicate rotor windings on a dc machine are subject to centrifugal forces ($m\omega^2R$) that can tear the machine apart at sufficient speeds.

Although the dc shunt motor has fairly constant speed (from zero loading to rated loading), the synchronous (ac) motor has essentially *perfect* speed regulation (constant speed). That *synchronous speed* is linked to the very well-regulated line frequency (60 Hz in the United States—by which you can run a clock). The largest machines (in size and power) now in operation are ac synchronous machines.

Finally, a very simple and obvious objection to dc motors is that most available electric power (other than batteries) is ac. (Actually, it was the foregoing disadvantages of dc and other factors that were responsible for ac power transmission, as discussed in Sec. 6.7.) So if a dc motor is to be run off electric lines, the sinusoidal (ac) voltages must first be converted to dc voltages. This introduces cost and inefficiency.

In this final chapter, the principles of ac motor operation, and to a lesser extent, ac generator operation are addressed in some detail. Specifically, the three main types of motor are examined: single-phase induction motors, three-phase induction motors, and three-phase synchronous motors. Crucial issues involved in the selection of motors are addressed—in particular, the dependence of developed torque on other machine parameters. After a short discussion of ac generators, the chapter closes with an introduction to a new variation on ac motors: the superconducting motor.

## 15.2 SINGLE-PHASE INDUCTION MOTOR

### 15.2.1 Principles of Operation

The obvious construction of an ac commutatorless machine would have a sinusoidal voltage connected to the field coil, as shown in Fig. 15.1a. Without the commutator or even slip rings, the rotor winding closes on itself, and functions similarly to the shorted secondary of a transformer. The only difference is that the secondary is on a rotatable core so that any torques that may be produced may produce rotation. Such torques are due to induced currents resulting from induced voltages generated when the rotor windings are exposed to $d\phi/dt$ presented by the primary windings.

Suppose in Fig. 15.1a, drawn for $t = 0$, the background $B$ field (and therefore $\phi$) is for this instant of time zero and about to become positive left to right. (Actually, the field is significantly radially directed, so terms such as "left to right" refer merely to net-flux directions.) Then $d\phi/dt$ is at a positive maximum at the horizontal rotor loop conductors because the derivative of a sinusoid has maximum absolute value when the sinusoid itself is zero.

The induced voltages in each coil will have polarity such that the induced flux opposes $d\phi/dt$ for the given coil. Thus by the right-hand rule, the currents in loops 1 and 2 will have directions as shown in Fig. 15.1b. But the current direction in conductor $a$ is opposite that in conductor $b$. By Ampère's force $B_{\text{APP}}\, i_{\text{IND}}\, L$ (again, "APP" means "applied"), where $L$ is the length of a conductor side, the torque on conductor $a$ (and $a'$) is clockwise, while simultaneously that on $b$ (and $b'$) is counterclockwise. These opposing torques cancel, as do those of all similar pairs on the rotor. Therefore, there is no net torque on the rotor at any time, and it will not begin to spin. It should be noted that there are fortuitous starting angles for which a net starting torque is sufficiently nonzero. However, the prospect of zero or near-zero starting torque roughly half the time is not acceptable.

Note that unlike dc motors, the torques in the ac *induction* motor depend on

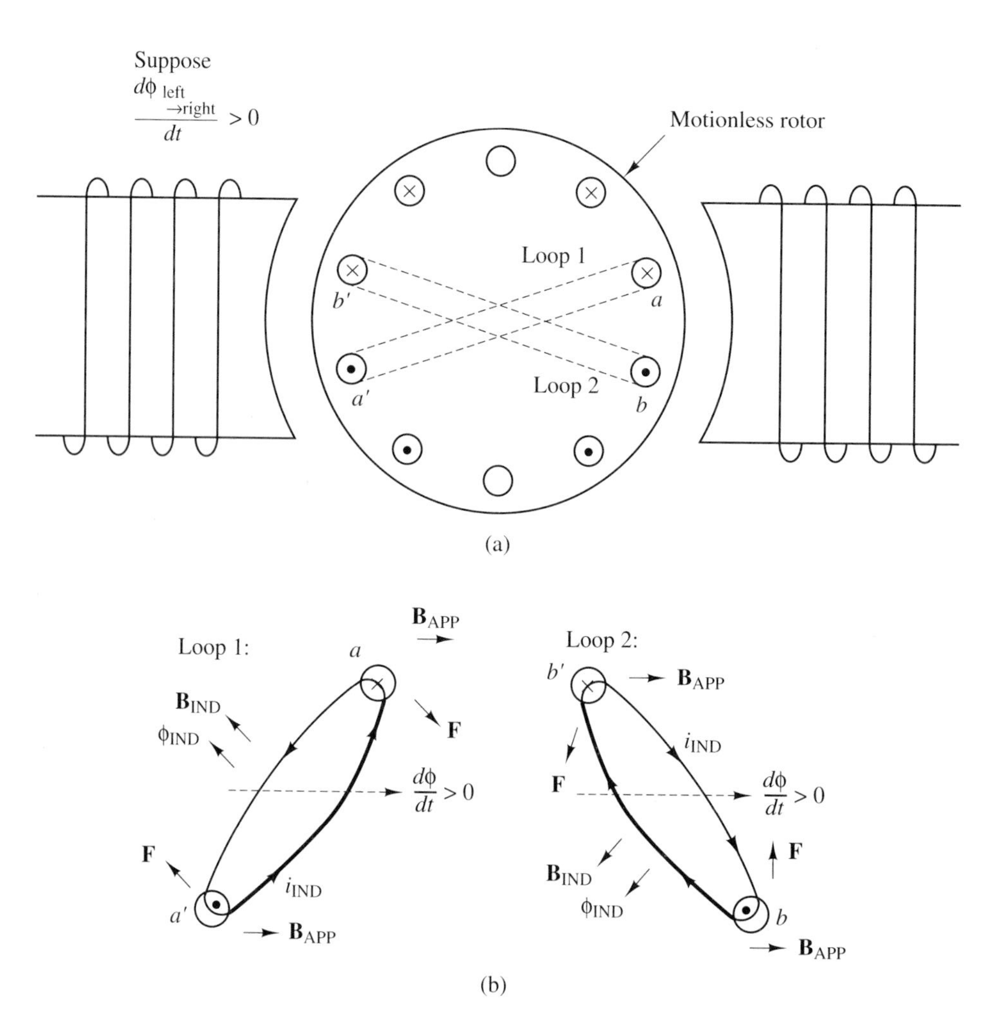


**Figure 15.1** (a) Single-phase ac motor. Two loops of the rotor winding are analyzed. (b) Loop 1 has a clockwise torque, while (c) loop 2 has a counterclockwise torque. The motor as shown does not self-start.

*induced* rather than applied currents. For dc motors, the applied currents to all rotor loops were "rigged" via the commutator so that the torques in all armature loops were in the same direction. In that case if the applied rotor current were ac and the field current dc, the torque direction would oscillate and the time-average torque would be zero.

Suppose that by some other means the rotor has been set into clockwise motion. Then not only is there a $d\phi/dt$ due to the sinusoidally varying $B_{\text{APP}}$, but there is a $d\phi/dt$ component due to the (externally initiated) motion of the rotor. The interaction of these two flux components with the induced currents produces net torque in the direction of the motion, so the motor will continue to spin in the same direction. If at any moment the torque falls to zero, the momentum will carry it past that point. A proper description of this interaction is, however, beyond the scope of this book.

This rather annoying property of the single-phase induction motor is inherently eliminated in the three-phase induction motor, described next. It has also been eliminated by a variety of design techniques applied to the single-phase machine. Old electric shavers used to have a thumbwheel which when spun in either direction would start it (the direction not being of great interest or consequence to the user). This technique was later cleverly disguised by an on–off "switch" that was mechanically linked to the spin-start mechanism!

In modern single-phase motors, the main idea is to create a magnetic asymmetry or imbalance, so that there is not a complete cancellation of torques at startup from any starting rotor angle. Often, an auxiliary winding having a reactance different from that of the main winding is placed 90° (mechanically) from the main winding. The auxiliary winding is connected at starting time and disconnected at near-operating speed (typically, 75% of synchronous speed) by the opening of a centrifugal switch. The auxiliary coil may be in series with a resistor or a capacitor to effect a near-90° electrical phase difference from the main winding field, a condition that will yield maximum starting torque.

In a capacitor-type motor, sometimes the centrifugal switch, prone to eventual failure, is removed. Or for best performance but higher construction cost, the switch is retained and different capacitors are used at start and at full load, each chosen for respective optimal operation at start and full-load conditions. These motors are called *split-phase motors.*

Alternatively, a thick short-circuited wire loop of one or a few turns is placed around the tip of one side of each pole. Induced currents in the loop cause the flux to sweep across the pole face much as would a rotating magnetic field. A small starting torque results. This type of motor, called a *shaded pole single-phase induction motor*, can be used only in very low-power applications (less than $\frac{1}{10}$ hp), a common example in the recent past being phonograph turntable motors.

### 15.2.2 Applications

The single-phase motor is the most common low-power (e.g., fractional horsepower) ac motor, and with the variety of designs satisfies an extensive range of application requirements for situations where only single-phase ac power is available. It may be found in air conditioners, refrigerators, electric shavers, centrifugal pumps, household oscillating fans, washing and ironing machines, cellar drain pumps, some home/yard work tools, furnace and ventilation blowers, gasoline pumps, hair driers, humidifiers, and many other devices.

## 15.3 THREE-PHASE INDUCTION MOTOR

### 15.3.1 Introduction

In general, three-phase motors have advantages over single-phase motors, such as those mentioned in Sec. 6.6. Specifically, the instantaneous power and therefore torque is constant, allowing smooth operation. Also, for a given output power, three-phase systems require lower line currents and therefore lower ohmic losses. Finally, three-phase motors have startup and run characteristics superior to ($\approx\frac{3}{2}$ as efficient as) similar-rated single-phase motors. It will be seen that for three-phase

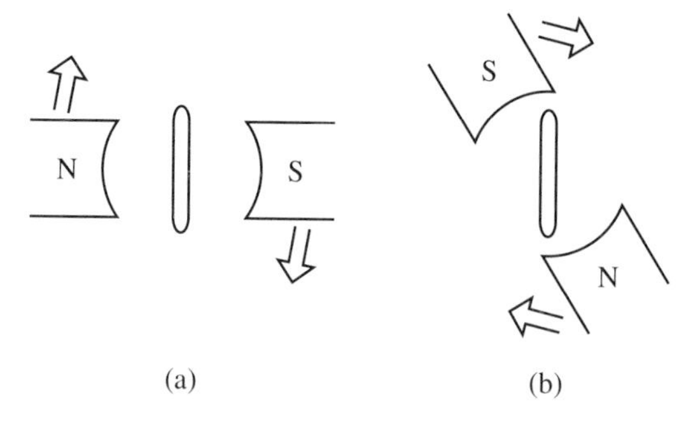


**Figure 15.2** The bar magnet field is rotated; two angular positions with respect to the loop are shown.

motors, the entire surface of the rotor is used for torque generation, which improves efficiency.

The three-phase induction motor is the most widely used motor for heavy industrial applications. With three phases of oscillating magnetic fields, it will be shown that a single constant-magnitude rotating field can be produced. Then the total (average) torque on the induction motor rotor will be derived. It will be shown that unlike the single-phase motor, the starting torque is nonzero. From the expression for the torque, the important torque–speed curve will be determined. It is the most useful curve for motor selection and the only curve derived in this introductory development.

### 15.3.2 Rotating Magnetic Field

First it will be appropriate to discuss the effects of a constant-magnitude rotating magnetic field on a single loop. For now, the rotating magnetic field is accomplished by physically rotating two bar magnets, as shown in Fig. 15.2. Define $\phi > 0$ to mean the left-to-right flux that passes through the loop. In Fig. 15.2 the flux is in the process of decreasing: $\phi_a > \phi_b$, so $d\phi/dt < 0$. The coil will consequently respond with a current to produce a $d\phi/dt > 0$ ("left-to-right" increase).

Figure 15.3 shows the situation of Fig. 15.2b in more detail where, for the moment, we assume the loop is motionless. The induced **B**, directed to oppose $d\phi/dt$, in this case aids the applied field. The right-hand rule determines the necessary current direction in the loop to produce the required induced **B**. As a result, by Ampère's force equation $\mathbf{F} = i_{\text{IND}}\boldsymbol{\ell} \times \mathbf{B}$ (in which "**B**" is the applied, dominant field $\mathbf{B}_{\text{APP}}$), the direction of the torque-producing forces on the conductors will be as shown in Fig. 15.3. Clearly, the greatest rotational force occurs when $\mathbf{B}_{\text{APP}}$ is vertical both because the strength of perpendicular **B** hitting the top and bottom loop conductors will be maximum, and because $d\phi/dt$ through the loop and therefore $i_{\text{IND}}$ will be maximum. Finally, torque = force × rotor radius vector. Thus the loop turns in the same direction as does the rotating $\mathbf{B}_{\text{APP}}$.

If now the coil were to turn as fast as $\mathbf{B}_{\text{APP}}$ (i.e., at *synchronous speed*), $d\phi/dt$ and therefore the induced current would be zero, which would result in zero net force or torque. Any friction or loading would immediately slow the rotor down. Consequently, the rotor will always spin at a speed slower than that of the rotating field $\mathbf{B}_{\text{APP}}$.

Obviously, it is rather inconvenient to have to spin the field magnets around; that might be harder than just to spin the rotor! Fortunately, as hinted above, there

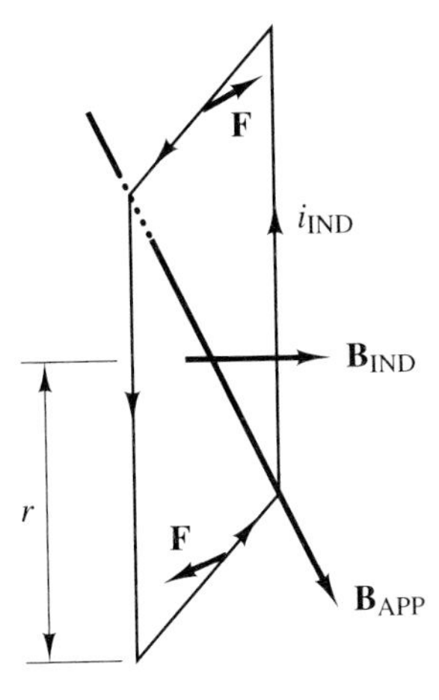


**Figure 15.3** Forces on the loop due to the rotating $B$ field tend to accelerate the loop clockwise, along with the direction of rotation of $\mathbf{B}_{\text{APP}}$.

is a way of producing a rotating $\mathbf{B}_{\text{APP}}$ *electrically*, with no physically moving parts. This is conveniently done using three electromagnets with currents $\frac{1}{3}(360°) = 120°$ out of phase in time with respect to each other and physically oriented 120° apart.

The situation is shown in Fig. 15.4. In Fig. 15.4a the three field windings are shown with actual current polarities for an assumed time $t = 0$. It is assumed that terminals $A^-$, $B^-$, and $C^-$ are electrically connected. For simplicity the rotor is not shown; it would be placed in the center of that diagram.

Figure 15.4b shows the current phasors for the three phases, each 120° apart from the others (in time). Note that the sum of the three current phasors is zero, so that the return line for the three-phase circuit ideally has zero current. Of course, this fact was proved in Sec. 6.6.2. But it should be noted that each flows in a separate

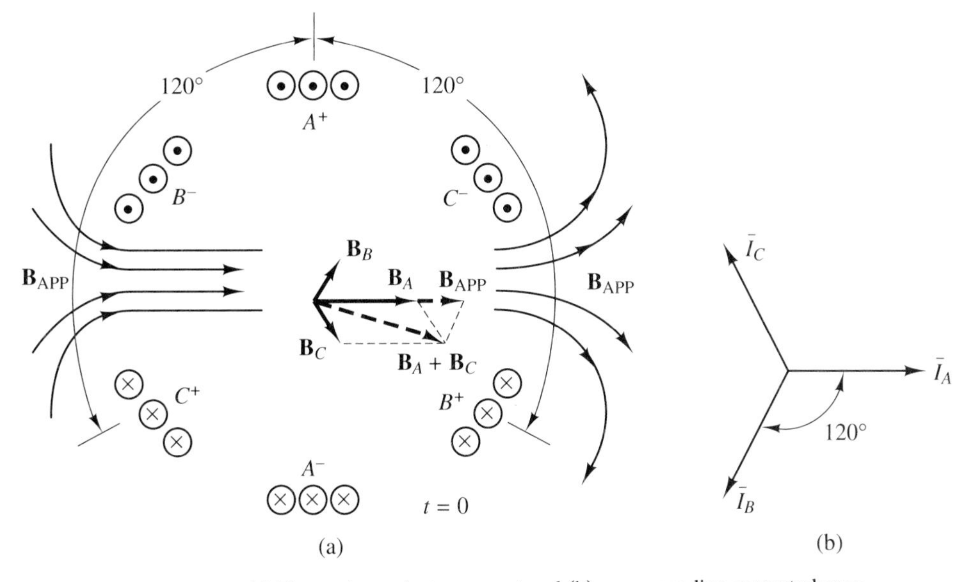


**Figure 15.4** (a) Three-phase electromagnet and (b) corresponding current phasor diagram, at time $t = 0$.

coil, producing its own magnetic field in a different direction. It is the vectorial sum of the magnetic fields that is of interest now.

As usual, the "horizontal" component would represent the actual real-valued current time function for a particular phasor. Thus at this time ($t = 0$), $i_A$ is large and positive and $i_B$ and $i_C$ are small and negative. In Fig. 15.4a, define positive current as entering "−" terminals and leaving the "+" terminals of each of the coils. These polarities are indicated respectively with the arrow heads · and tails × in Fig. 15.4a.

Using the right-hand rule for one coil at a time, the resulting flux density for each phase is determined as shown. Note that the **B**s for each phase are *vectors*, while the currents in Fig. 15.4b are *phasors*. The total $\mathbf{B}_{APP}$ is the vectorial sum of the three individual **B**s; at $t = 0$ it is seen to point to the right.

At a later time for which $\omega t = \pi/4$, the diagrams of Fig. 15.5 apply. From Fig. 15.5b it is seen that $i_A > 0$ (of medium magnitude), $i_B > 0$ (small), and $i_C < 0$ (of large magnitude). Consequently, the current directions are as shown with · and × in Fig. 15.5a. Again the individual **B**s are found by the right-hand rule and are vectorially summed to obtain $\mathbf{B}_{APP}$. It is found that $\mathbf{B}_{APP}$ now points 45° clockwise from its direction at $t = 0$. Thus, a rotating field has been synthesized. Incidentally, the rotating field completes one revolution for each cycle of the currents.

All that remains is to show that the magnitude of $\mathbf{B}_{APP}$ is constant in time. Clearly, just as $i_A(t)$, $i_B(t)$, and $i_C(t)$ are each 120° out of phase in time and are of the same magnitude, so will be $\mathbf{B}_A(t)$, $\mathbf{B}_B(t)$, and $\mathbf{B}_C(t)$. Furthermore, the magnetic fields are oriented in different (but fixed) directions, 120° = $2\pi/3$ radians apart. These directions are determined using the right-hand rule, which applies to *positive*

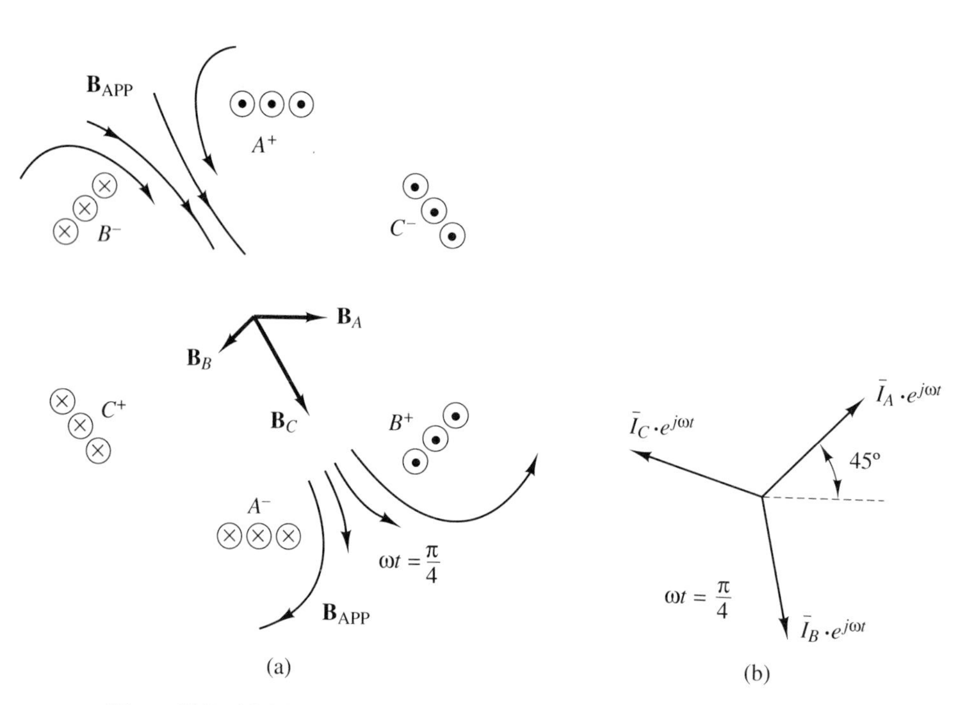


**Figure 15.5** (a) Three-phase electromagnet and (b) the corresponding current phasor diagram, at time such that $\omega t = \pi/4$.

current in each coil. (So, for example, $\mathbf{B}_C$ for $i_C > 0$ points in the direction opposite that shown in Figs. 15.4 and 15.5 because in those figures $i_C$ was negative.)

In the geometry of Figs. 15.4 and 15.5, let the right be the positive $x$-direction and up be the positive $y$-direction. Then

$$\begin{aligned}
\mathbf{B}_A(t) &= A\cos(\omega t)\hat{\mathbf{x}},\\
\mathbf{B}_B(t) &= A\cos\left(\omega t - \frac{2\pi}{3}\right)\left[\cos\left(\frac{4\pi}{3}\right)\hat{\mathbf{x}} + \sin\left(\frac{4\pi}{3}\right)\hat{\mathbf{y}}\right]\\
&= \frac{1}{2}A\cos\left(\omega t - \frac{2\pi}{3}\right)[-\hat{\mathbf{x}} - \sqrt{3}\,\hat{\mathbf{y}}],\\
\mathbf{B}_C(t) &= A\cos\left(\omega t - \frac{4\pi}{3}\right)\left[\cos\left(\frac{2\pi}{3}\right)\hat{\mathbf{x}} + \sin\left(\frac{2\pi}{3}\right)\hat{\mathbf{y}}\right]\\
&= \frac{1}{2}A\cos\left(\omega t - \frac{4\pi}{3}\right)[-\hat{\mathbf{x}} + \sqrt{3}\,\hat{\mathbf{y}}],
\end{aligned} \tag{15.1}$$

so that the total magnetic field is

$$\begin{aligned}
\mathbf{B}_{\text{APP}}(t) &= A\left[\left\{\cos(\omega t) + -\frac{1}{2}\cos\left(\omega t - \frac{2\pi}{3}\right) - \frac{1}{2}\cos\left(\omega t - \frac{4\pi}{3}\right)\right\}\hat{\mathbf{x}}\right.\\
&\quad\left. + \left\{-\frac{\sqrt{3}}{2}\cos\left(\omega t - \frac{2\pi}{3}\right) + \frac{\sqrt{3}}{2}\cos(\omega t - 4\pi/3)\right\}\hat{\mathbf{y}}\right]\\
&= A\left[\left\{\cos(\omega t) - \frac{1}{2}\left(-\frac{1}{2}\cos(\omega t) + \frac{\sqrt{3}}{2}\sin(\omega t) - \frac{1}{2}\cos(\omega t)\right.\right.\right.\\
&\quad\left.\left. - \frac{\sqrt{3}}{2}\sin(\omega t)\right)\right\}\hat{\mathbf{x}} + \frac{\sqrt{3}}{2}\left\{-\left(-\frac{1}{2}\cos(\omega t) + \frac{\sqrt{3}}{2}\sin(\omega t)\right)\right.\\
&\quad\left.\left. - \frac{1}{2}\cos(\omega t) - \frac{\sqrt{3}}{2}\sin(\omega t)\right\}\hat{\mathbf{y}}\right]\\
&= A\left[\left\{\cos(\omega t)\left(1 + \frac{1}{4} + \frac{1}{4}\right) + \sin(\omega t)\left(-\frac{\sqrt{3}}{4} + \frac{\sqrt{3}}{4}\right)\right\}\hat{\mathbf{x}}\right.\\
&\quad\left. + \frac{\sqrt{3}}{2}\left\{\cos(\omega t)\left(\frac{1}{2} - \frac{1}{2}\right) + \sin(\omega t)\left(-\frac{\sqrt{3}}{2} - \frac{\sqrt{3}}{2}\right)\right\}\hat{\mathbf{y}}\right]\\
&= 1.5A\,[\cos(\omega t)\hat{\mathbf{x}} - \sin(\omega t)\hat{\mathbf{y}}],
\end{aligned} \tag{15.2}$$

which is exactly a constant-magnitude rotating magnetic field. [It rotates clockwise, as dictated by the minus sign in front of $\sin(\omega t)\hat{\mathbf{y}}$; think of $t$ increasing just beyond zero to show this.]

### 15.3.3 Multiple-Pole Fields and the Rotor

To simplify discussions, let the three coils of each individual phase in Fig. 15.4 be represented by one coil, as in Fig. 15.6. Figure 15.6 and the previous discussion refer to a two-pole machine (i.e., a two-pole revolving $\mathbf{B}_{\text{APP}}$). Now suppose that all of the six stator coils are squeezed to take up only half the periphery, as shown in Fig. 15.7.

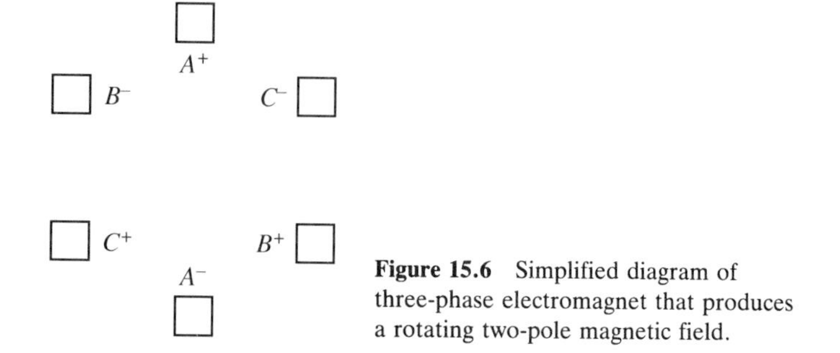


**Figure 15.6** Simplified diagram of three-phase electromagnet that produces a rotating two-pole magnetic field.

(This idea is suggested in Rosenblatt).* The total field $\mathbf{B}_{APP}$ still points at $t = 0$ from between $B^-$ and $C^+$ to between $C^-$ and $B^+$, but is of course now necessarily curved. Although $\mathbf{B}_{APP}$ still "rotates," it now traverses only 180° geometrically in a full (electrical) cycle, as opposed to 360° before.

This lopsided configuration can be made symmetrical again by adding three more coils linked to the original phases as shown in Fig. 15.8. When the **B** direction lines are drawn in, it is evident that the total $\mathbf{B}_{APP}$ is now a four-pole field rather than a two-pole field. As just discussed (regarding Fig. 15.7), the new field rotates only half as fast as the two-pole $\mathbf{B}_{APP}$ rotated. Again, this is because during an electrical 360° cycle, $\mathbf{B}_{APP}$ now rotates only halfway around the periphery.

In rpm, the geometric *synchronous frequency* $n_s$ is $60f/(4 \text{ poles}/2)$, where $f$ is the electrical frequency of phase voltages $f$ (in hertz). For a general number of poles $P$, this relation becomes

$$n_s = \frac{60f}{P/2} = \frac{120f}{P} \quad \text{rpm.} \tag{15.3}$$

It should be pointed out that while $n$ represents rotational frequency in rpm it is usually called the *speed.* The adjective "synchronous" will be clarified in Sec. 15.4, where the rotor spins at precisely $n_s$, the rotational speed of the rotating magnetic field.

As pointed out above, the rotor of an induction motor spins at a speed slower than the rotating field; it is therefore sometimes called an *asynchronous motor.* This type of motor is most often called a *three-phase induction motor* because, like the single-phase induction motor, its operation depends on the *induced* currents circu-

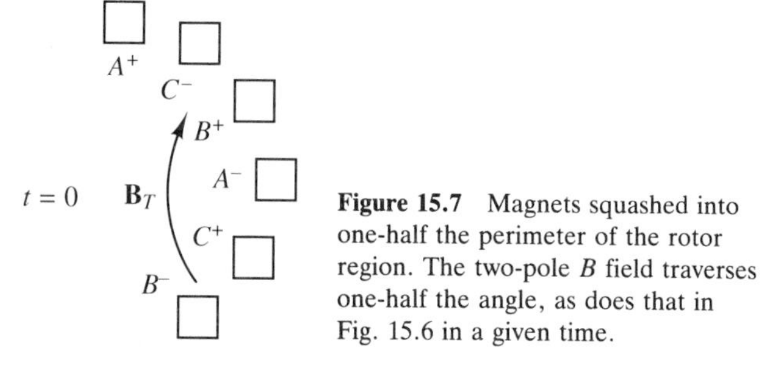


**Figure 15.7** Magnets squashed into one-half the perimeter of the rotor region. The two-pole $B$ field traverses one-half the angle, as does that in Fig. 15.6 in a given time.

*J. Rosenblatt, M. H. Friedman, *Direct and Alternating Current Machinery* (C. E. Merrill, Columbus, OH, 1984), p. 275.

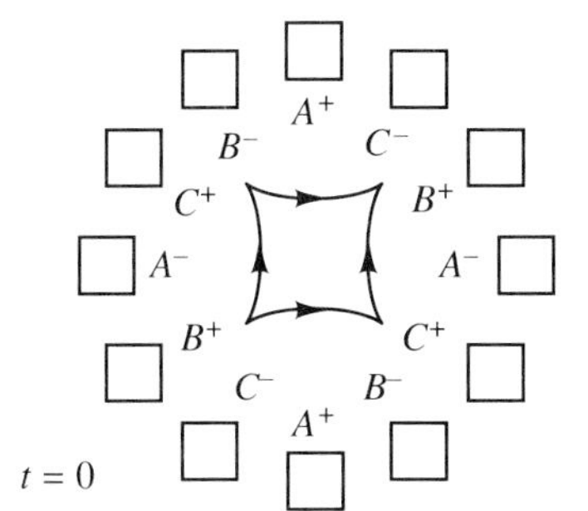


**Figure 15.8** With the addition of three more electromagnets, we have a four-pole electromagnet spinning at one-half the speed, as does the two-pole field in Fig. 15.6.

lating in the rotor (akin to the currents in a shorted transformer secondary coil). The rotor speed is thus a fraction of synchronous speed; this fraction is quantified by the *slip* parameter $s$:

$$n_{\text{rotor}} = n_r = (1 - s)n_s, \tag{15.4}$$

where $0 < s \leq 1$. Solving for $s$ gives

$$s = \frac{n_s - n_r}{n_s} = 1 - \frac{n_r}{n_s}. \tag{15.5}$$

Sometimes $s$ is expressed as a percentage; note the similarity to speed regulation of a dc machine—see Eq. (14.15). From Eq. (15.5) we see that at standstill ($n_r = 0$), the slip $s$ is unity, while if it were possible to rotate at synchronous speed, $s$ would be zero.

**Example 15.1**

A three-phase induction motor has four poles and operates at 60 Hz. At full load it spins at 1500 rev/s. What is the synchronous speed, and what is the slip at this load?

**Solution** By Eq. (15.3), $n_s = 120f/P = 120(60)/4 = 1800$ rpm. By Eq. (15.5), $s = 1 - n_r/n_s = 1 - 1500/1800 = 0.17 = 17\%$.

A brief word about the rotor. Instead of a single loop, as described in the beginning of this section, either a *squirrel-cage* or a *wound rotor* is used. Lloyd and Schwarz* claim that "the squirrel cage is, without doubt, the basic workhorse on which our civilization is based." Pretty strong words for such a humble rodent! Anyway, the basic construction, complete with squirrel, is as shown in Fig. 15.9. Note the connecting end rings that electrically connect the horizontal bars.

We can think of the squirrel cage as a variation on the form shown in Fig. 15.10, where there is no harm in electrically connecting the midpoints of the side conductors of all loops. In fact, the end rings of the squirrel cage reduce the required metal compared with that in Fig. 15.10. In any case, all the torque action takes place on the horizontal bars, through which the rotating magnetic field (and flux) pass. With these numerous loops spread around the rotor surface, the effect of a distributed winding (an increased total torque) is accomplished using a simple, rugged construction. Of course, in practice the rotor interior is filled with a magnetic core.

Alternatively, a distributed winding (wound rotor) may be constructed to

*B. J. Chalmers, ed., *Electrical Motor Handbook* (Butterworth, Boston, MA, 1988), p. 133.

**Figure 15.9** Squirrel-cage rotor.

facilitate speed and power factor control and other tasks beyond the scope of this book. Through slip rings, the ends of the winding may either be short-circuited or connected through a resistor to alter the motor behavior. However, these are more expensive and problem-prone than is the rugged squirrel cage.

Finally, a further brief word about multiple poles. Typically, the number of stator poles on a *synchronous generator* is specified by the number of poles on the rotor field. (Synchronous generators are merely synchronous motors run as generators.) But typically, the number of poles on induction and synchronous motors is stated as the number of poles on the stator. This is because the squirrel-cage rotor automatically accommodates however many poles $P$ the field produces in a $P$-pole $\mathbf{B}_{\text{IND}}$.

In fact, the number of poles on the stator and rotor fields *must be the same.* Figure 15.11 shows why by three examples. In Fig. 15.11a is shown a two-pole machine properly specified, in that the relation between rotor and stator poles is the same at $A^+$ as it is at $A^-$ (like poles). This is a consequence of the stator also having two poles. But in Fig. 15.11b, the rotor has four poles while the stator has only two; at $A^+$ the poles are alike, while at $A^-$ they are opposite. The result is that there is no net torque (motor) or generated voltage (generator). So that connection diagram is wrong. The proper four-pole configuration is shown in Fig. 15.11c, where alike poles are together at $A_1^+$, $A_1^-$, $A_2^+$, and $A_2^-$; in this situation net torque or voltage will be obtained as desired.

### 15.3.4 Developed Torque

For any motor to spin, there must be developed torque to overcome the loading and frictional torques. So top priority must be given to the ability to predict the values

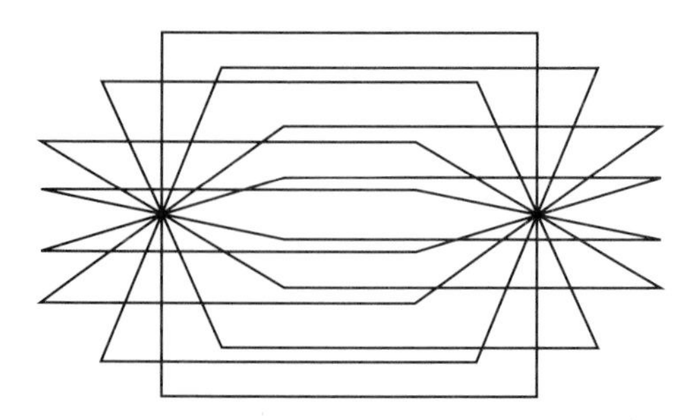

**Figure 15.10** Near-equivalent of squirrel-cage rotor in Fig. 15.9.

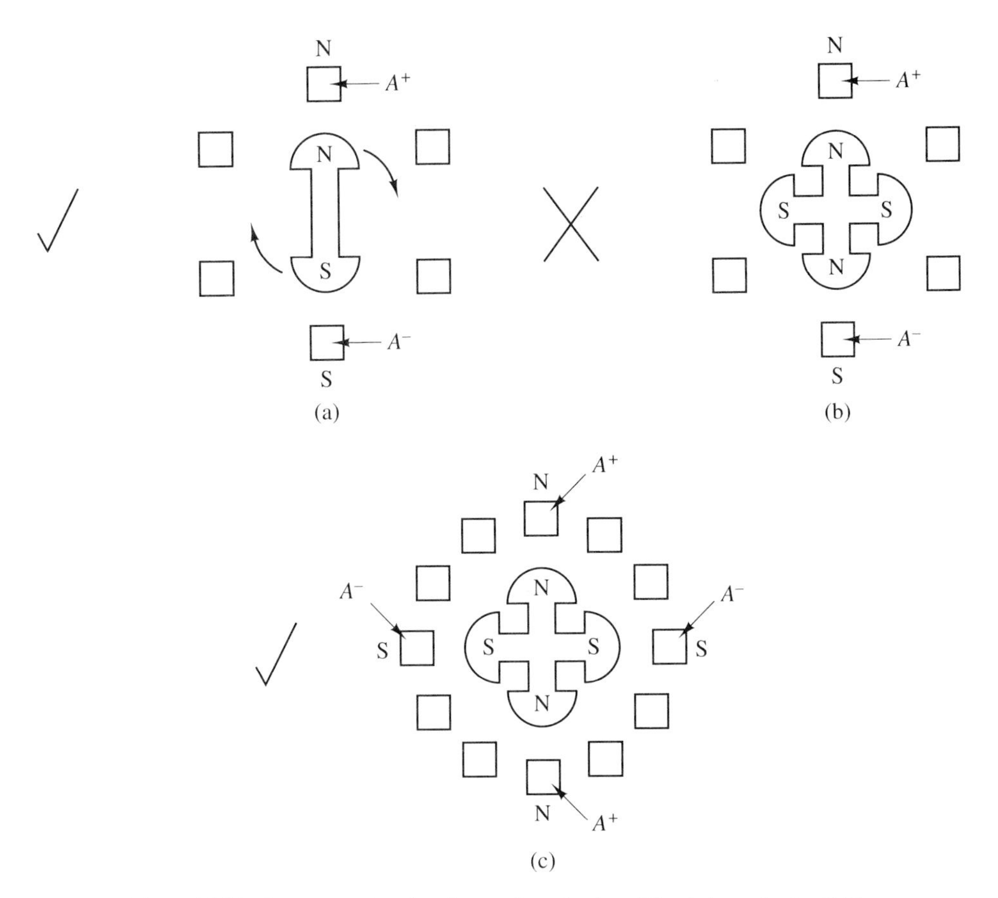


**Figure 15.11** Demonstration that the number of poles of the stator and rotor fields must be equal.

of torque at any speed. Clearly, the first calculation should be the starting torque ($n_r = 0$). Often the condition ($n_r = 0$) is called *standstill* or *locked rotor*. The implication in the latter term is the torque that would persist were the rotor prevented from spinning by a very large (infinite) load torque; that is, the rotor is locked.

Consider a two-pole machine. The final expression for $\mathbf{B}_{\text{APP}}$ in Eq. (15.2) was an idealized determination of the field at the center of the rotor. In a real motor, the distribution is such that the field (and therefore the flux) is *sinusoidally* distributed around the air gap while still rotating at speed $n_s$. (The more field windings, the better the sinusoidal assumption can be.)

The flux, like the magnitude of $\mathbf{B}_{\text{APP}}$, is temporally constant in its sinusoidal spatial form, but the entire distribution rotates around the rotor at speed $n_s$. Furthermore, this flux is usually thought of as a radially inward flux component. Recall that the generated voltage in a single loop is $e_g = d\phi/dt$. In this equation the flux is considered to be that passing *through the loop*. From previous discussions, $e_g$ is spatially 90° out of phase with $\phi$. Because $e_g$ is not in the same circuit as $v_{\text{APP}}$, we cannot call it a "back" emf. Thus in this chapter we use the subscripts $g$ or $G$ rather than $B$.

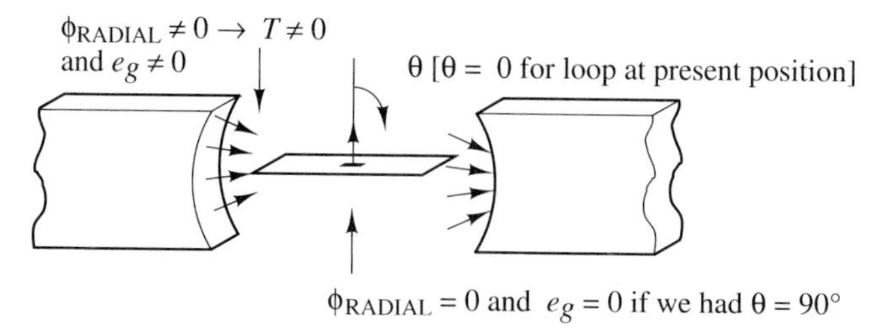


**Figure 15.12** Illustration that the generated voltage is in phase with the flux at the outer conductors.

Alternatively, if the rotor has lap or wave windings, the conductor on the right edge of the loop (for example) will be aligned with maximum flux at times when the center of the coil will be exposed to the angle of minimum flux; similar arguments could be made for the squirrel-cage rotor. But for the following analysis, it will be stated that the loop $e_g$ is seen to be spatially *in phase* with the flux.

Why? Remember that the purpose of the present discussion is to calculate the *torque*. In the context of torque calculations, it is convenient to redefine $\phi$ temporarily as the flux on the coil edge rather than the coil center. The reason for this is as follows.

The torque on a conductor is proportional to the $B$ field (and therefore the flux) and the current in the conductor. It is the flux *at the conductor* that produces the torque, not the flux that passes through the loop center and produces $e_g$ (see Fig. 15.12). With respect to *this* definition of $\phi$—the only reasonable definition for torque calculation—it can be said that $e_g$ is spatially *in phase* with $\phi$.

Recall that this is the same conclusion as was reached in Sec. 13.4.3, where we noted that the induced voltage is maximum when the right coil edge is passing under the north pole; the induced voltage was in phase with the flux at the flux-cutting conductor side.

The foregoing reasoning is intended to help answer questions such as "but $e_g = d\phi/dt$, so $e_g$ should be 90° from $\phi$ because $\phi$ is temporally sinusoidal."

Now Fig. 15.13 should make sense. It shows the split-open and unrolled rotor flux distribution, which is labeled both $\phi(\theta)$ (referenced to the coil edge) and $e_g(\theta)$. Here and in Fig. 15.12, $\theta$ is the angle between the normal of the loop and the vertical axis. Thus Fig. 15.12 represents the position $\theta = 0$. The entire sine wave travels to the "right" as time progresses and continually reenters the left side.

Any given loop has generated within it a sinusoidally time-varying $e_g(t)$, as just described. The frequency of the induced voltage is by Eq. (15.3) $n_s/60$ (for $P = 2$) because the rotor coils themselves are not rotating (locked rotor); that is, the relative speed between rotor and rotating magnetic field is $n_s$. For this frequency the coils are quite inductive; the impedance of the rotor is

$$\bar{Z}_r = R_r + jX_r = Z_r\angle\theta_r. \tag{15.6}$$

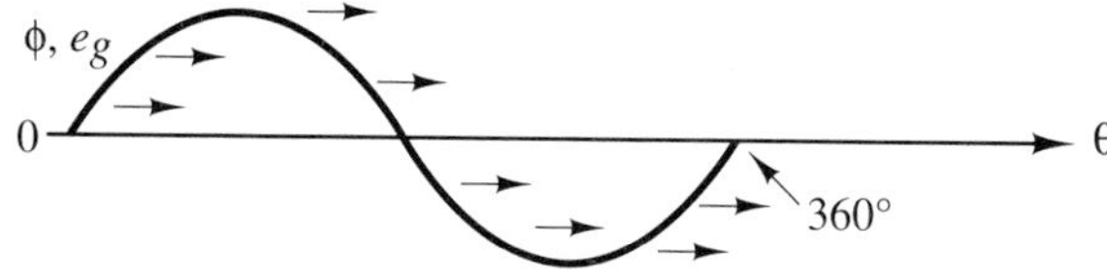


**Figure 15.13** Sinusoidal distribution of flux and generated voltage.

The angle $\theta_r$ may at locked rotor be quite large, on the order of 80°. Thus the induced rotor current lags substantially behind the induced voltage and therefore the flux at that conductor (because the induced voltage and therefore the flux are spatially in phase, as argued above).

These relations are illustrated in Fig. 15.14. Again, each point on the current curve represents the induced current in a coil conductor at that particular angle $\theta$. So the continuous current distribution shown is the ideal case of infinitely many conductors. (Recall that for dc analysis it was sufficient to consider only $R_r$, then called $R_A$; there was no phase lag, because in that case $f = 0$.)

The total torque on the motor will be the sum of all the torques on the individual conductors on the rotor. This is equivalent to an average torque averaged over $\theta$. But at any $\theta$,

$$T(\theta) = k_t\phi(\phi)i_A(\theta) = k_t\phi \cos(\theta)I_A \cos(\theta - \theta_r), \tag{15.7}$$

where $i_A(\theta)$ is shown lagging behind $\phi(\theta)$ by $\theta_r$, $\phi$ is the effective magnitude of $\phi(\theta)$, $I_A$ is the effective magnitude of $i_A(\theta)$, and $\theta = \omega_s t = n_s\,[(2\pi \text{ rad/rev})/(60 \text{ s/min})]\,t = \pi n_s t/30$ rad ($n$ is in rpm). The average of $T(\theta)$ over $\theta$ is merely the average of two sinusoids of the same frequency and differing phases, a calculation done in Sec. 6.3.2. That calculation came out to be the product of the effective magnitudes of the sinusoids times the cosine of the phase angle between them [see Eq. (6.11)]. Thus in this case,

$$T_{\text{avg}} = k_t\phi\, I_A \cos(\theta_r), \tag{15.8}$$

where

$$\theta_r = \tan^{-1}\!\left(\frac{\omega_{\text{rel}} L_r}{R_r}\right), \tag{15.9}$$

in which $\omega_{\text{rel}} = 2\pi f_{\text{rel}}$, where $f_{\text{rel}}$ is the electrical frequency (hertz) associated with the relative difference in speeds between the rotor and the rotating magnetic field. By Eq. (15.3), $\omega_{\text{rel}}$ is

$$\begin{aligned}\omega_{\text{rel}} &= 2\pi\left[P\frac{(n_s - n_r)}{120}\right] \\ &= \frac{\pi s n_s P}{60} \\ &= s\omega_s.\end{aligned} \tag{15.10}$$

So $\omega_{\text{rel}}$ is the relative radial frequency between the rotating field and the rotor. Note that $\omega_{\text{rel}} = \pi n_s P/60 = \omega_s$ for a locked rotor.

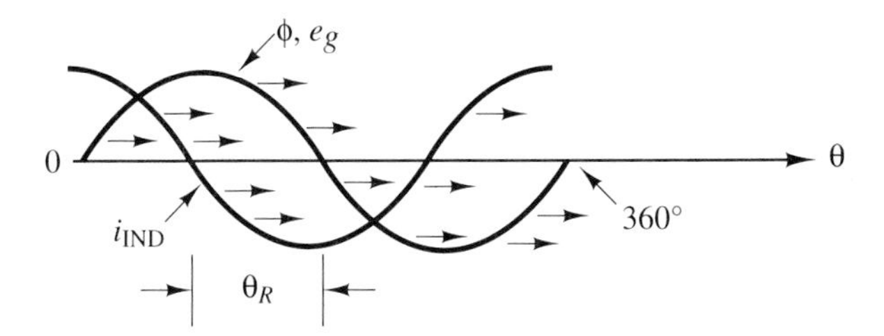


**Figure 15.14** Sinusoidal distribution of flux and generated voltage, and the lagging induced current.

It is helpful at this point to recall that the rotor winding closes on itself, so an equivalent circuit for the rotor would appear as in Fig. 15.15. The important result here is that (for a locked rotor)

$$I_A = \frac{E_G}{Z_r} = \frac{E_G}{[R_r^2 + (X_r^L)^2]^{1/2}}, \tag{15.11}$$

where if $X_r = \omega_{\text{rel}} L_r$, then $X_r^L = \omega_s L_r$ is the reactance of the locked rotor. (Keep in mind that $E_G$ now represents the effective magnitude of the phasor representation of the sinusoidal generated voltage. It is not the time-averaged generated voltage as it was in the study of dc machines.)

The effective value of the field current is proportional (via $Z_{\text{field}}$) to the effective applied voltage $V_{\text{APP}} = V_P$, where $V_P$ represents one of the three phase voltages. Therefore, the effective flux and the generated voltage are proportional to $V_P$. Finally, from the definition of $Z_r$, at the locked rotor condition

$$\cos(\theta_r) = \frac{R_r}{[R_r^2 + (X_r^L)^2]^{1/2}}. \tag{15.12}$$

Using the facts above and Eq. (15.11) for $I_A$ in Eq. (15.8) yields for the starting (locked rotor) torque

$$T = \frac{K_1 V_P^2 R_r}{R_r^2 + (X_r^L)^2}, \tag{15.13}$$

where $K_1$ is a machine constant.

After startup, assuming that the rotor is now spinning at some nonzero speed (nonunity slip), only a few changes need to be made to the expression for the average torque. Recall that now the relative frequency between the rotating magnetic field and the rotor is $\omega_{\text{rel}} = s\omega_s$ [Eq. (15.10)]. Consequently, $E_G$, which depends on $d\phi/dt$, is evidently proportional to $s$ as well as $V_P$ [because the amplitude of the derivative of $\cos(\omega t)$, $-\omega \sin(\omega t)$, is proportional to the frequency]. If the generated voltage at unity slip (locked rotor) is called $E_G^L$, then

$$E_{G,\text{ any slip}} = sE_G^L, \tag{15.14}$$

which reduces to $E_G^L$ for $s = 1$. Also at nonunity slip,

$$X_r = (s\omega_s)L_r = sX_r^L, \tag{15.15}$$

so that

$$T_{\text{any slip}} = \frac{K_2 V_P^2 s R_r}{R_r^2 + (sX_r^L)^2}, \tag{15.16}$$

where $K_2$ is a machine constant.

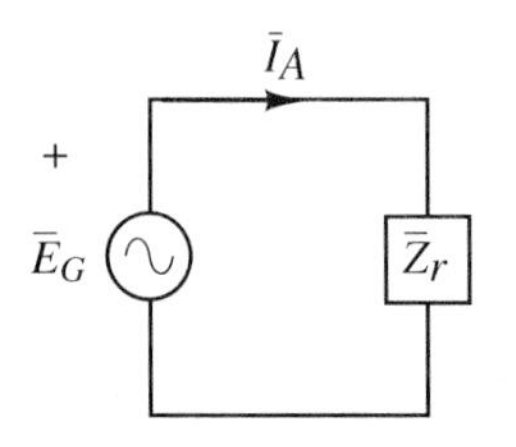


**Figure 15.15** Simple circuit model of induction motor rotor coil.

As stated above, $V_P$ is a phase voltage. In addition, $R_r$ and $X_r$ are "per phase" quantities, so terms such as $V_P^2 sR_r/[R_r^2 + (sX_r^2)^2]$ are proportional to *per phase* torques. The total torque would be three times any one phase torque; the factor 3 is embedded in $K_2$. Often the "per phase" designation is omitted for brevity.

**Example 15.2**

At locked rotor conditions, the generated voltage per phase in a three-phase induction motor is 100 V. The motor has six poles and runs at 60 Hz. If the speed at full load is 1000 rpm, what is the full-load generated voltage?

**Solution** The synchronous speed should usually be calculated first. It is $n_s = 120(60)/6 = 1200$ rpm. The slip is $s = 1 - n_r/n_s = 1 - 1000/1200 = 0.17$. Therefore, by Eq. (15.14), $E_G = 0.17(100) = 17$ V.

It is not only interesting but also important to find the conditions for maximum torque. Often in wound rotors $R_r$ can be made variable (by a potentiometer connected in series with the rotor winding via slip rings) and thus be adjusted to maximize torque. This may be particularly crucial at startup. In fact, it was already mentioned that at startup, $\theta_r$ is near 90°; the cosine of 90° is zero, so the torque would be near zero at startup. If the rotor resistance is increased, the *power factor* $\cos(\theta_r)$ can be increased. But it must also be noted that increasing $R_r$ increases the denominator of $T$, thereby tending to decrease $T$. A simple differentiation will determine

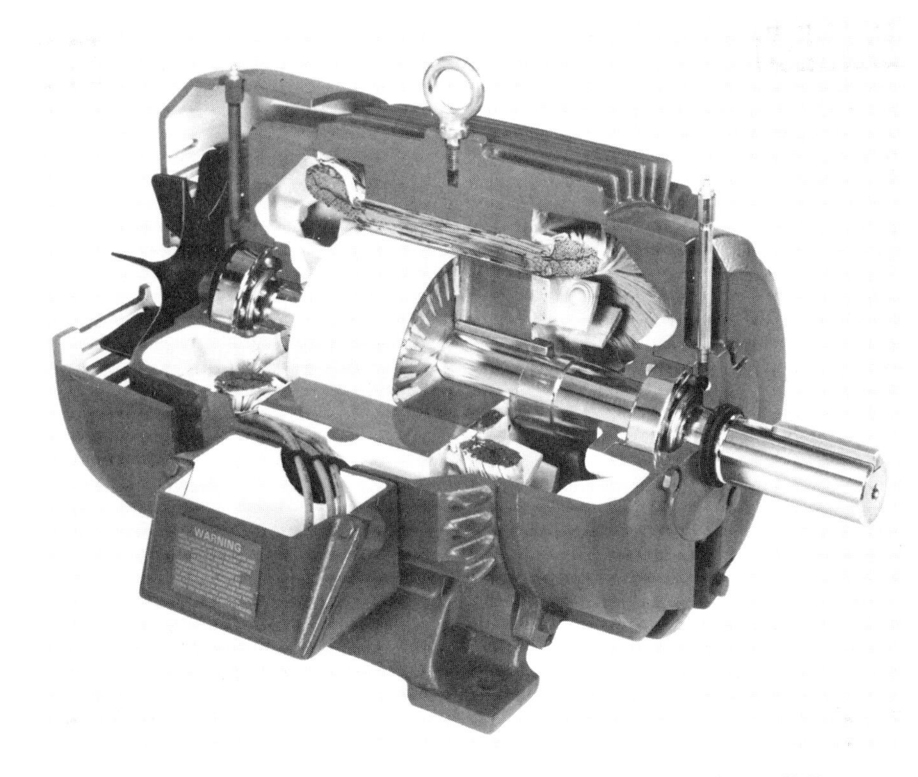


Three-phase squirrel cage induction motor. 25 to 30 hp, 4-pole, 1750 rpm, 60 Hz, 230/460 V. Used for material handling, fans, conveyors, pumps. *Source:* Reliance Electric.

the optimal value, just as it gave the maximum power transfer *matched load* result (in Sec. 4.11).

Consider first differentiation of $T$ with respect to $R_r$:

$$\frac{\partial T}{\partial R_r} = K_2 V_{PS}^2 \frac{R_r^2 + (sX_r^L)^2 - 2R_r^2}{[R_r^2 + (sX_r^L)^2]^2} = 0, \tag{15.17}$$

which yields

$$R_r = sX_r^L \tag{15.18}$$

as the optimal rotor resistance for maximum torque. A special case of this condition is that for unity slip (locked rotor)

$$R_r = X_r^L. \tag{15.19}$$

So for maximum torque, $R_r$ is adjusted (increased) by an external resistor to satisfy Eq. (15.19). Because $T$ depends in exactly the same way on $s$ as it does on $R_r$ (linear in the numerator and squared in one term of the denominator), differentiating with respect to $s$ will yield the same condition:

$$s_b = \frac{R_r}{X_r^L}, \tag{15.20}$$

which is known as the *breakdown slip*. This slip corresponds to the speed at which maximum torque is developed, assuming fixed $R_r$ and $X_r^L$.

When either the optimal rotor resistance or the breakdown torque is substituted into the expression for average torque, the same maximum torque value results:

$$\begin{aligned} T_{\max} &= \frac{K_2 s^2 X_r^L}{2(sX_r^L)^2} \\ &= \frac{K_2 V_P^2}{2X_r^L}, \end{aligned} \tag{15.21}$$

where the intermediate step was shown for the case of substituting in the optimal $R_r$.

Incidentally, if $n_r$ is very close to $n_s$, $s$ is small enough that the reactance $(sX_r^L)$ can be neglected in comparison with $R_r$, so that Eq. (15.16) becomes

$$T \approx \frac{K_2 V_P^2}{R_r} s. \tag{15.22}$$

This expression shows that for small $s$, the torque is proportional to $s$.

Finally, it should be noted that the mechanical power developed by the motor is

$$\begin{aligned} P_{\text{dev}} &= T\omega_r \\ &= T(1 - s)n_s \frac{\pi}{30}. \end{aligned} \tag{15.23}$$

A curve summarizing the torque–speed behavior derived above is shown in Fig. 15.16. For the case shown, the maximum torque occurs at a speed $n_b$ about three-

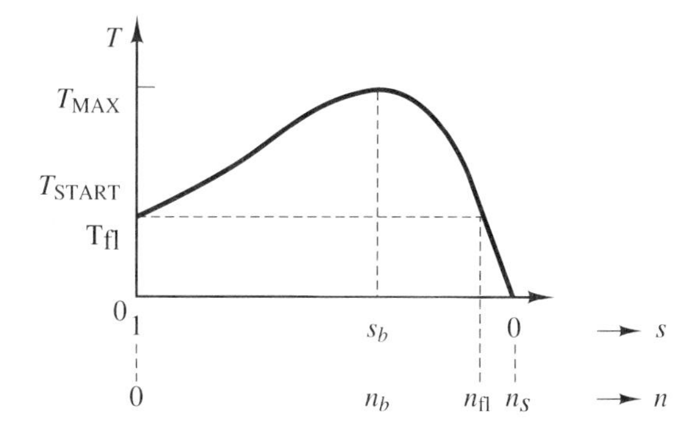


**Figure 15.16** Torque–speed characteristic of an induction motor.

fourths of the "full-load" or "rated" speed $n_{fl}$. Why not choose the rated full load to coincide with $n_b$? Because "breakdown" (also called "pullout") actually means that if that torque is exceeded, the motor will stop (break down); from Fig. 15.16 it is easily seen that the torque decreases for $n < n_b$.

Having $n_{fl} > n_b$ provides a buffer, so that if the load is increased, causing the motor to slow down a little, the increased torque will tend to bring the motor back up to rated speed. The linear torque–slip (torque–speed) relation for $s$ near zero is evident, as predicted by Eq. (15.22). Also, the torque is zero for $n_r = n_s$, as mentioned in Sec. 15.3.2. Note further that it is often the case that the starting torque exceeds or is close to the full-load torque; this makes starting relatively easy.

Variations on this torque–speed relation have been made possible by varying the size and shape of the squirrel-cage conductors. Four standard types have been defined by the National Electrical Manufacturers' Association (NEMA). Classes A and B have high efficiency at full load but very high starting currents, which are problematic from a power systems viewpoint. The only difference between B and A is that B has three-fourths of the starting current, at the expense of a lower power factor than A. Class C has a higher starting torque than A and B, but lower efficiency at full load.

Finally, Class D has a yet higher starting torque, but a monotonically declining torque with speed and consequent reduction in efficiency. These four classes are pictured in Fig. 15.17. There is also a NEMA class F that is more sharply peaked near full load and is good for loads that pulsate at rated speed.

**Example 15.3**

Consider a six-pole 60-Hz three-phase induction motor. The locked rotor reactance is 2 Ω and the rotor resistance is 0.5 Ω.

(**a**) At what speed does maximum torque occur?

(**b**) If that maximum torque is 40 N · m, what is the mechanical power delivered by the motor when spinning at speed $n_b$?

**Solution**

(**a**) First, $n_s = 120(60)/6 = 1200$ rpm. The breakdown slip $s_b = R_r/X_r^L = 0.5/2 = 0.25$, so that $n_b = n_s(1 - s_b) = 1200(1 - 0.25) = 900$ rpm.

(**b**) By Eq. (15.23), at breakdown $P_{dev} = T\omega_b = 40(900)\pi/30 = 3.8$ kW.

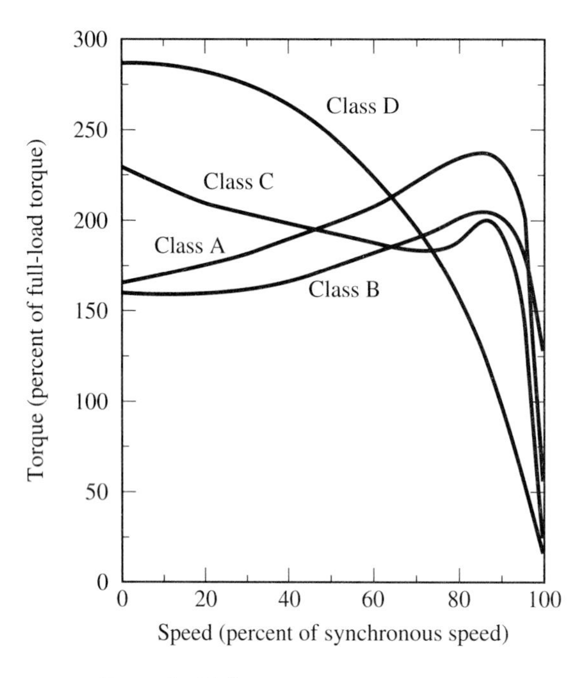


**Figure 15.17** NEMA classification of induction motor torque–speed characteristics. (*Source:* Vincent Del Toro, *Basic Electric Machinery* (Reprinted by permission of Prentice Hall, Englewood Cliffs, New Jersey, 1990), p. 208.)

**Example 15.4**

A six-pole wound-rotor induction motor has a full-load speed of 1000 rpm at 60 Hz. Suppose that $R_r = 0.5\ \Omega$, $X_r^L = 1.5\ \Omega$.

(**a**) At what speed will pullout occur for a "short-circuited" rotor?

(**b**) What resistance must be added to the rotor to achieve maximum starting torque?

(**c**) What is the full-load speed with the new resistor in place?

**Solution**

(**a**) "Short-circuited" here means no added resistance to the rotor; the exposed terminals accessible through slip rings are short-circuited. The synchronous speed is $n_s = 120(60)/6 = 1200$ rpm. Thus $n_b = n_s(1 - s_b) = 1200(1 - 0.5/1.5) = 800$ rpm.

(**b**) Now by Eq. (15.19), the total rotor resistance $R_r + R_{\text{add}}$ must equal $X_r^L$ for maximum torque to occur at startup. So $R_{\text{add}} = X_r^L - R_r = 1.5 - 0.5 = 1\ \Omega$.

(**c**) The trick here is to recognize that the full-load torque will remain the same, because it is equal to the applied mechanical load, which is independent of $R_r$. So by Eq. (15.22), if $R_r$ has been increased by the factor $1.5/0.5 = 3$, the slip must also increase over the original full-load slip by the factor 3. Consequently $s_{\text{fl, old}} = 1 - 1000/1200 = 0.17$, which multiplied by 3 gives $s_{\text{fl, new}} = 0.5$. Thus $n_{\text{fl, new}} = 1200(1 - 0.5) = 600$ rpm.

### 15.3.5 Applications

Squirrel-cage rotors predominate and are cheaper than wound rotors. Wound rotors are reserved for severe starting-torque conditions, such as some cranes and hoists. Squirrel-cage motors drive blowers, agitators, dumbwaiters, elevators, mixers, milling machines, refrigerators, electric cars, and many of the same uses as the higher-power applications of single-phase induction motors. For additional details on the general principles of motor selection, see Sec. 14.5.

## 15.4 SYNCHRONOUS MOTORS

### 15.4.1 Introduction

Induction motors are extremely valuable as ac motors. They are relatively inexpensive, at least for low to moderate power applications. For example, they do not require external field excitation other than the revolving stator field, and their starting and control devices are inexpensive. Also, there are possibilities for speed variation of induction motors (although not as convenient as dc motors).

But what if we want essentially perfect speed regulation? The induction motor will always slow down under loading ($s$ increases). Also, induction motors always have lagging power factors. It would be nice to have other motors whose power factors are leading in order to raise the overall power factor of the industrial plant in question. That would help reduce power bills.

Both of these advantages are present in the synchronous motor. In addition, synchronous motors have higher efficiency; for large high-power machines this overcomes the problem that they sometimes have a higher construction cost. In general, for lower speeds, synchronous motors end up being cheaper than induction motors except for very low-power applications (see Fig. 15.18).

### 15.4.2 Basic Principles

What is a synchronous motor? Essentially, it is a three-phase induction motor with a dc rotor field excitation that causes it to run at speed $n_s$ *independent of the load.* This rotor field may be created either by running slip rings out to a dc supply or by

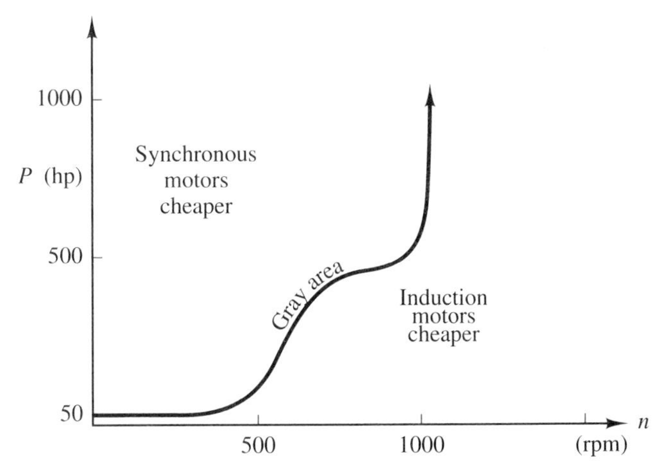


**Figure 15.18** Application suggestions for induction motors versus synchronous motors [Data from *Introduction to Electrical Machines,* by Harit Majmudar (Allyn and Bacon, Boston, 1969), p. 212.]

a self-excited dc generator situated on the same shaft as the rotor, thereby eliminating any slip rings.

The rotor may be either of the *salient* type (electromagnetic poles physically project) or *cylindrical* type (all windings on the periphery of a cylinder). Salient poles are cheaper to fabricate and are appropriate for low speed because they typically have many more poles than the cylindrical type. Thus by Eq. (15.3), a low electrical frequency is associated with a low mechanical speed. At high speeds, the centrifugal forces could rip apart salient poles, so the cylindrical type must be used in those applications. And the smaller number of poles allows the low frequency to be associated with a high mechanical speed. The stator rotating field is the same as that for a three-phase induction motor.

How does it work? The following argument is made by D. D. Stephen et al. in *Electric Motor Handbook.** Suppose that a wound-rotor induction motor is spinning at $n_r = (1 - s)n_s$. Also suppose now that a sinusoidal voltage of frequency $s\omega_s$ is applied across the rotor coil. Recall that $s\omega_s$ is the frequency of the existing induced currents in the rotor coil. Further suppose that the motor load is very light, so that $s$ is very small. Then $s\omega_s$ is small.

If the polarity of the new applied voltage is chosen so as to aid rotation, the rotor speed will become even closer to $n_s$ and $s$ will consequently further decrease. (In this process, the frequency of the voltage applied to the rotor was made to change continuously with the frequency of induced voltages; that is, the frequency was kept at $s\omega_s$ as $s$ changed.) Eventually, $n_r$ will equal $n_s$ and therefore $s$ will become zero, which means that dc is being applied to the rotor.

The purpose of the argument above is merely to show why it makes sense to apply a *dc* voltage to the excitation (rotor) coil to make $n_r = n_s$, even though three-phase sinusoidal voltages are applied to the stator. The major claim, however, is that now $n_r$ will *stay* equal to $n_s$, even as heavier loads are applied. That is, the rotor is "locked in" to the synchronous speed. (Stephen's argument above did not show this.) It is claimed that instead of the load slowing the motor down, the following behavior occurs. The stator (armature) current increases and the rotor lags at an angle $\alpha$ behind the rotating (stator) field, but the rotor spins *at the same speed*, $n_s$.

That will take a bit of explaining. Suppose first that the directions of D. D. Stephen described above are followed, and at least temporarily the rotor is spinning at $n_s$. (This would occur only for zero load.) Consequently, the excitation (rotor) field is perfectly aligned with the rotating (stator) field. Because of this (dc) rotating excitation field, by Faraday's law a sinusoidal voltage $e_g = d\phi_{\text{excit}}/dt$ will be induced in the stator coils, where $\phi_{\text{excit}}$ is the flux due to the dc rotor excitation field. Therefore, because the excitation flux is proportional to the field current, $E_G$ is proportional to $I_F$.

Notice that now the main induction (which will be shown to produce the lock-in torque) is in the stator. Therefore, for this mode of operation, the *stator* acts as and will be called the armature, while the *rotor* acts as the field.

Suppose that the machine has a two-pole rotor field rotating clockwise. Con-

*D. D. Stephen, E. H. Werninck, G. H. Rawcliffe in E. H. Werninck, ed., *Electric Motor Handbook* (McGraw-Hill, NY, 1978) p. 135.

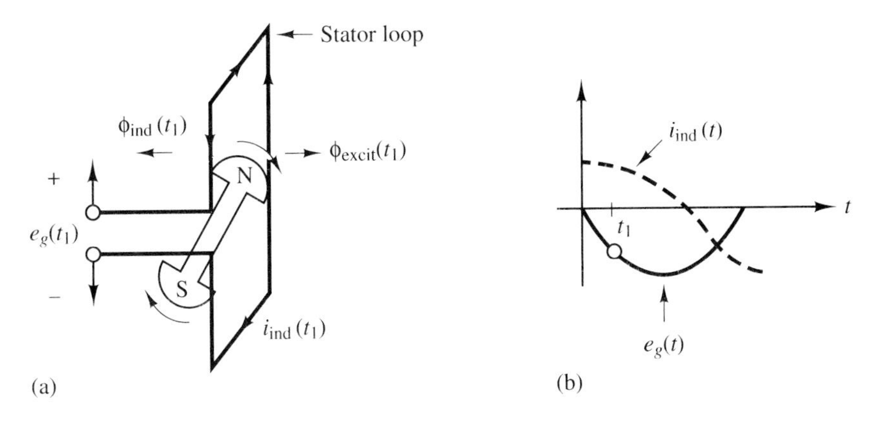


**Figure 15.19** Induced current, flux, and voltage produced in a loop exposed to a rotating magnetic field: (a) physical diagram; (b) induced voltage and current waveforms.

sider one loop of one phase of the stator through which $\phi_{excit}$ passes (see Fig. 15.19a). There are two components to the total $d\phi_{tot}/dt$ responsible for $e_g(t)$: The external (field) induction $d\phi_{excit}/dt$ and the usual self-induction $d\phi_{self}/dt$. Consider first $d\phi_{excit}/dt$. The arrows on the terminals of the loop indicate connections to the rest of the stator coil circuit. Define any flux $\phi$ passing through the loop to the right as "positive." Then at the moment $t_1$ pictured in Fig. 15.19a, $d\phi_{excit}(t_1)/dt > 0$. By Lenz's law, we may think of $e_g(t_1)$ as a voltage source producing an induced current $i_{ind}(t_1)$ having a $\mathbf{B}_{ind}(t_1)$ and therefore induced flux $\phi_{ind}(t_1)$ pointing left [i.e., opposing the positive $d\phi_{excit}(t_1)/dt$]. With the polarity of $e_g(t_1)$ as defined in Fig. 15.19, $e_g(t_1) < 0$. (Recall that positive current flows out of the positive terminal of a voltage source.) The waveforms $e_g(t)$ and $i_{ind}(t)$ are shown for a half-cycle in Fig. 15.19b.

Because we have assumed that $\mathbf{B}_{field}(t)$ is aligned with $\mathbf{B}_{stator}(t)$, we also have the self-induction situation at $t = t_1$ shown in Fig. 15.20a. The stator field, it will be recalled, is due to $v_{APP}(t)$. Using the right-hand rule, we may verify that the stator (armature) current $i_{APP}(t_1)$ is flowing in the correct direction in Fig. 15.20a at this

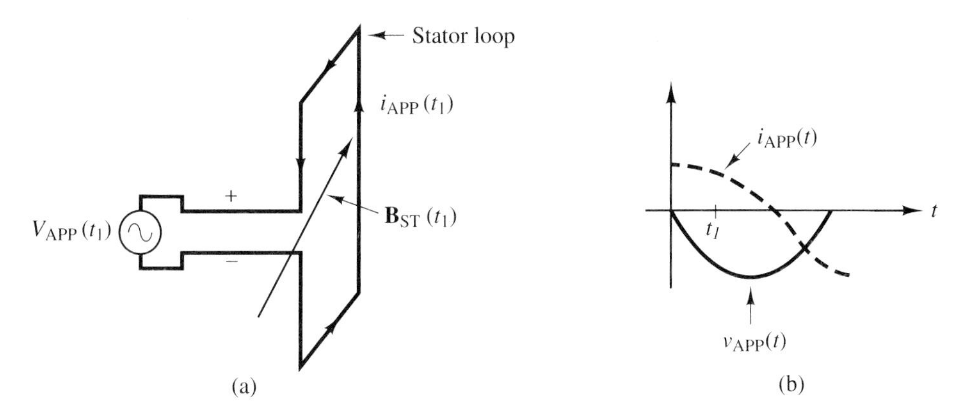


**Figure 15.20** Current, $B$, and voltage applied to a loop: (a) physical diagram; (b) applied voltage and current waveforms. Note difference in polarity conventions for $i_{APP}(t)$ in (a) and $i_{IND}(t)$ in Fig. 15.19a.

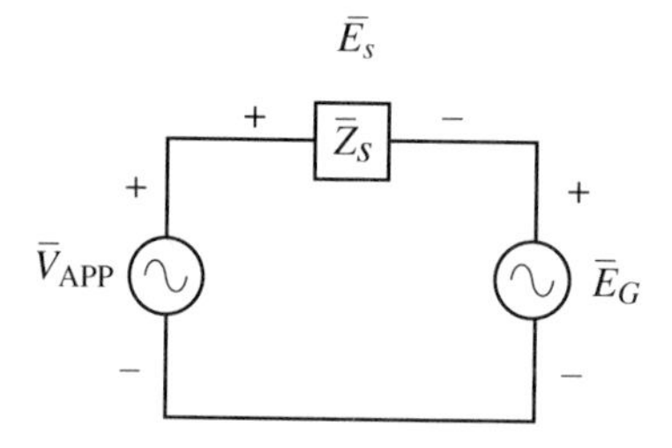


**Figure 15.21** Simple circuit model of stator winding in a synchronous motor. $\overline{V}_{\mathrm{APP}}$ is added to the model of Fig. 15.15 because of the additional $d\phi/dt$ from $v_{\mathrm{APP}}(t)$.

instant for this loop's contribution to $\mathbf{B}_{\mathrm{stator}}$, and its magnitude approaching its maximum. Therefore, as shown in Fig. 15.20b, $v_{\mathrm{APP}}(t_1)$, is indeed negative at this particular instant. Note that the same convention of polarity of $v_{\mathrm{APP}}$ is used as was used for $e_g$. We see in Fig. 15.20b that $e_g(t)$ is exactly in phase with $v_{\mathrm{APP}}(t)$ because $\mathbf{B}_{\mathrm{field}}(t)$ and $\mathbf{B}_{\mathrm{stator}}(t)$ are assumed to be in phase. However, its magnitude depends on that of $\mathbf{B}_{\mathrm{field}}$, so it may differ in magnitude from that of $v_{\mathrm{APP}}(t)$.

The two emf effects above are combined in the circuit model of Fig. 15.21. The phasor diagram for $\overline{E}_G$ and $\overline{V}_{\mathrm{APP}}$ for perfect alignment of $\mathbf{B}_{\mathrm{field}}$ with $\mathbf{B}_{\mathrm{stator}}$ as shown in Fig. 15.22a is now easily understood ($\overline{I}_A$ will be discussed momentarily). The phasor $\overline{V}_{\mathrm{APP}}$ is chosen as the reference, so it has zero phase angle by definition. The net (resultant) voltage that the stator coil experiences is

$$\overline{E}_S = \overline{V}_{\mathrm{APP}} - \overline{E}_G. \tag{15.24}$$

Thus we see that $e_g(t)$ effectively modifies—in this case reduces—the net $d\phi_{\mathrm{tot}}(t)/dt$, and therefore the net voltage $e_r(t)$, from what it would be without the field, namely the self-emf $v_{\mathrm{APP}}(t)$.

The impedance of the armature $\overline{Z}_S$ (stator, also called the *synchronous impedance*) is highly inductive and has an angle β very close to 90° ($\overline{Z}_S = Z_S e^{j\beta}$). Taking β to be 90° in this discussion and including the resultant armature current phasor $\overline{I}_A$ in the diagram yields the armature current shown in Fig. 15.22. It is the resultant voltage $\overline{E}_S$ that gives rise to the actual armature current $\overline{I}_A$; therefore, $\overline{I}_A$ lags an angle β behind $\overline{E}_S$, and $\overline{E}_S = \overline{I}_A\overline{Z}_S$.

The analysis above has shown that if the stator and rotor fields are perfectly aligned, $\overline{E}_G$ is in phase with $\overline{V}_{\mathrm{APP}}$. Now suppose that instead of perfect alignment, the rotor field is behind the impressed (stator) field by an angle α (see Fig. 15.23). This will occur for a nonzero load on the rotor. Then the phasor diagram would change to that shown in Fig. 15.24. The rotor field rotates at speed $n_r$, and the stator field rotates at speed $n_s$.

As an aside, there are $P/2$ "electrical" (N-S) cycles around the air gap; thus, an "electrical" angle of α corresponds to a "mechanical" angle of $\alpha/[P/2] = 2\alpha/P$. [Refer also to the arguments made concerning Eq. (15.3).]

Recall from previous discussions that $\overline{E}_G$ depends *only* on (is due to only) the

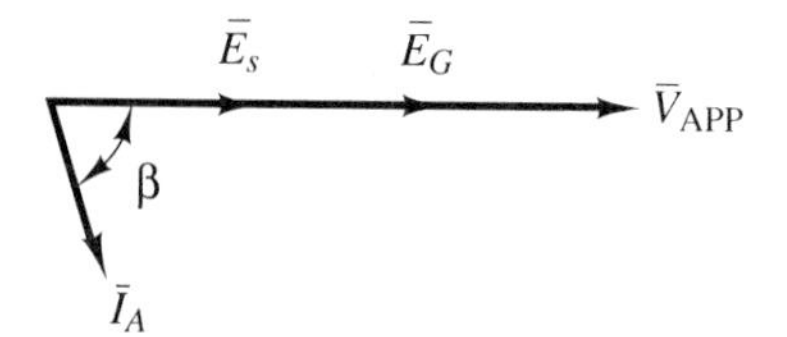


**Figure 15.22** No-load phasor diagram for applied, generated, and resultant stator voltages and current in a synchronous motor.

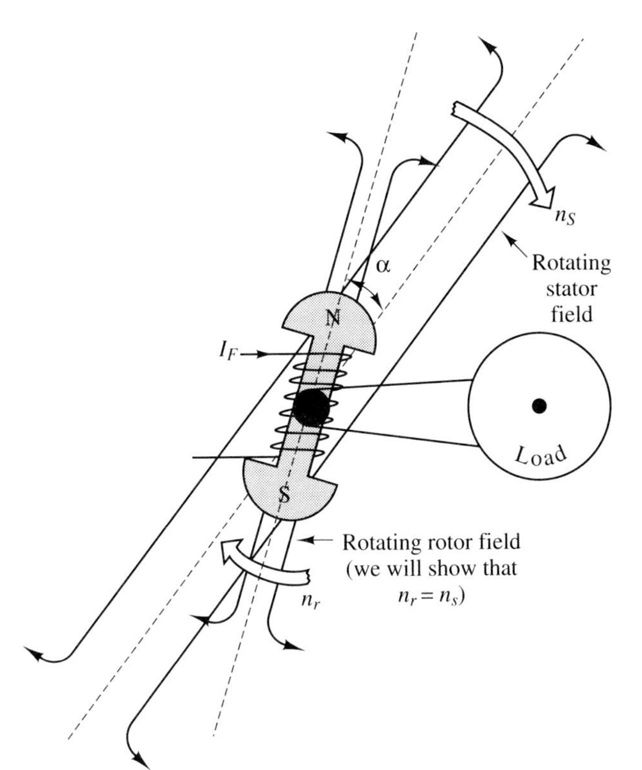


**Figure 15.23** A load on the rotor causes the rotor field to fall behind the stator field by an angle α.

dc rotor field $\overline{\phi}_{\text{excit}}$ and $\overline{B}_{\text{excit}}$ (and constant parameters). Consequently, if the rotor field is an angle α behind the stator field, $\overline{E}_G$ will correspondingly be an angle α behind its former position when the rotor and stator fields were aligned. This effect is shown in Fig. 15.24, where "behind" means clockwise, and $\overline{E}_G$ is to be compared with its former position in Fig. 15.22.

Notice in Fig. 15.24 the angle δ between $\overline{E}_S$ and $\overline{V}_{\text{APP}}$. What is its value? From Fig. 15.24 it is seen that

$$\begin{aligned}\overline{E}_S &= \overline{V}_{\text{APP}} - \overline{E}_G \\ &= V_{\text{APP}} - E_G \cos(\alpha) + jE_G \sin(\alpha) = E_S e^{j\delta},\end{aligned} \tag{15.25}$$

where

$$\begin{aligned}\delta &= \tan^{-1}\left[\frac{E_G \sin(\alpha)}{V_{\text{APP}} - E_G \cos(\alpha)}\right] \\ &= \tan^{-1}\left[\frac{\sin(\alpha)}{V_{\text{APP}}/E_G - \cos(\alpha)}\right]\end{aligned} \tag{15.26}$$

and

$$\begin{aligned}E_S &= \{[V_{\text{APP}} - E_G \cos(\alpha)]^2 + [E_G \sin(\alpha)]^2\}^{1/2} \\ &= V_{\text{APP}}\left\{\left[1 - \frac{E_G \cos(\alpha)}{V_{\text{APP}}}\right]^2 + \left(\frac{E_G \sin(\alpha)}{V_{\text{APP}}}\right)^2\right\}^{1/2}.\end{aligned} \tag{15.27}$$

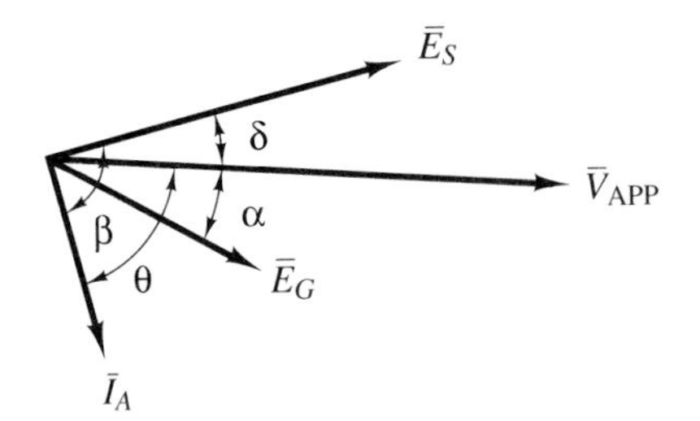


**Figure 15.24** Phasor diagram for applied, generated, and resultant stator voltages and current in a synchronous motor with a load applied.

The net armature current, as stated above, is determined by the complex Ohm's law as

$$\overline{I}_A = \frac{\overline{E}_S}{\overline{Z}_S} = \frac{E_S}{Z_S} e^{j(\delta-\beta)}. \tag{15.28}$$

Incidentally, this means that the power factor angle θ (the angle between $\overline{I}_A$ and $\overline{V}_{APP}$) is $\theta = \theta_v - \theta_i = 0 - (\delta - \beta)$, or

$$\theta = \beta - \delta. \tag{15.29}$$

**Example 15.5**

This problem concerns a 40-hp 400-V wye-connected synchronous motor. Suppose that currently $\alpha = 20°$; that is, the rotor lags behind the rotating stator field by 20°. Suppose also that the field current in the rotor is externally adjusted so that the generated voltage $\overline{E}_G$ is set to be equal in magnitude to $V_{APP}$. What is the resultant armature voltage (per phase), and at what angle is it with respect to the applied voltage phasor?

**Solution** For a wye-connected load, $V_{APP} = V_{phase} = V_{line}/\sqrt{3} = 400/\sqrt{3} = 231$ V. For the case $E_G = V_{APP}$, Eqs. (15.26) and (15.27) simplify and give for this example

$$\begin{aligned}\overline{E}_S &= V_{APP}\{[1 - \cos(\alpha)]^2 + \sin^2(\alpha)\}^{1/2} \\ &= V_{APP}\{2[1 - \cos(\alpha)]\}^{1/2} \\ &= 231\{2[1 - \cos(20°)]\}^{1/2} = 80.2 \text{ V}\end{aligned}$$

and

$$\begin{aligned}\delta &= \tan^{-1}\left[\frac{\sin(\alpha)}{1 - \cos(\alpha)}\right] \\ &= \tan^{-1}\left[\frac{\sin(20°)}{1 - \cos(20°)}\right] = 80°.\end{aligned}$$

**Example 15.6**

Suppose that the field excitation of the motor of Example 15.5 is increased so that $E_G$ becomes 300 V. The *synchronous resistance* is 0.5 Ω and the *synchronous reactance* is 6 Ω. Find (a) $E_S$, (b) δ, (c) $I_A$, (d) the power factor, and (e) the total input power to the motor.

**Solution**

**(a)** Using Eq. (15.27), $E_S = 231\{[1 - 300\cos(20°)/231]^2 + [300\sin(20°)/231]^2\}^{1/2} = 114.5$ V.

**(b)** Using Eq. (15.26), $\delta = \tan^{-1}\{\sin(20°)/[231/300 - \cos(20°)]\} = 116.4°$. Note that the answer for δ that we would obtain by blindly using a calculator would be $-63.6°$, which is wrong. The denominator of the inverse tangent argument is negative and the numerator is positive; this indicates that $E_S$ is in the second quadrant. Adding 180° to −63.6° gives the correct answer, 116.4°.

**(c)** First, the synchronous impedance is $\overline{Z} = [(0.5)^2 + 6^2]^{1/2} e^{j\tan^{-1}(6/0.5)} \approx 6e^{j85.2^\circ} = Z_S e^{j\beta}$. Then, by Eq. (15.28), $I_A = E_S/Z_S = 114.5/6 = 19.1$ A.

**(d)** By Eq. (15.29), $\theta = \beta - \delta = 85.2^\circ - 116.4^\circ = -31.2^\circ$. Because $\theta < 0$, the power factor is leading and is equal to $\cos(-31.2^\circ) = 0.86$.

**(e)** $P_{\text{in}} = 3V_{\text{APP}} I_A \cos(\theta) = 3(231)(16.4)(0.86) = 9.8$ kW [$V_{\text{APP}}$ and $I_A$ are both phase quantities; see Eq. (6.59)].

### 15.4.3 Developed Torque

In previous descriptions of induced torque, a background field from the stator coils produced forces on current-carrying rotor conductors according to Ampère's force. Now the "background field" is actually the rotating dc field from the rotor winding. The current-carrying conductors are the stator windings, whose current, $\overline{I}_A$, was just derived. The reader may object, "but the stator cannot move." However, the *underlying* concept is the mutual repulsion of the rotor field and the magnetic field surrounding the stator windings due to $\overline{I}_A$. By Newton's third law, equal and opposite forces act on the stator and rotor. The rotor, being free to turn, becomes the rotating member.

The excitation (rotor) flux $\phi_{\text{excit}}$ is spatially distributed sinusoidally, as is $\overline{I}_A$ in the stator windings. By the same arguments as in the discussion on induction motors, $\overline{E}_G$ is spatially in phase with $\overline{\phi}_{\text{excit}}$—now redefined as the flux at a stator *conductor*. But by Fig. 15.24, $\overline{E}_G$ is at an angle $\alpha - \theta$ away from $\overline{I}_A$. Thus the angle between $\overline{\phi}_{\text{excit}}$ and $\overline{I}_A$ is also $\alpha - \theta$. The expression for the total (spatial average) torque on the rotor is identical in form to that for the induction motor. Taking into account these observations and using Eqs. (15.27) and (15.28) gives

$$\begin{aligned} T &= K_3 I_A \phi_{\text{excit}} \cos(\alpha - \theta) \\ &= K_4 I_A I_F \cos(\alpha - \theta) \qquad (15.30) \\ &= K_5 \{[V_{\text{APP}} - E_G \cos(\alpha)]^2 + [E_G \sin(\alpha)]^2\}^{1/2} I_F \cos(\alpha - \theta), \end{aligned}$$

where the factor $1/Z_S$ is embedded within $K_5$ and $K_3$, $K_4$, $K_5$ are constants.

For what $\alpha$ is $T$ a maximum? The answer is $\alpha = \beta$ (as shown below), that is, when the angle between the rotor field and the stator field, $\alpha$, is equal to the impedance angle of the stator, $\beta$.

***Proof That Maximum Torque Occurs When $\alpha = \beta$.*** Recall that $\theta = \beta - \delta$, so that $\alpha - \theta = \alpha - \beta + \delta$. First some simplifications will be made on $T(\alpha)$. Then the derivative of $T$ with respect to $\alpha$ will be set to zero. Using Eq. (15.26) for $\delta$ in the expression above for $\alpha - \theta$ gives, in Eq. (15.30),

$$\begin{aligned} T = {} & K_5 \{[V_{\text{APP}} - E_G \cos(\alpha)]^2 + [E_G \sin(\alpha)]^2\}^{1/2} I_F \\ & \cdot \cos\left\{\alpha - \beta + \tan^{-1}\left[\frac{E_G \sin(\alpha)}{V_{\text{APP}} - E_G \cos(\alpha)}\right]\right\} \end{aligned}$$

$$= K_5\{[V_{APP} - E_G \cos(\alpha)]^2 + [E_G \sin(\alpha)]^2\}^{1/2} I_F$$

$$\cdot \left( \cos(\alpha - \beta) \cos\left\{ \tan^{-1}\left[ \frac{E_G \sin(\alpha)}{V_{APP} - E_G \cos(\alpha)} \right] \right\} - \sin(\alpha - \beta) \sin\left\{ \tan^{-1}\left[ \frac{E_G \sin(\alpha)}{V_{APP} - E_G \cos(\alpha)} \right] \right\} \right), \quad (15.31)$$

which, when using the triangle implied by the argument of the inverse tangent having vertical leg $E_G \sin(\alpha)$ and horizontal leg $V_{APP} - E_G \cos(\alpha)$ reduces to

$$T = K_5\{[V_{APP} - E_G \cos(\alpha)]^2 + [E_G \sin(\alpha)]^2\}^{1/2} I_F$$

$$\cdot \left( \cos(\alpha - \beta) \frac{V_{APP} - E_G \cos(\alpha)}{\{[V_{APP} - E_G \cos(\alpha)]^2 + [E_G \sin(\alpha)]^2\}^{1/2}} - \sin(\alpha - \beta) \frac{E_G \sin(\alpha)}{\{[V_{APP} - E_G \cos(\alpha)]^2 + [E_G \sin(\alpha)]^2\}^{1/2}} \right) \quad (15.32)$$

$$= K_5 I_F\{[V_{APP} - E_G \cos(\alpha)] \cos(\alpha - \beta) - E_G \sin(\alpha) \sin(\alpha - \beta)\}.$$

Finally, differentiating $T$ with respect to $\alpha$ gives

$$\frac{\partial T}{\partial \alpha} = K_5 I_F[E_G \sin(\alpha) \cos(\alpha - \beta) - (V_{APP} - E_G \cos(\alpha)) \sin(\alpha - \beta)$$

$$- E_G \cos(\alpha) \sin(\alpha - \beta) - E_G \sin(\alpha) \cos(\alpha - \beta)]$$

$$= K_5 I_F[E_G\{\cos(\alpha - \beta) \sin(\alpha) - \sin(\alpha - \beta) \cos(\alpha) \quad (15.33)$$

$$+ \sin(\alpha - \beta) \cos(\alpha) - \cos(\alpha - \beta) \sin(\alpha)\} - V_{APP} \sin(\alpha - \beta)]$$

$$= K_5 I_F V_{APP} \sin(\alpha - \beta),$$

which equals zero if and only if $\alpha = \beta$. Substituting $\alpha = \beta$ into Eq. (15.32) we find that the maximum torque, $K_5 I_F[V_{APP} - E_G \cos(\alpha)]$, is proportional to $I_F$ and also increases with $V_{APP}$.

The important points to remember from all this are that (1) the torque for any angle—particularly the maximum torque—is proportional to $I_F$ and (2) the maximum torque occurs for the angle $\alpha$ between the rotor and stator fields equal to the phase angle $\beta$ of $\bar{Z}_S$, which is close to 90°. The maximum torque is also roughly proportional to $V_{APP}$, assuming that $\beta \approx 90°$. Note that typically $V_{APP}$ is fixed, so the primary means by which the maximum torque can be increased (and therefore making it harder to pull the rotor out of synchrony) is by increasing $I_F$. We *cannot* increase the speed by increasing $I_F$, but only the maximum locking torque. Nor does $I_F$ increase under increased load; only $I_A$ increases, as will soon be demonstrated.

Suppose that initially the fields are in line: $\alpha = 0$. When the load is increased, naturally the motor initially begins to slow down ($T_{net} = dn/dt < 0$). This implies that $\alpha$ will increase and that will increase the developed torque toward its maximum at $\alpha = \beta$.

If $I_F$ is so small that the *maximum* torque happens to be less than the loading torque on the motor shaft, there will be no $\alpha$ permitting *lock-in* and the motor will slow down. To what speed will it reduce? First, it should be noted that a synchronous motor as described to this point is not self-starting; that is, the lock-in process above can work only if $n_r$ is already close to $n_s$. The time-averaged torque at $n_r = 0$ would be zero.

Therefore, typically, a squirrel-cage or other set of shorted windings is embedded within the rotor to bring the rotor to a speed sufficiently close to $n_s$ for the lock-in to be possible. These windings are called *amortisseur* and *damper windings.* ("Amort" means "not dead" and "damper" refers to the stabilizing effect these windings have in the case of unsmooth loading.)

With the damper windings, if lock-in does not occur for the given rotor field current, the rotor will slow down to the underlying induction motor speed. So unless the rotor field current is large enough, nothing particularly interesting happens—the motor merely operates as an induction motor. And without the squirrel cage, the motor would *stop* if the maximum torque were less than the load torque.

But if $I_F$ *is* sufficiently large (so that at least the maximum torque exceeds the load torque), lock-in *does* occur. In this case, $\alpha$ takes on the value ($<\beta$ in practice) for which the developed torque balances the load torque. As the load varies in time, $\alpha$ will vary, but the frequency of the rotor rotation maintains the synchronous value (unless the load torque exceeds that maximum torque at $\alpha = \beta$). If lock-in is temporarily lost, $\alpha$ will exceed $\beta$ and the rotor is said to "slip a pole." Assuming that the load surge that caused loss of lock-in is momentary, the rotor will eventually re-lock-in upon restoration of the normal (lower) load.

The maximum torque is sometimes called the *pull-out torque.* Its original meaning was the torque beyond which the rotor would "pull out" of synchronization with the rotating (stator) field. Below synchronous speed, the average torque due to the rotor excitation field is zero. In particular, as mentioned above, the starting torque is zero. For half a cycle the torque is, say, clockwise, but for the next half-cycle it is counterclockwise. This switching of direction thus happens (for $n_r = 0$ and $f = 60$ Hz) every $\frac{1}{120}$ s; the rotor never gets a chance to build up speed. When the squirrel cage brings $n_r$ sufficiently close to $n_s$, the torque oscillation frequency becomes extremely low, so that the torque, acting in one direction, can bring about lock-in.

It was originally stated that an increased mechanical load would increase not only $\alpha$ but also the stator current $I_A$. Note that $I_F$ is essentially fixed in value by the dc field source. That the stator current increases with load will now be shown. Suppose that $I_F$ and therefore $E_G$ are held fixed. As described above, $\alpha$ increases when the load is increased. The phasor diagram in Fig. 15.25 will indicate what happens to $\bar{E}_S$ and $\bar{I}_A$. Note in the following that because $Z_S$ is fixed, $I_A$ will increase or decrease in proportion to $E_S$. This discussion will also determine what happens to the power factor angle $\theta$.

Figure 15.25 shows phasor diagrams for two loads but fixed $|\bar{E}_G| = E_G$, where the load in Fig. 15.25b exceeds that in Fig. 15.25a, so that $\alpha_2 > \alpha_1$. By simple vector addition, clearly $E_{S,2} > E_{S,1}$; that is, the resultant emf increases for increased load. Consequently, $I_{A,2} > I_{A,1}$, as predicted.

This is an important point: The extra "oomph" to overcome an increased load comes *not* from the rotor field but from the stator. Of course, this is fortunate because it means that the major electrical power delivery is performed on the stationary member, not the rotating one, as it does in the dc motor. (This advantage was also present in the induction motor.) Also, $\theta_2 < \theta_1$; that is, the power factor improves for larger load. (The latter fact is true only for $|\theta_1|$ sufficiently large.)

Moreover, phasor diagram analysis shows that by using large field currents, the

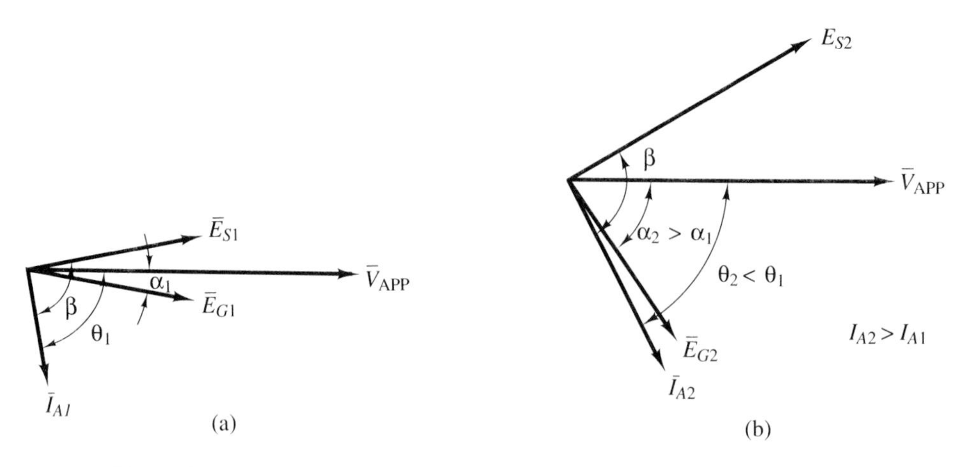


**Figure 15.25** Phasor diagrams for applied, generated, resultant stator voltages and stator (armature) current for two loads, that in (b) being greater than that in (a).

power factor of the machine can be made leading! In fact, we have already seen an instance of this in Example 15.6. Recall that $E_G$ is proportional only to $I_F$ and constants. Figure 15.26 shows three situations, dictated by the choice of magnitude of $\bar{I}_F$. The vectorial (parallelogram) subtraction of $\bar{E}_G$ from $\bar{V}_{APP}$, $\bar{E}_S$, can be at an angle away from $\bar{V}_{APP}$ less than β (Fig. 15.26a), equal to β (Fig. 15.26b), or greater than β (Fig. 15.26c). (In these diagrams we assume that $\beta = 90°$ for simplicity.)

These situations are termed, respectively, underexcitation, normal excitation, and overexcitation. The result is that θ is, respectively, positive, zero, or negative. Note that in all cases, the load was assumed to have a constant value. Consequently, $P_{in}$ and therefore $I_A \cos(\theta)$ ($V_{APP}$ is constant) are constants as well. From Fig. 15.26a to 15.26b to 15.26c, $I_F$ and therefore $E_G$ increase; the stronger dc field causes the torque angle α to decrease in that sequence.

A final brief note about power and torque. The developed torque and developed power are related by

$$T_{dev} = \frac{P_{dev}}{\omega_r}. \tag{15.34}$$

The developed power can be obtained from wattmeter measurements $W_1$ and $W_2$ and/or from other measurable quantities as

$$\begin{aligned} P_{dev} &= W_1 + W_2 - 3R_S I_A^2 \\ &= \sqrt{3}\, V_L I_L \cos(\theta) - 3R_S I_A^2, \end{aligned} \tag{15.35}$$

where, for example, in a Y-connected stator $I_L = I_A$ and $V_L = \sqrt{3}\, V_{APP}$ for each phase and the total developed power is $3V_{APP}I_A \cos(\theta) - 3R_S I_A^2$. The output power is

$$P_{out} = P_{dev} - P_{mech,core}, \tag{15.36}$$

where $P_{mech,core}$ is the developed power at no load to overcome frictional (mechanical) and magnetic (core) losses.

Also, it should be recalled that $\alpha = 0$ for zero (mechanical) load, ignoring

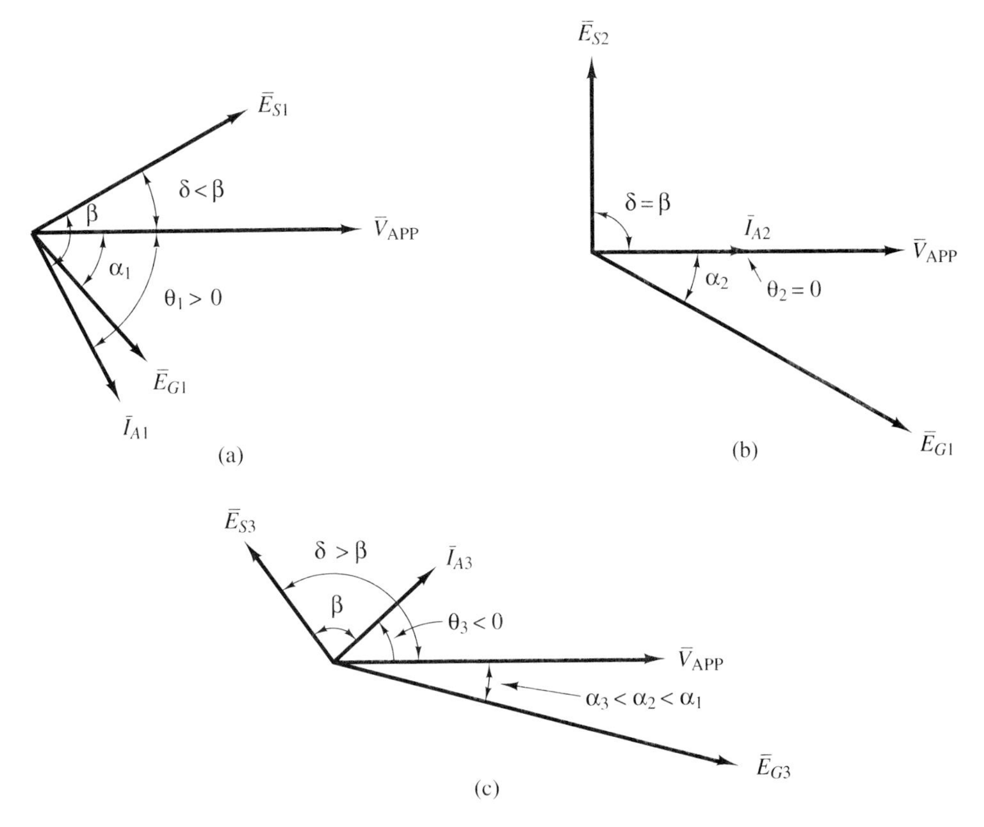


**Figure 15.26** Phasor diagram for three values of field current and thus generated voltage: (a) underexcitation; (b) normal excitation; (c) overexcitation.

friction. This does *not* in general imply that the armature current is zero, because there could still be a difference in the magnitudes of $\bar{E}_G$ and $\bar{V}_{APP}$ due to the dc exciting field, which determines $E_G$. [Of course, for generator operation, $I_A$ *will* be zero for no (electrical) load. In that case, $E_G$ will therefore be the open-circuit terminal voltage.]

An alternative expression for $P_{dev}$ can be derived as follows. Keep in mind that $E_G$ is independent of $\alpha$; it is a function of the dc field, not the mechanical load. We thus see from Fig. 15.24 that at no load ($\alpha = 0$, $\delta = 0$), $E_G = |V_{APP} - I_{A,nl}X_S|$ ignoring $R_S$. All of these quantities are measurable. (The value of $X_S$ may be determined by a procedure known as the synchronous impedance test, which involves open- and short-circuit measurements of the dynamo under generator conditions and synchronous speed.) By Eq. (15.25) we see that in Fig. 15.24, $E_G \sin(\alpha) = E_S \sin(\delta)$ because $\bar{V}_{APP}$ is horizontal. Under the assumption that $R_S$ is negligible compared with $X_S$ ($\beta = 90°$), this relation may be rewritten using Eqs. (15.28) and (15.29) as

$$E_G \sin(\alpha) = I_A \cos(\theta) X_S. \tag{15.37}$$

Consequently, the developed power, neglecting $R_S$ losses,

$$P_{dev} = 3V_{APP}I_A \cos(\theta)$$

may also be written as

$$P_{\text{dev}} = \frac{3V_{\text{APP}}E_G \sin(\alpha)}{X_S}. \tag{15.38}$$

Equation (15.38) is often cited in the context of generators but is equally valid for motors. In the context of generators and Eq. (15.38), $\alpha$ is known as the power angle (as opposed to the power factor angle) for an obvious reason, considering Eq. (15.38). This alternative view of $\alpha$ is more relevant for generator applications, where "torque angle" has less direct physical significance. But for constant $\omega$, it is clear that as $\alpha = 90°$ maximizes power in Eq. (15.38), it also maximizes torque. Thus the claim that maximum torque occurs for $\alpha = \beta$ [ $= 90°$ in Eq. (15.38)] is verified from a different viewpoint.

We have shown in Eq. (15.30) that the developed torque is proportional to $I_A I_F \cos(\alpha - \theta)$, so $P_{\text{dev}} = \omega_s K I_A I_F \cos(\alpha - \theta)$. Yet above we have shown that the developed power per phase is $P_{\text{dev}} = V_{\text{APP}} I_A \cos(\theta) - I_A^2 R_S = P_{\text{in}} - I_A^2 R_S$. Can both be true? To find out, let us decompose $\overline{E}_S$ into its components along and perpendicular to $\overline{I}_A$:

$$\overline{E}_S = \overline{I}_A R_S + j\overline{I}_A X_S. \tag{15.39}$$

But by Eq. (15.24), $\overline{E}_S = \overline{V}_{\text{APP}} - \overline{E}_G$. Taking the component of Eq. (15.24) along $\overline{I}_A$, and referring to Fig. 15.24, yields

$$I_A R_S = -E_G \cos(\theta - \alpha) + V_{\text{APP}} \cos(\theta), \tag{15.40}$$

or, equivalently,

$$E_G \cos(\alpha - \theta) = V_{\text{APP}} \cos(\theta) - I_A R_S. \tag{15.41}$$

Multiplying by $I_A$ and noting that $E_G$ is proportional to $I_F$ gives

$$K I_A I_F \cos(\alpha - \theta) = V_{\text{APP}} I_A \cos(\theta) - I_A^2 R_S = P_{\text{dev}}, \tag{15.42}$$

which is exactly what we sought to show. Indeed, dividing by $\omega_s$ gives the expression for $T_{\text{dev}}$ in Eq. (15.30).

**Example 15.7**

A 40-hp six-pole wye-connected synchronous motor operates off a 400-V 60-Hz bus. Suppose that at rated load the power factor is 0.5 leading (by design). Given that the rated line current is 30 A, $R_S = 0.1\ \Omega$, and $X_S = 5\ \Omega$, determine the following quantities at rated load: $\alpha$, $E_G$, $P_{\text{dev}}$, and $T_{\text{dev}}$.

**Solution** Some preliminary calculations are: $P_{\text{out}} = 40$ hp $= 29.8$ kW, $n_s = 120(60)/6 = 1200$ rpm, $V_{\text{APP}} = V_L/\sqrt{3} = 400/\sqrt{3} = 231$ V, $Z_S = [(0.1)^2 + 5^2]^{1/2} \approx 5\ \Omega$, $\beta = \tan^{-1}(5/0.1) = 88.9°$, $\theta = -\cos^{-1}(0.5) = -60°$, and $I_A = I_L = 30$ A (wye connection). We showed above that $E_G \sin(\alpha) = E_S \sin(\delta)$ or $\alpha = \sin^{-1}\{E_S \sin(\delta)/E_G\}$. Thus first $E_S$, $E_G$, and $\delta$ must be determined. $E_S = I_A Z_S = 30(5) = 150$ V. A little right triangle analysis of Fig. 15.24 ($\overline{E}_G$ the hypotenuse) and $\overline{E}_G = \overline{V}_{\text{APP}} - \overline{E}_S$ shows that $E_G = \{[V_{\text{APP}} \cos(\theta) - I_A R_S]^2 + [V_{\text{APP}} \sin(\theta) - I_A X_S]^2\}^{1/2}$. Thus $E_G = \{[231(0.5) - 30(0.1)]^2 + [231 \sin(-60°) - 30(5)]^2\}^{1/2} = 368$ V. Also, from Eq. (15.29), $\delta = 88.9° - (-60°) = 148.9°$. Therefore, $\alpha = \sin^{-1}[150 \sin(148.9°)/368] = 12°$. From Eq. (15.35), $P_{\text{dev}} = \sqrt{3}(400)(30)(0.5) - 3(0.1)(30)^2 = 10.1$ kW. Finally, $T_{\text{dev}} = P_{\text{dev}}/\omega_s = 10.1 \cdot 10^3/(1200\pi/30) = 80.4$ N·m.

### 15.4.4 Applications

Synchronous motors are useful in situations where relatively constant load torque may be expected and constant speed desired. If there are heavy, sudden loads, the motor will "slip poles" and remain out of synchrony, defeating the purpose of the machine. They are also not recommended for applications where there is to be a lot of starting and stopping, for again the motor will be operating out of synchrony with the rotating field. Specific applications include bandsaws, blowers, pumps, dc generator driving, fans, pulp grinders, and metal rolling mills.

They are also being tested for use in electric cars. There is a debate as yet unresolved concerning whether to use dc motors that have inexpensive control circuits or ac motors, which require advanced electronic control but may have performance advantages. Another common application is in power factor correction, as noted in Sec. 15.4.3; such motors are referred to as *synchronous condensers* because they function as capacitors.

One disadvantage of synchronous motors is their relatively low starting torque for a given size machine. Also, they are somewhat more expensive than induction motors for low-power applications, due to the dc field winding and connections added to a squirrel-cage rotor. However, they are very efficient for, cheaper in, and well suited to high-power applications. The crossover point between suggested use of induction motors versus synchronous motors has been estimated to be 1 hp/rpm; see also Fig. 15.18.

Naturally, the synchronous motor is useful for timekeeping as well as motors in tape recorders and phonograph turntables. In this case the excitation is single-phase, and the round rotor is a permanent magnet, which produces the rotor field (the dc-excited winding in our previous studies). These motors are called hysteresis motors. Similar techniques (e.g., shaded pole) are used to start hysteresis motors as are used for single-phase induction motors.

If the rotor material is instead salient-pole and not a permanent magnet, the operational principles resemble more those governing relays. These motors are called reluctance motors and have also been used for similar applications, including electric clocks.

## 15.5 AC GENERATORS

The terms *alternator* and *synchronous generator* are synonymous. As was the case for dc generators and motors, essentially the same dynamo can be used for either synchronous motor or synchronous generator action. The synchronous generator may differ from the synchronous motor in that the squirrel cage is no longer required; there is no issue of starting torque. However, the squirrel-cage "damper" winding is often retained, as it stabilizes the synchronization in the face of pulses in the prime mover torque.

Note that the frequency of the output voltages is directly tied to the mechanical frequency; in this sense the generator is "synchronous." Thus if precisely 60 Hz is required, the rotor must spin at constant speed determined by the number of poles. Detailed analyses of voltage regulation performance must include such issues as armature reaction variation with load and leakage flux effects.

Another factor is the air gap width. The larger it is, the better the voltage regulation but the larger the required excitation currents. An inductive load can cause drooping of the terminal voltage magnitude, while capacitive loading will result in a rise. Automatic voltage regulation circuitry is required to maintain high performance in the face of such effects. Unbalanced loading is another problem, which can cause overheating. As with any machine, losses such as friction, windage, cooper losses ($I^2R$), eddy currents, and hysteresis all contribute to reduction in efficiency.

Three-phase synchronous generators are the workhorses for electric power production on mass scales in power plants of all kinds. The three-phase $v_{APP}(t)$ is removed and the armature now supplies the output voltage. The dc field remains connected.

Another important use is as automobile alternators for recharging the car battery and supplying other electrical loads while driving or idling. Although the alternating voltages must be rectified for dc battery charging, the alternator is far more efficient, cheaper, and more reliable than an equivalent dc car generator (used long ago). The generated voltage varies widely with engine speed. A voltage regulator circuit uses switching transistors, diodes, and capacitors to provide a smooth voltage at an acceptable level. The field current is started by the battery, but is soon supplied by a fraction of the generated output voltage.

As discussed in Secs. 6.6 and 6.7, all three phases of electric power are transmitted long-distance. Individual customers may use either all three phases (industrial) or only one of the phases (homeowners), according to the requirements of the customer's equipment. The generators are invariably synchronous.

Induction generators can also be obtained by rotating an induction motor above synchronous speed. However, an induction generator is not self-exciting. An external sinusoidal voltage source must supply sinusoidal currents to set up the revolving field. This source determines the *synchronous speed.* Yet to act as a generator, the rotor must be spun faster than the synchronous speed. Therefore, unlike the synchronous generator, the induction generator output cannot straightforwardly be used to generate its own field. In fact, it really needs a synchronous generator to start it. Proposals to overcome this problem such as connecting capacitors to the terminals have had some success in applications where precise frequency is unimportant. Because of this inability to self-"start" and because of the great efficiency of synchronous generators, the induction generator is infrequently used.

## 15.6 SUPERCONDUCTING MACHINES

An exciting recent development in machines is the application of superconducting materials. These materials have zero resistance and can carry very large currents. As a result, the size and weight of a machine having a particular power rating can be reduced. Because of the necessity of cryogenic conditions (liquid nitrogen, 77 K), such machines are economical only for large machinery such as in power plants or other large conversion situations, such as large ship or high-speed rail propulsion.

Existing large machines are already efficient; it has been reported that an increase of only 0.5% is possible using superconductors. Nevertheless, for huge power generating systems, that gain translates to megawatts of power. For motors, it is the reduction in weight that is attractive (one-third the weight). Because extremely high currents and therefore fluxes (5 to 6 T) are obtained easily and efficiently, no inner iron core is necessary; therefore, a large fraction of the bulk and weight of the machine may be eliminated. (An outer core shield may be necessary to protect people from the huge magnetic fields.)

As noted previously, a large air gap translates to good voltage regulation in generator applications, so the superconducting generator inherently has good voltage regulation. It is unfortunate that the currents must be dc; alternating currents will heat up the conductors (by inductive heating), causing them no longer to be superconducting. Thus only the field winding, not the armature of, for example, a synchronous motor may be a superconducting coil. This necessitates bringing the supercoolant to the rotating field—a difficult design problem.

Still in its infancy, the area of superconducting machinery is progressing well. In Japan, 70- and 200-MVA generator prototypes are being designed. This area is also being investigated in the former Soviet Union. In Germany, a 1000-MVA machine using superconducting coils is being worked on. But hardware demonstrations remain to be performed.

## 15.7 SUMMARY

Disadvantages of the dc machine such as the vulnerability and expense of the commutator, the fact that the large power must be transferred to or from the rotating member, and the relative difficulty in obtaining high-power dc lines are factors that favor ac machines. Practical ac machines have none of these disadvantages. The synchronous motor has the additional advantage of precisely constant speed. The main advantages of dc motors are the ease of achieving speed control accurately over a wide range, and their high starting torque.

The most common household ac motor is the single-phase induction motor. It is also the most straightforward to understand except for the difficult problem of starting the motor. A number of techniques that are admittedly partly empirically designed have solved the starting problem. Numerous applications were listed; many more can be found in the home or office.

A number of advantages of using three-phase rather than single phase for higher-power applications were given, including the important fact that the instantaneous power into a balanced three-phase machine is constant. The three-phase induction and synchronous motors are the most widely used motors for heavy industrial applications. With three phases of oscillating magnetic fields produced by the stator windings, a single, constant-magnitude rotating magnetic field results. This field produces the driving torque in induction motors and the starting torque in synchronous motors.

An important and basic relation is that between the synchronous speed $n_s$, the electrical frequency $f$, and the chosen number of stator poles $P$: $n_s = 120f/P$ rpm. The synchronous speed is the speed at which the induction motor rotor would spin were there zero mechanical load. It is precisely the speed at which a synchronous motor spins.

In practice the induction motor rotor always spins at a speed less than $n_s$, quantified by the slip parameter $s$ as $n_{\text{rotor}} = (1 - s)n_s$. The rotor can either consist of a squirrel-cage set of shorted conducting bars or a wound rotor similar to that in dc machines. The advantage of the wound rotor is that its resistance can be controlled via an external resistance coupled to the rotor by slip rings.

The flux on the periphery of the rotor is sinusoidally distributed in three-phase machines. The generated voltage is found to be in phase with the flux at the conductor experiencing magnetic force, which is the flux that produces the rotational torque. Because of this fact, the angle between flux and current is the same as that between generated voltage and current: the rotor impedance angle.

Unlike the single-phase motor, the starting torque of a three-phase induction motor is nonzero. A general expression for the torque valid both for "locked rotor" and general slip was derived. From the expression for the torque, the important torque–speed curve was determined. The slip at which maximum torque occurs is called the breakdown slip. Finally, several examples of applications of three-phase induction motors were cited.

For high-power constant-speed applications requiring high efficiency, or if we desire to use a machine to raise the overall power factor of an industrial plant, the synchronous motor is the best choice. The synchronous motor is a three-phase induction motor with a dc rotor field excitation that causes it to run at speed $n_s$ independent of the load (within limits). This rotor field may be created either by running slip rings out to a dc supply or by a self-excited dc generator situated on the same shaft as the rotor, thereby eliminating any slip rings.

A squirrel-cage rotor is typically used to bring the motor up to near-synchronous speed. Then the dc field is applied, which fixes the rotor speed at synchronous speed. If a load is applied, the torque angle increases. It was shown that as a result of induced armature current in proportion to the torque angle, the developed torque increases. If a sufficiently strong dc field is applied, this developed torque will restore the speed back to synchronous speed. At this point, the load and developed torques cancel and steady-state operation results.

However, as noted, the power to accomplish this is drawn from the armature, not the rotating field. Thus the major power needs are supplied by the motionless stator, as is desired. Even if the load varies with time, only the balancing torque angle will vary—not the average speed, which is fixed at synchronous speed by the mechanism above.

If the load brings the torque angle $\alpha$ above its maximum (which occurs at the highly inductive stator impedance angle $\beta$), the motor will "slip a pole." Were this to continue, the motor would no longer be operating in synchrony with the applied rotating magnetic field and would just operate as an induction motor.

Some comments concerning power, torque, and applications were made. There is a very rough 1 hp/rpm crossover between recommendation of induction versus synchronous motors; other factors may easily override this rule of thumb.

Next, the ac generator was discussed. As was the case for dc generators and motors, essentially the same dynamo can be used for either ac motor or ac generator action. The synchronous generator may differ from the synchronous motor in that the squirrel cage is no longer required; there is no issue of starting torque. The required rotating field is produced by the spinning rotor dc electromagnet.

Three-phase synchronous generators are the workhorses for electric power production on mass scales in power plants of all kinds. Another important use is as automobile alternators for recharging the car battery while driving or idling. Induction generators are infrequently used because they are not self-exciting; they require an external source to supply the rotating excitation field.

As a final topic, superconducting machines were introduced. Superconducting materials have zero resistance and can carry very large currents with no $I^2R$ losses. As a result, the size and weight of a machine having a particular power rating can be reduced. Because of the necessity of cryogenic conditions, such machines are economical only for large machinery, such as in power plants, or other large conversion situations, such as large ship or high-speed rail propulsion. Although much further research is necessary to make them economical for practical use, they do show a potential for significant advantages.

## PROBLEMS

**15.1.** List five advantages of dc motors/generators.

**15.2.** List six disadvantages of dc motors/generators.

**15.3.** Draw an electrical schematic diagram of a single-phase induction motor.

**15.4.** Why is it likely that when a single-phase induction motor is started, it will not begin to spin; that is, why is the starting torque zero or near-zero?

**15.5.** Briefly discuss a few ways that the problem of starting a single-phase induction motor has been achieved.

**15.6.** Frequently, the capacitor and the auxiliary winding of single-phase motors are connected across the field winding, and are left connected all the time that the motor is on (as opposed to using centrifugal switching). Draw the circuit diagram of a single-phase capacitor induction motor as used in a household three-speed oscillating fan. Let the speeds be taps off the "auxiliary" winding.

**15.7.** In the circuit in Problem 15.6, an impedance bridge is used to measure the inductance and/or capacitance between various points in the circuit. It is found that with the power off, if the fan is *manually* spun (i.e., by hand) with increasing speed, the overall impedance changes from capacitive to purely resistive, to inductive. Expalin why this occurs.

**15.8.** Name four advantages of the three-phase induction motor over the single-phase induction motor.

**15.9.** Why does an induction motor rotor always spin at a speed below synchronous speed?

**15.10.** Suppose that in Figs. 15.4 and 15.5 we look instead at the time $\omega t = 5\pi/4$.

**(a)** Draw a phasor diagram of the three current phasors $\bar{I}_A$, $\bar{I}_B$, and $\bar{I}_C$, explicitly indicating the angles each makes with the horizontal (in degrees). What are the relative magnitudes and polarities of the three currents $i_A(t)$, $i_B(t)$, and $i_C(t)$?

**(b)** Draw a vector diagram of the corresponding magnetic field vectors from the three electromagnets, at the center of the rotor. In what direction is the total magnetic field pointing at this instant?

**15.11.** A 12-pole motor driven by a 60-Hz voltage source has a slip of 0.2. What are the synchronous and rotor speeds in rpm?

**15.12.** The generated voltage in an induction motor is $e_g = d\phi/dt$, so $e_g$ should be 90° out of phase with $\phi$, because $\phi$ is spatially and temporally sinusoidal. Why, then, do we say that $e_g$ is spatially in phase with $\phi$?

**15.13.** Defining $\phi$ as the flux on the right edge of a current-carrying loop, draw the flux and generated voltage distributions around the circumference of a four-pole induction motor. Draw the distributions at one particular time, and then at a slightly later instant of time.

**15.14.** What is the maximum synchronous speed for a three-phase machine driven at 60 Hz? If $f$ is fixed, what must be done to obtain a slow-spinning machine?

**15.15.** At standstill, what is the electrical frequency of currents induced in a $P$-pole induction motor rotor operated at $f$ hertz?

**15.16.** Suppose that the slip of a four-pole induction motor is 0.2 and that the excitation frequency is 60 Hz. Find the rotor impedance if the rotor resistance is 0.1 Ω and the locked rotor reactance is 0.8 Ω. Also find the locked rotor impedance.

**15.17.** Find the slips at which the per-phase torque of an induction motor is equal to half the maximum value. Interpret your results.

**15.18.** **(a)** From Eq. (15.23), determine the developed power at locked rotor and at synchronous speed (were that attainable).

**(b)** Reconcile your results with the fact that at locked rotor, $T$ can be substantial [see Eq. (15.13)] and that at synchronous speed, Eq. (15.23) becomes $T \cdot n_s\, \pi/30$.

**15.19.** At what slip $s_m$, is $P_{\text{dev}}$ a maximum? Express $s_m$ in terms of $s_b$, the breakdown slip. Is $s_m$ greater than or less than $s_b$? In other words, which is greater: the maximum-power speed or the maximum-torque speed? Prove your results mathematically, and calculate $s_m$ for $s_b = 0.3$. Make a rough sketch of $P_{dev}$ vs. $\omega_r$ for this case.

**15.20.** Determine whether a variable load whose opposing torque declines with rotational speed is more efficiently driven by a class A or a class D induction motor, and provide a reason.

**15.21.** Judging from Fig. 15.17, which type of induction motor has the highest ratio of starting torque to rated torque? What might be an advantage of a class B motor over a class A motor?

**15.22.** **(a)** Find the breakdown slip and speed of a four-pole 60-Hz three-phase induction motor whose rotor resistance is 0.6 Ω and whose rotor reactance is 0.5 Ω at a speed of 1200 rpm.

**(b)** If the developed power at breakdown speed is 10 kW, what is the breakdown torque?

**15.23.** A three-phase six-pole squirrel-cage induction motor operates on a line-to-line 60-Hz voltage of 440 V (rms). At full load, its output power is 50 hp, it has 80% efficiency, its slip is 0.06, and its power factor is 0.75 lagging. Determine at full load:

**(a)** The speed in rpm.

**(b)** The load torque in N · m.

**(c)** The average input power.

**(d)** The line current.

**15.24.** A four-pole 60-Hz three-phase induction motor has a total average input power of 40 kW. The speed is 1750 rpm, the efficiency is 80%, and the power factor is 0.70. Find the slip, losses, and output power in hp.

**15.25.** A four-pole 60-Hz squirrel-cage three-phase induction motor has an input line-to-line voltage of 440 V, a line current of 50 A, and a power factor of 0.80 (lagging). The motor provides a torque of 150 N · m to a load, at a slip of 5%. Find the efficiency of the motor.

**15.26.** A three-phase squirrel-cage induction motor is rated for 50 hp output at a speed of 1750 rpm. The three-phase 60-Hz source provides a line-to-line voltage of 260 V (rms). At rated output, the motor has a power factor of 0.70 and an efficiency of 85%.

**(a)** Find the output torque of the motor at rated output.

**(b)** Find the slip at rated output.

**(c)** How many poles does the motor have?

**(d)** Find the line current at rated output.

**15.27.** A three-phase six-pole squirrel-cage induction motor operating at full load has a 0.75 power factor, 80% efficiency, and runs at a slip of 3%. The 60-Hz source has a line-to-line voltage of 240 V and the line current is 40 A. Find:

**(a)** The output power in hp.

**(b)** The rotor speed in rpm.

***15.28.** A polyphase induction motor has a stator that is wound for four poles.

**(a)** At what speed will the field rotate if a 60-Hz supply is applied to the stator terminals?

**(b)** What will be the speed for a 50-Hz supply? Express your answer in rpm and rad/s.

**15.29.** What frequency must be applied to the stator of a four-pole polyphase induction motor in order to obtain a synchronous speed of 6000 rpm? (*Source:* Rosenblatt and Friedman)

**15.30.** A three-phase squirrel-cage induction motor develops a starting torque of 32 lb-ft with full voltage applied to the stator.

(**a**) What is the starting torque, in lb-ft, with 80% of rated voltage applied?

(**b**) What is the reduced-voltage starting torque in N · m? (*Source:* Rosenblatt and Friedman)

**15.31.** When rated voltage of 440 V is applied to the stator terminals of a polyphase induction motor, it develops a starting torque equal to 1.5 times the rated torque. At what voltage will the starting torque be exactly equal to the rated torque? (*Source:* Rosenblatt and Friedman)

**15.32.** A six-pole 60-Hz polyphase induction motor has a full-load speed of 1155 rpm.

(**a**) What is the slip in rpm?

(**b**) What is the slip in rad/s?

(**c**) What is the percent slip? (*Source:* Rosenblatt and Friedman)

**15.33.** A 60-Hz polyphase induction motor has an operating speed of 575 rpm.

(**a**) For how many poles is the stator wound?

(**b**) At this load, what is the slip? Express as a decimal. (*Source:* Rosenblatt and Friedman)

**15.34.** A 50-Hz polyphase induction motor has a no-load speed of 2990 rpm and a full-load speed of 2880 rpm. Calculate:

(**a**) The frequency of the rotor induced emf at the instant of starting

(**b**) The rotor frequency at no load and at full load. (*Source:* Rosenblatt and Friedman)

**15.35.** A six-pole 60-Hz induction motor has a no-load speed of 1196 rpm and a full-load speed of 1150 rpm. Suppose that at the instant of starting, the induced emf per phase in the rotor is 55 V. Now the motor is driven by another motor while its stator is excited.

(**a**) If it is driven at synchronous speed in the same direction as the stator field, what is the rotor induced emf per phase?

(**b**) If the direction of rotation remains the same, what is the rotor induced emf at speeds of 1204, 1250, and 2400 rpm?

(**c**) What is the rotor induced emf per phase when the rotor is driven at synchronous speed but opposite to the direction of the stator field? (*Source:* Rosenblatt and Friedman)

**15.36.** A polyphase induction motor has a resistance per phase of 0.10 Ω and a rotor reactance per phase at standstill of 0.40 Ω. If the induced emf per phase at standstill is 30 V, calculate:

(**a**) The rotor phase current at standstill.

(**b**) The rotor phase current at full load (slip is 5%).

(**c**) The rotor phase current at a slip of 10%. (*Source:* Rosenblatt and Friedman)

**15.37.** A two-pole 60-Hz polyphase induction motor has a full-load speed of 3450 rpm. Calculate the approximate speeds, in rpm and rad/s, at loads of $\frac{1}{4}$, $\frac{1}{2}$, $\frac{3}{4}$, and $1\frac{1}{4}$ times full load. (*Source:* Rosenblatt and Friedman)

**15.38.** A polyphase induction motor has a rotor resistance per phase of 0.20 Ω and a standstill reactance per phase of 1.0 Ω. At standstill, the rotor induced emf per phase is 70 V. Full-load slip is 3.5%. If the torque developed at full load is 30 lb-ft, calculate the torque developed at standstill. (*Source:* Rosenblatt and Friedman)

**15.39.** An eight-pole 60-Hz polyphase induction motor develops maximum torque at a speed of 73 rad/s. If the rotor resistance per phase is 0.50 Ω, calculate the rotor reactance per phase at standstill. (*Source:* Rosenblatt and Friedman)

**15.40.** An eight-pole 60-Hz polyphase wound-rotor induction motor has a full-load speed of 90 rad/s. If external resistance equal to the rotor resistance per phase is added, what will be the new full-load speed, in rad/s and rpm? (*Source:* Rosenblatt and Friedman)

**15.41.** A 10-pole 60-Hz wound-rotor induction motor has a full-load speed of 690 rpm. The rotor resistance per phase is 0.30 Ω, the standstill rotor reactance per phase is 1.10 Ω.

**(a)** At what speed will pull-out (maximum torque) occur for the short-circuited rotor?

**(b)** If maximum starting torque is to be obtained, what rotor resistance per phase must be added?

**(c)** What is the full-load speed with the added resistance? Express in both English and SI units. (*Source:* Rosenblatt and Friedman)

**15.42.** A 10-pole 60-Hz wound-rotor induction motor has a full-load speed of 690 rpm. The rotor resistance per phase is 0.30 Ω, the standstill rotor reactance per phase is 1.10 Ω. Also, the voltage per phase induced in the rotor at standstill is 65 V. Determine:

**(a)** The starting rotor-phase current with the rotor short-circuited.

**(b)** The rotor-phase current at rated load with the rotor short-circuited.

**(c)** The rotor-phase current at standstill when maximum torque is developed.

**(d)** The rotor-phase current at standstill when developing full-load torque at starting. (*Source:* Rosenblatt and Friedman)

**15.43.** A six-pole 60 Hz wound-rotor induction motor has a full load speed of 1000 rev/min. The rotor resistance per phase is 0.30 Ω, the locked rotor reactance per phase is 0.80 Ω.

**(a)** Calculate the breakdown speed for the case in which the rotor is short-circuited.

**(b)** What is the new full-load speed if a resistance is placed in series with the rotor winding such that the torque at zero speed equals that at 1000 rpm?

**(c)** Repeat Problem 15.42b and c for this motor, if $E_G^L = 75$ V.

**15.44.** A 22.5-kW (output) 60-Hz three-phase wound-rotor induction motor has a no-load speed of 125 rad/s and a full-load speed of 120 rad/s, and pulls out (has maximum torque) at 88 rad/s. The rotor resistance per phase is 0.25 Ω.

**(a)** What is the standstill rotor reactance per phase?

**(b)** What resistance per phase must be added to the rotor in order to obtain a full-load speed of 113 rad/s?

**(c)** What is the no-load speed with the added resistance?

**(d)** What is the new speed regulation?

**(e)** At what speed will maximum torque be developed with the added rotor resistance? (*Source:* Rosenblatt and Friedman)

**15.45.** A three-phase induction motor has $\overline{Z}_r^L = 0.316\,\Omega\angle 71.57°$. The machine is 2-pole and operates at 60 Hz. The full-load speed is 3400 rev/min. The induced emf at locked rotor is 40 V.

**(a)** What is the synchronous speed $n_s$ in rpm?

**(b)** What is the full-load slip (in decimal)?

**(c)** What is the rotor phase current at the locked rotor condition, and at the rotor speeds 2880 rpm and 3400 rpm?

**(d)** Does the rotor current increase or decrease with increasing slip? Explain your reasoning by analyzing an appropriate equation.

**15.46.** **(a)** The rotor current in an induction motor can be calculated from

$$I_R = s\frac{E_G^L}{[R_R^2 + (sX_R^L)^2]^{1/2}}$$

$$= \frac{E_G^L}{[(R_R/s)^2 + (X_R^L)^2]^{1/2}} \tag{P15.1}$$

The term $R_R/s$ can be rewritten as $R_R/s = R_R + R_R(1 - s)/s$. When this expression is multiplied by $I_R^2$, the result is

$$\frac{I_R^2 R_R}{s} = I_R^2 R_R + \frac{I_R^2 R_R (1 - s)}{s} \qquad \text{(P15.2)}$$

In Eq. (P15.2), the first term can be considered the rotor input power per phase, $P_{\text{in}}$, the second can be considered the rotor copper loss per phase, $P_{\text{copper}}$, while the third term is the rotor developed power, $P_{\text{dev}}$. Using these facts, find the developed torque in terms of $P_{\text{in}}$ and $\omega_s$, the radial synchronous speed.

**(b)** At locked conditions, a six-pole 50-Hz induction motor has a generated voltage of 200 V, while the rotor reactance is 10 Ω and the rotor resistance is 4 Ω. From Eqs. (P15.1) and (P15.2), find a simple expression for $P_{\text{in}}$ at "breakdown" in terms of known quantities. What is the numerical value in watts for this example? What is the maximum developed torque in N · m?

**15.47.** Name three possible advantages of synchronous motors over induction motors.

**15.48.** Suppose that the stator impedance of a synchronous motor is $\overline{Z}_s = 0.2\,\Omega + j4\,\Omega$. What is the angle of maximum torque? Suppose that it is found that $\alpha$ reaches an angle of 83°. Is and will the motor be slipping poles?

**15.49.** Is the developed torque of a synchronous motor nonzero at $\alpha = 0$? If so, what is its value, and is it relatively large or small?

**15.50.** For what value of $\alpha$ is $P_{\text{dev}}$ maximum for a synchronous motor? Compare this result with that for the maximum torque and power for an induction motor.

**15.51.** What will be the full-load speed of a 24-pole 50-Hz polyphase synchronous motor? Express in rpm and rad/s. (*Source:* Rosenblatt and Friedman)

**15.52.** A polyphase synchronous motor has a full-load speed of 250 rpm when a 25-Hz supply is applied to the stator. How many poles does it have? (*Source:* Rosenblatt and Friedman)

**15.53.** A 30-hp 440-V wye-connected synchronous motor is operating at loads that cause the rotor to lag the stator field by (i) 15, (ii) 20, and (iii) 25 electrical degrees. If the field excitation is adjusted to produce a generated phase voltage equal to the applied phase voltage:

**(a)** What is the resultant voltage per phase in the armature?

**(b)** What angle, δ, does it make with the applied phase voltage? (*Source:* Rosenblatt and Friedman)

**15.54.** A 30-hp 440-V wye-connected synchronous motor has an effective resistance per phase of 0.40 Ω and a synchronous reactance per phase of 6.5 Ω. The load causes a torque angle of 30°, and the field excitation is adjusted to give a generated voltage per phase of 320 V. Calculate:

**(a)** The resultant armature voltage per phase.

**(b)** The angle δ.

**(c)** The armature current.

**(d)** The power factor at which the motor operates.

**(e)** The total power input to the motor. (*Source:* Rosenblatt and Friedman)

**15.55.** A 30-hp 440-V wye-connected synchronous motor has an effective resistance per phase of 0.40 Ω and a synchronous reactance per phase of 6.5 Ω. The load is increased until pull-out (maximum torque) occurs. The field excitation is adjusted to give a generated voltage per phase of 320 V. The rated motor line current is 33 A.

**(a)** What is the torque angle?

**(b)** What is the resultant armature voltage per phase?

**(c)** Find the angle δ.

**(d)** What is the armature current?

**(e)** Determine the power factor and calculate the total power input.

**(f)** At pull-out (maximum torque), calculate the developed kilowatts and horsepower, and the developed torque in lb-ft and N · m. (*Source:* Rosenblatt and Friedman)

**15.56.** A 1500-hp four-pole 4160-V 60-Hz wye-connected synchronous motor has a synchronous impedance of 10 Ω and an effective armature resistance per phase of 0.80 Ω. Determine:

**(a)** The synchronous reactance per phase.

**(b)** The synchronous impedance angle.

**(c)** The power factor for the following operating condition: $E_G = 2840$ V and $\alpha = 35°$.

**(d)** The power and torque developed, in English and SI units for the condition defined in part (c). (*Source:* Rosenblatt and Friedman)

**15.57.** Repeat Problem 15.55 for the following data values: $P = 4$, $f = 60$ Hz, $Z_s = 10$ Ω, $R_s = 0.80$ Ω, $P_{out} = 1$ MW, $\alpha = 30°$, and $E_G = 2200$ V. Indicate whether the power factor is leading or lagging, and how you know this.

**15.58.** A four-pole 60-Hz wye-connected 88% efficiency three-phase synchronous motor has synchronous impedance magnitude 10 Ω, armature reactance 4 Ω, and 0.9 PF leading. The output power is 10 kW and the line voltage is 440 V. Find **(a)** $I_A$, **(b)** $\theta$, **(c)** $\delta$, **(d)** $E_G$, and **(e)** $\alpha$.

**(f)** Which answers would change if instead the motor were delta connected? Whichever answers would change, recalculate.

**15.59.** A 1500-hp four-pole 4160-V 60-Hz wye-connected synchronous motor has a synchronous impedance of 10 Ω and an effective armature resistance per phase of 0.80 Ω. The motor is operated with a load such that the torque angle is 22° and the generated voltage is 2400 V per phase. Calculate:

**(a)** The resultant voltage per phase.

**(b)** The armature current.

**(c)** The power factor.

**(d)** The total power input.

**(e)** The developed power and torque, in English and SI units. (*Source:* Rosenblatt and Friedman)

**15.60.** A 1500-hp four-pole 4160-V 60-Hz wye-connected synchronous motor has a synchronous impedance of 10 Ω and an effective armature resistance per phase of 0.80 Ω. The dc field current is increased until the generated phase voltage is 3200 V. For a torque angle of 35°, calculate:

**(a)** The resultant phase voltage.

**(b)** The armature current.

**(c)** The power factor.

**(d)** The power output, in kW and hp, assuming an ac efficiency of 87%.

**(e)** The power and torque developed, in English and SI units. (*Source:* Rosenblatt and Friedman)

**15.61.** A 75-hp 440-V eight-pole 60-Hz wye-connected synchronous motor operates at a torque angle of 30°. The armature synchronous impedance per phase is 2.0 Ω, and the effective resistance per phase is 0.15 Ω. For each generated phase voltage of 220, 280, and 340 V and a constant torque angle of 30°, calculate:

**(a)** The resultant phase voltage.

**(b)** The armature current.

**(c)** The power factor, stating whether it is leading or lagging. (*Source:* Rosenblatt and Friedman)

**15.62.** A 75-hp 440-V eight-pole 60-Hz wye-connected synchronous motor operates at a torque angle of 30°. The armature synchronous impedance per phase is 2.0 Ω, and the effective resistance per phase is 0.15 Ω. At pull-out, determine the armature current, kilowatts, and horsepower developed in this motor for generated voltages of (**a**) 220 V, (**b**) 280 V, and (**c**) 340 V. (*Source:* Rosenblatt and Friedman)

**15.63.** A 60-hp 550-V four-pole 50-Hz wye-connected synchronous motor is designed to operate at an 80% leading power factor at rated load. Rated line current is 51 A. The armature resistance per phase is 0.30 Ω, and the synchronous reactance per phase is 4.2 Ω. Determine:

(**a**) The torque angle and the generated voltage.

(**b**) The power and torque developed, in English and SI units. (*Source:* Rosenblatt and Friedman)

**15.64.** A 60-hp 550-V four-pole 50-Hz wye-connected synchronous motor is designed to operate at unity power factor at rated load. Rated line current is 51 A. The armature resistance per phase is 0.30 Ω, and the synchronous reactance per phase is 4.2 Ω. The excitation producing full load, unity power factor is maintained while the motor is loaded until pull-out occurs. Determine:

(**a**) The line current.

(**b**) The power factor.

(**c**) The developed power and torque, in English and SI units.

(**d**) The ratio of pull-out torque to rated torque. (*Source:* Rosenblatt and Friedman)

**15.65.** A 2500-hp 23,000-V 12-pole 60-Hz wye-connected synchronous motor is designed to operate at 0.80 leading power factor. The effective armature resistance per phase is 6.0 Ω, and the synchronous reactance per phase is 100 Ω. The full-load ac efficiency is 0.90. Suppose that the load is increased while maintaining the same generated voltage.

(**a**) At what torque angle will the motor operate at unity power factor?

(**b**) What is the armature current?

(**c**) Find the developed kilowatts and horsepower. (*Source:* Rosenblatt and Friedman)

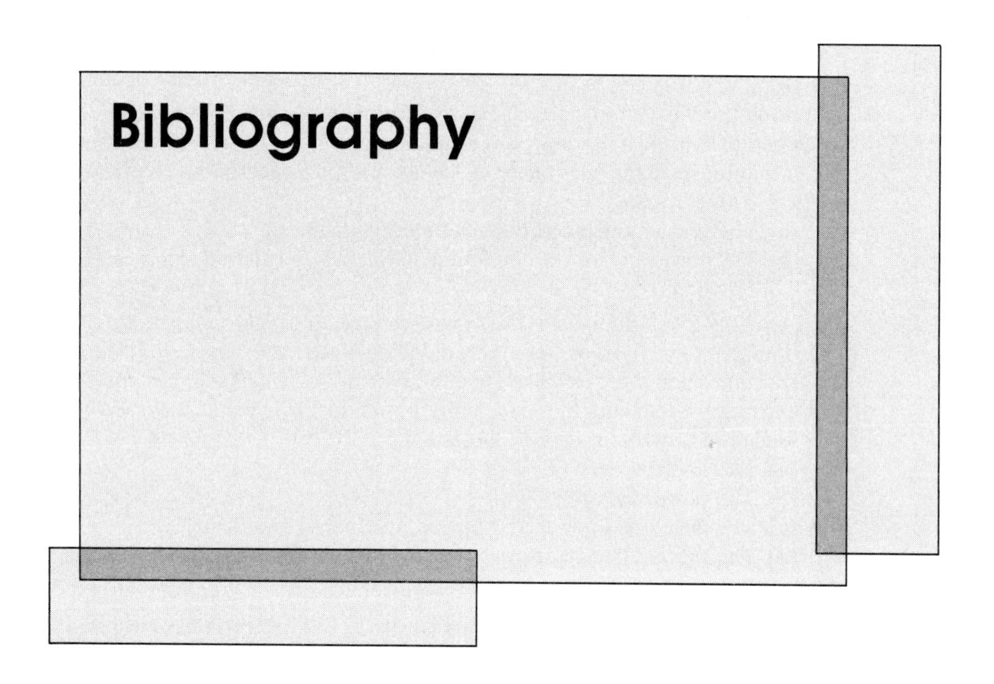


## Chapters 2 and 3

BORDEAU, S. P., *Volts to Hertz: The Rise of Electricity,* Burgess, Minneapolis, MN, 1982.

CARTER, W., *The Electromagnetic Field in Its Engineering Aspects,* American Elsevier, New York, 1967.

CHEN, D. K., *Field and Wave Electromagnetics,* Addison-Wesley, Reading, MA, 1989.

CHRISTOPOULOS, C., *An Introduction to Applied Electromagnetism,* Wiley, New York, 1990.

CRAVEN, A. T., *Electricity and Magnetism for Electrical Engineers,* Addison-Wesley, Reading, MA, 1962.

CUNNINGHAM, D. R., and J. A. STULLER, *Basic Circuit Analysis,* Houghton Mifflin, Boston, 1991.

ELLIOTT, R. S., *Electromagnetics,* McGraw-Hill, New York, 1966.

FRANK, N. H., *Introduction to Electricity and Optics,* McGraw-Hill, New York, 1940.

HALLIDAY, D., and R. RESNICK, *Physics* (Vols. 1 and 2), Wiley, New York, 1978.

KRAUS, J. D., *Electromagnetics,* McGraw-Hill, New York, 1992.

KRAUS, J. D., and K. R. CARVER, *Electromagnetics,* McGraw-Hill, New York, 1973.

LORRAIN, P., D. R. CORSON, and F. LORRAIN, *Electromagnetic Fields and Waves, Including Electric Circuits,* W. H. Freeman, New York, 1988.

MAGID, L. M., *Electromagnetic Fields, Energy, and Waves,* Wiley, New York, 1972.

MARTIN, T. L., *Physical Basis for Electrical Engineering,* Prentice Hall, Englewood Cliffs, NJ, 1957.

MATSCH, L. W., *Capacitors, Magnetic Circuits, and Transformers,* Prentice Hall, Englewood Cliffs, NJ, 1964.

McKenzie, A. E. E., *A Second Course of Electricity,* Cambridge University Press, Cambridge, 1973.

Meyer, H. W., *A History of Electricity and Magnetism,* MIT Press, Cambridge, MA, 1971.

National Electrical Manufacturers Association, *A Chronological History of Electrical Development from 600 B.C.,* NEMA, New York, 1946.

Neff, H. P., Jr., *Introductory Electromagnetics,* Wiley, New York, 1991.

Nussbaum, A., *Electromagnetic Theory for Engineers and Scientists,* Prentice Hall, Englewood Cliffs, NJ, 1965.

Page, L., and N. I. Adams, Jr., *Principles of Electricity: An Intermediate Text in Electricity and Magnetism,* D. Van Nostrand, Princeton, NJ, 1969.

Plonsley, R., and R. E. Collin, *Principles and Applications of Electromagnetic Fields,* McGraw-Hill, New York, 1961.

Pugh, E. M., and E. W. Pugh, *Principles of Electricity and Magnetism,* Addision-Wesley, Reading, MA, 1970.

Purcell, E. M., *Berkeley Physics Course,* Vol. 2: *Electricity and Magnetism,* McGraw-Hill, New York, 1965.

Ramo, S., J. R. Whinnery, and T. Van Duzer, *Fields and Waves in Communications Electronics,* Wiley, New York, 1984.

Ryder, J. D., and D. G. Fink, *Engineers and Electrons: A Century of Electrical Progress,* IEEE Press, New York, 1984.

Schwarz, S. E., *Electromagnetics for Engineers,* Saunders College, Philadelphia, PA, 1990.

Starling, S. G., *Electricity and Magnetism for Degree Students,* Longman, New York, 1953.

Timbie, W. H., and A. Kusko, *Elements of Electricity,* Wiley, New York, 1953.

Tipler, P. A., *Physics,* Worth, New York, 1976.

Zilberman, G., *Electricity and Magnetism,* Mir, Moscow, 1973.

## Chapters 4 to 6

Allocca, J. A., and H. E. Levenson, *Electrical and Electronic Safety,* Reston/Prentice Hall, Reston, VA, 1982.

Bell, D. A., *Fundamentals of Electric Circuits,* Prentice Hall, Englewood Cliffs, NJ, 1988.

Blalock, G. C., *Principles of Electrical Engineering,* McGraw-Hill, New York, 1950.

Bobrow, L. S., *Elementary Linear Circuit Analysis,* Holt, Rinehart and Winston, New York, 1987.

Budak, A., *Circuit Theory Fundamentals and Applications,* Prentice Hall, Englewood Cliffs, NJ, 1987.

Carter, G. W., *Techniques of Circuit Analysis,* Cambridge University Press, Cambridge, 1972.

Churchill, R. V., and J. W. Brown, *Fourier Series and Boundary Value Problems,* 4th ed., McGraw-Hill, New York, 1987.

Cook, A. L., and C. C. Carr, *Elements of Electrical Engineering,* Wiley, New York, 1954.

Cooper, W. F., *Electrical Safety Engineering,* Newnes-Butterworths, Boston, 1979.

Cunningham, D. R., and J. A. Stuller, *Basic Circuit Analysis,* Houghton Mifflin, Boston, 1991.

Dawes, C. L., *A Course in Electrical Engineering* (Vols. I and II), McGraw-Hill, New York, 1937.

DE JONG, H. C. J., *AC Motor Design,* Springer-Verlag, New York, 1989.

EE Staff, MIT, *Magnetic Circuits and Transformers,* MIT Press, Cambridge, MA, 1943.

FLOYD, T. L., *Principles of Electric Circuits,* Charles E. Merrill, Columbus, OH, 1985.

GABEL, R. A., and R. A. ROBERTS, *Signals and Linear Systems,* Wiley, New York, 1980.

GIBBS, J. B., *Transformer Principles and Practice,* McGraw-Hill, New York, 1937.

GLOVER, J. D., and M. SARMA, *Power System Analysis and Design,* PWS, Boston, 1987.

GONEN, T., *Power Distribution System Engineering,* McGraw-Hill, New York, 1986.

GREENWALD, E. K., ed., *Electrical Hazards and Accidents: Their Cause and Prevention,* Van Nostrand Reinhold, New York, 1991.

GROSS, C. A., *Power System Analysis,* Wiley, New York, 1979.

GUILLEMIN, E. A., *The Mathematics of Circuit Analysis,* Wiley/MIT Press, New York, 1949.

HAASE, H., *Electrostatic Hazards: Their Evaluation and Control,* Verlag Chemie, New York, 1977.

HAYT, W. H., JR., and J. E. KEMMERLY, *Engineering Circuit Analysis,* McGraw-Hill, New York, 1986.

HUELSMAN, L. P., *Basic Circuit Theory,* Prentice Hall, Englewood Cliffs, NJ, 1991.

JOHNSON, D. E., J. R. JOHNSON, and J. L. HILBURN, *Electric Circuit Analysis,* Prentice Hall, Englewood Cliffs, NJ, 1992.

KALE, C. O., *Industrial Circuits and Automated Manufacturing,* Saunders College, Philadelphia, PA, 1989.

KELLEY, M. C., and B. NICHOLS, *Introductory Linear Electrical Circuits and Electronics,* Wiley, New York, 1988.

KIRWIN, G. J., and S. E. GRODZINSKY, *Basic Circuit Analysis,* Houghton Mifflin, Boston, 1980.

KRAUS, A. D., *Circuit Analysis,* West, St. Paul, MN, 1991.

LAYNE, K., *Automobile Electronics and Basic Electrical Systems,* Wiley, New York, 1990.

LOWDON, E., *Practical Transformer Design Handbook,* TAB Books, Blue Ridge Summit, PA, 1989.

MARSDEN, J. E., *Basic Complex Analysis,* W. H. Freeman, San Francisco, 1973.

NADON, J. M., B. J. GELMINE, and E. D. MCLAUGHLIN, *Industrial Electricity,* Delmar, Albany, New York, 1989.

NILSSON, J. W., *Electric Circuits,* Addison-Wesley, Reading, MA, 1990.

PANSINI, A. J., *Electrical Transformers and Power Equipment,* Prentice Hall, Englewood Cliffs, NJ, 1988.

PATRICK, D. R., and S. W. FARDO, *Electricity and Electronics: A Survey,* Prentice Hall, Englewood Cliffs, NJ, 1990.

PRICE, J. F., ed., *Fourier Techniques and Applications,* Plenum Press, New York, 1985.

REDDING, R. J., *Intrinsic Safety: The Safe Use of Electronics in Hazardous Locations,* McGraw-Hill, New York, 1971.

RICHES, B. E., *Electric Circuit Theory: A Computer Illustrated Text,* IOP, Bristol, England, 1989.

SANTINI, A., *Automotive Electricity and Electronics,* Delmar, Albany, New York, 1988.

SCHULER, C. A., and R. J. FOWLER, *Basic Electricity and Electronics,* McGraw-Hill, New York, 1988.

SOLIMAN, S. S., and M. D. SRINATH, *Continuous and Discrete Signals and Systems,* Prentice Hall, Englewood Cliffs, NJ, 1990.

STEVENSON, W. D., *Elements of Power System Analysis,* McGraw-Hill, New York, 1982.

TUTTLE, D. F., JR., *Circuits,* McGraw-Hill, New York, 1977.
WEAVER, H. J., *Applications of Discrete and Continuous Fourier Analysis,* Wiley, New York, 1983.
WINBURN, D. C., *Practical Electrical Safety,* Marcel Dekker, New York, 1988.

## Chapter 7

ANGELO, E. J., *Electronic Circuits,* McGraw-Hill, New York, 1964.
ANKRUM, P. D., *Semiconductor Electronics,* Prentice Hall, Englewood Cliffs, NJ, 1971.
BOYLESTAD, R., and L. NASHELSKY, *Electronic Devices and Circuit Theory,* Prentice Hall, Englewood Cliffs, NJ, 1978.
BROPHY, J. J., *Basic Electronics for Scientists,* McGraw-Hill, New York, 1990.
BUSCOMBE, C. G., *Television and Video Systems: Operation, Maintenance, Troubleshooting, and Repair,* Prentice Hall, Englewood Cliffs, NJ, 1990.
BYERS, T. J., *Electronic Test Equipment,* McGraw-Hill, New York, 1987.
CHIRLIAN, P. M., *Analysis and Design of Integrated Electronic Circuits,* Harper & Row, New York, 1987.
DOWDING, B., *Principles of Electronics,* Prentice Hall, Englewood Cliffs, NJ, 1988.
DOYLE, J. M., *Pulse Fundamentals,* Prentice Hall, Englewood Cliffs, NJ, 1973.
DUNSTER, D. F., *Semiconductors for Engineers,* Business Books, London, 1969.
EDGERTON, H. E., *Electronic Flash, Strobe,* McGraw-Hill, New York, 1970.
FARLEY, F. J. M., *Elements of Pulse Circuits,* Wiley, New York, 1964.
FLOYD, T. L., *Electronic Devices,* Charles E. Merrill, Columbus, OH, 1988.
FREDERIKSEN, T. M., *Intuitive Analog Electronics,* McGraw-Hill, New York, 1989.
GIBBONS, J. F., *Semiconductor Electronics,* McGraw-Hill, New York, 1966.
GRAY, P. E., and C. L. SEARLE, *Electronic Principles,* Wiley, New York, 1969.
HARRIS, D. J., and P. N. ROBSON, *The Physical Basis of Electronics,* Pergamon Press, Elmsford, New York, 1974.
HNATEK, E. R., *Design of Solid State Power Amplifiers,* Van Nostrand Reinhold, New York, 1981.
HODGES, D. A., and H. G. JACKSON, *Analysis and Design of Digital Integrated Circuits,* McGraw-Hill, New York, 1983.
HOROWITZ, P., and W. HILL, *The Art of Electronics,* Cambridge University Press, New York, 1989.
HOROWITZ, M., and T. HORN, *How to Design Solid-State Circuits,* TAB Books, Blue Ridge Summit, PA, 1988.
KETCHUM, D. J., and E. C. ALVAREZ, *Pulse and Switching Circuits,* McGraw-Hill, New York, 1965.
LITTAUER, R., *Pulse Electronics,* McGraw-Hill, New York, 1965.
LURCH, E. N., *Fundamentals of Electronics,* Wiley, New York, 1981.
MILLMAN, J., and A. GRABEL, *Microelectronics,* McGraw-Hill, New York, 1987.
MILLMAN, J., and C. C. HALKIAS, *Electronic Devices and Circuits,* McGraw-Hill, New York, 1967.
MILLMAN, J., and H. TAUB, *Pulse and Digital Circuits,* McGraw-Hill, New York, 1956.
MITCHELL, F. H., JR., and F. H. MITCHELL, SR., *Introduction to Electronics Design,* Prentice Hall, Englewood Cliffs, NJ, 1992.

MORTON, R. A., ed., *Photography for the Scientist,* Academic Press, New York, 1984.

NEUDECK, G. W., *The PN Junction Diode,* Addison-Wesley, Reading, MA, 1988.

OVERHAGE, C. F., ed., *The Age of Electronics,* McGraw-Hill, New York, 1962.

PAYNTER, R. T., *Introductory Electronic Devices and Circuits,* Prentice Hall, Englewood Cliffs, NJ, 1991.

PRENTISS, S., *Modern Television: Service and Repair,* Prentice Hall, Englewood Cliffs, NJ, 1989.

ROULSTON, D. J., *Bipolar Semiconductor Devices,* McGraw-Hill, New York, 1990.

RYDER, J. D., *Electronic Fundamentals and Applications,* Prentice Hall, Englewood Cliffs, NJ, 1976.

SAVANT, C. J., JR., M. S. RODEN, and G. L. CARPENTER, *Electronic Circuit Design: An Engineering Approach,* Benjamin-Cummings, Menlo Park, CA, 1987.

SEDRA, A. S., and K. C. SMITH, *Microelectronic Circuits,* Saunders College, Philadelphia, PA, 1991.

Semiconductor Electronics Education Committee, *Elementary Circuit Properties of Transistors,* Wiley, New York, 1964.

SENTURIA, S. D., and B. D. WEDLOCK, *Electronic Circuits and Applications,* Wiley, New York, 1975.

SEVIN, L. J., JR., *Field-Effect Transistors,* McGraw-Hill, New York, 1965.

SEYMOUR, J., *Electronic Devices and Components,* Longman, Harlow, England, 1981.

SHAW, D. F., *An Introduction to Electronics,* Longman, London, 1970.

SHIVE, J. N., *The Properties, Physics, and Design of Semiconductor Devices,* D. Van Nostrand, Princeton, NJ, 1959.

SHOCKLEY, W., *Electrons and Holes in Semiconductors, with Applications to Transistor Electronics,* Van Nostrand, New York, 1950.

SMITH, R. A., *Semiconductors,* Cambridge University Press, Cambridge, 1961.

SMITH, R. J., and R. C. DORF, *Circuits, Devices and Systems,* Wiley, New York, 1992.

SPARKES, J. J., *Junction Transistors,* Pergamon Press, Elmsford, New York, 1966.

STANLEY, W. D., *Electronic Devices: Circuits and Applications,* Prentice Hall, Englewood Cliffs, NJ, 1989.

STREETMAN, B. G., *Solid State Electronic Devices,* 3rd ed., Prentice Hall, Englewood Cliffs, NJ, 1990.

TOWERS, T. D., *Elements of Transistor Pulse Circuits,* Newnes-Butterworths, London, 1974.

UMAN, M. F., *Introduction to the Physics of Electronics,* Prentice Hall, Englewood Cliffs, NJ, 1974.

WILLIAMS, R. A., *Communication Systems Analysis and Design,* Prentice Hall, Englewood Cliffs, NJ, 1987.

ZAMBUTO, M., *Semiconductor Devices,* McGraw-Hill, New York, 1989.

ZEINES, B., *Principles of Applied Electronics,* Wiley, New York, 1963.

## Chapter 8

BARNA, A., *Operational Amplifiers,* Wiley, New York, 1971.

BELL, D. A., *Operational Amplifiers: Applications, Troubleshooting, and Design,* Prentice Hall, Englewood Cliffs, NJ, 1990.

CARR, J. J., *Integrated Electronics: Operational Amplifiers and Linear ICs with Applications,* Harcourt Brace Jovanovich, New York, 1990.

CLAYTON, G. B., *Operational Amplifiers,* Butterworths, London, 1971.

COUGHLIN, R. F., and F. F. DRISCOLL, *Operational Amplifiers and Linear Integrated Circuits,* Prentice Hall, Englewood Cliffs, NJ, 1991.

COUGHLIN, R. F., and R. S. VILLANUCCI, *Introductory Operational Amplifiers and Linear ICs,* Prentice Hall, Englewood Cliffs, NJ, 1990.

FAULKENBERRY, L. M., *An Introduction to Operational Amplifiers with Linear IC Applications,* Wiley, New York, 1982.

FREDERIKSEN, T. M., *Intuitive Operational Amplifiers,* McGraw-Hill, New York, 1988.

GRAEME, J. G./Burr Brown Corp., *Designing with Operational Amplifiers: Applications Alternatives,* McGraw-Hill, New York, 1977.

HONEYCUTT, R. A., *Op Amps and Linear Integrated Circuits,* Delmar, Albany, New York, 1988.

IRVINE, R. G., *Operational Amplifier Characteristics and Applications,* Prentice Hall, Englewood Cliffs, NJ, 1987.

JOHNSON, D. E., and V. JAYAKUMAR, *Operational Amplifier Circuits: Design and Application,* Prentice Hall, Englewood Cliffs, NJ, 1982.

LENK, J. D., *Manual for Integrated Circuit Users,* Reston, Reston, VA, 1973.

SOCLOF, S., *Design and Applications of Analog Integrated Circuits,* Prentice Hall, Englewood Cliffs, NJ, 1991.

STANLEY, W. D., *Operational Amplifiers with Linear Integrated Circuits,* Charles E. Merrill, Columbus, OH, 1989.

TRAISTER, R. J., *Operational Amplifier Circuit Manual,* Academic Press, San Diego, CA, 1989.

## Chapter 9

American Society for Testing and Materials, *Instrumentation for Monitoring Air Quality,* ASTM Technical Publications, Philadelphia, PA, 1974.

ASTON, R., *Principles of Biomedical Instrumentation and Measurement,* Charles E. Merrill, Columbus, OH, 1990.

BEAUCHAMP, K., and C. YUEN, *Data Acquisition for Signal Analysis,* Allen & Unwin, Boston, 1980.

BECKWITH, T. G., and R. D. MARANGONI, *Mechanical Measurements,* Addison-Wesley, Reading, MA, 1990.

BORWICK, J., *Microphones: Technology and Technique,* Focal Press, London, 1990.

BRINDLEY, K., *Sensors and Transducers,* Heinemann Newnes, London, 1988.

CAMRAS, M., *Magnetic Tape Recording,* Van Nostrand Reinhold, New York, 1985.

GILLUM, D. R., *Industrial Pressure Measurement,* Instrument Society of America, Research Triangle Park, NC, 1982.

HARRINGTON, J., *Automated Process Control Electronics,* Delmar, Albany, New York, 1989.

HUMPHRIES, J. T., and L. P. SHEETS, *Industrial Electronics,* Delmar, Albany, New York, 1989.

JACOB, J. M., *Industrial Control Electronics,* Prentice Hall, Englewood Cliffs, NJ, 1988.

JONES, L. D., and A. F. CHIN, *Electronic Instruments and Measurements,* Prentice Hall, Englewood Cliffs, NJ, 1991.

MANSFIELD, P. H., *Electrical Transducers for Industrial Measurement,* Butterworths, London, 1973.

MEE, C. D., and E. D. DANIEL, eds., *Magnetic Recording,* McGraw-Hill, New York, 1987.

NORTON, H. N., *Handbook of Transducers,* Prentice Hall, Englewood Cliffs, NJ, 1989.

OLIVER, F. J., *Practical Instrumentation Transducers,* Hayden, Rochelle Park, NJ, 1971.

Omega Engineering, *Flow and Level, Pressure and Strain, Temperature, Ph and Conductivity Handbooks and Encyclopedias,* Omega Engineering, Inc., Stamford, CT, 1991 (yearly).

PUTTEN, A. F., *Electronic Measurement Systems,* Prentice Hall, Englewood Cliffs, NJ, 1988.

RUNSTEIN, R. E., *Modern Recording Techniques,* Howard W. Sams, Indianapolis, IN, 1974.

SHEINGOLD, D. H., ed., *Transducer Interfacing Handbook: A Guide to Analog Signal Conditioning,* Analog Devices, Norwood, MA, 1981.

STROBEL, H. A., *Chemical Instrumentation,* Addison-Wesley, Reading, MA, 1973.

SYDENHAM, P., *Basic Electronics for Instrumentation,* Instrument Society of America, Research Triangle Park, NC, 1982.

TRIETLEY, H. L., *Transducers in Mechanical and Electronic Design,* Marcel Dekker, New York, 1986.

TSE, F. S., and I. E. MORSE, *Measurement and Instrumentation in Engineering,* Marcel Dekker, New York, 1989.

WHITE, R. M., ed., *Introduction to Magnetic Recording,* IEEE Press, New York, 1984.

WILMSHURST, T. H., *Signal Recovery from Noise in Electronic Instrumentation,* Adam Hilger, New York, 1990.

WOBSCHALL, D., *Circuit Design for Electronic Instrumentation,* McGraw-Hill, New York, 1987.

WOOLVET, G. A., *Transducers in Digital Systems,* Peter Peregrinus (Institution of Electrical Engineers), New York, 1979.

WORAM, J. M., *Sound Recording Handbook,* Howard W. Sams, Indianapolis, IN, 1989.

### Chapters 10 and 11

ADAMSON, T. A., *Digital Systems, Logic, and Applications,* Delmar, Albany, New York, 1989.

BECHER, W. D., *Logical Design Using Integrated Circuits,* Hayden, Rochelle Park, NJ, 1977.

CAVANAGH, J. F., *Digital Computer Arithmetic: Design and Implementation,* McGraw-Hill, New York, 1984.

COMER, D. J., *Digital Logic and State Machine Design,* Saunders College, Philadelphia, PA, 1990.

DAVIO, M., J. P. DESCHAMPS, and A. THAYSE, *Digital Systems, with Algorithm Implementation,* Wiley, New York, 1983.

FLOYD, T. L., *Digital Fundamentals,* Charles E. Merrill, Columbus, OH, 1982.

FREDERIKSEN, T. M., *Intuitive CMOS Electronics: The Revolution in VLSI Processing, Packaging, and Design,* McGraw-Hill, New York, 1989.

HEISERMAN, P. L., *Surface-Mount Devices: Troubleshooting and Repair,* Prentice Hall, Englewood Cliffs, NJ, 1990.

HILL, F. J., and G. R. PETERSON, *Digital Logic and Microprocessors,* Wiley, New York, 1984.

ISMAIL, A. R., and V. M. ROONEY, *Digital Concepts and Applications,* Saunders College, Philadelphia, PA, 1990.

KERSHAW, J. D., *Digital Electronics: Logic and Systems,* Duxbury Press, North Scituate, MA, 1976.

KLEITZ, W., *Digital and Microprocessor Fundamentals,* Prentice Hall, Englewood Cliffs, NJ, 1990.

PIPPENGER, D. E., and E. J. TOBABEN, *Linear and Interface Circuits and Applications,* McGraw-Hill, New York, 1988.

PUTMAN, B. W., *Digital Electronics: Theory, Applications, and Troubleshooting,* Prentice Hall, Englewood Cliffs, NJ, 1986.

ROTH, C. H., JR., *Findamentals of Logic Design,* West, St. Paul, MN, 1985.

SANDIGE, R. S., *Modern Digital Design,* McGraw-Hill, New York, 1990.

SCOTT, N. R., *Computer Number Systems and Arithmetic,* Prentice Hall, Englewood Cliffs, NJ, 1985.

TAUB, H., *Digital Circuits and Microprocessors,* McGraw-Hill, New York, 1982.

TAUB, H., and D. SCHILLING, *Digital Integrated Electronics,* McGraw-Hill, New York, 1977.

THIJSSEN, A. P., H. A. VINK, and C. H. EVERSDIJK, *Digital Techniques: From Problem to Circuit,* Edward Arnold, London, 1989.

TRIEBEL, W. A., *Integrated Digital Electronics,* Prentice Hall, Englewood Cliffs, NJ, 1985.

WAKERLY, J. F., *Digital Design Principles and Practice,* Prentice Hall, Englewood Cliffs, NJ, 1990.

## Chapter 12

AUSLANDER, D. M., and P. SAGUES, *Microprocessors for Measurement and Control,* Osborne/McGraw-Hill, Berkeley, CA, 1981.

AYALA, K. J., *The 8051 Microcontroller: Architecture, Programming, and Applications,* West, St. Paul, MN, 1991.

BARNEY, G. C., *Intelligent Instrumentation: Microprocessor Applications in Measurement and Control,* Prentice Hall, Englewood Cliffs, NJ, 1988.

BIBBERO, R. J., *Microprocessors in Industrial Control,* Instrument Society of America, Research Triangle Park, NC, 1982.

BREY, B. B., *Microprocessors and Peripherals: Hardware, Software, Interfacing, and Applications,* Charles E. Merrill, Columbus, OH, 1988.

BROGAN, W. L., *Modern Control Theory,* Prentice Hall, Englewood Cliffs, NJ, 1985.

CAVE, F. E., and D. L. TERRELL, *Digital Technology with Microprocessors,* Reston, Reston, VA, 1981.

FULCHER, J., *An Introduction to Microcomputer Systems: Architecture and Interfacing,* Addison-Wesley, Reading MA, 1989.

GILMORE, C. M., *Microprocessors: Principles and Applications,* McGraw-Hill, New York, 1989.

GROSSBLATT, R., *The 8088 Project Book,* Tab Books, Blue Ridge Summit, PA, 1989.

HALL, D. V., *Microprocessors and Interfacing: Programming and Hardware,* McGraw-Hill, New York, 1986.

KING, D. H., *Computerized Engine Controls,* Delmar, Albany, New York, 1989.

KNOWLES, D., *Automotive Emission Control and Computer Systems,* Prentice Hall, Englewood Cliffs, NJ, 1989.

PASAHOW, E. J., *Microprocessor Technology and Microcomputers,* McGraw-Hill, New York, 1988.

SLATER, M., *Microprocessor-Based Design,* Prentice Hall, Englewood Cliffs, NJ, 1989.

STECKHAHN, A. D., and J. DEN OTTER, *Industrial Applications for Microprocessors,* Reston, Reston, VA, 1982.

STOUT, D. F., editor-in-chief, *Microprocessor Applications Handbook,* McGraw-Hill, New York, 1982.

UFFENBECK, J., *Microcomputers and Microprocessors,* Prentice Hall, Englewood Cliffs, NJ, 1991.

WARD, S. A., and R. H. HALSTEAD, *Computation Structures,* MIT Press, Cambridge, MA, 1990.

YUEN, C. K., K. G. BEAUCHAMP, and D. FRASER, *Microprocessor Systems in Signal Processing,* Academic Press, New York, 1989.

ZAKS, R., *From Chips to Systems: An Introduction to Microprocessors,* SYBEX, Berkeley, CA, 1981.

## Chapters 13 to 15

CHALMERS, B. J., ed., *Electric Motor Handbook,* Butterworth, Boston, MA, 1988.

CHAPMAN, S. J., *Electric Machinery Fundamentals,* McGraw-Hill, New York, 1991.

DEL TORO, V., *Basic Electric Machines,* Prentice Hall, Englewood Cliffs, NJ, 1990.

DEL TORO, V., *Electric Machines and Power Systems,* Prentice Hall, Englewood Cliffs, NJ, 1985.

DUFFIN, D. J., *Generators and Motors and Their Applications,* McGraw-Hill, New York, 1947 (large applications chart on pp. 144–151).

EMANNUEL, P., *Motors, Generators, Transformers, and Energy,* Prentice Hall, Englewood Cliffs, NJ, 1985.

FITZGERALD, A. E., C. KINGSLEY, JR., and S. D. UMANS, *Electric Machinery,* McGraw-Hill, New York, 1990.

HINDMARCH, H., *Electrical Machines and their Applications,* 4th ed. Pergamon Press, Elmsford, NY, 1984.

HUBERT, C. I. *Electric Machines: Theory, Operation, Applications, Adjustment, and Control,* Charles E. Merrill, Columbus, OH, 1991.

JORDON, H. E., *Energy Efficient Electric Motors and Their Application,* Van Nostrand Reinhold, New York, 1983.

KOSOW, I. L., *Electric Machinery and Transformers,* Prentice Hall, Englewood Cliffs, NJ, 1991.

KRAUSE, P. C., *Analysis of Electric Machinery,* McGraw-Hill, New York, 1986.

LOEW, E. A., and F. R. BERGSETH, *Direct and Alternating Currents: Theory and Machinery,* McGraw-Hill, New York, 1954.

MABLEKOS, V. E., *Electric Machine Theory for Power Engineers,* Harper & Row, New York, 1980.

MAJMUDAR, H., *Electromechanical Energy Converters,* Allyn and Bacon, Boston, 1965.

MAJMUDAR, H., *Introduction to Electrical Machines,* Allyn and Bacon, Boston, 1969.

MATSCH, L. W., *Electromagnetic and Electromechanical Machines,* IEP, New York, 1977.

MCPHERSON, G., and R. D. LARAMORE, *An Introduction to Electrical Machines and Transformers,* Wiley, New York, 1990.

NASAR, S. A., and I. BOLDEA, *Electric Machines: Steady-State Operation,* Hemisphere, New York, 1990.

RAMSHAW, R., and R. G. VAN HEESWIJK, *Energy Conversion: Electric Motors and Generators,* Saunders College, Philadelphia, PA, 1990.

RICHARDSON, D. V., *Rotating Electric Machinery and Transformer Technology,* Reston, Reston, VA, 1982.

ROSENBLATT, J., and M. H. FRIEDMAN, *Direct and Alternating Current Machinery,* Charles E. Merrill, Columbus, OH, 1984.

SISKIND, C. S., *Direct-Current Machinery,* McGraw-Hill, New York, 1952.

SISKIND, C. S., *Induction Motors: Single-Phase and Polyphase,* McGraw-Hill, New York, 1958.

SMEATON, R. W., editor-in-chief, *Motor Applications and Maintenance Handbook,* McGraw-Hill, New York, 1969.

WERNINCK, E. H., ed., *Electric Motor Handbook,* McGraw-Hill, New York, 1978.

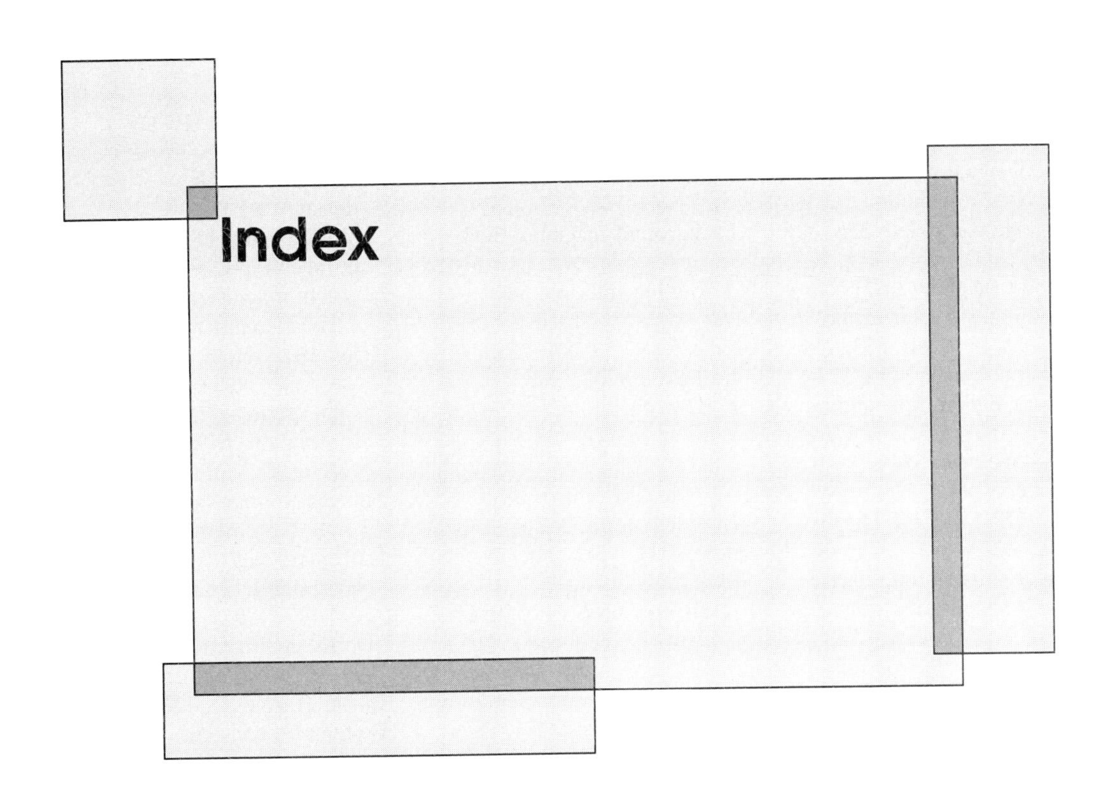

# Index

## A

## C

## D

## E

**F**

G



H

**Q**



**R**

## S

## T

## U



## V

W



X



Z